First International Congress on Environmental Geotechnics

July 10-15, 1994
Edmonton, Alberta
Canada

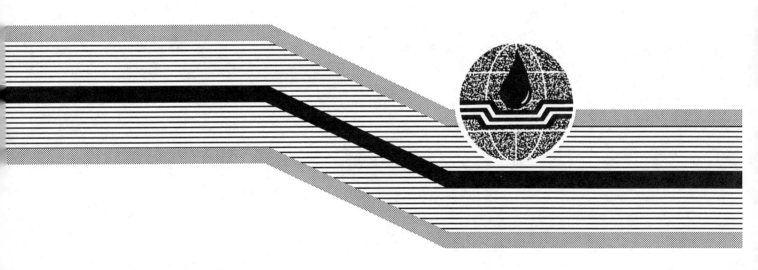

PROCEEDINGS OF THE FIRST INTERNATIONAL CONGRESS ON ENVIRONMENTAL GEOTECHNICS

Sponsored by the International Society for Soil Mechanics and Foundation Engineering and Canadian Geotechnical Society

Edmonton 11-15 July 1994

EDITOR
Technical Program Committee of 1st ICEG
W. David Carrier, III, Chairman

STEERING COMMITTEE

ISSMFE
Prof. N.R. Morgenstern, Past President
Dr. R.H.G. Parry, Secretary General
Prof. H.L. Jessberger, Germany
Dr. Z.-C. Moh, Taiwan

CGS
Mr. J. Seychuk, President
Mr. A.G. Stermac, Director General
Prof. J. Graham, VP-Technical
Mr. P. Wu, VP-Finance

CONFERENCE CO-CHAIRMEN
Prof. J.D. Nelson
Prof. J.H. Troncoso
Prof. R.N. Yong

ORGANIZING COMMITTEE

Dr. W.D. Carrier, III, Technical Program
Prof. J.D. Nelson, Environmental control, ISSMFE TC-5
Prof. D.C. Sego, Arrangements
Prof. J.H. Troncoso, Mines Waste, ISSMFE TC-7
Prof. R.N. Yong, Geoenvironment Division, CGS

TECHNICAL PROGRAM COMMITTEE

Y.B. Acar	J.D. Nelson
D.W. Airy	M.J. O'Connor
M. Aubertin	R.H.G. Parry
J.E. Baumann	H.G. Poulos
J.R. Booker	R.M. Quigley
W.D. Carrier, III	J. Saarela
R.P. Chapuis	D.C. Sego
R.G. Clark	C.D. Shackelford
A.W. Clifton	B. Soyez
L.G. de Mello	R.S. Tenove
K. Jha	J.H. Troncoso
M. Kamon	M.T. Tumay
P. Klablena	M. Usmen
R.C. Lo	S.G. Vick
T.A. Lundgren	J.P. Welsh
M. Manassero	R.N. Yong
Z.-C. Moh	Yudhbir
N.R. Morgenstern	T F Zimmie

ACKNOWLEDGEMENTS

The Organizing Committee wishes to express its great appreciation to L.K. Jarrah and C.D. Manuel, both of Bromwell & Carrier, Inc.

Printed and bound in Canada

Published 1994

Published and sold by:
BiTech Publishers Ltd.
173 - 11860 Hammersmith Way
Richmond, British Columbia
Canada V7A 5G1

ISBN 0 921095 32 5

Table of Contents

DREDGING

Plenary Speakers

E.W. Brand, J.B. Massey, P.G.D. Whiteside
    Environmental Aspects of Sand Dredging and Mud Disposal in Hong Kong 1

M.A. Viergever, F.A. Westrate, M. Loxham
    Reclamation of Dredged Materials in Europe 11

Keynote Speaker

Marian Poindexter Rollings
    Geotechnical Considerations in Dredged Material Management 21

Submitted Papers

P.K. Egyir, D.C. Sego
    Characteristics of a Hydraulic Fill Beach 33

David P. Knox, Rosemary A. Najjar
    Solidification and Stabilization of Lagoon Sludge and Dredged Spoils 41

Takuya Ogino, Toshiyoshi Goto, Koji Kataoka, Minoru Kuroda
    Utilization of Stabilized Dredged Waste for Construction Material 49

Naresh C. Samtani, Vahan Tanal, Joe Wang, Anthony R. Lancellotti
    Effect of Lime Admixtures on Contaminated Dredged Sediments 57

A. Suzuki, Y. Kitazono, S. Maruyama, Y. Hayashi, J. Yang
    On the Influence of Ca^{2+} Leached out of the Light Weight Stabilized Soils 65

T. Yamanouchi, T. Ishibashi, M. Takagi, T. Tashima
    **Environmentally Acceptable Reclamation using Materials Dredged
from Adjoining Seabed** 71

INDUSTRIAL

Plenary Speakers

John A. Cherry
    In Situ Control of Contaminated Groundwater 77

A.E. Fair, D.E. Sheeran
    Assessment of Oil Sand Fine Tails Reclamation Strategies 79

F. Schlosser, B. Soyez, M. Wojnarowicz
    Aspects Geotechniques de la Gestion des Dechets Industriels en France 91

Keynote Speakers

Robert Hinchee
    State-of-the-Art of In Situ Remediation of Hydrocarbon Contamination 101

M. Manassero, C.D. Shackelford
    Classification of Industrial Wastes for Re-use and Landfilling 103

Submitted Papers

I. Alimi-Ichola, O. Bentoumi, G. Didier
    Analyse des Transferts d'Eau Dans un Materiau d'Etancheite de Fond de Decharge Evaluation de la Permeabilite en Regime Transitoire 115

P. Amann, F.T. Madsen, L. Martinenghi
    Waste Immobilisation, Soil Liner and Slurry Wall Material Research in Switzerland . 121

Cem B. Avci, Daniel Bodine, Erol Güler
    Effects of Clay Swelling on Permeability Calculations Obtained from Sealed Double Ring Infiltrometer Tests 127

N.P. Betelev
    A New Method for Determination of Organic Carbon Content in Soils and Rocks to Estimate Pollutant Levels 133

N.P. Betelev, B.I. Kulachkin, Ya M. Kislyakov
    Dependence between Uranium and Organic Matter in Soils, Rocks and in Uranium Deposits in Connection with Geotechnical Problems 137

Brigitte Boldt-Leppin, Paul Kozicki, Moir D. Haug, John Kozicki
    Use of Organophilic Clay to Control Seepage from Underground Gasoline Storage Tanks 141

J.R. Booker, C.J. Leo
    A Boundary Element Method for Analysis of Contaminant Transport in Heterogeneous Porous Media 147

R.G. Campanella, M.P. Davies, T.J. Boyd, J.L. Everard
    Geoenvironmental Subsurface Site Characterization Using In-Situ Soil Testing Methods 153

Wah-Tak Chan, Daniel C. Hsu
    Design and Construction of Soil-Bentonite Slurry Wall Helen Kramer Superfund Site . 161

R.G. Clark, J.A. Scarrow, R.W. Skinner
    Safety Considerations Specific to the Investigation of Landfills and Contaminated Land 167

Brian Cooke, Jason Fong
    Apparatus for Measurement of Molecular Diffusion Coefficients in Partially Saturated Soils 173

Henry S. Crawford, Kevin Parks
    Seeking Optimal Containment Design at Creosote-Contaminated Site Through Cost-Benefit Analysis 179

J. Crosnier, C. Delolme
    Le Controle de la Charge Bacteriologique des Lixiviats de Decharge par les Sols: Influence des Parametres Physico-Chimiques des Lixivats de Decharge sur la Migration Bacterienna a Travers un Sol Sature 185

Raghava Dasika, James Atwater
    A Review of Clogging Processes in Porous Media as Related to Wastewater
    Collection and Drainage 187

Greg H. Deaver, Carol S. Mowder
    Soil Vapor Extraction Below a Building 193

Greg H. Deaver, Robert J. Tworkowski
    Application of Air Sparging to 1,1,1-Trichloroethane Contaminated Soil -
    A Success Story 201

Maurice B. Dusseault, Roman A. Bilak, Leo Rothenburg
    Slurry Injection Disposal of Granular Solid Wastes 209

A. Esnault, F. Dufournet-Bourgeois
    Ecrans Etanches Destinés au Confinement de Sites Pollués:Durabilité
    des Matériaux Constitutifs 215

Marco Favaretti, Nicola Moraci, Paolo Previatello
    Effects of Leachate on the Hydraulic and Mechanical Behaviour of Clay Liners 221

Jean-Marie Fleureau, Said Taibi
    A New Apparatus for the Measurement of Water-Air Permeabilities 227

Mark Gemperline
    Surface Sampling to Detect Unacceptable Human Health Risk 233

Michio Gomyoh, Hideo Hanzawa, Tej Pradhan
    Investigations and Measurements Associated with Environmental Improvement
    of Inner Bay Areas by Covering Polluted Soft Mud with a Thin Sand Layer 241

V.I. Grebenets, A.B. Lolaev, D.B. Fedoseev, V.A. Savchenko
    Geotechnical Aspects of Environmental Violations in Cryolitic Zone 247

Paul G. Hansen, G. Robert Crotty
    Use of Geomembranes as Vertical Barrier Liners for Containment on the
    North Slope of Alaska 255

Musatapha Hellou, Mohamed El Yazidi
    Visualisation du déplacement de polluants liquides dans un milieu poreux 261

Satoshi Imamura, Masanori Shimomura, Toru Sueoka
    Migration Characteristics and Remediation Efficiency at Organic Compounds
    Contamination Sites of Volcanic Ash Soil Layer in Japan 267

Hilary I. Inyang
    A Weibull-based Reliability Analysis of Waste Containment Systems 273

Hank H. Jong, Nestor O. Acedera, Todd F. Battey, Roger R. Batra
    Application of Cone Penetration Test for a Background Sampling Program 279

Masashi Kamon, Takeshi Katsumi
    Potential Utilization of Waste Rock Powder 287

Raj P. Khera
    Calcium Bentonite and Fly Ash in Slurry Wall Construction Materials 293

Suresh R. Kikkeri, Edward P. Hagarty, Jimmy L. Wilcher, John Bowman
  Site Characterization and Remediation of Hydrocarbon Contaminated Area 299

Paul Kozicki
  Remediation of a Chemical Landfill at the University of Saskatchewan, Saskatoon, SK   305

G. Lefebvre, F. Burnotte, P. Rosenberg
  Contrôle Électrocinétique des Écoulements Souterrains 313

P.C. Lim, S.L. Barbour, D.G. Fredlund
  Laboratory Determination of Diffusion and Adsorption Coefficients of
  Inorganic Chemicals for Unsaturated Soil 319

P.L. Martin, G. Vaciago, N.R. Somers, H.T. Burbidge
  Additional Lagooning on Stage I of the Gale Common Ash Disposal Scheme:
  Investigation, Design and Preparatory Works 325

Godwin N. Nnad
  Contamination Assessment and Remediation Approaches in North America 331

Paul M. Przygocki, L. Alan Johnston, Walter Kosinski
  Containment System of an Industrial Wastewater Lagoon, including Surcharge
  Loading of Sediments for Cap Support 337

P.B. Rainbow, R.G. Clark, J.H. Donovan
  The Statistical Analysis of Leachate and Water Quality Monitoring Data
  at a Large Landfill Site 343

R. Kerry Rowe, M.J. Fraser
  Composite Liners as Barriers: Critical Considerations 349

Martin Sanchez, Alain Grovel
  Sedimentation sur un Fond Permeable 355

Kohei Sawa, Masashi Kamon, Seishi Tomohisa, Nagahide Naito
  Waste Oil Hardening Treatment by Industrial Waste Materials 361

Markus Sjöholm, Tapio Strandberg
  Using Glacial Till as Liner Material for a Waste Disposal 367

Mohd R. Taha, Yalcin B. Acar, Robert J. Gale, Mark E. Zappi
  Surfactant Enhanced Electrokinetic Remediation of NAPLs in Soils 373

S. Thevanayagam, J. Wang
  Flow Behavior During Electrokinetic Soil Decontamination 379

Matthias Vogler, Helmut Dörr, Rolf Katzenbach, Norbert Molitor
  Quality and Efficiency Control During Soil Air Sampling and Remediation 387

T. Winiarski, J. Cavalcante Rocha, G. Didier
  Effet de Trois Lixiviats de Decharges d'Ordures Menageres sur le Gonflement
  de Quatre Bentonites 393

Kazuya Yasuhara, Masayuki Hyodo, Kazutoshi Hirao, Sumio Horiuchi
  Liquefaction Characteristics of Coal Fly Ash as a Reclamation Material 410

M. Yudhbir, M. Kumar, S. Kumar, A. Kumar
   A Novel Concept of Design of an Ash Pond 409

MINING

Plenary Speaker

Richard R. Davidson
   Seepage Patterns in Tailings Dams: Implications for Operation and Remediation 415

Keynote Speaker

G.E. Blight
   Environmentally Acceptable Tailings Dams 417

Submitted Papers

M. Aubertin, R.P. Chapuis, M. Aachib, J.F. Ricard, L. Tremblay, B. Bussière
   Cover Technology for Acidic Tailings: Hydrogeological Properties of Milling Wastes used as Capillary Barrier 427

R.F. Azevedo, M.C.M. Alves, T.M.P. de Campos
   Numerical Analysis of the Sedimentation and Consolidation of a Neutralized Red Mud 433

S.E.T. Bullock, F.G. Bell
   Ground and Surface Water Pollution at a Tin Mine in Transvaal, South Africa 441

G. Calabresi, V. Pane, S. Rampello, O. Bianco
   Geotechnical Problems in Construction over a Thick Layer of a Mine Waste 449

B.C. Davidson, M.B. Dusseault, R.J. Demers
   Solution Cavern Disposal of Solvay Process Waste 455

Tácio M.P. de Campos, Maria Cristina M. Alves, Roberto F. Azevedo,
   Laboratory Settling and Consolidation of a Neutralized Red Mud 461

Euler De Souza
   Engineering and Environmental Considerations in Salt Tailings Management and Disposal Underground 467

W.A. Ericson, C. Winkler III, M.E. Plaskett
   Comparison of Fine Tailings Properties and Disposal Methods for the Alumina, Phosphate, and Oil Sands Industries 475

Martin Fahey, Yoshimasa Fujiyasu
   The Influence of Evaporation on the Consolidation Behaviour of Gold Tailings 481

A.B. Fourie, B. Wrench
   Use of Dynamic Compaction for the Rehabilitation of a Mining Backfill Area 487

Dieter D. Genske, Jean Thein
   Recycling Derelict Land 493

T. Germanov, V. Kostov
   Determination of Tailings Properties for Seismic Response Analysis of a Tailings Dam . 499

J. Hadjigeorgiou, R. Poulin, K. Aref, T. Boyd
    An Assessment of the Co-mingled Waste Disposal Method for ARD Control 505

Peter C. Lighthall, Harvey N. McLeod, Serena Domvile
    A Review of Mine Tailings Management Practises for Closure 513

Robert C. Lo, Edward R.F. Lord, Julian M. Coward
    Oil-Sands Fine Tailings Reclamation Involving Sand-Fines Mixtures 519

L.J. Potter, C. Savvidou, R.E. Gibson
    Consolidation and Pollutant Transport Associated with
    Slurried Mineral Waste Disposal 525

S. Rampello, V. Pane, G. Calabresi, O. Bianco
    Geotechnical Characterization of a Clayey Mine Waste 531

Krishna R. Reddy, Jeffrey C. Schuh
    Computer Modeling to Define the Extent of Groundwater Contamination at a
    Coal Refuse Disposal Facility 539

Yakov M. Reznik
    Water Resisting Barriers at Coal Mining Activity Sites 545

Pechka Stoeva, Paoulin Zlatanov
    Estimation de la Stabilite des Terrils Interieurs sur la Base de la Theorie du Risque .. 551

Nagula N. Suthaker, J.Don Scott
    Large Scale Consolidation Testing of Oil Sand Fine Tails 557

Gareth E. Swarbrick
    The Use of Small Scale Experiments to Predict Desiccation of Tailings 563

M.B. Szymanski, H.L. MacPhie
    Approach to Closure Designs for Mine Sites with Severe AMD 569

MUNICIPAL

Plenary Speaker

Hans L. Jessberger
    Remedial Works in a Highly Industrialized Region 577

Keynote Speaker

Robert M. Quigley
    Municipal Solid Waste Landfilling 589

Submitted Papers

A. Al-Tabbaa, S. Walsh
    Geotechnical Properties of a Clay Contaminated with an Organic Chemical 599

Olivier Artières, Philippe Delmas, Michel Correnoz, Odile Oberti
    Stability of Wastes on Lining Systems: Analysis and Design 605

C. Bernhard, J.P. Gourc, I. Le Tellier, Y. Matichard
   Compared Performances of Bottom and Slope Lining Systems of Municipal Landfills . 611

Brent A. Black, Abdul Shakoor
   A Geotechnical Investigation of Soil-Tire Mixtures for Engineering Applications 617

R.G. Clark
   Landfill Gas Protection Measures for Industrial Developments 625

C. Coulet, M. Boudissa, L. Curtil
   Remblais Alleges en Dechets de Matieres Plastiques 631

Francisco Casanova de O. e Castro, Mauricio Ehrlich, Maria Claudia Barbosa,
Marcio S.S. de Almeida
   Site Investigation for Rehabilitation of a Municipal Landfill 637

O. Del Greco, C. Oggeri
   Shear Resistance Tests on Solid Municipal Wastes 643

Eduardo C. do Val, Luiz Antoniutti Neto
   A New Piezometer for Sanitary Landfills 651

M. Ehrlich, M.S.S. Almeida, M.C. Barbosa
   Pollution Control of Gramacho Municipal Landfill 657

Richard B. Erickson, Echol E. Cook, Braja M. Das
   Aerobic Biological Clogging of Soil/Geotextile/Geonet Filter in
   Leachate Collection Systems 665

F. Federico
   Numerical Analysis of Effectiveness of Imperfect Underground Barriers 671

Angelo Lucio Garassino, Massimo Giambastiani, Bruno Salesi
   Design and Performances of a Clay Liner for Containment and Impermeabilization
   of an Aeration Waste Water Plant in Ghana 677

S.W. Hong, Y.S. Jang, H.I. Jeong
   A Study on the Characteristics of Marine Clay and Admixed Liners 683

Rosario Iturbe, Ana E. Silva
   Behavior of Bordo Poniente Landfill in Texcoco Zone, Near Mexico City 689

Y.S. Jang, K.Y. Lee, S.S. Kim, G.S. Lee
   Numerical Modelling of Contaminant Migration in two Municipal Landfill Sites 695

Mir Fazlul Karim, Mohammed Jamal Haider
   Impact of Geo-Engineering Mapping on Waste Management in
   Greater Dhaka City, Bangladesh 701

Edward Kavazanjian, Jr., Michael S. Snow, Neven Matasovic, Chaim J. Poran, Takenori Satoh
   Non-Intrusive Rayleigh Wave Investigations at Solid Waste Landfills 707

Paul Kozicki, Steven Harty, John P. Kozicki
   Design and Construction of Soil-Bentonite Liners and Two Case Histories 713

M. Miyake, M. Wada, A. Maruyama, H. Iwatani
  **Geotechnical Characteristics of Refuse Ash Dumped in Phoenix Project
  in the Amagasaki Offshore Site** 721

Vince O'Shaughnessy, Vinod K. Garga
  **Assessment of Hydrogeological and Contaminant Transport Properties of a
  Fractured Clay in Eastern Ontario** 729

Issa S. Oweis, Edward M. Zamiskie, M. Golam Kabir
  Evaluation of Soil Bentonite Slurry Wall by Piezocone 735

D.K. Papademetriou, G.A. Parisopoulos, G.T. Dounias
  Deep Subsurface Wastewater Injection in Ioannina, Greece 743

J. Perrin, J.P. Khizardjian, C. Coulet
  Reemploi de Pneumatiques Usages en Remblais Alleges 749

Tej B.S. Pradhan, Goro Imai, Masami Uchiyama
  Mechanical Properties of Stabilized Light Soil Using Expanded Polystyrene 755

M.H. Rahman
  Environmental Impact Assessment of Greater Dhaka City Flood Protection Structures 761

J.D. Rakotondramanitra, C. Coulet, R. Azzouz
  Etudes et Applications du Renforcements des Sols par le Procede Plasterre 767

Jouko Saarela
  Aftercare Research of Closed Sanitary Landfills 773

Hiroyuki Sakai, Osamu Murata, Hisashi Tarumi
  Sensitive and Simplified Methods for Monitoring of Chemical Species in Ground ... 781

Charles D. Shackelford, Chung-Kan Chang, Te-Fu Chiu
  The Capillary Barrier Effect in Unsaturated Flow Through Soil Barriers 789

A.J. Sillito, F.G. Bell
  **A Preliminary Survey of Leachate Development in Some Landfills
  in the Greater Durban Area** 795

Robert E. Smiley
  Replacement Wetland Emphasizes Biodiversity 801

K.C. Sohn, S. Lee
  A Method for Prediction of Long Term Settlement of Sanitary Landfill 807
H. Uehara, H. Hara
  Experimental Studies on Prevention of Reddish Soils Contamination of the Sea 813

Mark E. Unruh, Richard P. Kraft
  Siting a Municipal Solid Waste Landfill - A Case Study 819

NUCLEAR

Plenary Speakers

Carl P. Gertz
 Status of the Yucca Mountain Site Characterization Program 827

Thomas A. Shepherd, Louis L. Miller, Robert L. Medlock
 UMTRCA Regulations - Evolution of the Technical Basis 829

Keynote Speaker

Vern Rogers
 Present Trends in Nuclear Waste Disposal 837

Submitted Papers

E.E. Alonso, A. Gens, A. Lloret, C.H. Delahaye
 Analysis of a Compacted Clay Barrier in a Radioactive Waste Disposal Scheme 847

Maurice B. Dusseault, J. Carlos Santamarina, Christine Trentesaux,
Patrick Lebon, Eduardo E. Alonso
 Granular Halite Creep and Compaction for Backfilling 855

M. Gandais, F. Dufournet-Bourgeois, A. Esnault, J.F. Ouvry
 Scellement de Forage sur les Sites de Stockage de Déchets Radioactifs 861

Jozef Hulla, Július Plsko, Mohamed Taha
 Influence of Nuclear Power Stations on the Groundwater 867

Hideo Komine, Nobuhide Ogata
 Evaluation of Swelling Characteristics of Compacted Bentonite 873

B.I. Kulachkin, V.A. Ilyichev, V.S. Yamschikov, A.I. Radkevich, V.I. Sheinin,
L.R. Stavnitser, I.P. Korenkov, O.G. Polsky, V.F. Kirillov, V.M. Pankov, A.P. Van den Berg
 Ground Testings of Radon as a Source of Ecological Danger 879

Lyesse Laloui, Hormoz Modaressi
 Effets de la Thermo-Plasticite des Argiles sur le Comportement des Puits de Stockage . 881

H.K. Mittal, N. Holl, S. Donald
 Geoenvironmental Design of a Uranium Mill Tailings Facility in
 Northern Saskatchewan 887

S. Olivella, A. Gens, J. Carrera, E.E. Alonso
 A Model for Coupled Deformation and Non-isothermal Multiphase Flow
 Through Saline Media. Application to Borehole Seal Behaviour 895

J-C. Robinet, A. Rohbaoui, F. Plas
 A Thermo-Elastoplastic Model with Thermal Hardening for Saturated Clay Barriers . 901

J.C. Robinet, M. Rhattas, F. Plas, J.M. Palut
 Modélisation des Transferts de Masse dans les Argiles à Faible Porosité.
 Application au Stockage des Déchets Radioactifs 909

Clint Strachan, Claude Olenick
**Design and Construction of the Closure Plan for the Conquista Uranium
Mill Tailings Impoundment** 915

Phi Oanh Tran Duc, Hiroya Komada, Takao Endo, Michihiko Hironaga, Koichi Taniguchi
Strength and Deformation of Cement-Bentonite Underground Filling Materials 921

PERORATIONS

Plenary Speakers

Joseph D. Ben Dak
Perspectives of Sustainable Human Development in Environmental Geotechnics 927

W.F. Brumund
**Comparing the Practice of Environmental Geotechnics in Various
Regulatory Environments** 929

Andrew Lord
Redevelopment of Contaminated Sites in the UK 937

William J.C. Meynink
Tailings Disposal in an Equatorial Environment 949

James K. Mitchell
Physical Barriers for Waste Containment 951

N.R. Morgenstern
The Observational Method in Environmental Geotechnics 963

Andrew Steer
World Development Report on Environment and Development 977

Wendell D. Weart
Geotechnical Studies for the Waste Isolation Pilot Plant Repository 979

Appendix 1

Charles D. Shackelford
Report of Technical Committee on Environmental Control (TC5) 981

Appendix 2
Author Index 1007

Appendix 3
Country Index 1013

Dredging

Environmental Aspects of Sand Dredging and Mud Disposal in Hong Kong

E.W. Brand, J.B. Massey, P.G.D. Whiteside
Civil Engineering Department, Hong Kong Government

Abstract : Hong Kong has a massive ongoing programme of reclamations, for which more than 400Mm^3 of marine fill is required this decade. In addition, about 315Mm^3 of mud requires marine disposal, of which 26Mm^3 is highly contaminated. Centralized control of fill resources and related matters has resulted in a successful programme of exploration for marine sand, as well as satisfactory solutions to the problems of mud disposal. Environmental Assessments are undertaken separately for all reclamation, dredging and disposal sites, often with extensive longterm monitoring of environmental impacts. Surveys completed so far indicate that plumes of suspended sediment migrating away from sand dredging areas probably have less effect on marine ecology than do natural disturbances. Of particular note is the disposal of highly contaminated mud into specially dredged seabed pits, each of which is provided with an engineered cap composed of two metres of clean mud over a one-metre sand layer. Detailed analysis has shown that this will effectively provide longterm isolation of the contaminated material from the environment.

Introduction

The Territory of Hong Kong, with a land area of only 1,050 square kilometres and a population of less than six million, has been phenomenally successful economically by any standards. According to GATT statistics, it ranks as the world's tenth trading entity (fifth if the European Union is counted as one), although only 89th in terms of population. The GDP per capita of about US$18,500 is amongst the highest in the world. To ensure the longterm maintenance of economic prosperity, there is continuing high expenditure on infrastructure and other public works. Of particular note is the ongoing implementation of the huge infrastructure projects emanating from Hong Kong's Port & Airport Development Strategy (PADS), which together are estimated to cost about US$28 billion, a significant proportion of which will be privately financed. With much of the Territory's land being undevelopable because of its steep terrain, most major projects involve the formation of land by marine reclamation.

Hong Kong has been forming new land by reclamation for over a hundred years. The rate of land formation, however, has gradually increased, and this has meant a corresponding increase in the rate of use of fill material, as illustrated in Figure 1. From the last century to the 1960s, fill material was placed in reclamations at the rate of about 1Mm^3 per year, and from the late 1960s to the late 1980s, fill placement averaged 15Mm^3 per year. Now, with the extensive reclamations well underway for the new airport and related infrastructure development, fill is being placed at a rate of nearly 70Mm^3 per year. The fill supply and demand have been discussed by Brand & Whiteside (1990), Brand (1992) and Ooms et al (1994).

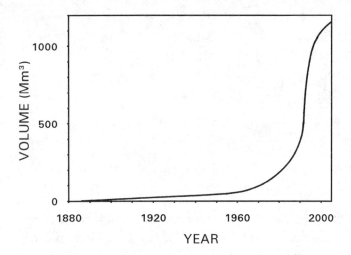

Figure 1 Cumulative fill usage in Hong Kong

The total demand for fill over the decade 1990 to 2000 is about 650Mm³, of which about 260Mm³ has already been used. The annual requirements are shown in Figure 2.

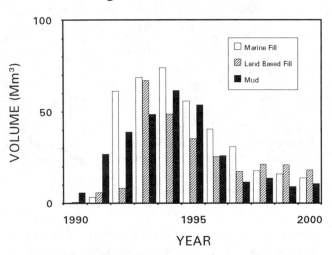

Figure 2 Annual requirements for fill and mud disposal in Hong Kong

Until a few years ago, most fill material was excavated from hillsides, whereas the present rate of reclamation can only be achieved by the use of marine-dredged sand won with high production dredgers. The use of marine rather than land borrow areas is also considerably more environmentally acceptable in densely populated Hong Kong, and the maximum possible reliance is therefore being placed on marine sources of fill.

A major portion of the 'demand' for land-based fill shown in Figure 2, totalling about 270Mm³, results from an inadequate supply of marine fill, and much of this will be changed to marine fill if additional marine sources can be located by the ongoing programme of exploration (see below).

A consequence of the historical development of large areas of reclamation in Hong Kong is that the latest reclamations are now far seaward of the original natural coastline, where the deposits of marine mud were non-existent or thin. In the areas now being reclaimed, the seabed is covered by marine mud that is commonly up to 10 metres thick, and the new Lantau Port reclamations are being designed for up to 20 metres of mud. For various reasons, mostly related to construction programme constraints, the mud has had to be completely removed from beneath a number of large reclamations in the recent past, including those for the new Chek Lap Kok airport and several container terminals. In addition, considerable quantities of mud need to be dredged to maintain existing shipping channels and to create new ones. As shown in Figure 2, a total of almost 315Mm³ of mud needs to be disposed of this decade. This major disposal problem (Hong Kong Government, 1993, 1994) is having to be solved within Hong Kong waters, which are less than 1,800 square kilometres in size. The disposal process is also complicated by the fact that much of the seabed mud in the urban harbour area is highly contaminated because of past industrial discharges at inshore outfalls, and this means that about 26Mm³ of the mud to be disposed of requires special disposal facilities.

In planning and undertaking the reclamations of the 1990s, Hong Kong has therefore had to deal with the related problems of locating sources of large quantities of suitable fill, of high production marine sand dredging operations, and of the marine disposal of large quantities of both clean and contaminated mud, all in a very limited sea area. The projects now underway are occupying almost 60% of the world's fleet of large trailer suction hopper dredgers. The extent of the dredging and reclamation activity in Hong Kong can be judged from Figure 3. The great importance of fill to Hong Kong's programme of infrastructure development has necessitated central control of this strategic resource and related issues, and the Hong Kong Government established the Fill Management Committee (FMC) in 1989 for this purpose.

Environmental Assessment Process

In common with a growing number of countries, Hong Kong has a formalised procedure under which major Government projects are subjected to a progressive sequence of environmental assessment (EA) at the planning stage. Most reclamations are constructed by the Government, but some major ones (e.g. container terminals) are planned by Government prior to being implemented by the private sector. Although EAs are not yet a statutory requirement, all major private projects are subject to statutory approvals of one kind or another, and this ensures that EA procedures are in fact implemented.

The EA process in Hong Kong commences with an Environmental Review, normally undertaken by the Government's Environmental Protection Department (EPD), which provides a brief overview of the proposed works and identifies any areas of potential adverse impact. If these potential impacts are perceived as being sufficiently severe, a detailed Environmental Impact Assessment (EIA) is undertaken, with attention being focused on the key issues. Separate EIAs are undertaken for works at the reclamation site (not discussed further in this paper), at the marine borrow areas, and at the marine mud disposal sites. Finalized EIA reports are now routinely placed in the public domain. In addition to the EA process, the laws of Hong Kong require that, before any works can be commenced that affect the foreshore or seabed, there has to be public notification of the proposal (through publication in the Government Gazette), to enable any interested or affected parties to lodge an objection or to claim compensation. This process protects the interests of directly affected parties such as fishermen, cargo handlers and waterfront operators; it also enables interested groups to lodge objections on environmental grounds.

A concern that has figured prominently in the Hong Kong press is the effect that sand dredging operations are having on the fishing industry, including both capture fisheries and mariculture farms. The assessment of the commercial disbenefits resulting from dredging has been particularly difficult. The situation with regard to the capture fisheries is complicated by the fact that the pattern of fishing in Hong Kong's territorial waters has been changing. More use is now being made of larger better-equipped offshore vessels, and less use is being made of small boats that are restricted to the near-shore area. Furthermore, past overfishing has probably affected near-shore catches, and recent reduced catches cannot therefore be attributed directly to dredging operations. The mariculture situation is also complex because of the tendency in Hong Kong to use fish farms not only as breeding and raising grounds for fish, but as holding areas for imported semi-mature fish from areas of warmer clearer water elsewhere in Southeast Asia. In areas set aside for sand dredging, the Government at present pays ex-gratia allowances to capture fishermen to cover a fixed period of disturbance. For the fish farmers affected by dredging, there is a package of compensation options which, in part, is related to measured increases in suspended sediment levels.

Sand Dredging

The locations of the main marine sand resources in Hong Kong are shown in Figure 3. Most of these have been discovered since 1987, when the Geotechnical Engineering Office of the

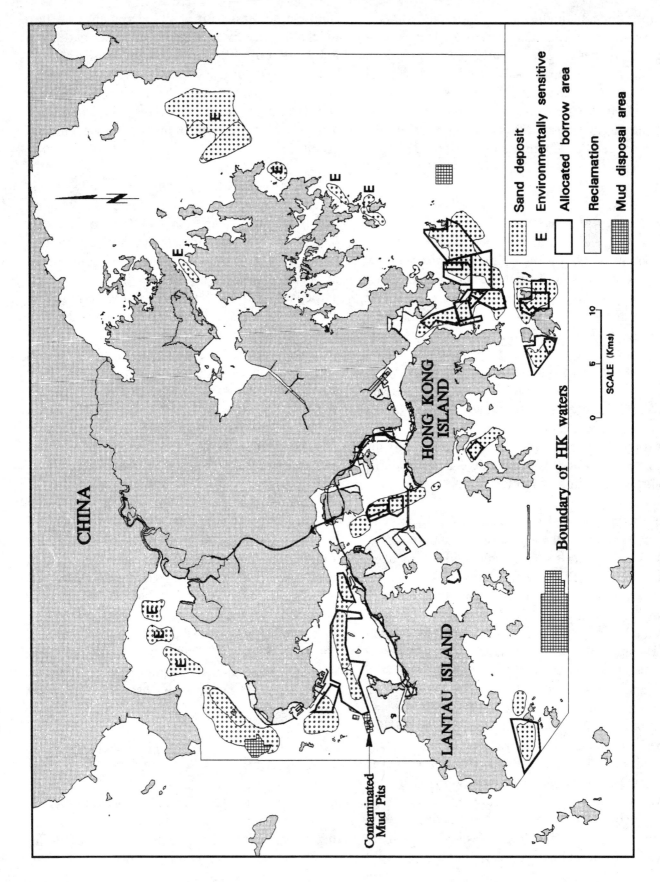

Figure 3 Current reclamations, marine borrow areas, and mud disposal sites in Hong Kong

Government's Civil Engineering Department commenced its programme of offshore exploration for sand. Details of the early stages of this work are contained in the publication edited by Whiteside & Wragge-Morley (1988), and the more recent stages have been described by Whiteside & Massey (1992).

Hong Kong's drowned landscape, with its typical ria coastline, is the result of continued submergence, most recently by the postglacial sealevel rise of Holocene times. The offshore sand resources are largely alluvial deposits laid down in the systematic but complex network of now-submerged river channels. In some areas, the early Holocene seas reworked these river sands into low asymmetrical banks that migrated away from the original river courses. As the sealevel continued to rise, a swath of marine mud was deposited on top of the alluvial deposits, including the sand. The thickness of the mud cover varies, being thinnest in constricted areas where the tidal currents are swiftest. The economic sand therefore tends to be along the pre-Holocene drainage channels, in places where modern tidal currents have restricted the deposition of mud.

The Government's exploration programme has so far identified $750 Mm^3$ of sand in Hong Kong waters, but the exploitation of some $420 Mm^3$ of this has been found from Environmental Assessments to be unacceptable. These environmentally sensitive marine borrow areas are indicated in Figure 3. It became clear at the Environmental Review stage that the dredging of some sand resources in the northeast and northwest of Hong Kong could result in significant pollution if conventional means of dredging were used, because the sands were covered by polluted overburden that would first have to be removed. At this time too, it was judged that a series of small sand bodies close to the eastern coast of Hong Kong, where there are popular unspoilt beaches, should also be left unworked.

The large sand resource (about $230 Mm^3$) in Mirs Bay to the northeast of Hong Kong (Figure 3) was initially judged to have potential for sand winning even though there were sensitive receivers nearby. A detailed geological investigation was undertaken to delineate and characterise the sand, and then a full EIA was undertaken. The EIA concentrated on the effects of suspended sediment on the marine ecology, especially the coral habitats, which are located at a popular diving venue. On the basis of the results of the EIA, the Fill Management Committee concluded that the impacts of the Mirs Bay dredging would not be acceptable, and the proposal to dredge in this area has therefore never been implemented.

The likely environmental impacts of dredging the other sand resources shown in Figure 3 have also been fully considered, and in each case the assessed impacts have been found to be acceptable. The detailed EIA undertaken for the potential borrow area (about $60 Mm^3$) in the East Lamma Channel, which is located within a few kilometres of some of Hong Kong's most popular bathing beaches, showed that dredging in this area would be marginally acceptable. Sand dredging in this area will therefore only proceed on the basis of a pilot exercise at the northern end to verify the accuracy of the predictions made by the hydraulic and plume modelling carried out during the EIA.

As part of the EA studies for each potential marine borrow area, the likely impacts of suspended sediment plumes within the dredging area are examined, but no examination is made of the potential impacts at distant locations caused by the cumulative effects from a number of different dredging operations. On the basis of experience from elsewhere in the world, it has been assumed in Hong Kong that, while the effects of overflowing during dredging can be very significant in the immediate vicinity of the borrow area, these effects rapidly diminish away from the vicinity of the dredging. Nevertheless, in view of the great concentration of dredging operations in Hong Kong, surveys are being undertaken throughout local waters to verify this assumption by measuring the effects on both suspended sediment levels and on the accumulated deposited sediment on the seabed.

The surveys of suspended sediment employ a number of techniques, data from which are being collated to provide an overall picture for the Territory. Satellite imagery (SPOT), and high-level fixed-wing and lower-level helicopter colour photography provide a framework for assessing the results of more localised direct measurements. These direct measurements are made by water sampling, turbidity meters and Acoustic Doppler Current Profilers (ADCP) used in backscatter mode. Although the integration of these different techniques has not yet been completed, the general pattern emerging confirms that sediment plumes from sand dredging operations do decay rapidly with distance, mostly within the confines of the dredging area itself, and that only rarely do visible remnants persist much beyond a kilometre from the dredger. The surveys have shown that, although relatively low sediment levels of 10 to 20mg per litre are detectable visually, these are not damaging to the environment. Most of the sand dredged so far has been a kilometre or more away from land. In one area, however, the course of the pre-Holocene river channel is within a few hundred metres of the present coastline, and in winning this sand, sufficient suspended sediment reached the coast to smother some of the hard corals and a smaller proportion of the soft corals.

Surveys have also been undertaken to establish the nature of the seabed ecology and the effect that dredging-related sedimentation is having on this. The techniques used are traditional grab sampling and the more innovative REMOTS seabed camera system (Rhoads & Germano, 1982). The seabed in Hong Kong is fairly dynamic, being affected by frequent high inputs of sediment and energy related to the Pearl River, typhoons and localised high tidal currents. Although the surveys are not yet complete, it is already apparent that the effects of dredging on the health of seabed life away from the dredging area are significantly less than the effects of the natural variations.

As Hong Kong's more centrally-placed near-shore sand deposits are becoming worked out, sand dredging is having to move further offshore, albeit only a few kilometres. This means that detectable plumes and their environmental impacts are tending to decrease in significance.

Disposal of Uncontaminated Mud

Hong Kong has for many years disposed of unwanted excavated inert material at one of three marine disposal sites, but only the two shown in Figure 3 are now used. The disposal area to the south of Hong Kong was greatly increased in size in 1991 to accommodate the large volumes of material needing disposal in the 1990s. It was at the time of this extension that environmental considerations of marine dumping sites became of greater concern than previously.

Surveys made at the disposal sites by means of Acoustic Doppler Current Profilers and turbidity meters have indicated that, during and after normal dumping operations by either trailer dredgers or barges carrying grab-dredged material, very little sediment is released into suspension. Indeed, there is almost no sea surface expression of dumping. Where the effects are significant, however, is in the near-bed zone, where dense suspensions and fluid mud gradually move down slope. The grab-dredged mud, being close to its insitu density, tends to remain in gentle mounds, with slopes of about 1 in 20 which, by wave attrition and current scour, slowly reduce in height. The trailer-dredged mud, being essentially a lumpy slurry, is inherently less stable, commonly lying at slopes of 1 in 100. In times of severe typhoons, wave-induced instability results in considerable lateral spreading of this weaker material.

The REMOTS seabed camera has also been used at the uncontaminated mud disposal sites, where surveys have so far revealed that, even close to active dumping, the seabed is quickly recolonised by the organisms typical of early trophic levels.

Disposal of Contaminated Mud

Many decades of uncontrolled disposal of industrially-polluted wastes has resulted in much of Hong Kong's central harbour seabed being contaminated with toxins, particularly metallic radicals. Large amounts of copper, chromium and other metals associated with industrial processes such as electro-plating have accumulated in the upper layers of marine mud. Together with organic pollutants (PCB, PAH, antifouling TBT, etc) and large volumes of raw sewage, these toxins have been mixed deeper into the seabed by ships' anchors, resulting in a layer of highly contaminated mud, commonly three or more metres thick over much of the areas now being reclaimed.

In 1991, a study was completed of the nature and extent of the mud contamination in Hong Kong waters. At a very early stage in the study, it was apparent that some form of marine disposal was going to be the only practical solution for coping with the large annual volumes of contaminated mud being generated by reclamation works. The study concluded that, if the highly contaminated mud could not be left in place beneath new reclamations, then the best and most practical disposal solution was to place it in seabed pits to be capped with inert material. Having considered practices and standards adopted elsewhere in the world, the Environmental Protection Department of the Hong Kong Government specified criteria to differentiate between mud that could be safely disposed of at open marine disposal sites and mud that required special disposal facilities to isolate contaminants from the marine environment. Table 1 lists these criteria, and Figure 4 shows the estimated annual volumes of contaminated material that require disposal. It should be noted that, while the contamination is assessed in terms of metals only, the distribution of most other pollutants is very similar to that of the metallic pollutants. For this reason, the relatively simpler total metal laboratory testing provides a reliable guide to overall contamination.

Table 1 EPD's contamination criteria for marine mud

Metal	Cd	Cr	Cu	Hg	Ni	Pb	Zn
ppm dry wt	>1.5	>80	>65	>1.0	>40	>75	>200

Having decided to dispose of the contaminated mud in seabed pits, the Government undertook an experimental disposal operation using uncontaminated mud to determine whether empty seabed pits that had been used for sand extraction could be used for contaminated mud.

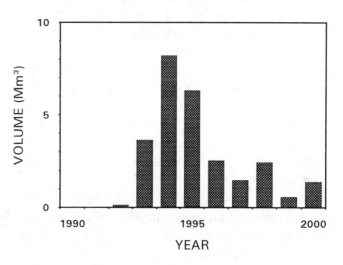

Figure 4 Annual volumes of contaminated mud for disposal

Unfortunately, the location proved unsuitable because of unacceptable levels of mud loss during placing. The relatively deep water (about 20 metres) and high tidal currents combined to give losses of up to 10%. Simple bottom-dumping and pipe-discharge methods both resulted in similar losses of material. It was therefore clear that a shallow-water site with relatively slow tidal currents would be required. In consequence, an area of seabed with a water depth of only five to six metres was selected, where hydraulic modelling had indicated currents generally below 0.5 metres per second. At this location, shown in Figure 3, the first Contaminated Mud Pit (CMP) was dredged in 1992.

In the area of seabed now designated for contaminated mud disposal, there are currently five CMPs, and plans are in place for three more. These pits are expected to accommodate contaminated material up to the end of the decade, by which time the main disposal problem will have passed. The design of the CMP facilities has been under constant review since disposal commenced in late 1992, but the overall scheme remains essentially as initially planned. Figure 5 shows schematically the design of a typical CMP. The pits are dredged as deep as can be readily achieved; in practice, this means to the base of the soft Holocene (post-glacial) marine mud deposits, commonly about 15 metres below the seabed.

As part of the initial design process, induced bed shear stresses were calculated for return periods of 10 per year, one per year, one per 10 years, and one per 100 years, and these were used to assess the potential for remobilisation of mud at various depths within the pits. A range of mud properties was adopted to cover the different states of the spoil material deposited in the CMPs. Grab-dredged mud commonly comprises more than 90% by mass of material that is close to its insitu density ($1.3t/m^3$ and over), together with less than 10% of fluid mud with a density of usually less than $1.1t/m^3$. Field investigations of mud deposited in a pit helped to confirm these parameters. The fluid portion of the mud is most critical to the potential for remobilisation by storm waves; the higher the filling level of the CMP, the greater the potential for remobilisation. The analysis indicated that there was no critical level at which the remobilisation risk increases rapidly and so, with due regard for the need to allow sufficient space for a capping layer, a maximum filling level of three metres below seabed was selected. This level, which is approximately nine metres below the sea surface, corresponds to the level at which there is little or no resuspension of fluid mud for a storm wave with a one-year return period. For full details of the analytical treatment, reference should be made to the paper by Premchitt et al (1993).

Sampling and testing of the Hong Kong harbour muds showed that the contaminant build-ups were heavily concentrated around drainage outlets, indicating that the metals had become attached to sediments close to the exit points into the sea. This, together with the fact that there was minimal change in salinity and little opportunity for oxidation during the dredging and disposal process, led to the expectation that there was unlikely to be significant release of metallic radicals into the pore water during consolidation of the mud in the CMPs. In designing the CMP cap, therefore, the main consideration was simply to ensure that natural longterm scouring would not expose the contaminated material.

Geological evidence from seismic boomer surveys suggested that the deepest (presumably storm-related) scour in the area extended about one metre below the seabed. In view of this, and with due consideration for other cap studies (Brannon et al, 1985; Gunnison et al, 1987), a two-metre thick mud cap was chosen. In order to facilitate placement of the mud cap onto the semi-fluid contaminated mud, a one metre thick layer of sand is first sprinkled onto the surface of the contaminated mud; this sinks differentially into the surface, and so thickens and strengthens the mud surface prior to placement of the clean capping mud. The full cap design, therefore, comprises a one-metre thick sand layer followed by two metres or more of uncontaminated mud to reinstate the seabed to its original level (Figure 5).

Each CMP is operated by the Government as a multi-user facility for several contractors. To ensure that the contractors comply with the CMP's strict conditions of use, and to ensure that each pit is filled evenly, a 24-hour on-site management team is deployed to position barges, to record users, etc. The contaminated mud is usually deposited into the pit from bottom-opening barges or dredgers. At present, a silt curtain is deployed on an experimental basis to help restrict any plume of suspended sediment around individual dumping barges, and the effectiveness of this is being carefully monitored.

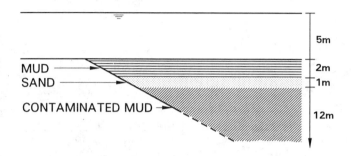

Figure 5 Schematic section of contaminated mud pit

Because of the novel nature of Hong Kong's disposal arrangements for contaminated material, a comprehensive programme of water, sediment and biota sampling is carried out independently of the actual disposal operations. The results of this monitoring programme are fed back continuously to the Environmental Protection Department, which is the regulatory authority for all disposal facilities. Interpretation of the measurements takes account of the fact that the CMP area is affected by large diurnal and seasonal fluctuations in the natural suspended sediment load coming from the Pearl River. To date, however, there has been no measured trend towards degradation of the marine environment. The public, who had expressed strong concern during the planning and implementation of the CMP scheme, are regularly informed of the progress of the work and of the results of the environmental monitoring.

Conclusions

Hong Kong's extensive amount of reclamation being undertaken at an unprecedented rate in such a small area of water has brought into focus a range of environmental problems related to sand dredging and mud disposal. Particularly noteworthy is the solution adopted for the marine disposal of the highly contaminated mud. Although the naturally dynamic marine environment around Hong Kong complicates studies of suspended sediment and trace pollutants, the ongoing intensive programme of measurements, using a wide variety of direct and indirect techniques, has so far indicated that the environmental impacts are within acceptable limits.

References

Brand, E.W. (1992). Fill Management for the Airport Core Programme. Hong Kong Engineer, vol. 20, no.8, pp 19-21.

Brand, E.W. & Whiteside, P.G.D. (1990). Hong Kong's fill resources for the 1990s. The Hong Kong Quarrying Industry, 1990-2000, edited by P. Fowler & Q.G. Earle, pp 101-112. Institute of Quarrying, Hong Kong.

Brannon, J.M., Hoeppel, R.E., Sturgis, T.C. Smith, I. & Gunnison, D. (1985). Effectiveness of Capping in Isolating Contaminated Dredged Material from Biota and Overlying Water. Technical Report D-85-10, US Army Waterways Experiment Station, Vicksburg, Mississippi.

Gunnison, D., Brannon, J.M. Sturgis, T.C. & Smith, I. (1987). Development of a Simplified Column Test for Evaluation of Thickness of Capping Material Required to Isolate Contaminated Dredged Material. Miscellaneous Paper D-87-2, US Army Waterways Experiment Station, Vicksburg, Mississippi.

Hong Kong Government (1993). Disposal of dredged materials. Report to the 16th

Consultative Meeting of the London Convention on the Prevention of Marine Pollution by Dumping of Wastes, International Maritime Organization, London, paper no. LC 16/Inf, 5 p.

Hong Kong Government (1994). Marine disposal of dredged material in Hong Kong. Report to the 17th Scientific Group Meeting of the London Convention on the Prevention of Marine Pollution by Dumping of Wastes, International Maritime Organization, London, in press.

Ooms, K., Woods, N. & Whiteside, P. (1994). Marine sand dredging : key to the development of Hong Kong. Terra et Aqua (Netherlands), no. 54, in press.

Premchitt, J., Rodger, J.G., Evans, N.C., Ip, K.L. & Dearnley, M.P. (1993). Geotechnical aspects of contaminated mud dredging and disposal in Hong Kong. Proceedings of the Seminar on Geotechnics and the Environment, Hong Kong, pp 15-24.

Rhoads, D.C. and Germano, J.D. (1982). Characterisation of benthic processes using sediment profile imaging : An efficient method of remote ecological monitoring of the seafloor (REMOTES TM System). Marine Ecology Progress Series, vol. 8, pp 115-128.

Whiteside, P.G.D. & Massey, J.B. (1992). Strategy for exploration of Hong Kong's offshore sand resources. Proceedings of the International Conference on the Pearl River Estuary in the Surrounding Area of Macao, Macao, vol. 1, pp 273-281.

Whiteside, P.G.D. & Wragge-Morley, N. (Editors) (1988). Marine Sand and Gravel Resources of Hong Kong. Geological Society of Hong Kong, 221 p.

Reclamation of Dredged Materials in Europe

M.A. Viergever, F.A. Westrate, M. Loxham
Delft Geotechnics, The Netherlands

Abstract

Due to the sedimentation in delta areas of suspended minerals from the hinterland, vaste amounts of solids accumulate in rivers and harbours situated in these delta regions. As rivers and harbours offer transport and commercial facilities of economical importance, these should be kept at depth by regular dredging operations, where the dredged materials have to be disposed. Economical application of the spoil at this very moment is restricted due to its unsuitable geotechnical properties, for example its very high water content, and its contamination levels. As most of the dredged spoil is polluted by a wide diversity of contaminants, due to industrial, agricultural and other human activities, the presence in situ and the disposal of dredged materials will cause a serious thread for human health and the ecosystem. As long as sufficient countermeasures at the contamination sources will lack, the presence and disposal of dredged spoil will be an environmental problem for several decades if no other measures are taken.
For this in several European countries programs have started to overcome this problem, for which two main strategies are available. The first one is to upgrade the geotechnical and environmental properties to levels allowing it's use, for example by cleaning or by regrading. Secondly, the end uses of the material can be so adjusted or engineered that modification of the spoil is not necessary to attain a reasonable economic use value from it. Landfill with pollution control, based on the fail safe principle, and direct use as wetland material support are examples here.
These programs deal with subjects as the chemical behaviour of the contaminants present in the sludge, in order to get insight into the transport phenomena of the pollutants as they are disposed in specific sites, into the interaction with engineered barrier systems and into potential clean-up processes. Besides this attention is paid to sampling- and in-situ measuringtechniques establishing the character and extent of contamination and the geotechnical properties, as for monitoring purposes. As the geotechnical properties of the material are of importance, wether for reusing or for disposal, and to restrict the volumes of process water, that will be contaminated during dredging, dredging-, dewatering-, consolidation- and separationtechniques are studied as well.
In this paper an overview will be given on these subjects based on some case studies for the European situation.

Present state

In Europe, due to the sedimentation in delta areas of suspended minerals from the hinterland, vaste amounts of solids accumulate in rivers and harbours situated in these delta regions. In the Netherlands more then 45 million cubic meters have to be removed annualy. For Europe these amounts are 10 to 20 times as much, from which 25 to 30 % is contaminated, however only smaller quantities of it are severely polluted.

The contamination changes with time. Due to regulations and legislation certain types of contamination are no longer brought into the system, unless in upstream deposited sediments the contamination is still present and will be released for many years ahead.

Certain types of contamination may be used as tracer, and with knowledge of deposition patterns areas with high concentrations of contaminants can be located. Deposition patterns depend on currents and geographical lay out. Only with adequate investigation can a good plan for removal of the contaminated soil be made. An example of the outcome of site investigations is the pilot project near Elburg in The Netherlands.

The estimation of the amount of sludge to be removed was 15.10^3 cubic meters, based on the first in situ investigations. After more thorough investigation and additional lab tests this amount appeared to be 40.10^3 cubic meters. The costs of the final investigations were only 30% of the costs of the first investigations.
It shows that it is important to choose the level of investigation that makes it possible to judge the type of contamination and its extent.

Site investigation

The techniques used in site investigations are evolving. Undisturbed or anearobic sampling of sludges is difficult. Samplers are available now in which in situ back pressure can be applied when the sample is taken and lifted to the surface. Specially, evaporation of gas from the sample is prohibited in this way, lit [a]. Other probes are available or still in development. Table 1 indicates the present state of development. Measurement in sludges is in most cases not yet available, but may be operational within a few years.

TABLE 1 -- Present and future available in situ techniques

type of contamination	present testmethod

chlorated hydrocarbons
polycyclic hydrocarbons
fossil fuels
macro chemical parameters:
    ph
    redox potential
    electrical conductivity
    Ka, Ca, Na, Cl, F, P, NO_3
gas in non saturated soils
	chemical analysis on site
	hydrocarbon probe lit [b]
	hydrocarbon probe lit [b]
	chemical sounding probe,
	measuring pore water at in-
	termediate depth lit [c]
	
	gas probe lit [d]

With regression analysis expensive and difficult measurements can be minimised by using correlations. Local circumstances and sources of contamination must be known and a correlation can be possible, lit [e].

Another method to get more information about the areal spreading of contaminants is the use of acoustic methods as an additional tool. Based on local borings with analysed samples the measuring of changes in concentration of contaminants is possible, depending on water depth and gas content of the sludge.

Clean up strategies

Measures taken at this moment can be classified as:
- no measures, zero option;
- disposal in depots;
- treatment.

The present state of the art of these four categories will be given in this article.

Zero option

Taking no measures has two different options. The first one is really doing nothing and keep the sludge in place. This solution is only possible when the location is not subjected to navigational or discharge conditions. Secondly this solution depends on the exchange of the contamination with the surface water or the groundwater underneath the sludge. This solution is preferred for sludges with only minor contaminants or contaminants with a degradation time shorter then the time to leach out.

The second option of taking no measures is dredging and dumping the sludge without restrictions. With the present legislation this method is only suitable for uncontaminated sludges or sludges with minor contaminations. An example of this type of treatment is the dumping of large quantities of sludge from the Rotterdam harbour into the North Sea. Dumping is only allowed when no other suitable disposal option is available and when it can be shown that no harm is done to the marine environment. This means that the amount and concentration of materials that could effect the ecological environment are closely restricted. The total amount of contamination may not exceed the load in the year 1988.

The allowable concentration is based on the effects on the ecosystem at several target areas. The allowable effect on each key bio species depends on its present situation.

TABLE 2 -- European depots

Depot	Size [m^3]	Depth [m below MSL]	Height [m above MSL]	Degree of contamination
Slufter Rotterdam	$90 * 10^6$	28	22	medium
Papegaaiebek Rotterdam	$90 * 10^4$	- 4,50	+ 8,80	heavy
Metha Bremen	$12 * 10^6$	- 5	25	medium
Zelzate Antwerpen	$12 * 10^5$	18	3	heavy

For a species which has already strongly declined in relation with the reference year 1930, the maximum allowable decrease should be lower than for a species which is nowadays at an acceptable level, lit [f].

Disposal in depots

Present solutions are found in design and construction of large scale depots. The material is concentrated in areas where it can be stored and separated from the environment and the effects on surrounding areas can be easily monitored. Table 2 shows the contents and the dimensions of some large European depots.

Sludges contain large quantities of sand. Contamination is mainly attached to the clay and silt particles. The amount of sludge to be stored can be minimised by separating sand and silt or clay particles.

The city of Rotterdam, lit [g], conducted experiments with sedimentation basins. Sandy particles settle near the outlet point of the transport pipe, while the clayey and silty sediments settle near the end of the basin. The economic break even point of the extra sedimentation before dumping in the depot depends largely on the amount of sand in the sludge. In this case the break even point was reached at 50 % particles < 63 μm.

By separating sand from the silt and clay the total amount of contamination will be concentrated in a smaller volume. The total load of contaminants inside the depot will increase in this way. Secondly the remaining sludge contains more clay and silt particles. This lowers the permeability and thus the rate of consolidation will be much lower. However in this way the storage capacity increases and the depot can be used over a longer period. As in this way the total contaminant load increases, the contaminant emission into the surroundings may take a longer time, but due to the decreased permeability the emission fluxes may decrease, resulting in a lower concentration in the surrounding, surface- and/or groundwater. A thourough study of this phenomena is needed to predict the net outcome.

In lit [h] the results of comparitive calculations with respect to consolidation of sludges were published. The consolidation rate is calculated for the disposal side of Rotterdam Slufter. The rate of consolidation is of importance to predict the additional storage capacity due to volume reduction during the filling period. Also important is the amount of expelled water to the top and the bottom of the disposal site. This water will contain contaminants and must be managed. In the long term the behaviour of the site is important in the planning for re-use.

Dredged sludge with a water content of 90 % will consolidate by its own weight or by overburden to a material with a water content in the order of 50 % and large strains are involved in this process. Furthermore the mechanical properties of the material will change considerably.

Two different mathematical models have been used to calculate the present condition. The first computer model SlIB, developped by the municipality of Rotterdam, is based on the Terzaghi model with a non-linear stress strain relationship and variable soil properties. It was validated with large strain self weight column tests. The second model, FSCONSBAG, developped by Delft Geotechnics for the Dutch Ministry of Transport and Waterways, is based on the Finite Strain Theory as formulated by Gibson in 1981. This model was also validated with lab-tests.

The Slufter is at present filled with 20 m of sludge, with an average sedimentation density of 1.140 kg/dm^3. The results of the computer models differ from the measured values. The models predict less subsidence. From the results it is clear that it is very difficult to measure accurately the density in situ and the top surface of the sludge under water. Furthermore it became clear that measuring soil parameters at very low stress levels in this first part of the consolidation process requires more attention. Finally, the effect of gas in the sludge on the consolidation process requiers more study.

In the Metha project at Bremen a two stage cyclone is used to separate the sand from the silt. The silt is compacted by using anionic and cationic flocculants combined with lamelar thickeners and belt presses. After that the compacted slurry is transported and dumped in the depot. The sand is partly used for layering between the compacted sludge to provide a path for the consolidation water, lit [i].

Much attention has been paid to the aspects of leaching, lit [j], and advection and diffusion of the leachate. The fluxes of leaching contaminants, due to advection are determined by the flow rate and the concentration of dissolved contaminants. The diffusion rate is determined by the difference in concentration inside and outside the sludge and the resistance against transport (diffusion coefficient, tortuosity and adsorption processes at the interface).

Restriction of flow rate can be achieved by lowering the hydraulic gradients. This is a matter of design. Depots below mean sea level or below the groundwater table restrict the hydraulic gradient whereas depots above mean sea level or above the groundwater table will cause additional dissipation of contaminated water. The water table inside the depot may even be lowered to a level lower then sea level and/or groundwater level to create a flow into the disposal site rather than to the environment. This measure must be taken as long as the depot exists and its effect must be guaranteed to prevent leakage in the long term. This is an example of design using fail safe principles.

When the emission contribution due to advection is strongly reduced or even reduced to zero, the emission of contaminants into the surroundings is only determined by diffusion processes. To decrease the diffusion rate, the advective flow direction can be reversed or measures can be taken to reduce the effective diffusion coefficient. The effective diffusion coefficient is the molecular diffusion coefficient in water for a species, corrected for the tortuosity of the material to diffuse through and for the porosity. The tortuosity is a measure for the pathlength through the material. Tortuosity and porosity can be related to each other leading to an expression for the effective diffusion coefficient $D_{eff}=D_{mol}*n^2$, where n is the porosity of the related material lit, [m].

D_{mol} = molecular diffusion coefficient [m^2/s]
D_{eff} = effective diffusion coefficient [m^2/s]
n = porosity [-]

Containment systems are often used to isolate the contaminants from the surroundings. Barriers may be rigid or flexible, and may have a heigh or a low barrier capacity. Rigid barriers, such as concrete slabs, will have cracks at certain points, but also flexible barriers could have gaps. Around these cracks or gaps the contamination will be leached out and the difference in concentration will be spreaded over a growing area around the crack or gap. With decreasing difference in concentration the diffusion will also decrease. The leaching will be restricted to a local exhausting of the disposal side.

Continuous barriers of high integrity resulting in a reduced effective diffusion coefficient, due to its strongly reduced porosity will reduce the diffusion considerably. In that case the concentration of contaminants on the deposit side of the barrier remains high and the emission by diffusion will continue for a long time but with strongly reduced emission fluxes. The effect of small or large diffusion coefficients and thus the use of a liner strongly depends on the allowable emission fluxes or developing concentrations outside of the site.

As liner the following materials could be used:
- inert materials such as concrete, glass, shale;

- self healing materials [1];
- materials that adsorb contaminants;
  These type of materials will actually only retard the diffusion process for a period until the liner is loaded to a certain concentration lit [1].
- plastic liners;
  The common plastics are apolar plastics with a very low porosity resulting in a strongly reduced effective diffusion coefficient. These plastics may also adsorb apolar (organic) contaminants, and may thus retard the emission by diffusion as well. A combination of polar and apolar plastic liners will reduce and retard the diffusion, for organic contaminants as well as for heavy metals. In table 3 the cost indications are given fora dredged sludge deposit, designed as an island inside a lake, with different options:

TABLE 3 -- Cost indications for a dredged sludge depot

	advection or diffusion	costs
no measures	100 %	
storing in depot	5 %	100 %
no hydraulic gradients	1 %	110 %
additional sealings	0,1 %	300 %

In the design and selection of sites for disposal of contaminated dredged materials usually the short term impacts on the environment are considered. However due to loss of institutional control, that might be expected in the long term, the long term behaviour is probably of more importance and attention should be paid to this. It means that release processes, like erosion, physical intrusion or removal of material and direct uptake into the ecosystem, all processes that can be controlled as long as institutional control is present, have to be considered, with respect to the selection and design of permanent disposal sites, lit.[k].

Treatment

At this moment there are three major groups of clean-up techniques available for contaminated soils and which have found some application in the treatment of dredging spoil.

1. Chemical treatment
2. Thermal treatment
3. Biological treatment

ad 1. The chemical clean-up techniques can be divided in two main techniques, the oxidation of organic contaminants and the extraction of inorganic and/or organic contaminants.
The oxidation of organic contaminants can be realised by mixing the contaminated soil material with adequate oxidising agents, like peroxides. If however

as well as organic, as inorganic contaminants are present this method only may not be sufficiënt and the oxidising technique should be followed by extraction techniques.

The extraction technique consists of three major treatment steps:
a. Mixing of the contaminated soil with the extracting agent
b. Separation of the extracting agent and the clean soil particles
c. Treatment of the extracting agent.

Within this technique two removal mechanisms can be distinguished, one is aimed at the dissolution of contaminants into the extracting agent, the other is aimed at the dispersion of polluted particles into the extracting agent.

The oxidation and extraction techniques, can be used for clean-up of excavated material and for in situ treatment, with the remark that the extraction technique, based on the dispersion of polluted particals is not suited for in situ clean-up.

Up till now the chemical clean-up techniques are only suitable for sandy soil materials with a clay and humic content of less than 10 to 15 %, lit [n].

The developments for the clean-up of contaminated sludges by the men-tioned chemical treatment techniques is still ongoing, lit [o].

ad 2. Thermal treatment
This clean-up technique is frequently used for the sanitation of soil materials contaminated with organics. Organic contaminants are volatilised and/or destroyed into volatile components.

Afterwards the volatiles are completely destroyed in a separate incineration step. Treatment of contaminated sludge in a rotating kiln, with a burner for directly heating the sludge is the most frequently used method. Other heating methods, as fluïdised bed heating and the infrared thermal treatment system can be used as well.

This method is suitable for the removal of all types of organic contaminants, but the removal depends strongly on the temperatures, that can be applied. In reality this method is mainly applied on soils contaminated with non-halogenated organics, lit [n].

Thermal treatment of polluted dredged material is however not a common practice as the dredged spoil is often contaminated with both organic and inorganic contaminants. Most of the inorganic contaminants, heavy metals, can not be removed by heating.

Besides that, the energy consumption for cleaning this material is very high due to the high water content of dredged material, lit [o].

Thermal immobilisation, by melting or sintering of sediment particles, is not a common practice in Europe because of the high associated costs, lit [o].

ad 3. Biological treatment
Biological treatment of soil materials, contaminated with organic pollutants is common practice in The Netherlands, especially for soils contaminated with oil products. The principle of biological clean-up is based on forced ventilation combined with enhanced biodegradation, in which manipulated or naturally existing microörganisms convert the contaminants into water and carbon dioxide. This method can be applied

on excavated soil material, in land farming, as well as in situ. Restrictions at this moment however are the permeability of the soil material for air and water and the presence of non-degradable organics and heavy metals. The standards to be met by the techniques are becoming more and more stringent and it is unclear as to the long term viability of this method. All these factors restrict the application of biodegradation for polluted dredged material. Further more to date only laboratory tests on biodegradability are available, lit [n, o].

Literature

[a] Opstal, A.T. and A.F. Van Tol;
A new sludge-sampling system; CATS II congress, Antwerp 1993;

[b] Kunst, D.J., Olie J.J.;
The development of in situ detection of oil and related compounds (in Dutch): Land en Water Jan. 1994;

[c] Olie, J.J., Visser, W., The use of in situ measurement techniques for soil pollution problems; Fourth International TNO/KFK Conference On Contaminated Soil, Berlin, 1993

[d] Olie, J.J.; In situ detection of migration paths in aquifers from landfills by chemo probe; Second International Conference On Site Analysis and Field Portable Instrumentation, Houston, Jan. 1994;

[e] Tack, F., A. Demeyer and M. Verloo;
Using cluster analysis and multiple regression techniques for a more efficient characterization of sediments;
CATS II congress, Antwerp 1993;

[f] Wieriks, J.P., and C.J.Otten;
Expected environmental effects of the dumping of dredged material from the port of Rotterdam at dumping site Loswal Noord;
CATS II congress, Antwerp 1993;

[g] Deibel, I.K., and J.W. Zwakhals;
Separation from sand out of sludge; a large scale test; CATS II congress, Antwerp 1993;

[h] Elprama, R., A.F. van Tol, G. Greeuw and B. B. Thorborg;

Validation of consolidation models for sludge disposal sides; CATS II congress, Antwerp 1993;

[i] Detzner H.D.;
Mechanical treatment of the dredged material from the Hamburg harbour; CATS II congress, Antwerp 1993;

[j] Doef, M.R van der, and G.E. Kamerling;
Minimization of contamination of groundwater caused by disposal sites for contaminated dredged material; CATS II congress, Antwerp 1993;

[k] Loxham, M., Weststrate, F.A.; Fail safe criteria for the environmental impact design of large scale landfills for contaminated dredging spoil; WODCON XII, Orlando, 1988

[l] Loxham M., Taat J., Eede, E. van den, Silence P.,
The theory and practice of chemically active liners for dredged spoil disposal sites;
Cat II Congress, Antwerp 1993

[m] Bear J.; Hydraulics of Groundwater, McGraw Hill Int. Book Company, New York.

[n] Rulkens W.H., Grotenhuis J.T.C., Soczó E.R.;
Remediation of contaminated soil: State of the art and desirable future developments. Fourth International KfK/TNO Conference on Contaminated Soil, 1993, Berlin.

[o] Stokman G.N.M., Bruggeman W.A.; Remediation of contaminated sediments, developments in The Netherlands. CATS II Congress, Characterisation and treatment of contaminated dredged materials, 1993, Antwerp.

Geotechnical Considerations in Dredged Material Management

Marian Poindexter Rollings
USAE Waterways Experiment Station, Vicksburg, Mississippi, USA

ABSTRACT

For international commerce to continue, dredging must be performed to create and maintain navigable waterways in industrialized nations. The method of dredging and type of disposal facility used will affect the initial dredged material properties and will have a pronounced effect on long-term material behavior. The presence of contaminants in sediments can affect the design and future use of disposal facilities as well as potential offsite use of the material itself. Geotechnical engineering properties of many disposal facilities can be improved through intelligent site management allowing various ultimate-use options such as environmental enhancement, parks, agriculture, ports, airports, and industrial complexes.

Background

The dredged material "waste stream" defies categorization more so than most other waste streams considered at this Congress. Dredged material is not regulated as a waste by current environmental laws in the United States, e.g., Resource Conservation and Recovery Act of 1976. The confined disposal facilities (CDFs) in which much dredged material is placed are neither conventional wastewater treatment facilities nor conventional solids-handling facilities but they must be operated to incorporate some of the features of each. Dredged material is typically a fertile geotechnical material that can be employed for many beneficial uses in addition to those normally considered by engineers. Thus, to treat dredged material solely as a waste material is to ignore its value as a natural resource.

This view of dredged material as a resource, not a waste product, was not always held. As recently as the early 1970's, dredged material was often referred to as "dredge spoil," and the material was characterized as either "mud or sand." The material was often placed in "disposal pits" which dotted the landscape along the navigable waters of the world and were left to the wading birds, water fowl, snakes, and mosquitos. Historically, dredged material has found use as a construction fill material (e.g., The Netherlands' numerous land development projects, San Francisco International Airport, Port of Melbourne), but this was the exception not the rule. However as environmental awareness increased, environmentally sound disposal procedures became mandatory, and the value of dredged material as a geotechnical resource could no longer be ignored.

During the 1970s a number of dredging research programs (e.g., sponsored in the U.S. by federal government and in The Netherlands by a consortium of industry, academia, and government) made significant strides in our views and knowledge of dredged material handling, disposal, use, and management. We now have sophisticated procedures for engineering and managing dredged material disposal and use/reuse; the adverse effect of dredged material disposal on the environment has been minimized and, in many cases, turned into a positive effect. Worldwide, we have the knowledge to use this material as a valuable resource for construction site development, agricultural applications, upland and wetland creation, beach nourishment, and habitat development; and when properly engineered, we can safely handle the challenge of dealing with contaminated sediments. However, knowledge is not always translated into practice.

Huge quantities of geotechnical materials are dredged annually around the world. In the United States alone, the annual dredging conducted or permitted by the U.S. Army Corps of Engineers amounts to approximately 500 million cu yd. This material includes new work and maintenance dredging in both salt- and freshwater environments and may range in type from very soft clays to rock. Some of the materials contain dissolved or adsorbed contaminants, although a large percentage of the material (about 90% worldwide) is considered uncontaminated or "clean" [5] and may be placed in a wide variety of locations. In engineering terms, dredged material may be placed in upland, nearshore, or aqueous disposal sites, whereas in regulatory terms, disposal options are termed open-water disposal, confined (diked) disposal, and beneficial uses [32]. Typically material deposited in aqueous sites remains permanently in these sites, unaltered further by human activity, while that material in upland and some nearshore (diked) sites may be managed to improve material properties for subsequent beneficial use of either the site or the material.

When dredging projects are contemplated, many considerations other than purely geotechnical ones normally take front stage. These include such problems as real estate availability and liability, site configuration and access, proximity to sensitive ecological environments, sediment contamination levels, limited disposal capacity, and optimization of long-term (50 years or more) disposal plans. Many of these issues are settled by the legal and administrative staffs of involved agencies with some public involvement. Far too often, geotechnical and environmental engineers have minimal impact on this process.

Material Properties

The variation in dredged material properties (both physical and chemical) is much greater than that encountered in other waste streams considered at this Congress. Engineering properties of dredged material vary not only by insitu geographic location but also with type of dredge used, and with time as sedimentation, consolidation, and desiccation occur. Table 1 illustrates the types of material and initial conditions typically encountered immediately after dredging and disposal at aqueous sites [18]; void ratios reported would be similar to those encountered in confined sites. The type of dredge used will be dictated primarily by the properties and quantities of geotechnical material to be excavated [10], although other factors such as dredge availability and production rate, dredging and disposal site locations and conditions, contamination level of sediments, and cost [32] must be considered. Variations in dredged material volume over time may be significant as illustrated in Table 2 for selected sediments that were hydraulically dredged. Typical properties of dredged material placed within CDFs are given in Table 3; properties of other recognized soft soils are provided for comparison [21]. The high water contents in dredged material deposits indicate high compressibility as has been evidenced in numerous field monitoring projects and laboratory testing programs conducted around the world. The comparative range of dredged material compressibility is illustrated in Figure 1. Associated with high compressibility and settlement is high but variable permeability.

Engineering Characteristics

As CDFs are managed to maximize storage capacity (or air space), sedimentation, self-weight consolidation, desiccation, and additional primary consolidation caused by dried crust surcharge loading typically occur in dredged material deposits. The processes of sedimentation and consolidation have been investigated, described theoretically, and modeled analytically [16, 22, 2, 30, 18]. The formation of a desiccation crust has been documented, studied, and modeled empirically although the mechanisms of crust formation have not received much attention. Dredged material drying, cracking, and shrinkage leading to formation of a desiccation crust should be fully describable using the principles of mechanics of partially saturated soils, specifically considering soil suction. Chemical composition of the mixing water is a critical factor to be included in any such analysis.

Two areas of characterizing sediments for dredging have proven particularly troublesome. These are definition of the "bottom" of a waterbody and prediction of

TABLE 1--Sediment Type and Conditions Associated with Dredge Type

Type of Dredge	Sediment Type	Range of Void Ratio	Side Slopes (if Mounded)
Hydraulic Pipeline Cutterhead	clay, silt, sand, gravel, soft rock	7-18	0.36V:100H (max) for fine-grained soil; otherwise, angle of repose
Dustpan	sand, gravel	same as insitu or slightly higher	Discrete mound
Hopper	clay, silt	15-18	0.5V:100H (max)
Dipper	stiff clay, rock, shell, coral	same as insitu	Discrete mound
Ladder	coarse-grained material	same as insitu	Discrete mound
Clamshell			
Bucket	clay, silt, sand	same as insitu	1.0V:100H to 3.5V:100H
Orange-peel	rock	same as insitu	Discrete mound

TABLE 2--Typical Changes in Volume of Hydraulically Dredged Material Over Time

Location	Material Type	Property	Time of Measurement			
			Insitu	At Deposition	180 Days	720 Days
Mobile, AL (w/o desiccation)	CH	void ratio	5.65	10.00	7.98	4.16
		volume	0.60	1.00	0.82	0.47
Mobile, AL (w/ desiccation)	CH	void ratio	5.65	10.00	7.98	2.90
		volume	0.60	1.00	0.82	0.35
Charleston, SC	CH	void ratio	na	12.15	7.61	6.82
		volume	na	1.00	0.65	0.59
Norfolk, VA	CH	void ratio	6.86	10.50	6.04	6.04
		volume	0.68	1.00	0.61	0.61
Bridgeport, CT	OH	void ratio	6.47	8.50	5.89	5.43
		volume	0.79	1.00	0.72	0.68
Indiana Harbor, IL	CH	void ratio	2.37	10.00	3.88	3.53
		volume	0.33	1.00	0.44	0.41
Seattle, WA (clamshell dredged w/ 120# surcharge)	CH	void ratio	2.25	2.60	1.96	1.96
		volume	0.90	1.00	0.82	0.82

TABLE 3--Typical Soft Soil Properties [21,1]

Material Type	Engineering Properties				
	w_c, %	G_s	LL	PI	Shear Strength (kPa)
Dredged Material	50-400	2.4-2.8	0-270	0-185	0-50
Marsh/organic soil	100-500	1.7-2.5	50-200	100-165	<4
Peat	100-1,800	1.5-1.75	250-500	150-400	<4
Phosphatic clay	300-1,000	2.5-2.9	76-245	45-175	<1
London clay	20-40	2.71	65-95	35-65	50-500
Norwegian marine clay	18-45	2.77	20-44	4-23	3-50
Boston blue clay	32-42	2.78	40-52	18-32	40-80
Mexico City clay	100-550	2.35	150-500	100-400	25-175
Coode Island silt, Australia	23-130	2.4	37-90	20-53	7-120
Alumina red muds	55-65	2.8-3.3	42-46	7-39	na
Pulp & paper wastes	210-265	1.9-2.3	70-413	40-380	<7
Fly ash, type F	~50	2.1-2.5	--	NP	pozzolanic

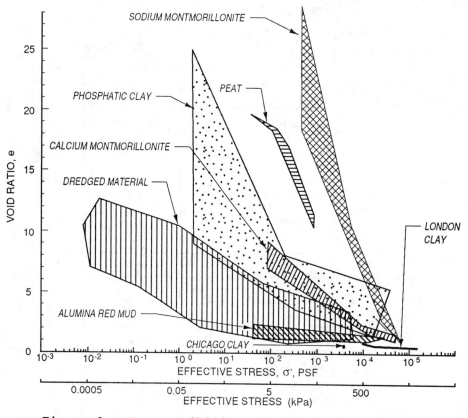

Figure 1. Compressibility of Various Soft Soils.

clay ball formation. Definition of the bottom for navigation and dredging purposes has been difficult to establish. At some point as the material near the bottom of the waterbody changes from water to "fluff" to fluid mud to a soil with some minimal firmness, the datum at which ships can no longer navigate, and thus the "bottom," is reached. As the result of current research in Europe and the Americas, the point of this transition is often defined as a density of 1.2 g/cc [24]. Techniques are being developed for continuous pre- and post-dredging characterization of the bottom below the 1.2 g/cc contour for purposes of determining material type and consistency, identifying the channel navigation prism, aiding in dredge selection, and calculating dredged volume. When dredging for new projects is performed in some clays, balls of clay are formed and are expelled from the dredge pipeline. The formation of clay balls has not been predictable a priori. However, current research is indicating that the Atterberg limits can be used as an indicator of clay ball formation and durability [E. C. McNair, USAEWES, Vicksburg, MS, personal communication, 1993]. When on-going work is completed in 1994, it is hoped that better material characterization data can be provided to dredging contractors, thus minimizing contractor claims for additional compensation caused by "changed conditions." To facilitate incorporation of the new technology, a computerized expert system is being developed that is to solicit needed geotechnical input data from the user after which it will determine likelihood of clay ball formation and sediment bulking factors for sizing containment areas.

Bulking factors have been the dredging industry's traditional and time-honored method of dealing with sediment volume change between in situ and final placement, but modern geotechnical research [e.g., 16, 2, 18, 11] has developed far more accurate and comprehensive approaches that makes the use of bulking factors for geotechnical analysis as outdated as the old empirical pile formulas were over a quarter century ago [25]. Inclusion of "bulking factors" in geotechnical research of the 1990s is a major step backwards in this author's opinion. In this era of sophisticated laboratory testing and analytical modeling (summarized by Townsend and McVay [27]), the absurdity of using bulking factors is illustrated in this example from an actual project in the United States. On a large construction project, a $60 million disposal facility was designed to retain 600,000 cu yd of highly contaminated soft bottom sediments, 650,000 cu yd of organic clayey-sandy silt, 1,200,000 cu yd of gravelly sands, and 950,000 cu yd of stiff to hard clay/clayey silt. "Different bulking factors were used to predict the increased volume of these materials....The increased volumes....were used to estimate the required size of the disposal facility" (quoted from a project report). Would it not be immensely more rational to conduct laboratory tests and perform analytical modeling to calculate the actual disposal site requirements? The cost of determining actual material behavior would have been only a very small fraction of the cost of the disposal facility. That cost, and probably more, could have been saved in modified CDF design, construction and site management.

Contaminants

Although dredged materials are simply composites of geotechnical materials naturally occurring in the earth's crust, they sometimes contain contaminants which may come from agricultural, urban, and/or industrial sources; types of contaminants will depend upon the insitu location of the sediments. In agricultural settings, the primary contaminants will normally be fertilizers and pesticides. Typical contaminants in industrialized harbors include toxic metals, organohalogens (i.e., PCBs), petrochemical byproducts, excess nutrients, and harmful microbes. Levels of contaminants vary widely, from below detection limits to extremely high levels. Table 4 shows contaminant concentrations in two U.S. dredged materials compared to USEPA maximum limits for digested sewage sludges (because no

TABLE 4 -- Contaminant Levels in Two Sediments (mg/kg dry material)

Contaminant	Indiana Harbor	Lake Michigan	EPA Max
Total PCB	7.39	0.41	
Total PAH	3229	9.4	
Cadmium	20	0.4	15
Copper	ND	43	1000
Lead	879	43	1000
Mercury	0.5	<0.1	10
Zinc	4125	98	2000

TABLE 5 -- Heavy Metals Content of German Sediments [20]

Metal	Range (mg/kg dry material)
Arsenic	50 - 150
Lead	150 - 250
Cadmium	7 - 25
Chromium	150 - 300
Copper	250 - 600
Manganese	1200 - 1800
Nickel	50 - 100
Mercury	10 - 20
Zinc	1500 - 2500

specific limits have been set for dredged materials). For comparison, the ranges of heavy metals content for 30 German sediments are shown in Table 5. Much required dredging occurs in industrialized waterways, so disposal of these potentially contaminated sediments requires chemical and biological testing prior to dredging and, if contaminants are present, special dredging and post-dredging handling and management techniques.

Management of Placement Areas

Efficient geotechnical management of dredged material placement sites requires pre-dredging predictive modeling and active post-dredging field operations to assess material behavior and facilitate site improvement. Prior to dredging, predictions should be made of dredged material surface elevation over time, accounting for projected placement rates and periods and consolidation and/or desiccation of the dredged material [30, 18, 12]. Predictions should be made for all dredged material placement sites. Common practice has been to model surface elevation over time for CDFs, but little effort has been made to predict a priori the height, shape, etc. of subaqueous mounds after initial placement. In a recent study, coastal engineers who are accustomed to working only with sands measured a mound height loss of 5 ft and promptly assumed such loss was caused entirely by erosion (a slope failure was ruled out when subbottom profiling did not reveal any displaced material), yet later calculations indicated that consolidation of the fine-grained material actually in the mound could easily have accounted for 100% of the observed height loss. The potential effect of consolidation on subaqueous mounds has been illustrated for several very different mounds (contaminated material a)uncapped, b)capped with sand, and c)capped with silt); the maximum reduction in mound volume for these cases was about 67% resulting from *consolidation only* in the mounded materials and foundation soils [18].

Because of the large strains involved in consolidation of very soft fine-grained materials (typified by loss of as much as 50% of initial deposit thickness from self-weight consolidation alone), general Terzaghi 1-D consolidation is totally inadequate to handle the problem. Instead, a finite (large) strain consolidation theory provides the proper analytical approach. The most general and least restrictive of the many one-dimensional primary consolidation formulations was developed by Gibson, England, and Hussey [6] and has been expanded and applied to various dredged material and soft soil analyses [7, 2, 26, 18, 27]. These models include consideration of primary consolidation and desiccation crust formation. Secondary compression (which may be significant in these thick, soft deposits) is not explicitly considered but is implicitly included in the empirical desiccation model. The most recent version of this finite strain consolidation/desiccation model, including typical input values, is included in the personal computer version of the Corps of Engineers' Automated Dredging and Disposal Area Management System (ADDAMS), an assortment of dredging related computer programs that can assist in management of dredging and disposal operations [23].

Material usable for construction projects or other non-slurry applications is most likely to come from upland sites because dredged material placed here is most easily managed to improve its engineering properties. Several cost-effective field techniques have been identified for management of CDFs receiving clean material including: 1)location of inlets and outlets to facilitate establishment of a gently sloping dredged material surface, 2)placement of thin lifts of material (3-5 ft initial thickness), 3)surface trenching to enhance drainage (at site perimeter and then internally using low ground-pressure ditchers), 4)removal of dried material for access road or dike improvement or for sale/use offsite, and 5)compartmentalization of large sites so they can be operated as outlined in 1) through 4) above. Evidence of the effectiveness of this approach is shown in Figure 2 where three individual lifts of dewatered material can be seen on the side slope of a 5 ft deep trench. The highly plastic clay (CH) material in this site had an initial void ratio of about 12.2% upon placement while at the time of the picture (18 months later) it had dried sufficiently to permit operation of large conventional earthmoving equipment to remove dried surface crust.

Figure 2. Dewatered material at Daniel Island in Charleston, SC

The effectiveness of using analytical modeling to plan management practices to maximize placement volume is illustrated by the Corps of Engineers' operation of the Craney Island CDF in Norfolk, VA. This 2500 acre site receives about 5 million cu yd of dredged material annually, and for many years the slurry was simply dumped into the site. The available disposal volume was decreasing at the site more rapidly than hoped, so a series of analytical modeling studies by WES formed the basis of a plan to divide the site into 3 compartments [17]. Each compartment then received material for one year and was allowed to consolidate and desiccate for two years as the other compartments were actively used. The understanding and modeling of the geotechnical processes involved in dredged material disposal allowed the useful site life to be increased by 17 years (doubling the remaining life over original operating practices) without considering the potential effect of removing dried material [19].

In situations where contaminated dredged material is involved, other management techniques may be necessary to protect the environment. When contaminated sediments are placed in CDFs, the potential migration pathways through which contaminants may exit the site must be assessed and appropriate management techniques employed to prevent or minimize this movement. The pathways of concern at upland CDFs include effluent and precipitation runoff discharges, leachate movement into groundwater, volatilization to the atmosphere, and direct plant and animal uptake. Migration pathways for nearshore CDFs include all previously mentioned pathways as well as soluble convection and diffusion and groundwater seepage through the dikes. Methods to address migration pathways are presented in USACE/USEPA [32]. Research has consistently found that the geochemical environment of a placement site greatly affects the availability of contaminants to the environment, with many being less available in wet environments.

Management of aqueous and many nearshore placement sites may be as economically costly as upland CDFs although the post-dredging management effort may be much less. At aqueous sites, pre-dredging management typically involves proper site selection to provide the desired type of site (dispersive or non-dispersive [31]), operational modifications (e.g., change dredging method to reduce water-column dispersion), use of subaqueous discharge points or diffusers, subaqueous lateral confinement of material, and thin-layer placement (12 inches or less). When contaminated sediments are involved, selection and placement of appropriate (clean) capping material will be necessary whether level bottom capping or contained aquatic disposal is used [32]. For either clean or contaminated materials, long-term monitoring is necessary to follow the deposit's behavior and fate and to establish the effectiveness of other management actions. If habitat creation is desired in nearshore sites, management consists of monitoring watertable/tidal elevations in relation to the material surface and adjusting either the surface elevation (by addition or removal of material) or the water levels (with control structures) to ensure proper long-term elevations for development or sustainment of the desired habitat type. Management of nearshore sites where the dredged material emerges sufficiently above the water table may be managed as upland sites. Placement of contaminated material (which would not normally be used for habitat development) in a nearshore site would require the same control options to be considered as upland CDFs.

Soil/Site Improvement

Adjacent to many navigable waters, land available for development is non-existent or extremely expensive; thus filled CDFs can provide a needed resource if the material contained in the facility has sufficient strength, or bearing capacity, to support the potential uses. In most cases some type of site improvement can be used to bring the dredged material to an acceptable engineering condition.

Numerous methods of ground improvement have been developed through the years, dating possibly from early Chinese [13] or Roman times [8]. In recent years significant effort has gone into development of sophisticated soil improvement techniques [14] with the theoretical base provided by geotechnical engineers and the equipment and construction techniques provided by specialty contractors. Many of these approaches have application to dredged material deposits (Table 6). The relative cost of a site improvement technique will vary depending upon site-specific conditions and location; Swedish experience on relative costs of soft soil improvement for various depths of deposit are shown in Figure 3.

Improvement of a material placement site typically means reducing the water content, consolidating or densifying the material, and increasing the shear

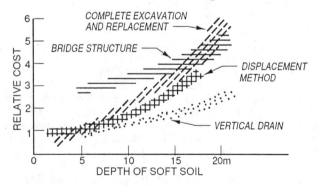

Figure 3. Relative Cost of Several Site Improvement Techniques for Different Depths of Soft Soil [15]

strength of the deposit. Because sediment placed in a CDF is typically deposited by hydraulic dredge (although other methods such as clamshell dredging may be used), the material is deposited at a high and/or variable water content. If a CDF is to be used for any purpose other than storage of dredged material or if the material itself is to be used, then some improvement technique must be employed to bring the material properties to a usable condition.

Unless intensive management has been practiced at a CDF, the material will typically be in a very wet, soft condition (at depth, if not throughout the deposit). Thus some type of remedial site improvement may be necessary before the site can be accessed for application of more conventional techniques. The U.S. Army Corps of Engineers and the phosphate mining industry have used progressive surface trenching (initially shallow trenches that are progressively deepened as the material strengthens adequately for trenches to stand open) which provides sufficient drying and strength gain to allow site access after 1 to 2 drying seasons. Then other site improvement techniques may be employed.

Site improvement techniques are not commonly used at dredged material placement sites unless the site will be used for port development, airport expansion, or some similar industrial purpose from which the initial financial investment for site improvement will be recouped. In this case, the most likely techniques to be utilized on fine-grained dredged materials are vertical strip drains and surcharge loading; for coarse-grained deposits, dynamic compaction may prove useful. In fine-grained materials, some type of vertical drain is often needed in conjunction with a surcharge load because many deposits have an initially low permeability while others will consolidate rapidly near the drainage boundaries forming a cake of low permeability material that will inhibit further drainage. In either case, the low permeability material may prevent the surcharge from causing sufficient consolidation within the time-frame of interest, but use of a vertical drain can speed drainage and consolidation. For example, vertical strip drains and surcharge loading were used successfully in combination in a portion of the disposal site used to contain the 3.5 million cubic yards of material removed from the Baltimore, Maryland Harbor during

TABLE 6 -- Soil Improvement Techniques for Dredged Material [21,14]

Method	Most Suitable Soil Conditions/Types	Maximum Effective Treatment Depth	Economical Size of Treated Area
Surface Trenching	Soft, fine-grained	2-3 m	>50,000 m^2
Preload, Surcharge	Fine-grained	--	>1000 m^2
Vertical Drains	Fine-grained	30 m	Small to moderate
Deep Dynamic Compaction	Cohesionless	15-18 m	>3400 m^2
Vibro- Compaction	Cohesionless	27 m	>1000 m^2
Vibro-Replacement	loose sandy; partly saturated clayey soils; loess	18 m	Small to moderate

construction of the Fort McHenry (I-95) Tunnel. Use of the geosynthetic drains in conjunction with the surcharge loading decreased the time required for completion of primary consolidation from 20 years to 9 months [Thomas Shafer, STV Lyons, Baltimore, MD, personal communication, 1987]. The site was then developed as the Seagirt Marine Terminal, a container port facility subjected to extremely heavy surface loads. In another example, dynamic compaction was used successfully in the 1980s to densify a silty sand dredged material deposit at the Port of Columbus, Mississippi. As land values continue to increase, especially in coastal and port areas, and as more marginal lands must be developed, the use of site improvement techniques in dredged material containment areas will continue to increase.

Beneficial Uses

Although historically dredged material has been used as fill for development of ports, urban areas, industrial sites, parks, and beach nourishment, only since the early 1970s have natural resource beneficial uses of the material (e.g., development or rehabilitation of wetlands, wildlife islands, and nesting beaches) come to the fore. The popularity and acceptance of using dredged material for natural resource enhancement has resulted in consideration of beneficial uses as an option to traditional disposal. In fact in the United States, law (e.g., the Code of Federal Regulations, Title 33, Part 337.9) now requires that dredging organizations give full and equal consideration to all practicable alternatives, including beneficial uses of dredged material.

Based on the functional use of dredged material, beneficial uses include, in addition to the traditional engineering uses: habitat development, beach nourishment, aquaculture, parks and recreation, agriculture (including forestry and horticulture), strip mine reclamation, solid waste management, shoreline stabilization, and erosion control [29]. Proposals to chemically or thermally treat dredged material to form lightweight, small aggregate have been advanced but no large scale operations have been implemented.

For each beneficial use, the physical, chemical, and logistical (e.g., size and timing of project and site accessibility) compatability between the material to be dredged and the intended use must be evaluated [32]. For example, sandy material must be used for island creation if the intended purpose is nesting habitat for certain threatened or endangered colonial-nesting birds, although species simply need a secluded site (Figure 4). Monitoring of both environmental and engineering aspects is another essential element on each beneficial use project; it is needed to determine the success of the project, possible need for rehabilitation, and how the site blends with the surrounding environment [9]. Guidelines for and examples of use of dredged

Figure 4. Endangered Brown Pelicans nesting on Galliard Island CDF, their first nesting in Alabama this century [29].

material for beneficial uses are presented in USACE [29]. Dredged material is a valuable resource - not a waste product, and it should be used beneficially as often as possible. This is simply good economics and sound environmental practice.

Risk Assessment

Risk assessment in geotechnical engineering nowadays implies use of probabilistic techniques to determine the risk of failure, or conversely, the reliability of a system to perform as intended [3, 34, 33], while USEPA environmental documents requiring "risk assessments" refer to human health- and environment-related assessments and do not include any probabilistic analyses.

Performance of a probability-based risk assessment requires some level of knowledge of the certainty of parameters to be evaluated. Many factors associated with dredging and dredged material placement are imprecisely known. Historically, the placement facilities have not been engineered. Quantification and (USCS) classification of sediments to be dredged have not been accurately and routinely done. Thus probabilistic risk assessments may be difficult or impossible to conduct. Environmental/health risk analyses recognize uncertainty and address it only qualitatively, so these methods

should be considered as deterministic [33]. Because it is important to use a consistent procedure throughout an analysis, it may often be necessary to use deterministic approaches when assessing risks associated with dredged material.

The consequences of failure must also be included in a risk analysis. The consequences of failure of a CDF dike, for example, may be a degradation of adjacent habitat, refilling of the navigation channel with slurry, contamination of a large "clean" area, or loss of needed fill for a construction project. Some method of quantification of the consequences of failure (e.g., monetary, lives lost, environmental effect, etc.) must be selected (Figure 5), and the level of acceptable risk must be determined.

Dredged material risk analyses will probably indicate an abnormally high level of risk when compared to others in geotechnical engineering. When an engineered product like an airfield pavement is typically designed for 50% reliability (and therefore 50% risk of premature failure), it is not unthinkable that dredged material placement may have a much higher risk associated with it. However this does not imply that risk assessments should not be conducted. Instead, "the use of *imperfect knowledge*, guided by judgment and experience, to estimate the probable ranges for all pertinent quantities that enter into the

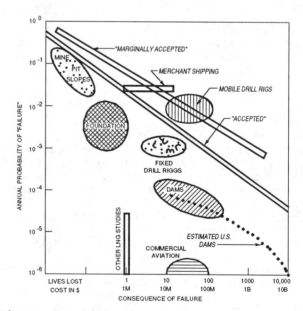

Figure 5. Risks for Selected Engineering Projects [34].

solution of the problem" [3] should lead to a reasonable, and continuously improving, risk assessment capability.

Conclusions

Significant progress has been made in characterizing dredged material physically and chemically. If these existing engineering techniques are applied on a consistent and thorough basis, better planning, design, and performance of placement facilities can be expected. Broader application of site improvement techniques will permit more beneficial use of dredged material and its placement sites. Application of environmental and geotechnical engineering techniques will allow better assessment of the risks associated with various dredged material placement options.

The challenge of the future is to implement what we know today but we currently ignore for economic reasons or intellectual laziness and to take full advantage of the rapid advancement of our knowledge in geotechnical engineering, environmental engineering, and risk assessment concepts. Unfortunately development of such things in current research programs show little indication of making significant progress in converting knowledge into practice.

Acknowledgments

The opinions expressed are those of the author alone, although much of the data reported here was collected by the author while working on various dredging research programs of the US Army Corps of Engineers. This paper is published with permission of the Office, Chief of Engineers. In addition to the in-house reviewers, the author extends her appreciation for reviewing this paper to: Dr. Robert L. Lytton, Dr. C. C. Mathewson, Mr. Thomas Patin, Dr. Robert L. Schiffman, and Dr. Frank C. Townsend.

References

1. Bromwell, L. G. 1982. "Evaluation of Alternative Processes for Disposal of Fine-grained Waste Materials," Seminar on Consolidation Behavior of Fine-Grained Waste Materials, Denver, CO.

2. Cargill, K. W. 1985. "Mathematical Model of the Consolidation/Desiccation Process in Dredged Material," Technical Report D-85-4, USAEWES, Vicksburg, MS.

3. Casagrande, A. 1965. "Role of the 'Calculated Risk' in Earthwork and Foundation Engineering," Journal of the Soil Mechanics Division, ASCE, (91)4, New York, NY.

4. Dredging Research Program. 1988."Dredging Research Program," Information Exchange Bulletin, Vol. DRP-88-1, August 1988, USAEWES, Vicksburg, MS.

5. Engler, R. M. 1992. "The Role of London Dumping Convention in Regulating Dredged Material Disposal: Past Development and Future Issues," *Proceedings*, Ports '92, U.S. Section, PIANC, Seattle, WA.

6. Gibson, R. E., G. L. England, and M. J. L. Hussey. 1967. "The Theory of One-Dimensional Consolidation of Saturated Clays, I. Finite Non-Linear Consolidation of Thin Homogeneous Layers," Geotechnique, (17)13.

7. Gibson, R. E., R. L. Schiffman, and K. W. Cargill. 1981. "The Theory of One-Dimensional Consolidation of Saturated Clays, II. Finite Non-Linear Consolidation of Thick Homogeneous Layers," Canadian Geotechnical Journal, 18.

8. Kerisel, J. 1985. "The History of Geotechnical Engineering Up Until 1700," *Proceedings*, 11th International Conference on Soil Mechanics and Foundation Engineering, San Francisco, CA.

9. Landin, M. C. 1992. "Beneficial Uses of Dredged Material Projects: How, When, Where, and What to Monitor, and Why It Matters," *Proceedings*, Ports '92, U.S. Section, PIANC, Seattle, WA.

10. Leussen, W. van and J. D. Nieuwenhuis. 1984. "Soil Mechanics Aspects of Dredging," Geotechnique, (34)3.

11. Martinez, R. E., Bloomquist, D., McVay, M. C., and Townsend, F. C. 1988. "Consolidation Properties of Slurried Soils," *Proceedings*, Soil Properties Evaluation from Centrifugal Models and Field Performance, ASCE National Convention, Nashville, TN.

12. McVay, M. C., Townsend, F. C., and Bloomquist, D. 1989. "Quiescent Consolidation of Phosphatic Waste Clays," ASCE Journal of Geotechnical Engineering

112(11), New York.

13. Menard, L. and Y. Broise. 1976. Theoretical and Practical Aspects of Dynamic Consolidation," *Proceedings*, Ground Treatment by Deep Compaction, Institute of Civil Engineers, London, UK.

14. Moseley, M. P., Ed. 1993. *Ground Improvement*, Blackie Academic and Professional, CRC Press, Boca Raton, FL.

15. Organization for Economic Cooperation and Development. 1979. *Construction of Roads on Compressible Soils*, Paris.

16. Palermo, M. R., R. L. Montgomery, and M. E. Poindexter. 1978. "Guidelines for Designing, Operating, and Managing Dredged Material Containment Areas," Technical Report DS-78-10, USAEWES, Vicksburg, MS.

17. Palermo, M. R., F. D. Shields, and D. F. Hayes. 1981. "Development of a Management Plan for Craney Island Disposal Area," Technical Report EL-81-11, USAEWES, Vicksburg, MS.

18. Poindexter, M. E. 1988. "Behavior of Subaqueous Sediment Mounds: Effect on Disposal Site Capacity," PhD Dissertation, Texas A&M University, College Station, TX.

19. Poindexter-Rollings, M. E. 1989. "Storage Capacity Evaluation for Craney Island Expansion Alternatives," Technical Report EL-89-6, USAEWES, Vicksburg, MS.

20. Rizkallah, V. 1987. "Geotechnical Properties of Polluted Dredged Material," *Proceedings*, Geotechnical Practice in Waste Disposal '87, ASCE, New York, NY.

21. Rollings. R. S., M. E. Poindexter, and K. G. Sharp. 1988. "Heavy Load Pavements on Soft Soils," *Proceedings*, 14th Australian Road Research Board Conference, (14)5, pp.219-231, Canberra, Australia.

22. Schiffman, R. L., V. Pane, and R. E. Gibson. 1984. The Theory of One-Dimensional Consolidation of Saturated Clays: IV. An Overview of Nonlinear Finite Strain Sedimentation and Consolidation," *Proceedings*, ASCE Sedimentation/Consolidation Models, San Francisco, CA.

23. Schroeder, P. R. and M. R. Palermo. 1990. "The Automated Dredging and Disposal Alternatives Management System," EEDP Technical Note 06-12, USAEWES, Vicksburg, MS.

24. Teeter, A. M. 1991. "Navigable Depth Concept for Channels with Fine-Grained Sediment," Information Exchange Bulletin DRP-91-4, USAEWES, Vicksburg, MS.

25. Terzaghi, K. and R. B. Peck. 1967. *Soil Mechanics in Engineering Practice*, John Wiley and Sons, New York, NY.

26. Townsend, F. C. 1987. "Clay Waste Pond Reclamation by Sand/Clay Mix or Capping," ASCE Journal of Geotechnical Engineering 115(11), New York.

27. Townsend, F. C. and McVay, M. C. 1990. "State-of-the-Art: Large Strain Consolidation Predictions," ASCE Journal of Geotechnical Engineering 116(2), New York.

28. U.S. Army Corps of Engineers (USACE). 1983. "Dredging and Dredged Material Disposal," Engineer Manual 1110-2-50-25, Office, Chief of Engineers, Washington, DC.

29. USACE. 1986. "Beneficial Uses of Dredged Material," Engineer Manual 1110-2-5026, Office, Chief of Engineers, Washington, D.C.

30. USACE. 1987. "Confined Disposal of Dredged Material," Engineer Manual 1110-2-50-27, Office, Chief of Engineers, Washington, DC.

31. USACE/U.S. Environmental Protection Agency (USEPA). 1992. "Evaluating Environmental Effects of Dredged Material Management Alternatives - A Technical Framework," EPA/842-B-92-008, USEPA, Washington, D.C.

32. USACE/USEPA. 1984. "General Approach to Designation Studies of Ocean Dredged Material Disposal Sites," US Army Engineer Water Resources Support Center, Ft. Belvoir, VA.

33. Van Zyl, D. 1987. "Health Risk Assessment and Geotechnical Perspective," *Proceedings*, Geotechnical Practice for Waste Disposal '87, ASCE Specialty Conference, Ann Arbor, MI.

34. Whitman, R. V. 1986. "Evaluating Calculated Risk in Geotechnical Engineering," Terzaghi Lectures 1974-1982, ASCE Geotechnical Special Publication No. 1, New York, NY.

Characteristics of a Hydraulic Fill Beach

P.K. Egyir, D.C. Sego
University of Alberta, Department of Civil Engineering, Edmonton, Alberta, Canada

Abstract

In the mining industry, hydraulic placement is used extensively to construct tailings embankments. These fills are often considered to have low shear strength, low relative density and high susceptibility to liquefaction. A research project carried out at the Brenda Mines abandoned emergency depositional area revealed that the single point discharge of the tailings slurry deposited material with relative densities comparable to those attained via cell construction used on the main tailings dam [1]. Consolidated undrained triaxial tests results from specimens obtained from undisturbed samples showed a consistent dilative response which suggest that the hydraulic deposit would resist both static and dynamic flow deformation or liquefaction.

Introduction

The Brenda Mines was an open-pit copper-molybdenum operation located at (49° 52' N, 119° 58' W) about 225 km east northeast of Vancouver and 40 km northwest of the City of Kelowna in the south central British Columbia (Figure 1).

The ore body occurs in a relatively homogeneous quartz diorite rockmass of Jurassic age [2]. The mill was designed to process about 22,000 tonnes of low grade ore per day for 20 years to produce copper and molybdenum concentrates. However between January 1988 and May 1990, an average of 32,000 tonnes of ore was processed daily.

Processing consisted of crushing, grinding and a flotation extraction. Fresh water for the milling process was obtained by diverting surface flows to storage in Peachland Lake. Reclaimed water was recycled and also stored in a separate basin for reuse. An estimated 178 to 203 million tonnes of tailings were to be produced [3] which required disposal. The tailings consist of a relatively uniform angular sand comprising of 60% feldspar, 30% quartz and 10% trace minerals [4].

Due to the steep longitudinal and lateral gradients of the adjacent valley containing the tailings disposal facilities, a large tailings dam was constructed to provide sufficient storage volume for the tailings in the valley. Various methods of depositing hydraulic fill were used during the mine operation to construct it to a final height of 138 m by 1986. These methods included single and double-stage cycloning and hydraulic cell construction to separate the coarse fraction from the fines. The fines were spigotted or allowed to drain from the crest to the storage area behind the dam.

During construction shutdowns, the tailings were discharged to an emergency depositional area to ensure uninterrupted mining and milling operations. The beach was formed by sporadically discharging tailings from the crest of the adjacent mountain into a steep walled narrow valley. The tailings was transported via a 600 mm diameter pipe and discharged from a height of about 10 m. After completion of the dam, subsequent tailings were deposited in the upstream reservoir. Figure 2 is an aerial view showing the layout of the tailings disposal facilities at the Brenda Mines.

This paper outlines the findings of a study undertaken to evaluate the characteristics of the as-placed tailings in the emergency deposition area. The study was part of an ongoing evaluation of the resistance of mine tailings deposited by hydraulic fill methods to flow liquefaction.

Sampling and Test Program

Disturbed and undisturbed samples were retrieved at different locations along the emergency beach. The sampling locations followed a grid pattern and were identified by their distance from the point of discharge to the beach. At each transverse section, samples were taken at centreline and to the right and left of it.

At each sample point, disturbed samples from various depths were obtained to establish the particle grain size variation. The profiles were from 0 to 120 cm with samples taken at 30 cm intervals. Undisturbed samples were obtained adjacent to these sample locations by removing the surface wind blown material and then the next 100 mm layer was gently excavated by hand. Then a 100 mm diameter sample tube was gently pushed 200 mm into the undisturbed deposit while excavating around it. Carbon dioxide pellets (dry ice) were placed around the base and allowed to freeze the sample from the base upward.

The sample tubes were retrieved and wrapped in double plastic bags. Then they were stored in a refrigerator containing dry ice for subsequent transportation to a laboratory at the University of Alberta, where they were stored in a temperature-controlled cold room until required for testing.

Index and stress-strain response were determined to evaluate the geotechnical properties of the tailings deposit. The tests were performed following standard ASTM procedures [5].

Disturbed bed samples were used for grain size analysis and specific gravity tests to evaluate if variation occurred along the beach. These samples were also used to determine the Atterberg Limits and the maximum and minimum dry densities.

The in-situ bulk density was determined using the frozen samples. The in-situ moisture contents allowed dry density to be calculated.

A series of consolidated undrained triaxial compression tests was performed on specimens prepared from the undisturbed frozen samples. The field samples were cut and machined to 76 to 90 mm long and 38 mm diameter specimens.

To prevent thawing and collapsing prior to the application of load, the specimen was mounted in the triaxial cell in the $-10°C$ cold room. The triaxial apparatus, the rubber membranes and the water used as the cell fluid were allowed to cool before the specimens were mounted. A small quantity of antifreeze kept the water from freezing while in the cold room.

A confining pressure of 50 kPa was applied and the specimen was allowed to thaw. The specimens were saturated by percolating water through them from an elevated reservoir. After forcing out the air, the specimens had a back pressure applied to saturate them and then they were isotropically consolidated. The samples were consolidated to an effective stress eight times greater than the estimated in-situ effective stresses at the sampling depth. After consolidation, the drainage valve was closed and the specimens were axially loaded at a constant vertical displacement rate until at least 20% vertical strain was achieved.

Test Results

Figure 3 presents the areal variation in the particle grain size along the emergency beach. For a distance of 1200 m from the point of discharge, the surface deposits varied within a relatively narrow range. However, material close to the disposal pond (CH 1+200 to CH 1+400) varied significantly in its grain size. Little variation in the grain size with depth occurred except at CH 1+200 where more silt was deposited at the 120 cm depth (Figure 4). All the tailings tested as nonplastic with an overall average specific gravity, G_S of 2.74.

The variation of the in-situ dry density along the beach is presented in Figure 5. The in-situ bulk density ranged from 1.51 to 1.92 Mg/m^3 and dry density from 1.33 to 1.62 Mg/m^3 respectively. The loosest material was deposited near the discharge and close to the pond with the majority of the deposit achieving greater than 60% relative density.

Typical stress strain curves, stress paths and pore pressure change plots are presented for two of the test specimens in Figures 5 and 6. These two specimens had relative densities of approximately 30% but contained much different amount of fines. The index properties of all the specimens tested are summarized in Table 1.

Figures 6a and 7a show that the tailings exhibit a post peak strain hardening characteristics. Failure strengths were reached at axial strains ranging from 9 to 23%. Strain localization resulting in nonhomogeneous deformation occurred in the samples after they reached their steady state. Results on undisturbed samples reported by Klohn [6] show similar large strains prior to failure. However, triaxial tests obtained using reconstituted samples indicated that the peak strengths were reached at were reached at relatively low axial strains [4; 6; 7]. These differences may be attributed to differences in the particle arrangement (fabric) developed within the undisturbed and reconstituted tailings test specimens.

Figures 6b and 7b show a decrease followed an initial increase in pore pressure resulting in the pore pressure becoming negative as the dilative material behavior becomes pronounced. The effective stress paths (Figures 6c and 7c) for the specimens indicate a consistent dilative response up to the peak resistance. The specimens continued to gain in strength with increase shearing until the steady state line was reached thereafter the samples began to fail.

Discussion of Results

The particle grain size distribution curves show that the tailings slurry behaved as a segregating mixture depositing predominantly medium to fine sand at the upstream portion of the beach. The fine grained sandy silt fractions were transported and deposited further downstream close to the pond. The fines may have been initially deposited below the previous pond level but is presently exposed due to the subsequent lowering of the pond level due to its transfer into the mined out pit [8].

An overall slight reduction in grain size with distance is observed in Figure 3. The D_{60}/D_{10} and D_{90}/D_{10} were greater than 2.5 and 5 respectively and suggested that hydraulic sorting was possible [9]. Hydraulic sorting may have resulted from the relatively high flow rate ($0.65 \text{ m}^3/\text{s}$) and the low slurry concentration (30% solids by weight) at which the tailings were discharged to the emergency beach.

The configuration of the valley containing the beach suggest that it is probable that the tailings were deposited under a higher velocity in the upstream area and it decreased significantly as the beach broadened allowing the flowing sand to spread laterally. Thus, the specific flow rate decreased with increasing distance along the beach, a condition which enhances the segregation of the flowing material [9].

The in-situ densities of the emergency beach are comparable with densities reported after cycloning and deposition and after hydraulic cell construction was used during the construction of the main dam. Hydraulic fills constructed using high flow rates and low slurry concentrations are known to produce higher density by creating a high energy environment associated with a low rate of deposition, but it may also cause a density decrease by accentuating hydraulic sorting which causes a uniform material to be deposited [9].

The emergency beach tailings have relative density greater than 60% for most of its length. Mittal [3] suggests that the quality of granular fills used in the construction of tailings embankment is controlled by specifying a minimum in-situ relative density of 60%. The relative density of the emergency beach tailings varied from 28 to 73% with an overall average value of 52%. Results from static CPT carried out on the main dam and contained in Klohn's Report [6] indicate that the relative density of the sandfill ranged from 40 to 80%.

Figure 5 shows that the sample retrieved from CH 0+800 c/L attained a relative density of about 73%. This high relative density may be attributed to compaction provided by construction traffic which used an access road that ran temporarily across the emergency beach and close to this location as shown in figure 2. However, samples which were taken from two other locations far from the access road attained relative densities greater than 60%

The loose sand deposit which was found close to the point of discharge (figure 5) may have resulted from material deposited downslope of a fanhead entrenched channel. In the study of alluvial fans entrenchment of channels near the apex of the fan is a common and hydraulically important characteristics [10]. Channel entrenchment occurs when erosion rather than deposition takes place at the apex.

The configuration of the valley containing the emergency beach, the sporadic high rate at which the tailings slurry was discharged to the beach and the subsequent high energy flow of the tailings suggest that the formation of fanhead entrenchment was possible in the narrow upstream section.

The low relative density of the material near the pond suggests that the tailings were initially deposited below water. In addition, lowering of the pond water level also may have led to an artificial oversteepening of the previously underwater slope and disturbing the already loose deposit [9].

The dilative mechanical behavior of the samples at the higher relative density from the triaxial tests is consistent with the dense state at which the tailings were deposited. However, the tailings which were deposited in a loose state were expected to show a contractive response during loading but as shown in Figures 6 and 7 they showed a consistent dilative response.

The test shown in Figure 6 contains about 7% non-plastic fines. At such low fines content, the undrained behaviour is expected to be purely contractive [4]. The Brenda tailings consists of predominantly pulverized, angular sand. The dilative response of the specimen in Figure 6 may be attributed to the inparticle structure of the deposited tailings. The end of consolidation void ratios of the specimens indicate that they were consolidated to dense states prior to shearing. Results reported by Klohn Leonoff Ltd (1984) [6] show a similar dilative response in a fine, predominantly non stratified cycloned sand containing about eight per cent non-plastic fines.

Kuerbis *et. al.* [11] reporting on Brenda tailings specimen containing up to 40% silt established that the dilative behaviour of the tailings sand under undrained monotonic triaxial compression is enhanced when the test samples have increasing amounts of silt. The specimen in Figure 7 contains about 69% non-plastic fines. The dilative behavior therefore confirms these findings.

Figure 8 is a plot of the steady state line for the Brenda cycloned sand obtained by substituting values for the state boundary constants Γ and λ into equation 1 [12].

$$e = \Gamma - \lambda \ln p' \qquad (1)$$

The insitu and the end of consolidation void ratios of the specimens are respectively plotted in figure 8. The results show that with the exception of the sample retrieved from CH 1+400 c/L, all the other specimens were dense of the steady state line.

A modified Mohr plot obtained by plotting the peak values of the mean shear stress versus the mean normal stress shows that the effective friction angle ϕ' of the tailings sample ranges from 35° to 39°. The high ϕ' value results from the high degree of particle angularity of the tailings sand found on the beach. Table 2 compares the properties of the Brenda tailings as reported by various investigators.

Conclusion

Index and triaxial test results of the total tailings presented in this paper show that the single point gravity discharge of tailings slurry at the emergency area yielded suitable fill material comparable with the cycloned sand utilized in the construction of the Brenda Mines tailings dam. It is interesting to note that no design or little construction effort was applied to the deposition at the emergency area and yet a stable hydraulic fill developed [8].

Acknowledgements

Financial support was provided by the Natural Science and Engineering Research Council of Canada. The authors also wish to thank Brenda Mines Ltd. for access to the mine site, the tailings facility and for furnishing pertinent data and information. The technical assistance of Gerry Cyre and Steve Gamble in the field sampling and the laboratory programs is greatly appreciated.

References

[1] Lighthall, P.E., Watts, B.D., and Rice, S., 1989. Deposition methods for construction of hydraulic fill tailings dams. Geotechnical Aspects of Tailings Disposal and Acid Mine Drainage, The Vancouver Geotechnical Society.
[2] Soregaroli, A.E. 1974. Geology of the Brenda copper-molybdenum deposit in British Columbia. The Canadian Institution of Mining and Metallurgy (CIM) Bull. October, pp 76-83.
[3] Mittal, H.K., 1974. Design and performance of tailings dams. Ph.D. Thesis, Department of Civil Engineering, University of Alberta, Edmonton.
[4] Pitman, T.D., 1993. Effect of fines and gradation on the collapse surface of a loose saturated soil. MSc. Thesis, Department of Civil Engineering, University of Alberta, Edmonton.
[5] ASTM, 1988. Annual Book of ASTM Standards, Section 4, Construction, Soil and Rock Building Stones, Geotextiles, **04.08**.
[6] Klohn Leonoff Consulting Engineers, 1984. Seismic investigations for Brenda Main Sand Dam, Peachland, British Columbia, **III**-Laboratory Investigation.
[7] Kuerbis, R.H., 1989. The effect of gradation and fines content on the undrained loading response of sand. MSc. Thesis, Department of Civil Engineering, University of British Columbia, Vancouver.
[8] Brown, R., 1992. Personal Communications.
[9] Küpper, A.M., 1991. Design of hydraulic fill. Ph.D. Thesis, Department of Civil Engineering, University of Alberta, Edmonton.
[10] French, R.H. 1987. Hydraulic processes on alluvial fans. Developments in Water Science 31, pp. 62-65.
[11] Kuerbis, R.H., Negussey and Vaid Y.P., 1988. Effect of gradation and fines content on the undrained response of sand. Geotechnical Special Publication (21). Edited by D.J.A.Van Zyl and S.G. Vick, pp. 330-345.
[12] Sasitharan, S. 1993. Collapse behavior of very loose sand. Ph.D. Thesis, Department of Civil Engineering, University of Alberta, Edmonton.

TABLE 1--
Index Properties of Emergency Beach Tailings Sample

Sample	C_u	e_o	e_c	fines (%)	γ_b	γ_d (Mg/m^3)		
						In-situ	Min.	Max.
0+400	2.92	0.881	0.693	6.00	1.626	1.457	1.357	1.741
0+800	2.46	0.690	0.520	5.93	1.920	1.618	1.326	1.764
1+000	3.07	0.752	0.742	8.00	1.792	1.564	1.327	1.761
1+200	2.89	0.953	0.552	64.2	1.507	1.403	1.183	1.562
1+400	3.10	1.064	0.970	69.9	1.796	1.328	1.229	1.679

TABLE 2--
Properties of Brenda Tailings Deposits

Location of tailings	Dry density (Mg/m^3)		Reference
	range	mean	
Initial cycloned sand	1.414-1.529	1.455	[6]
Hydraulic cell construction:			
Dozer compacted sand	1.470-1.550	1.520	[1]
Sand unaffected by dozer	1.410-1.530	1.460	[1]
Emergency beach	1.328-1.618	1.470	Current study

Figure 1. Map Showing the Location of the Brenda Mine Ltd.

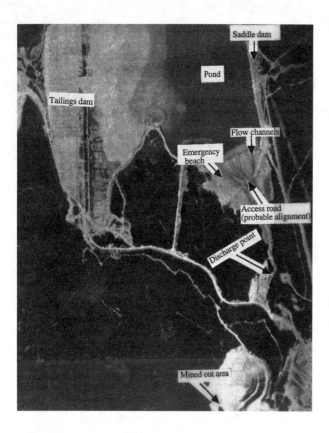

Figure 2. Aerial View of the Brenda Mines Site

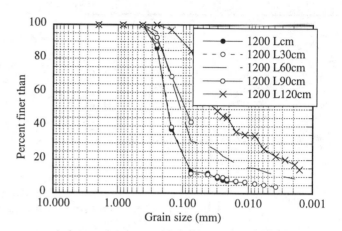

Figure 4. Vertical Distribution of Grain Sizes

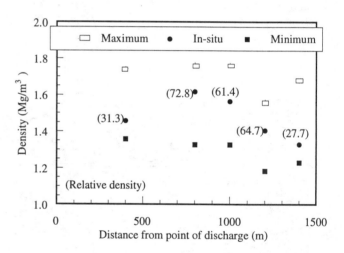

Figure 5. Density of Emergency Beach Tailings

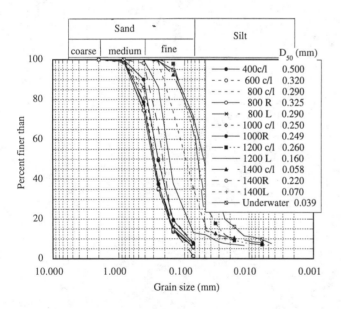

Figure 3. Areal Distribution of Grain Sizes

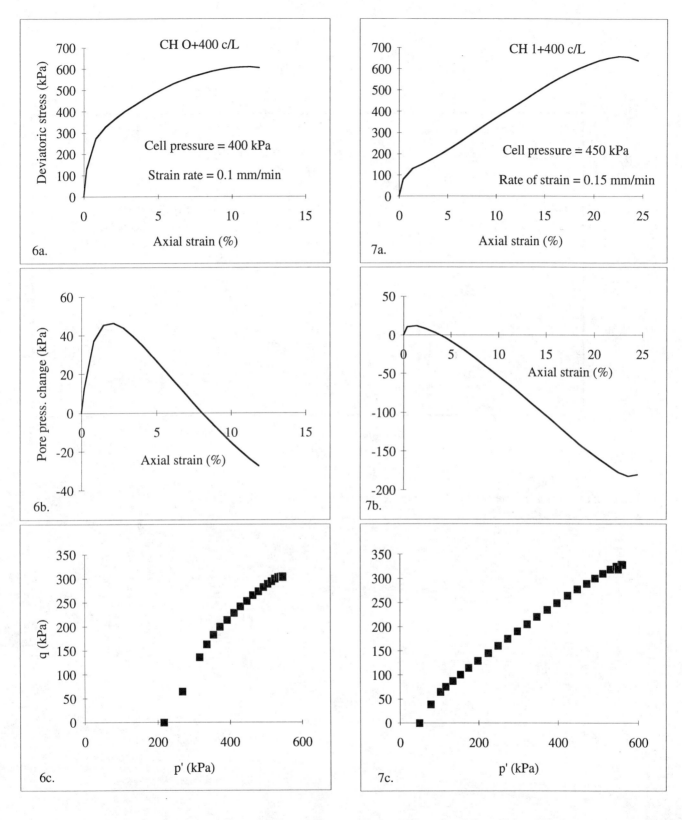

Figure 6. Results of Consolidated Undrained Triaxial Test

Figure 7. Results of Consolidated Undrained Triaxial Test

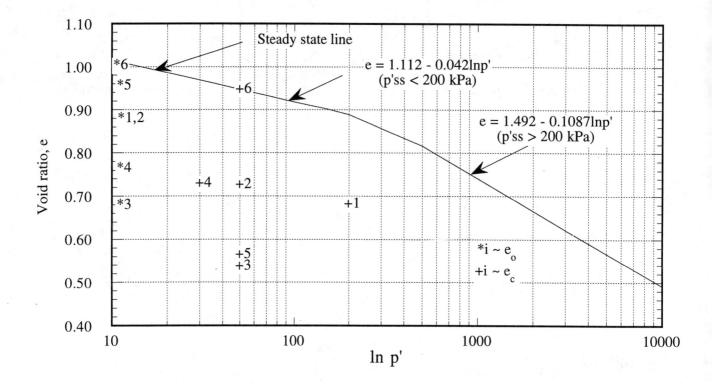

Figure 8. Steady State Line for the Brenda Cycloned Sand in Relation to Void Ratio of Emergency Beach Specimens

Solidification and Stabilization of Lagoon Sludge and Dredged Spoils

David P. Knox, P.E., Rosemary A. Najjar, P.E.
Camp Dresser and McKee Inc., Cambridge, Massachusetts, USA

ABSTRACT

A solidification/stabilization (S/S) program was developed to evaluate treatment options for 40,000 cubic yards of lagoon sludge, dredged spoils, and contaminated soils at an aluminum production facility in Massena, New York. The fine-grained waste material was generated from a settling basin for carbon fines from a wet scrubber process. The waste sludge, spoil and contaminated soils will be excavated, stabilized and/or solidified, and placed within an on-site RCRA (Resource Conservation and Recovery Act) and TSCA (Toxic Substance Control Act) permitted landfill. An increase in the bearing strength of these very soft sludge and spoil materials is required prior to permanent disposal in the on-site landfill.

Initially, this S/S program considered a wide range of additive and stabilization alternatives. The program was "phased" to focus in and narrow down the retained options as each stage was completed. The material types selected for treatment included a representative sample of the wetter sludge (30 percent solids content) and a sample of the drier sludge (50 percent solids content). The additives selected for study included cement, cement kiln dust, quick lime, hydrated lime, lime kiln dust, fly ash, and a mixture of hydrated lime/fly ash/cement. Strength tests were performed on initial cure and remolded samples to evaluate the effect of disturbance and sensitivity on solidified material. A correlation was also developed between field penetrometer and laboratory unconfined compressive or triaxial strengths to be used for field evaluation of treatment effectiveness during full-scale remediation.

The results of this program will enable the field engineer to monitor the efficiency of the S/S process in the field through simple index testing for water content, grain size and percent additive, and through field index strength measurements.

Introduction

The intent of this paper is to: 1) summarize laboratory test results for seven trial additives for an ongoing solidification/stabilization (S/S) program; 2) to screen the selected trial additives for effective bearing strength results and related treatment cost; 3) to recommend an additive(s) and percent dosage ranges for final mix design for handling and treatment of excavated waste materials; 4) to develop a correlation between field index and laboratory strength tests to monitor treatment in the field; and 5) to address disturbance effects on long-term remolded strength caused by movement and placement after initial curing in the on-site secure landfill.

Background

Solidification options included the mixing of pozzolanic additives to two waste material types, representing the approximate lower and upper range of in-situ solids content (30 and 50 percent solids). The intent was to evaluate improvement of the waste material characteristics by adding pozzolanics that would provide some measure of chemical bonding. Time effects and remolding/sensitivity were evaluated since pozzolanic materials are curing time dependent and can be affected by disturbance during (and after) the curing process.

Stabilization options included the addition of on-site contaminated granular soil to the soft, waste material to improve its bearing strength and handling characteristics. The intent was to evaluate the improvement of the waste material's characteristics by the physical modification of the material grain size distribution rather than by pozzolanic additives.

Design Criteria

The minimum criteria used as a basis for the initial screening and retaining of additives was:

- Landfill bearing strength capacity [long-term (stability) and short-term (placement)]

- Free-water content

No leachability criteria were established for the waste. The chemical quality of potential leachate generated within the on-site RCRA/TSCA landfill is not a design consideration of the S/S program.

A minimum compressive bearing strength of 16 psi (1.2 tsf) for waste material with a unit weight of 120 pounds per cubic foot (pcf) is required for long-term, in-situ strength. This design bearing strength is based on stability design considerations for the completed landfill cell. Bearing strength is equivalent to unconfined compressive strength.

The minimum bearing strength for placement operations within the landfill was 6 psi (0.45 tsf), allowing a reduction in strength for disturbance. This strength is based on construction equipment with low tire or track pressures.

Figure 1 illustrates the impact of curing on initial strength, the effects of disturbance on remolded strength, and the re-curing impacts on residual (re-cured) strength. Upon initial mixing, the strength of the solidified material is increased through a pozzolanic reaction and is therefore curing-time dependent. Only materials that achieved the minimum strength requirement for long-term landfilling (16 psi) using the initial (undisturbed) strengths were retained as part of the initial screening process.

Upon handling, the material is disturbed. Disturbance breaks the pozzolanic bonds and reduces the strength of the solidified material. This reduction in strength is defined as the sensitivity of the material. This disturbed strength must be above the placement strength of 6 psi. Material must be allowed to initially cure such that disturbance effects will still allow a placement strength above 6 psi. The disturbed strength of the material may be below the long-term strength (16 psi). After placement, however, the material will rebound by re-curing as indicated on Figure 1 to above the long-term criteria within the cell.

Direct landfilling (i.e. no initial curing) may be possible for waste stabilized with granular soil. For this case, the placement bearing strength of 6 psi may become the deciding factor since no curing would be needed.

During full-scale treatment both disturbed and residual bearing strength can be estimated by field strength index testing (e.g., pocket penetrometer, torvane, shear vane). Correlation of strength versus time between field index testing and laboratory triaxial testing will be developed. The correlation will be statistically valid with an acceptable level of confidence.

The other landfill design criterion is that no free-water be present in the solidified or stabilized materials following placement within the landfill.

Additional design considerations evaluated during additive screening included chemical interferences, durability, volume/density of final waste mixture, total costs of material handling and treatment, and health and safety in material handling.

Waste Characterization

Samples of waste material were collected from the site. Testing was performed to measure the waste material physical characteristics and existing in-situ strength properties, and to evaluate the variation in the material properties for the large volumes and areal extent of waste materials. The results of physical testing are:

- The solids content for the waste material varied from approximately 15 to 30 percent for wet lagoon sludge, up to 50 percent for drier dredge spoil. Water contents ranged up to 200 to 300 percent.

- The range of fines content (percent passing No. 200 sieve) for the waste material varied up to 100 percent for the sludge and spoil.

- The organic content range for the waste material varied up to approximately 62 percent for the lagoon sludge.

- The Plasticity Index for all of the waste material varied up to 127 for the lagoon sludge.

The two material types used for the S/S program included a 30 percent bulk solids sludge and a 50 percent bulk solids dredge spoil. These represent the bounds of material at this site, with the 30 percent comprising the bulk of the material to be treated. Consolidated undrained triaxial tests indicated an in-situ strength of 4 psi for the 50 percent solids material and no measured strength for the 30 percent material. Approximately 37,000 cubic yards of 30 percent waste are present, with another 3,000 cubic yards of the 50 percent material on-site.

Additive Trial Mix Dosages

Solidification additives evaluated for this study were: cement, cement kiln dust, lime, quick lime, lime kiln dust, fly ash, and lime cement fly ash (1 part to 6 parts to 13 parts).

Solidification was initially studied by adding 5, 10 and 20 percent additive by wet weight volume (percent of waste material's wet weight volume) for the 30 percent bulk solids material and 2, 5 and 10 percent additive by wet weight volume for the 50 percent bulk solids material. Stabilization (by adding granular soil) was evaluated by the addition of 10, 20 and 30 percent sand by wet weight volume.

Following the initial assessment of percent additive versus bearing strength, additives were eliminated from further consideration. Additional trial mix dosages were then selected at intermittent percentages for the retained additive to develop a design curve of percent additive versus bulk solids content (and water content).

Result of Testing

For the range of initial additive dosages for the 30 percent bulk solid waste, only cement achieved the 16 psi long-term bearing strength for initial curing. It achieved a strength of 20 psi at 10 percent additive. All other additives required over 20 percent additive by wet weight volume. This is shown in Figure 2.

For the range of initial additive dosages for the 50 percent bulk solids waste, cement, lime, quick lime and the combination of lime-cement-fly ash achieved the long-term bearing strength. Bearing strengths ranged up to 70 psi, as shown in Figure 3.

Granular soils alone did not provide any significant strength improvement until a bulk solids content of 70 to 80 percent was achieved. The amount of soil needed added so much to the waste volume (50 to 1,500 percent) that it was not economical.

A third material type was formed by adding granular sand to the 30 percent material, resulting in a stabilized 50 percent material. The purpose was to evaluate if "bulking" with sand could be cost-effective by reducing additive dosage. There was a reduction in percent additive sand and additional handling step made treating with additive alone more cost-effective.

Correlation of laboratory to field index tests indicated the pocket penetrometer could accurately predict measured laboratory strength. Disturbance of the cured samples resulted in remolded strength from 50 percent to 100 percent of the initial cure strength. Sensitivity measurements (the ratio of initial cured strength to disturbed strength) ranged from 1 to 17 with the more "brittle" cement-based additives being more sensitive.

The additives were compared for percent solids and long-term and short-term strengths. Table 1 summarizes the predicted additive dosages.

Based on additive material costs, cement and lime kiln dust were retained for further mix dosage testing. Percent additive and bulk solids contents were varied and the effects of disturbance and remolding were studied for both additives. Free water was also studied for the range of additive and bulk solids.

The additional mix testing yielded the following results:

- Cement produced the necessary strength to meet the short-term (6 psi) strength criteria for placement and the long-term (16 psi) design strength criteria for stability. Lime kiln dust was eliminated from further analysis because it was more costly to achieve the same strength increases.

- The long-term (recured) strength of 16 psi can be attained for the range of bulk solids more readily than can the short-term strength of 6 psi. The short-term strength is measured immediately upon disturbance after an initial curing period. Therefore, the short-term placement (6 psi) will govern additive dosage. The treated soil can attain long-term, recured strength within approximately 14 days of disturbance if the short-term strength is attained, it is anticipated that long-term strength can also be attained as the material recures in the landfill.

- Unconsolidated undrained (UU) triaxial tests are representative of short-term strength where no consolidation, recuring, or drainage will improve soil strength. The short-term, residual friction angle was measured by UU triaxial testing to be approximately 0 degrees treated with any percent dosage additive. Therefore, friction contributes very little to the short-term strength of the treated waste material and cohesion is the governing factor for remolded samples. This is consistent with the very fine-grained nature of the waste and the fine-grained physical composition of cement. The UU triaxial tests were performed at one confining pressure (2 psi) to represent placement conditions where material will be placed in lifts from 1 to 3 feet high.

- Consolidation Undrained (CU) triaxial testing was performed to represent long-term strength of the material where consolidation and recuring will increase the treated waste strength. The long-term friction angle was measured by CU testing to be approximately 15 to 20 degrees for the 50 percent bulk solids. Therefore, friction contributes significantly to long-term strength of material. This occurs due to the chemical bonding of the additive rather than the physical composition of the waste material.

- Water content and free-water is a major consideration in the placement of the treated waste in the landfill. Based on the results, the raw (not bulked with soil) 30 percent solids samples with no cement additive failed the EPA Paint Filter Test (SW846 9095), and the raw 50 percent solids samples with no additive passed the test. Therefore, a raw bulk solids content between 30 and 50 percent is required to meet free-liquid design criteria. The raw 30 percent bulk solids material with 10 percent cement additive passed the Paint Filter test. Therefore, additive must be added for bulk solids below 50 percent.

However, increasing the bulk solids content to 50 percent by bulking the waste with soil may not result in the material passing the Paint Filter test. Their addition to the sludge may modify the grain size distribution such that the soil will not retain pore water. The required upper limit of percent bulk solids by bulking with soil and no additive that is required to retain pore water is 80 percent.

- The pocket penetrometer has the highest and most consistent statistical correlation with lab triaxial testing. Therefore, the pocket penetrometer remains the recommended instrument for field index testing. Penetrometer strengths are on the order of twice that measured in the lab. This instrument should be used to make field decisions regarding adequacy of treatment dosage during full-scale treatment operations.

- Long-term compatibility of the cement additive and the contaminants with the sludge and spoil waste was evaluated. There appears to be no decrease in strength after 150 days of study.

Conclusions and Recommendations

Cement was selected as the additive to solidify this soft lagoon sludge and dredge spoil. The percent dosage will range from 40 percent (at 30 percent bulk solids) to 20 percent (at 50 percent bulk solids). The dosage is governed by attaining the short-term bearing strength of 6 psi. Free-water will not exist at these additive dosages. Pocket penetrometer testing will be used to monitor initial cure strengths and verify that the material was adequately treated with cement.

Table 1 - Summary of Additive Dosages for Short-Term and Long-Term Strengths

Additive	Short-Term (6 psi) % Dosage	Long-Term (16 psi) % Dosage	Approximate Design Dosage Used for Costing	Probable Additive Range
30% Solids				
Cement	12	12	12	10-15
Cement Kiln Dust	41	41	41	35-45
50% Solids				
Cement	5	6	6	4-8
Cement Kiln Dust	15	15	15	10-20
Lime Kiln Dust	18	18	18	15-25
Quick Lime	5	70	70	65-75
Lime-Cement-Fly Ash	20	20	20	15-25
50% Solids with Stabilization by Soils				
Cement	2	2	2	1-5
Cement Kiln Dust	10	10	10	8-12
Lime Kiln Dust	8	8	8	5-10
Quick Lime	7	7	7	5-10
Lime-Cement-Fly Ash	6	6	6	4-8

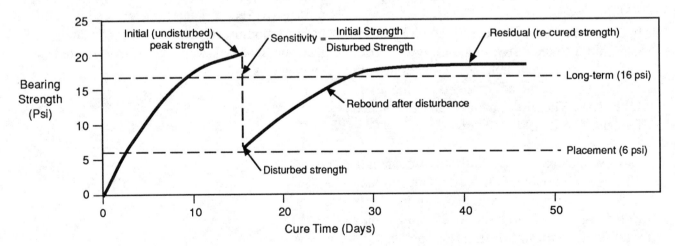

FIGURE 1 - Bearing Strength vs. Cure Time and Disturbance Effects

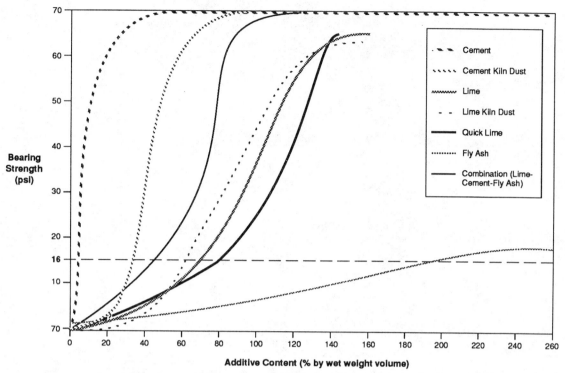

FIGURE 2 - 30% Solids Bearing Strength vs. Percent Additive

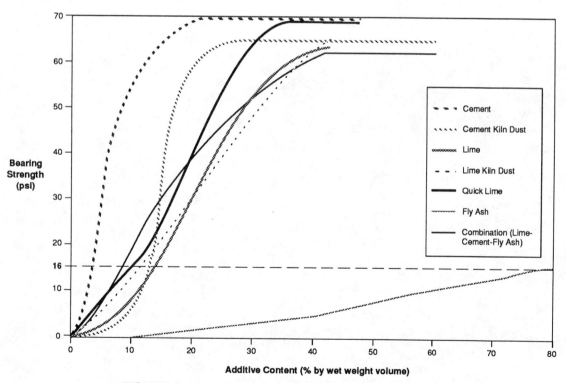

FIGURE 3 - 50% Solids Bearing Strength vs. Percent Additive

Utilization of Stabilized Dredged Waste for Construction Material

Takuya Ogino
Chuken Consultant Co., Ltd.

Toshiyoshi Goto
Osaka Cement Co., Ltd.

Koji Kataoka
Chuken Consultant Co., Ltd.

Minoru Kuroda
Estec Co., Ltd.

ABSTRACT

In Japan, wastes from construction works in urban area, excavated soil, dredged deposit and crushed concrete etc., were treated as reclamation material for the disposal area near the coast. Because of rapid increase of quantity of such waste with increase of construction works, it is becoming difficult to construct new disposal areas because of environmental problems in late 1980s, methods to decrease and utilize the wastes have been expected.

Also in Tokyo metropolis, wastes from construction works have increased rapidly and disposal area are lacking. In reconstruction project of an embankment of Sumida rever flowing down center area of the city, a river bed deposit should be dredged for improvement of the river. So recycling of such deposit was strongly desired. In this paper, laboratory test, practical application of utilization of dredged deposit for renewal project of river levee of Sumida and executed result are described.

INTRODUCTION

Present and planned standard cross section of the embankment of Sumida river are shown in Fig. 1.

To construct the river side terrace in front of existing concrete revetment, they planed constructing double wall cofferdam with sheet piles and filled it

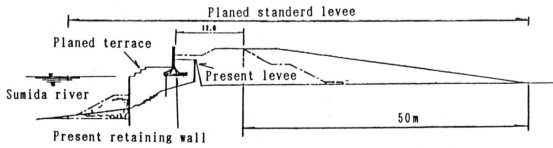

Fig. 1 Present and planed standard cross section of the embankment

with sand. Higher embankment will be constructed behind concrete revetment and excavation of the river bed will be carried out.

We proposed using the deposit stabilized with geo-cement instead of sand to prevent destruction of the nature by mining sand at mountain area.

Design

The design strength of treated deposit has been determined, based on either the cohesion or allowable bearing capacity. As a result, unconfined compressive strength have to be more than 1.0 kgf/cm^2.

In this condition, we analyzed structural stability. The stability of embankment and terrace was perfect. Because of the strength ratio of site mixing to laboratory mixing test is considered 0.5 [1] and strength reduction caused by placing under water was assumed as 50%, strength needed in laboratory mixing test was determined as 4.0 kgf/cm^2.

PROPERTIES OF DEPOSITS

To investigate a properties of deposit at the river bed, its sampling was carried out at 8 point along the river. Properties of the deposits are shown in Table 1.

Table 1 Properties of deposits

No.	Water content (%)	Wet unit weight (g/cm^3)	specific gravity	organic matter (%)	grain under 74μm (%)	soil classification
1	104.1	1.422	2.62	10.1	87.8	MH
2	69.2	1.534	2.63	8.3	44.1	SM
3	129.6	1.335	2.57	12.0	79.3	MH
4	114.1	1.376	2.53	13.8	98.2	MH
5	111.4	1.427	2.63	11.3	63.2	MH
6	117.6	1.391	2.57	13.1	68.9	MH
7	58.8	1.649	2.65	8.0	50.6	MH
8	85.7	1.580	2.63	9.1	45.1	SM

Water content, wet unit weight, organic mater content ranged from 58.8% to 129.6%, from 1.335 g/cm^3 to 1.649 g/cm^3 and from 8.0 % to 13.8 %, respectively.

LABORATORY MIXING TEST

Laboratory mixing test was carried out with three types of cement for every eight deposits to investigate the properties of the treated deposits.

Used cements were normal Portland cement(NPC), blast furnace slug cement (BSC) and geo-cement(GEC) which was blended for stabilizing soils containing much organic matter. Chemical components of the cements are shown in Table 2.

Table 2 Chemical components of cements

Type	igloss	SiO_2	CaO	Al_2O_3	SO_3
NPC	1.0	21.6	64.4	5.5	1.8
BSC	1.0	25.9	54.5	8.5	2.0
GEC	0.7	21.7	59.0	6.3	6.1

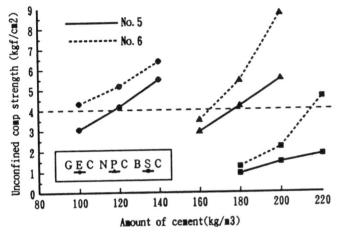

Fig. 2 Result of Laboratory mixing test

Some of the results are shown in Fig. 2. To obtain a expected unconfined compressive strength for every treated deposit, needed amount of each cement was as follow,

   normal poltland cement   : 220 kg/m3
   blast furnace slug cement : >220 kg/m3
   geo-cement   : 120 kg/m3

From the economical aspect and not increase the volume of treated deposit, we selected to use geo-cement.

MODEL TEST

To get a expected quality, suitable placing method should be chosen. Laboratory test was carried out to investigate the effect of placing condition and behavior of placed treated deposit in the water. The deposit used for the test was got from one of sites, and properties are shown in Table 3.

Table 3 Properties of used deposit

Water content (%)	Wet unit weight (g/cm3)	specific gravity	organic mater content (%)	grain under 74 μm (%)	Liquid limit (%)	soil classification
94.2	1.421	2.57	14.0	88.0	116.0	MH

Water content of the deposit was 94.2 % but was below its liquid limit. As a pumpability of the deposit seemed very low, tests were carried out with deposit whose water content were adjusted 94 %(natural), 113%, 133%. Geo-cement were used for stabilization of the deposit at rate of 120 kg/m3. Each sample was mixed in soil mixer for 3 minutes. As a index for pumpability, slump value is often used for concrete, we employed the value for the deposit with and without addition of geo-cement. Measured result are shown Fig. 3.

Slump value is decreased 5 to 12 cm with addition of geo-cement. From the consideration of a pumpability, water content of the deposit should be controlled 120% or more. To investigate the effect of placing method of treated deposit into the water, specimen were

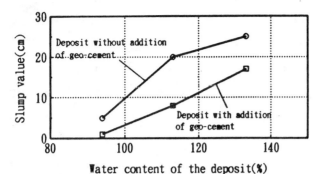

Fig. 3 Slump value of deposit

made in 3 ways, first case was placing the treated sample into molds (diameter of 5 cm, height of 10 cm) filled with water through a pipe(diameter of 4 cm) reached to the bottom of them, second case was placing into larger molds(diameter of 15 cm, height of 22 cm) continuously by the same way and specimen were made by thin wall sampler, and the last case was placing into the acrylic container(30×20×25 cm) filled with water through the pipe. Placing work was divided into 3 steps at one side and placed deposit was 2 litter at one step. Placing work was done on next day at another side in the same step. Specimen were sampled with thin wall sampler at 3 points. Specimen were also made by pouring into mold in air. Unconfined compression test were carried out at age of 7 days. Results were shown in Fig. 4, and figures on the columns show strength ratio to the specimens made in air. Ratio were about 0.7-1.0 in first case, 0.5-0.64 in second case and 0.45-0.59 in last case. The ratio is higher in case of higher water content and higher flow ability. In last case, there was not a clear construction joint and decrease of strength of treated deposit by the construction joint seemed relatively small.

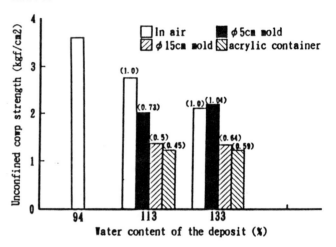

Fig. 4 Effect of placing condition on the strength

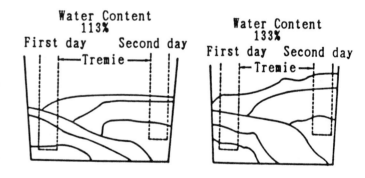

Fig. 5 Side view of the container

In the last case, flowing nature of the treated deposit in the water was observed. Fig. 5 shows side view of the container. Treated deposit which has low flowability flow horizontally and gradient of slope of flowed deposit was approximately 1:2, and the deposit which has high flowability flowed vertically.

EXECUTION

Fig. 6 shows the typical cross section of the embankment and terrace of the river being constructed. The terrace were constructed as follow; bottom ground of river at the side of existing concrete revetment was improved by deep mixing method, sheet piles were driven, then excavation of the river bed was carried out. Dredged deposit was mixed with geo-cement and placed into the river between sheet piles. Rubble stones were placed in front of the sheet piles.

The deposit was treated as to obtain unconfined compressive strength of 1.0 kgf/cm^2.

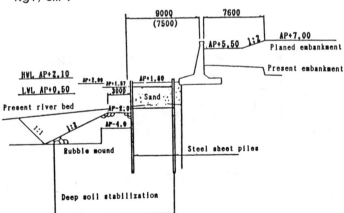

Fig. 6 Typical cross section of the embankment and terrace

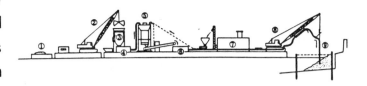

① Dredged deposit ② Clamshell ③ Vibrating screen
④ Pump ⑤ Mixing plant ⑥ Concrete pump
⑦ Cement silo ⑧ Crawler crane ⑨ Tremie

Fig. 7 Execution flow of treatment of dredged river bed deposit

Outline of execution flow of treatment of dredged river bed deposit is shown in Fig. 7. The deposit was dredged by backhoe and transported by barge and water content was adjusted in it by putting water and throw into vibrating screen and impurities over 70 mm were screened. The deposit was taken out constantly from the feeder under the screen, and moved to mixing plant by concrete pump. Geo-cement was added in the powdered state by air at rate of 120 kg per 1 m^3 of deposit and mixed with it in the plant. Capacity of the plant is 50 m^3 per hour. Treated deposit was placed into the double wall cofferdam by concrete pump through 8 inches pipeline and tremie. Tremie was moved by crane and a tip of it put always into the placed treated deposit to prevent mixed with water.

Sampling of dredged deposit for testing was done twice a day and water content, unit weight, grain size, vane shear strength, ignition loss and pH were measured. Results are shown in Table 3. Properties of the deposit ranged very wide.

Table 3 Properties of the deposit

Water content (%)	92.6-146.2
Wet unit weight (g/cm³)	1.303-1.447
Fine-grained size (%)	57.1-86.3
Vane shear strength (kgf/cm²)	$(0.73-2.11) \times 10^{-2}$
Ignition loss (%)	9.7-25.0
pH	7.09-7.87

To check the quality of treatment, treated deposit sample was put into the mold(diameter of 5 cm, height of 10 cm) and cured on the barge. Unconfined compressive strength was measured at age of 28 days and results were shown in Fig. 8.

Average strength was 2.29 kgf/cm². Strength distributed relatively wide because the properties of deposit

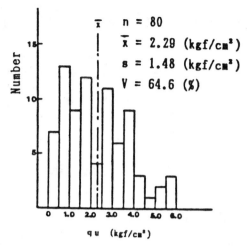

Fig. 8 Unconfined compressive strength

especially ignition loss and grain size differ very much. The reason why being the specimen below the expected strength seems that a curing temperature was 0 to 6 °C in winter season. The temperature of the water was 7 to 8 °C.

PROPERTIES OF TREATED DEPOSIT

About 3 weeks after filling work, bored core were sampled with double tube sampler of diameter of 86 mm and hand

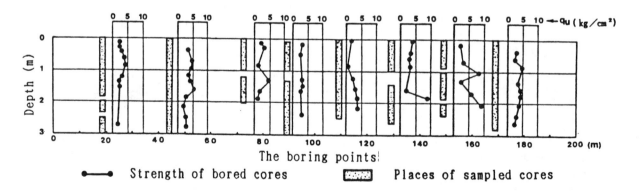

Fig. 9 Depth of the cores obtaind and unconfined compressive streength

held boring machine. The cores were aped as 70 mm in diameter and 140 mm in height and unconfined compression test was carried out for them at age of 28 days. Depth of the cores obtained and unconfined compressive strength are shown in Fig. 9, and distribution of the strength is shown in Fig. 10.

slope of flowed deposit was 1:2 and nearly equal to the result in model test in laboratory.

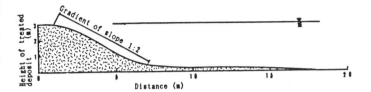

Fig. 11 Cross section of the placed treated deposit

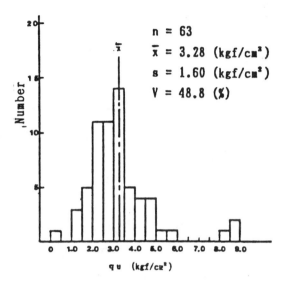

Fig. 10 Unconfined compressive strength

Core recovery was 82.5% and average strength was 3.29 kgf/cm² which is higher than cured specimen on the barge and coefficient of validity was rather small. This shows treated deposit was placed in good condition and curing temperature was higher.

FLOWABILITY OF TREATED DEPOSIT

Cross section of the placed treated deposit was shown in Fig. 11. Gradient of

DISPLACEMENT OF THE SHEET PILE

Displacement of the sheet pile was measured at a point before and after placing the treated deposit between the piles. Result was shown in Fig. 12. Displacement of the sheet pile was 13 mm right after placing the deposit, but was increasing as the time and reached 35 mm in 1 month and tend to stop.

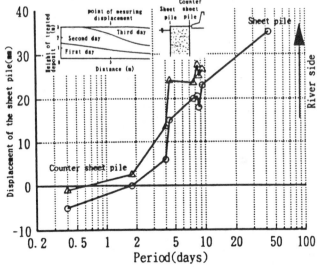

Fig. 12 Displacement of the sheet piles

CONCLUSION

Utilizing the dredged deposit from the river bed, the deposit was mixed with geo-cement in the plant and placed between the sheet piles instead of sandy soil to construct the terrace in front of the concrete river levee. Treatment and placement of dredged deposit was carried out successfully, and new space with amenity was created along the river. Utilization of the deposit has been continued for 4 years and recycled deposit become about 400,000 m^3. More and more dredged deposit should be disposed in the near future, another way of utilizing the deposit must be considered. For the purpose, in the project, using the treated deposit for embankment material or placing into the certain shaped geotextile bag in front of sheet piles instead of rubble stone are considered.

Recently, preventing environment from distraction by disposing construction wastes, some project are carrying out under the initiative of Ministry of construction and Ministry of transportation in Japan.

Some guides for utilizing the wastes from construction works will be made in near future.

ACKNOWLEDGMENT

Authors would like to our grateful thanks to Mr. H. Hayashi of Tokyo Metropolitan Government.

REFERENCES

1) Japan Cement Association, Soil Stabilization Manual with Geo-cement, pp52, 1985

2) T. Goto, T. Ogino, Model test on underwater placement of treated mud with geo-cement, 25th Japan National Conference on Soil Mechanics and Foundation Engineering, 1990

3) T. Goto, T. Ogino, Method of underwater placement and quality of treated dredged mud with geo-cement, 26th Japan National Conference on Soil Mechanics and Foundation Engineering, 1991

4) M. Ito, Y. Ito, H. Minaguch, Renewal project of Sumida river with geo-cement, pp75-78, Technical conference on construction in Tokyo, 1990

Effect of Lime Admixtures on Contaminated Dredged Sediments

Naresh C. Samtani, Vahan Tanal, Joe Wang
Parsons Brinckerhoff Quade & Douglas, Inc., New York, New York, USA

Anthony R. Lancellotti
Bechtel/Parsons Brinckerhoff, Boston, USA

Abstract: Field sampling and laboratory testing was conducted to investigate stabilization and solidification effects of lime, fly ash, kiln dust and portland cement admixtures on contaminated dredged harbor sediments. The testing was undertaken for the Central Artery Third Harbor Tunnel crossing in Boston where 89,000 cubic yards of contaminated sediments were dredged for the excavation of the immersed tube tunnel trench. The goals were to design a method to stabilize the contaminants and to rapidly solidify the dredged sediments so that their disposal site could be reclaimed for future development. Samples of the contaminated sediments were mixed with various proportions of cement, lime, fly ash and kiln dust. Laboratory tests were conducted on virgin samples and on mixed samples at various curing conditions to investigate both the geotechnical properties and the leachability of contaminants. The laboratory test results indicated that lime reacts with the sediments producing a pozzolanic reaction which significantly improves the compressive strength of the sediments. The curing temperature plays a critical role in the pozzolanic reaction. The leach test results indicated that the sediments were completely stabilized in the sense that no detectable level of contaminants were observed. The sampling and testing are described and the results of the laboratory tests are presented. Additionally, some examples where these procedures have been used are cited.

Introduction

The construction of the Central Artery Third Harbor Tunnel (THT) involved the excavation of a subaqueous trench to accommodate twelve immersed tubes across the harbor. Approximately 1.1 million yd^3 of material were dredged/excavated from the THT alignment which is across the main channel of Boston Inner Harbor. Approximately 89,000 yd^3 of shallow marine sediments, within 5 feet of the harbor bottom, were contaminated and had to be disposed of in a manner to meet the Massachusetts Department of Environmental Protection (MDEP) regulations for disposal.

A search for a suitable inland disposal site converged on Governors Island, where Boston's Logan Airport is located. To satisfy MDEP and Massport, the operator of the airport, the dredged material had to be stabilized and placed in a totally contained and capped environment. Furthermore, it was required to quickly solidify the material so that, a) additional land excavated material could be temporarily stored 15 feet high over the capped sediments, and b) the site could support a planned construction of airport appurtenant structures. It was estimated that the solidified sediments should have a laboratory unconfined compressive strength of 30 psi to support these future constructions.

Bechtel/Parsons Brinckerhoff (B/PB) embarked upon a sampling and testing program to design a process to chemically stabilize and solidify the contaminated marine sediments. As a first step, a literature search was undertaken to review similar case histories and experimental work by others on stabilization and solidification of contaminated dredged sediments.

Stabilization/Solidification (Immobilization) Technology

Stabilization/solidification is a state-of-the-art technology for the treatment and disposal of contaminated dredge materials. The immobilization techniques, stabilization and solidification, have two different goals. Stabilization attempts to reduce the solubility or chemical reactivity of the contaminated material. Solidification is the process of eliminating the free water in a semisolid by hydration with a setting agent(s). Typical setting agents include portland cement, lime, fly ash, kiln dust, slag, and combinations of these materials. Co-additives such as bentonite, soluble silicates, and sorbents are sometimes used with the setting agents to give special properties to the final product. In addition to physical stabilization which immobilizes contaminants, the admixtures may also chemically stabilize hazardous constituents such that leachability is eliminated or substantially reduced. The immobilization processes are usually formulated to minimize the solubility of metals by controlling the pH and alkalinity.

The stabilization/solidification technology has been successfully used to stabilize dredge spoils in addition to scums, sludges, and oil wastes. The quantities of these fluid spoils treated vary from small quantities used in laboratory to large quantities, e.g., 45850 yd^3 in Hama River and 157200 yd^3 in Tagonoura Harbor in Japan (Kita and Kubo 1983). The primary thrust of the Japanese research has been the improvement of handling and enhancing bearing capacities of the sediments for filling to improve or reclaim land, with minor emphasis on stabilization of contaminants. In similar practice in the United States, such techniques have been used almost exclusively in oil and gas drilling operations and on wastes from power plants, sewage treatment, chemical manufacturing, and the nuclear industry. A review of thirty three cases of contaminated sediment remediation in the United States by JRB Associates (1984) revealed only two cases, the Upper Hudson River in New York and Waukegan Harbor in Illinois, where stabilization was among the remediation techniques evaluated, but no cases of actual use were identified.

Most immobilization systems being marketed in the United States are proprietary processes involving addition of absorbents and solidifying agents to a waste. The process is often changed to accommodate specific wastes. In general, the waste immobilization systems that have potentially useful applications in dredge material disposal are, (a) Lime-flyash pozzolan processes, (b) Portland cement-pozzolan processes and (c) Sorption. Other technologies such as thermoplastic microencapsulation, macroencapsulation, and fusing waste to a vitreous mass or using self-cementing material are either too specialized or not sufficiently field applicable to be used at present (Environmental Laboratory 1980).

Depending upon the type of sediment and project requirements, the type and amount of admixture varies. Cement, lime and flyash are the most commonly used admixtures.

Sampling

Channel bottom sediments were collected at six locations along the centerline of the THT alignment where sediment samples were previously tested for contaminants. Sampling was done using a gravity corer and a clamshell. Twenty five gallons of sample was obtained from each of the six designated locations. Tube samples were obtained by a 3 inch diameter and 5 foot long gravity corer. Fifteen gallons of bottom sediments from each sampling location were collected using the gravity corer. The remaining ten gallons from each sampling location were collected by the clamshell.

The purpose of taking the clamshell samples was to simulate the actual dredging operations. These samples were retained in the solid to liquid ratio as obtained by the clamshell bucket, homogenized on the barge, and placed and sealed in five-gallon buckets. As the clamshell was 2.5 feet high, these samples represented the top 2.5 feet of the harbor sediments.

The aim of the tube samples was to collect sediments from the top five feet of the harbor bottom. Each core extended from the surface of the harbor to the native sediments below. As the tube samples were primarily meant to collect sediment solids, the water was decanted before sealing the tubes.

Properties of Raw Sediments

Laboratory tests for contaminant levels and for geotechnical properties were conducted on raw sediments. These are presented below:

TABLE 1 -- Contaminant Levels

Chemical	Range (ppm)
Arsenic (As)	8.5 - 49
Cadmium (Cd)	2.3 - 14
Chromium (Cr)	220 - 653
Copper (Cu)	200 - 649
Lead (Pb)	130 - 337
Mercury (Hg)	0.8 - 6.8
Nickel (Ni)	31 - 106
PCB's	0.2 - 0.4
Vanadium (V)	153 - 226
Zinc (Zn)	230 - 410

Based on the above contaminant levels, the dredge materials were classified as unsuitable for ocean disposal according to Massachusetts regulations (314 CMR 9.00).

TABLE 2 -- Geotechnical Properties

Property	Range
pH	7-8
Total Organic Content (TOC), %	4-10
Fines (<#200 sieve) content, %	70-95
Water Content, %	
- Clamshell samples	100-140
- Tube samples	60-140
Liquid limit, %	
- Clamshell samples	55-80
- Tube samples	40-65
Plastic limit, %	
- Clamshell samples	30-45
- Tube Samples	20-35

The water content and the Atterberg limits of the clamshell samples were generally higher than the tube samples as they represented the shallower softer sediments. To test representative sediments a composite sample was prepared by mixing samples obtained from three locations where the TOC, fines content and Plasticity Index (PI) test values were the highest. The average specific gravity of solids was determined as 2.52.

Stabilization/Solidification tests

The tests were performed with powdered quicklime and cement admixtures. Additional tests were conducted by adding pozzolans such as fly ash and kiln-dust. Mixes prepared with various proportions of admixtures were tested for pH, unconfined compressive strength, compaction, torvane shear strength, California Bearing Ratio (CBR), workability, Atterberg limits and contaminant levels. The results are presented in the following paragraphs. All admixture percentages herein are on a dry weight basis unless otherwise noted.

pH

Lime is alkaline with a pH of 12.4 at 25°C. Since a high alkaline condition is required for the formation of calcium silicates, the lowest lime content that will produce a pH of 12.4 is the minimum amount required to stabilize the soils (Arman and Munfakh 1970). The pH of the soil-lime mixes immediately after mixing was 12.3 and 12.41 for the 4 percent and 8 percent mixes respectively and increased slightly with time. The pH of the soil-lime mixes was greater than the pH of the soil-cement mixes.

Strength of uncompacted air cured mixes

Tests were conducted on air cured as well as moist-cured uncompacted mixes. Air curing of samples at room temperature resulted in considerable desiccation, carbonation, and shrinking of samples. The unconfined compressive strength of such samples was greater than 150 psi, a result of desiccation of the sample rather than a pozzolanic reaction. In contrast, samples cured at 35°F remained in a moist state and yielded very low torvane shear strengths of the order of 0.03 to 0.06 psi. These tests indicated that lime reaction does not occur in such low temperatures.

Moisture-Density Relations

The following moisture density relationships were obtained,

* for 8 percent, by wet weight, soil lime mix
- maximum dry density of 76.7 pcf
- optimum moisture content of 35.4 percent.

* for 8 percent, by dry weight, soil lime mix
- maximum dry density of 82.6 pcf
- optimum moisture content of 32.2 percent.

Figure 1 shows CBR values and torvane shear strength values plotted on the moisture density curve for a mix of 8 percent lime by dry weight. The CBR value drops sharply on the wet side of the optimum moisture content (OMC). This indicated that the mix may not be able to support compaction equipment if the moisture content were higher than the optimum.

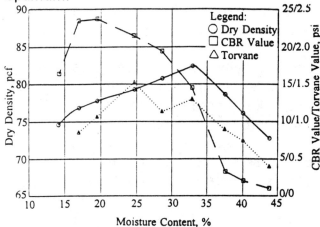

Figure 1 Relationship between the moisture density curve, CBR and torvane shear strength.

Strength of compacted moist cured mixes

Figure 2 shows the variation of unconfined compressive strength with curing time for different admixtures. The unconfined compressive strength data points in Figure 2 represent the average of the three tests done for each case on mixes compacted to 90 percent of their standard Proctor maximum density. The 28-day strengths were obtained from mixes which were cured for 48 hours at an temperature of 120°F.

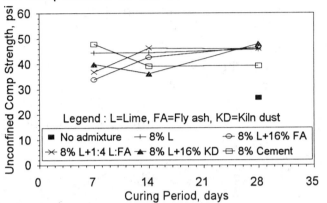

Figure 2 Effect of admixtures and curing periods

The following conclusions were drawn from Figure 2.

* Strength gain for all cases ranges from 35 to 50 psi; this satisfies the strength criterion of 30 psi.

* Although lime reacts quickly, the rate of strength gain decreased with time.

* Lime with additional pozzolans shows a weaker initial reaction compared to the use of lime alone. However, the long term strength for lime alone or lime with admixtures is virtually the same. Thus, the additional pozzolans do not appear to contribute much to strength gain.

* If the water content of the sediment is lowered to the optimum moisture content and if it is compacted to about 90 percent of its maximum dry density, an unconfined compressive strength of 25 to 28 psi may be achieved without any admixtures, and 40 to 50 psi may be achieved with 8 percent lime.

Effect of admixture quantities on moisture content (workability)

To reduce the initial moisture content and increase workability, different percentages of flyash and kiln-dust, in addition to a fixed percentage (9.6%) of lime, were added to the sediment on a dry weight basis. The moisture contents were measured at various times following the mixing. The mixes were prepared and kept at temperatures close to 35°F in an attempt to simulate the conditions during actual full scale operations. All the reduction in moisture content occurred immediately after mixing; further curing did not affect the moisture content because at such low temperatures pozzolanic reaction does not take place.

Figure 3 shows the variation of moisture content with increasing quantities of flyash and kiln-dust. Flyash and kiln-dust cause virtually the same decrease in moisture content, and even an addition of 50 percent of admixture is not sufficient to reach the optimum moisture.

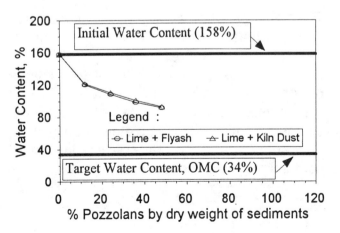

Figure 3 Effect of admixtures on moisture content of mixes

Atterberg limits

Mixes with three different percentages (4, 8 and 15) of lime were prepared and subjected to different curing temperatures (35°F, room and 120°F) for 48 hours. Atterberg limit determinations made on the cured mixes showed that the PI varies from 7 to 12 regardless of the quantity of lime or the curing temperature; also, the PI did not vary significantly beyond the initial reduction immediately after mixing.

Leach Tests

Samples for leach tests were prepared as follows:

i) 10 percent lime only
ii) 10 percent lime plus 30 percent flyash
iii) 10 percent lime plus 30 percent kiln-dust

The mixes were prepared from tube samples obtained at the sampling locations where the concentration of contaminants were the highest. The mixes were prepared to simulate 28-day curing by accelerating cure for 48 hours at 120°F. The cured specimens were then crushed to pass the 3/8" sieve opening.

The samples were analyzed for metals and semivolatiles. Both Total Contaminant Leachate Procedure Tests (TCLP) and Sequential Batch Leach Tests (SBLT) were performed on each sample. None of the metals listed in Table 1 were detected. Thus, the sediments were completely stabilized in the sense that no detectable levels of contaminants were observed.

Conclusions

Based on results of sampling and testing, the following conclusions were made:

1) Stabilization of contaminants detected in the marine sediments can be achieved by addition of quicklime.

2) Solidification of the marine sediments can be achieved by using quicklime. On a dry weight basis, about 8 percent quicklime is needed to achieve a pH of 12.4 and the required strength of 30 psi at room temperature.

3) The primary controlling parameters for a successful mix in the field are, (a) uniform mixing, (b) effective compaction, and (c) temperature greater than 40°F during curing of mix.

4) Since the initial water content of the sediments is very high, it is necessary to either add sorbents or air-dry the mixed sediments prior to compaction. Dry soil, lime-kiln dust or flyash can serve as sorbents.

The full scale field operation was performed in 1992 during the construction of the THT immersed tube trench. The contaminated sediments were dredged by clamshell, mixed with quicklime, and contained in a lined and capped site on Governors Island.

Acknowledgements

The authors thank Peter Zuk, Project Director, the Massachusetts Highway Department and the Federal Highway Administration for permission to publish this paper. Marshall

Thompson of the University of Illinois and Ara Arman of Woodward-Clyde Consultants assisted the authors in developing the laboratory testing program and interpreting the test results.

References

Arman, A., and Munfakh, G. A. 1970. "Stabilization of Organic Soils with Lime", Engineering Research Bulletin No. 103, Division of Engineering Research, Louisiana State University.

Environmental Laboratory. 1980. "Guide to the Disposal of Chemically Stabilized and Solidified Waste", SW-872, US Environmental Protection Agency, Washington, DC, 126 p.

Kita, D. and Kubo, H. 1983. "Several Solidified Sediment Examples", Proc. 7th Annual US-Japan Experts' Meeting, US Army Corps of Engineers, Fort Belvoir, Virginia.

JRB Associates. 1984. "Removal and Mitigation of Sediments Contaminated with Hazardous Substances", Draft Report, Municipal Environmental Research Laboratory, Office of Research and Development, US Environmental Protection Agency, Cincinnati, Ohio.

Massachusetts Department of Environmental Protection Regulations. 314 CMR 9.00. "Certification for Dredging, Dredged Material Disposal, and Filling in Waters".

On the Influence of Ca^{2+} Leached out of the Light Weight Stabilized Soils

A. Suzuki, Y. Kitazono, S. Maruyama
Dept. Of Civil and Environmental Engineering, Kumamoto University, Japan

Y. Hayashi
Dept. Of Asahi Kasei Inc., Japan

J. Yang
Dept. Of Civil and Environmental Engineering, Kumamoto University, Japan

SYNOPSIS: The light weight stabilization aims to stabilize the dredged-reclaimed soft grounds by strengthening and lightening ground surface to decrease the net load on the lower soft layers. In this stabilization, the surface part of reclaimed soft ground is stabilized by cement milk with fine air bubbles called "air cement milk". It was feared that leaching out of Ca^{2+} could cause loss of strength of the stabilized soil and pollution to the neighboring environment. It was found from our experimental research in laboratory and field that the effects of leaching out of Ca^{2+} on the strength of the stabilized soil and the neighboring environment of the stabilized area were small.

1 Introduction

In port and harbor areas, the sea routes are kept open by dredging, and the dredged soils are usually used in reclamation. The dredged-reclaimed soils make very soft ground, and soil improvement is necessary in advance of construction of structures. The light weight stabilization aims to stabilize the dredged-reclaimed soft ground by strengthening and lightening ground surface to decrease the net load on the lower soft layers. In this stabilization, the surface part of reclaimed soft ground is stabilized by cement milk with fine air bubbles called "air cement milk".

The porous structure of the light weight stabilized soils were considered to suffer from leaching out of Ca^{2+}. It was feared that leaching of Ca^{2+} could cause loss of strength of the stabilized soil and pollution to the neighboring environment. Our research was performed experimentally on the above problems in our laboratory and in the reclaiming site of Kumamoto port.

2 Samples and Field Test Site

Figure 1 shows the sampling points and the field test site. Four kinds of samples (SM, ML, MH, MH2) were taken in Kumamoto port and another one in a retarding basin in Ariake reclaimed land (the inner part of Ariake sea). In the former samples, two kinds of samples (ML, MH2) were taken in reclaimed site. The field test were performed in a part of reclaimed site where the sample, MH2 were taken.

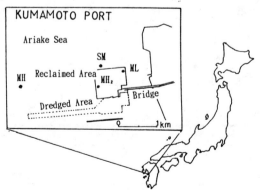

Fig.1 Location of Kumamoto port

3 Soil Samples and Air Cement Milk

Physical characteristics of samples shown in Fig.2 and table 1 were classified according to Japanese engineering classification system [1] as follows.

SM   : silty sand
ML   : low liquid limit silt
MH, MH2 : high liquid limit silt
CH   : clay

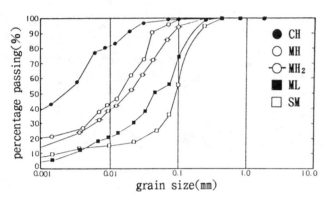

Fig.2 Grain size accumulation

TABLE 1 Physical Caracteristics of Samples

	S M	M L	M H	M H 2	C H
$\omega n(\%)$	51.0	64.6	111	68.0	180
$\omega L(\%)$	N.P.	N.P.	83.2	59.2	135
$\omega P(\%)$	N.P.	N.P.	41.7	37.1	62.5
I P	N.P.	N.P.	41.5	22.1	72.7

Clay minerals in these samples were mainly kaolinite and illite.

Two kinds of air cement milk (A.C.M) different in proportion of air contained, A and B, were prepared depending on samples as follows.

A : for SM, ML, MH, MH2
B : for CH

TABLE 2 Composition of air cement milk

	cement (g/l)	water(g/l)	density(g/cm³)
A	219.2	219.2	0.46
B	128.1	128.1	0.28

4 Laboratory Experiment

4.1 Research Method

Specimens for unconfined compression test were prepared in different addition ratio of cement for each kind of soil as shown in table 3.

TABLE 3 Addition ratio of cement for samples

Sample	Add. Rat. of Cement (%)					
	10	15	20	22.5	25	30
CH	○	○	○			
MH			○		○	○
ML			○	○	○	○
SM		○	○		○	

Specimens in each case of table 3 were separated to three different groups in curing conditions.
The first group, denoted by 1-6, were cured for 7 days, in air for the first day (submg. 1 in Fig. 3) and in water for the other 6 days. The second group, denoted by 6-1, were cured for 7 days, in air for 6 days at first (submg. 2 in Fig. 3) and in water for the other one day. The third one, denoted by 6-22, were cured for 28 days, in air for 6 days at first (submg. 2 in Fig.3) and in water for the other 22 days. In curing in water, amount of Ca^{2+} leached out from a specimen would change depending on the ratio of volume of water to the surface area of a specimen. The ratio were defined as 'equivalent water depth'. The water depths of 3 cm, 6 cm, 10 cm and 20 cm, were selected in this time. After the above curing, specimens were subjected to unconfined compression test and Ca^{2+} in the curing water was examined [2]. Also, leaching test on Ca^{2+} remained at points of different distance from center of each specimen was practiced after unconfined compression test, where samples through 2 mm sieve were stirred in water for 1 minute.

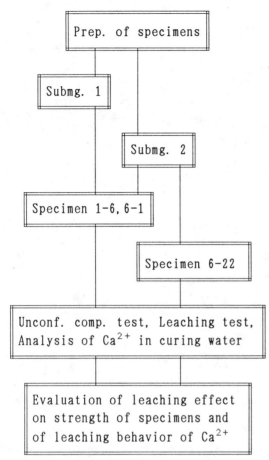

Fig.3 Flow of laboratory experiment

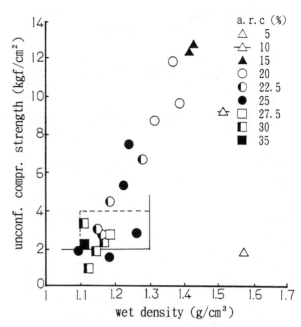

Fig.4 Change of wet density and unconfined compression strength due to addition ratio of cement

4.2 Wet Density and Unconfined Compression Strength

The wet density decreased with increasing of addition ratio of cement for a composition of air cement milk but unconfined compression strength had a peak value at a wet density for each sample. Figure 4 shows a example of sample ML, where the aimed area is enclosed by flame in this stabilization and a.r.c. means addition ratio of cement.

4.3 Leaching out of Ca^{2+} during Water Curing

Figure 5 shows the relation between equivalent depth and amount of Ca^{2+} leached during water curing of 6 days after aerial curing of one day. This figure gives the next results.

1) Amount of Ca^{2+} leached increased with equivalent depth for small value of the depth but the amount changed little for the depth more than 10 cm for each sample.
2) The finer sample soil was, the less was Ca^{2+} leached.

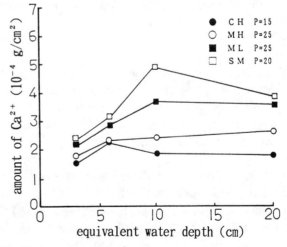

Fig.5 Amount of Ca^{2+} leached versus equivalent water depth

Figure 6 shows the relation between amount of Ca^{2+} leached during water curing in 10 cm of equivalent depth and addition ratio of cement.

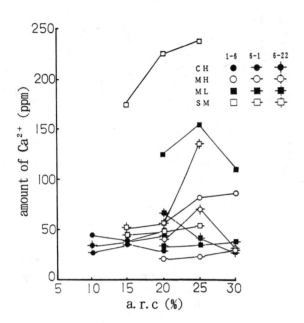

Fig.6 Amount of Ca^{2+} leached versus addition ratio of cement

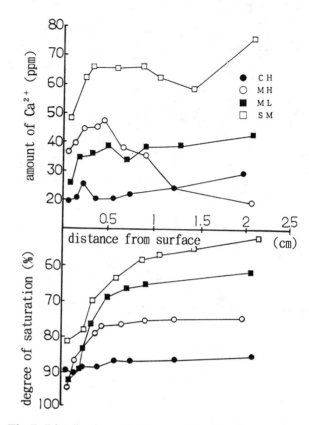

Fig.7 Distribution of degree of saturation and amount of Ca^{2+} leached in specimens after curing (6-22)

The following results is found in this figure.

1) Amount of Ca^{2+} leached increased with addition ratio of cement. The tendency was more considerable in sandy sample (SM) than in other samples.

2) The amount was particularly large in specimens of short aerial curing (1-6). The difference between the amounts of specimen (6-1) and specimen (6-22) was little except specimen (6-22) of 25% of addition ratio of cement of the sample, SM.

Figure 7 displays radial distribution of degree of saturation and amount of Ca^{2+} leached after unconfined compression test in specimens cured in the condition of 6-22, where axis of abscissa presents distance from surface of the specimens, ordinate presents amount of Ca^{2+} leached in upper part and degree of saturation in lower part each.

The following tendencies were found in this figure.

1) Degree of saturation was the highest at the surface and decrease toward the center, particularly in steep gradient within the distance less than 0.5 cm from surface except the sample CH whose degree of saturation was very high in the whole points of the specimen.

2) Amount of Ca^{2+} leached was the least at the surface conversely and increased to the center, particularly in steep gradient comparably to the above tendency except the samples CH and MH.

These tendencies mean that most of absorption of water and leaching of Ca^{2+} during water curing for 22 days occurred in the thin surface layer less than 0.5 cm.

4.4 Leaching Effect of Ca^{2+} on the Strength of Stabilized Soils

Let us examine the leaching effect of Ca^{2+} on unconfined compression strength of stabilized specimens in table 4. In this table, it is found that the specimens cured in the condition (1-6) were about from 15 to 30 % weaker than the ones cured in the condition(6-1) but the specimens cured in the condition (6-22) were from 35 to 58 % stronger than the ones cured in the condition (6-1) and these difference in strength due to curing condition was remarkable in the sample, SM.

TABLE 4   Difference of unconfined compression strength(qu) due to curing condition

SAMPLE	$\dfrac{qu(1-6)}{qu(6-1)}$	$\dfrac{qu(6-22)}{qu(6-1)}$
CH	0.84	1.39
MH	0.96	1.35
ML	0.88	1.53
SM	0.72	1.58

5 Field Experiment

5.1 Outline of Experiment

Field experiment was performed at the point indicated by MH2 in fig.1 to confirm the effect of stabilization with air cement milk and to investigate the influence on the neighborhood of the stabilized area.

In advance of stabilization, shear strength distribution were obtained insitu by vane shear test up to the depth of 4.2 m, at intervals of 0.2 cm up to the depth of 2.2 m and 0.5 m in the deeper part. Shear strength at each depth up to 2.2 m was less than 0.6 tf/m² except 0.74 tf/m² at the depth of 0.2 m.

Stabilization practiced by the system developed for this research whose main parts consist of a twin high speed mortal mixer (300 rpm), stabilizing machine with auger and mixing bar at the tip of mixing rod to mix air cement milk into the soft ground, and a mortal pump.

Addition ratio of cement was decided from laboratory test on the sample, MH2 in advance so that the result in the site will reach the target in fig.4.

The size of stabilized area was 300cm×300cm in spread as shown in fig. 8 and 120 cm in depth. Sampling was performed by Denison type sampler at 5 points, No.1 - No.5 in the figure, one month after stabilizing to confirm the effect.
Samples of only 2 points, No. 4 and No. 5 obtained but the sampling at the other points did not succeed. Then homogeneity of wet density and sounding indices were checked at the points near the above sampling points in the test pit dug in the stabilized area.

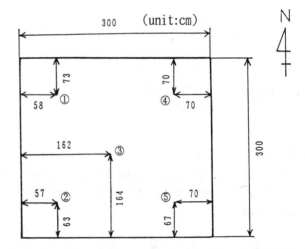

Fig.8 Location of investigation points in stabilized area

5.2 Result of Experiment

Effect of stabilization was investigated on wet density and unconfined compression strength of the samples of points of No. 4 and No. 5 comparing the results of laboratory test on the sample, MH, in fig.9 where laboratory test results were presented by the open marks with different addition ratio of cement. It is found that the field stabilization was effective but distant from the target area set up in fig.4. The reason of the result was thought that the actual amount of air cement milk mixed into the ground was about 30% smaller than the required one, because of loss of pressure head in transporting hose due to high compressibility of air cement milk.

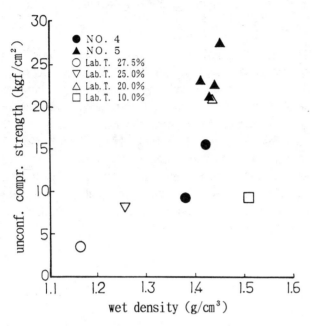

Fig.9 Comparison of result between laboratory and field experiments

The distribution of wet density in the test pit is shown in fig. 10. The homogeneity of stabilization effect is understood to be almost kept among each point except No. 5.

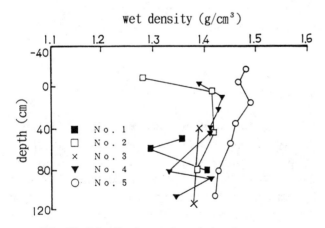

Fig.10 Distribution of wet density

Almost no influence is recognized on the environment of neighborhood of the stabilized area in terms of Ca^{2+} as shown in table 5 comparing data after stabilization with the one before.

TABLE 5 Result of analysis Ca^{2+} leached from soils of a neighboring point

(ppm)

	Ratio of Water to Soil*			
	5	10	20	50
Before	66.4	32.5	19.9	9.03
After	47.0	23.0	12.6	7.22

* mass ratio of water to dry density of soil

6 Concluding Remarks

The following result were obtained from this research.
1) The higher the air cement milk ratio was added in stabilization, the lower density and the more were the amount of Ca^{2+} leached from the stabilized soil. The strength of the specimens, however, showed a peek at a certain ratio of that.
2) The earlier the stabilized soil contacted with water, the more was Ca^{2+} leached.
3) The stabilized soil with fine grain size distribution were susceptible to leaching of Ca^{2+}.
4) Loss in strength was recorded in the stabilized specimens subjected to leaching in younger age. Loss in strength was small in specimens cured more than 6 days in air.
5) There was no significant increase in Ca^{2+} ion concentration at neighboring points to the stabilized area.

References

[1] JSSMF, JSF M 111-1990
[2] Hokkaido Chpt. of JSAC. water, pp. 161-165, 1971

Environmentally Acceptable Reclamation using Materials Dredged from Adjoining Seabed

T. Yamanouchi
Kyushu Sangyo University, Fukuoka

T. Ishibashi, M. Takagi
Bureau of Port and Harbor Construction, Fukuoka City, Fukuoka

T. Tashima
Nihon-Chiken Co., Ltd., Fukuoka, Japan

ABSTRACT Bureau of Port and Harbor Construction, Fukuoka City, Kyushu, Japan, carried out from 1982 to 1986 coastal reclamation works in the seaside area. The bay had a total area of 254.7 ha divided into four sections for the reclamation. Materials used for the reclamation were silty clay and sandy soil which were dredged up from the seabed by using grab-type excavators and the purchased gravelly soil was used for the top layer of reclaimed deposit of the land mass. As the area adjacent or behind the reclamation is inhabited by residential houses and schools, numerous strict environmental preservation measures such as measures against sea water pollution, air pollution, measures against noise and vibration, and pollution of neighboring seashores and beaches are taken at every stage of works. In order to meet such requirements as several geotechnical countermeasures were adopted and various environmental protection standards were observed during the works, and checked if the measured site-data did not exceed the limit values authorized by the Government as standard regulations at all stages of works. At present the lands are being rigorously used as lands for residential, public, cultural and sport facilities including a marine one especially at the artificial beach areas.

1. ENVIRONMENTAL CONDITION OF ADJOINING AREAS BEFORE RECLAMATION

The environmental condition of the areas around the reclamation site is fair with a quiet atmosphere. The reclamation site adjoins the sub-downtown center of Fukuoka City, and the area toward the reclamation seaside is densely populated with the accumulation of residential quarters and schools and a port of seaside land is being used as a public recreation center.

Without detriment to such an environmentally fair condition precautionary measures are taken to preserve the environmental aspects of the area in constructing a new seaside land and also to exert efforts toward the improvement of city environment in harmony with the existing nature.

Prior to the construction works, a plan for monitoring of the environment is laid, and the monitoring items and their standard values are set up.

Moreover, when an emergency case is predicted, or occurs as expected, a coordination or information sharing organization is envisaged to propose and locate new necessary facilities.

Locations of the points that are to be environmentally monitored are described in Fig. 1 and the monitoring items and their standard values are stated in Table 1.

2. MEASURES AND METHODS OF ENVIRONMENTAL PRESERVATION

2.1 Methods and Countermeasures Against Water Pollution

As to the measures against water pollution, some countermeasures against the

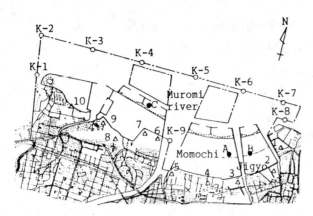

Environmental monitoring points
○ Basic water-quality monitoring points
● Excess-water monitoring points
△ Noise and vibration monitoring points
× Air-quality monitoring points

Fig 1. Location of Monitoring Points

section through which earth moving barges can pass. The protection blanket is laid to go down to the depth of HWL or seabed as shown in Fig. 2 with an aim to ensure complete protection from turbidity dispersion.

Moreover during the excavation by grab-type dredgers, the anti-turbidity blanket is set up; the grab-type dredger operation is performed within the blanket enclosure. Thus the sea water is protected against turbidity dispersion.

2) Reclamation Works

Construction works, placing of fill

Table 1. ENVIRONMENTAL OBSERVATIONS AND THEIR STANDARDS

Item	Observation name	Observation site	Observation items	Standard values
Water quality	Basic monitoring	9 sites (K1 to K9)	SS turbidity, clarity, daily-life environmental items (pH, COD etc.) and poisonous items (cadmium, cyanogen stc.)	Density of SS < 10 mg/l resulting from construction works
	Auxiliary monitoring	Flow direction of dredger ship	SS turbidity, clarity during dredging	
	Out-flow water monitoring	Out-flow water treatment facilities	SS, poisonous items	SS < 50 mg/l Poisonous items: not detected
Noise and vibration	Noise, vibration	10 sites (1 to 10)	Level of noise and level of vibration	Noise: Pile driving < 85 phon Other works < 60 " Vibration: Pile driving < 75 dB
Air	Air quality	On roof top of building	SO_2, NO_2, wind direction, wind velocity	Both SO_2, NO_2 < 40 ppm
Wild birds	Wild birds	Mouth of the Muromi river	Kinds of wild birds and habitat condition	

wide-spread of turbidity are taken for all cases of sea water turbidity that resulted from construction works.

1) Dredger Operation

The dredged site is surrounded by turbidity protection blankets on four directions leaving only the river-mouth

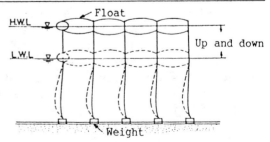

Fig. 2. Turbidity Protection Blanket

materials were carried out starting from No. 1 Construction Zone where the outer-perimeter revetment and temporary revetment had been completed. Clayey soil was damped directly from the open-base type barges. Such a "direct damping" was considered to be a protection method against muddy-water dispersion as the extraction of clay by grabs leads to less amount of water content and less agitation of clayey soil.

Pumping ships were employed as shown in Fig. 3 for reclamation at the area of more than -1.0 m where the open-base barges were not able to operate. Hence the area had a very high water turbidity. Thus the area was protected against the turbidity dispersion by a temporary wall enclosure and surplus clear water was

The area higher than DL -1.00 m had an extremely low water content. Here a floating conveyor system ("FCS") of reclamation was adopted as shown in Fig. 4. This method was capable of spreading sandy soil in homogeneous layers on soft clayey soil without the occurrence of slip-failure and lateral movement of the combined mass of soil.

With the adoption of that method the water content of soil can be extremely reduced as compared with the pumping method and the clayey soil can be fully contained. The area above the level +2.5 m is covered, in a similar way with No.1 Construction Zone, by a good quality soil. Consequently the reclaimed land has a composition as shown in Fig. 5.

3) Treatment of Surplus Water

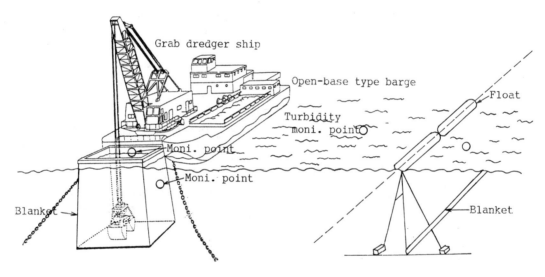

Fig. 3. Grab Dredger Assembly

drained out after sedimenting the suspended particles by flocculation process in a treatment pond.

The area higher than DL +2.5 m is covered with a good-quality gravelly soil and planted with grass as a part of land beautification measures. At No. 2 Construction Zone clayey soils are distributed around the areas 2 to 3 m below sea level. On this area the open-base barges were employed for direct samping of fill materials (clayey soil) up to the depth of DL -1.00 m.

The pump dredging method accomplishes the reclamation by transporting a large quantity of sand and sea water. At that time the turbid polluted water that resulted in is drained out of the reclamation area. It is required to drain out this water after it is treated to the specified water quality. As a countermeasure to this effect a treatment facility for the surplus water is provided in the reclamation area to enhance the sedimentation of soil grains.

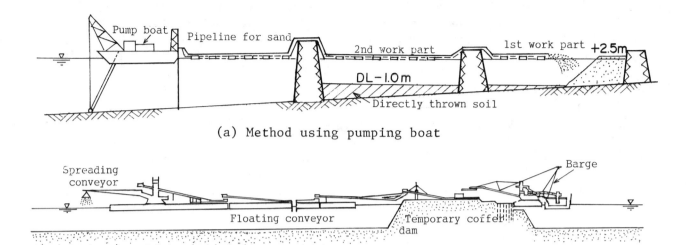

(a) Method using pumping boat

(b) Method by means of "floating conveyor system"

Fig. 4. Two Kinds of Reclamation Methods

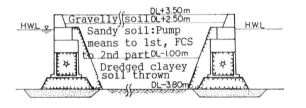

Fig. 5. Section of Reclaimed Land

The treatment method is as shown in Fig. 6. The turbid polluted water from the reclamation site is led to a high-speed agitating tank where an inorganic flocculation agent is introduced to make the suspended materials settle down. And then the treated water is drained to a settlement pond after passing through a slow-speed agitating tank. The clear water is pumped out of the reclamation area into the sea. The flocculation agent that was used here is not injurious to living creatures and the agent was determined on the basis of the results of flocculation tests conducted with several flocculation agents.

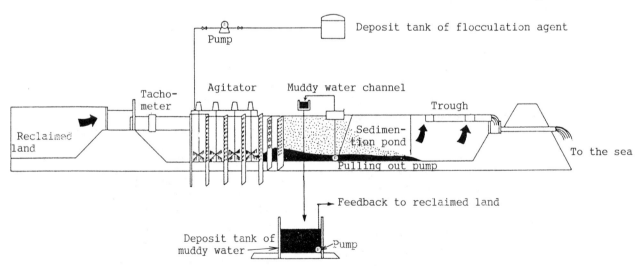

Fig. 6. Excess-Water Treatment Facility

2.2 Other Methods of Countermeasures

Countermeasures against noise and vibration (Figs. 7 and 8), air pollution and dust, and the preservation on birds and marine life were also respectively adopted during the construction works.

3. RESULTS FROM ENVIRONMENTAL PRESERVATION STUDY

With an objective of monitoring the environmental changes during construction environmental studies were conducted at the fixed points at every fixed interval time.

Construction works were started in April of 1982 simultaneously at four construction zones. The dredging and reclamation period differ from construction zone, and the dredging and reclamation associated with operation on the seaside were finished in the year of 1985. Such preparation works as revetments were completed on the two sections (Jigyo and Momochi) and other two sections in 1988 (Fig. 9).

The results of water-quality monitoring by SS-concentration degree are summarized below.

Fig. 7. Noise-Proofing Dredger Ship

Fig. 8. Pipeline Covered by Noise Absorber

Fig. 9. Recent Aerial View of the West Land (Jigyo and Momochi) (Mar. 1993)

Table 2. RESULTS OF WATER-QUALITY OBSERVATION UNIT (in mg/l)

Period	Year	Observation number								
		K-1	K-2	K-3	K-4	K-5	K-6	K-7	K-8	K-9
Before construction	1981	8	8	7	8	9	10	11	11	11
During construction	1982	10	8	9	9	9	10	9	10	14
	1983	11	11	12	13	14	14	12	13	13
	1984	7	7	8	8	9	10	9	10	10
	1985	8	8	8	8	9	10	10	11	10

The values are the averaged ones of observation (it is the average of monthly and the that of water depths of 0.5 m and 2.0 m). Locations are shown in Fig. 1.

When the results of daily observations at a fixed point are expressed in terms of annual average values, they will be as shown in Table 2. At some places the SS-concentration is compared with the value before construction, higher by 3 to 5 mg/l especially the first and second construction years. The time is the most active one for dredging and reclamation of the whole area, but the values do not change and they are almost the same as the values before construction, and the values do not go beyond the specified values.

4. CONCLUSION

Upon consideration of the fact that the reclamation works involve the area adjacent to the city's residential quarters and the bay itself is a sanctuary for birds and wild life, the reclamation works were carried out taking precautionary considerate measures for water quality preservation, protection against noise and vibration and air pollution placing emphasis on general environment and daily-life environment. Especially for the water-quality preservation, the countermeasures such as setting up of blanket for containing turbidity and pollution during dredging, prevention of clayey soil dispersion during reclamation and prevention of sea water pollution by setting up treatment ponds for excess water that comes out from reclamation were taken.

As a result of countermeasures the impact on the residents is negligible and changes hardly takes place with kinds and number of birds and wild life. Thus it is to be concluded that the reclamation works were successfully completed.

APPENDIX A LAWS AND REGULATIONS RELEVANT TO THE ENVIRONMENTAL PRESERVATION

Bureau of Port and Harbor, Fukuoka City, and Hakata Port Development Co., Ltd. (1982). Note on the countermeasure against pollution during the water break works, 10 pp. (in Japanese).

Ditto (1983). Environmental preservation countermeasures during the reclamation works, 9 pp. (ditto).

Ditto (1987). General analytical report on the Jigyo and Momochi reclamation grounds, 230 pp. (ditto).

Japan Harbor Assoc. (1987). Future development plan of Hakata Harbor, 19 pp. (ditto).

Japanese Government (1971). Law on the noise control, Law No. 88, Rev. (ditto).

Ditto (1974). Law on the air pollution prevention, Law No. 65. (ditto).

Ditto (1976). Law on the vibration control, Law No. 64. (ditto).

Industrial

In Situ Control of Contaminated Groundwater

John A. Cherry
University of Waterloo, Waterloo, Ontario, Canada

Final manuscript not received in time for inclusion in the *Proceedings*. For additional information, please contact:

        Prof. John A. Cherry
        Waterloo Centre for Groundwater Research
        University of Waterloo
        Waterloo, Ontario NL2 3G1
        CANADA
        Voice: +519-885-1211
        Fax: +519-746-5644

Assessment of Oil Sand Fine Tails Reclamation Strategies

A.E. Fair
Syncrude Canada Ltd., Fort McMurray, Alberta, Canada

D.E. Sheeran
Suncor Oil Sands Group, Fort McMurray, Alberta, Canada

Abstract

The Athabasca Oil Sand Deposit, located in northern Alberta, is the largest of Alberta's four oil sands deposits. It contains reserves of some one trillion barrels of bitumen in place. The technology currently utilized to extract the bitumen from the oil sands utilizes a water-based extraction process which results in the generation of large volumes of coarse and fine tails. The properties of the fine tails are such that they consolidate very slowly and essentially remain as a fluid indefinitely. The development of a suitable reclamation plan for these fluid fine tails is consequently a significant challenge. The absence of an acceptable reclamation scheme for the fine tails could limit future development of the oil sands using current extraction techniques.

Extensive work, aimed at understanding the behavior of the fine tails and developing reclamation options has been conducted, and is currently underway. Much of the work is conducted using a collaborative approach involving industry, provincial and national research groups, academic institutions, government regulatory bodies and private research groups. The work is conducted by multi-disciplinary teams who are able to bring various perspectives to the problem.

This paper is intended to describe an approach which is being utilized to assess various potential oil sand fine tails reclamation strategies. The approach is based on utilizing probabilistic methods to quantify various input parameters. The application of the probability assessment methodology to the oil sand fine tails differs from other probability assessments in that the methodology is being used as a tool to facilitate understanding and to consolidate a large number of complicated issues so that all stakeholders can have input into the assessment.

Introduction

The Athabasca Oil Sands deposit contains approximately one trillion barrels of bitumen in place. Based on surface mineable areas only, the recoverable reserves are estimated at 33 billion barrels of bitumen. Figure 1 shows the location of the Athabasca oil sands relative to the other oil sand deposits in Alberta. Commercial operations in the Athabasca oil sands currently produce over 15 million cubic metres (90 million barrels) of synthetic crude oil per year. This represents about 16% of Canada's total petroleum needs. The projected decline in Alberta's conventional oil production will result in a need to continue the development of the oil sands in order to limit Canada's need for foreign oil imports.

Two companies are currently commercially developing the Athabasca oil sands. They are Syncrude Canada Ltd. and Suncor Oil Sands Group. Suncor began operations in 1967 and currently produces about 5 million cubic

Figure 1 Distribution of Alberta Oil Sands

metres of synthetic crude oil (SCO) per year. Syncrude began its operations in 1978 and currently produces approximately 10 million cubic metres of SCO per year. Both companies utilize surface mining methods to mine the oil sand ore and then subsequently transport it to the Extraction Plant.

Methods Used for Bitumen Extraction

Both Syncrude and Suncor also use hot water-based extraction techniques to recover the bitumen. The hot water extraction process generates a tailings stream which is subsequently pumped to a tailings storage area. Soon after start-up of the Suncor operations in 1967, it became apparent that the extraction process also resulted in the generation of fine tails. This was not initially anticipated. Upon discharge into the tailings storage area these fine tails segregate from the coarse tails and flow to the central portion of the tailings storage area. Figure 2 shows a schematic cross-section of a tailings storage area illustrating the accumulation of the fluid fine tails within the tailings settling basin. Fine tails originate in the colloidal size mineral particles (primarily clays) of the oil sands. They are mined and processed along with the oil sand. During processing, the clays are broken up and suspended in the process water by the combined action of the hot water and the mechanical energy input of the initial slurry preparation step. The hot water extraction process requires large quantities of water, as a

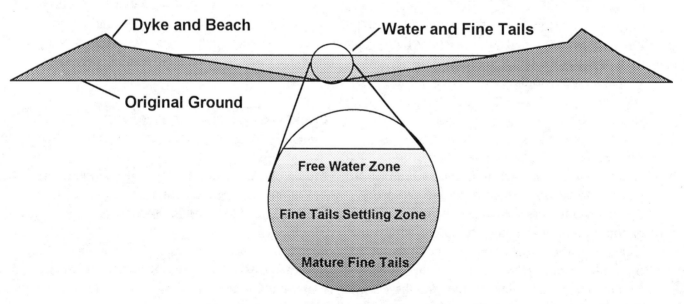

Figure 2 Tailings Settling Basin

result, large quantities of fluid tailings are also produced. Every cubic meter of synthetic crude oil produced results in about 2 cubic metres of fluid fine tails. Today over 350 million cubic metres of fluid fine tails have accumulated at the Syncrude and Suncor sites with about another 30 million cubic metres being produced annually. The coarse solids (i.e., greater than 22 microns) settle relatively quickly and are used to construct containment dykes and beaches. About 40 to 50% of the fine tails fraction (i.e., less than 22 microns) of the oil sands solids enter the tailings settling basin as a thin slurry where they are allowed to settle. However, consolidation of the fine tails is very slow and much of the material is essentially non-settling.

Fine tails management is a key lease development activity in that the nature of the material (i.e., its fluid state and very slow consolidation rate) make it more difficult to handle. Interim storage is provided by large tailings settling basins. Long term storage of fluid fine tails will require containment below original ground elevation.

Fine Tails Characteristics

The mean particle size of the fine tails is between 5 and 10 microns. The average solids content of mature fine tails is about 33% (g/100g) which equates to an average void ratio (volume of voids/volume of solids) of greater than 5. The void ratios of the fine tails range from more than 25 in the initial tailings slurry that enters the settling basin to less than 2 in the deepest portions of the settling basin. The permeability of mature fine tails is very low; laboratory measurements give results within a range of 10E-6 to 10E-9 m/s.

Fine tails display a gel-like or thixotropic character which tends to explain some of their unique properties. This gel-like character can be temporarily broken by shearing but when left alone, the gel-like structure will quickly reestablish itself. This gel-like structure causes fine tails to behave as a non-Newtonian fluid (i.e., its rheological properties change as a function of the shear rate). The rheological properties of the fine tails influence their rate of development and composition.

The water fraction of fine tails exhibit some short term toxicity to aquatic organisms. Toxicity of fine tails is generally measured in terms of specific tests in which the rate of survival of certain organisms is measured over a specific period of exposure to the fine tails. Typically the results are reported in terms of a lethal (LC) or effective (EC) concentration. Various bioassays for lethal response, viability and quality have been developed using various organisms (e.g., fish zoo plankton, larvae) to determine short term (acute) and long term (chronic) effects of the fine tails. In a whole sample of fine tails (i.e., water and solids) toxicity can also result from the physical impact of the high turbidity of the fine tails in the storage areas and from the fouling of the gills by unrecovered hydrocarbons.

The toxicity of tailings waters, overlying the fine tails, rapidly decreases with time. When tailings water is separated from the tailings settling basin water and maintained under aerobic conditions, a rapid decrease in toxicity is measured, such that after three to nine months, the toxicity is essentially reduced to zero.

Collaborative Approach to Fine Tails Reclamation

Both Syncrude and Suncor have invested large amounts of time and money into developing a better understanding of fine tails over the years. The work has been primarily aimed at developing techniques to manage and eventually dispose of the fine tails. During the late 1960's, soon after start-up, Suncor recognize that fine tails had many unique properties which would require special handing. As a result, Suncor initiated the early studies into the nature of fine tails. These studies concentrated on both mineralogical and surface chemistry investigation as well as mechanical methods for accelerated dewatering of the fine tails. Syncrude has also conducted extensive research into the nature of fine tails. Building on Suncor's knowledge that fine tails would be a by-product of the extraction operations, Syncrude began work prior to start-up on developing an understanding of fine tails.

Much of the early work was based on the assumption that a "magic bullet" existed which would cause the non-settling fine tails to consolidate very rapidly. If such a solution could be found it would be relatively easy to provide adequate containment for the remaining solids portion of the fine tails. However, after many years of trying researchers began to acknowledge that it was not going to be possible to find the magic solution, at least not in a time frame necessary to implement into the existing operations. At this point, the research shifted to focus on techniques aimed at managing the fine tails in its fluid form.

In order to manage the fluid fine tails, it was necessary to develop techniques for integrating them back into the mined ore areas. This led to the development of the "water-capping" technique which is described in more detail later in this paper. In simple terms, the water-capping technique is aimed at developing a "live lake" over fluid fine tails. The overlying lake acts to isolate the fine tails and "receives" the pore waters being released from the fine tails, which are eventually allowed to be discharged to natural waterways.

As both Syncrude and Suncor began to focus in on this water-capping technique, it also became apparent that there was value in sharing information and conducing joint research in common areas of interest. In addition, the magnitude of the fine tails problem had necessitated the involvement of many others. Several consulting and university personnel had been engaged by both Syncrude and Suncor to assist with the fine tails research work. As a result, the Fine Tails Fundamental Consortium was formed in 1989 with the objective of coordinating research aimed at obtaining a better understanding of the mechanisms controlling the properties and behavior of fine tails. The Consortium recognized the value of multi-discipline teams to enhance the level of understanding. The water-capping technique alone requires the integration of numerous different areas of study such as colloidal scientists, geologists, hydrogeologists, hydraulic engineers, geotechnical engineers, biologists, chemists, etc. It was felt that an improved understanding of fine tails fundamentals would lead to the development of technologies which could alleviate the fine tails accumulation. The Consortium consists of Syncrude, Suncor, AOSTRA (now part of Alberta Energy Oil Sands and Research Division), CANMET, OSLO, ARC, NRC, Alberta Energy, and Environment Canada. In addition, contract work by numerous university and private scientists has been funded by the Consortium.

In order to provide the necessary data base for the Consortium to conduct its work, both Syncrude and Suncor agreed to make all their relevant information available. Initially the information that was placed in public files tended to revolve around the fundamental properties of the fine tails. In more recent years, both Suncor and Syncrude have agreed to share all their respective data on various aspects of fine tails. This represents millions and millions of dollars and over two decades of research into the behaviour of fine tails. This collaborative approach to developing both fundamental knowledge and fine tails management techniques, is seen as a major step forward in terms of conducting much more efficient and effective research and development work on fine tails. In the past several years since this collaborative approach has been implemented, several fine tails management techniques have been brought forward without the encumbrance of ownership due to patents, etc. Many of these have been presented in the form of technical papers at conferences sponsored by the Fine Tails Fundamental Consortium.

The collaborative approach to fine tails research has also been extended to include the government regulatory groups. The regulatory staff freely participate in the various Fine Tails Fundamental Consortium forums and have equal access to data. As part of both Syncrude and Suncor's operating permits, an annual workshop is held in conjunction with both industry and government regulatory groups to review and discuss progress on fine tails research work over the previous year and the workplans for the upcoming year. The regulatory groups are also provided with extensive detail on fine tails research as part of various submissions, including Conservation and Reclamation Plans and as part of the supporting documentation for applications to the Energy Resources Conservation Board (ERCB) which is now part of the Alberta Energy Oil Sands and Research Division.

On a broader front, this type of collaborative approach to research in the oil sands has led to the formation of an organization referred to as the Canadian Oil Sands Network for Research and Development (CONRAD). CONRAD is a multi-stakeholder research and development consortium, focused on oil sands and heavy oil, with a vision based on using a cooperative approach to develop new perspectives, challenge old assumptions and ensure that the research is targeted to support the industry in the future.

Fine Tails Reclamation Options

The large volume and non-settling character of fine tails together represent a significant challenge in terms of developing reclamation techniques that provide a suitable long term disposition of the fine tails. A wide range of reclamation options are being actively pursued to manage the fine tails. The approaches can be broadly classified as:

- incorporation of the mature fine tails into a wet landscape,
- incorporation of fine tails into a dry landscape, and
- reduction of the volume of fluid fine tails.

<u>Water-Capping in a Wet Landscape</u>

Current mining and reclamation plans at both Syncrude and Suncor involve the placement of fluid mature fine tails in the mined out pit areas where they will be capped with a water layers. This approach relies on large land mass structures to provide containment for these artificial lakes along with a surface drainage pathway into the natural adjacent waterways. The technique is aimed at developing a stable, maintenance free, aquatic reclamation site which allows for the development of a self-sustaining, aquatic ecosystem similar to natural local lakes. Natural water flow through the system is provided to stabilize the lake surface elevation and the thickness of the water cap layer. Surface water drainage from the lakes will pass through a marsh which acts as a buffer before passing into open waterways. The system is intended to ultimately provide a useful wildlife habitat, such as wetland for migrating water fowl. Figure 3 shows a schematic illustration of this technique, as well as the dry landscape approach.

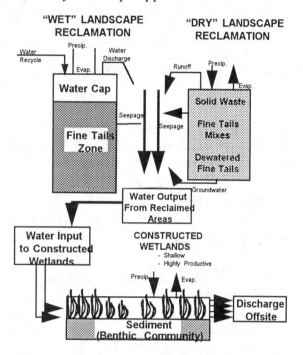

Figure 3: Fine Tails Reclamation Strategies

Both Syncrude and Suncor recognize that prior to accepting this reclamation approach for fine tails, several important questions need to be answered. Much of the research in this area of fine tails management has focused on these questions. They include:

- depth of capping water required to prevent mixing with the fine tails zone,
- impact of water depth on the long term stratification of the capping layer,
- impact of water released during densification of fine tails on the quality of the surface zone,
- potential for gas formation in the fine tails,
- rate of biological colonization and level of productivity achieved,
- role of the surrounding watershed on the depth and quality of the surface zone,
- role of littoral zone development,
- hydrogeology of the in-pit storage area, and
- potential for build-up of toxicity in the water layer in contact with the fine tails.

Extensive research and development work including analytical, laboratory and field work has been conducted over the years in order to advance the understanding associated with the physical, chemical and biological aspects of the water capping approach to managing fine tails.

Fine Tails in a Dry Landscape

Advocates of the dry landscape approach argue that, if the fine tails were disposed of as a solid, the mined-out pits could be utilized for storage of greater volumes of overburden and coarse tailings, which are otherwise often stored some distance away, resulting in higher materials handling costs. In addition, they argue that the solid disposal form has inherently less risk in terms of long term stability of the containment structures required to store the fluid fine tails. As a result of these issues, extensive research has also been conducted in order to develop alternate fine tails management techniques that focus on achieving a solid land mass. The work has primarily concentrated on combining the fine tails with other waste streams, such as overburden or the coarse tails, although several other mechanically based or chemically based approaches have also been proposed. We will limit our discussion to two techniques which Syncrude and Suncor feel hold the most potential.

The first approach involves increasing the fine tails solids component in the coarse tails deposit. In the current tailings disposal method, about 4 to 5% (by weight of dry solids) of the coarse tails deposits consist of less than 22 microns sized solids, which are generally considered to represent the main component of the fine tails. Laboratory samples have been tested to determine the engineering properties of various sand-fine tails deposits with fines content ranging from 5 to 20%. The results indicate that the fines content can be significantly increased to somewhere in excess of 10% without jeopardizing the overall stability of the deposit.

Two methods have been put forward to increase the fine solids contents of the sand deposits. The first involves replacing the coarse tailings pipeline make-up water with mature fine tails at the Extraction Plant (referred to as "spiking") so that there are more fine tails in the pore spaces of the saturated sand deposits. The second method involves adding lime and/or acid to coagulate the fines particles to produce a non-segregating fines particles to produce a non-segregating tailings mix where the fine and coarse solids do not separate upon discharge from the tailings pipeline. Comprehensive laboratory and field work has been conducted by both Syncrude and Suncor in order to progress and develop these methods. Syncrude is currently planning to implement the first method (i.e., spiking) by 1996, while Suncor is giving serious consideration to implementing the second, non-segregating tails approach at some point in the future.

The second approach to achieving a solid landscape involves mixing the fine tails with overburden. The Clearwater Formation clay shale overburden, found mainly at the Syncrude site, is very dry and has an extremely high plasticity range. By mixing the fine tails with an appropriate amount of clay shale, a soft clay deposit can be produced. This material can be incorporated within stable overburden dumps or contained in the mined-out pits. Various methods have been experimented with to mix the overburden and fine tails. Current efforts at Syncrude are focused on using a mixing device, referred to as a cyclofeeder, which creates a slurry allowing the fine tails-overburden mixture to be pipelined to a disposal site.

Reduction in Fine Tails Volume

Considerable work has also been undertaken in order to develop techniques which reduce the volume of fluid fine tails. It typically takes about 10 years for fine tails to mature to a state where it reaches a solid content of about 30% at which point the fluid is still 85% water by volume. Further dewatering or consolidation to anything resembling a soft clay (i.e., 67% solids by weight, 57% water by volume) would take a very long time.

Utilization of natural freeze-thaw cycles to increase the solids content of the fine tails beyond 30% is currently being evaluated by both Syncrude and Suncor as a method of reducing the ultimate volume of fluid fine tails. Results to date have demonstrated that the volume of fluid fine tails can be reduced by up to 50% over a single freeze-thaw cycle. Repeated application of annual freeze-thaw cycles would improve this number. Evaporation of water from thin layers of fine tails has also been experimented with. The evaporation of the water from the fine tails transforms the fluid fine tails into a soft clay. Additional studies undertaken at Syncrude have shown that adding these more dense fine tails to tailings sand improves the soil texture, increases the moisture retention and plant nutrient holding capacity, and consequently improves overall revegetation performance.

Figure 4 provides a schematic summary of the various reclamation options currently being considered.

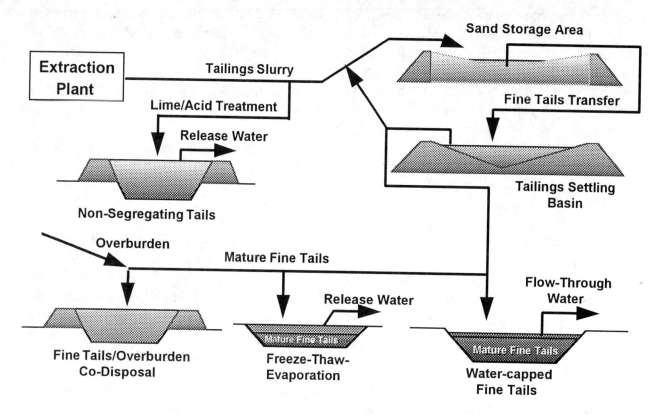

Figure 4 Potential Fine Tails Reclamation Options

Fine Tails Performance Assessment

The debate between the advocates of the fluid landscape approach to fine tails management versus the advocates of the solid landscape approach to fine tails management has resulted in numerous discussions which are often very subjective and difficult to quantify in terms of costs and/or potential liabilities. in order to provide more definitive answers to some of these subjective questions both Suncor and Syncrude recently considered the use of a performance assessment approach to evaluating various methods for fine tails management. This technique allows for a comparative assessment of fine tails reclamation alternatives on a more quantitative basis as compared to the various subjective type of arguments typically used today.

Before we explore the technique, it is important to understand what the underlying objective is for any fine tails reclamation scheme. Overall, the objective is to produce a landscape which is stable, productive and self-sustaining. The final landscape must meet the following criteria:

- it must be geotechnically stable and non-erosive,
- have a productive capability at or above the pre-disturbance landscape, and
- be free of any need for long term maintenance requirements.

In addition, the resulting environment must develop into an ecosystem that is:
- self-sustaining, and
- diverse and self-supporting with natural communities of plants and animals.

In a recent preliminary joint performance

assessment study, various generic landscape units were identified which were considered representative of the components of the fluid and/or solid landscapes referred to earlier. The assessment focused on several different factors as measures of the relative merits of a particular landscape unit associated with the fluid or solid landscape. The factors included:
- engineering risk,
- ecological risk,
- human health risk,
- financial factors, and
- public and regulatory acceptance.

The various landscape units were assessed against the different factors using a probabilistic method to estimate the hazards and derive the risks. The method used also allowed for the inclusion of qualitative data by means of expect professional judgment. Although this input is not directly quantifiable, it was captured and incorporated in the risk analysis through the use of subjective probability techniques. Figure 5 schematically shows how the input parameters were considered in probabilistic terms for the water-capping technique.

The output from the performance assessment is not intended to provide the final word as to which fine tails reclamation scheme should or should not be implemented, but rather to provide input into the decision. Issues such as the geotechnical design requirements associated with a containment structure for fluid fine tails which must function for an essentially indefinite period of time are unprecedented in terms of current practice. Likewise, the issue of potential release of contaminants over a very long period of time at the huge scale of the oil sand operations is also largely unprecedented. Figure 6 shows a typical probability curve for a fine tails reclamation input parameter which can be assessed against other scenarios in relative terms.

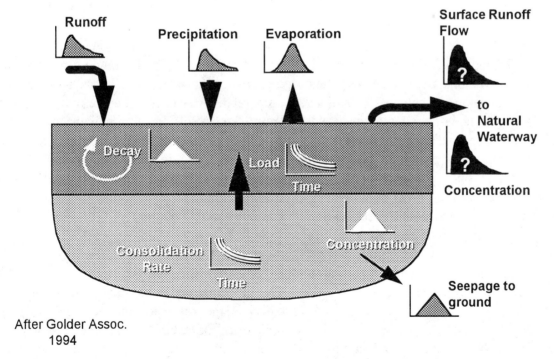

After Golder Assoc. 1994

Figure 5 Probabilistic Charaterization of Wet Landscape Input Parameters

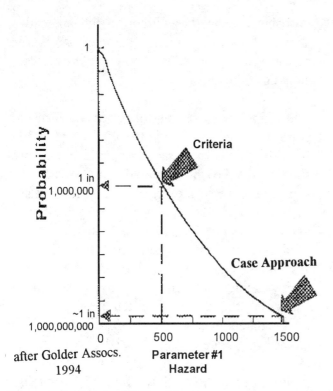

Figure 6 Probability Assessment for a Reclamation Option versus a Particular Hazard

Conclusions

The issue of oil sands fine tails reclamation is a very real one. Millions of dollars are spent every year to develop acceptable techniques for their disposal. The huge volume which will ultimately exceed one billion cubic meters for the current Suncor and Syncrude operations combined, results in significant costs to provide the necessary long term containment. The absence of an acceptable reclamation scheme for the fine tails may limit future development of the oil sands using current extraction techniques. Extensive research work has shown that there are no "magic solutions".

In the end, it is likely that a combination of techniques will need to be adopted in order to manage the fine tails. If one attempts to utilize only one technique, the marginal costs to reduce additional volumes of fine tails, using the same technique generally tend to increase. Consequently, it is more efficient to utilize a collection of techniques and optimize the use of each to achieve an overall optimal approach to fine tails management which considers all factors including cost, long term liability, and environmental acceptability.

The issue then becomes, "How do we make decisions which take into account both environment and cost considerations?" Syncrude and Suncor are proposing that a performance assessment approach be adopted as a tool to facilitate a better understanding of the issue. This probability based approach allows:

- the quantification of both tangible and less tangible risks,
- all stakeholders can provide input,
- sensitivity analysis can be conducted, and
- a guide for future research by identification of key parameters can be obtained.

The performance assessment approach provides a level of understanding for fine tails reclamation activities which can be compared on a relative basis to other activities, such as your chances of developing cancer or being struck by lightning. This insight allows the various stakeholders to assess the relative merits of different fine tails reclamation schemes and to consequently decide what strategy should be adopted.

Acknowledgments

The information upon which this paper is based is the result of the efforts of a large number of individuals throughout the industry. Specific thanks are expressed to both Syncrude and Suncor for their support and approval to publish this paper.

References

- Caughill, D.L., Morgenstern, N.R. and Scott, J.D. 1993. Geotechnics of Non-segregating Oil Sand Tailings. Canadian Geotech. J.30, 801-811.

- Cuddy, G., Mimura, W. and Lahaie, R., 1993. Spiking Tailings as a Method to Increase Fine Tails Retention in Beach Deposits. Proc. Oil Sands - Our Petroleum Future Conference, Edmonton, Alta. Paper F21.

- Gully, J. and MacKinnon, M., 1993. Fine Tails Reclamation Utilizing a Wet Landscape Approach. Proc. Oil Sands - Our Petroleum Future Conference, Edmonton, Alta. Paper F23.

- Lord, E.R., Mimura, D.W., and Scott J.D., 1991. Disposal of Oil Sand Fine Tails (Sludge) in Overburden Waste Dumps. Proc. of First Canadian Conference On Environmental Geotechnics, Montreal, Quebec.

- Lord, E.R. and Lahaie, R., 1992. Research Directions for Solid Tailings Disposal at the Syncrude Oil Sands Mine. Proc. Second International Conference on Environmental Issues and Management of Waste in Energy and Mineral Production, Calgary, Alta. 1229 - 1239.

- Lord, E.R. and Isaac, B.A., 1989. Geotechnical Investigation of Dredged Overburden at the Syncrude Oil Sands Mine in Northern Alberta, Canada. Can. Geotech. J. 26, 132 - 153.

- MacKinnon, M. and Sethi, A., 1993. A Comparison of the Physical and Chemical Properties of the Tailings Ponds at the Syncrude and Suncor Oil Sands Plants. Proc. Oil Sands - Our Petroleum Future Conference, Edmonton, Alta. Paper F2.

- MacKinnon, M. and Boerger, H., 1991. Assessment of a Wet Landscape Option for Disposal of Fine Tails Sludge From Oil Sands Sludge Clays. Proc. Petroleum Society CIM and AOSTRA Technical Conference, Banff. Paper 91-124.

- Martin, R.W. and Klym, D.J., 1991. Sludge Storage at Lease Abandonment. Proc. Petroleum Society CIM and AOSTRA Technical Conference, Banff. Paper 91-131.

- Mimura, D.W. and Lord, E.R., 1991. Oil Sand Fine Tails Absorption into Overburden Clay Shales - A Dry Landscape Alternative. Proc. Petroleum Society CIM and AOSTRA Technical Conference, Banff. Paper 91-128.

- Morgenstern, N.R., Fair, A.E. and McRoberts, E.C., 1988. Geotechnical Engineering Beyond Soil Mechanics - A Case Study. Can. Geotech. J.25, 637 - 661.

- Scott, J.D., Liu, Y. and Caughill, D.L., 1993 Fine Tails Disposal Utilizing Non-segregating Mixes. Proc. Oil Sands -- Our Petroleum Future Conference, Edmonton, Alta. Paper F18.

- Sheeran, D., Burns, R. and Gaston, L., 1991. Pond Dynamics and Sludge Handling Logistics. Proc. Petroleum Society CIM and AOSTRA Technical Conference, Banff. Paper 91-116.

- Sheeran, D., 1993. An Improved Understanding of Fine tailings Structure and Behavior. Proc. Oil Sands - Our Petroleum Future Conference, Edmonton, Alta. Paper F1.

- Syncrude Canada Ltd., 1992. Application To The Energy Resources Conservation Board for Continued Improvement and Development Of The Syncrude Mildred Lake Operation.

- Yano, L.T. and Fair, A.E., 1988. Overview of Hydraulic Construction Techniques Utilized at the Syncrude Canada Ltd. Tailings Pond. Pro. of Conference on Hydraulic Fill Structures, Fort Collins, U.S.A., 971-986.

Aspects Geotechniques de la Gestion des Dechets Industriels en France

F. Schlosser
Professeur à l'Ecole Nationale des Ponts et Chaussées, Président de Terrasol, France

B. Soyez
Laboratoire Central des Ponts et Chaussées, Paris, France,

M. Wojnarowicz
Terrasol, France

Résumé : La France produit environ 150 millions de tonnes de déchets industriels par an, dont 7 millions de tonnes de déchets spéciaux mis en décharge dans des centres d'enfouissement technique contrôlés, avec ou sans traitement préalable. L'utilisation dans le domaine routier des laitiers de haut fourneau, des schistes houillers et des cendres volantes a été beaucoup développée, mais certains remblais en scories ont conduit à des désordres notables par suite de gonflements et quelques cas typiques sont exposés. On présente également un cas de pollution du sous-sol par l'utilisation de matériaux de remblai provenant du traitement du minerai de chrome. Par ailleurs l'utilisation des pneus usagés a trouvé des applications intéressantes sur le plan géotechnique. Les boues industrielles et de lavage conduisent à des problèmes spécifiques dont on donne deux exemples. Enfin, parmi les nombreuses décharges de déchets spéciaux, certaines ont posé des problèmes de pollution et ont dû être réhabilitées: un cas typique est présenté.

1. INTRODUCTION

Comme tout pays industrialisé, la France est confrontée aux problèmes de la gestion des déchets, qu'ils proviennent de l'activité des entreprises ou de la consommation des individus. Pour les déchets industriels, la France doit par ailleurs faire face, dans le Nord et l'Est du pays, aux conséquences de la présence de quantités importantes de déchets inertes ou potentiellement à risques, qui sont l'héritage d'un passé industriel parfois centenaire.

Dans le domaine de l'environnement, la France s'est doté de textes juridiques et réglementaires, ainsi que des services et structures administratives nécessaires à leur application. La géotechnique y a une part importante en particulier dans le stockage et la valorisation des déchets, comme dans l'Aménagement du Territoire.

L'objet de cette communication est de présenter, à l'occasion de ce congrès international sur la géotechnique et l'environnement, quelques cas particulièrement représentatifs de l'expérience française en matière de géotechnique appliquée au domaine des déchets et sous-produits industriels.

Après une rapide description du contexte des déchets industriels en France, on abordera successivement les sujets suivants :

- La réutilisation en remblai des déchets et sous-produits industriels.
- La géotechnique des déchets boueux, qui constituent le tiers du tonnage des déchets industriels produits chaque année en France.
- Les aspects géotechniques propres au stockage en décharge des déchets industriels spéciaux solides.

2. LES DECHETS INDUSTRIELS EN FRANCE.

2.1 Aspects réglementaires

En France, comme dans de nombreux autres pays, tous les déchets font l'objet de textes législatifs et réglementaires. Il s'y ajoute également des textes de la Communauté Européenne, qui donnent des règles supplémentaires en matière de gestion de déchets (AGTHM, 1992). Le gouvernement français a présenté au Parlement, en 1990, les objectifs d'un programme national de maîtrise des déchets orienté autour de quatre axes principaux :

1) limiter la production de déchets en agissant sur les procédés de fabrication industrielle et sur les modes de consommation;
2) connaître et contrôler les mouvements de déchets;
3) assurer, lorsque cela est possible, la valorisation des déchets ou leur destruction;
4) effectuer, dans de bonnes conditions, la mise en décharge des déchets résiduels ou ultimes, résultant notamment de l'incinération.

En France, les déchets industriels, dont ne font pas partie les déchets radioactifs qui font l'objet de traitements et de stockages particuliers, sont classés de la façon suivante.

1) Les déchets inertes. Ils sont pour la plupart constitués par des déblais et gravats de démolition, ainsi que par les résidus minéraux des industries d'extraction, des industries sidérurgiques et des industries de fabrication de matériaux de construction. Ils peuvent être soit utilisés en remblais, soit réutilisés dans le secteur du bâtiment et des travaux publics. Actuellement, des études sont entreprises dans le but de les recycler dans l'industrie des agrégats.

Ces déchets sont normalement placés dans des décharges, dites de "classe III", qui ne comportent aucune spécification vis à vis de la perméabilité de leurs parois et/ou du sous-sol.

2) Les déchets banals. Le terme "banal" est issu du fait que ces déchets peuvent être incinérés comme les ordures ménagères. En effet, cette catégorie regroupe des déchets constitués de papiers, cartons, plastiques, bois, emballages, verres, matières organiques... La collecte sélective, les déchetteries permettent de valoriser ou de recycler tout ou partie de ces déchets, l'élimination en étant la phase ultime. Les déchets banals sont, comme les déchets ménagers, placés dans des décharges, dites de "classe II", dont le principe était d'avoir des parois ou un terrain de confinement "semi-perméables". Les réflexions en cours, liées à la mise en place d'une réglementation européenne, feront vraisemblablement évoluer la sécurité de ces décharges vers une conception de site étanche.

3) Les déchets spéciaux. Certains déchets industriels peuvent être des générateurs potentiels de nuisances. Ces déchets, appelés déchets spéciaux, font l'objet d'un contrôle particulier à tous les niveaux : production, stockage, transport, prétraitement et élimination. Ces déchets sont les suivants: - déchets organiques solides ou liquides (hydrocarbures, goudrons, solvants et huiles usagées, boues de peinture, sous produits de l'industrie chimique, etc); - déchets minéraux liquides ou semi-liquides (bains de traitement de surface, acides de décapage, etc); - déchets minéraux solides (sables de fonderie, scories, sels de trempe cyanurés, cendres volantes, résidus de traitement des fumées, etc). Ils sont placés dans des décharges spécifiques, dites de "classe I", dont l'étanchéité doit être assurée de telle manière qu'il n'y ait aucun risque de pollution pour le voisinage.

Alors que jusqu'à présent la mise en décharge de ces déchets spéciaux était considérée comme une des solutions possibles pour leur élimination, on considère de plus en plus que leur traitement est devenu nécessaire. Il reste, après traitement, un déchet ultime ou résidu ultime, qui doit être mis en décharge sans ou avec traitement (vitrification, solidification, etc) et qui est potentiellement polluant car il contient notamment des métaux lourds.

2.2. Importance et nature des déchets industriels en France

La France produit en moyenne chaque année un peu moins de 600 millions de tonnes de déchets:

- 30 millions de tonnes de déchets urbains ;
- 150 millions de tonnes de déchets industriels;
- 400 millions de tonnes de déchets produits ou recyclés dans l'agriculture et les industries agroalimentaires.

Les 150 millions de tonnes de déchets industriels comprennent :

- 100 millions de tonnes de déchets inertes (déblais, gravats, etc.);
- 40 millions de tonnes de déchets banals, assimilables aux ordures ménagères et relevant du même traitement;
- 7 millions de tonnes de déchets spéciaux nécessitant des traitements particuliers.

Les déchets industriels spéciaux doivent, avant d'être traités, être identifiés et analysés. Leur traitement peut être de nature chimique pour les déchets liquides, ou de nature physique afin de séparer les phases liquides et solide. L'incinération peut s'appliquer à presque tous les déchets organiques et dérivés hydrocarbonés ; elle est irremplaçable pour éliminer de façon sûre certains déchets très nocifs. Elle s'opère pour l'essentiel dans des unités industrielles complexes, où les déchets ultimes sont dans certains cas valorisés. Enfin, quelques 600 000 tonnes de déchets nocifs ou dangereux sont stockés en France dans 11 centres d'enfouissement technique (décharges contrôlées dites de classe I) répondant à des normes réglementaires sévères. Il faut ajouter en outre que les industriels producteurs éliminent directement dans des décharges internes environ la moitié de leurs déchets contenant des éléments nocifs.

La France traite également des déchets en provenance d'autres pays. Elle a ainsi reçu, en 1991, 630 000 tonnes de déchets industriels, en provenance pour l'essentiel d'Allemagne, de Belgique, des Pays-Bas et de Suisse. Ces déchets sont allés pour 27 % d'entre eux en décharges de classe II (machefers et cendres), pour 44 % en incinération, pour 12 % en prétraitement, pour 5 % en décharges de déchets spéciaux (décharges de classe I) et 13 % d'entre eux ont été valorisés. En outre, plus d'un million de tonnes de déchets ménagers ont été importées en 1991, pour la quasi totalité en provenance d'Allemagne.

De son côté, la France exporte 3 000 tonnes de déchets dangereux (arsenic, mercure...), qui sont enfouis dans les mines de sel d'Allemagne, et 17 000 tonnes de déchets industriels qui sont valorisés.

3. HISTORIQUE DE LA REUTILISATION DES DECHETS ET DES SOUS-PRODUITS INDUSTRIELS

3.1. L'expérience des Laboratoires des Ponts et Chaussées de 1975 à 1988

L'idée d'utiliser tous les produits d'une fabrication et donc d'en récupérer et d'en valoriser les sous-produits et les déchets n'est pas neuve. Ainsi dès la fin du siècle dernier l'industrie du fer et du charbon produisaient de grandes quantités de déchets et de sous-produits, dont la réutilisation a donné lieu à des premières applications en génie civil. J.H. Colombel (1992) donne les premières dates de réutilisation des laitiers de haut fourneau, des schistes houillers et des cendres volantes en France:

1. Emploi du laitier granulé en cimenterie : 1800
2. Emploi du laitier concassé en remblai de chemin de fer (ballast) : 1885
3. Emploi du laitier de haut fourneau granulé comme liant : 1950/55
4. Utilisation des schistes houilliers noirs et rouges : 1950
5. Utilisation des cendres volantes de houilles dans les chaussées : 1960/65

Si les premières demandes dans le domaine du génie civil ont concerné les agrégats, ce sont plus tard des liants qu'il a fallu trouver, puis des additifs de plus en plus performants. La réutilisation des déchets provenant de la sidérurgie et des centrales thermiques s'est faite progressivement depuis une cinquantaine d'années, mais a connu une très forte accélération depuis 20 ans, posant des problèmes d'ordre notamment écologique.

Dans le domaine routier, les Laboratoires des Ponts et Chaussées ont joué un rôle déterminant dans la mise au point des techniques de réutilisation des laitiers de haut fourneau, des schistes houillers et des cendres volantes dans les couches de chaussée.

Il faut en particulier signaler la mise au point par E. Prandi (1965) des graves-laitiers, puis des sables-laitiers. Ces matériaux ont constitué une nouvelle gamme de matériaux très performants, dans la mesure où ils se sont révélés plus déformables et moins sensibles au retrait que ne le sont les matériaux à base de ciment. La grave-laitier a été utilisé en grand pour plusieurs centaines de kilomètres de remise en état du réseau routier français à partir du milieu des années 1960. Elle s'est révélée un matériau économique, facile de mise en oeuvre et résistant à la fatigue.

A cette époque des années 60, des quantités énormes de déchets industriels étaient produites en France et absolument non réutilisées. Ainsi 5 à 7 millions de tonnes de phosphogypses provenant de la fabrication d'acide phosphorique étaient évacuées dans la Seine et dans la mer chaque année. Les usines d'incinération d'ordures ménagères produisaient des centaines de milliers de tonnes de mâchefers et de cendres non réutilisés. Des centaines de milliers de pneus étaient brûlés chaque année et les boues de lavage des carrières étaient systématiquement rejetées dans les rivières.

En 1976, le réseau des Laboratoires des Ponts et Chaussées a créé un groupe de recherches sur les déchets industriels pour détecter les besoins et mettre au point des matériaux et procédés nouveaux.

Ce groupe a en particulier étudié (1976-1988) la faisabilité du réemploi des déchets d'industries chimiques (phosphogypses, polyéthylènes), des déchets miniers (schistes houillers, déchets des mines de potasse), des déchets sidérurgiques (scories d'aciérie, sables de fonderie) et des déchets de consommation (pneumatiques, plastiques, mâchefers d'incinération, etc.).

Les résultats des travaux et des recherches entreprises ont été en partie publiées dans les compte-rendus du Colloque International sur l'Utilisation des Sous-produits et Déchets dans le Génie Civil (ENPC, Paris, Novembre 1978) et de la Journée du Bourget du LCPC (4 et 5 Mai 1983).

3.2. Quelques exemples de mauvaises réutilisations de déchets en remblai

3.2.1. Caractéristiques et aspects géotechniques

A côté de tous les cas où la réutilisation en remblai de déchets industriels n'a posé aucun problème, il paraît intéressant et pédagogique pour l'ingénieur, d'analyser quelques exemples d'échecs. Pour la seule région Nord de la France, une enquête, réalisée en 1986, a mis en évidence cinq cas de désordres typiques, imputables en particulier à l'utilisation de déchets d'usines sidérurgiques ou métallurgiques. Au niveau géotechnique, les phénomènes en cause ont été des gonflements provoqués par la présence, au sein des scories, de chaux vive, de magnésie ou de silicate bicalcique.

Trois de ces cas sont présentés ci-après. Les deux premiers concernent des désordres ayant eu des conséquences à court terme sur la tenue mécanique des ouvrages (entre 6 mois et 1 an après la mise en service). Le troisième cas correspond à un problème de pollution de la nappe phréatique et des sols apparu 20 ans après la mise en service de l'ouvrage.

3.2.2. Exemple d'une rocade urbaine

Il s'agit d'une autoroute urbaine construite principalement en remblai de 3 à 10m de hauteur et comportant de nombreux ouvrages de franchissement. Les problèmes ont été liés à l'utilisation en remblai de 1974 à 1975 d'un tout-venant de crassier d'usines sidérurgiques.

Les premiers désordres sont apparus sur la chaussée, 6 mois à 1 an après la construction, et il s'est par la suite avéré qu'ils étaient généralisés et s'étendaient sur une grande partie de la chaussée et des ouvrages d'art. Les désordres, dans leur ensemble, ont été les suivants (Figure 1):

- sur la chaussée : fissurations, forts gonflements localisés, déformations des profils en long et en travers;

- sur les ouvrages d'art : rupture du parement d'un mur en terre armée; poussées lentes et continues du remblai sur les ouvrages des passages inférieurs (culées et pieux de fondation) avec des déplacements horizontaux atteignant 30 cm. Plus de 10 ans après, l'évolution de ces mouvements apparaissait encore linéaire.

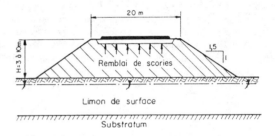

a) Gonflement des chaussées

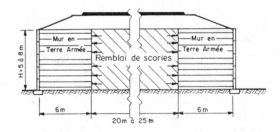

b) Mur en Terre Armée désorganisé

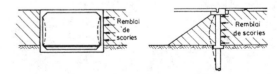

c) Poussée sur les culées des ouvrages

Figure 1 : Désordres sur la rocade urbaine Sud de Dunkerque

L'ensemble de ces désordres a été dû aux gonflements du matériau

L'ensemble de ces désordres a été dû aux gonflements du matériau de remblai. Parmi les divers produits susceptibles de gonfler et présents dans les remblais de crassiers, il faut mentionner les nodules de chaux vive au sein des blocs de scories. Comme l'a expliqué E.Prandi à propos de plusieurs cas de désordres analogues, l'humidité progresse lentement au sein des blocs de scories peu perméables et provoque leur cassure par gonflement lorsque les nodules de chaux vive (ou de magnésie) sont atteints. Il en résulte un gonflement du matériau par expansion et foisonnement. Pour cette raison et afin d'obtenir un matériau de remblai inerte, les scories doivent être laissées en dépôt à l'air libre et soumises aux intempéries pendant plusieurs années avant d'être utilisées.

Parmi les solutions de réparation, il faut citer, outre les réfections des chaussées, l'interposition entre les parois des ouvrages d'art et le remblai d'un amortisseur en polystyrène expansé d'un mètre d'épaisseur, la réfection complète des pieux, colonnes et chevêtres des ponts ainsi qu'une purge importante du matériau de remblai et son remplacement par du sable.

3.2.3. Exemple d'une piscine à structure sensible

Il s'agit d'une piscine du type "Tournesol" construite dans le cadre d'un programme d'équipement portant sur 1000 piscines sur l'ensemble du territoire français. Ce type d'ouvrage présente la particularité d'être modulable en fonction des conditions climatiques saisonnières et de pouvoir être ouvert ou fermé par coulissement d'une partie du toit sur une longrine périphérique. Il en résulte bien sûr qu'un tel ouvrage est très sensible aux tassements différentiels du sol de fondation. Dans le cas présent, la piscine fut fondée sur un remblai rapporté, constitué de cendres volantes et de scories sur une épaisseur moyenne de 1 m, puis de matériau de crassier sur 1 m d'épaisseur également (Figure 2).

Les premiers désordres sont apparus début 1978, soit moins de 6 mois après la construction. Des mouvements différentiels, provoqués par un gonflement du remblai se sont produits entre le dallage, le bassin et la galerie technique et les fondations périphérique. Les mouvements les plus importants ont été observés dans la zone où l'épaisseur de remblai était la plus importante. Dans la galerie technique, furent observées des fissures et des fentes verticales dans le mur circulaire.

Les causes des gonflements, qui n'ont concerné que le mètre inférieur des matériaux de crassier, résultent d'une part de la présence d'éléments susceptibles de gonfler avec l'eau (chaux libre, magnésie libre, ferrite) dans ces matériaux, d'autre part d'une conception qui n'a pas tenu compte de la présence de la nappe phréatique à faible profondeur.

L'évolution des gonflements a connu deux phases distinctes. Dans un premier temps, il y a eu stabilisation apparente au bout de deux ans après la construction et la réparation des désordres. Malheureusement quelques années après ces réparations, les gonflements et les désordres ont repris et ne sont pas encore à l'heure actuelle stabilisés. Une deuxième série de travaux de réparations a dû être effectuée.

3.2.4 La pollution par le chrome du sous-sol de l'Autoroute A22
(Dujardin et al., 1991 ; Van Laethem et Legrand, 1993)

L'autoroute A22, au nord de Lille, longue de 18 km et construite de 1969 à 1971, a permis la jonction des réseaux autoroutiers français et belge. Elle a été réalisée en remblai, sur un sol de fondation assez plat, constitué de sables fins et de limons. Il a été notamment utilisé 450 000 tonnes de charrées, un sous-produit du traitement du minerai de chrome.

Les charrées proviennent du procédé de fabrication du chrome à partir du minerai de chromite et des carbonates de sodium et de magnésium. Une fois le mélange calciné au four tournant, le bichromate de soude est extrait par lavage à l'eau. Les stériles ainsi obtenus sont recyclés au four pour moitié, l'autre moitié étant mise en dépôt et constituant les charrées. Une partie des charrées ainsi stockée a donc été réutilisée en remblai pour la construction de l'autoroute car les caractéristiques physiques mécaniques de ces déchets, qui se présentes sous la forme de sables, sont bonnes.

En 1987, dix-sept ans après la mise en service de l'autoroute, il fut constaté que l'herbe, sur certains talus de l'autoroute, avait du mal à pousser et qu'il y avait présence d'efflorescences jaunâtres. Une étude spécifique, effectuée sur le matériau de remblai et le sous-sol, a révélé qu'il s'agissait d'une pollution résultant de l'entraînement des chromates des charrées (sels alcalins de chrome hexavalent) par les eaux de pluie. Le terre-plein central de l'autoroute, très large, aurait été une source privilégiée d'infiltration des eaux de pluie dans le remblai, provoquant une lixivation, pendant 17 ans, des sables chromifères présents au cœur du remblai (figure 3).

La diffusion latérale du halo de pollution autour de l'ouvrage n'a heureusement concerné que le sol en surface. En effet, par suite de la faible perméabilité des limons et des argiles sous-jacents, la nappe phréatique profonde et exploitée n'a pas été atteinte. La pollution n'apparaissait que d'un seul côté des remblais et sur une distance limitée, en dehors de l'autoroute, en raison de la faible perméabilité des sols de fondation.

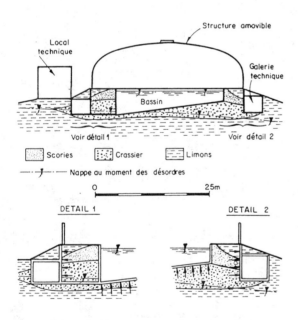

Figure 2 : Coupe de la piscine ayant subi des désordres par gonflement des remblais

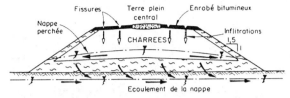

Figure 3 : Pollution par le corps de remblai en charrées (chromates)

Cette pollution fut d'autant plus anormale, qu'à l'époque des travaux de construction de l'autoroute, les risques liés à la présence du chrome hexavalent, soluble, dans le corps de remblai étaient connus et avaient été pris en compte. Cette toxicité potentielle avait justifié des précautions particulières, au moins en ce qui concerne la structure des remblais : les charrées ne devaient être utilisées qu'au coeur des remblais avec une étanchéité à la fois à la base des remblais à l'interface sol-remblai et latéralement en bord des remblais. Cette étanchéité fut réalisée par apport de matériaux fins compactés. Malheureusement, la conception du terre-plein central fut déficiente et n'a pas assuré une totale imperméabilisation. En outre, les infiltrations dûes au viellissement général des chaussées et des structures d'assainissement de l'ouvrage avaient été sous-estimées, voire ignorées.

Une nouvelle étanchéité, réalisée et expérimentée en 1991, a concerné tant le terre-plein central (membrane bitumineuse recouverte d'un matériau granulaire drainant), que les chaussées (enrobé dense) et les talus (membrane bitumineuse recouverte d'une géogrille pour l'accrochage de la terre végétale). Elle s'est accompagnée de la pose de glissières en béton et d'une remise en état du réseau de collecte des eaux pluviales (figure 4).

En dépit de difficultés de réalisation et d'un coût assez élevé, les résultats de cette solution ont cependant été positifs, car les eaux, prélevées en aval dans les écoulements de surface, ont montré une très nette et rapide diminution dans le temps de la concentration en chromates.

En plus d'une tentative de rabattement de la nappe perchée du remblai par pointes filtrantes, d'autres techniques sont actuellement expérimentées ou envisagées sur certaines autres parties de l'autoroute, notamment la réalisation d'un barrage filtrant à l'aide d'une tranchée de 5 à 6 m de profondeur située à l'aval de la source de pollution et le traitement des charrées au sulfate de fer pour neutraliser les chromates, par des matériels et des techniques analogues à ceux utilisés pour le traitement des sols fins par liant hydraulique.

3.3 Utilisation des pneumatiques usagés en géotechnique : le Pneusol

Dès le début des années 80, Nguyen Thanh Long (1985) a développé des recherches pour étudier les possibilités d'utilisation de pneus usagés placés au sein d'un remblai. Les pneus, ou éléments de pneus, peuvent être utilisés soit comme éléments de renforcement par liaisonnement des pneus (ou éléments de pneus) les uns aux autres pour constituer des lits horizontaux de renforcement, soit comme éléments compressibles pour constituer avec le sol de remblai un matériau pouvant absorber facilement des déformations statiques ou dynamiques (Figure 5). Si la première utilisation est intéressante, car elle permet de construire des ouvrages en sol renforcé à un très faible coût (Figure 6), la seconde utilisation présente en revanche un caractère innovant tout à fait notable. En effet le matériau constitué par des lits de pneus horizontaux placés au sein d'un remblai, et dénommé Pneusol, présente dans le sens vertical une grande compressibilité. Les ordres de grandeur des modules

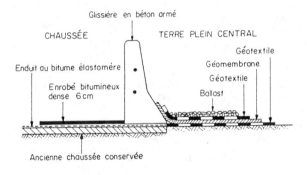

a) Etanchéité du terre-plein central et des chaussées

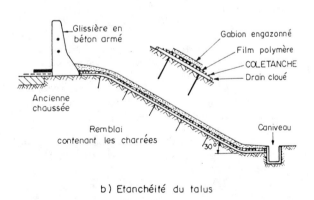

b) Etanchéité du talus

Figure 4 : Dispositifs d'étanchéité mis en oeuvre

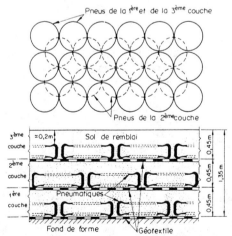

Figure 5 : Utilisation du Pneusol comme matériau compressible

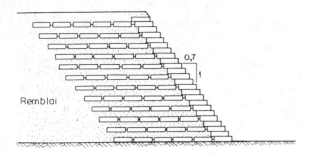

Figure 6 : Mur en Pneusol

statiques et dynamiques du Pneusol et du remblai seul sont les suivants, dans le cas d'un matériau Pneusol avec remblai en sable (Nguyen Thanh Long, 1993) :

	Sable	Pneusol
E_{sta}	4 MPa	1,0 à 1,3 MPa
E_{dyn}	30 MPa	7 à 10 MPa

Tableau 1 : Modules statiques et dynamiques d'un sable et du Pneusol (Nguyen Thanh Long, 1993)

Cette propriété a trouvé une application dans la réduction des pressions s'exerçant sur des conduites ou des buses enterrées dans des remblais. (Doan Tu Ho et al., 1982). Le principe est de placer au-dessus de la buse, qui constitue dans le remblai un point dur sur lequel se concentrent les contraintes verticales (effet Marston), une hauteur H de Pneusol, de module E, de façon à obtenir un plan d'égal tassement aussi près que possible de la génératrice supérieure de la buse (Figure 7). Si on note E_o et H_o le module et la hauteur des remblais latéraux situés entre le niveau d'assise de la buse et le plan d'égal tassement, il vient dans le cas d'une buse supposée très rigide :

$$\frac{H}{E} = \frac{Ho}{Eo} = \frac{H-Ho}{E-Eo} = \frac{D}{E-Eo}$$

ce qui permet de déterminer la hauteur H de Pneusol à placer au-dessus de la buse (Nguyen Thanh Long, 1985). Cette application du Pneusol, appelée répartiteur de contraintes, s'est bien développée durant ces dernières années par suite de son intérêt économique et de sa facilité d'utilisation.

Actuellement, plus de 250 ouvrages en Pneusol ont été construits en France dans des domaines très divers du génie civil. On peut les classer de la façon suivante :

1) Murs de soutènement, y compris les raidissements de pentes.

2) Remblais légers.

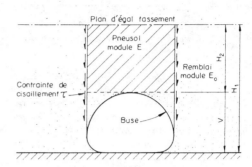

Figure 7 : Utilisation du Pneusol comme réducteur de pression

3) Répartiteurs de contraintes au-dessus des conduites en béton.

4) Murs absorbeurs d'énergie, notamment pour arrêter les chutes de blocs en montagne.

5) Réducteurs de poussée derrière les ouvrages de soutènement et les culées.

6) Protection des pentes et des berges.

L'absorption d'énergie, comme l'arrêt de blocs, l'atténuation du bruit ou la réduction des vibrations, constitue un domaine très prometteur comme ont pu le montrer les premiers ouvrages réalisés.

4. ASPECTS GEOTECHNIQUES DES DECHETS ET RESIDUS SEMI-LIQUIDES.

4.1 Généralités géotechniques et environnementales

Sur les plans géotechnique et environnemental les problèmes liés à la décantation et au stockage des déchets industriels semi-liquides dans des bassins sont les suivantes :

1) Eviter toute percolation des liquides, surtout lorsqu'il s'agit de résidus polluants, à travers le sol de fondation. L'utilisation de géomembranes et de techniques d'étanchéité propres aux stockages sont recommandés.

2) Contrôler la collecte des liquides récupérés, tant en surface du bassin que par les drains de fond.

3) Assurer la stabilité des digues tout au long de l'exploitation du bassin de sédimentation, en tenant compte à la fois de la consolidation des boues, des écoulements de liquide et des surélévations éventuelles des digues.

4) Contrôler et assurer l'étanchéité du stockage après l'exploitation du bassin de décantation.

Pour illustrer ces aspects, deux cas relatifs à des bassins de stockage de boues sont présentés ci-après.

4.2 Etude de la stabilité des digues de bassins de décantation des usines Solvay.

L'usine de Tavaux (Jura) de la Société Solvay stocke ses déchets de fabrication de la soude, à partir du chlorure de sodium, dans plusieurs bassins d'environ vingt hectares de superficie chacun. Ces résidus basiques, fortement chargés en chlorures, se présentent sous la forme d'une boue blanchâtre très liquide, appelée "blanc Solvay".

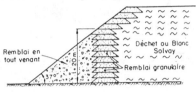

a) Digue périphérique

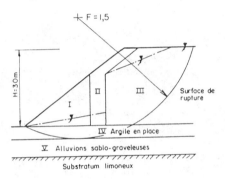

b) Première analyse de la stabilité

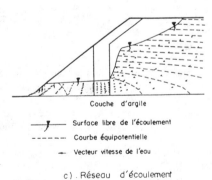

c). Réseau d'écoulement

Figure 8 : Stabilité des digues de décantation de bassins des Usines Solvay (déchets de la production de soude)

Après décantation, cette dernière acquiert une consistance pâteuse, puis solide. Actuellement, près de 12 millions de m^3 de ce déchet industriel sont stockés dans quatre bassins de 18 à 20 m de hauteur.

Le souhait d'augmenter les niveaux des bassins actuels, conjugué au projet de création de deux nouveaux bassins pour lesquels il est envisagé d'atteindre des hauteurs de 30 à 40 m, voire plus, ont nécessité une reconnaissance géotechnique préalable et une étude de stabilité.

Les digues périphériques (figure 8a) sont constituées d'un empilement successif de petites digues de 2,50 m de hauteur, dont la fonction initiale est de contenir la boue déversée et de faciliter l'évacuation de l'eau. Par la suite, lorsque le "blanc" est suffisamment consolidé pour permettre la surélévation de la digue, le rôle essentiel de celle-ci est d'assurer le rabattement de la nappe du bassin. Ces digues sont construites en concassé calcaire, assez propre, pour la partie intérieure, et en matériaux plus pollués (découverte de carrière) pour la partie extérieure.

Une première étude a montré que la stabilité de cet ouvrage était assurée (F ~ 1,5) jusqu'à 30 m de hauteur, à condition que la nappe du bassin soit complètement rabattue par la digue (Figure 8b).

Des mesures piézométriques effectuées sur les anciennes digues, ainsi que l'observation de niveaux de résurgences situées nettement au-dessus du pied des digues, ont conduit à prendre en compte l'hypothèse d'une nappe partiellement rabattue. La prise en compte de l'anisotropie de perméabilité des différents matériaux a permis d'établir un réseau d'écoulement compatible avec les constatations faites (Figure 8c).

De nouveaux calculs de stabilité ont alors été effectués, dans lesquels il a été considéré une cohésion du "blanc" variable avec la profondeur. Ces calculs ont permis de définir une nouvelle géométrie de digue pour atteindre une hauteur de 30 m dans l'hypothèse d'une nappe de bassin partiellement rabattue.

4.3 Franchissement d'un bassin de décantation par une autoroute en construction. (Delmas et al, 1986)

Lors de sa construction, l'autoroute A 26, entre Calais et Reims, traversait localement, au nord de Laon, un bassin de décantation de boues de lavage de betteraves sucrières. Le profil en long du projet plaçait la chaussée à 70 cm au-dessus du niveau des déchets, dont l'épaisseur moyenne était de 6 m dans l'axe de l'autoroute (figure 6a).

Ces déchets de lavage ressemblent à des limons plus ou moins sableux, voire à des vases. Delmas et al. (1986) ont présenté un tableau exhaustif des caractéristiques mécaniques de ces matériaux. A titre indicatif, au centre du bassin, la cohésion non drainée c_u variait entre 7 et 19 kPa à 0,6 m de profondeur et entre 28 et 36 kPa à 6,9 m.

Les différentes solutions, envisagées pour le franchissement de cette zone de déchets, étaient la construction d'un viaduc "rasant", l'édification d'un remblai sur pieux ou un remblai construit par poinçonnement partiel des déchets. La réalisation d'une piste de chantier montra qu'un remblai était envisageable et un remblai d'essai, renforcé par géotextiles, démontra alors la faisabilité d'une solution par poinçonnement. Le projet retenu (Figure 9b) consistait en : - un poinçonnement partiel, par le remblai, des déchets du bassin; - un préchargement assurant une consolidation partielle des déchets restant en place ; -un déchargement du remblai jusqu'à une cote permettant la réalisation d'un remblai léger supportant la chaussée à son niveau définitif.

Le dimensionnement du projet fut jugé délicat et, à l'initiative du maître-d'oeuvre, il fut décidé de réaliser un remblai expérimental. Cet ouvrage fut construit dans la partie centrale du bassin, a priori plus compressible et relativement homogène d'après la reconnaissance. Sur la base d'un calcul de force portante pour un remblai de longueur infinie, la hauteur de remblai, à mettre en oeuvre pour poinçonner le sol fut estimée à 5 m.

Parallèlement, un autre ouvrage expérimental en vraie grandeur fut réalisé à l'extérieur de l'emprise du bassin, afin de tester la mise en place du remblai en matériau léger. Ce matériau, à base de polypropylène et dénommé Nidaplast (Figure 9a), est un matériau alvéolaire en nid d'abeille ($\gamma = 5$ kN/m^3). Il fut choisi de préférence à un polystyrène expansé, plus traditionnel à l'époque, pour éviter la poussée d'Archimède due à une éventuelle remontée du niveau de la nappe dans le bassin.

Les travaux de réalisation ont comporté par ailleurs les éléments caractéristiques suivants : - utilisation de nappes de géotextile dans le remblai, pour le rigidifier et faciliter son poinçonnement ; - réalisation de tranchées drainantes de 1,50 m de profondeur, de part et d'autre du remblai et après poinçonnement, pour faire chuter les pressions interstitielles ; - mise en oeuvre, lors de la réalisation du remblai en matériau léger, de deux nappes de géotextile, au-dessous et au-dessus des blocs de Nidaplast, pour les solidariser et empêcher leur contamination par les fines.

Cette solution de franchissement, originale et moins onéreuse qu'un ouvrage d'art, a donné toute satisfaction depuis la mise en service de l'autoroute.

5. STOCKAGE DE DECHETS INDUSTRIELS SPECIAUX "SOLIDES"

5.1. Aspects géotechniques de la réglementation : l'introduction des géomembranes

L'emploi des géomembranes en France dans les décharges contrôlées de déchets industriels spéciaux (décharges de classe I) et de déchets industriels autres et ménagers (classe II et III) est, à la différence de nombreux autres pays, encore peu répandu pour des raisons liées à la réglementation fixant les modalités d'exploitation de ces sites. Ces textes, dans leur première version reposaient en effet sur le concept d'étanchéité naturelle, avec la présence au contact des déchets d'au moins 5 m d'un sol présentant une perméabilité en place respectivement inférieure ou égale à 1.10^{-9} m/s et 1.10^{-6} m/s pour les décharges des classes I et II. Sous l'influence de nombreux facteurs techniques, économiques et sociologiques, le Ministère français de l'Environnement a entrepris de façon concertée la révision des règles d'exploitation des décharges françaises. Cette action s'est traduite par la publication en mars 93 de deux arrêtés relatifs aux décharges de classe I, traitant de manière séparée les installations nouvelles et les installations existantes.

Il est d'une part stipulé que ne seront plus admis en décharge, "que des résidus ultimes, résultant du traitement des déchets ou de la dépollution ayant permis d'en extraire la part valorisable ou d'en réduire la nocivité, ou des résidus ne pouvant faire l'objet d'un traitement préalable à un coût économiquement acceptable. Ces déchets seront essentiellement solides, minéraux, avec un potentiel polluant constitué de métaux lourds peu mobilisables. Ils seront très peu réactifs, très peu évolutifs, très peu solubles".

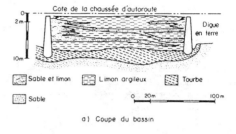

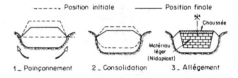

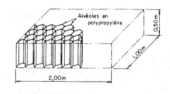

Figure 9 : *Remblai d'autoroute construit sur un bassin de boues de lavage de betteraves*

D'autre part, pour la conception et la construction de décharges nouvelles, il est imposé :

- "un niveau de sécurité passive représenté par un sol en place ou rapporté, d'épaisseur minimum 5 m, dont le coefficient de perméabilité k soit inférieur ou égal à 1.10^{-9} m/s, sur le fond et sur les flancs de la décharge ;

- un niveau complémentaire de "sécurité active" constitué d'une géomembrane et d'une couche drainante en matériau granulaire, dans laquelle soit intégré un réseau de drains relié à un collecteur principal. Toutes ces canalisations doivent être vidéo-inspectables et éventuellement "réparables" à partir d'une galerie de visite.

On notera enfin, que les procédures, concernant les décharges existantes ou à créer, imposent la réalisation d'une couverture finale multicouche dont la partie étanche est elle-même composite (une géomembrane et une couche de sol de 1m d'épaisseur caractérisé par une perméabilité de 1.10^{-9} m/s).

Un projet d'arrêté, relatif cette fois aux décharges de classe II, a commencé à être élaboré en 1992. Bien qu'il soit encore trop tôt pour annoncer quelles seront les dispositions prévues à cette occasion, on peut d'ores et déjà se demander quelle sera l'influence sur ce texte des récentes réflexions de la Communauté Européenne, qui envisage de ne plus faire de différence entre les résidus ménagers et les résidus industriels, au niveau du critère minimal d'étanchéité requis pour les deux types de décharges (3 m de matériau de perméabilité inférieure à 10^{-9} m/s).

5.2. Un exemple de réhabilitation de décharge industrielle

Parmi toutes les décharges en place, ou encore en service, en France, quelques unes sont à l'heure actuelle dans des situations critiques, vis à vis du risque de pollution qu'elles font courir à leur environnement. Au-delà des problèmes posés par des sites industriels anciens, il faut malheureusement constater que parmi les urgences actuelles figurent des installations récentes que l'on peut scinder en deux groupes : - les installations pour lesquelles aucune précaution particulière n'avait jamais été prise ; - celles pour lesquelles toutes les précautions, requises à l'époque de leur création, avaient été apparemment prises.

Parmi ces dernières, on peut citer le cas d'une décharge de classe I de déchets industriels spéciaux, située dans le centre de la France, ouverte au début des années 80. La décharge fut utilisée pendant dix ans, principalement pour des déchets industriels chimiques. Placée dans le site d'une ancienne carrière d'argile, elle fut construite sans chercher à imperméabiliser le fond et les côtés. Les déchets étaient placés dans des cellules, remplies les unes après les autres. Chaque cellule, une fois remplie, était recouverte d'argile compactée pour empêcher les infiltrations des eaux de pluie. Il était prévu de construire une couverture très imperméable constituée de géomembranes et de couches d'argile, une fois la décharge complètement remplie.

A la fin des années 80, les autorités ont brusquement décidé, pour des raisons de sécurité, de fermer la décharge. Les deux raisons principales étaient les suivantes :

- la décharge produisait de grandes quantités de gaz toxiques, nauséabonds et dangereux pour la population, à des kilomètres à la ronde;

- les analyses des eaux des lacs, des rivières et des nappes phréatiques autour du site montraient une pollution significative.

Il fut établi par la suite que ces deux types de pollution étaient liées à la mauvaise qualité du sous-sol. En effet les couches d'argile étaient faillées et comportaient des veines de grès; il en résultait une assez forte valeur du coefficient de perméabilité global, comparé au faible coefficient de perméabilité de l'argile intacte. Cet aspect fondamental n'avait pas été vu lors du choix du site. Lorsqu'une cellule était remplie et couverte, la teneur en eau des déchets augmentait par suite d'infiltrations par les côtés et par l'argile de couverture. Il s'ensuivait une intense biodégradation des déchets, avec la production de méthane et de gaz toxiques qui s'échappaient par la couverture. En outre des lixiviats très polluants s'infiltraient dans le sous-sol.

Les solutions qui ont été étudiées et celle qui a été retenue pour la réhabilitation de la décharge constituent des exemples intéressants vis à vis d'autres désordres de décharges industrielles.

Quatre solutions furent examinées : 1) l'excavation puis l'incinération des déchets fut une solution rejetée, car les dangers liés à une excavation étaient trop grands, notamment le danger d'enflammation de la décharge; 2) la stabilisation des déchets par un mélange, en place, avec de la chaux fut également rejetée, car le procédé apparaissait dangereux, difficile à mettre en oeuvre et peu fiable ; 3) le confinement, à la fois en tête de la décharge par une couverture totalement étanche et latéralement par des parois moulées profondes empêchant toute infiltration d'eau et toute sortie de lixiviat, bien que faisable, fut également rejeté car il ne diminuait pas la teneur en eau trop élevée de la décharge et sa pérennité n'était pas assurée ; 4) le confinement, associé à tout un réseau de drainage par le fond et à un contrôle de l'évolution biochimique des déchets, a été la solution finalement retenue.

L'originalité de cette dernière solution était de réaliser, en plus d'un confinement par une couverture étanche en tête et par des parois moulées étanches sur le pourtour, un tunnel visitable sous la décharge. Ce tunnel permettait la mise en place de tout un réseau de drains destiné à extraire les lixiviats, éventuellement les gaz, à suivre l'évolution de la décharge et à pouvoir, si nécessaire, agir sur cette évolution en introduisant dans la décharge des éléments actifs (bactéries, fluides, etc).

La figure 10 montre une coupe de la décharge avec le principe de la réhabilitation. L'ouvrage principal et original est le tunnel à construire à 10 m de profondeur sous le fond de la décharge et dans son axe. Avec deux accès de part et d'autre de la décharge, ce tunnel, de 3,5 m de diamètre fini, doit être excavé dans un terrain très pollué et c'est la raison pour laquelle il a été décidé de le réaliser avec un tunnelier fermé à pression de terre, qui permet une plus grande automatisation et un moindre risque pour le personnel de l'entreprise qui doit construire ce tunnel. Il est prévu de construire le soutènement du tunnel par mise en place, avec un érecteur, d'éléments préfabriqués en béton. Des injections doivent permettre d'obtenir une complète étanchéité du soutènement.

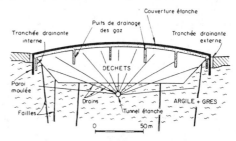

Figure 10 : Solution de réhabilitation d'une décharge industrielle par contrôle de l'évolution des décharges

Une fois le tunnel réalisé, tout un réseau de drains, à fonctions multiples, doit être foré à partir du tunnel. Tout d'abord des drains au-dessus du tunnel, avec une répartition à peu près homogène dans l'ensemble de la masse des déchets. Ces drains doivent permettre une collecte des lixiviats, éventuellement des gaz émis, en même temps qu'un contrôle et, si nécessaire, une modification de l'évolution biochimique des déchets. D'autres drains, forés à la fois horizontalement et au besoin plus en profondeur, doivent assurer le traitement et le contrôle du sous-sol pollué.

La décharge doit être ceinturée par une paroi étanche verticale en béton plastique de 20 à 25 m de profondeur, elle-même raccordée à une couverture étanche et souple. Sous cette couverture un réseau interne doit collecter les gaz produits pour les incinérer avec ceux drainés par la base dans le tunnel. A l'extérieur et à l'intérieur de la paroi moulée, deux tranchées drainantes doivent empêcher à la fois l'arrivée latérale d'eau dans la décharge et l'infiltration latérale de lixiviats vers le sous-sol extérieur à la décharge.

La solution retenue pour réhabiliter cette décharge industrielle a ainsi trois objectifs : 1) le contrôle de l'évolution de la décharge dans sa masse et à sa base, 2) le confinement de la décharge en périphérie et en tête, 3) le traitement des lixiviats et des gaz.

6. CONCLUSIONS

Les quelques exemples précédents illustrent les principaux problèmes posés par les 150 millions de tonnes de déchets industriels produits chaque année en France.

La principale réutilisation de ces déchets concerne le domaine routier avec les couches de chaussée, mais les scories sont assez largement utilisées en remblai et provoquent parfois des désordres par suite de gonflements. L'expérience montre par ailleurs qu'il faut être très prudent dans l'utilisation de matériaux polluants, en corps de remblai, car l'étanchéité n'est jamais complètement assurée.

L'utilisation des pneus usagés est un domaine géotechnique en pleine expansion.

Les stockages de déchets industriels boueux et non polluants posent des problèmes spécifiques, mais purement géotechniques.

Par contre l'enfouissement des déchets industriels spéciaux montre qu'il convient d'accorder une attention toute particulière à l'étanchéité des parois latérales et des fonds de décharges ainsi qu'à sa pérennité. L'expérience de désordres survenus sur une décharge ayant quinze ans d'âge montre que la réhabilitation nécessite non seulement un confinement, mais surtout un contrôle de l'évolution biochimique des déchets. Pour ce faire la construction d'un tunnel sous la décharge, permettant la réalisation de tout un réseau de drains, s'avère une solution intéressante.

Parmi les évolutions importantes dans la gestion des déchets industriels, il convient de souligner que la simple mise en décharge des déchets spéciaux n'est plus considérée comme une solution d'élimination de ces déchets et que leur traitement préalable s'est avéré nécessaire. Ce sont les résidus ou déchets ultimes qui sont alors mis en décharge et qui peuvent eux-même faire l'objet de traitement (vitrification, solidification,...) avant enfouissement.

On voit ainsi apparaître dans la gestion des déchets et à tous les niveaux de la conception, la notion de barrières de sécurité multiples.

REMERCIEMENTS

Les auteurs tiennent à exprimer ici leur reconnaissance aux ingénieurs responsables de projets, notamment de ceux qui ont subi des désordres, pour leur autorisation à publier les résultats et les enseignements à tirer des difficultés rencontrées.

BIBLIOGRAPHIE

ADEME, (1993). Rapport d'étude. Inventaire national des flux de déchets industriels nécessitant un traitement spécial.

AGHTM, (1992). Commission "Déchets et propreté". T.S.M. L'EAU, Mars 1992 n° 3 bis, pp. 1 - 22.

CETE NORD ET PAS DE CALAIS (1986). Déchets sidérurgiques et métallurgiques. Rapport interne du Laboratoire Régional des Ponts et Chaussées du Nord et du Pas de Calais. F.A.E.R. n° 1.36.09.4. 30 Décembre 1986.

COLOMBEL J.H., (1992). Les recherches pour la réutilisation des déchets et sous-produits en génie civil. Historique. Stage sur la Valorisation des déchets et sous-produits dans les travaux de Génie Civil. Ecole Nationale des Ponts et Chaussées. Paris 15-17 Décembre 1992.

DELMAS P., SOYEZ B., (1986). Poinçonnement expérimental d'un remblai renforcé par géotextiles sur une décharge industrielle. Third International Conference on Geotextiles, Vienna, Austria, pp. 223-227.

DOAN TU HO, NGUYEN THANH LONG, PIAU J.M., VEZOLE P. (1992). Pneusol répartiteur de contraintes. Colloque International Géotechnique et Informatique. Ecole Nationale des Ponts et Chaussées. Paris, 19 Septembre 1992.

DUJARDIN T., HERMENT R., LAUREAU D., BIENAIME C., COQUILLE J.P., (1991). Imperméabilisation des remblais de l'autoroute A22. Revue générale des Routes et des Aérodromes, n°683, Mars 1991, pp. 1-8.

ENPC-LCPC ,(1978). Colloque International sur l'utilisation des sous-produits et déchets dans le Génie Civil. Ecole Nationale des Ponts et Chaussées. Paris 28-30 Novembre 1978.

LCPC, (1983). Valorisation et élimination des déchets et sous-produits industriels et urbains. Journées de Bilan de l'Action de Recherche n° 36. Laboratoire Régional du Bourget. 4 et 5 Mai 1983. Laboratoire Central des Ponts et Chaussées.

LCPC, (1985). Rapport Général d'Activité du Laboratoire Central des Ponts et Chaussées, 1985, pp. 97-98.

NGUYEN THANH LONG (1985). Le Pneusol. Rapport de recherche n°7 des Laboratoires des Ponts et Chaussées. LCPC Paris. Juillet 1985.

NGUYEN THANH LONG (1985). Le Pneusol : réalisations. Colloque Innovations dans les Techniques de la Route. Paris n° 16/17.

NGUYEN THANH LONG (1993). Le Pneusol : recherches, réalisations, perspectives. Thèse de Doctorat. Institut Natioanal des Sciences Appliquées de Lyon. Décembre 1993.

PRANDI E., (1965). Traitement au laitier granulé des matériaux routiers. Bulletin de Liaison des Laboratoires Routiers. Spécial C., LCPC. Mai 1965. pp. 72 à 119.

VAN LAETHEM F., LEGRAND J., (1994). Impact des remblais polluants de l'autoroute A22 au Nord de Lille. Recherche de solutions. Bulletin de Liaison des Laboratoires des Ponts et Chaussées (à paraître).

State-of-the-Art of In Situ Remediation of Hydrocarbon Contamination

Robert Hinchee
Battelle Institute, Columbus, Ohio, USA

Final manuscript not received in time for inclusion in the *Proceedings*. For additional information, please contact:

    Dr. Robert Hinchee
    Battelle Institute
    505 King Avenue
    Columbus, OH 43201-2693
    USA
    Voice: +614-424-6424
    Fax: +614-424-5263

Classification of Industrial Wastes for Re-use and Landfilling

M. Manassero
University of Ancona, Italy,

C.D. Shackelford
Colorado State University, USA

ABSTRACT : Basic regulations and recommendations for industrial wastes (IW) by most industrialized countries take into account classification and characterization procedures which consider the waste materials from a chemical and toxic viewpoint thereby implying that only tests such as leachability and chemical analyses on the pore liquid and on the finer particle fraction are needed. In general, no detailed regulations exist to classify the industrial wastes from a mechanical and hydraulic standpoint taking into account the final destination and/or re-use. Moreover, when an IW is landfilled the compatibility of its chemical composition with the mineral sealing layer that will be used in the depository should be evaluated. In this paper, some observations and suggestions about the geotechnical contribution to classifying IW in terms of mechanical and hydraulic properties and chemical compatibility with containment mineral barriers are proposed. After a general outline of the topic, some examples from the literature and from the authors' experiences will be presented to support the proposed classification approaches.

Introduction

Final destination, containment systems and placement procedures for by-products and wastes from industrial activities should be determined considering both the chemical composition and the mechanical behaviour of the wastes. However, national and local regulations as well as classification systems of most of the industrialized countries are based only on the chemical composition and toxicity levels of wastes and related leachates. In spite of this trend, an appropriate knowledge of the mechanical behaviour of both the industrial wastes and the by-products is of primary importance for re-use and landfilling in terms of improving or optimizing (1) the placement and compaction procedures, (2) the strength and deformation behaviour, (3) the stability of the waste body, (4) the safety during the landfilling activities, (5) the volume capacity of the landfill, and (6) the final closure and reclamation of the storage area. Conventional laboratory and in situ tests commonly used in geotechnical engineering to evaluate the mechanical behaviour of particulate materials can also be adopted to assess the mechanical behaviour of many soil-like industrial solid wastes (ETC8, 1993). Laboratory test programs to characterize the mechanical behaviour of industrial solid wastes are suggested and discussed in this paper. Moreover, some examples of results from laboratory tests performed on industrial wastes (IW) and by-products are presented.

Compacted clay liners and cutoff slurry walls backfilled with clay minerals typically are used as components of containment barrier systems for toxic wastes relying, in particular, on their long-term performances, attenuation and buffer capacities. However, existing classification systems for industrial wastes and waste disposal regulations generally do not account specifically for the interaction phenomena between the mineral sealing layers and the chemical compounds in the wastes. The importance of considering the waste-soil interactions with respect to containment barrier performance in terms of hydraulic conductivity is outlined and a general scheme related to some typical pollutant compounds included in IW leachate and to the features of clay minerals is presented. This scheme can be used to obtain a preliminary assessment of the performances of various types of clay mineral barriers as well as to optimize the choice and the design of these barriers with respect to specific types of IW.

Mechanical behaviour of wastes

According to the general regulations of the countries of the European Economic Community, wastes are classified in general terms as: (1) Inert Wastes (ITW), (2) Municipal Wastes (MW), and (3) Industrial Wastes (IW). The IW are generally sub-classified into two or three groups which, disregarding the local terminology, can be defined as dangerous materials of low, average and high toxicity. Each of these classes must be stored in specific landfills with increasing degree of safety which means an increasing cost of the liner system. The only admission requirement, concerning the mechanical properties for landfilling solid IW, is that the water content be lower than an allowable maximum limit (e.g., $w \leq 230\%$ in Italy; $w \leq 180\%$ in France). Referring to the water content it is possible to group IW into five classes taking into account the main waste stream as shown in Table 1.

TABLE 1--Preliminary classification system

(a) residues from incineration processes (fly and bottom ash from coal power plant and from incineration of municipal and industrial wastes, flue gas ash from dry treatments, boiler slag, etc.)
(b) residues from metallurgical industry processes (steel slag, blast furnace slag, foundry sands etc.)
(c) residues from construction, oil industries, subsoil treatments and investigations (construction debris, asbestos from removal of isolation, polluted soils, drilling sludges, etc.)
(d) dry or quasi-dry residues from physico- chemical treatments (dust from air treatments, oxides, salts from chemical, metallurgical, pharmaceutical and other industries. etc.)
(e) residues from waste liquid, and gas treatment plants (slurries from sewage water treatments, slurries and ion exchange resins from waste water and gas treatments of painting, leather, paper, agricultural, metallurgical, mechanical, chemical, pharmaceutical and other industries, flue gas desulphurization treatments of power plants, etc.)

IW classes (a) and (b) are characterized by source processes in a wide range of grain size distribution for dry particulate material. In general, this material needs to be wetted for transportation and for provisional and final storage. On the contrary, class (e) includes fine grained wastes with high water contents near their liquid limits (i.e. sludges and muds). Classes (c) and (d) include a wide range of wastes both in terms of grain size distribution and water content. A preliminary classification of the considered IW using Table 1 allows one to focus the possibility of re-use, the need of pre-treatments or field-treatments, and the probability of stability and settlement problems. Class (e) IW are generated with very high water contents (w > 200 to 300%) that is usually afterwards reduced with mechanical treatments (e.g., filter press and centrifuge). Sometimes it is possible to observe this mechanical overconsolidation through the behaviour of waste body in the landfill (Belfiore et al., 1990). Besides the mechanical consolidation, chemico-physical treatments also are used to improve the stress-strain behaviour and to reduce the pollutants mobility of some IW. The chemico-physical treatments can be grouped into three main classes: (1) stabilization with cement based mixtures; (2) stabilization with lime based mixtures; (3) stabilization with organic polymers and thermo plastic additives. Bentonite, lime, and cement also are used often to reduce hydraulic conductivity of by-products such as fly ashes or steel slags which can be re-used in some cases as sealing materials (Bowders et al., 1987; Fioretti et al., 1993).
Another important distinction among the different IW streams pertains to soil-like and non-soil-like materials (ETC8, 1993). Soil-like IW are defined as particulate materials for which soil mechanics principles are applicable (i.e. sludges, ashes, excavation materials, etc.). Non-soil-like IW include materials for which soil mechanics principles are not directly applicable (i.e. plastic bags filled with ashes, drums, wood or metal boxes, etc.).
The major steps which should be considered in planning

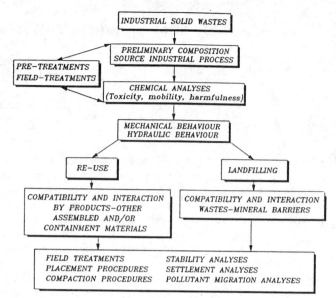

FIG.1 : Some of the major steps to be considered in landfill design and re-use of industrial wastes.

and designing landfilling and re-use of IW are shown in Fig. 1. After the proposed preliminary classification (Table 1), the mechanical behaviour of IW must be examined in more detail. The basic mechanical features of IW can be defined with respect to the aspects reported in Table 2 which refer to the behaviour of a multiphase lightly cemented particulate medium. The aspects of the IW mechanical behaviour listed in Table 2 are detailed and discussed in the following showing examples when available.

TABLE 2--Basic aspects of IW mechanical behaviour

(a) shear strength components: (1) pure friction; (2) locking or dilatant behaviour; (3) osmotic and matric suction; (4) cementation; (5) fibers or rigid inclusions reinforcement
(b) components of deformability for external stress variation: (1) intrinsic deformability and/or structural collapse of solid particles and rigid bodies; (2) deformability of solid skeleton
(c) strain components: (1) collapsible solid skeleton; (2) positive or negative changes in pore pressure; (3) changes in pore liquid content (saturation degree); (4) changes in chemical composition of pore liquid (osmotic suction) (5) biological and/or physico-chemical degradation; (6) ravelling (movement of fines into large voids); (7) viscosity of the solid skeleton or single solid particles or inclusions

Shear strength

Most IW typically shows a stress-strain behaviour of structured and lightly cemented elasto-plastic particulate materials. A qualitative model to illustrate this kind of behaviour is depicted in Fig. 2. The main features of the proposed model are as follows.

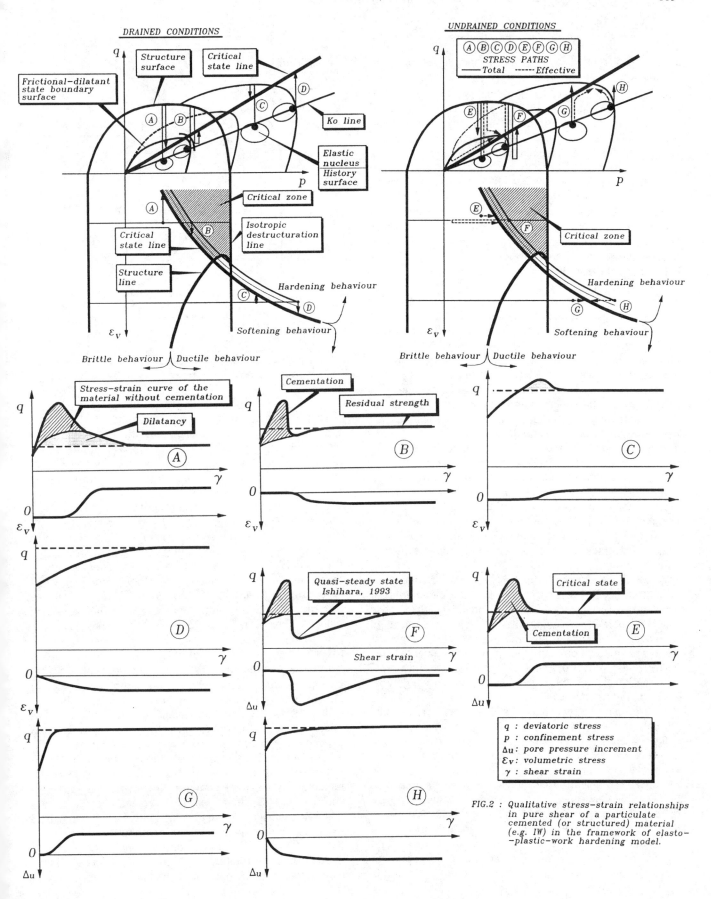

FIG.2 : Qualitative stress-strain relationships in pure shear of a particulate cemented (or structured) material (e.g. IW) in the framework of elasto-plastic-work hardening model.

- There are two basic boundary surfaces plotted in the p - q and in the p-ε_v planes (where p is the isotropic confinement stress, q is the deviatoric stress and ε_v is the volumetric strain). The first (1) surface describes the limit states of a particulate, frictional, non cemented, non structured (i.e. remoulded) material. This "frictional-dilatant state boundary surface" expands for increasing density values of the considered material (i.e. decreasing specific volume). The second surface (2) is a "structure surface" that describes the limit states when the particulate material is cemented or shows a structure that gives the undisturbed material a higher strength (at low confining stresses) in comparison with the remoulded material of the same density. This last surface, as a first tentative hypothesis, can be considered not significantly dependent on the material density.
- The p-ε_v plane is divided into four main zones by the projections of: (1) the well known critical state line and (2) the structure line that presents the intersection line between the structure surface and the friction-dilatant state boundary surface. The four main kinds of stress-strain behaviour of the considered material can be obtained via the initial state parameters and, therefore are located inside one of the aforementioned zones acknowledged as brittle-softening zone (stress-paths A and E), brittle-hardening zone (stress-paths B and F), ductile-softening zone (stress-paths C and G) and ductile-hardening zones (stress-paths D and H).

This proposed qualitative model can take into account the basic contributions (see stress paths of Fig. 2) to the peak and residual strengths and the related stress-strain behaviour of the considered material via: (1) the critical state surface for the pure friction contribution to the shear strength, (2) the frictional-dilatant state boundary surface for the locking (dilatancy) contribution and (3) the structure surface for the cementation and/or structure contribution.

For dry fine-grained IW, partially saturated conditions very often occur in the field. Due to the requirements for transportation, placement and compaction, dry ashes must be partially wetted without reaching saturation. Moreover, wastes and industrial by-products are almost always isolated to prevent leakage of contaminants. This isolation also prevents water infiltration which maintains, in some cases, the initial partially saturation conditions. There are some models (Toll, 1990; Alonso et al., 1991) that are able to describe the behaviour of partially saturated granular materials, including the swelling and collapsing behaviour when water content increases, and can be considered a useful tool to model the behaviour of some IW. Within the proposed qualitative model, the effect of partial saturation and suction are sketched in Fig. 3 which refers to the indications of Alonso et al. (1990); Toll (1990); Fredlund and Rahardjo (1993).

Based on Figures 2 and 3, the importance of quantifying the stress-strain behaviour when an IW is taken into consideration for re-use and landfilling in terms of stability and reduced deformations can be outlined. In

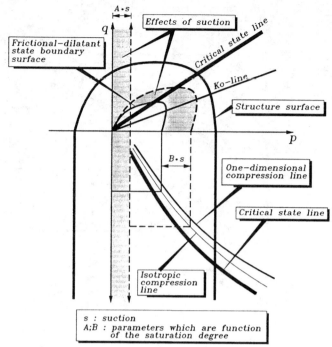

FIG.3 : Effects of partial saturation (suction).

particular Fig. 2 (stress path F) shows the possibility of a dramatic strength decrease due to the brittle-hardening behaviour in undrained conditions. Therefore, when stability or bearing capacity analyses must be carried out, the safety factors must be calibrated paying attention to the following considerations.

(1) Pure friction is a permanent contribution to the shear strength with respect to external scenarios and boundary conditions (i.e. water content, strain field, etc.) which directly depends on the effective confinement stress.

(2) Dilatancy (locking) contributes to the strength at low to average strains, while softening behaviour follows at large strains. Referring to conventional stability analyses a partial reduction factor $f_d \cong 0.5$ is proposed for reducing this contribution to the shear strength over the pure friction.

(3) The osmotic and matric suction give a certain contribution to the strength, but this component suddenly disappears due to changes in water content (see Fig. 4) i.e. it is strongly dependent on weather conditions, isolation systems, and permeability of the IW itself. In many cases, this contribution in the stability or bearing capacity calculations should not be considered.

(4) Cementation can provide an important contribution to the strength of the material, but it is important to check its durability. Slightly cemented or structured materials can be very sensitive and brittle (see. Fig. 5). In the stability analyses a reduction factor, $f_c \leq 0.3$, should be used for cementation strength contribution overcoming the friction and dilatant contributions. The outlined critical zone (brittle-hardening area) in terms of state parameters (Fig. 2) should be avoided, if possible, both in

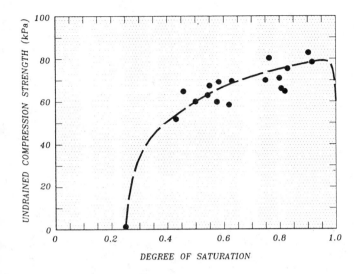

FIG.4 : Unconfined compression strength vs. saturation of a fly ash (after ROHDE et al., 1992)

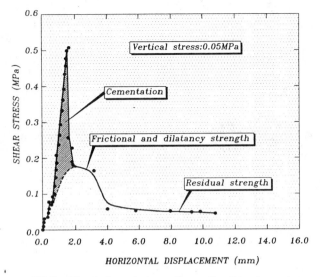

FIG.5 : Direct shear test results on the fly ash coming from the Brindisi (Italy) power plant.

landfillings and structural re-uses dealing with slightly cemented and saturated IW.

(5) Particulate fibrous materials can show a high friction angle, cohesion intercept, and hardening at large strains, which provide, in general, good engineering behaviour. In this case, an evaluation of the long-term stability of the tensile resistance of the fibers is important. It is very difficult to take into account the effects of inclusions on settlements and stability, even though they are very often present, in particular, in non-soil-like IW. At the present, it is not possible to apply general indications; therefore, each case must be approached with a specific analysis always paying attention to the structural stability of the inclusions, such as drums and packed wastes.

The partial reduction factors previously suggested should be applied to the different shear strength components so that the stability or bearing capacity calculations referring to perfectly plastic materials can be carried out using the reduced total strength. The overall stability and/or bearing capacity safety factors suggested by the local regulations must be applied afterwards.

The following laboratory test procedure is suggested for providing, in practice, the main features of IW shear strength with particular reference to the location of the structure surface, the frictional-dilatant state boundary surface and the critical state surface.

• Large shear strain tests (i.e. ring shear or direct shear with repeating shearing) on remoulded, fully saturated samples. This test series allows one to locate the critical state line and the residual strength (i.e. the failure envelope at very large strains).

• Consolidated-drained and consolidated-undrained triaxial tests on remoulded overconsolidated or dynamically compacted samples in fully saturated conditions. This test series allows one to locate the frictional-dilatant state boundary surface of the considered material without taking into account cementation effects.

• Consolidated-drained triaxial tests on remoulded partially saturated samples at different water contents and densities, possibly measuring negative pore water pressure (suction). This test series allows one to quantify the effect of matric suction on shear strength.

• The same test described at the previous point should be repeated changing the concentrations in the pore water of the typical chemical compounds expected in the considered material. This test series allows one to quantify the effect of osmotic suction on shear strength.

• Oedometer tests on the considered material starting from an unsaturated sample and, thereafter, allowing saturation at constant vertical stresses using different concentrations of expected chemical compounds. The test can be completed afterwards by increasing vertical stresses. This test series allows one to exhibit shrinkage or swelling behaviour of the considered IW.

• Unconfined compression tests and/or triaxial tests on undisturbed and cured samples to exhibit the contribution of cementation and/or structure to the shear strength. An example of consolidated-undrained triaxial tests on a cemented IW is shown in Fig. 6.

By means of this test series, it is possible to locate the different state surfaces and state areas in the p-ε_v plane paying particular attention to the critical zones to be avoided in the final placements. The in situ compaction procedures should be established in a way, if possible, to be in the ductile hardening zone with respect to the structured material (see Fig. 2). The brittle-hardening area must be avoided in particular when fully saturated conditions and undrained behaviour are expected.

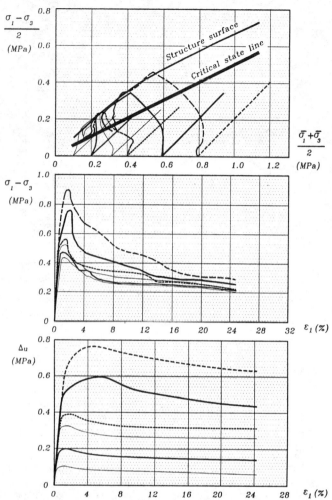

FIG.6 : Consolidated-undrained triaxial tests on a polluted cement-bentonite mixture.

Deformability and Strains

The deformability of IW due to external stresses must be investigated, looking at soil-like wastes, in terms of solid skeleton deformability. For this kind of material the expected stress-strain behaviour can be simulated with traditional soil mechanic models. As far as the non-soil-like IW are concerned, it is necessary to account for the deformability and possible structural collapse of the single solid particles or bodies (i.e. drums, bags, etc.). In this case, the stress-strain models are very complex; otherwise, settlements can be evaluated only via empirical considerations based on experience.

Besides the variation in external stresses, the following phenomena also can cause deformation and settlements of IW, in particular, referring to soil-like materials:

- Changes in pore water pressure reflect consolidation settlements with time. The resulting time-settlement curve is a function of the permeability and deformability of the solid skeleton.

- Changes in pore water (liquid) content (i.e. saturation) cause variation in matric suction and, thereafter, shrinkage or swelling.

- Chemical composition of pore liquid can play a very important role not only in terms of pollutant mobility and migration but also in terms of osmotic suction (Fredlund and Rahardjo, 1993). Therefore, it is possible that settlement or heave occurs only due to change in chemical composition of the pore liquid. Figure 7 shows the deformations of a polluted soil due to variations of chemical composition in the pore liquid.

- Biological or physico-chemical degradation can cause volumetric strains and settlements. This phenomenon is well known as far as municipal wastes are concerned. Some wastes from paper and leather industries also can show this kind of behaviour. For example, a settlement representing 10% of the total height of the Casa Carraia landfill containing IW from leather industry has been attributed to degradation of the organic matter content (Belfiore et al., 1990).

- The movement of the finer material due to erosion phenomena can cause settlements in IW having non-uniform grain size distributions. In particular, non-soil-like materials can be very sensitive to this kind of settlement when intervoids between drums or bags are filled with soil. Ravelling settlements typically are long-term settlements non uniformly distributed with time.

- Secondary settlements due to IW viscosity is a typical mechanical behaviour of fine grained wastes. In general, the secondary viscous settlements are the only deformations besides ravelling and degradation that influence the long-term behaviour and, therefore, the re-use of the landfill areas.

An example for fly ash

A typical well known industrial by-product, fly ash (FA), which is very often used in embankments construction

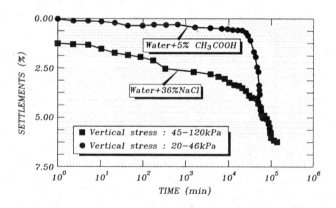

FIG.7 : Effect of NaCl and CH_3COOH solution on the settlements of a polluted peat tested in the oedometric cell. (after CASCINI and DI MAIO, 1993)

and as structural fill is classified based on the suggested procedure. The FA represents many of the features pointed out in the previous paragraphs. Following Table 1, fly ash is a class (a) material, i.e. a waste produced in dry conditions. Shear strength of FA is based on (see Fig. 5).

- Pure friction with a constant volume friction angle φ_{cv} (at initial state) ranging around 30° to 35° as reported by many authors (Geot. Dept. of Helsinki, 1983; McLaren and Di Gioia, 1987; Chen et al., 1992; Rohde et al., 1992).
- Dilatant behaviour (for compacted material) which leads to a cohesion intercept ranging around 10 to 100 KPa and a friction angle of 32° to 40°, referring to the linearized failure envelope relative to the frictional-dilatant state boundary surface in the considered range of effective confining stresses 0-500 KPa.
- The pozzolanic activity gives a contribution, in terms of true cohesion, to the shear strength. This contribution is very variable depending on the coal fuel type and on the lime and cement contents which are added in some cases during desulphurization processes or during field treatments. Peak unconfined compression strength higher than 1 MPa can be obtained easily with 3 to 5% by weight of additive (Bowders et al., 1987).
- Rohde et al. (1992) have carried out an interesting study on the influence of the degree of saturation on the FA strength (see Fig. 4). In non pozzolanic FA, the effect of suction can give apparent intercept cohesion of about 30 to 40 KPa which disappears when the material is fully wetted or completely dried.

As far as deformability and strains are concerned the following points may be outlined.

- Deformability under external stresses of FA is very low referring, in particular, to compacted materials. Particle collapse does not occur in the normal range of confining stresses. Considering the general trend of the grain size distribution, no strains are due to ravelling and collapsible skeleton if the FA has been properly compacted. Viscous behaviour is not a typical behaviour of FA.
- Consolidation settlements due to changes in pore pressure occur in a short time due to a good combination of permeabiltiy and deformability which gives rather high consolidation coefficients.
- On the other hand, attention must be paid to changes in the chemical composition and in the pore water content that can induce settlements due to yield of the material caused by the loss of strength when suction disappears (Rohde et al., 1992).

Waste-soil interactions and compability

In most cases, relatively low ($\leq 1 \times 10^{-9}$ m/s) hydraulic conductivity values are required for soils typically used as components of barrier systems (e.g., compacted soil liners and cutoff slurry walls) for containment of toxic industrial wastes. In other cases, when low polluted by-products are used as structural fills, confinement soil layers are provided to avoid pollutant migration and/or water infiltration. While generally not clearly specified, the implication of most environmental regulations is that the requirement for a low hydraulic conductivity for a soil barrier material refers to the measurement of a saturated hydraulic conductivity based on permeation with water. The reference to a saturated hydraulic conductivity is sound on the basis that the unsaturated hydraulic conductivity will be lower and, therefore, less conservative, all other conditions being equal. However, the use of water as the permeant to satisfy the regulatory requirements with respect to hydraulic conductivity of the barrier material is seldom appropriate since containment of toxic wastes (not water) generally is the goal of the barrier.

Since low hydraulic conductivity values typically are associated only with clay soils, only soils with a significant clay mineral content usually are considered for use in waste containment barriers. With respect to the requirement for achieving a low hydraulic conductivity (e.g., $k \leq 10^{-9}$ m/s), the same properties which make clay mineral soils desirable from the standpoint of an ability to achieve relatively low hydraulic conductivity values when permeated with water also typically make the same clay mineral soils more susceptible to adverse interactions with waste liquids. For example, the relatively small particles sizes and large surface areas associated with the clay mineral montmorillonite result in relatively large swelling potentials for clay soils containing significant amounts of montmorillonite, such as bentonites. Due to the large swelling potential of bentonites, in general, and sodium bentonites, in particular, permeation with water typically results in hydraulic conductivity values less than 10^{-9} m/s. As a result, bentonites commonly are used as admixture constituents in compacted soil liners and in backfill for slurry cutoff walls (e.g., Alther, 1987; Evans, 1993). However, the same properties of bentonites which result in swelling tendencies upon exposure to water also result in shrinkage tendencies upon exposure to organic solvents. As a result, permeation of clay soils containing large amounts of bentonite with organic solvents often has resulted in large increases in hydraulic conductivity relative to water (e.g., see Alther et al., 1985; and Ryan, 1987). Based on this knowledge, current waste containment regulations which do not account specifically for the interaction phenomena between mineral sealing layers and the chemical compounds in the wastes may be deficient. Therefore, any classification of wastes for re-use and landfilling should consider the potential ramifications resulting from adverse interactions between the mineral sealing layers and the waste.

Mineral Sealing Layers

The clay content of clay soils is evaluated in terms of particle size and mineralogy. The Atterberg limits commonly are used to characterize the mineralogical behaviour of clay soils (e.g., see Mitchell, 1993) in the absence of direct analysis of the mineralogical composition of clay soil (e.g., by scanning electron microscopy, x-ray diffraction, etc.). The particle size distribution of fine-grained soils typically is characterized by use of a hydrometer. A useful parameter which combines the influence of both the mineralogy and the particle size of the clay soil is the activity, A, or

$$A = \frac{PI(\%)}{\text{clay fraction}\% \leq 2\mu m} \qquad (1)$$

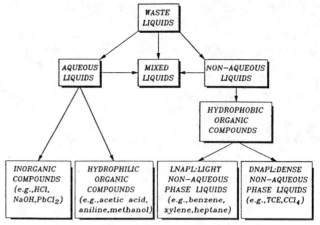

FIG.8 : Classification of waste liquids. (after SHACKELFORD, 1994)

where PI is the plasticity index, or difference between the liquid limit (LL) and the plastic limit (PL), and the clay fraction $\% < 2\mu m$ represents the amount of clay-sized particles in the soil. In general, the greater the activity the soil, the more susceptible the soil to adverse waste-soil interactions. Although there are numerous other compositional factors of clay soils which can have a significant effect on waste-soil interactions (e.g., pH, electrical conductance, cation exchange capacity, carbonate content, organic matter content, specific surface area, metal oxide content, exchangeable cations, soluble salts, and sodium adsorption ratio), the activity of the soil is easily measured and, as a first approximation, may be used to evaluate the susceptibility of the clay soil to adverse waste-soil interactions.

Waste Liquids

As indicated in Fig. 8, waste liquids or leachates with the potential to affect adversely the hydraulic conductivity of soils used for waste containment barriers may be grouped into three categories: (1) aqueous solutions or liquids containing miscible contaminants in water, such as inorganic compounds and hydrophilic organic compounds; (2) non aqueous liquids consisting of hydrophobic compounds (e.g., LNAPLs and DNAPLs); and (3) mixtures of both aqueous and non-aqueous liquids. Although hydrophobic organic compounds are considered immiscible in water, most hydrophobic organic compounds are slightly soluble in water.

A comparison of the toxic and hazardous waste sources with the waste liquid categories outlined in Fig. 8 would reveal that most waste liquids fall within the mixed liquid category. However, since there are an infinite number of mixtures which could be evaluated to determine the effects of waste liquids on the hydraulic conductivity of clay soils, the vast majority of studies performed to date have involved simple aqueous solutions or non-aqueous liquids. While limited in terms of practical applications, these studies have provided insight into the waste-soil interactions which can alter significantly the hydraulic conductivity of clay soils.

Waste-Soil Interactions

The effects of various waste liquids on the hydraulic conductivity of different clay soils commonly used in waste containment barriers based on the results of numerous studies have been summarized by Mitchell and Madsen (1987) and Dragun (1988). In addition, a comprehensive evaluation of the potential waste-soil interactions which can alter hydraulic conductivity is presented by Shackelford (1994). A summary of these findings is provided below.

- Electrolyte solutions primarily influence the hydraulic conductivity of clay soils through changes in the microfabric of the clay soils;
- organic solvents with dielectric constants (ε) significantly lower than that of water ($\varepsilon < 80.4$ @ 20°C) tend to result in shrinkage of the clay soil matrix, cracking, and large increases in hydraulic conductivity; and
- the pH of aqueous solutions must be either very low (pH$\leq$2) or very high (pH$\geq$12) to result in significant changes in hydraulic conductivity.

The microfabric of clay soils is the smallest level of fabric and refers to the arrangement of individual particles into small aggregations (Mitchell and Madsen, 1987). Flocculated microfabrics are characterized by large micropores relative to dispersed microfabrics and, therefore, relatively larger hydraulic conductivities are associated with flocculated microfabrics. Due to the similarities between colloidal particles and clay particles, the influence of electrolyte solutions on the fabric of clay soils generally is assumed to be described by the Gouy-Chapman theory from colloidal science. Based on this theory, increases in electrolyte concentration and/or cation valence, and/or decreases in dielectric constant and/or pH result in the formation of flocculated microfabrics and an increase in hydraulic conductivity (e.g., see Mitchell, 1993). These factors are more significant for high activity clays (e.g., Na-bentonites) and for clays under low confining stresses such as in the case of clay slurry filter cakes formed during construction of slurry walls (e.g., McNeal and Coleman, 1966; and Alther et al., 1985).

Numerous studies (see Mitchell and Madsen, 1987; Dragun, 1988; and Shackelford, 1994) have indicated that permeation of clay soils with organic compounds with dielectric constants significantly less than that of water have resulted in shrinkage of the clay with concomitant large increases in the hydraulic conductivity. However, there also is significant evidence indicating that the only pure or relatively high concentrations (≥ 80%, by volume) will result in large increases in hydraulic conductivity since dilute mixtures of organic compounds result in larger dielectric constants which, in the limit, approach that of water (e.g., see Bowders and Daniel, 1986; Daniel et al., 1988; and Sai and Anderson, 1991).

Three mechanisms may contribute to an increase in the hydraulic conductivity of clay soils upon permeation with acid permeants (Shackelford, 1994): (1) flocculation of the clay (microfabric effect); (2) dissolution of the clay minerals (e.g., aluminosilicates); and (3) dissolution of other minerals (e.g., $CaCO_3$) in the clay soil. Dissolution and piping of the clay minerals leads to increases in hydraulic conductivity. Dissolution of carbonates initially leads to buffering, re-precipitation, pore clogging, and a decrease hydraulic conductivity. Depletion of the buffering capacity leads to a decrease in pH, dissolution of constituents, and a possible increase in hydraulic conductivity.

General Framework for Compatibility

In terms of classifying IW for re-use and landfilling based on the potential for adverse waste-soil interactions, a compatibility index, I_c, might be defined as follows:

$$I_c = \prod_{i=1}^{n} \mu_i \quad (2)$$

where $\prod$ is the multiplier symbol (= $\mu_1 \cdot ... \cdot \mu_n$), μ_i represents an index factor associated with a soil and/or a waste liquid property indicating a potential for an adverse interaction between the waste and the soil, and n represents the number of index factors to be considered. A convenient scale should be chosen for I_c to provide an indication of the combinations of wastes and soils which might warrant concern and/or further investigation. For example, a scale of $1 \leq I_c \leq 100$ might be chosen such that an I_c value of one would represent the best case (i.e., small potential for adverse waste-soil interactions) whereas an I_c value of 100 would represent a worst case. Since the effects of inorganic compounds and organic compounds on the hydraulic conductivity of clay soils typically are evaluated separately, a separate I_c parameter can be defined for inorganic and organic compounds.

For example, consider the effect of organic compounds on mineral sealing layers. Based on the previous discussion, the major soil property indicating a potential for swelling or shrinking of the soil upon exposure to organic compounds is the soil activity (A). The major waste liquid properties associated with adverse interactions with mineral sealing layers are the dielectric constant (ε), the concentration (c), and the density (ρ) of the liquid. As a first approximation, these four properties might be considered in the development of a suitable index parameter.

With respect to the four parameters (A, ε, c and ρ), a general form of the factor, μ_i, may be defined as follows:

$$\mu_i = b_i (P_r)_i^{m_i} \quad (3)$$

where $(P_r)_i$ is a relative (dimensionless) parameter which reflects the potential influence of property i, m_i is the slope of a log-log plot of μ_i versus $(P_r)_i$, and b_i is the y intercept. In order to determine the correct form of the index factor from Eq. (3) for each property, the relative effect of each property on the hydraulic conductivity of the mineral sealing layer must be evaluated as well as the potential maximum and minimum values of $(P_r)_i$ expected for each property.

As described previously, an increase in A and c and a decrease in ε, relative to water (ε_w), results in an increase in the potential for an adverse waste-soil interaction with respect to hydraulic conductivity. Also, it might be argued that a DNAPL poses a greater threat to the integrity of a clay liner than does a LNAPL, since a DNAPL can migrate by gravity through a large crack or fracture in a mineral sealing layer whereas a LNAPL will tend to float on the surface of a waste pond. Therefore, the relative parameters, $(P_r)_i$, may be defined as shown in Table 3.

With respect to the maximum and minimum values of the relative parameters, the activity of the common clay minerals in engineering practice tends to range from a minimum value of 0.1 for halloysite ($4H_2O$) to a maximum value of 7 for the smectites (Mitchell, 1993). As shown in Table 4, the dielectric constants,

TABLE 3--Limits on Relative Parameters, $(P_r)_i$, for the Index Factors, μ_i, for Organic Compounds

Property	Symbol	Index Factor	Relative Parameter* P_r	Range in P_r Values	
				Minimum Value	Maximum Value
Soil Activity	A	μ_1	A/A_{min}	1	70
Dielectric Constant	ε	μ_2	$\varepsilon_w/\varepsilon$	1	80.4
Concentration	c	μ_3	c/c_s	0.1	1**
Liquid Density	ρ	μ_4	ρ/ρ_w	0.158	1.625

*
A_{min} = minimum activity of clay soil (=0.1);
ε_w = dielectric constant of water (=80.4);
c_s = solubility limit of compound (given in Table 4);
ρ_w = density of water (given in Table 4).

** For c = ∞, use c/c_s = 1.; for c/c_s < 0.1, use 0.1

TABLE 4--Properties of Some Organic Compounds and Water (data from Griffin and Roy 1985, and Shackelford 1994)

Compound name	Compound Formula	Dielectric Constant ε	Solubility Limit c_s (mg/kg)	Density ρ (kg/m^3)
acetic acid	$C_2H_4O_2$	6.15[a]	∞	1049.2[b]
acetone	CH_3COCH_3	20.7[b]	∞	789.9[a]
aniline	$C_6H_5NH_2$	6.89[a]	36.110[b]	1021.7[a]
benzene	C_6H_6	2.274[b]	700[b]	876.5[a]
n-butyl alcohol	$CH_3(CH_2)_3OH$	17.7[b]	71.080[c]	809.8[a]
carbon disulfide	CS_2	2.64[a]	2946[a]	1263.2[a]
carbon tetrachloride	CCl_4	2.21[b]	800[b]	1594[b]
chlorobenzene	C_6H_5Cl	5.65[b]	450[c]	1105.8[a]
m-cresol	$C_6H_4OHCH_3$	11.8[b]	21.840[a]	1027.3[a]
o-cresol	$C_6H_4OHCH_3$	11.5[b]	24.550[a]	1033.6[a]
p-cresol	$C_6H_4OHCH_3$	9.9[b]	19.440[a]	1017.8[a]
cyclohexane	C_6H_{12}	2.105[b]	<300[b]	778.5[a]
cyclohexanone	$CH_2(CH_2)_4CO$	18.3[b]	50.200[c]	947.8[a]
diethyl ether	$C_2H_5OC_2H_5$	4.2[b]	60.230[b]	713.8[a]
1.4 - dioxane	$C_4H_8O_2$	2.209[b]	∞	1033.7[a]
ethyl acetate	$CH_3CO_2C_2H_5$	6.02[b]	85.470[a]	900.3[a]
ethylbenzene	$C_6H_5CH_2CH_3$	2.41[b]	150[a]	867.0[a]
ethylene glycol	$C_2H_6O_2$	38.66[a]	∞	1108.8[a]
heptane	C_7H_{16}	1.0[a]	<300[b]	693.7[a]
isobutyl alcohol	$(CH_3)CHCH_2OH$	17.9[b]	87.170[a]	801.8[a]
methanol	CH_3OH	32.6[b]	∞	791.4[a]
methylene chloride	CH_2Cl_2	9.1[b]	13.230[a]	1326.6[a]
methyl ethyl ketone	$CH_3CH_2COCH_3$	18.5[b]	270.540[a]	805.4[a]
methyl isobuthyl ketone	$C_4H_9COCH_3$	15[b]	19.040[a]	797.8[a]
nitrobenzene	$C_6H_5NO_2$	34.8[b]	2.000[a]	1204[b]
o-dichloro-benzene	$C_6H_4Cl_2$	6.83[b]	150[a]	1304.8[a]
phenol	C_6H_5OH	13.13[a]	86.250[b]	157.6[a]
pyridine	C_5H_5N	12.3[b]	∞	981.8[a]
tetrachloroethylene	C_2Cl_4	2.35[b]	150[a]	1620[b]
trichloroethylene	C_2HCl_3	3.42[b]	1.000[a]	1464.2[a]
1,1,1-trichloro ethane	CH_3CCl_3	7.1[d]	700[a]	1339.0[a]
toluene	$C_6H_5CH_3$	2.44[b]	500[a]	866.9[a]
xylene:	$C_6H_4(CH_3)_2$		175[a]	
m-xylene		2.44[b]		864.2[a]
o-xylene		2.57[b]		880.2[a]
p-xylene		2.27[b]		861.1[a]
water	H_2O	80.4[a] ; 78.5[b]	1 x 10^6	998[a] ; 997[b]

[a] @20°C; [b] @25°C; [c] @30°C; [d] @0°C

solubilities, and densities of some common organic solvents vary widely. For dielectric constant, a value of one for heptane may be assumed as a lower limit, whereas ε = 80.4 at 20°C for water may be taken as the upper limit. The lowest density of the organic compounds listed in Table 4 is for phenol (ρ = 157.6 kg/m^3) whereas the highest density listed in Table 4 is for pyridine (ρ = 1620 kg/m^3). The solubilities, c_s, range from 150 mg/kg for ethylbenzene and tetrachlorethylene to infinity for the completely miscible organic compounds, such as acetic acid. As a working hypothesis, minimum and maximum relative solubility, c/c_s, values of 0.1 and 1.0 are assumed. Based on these ranges in parameters, the limiting values of $(P_r)_i$ are shown in Table 3. It should be noted that the limiting values of $(P_r)_i$ should be specified with respect to a reference temperature, such as 20°C, since the properties of the organic compounds in Table 3 are functions of temperature. However, the difference in the limiting values of $(P_r)_i$ will be insignificant if either 20°C or 25°C are used as reference temperatures.

If the range of I_c values is taken to be from 1 to 100, and equal weight is given to each index factor, then the possible range of values for each index factor is given by

$$1 \leq \mu_i \leq 10^{2/n} \qquad (4)$$

In the current scenario, n=4, so $1 \leq \mu_i \leq 3.1623$. Based on this possible range in μ_i values and the limiting values of $(P_r)_i$ listed in Table 3, the correction factors may be written as follows:

$$\mu_1 = (10A)^{0.271}; \quad \mu_2 = \left(\frac{80.4}{\varepsilon}\right)^{0.262};$$

$$\mu_3 = \left(\frac{10c}{c_s}\right)^{0.5} ; \quad \mu_4 = \left(\frac{6.329\rho}{\rho_w}\right)^{0.494} \quad (5)$$

where ρ_w is the density of water (see Table 4) and all other parameters are as previously defined. Therefore, the equation for the compatibility index, I_c, is determined by combining Eqs. (2) and (5), or:

$$I_c = (10A)^{0.271} \cdot \left(\frac{80.4}{\varepsilon}\right)^{0.262} \cdot \left(\frac{10c}{c_s}\right)^{0.5} \cdot (0.00634\rho)^{0.494} \quad (6)$$

where the density of water has been taken as 998 kg/m^3. A check of Eq. (6) using the best case set of values for the parameters ($A = 0.1$, $\varepsilon = 80.4$, $c/c_s = 0.1$, and $\rho = 157.6$ kg/m^3) results in the desired minimum value of 1 for I_c, whereas the worst case set of values for the parameters ($A = 7$, $\varepsilon = 1$, $c/c_s = 1.0$, and $\rho = 1620$ kg/m^3) results in the desired maximum value of 100 for I_c.

In order to determine a reasonable maximum value of I_c in terms of the suitability of a particular combination of a waste liquid and a mineral sealing layer, representative values for the parameters A, ε, c/c_s, and ρ must be established. As a first approximation, the activities of the most reactive soils, such as Na-bentonites, typically are around 4. The largest pure-phase dielectric constant for the organic compounds listed in Table 4 is 38.66 for ethylene glycol. As previously mentioned, only relative concentrations of organic compounds in excess of about 80% have been shown to affect significantly the hydraulic conductivity of clay soils; therefore, $c/c_s = 0.8$ is a reasonable first approximation. Finally, the separation between a LNAPL and a DNAPL is the density of water; therefore, $\rho = \rho_w = 998$ kg/m^3 seems reasonable as a cutoff value. Based on these values of A, ε, c/c_s, and ρ, the maximum value of I_c in terms of the suitability between waste liquid and mineral sealing layer for classification purposes is approximately 23. Therefore, I_c values ≤ 23 would be considered acceptable, whereas an I_c value greater than 23 would indicate the need for a more suitable combination of parameters, such as the use of a lower activity clay and/or treatment of the waste stream.

For example, consider the proposed use of a smectitic clay soil ($A = 3$) as a liner material for storage of a pure carbon tetrachloride (CCl_4) solution. Based on the activity of the soil, and the properties of CCl_4 in Table 4, the I_c classification is 64, which is greater than 23. Therefore, either a lower activity clay is required and/or the waste stream must be treated. Since the value of I_c of 64 is significantly greater that the cutoff value of 23, a lower activity soil probably will not solve the problem and treatment or stabilization of the waste stream probably would be required. Although dilution of the waste stream, which results in an increase in ε, a decrease in ρ, and a decrease in c/c_s, would help to reduce I_c, dilution probably is not an acceptable option since the volume of the waste stream will increase significantly.

Limitations of General Classification

The general framework for classifying IW for landfilling in terms of an index parameter, I_c, as outlined above can be used to provide a preliminary assessment of the various types of clay mineral barriers as well as to optimize the choice and design of clay mineral barriers in terms of a particular type of IW. However, the formulation of the I_c index for organic compounds as outlined above should be considered only as a first approximation. An evaluation of the compatibility index, I_c, is required before wide-spread use of the index is invoked. Also, an extension of this analysis may be made by considering other soil and/or liquid properties as well as other weighting factors. In addition, a separate compatibility index which accounts for both the ionic strength and pH of the solution should be developed for inorganic solutions. Finally, appropriate adjustements must be made to the index factors for mixtures of one or more compounds.

Final remarks

Two general schemes have been attempted and proposed for classifying IW from a geotechnical point of view.
The first scheme deals with the mechanical behaviour referring to the elasto-plastic-work hardening models and related tests in order to split the different contributions to the strength and deformability of solid IW. This procedure allows one to be aware of the short and long-term behaviour of the material and the safety margins in terms of stability and settlements to be expected when the considered IW are landfilled or re-used as structural fills.
The second proposed classification scheme outlines the compatibility assessment between some solid and liquid IW and mineral barriers via a compatibility index which allows to evaluate the possibility to use a given fine grained soil as a sealing barrier for a given IW. Moreover, it is possible to optimize mixtures and to design in detail a liner taking into account, in particular, the long-term performance.

Both classification methods represent non-comprehensive and non-validated first attempts. Future use, modifications, and associated research will contribute to enhance the geotechnical approach to this complex materials.

References

Alonso E.E., Gens A. and Josa A., 1990, "A Constitutive Model for Partially Saturated Soils", Geotechnique 40, no. 3, pp. 405-430, The Institution of Civil Engineers, London.

Alther G.R., 1987, "The Qualifications of Bentonite as a Soil Sealant", Engineering Geology, Vol. 23, pp. 177-191.

Alther G.R., Evans J.C., Fang H-Y, and Witmer K., 1985, "Influence of Organic Permeants upon the Permeability of Bentonite", Hydraulic Barriers in Soil and Rock, ASTM STP 874, A.I. Johnson, R.K. Frobel, N.J. Cavalli, and C.B. Petterson, Eds. ASTM, Philadelphia, pp. 64-73.

Belfiore F., Manassero M. and Viola C., 1990, "Geotechnical Analysis of Some Industrial Sludges", Geotechnics of Waste Fills - Theory and Practice, ASTM STP 1070, Arvid Landva, G. David Knowles, Editors, American Society for Testing and Materials, Philadelphia.

Bowders J.J., Jr. and Daniel D.E., 1986, "Hydraulic Conductivity of Compacted Clay to Dilute Organic Compounds", Journal of Geotechnical Engineering, ASCE, Vol. 113, No. 12, pp. 1432-1448.

Bowders J.J., Usmen M.A. and Gidley J.S., 1987, "Stabilized Fly Ash for Use as Low-Permeability Barriers", Geotechnical Practice for Waste Disposal 87, GSP no. 13, R.D. Woods Editor, ASCE, pp. 320-333.

Cascini L. and Di Maio C., 1993, "Subsidence Phenomena in the Sarno Village", National Research Council. Proceedings of the Annual Meeting of the Geotechnical Researchers, Roma, November (in Italian).

Chen Y.J., Zhu Y. and Shi F.X., 1992, "Air Pollution Prevention During Fly Ash Disposal", Geotechnical Engineering, vol. 23, no. 2, December, Asian Institute of Technology, pp. 1-10.

Daniel D.E., Liljestrand H.M., Broderick G.P. and Bowders J.J., Jr., 1988, "Interaction of Earthen Liner Materials with Industrial Waste Leachate", Hazardous Waste and Hazardous Materials, Vol. 5, No. 2, pp. 93-107.

Dragun J., 1988, "The Soil Chemistry of Hazardous Materials". The Hazardous Materials Control Research Institute, Silver Spring, Maryland.

ETC.8-European Technical Committee n. 8, 1993, "Geotechnics of Landfill Design and Remedial Works. Technical Recommendations. 2nd Edition. Ernst and Sohn. Berlin.

Evans J.C., 1993, "Vertical Cutoff Walls", Chapter 17, Geotechnical Practice for Waste Disposal, D.E. Daniel, Ed., Chapman and Hall, London, pp. 430-454.

Fioretti A., Ghionna V.N., Manassero M. and Pedroni S., 1993, "Geotechnical Characterization of Steel Slags for Landfills Construction in a Quarry Blasts Area", Proceedings of Int. Conf. on Environmental and Geotechnics ENPC, Paris 6-8 April, pp. 193-200.

Fredlund D.G. and Rahardjo H., 1993, "Soil Mechanics for Unsaturated Soils". John Wiley and Sons Inc., New York. 517 pp.

Geotechical Department of the City of Helsinki, 1983, "The Utilization of Coal Ash in Earth Works. Technical Guidelines", Bulletin no. 33, pp. 32.

Griffin R.A. and Roy W.R., 1985, "Interaction of Organic Solvents with Saturated Soil-Water Systems", Open File Report No. 3, Environmental Institute for Waste Management Studies, University of Alabama, Tuscaloosa, 86 pp.

Ishihara K., 1993, "Liquefaction and Flow Failure During Earthquakes", Geotechnique 43, no. 3, pp. 351-415, The Institution of Civil Engineering, London.

McLaren R.J. and Di Gioia A.M., 1987, "The Typical Engineering Properties of Fly Ash", Geotechnical Practice for Waste Disposal 87, GSP no. 13, R.D. Woods Editor, ASCE, pp. 683-697.

McNeal B.L. and Coleman N.T., 1966, "Effect of Solution Composition on Soil Hydraulic Conductivity", Proceedings, Soil Science Society of America, Vol. 30, pp. 308-312.

Mitchell J.K., 1993, Fundamentals of Soil Behaviour, 2nd Edition, John Wiley and Sons, Inc., New York, 437 pp.

Mitchell J.K. and Madsen F.T., 1987, "Chemical Effects on Clay Hydraulic Conductivity", Proceedings, Geotechnical Practice for Waste Disposal '87, R.D. Woods, Ed., Geotechnical Special Publication No. 13, ASCE, pp. 87-116.

Rhode H.L., Martin J.P. and Cheng S.C., 1992, "Effects of Drainage Conditions on Subgrade and SLope Stability of Fly Ash Embankments", Proceedings of the Mediterranean Conf. on Environmental Geotechnology, Cesme, Turkey, 25-27 May, pp. 445-452.

Ryan C.R., 1987, "Vertical Barriers in Soil for Pollution Containment", Proceedings, Geotechnical Practice for Waste Disposal '87, R.D. Woods, Ed., Geotechnical Special Publication No. 13, ASCE, pp. 182-204.

Sai J.O. and Anderson D.C., 1991, "Long-Term Effect of an Aqueous Leachate on the Permeability of a Compacted Clay Liner", Hazardous Waste and Hazardous Materials, Vol. 8, No. 4, pp. 303-312.

Shackelford C.D., 1994, "Waste-Soil Interactions that Alter Hydraulic Conductivity", Hydraulic Conductivity and Waste Contaminant Transport in Soils, ASTM STP 1142, D.E. Daniel and S.J. Trautwein, Eds. ASTM, Philadelphia, PA (in press).

Toll D.G., 1990, "A Framework for Unsaturated Soil Behaviour", Geotechnique 40, no. 1, pp. 31-44, The Institution of Civil Engineers, London.

Analyse des Transferts d'Eau Dans un Materiau d'Etancheite de Fond de Decharge Evaluation de la Permeabilite en Regime Transitoire

I. Alimi-Ichola, O. Bentoumi, G. Didier
Laboratoire Géotechnique INSA-LYON, France

RESUME : L'étude du transfert d'eau dans une couche d'argile de Gault compactée nous montre que l'évolution de la conductivité et de la diffusivité avec la teneur en eau volumique dépend de la charge hydraulique.
La méthode du profil instantané et la méthode de la courbe de rétention donnent des résultats divergents.
Nous expliquons cette divergence par le fait que l'hypothèse d'un milieu semi-infini n'est pas valable et par le fait que le potentiel de rétention ralenti le mouvement par convection de l'eau. La prédiction de l'épaisseur d'une couche d'étanchéité doit tenir compte de cet aspect.

INTRODUCTION

Lors de l'extension du site de décharge de Montreuil sur Barse (France), il a été réalisé quatre casiers afin de tester plusieurs systèmes d'étanchéité de fond de décharge. Il est actuellement recommandé d'utiliser les matériaux pouvant assurer un coefficient de perméabilité de 10^{-9} m/s. La mesure in situ d'une telle perméabilité pose actuellement des problèmes aussi bien au niveau de la technique de mesure que de la durée de l'essai.

Les matériaux retenus pour réaliser l'étanchéité des fonds des casiers sont : l'argile noire de Gault compacté sur une épaisseur de 0.6m, une géomembrane, et un composite bentonitique. Ces matériaux sont mis en place dans des casiers de dimensions 30m x 30m et présentant une pente de 1/1 sur deux côtés et de 2/1 sur les deux autres jusqu'au sommet situé à 5.5m du fond. Un drain est réalisé au point de convergence des plans inclinés constituant le fond des casiers. Les lixiviats collectés sont acheminés vers une bâche de reprise située au point le plus bas du casier. Le fond est recouvert d'une couche de gravier 20/40 sur une hauteur de 0.30m.

En vue de la comparaison des différents systèmes d'étanchéité, des systèmes de détection et de collecte des fuites ont été installés. Un test est envisagé sous un mètre d'eau. Nous nous sommes proposés d'étudier le transfert d'eau dans la couche d'argile compactée lors d'une telle mise en charge hydraulique.

DISPOSITIF EXPERIMENTAL

Les caractéristiques de l'argile sont rassemblées sur le tableau 1. L'essai de transfert est réalisé sur une colonne composée de plusieurs anneaux comme le montre la figure 1. Trois disques de papiers filtre saturés d'eau sont intercalés entre les anneaux. L'échantillon est mis en place dans chaque anneau à la teneur en eau et la densité requises. Nous avons choisi une densité et une teneur en eau initiales proches des conditions du sol compacté en place ($\gamma_d = 17.23$ KN/m^3, w=17.4%). Une fois le montage des anneaux effectué, la colonne est emballée pour attendre l'équilibre de potentiel entre le sol et le papier filtre avant les essais d'infiltration (une dizaine de jours).

Tableau 1. Caractéristiques physique.

w_{opt} (%)	γ_{dopt} KN/m^3	γ_s KN/m^3	w_l (%)	I_p (%)	w_{nat} (%)
17.5	17.7	26.5	40	21	15.1

w_{nat} : teneur en eau naturelle.

L'essai est réalisé sous une charge hydraulique de 1m d'eau. La quantité d'eau infiltré est détectée à l'aide d'un capteur de contrôle du niveau d'eau. Le signal du capteur est transmis à une table traçante.

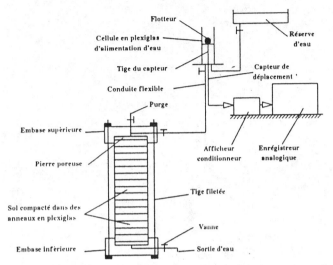

Figure 1 : Schéma du dispositif expérimental.

La courbe de rétention de l'argile utilisée pour l'analyse du transfert, a été déterminée par trois méthodes différentes afin de couvrir une grande plage de potentiel de succion :

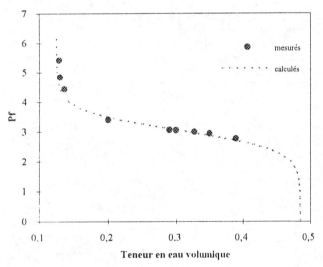

Figure 2 : Courbe de rétention.

- méthode de la pression d'air pour un PF de 1 à 2.69
- tensiomètre électrique pour un PF de 2 à 3.3
- méthode de la vapeur saturante pour un PF de 4.45 à 6.2

Chaque point expérimental de la figure 2 est obtenu après une mise en équilibre d'une dizaine de jours. Cet équilibre est atteint par imbibition. Le PF ainsi obtenu représente la succion du sol à la teneur en eau d'équilibre. Elle est composée de la succion matricielle et de la succion osmotique.

Le coefficient de perméabilité de l'argile saturée est déterminé à l'aide d'un perméamètre à charge constante. Les essais de perméabilité ont été réalisés pour plusieurs charges hydrauliques. Nous donnons sur la figure 3, les courbes d'évolution du coefficient avec le temps. La valeur atteinte est de l'ordre de $2.5 \cdot 10^{-10}$ m/s.

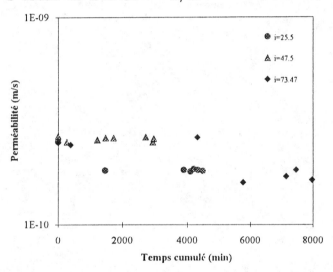

Figure 3 : Evolution de K_s.

RESULTATS D'ESSAIS

Le suivi de la quantité d'eau infiltrée au cours du temps nous permet de tracer la courbe d'infiltration cumulée afin de vérifier la relation proposée par Philip pour décrire le phénomène d'infiltration. L'infiltration cumulée I est le volume d'eau infiltré par unité de surface ; d'après les études de Philip, il vérifie la relation :
$I = S\sqrt{t} + At$. La sorptivité S [elle exprime la capacité plus ou moins grande que possède le sol à absorber l'eau par capillarité (Philip, 1969)] nous permet de calculer la valeur moyenne de la diffusivité de l'eau dans

le sol, tandis que A permet d'estimer le coefficient de perméabilité du sol. La pesée périodique des anneaux et du papier filtre nous permet de suivre la progression de l'humidité. Ce processus de mesure est adopté à cause de la lenteur du phénomène d'infiltration. Les arrêts successifs ont apparemment peu d'influence car nous n'avons pas observé des sauts sur la courbe d'infiltration. Grâce à la courbe de rétention du papier filtre utilisé, nous pouvons tracer les courbes de distribution de la pression effective de l'eau dans le sol avec la profondeur, pour un temps donné. En effet, nous avons remarqué qu'il a peu de différence entre le potentiel donné par le papier filtre et le potentiel correspondant à la teneur en eau du sol pour la gamme de teneur en considérée. L'écart observé ne dépasse pas 3cm d'eau. Les deux familles de courbes vont nous permettre d'utiliser la méthode du profil instantané pour déterminer les relations entre la conductivité, la diffusivité de l'eau et la teneur en eau volumique du sol.

ANALYSE DU PHENOMÈNE DE TRANSFERT

Le transfert de l'eau dans un sol non saturé est décrit par l'équation de Richards. Cette équation dérive de la loi de Darcy généralisée où la conductivité hydraulique et la diffusivité ne dépendent que de la teneur en eau volumique. Différents auteurs ont proposé des solutions de cette équation pour l'interprétation des résultats expérimentaux. Pour utiliser la proposition de Philip citée ci-dessus, nous effectuons la représentation des résultats du flux entrant dans le diagramme $(I,\sqrt{t})$ ou dans le diagramme $(I/\sqrt{t},\sqrt{t})$ pour obtenir les coefficients S et A.

Nous remarquons que l'évolution du flux entrant est bien représentée par une droite dans le diagramme $(I,\sqrt{t})$ (fig. 4) pour les premiers instants de l'infiltration, alors que les points sont dispersés lorsqu'on les représente dans le diagramme $(I/\sqrt{t},\sqrt{t})$ (fig. 5).

D'après la représentation de la figure 5 et l'allure de la courbe d'infiltration de la figure 6, l'infiltration de l'eau dans la colonne se caractérise par deux régimes différents correspondant à deux vitesses d'infiltration différentes. La valeur de A qui traduit l'influence de la gravité est donc négligeable au cours des premières heures d'infiltration. D'après la théorie d'infiltration de Philip, l'écoulement se comporte alors comme un écoulement horizontal. Le changement de vitesse traduit l'influence différée de la charge hydraulique extérieure imposée.

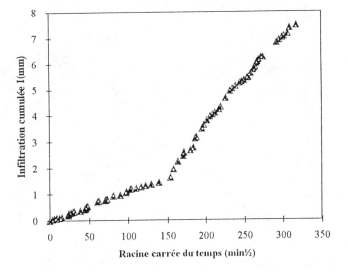

Figure 4 : Evolution de I avec la racine carrée du temps.

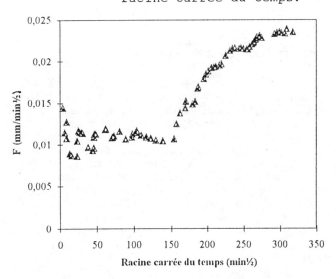

Figure 5 : Evolution du terme $F = I/\sqrt{t}$ avec la racine carrée du temps.

L'analyse des profils hydriques et des distributions de la pression effective dans l'épaisseur de la colonne montre que le front d'humidité progresse dans l'échantillon tout en se déformant. Le profil à l'instant t ne se déduit pas, par homothétie, du profil de l'instant précédent. (fig. 9). Nous ne pouvons donc pas décrire les profils hydriques en utilisant la transformation de Boltzmann, solution de l'équation de Richards, en écrivant : $z(\theta, t) = \eta(\theta)\sqrt{t}$, et nous ne pouvons pas appliquer l'hypothèse d'un milieu semi-infini pour analyser ce transfert.

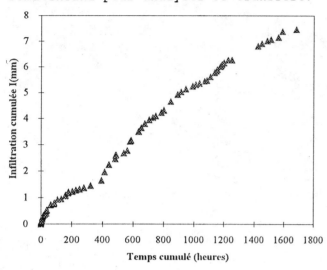

Figure 6: Courbe d'infiltration mesurée.

DETERMINATION DE K ET DE D

La connaissance de la conductivité hydraulique K et de la diffusivité D est nécessaire pour résoudre l'équation de Richards, déterminer les profils hydriques dans l'épaisseur de la couche et pour évaluer le flux entrant au cours du temps. Nous utilisons deux méthodes ; la première est basée sur le modèle de Mualem qui utilise la courbe de rétention du sol et la deuxième utilise le flux qui traverse une section donnée sous la différence de pression effective de l'eau de part et d'autre de cette section.

MODELE DE MUALEM

Le modèle de Mualem utilise la représentation de la courbe de rétention sous la forme (Van Genuchten, 1980) :

$$\frac{\theta - \theta_r}{\theta_s - \theta_r} = \left[\frac{1}{1+(\alpha h)^n}\right]^m \quad (1)$$

Les constantes θ_r (teneur en eau résiduelle), θ_s (teneur en eau de saturation correspondant à h=0), m et n sont estimées à l'aide des points expérimentaux de la figure 2. Les valeurs obtenues pour ces différents paramètres sont rassemblées dans le tableau 2.

Nous présentons sur cette figure, en pointillé, l'expression (1) correspondant aux valeurs des paramètres indiquées sur le tableau 2. Cette courbe définit la pression effective de l'eau dans le sol pour une teneur en eau volumique θ. Elle définit le potentiel énergétique de rétention de l'argile de Gault. La conductivité hydraulique s'écrit alors :

$$K = K_s\left[\frac{1}{1+(\alpha h)^n}\right] \quad (2)$$

Ks étant la perméabilité à saturation de l'argile compactée.

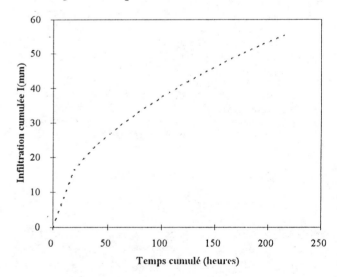

Figure 7 : Courbe d'infiltration calculée.

Tableau 2. Paramètres du modèle.

θ_r	θ_s	m	n	α	K_s
0.125	0.485	1	1.33	4.10^{-4}	9.10^{-5}

K en (cm/h) et h en (cm).

La diffusivité D est donnée par : $D = K \frac{\partial \theta}{\partial h}$. Connaissant les expressions de K et de D, nous utilisons le développement :

$$z(\theta, t) = \sum_{m=1}^{M} f_m(\theta) t^{m/2} \quad (3)$$

proposé par Philip (Vauclin et al., 1979), pour calculer le profil hydrique à travers la couche d'argile, et l'évolution de l'infiltration I avec le temps. Les résultats de ce calcul sont présentés sur les figures 7 et 8.

METHODE DU PROFIL INSTANTANE

Les différentes étapes de cette méthode ont été décrites et utilisée par plusieurs auteurs (Klute, 1972). Elle repose sur la mesure simultanée du flux et de la pression effective de l'eau dans une section, au cours de l'écoulement. La variation de la pression effective de l'eau, d'un point à un autre de la couche de sol, est supposée représenter la variation du potentiel de rétention, cause du mouvement de l'eau dans le sol.

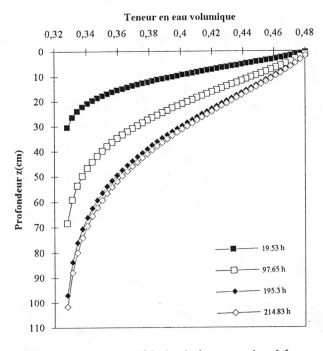

Figure 8 : Profil hydrique calculé.

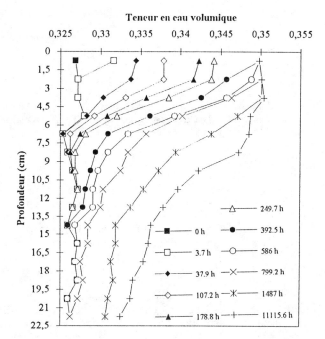

Figure 9 : Profil hydrique mesuré.

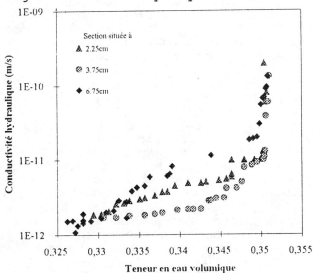

Figure 10 : Evolution de la conductivité hydraulique avec θ.

Nous utilisons les profils représentés sur la figure 9 pour effectuer ce calcul. Les valeurs de la conductivité hydraulique calculées dans une section par cette méthode, sont rassemblées sur la figure 11. Nous constatons une lente évolution de la conductivité avec la teneur en eau aux premiers instants de l'infiltration et une croissance plus rapide lorsqu'on approche la teneur en eau d'équilibre dans la section de calcul. Cette montée de la conductivité se produit lorsque le degré de saturation dépasse 93%. On observe une fluctuation de la conductivité avec la profondeur pour des degrés de saturation plus faibles. Les différentes courbes convergent vers la valeur de Ks obtenue au perméamètre à charge constante. Cette valeur est atteinte presque instantanément lorsqu'on approche la saturation.

CONCLUSION

Lorsqu'on compare les courbes d'infiltration des figures 6 et 7, on remarque que la méthode basée sur la courbe de rétention surestime la quantité d'eau infiltrée au cours du temps. Le profil hydrique calculé par la première méthode est différent du profil observé. D'après le modèle de Philip, le front d'humidité avance plus vite. Ce résultat est justifié par les coefficients du modèle de Mualem qui impliquent une relation linéaire entre la conductivité K et la teneur en eau volumique θ. Cette allure de l'évolution de la conductivité est loin de celle donnée par la méthode du profil instantané représentée sur la figure 10.

Au vu de ces résultats, deux remarques s'imposent. La première est que le flux entrant n'est pas sensible à la faible charge imposée les premières heures de l'infiltration, alors que celle-ci est prise en compte dans l'équation (3) dès le début, puisque nous avons utilisé quatre termes, les trois derniers prenant en compte l'effet gravitaire.

La deuxième remarque est le fait que le potentiel de rétention du sol n'est pas nul lorsque l'échantillon compacté est saturé. Ceci nous fait penser à une rétention de l'eau que doit vaincre la charge hydraulique pour obtenir une circulation convective.

Nous montrons sur la figure 3, l'évolution de K_S avec le temps. On remarque que le gradient imposé a peu d'influence sur la valeur de K_S. Il semble qu'à un gradient imposé faible correspond un K_S plus petit. Mais ceci dépend de la durée de l'écoulement. Nous attribuons ce comportement au phénomène de rétention que ne prend pas en compte le modèle de Mualem lorsque le sol est saturé. Nous continuons à étudier la mesure de la perméabilité sous faible gradient. Nous devons rechercher des modèles qui permettent d'obtenir les caractéristiques du régime permanent à partir de celles obtenues en régime transitoire.

Notre étude montre que l'épaisseur de 0.6m donnée à la couche est plus que suffisante, vu le système de drainage placé au dessus de la couche, même si on considère le cas le plus pessimiste donné par la première méthode. Au bout de 10 jours d'infiltration sous charge constante, la teneur en eau volumique à 0.60m n'a varié que de 3%.

REFERENCES

BENTOUMI O. et **ALIMI-ICHOLA I.** (1993). Transfert d'eau dans un sol non saturé. Influence de la teneur en eau et de la densité sèche initiales. 6ème colloque Franco-Polonais de mécanique des sols. pp. 351-358.

KLUTE A. (1972). The determination of the hydraulic conductivity and diffusivity of unsaturated soils. Soil Sci. Am. 113, pp. 264-276.

PHILIP J. (1969). The theory of infiltration. Adv. in hydrosc. 5, pp. 215-305.

VAN GENUCHTEN M.Th. (1980). A Closed-form equation for predicting the hydraulic conductivity of unsaturated soil. Soil Sci. Soc. Am. J. Vol. 44, pp. 892-898.

VAUCLIN M., HAVERKAMP R. et **VACHAUD G.** (1979). Résolution numérique d'une équation de diffusion non linéaire. Application à l'infiltration de l'eau dans les sols non saturés. Presse universitaire de Grenoble. 183p.

Waste Immobilisation, Soil Liner and Slurry Wall Material Research in Switzerland

P. Amann, F.T. Madsen, L. Martinenghi
Swiss Federal Institute of Technology of Zurich, Geotechnical Engineering Division, Zürich

Abstract: The Swiss concept for waste disposal defines three types of landfills, namely: for inert materials requiring no treatment, for materials requiring special treatment such as immobilisation of heavy metals, and for materials containing degradable substances [12]. At the present time, more than 90% of all municipal waste is incinerated, yielding more than 60'000 tons of fly ash and filter residue per year. Therefore, the final disposal of fly ash and slag is a major concern in Switzerland. Such materials are commonly immobilised with cement, however, research carried out at the Swiss Federal Institute of Technology of Zurich has shown that clay can be used effectively to absorb the heavy metals in the fly ash [8]. Further research into the effects of organophilic clays to enhance sealing capacities of soil liners has also revealed promising heavy metal adsorption capacities [11]. Coal incineration fly ash has also been found to decrease the permeability of vertical cut-off walls used to encapsulate waste deposits, thus enabling a waste product to be used as a construction material [3].

Introduction

The Swiss Federal Institute of Technology in Zurich (ETHZ) is one of two engineering universities in Switzerland. The Geotechnical Engineering Division (IGT), part of the Civil Engineering Department (BAUM), concentrates on geoenvironmental problems, and is composed of three separate laboratories: soil mechanics, rock mechanics, and clay mineralogy.

Conductivity and diffusion properties are decisive for barrier waterproofing capacities, as are the material's ability to absorb organic substances or heavy metals and to resist chemical influences. Mineral barriers, constructed from natural materials, must also respond to mechanical demands. Barrier deformability is critical, as differential settlements caused by waste load may induce cracking if the barrier is too rigid, resulting in leaks.

The IGT has undertaken research directed at finding solutions to these problems.

Waste immobilisation

In Switzerland, the incineration of municipal solid waste (MSW) has become a primary waste reduction procedure. Through incineration, 1 kg of MSW is converted to approximately 700 g of gas, 270 g of slag and 30 g of filter ash. As filter ash contains leachable heavy metals, this material must be treated before being placed in a landfill.

A complete characterisation of fly ash and filter residue using X-ray image analysis techniques in an electron microprobe was carried out in order to understand the potential leaching by percolation water [10]. The crystalline materials in the ashes were revealed by X-ray powder diffraction techniques to be gypsum, anhydrite, quartz, halite, sylvite, calcite and ettringite. Microchemical analysis techniques have shown that the bulk of the Pb and most of the other heavy metals are associated with silicate glasses of varying composition.

Traditionally, fly ash has been immobilised with cement. Research has been carried out to investigate the possibilities of immobilisation using clay

TABLE 1--Mineralogical and physical data for clays and fly ash used [9]

parameter	Montigel (M)	Opalinus clay (Op)	Arizona clay (A)	Pfungen clay (Pf)	Fresh fly ash (QTR,f)	Washed fly ash (QTR, w)
smectite %	66	-	90	10	portlandite	portlandite
ml illite/smectite	n.d.	20	-	-	quartz	quartz
illite %	-	15	-	5-10	syngenite	calcite
chlorite %	-	10	-	5	sylvite	ettringite
kaolinite %	2	10	-	5	halite	magnetite
quartz %	8.3	30	10	10-20	calcite	-
mica %	12-15	-	-	-	anhydrite	-
feldspar %	2-4	-	-	10	mahnetite	-
carbonate %	3.8	10	-	43	-	-
accessories %	2-3	5	-	<5	-	-
organic carbon %	0.03	-	-	-	-	-
CEC mmol/100g	62	12	120	10	-	-
BET-surface m^2/g	72	36	97	17	13	31
γ_s kN/m^3	27.7	26.6	25.9	27.5	26.5	24.1
water content %	17	2.7	n.d.	n.d.	2	80
> 63 µm	8.0	2.0	0.1	n.d.	13.0	14.0
< 20 µm	89.0	89.0	97.0	n.d.	87.0	86.0
< 2 µm	77.5	57.0	88.0	n.d.	15.2	15.1

[8, 9]. One type of fly ash and four types of clay were used, and their data are listed in Table 1. Four different tests were carried out to determine the behaviour of the test materials: leaching, compaction, deformation and shearing.

Leaching tests with percolation columns were used to study the adsorption capacity of fly ash-clay mixtures. Test series were run using 20, 30 and 40% weight of clay mixed with the filter ash for the four clay types. One percolation cycle corresponded to filling the column first with test liquid and then with distilled water for the same fly ash-clay test sample. This procedure yielded information concerning leaching evolution. All of the eluated solutions were then analysed by Atomic Adsorption Spectroscopy (AAS) to determine the concentration of heavy metals in solution. Figure 1 shows the immobilisation of Zn for the four different clays. It may be seen that the immobilisation is dependent on both the type and the percentage of the clay added to the fly ash.

Standard and modified Proctor tests showed that the fly ash-clay mixtures may be compacted with standard procedures, even for high clay contents. Permeability was seen to depend both on clay content and compaction energy.

Consolidation tests indicated swelling of the fly ash-clay mixtures, due both to the presence of anhydrite and to the intercrystalline swelling of the clay minerals. Pure fly ash showed virtually no swelling behaviour.

Ring shear tests were carried out on mixtures of fly ash with both 30% Montigel and 30% Opalinus clay, respectively. Friction angle test results are given in Table 2.

TABLE 2--Ring shear test results on fly ash-clay mixtures [9]

	Fly ash	30% Montigel added	30% Opalinus clay added
ϕ' max	43.0	33.1	39.1
ϕ' min	40.2	30.9	34.3

The Swiss National Research Fund will soon begin funding of a new project in which the immobilisation of slag will be investigated. In this project the research work will be divided among the Swiss Federal Institute of Technology and industrial partners. The research program will also treat the effect of residual waste on both plant and animal cells.

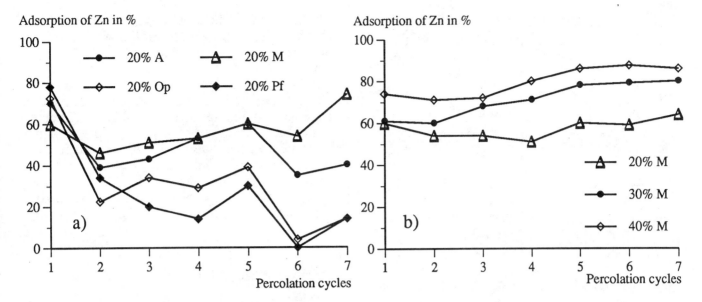

FIGURE 1--Adsorption of Zn ions for a) Arizona clay (A), Montigel (M), Opalinus clay (Op) and Pfungen clay (Pf) using the same quantity (20 % wt. addition) and b) Montigel using 20 % wt., 30 % wt. and 40 % wt. addition to the fly ash [9]

Soil liners

The hydraulic conductivity and its susceptibility to changes with time or exposure to chemicals are the major factors in the selection of clay for use in waste containment barriers. Concentrated organic substances have been found to influence the sealing capacity of clays adversely [7]. The adsorption behaviour of organophilic bentonites in contact with aqueous solutions of organic compounds and contaminant retention in liner material improved with organobentonites was extensively studied by percolation, permeability and diffusion tests [11].

Organophilic bentonites with a broad adsorption ability are needed for use in soil liner systems, as the composition and concentration of leakage water vary with time. The following bentonites were used in this study:

Tixosorb and **Tixosorb VZ**: dry-fabricated Bavarian calcium bentonites with 50% organic cation exchange capacity (CEC).
Tixogel VP and **Tixogel VZ**: wet-fabricated sodium bentonites from Wyoming with 100% organic cation exchange capacity.
Viscogel B4: a 50/50 mixture of Turkish and Wyoming sodium bentonites with 100% organic cation exchange capacity.
Montigel: a Bavarian unmodified calcium bentonite. This nonorganophilic bentonite was used as a control substance.
Phenol, aniline, nitroethane, diethyl ketone, ethoxy acetic acid, maleic acid and hexadecyl pyridunium bromide were used as test substances.

Percolation tests revealed that the adsorption of organic liquid compounds by organophilic bentonite depends on the type of organic cations exchanged by the bentonite, and the ion exchange capacity of the bentonite [11]. Table 3 lists which of the bentonites were found most suitable for the test substances.

Permeability of silty sand (70% sand, 25% silt and 5% clay), improved with varying percentages of the different test bentonites, to both water and phenol solutions was tested using an oedometric cell. Samples containing 5, 10 and 20% of the bentonites were tested. Permeability decreases of between 1 and 3 orders of magnitude were observed, with the lowest values being for all percentages of the Montigel. Increased bentonite led to decreased permeability in some, but not all cases. Detailed results of these tests may be found in the literature [11].

TABLE 3--Overview of adsorption test substances and corresponding bentonites

Test substance	Suitable bentonite
Phenol	Tixogel VP
Aniline	Tixogel VP or Viscogel B4
Nitroethane	Tixosorb
Diethyl ketone	Tixogel VP or Viscogel B4
Ethoxy acetic acid	Tixosorb VZ or Tixogel VZ
Maleic acid	Tixogel VZ
Hexadecyl pyridunium bromide	Tixogel VP, Tixogel VP or Viscogel B4

Proctor-compacted samples of the silty sand improved with bentonites of various organic adsorption capacity were used in order to carry out diffusion tests with phenol test solutions. Figure 2 presents the diffusion test results. The test solutions were 10'000 ppm phenol, and the total test times are indicated in parentheses. The treated samples (especially Tixosorb and Tixogel VP) showed higher phenol concentrations for a given penetration depth than the untreated soil. A considerably increased phenol retention was observed for these treated samples, as the diffusion transport mechanisms were hindered in the presence of the organic bentonites.

The mechanical behaviour of clay liners is currently the subject of a research project involving both laboratory investigations and field results from an instrumented landfill in Germany [1, 2]. Small-scale analogy tests using foam rubber to simulate both the clay liner and the waste material have yielded qualitative results in good agreement with field measurements of pressure and displacement in the clay liner. This correlation will be further tested in large-scale instrumented laboratory tests, and simulated by F.E. calculations, in order to define deformation mechanisms and failure criteria for such liners.

Slurry wall materials

Different sealing materials were tested in order to improve the encapsulation of old waste deposits using slurry walls [3, 4]. Active clay material (Opalite) was used in order to improve long-term behaviour and to increase the retention of polluted liquids. Of the many mixtures tested in this research project, results from two only will be presented in this paper. Table 4 gives the make-up of these mixtures (A and B).

Permeability tests using both tap water and simulated leachate were carried out on samples of test Mixtures A and B. The results of these tests in Figure 3 show a much lower permeability with both clean and contaminated water for Mixture A than for Mixture B.

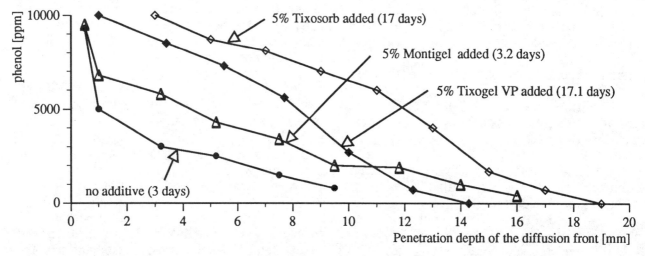

FIGURE 2--Effect of added bentonite on the diffusion of phenol in silty sand [2, 11]

The collected percolates were then analysed chemically in order to determine calcium concentration. Lower calcium concentrations and a smaller decrease of calcium content with time for Mixture A for both clean and contaminated water are seen in Figure 4. A slightly modified version of Mixture A has been chosen for the slurry walls at a hazardous waste site in Germany [13].

TABLE 4--Test mixtures for slurry walls

Compound	Mixture A [kg/m^3]	Mixture B [kg/m^3]
Clay	491	638
Portland cement	368	279
Coal incineration fly ash	123	0
Water	654	678

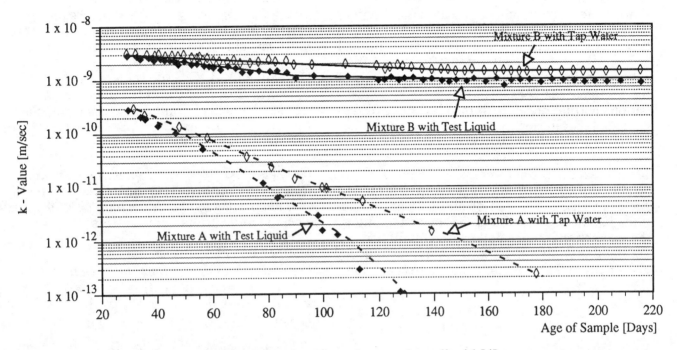

FIGURE 3--k-values of Mixtures A and B as a function of age and test liquid [4]

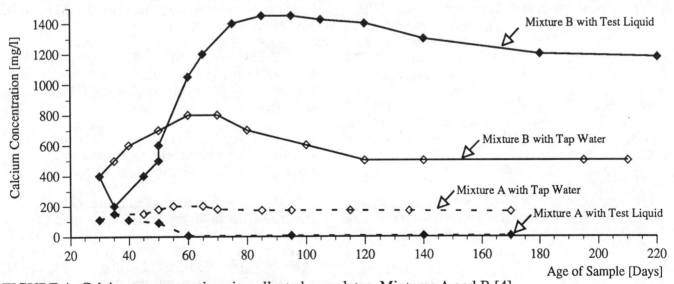

FIGURE 4--Calcium concentrations in collected percolates, Mixtures A and B [4]

Summary and Conclusion

In Switzerland, research and industry strive to attain both short and long term landfill quality through adequate legislation [12]. At the Geotechnical Engineering Division at the Swiss Federal Institute of Technology of Zurich, geoenvironmental research is carried out in accordance with Swiss legislation in order to promote development. Such research has mainly concentrated on the adsorption and deformation behaviour of barrier materials, with additional attention to the immobilisation of slag and filter ash resulting from incineration. A major part of the research has consisted of laboratory investigations carried out in both the clay and soil mechanics laboratories. Some additional work has taken the form of field tests.

In this Paper, a brief summary of relevant research work carried out over the past year has been presented. The aim of all of the investigations has been to facilitate the understanding of the interactions among pollutants, landfill barriers and in situ ground in order to envisage coherent and environmentally sound waste disposal solutions.

References

[1] Amann, P. and Hertweck, M. (1992), *Untersuchung und Anwendung von Steilwandbarrieren für Deponien in Steinbrüchen*, XLI. Geomechanik - Kolloquium, Salzburg, 8-9 October (in German)

[2] Amann, P. and Martinenghi, L. (1993), *Geotechnical design criteria for landfills and waste disposal sites*, Joint CSCE-ASCE National Conference on Environmental Engineering, July 12-14, Montreal, Canada

[3] Hermanns, R. (1992), *Sicherung von Altlasten mit vertikalen mineralischen Barrieresystemen im Zweiphasen-Schlitzwandverfahren*, Dissertation No. 9833, ETH Zürich (in German)

[4] Hermanns, R. (1993), *Waste deposit encapsulation using vertical barriers*, Joint CSCE-ASCE National Conference on Environmental Engineering, July 12-14, Montreal, Canada

[5] Kruse, K. (1992), *Die Adsorption von Schwermetalleionen an verschiedenen Tonen*, Dissertation No. 9737, ETH Zürich (in German)

[6] Kruse, K. und Stockmeyer, M. (1991), *The adsorption of heavy metals and organic pollutants by bentonites of various organophilic covering*, Clay and Minerals 26, pp. 431-434

[7] Mitchell, J.K. and Madsen, F.T., (1987), *Chemical Effects of Clay Hydraulic Conductivity*, ASCE Specialty Conference on Geotechnical Practice for Waste Disposal, Ann Arbor, Michigan, June 14-17

[8] Plüss, A., (1992), *Charakterisierung von Rauchgasreinigungsrückständen aus Kehrichtverbrennungsanlagen und deren Immobilisierung mit Tonmineralien*, Dissertation No. 9824, ETH Zürich (in German)

[9] Plüss, A., (1993), *Immobilization and stabilization of fly ash with clay minerals*, Joint CSCE-ASCE National Conference on Environmental Engineering, July 12-14, Montreal, Canada

[10] Plüss, A. and Ferrell, R. (1991), *Characterization of lead and other heavy metals in fly ash from municipal waste incinerators*, Hazardous Waste and Hazardous Materials, Vol. 8, Nr. 4, p. 275-292

[11] Stockmeyer, M. (1992), *Organophile Bentonite als Komoponente in Deponiebarriere-Systemen*, Dissertation No. 9740, ETH Zürich (in German)

[12] TVA (1990), *Technische Verordung über Abfälle*, Schweizerische Bundesrat, Bern (available in German, French and Italian)

[13] Hollenweger, R. and Martinenghi, L., *New Developments in Slurry Wall Construction for the Remediation of Contaminated Sites*, ASTM Symposium on Dredging, Remediation and Contaminated Sediments, June 23-24, 1994, Montreal, Canada (in preparation)

Effects of Clay Swelling on Permeability Calculations Obtained from Sealed Double Ring Infiltrometer Tests

Cem B. Avci
Associate Professor. Soil Environment Protection and Remediation Research Center & Department of Civil Engineering, Bogaziçi University, Istanbul, Turkey

Daniel Bodine
Associate, Woodward-Clyde Consultants, Chicago, Illinois USA

Erol Güler
Professor. Soil Environment Protection and Remediation Research Center & Department of Civil Engineering, Bogaziçi, University, Istanbul, Turkey

Abstract: The sealed double ring infiltrometer (SDRI) test is one of the in situ testing techniques available to estimate the hydraulic conductivity of compacted earth liners constructed as part of containment systems. It is known that swelling of the earth material during SDRI testing can cause overestimation of the saturated hydraulic conductivity and ultimately result in longer field testing periods. This study was performed to review the effects of swelling on the permeability calculations in SDRI tests. The governing equations for water movement in unsaturated medium were revised to take into account the deformation of the media; the revised equations were subsequently solved using numerical methods. Field data obtained from an SDRI test were analyzed to estimate the effect of swelling on the prediction of the saturated hydraulic conductivity.

Introduction

Earth liners represent a key component of containment systems for minimizing the environmental impact of leachate generated in nonhazardous and hazardous waste landfills. Environmental regulations require the hydraulic conductivity of constructed earth liners to be at most 1×10^{-7} cm/s; this requirement has led to extensive research for the identification of the variables affecting the hydraulic conductivity property of earth liners [1]. Material properties, gradation, plasticity index, lift thicknesses during construction, maximum clod size, degree of compaction, compaction moisture content, scarification between lifts and compaction equipment used [2] have been determined to affect the value of the hydraulic conductivity.

The trend in performance evaluation of liner construction techniques has lately been the use of in situ testing methods [3]. The sealed double ring infiltrometer (SDRI) test represents one of the available in situ methods for measuring the hydraulic conductivity of earth liners. The test is performed on test pads prepared with the proposed construction techniques. The test set up allows for the measurement of the infiltration rates on top of the liner; tensiometers can be used to detect the movement of the wetting front and swell gages can record the total swell occurrence during the testing phase.

The saturated hydraulic conductivity of the earth liner can be estimated using analytical methods as well as numerical methods [4]. The saturated hydraulic conductivity of the liner is often calculated using the Green-Ampt Infiltration model [5] formulated as:

$$K_s = i \left[1 + \frac{h + \psi_c}{L_f} \right]^{-1} \quad (1)$$

where i is the infiltration rate, ψ_c is the matric potential, h is the ponding depth, L_f is the depth to the wetting front and K_s is the saturated hydraulic conductivity. The saturated hydraulic conductivity can also be determined using numerical methods [6] which solve the nonlinear differential equation governing water movement in unsaturated domains. The governing equation for one dimensional unsaturated flow in the vertical direction can be written [7] as:

$$\frac{\partial \psi}{\partial t} \frac{d\theta}{dz} - \frac{\partial}{\partial z}\left[K(\psi)\frac{\partial \psi}{\partial z} + K(\psi) \right] = 0 \quad (2)$$

in which $\psi = \phi - z$ is the matric potential or capillary pressure head, where ϕ is the piezometric head; θ is the volumetric moisture content; $K(\psi) = K_r(\psi) K_s$ is the unsaturated hydraulic conductivity, where $K_r(\psi)$ is relative hydraulic conductivity, K_s is the saturated hydraulic conductivity; z is the vertical coordinate, positive upward and t is time.

One factor which affects the infiltration rates measured during SDRI tests is the swelling of the compacted material [8]. Infiltration rates being larger in swelling soils than nonswelling soils lead to longer SDRI testing periods for swelling soils than in nonswelling soils.

The present study investigates the swell occurrence during SDRI testing and its effects on the measured infiltration rates on top of the liner. The analysis was performed by modifying equation (2) to account for deformations in the flow domain when water movement occurs. The governing equations were subsequently solved using finite difference techniques. The numerical model was applied to SDRI data collected during the performance testing of an earth liner test pad.

Problem formulation

The mathematical theory of water movement in swelling unsaturated soils was derived by Philip [9] using a Lagrangian Coordinate system. The relationship between the deforming material coordinate m and the vertical system z is:

$$\frac{dm}{dz} = (1 + e)^{-1} \quad (3)$$

where $e(\theta)$ represents the void ratio. The governing equation (2) formulated in terms of the moisture ratio ϑ defined as:

$$\vartheta = (1 + e)\theta \quad (4)$$

yields:

$$\frac{\partial \vartheta}{\partial t} = \frac{\partial}{\partial m}\left[\frac{K}{1+e}\left[\frac{d\psi}{d\vartheta} + P(O)\frac{\partial^2 e}{\partial \vartheta^2} \right] \frac{\partial \vartheta}{\partial m} \right]$$

$$- \frac{\partial}{\partial m}\left[K \left[1 - \gamma \frac{de}{d\vartheta} - \frac{d^2 e/d\vartheta^2}{1+e} \int_0^m \gamma(1+e) dm \frac{\partial \vartheta}{\partial m} \right] \right] \quad (5)$$

where $P(O)$ represents the potential head at the top of the liner and γ is the unit weight of the soil. The void ratio dependence on the moisture ratio can be taken to be linear within the range being considered for SDRI testing. The above equation then becomes:

$$\frac{\partial \vartheta}{\partial t} = \frac{\partial}{\partial m}\left[\frac{K}{1+e}\frac{d\psi}{dm} - K\left[1 - \gamma \frac{de}{d\vartheta}\right]\right] \quad (6)$$

subject to the following initial and boundary conditions:

$$t=0, m>0, \vartheta = \vartheta_i$$
$$t>0, m=0, \vartheta = \vartheta_o$$

Equation (6) with the appropriate conditions were simulated using an implicit finite difference scheme that was based on the SOILINER numerical model [10]. The numerical model yields the distribution of the moisture ratio ϑ with respect to the material coordinate m. The moisture content distribution with respect to time and the vertical coordinates can then be obtained from equation (3) and equation (4), respectively.

Case study

An SDRI test was performed at a site in Indiana where a clay liner was proposed to be built as part of a final cover system for a waste containment unit. A 90 cm thick test pad with a 250 m² surface area was prepared for the SDRI testing. Table 1 lists the results of the geotechnical tests performed on the Shelby tube samples collected from each of the five lifts which made up the test pad thickness; the geotechnical test results met the project specifications listed in Table 1.

TABLE 1

Geotechnical Parameters

Parameter	Average	Specified
Water Content %	17.7	≥ 13.9
Liquid Limit %	35	≥ 30
Plastic Limit %	16	
Plasticity Index	19	≥ 15
Fines Control %	88	≥ 60
Dry Density kN/m³	17.9	
K cm/s	1.5×10^{-8}	$\leq 1 \times 10^{-7}$
Percent Compaction % (Standard Proctor)	94.7	≥ 90

The SDRI testing was conducted for a period of 73 days where measurements were obtained for the infiltration rates, suction head from the tensiometers installed at 15 cm, 30 cm and 45 cm depths and the heave of the inner ring area.

Figure 1 shows the infiltration rates during the testing period. Tensiometer readings showed that the wetting front reached depths of 15 cm and 30 cm in 13 days and 54 days, respectively. The total swell of the inner ring area at the end of 73 days was measured to be 0.35 cm. The hydraulic conductivity value calculated using equation (1) with the infiltration rates measured at the time when the wetting front reached the tensiometers was estimated at 6×10^{-8} cm/s. This result indicated that the length of the test was adequate to establish that the test pad met the hydraulic conductivity criteria ($\leq 1 \times 10^{-7}$ cm/s) and, therefore, a complete wetting of the liner was not necessary. It should be noted that this hydraulic conductivity value is conservative since the suction head at the wetting front was taken as zero in the use of equation (1).

Analysis results

The field data was analyzed and used as input into the numerical model. The porosity and the specific gravity of the clay material were selected as 0.35 and 2.67, respectively. The initial volumetric moisture content was calculated to be 0.32 based on 92% saturation of the clay liner. The $\psi(\theta)$ relationship was based on the relationship proposed by Haverkamp et al. [11] and was derived for heavy clays. An initial suction head of 20 cm was selected throughout the clay liner to achieve the 92% initial saturation condition. The numerical simulations were performed on a uniform grid with 90 nodes and a 1 cm spacing.

The saturated hydraulic conductivity K_s and the swell factor $\alpha = de/d\theta$ were the two variables which were used to adjust and compare the numerical results with the collected field data. The initial simulations neglected the swell phenomenon ($\alpha = 0$) in order to determine the saturated hydraulic conductivity which closely simulated field conditions. The best numerical simulations were obtained using $K_s = 5 \times 10^{-8}$ cm/s. Figure 1 shows the numerical prediction of the infiltration rates at the top of the clay liner and the data collected during the testing period.

The results of the simulations reaffirmed once more that the test pad construction techniques and the selected clay material satisfied the hydraulic conductivity performance criteria ($K_s < 1 \times 10^{-7}$ cm/s). This analysis was again considered to be on the conservative side since, despite taking into account the suction at the wetting front, the swelling of the clay material would increase the measured infiltration rates which would, in turn, yield a higher value of K_s than the actual value existing in the clay material.

A sensitivity analysis was performed using a value of $K_s = 5 \times 10^{-8}$ cm/s and various values of α; the results are shown in Figure 1. The swelling phenomenon was noted to affect the infiltration rates at the top of the clay liner. The infiltration rate curves increased with increasing values of α; at the same time, the shape of these curves tended to be flatter toward the latter part of the simulations. These results showed that the predictions of the saturated hydraulic conductivity would be conservative if the swelling phenomenon was neglected in the analysis.

The numerical model was used to estimate the actual value of the saturated hydraulic conductivity by taking into account the swelling phenomenon. For this, various values for K_s and α were used to fit the infiltration curve data, the wetting front movement and the total swell which occurred at the end of 73 days. The best match between these results and field data was obtained using a value of $K_s = 3.5 \times 10^{-8}$ cm/s, and $\alpha = 0.4$. The infiltration rate and the wetting front predictions are shown in Figures 1 and 2, respectively. The total swell occurring in the inner ring area was simulated to be 0.29 cm at the end of 73 days.

Summary and conclusions

The water movement occurring in an unsaturated and deforming media was simulated using an implicit finite difference model. The aim of the study was to review the influence of the swelling phenomenon on the infiltration rates measured during SDRI tests. Numerical simulations were performed using field data obtained from an SDRI test. A saturated hydraulic conductivity of 5×10^{-8} cm/s was predicted if swelling was neglected; a hydraulic conductivity of 3.5×10^{-8} cm/s and an 0.29 cm total swelling occurrence was found to match the field data when swelling was considered.

Swelling of the material during the SDRI test causes an increase in the infiltration rates measured at the top of the liner. Estimates of the saturated hydraulic conductivity are, therefore, on the conservative side if the swelling phenomenon is not taken into account in the analysis. The overestimation of the hydraulic conductivity may lead to prolonged test duration periods or, in the extreme case, may lead to rejection of the test pad if the required hydraulic conductivity criteria is not satisfied.

References

1. Daniel, D.E. and Trautwein, S.J, 1985, Field Permeability Tests for Earthen Liners, Use of Insitu Tests in Geotechnical Engineering, ASCE, S.P. Clemence, pp. 146-160.

2. Mitchell, J.K. and Jaber, M., 1990, Waste Containment Systems: Construction, Regulation and Performance, Geotechnical Special Publication, No. 26, pp. 84-105.

3. Trautwein, S.J. and Williams, C.E., 1990, Waste Containment Systems: Construction, Regulation and Performance, Geotechnical Special Publication, No. 26, pp. 30-52.

4. Goldman, L.J., Kingsbur, G.L., Northeim, C.M. and Truesday, R.S., Draft Design, Construction, Maintenance and Evaluation of Clay Liners for Hazardous Waste Facilities, EPA/68-03-3149-1-2, Technical Resources Document, Research Triangle Institute, NC, 1985.

5. Green, W.H. and Ampt. G.A., 1911, "Studies on Soil Physics: I Flow of Air and Water through Soils", Journal of Agricultural Science, V. 4. No. 1, pp. 1-24.

6. GCA Corporation, 1984, Procedures for Modelling Flow through Clay Liners to Determine the Required Liner Thickness, U.S.EPA Document EPA/530-SW-84-001.

7. Richards, L.A., 1931, "Capillary Conduction of Liquids through Porous Mediums", Physics, 1, pp. 318-333.

8. Trautwein, 1989, Installation and Operation Instructions for the Sealed Double Ring Infiltrometer, Trautwein Soil Testing Equipment Co. Houston, Tex.

9. Phillip, J.R., 1969, Hydrostatics and Hydrodynamics in Swelling Soils, Water Resources Research, Vol. 3, No. 5, pp. 1070-1077.

10. Haverkamp, R., Vauclin, M., Toma, J., Wierenga, P.J., and Vachaud, G., "A Comparison of Numerical Simulation Models for One-Dimensional Infiltration", Soil Science Society of America Journal, 41, 285-294, 1977.

11. U.S. EPA, 1986, Soiliner Model- Documentation and User's Guide, U.S.EPA Document, EPA/530-SW-86-006a.

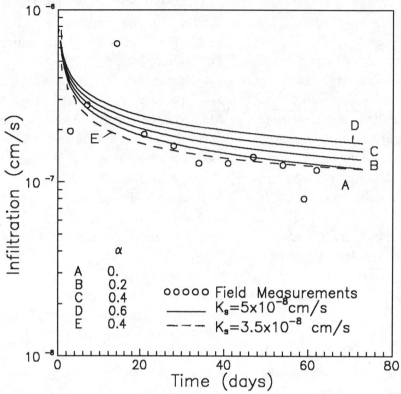

Figure 1. Infiltration rates at the top of the clay liner

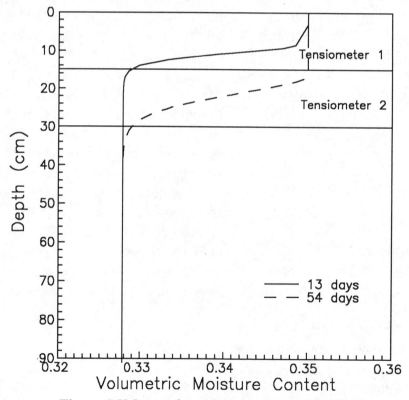

Figure 2 Volumetric moisture content distribution

A New Method for Determination of Organic Carbon Content in Soils and Rocks to Estimate Pollutant Levels

N.P. Betelev
Research Institute of Bases and Underground Structures (NIIOSP), Moscow, Russia

ABSTRACT The data of organic carbon (C_{org}) content in soils and rocks can be used in the estimation of their oil pollution, which is determined by the comparison of C_{org}-content in polluted and non-polluted sections. The data of organic matter (OM) content in soils and rocks allow to estimate the possibility of radio-active and other harmful substances accumulation in them because OM promotes such accumulation. The new method for determination of C_{org}-content in soils and rocks by dry combustion in oxygen or air stream is proposed. C_{org}-content is determined by the quantity of carbon dioxide CO_2 that is produced from combustion of OM. The use of the catalyst of oxidation allows to decrease the temperature of calcination to 500°C instead of the temperature of 1000°C formerly employed. Natural carbonates calcite $CaCO_3$ and dolomite $CaMg(CO_3)_2$ are not decomposed with CO_2 generation being calcinated at the temperature of 500°C. It gives the possibility not to remove these carbonates from samples before analysis. Due to this, the procedure of analysis is more simple than that being carried out at the temperature of 1000°C. The analysis is carried out on various types of gas-analysers, CHN-analysers and instruments for micro- analysis.

OM-content in soils and rocks is commonly determined from C_{org}-content. The data of C_{org}-content in natural objects are important for ecological investigations, in estimation of oil pollution, for instance. The data of OM-content in soils and rocks allow to estimate the possibility of accumulation of radio-active and other harmful substances because OM promotes such accumulation. The information of OM-content in natural objects can be used for determination of regions where accumulation of radio-active and other harmful substances has been the most probable when these substances has been dispersed from the sources of pollution.

There are several methods for C_{org}-content determination. We attempted to improve the method of determination by dry combustion (1).

The method of C_{org}-content determination by dry combustion at the temperature of 1000°C in oxygen stream had been known before we started our investigations. The C_{org}-content is usually determined by the quantity of the carbon dioxide CO_2 that has been produced from combustion of OM. The temperature of 1000°C has been taken because at lower temperature the OM of the sample can not burn completely with CO_2 generation. Since the carbonates (for example calcite $CaCO_3$) are decomposed with CO_2 generation at this temperature it is necessary to remove the carbonates from the samples before combustion. The decarbonization by the action of acid is a labour-consuming operation that demolishes some part of OM in sample. The loss of OM is especially high (up to 40%) in the case of low degree of metamorphism of OM in soils and recent sediments. It is useful to lower the temperature of combustion to the level in which carbonates will not decompose and avoid the acid decarbonization.

Our experiments showed that the use of the catalyst of oxidation three cobalt tetroxide Co_3O_4 allowed to reach complete oxidation of OM with CO_2 generation at the temperature of 500°C. The most common natural carbonates calcite $CaCO_3$ and dolomite $CaMg(CO_3)_2$ are not decomposed with CO_2 generation at the temperature of 500°C hence it is not necessary to remove these carbonates by the action of acid. The experiments showed that calcite started to decompose with CO_2 generation at the temperature of 635°C, and dolomite- at the temperature of 575°C. These two carbonates are not decomposed at the temperature of 500°C even when calcination lasts 5 hours.

There are some carbonates in nature that begin to decompose with CO_2 generation at the temperature of 500°C: siderite $FeCO_3$, magnesite $MgCO_3$ and rhodochrosite $MnCO_3$. Magnesite and rhodochrosite are found in soils and rocks very seldom and in small amounts (less than 1%) and can not influence the results of C_{org} determination seriously. The siderite occurs in soils and rocks seldom enough and in the amounts no more than 3%. Our special experiments demonstrated that the presence of 3% of siderite in the samples being analysed resulted in error no more than 3% in C_{org}-content determination that is permissible for a analytic method. Siderite ores and concretions containing more than 3% of siderite can not be analysed by the proposed method.

The action of the catalyst of oxidation is based on the activity of oxygen of crystalline lattice (2):

$$2Co_3O_4 \rightleftarrows 6CoO + O_2$$

We analysed hundreds of samples of various genesis by this and some other methods and got similar results. Some of these results are given in table 1.

Oxygen can be used in special laboratory rooms only and it is explosive in the presence of oil, that is why it is reasonable to use air instead of oxygen for dry combustion. Our experiments showed that air in the presence of the catalyst Co_3O_4 can be used for combustion instead of oxygen. The results achieved with the use of air are close to that got with using oxygen for dry combustion (table 2).

The coefficients K_1, K_2, K_3, K_4 were given to estimate and compare the influence of oxygen, air and catalyst in dry combustion. These coefficients represent ratious of percentages of C_{org}-contents determined in various conditions (in oxygen or air stream, with the catalyst Co_3O_4 and without the catalyst):

$$K_1 = \frac{C_{org} \text{ det. in oxygen with } Co_3O_4}{C_{org} \text{ det. in air with } Co_3O_4},$$

TABLE 1 C_{org}-content determined from various methods of analysis

type of sample	C_{org}-content, %			
	determined from dry comb. at 500°C with catalyst Co_3O_4	det. V	from wet comb. by the Knopp method	det. oxidimetrically by the Tyurin method
silt	1.54	0.6	1.53	1.52
silt	4.26	1.9	4.42	4.22
silt	2.63	1.0	2.58	2.56
silt	4.59	2.5	4.45	4.56
clay	2.25	1.3	2.30	2.10
clay	2.54	1.1	2.60	2.38
clay	1.41	2.6	1.42	1.42
clay	0.55	3.6	0.55	0.61

V-coefficient of variation,%

$$K_2 = \frac{C_{org} \text{ det. in oxyg. without cat.}}{C_{org} \text{ det. in air without cat.}},$$

$$K_3 = \frac{C_{org} \text{ det. in oxyg. with } Co_3O_4}{C_{org} \text{ det. in oxyg. without cat.}},$$

$$K_4 = \frac{C_{org} \text{ det. in air with } Co_3O_4}{C_{org} \text{ det. in air without cat.}}.$$

Arithmetical mean values of coefficients K_1, K_2, K_3, K_4 for all samples being analysed show that presence or absence of the catalyst affects the results of C_{org} determination stronger (arithmetical mean values of coefficients K_3 and K_4 are respectively 1.204 and 1.247) than the use for combustion of oxygen or air (arithmetical mean values of coefficients K_1 and K_2 are respectively 1.027 and 1.064). Results of the analyses in oxygen and air stream are given in table 2.

The use of the coefficient K_1 allows to calculate C_{org}-content for analysis in oxygen stream by C_{org}-content received in air stream. Our experiments show that arithmetical mean values of the coefficient K_1 are 1.032 and 1.014 respectively for non-polluted and oily polluted samples (see table 2).

The analysis by the method proposed is carried out with gas-analysers, CHN-analysers and instruments for micro-analysis in which C_{org}-content is determined by dry combustion. For this analysis we used the gas-analyser КГА-4-2 (ГОУ-1) made in Russia. C_{org}-content was calculated by the quantity of carbon dioxide CO_2 volumetrically found. The error of determination was 3 relative %. The period of combustion in oxygen and air stream was 25-35 min.

References

1. N.P.Betelev. Determination of organic carbon content in silts and rocks by calcination in oxygen stream without preliminary removal of carbonates. Journ. Litologiya i poleznyie iscopaemyie, 1981, No. 5, p. 152-159.

2. M.Večeřa. The study of some basic reactions used in organic elemental analysis. Microchemical Journ., 1966, vol. 1-4, p. 250-259.

TABLE 2 C_{org}-content determined from dry combustion at 500°C in oxygen and air stream

type of sample	C_{org}-content determined in various cond., %								coefficients			
	in oxygen stream				in air stream				K_1	K_2	K_3	K_4
	with Co_3O_4	V	without Co_3O_4	V	with Co_3O_4	V	without Co_3O_4	V				
peat	28.70	3.0	25.85	1.7	28.25	3.0	20.45	4.0	1.016	1.264	1.110	1.381
peat polluted by oil	56.66	4.0	39.25	2.5	56.25	2.5	36.60	3.5	1.007	1.072	1.444	1.537
soil	5.10	4.0	4.26	4.0	4.97	3.0	4.03	3.0	1.026	1.044	1.197	1.218
soil polluted by oil	7.88	4.0	4.49	1.0	7.36	3.0	4.45	3.5	1.071	1.007	1.755	1.650
soil	2.46	2.0	2.07	2.0	2.42	2.0	2.12	4.0	1.017	0.976	1.188	1.142
soil polluted by oil	12.75	4.0	7.54	3.5	12.85	4.0	7.31	4.0	0.992	1.031	1.691	1.758
soil	0.34	2.0	0.34	2.0	0.34	1.0	0.26	1.0	1.000	1.308	1.000	1.308
soil polluted by oil	1.60	4.0	1.49	3.5	1.56	4.0	1.39	3.5	1.026	1.072	1.074	1.122
marine sediment	3.66	1.0	3.20	1.0	3.47	4.0	3.06	3.5	1.055	1.046	1.144	1.134
marine sediment polluted by oil	6.82	3.5	6.28	2.5	6.99	4.0	6.20	3.5	0.976	1.013	1.086	1.127
soil	4.34	2.0	3.68	1.4	4.02	0.2	3.40	0.6	1.080	1.082	1.179	1.182
soil	7.31	0.4	6.48	0.3	7.22	0.7	6.30	0.0	1.012	1.029	1.128	1.146
peat	45.40	0.4	41.50	3.8	44.10	0.2	38.40	3.4	1.029	1.081	1.094	1.148
peaty soil	2.61	0.8	2.44	1.2	2.55	0.4	2.30	1.7	1.024	1.061	1.070	1.109
silt	2.96	1.0	2.76	1.0	2.83	4.0	2.54	4.0	1.046	1.087	1.072	1.114
soil	4.14	4.0	3.72	3.5	3.98	2.0	3.48	4.0	1.040	1.069	1.113	1.144
soil	4.95	0.0	4.29	3.4	4.87	3.0	4.24	1.0	1.016	1.012	1.154	1.149

arithmetical mean values of coefficients:
for all samples being analysed (19 samples)   1.027 1.061 1.204 1.241
for samples without oil pollution (14 sampl.)   1.032 1.070 1.118 1.158
for samples polluted by oil (5 samples)   1.014 1.039 1.410 1.439

V-coefficient of variation, %

Dependence between Uranium and Organic Matter in Soils, Rocks and in Uranium Deposits in Connection with Geotechnical Problems

N.P. Betelev, B.I. Kulachkin
Research Institute of Bases and Underground Structures (NIIOSP), Moscow, Russia

Ya. M. Kislyakov
All-Russian Institute of Mineral Resources (VIMS), Moscow, Russia

ABSTRACT The content of radio-active elements in soils and rocks is connected closely to the content and composition of organic matter. Uranium is the most wide spread radio-active element. Many scientists have noted the close correlation between the content of uranium and organic matter of various types (humous and sapropelic). Transportation of uranium depends on hydrogeological situation and on microbiological processes. The conditions of formation of high concentrations of uranium are:
1) Presence of small amounts of dispersed uranium in soils and rocks.
2) Possible transportation of uranium. As the combinations of uranium are well dissolved in water so water is the most favourable medium for transportation of uranium.
3) Concentration of uranium by the action of organic matter.

In the case of the favourable combination of these conditions high concentrations of uranium can appear in soils and rocks. These high concentrations of uranium can be harmful for the health of people.

The organic matter of bitumens, brown coals and peats actively concentrates uranium ores and minerals in many deposits. The correlation between uranium and organic matter is the most evident in uranium deposits of infiltration type which are located in gray and coal-bearing sedimentary rocks in the regions of wedging of zones where oxidation and limonitization appear. The deposits of such type are called exogenic-epigenetic. There are many uranium deposits of such type on the Turan plate in Kazakhstan and Uzbekistan (1). These deposits are also rich in Se, Mo, Re, V and some other elements.

There is direct correlation between contents of uranium and organic matter in "young" deposits in the Kizil-Kum and Chu-Sarysui regions. The process of uranium concentration continues here till now in gray sandy horizons. The uranium mineralization is connected to brown-coal deposits in adjacent regions also.

The main role in exogenic-epigenetic accumulation of uranium belongs to microbiologically-active organic matter of beginning stages of metamorphism (stages B_1 and B_2). These types of organic matter are usual to peats, lignites and brown coals. There are no exogenic-epigenetic uranium concentrations connected to coals of high degree of metamorphism and especially to anthracite coals. But uranium mineralization can be connected to highly metamorphosed coal remains in case of hydrothermal alteration (the Strelzov deposit in Baikal region).

The content of uranium in uranium-bearing gray terrigenous rocks is 1-2 orders higher than in alumosilicate material. High concentrations of uranium present in dispersed coal pigment and in plant remains. There is correlation between concentration of uranium and degree of oxidation of organic matter. Degree of oxidation of organic matter is reflected in increase of content of oxygen and decrease of carbon and hydrogen contents. The ratios of (O+N+S)/H and (O+N+S)/C are the most sensitive indices of degree of oxidation. The process of oxidation is followed by structural-molecular reconstruction of organic matter. This reconstruction can be measured by technical analysis of coals and by thermoanalytical, infra-red, spectrometry and electronic microscope methods of investigation (2).

One of two types of correlation between degree of oxidation of organic matter and content of uranium prevails in each particular deposit. The direct correlation is peculiar to deposits in permeable sandy layers. The anaerobic bacteria oxidize brown coal matter in zones of recent accumulation of uranium in the deposits of infiltration type. The aerobic bacteria oxidize the matter in zones of limonitization. The bacteria use for their vital activity the aliphatic part of molecules of organic matter. This part of molecules is regenerated in the process of oxidation at the expense of aromatic nuclei of organic matter (3). The rise of degree of oxidation of organic matter is followed by concentration of uranium. But content of uranium begins to decrease in that part of oxidation zone where the most oxidized organic matter is located.

The reverse correlation between degree of oxidation of organic matter and content of uranium is observed in uranium-coal deposits. Low residual uranium concentrations are present in strongly oxidized and limonitized brown coals at the contacts with permeable oxidized sandstones. The concentration of uranium increases when degree of oxidation decreases. The maximum concentrations of uranium appear at a distance of 1-2 metres from the contact with oxidized sandstone. Very weak oxidation of organic matter in brown coals with maximum contents of uranium is masked by radio-active action of uranium.

High concentration of uranium in brown-coal matter is brought about by two causes:
1) by the decrease of Eh potential by the action of organic matter. At the same time uranium changes its form from U-6 to U-4 and uranium minerals uraninite and coffinite appear. These minerals are placed in coal remains and in containing rocks. This cause determines formation of the most part of so called "sandy" uranium deposits;
2) by the sorption of uranium by brown-coal matter without generation of new uranium minerals. This process is usual to recent superficial uranium deposits and to containing uranium peat-beds.

The two types of geochemical barriers correspond to these two causes. These barriers have been well studied in uranium-coal deposits. The main role in concentrating of uranium in brown coals belongs to the reduction barrier. This barrier is responsible for accumulation of uranium minerals and many accompanying elements (Se, Mo, etc.). The sorption barrier plays a secondary role. This barrier is placed in limonitized brown coals. The abundance of carboxil groups (COOH) in strongly oxidized organic matter generates this barrier. The presence of carboxil groups is cause of acid reaction of brown coals. The coals acquire the properties of natural ionites.

The properties and quality of underground water are closely connected with adsorption capacity of organic matter because organic matter concentrates and preserves uranium and accompanying elements when the elements are released and transported during the process of weathering. For example underground water containing oxygen and uranium becomes undrinkable because of a high concentration of selenium. On the other hand underground water can get a lot of radionucleides in the uranium mineralization zone. But such water is cleaned of selenium and radionucleides when this water filters through gray sand, which contains organic matter.

The use of brown coals and peats which contain uranium as a fuel can lead to dangerous radioactive and chemical contamination. Disturbance of the natural state of uranium in "sandy" deposits by the mining process of borehole underground leaching gives rise to difficult problems of cleaning residual solutions. These solutions, if not cleaned, contaminate underground water with sulphates and various heavy metals.

References

1. Shchetochkin V.N., Kislyakov Ya.M. Exogenic-epigenetic uranium deposits of the Kizilkums and adjacent regions, journ. Geology of ore deposits, 1993, vol. 35, no. 3, p. 222-245.

2. Kislyakov Ya.M., Urmanova A.M., Uspenskii V.A., Dubinchuk V.T., Solntseva L.P. Oxidizing conversions of the coal material of infiltrational deposits, journ. Geology of ore deposits, 1990, no. 6, p. 56-70.

3. Kislyakov Ya.M., Urmanova A.M., Shugina G.A. Correlation of thermoanalytical properties of the coal material and microbiological communities at exogenic- epigenetic uranium deposits, Nakoplenie i preobrazovanie organicheskogo veshchestva sovremennykh i iskopaemykh osadkov, Nauka, Moscow, 1990, p. 162-181.

Use of Organophilic Clay to Control Seepage from Underground Gasoline Storage Tanks

Brigitte Boldt-Leppin, Paul Kozicki, Moir D. Haug, John Kozicki
Ground Engineering Ltd., Regina, Saskatchewan, Canada

Abstract

A two phase laboratory test program was conducted to evaluate the performance of organophilic clay in controlling seepage of gasoline. In the first phase, different percentages of organophilic clay (15, 20 and 25 %) were blended with pea gravel and compacted into a fixed wall permeameter. Test specimens were prepared with both powdered and granular organophilic-clay. These test specimens were first saturated with water to simulate anticipated field conditions and then permeated with gasoline. In the second phase, 20 and 25% powdered organoclay concentrations were blended with pea gravel, moisture conditioned and compacted into 150 mm diameter Proctor molds. The extruded samples were tested in specially designed stainless steel triaxial permeameters capable of measuring hydraulic conductivity and sample volume change during permeations. The specimens were confined under an effective stress of 20.7 kPa and permeated with water using hydraulic gradients of less than 20. After reaching equilibrium conductions, these samples were permeated with gasoline. The results of this program demonstrated that organophilic clay-sand mixtures significantly restrict the flow of gasoline. Hydraulic conductivities in water were in the 1×10^{-5} cm/s range. In gasoline, the samples swelled (9 to 10 %) and the hydraulic conductivity dropped to approximately 1×10^{-8} cm/s after 30 days of testing.

Introduction

The results of a laboratory test program demonstrated that organophilic clay-pea gravel mixtures provide an effective barrier to gasoline seepage. The containment of gasoline spills is an important aspect of environmental protection. In many cities, explosions and fires caused by spills of gasoline have resulted in loss of life and extensive damage to property. As a result, considerable effort has been directed towards the development of measures to ensure that gasoline spills be contained and not allowed to seep into the groundwater system where they can rapidly spread.

Over the past decade numerous containment strategies have been developed involving both the use of synthetic and natural materials. One of the most common materials is compacted fine grained soil, which swell (hydrate) and decrease in hydraulic conductivity during the permeation with water. Unfortunately, this same hydration process does not occur when the permeating fluid is a hydrocarbon such as gasoline. However, specially designed organophilic clays originally developed for drilling through salt formations with diesel fuel slurries, do swell in the presence of hydrocarbons.

The objective of this study was to conduct a laboratory test program to determine if organophilic clays could be used as a constituent in soil liners intended to contain gasoline. A secondary objective was to investigate the percentage of organophilic clay required and whether the organophilic clay would perform better in a powdered or granular form. The hydraulic conductivities measured in this test program were consistently below the 1×10^{-7} cm/s range required for most gasoline containment facilities.

Background

In recent years there has been an increased awareness and concern of Canadian and U.S. citizens about the state of their environment. This awareness and concern reflects the strong demands that government and industry examine their activities to prevent further environmental damage. The problem of groundwater contamination as a result of leaking underground fuel storage tanks or pipelines became evident in the early 1970's (Matis, 1971). The post war automobile boom let to the installation of thousands of underground storage tanks (USTs) during the 1950's and 1960's, mainly at gasoline service stations. In 1989, the U.S. Environmental Protection Agency (EPA) estimated a total of 2.5 to 5 million USTs in the United States, containing mainly hazardous substances (Metelski and Anderson, 1989) with over 95% storing petroleum products.

The majority of these tanks were produced out of carbon steel (Metelski and Anderson, 1989), with a limited life span of 20 to 25 years. Consequently the oldest ones have now, after more than 30 years, exceeded their design life and corrosion related leaks are on the increase (Woods and Webster, 1984). Butts (1977) estimates, that less than 25% of the leaks are actually reported and that in most cases the stated losses are substantially less than the actual losses, and thus the spilled volume by far exceeds the highly publicized spills of crude oil.

According to Statistics Canada (1993), the reported losses (losses and adjustments) of gasoline were 251,780 m^3, or 0.76 % of total domestic sales.

Theoretical Considerations

Low diffusion speeds and permeability, a high adsorption capacity and plasticity make natural clays appropriate materials for the sealing of waste deposits, in order to prevent or retard the infiltration of chemicals into the groundwater. However, the hydraulic conductivity of a natural as well as a processed clay can be affected by physical changes in the structure of the clay, chemically influenced changes, or biological processes (Mitchell and Madsen, 1987). Pure, organic liquids or concentrated organic chemicals have been shown to cause large increases in the hydraulic conductivity of compacted clay barriers and in slurry wall backfill material. The chemical process changes associated with non aqueous phase liquids (NAPL) are also known to reduce or destroy the plasticity of the clay, and to alter the fabric of compacted clay soils. Generally, organic chemicals change the hydraulic conductivity of clay soils by altering the electrical double layer that surrounds the clay platelets.

In order to evaluate the effect of organic chemicals on unmodified clays, a research project was started in 1941, which resulted in the development of organically modified clays by John W. Jordan (1949). In the environmental field this research persists in the application of organophilic clays in the removal of pollutants from industrial waste waters, in the removal of organic solvents from exhaust air, and its integration as a waste barrier material (Boyd et al., 1991).

Organophilic clays are prepared by exchanging the original metallic counterions of a natural bentonite with organic alkyl chains bearing cations, which then have the ability to adsorb other organic compounds (Stockmeyer, 1992). By replacing the sodium, calcium and magnesium ions on the clay's surface with an organic cation, the clay is converted from a hydrophilic to an organophilic clay (Fig. 1). The adsorpiton capacity of the modified clay is dependant on the interactions of adsorbent-adsorbate and adsorbent-solvent.

The use of organoclays in the cleanup of spills of organic chemicals on land, in water, as well as from the gaseous phase is studied world wide. To examine the possible use of a commercially available organophilic clay as a secondary seepage control barrier for USTs containing gasoline, a laboratory testing program was financed by the National Research Council of Canada (NRC, 1993).

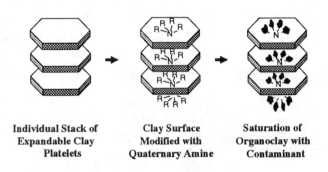

Fig. 1: Structure of Organophilic Clay Platelets

Laboratory Test Program

The phase I laboratory testing program examined the hydraulic conductivity of six (6) liner materials prepared from an admixture of pea gravel and 15%, 20%, and 25% of powdered organophilic clay (PT-1, supplied by Biomin Inc.) and with the same percentages of the granular form of the organoclay.

Preparation of Test Samples

The pea gravel and organophilic clay samples were compacted to a dry density of 95% of maximum. The samples were then transferred from the Proctor mold to the permeability mold. A slurry of bentonite and organophilic clay was used as a sealer around the sides of the mold to insure no leakage occurred between sample and mold. The sample was saturated with water for 2 hours, using a backpressure of 207 kPa. After saturation, the samples were permeated with gasoline at a constant head of 13 kPa for approximately 30 days.

The phase II test program was conducted on powdered 20 and 25 % organophilic clay-pea gravel samples. These 152 mm diameter samples were prepared by moisture conditioning and compacting the mixtures into Proctor molds. The extruded 60 mm high samples were then placed in specially designed stainless steel permeameters (Fig. 2). The sides of the sample were covered with a gasoline resistant paste, and the samples were sandwiched between two porous stones and placed in the flexible wall permeameter. The lines of the permeameter were connected to a vacuum source, and a vacuum was gradually applied to the specimen and the cell. The vacuum was left on for a minimum of four hours to remove air from the specimen.

After vacuuming, the lines were connected to the water reservoirs. The inflow, outflow and confining pressures were set at 206.8, 206.8 and 220.6 kPa, respectively. These pressures were maintained for a minimum of 24 hours to ensure saturation. Once saturation was confirmed, the cell and inflow pressures were increased to 234.4 and 220.6 kPa, respectively, and the test started.

Test Results

The results of the preliminary test program are presented in Table 1. The flow rates during the fixed wall constant head permeability testing were found to be variable and highly dependent on sample preparation, temperature and other factors. The hydraulic conductivities varied from 3.2×10^{-9} cm/s to 1.2×10^{-7} cm/s for specimens prepared with 15 % granular organophilic clay

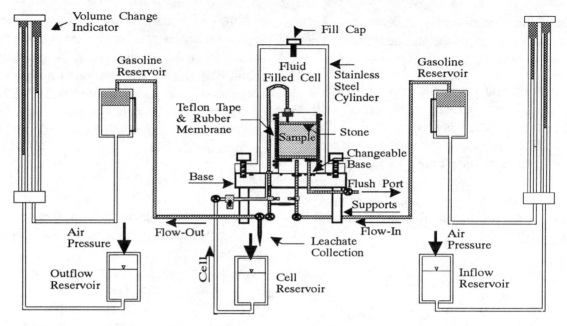

Fig. 2: Triaxial Permeameter System

TABLE 1 -- Phase I Test Results

DESCRIPTION OF MATERIAL	DRY DENSITY (Mg/cu.m.)	STD. PROCTOR DENSITY (Mg/cu.m.)	PRESSURE RANGE (kPa)	ELAPSED TIME (min / days)	HYDRAULIC CONDUCTIVITY (cm/sec)
Pea Gravel admixed with 15% powdered O-Clay	1.764	1.845	13.78 / 34.45	47791 / 33.2	1.50 E-08
Pea Gravel admixed with 15% granular O-Clay	1.737	1.825	13.78 / 34.45	65405 / 45.4	3.17 E-09
Pea Gravel admixed with 20% powdered O-Clay	1.616	1.709	13.78 / 34.45	34215 / 23.7	1.83 E-08
Pea Gravel admixed with 20% granular O-Clay	1.729	1.814	13.78 / 34.45	43110 / 29.9	1.57 E-08
Pea Gravel admixed with 25% powdered O-Clay	1.580	1.518	13.78 / 34.45	50510 / 35.1	1.18 E-07
Pea Gravel admixed with 25% granular O-Clay	1.670	1.777	13.78 / 34.45	50510 / 35.1	No Outflow

and 25 % powdered organophilic clay respectively. In the case of the 25 % granular organophilic clay specimen, after more than 35 days of testing, no flow out of the sample was measured. Of the six tests conducted using this material, only one resulted in a hydraulic conductivity above 1×10^{-7} cm/s.

The results of the phase II test program gave considerably more reliable results. Figure 3 shows the change in hydraulic conductivity with time for the 20 % organophilic clay-pea gravel specimen permeated with water. This specimen was prepared with 10.2 % molding water and compacted to a dry density of 1.79 Mg/m^3. The hydraulic conductivity of this mixture was 7.3×10^{-5} cm/s after 1371 minutes of testing. The hydraulic conductivity of this specimen was increasing at the end of the test. Figure 4 shows how the hydraulic conductivity changed when the

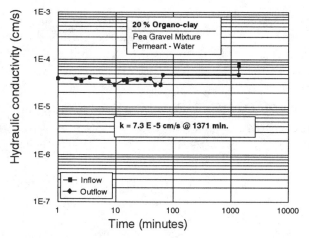

Fig. 3: k Vs. time for 20 % organophilic clay-pea gravel specimen with water

permeant was switched to gasoline. In this figure the cumulative (based on total flow divided by the duration of the test to that point in time) hydraulic conductivity decreased approximately two orders of magnitude. At the end of the test the cumulative flow in and out hydraulic conductivities were 3.12×10^{-8} cm/s and 1.06×10^{-8} cm/s respectively. The higher inflow hydraulic conductivity was due to swelling of the specimen. Figure 5 shows the volume change which took place during testing, indicating that this sample swelled by approximately 100.60 ml. However, Fig. 6 shows that at the end of the test the rate of increase in sample volume was slowing, indicating that the specimen was approaching equilibrium hydraulic conditions.

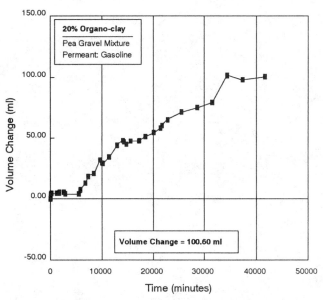

Fig. 5: Volume change during k-testing for the 20% organophilic clay specimen in gasoline

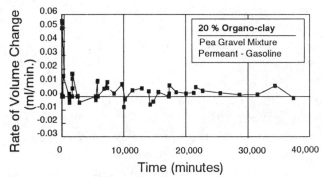

Fig. 6: Rate of volume change during k-testing for the 20 % specimen in gasoline

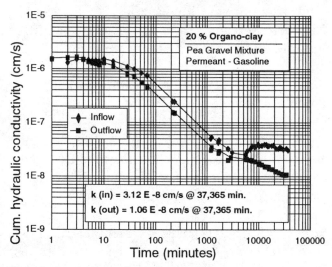

Fig. 4: k vs. time for 20 % organophilic clay-pea gravel specimen with gasoline

Figure 7 show the hydraulic conductivity versus time plot for the 25 % organophilic clay-pea gravel sample. This sample was prepared with a molding water content of 11.1 % and compacted to a dry density of 1.76 Mg/m^3. It had a hydraulic conductivity of 8.5×10^{-6} cm/s after 1416 minutes of testing. In gasoline, this value dropped sharply to 3.4×10^{-8} cm/s for the inflow and 1.06×10^{-8} cm/s sec for the outflow, after 37,368 minutes of testing. The volume change of this specimen was + 111.22 ml during testing, indicating an increased swelling capacity over the 20 % organophilic clay specimen.

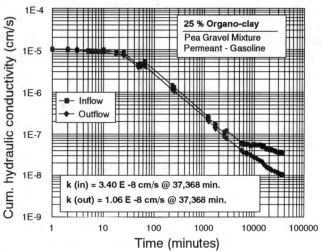

Fig. 7: k vs. time for 25 % organophilic clay-pea gravel specimen with gasoline

Summary and Conclusions

The results of this test program show that organophilic clay can be mixed with pea gravel to construct a low permeability barrier to contain gasoline. Organophilic clay was found to swell and hydrate during permeation with gasoline, even though the organophilic clay had initially been mixed with water during placement. The preliminary test program also found that granular organophilic clay produced similar results to those of powdered organophilic clay. The results of this program suggest that the incorporation of organophilic clay into traditional soil liners may be an effective technique to contain gasoline seepage from underground storage tanks.

References

Boyd, S.A., Jaynes, W.F., and Ross, B.S., 1991, Immobilization of Organic Contaminants by Organo-Clays: Application to Soil Restoration and Hazardous Waste Containment. in: R.A. Baker (ed.), Organic Substances and Sediments in Water, Lewis Publ., Celsen, Mich., pp.181-200.

Butts, E.O., 1977, Detecting Leaks in Pipelines and Storage Tanks, Engineering Journal, Vol. 60, No. 3, pp. 45-47.

Jordan, J.W., 1949, Organophilic Bentonites, I, Swelling in Organic Liquids, J. Phys. Colloid. Chem., Vol. 53.

Matis, J.R., 1971, Petroleum Contamination of Groundwater in Maryland, Groundwater, Vol. 9, No. 6, pp. 57-61.

Metelski, J.J., and Anderson, M.R., 1989, Managing Underground Storage Tanks, The Hazardous Waste Handbook Series, Executive Enterprises Publications, Corp. Inc., N.

Mitchell, J.K., and Madsen, F.T., 1987, Chemical Effects on Clay Hydraulic Conductivity, in: Geotechnical Practice for Waste Disposal, ASCE Speciality Conference, N. Y.

NRC, 1993, Use of Organophilic Clay to Control Seepage From Underground Gasoline Storage Tanks, by: Ground Engineering, Regina (Sask.), National Research Council of Canada, IRAP Project No. 31322P.

Statistics Canada, 1993, Refined Petroleum Products Catalogue, No. 45-004.

Stockmeyer, M.R., 1992. Organophile Bentonite als Komponente in Deponiebarriere-Systemen, Diss. Nr. 9740, Swiss Federal Institute of Technology (ETH) Zuerich, Switzerland.

Woods Jr., R.H., and Webster, D.E., 1984, Underground Storage Tanks: Problems, Technology and Trends, Pollution Engineering, 16, No. 7.

A Boundary Element Method for Analysis of Contaminant Transport in Heterogeneous Porous Media

J.R. Booker, C.J. Leo
School of Civil and Mining Engineering, University of Sydney, Australia

Abstract: In more recent years, the use of boundary element method has rapidly gained popularity in engineering analysis. It is shown in this paper how through the use of a series of transform, the advection-dispersion equation of contaminant transport is reduced into a form which facilitates the formulation of a boundary integral equation. The technique also uses conventional boundary element techniques to deal with heterogenity in the porous media by approximating the domain with a finite number of homogeneous zones. The method is particularly useful for assessing the effectiveness of liner configurations around waste repositories.

1 Introduction

Waste storage and disposal are sensitive undertakings which constitute a potential risk to the environment. For this reason, many regulatory authorities nowadays will require the design of waste disposal or storage facilities to comply with specifications limiting the migration of the stored waste to an acceptable level. Hence, it is usually necessary to incorporate in the design the use of low permeability liner system acting as barriers. The characteristics of these barrier liners must be: (1) compatible with the stored substances (2) of sufficient strength and thickness to resist damage during installation and operation (3) constructed of materials which will limit groundwater seepage and delay the migration of the contaminants from the waste disposal facility. There are many forms of liners in use today and these include compacted clay layers, sand-bentonite mixtures, asphaltic sealants, synthetic membranes or a combination thereof. The design of liners is a complex task and of great importance as the groundwater advection and contamination are relatively slow, thus the ramifications of the design will often not be known until many years later.

In addition to the engineered layers found in the liner system, natural soil formation is often either layered, in which the bedding planes are nearly parallel, or consist of zones of relatively homogeneous properties. If the underlying soil layer is composed of low permeability clay and the magnitude of the groundwater advection is low, molecular diffusion will be the dominant transport process. However, in sandy aquifers where the magnitude of the groundwater advection is high, it is the groundwater advection and mechnical dispersion which are the dominant transport processes.

Given the inherent variability of the soil-liner system and natural soil, the waste facility in many situations is acting in a heterogeneous geologic environment. Earlier, Rowe and Booker presented techniques for accurate and computationally efficient determination of 1-D [1], 2-D [2] and 3-D [3] contaminant migration emanating from a surface landfill into layered soil of varying properties and groundwater conditions. The object of this paper is to develop an alternative two-dimensional boundary integral equation technique with ability to simulate embedded and buried repositories in heterogeneous porous media by dividing up the domain into homogeneous isotropic zones. This work also represents an extension of the boundary integral equation method presented by Rahman, Booker and Hwang [4] and Leo and Booker [5] for a homogeneous isotropic porous medium.

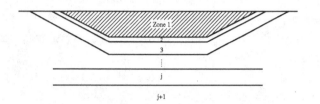

Figure 1: Schematic of typical landfill

2 Theory

2.1 Governing Equations

A typical waster repository situated in 2 dimensional heterogeneous porous media which may be divided up into a finite number of homogeneous isotropic zones is shown schematically in Fig. 1. In each zone j, the equation of contaminant transport under steady uniform groundwater flow field for a single species of contaminant in the global natural coordinate system is given by,

$$n_j(\boldsymbol{D}_j\boldsymbol{\nabla}).\boldsymbol{\nabla}c - n_j\boldsymbol{\nabla}.\boldsymbol{V}_j c = n_j\frac{\partial c}{\partial t} + g_j \quad (1)$$

where c denotes the concentration of the contaminant, n_j the porosity of zone j, $\boldsymbol{\nabla}^T = (\partial/\partial x, \partial/\partial z)$, $\boldsymbol{V}_j$ is the vector of groundwater advection in the jth zone, g_j is the rate at which the contaminant is lost from the groundwater due to adsorption onto the soil skeleton, radioactive decay and biodegradation.

$\boldsymbol{D}_j$ is the tensor of hydrodynamic dispersion and for isotropic homogeneous soil its components are defined by [6],

$$D_{kl_j} = (D_{o_j} + \alpha_{T_j}V_j)\delta_{kl} + (\alpha_{L_j} - \alpha_{T_j})\frac{V_{k_j}V_{l_j}}{V_j} \quad (2)$$

where the indices k, l range over the set (x, z), D_{o_j} is the coefficient of molecular diffusion of the jth zone; $\alpha_{T_j}, \alpha_{L_j}$ are the dispersivities in the transverse and longitudinal directions, δ_{kl} is kronecker's delta; V_{k_j}, V_{l_j} are the magnitude of the components of groundwater velocity, V_j is the magnitude of the groundwater velocity.

The quantity g_j can be thought of as the sum of three components, g_{a_j} (due to adsorption), g_{r_j} (due to radioactive decay) and g_{b_j} (due to biodegradation) i.e.,

$$g_j = g_{a_j} + g_{r_j} + g_{b_j} \quad (3)$$

The rate of adsorption onto the soil skeleton can under many circumstances be expressed in the form:

$$g_{a_j} = \rho_j K_{d_j}\frac{\partial c}{\partial t} \quad (4)$$

where ρ_j is the dry density of the soil, K_{d_j} is the partitioning or distribution coefficient

The ratio at which contaminant is lost due to both biodegradation and radioactive decay is usually proportional to the concentration, thus

$$g_{r_j} = \gamma_{r_j}(n_j + \rho_j K_{d_j})c \quad (5)$$

$$g_{b_j} = \gamma_{b_j}(n_j + \rho_j K_{d_j})c \quad (6)$$

2.2 Transformed Equations

Equation (1) which is represented in the global $x - z$ coordinate system can be transformed to a local $x - z$ system by a simple rotation of the coordinate axes using the relation defined by,

$$\begin{bmatrix} x \\ z \end{bmatrix} = \begin{bmatrix} \cos\theta & \sin\theta \\ -\sin\theta & \cos\theta \end{bmatrix} \begin{bmatrix} x \\ z \end{bmatrix} \quad (7)$$

where $\theta = \tan^{-1} V_{x_j}/V_{z_j}$, so that the groundwater advection is parallel to the local z axis. It then yields,

$$n_j D_{xx_j}\frac{\partial^2 c}{\partial x^2} + n_j D_{zz_j}\frac{\partial^2 c}{\partial z^2} - n_j V_{z_j}\frac{\partial c}{\partial z} = n_j \frac{\partial c}{\partial t} + g_j \quad (8)$$

D_{xx_j}, D_{zz_j} are the components of hydrodynamic dispersion in the local $x - z$ space which may be deduced from equation (2) to be,

$$D_{xx_j} = D_{o_j} + \alpha_{T_j} V_j$$

$$D_{zz_j} = D_{o_j} + \alpha_{L_j} V_j$$

In passing it is noted that V_j is now the magnitude of V_{z_j}. Next, it is convenient to introduce the following coordinate transformation:

$$x = uX$$

$$z = wZ$$

where,

$$u = \left(\frac{D_{xx_j}}{D_j}\right)^{\frac{1}{2}}$$

$$w = \left(\frac{D_{zz_j}}{D_j}\right)^{\frac{1}{2}}$$

and

$$D_j = (D_{xx_j} D_{zz_j})^{\frac{1}{2}}$$

then equation (8) leads to,

$$\frac{\partial^2 c}{\partial X^2} + \frac{\partial^2 c}{\partial Z^2} - \frac{V_{Z_j}}{D_j^*}\frac{\partial c}{\partial Z} = \frac{1}{D_j^*}(\frac{\partial c}{\partial t} + \gamma_j c) \quad (9)$$

where $D_j^* = D_j/(1 + \rho_j K_{d_j}/n_j)$, $V_Z = uV_z$ or V_z/w. It is seen that in the coordinate transformed space, the coefficient of hydrodynamic dispersion is rendered 'isotropic' so that it may be represented by an equivalent coefficient D_j, while V_{z_j} is transformed to V_{Z_j}.

Introducing the Laplace transform,

$$\bar{c} = \int_0^\infty c e^{-st} dt \quad (10)$$

it is found that,

$$\frac{\partial^2 \bar{c}}{\partial X^2} + \frac{\partial^2 \bar{c}}{\partial Z^2} - \frac{V_{Z_j}}{D_j}\frac{\partial \bar{c}}{\partial Z} = (\frac{s + \gamma_j}{D_j^*})(\bar{c} - \frac{c_{o_j}}{s + \gamma_j}) \quad (11)$$

where c_{o_j} is the initial concentration in zone j. The terms due to advection in equation (11) can be eliminated by introducing the variable $\bar{c}^*$ defined by the relationship,

$$\bar{c} - \frac{c_{o_j}}{s + \gamma_j} = \bar{c}^* e^{\kappa_j Z} \quad (12)$$

where,

$$\kappa_j = \frac{V_{jZ}}{2D_j}$$

thus,

$$\frac{\partial^2 \bar{c}^*}{\partial X^2} + \frac{\partial^2 \bar{c}^*}{\partial Z^2} + = (\frac{s + \gamma_j}{D_j^*} + \kappa_j^2)\bar{c}^* \quad (13)$$

leading to the mathematically convenient form,

$$\nabla^2 \bar{c}^* = \mu_j^2 \bar{c}^* \quad (14)$$

where $\mu_j = \sqrt{(\frac{s+\gamma_j}{D_j^*} + \kappa_j^2)}$, denotes the branch with the positive real part.

3 Boundary Integral Equation Formulation

In the usual manner Green's theorem can be used to reformulate equation (14) to the form of a boundary integral equation:

$$\epsilon(\boldsymbol{r}_o)\bar{c}(\boldsymbol{r}_o) = \int_\Gamma (\bar{c}^\sharp \frac{\partial \bar{c}}{\partial N} - \bar{c}\frac{\partial \bar{c}^\sharp}{\partial N}) d\Gamma \quad (15)$$

where Γ represents the boundary to the domain, the fundamental solution $\bar{c}^\sharp = \frac{1}{2\pi}K_0(\mu r^\sharp)$, K_0 is the modified Bessel function of zero order, $r^\sharp$ is the distance of the point of application of the point of disturbance from the field point, $\partial/\partial N$ is the gradient with respect to the boundary normal, and $\epsilon(\boldsymbol{r}_o)$ is given by,

$$\epsilon(\boldsymbol{r}_o) = \begin{cases} 1 & \text{if } \boldsymbol{r}_o \text{ is within domain } \Omega \\ 0 & \text{if } \boldsymbol{r}_o \text{ is outside domain } \Omega \\ \frac{1}{2} & \boldsymbol{r}_o \text{ lies on a smooth boundary or} \\ & \text{the subtended angle} \div 2\pi \text{ if the} \\ & \text{boundary is not smooth} \end{cases} \quad (16)$$

Using conventional Boundary element techniques, it is found that an approximating set of algebraic equations to equation (15) is,

$$\boldsymbol{H}_j^* \bar{c}^* = \boldsymbol{G}_j^* \frac{\partial \bar{c}^*}{\partial N} \quad (17)$$

where $\bar{c}^*, \partial \bar{c}^*/\partial N$ are the vectors of nodal values of $\bar{c}^*$ and $\partial \bar{c}^*/\partial N$. For constant elements, the component h_{pq}^* of the matrix $\boldsymbol{H}_j^*$ is given by,

$$h_{pq}^* = \int_{\Gamma_q} \frac{\partial \bar{c}^\sharp}{\partial N} d\Gamma_q \quad if \quad p \neq q$$

$$h_{pq}^* = \frac{1}{2} \quad if \quad p = q$$

where Γ_q is the boundary of element q and the point disturbance has been assumed to act at the mid-point of element p. $\boldsymbol{G}_j^*$ is the matrix in which the component g_{pq}^* is given by,

$$g_{pq}^* = \int_{\Gamma_q} \bar{c}^\sharp d\Gamma_q$$

The matrices in equation (17) needs to be assembled into a global system involving boundary elements in every zone. Before proceeding, it is convenient to 'undo' the transformation of (12) as well as the coordinate rotation, stretching or compression. The inversion of the concentration $\bar{c}^*$ to $\bar{c}$ presents no difficulty and it follows immediately from 'undoing' equation (12). The normal derivative is, however, less straightforward and some care is necessary. From equation (12), it follows that,

$$\frac{\partial \bar{c}}{\partial N} = \left(\frac{\partial \bar{c}^*}{\partial N} + N_Z^1 \kappa_j \bar{c}^*\right) e^{\kappa_j Z}$$

where N_Z^1 is the $Z-$component of the directional cosines of the normal in the transformed coordinate space. To invert the normal derivative $\partial \bar{c}/\partial N$ back to $\partial \bar{c}/\partial n$ in the natural global coordinate space an additional step is required. In this paper, rather than the normal derivative, it was found convenient to 'undo' an equivalent quantity, the normal diffusion flux, defined as,

$$\bar{q}_N = -D_j \frac{\partial \bar{c}}{\partial N}$$

Now, it may be shown that,

$$\bar{q}_x = -n_x^1 D_{xx_j} \frac{\partial c}{\partial x} = \epsilon \bar{q}_X$$

$$\bar{q}_z = -n_z^1 D_{zz_j} \frac{\partial c}{\partial z} = \frac{\bar{q}_Z}{\epsilon}$$

where,

$$\epsilon = \left(\frac{D_{xx_j}}{D_{zz_j}}\right)^{\frac{1}{4}}$$

and n_x^1, n_z^1 are the directional cosines of the normal in the local coordinate space. After some mathematical manipulation, it is found that $\bar{q}_n$ in the local space is related to $\bar{q}_N$ by,

$$\bar{q}_n = \left(\epsilon N_X^1 n_x^1 + \frac{1}{\epsilon} N_Z^1 n_z^1\right) \bar{q}_N \quad (18)$$

Moreover, it can be shown that $\bar{q}_n$ is invariant under rotation, i.e.

$$\bar{q}_n = \bar{q}_n$$

After undoing all the transformations which in essence involve some algebraic operations on the components of the matrices H_j^* and G_j^*, equation (17) may then be written as,

$$H_j(\bar{c} - \bar{c}_{o_j}) = G_j \bar{q}_n \quad (19)$$

where $\bar{c}_{o_j} = [c_{o_j}/(s+\gamma_j), \ldots, c_{o_j}/(s+\gamma_j)]^T$, is the boundary element vector of the Laplace transformed initial concentration in zone j.

The matrices G_j and H_j of each zone will be assembled into the global G and H matrices by invoking the compatibility of the concentrations and the contaminant fluxes at the interface boundaries of the zones so as to yield a global system of equation represented by (see e.g. [7]),

$$H\bar{c} = G\bar{q}_n \quad (20)$$

The boundary value problem will be well posed if either the concentrations or the normal gradients (hence the diffusion flux), or linear combinations of both are known on the external boundary. If the unknowns are moved to one side of equation (20), the resulting system of equations may then be solved to yield the values of the unknowns. It is clear that once concentrations and the gradients are known on the external boundaries and interfaces, the concentration at any internal point within any zone may then be found using equation (15). The solutions in the time domain are obtained by applying the efficient inversion algorithm developed by Talbot [8].

4 Application

In the following analysis, the boundary element technique developed above will be used to assess the design of a hypothetical sanitary landfill situated in a clay stratum of 8 m and further underlain by a 2 m deep confined aquifer. Beneath the aquifer is a deep layer of largely impermeable soil. The landfill is long with a base width of 50 m, depth of 4 m and side slopes of 3H:1V. It is represented as a porous medium with a finite diffusion coefficient value of 0.5 m²/a and initial concentration of 1000 mg/l uniformly throughout. The contaminant plume

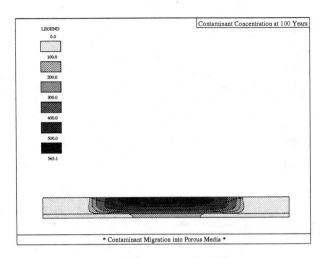

Figure 2: Contaminant contours at 100 years when no clay liner is used

will be assumed to be the concentration contour of 100 mg/l and the contaminant species is assumed to be conservative. The ground surface and the far field boundaries at infinity are specified as having zero concentration throughout the period of simulation. For simplicity, it has been assumed that a total head difference of about 1 m exist between the base of the landfill and the top of the aquifer, the hydraulic conductivity of the soil in the aquitard is of the order of 10^{-7} cm/s, thus yielding a Darcy velocity of about 0.02 m/a in the z (downwards) direction. The molecular diffusion of the aquitard is 0.05 m²/a, the transversal dispersivity is 0.1 m, the longitudinal dispersivity is 1 m, the porosity is 0.4. As for the aquifer the respective parameters are 1 m²/a, 1 m, 10 m, 0.35. The seepage velocity in the aquifer is assumed to be 8 m/a in the x direction.

If no engineered liner is used, the natural aquitard itself is clearly not sufficient to prevent contamination of the aquifer within 100 years as shown in Fig. 2. However if a 1 m thick clay barrier with adsorptive properties ($\rho K_d = 10$) and a leachate collection system are introduced, the combined system is sufficient to prevent aquifer contamination within 100 years (shown in Fig. 3).

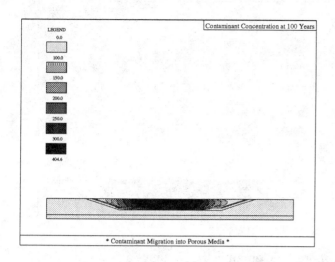

Figure 3: Contaminant contours at 100 years when clay liner is used

5 Conclusion

A boundary element method is developed in this paper for analysing contaminant migration in heterogeneous porous media. The method is useful for assessing the design of liner barriers surrounding waste facilities.

5.1 References

1. Rowe, R.K. and Booker, J.R. (1985),'1-D pollutant migration is soils of finite depth', *Journal of Geotechnical Engineering, ASCE,* Vol. 111, 4, pp. 479-499.

2. Rowe, R.K. and Booker, J.R. (1985), 'Two-dimensional pollutant migration in soils of finite depth', *Canadian Geotechnical Journal*, Vol. 22, p429-436.

3. Rowe, R.K. and Booker, J.R. (1986), 'A Finite Layer Technique for Solving General 3-D Pollutant Migration Problems', *Geotechnique*, Vol. 36, p205-214.

4. Rahman, M.S., Booker, J.R. and Hwang, C.W. (1989), 'A boundary integral formulation for modelling pollutant migration in groundwater', *Proceedings of the International Conference on 'Solving Groundwater Problems with Models', Indianapolis.*

5. Leo, C.J. and Booker, J.R. (1992), 'Boundary element analysis of contaminant transport from waste repositories', *Proceedings of the International Conference. on Computational Methods in Engineering, Singapore.*

6. Bear, J. (1972), 'Dynamics of Fluids in Porous Media', *Elsevier, New York.*

7. Brebbia, C.A. and Dominguez, J. (1989), 'Boundary Element – An Introductory Course', *Computational Mechanics Publ.*

8. Talbot, A. (1979), 'The Accurate Numerical Integration of Laplace Transforms ', *Journal Inst. Mathematical Applications* , Vol. 23, p97-120.

Geoenvironmental Subsurface Site Characterization Using In-Situ Soil Testing Methods

R.G. Campanella, M.P. Davies, T.J. Boyd, J.L. Everard
In-Situ Testing Group, Civil Engineering Department, The University of British Columbia, Vancouver, B.C., Canada

SYNOPSIS Geoenvironmental site characterization refers to the surficial and subsurficial representation that approximates the actual in-situ conditions. Surficial characterization, although often not straightforward, does allow visual appraisal of essentially all of the materials involved. Subsurface characterization, on the other hand, rarely allows even a micro-percentage appraisal. From a statistical significance standpoint, subsurface characterizations are almost always inadequate and therefore large assumptions based on the best possible information are required.

In-Situ testing methods, such as the piezometer cone penetration test, often provide the best possible information at selected locations for stratigraphic, geotechnical, hydrogeological and some specific environmental parameters on a specific and/or screening basis. This paper presents a review of the currently available in-situ methods for Geoenvironmental site characterization. The methods reviewed include conventional drilling techniques, piezometer cone penetration test technology and discrete water sampling systems. The advantages and disadvantages of using each methodology are presented. Included in the advantages and disadvantages are comparisons of testing cost, data reliability and regulatory acceptance. Brief case examples are introduced to demonstrate the preferred technologies. The paper concludes with an overview of current research including recent advances in self-grouting technology for cross-contamination concerns.

INTRODUCTION

Geoenvironmental is a relatively new term in the geoscience field. There are currently several interpretations of meaning for Geoenvironmental and the definition used herein is defined as:

> "the field of study that links geological, geotechnical and environmental engineering and engineering sciences to form an area of interest that includes all physical and environmental concerns within geological media."

Groundwater contamination is an excellent example of an area of interest within the scope of the GeoEnvironmetal field. With the basic definition established, Geoenvironmental subsurface site characterization can then be defined as:

> "the formulation of the physical, chemical and/or biological information for the site that allows <u>adequate</u> three-dimensional representation for the prevailing subsurface conditions and the engineering requirements of the data."

Modern in-situ testing instruments provide the best current technology for developing this three-dimensional representation for many site conditions. This paper presents an overview of the currently available in-situ methods for site characterization. In-situ testing embodies all of the prevailing field conditions on material behaviour whereas laboratory testing, usually on disturbed samples, attempts to estimate these many factors. The value of a given in-situ test is directly related to how much non-quantifiable disturbance is caused during tool insertion; e.g., the degree of data repeatability.

GEOENVIRONMENTAL SITE CHARACTERIZATION

As will be shown in the following sections, what makes modern in-situ testing methods attractive for Geoenvironmental site characterization is that accurate, repeatable, direct results can be obtained in a fast, cost-effective manner. Traditional methods of in-situ site characterization, like conventional drilling and sampling, cannot match the speed, accuracy or economy achieved by the tools described in this paper. However, in cemented and fractured sedimentary formations and in compact, thick gravel layers (where penetration tools are restricted) drilling and coring likely have an advantage. Surface geophysical methods are the least disruptive but they lack ground-truthing capability and therefore the results are often difficult to interpret.

Modern in-situ testing also provides a huge advantage over conventional drilling in that soil cuttings are not produced. On many contaminated sites, the cost of preparing for, handling and disposing of, these cuttings can be extremely high and tends to limit the amount of necessary subsurface investigation ultimately carried out.

The key items that must be evaluated in order to formulate the subsurface characterization are:
- the stratigraphic profile;
- the relative geotechnical properties of each stratigraphic unit;
- the hydraulic properties of aquifer and aquitard materials; and
- the nature of the pore fluid/gases that are present in the materials.

Therefore, the assessment of any given in-situ technique must include an evaluation of whether it will provide the requisite items and how comprehensive and accurate the results provided will be.

IN-SITU TESTING

As defined above, in-situ testing involves evaluating material properties "in-place"; where physical and biochemical stresses are evaluated as is. Conventional drilling and discrete sampling represents a *quasi* in-situ test method although laboratory evaluations are typically required for samples retrieved. However, drilling is included in the definition for this paper to allow a comparison of the current industry standard with the more sophisticated tools and evaluation procedures available. On the other hand, surface geophysical methods have not been included as most of these approaches are in the early stages with respect to assessing environmental parameters. These geophysical techniques are acknowledged as having great promise, particularly for site screening studies.

Conventional Drilling and Discrete Sampling

The most common manner in which Geoenvironmental site characterization is carried out remains conventional drilling with discrete sampling. The large experience base and the retrieval of a physical sample are important advantages of the method, whereas lack of standards, relatively high cost and questionable sample quality represent the key disadvantages.

Drilling with reasonable, albeit far from continuous, sampling frequency will result in a production rate for a 30 metre (100 ft) hole of from 10 to 20 hours and more. The actual production rate will depend upon soil conditions, equipment and operator quality, the method of sampling used and decontamination and grouting requirements. Although specifying conventional drilling on a metreage basis is never recommended for obvious reasons, we can apply a generic $150 hourly rate with modest consumables to provide an approximate $50 to $100 per metre for this discrete sampling.

Once a sample is obtained, its quality must be assessed. Table 1 presents the custody steps and sources of error in taking a sample from its in-situ condition to the laboratory. Table 1 does not include the errors introduced by the laboratory itself, such as sample disturbance, loss of in-situ stresses, and chemical alterations, etc.

TABLE 1 – Steps in Geoenvironmental sampling and their related sources of error

	Step	Sources of error
	In-Situ Condition	---------
1.	Establishing a sample location	Improper borehole completion/placement; poor location choice; poor material choice
2.	Sample collection	Sampling mechanism/procedural bias; operator error
3.	Sample retrieval	Sampling mechanism/procedural bias; sample exposure; degassing; oxygenation; cross-contamination
4.	Preservation/storage	Handling/labeling errors; temperature, etc. control
5.	Transportation	Delay; sample loss

Drilling and sampling is therefore shown to be relatively expensive and provide potentially suspect reliability on a non-continuous basis. However, due to its popular use, regulatory acceptance has traditionally been high although recent trends in data quality management and geostatistical evaluations are tending to make it more difficult to convince all regulators that data from drilling and sampling programs are either comprehensive or accurate enough.

Piezometer Cone Penetration Test

The piezometer cone penetration test (CPTU) is increasingly being used by practicing engineers as an effective means of geotechnical classification. The method provides a fast, economical, and accurate means of delineating soil stratigraphy and determining geotechnical parameters. The cone has a standard 10 cm^2, 60° conical tip, a friction sleeve with an area of 150 cm^2 and one or more pore pressure sensors. The CPTU measures tip resistance (q_c), friction sleeve stress (f_s) and pore pressure at up to three locations on the cone. All channels are continuously monitored and reported at either typically 25 or 50 mm intervals, thus providing essentially continuous in-situ data sampling. There are extensive relationships available between the various CPTU channels, and combinations thereof, that provide soil behaviour type (often equivalent to stratigraphy), geotechnical strength parameters and hydraulic parameters. Temperature (t) and instrument inclination (i) are usually measured simultaneously. In addition, a seismic model of the cone has been developed that allows the measurement of low strain dynamic properties of the soil.

The CPTU is generally regarded as the most effective and cost efficient tool for stratigraphic logging of soils (Robertson et al., 1986). The cone is pushed into the ground at a constant 2 cm per second (or about 1 metre per minute) by a hydraulic pushing source; often a drill rig or a specially outfitted cone truck or penetration vehicle. As the cone is advanced, pore pressures are generated around the cone tip and sleeve. The measurement of the excess pore pressures which are generated during penetration, and their subsequent dissipation, provide insight into the soil type and its hydraulic parameters.

The CPTU can and should be specified on a metreage basis. The rigorous ASTM and International standards ensure high repeatability and reliability, thus eliminating traditional concerns such as the speed of testing being inversely related to data quality. Although there are regional variances, CPTU soundings are typically carried out for between $20 and $30 per metre for the continuous data and plots provided. Due to its high standards and operator requirements, engineering supervision to the degree required for conventional drilling is not necessary. Its continuous data, speed, relatively low cost and excellent repeatability are its main advantages. The main disadvantage is the lack of physical samples but for screening assessments, confirmatory investigations or routine monitoring, the requirement for a physical sample is becoming much less important as confidence in the method grows. Regulatory acceptance is variable with the knowledge of the regulator being the key discriminator as to the success of approval. The trend, however, is one of acceptance with the provision of at least an occasional confirmatory/collaborative borehole; an approach with which we fully concur. The recent trend in some jurisdictions of requiring full testing and appropriate disposal of drill cuttings represents a significant advantage of the CPTU.

Resistivity Piezometer Cone Penetration Test

The resistivity piezometer cone penetration test (RCPTU) is a recent modification of the standard piezocone test (CPTU). The ability to measure the resistance to current flow in saturated soils on a continuous basis is extremely valuable due to the large effects that dissolved and free product constituents have on soil resistivity (conductivity).

Table 2 presents typical values of bulk soil resistivity measurements with the RCPTU and corresponding measurements of pore fluid resistivity. Since conductivity is the reciprocal of resistivity it is easy to convert according to:

$$\text{Conductivity}(\mu S/cm) = 10{,}000 \div \text{Resistivity}(\Omega\text{-m}) \qquad (1)$$

Table 2 also gives corresponding values in units of conductivity due to their popular use in the geochemical field. The results in Table 2 show that the range of resistivity (conductivity) values is very large from about one (10000) to about 700 Ω-m (14 μS/cm) and is very sensitive to both soluble salts and low solubility organic contaminants.

The RCPTU has the advantage of providing continuous resistivity measurements (which directly correlate on a site specific basis with contaminant concentration) in combination with all of the traditional CPTU data. The current electrode separation used by researchers and practitioners varies from about 10 mm to over 150 mm providing different depths of lateral penetration. The more advanced tools have several electrode spacings so that both profiling and sounding resistivity tests are carried out simultaneously. Figure 1 shows a schematic representation of one of UBC's RCPTU tools. Campanella and Weemes (1990) provide a complete summary of the RCPTU and its development.

The RCPTU meets all of the key items identified for formulating a subsurface site characterization. The actual quantification of non-background geochemical processes requires site specific calibration if more than "screening" evaluations are intended. This site specific calibration involves accurate pore fluid-gas sampling which is described in the following section. The costs per metre of RCPTU programs is essentially the same as for the CPTU with an allowance of an additional 10% for data reduction. Regulatory approval is variable due to the relative newness of the tool but in our experience, once the technology is demonstrated, the acceptance is typically rapid.

TABLE 2 -- Summary of typical resistivity (conductivity) measurements of bulk soil mixtures and pore fluid (UBC experience)

Material type	Bulk Resistivity ρ_b, Ω-m	Fluid Resistivity ρ_f, Ω-m	Bulk Conductivity μS/cm	Fluid Conductivity, μS/cm
Sea water	---	0.2	---	50000
Drinking water	---	>15	---	<665
McDonald Farm clay	1.5	0.3	6700	33300
Laing Bridge site clay	20	7	500	1430
Colebrook site clay	25	18.2	400	550
401 @ 232 Ave clay	8	---	1250	---
Strong Pit clay	35	---	285	---
Kidd 2 site clay	14	12.5	715	800
McDonald Farm site sand	5-20	1.5-6	2000-500	6700-1670
Laing Bridge site sand	5-40	1.5-10	2000-250	6700-1000
Colebrook site sand	70		143	
Strong Pit site sand	115		89	
Kidd 2 site sand	1.5-40	0.5-21	6700-225	20000-475
Typical landfill leachate	1-30	.5-10	10000-330	20000-1000
Mine tailings site sand with acid drainage leachate	1-40	2-27	10000-250	5000-370
Mine tailings site sand without acid drainage	70-100	15-50	145-100	665-200
Industry site-inorganic contaminants in sand	0.5-1.5	0.3-0.5	20000-6500	33000-20000
100% ethylene dichloride (ED)	---	20,400	---	0.5
50% ED/50% water in sand	700	---	14	---
17% ED/83% water in sand	275	---	36	---
Industry site - organic contaminants in sand	125	---	80	---
BC Place Parcel 2, PAHs (coal gas plant)	200-300	---	50-33	---
BC Place Parcel 2 (wood waste)	300-600	---	33-66	---

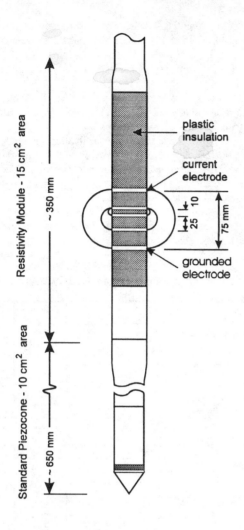

Fig. 1 UBC Resistivity Piezocone (RCPTU)

In-Situ Water Sampling

The main goal of any water sampling program is to rapidly and economically collect representative samples of the in-situ regime. Conventionally, bailers and pumps are used to obtain samples from drilled or installed wells or piezometers. To obtain high levels of sample integrity with conventional methods, costly and time consuming installations and procedures are typically required.

In recent years there have been significant developments in in-situ discrete depth water samplers. These samplers, such as the BAT Enviroprobe and QED Hydropunch II, minimize purging and exposure to the sample, allow for rapid, high quality, representative samples and are easy and economical to use (Blegen et al, 1988, Zemo et al, 1992).

A key advantage of these discrete depth samplers is that they can be used effectively in screening studies to better delineate contaminant and thus optimize the location(s) of permanent monitoring wells. Regulatory approval of these new in-situ sampling methods has been widespread and rapid. As use and experience increases, the methods will likely become industry standard for appropriate sites.

Figure 2 shows a sketch of the rapid sampling probe developed at UBC which uses the BAT groundwater system (Torstensson, 1984). This 50 mm diameter probe with a central HDP filter can be pushed on its own or be pushed down the 44 mm diameter hole previously made by the RCPTU. The probe is pushed to a specific depth to obtain a pore water sample which is then chemically analyzed to provide site specific correlations of contaminant concentrations with bulk resistivity measurements. The preceding and/or adjacent RCPTU sounding provides soil classification information to precisely indicate the depth where high hydraulic conductivity zones exist so rapid water sampling can be achieved. A complete RCPTU sounding to 30m depth followed by 10 or more water samples (including a purge sample at each depth) can be completed in one working day.

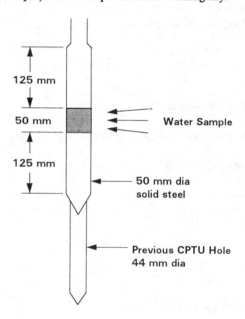

Fig. 2 UBC Modified BAT Pore Water Sampler

CASE EXAMPLES

Organic Contaminants: Heavy Oils

The RCTPU and discrete depth water sampling equipment were used to confirm and augment data obtained through a conventional drilling and sampling program at a contaminated site in the Lower Mainland, B.C. The aim behind the program was to demonstrate the capabilities and limitations of using this technology as a screening tool on organically contaminated sites. The main organic contaminant was creosote from timber treatment.

The field program consisted of 9 RCPTU soundings (3 were used as water sampling holes) located to make optimum use of existing information. Figure 3 shows a typical RCPTU profile of cone tip resistance, pore pressure, friction ratio and resistivity with depth. Contrasts in bulk resistivity showing elevated values (due to the insulating quality of the organic contaminants) compared to "background" values are indicated by the shaded zones in Fig. 3. The changes in bulk resistivity are dominated by the presence of contaminant within the pore fluid which was verified by discrete depth water sampling and lab testing. "Spikes" in the bulk resistivity profile, as from 15 to 19 metres, indicate the presence of free product. The free product was verified by monitoring well sampling.

et al (1990). In contaminated zones, bulk resistivities varied between 0.5 and 1.5 Ω-m (20,000-6500µS/cm). Pore fluid resistivities within contaminated zones varied between 0.3 and 0.5 Ω-m (30,000-20,000µS/cm). Contaminated zones and their direction and rate of movement were subsequently identified through on-going RCPTU studies..

Acid Rock Drainage (ARD) in Mine Tailings - Eastern Canada

A field program was conducted at a sulphide tailings impoundment in eastern Canada at which a well-documented case of ARD is occurring. The field program consisted of more than thirty RCPTU soundings with calibrative discrete groundwater sampling using the UBC modified BAT System, previously described, at selected holes. Other high quality samples were obtained from established monitoring well locations.

A typical RCPTU sounding in acid generating tailings is shown in Fig. 4. Note that the tailings below 5 m depth are very mixed and stratified as indicated by the cyclic nature of the cone resistance and friction ratio with depth and are fairly fine and quite loose. However the soil to 5 m depth is a fairly uniform fine sand with little fines as shown by the consistent value of 1% and less for friction ratio. The stratified nature of the deposit is typical for hydraulically deposited tailings.

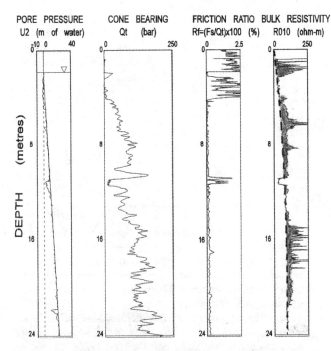

Fig. 3 RCPTU Sounding: Organic Contamination by Heavy Oils

Industrial Inorganic Pollutants - Western Canada

The RCTPU and the discrete depth water sampling equipment were used to characterize an industrial site in Western Canada with respect to the geological, hydrogeological and geochemical setting. The purpose of the investigation was to delineate the plume of inorganic contaminants (sulphates, chlorides, and phosphates) and to speculate on the potential for offsite migration. RCPTU and discrete depth water sampling were chosen for their technical and cost advantages.

Geotechnical and hydrogeological parameters (e.g., soil type, relative density, hydraulic conductivity, etc.) were readily obtained from the investigation. Chemical analyses conducted on groundwater samples correlated reasonably well with bulk resistivity contours. A trend between total dissolved solids (TDS) and bulk resistivity was noted to be similar to Ebraheem

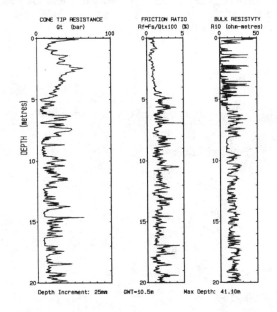

Fig. 4 RCPTU Sounding: Acid Generating Mine Tailings

The extremely low bulk resistivity measurements of less than 1 Ω-m (or bulk conductivity more than 10,000 µS/cm) suggest that from ground surface to about 5 m in depth there is very high accumulation of ions in solution. This is a zone of high capillary pore water tension in fairly fine sands where saturation levels are high (causing lower resistivity), but, more important, this is a zone of high oxidation which gives rise to sulphate and acid generation which is the main cause for the very low resistivities. Note that from 5 m to well below the water table (10.5 m) the lowest bulk resistivity values are about 10 Ω-m indicating a dilution of dissolved solids compared to the top 5 m.

The results of resistivity logging and pore water are summarized in Fig. 5 where bulk resistivity is plotted against fluid resistivity at all depths where discrete water samples were taken. The linear relationship in Fig. 5 gives a ratio of bulk to fluid resistivity (called Formation Factor) of about 2 in these tailings. Resistivity of soil particles is much higher than that of the pore fluid, and therefore it is expected that the bulk resistivity values would be significantly higher than that of the pore water. Typically the Formation Factor for clean quartz sands is between 2 and 5, but can be much lower with clay minerals and particles with low resistivity surfaces. The agreement between the data in Fig. 5 demonstrate the effectiveness of the RCPTU in identifying zones of lower resistivity and potential groundwater contamination in mine tailings which contain mixed and often conductive fine grains. Chemical testing has been conducted to determine sulphate and metal levels in the water samples. The results of the sulphate tests are presented in Fig. 6 which shows a good correlation between sulphate concentration in mg/l and bulk conductivity in μS/cm. With this average correlation the concentration of sulphates in the pore water can be mapped, three-dimensionally, using the bulk resistivity measurements from the rapidly performed RCPTU soundings. This work is ongoing and promising results are indicated.

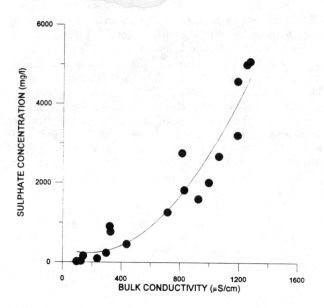

Fig. 6 Sulphate Concentration versus Bulk Conductivity: Mine Tailings

CURRENT RESEARCH

Our current priorities include the following:

1. To develop improved methods of rapid water sampling for screening studies.
2. To develop new techniques to measure the hydraulic conductivity, K, of coarse soils in-situ and to correlate measurements with CPTU parameters to allow rapid estimates of K vs. depth.
3. To develop a grouting and decontamination system for the cone penetrometer. This system must easily and economically grout and seal the cone hole to prevent potential cross contamination occurring. Conventionally, the hole is grouted after the cone is removed and requires a special grouting pipe to be pushed back down the cone hole and withdraw slowly while introducing grout. This procedure takes at least as long as the RCPTU sounding itself and is therefore time consuming and requires extra equipment, thereby increasing the cost substantially. It is therefore desirable to develop a 'self-grouting' system.
4. To evaluate and develop other geophysical techniques applied to cone penetration technology, such as the induced polarization technique and broad band radar.
5. To evaluate and apply other sensor technology, especially in the vadose zone, in the detection and quantification of DNAPLs and LNAPLs.
6. To investigate the ability of the RCPTU to determine the degree of saturation in the vadose zone.
7. The importation, application and use of high density, in-situ measurements of contamination levels directly into SQL databases, data worth models and contaminant transport models.

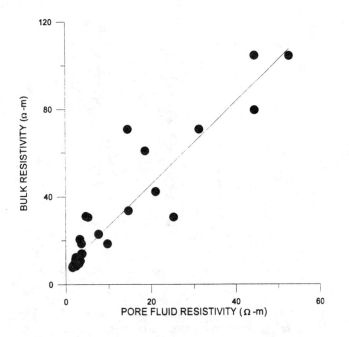

Fig. 5 Bulk versus Fluid Resistivity for Mine Tailings

Specific groundwater contamination problems currently under investigation at appropriate sites include:

1. Hydrocarbon detection and mapping: DNAPLs and LNAPLs.
2. Assessing the accumulation of nitrates and pesticides due to excessive agricultural activities.
3. Mine waste facilities where we are developing rapid methods of detection, mapping and characterization of metals and sulphates in mine tailings and waste dumps coupled with characterizing acid generation and drainage mechanisms.

All of our current research thrusts are related and applied to industrial and public needs concerning rapid and economical screening methods to measure geotechnical parameters and to assess groundwater quality.

SUMMARY

The use of a rapidly deployed logging tool like the Resistivity Piezocone (RCPTU) to identify the stratigraphy and estimate groundwater flow parameters to guide the location for specific depth pore water sampling have been presented and described. The general case examples relating to both organic and inorganic contamination demonstrate the application of this penetration technology. The results that these methods provide give a better definition of the lateral and vertical extent of the stratigraphy and contaminant zones than the conventional methods and at substantially lower cost. Their application can be either in a screening or specific evaluation capacity dependent upon whether global or site specific relationships are developed and used.

REFERENCES

Bear, J. and Verruijt, A., (1987). "Modelling groundwater flow and pollution", D. Reidel Publishing Co., Dordrecht, pp. 196-198.

Blegen, R.P., Hess, J.W. and Denne, J.E., (1988), "Field Comparison of Ground-Water Sampling Devices", North West Waterworks Assoc., 2nd Annual Outdoor Action Conference, Los Vegas, Nevada, 23 pgs.

Campanella, R.G. and Weemes, I., (1990). "Development and use of an electrical resistivity cone for groundwater contamination studies", Canadian Geotechnical Journal, **27**:5, pp. 557-567. Oct.

Campanella, R.G., Davies, M.P. and Boyd, T.J., (1993). "The use of in-situ testing to characterize contaminated soil and groundwater systems", Second International Joint ASCE-CSCE Conference on Environmental Engineering, Publ. by Geotechnical Research Centre, McGill Univ., Montreal, Vol.2, pp. 1497-1505.

Ebraheem, A.M., Hambeurger, M.W., Bayless, and Krothe, N.C., (1990). "A Study of Acid Mine Drainage Using Earth Resistivity Measurements", Groundwater, Vol. 28, No. 3, pp. 361-368, May

Kokan, M.J., (1990). "Evaluation of resistivity cone penetrometer in studying groundwater quality", unpublished B.A.Sc. Thesis in Geological Engineering, University of British Columbia.

Robertson, P.K., Campanella, R.G., Gillespie, D.J. and Grieg, J., (1986). "Use of piezometer cone data", Proc. of In-Situ-'86, ASCE, Geotechnical Special Publication No. 6., pp. 1263-1280.

Telford, W.M., Geldart, L.P., Sheriff, R.E., and Keys, D.A., (1976). "Applied Geophysics". Cambridge University Press. Cambridge, pp. 442-457.

Torstensson, B-A., (1984), "A new system for ground water monitoring", Ground Water Monitoring Review, Fall, pp. 131-138.

Zemo, D.A., Pierce, Y.G. and Gallinatte, J.D., (1992), "Cone Penetrometer Testing and Discrete-Depth Groundwater Sampling Techniques: A Cost-Effective Method of Site Characterization in a Multiple-Aquifer Setting", Proc. 6th Outdoor Action Conference, National Ground Water Assoc., May, Las Vegas, Nevada.

Design and Construction of Soil-Bentonite Slurry Wall Helen Kramer Superfund Site

Wah-Tak Chan, Daniel C. Hsu
IT Corporation, Monroeville, Pennsylvania, USA

Abstract

Design and construction of a soil-bentonite slurry wall (SB wall) has been successfully completed for the Helen Kramer Superfund site (HK) in Mantua Township, New Jersey, U.S.A. A comprehensive laboratory testing and design program was performed to develop a reasonable and practical SB wall backfill mix. A field demonstration was conducted to observe the mixing technique and to verify the laboratory results. Based upon our past experience and the information gained from this program, a technical solution was developed that emphasized the delicate material balance of bentonite content, clay fines, and water content. This finding was applied to the construction of the HK SB wall. This paper presents the data and the relationship of permeability and fines, slump and water content, and observed workability.

1.0 Introduction

The 413,000-square-foot SB wall is approximately 8,400 linear feet long, 3 feet wide, and completely encircles the 77-acre landfill area. The SB wall extends from the work platform and is keyed 3 feet into an impervious layer. The depth of the SB wall ranges from 20 to 73 feet and averages 48 feet. This SB wall project consisted of laboratory testing, mix design, field demonstration, and construction.

Because permeability is linked to a delicate balance between bentonite content, clay fines, and water content, many improper SB wall specifications may cause an improper selection of a bentonite and backfill mix design for the intended objectives. A too-rich bentonite content may cause excessive backfill shrinkage and consolidation; a too-lean bentonite content may not provide the required permeability. The bentonite requirement often depends upon the content of clay fines. The higher clay fines percentages generally require the least bentonite content. In practice, a minimum requirement of clay fines and bentonite has been established to provide a general guidance for the designer (D'Appolonia, 1980). The following describes the experience and lessons learned from this project.

2.0 Laboratory Testing and Mix Design

The laboratory program included the testing of bentonite, water, common soil, clayey soil, and backfill mix. The backfill mix specimen preparation is described below.

- The first one percent of bentonite was added in slurry form to the well-blended soil mixture of common soil and clayey soil. The slurry was hydrated overnight prior to mixing.

- The remaining percentages (1, 3, 5, 7, and 9) of bentonite in dry powder form were then added into the mixture.

- Additional water, if needed, was added to the mixture until the mix was workable.

- Prepared specimens were tested for permeability using the triaxial-cell back pressure method. The specimens were first consolidated at an average cell pressure of 7.5 pounds per square inch (psi). After primary consolidation was complete, the permeability tests were conducted under a hydraulic head of approximately 5 psi and an average hydraulic gradient of approximately 35.

One of the most important ingredients for the soil-bentonite backfill material is a minimum of 20 percent fines, preferably plastic fines (EPA, 1984) (D'Appolonia, 1980). The common soil had 19 to 21 percent fines passing the No. 200 sieve and was classified as SC-SM. Extra plastic fines were provided by adding clayey soil, CH, to the backfill mix. The clayey soil had 79 to 91 percent fines passing the No. 200 sieve. The liquid limit and plasticity index of the clayey soil were 60 and 40, respectively.

The laboratory test results for the soil-bentonite backfill mix are summarized in Table 1. As shown in Figure 1, the permeability of the backfill mix decreased approximately from 4×10^{-8} to 1×10^{-8} centimeters per second (cm/s) when the bentonite content increased from 2 to 10 percent.

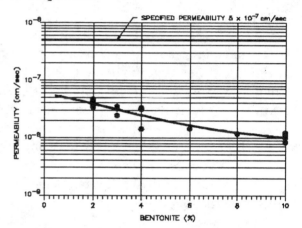

Figure 1 Permeability versus % bentonite for soil-bentonite backfill

This demonstrates that increased bentonite content will not significantly decrease permeability for the backfill mixes tested. The results of these permeability tests indicate that the soil-bentonite backfill with a 2 percent bentonite by dry weight of the mixture exceeds the specified permeability of 5×10^{-7} cm/s by more than one order of magnitude.

TABLE 1 -- Summary of laboratory testing and backfill mix[a] design

Sample No.	Bentonite Content (%)	Fines Passing No. 200 Sieve (%)	Water Content (%)	Permeability (× 10^{-8} cm/s)
1 - 10	2	37.3	50.2 - 53.8	3.41 - 4.54
11 - 13	3	38.2	53.7 - 55.0	2.58 - 3.42
14 - 16	4	39.2	59.4 - 78.2	1.42 - 3.38
17	6	41.1	52.0	1.59
18	8	42.9	96.8	1.20
19 - 28	10	48.8	72.6 - 107.4	0.85 - 1.29

[a] Soil mixture consists of 80% common soil and 20% clayey soil.

The workability of the backfill mix, as observed, was found to be dependent on the slump and the contents of water, bentonite, and clay fines. As shown in Figure 2, the slump increases as the water content increases. The clay fines tend to increase water content at the same slump.

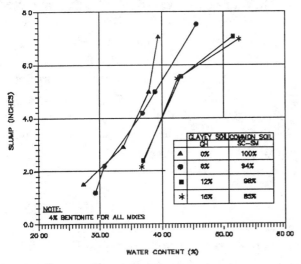

Figure 2 Slump versus water content for soil-bentonite backfill

To further confirm the laboratory results, a field demonstration program, as described in the following sections, was performed.

3.0 Field Demonstration

The technical approach for this demonstration consisted of slurry blending and soil-bentonite backfill mixing in the field using the construction method and equipment similar to that used in actual construction. A field demonstration program was conducted to evaluate the physical behavior and workability of both the design mix with 10 percent bentonite contents and the leaner backfill mix.

The field demonstration backfill mix logically began with a lesser bentonite content, then the dosage of bentonite was gradually increased at 3 percent intervals. This procedure allowed time for observation of the mixing and blending of the backfill mix, for performing slump and unit weight tests, and for obtaining samples for laboratory testing.

Backfill mix preparation began with a well-blended mixture of 20 percent clayey soil (CH) and 80 percent common soil (SC-SM). A hydrated slurry with 10 percent bentonite content was added to the mixture to achieve a 4 percent total bentonite content in the backfill mix. The moisture content, slump test, and pH were determined in the field. To evaluate the behavior of a rich bentonite backfill mix, additional dry bentonite power was added to this backfill at 3 percent intervals. As shown in Table 2, total bentonite contents of 4, 7, and 10 percent were evaluated. The slump decreased as the dry bentonite was added to the backfill mix. The 10 percent mixture (Mix No. 3) was so rigid that it showed a slump less than one inch. Additional water was introduced into this mixture to bring Mix No. 4 to a target slump of 3 inches or more. The slump increased from less than 1 inch to 3.5 inches when 22 additional gallons of water were blended into the mix. The water content also increased from 46 to 63.2 percent.

To evaluate the leaner bentonite mix, the above procedure was repeated for the hydrated bentonite slurry with 5 percent bentonite content and dry bentonite to achieve a 4 percent total bentonite content shown in Mixes 5 and 6 in Table 2.

TABLE 2 -- Properties of soil-bentonite backfill mixes for field demonstration program

Mix	Unit Weight (pcf)	Water Content (%)	Bentonite Content (%)		Slump inches
			Slurry Form	Dry Powder	
1	106.0	48.9	4	0	2-1/2
2	105.0	47.8	4	3	<2
3	99.0	46.0	4	6	<1
4	99.0	63.2	4	6	3-1/2[a]
5	106.0	44.6	1.7	2	3-1/8
6	107.5	48.8	2	2	4-1/4

[a]Water was added to Mix 3 to provide a 3.5-inch slump.

Subsequently, additional geotechnical testing for sieve analysis, water content, and unit weights was performed by the field laboratory personnel.

The results of this field demonstration program confirmed our earlier findings in the laboratory testing program, which are consistent with our past experiences. Primarily, the field demonstration program provided an opportunity to observe the mixing process and the behavior of workability with respect to slump, and to the delicate material balance between bentonite content, clay fines, and water content.

The program revealed that the following factors affect the backfill mix design:

- For backfill mix samples with the same slump, the increased bentonite content tends to increase water content. A phenomenon in which bentonite enhances the capability of the soil material to retain extra water content is known to have occurred.

- Dry bentonite powder added to the bentonite slurry premixed backfill decreases slump, reduces workability, and traps air pockets. As observed, the hydration of dry bentonite continued to reduce the slump for hours after the mixing. It is preferable to add the bentonite contents in premixed and hydrated slurry form, provided the water content is not excessive.

4.0 Construction

Based on the results of field demonstration programs, the following backfill mix was used for construction:

- Twenty percent off-site clayey soil (CH)

- Eighty percent off-site common soil (SC-SM)

- Two percent dry bentonite powder

- An adequate amount of 40 Marsh funnel second viscosity bentonite slurry to achieve a 4-inch slump.

The excavation of the SB wall was conducted from a work platform constructed of clayey material to minimize sloughing of the sidewall from wave action. The work platform had a maximum slope of one percent along the alignment to maintain the slurry level with the upper three feet of the platform. In order to maintain this maximum slope, two benches were constructed along the alignment to step the SB wall to its planned configuration.

Bentonite slurry was prepared in a centralized area situated in a remote clean zone. This area consisted of a water-holding pond, a primary slurry mixing pond, and a large storage pond for fully-hydrated slurry.

The slurry mixing plant included a high-velocity/high-pressure venturi jet mixer with a high-head centrifugal pump for mixing the powdered bentonite and water. To allow hydration to occur and to maximize the contact of the raw bentonite product and water, a set of agitator and circulation pumps was used to continuously circulate the slurry mixture. A second set of agitator and circulation pumps was provided in the larger storage pond to supply a fully-hydrated slurry of consistent quality to the trench and to the remote backfill mixing operation.

A Caterpillar 245 excavator with a 55-foot stick was used for excavation up to 40 feet deep, and a Koering 1266 excavator with an 86-foot stick was used for deeper excavation. The material excavated from the trench was considered to be contaminated material and deposited inside the landfill.

Approximately 80 percent common borrow and 20 percent clay materials were placed on a concrete platform by bucket counts from stockpiles. This material was then sluiced with bentonite slurry as dry bentonite was added. Dry bentonite was emptied evenly into the mixture via bulk bags suspended by a crane or loader bucket; discharge from the bags was conducted in a manner which minimized the creation of dust. The backfill was thoroughly mixed utilizing a Caterpillar D-4 LGP Dozer.

Prepared soil-bentonite backfill was loaded into trucks from two 100-cubic-yard (cy) transition storage boxes and transported to the trench location. Initial placement was accomplished using lead-in trenches constructed on 2H:1V slopes, eliminating the entrapment of slurry and the potential development of "windows."

Backfill was continuously built up on the lead-in slope until a natural angle of repose was achieved and the backfill surface was above the slurry in the trench. The toe of the backfill and the face of the excavation in the trench remained separated at a minimum distance of 30 feet to a maximum distance of 100 feet.

Table 3 and the following paragraphs present the field test results obtained during the SB wall construction.

A field permeability test was conducted prior to backfill placement (API 13-B, 1985). Mix from the batch mixing pad was prepared and placed into a rigid wall filter press and tested at 2 psi pressure until the flow was in a steady-state condition. Usually, it took about 2 to 3 days. These test results were used as an indicator during the field operation. Fifty-two rigid wall permeability tests were conducted with the coefficient of permeability ranging from 1.2×10^{-8} to 4.9×10^{-7} cm/s (averaging 1.9×10^{-7} cm/s), and with a standard deviation of 1.3×10^{-7} cm/s, (less than the specified 5×10^{-7} cm/s). The majority of the results were approximately 2×10^{-7} cm/s.

Fourteen additional laboratory flexible wall permeability tests were performed on the samples prepared during the slurry wall installation utilizing the triaxial cell method under back pressure (USACE 70). The coefficient of permeability ranged from 7.3×10^{-9} to 1.6×10^{-8} cm/s and averaged 1.3×10^{-8} cm/s, with a standard deviation of 2.5×10^{-9} cm/s. These flexible wall permeabilities are approximately one-half to one order of magnitude less than the rigid wall permeabilities.

Confirmation of the test results of the above design parameters revealed that the modified

TABLE 3 -- Statistical data for SB wall construction

	Bentonite Slurry			Soil-Bentonite Backfill						Permeability	
	Bentonite Content	API Viscosity (sec)	Unit Weight (pcf)	Bentonite Content	Wet Weight (pcf)	Water Content	Dry Weight (pcf)	Slump (inch)	Passing No. 200	API (cm/s)	Flexible (cm/s)
Number Tested	125	124	117	123	58	58	58	123	58	52	14
Maximum	6.87%	60	66.5	8.70%	123.3	73.0%	85.0	6.5	62.3%	4.9E-07	1.6E-08
Minimum	5.40%	41	64.5	4.40%	77.0	44.0%	49.0	3.0	33.3%	1.2E-08	7.3E-09
Average	6.77%	47	65.3	5.15%	106.9	54.8%	69.4	4.1	41.6%	1.9E-07	1.3E-08
Std. Dev.	0.237%	3.8	0.363	0.369%	8.92	8.482%	7.88	0.83	5.56%	1.3E-07	2.5E-09

backfill design mix could adequately provide the 5×10^{-7} cm/s slurry wall permeability, while reducing unnecessary bentonite, improving the workability, and eliminating the uncertainty, (i.e., trapped air pocket potential, delayed hydration, excessive postconstruction settlement, or backfill consolidation).

5.0 Conclusion

The technical solution presented in this paper was based on the results of project experience at HK which yielded the following conclusions:

- For a soil-bentonite backfill mix with a high clay content soil, increasing bentonite content results in a little to insignificant decrease in permeability.

- For samples having the same slump values, those with a rich bentonite content will have a higher water content.

- As observed in the field demonstration, adding large quantities of dry bentonite powder to the backfill mix reduces the slump; consequently, these quantities of bentonite powder will decrease the workability of the mix, and trapped air pockets may form while backfilling the slurry trench.

- By observation, the greater the amount of dry bentonite in the backfill, the greater the possibility of decreased slump by the time the mixture reaches the trench, workability is also reduced.

- A portion of the required bentonite content should be premixed and hydrated in slurry form prior to mixing with soil, and the remaining bentonite can be added in the form of dry powder.

References

API, 1985, "Standard Procedure for Testing Drilling Fluids," *API Recommended Practice*, API RP13B, American Petroleum Institute, Dallas, Texas.

API, 1986, "API Specification for Oil-Well Drilling Fluid Materials," *API, Std 13A, American Petroleum Institute, Dallas, Texas.*

D'Appolonia, D. J., 1980, "Soil-Bentonite Slurry Trench Cutoffs," *Journal of the Geotechnical Engineering Division*, ASCE, Vol. 106, No. GT4, pp. 399-417.

EPA, 1984, *Slurry Trench Construction for Pollution Migration Control, EPA-540/2-84-001.*

U.S. Army Corps of Engineers, 1970, *Laboratory Soil Testing, Engineer Manual, EM 1110-2-1906, Appendix VII.*

Safety Considerations Specific to the Investigation of Landfills and Contaminated Land

R.G. Clark
C L Associates, United Kingdom

J.A. Scarrow
Soil Mechanics Limited

R.W. Skinner
Foundation and Exploration Services Limited

Abstract

A colour coding system of site categorisation for landfills and contaminated land is presented and directly related to the basic levels of protective clothing and safety equipment that are required for each category. The importance of preparing a safety plan for each site is emphasised and guidance is given on the safety aspects relating to site investigation operations, gas detection and the handling of contaminated and potentially hazardous samples and spoil.

Introduction

There is now an increasing requirement for investigations to be carried out on sites which could pose a health hazard to the personnel who carry out the investigations. On some sites the risks are perhaps small but on others there could potentially be serious consequences of even very limited exposure to certain contaminants. Indeed, some people may already be suffering from complaints that as yet have not been identified as originating from exposure to hazardous contaminants in the ground.

In spite of the growing awareness of the risks involved and the need to identify as far as possible those sites that need particular precautions there is still no nationally accepted system in the UK of categorising sites in accordance with their known or suspected risk and the consequent level of protection that is required.

This paper presents a simple site categorisation system based on colour coding which was developed on behalf of the British Drilling Association. The implementation of this system could provide a significant step towards a coordinated approach to the safe investigation of potentially contaminated sites.

Categorisation System

A system of categorising sites in respect of their level of risk is clearly needed. This can then be used to indicate the safety and protective precautions required and enable staff on site and in the testing laboratory to understand the potential hazards that may exist.

Table 1 presents a colour coding system for categorising sites in terms of the nature of materials or substances that may be found on them. This can be applied to both landfills and contaminated land. Rather than try to list every material or substance that could potentially be found on a site, the intention is to provide an indication of the types of materials that fall into each of the three categories.

"Green" is designated the least hazardous category and applies to substantially inert materials. It may be necessary to put some apparently clean sites under this category if there is any uncertainty regarding the previous use of the site. A "yellow" site is one which contains substances which are unlikely to cause serious impairment to health but nevertheless constitutes some risk. A "red" categorisation should always be used for all sites that could potentially endanger health and applies to all substances that could subject persons to risk of death, injury or impairment of health.

A desk study concentrating on the previous use and history of the site should be carried out before any physical investigation work starts on any site where there is a possibility of contaminants being present. It is on the basis of the results of this desk study that the appropriate colour site categorisation should be allocated. However, the desk study may not detect whether indiscriminate dumping etc has taken place

TABLE I -- Site Categorisation

Category	Broad Description of Contaminants
GREEN	Subsoil, Topsoil, Hardcore, Bricks, Stone, Concrete, Clay, Excavated Road Materials, Glass, Ceramics, Abrasives, etc.
	Wood, Paper, Cardboard, Plastics, Metals, Wool, Cork, Ash, Clinker, Cement, etc.
	NOTE : There is a possibility that bonded asbestos could be present in otherwise inert areas.
YELLOW	Waste Food, Vegetable Matter, Floor Sweepings, Household Waste, Animal Carcasses, Sewage Sludge, Trees, Bushes, Garden Waste, Leather etc.
	Rubber and Latex, Tyres, Epoxy Resin, Electrical Fittings, Soaps, Cosmetics, Non-Toxic Metal and Organic Compounds, Tar, Pitch, Bitumen, Solidified Wastes, Dye Stuffs, Fuel Ash, Silica Dust, etc.
RED	All substances that could subject persons and animals to risk of death, injury or impairment of health.
	Wide range of Chemicals, Toxic Metal and Organic Compounds, etc. Pharmaceutical and Veterinary Wastes, Phenols, Medical Products, Solvents, Beryllium, Micro-organisms, Asbestos, Thiocyanates, Cyanides, etc.
	Hydrocarbons, Peroxides, Chlorates, Flammable and Explosive Materials. Materials that are particularly corrosive or carcenogenic, etc.

on a particular landfill or contaminated site and therefore the categorisation given in Table I should be treated as a guide only. Nevertheless, the implementation of this system can provide a workable framework in which to decide the level of basic protection that is required for a particular site.

There are some sites (both landfills and contaminated land) where initially there may be very little information regarding whether hazardous materials could be present. In these situations consideration should be given to first carrying out a limited probing and sampling exercise under a temporary "red" designation. The site can then be categorised prior to the full site investigation. In those situations where either a desk study is not carried out or the preliminary investigation is inconclusive then the "red" categorisation should automatically continue for the full investigation.

Safety Plan

A written Safety Plan for the project should be prepared prior to the commencement of work on site. In the first instance this should state the categorisation of the site and give an indication of the reliability of the available information on which the categorisation has been based.

The Plan should state the nature of the known or suspected hazards and give guidance on how best to recognise the contaminants involved and the effects of these contaminants on humans. Where appropriate, standard reference sheets can be included on certain chemicals or other substances. In the case of landfills, advice should be given on diseases associated with these sites such as Leptospirosis (Weils' disease) and tetanus. Likewise, advice should be given on the health hazards associated with particular contaminants in respect of other sites.

The document should state the telephone numbers and locations of emergency services on and off site including the nearest casualty department, together with the location and details of the availability of the nearest clean water supply. The safety equipment that will be required for the project (see below) should be listed and an indication given of the circumstances in which each appliance should be used. The person who is to be responsible for the safety and protection equipment should be stated

TABLE 2 -- Site Designation/Site Safety Equipment

Item	Site Designation		
	Green	Yellow	Red
Personal Protective Equipment			
Hard Hat	*	*	*
Eye Protection		*	*
Face Shield		*	*
Hand Protection	*	*	*
Overalls	*	*	
Disposable Overalls			*
Waterproofs	*	*	
Disposable Waterproofs			*
Industrial Boots	*	*	*
Wellington Boots with Sole and Toe Protections	*	*	*
Respiratory Equipment		*	*
Site Equipment/Services			
Mobile telephone (outside contaminated area)		*	*
Ropes, Cones and Barriers			*
Safety/Warning Signs	*	*	*
Clean water Supply	*	*	*
Changing Room/Washing Facilities		*	*
Decontamination Unit/Washing Facilities			*
Emergency Equipment			
Fire Extinguisher	*	*	*
Fire blanket	*	*	*
First Aid Kit	*	*	*

together with a schedule of any calibration or maintenance checks that are required.

Strict instructions should be included regarding such matters as reporting any illness, incident or injury and the actions to be taken should any suspicious substances or obnoxious gases or fumes be encountered. Guidance should also be given on safety/first aid procedures to be adopted.

There may be a number of other matters that are considered appropriate for inclusion in the Safety Plan for a particular site, for example the biological and bacteriological risks associated with a site, for example sites where anthrax may be present or where there have been old biological processing plants or sewage works. Sites where explosives or radioactive materials may be present would be other examples of specific hazards and risks.

All personnel working on the site should be given an induction on the contents of the Safety Plan and should acknowledge receipt of a copy. Where appropriate the Safety Plan may need to be modified during the course of the work and reissued, particularly if it is decided to change the categorisation of the site.

Safety Protection and Equipment

Once a site has been categorised in terms of the degree of hazard and consequently risk (and as part of the preparation of the Safety Plan), there is a need to determine the appropriate level of protective clothing and equipment that will be required. Table 2 provides a suggested guide to the minimum protection with different requirements applying to each colour category.

There are obviously certain items of protection that are required even for a "green" category site. It should always be borne in mind that after a site has been categorised following a desk study or initial probing exercise the unexpected may still be found.

A decontamination unit should be provided on all "red" category sites. This unit should consist of three distinct parts. On first entering the unit from the contaminated area there should be facilities for the removal and storage of contaminated overalls, footwear, safety equipment, etc. This should be separated from the second part of the unit which should contain washing facilities including showers in cases where full body washing is a specified requirement of the Safety Plan. The final part of the unit should be for the storage of the clothing and footwear that will be worn when leaving the site. A boot wash should be provided adjacent to the decontamination unit.

Under no circumstances should overalls or other clothing used on a contaminated site be taken offsite (for example taken home) apart from under special transport/packaged conditions. Such clothing should be dealt with by professional cleaners who have been given prior information on the nature of any contaminants.

Site Operations

The hazards associated with the investigation of contaminated or potentially contaminated sites should not be ignored or underestimated. All too often in the past such investigations have been approached on the basis that no additional precautions are required compared to the investigation of a clean site. This attitude has undoubtedly put the personnel who are involved in the investigation at serious risk.

Training is an important requirement to ensure that those who are involved are fully conversant with the hazards likely to be encountered and are able to implement appropriate measures to ensure that the operations are safe. As part of this training, supervisory staff and key operatives should receive formal instruction in basic first aid.

In the case of "red" category sites a qualified environmental specialist should be present on site at all times when investigation work is being carried out. An environmental specialist may also be required on "yellow" category sites if there is doubt about the accuracy of information on which the categorisation has been based. The specialist should have suitable training and experience to enable them to recognise by sight, smell or in situ test (where this is possible) any unidentified hazards and then be able to specify consequent changes to the working procedures.

In all cases the environmental specialist should have the executive authority to initiate further analytical studies at any time during the investigation to ensure that the hazard of any unidentified substances is accurately and quickly identified.

The mere nature of carrying out investigations means that the unexpected can be encountered regardless of the initial colour categorisation allocated to the site. Exposure of people to unidentified substances or obnoxious fumes and gases must be kept to the absolute minimum. If unidentified substances are encountered then the investigation work should be suspended until such time as an environmental specialist can inspect the site and determine what the contaminant is and whether a hazard exists. In the case of "green" and "yellow" category sites consideration will need to be given to whether they should be recategorised as "red" sites.

Strict personal hygiene should be promoted at all times. It should be borne in mind that contaminants can enter the body by skin penetration, skin absorption, ingestion or inhalation. Smoking can result in the ingestion of contaminants and therefore all "yellow" and "red" category sites should be designated No Smoking areas. The consumption of drink or food should not be permitted within the contaminated area and separate messing facilities should be provided at a clean location such that personnel can first remove protective clothing, gloves, boots etc and wash either within the contaminated area or within a decontamination unit prior to entering the messing facilities.

In the case of landfills (or any other sites where there is a potential for gas to be generated) huts, temporary offices etc should not be located where the infiltration of gas is a possibility.

Spark arrestors and automatic air intake shutdown valves should also be provided on all combustion engines operating on or near landfills.

The investigation of a contaminated site must not constitute a hazard to the general public. Sites should be made secure at all times, day and night, if hazardous materials are found or are suspected as being present. When the investigation is complete the site must be left in a satisfactory and safe condition. This means that boreholes and trial pits must be satisfactorily backfilled and all contaminated spoil must be removed from the surface of the site. Under no circumstances should contaminants or otherwise hazardous materials be left exposed on the surface. If this were allowed to occur then those responsible could be liable to prosecution.

Gas Detection

Gas monitoring equipment may be required on all three categories of site particularly where biodegradable materials are present. In many cases, particularly when investigating landfills, high concentrations of gas under pressure can be encountered. Indeed, it has been known for the gas emission from a borehole to push the drilling tools out of the hole without an explosion taking place.

Operations should stop when methane concentrations at the surface exceed an acceptable limit. In many publications and guidance documents an acceptable limit of 1% by volume has been quoted. However, this relates to potentially confined conditions and therefore, provided the work is being carried out in the open, it is suggested that a limit of 3% is likely to be more practical and still be acceptable. Work should not recommence until either natural or artificial ventilation has reduced concentrations to below this limit or purging has been carried out.

When carrying out work in the open, monitoring for carbon dioxide is not essential from a safety point of view because dilution of any emission from the ground quickly occurs. However, if working in a confined space cannot be avoided then monitoring for carbon dioxide will be necessary.

If gas is likely to be a problem on a particular site then monitoring checks should be carried out at regular intervals during drilling or the excavation of trial pits. An appropriate interval during drilling would be every 0.5 to 1.0 m of depth and immediately following any occurrence which would suggest gas emission. Gas monitoring should also be carried out in any partially completed boreholes at the start of each shift prior to the recommencement of work.

On certain sites such as old paper mills or chemical works hydrogen sulphide, which is flammable between the limits of 4.3 and 45.5% by volume in air, may need to be monitored. Detection of hydrocarbon vapours will also be a requirement when investigating filling stations, tank farms, pipelines and chemical works etc. Petroleum vapours are particularly hazardous and are flammable between 1 and 7% by volume in air.

It is strongly recommended that, in addition to the use of monitoring instruments, the drillers should also wear audible alarms, particularly in respect of some of the more dangerous emissions.

Contaminated Samples and Spoil

Arrangements will need to be made for the disposal of contaminated spoil well in advance of the work commencing. The spoil will need to be taken to a waste disposal facility which is licenced to receive the waste in question. Contact should be made with the appropriate Waste Regulatory Authority who will give an indication of those waste disposal facilities in the area that are licenced to receive such materials.

The use of a licenced waste disposal carrier would be one way of transporting the spoil for disposal. Alternatively the site investigation specialist may wish to transport the spoil using his own transport and in this situation the specialist may need to be licenced as a waste carrier.

Samples obtained from a contaminated site should be clearly labelled and colour coded in accordance with the site categorisation. Hazardous samples must be suitably contained to ensure that no leakage or spillage can occur. In order to protect personnel at the testing laboratory, notice both verbally and in writing should be given prior to the samples being despatched from the site.

In conjunction with the Safety Plan all organisations involved in the investigation of landfills or contaminated land should operate Health Surveillance Schemes to ensure regular monitoring and updating of inoculations, etc.

Concluding Remarks

There is an urgent need for stricter safety measures to be applied to the investigation of landfills and contaminated land. On some projects appropriate procedures are being followed but there are many others where the awareness of the hazards involved is sadly lacking. This paper addresses a number of the safety aspects which should be particularly noted and puts forward a categorisation system. Further detailed information can be obtained from the British Drilling Association publication entitled "Guidance Notes for the Safe Drilling of Landfills and Contaminated Land" which also includes a proforma which can be used to record and transfer information about previous use and possible contaminants prior to the start of an investigation.

Acknowledgement

The authors wish to thank the British Drilling Association for permission to publish this paper.

Apparatus for Measurement of Molecular Diffusion Coefficients in Partially Saturated Soils

Brian Cooke, Jason Fong
Geomechanics Group, Department of Civil Engineering, University of Western Australia, Nedlands, WA

ABSTRACT: As a soil desaturates, the hydraulic conductivity decreases rapidly due to the increased tortuosity of the flow paths. Generally when analysing dissolved contaminant transport, lower hydraulic conductivities indicate an increase in the relative importance of molecular diffusion as a transport mechanism. However, for partially saturated soils the molecular diffusion coefficient (D^*) also decreases with saturation, again because of the increased tortuosity of the path followed by the diffusing molecules. To date there has been little research into the nature of the relationship between water content (θ) and diffusion coefficient. It is apparent that if D^* decreases with decreasing saturation by a number of orders of magnitude, as does k, then soils with a low degree of saturation will effectively block contaminant migration. If, on the other hand, there is only a minor decrease in D^*, then molecular diffusion will become an important transport mechanism as a soil desaturates.

This paper describes an apparatus developed to determine the $D^*(\theta)$ relationship. By measuring concentration of a salt tracer at the down gradient side of a partially saturated soil sample subject to pure diffusion, it is possible to calculate the diffusion coefficient of the soil. By repeating this procedure at a number of different capillary pressure (P_c) levels, the complete $D^*(\theta)$ (or $D^*(P_c)$) relationship can be defined.

INTRODUCTION

Solute transport in porous media is generally described by the advective-dispersive equation (ADE), which combines the transport due to Darcian flow (advection) with the spreading of the solute front due to mixing in the pores and molecular diffusion. For a conservative contaminant in a flow system with no sources or sinks, the general form of the ADE is:

$$\frac{\partial \theta C}{\partial t} = \nabla \cdot [(\alpha v + D^*)\theta \nabla C - \theta C v] \quad (1)$$

where C is the concentration of the dissolved chemical in the pore water, θ is the water content of the soil on a volumetric basis, v is the interstitial fluid velocity, α is the dispersivity of the porous medium, and D^* is the coefficient of molecular diffusion of the chemical species in the soil.

It can be seen that there are three components to the transport process: molecular diffusion, and the two velocity dependent phenomena of advection and mechanical dispersion. It follows that for very low velocity flow regimes, such as in clay soils or under very low hydraulic gradients, molecular diffusion takes on relatively greater importance. As flow velocity tends towards zero, diffusion becomes the only transport mechanism - in the limit, the ADE reduces to Fick's second law:

$$\frac{\partial \theta C}{\partial t} = \nabla \cdot [D^* \theta \nabla C] \qquad (2)$$

Partially saturated soils present a flow regime where flow velocities are very low, due to the rapid decrease in hydraulic conductivity with decreasing saturation. Under these conditions, the ADE is still valid, but with the proviso that D^* is a function of degree of saturation (Bear and Verrijuit, 1987). Thus solution of the ADE requires knowledge of the function $D^*(\theta)$.

Both the molecular diffusion coefficient and the hydraulic conductivity are sometimes expressed as a function of the porous medium termed "tortuosity", and the decrease in k and D^* with decreasing saturation is attributed to an increase in the tortuosity of the transport pathways. According to Bear (1972), the diffusion coefficient may be expressed as a function of tortuosity alone:

$$D^* = TD \qquad (3)$$

where T is the medium tortuosity tensor and D is the coefficient of diffusion of the particular chemical species in pure water. (Note that the mathematical sense of tortuosity is opposite to the grammatical sense - that is, as the medium becomes more tortuous, the numerical values of the components of T decrease.)

Using the same T as contained in the definition of D^*, hydraulic conductivity can be expressed as:

$$k = nBT \qquad (4)$$

where B is the medium's conductance, which takes into account the velocity distribution across the flow tubes, and varies as the square of the pore radius. Thus it would appear that although both k and D^* decrease with decreasing saturation, k should decrease much more rapidly due to the effect of viscous drag caused by the decreasing radii of the pores available for flow. If this is the case, then the relative importance of diffusion as a transport mechanism should increase with decreasing saturation.

There have been sporadic attempts in the soil physics area to measure the relationships between D^* and θ (e.g., Kemper and van Schaik, 1966). The techniques used generally require that two soil plugs be prepared to the required moisture content, one using pure water and the other using a tracer solution. The soil plugs are then brought together in a sealed container to permit diffusion of the tracer between them. After a set time the combined soil plug is sectioned and analysed for distribution of the tracer. The drawbacks to this procedure include difficulty in preparing samples to a specified moisture content, problems with ensuring proper contact between the soil plugs, and analytical problems due to uncertainty about boundary conditions.

The procedure outlined below uses standard preparation techniques to compact all samples in a saturated state. Moisture content is controlled by precise control of capillary pressure, which is related to water content through a known pressure - saturation function. The boundary and initial conditions are clearly defined, permitting application of a known analytical solution to the diffusion equation to determine the required diffusion coefficient.

APPARATUS

The apparatus developed for this experimental program is designed to use the time lag method of determination of D^*, as described by Yong *et* *al.* (1992). This requires that a resevoir be maintained at a constant concentration at the high concentration end of the cell, and a resevoir at the downstream end maintain a constant (nominally zero) concentration by means of continuous flushing, permitting the eventual establishment of a constant concentration gradient through the soil sample.

The main parts of the apparatus, described briefly below, are shown schematically in Figure 1 and photographically in Figure 2. The soil sample is contained within a brass collar, 30mm long by 50mm diameter, which is equipped with an air pressure inlet to provide the pressure for desaturating the soil. This collar fits snugly into lucite end pieces and flush against the porous plates, with two pairs of O-rings to prevent leakage from the resevoir. The ceramic plates are available in various air entry pressures; 1 bar plates were used for the current project. The two end pieces are tightly held against the brass collar by means of three bolts which pass through the lucite and are tightened by wing nuts.

The resevoirs thread directly into the end pieces to fit snugly against the ceramic plates and O-rings. The upstream resevoir has a threaded plug to permit addition of salt, and withdrawal of samples for monitoring of salt concentration. A similar opening in the downstream resevoir is used as the inlet for the flushing water from a peristaltic pump. The downstream resevoir has an additional opening which serves as the outlet through which the flushing solution drains.

As the purpose of the apparatus is to measure diffusion, it is important to ensure that this is the only transport mechanism available. Therefore the apparatus was designed to eliminate any hydraulic gradient which would result in advective transport. This was achieved by means of a standpipe hydraulicly connected to both resevoirs. To prevent the chemical tracer from "short circuiting" the soil by diffusing through the standpipe system, the opening into the upgradient resevoir was covered with a flexible latex membrane.

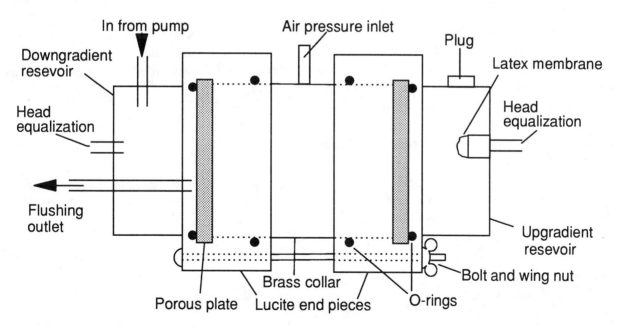

Figure 1. Schematic diagram of diffusion apparatus.

Figure 2. Photograph of disassembled apparatus.

TESTING PROCEDURE

The testing procedure comprises two phases: a desaturation phase and a diffusion phase. The soil used for the testing program was a silica flour for which the pressure-saturation and pressure-hydraulic conductivity relationships were already known. At the beginning of a test, a saturated sample is prepared in the brass collar by raining the soil into the water-filled collar on a vibrating table. The collar and sample are then placed into the end pieces and the bolts tightened. The resevoirs are filled with deionized water, and the peristaltic pump started, with the effluent from the flushing process directed to waste. Flushing is not a necessary part of the desaturation process; it is carried out to ensure that there is no change in hydraulic conditions when the diffusion phase begins.

Air pressure is applied to the soil sample, and it is allowed between one and three days to desaturate, with longer times being allotted to higher pressure levels. Water can flow out of the soil through the porous plates to the resevoirs; the capillary pressure is the difference between the applied air pressure and the pressure indicated by the common standpipe.

At the completion of the desaturation phase, the deionised water in the upstream resevoir is replaced with a saturated sodium chloride solution, with additional sodium chloride added to the resevoir to ensure that saturation is maintained. Thereafter, the discharge from the downstream resevoir flushing system is directed to a fraction collector for determination of salt content, which is carried out using a specific ion analyser. After experimentation with a range of peristaltic pump flow rates it was found that a rate of 2.3ml/min gives the optimum combination of maintaining a low concentration in the resevoir and reducing the number of effluent samples to be analyzed.

At the end of each test the soil sample is removed from the apparatus and a moisture content determination carried out.

Prior to the soil tests, it was necessary to determine the diffusion coefficient through the porous plates (D^*_{PP}) in order that subsequent calculations could be corrected for the presence of the plates. This was accomplished by running a test with the brass collar full of deionized water. The calculated experimentally determined diffusion coefficient was then corrected using the known coefficient for water, in a manner similar to that described in the analysis section below.

ANALYSIS

The analysis of the results is based on the time lag method, described in Yong *et al.* (1992). Cumulative flux is plotted against time, and the constant slope portion of the graph is projected down to the time axis, with the axis intercept being the time lag. A typical plot is shown in Figure 3.

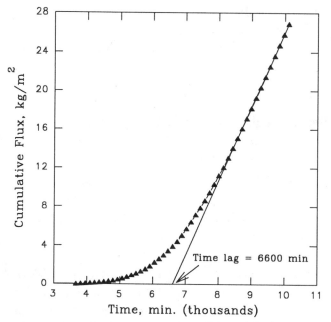

Figure 3. Plot of cumulative flux with time.

Once the time lag is determined, diffusion coefficient is calculated from:

$$D_B^* = \frac{L^2}{6T_L} \quad (5)$$

where T_L is the time lag, L is the diffusion length (soil sample plus porous plates), and D_B^* is the bulk diffusion coefficient. For a derivation of the time lag solution, the reader is directed to Crank (1975).

The soil diffusion coefficient, D_S^*, is calculated by correcting the bulk diffusion coefficient for the presence of the porous plates by assuming that it is a weighted geometric mean:

$$\frac{L_S}{D_S^*} = \frac{L_{PP}}{D_{PP}^*} + \frac{L_B}{D_B^*} \quad (6)$$

where L_S is the length of the soil sample, L_{PP} is the combined thickness of the two porous plates, and other terms are as defined previously.

RESULTS

Following an initial period of experimentation with and fine tuning of the apparatus, a satisfactory procedure has been established which has been found to produce reasonable results. The moisture content of the soil samples after testing agree well with the results of pressure-saturation measurements from standard pressure plate extraction tests. Although diffusion coefficients have to date been determined only for the saturated soil and three pressure levels (testing times range between one and three weeks for each pressure increment), it appears that the decrease in D^* is slower than the decrease in k, as suspected. The results available to date are plotted in Figure 4 as plots of D_r^*, which is the diffusion coefficient relative to the value at saturation, against pressure head. Also plotted on Figure 4 is relative hydraulic conductivity, k_r, which is defined similarly. Figure 5 plots the same two quantities against volumetric moisture content.

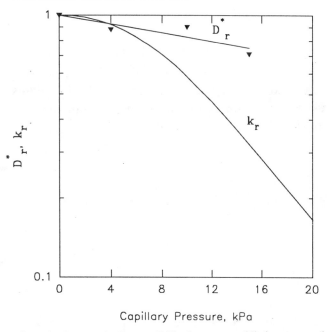

Figure 4. Relative diffusion coefficient and relative hydraulic conducivity as functions of capillary pressure.

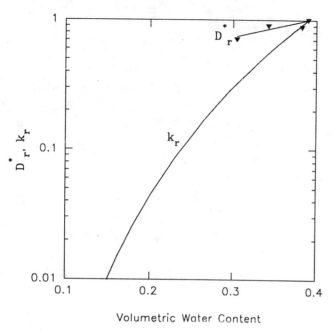

Figure 5. Relative diffusion coefficient and relative hydraulic conductivity as functions of moisture content.

CONCLUSIONS

An apparatus has been developed to measure the molecular diffusion coefficient of partially saturated soils. The equipment and procedures work well, and preliminary results indicate that the $D^*(\theta)$ relationship is of the anticipated form.

REFERENCES

Bear, J. (1972). *Dynamics of fluids in porous media*. American Elsevier Inc., New York, N.Y.

Bear, J. and A. Verruijt (1987). *Modeling Groundwater Flow and Pollution*. D. Reidel Publishing Company, Dordrecht, Holland.

Crank, J. (1975). *The Mathematics of Diffusion*. Clarendon Press, Oxford.

Kemper, W.D. and J.C. van Schaik (1966). "Diffusion of salts in clay-water systems." *Soil Science Society of America Proceedings,* v30, 534-540.

Yong, R.N., A.M.O. Mohamed, and B.P. Warkentin (1992). *Principles of Contaminant Transport in Soils*. Elsevier, Amsterdam.

Seeking Optimal Containment Design at Creosote-Contaminated Site Through Cost-Benefit Analysis

Henry S. Crawford, P.Eng.
Thurber Engineering Ltd., Calgary, Alberta, Canada

Kevin Parks, P.Geol.
University of Calgary (formerly of CH2M Hill Engineering Ltd.) Calgary, Alberta, Canada

Abstract

The site of a former wood treatment facility, now redeveloped within an active urban environment, has been discharging creosote to a major river that parallels the site. A cost-benefit analysis of potential containment systems has been completed considering both capital and operating costs. The recommended design incorporates partial containment by a slurry trench wall with groundwater control within the contaminated area and upgradient of the contaminated area. The system is to be built in a modular fashion such that the performance of each phase can be monitored prior to implementing subsequent phases.

Introduction

Full remediation of a contaminated site is often impractical, if not impossible due to costs or other constraints. In many instances owners of contaminated sites have no option but to implement some form of containment system to eliminate or reduce off-site migration of the contaminants. Because construction and operation of a containment system can be expensive, there is considerable incentive to optimize the system to minimize the initial capital costs and long term operating costs.

This paper presents the preliminary design and cost-benefit analyses undertaken to optimize containment at a former wood treatment facility located in Calgary, Alberta, Canada.

Site Description

The project site is located on a flood plain terrace along the south bank of the Bow River, on the western edge of downtown Calgary, as shown on Figure 1. The site is a former wood treating facility which operated between 1924 and 1962. It was reclaimed in the mid 1960's and now accommodates a

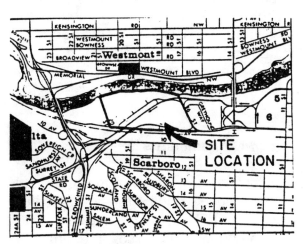

Figure 1 - Site Location

major freeway, smaller access roads and two car dealerships.

Overburden sediments at the site are up to 8 m thick and typically consist of coarse-grained fill material overlying alluvial sand and gravel. The underlying bedrock consists predominantly of siltstone with layers of interbedded sandstone and claystone. The bedrock surface generally slopes towards the river at a grade of about 3%, with some local depressions or secondary channels subperpendicular to the main bedrock channel. Statistical analysis of Rock Quality Designations (RQD's) of collected core samples, indicates the upper 2 m of bedrock are more fractured than the material at depth.

Groundwater within the surficial sediments is generally at 4 to 7 m below ground surface and flows northeast across the site to the Bow River. Horizontal gradients within the overburden sediments typically range between 0.005 and 0.013, with the steeper gradients occurring nearer the river. Vertical upward gradients in the range of 0.004 were measured in the bedrock near the river. Based on in-situ permeability tests conducted by others, the hydraulic conductivity of the overburden material is estimated to range between 10^0 and 10^{-3} cm/s, with the more permeable material occurring nearer the river. Hydraulic conductivity of the bedrock is estimated to range from 10^{-3} to 10^{-6} cm/s with the higher permeability occurring in the upper 2 m.

Extent of Contaminants

The contaminants at the site consist primarily of coal tar creosote and petroleum based oils used to dilute the creosote. Pentachlorophenol is also present, but represents less than 0.5% of the mixture. The contaminants are present in the form of dense and light non-aqueous phase liquids (DNAPLs and LNAPLs), dissolved contaminants in groundwater, and as sorbed contaminants on soils. The total mass of contaminant is estimated to be between 3,000 and 7,000 tonnes, with the main pool covering an area of about 20,000 m^2. Some creosote can be found in the river gravels opposite and immediately downstream of the site and in bedrock fractures on the opposite side of the river.

Multi-phase modelling studies, performed by the Alberta Research Council, suggest that approximately 70 tonnes of contaminant moves off-site each year, with a portion of this getting into the Bow River in the form of DNAPL and dissolved phase. The modelling also suggests that most of the contaminant enters the river via the overburden sediments with only a small amount travelling through the bedrock. It was recognized, however, that the modelling studies were very sensitive to a number of parameters and the calculated amounts from each source (eg. DNAPL through overburden, dissolved phase through overburden, etc.) were treated as guidelines only.

Design Constraints

The objective of the containment system was to significantly reduce contaminant loadings from the site to the Bow River, and to be compatible with possible future remediation of the site.

Due to uncertainties in the existing contaminant loading conditions, it was established as a design constraint that the containment system must serve to reduce all sources of loading. It would not acceptable for one source of loading to decrease while another source increased. The design also had to be compatible with a wide range of possible soil and groundwater conditions.

Components of Containment System

To meet the design objectives stated above, the containment system could be composed of one or more of the following components:

- a physical barrier;
- groundwater control within contaminated area, requiring water treatment;
- groundwater control upstream of contaminated area to divert clean water away from the contaminated area.

Inclusion of a product recovery system was deemed desirable but not within the scope of this assignment.

A comprehensive evaluation of each component and the different technologies available was completed and it was established that the preferred system would include a physical barrier in the form of a soil bentonite slurry trench with groundwater control provided by vertical pumping wells.

Containment systems considered, included:

- a simple, straight-line physical barrier along the downstream side of the contaminant pool;
- singular groundwater control around the perimeter of the site, with no physical barrier;
- partial enclosure of the site with an 'L-shaped' or 'U-shaped' physical barrier with and without groundwater control;
- full enclosure of the site by a physical barrier with and without groundwater control.

In total, more than 20 different containment configurations were considered. Schematic illustrations of typical configurations are shown on Figure 2.

To permit a rational evaluation of each potential system, it was considered prudent to develop a cost-benefit relationship between each of the candidate systems, as described in the following sections.

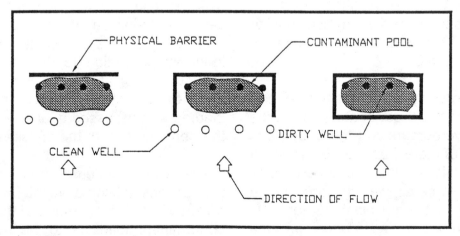

Figure 2 - Examples of Straight Line Barrier, U-Shaped Barrier and Full Enclosure

Groundwater Modelling

To quantify the performance of each system, relatively simple groundwater modelling studies were undertaken using two commercially available software packages.

The finite difference program, FLOWPATH, was used to model the impact of each candidate system on the natural groundwater flow in the overburden. Analyses were performed using a range of hydraulic conductivities and groundwater recharge coefficients considered appropriate for the site.

Output from the analyses included: total flux to the river from the contaminated area; volume of "dirty" groundwater pumped from within the contaminated area; volume of "clean" groundwater pumped from outside the contaminated area, and; total flux diverted away from the site.

The relative effectiveness of each system was assessed based on the total volume of groundwater which passed through the contaminated zone and into the river.

Additional groundwater modelling studies were undertaken using the finite element program, SEEP/W, to establish the depth of penetration required by the physical barrier and the impact of groundwater mounding up-gradient of the barrier.

Based on these analyses, it was established that the barrier should penetrate through the full depth of overburden and the upper 2 m (approximate) of more fractured bedrock. It was also established that no groundwater mounding should be allowed to occur on the upstream side of the physical barrier. By preventing groundwater mounding, the hydraulic gradient across the barrier becomes zero or negative, resulting in zero or reversed flow across the barrier, even if the barrier is not 100% effective. This provides a level of redundancy to the containment system which is desirable.

Cost-Benefit Analysis

A cost benefit analysis was undertaken to compare the relative cost and anticipated benefits of the different containment systems. The total cost of each system was estimated based on the physical dimensions of the system and the estimated unit cost of each component. Costs for pumping and water treatment were based on unit rates and the estimated pumping rates obtained from the groundwater modelling studies. Lump sum amounts were provided for items such as mobilization, crossing of utilities and roadways, instrumentation and monitoring, site cleanup, etc. Costs of each system were converted to their Net Present Value (NPV) based on a 25 year design life and a discount factor of 3% applied to operating costs.

The benefits derived from the containment system were estimated based on the degree to which the contaminant loading to the river was reduced, as compared to the do-nothing case. Reduction in the loading of the dissolved phase component was calculated in direct proportion to the reduced rate of groundwater flow through the contaminated area to the river. Reduction in DNAPL loading was assessed qualitatively, based on the configuration of the physical barrier.

These analyses demonstrated that most of the proposed containment systems were effective with respect to reducing contaminant loadings to the river from the main contaminant plume. However, some systems were rejected because they did not meet the design criteria of not causing an increase in contaminant loading downstream of the main contaminant plume. Other systems were rejected because they did not meet the design criteria of no groundwater mounding.

Table 1 presents a summary of the preferred configurations for the range of hydraulic parameters investigated. It shows that the optimum design will ultimately depend on the in-situ hydraulic parameters. For example, a configuration with full enclosure changes from being the most expensive to becoming the least expensive if the hydraulic conductivity increases by one order of magnitude. Conversely, at lower hydraulic conductivities, a partially enclosed system, with a U-shaped physical barrier and groundwater control, is more cost effective.

The optimum design established for this site, based on the best estimate of the actual in-situ hydraulic parameters, consisted of an open, U-shaped barrier with groundwater pumping wells maintained within the contaminated zone, immediately upstream of the barrier ("dirty wells"), and interceptor wells maintained upstream of the contaminated zone ("clean wells"), to intercept clean water before it enters the contaminated area. Figure 3 shows the preliminary layout of the proposed system.

Advantage of Modular Construction

Due to the uncertainty in the in-situ hydraulic parameters at the site, and in the actual loading rates to the river, it was recommended that the containment system be built in stages over a period of years such that the performance of each stage, including operating costs, could be monitored and the design modified if appropriate. This was possible at this particular site as the risk analyses, undertaken by others, had demonstrated that there was not a high risk to the public if full containment was not implemented immediately.

The preliminary recommended sequence of construction was:

- construct a straight line barrier with west wing wall (Segment A-B-C on Figure 3);
- add groundwater control within the contaminated area (dirty wells);
- add the east wing (Segment C-D-E);
- add upstream groundwater control ("clean wells").

The final configuration would depend upon actual in-situ hydraulic parameters as determined from the observed performance.

Acknowledgements

The authors gratefully acknowledge the Waste Management Division of Alberta Environmental Protection for granting permission to publish this paper.

TABLE 1 -- Preferred Containment Options

Scheme No.	Barrier Configuration	Upstream "Clean" Wells	"Dirty" Wells	Capital Cost (millions)	O&M Cost (thousands)	NPV (millions)
$K_H = 1 \times 10^{-2}$ cm/s						
18	Open "U-Shaped"	yes	yes	$2.80	$25.9	$3.25
29	Open "L-Shaped"	no	yes	$2.82	$56.6	$3.81
7	Open "U-Shaped"	no	yes	$3.09	$55.4	$4.05
22	Closed	yes	yes	$4.15	$27.4	$4.63
$K_H = 3 \times 10^{-2}$ cm/s						
18	Open "U-Shaped"	yes	yes	$2.90	$30.9	$3.43
22	Closed	yes	yes	$3.96	$19.7	$4.31
29	Open "L-Shaped"	no	yes	$3.19	$99.1	$4.91
7	Open "L-Shaped"	no	yes	$3.47	$100.4	$5.22
$K_H = 1 \times 10^{-1}$ cm/s						
22	Closed	yes	yes	$4.10	$24.9	$4.54
18	Open "U-Shaped"	yes	yes	$3.47	$78.8	$4.84
29	Open "L-Shaped"	no	yes	$3.88	$206.6	$7.47
7	Open "U-Shaped"	no	yes	$4.22	$217.5	$8.00

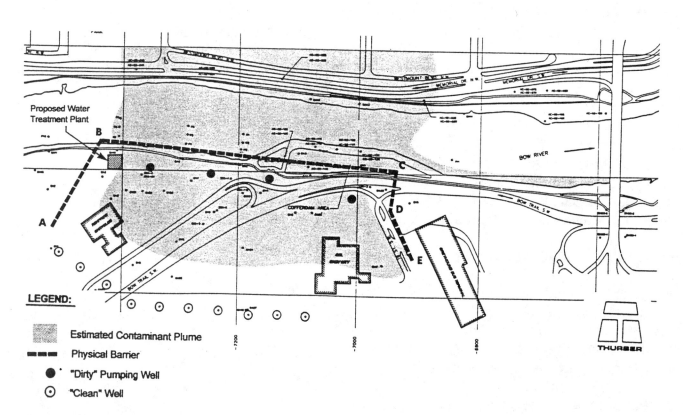

Figure 3 - Site Plan Showing Recommended Containment Alternative

Le Controle de la Charge Bacteriologique des Lixiviats de Decharge par les Sols: Influence des Parametres Physico-Chimiques des Lixivats de Decharge sur la Migration Bacterienna a Travers un Sol Sature

J. Crosnier, C. Delolme
Ecole Nationale Des Travaux Publics De L'etat, France

Final manuscript not received in time for inclusion in the *Proceedings*. For additional information, please contact:

> Prof. Jerome Crosnier
> Ecole Nationale Des Travaux Publics De L'etat
> Laboratoire des Sciences de l'Environment
> Rue Maurice Audin
> 69518 Vaulx-En Velin Cedex
> FRANCE
> Voice: +33-72-04-70-42
> Fax: +33-72-04-62-54

A Review of Clogging Processes in Porous Media as Related to Wastewater Collection and Drainage

Raghava Dasika, James Atwater
Department of Civil Engineering, University of British Columbia, Vancouver, B.C. Canada

ABSTRACT:

Clogging by the accumulation of solids within porous media can result from physical, microbial and chemical processes. When this occurs in engineered or natural collection and drainage structures, the resultant loss in permeability can reduce the system effectiveness, and also lead to potentially expensive and difficult remediation requirements. A review of available literature on each of the various processes provides some insight into the likely factors leading to clogging, and enables some prediction of clogging susceptibility. Such knowledge is useful at the design stage, but the long term extent and effects of gradual accumulations are currently difficult to predict.

INTRODUCTION:

The objective of this paper is to present a summary review of the various clogging processes that can effect the performance of engineered or natural porous media that are used for the collection and drainage of wastewaters. The following discussion aims to identify the conditions under which these processes individually, or in combination, result in significant impacts on collection and drainage systems.

Engineered or natural porous media that are used for the collection and drainage of wastewaters, or turbid runoff waters, can be susceptible to clogging as a result of time dependent accumulations of solid matter. Examples of engineered collection and drainage systems include sand and gravel trenches, composite geonet structures, and slotted pipe. Collection and drainage through natural soils, can occur, for example, below basins used for land treatment and recharge of wastewaters.

For the geo-environmental engineer, the clogging of critical structures such as landfill leachate leak-detection and collection systems is undesirable. Recent research (Fannin et al, 1993; Noyon, 1994) has found that clogging of composite geonet drainage systems during in-plane leachate flow can lead to rapid and significant losses in permeability. Given the increasing trend towards the management of land discharged wastes, and the minimization of contamination of subsurface regions, findings such as these raise serious questions about the long term performance of leachate collection and drainage systems. However, findings such as these are relatively few in number, and there is currently no unified criteria or approach available for the geo-environmental engineer concerned with the drainage of wastewaters through engineered or natural porous media, and the long term performance of drainage systems. Bass et al (1983) performed an early investigation of the potential for clogging of landfill leachate collection systems, and concluded on the basis of qualitative failure mode analyses that the potential for clogging was greater within hazardous waste landfills. These authors considered also that preventative methods such as redundancy in design, monitoring and periodic maintenance were preferable to remedial design, given the difficulties inherent in repairing buried leachate collection systems that are clogged.

A review of related literature shows that clogging processes as experienced during wastewater flow through porous media have been the subject of research over the past several decades. A representative list of references of this past research has been presented recently by Vandevivere & Baveye (1992). However, much of this early work is limited to investigation of the clogging process under one-dimensional gravity flow conditions through sands. Furthermore, despite the seemingly extensive nature of the available body of knowledge there is still no direct way by which the potential for, and the long term effects of, clogging may be quantified on the basis of the wastewater and porous media characteristics alone. This is perhaps not too surprising, given the highly complex and potentially inter-related nature of the physical, microbial and chemical processes that can result in clogging. The wide range in wastewater characteristics and also the types of collection and drainage systems also adds to the complexity of the problem. As a result of this, clogging potential evaluation is currently performed using empirical guidelines. For example, the use of a water quality criteria has been proposed (Nakayama & Bucks, 1991) for evaluation of the clogging hazard associated with drip irrigation systems using wastewater. This criteria is shown on Table 1.

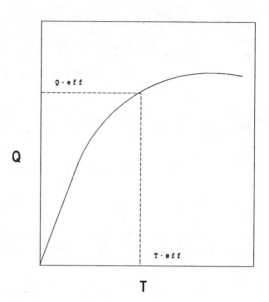

Figure 1: Conceptual trend of Cumulative Flow versus Time for gradual clogging

Clogging Factors	Hazard Rating		
	Minor	Moderate	Severe
Physical (mg/l)			
Suspended solids	< 50	50 - 100	> 100
Chemical (mg/l)			
pH	< 7.0	7.0 - 8.0	> 8.0
Dissolved solids	< 500	500 - 2,000	> 2,000
Manganese	< 0.1	0.1 - 1.5	> 1.5
Total Iron	< 0.2	0.2 - 1.5	> 1.5
Hydrogen Sulfide	< 0.2	0.2 - 2.0	> 2.0
Biological (no/ml)			
Bacterial number	< 10,000	10,000 - 50,000	> 50,000

Table 1: Wastewater quality criteria for evaluation of dripper irrigation clogging hazard (Nakayama & Bucks, 1991)

Figure 1 shows conceptually the cumulative flow volume, Q, versus time, T, trend that is affected on gradual clogging. Ideally, where clogging is expected, the prediction of the Q-T trend for any given set of conditions can provide the designer with values for the effective life time (T_{eff}) and effective total drainage volume (Q_{eff}) of a proposed drain system. As shown on Figure 1, T_{eff} and Q_{eff} would be selected on the basis of a minimum desired drainage rate.

It has been assumed, for the purposes of this review, that the porous media is physically inert or, more specifically, is non-deformable and does not contain structurally significant amounts of clay (the subject of wastewater interaction with clays, and the resultant impacts on hydraulic conductivity, has been extensively reviewed in the geo-environmental arena and is beyond the scope of this review).

PHYSICAL PROCESSES:

Filtration and accumulation of suspended matter during wastewater flow can lead to total clogging of drainage media. Suspended matter present in wastewaters may accumulate within drainage porous media by either simply settling out (sedimentation) during flow; by being strained at small pore openings; or by adsorption onto the drainage surfaces. The location of accumulation will depend upon the suspended particle size distribution as well as the pore size distribution within the porous drainage media. Other factors that will control the rate and extent of clogging due to suspended matter include: suspended solids composition (organic or inorganic?); drainage flow volumes, flow rate and direction, and periodicity. For colloid and clay suspensions the wastewater chemistry and the surface activity of the porous media will also play a role in the extent of accumulation of these particles within the media. For example, the formation of flocs under favourable physico-chemical conditions will result in a more rapid filtration of suspended matter.

A relatively comprehensive literature review on particle transport through porous media was presented by McDowell-Boyer et al (1986), though in conclusion they noted a paucity of predictive models relating particle accumulation and permeability reduction (or clogging). A considerable amount of research is currently being performed on the mechanisms of particle capture by porous media, though this work is primarily focused on the movement and fate of colloidal sized particles through granular media (Elimelech, 1992; Tipping et al, 1993). Though the mechanisms of gravitational settling and straining may appear obvious during the consideration of physical clogging, the theoretical quantification of the rate and extent of clogging under a given set of conditions is still not possible.

Available empirical knowledge can be used to estimate conditions under which suspended particles will not penetrate through granular drainage media. However, it is important to keep in mind that the effect of suspended particle accumulations on top of, or within, drain media can lead to a change in the pore size distribution in addition to a decrease in porosity, leading to a variation in pore velocities. Added to this is the difference in behaviour between organic and inorganic (eg. silica) suspended particles. Organic matter is susceptible to deformation and compaction. This characteristic of organic matter results in accumulations that can be highly impermeable in comparison with accumulations of inorganic and inert particles. This relationship between suspended matter accumulation, porosity and flow velocities can result in a spatially and temporally varying condition within the drain.

Soil and rock filter/drain evaluations, as traditionally performed by geotechnical engineers, are aimed at minimizing the migration of finer particles through the coarse filter (or drainage) media. The findings of Sherard et al (1984a,b), obtained from an extensive series of laboratory testing on sand and gravel filter materials, suggest that clogging of granular filters may occur under these conditions for suspensions of a) silts and sands ($d_{85} = 0.1$ to 2mm) if $d_{85} > 0.10 D_{15}$; b) sandy silts or clays ($d_{85} = 0.1$ to 0.5mm) if $d_{85} > 0.2 D_{15}$; c) fine grained clays ($d_{85} = 0.03$ to 0.1mm) if $D_{15} < 0.5$mm; d) fine grained silts of low cohesion ($d_{85} = 0.03$ to 0.1mm) if $D_{15} < 0.3$mm; and e) "exceptionally fine soils" ($d_{85} < 0.02$mm) if $D_{15} < 0.2$mm. In these criteria d_{85} is the sieve size through which 85% by weight of the suspended particles will pass, and D_{15} is the sieve size through which 15% by weight of the drainage material will pass. Clearly, in these cases the extent and rate of clogging will also be dependent upon the concentration of suspended solids.

Early research performed by Berend (1967) and Behnke (1969) on the flow of turbid runoff waters and other wastewaters through sand, as related to recharge spreading grounds, found that clogging under these conditions resulted primarily due to surface sealing. On the basis of experimental evidence, Behnke concluded that the clogging mechanism was composed of two processes: for suspended matter having a range of particle sizes "gravitational grading" (settling) initiates clogging at and below the infiltration surface, and as the deposited layer becomes progressively more graded the uppermost pores become small enough to "strain" out most of the remaining particles. Both Berend and Behnke provide empirical equations to predict the Q-T curves as applicable to clogging of spreading grounds during recharge using a variety of wastewaters, though their research was performed on fine to medium sand deposits. These findings show that complete clogging and permeability reduction of recharge basins in fine to medium sands can occur within less than a week with suspended solid concentrations as low as 50ppm (by weight). Similar work performed by Okubo & Matsumoto (1983) lead to the conclusion that long term flow through medium sand could be maintained as long as the suspended solids concentration in the wastewater is less than 2ppm (by weight).

There is little information available on the accumulation of suspended matter within non-granular systems such as geonets or pipe. Investigations by Adin & Sacks (1991) of dripper performance, during irrigation using wastewater, showed that sediment accumulation occurred within the bottom of the emitter tubing as a result of quiescent settling conditions developing behind corrugations and other obstructions to flow. Such settling would also be expected during in-plane flow through geonet collection systems, and during low velocity horizontal flow through granular media.

MICROBIOLOGICAL PROCESSES:

Drainage media exposed to wastewaters can offer ideal conditions for microbial growth and activity. The surfaces of the drainage media offer attachment opportunities for microbes present in the wastewater and also for microbes which are ubiquitous in the environment. Once established in this manner, the microbial population is free to utilize the wide variety of degradable waste constituents in order to satisfy it's energy and nutrient requirements.

Available knowledge on microbial clogging of porous media includes experiences in the use of wastewater recharge basins and injection wells (eg. Huisman & Olsthoorn, 1983); slow sand filtration (Logsdon, 1991); groundwater pumping wells (Cullimore, 1986); subsurface drains beneath agricultural lands (Ford, 1979, 1982); and anaerobic filters for wastewater treatment (Ehlinger et al, 1987).

Where observed in the case of wastewater recharge basins, injection wells, and slow sand filters, clogging has been typically associated with the formation of a thin bacterial mat at the infiltration surface. Mat formation may be considered to result from the initial attachment of bacteria followed by subsequent growth and activity. This growth and activity is greatest at the infiltrating surface where the inflowing concentrations of nutrients and other substrates, including dissolved oxygen, are greatest. The depletion of these growth factors across the infiltration surface subsequently reduces the clogging extent with depth. The formation of a continuous biomat at the infiltration surface is clearly more likely where the drainage media contains small pores that can be rapidly blocked and sealed. The rate and extent of clogging by such mat formation can be readily demonstrated by continuously infiltrating tap water through a column of fine sand. Even under these relatively "clean" conditions, it is found that formation of a thin (approx 1mm) biomat and measurable head losses across the infiltration surface are affected within one to two days. It should be noted that slow sand filtration techniques rely on the formation of the biomat (or "Schmutzdecke"), as it enhances the capture of suspended matter, and removal of dissolved substances from the water being treated. Periodic drying and scraping of the infiltration surface form part of slow sand filter operation.

In the case of pumping wells and subsurface agricultural drains, the microbial clogging process arises due to the development of an anaerobic-aerobic interface. In pumping wells this interface occurs at the well screen where the (presumably) anaerobic groundwater mixes with the aerobic well water. In agricultural drains this transition arises as a result of the anaerobic seepage water infiltrating the aerobic (due to air flow) environment within the drain. Clogging due to microbial growth and the release of extracellular polysaccharide ("slime") in these environments is enhanced by the aggregation of the slime with oxides of iron and manganese that also form at the interface. Reduced iron (Fe^{2+}) and manganese (Mn^{2+}) may be naturally present in groundwater, or formed by microbially mediated reduction on seepage through anaerobic zones. In the case of iron-precipitation, the slime-precipitate accumulation is commonly known as "ochre", due to its distinctive red-orange colour.

Ford (1982) found, from a relatively extensive survey of agricultural drain performance, that no ochre was detected in drains where Fe^{2+} levels in the surrounding groundwater were less than 0.5ppm. Severe ochre problems were encountered at sites where the Fe^{2+} levels were above 2.5ppm. In experimental studies, Ford (1979) found that significant ochre clogging occurred within 7 days in 5cm diameter corrugated polyethylene tubes receiving continuous injections of 2ppm Fe^{2+} solution. Of particular interest is the finding from these studies that chemically precipitated iron oxide deposits, in the absence of microbial activity, were more porous than the microbially enhanced and slime encapsulated oxide deposits (ochre). These findings related to the critical clogging Fe content are comparable to the clogging hazard criteria presented for total iron in Table 1.

An important aspect of the aerobic-anaerobic interface, as a location of greatest microbial activity, is that it provides an environment where anaerobes and aerobes can both flourish. The likelihood of detrimental clogging would presumably be increased in a collection and drainage system receiving oxygenated wastewaters (or where the system is partly open to the atmosphere) and which also provides conditions for the development of anaerobic zones.

Microbial growth kinetics provide a quantitative means by which the potential extent of microbial activity may be predicted. Principles associated with the biological treatment of wastewater can be applied for the prediction of microbial activity during wastewater flow through porous media, though the correlation between microbial numbers and potential clogging may not be as straightforward as implied in Table 1. In addition to slime formation and it's co-deposition with chemical precipitates, clogging as a result of microbial activity may also occur simply by the accumulation of microbial cells at pore throats. The controlling factor associated with microbial growth may be generally considered to be the availability of degradable carbon, and Okubo & Matsumoto (1983), from their studies on wastewater flow through medium sand, concluded that clogging will occur at soluble organic carbon (SOC) levels greater than about 10mg/l.

CHEMICAL PROCESSES:

The formation and accumulation of precipitates, as a result of disequilibrium conditions being present within a collection and drainage system, can result in a gradual reduction in the drainage pore volume. The precipitates most commonly associated with clogging are calcium carbonate, and hydroxides of iron and manganese. As already noted, chemical precipitation can be enhanced by microbial activity, resulting in a greater extent of clogging. For example, the reduction of sulfate to sulfide and the subsequent precipitation of iron sulfide is a microbially mediated process.

There is some empirical evidence relating chemical precipitation as a sole contributor to clogging of wastewater collection systems. As noted in the preceding sections, microbial enhancement of precipitate formation and the combined role of biomass in porosity reduction can enhance the role of chemical precipitation in clogging. However, Bass et al (1983), report of an exhumed gravel drain that received landfill leachate over a period of nine years, and which displayed significant cementation by calcium, iron and manganese precipitates. Similarly, Nakayama & Bucks (1985) reported that clogging of trickle irrigation emitters by calcium carbonate precipitates is encountered where the irrigation water has a combination of high pH, high calcium and carbonate concentrations, and excessively fluctuating temperatures.

Booram et al (1975) reported that significant quantities of magnesium ammonium phosphate ($MgNH_4PO_4.6H_2O$) were deposited in metal pump and pipe components used to recycle liquid from an anaerobic lagoon during the treatment of livestock wastes. In this case, the average concentrations of the various constituents included 67mg/l calcium; 47mg/l magnesium; 407mg/l ammonia-N; and 75mg/l phosphorous. These authors noted that precipitate deposition was greater on metal surfaces than on plastic surfaces.

Precipitate formation is dependent on the pH and redox conditions as well as the relative quantity and ratio of the various dissolved species in the wastewater. High temperatures can lead to precipitation by simple evaporation of water, in addition to increasing reaction rates. As suggested by the criteria in Table 1, the likelihood of precipitation increases with the amount of dissolved ionic species in the wastewater. The likelihood of precipitate formation can be evaluated given knowledge of the various species and quantities of dissolved constituents present, and using equilibrium calculations (Benefield et al, 1982). Bass (1983) notes the use of an "Incrustation Potential Ratio" (IPR) for the evaluation of calcium carbonate precipitation:

$$IPR = \frac{(Total\ Alkalinity)(Hardness)}{10.3 \times 10^{(11-pH)}}$$

where total alkalinity and hardness are given in ppm $CaCO_3$. If IPR is less than 1.0 then theoretically no calcium carbonate should precipitate.

Though equilibrium calculations can be used to predict the likelihood of precipitation, they do not provide any indication on the rate of precipitation. Precipitation rates are typically very slow, and can also be effected by environmental and physical conditions, such as flow rates, and the presence of any suspended matter which can provide nucleation opportunities during crystallization. Even if chemical precipitation occurs, the slow rate of deposition, combined with the relatively small volume occupied by the precipitate, may make this process generally less significant towards short term clogging - in comparison with physical and microbial processes.

Computational tools for multiple species chemical equilibrium evaluations exist, however the prediction of precipitation during complex wastewater flow through porous media is an aspect that is still currently under research (Liu & Narasimhan, 1989).

DISCUSSION:

Consideration of the fundamental processes that can result in the clogging of porous media provides some insight into the potential interactions between wastewaters and drainage media, and the likelihood of clogging. It has been noted that microbial activity can increase the clogging process by enhancing the capture and accumulation of suspended solids as well as chemical precipitates. This inter-relationship between the various clogging mechanisms can essentially be considered to result in an equivalent single clogging mechanism, and which perhaps is currently best predicted by the use of a total water quality criteria such as that shown in Table 1.

The nature of the wastewater and in particular the geometry and physical makeup of the collection and drainage system will both control the likelihood of any clogging tendency during the lifetime of the drainage system. The currently increasing reliance on engineered materials for the construction of drainage systems offers an opportunity for improved control on the long term performance of the systems. However, there is a very limited knowledge base available at the present time on the long-term interaction between wastewaters and engineered drainage systems. Significant benefits can be realised by detailed and long-term monitoring of systems currently in operation, and by the provision of monitoring during the design of such systems. Given the apparent complexity associated with the interaction of wastewaters and drainage media, it is likely that geo-environmental engineers concerned about clogging will continue to rely on empirical findings for some time to come.

REFERENCES:

Adin, A. & Sacks, M. (1991), "Dripper-clogging factors in wastewater irrigation", ASCE Journal of Irrigation and Drainage Engineering, 117(6), pp 813-826.

Bass, J.M., Ehrenfeld, J.R. & Valentine, J.N. (1983), "Potential clogging of landfill drainage systems", USEPA Report No. EPA-600/2-83-109.

Bass, J.M. (1984), "Avoiding failure of leachate collection systems at hazardous waste landfills", Report for Sept 83-Sept 84, USEPA Report No. EPA/600/D-84/210.

Behnke, J.J. (1969), "Clogging in surface spreading operations for artificial groundwater recharge", Water Resources Research, 5(4), pp 870-876.

Benefield, L.D., Judkins, J.F. & Weand, B.L. (1982), "Process chemistry for water & wastewater treatment", Prentice-Hall Inc.

Berend, J.E. (1967), "An analytical approach to the clogging effect of suspended matter", Bulletin Int. Assoc. of Scientific Hydrology, 12(2), pp 42-55.

Booram, C.V., Smith, R.J., & Hazen, T.E. (1975), "Crystalline phosphate precipitation from anaerobic animal waste treatment lagoon liquors", Transactions of the American Society of Agricultural Engineers, 18, pp 340-343.

Cullimore, D.R. ed (1986) "Proceedings of the 1986 International Symposium on Biofouled Aquifers: Prevention & Restoration", American Water Resources Association.

Ehlinger, P., Audic, J.M., Verrier, D. & Faup, G.M. (1987), "The influence of carbon source on microbial clogging in an anaerobic filter", Water, Science & Technology, 19, pp 261-273.

Elimelech, M. (1992), "Predicting collision efficiencies of colloidal particles in porous media", Water Research, 26(1), pp 1-8.

Fannin, R.J. et al (1993), "Laboratory measurement of the in-plane flow capacity of geonets", Proc. 1993 Joint CSCE-ASCE National Conference on Environmental Engineering, Montreal, Canada.

Ford, H.W. (1979), "Characteristics of slime and ochre in drainage and irrigation systems", Trans. American Society of Agricultural Engineers, 22(5), pp 1093-1096.

Ford, H.W. (1982), "Estimating the potential for ochre clogging before installing drains", Trans. ASAE, 25(6), pp 1597-1602.

Huisman, L. & Olsthoorn, T.N. (1983), "Artificial groundwater recharge", Pitman.

Liu, C.W. & Narasimhan, T.N. (1989), "Redox controlled multiple species reactive chemical transport: 1. model development", Water Resources Research, 25(5), pp 869-882.

McDowell-Boyer, L.M., Hunt, J.R. & Sitar, N. (1986), "Particle transport in porous media", Water Resources Research, 22(13), pp 1900-1921.

Nakayama, F.S. & Bucks, D.A. (1985), "Temperature effect on calcium carbonate precipitation clogging of trickle emitters", in Proc. 3rd Int. Drip/Trickle Irrigation Congress, ASAE, pp 45-49.

Nakayama, F.S. & Bucks, D.A. (1991), "Water quality in drip/trickle irrigation: A review", Irrigation Science, Vol 12, pp 187-192.

Noyon, M. (1994), "The effects of biological and chemical clogging on landfill leachate detection and collection systems", unpublished MASc thesis, Department of Civil Engineering, University of British Columbia, Canada.

Okubo, T. & Matsumoto, J. (1983), "Biological clogging of sand and changes in organic constituents during artificial recharge", Water Research, 17(7), pp 813-821.

Rebhun, M. & Schwarz, J. (1968), "Clogging and contamination processes in recharge wells", Water Resources Research, 4(6), pp 1207-1217.

Sherard, J.L., Dunnigan, L.P. & Talbot, J.R. (1984a), "Basic properties of sand and gravel filters", ASCE Journal of Geotechnical Engineering, 110(6), pp 684-700.

Sherard, J.L., Dunnigan, L.P. & Talbot, J.R. (1984b), "Filters for silts and clays", ASCE Journal of Geotechnical Engineering, 110(6), pp 701-718.

Tipping, E., Thompson, D.W., Woof, C. & Longworth, G. (1993), "Transport of haematite and silica colloids through sand columns eluted with artificial groundwaters", Environmental Technology, 14, pp 367-372.

Vandevivere, P. & Baveye, P. (1992), "Effect of extracellular polymers on the saturated hydraulic conductivity of sand columns", Applied & Environmental Microbiology, 58(5), pp 1690-1698.

Soil Vapor Extraction Below a Building

Greg H. Deaver, Carol S. Mowder
Dames & Moore, Inc., Bethesda, Maryland, USA

Results from a RCRA Facility Investigation (RFI) indicated that soil at a chemical manufacturing facility contained elevated levels of volatile organic compounds (VOCs), semivolatile organic compounds (SVOCs), pesticides, polychlorinated biphenyls (PCBs), and metals. The impacted area was the site of a proposed research building which was to be constructed immediately. SVOCs, pesticides, PCBs, and metals exceeded State/Federal criteria for incidental ingestion of surface soil; metals exceeded State criteria for inhalation of surface soil as dust; and VOCs exceeded State criteria for groundwater protection.

Although excavation and offsite treatment or landfilling was an immediate solution to the problem, costs associated with this activity were estimated to be $6,750,000. It was determined that the building foundation, once constructed, would act as a cap and preclude exposure to contaminants via inhalation and ingestion. However, because the groundwater at the site is relatively shallow (less than 8 feet below grade), the presence of VOCs could still potentially impact the groundwater. To meet the tight construction schedule, a soil vapor extraction system (SVES) was designed and integrated into the building foundation plans. This design allowed the client to proceed with building construction and gave them the ability to remediate soil below the building at a later date. It is estimated that the total maximum cost for the SVES, including design, permitting, installation, and operation, will be less than $650,000.

Introduction

Soil vapor extraction is a method used for in-situ remediation of soil contaminated with VOCs and some SVOCs. It is typically used as an alternative to excavation and treatment/landfilling because of the significant cost savings that can be incurred. The SVES discussed in this paper not only resulted in significant cost savings, but allowed the client to proceed with building construction and gave them the ability to remediate soil below the building at a future date, if necessary.

Background

A new research and development (R&D) complex, consisting of several buildings, was planned to be constructed at a chemical manufacturing facility. The buildings were to be a key component in the company's extensive R&D program which will continue well into the next century.

As part of the RFI conducted at the manufacturing facility, samples were collected from the site of the proposed R&D complex and analyzed for VOCs, SVOCs, pesticides, PCBs, and metals. Analytical results were compared with five sets of relevant and appropriate action levels which are protective of human health and the environment. The five action levels included:

- State Criteria for Grossly Contaminated Soils,

- EPA Incidental Ingestion of Surface Soils,

- State Criteria for Incidental Ingestion of Surface Soils,

- State Criteria for Inhalation of Surface Soils as Dust, and

- State Criteria for Soils Concentrations for the Protection of Groundwater for Class II-A Aquifers.

In addition, the metals concentrations detected were compared with concentrations of metals normally found in soils of the eastern United States. The results of the comparisons were used to identify areas of potential concern. The primary exceedances of criteria were as follows:

- Ingestion (SVOCs; pesticides; PCBs; metals)

- Inhalation (metals)

- Protection of Groundwater (VOCs).

A Response Action Plan was prepared to demonstrate that--with some mitigation and remedial measures--construction of the research building would not interfere with any future RFI investigations or remedial actions that may become necessary as part of the RCRA Corrective Action activities at the facility. Because areas outside of the proposed building footprint will be accessible for remedial activities at a future date, if necessary, the Remedial Action Plan focused on soil contamination within the building footprint.

To not interfere with a tight construction schedule, a decision had to be made regarding what would be done with the contaminated soil in a timely manner. One option was offsite thermal treatment and/or disposal at a RCRA-permitted facility. Although this was a technically feasible option that would provide immediate results, it was cost prohibitive; costs associated with this activity were estimated to be $6,750,000.

As previously mentioned, criteria exceeded for the proposed research building site included (1) EPA and State Criteria for Ingestion of Surface Soils, (2) State Criteria for Inhalation of Surface Soils as Dust, and (3) State Criteria for Protection of Groundwater. Because the building, once constructed, would act as a cap and preclude exposure via ingestion and inhalation, the only criteria of concern for soil contamination below the building is the State Criteria for Protection of Groundwater. Although the building foundation would reduce infiltration through the contaminated soil, the water table at this site is relatively shallow (less than 8 feet deep) and contaminated soil could still act as a source of groundwater contamination during periods of water table fluctuation.

The soil at the site consisted of silty sand and some construction debris was present in the soil. The estimated permeability of the soil was 5×10^{-4} centimeters per second (cm/s).

A pilot study was conducted at the site to determine if soil vapor extraction was a viable option for soils remediation. The results of the 2-day pilot study indicated that a 15-foot radius of influence could be obtained (greater radii of influence were also observed, but the minimum distance was used for design purposes).

System Design

In the conceptual design phase of the project, a trench-like SVES was proposed to be installed below the future buildings and an existing slab which was to be built upon (Figure 1). However, to minimize the amount of excavated soil -- which would also minimize the costs associated with disposing of the excavated soil -- it was decided that vertically-oriented wells would be connected to horizontal manifolds (Figure 2). Construction designs for the proposed building were reviewed during the SVES design to ensure that the SVES would not conflict with underground utilities and the 250+ piles used to support the building. The proposed SVES layout was modified several times due to changes in the building foundation plans. Because the overall tone of the project was one of "construction" and not "soils remediation," close contact with the construction management company and the architectural firm on the project was vital to make sure that all changes in the building design were communicated to the SVES designers.

The final design involved 56 extraction wells below the proposed research building and the existing slab. The SVES below the existing slab is slightly different than that below the research building in that the manifold lines for the existing slab SVES lay between the existing slab and the new slab foundation (Figure 3). The SVES design did not require any changes to the architects' design for the building foundation.

Installation

The SVES was installed at the same time as the building foundation piles. It was recommended to the client that, because the company that designed the SVES would not be installing the system, the individual wells and manifold lines should be tested and approved by the SVES designers prior to continuing with building construction. The various segments of the system (i.e., individual wells and manifolds) were tested after each segment was installed; several wells required replacement because they were "off spec." Although this activity temporarily delayed completion of SVES installation, the design specifications needed to be adhered to in order for the system to operate, if required, properly. The reinstallation of the wells did not impact the overall building construction schedule. Construction is nearing completion.

Operation

It is currently unknown whether the client will be required by regulatory authorities to operate the SVES. If SVES operation is required, the associated equipment (e.g., blower, knock-out pot, activated carbon

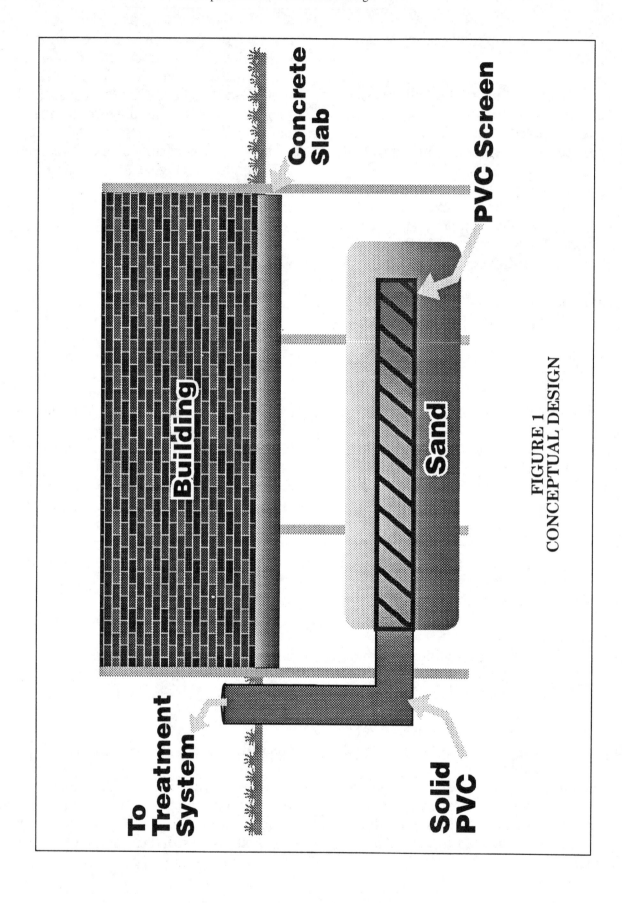

FIGURE 1
CONCEPTUAL DESIGN

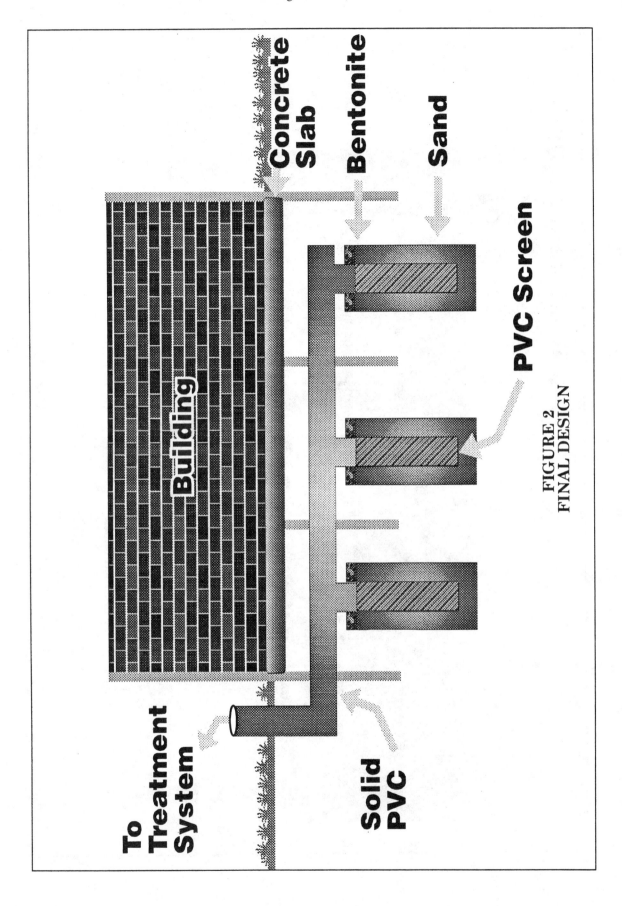

FIGURE 2
FINAL DESIGN

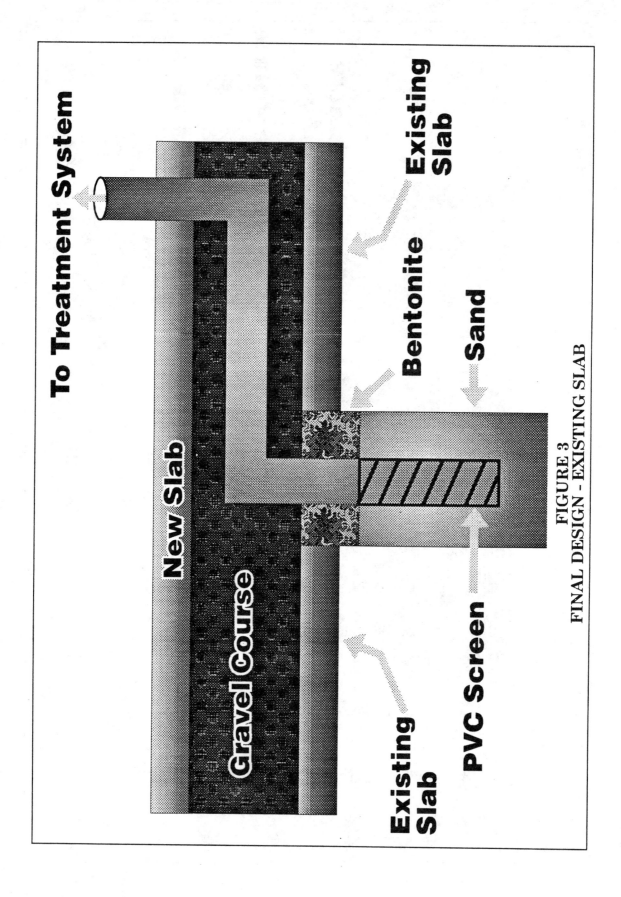

FIGURE 3
FINAL DESIGN - EXISTING SLAB

units, etc.) has been specified and can be purchased. This ancillary equipment, however, will not be purchased until system operation is mandated.

Cost

In a worst-case scenario, the SVES will require operation. In this instance, it is estimated that the total SVES cost, including design, permitting, installation, and operation, will be less than $650,000 -- which saved the client over $6,100,000 in remediation costs (as opposed to excavating and treating/disposing of the soil), while still meeting the tight construction schedule.

Conclusion

The SVES installed below the R&D buildings not only saved the client money (by not having to excavate and treat the contaminated soil that existed within the building foundation boundaries), but it also allowed them to keep their construction plans on schedule. If it is determined that the client does not have to operate the system, the client will have only invested approximately $200,000 in the project.

Application of Air Sparging to 1,1,1-Trichloroethane Contaminated Soil - A Success Story

Greg H. Deaver, Robert J. Tworkowski
Dames & Moore, Inc., Bethesda, Maryland, USA

Innovative remedial technologies have recently been developed in the hazardous waste remediation field which allow more cost-effective and rapid remediation. Dames & Moore has successfully applied one such technology - air sparging - with significant cost savings by remediating soils adjacent to and under a structure.

Specifically, as part of a State Superfund environmental investigation, Dames & Moore was retained by a Canadian-based client to perform a Remedial Investigation/Feasibility Study (RI/FS) of a release of 1,1,1-Trichloroethane (1,1,1-TCA) from an adhesives manufacturing facility.

An important component of the RI phase consisted of an immediate response program which utilized a portable soil vapor extraction (SVE) system. This system was designed to eradicate the residual "source" of 1,1,1-TCA adsorbed to soils in the unsaturated zone so that the continued degradation of the underlying groundwater quality would be minimized while the phased RI program was being conducted. These soils were located immediately adjacent to and under the process building. The SVE system was selected due to its applicability to remediating volatile contaminated soils and since the soils could be remediated without impacting the structures integrity.

Following a 6-month SVE program, a confirmatory soil boring program was conducted. Collected soil samples indicated that an order of magnitude decline was experienced in 1,1,1-TCA concentration in the impacted unsaturated zone.

However, the confirmatory soil boring program also indicated that elevated residual concentrations of 1,1,1-TCA were still adsorbed to soils in the upper portion of the saturated zone (including the capillary fringe).

In an attempt to flush this zone, an air sparging system was designed and installed. Air sparging, also referred to as in-situ air stripping, in-situ volatilization, and enhanced volatilization, is an innovative treatment technology which is utilized to remove volatile organic compounds (such as 1,1,1-TCA) from the subsurface saturated zone (groundwater).

Confirmatory soil samples collected immediately following the air sparging program indicated an order of magnitude decline in 1,1,1-TCA concentration in the impacted zone. Subsequent groundwater sampling indicated that the 1,1,1-TCA plume was no longer emanating from the source area and was detached. The impacted soils were no longer contributing as a source of 1,1,1-TCA, with the potential for degrading the underlying groundwater quality.

Due to this immediate remedial response initiative (SVE and air sparging program) exercised in the initial phase of this site investigation, the continual source impact was minimized. As a result of this successful source removal and the resulting reduction of 1,1,1-TCA concentrations within the groundwater plume, the State is currently considering a remedial alternative for this site which incorporates institutional controls along with long-term groundwater and surface water monitoring. This remedial alternative selection may be one of the first of its kind implemented as part of this State's Superfund program.

As a result of the immediate remedial response initiative, a less intrusive remedial alternative may be more appropriate and accepted by the State. Overall, the cost associated with the institutional controls along with the long-term monitoring will be significantly less than alternative remedial approaches originally discussed for this site prior to conducting the successful immediate remedial response program.

Overview

In 1990, Dames & Moore was informed by a Canadian-based client that a release of the solvent 1,1,1-Trichloroethane (1,1,1-TCA) had occurred at one of its U.S. adhesives manufacturing facilities. Initial sampling of neighboring residential drinking water wells by the State of Indiana Department of the Environment (IDEM) detected concentrations of 1,1,1-TCA and its degradation byproduct 1,1, Dichloroethene (1,1-DCE) in several of the samples collected. Just to the east of the facility, one of the residential wells contained concentrations of 56,000 μg/l 1,1,1-TCA and 4,000 μg/l 1,1,-DCE.

The occupants of the affected residences were immediately advised to discontinue use of the well water and bottled water was supplied to them through IDEM Emergency Procurement. Later, the impacted residents were converted to the municipal water supply system.

Characteristics of 1,1,1-Trichloroethane

1,1,1-Trichloroethane, also known as methyl chloroform, is a colorless, non-flammable, light-weight chlorinated hydrocarbon with an odor similar to that of chloroform. Commonly used as a metal cleaning fluid, degreasing agent, and aerosol-propellant, 1,1,1-TCA is a narcotic, which depresses the central nervous system. Acute exposure causes dizziness, loss of coordination, drowsiness, increased reaction time, unconsciousness, and eventually death (Boyer, Ahlert, and Lossen, 1987). With a specific gravity of 1.325, 1,1,1-TCA is heavier than water and would, therefore, have a tendency to migrate vertically within an aquifer system. This property is characteristic of dense non-aqueous phase liquids (DNAPLs).

Movements of DNAPLs in the subsurface is governed by pore pressure produced by capillary forces and gravity. Sufficient capillary pressure must be maintained within the soil pores for DNAPLs to sink downward in the saturated zone. This pressure is directly related to the magnitude of release (downward mass of the wetting front) as well as the pore size. In general, the pressure required increases in direct relation to decreasing pore sizes.

Initial Site Assessment

The spill event was determined to occur while filling a batch mixing tank with the 1,1,1-TCA solvent during which the tank overflowed for a period of time. The 1,1,1-TCA flowed along the cement floor slab and exited the building at the floor/wall contact at the north end of the building where it infiltrated into the adjacent soil. As part of the initial site evaluation, Dames & Moore conducted a soil gas survey at the site in an attempt to locate potential 1,1,1-TCA source areas as well as delineate plumes of groundwater contamination. A sampling grid of 66 sampling points with a spacing of approximately 50 feet was laid out over the study area. Based upon the analytical data generated from the soil gas sampling program, Dames & Moore conducted a subsurface soil boring program in the source area suspected of being impacted by the 1,1,1-TCA release.

At the source area (the north end of the building), groundwater was encountered at 12 feet below ground surface (bgs) and the impacted soils were found to extend from 4 to 18 feet bgs. Concentrations of 1,1,1-TCA as high as 190 $\mu g/kg$ were detected in soils above the water table (unsaturated zone). Concentrations of 1,1,1-TCA in soil increased to 5,500 $\mu g/kg$ in the water table (saturated zone). In general, concentrations of 1,1,1-TCA increased with depth, reaching maximum concentrations near the top of the water table (i.e., 12 to 14 feet below ground surface), and then decreased with depth in the saturated zone. These findings indicated that 1,1,1-TCA impacted soils were limited in vertical extent to the unsaturated zone and the upper few feet of the saturated zone. Based upon preliminary soil boring data, the areal extent of impacted soil appeared to decrease with increasing distance from the location of the spill and was limited to an area approximately 20 feet by 20 feet adjacent to the north side of the building. The groundwater in this area was found to contain approximately 2,000 $\mu g/l$ 1,1,1-TCA, approximately 10 times the allowable drinking water standard of 200 $\mu g/l$.

Soil Vacuum Extraction

Due to the accessibility/structural constraints imposed by potentially removing the impacted soil from the area adjacent to and beneath the building foundation (as well as the concerns of associated cost and potential longterm liabilities associated with disposal), Dames & Moore mobilized a soil vacuum extraction system (SVES) to the site as an initial immediate remedial response treatment measure. The objective of the system was to remove the residual "source" 1,1,1-TCA adsorbed to the sandy soil in the unsaturated zone at the spill site so that its potential for migration to groundwater would be minimized. Soil vacuum extraction, also referred to as soil vapor extraction, vacuum extraction, soil venting aeration, in situ volatilization and enhanced volatilization is a technique commonly used for the remediation of soils impacted by VOCs such as 1,1,1-TCA. Over the past several years, SVE technology has received increasing acceptance because of its demonstrated effectiveness in removing volatile compounds from impacted soil, its relatively low cost, and its apparent simplistic design and operation. This technique induces a negative pressure gradient in subsurface unsaturated soils, thus creating air movement through the soil pore space. Entrapped/adsorbed VOCs will volatilize in the moving air stream and are removed from the subsurface through an individual vacuum extraction well or series of wells that are screened in the impacted soils. The effectiveness of soil vacuum extraction is dependent upon several site-specific conditions which include the soil permeability and contaminant volatility. Generally, the higher the soil permeability and contaminant volatility the greater the probability that the application of SVE will be successful.

As part of the field investigation, a monitoring well was installed in the area of the 1,1,1-TCA release. The objective of the placement of this monitoring well was to document the groundwater quality at the 'source' location. In anticipation of implementing an SVE program at this 'source' location, the construction of the monitoring well was modified by increasing the screen length so that it extended into the impacted unsaturated zone. This extension would allow an induced vacuum to be transmitted into this zone.

In addition to the placement of the vacuum extraction well, a passive air inlet well was installed at the periphery of the impacted area. The air inlet well was designed to control the flow path of the vapors being extracted, resulting in more efficient contaminant removal. The air inlet well was open to the atmosphere, allowing air to be drawn passively into the soil from the surface. The air inlet well also acted as a vapor barrier, where the radius of influence induced by the SVE system did not propagate beyond the air inlet well in that vicinity. This allows the SVE system to give intensive treatment of a specific smaller area rather than less intensive treatment of a larger area.

To minimize the infiltration of precipitation that would potentially enhance the migration of the adsorbed 1,1,1-TCA from the soil to the groundwater, an impermeable tarp was placed over the impacted area. Roof drains from the facility were also redirected to minimize the infiltration of precipitation.

The system was composed of a 150-standard cubic foot per minute (scfm) vacuum pump that was connected to a knock-out drum designed to separate entrained sediment and groundwater from the vapor stream. Although it was not the intent of the system to generate water, any resulting entrained water was treated by passing it through two granular-activated carbon canisters in series. The 1,1,1-TCA air discharge rate was calculated based upon a concentration of 100 ppm, which is equal to twice the highest soil gas survey concentration obtained at the site. Based upon a flow rate of 150 scfm and a 24-hour-per-day operation time, it was calculated that a maximum 7.9 lbs. of 1,1,1-TCA would be removed from the subsurface each day. This 'worst-case' concentration was below the maximum daily discharge rate allowed by the State Office of Air Management. Thus, no vapor treatment devices were required and the vapor was discharged directly to the atmosphere.

When the system was brought on line, it induced a vacuum of 7 inches of mercury to the extraction well. Initially, the vacuum induced in the subsurface resulted in groundwater mounding in the vicinity of the vacuum extraction well. As a result, groundwater was being entrained in the vapor stream and collected in the knock-out drum. Because the intent of the SVE system was to treat the vapors from the unsaturated zone and not to treat groundwater, a packer was designed and set in the extraction well (at a depth of approximately 10 feet below ground surface) to minimize groundwater withdrawal. The packer design directed the vacuum horizontally into the unsaturated zone, minimizing the vertical influence on the groundwater table.

The SVES was operated at this site for approximately a 6-month period (July through December 1991). A confirmatory soil program (conducted in December 1991) was performed to evaluate the effectiveness of the SVES. Of the six soil borings advanced in the spill area (unsaturated zone), four did not encounter detectable concentrations of 1,1,1-TCA. The residual concentrations detected in the unsaturated soils in the remaining two borings were 10 μg/kg and 6 μg/kg. Over an order of magnitude (190 μg/kg → 10 μg/kg) decline was experienced as a result of the SVES system operation.

However, soil samples collected from the upper portion of the saturated zone (including the capillary fringe) still indicated residual 1,1,1-TCA concentrations as high as 310 μg/kg. Apparently, the adsorptive properties of the residual 1,1,1-TCA with soil retarded it from being flushed into the underlying groundwater system and the adsorbed 1,1,1-TCA was acting as a continual emitting source degrading the groundwater quality.

At sites with similar situations, it has been found that the groundwater table can be depressed by pumping to expose the impacted soil so that a vacuum extraction program can effectively remediate the formerly saturated soils.

Unfortunately, the data generated from a pump test performed at the site indicated that a sustained pumping rate of approximately 50 to 100 gpm was only capable of a 2-foot drawdown in the water table. Thus, the hydrogeologic conditions at the site did not permit the water table depression scenario without the induction of a large-scale program. As an alternative, an innovative remedial approach referred to air sparging was evaluated and applied to the site.

Air Sparging

In an attempt to flush the zone containing the residual 1,1,1-TCA, an air sparging system was designed and installed. Air sparging, also referred to as in-situ air stripping and in-situ volatilization, is an alternative treatment technology which is utilized to remove volatile organic compounds (such as 1,1,1-TCA) from the subsurface saturated zone (groundwater). The technology introduces contaminant-free air into an impacted aquifer system, inducing contaminants to transfer from subsurface soil and groundwater into sparged air bubbles. The air bubbles are then transported into soil pore spaces in the unsaturated zone where they can be captured by a SVES. Air sparging systems must operate in tandem with SVES to capture the volatile compounds stripped from the saturated zone. Using air sparging without accompanying SVE can create a net positive subsurface pressure extending contaminant migration to as-yet-unaffected areas. The effectiveness of a combined SVE/air sparging system results from two mechanisms: contaminant mass transport and biodegradation. In both remedial mechanisms, oxygen transport in the saturated and unsaturated zone plays a key role.

The air sparging program consisted of installing well points into the uppermost portion of the saturated zone where the impacted soils were detected. An air compressor capable of delivering air at a rate of approximately 10 to 50 cfm was hooked up to the well points and air was injected into the targeted zone. The bubbling effect created by the forced air enhanced volatilization of the adsorbed 1,1,1-TCA, which was mobilized into the unsaturated zone. The vacuum induced by the SVE system then removed the volatiles from the unsaturated zone. A minimum 5-foot radius of influence was assumed with each well point. Based upon that assumption and a review of the analytical data generated by the previous confirmatory soil sampling programs, five well point locations were identified. The well points consisted of 5 feet of screen, which was situated from 20 to 25 feet below ground surface. Compressed air was applied to each well point location for a period of 4 to 6 hours.

Confirmatory soil boring samples collected immediately following the program indicated an order of magnitude decline in 1,1,1-TCA concentrations in the impacted zone. Of the four soil samples submitted for analysis, three contained 1,1,1-TCA concentrations below 10 $\mu g/kg$ (non-detect, 2 $\mu g/kg$, and 9 $\mu g/kg$). The fourth soil sample contained a concentration of 76 $\mu g/kg$.

Subsequent groundwater sampling indicated that the 1,1,1-TCA plume was no longer emanating from the "source" area and was detached, migrating in an easterly direction. The impacted soils were no longer acting as a source of 1,1,1-TCA, continually degrading the underlying groundwater quality.

Since the source of the 1,1,1-TCA has been remediated, the natural processes such as dispersion, sorption, dilution, and biodegradation have reduced the 1,1,1-TCA concentration in the groundwater plume from a maximum 2,000 $\mu g/l$ to approximately 200 $\mu g/l$.

Regulatory Concern

A primary concern in a hazardous waste site investigation is source control and source removal. Due to the immediate remedial response (vacuum extraction and air sparging programs) exercised in the initial phases of this site investigation, the source was remediated in-place and the impact due to continuing releases to groundwater minimized. As a result of the successful source removal and the resulting reduction of 1,1,1-TCA concentrations within the groundwater plume, the State of Indiana is currently reviewing a remedial alternative for the site which incorporates institutional controls along with long-term groundwater and surface water monitoring. This remedial alternative may be one of the first of its kind implemented at a State of Indiana Superfund site. This remedial alternative appears to be an appropriate remedial response measure. For in this case, the 1,1,1-TCA concentrations in the groundwater will be naturally reduced and ultimately discharged to a surface water body where residual concentrations of 1,1,1-TCA will naturally volatilize to the atmosphere.

Cost Considerations

As a result of the immediate remedial response activities implemented at the site, a less intrusive long term remedial alternative appears to be more appropriate and accepted by the State of Indiana. Overall, the initial costs associated with in-situ soil treatment and immediate response in conjunction with the long term institutional controls along with long term monitoring are significantly less than the excavation and offsite soil disposal; and construction, operation and maintenance of a groundwater pump and treatment unit which could be a viable alternative if the continual emitting source remained.

In addition, soils adjacent to and beneath the building foundation were remediated in place without having to jeopardize the integrity of the foundation or by affecting the manufacturing process conducted within the building.

Conclusion

As shown in this case history, immediate response activities at spill sites which control or remove sources can minimize the long-term impact to groundwater and allow the incorporation of less costly alternatives to traditional groundwater pump and treat systems. Additionally, the correct application of in-situ innovative technologies can prove to be less costly and less intrusive alternatives to soil excavation and offsite disposal.

Slurry Injection Disposal of Granular Solid Wastes

Maurice B. Dusseault, PhD., PEng., Roman A. Bilak, MSc., Leo Rothenburg, PhD., PEng.
Terralog Technologies Inc., Calgary, Alberta, Canada

ABSTRACT

Slurry fracturing was successfully used in Saskatchewan for large volume disposal of an inert, low-toxicity fine-grained oily quartzose sand. This trial shows that novel waste disposal methods can be extended to civil wastes with reasonable cost and low risk; all technical factors are clearly favourable. Constraints on geological parameters of suitable sites are presented and discussed, a new sensitivity analysis of surface uplift is presented, the case history is described, and the critical issue of hydrogeological factors is addressed. We conclude that there are no technical barriers to the implementation of this approach.

1 Introduction

Injection of a water-solids slurry under hydraulic fracture conditions into permeable, porous strata at depth using oil-field technology is proposed for large volume, low-toxicity, inert, granular terminal wastes disposal. Large volumes means that each injection well will accept 3×10^4-$3 \times 10^5 m^3$ of solid waste (volume *in situ*). Low toxicity is a quantitative issue for regulatory agencies to define. Possible examples include flue gas desulphurization sludges, clinker, ash and fly ash, foundry sands, non-reusable plastics and composites, and HC-contaminated soils. Inert means no decomposition or gas generation, and minimal chemical reactivity with strata or other wastes; inert does not necessarily mean insoluble. Granular means that the solid waste can be prepared as a particulate medium to be slurried in a liquid stream for injection. Terminal wastes are those remaining after reducing, reusing, recycling, and rehabilitating have been economically implemented.

2 Precedents and Fracturing

Cement or clay slurry grouting at pressures high enough to part or lift the rocks ($p_{inj} > \sigma_3$), has been used to seal dam foundations and improve rock properties for many decades (Franklin and Dusseault, 1991). For at least five decades, clear waste water disposal in deep porous permeable strata has been practiced in all industrial countries; it is a safe means of noxious aqueous stream disposal (Flak and Brown, 1988). In the 1950's and 60's, nuclear waste disposal by grouting cement slurries into shallow impermeable rocks was tried and abandoned in the United States. More recently, due to environmental concerns in the North Sea, drilling muds and wastes have been injected into shales at fracture pressures (Willson et al., 1993).

Since 1948, the oil industry has been using hydraulic fracturing to enhance productivity by increasing well drainage radius, or to introduce chemicals or thermal energy. Gases, liquids, or slurries are injected at high bottomhole pressures (p_{inj}) to part the strata normal to the least compressive stress (σ_3). If σ_3 is horizontal, fractures are initially vertical; if σ_3 is vertical, fractures are quasi-horizontal (Figure 1). Slurries are mixed at surface and injected down 60-95 mm diameter tubing to depths of 100-7,000 m. During placement, solids remain suspended during injecting through many 10-20 mm casing perforation holes. Viscosifying agents maintain the suspension and reduce fluid loss, and fluid volumes as high as 10^4-$10^5 m^3$ with up to 10% by volume of sand are injected over periods of hours or days. Recently, through use of better carrier fluids, this technology has progressed to where 30% by volume slurries of 250-600 μm sand or larger bauxite beads can be forced into strata at depths exceeding 4000 m at rates in excess of 10 m³/min. Hydraulic fracture technology is mature, safe, relatively well-understood (Gidley et al., 1989), and ideally suited for the civil waste disposal industry (Arnould et al., 1993).

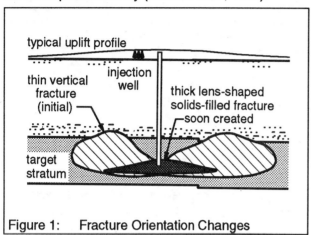

Figure 1: Fracture Orientation Changes

For waste injection, however, the goals are somewhat different; wastes should reside close to the wellbore, and viscosifying or water-loss control agents are not desirable. Also, the wellbore completion can be different than conventional oil field approaches: two possible designs are sketched in Figure 2.

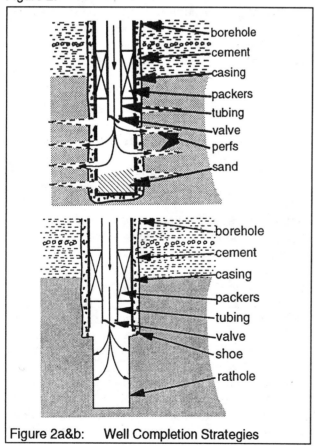

Figure 2a&b: Well Completion Strategies

3 Choice of Geological Strata

Impermeable strata have disadvantages as disposal targets; the water must travel far to dissipate pressures, causing fracture plugging and stratum pressurization. Placement control is lost, and pressured zones can lead to induced seismicity and well impairment (reference). An ideal site for slurried waste fracture injection will have the following major geological conditions (Figure 3):

a. 100-1500 m depth, depending on toxicity.
b. Flat-lying laterally continuous sediments.
c. Dominantly sand-shale lithostratigraphy, with thick ductile shales overlying target beds.
d. Thick (>7.5 m), porous (>25%), permeable (>1-2 Darcy) but weak target stratum.
e. Hydrogeologically isolated from potable waters (in a region of deep horizontal flow).
f. No large faults, oil reservoirs, soluble salts (unless the fluid is brine), and no other potentially valuable resources.

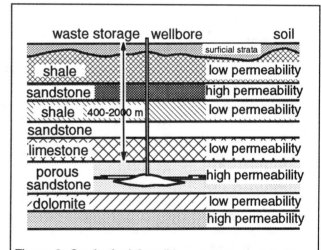

Figure 3: Geological Conditions for Waste Injection

These conditions can be easily modified if wastes and carrier water are totally non-toxic. Depth depends on hydrogeological conditions and displacement limits at surface; in general, shallower sites permit more modest injection pressures and rates. High porosity implies large carrier fluid storage capacity, but we point out that liquid volumes in our approach are 10 times less than for typical waste water injection wells. High permeability, thickness, and continuity mean that no permanent pressure build-up is possible; flat-lying strata assure horizontal groundwater flow; and ductile, thick shale cap rock implies that injection strains can be easily accommodated without damage, and the shales provide a secure barrier to upward fluid flow.

4 Confining Stresses and Pressures

If σ_3 is initially horizontal at injection depths, as in younger sedimentary basins with modest erosion, fractures are initially vertical, but large volume solids injection soon alters stress orientations, generating horizontal fractures. This occurs under rapid bleed-off as vertical fracture aperture increases to accommodate the solids being injected (Dusseault and Simmons, 1982), thus lateral stresses increase until σ_3 becomes σ_v. Orientation will permanently become near-horizontal. This is observed in all large-volume injection processes in Alberta and Saskatchewan oil sands and heavy oil reservoirs (Dusseault, 1993). It is also observed when grouting underneath dams; parting may initially be vertical, but continued grouting leads to pressures greater than σ_v, indicating horizontal fracturing. Thus, any solid waste injection project at the recommended depths will generate a dominantly horizontal injection plane geometry, and rapid fluid bleed-off rates will insure that the solids remain near the wellbore.

Horizontal fracturing means that solids remain in the target stratum; once fractures are horizontal, p_{inj} stabilizes at $1.15\text{-}1.25 \cdot \sigma_v$. Also, the storage capacity for a horizontal fracture is not limited by pressure or stratum thickness; rather, it is limited by surface effects and regulatory aspects. Accounting for injection slurry density and frictional losses, surface injection pressures are generally 40% lower than p_{inj}. For example, at -750 m near Edmonton, $p_{inj} \sim 20$ MPa, and $p_{surf} \sim 12$ MPa.

Will pressures in the target stratum be permanently increased? No, given the geological conditions outlined above. The permeable stratum bleeds off injection fluids rapidly, wastes are permeable and cannot trap fluid pressure for long, even if they contain fine-grained clay minerals. At -750 m, typical pore pressures are 6-7 MPa, and $\sigma_v \sim 17$ MPa; once injection ceases, effective stresses of ~10 MPa will consolidate clay-rich material rapidly. Also, slurries can be "designed" to contain proportions of fine- and coarse-grained materials so as to increase consolidation rate.

The carrier fluid (water) leaves the injection area by displacing pore waters. Water use is ~5-7 times the absolute volume of the solid wastes. For example, if 100 m³ of oil-contaminated dry sand in a pit at 40% porosity must be disposed of, the mineral volume is 60 m³; with a 15-20% by volume slurry, water requirement is 300-450 m³. Thus, for an facility licensed to accept 100,000 m³ of solid waste at an *in situ* porosity of 25%, about 500,000 m³ of water are required. With a 10 m stratum at porosity = 30%, a 250 m radius zone around the well will have pore fluids displaced by injection water. For carrier water composition, deep well injection standards are suggested, with relaxed requirements for water clarity.

Once emplacement is completed and the well abandoned, solids are subjected to the full σ_v and p_o which existed before injection ($\sigma_v \sim 17$ MPa and $p_o = 7$ MPa for the example above). Stress analysis suggests that $\sigma_v]_{max}$ in the wellbore region may approach $1.15\sigma_v$; $\sigma_h]_{max}$ may be 30-40% higher than this. Wastes are thus entombed permanently, immobilized, and incapable of being transported. Also, because initial p_o is recovered rapidly, there is no tendency for injected liquids to discover new flow paths; they become part of the fluids that were originally flowing through the stratum.

5 Uplift

Carrier fluids are dissipated in the stratum, but solids are left behind as a volume that did not exist before injection. What are the consequences of this? First, although there may be local fracturing and plastic effects in the region of the wellbore, the overburden behaves largely as an elastic beam, flexing upwards under the effect of the injected solids. Figure 4 shows surface uplifts calculated for a range of volumes and depths. For example, for $\Delta V = 10,000$ m³ at -500 m, maximum surface uplift of ~80 mm is expected. Because the precise injected body geometry is uncertain, an assumption had to be made; however, analyses show that surface shape is a weak function of injected body shape (Wang *et al.*, 1994). Though uplift magnitude is linearly linked to ΔV, the subsidence bowl shape is only mildly dependent on geometric details of the ΔV zones. Nevertheless, as demonstrated by surface displacement analysis (Dusseault *et al.*, 1993), these weak but consistent dependencies can be used to monitor injection processes.

Figure 5 shows maximum slopes for the cases in Figure 4. The maximum slope occurs at a distance of about $0.45 \cdot Z$, where Z is depth to the zone. Maximum slopes are < 0.1%, acceptable for most existing structures in areas where such technology will be implemented. Slopes can be reduced if waste injection is through a number of wells spaced at about 0.5-0.75·Z, with outer wells accepting less than inner wells. Once injection ceases, the ground is stable; there is no continued subsidence risk associated with waste injection after it ceases.

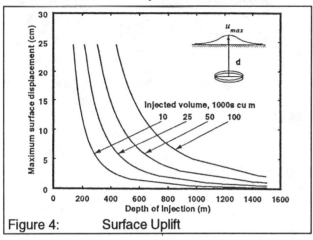

Figure 4:   Surface Uplift

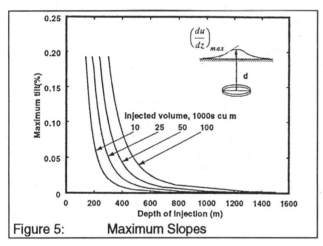

Figure 5:   Maximum Slopes

6 Saskatchewan Case History

Over 30 months, Mobil Oil Canada Ltd. of Calgary, Alberta disposed of 9500 m³ of oil-contaminated sand produced along with heavy oil during reservoir exploitation in cohesionless sandstones (Dusseault, 1993). The sand was disposed by slurrying and injecting into a porous, permeable, stratum at a depth of 655-690 m. The target stratum is a laterally extensive 35 m thick, 30% porosity, 3-7 Darcy permeability quartzose sandstone (very dense sand) in the Lloydminster region of Alberta and Saskatchewan (Dina Formation, Lower Cretaceous age). Sand was slurried into waste process water (produced along with the oil) and injected at pressures of from $1.15\text{-}1.30\cdot\sigma_v$, at -675 m. The sand injected is mainly 80-140 microns in grain size; the Dina Formation is coarser-grained, on the order of 200-1000 microns (i.e.: up to a millimetre diameter).

In situ, emplaced sand permeability is likely about 2-4 Darcy at 33-35% porosity.

Sand injection episodes were carried out regularly, usually over periods of a few hours, daily or several times a week. The well was operated successfully without blockage or increases in injection pressures, and fracturing conditions were reinitiated readily at the beginning of each injection episode. Back pressures during injection and decaying pressures in the reservoir after injection were monitored. It is clear that the sand did not go far from the wellbore, and the reservoir was not impaired with respect to its ability to accept sand, transmit fluids, and drain off excess pressures generated by the injection process. We believe that the injected sand went no further than 35-40 m from the well forming a lenticular body.

The sand stayed near the wellbore, but what about fluids injected with the solids, approximately 60,000-80,000 m³? Figure 6 shows site stratigraphy. Assuming 50% displacement of the stratum height, injected water went no further than 75-80 m from the well. Nevertheless, this is a small-scale operation, and communication with surface waters must be addressed within the recommended context: the potential use of slurry fracture injection as a large-scale disposal approach for terminal wastes.

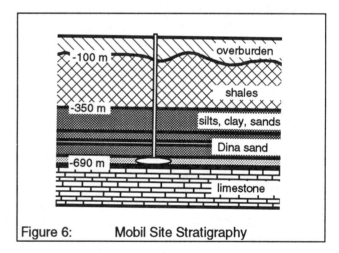

Figure 6:   Mobil Site Stratigraphy

7 Hydrogeological Conditions

Hydrogeological and lithostratigraphic conditions at the Mobil site are excellent. (Fortunately, they are neither unique nor local; they can be found in many sedimentary basin environments.)

a. There are extensive, unfractured and unfaulted, thick (250 m) low-permeability (nano-Darcy) shales above the site (Colorado Group clay-shales). This barrier is continuous for hundreds of kilometres in all directions (Figure 7).

b. Geochemical studies show that mixing of water above and below these strata has not occurred by vertical migration. Deep groundwater systems have no connection with surficial strata over great distances.

c. Regional fluid flow is horizontal from the southwest to the northeast, but at a slow rate: fluid age of 20-50 million years is typical, and exit points for these fluids is several hundred kilometres to the northeast, under thick surficial sediments (Figure 7).

d. Injected fluid is brackish whereas surface waters are of density 1.0. Energy required to force flow to surface is large, and no such forces can exist in this environment.

e. Considering rates at which interaction with surface waters could ever occur, dilution factors from annual rainfall would reduce any salinity to low levels.

f. Available pore volumes in the sediments are huge: In a 1000 m thick, 100 km^2 area, there is 25 km^3 of saturated pore space. Likely volumes are minuscule compared to these figures.

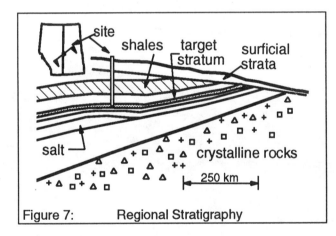

Figure 7: Regional Stratigraphy

In conclusion, interaction with surface waters is not considered an issue, on the condition that the well is not breached at higher elevations, and high abandonment standards are used.

8 Discussion

A suitable technology for terminal waste should meet the following requirements:

a. The site must have a minuscule probability of interacting negatively with the biosphere;
b. The disposal technology should be straightforward and flexible in its capacity to handle the waste materials;
c. The approach for permanent disposal must carry society's acceptance;
d. Procedures for waste transporting, handling, and disposal must be safe and healthy for workers and nearby communities;
e. Disposal sites must be permanent, must not impair current and future surface land use, and must not require permanent post-maintenance or treatment; and,
f. The price of the disposal method must allow it to operate economically.

The proposed technology meets these requirements, but acceptance by society and regulatory agencies needs open discussion of the technology and procedures involved. Irrational criticisms and highly improbable suppositions must be addressed openly. A criticism we have met is: "Regulatory agencies are not likely to approve of this, and there is a reluctance to permit deep injection." Such attitudes must be reversed; this can be helped by quantitative risk analyses based on rational premises. The reduced risk that accompanies deep geological entombment must be recognised and used for environmental benefits.

To place the discussion into context, consider the issue of landfill security (Arnould et al., 1993). Landfills suffer problems with respect to leaks and breaching, and they cause a permanent land quality reduction. Maintenance by barrier wells and by other means may be required indefinitely. Clay barriers have proved highly unreliable, and a single hole in a plastic or rubber liner is enough to contaminate surface waters that are consumed by people. Plastic and rubber liner integrity for thousands of years cannot be guaranteed, and breaching is likely. Clearly, surface landfills are inherently unstable and meet few of the requirements listed above (Figure 8).

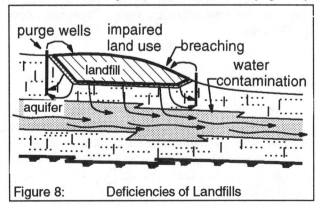

Figure 8: Deficiencies of Landfills

Slurry fracture placement of non-toxic or low-toxicity wastes in deep geological disposal sites carries environmental risks orders of magnitude less than those for surface placement. Given proper procedures and the right geological conditions, no biosphere interaction for periods in excess of 10^5yr can be expected. If interaction occurs, flow rates are so slow that natural dilution will mitigate unacceptable concentrations. Furthermore, placement of highly toxic wastes or liquids is not recommended, thus groundwaters that flow through the wastes will at most dissolve a few extra species, and quality is not likely to be affected because deep waters are always non-potable and saline. Site security is high, maintenance minimal, the technology exists and is demonstrable, and economics are favourable compared with landfill costs. If costs are low, reducing the disincentive to generate wastes, a surtax directly related to toxicity ranking could be used.

9 Conclusions

Oil field wastes have been successfully disposed of by fracture injection of a slurry into porous permeable sandstones of intermediate depth. The same approach promises to be a viable means of disposing low-toxicity granular civil wastes. The approach proposed carries an environmental security far in excess of surface landfills, does not impair long-term land use, and will be suitable for many types of wastes for which acceptable waste disposal technology is problematic and expensive.

10 Acknowledgments

Mobil Oil of Canada had the vision to try an idea that seemed a bit far-fetched. The Alberta Oil Sands Technology and Research Authority has helped all the writers during their endeavours, as has the Natural Sciences and Engineering Research Council of Canada.

11 References

[1] Arnould, M., Barrès, M. and Côme, B., (eds.), Geology and Confinement of Toxic Wastes. Balkema, Rotterdam, 605 p. (1993).

[2] Dusseault, M.B., Cold Production and Enhanced Oil Recovery. CIM J. Can. Pet. Tech., in press (1993).

[3] Dusseault, M.B., Bilak, R.A. and Rothenburg, L., Inversion of surface displacements to monitor *in situ* processes. Int. J. Rock Mech., Mining Sci., and Geomech. Abst., 4 p., at the press (1993).

[4] Dusseault, M.B. and Simmons, J.V., Injection-induced stress and fracture orientation changes. Can. Geot. J. 19, 4, 483-493 (1982).

[5] Flak, L.H. and Brown, J. Case history of ultra-deep disposal well in western Colorado. Proc. Int. Assoc. Drilling Contractors and SPE Drilling Conf., SPE, Richardson TX, pp. 381-394 (1988).

[6] Franklin, J.A. and Dusseault, M.B., Rock Engineering Applications. McGraw-Hill Inc., New York, 431 p. (1991).

[7] Gidley, J.L., Holditch, S.A., Nierode, D.E. and Veatch, R.W., Recent Advances in Hydraulic Fracturing. Monograph Series, Soc. of Pet. Engineers (SPE), Richardson, TX (1989).

[8] Wang, G., Dusseault, M.B., Pindera, J.T. and Rothenburg, L., Influence of sub-surface fractures on surface deformation of an elastic half-space. Int. J. Anal. and Num. Meth. in Geomech., at the press (1994).

[9] Willson, S.M., Rylance, M. and Last, N.C., Fracture mechanics issues relating to cuttings re-injection at shallow depth. SPE/IADC Drilling Conf. Proc., SPE #25756 (1993).

Ecrans Etanches Destinés au Confinement de Sites Pollués: Durabilité des Matériaux Constitutifs

A. Esnault, F. Dufournet-Bourgeois
S.I.F. Entreprise BACHY, France

Résumé : Le confinement de sites pollués implique une nouvelle approche au niveau de la conception et de la réalisation des parois d'étanchéité. Les matériaux doivent pouvoir résister à l'agression éventuelle de différents produits chimiques. Cette notion n'avait été jusqu'alors que très peu abordée dans le domaine de l'hydraulique. Des phénomènes de dégradation sont mis en évidence par des mesures portant sur des contrôles de pérennité par percolixiviation, perméabilité et stabilité mécanique. Ces essais montrent le besoin impératif de formulations nouvelles. La présentation fait le point sur les résultats d'une étude expérimentale en laboratoire destinée à mettre en évidence le comportement physico-chimique de matériaux soumis à des environnements agressifs.

Abstract : Containment of polluted sites involves an original design to realize watertight cut off. Components are required to be capable of resisting the contaminants. This approach constitutes a new way of research in the hydraulic field. It is possible to find a very vast range of soil contamination. Their action can be disastrous to classical slurries. Several laboratory tests have been carried out to determine durability, permeability and mechanical behaviour of slurries. The article presents the latest developments on the formulation of slurries.

1. Introduction

Une technique de réhabilitation de sites pollués consiste à mettre en place un confinement vertical sous la forme d'une paroi moulée dans le sol.
L'écran étanche ainsi constitué s'oppose au passage de fluide à l'extérieur de la zone contaminée.

Dans le cas présent, le fluide est le lixiviat ou le gaz émanant du dépôt. En plus des caractéristiques hydrauliques et mécaniques auxquelles la paroi aura à répondre, elle devra offrir une durabilité suffisante au contact des différents produits chimiques présents pour garantir la pérennité de l'ouvrage. Ceci implique une nouvelle approche au niveau de la conception des écrans verticaux d'étanchéité, et plus particulièrement au niveau du choix des matériaux constitutifs du coulis comme le ciment et l'argile.

2. Le Confinement de Sites Pollués

Certains produits chimiques, comme les sulfates ou les acides, sont connus pour leurs actions néfastes sur les ciments et leur mode d'attaque est bien documenté. Les transports des produits chimiques à l'intérieur du coulis s'effectuent par diffusion ou capillarité juxtaposées à la percolation.

Globalement, la dégradation du coulis peut se manifester de deux façons (Bodosci et al, 1988):

- expansion accompagnée en général d'une chute de résistance mécanique;

- dégradation par perte de poids accompagnée ou non d'une chute de résistance.

Ces deux phénomènes ont une conséquence sur la perméabilité du mélange et l'efficacité du système en tant qu'écran peut être remise en cause. Il est donc nécessaire de procéder à des tests de durabilité sur les coulis destinés à la réalisation d'écrans d'étanchéité en complément des études mécaniques et hydrauliques couramment pratiquées.

Une étude portant sur le comportement physico-chimique de coulis en contact avec différents produits chimiques est en cours depuis plusieurs années afin de vérifier leur durabilité au laboratoire Bachy à Paris. L'effet de la diffusion de ces produits à l'intérieur des coulis est estimé par rapport aux variations pondérales et aux variations de résistances mécaniques obtenues sur des échantillons placés en immersion dans des solutions tests. Ces essais sont complétés par des mesures de perméabilité et des études de lixiviation par percolation visant à identifier rapidement les processus de dégradation et confirmer la validité des résultats des essais de diffusion.

3. Approche Expérimentale

3.1. Produits Chimiques Testés

Les produits les plus souvent rencontrés dans des cas de pollution sont représentés par les hydrocarbures, les métaux lourds, les solvants, les acides, etc...et tous à des concentrations extrêmement variées. Les décharges, quant à elles, génèrent des lixiviats évolutifs de compositions variables. Il est donc impossible de dresser une liste exhaustive de produits chimiques suceptibles d'être rencontrés. L'approche consiste donc à sélectionner des solutions les plus représentatives des grandes familles chimiques. Elles correspondent aux groupes suivants :
- les acides minéraux et organiques : acide sulfurique, chlorhydrique, fluorhydrique, acétique;
- les bases minérales et organiques : soude, aniline;
- les solutions salines : sulfate, chlorure, nitrate;
- les métaux lourds : cuivre, nickel, cobalt;
- les hydrocarbures et solvants : heptane, benzène, ethylbenzène, chlorobenzène, naphtalène.

Le choix de leur concentration est effectué sur la base de la norme NFP18011 sur la classification des environnements agressifs pour les bétons en prenant, quand elle y figure, la limite supérieure des concentrations acceptables pour la composition testée. Des solutions assez concentrées sont aussi utilisées de façon à vérifier rapidement la compatibilité du ciment en accélérant le processus éventuel de dégradation. L'expérience prouve aussi que des cas isolés de pollution due à des produits purs ou très concentrés existent (cas des acides par exemple). Des mélanges sont aussi reproduits sur la base de compositions trouvées sur des sites pollués. Ils correspondent à des pollutions organiques type BTEX (Davis et al, 1984), un mélange type créosote (Arvin et al, 1991) et trois lixiviats de synthèse représentatifs de décharges de déchets domestiques (Colin, 1985) et de déchets industriels (Esnault, 1992).

3.2. Coulis Testés

Toutes les compositions testées sont à base de ciment de laitier additionné d'argile et d'un plastifiant réducteur de filtrat. Plusieurs types d'argile sont utilisés avec addition de matériaux pouzzolaniques.

La composition de base servant de témoin au test correspond à la composition N 1. Les autres coulis possèdent des caractéristiques d'ouvrabilité et de stabilité au moins égales à celles du témoin. Leur perméabilité est améliorée pour atteindre des valeurs inférieures à 10^{-9} m/s. Leurs caractéristiques sont illustrées dans le tableau suivant.

Tab. 1 : Caractéristiques des coulis testés.

COULIS	Viscosité Marsh sec.	% Décan. à 2h	Filtrat 7'30 7b cc	Rc 28j MPa	k 28j m/s
1	37	0	200	0,5	$5*10^{-8}$
2	35	0	116	0,7	$6*10^{-9}$
3	44	1	164	0,8	$6*10^{-10}$
4	37	0	130	0,9	$3*10^{-10}$

3.3. Test de Diffusion

L'essai consiste à mesurer les variations pondérales de coulis placés en immersion dans différentes solutions de produits chimiques. Le protocole d'essai est inspiré de la norme ASTM C267-82. Des échantillons de diamètre 4 cm et d'épaisseur 2 cm sont placés à l'âge de 7 et 28 jours dans une cellule hermétiquement fermée et contenant 300 cm^3 de solution test. L'ensemble est maintenu à une température constante de 18 +/- 2°C. Les pesées et le renouvellement des solutions sont effectués à intervalles réguliers. Les essais sont poursuivis jusqu'à ce que les variations pondérales se stabilisent. Les échantillons destinés aux mesures de résistance mécanique sont stockés dans les mêmes conditions. Ils ont une hauteur et un diamètre de 40 mm.

3.4. Test de Lixiviation par Percolation

Il est réalisé en cellule triaxiale pouvant travailler à des gradients de l'ordre de 1500 et résister aux divers produits chimiques agressifs. Le test permet de calculer le coefficient de conductivité en fonction du fluide utilisé.
L'analyse du liquide de lixiviation est réalisée à l'aide d'un spectromètre d'absorption atomique Perkim Elmer 3100 pour la phase minérale et par des techniques analytiques spécifiques aux produits organiques testés (colorimétrie ou chromatographie). Les analyses effectuées par spectrométrie portent sur les éléments constitutifs lixiviables du ciment, principalement le calcium, le silicium, le potassium, le sodium, le magnésium et l'aluminium.

4. Résultats des Tests de Diffusion

4.1. Essais d'Immersion

Les résultats des mesures de variations pondérales sont interprétés par rapport aux comportements de coulis stockés dans de l'eau dans les mêmes conditions d'essais.

Une première synthèse permet de dégager certaines tendances au niveau du comportement des coulis en fonction de leur composition de base. Les chiffres de variations pondérales obtenus ne sont pas à prendre en termes de valeur absolue mais plutôt comme indice de comportement (Tab.2)

L'ensemble des coulis stockés dans l'eau subissent des variations pondérales dP de +/- 2% après 600 jours d'immersion. La précision de la mesure au niveau de la pesée est de l'ordre de +/-1 %.

4.2. Résistance Mécanique

L'effet de la diffusion sur la résistance mécanique des coulis est étudié de façon à corréler les variations pondérales avec les variations des propriétés mécaniques en compression simple.
Une classification à base d'indices est adoptée de la manière suivante :
- indice Rc = Rc immersion dans la solution test / Rc immersion dans l'eau;
- indice stabilité pondérale ST = indice arbitraire determiné à partir des variations pondérales présentées tableau 2 :
A = 1 B= 0,8 C = 0,6 D = 0,4 E= 0,2 F = 0.

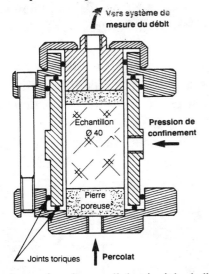

Fig. 1: Représentation d'une cellule triaxiale de lixiviation.

Tab. 2 : Résultats des essais d'immersion :
Test de durabilité sur coulis de ciment.

PRODUITS	Ech.1	Ech.2	Ech.3	Ech.4	PRODUITS	Ech.1	Ech.2	Ech.3	Ech.4
ACIDES :					FONCTIONS SALINES :				
Acide sulfurique pH2	F	F	F	F	NH4NO3 25 mmol/l	C	D	B	D
Acide chlorhydrique pH2	F	E	E	E	NH4Cl 7 mmol/l	C	B	C	A
Acide fluorhydrique pH2	F	F	E	D	MgSO4 330 mmol/l	A	B+	D	A
Acide acétique pH4	D	E	B	C	Na2SO4 77 mmol/l	B	A	A+	A+
BASES :					Li2SO4 250 mmol/l	A	A+	A+	A+
ANILINE pH 9	C+	A+	B+	B+	CaSO4 250 mmol/l	A+	A+	A+	A+
SOUDE pH 14	B+	A+	A+	A+	CuSO4 250 mmol/l	F+	F+	F+	F+
LIXIVIATS :					K2SO4 250 mmol/l	A+-	A+	C+	C+
Type PB pH 6,5	B+	C+	B+	C+	CoSO4 250 mmol/l	F+	F+	F+	F+
Type IRH pH 5	E	C	B-+	B-+	BeSO4 250 mmol/l	E+	B+-	A+-	A+-
Type Bachy pH 6,8	C	D	B	A	NiSO4 250 mmol/l	F+	F+	F+	F+
FONCTIONS OXYGENEES :					HYDROCARBURES-SOLVANTS :				
METHANOL PUR	E	E	E	E	HEPTANE	B	E	A	A
METHANOL 1/10	A	A	A	A	BENZENE	A	D	A	A
ACETONE PUR	E	E	E	E	CHLOROBENZENE	E	B	A	A
ACETONE 1/10	A	A	A	A	GASOIL	A	A	A	A
FORMALDEHYDE 38%	C+	C+	B+	B+	NAPHTALENE	A	A	A	A
PHENOL 1g/L	C	B	B	C	TYPE CREOSOTES	A	A	A	A
PHENOL 10g/l	C	E	B	C	TYPE BTEX	A	A	A	A
					ETHYLBENZENE	A	A	A	A

La classification suivante a été adoptée pour différencier les comportements observés :

- dP - 2 %   = A
- dP - 2 à 4 % = B
- dP - 4 à 6 % = C
- dP - 6 à 8 % = D
- dP - 8 à 15 % = E
- dP > - 15 %   = F

l'indice + indique un gain de poids, considéré aussi préjudiciable qu'une perte de poids.

Les deux indices se comparent ainsi par rapport à un niveau 1, ils n'évoluent pas forcément dans le même sens et ne sont pas toujours proportionnels. Un résultat est illustré figure 2.

Ces résultats mettent en évidence la nécessité de procéder à des tests dynamiques de lixiviation en complément des essais statiques de diffusion.

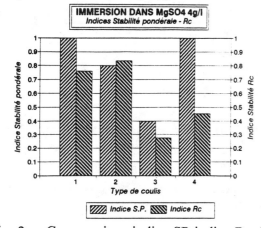

Fig. 2 : Comparaison indice SP-indice Rc dans du sulfate de magnésium à 4 g/l.

La présentation qui suit décrit les résultats expérimentaux d'une étude portant sur la détermination des effets d'un lixiviat type créosote sur un coulis témoin.

5. Etude sur Lixiviat Type Créosote

Le coulis testé correspond à la formulation 1 du tableau 1. Le lixiviat de synthèse type créosote a été retenu car il est représentatif d'une catégorie de sites industriels à fort potentiel de pollution organique. Il est constitué d'un mélange de phénols, HAP, BTEX, NSO, et NH4.

5.1. Résultats des Essais d'Immersion : Variation Pondérale et Résistance Mécanique

Les variations pondérales mesurées après 150 jours d'immersion ne sont pas significatives d'une dégradation du coulis dans les conditions du test (tableau 2).

Les échantillons placés en immersion dans des solutions renouvellées (R) et non renouvellées (NR) présentent après 90 jours les résistances mécaniques suivantes :

COULIS TEMOIN	
Eau R	6,2 bar
Eau NR	7,5 bar
Créosote R	6,6 bar
Créosote NR	6,4 bar

Il n'apparait pas de différence significative quelles que soient les solutions d'immersion ou les conditions du test.

5.2. Résultats des Essais de Percolixiviation

5.2.1. Concentrations du fluide de percolation après lixiviation

Trois types de comportements se dégagent des modes de variations des concentrations en fonction du temps :

- Type 1 : la concentration maximale est atteinte dès le premier jour du test et décroit en fonction du temps en se rapprochant d'un niveau d'équilibre.

- Type 2 : la concentration passe par un maximum puis décroit en fonction du temps.

- Type 3 : la concentration croit en fonction du temps sans atteindre un niveau d'équilibre.

Le type de comportement spécifique à un élément est commun pour l'eau et le lixiviat type créosote. Le comportement de type 1 est observé pour le potassium et le sodium.

Ceci s'explique par la grande solubilité des deux cations qui sont des traceurs de type non réactif.

Le comportement de type 2 est observé pour le calcium et l'aluminium. Les deux éléments sont de solubilité moindre et les concentrations diminuent quand la partie solubilisable de l'échantillon décroit.

Le comportement de type 3 est observé pour la silice très peu soluble et pour laquelle l'équilibre n'est pas atteint.

5.2.2. Différence de comportement entre les fluides de percolation

Après une durée égale de percolation, on remarque que le nombre de renouvellement du volume des pores (RNV) est supérieur dans le cas des créosotes.
En effet, après 121 jours de percolation, le volume des pores a été renouvellé 160 fois dans le cas de l'eau et 390 fois dans le cas des créosotes. La conductivité hydraulique de l'échantillon est affectée par la percolation des créosotes :

- K Eau = 0.6 * E-08 m/s

- K Créosotes = 1.3 * E-08 m/s

Tab. 3 : Essais de percolixiviation, coulis témoin.

			COULIS TEMOIN									
EAU												
Temps (jours)	RNV Vol. pores	Conductivité k*E-08 m/s	Quantité cumulée élément lixivié mg					pH	Concentration cumulée élément lixivié mg/l				
			Ca	Al	K	Na	Si		Ca	Al	K	Na	Si
1	3,5	1,2	14,6	0,3	17,4	56,7	0,1	11,91	91,65	1,88	109,23	355,93	0,63
3	9,1	0,9	48,7	1	18,1	58	0,1	11,14	119,57	2,46	44,44	142,4	0,25
7	16,4	0,6	62,6	2,1	18,7	60,3	0,2	11,27	84,82	2,85	25,34	81,71	0,27
14	24,4	0,5	107,4	4	18,7	61,6	0,6	11,27	98,23	3,66	17,1	56,34	0,55
28	39,2	0,4	157,3	7,9	18,7	67,9	1,9	11,16	89,47	4,49	10,64	38,62	1,08
63	75,8	0,4	169,8	12,6	18,7	81,8	7,3	10,33	49,9	3,7	5,5	24,04	2,15
91	112,7	0,4	190,8	16,4	20,3	108,9	12,3	9,41	37,72	3,24	4,01	21,53	2,43
121	160,3	0,6	221,3	19,8	22,2	135,6	38,8	7,71	30,77	2,75	3,09	18,85	5,39
CREOSOTE													
1	3,6	1,2	16,2	0,4	17,8	47,9	0,2	12,25	100,19	2,47	110,08	296,23	1,24
3	9,4	0,9	63,3	0,8	20,4	52,7	0,5	11,54	150,54	1,9	48,51	125,33	1,19
7	18,1	0,7	130,5	2,2	25,4	57	0,8	11,65	160,32	2,7	31,2	70,02	0,98
14	28,6	0,6	222,9	4	27,6	59,5	1,6	11,58	173,36	3,11	21,47	46,27	1,24
28	56,3	0,6	393,6	6,6	27,6	70,7	4	10,66	155,73	2,61	10,92	27,97	1,58
63	148,4	1	425,4	8,5	27,6	98,9	35	9,58	63,88	1,28	4,14	14,85	5,26
91	248,1	1,6	509,3	8,6	29,8	197	122,8	9,41	39,94	0,67	2,34	15,45	9,63
121	390	1,3	535,9	11,4	57,4	255,4	186,4	9,54	30,62	0,65	3,28	14,59	10,65

Par ailleurs, l'observation visuelle des deux échantillons à la sortie des cellules de lixiviation a permis de constater une dégradation des coulis par lessivage : perte de couleur, fragilité. Ce phénomène est plus marqué dans le cas des créosotes.

5.2.3. Comparaison des quantités lixiviées

Après le premier jour de percolation, le coefficient de RNV est sensiblement identique pour l'eau et les créosotes, respectivement 3,5 et 3,6. Les quantités lixiviées sont alors similaires pour les deux fluides de percolation. La différence de comportement des coulis entre eau et créosote apparait dès le troisième jour de l'expérience puisqu'elle est liée à la fréquence de renouvellement de l'eau des pores. Cependant, pour des RNV identiques, ce sont les ions calcium et silicium qui sont les plus sensibles à la solubilisation par les créosotes.

6. Conclusion

Cet article ne présente que les résultats obtenus sur un couple Coulis/Solution de percolation.

Les essais dynamiques de lixiviation par percolation, contrairement aux essais statiques en immersion, font apparaitre des différences de comportement entre les coulis en fonction de la solution utilisée.

Les cinétiques de dissolution des coulis varient en fonction du type de coulis, de sa perméabilité et de la nature de la solution de percolation.

La détermination de la composition des matériaux à mettre en oeuvre passe désormais par la réalisation en laboratoire d'essais sévères de détermination de pérennité.
Les résultats montrent que les deux tests, percolixiviation et immersion, doivent être réalisés simultanément car l'action des produits chimiques procèdent de processus différents dans les deux cas.

Les conséquences sur la stabilité mécanique et sur les variations éventuelles du coefficient de perméabilité diffèrent selon le mode de transport des produits.

Cette nouvelle approche de la détermination de la composition des coulis a été adoptée systématiquement pour la réalisation de plusieurs coupures d'étanchéité destinées à prévenir de l'extension de pollutions potentielles.

Références

BODOSCI, A., BOWERS, M.T. AND SHERER, R. (1988) Reactivity of various grouts to hazardous wastes and leachates. EPA/600/2-88/021.

DAVIS, K.E. AND HERRING, M.C. (1984) Laboratory evaluation of slurry wall materials of construction to prevent contamination of groundwater from organic constituents, Proc. of 7th National Ground water quality symposium, Las Vegas, pp. 1-21.

ARVIN, E. AND FLYVBJERG, J. (1991) Groundwater pollution arising from the disposal of creosote waste, Proc. IWEM Groundwater pollution of aquifer protection in Europe, Paris, pp 8/1-8/12.

COLIN, F. (1985) Etude des mécanismes de la génèse des lixiviats Inventaire et examen critique des tests de laboratoire, CCE et Ministère de l'environnement, pp 61-88.

DUPLAINE, H., ESNAULT, A. AND DUFOURNET-BOURGEOIS, F. (1993) Le confinement de sites contaminés, Géoconfine 1993, Montpellier, pp. 183-188.

ESNAULT, A. (1992) Adaptation et évolution des techniques de traitement de sol en matière de protection de l'environnement, Rev. Fran. Géot., No 60, pp 27-40.

Effects of Leachate on the Hydraulic and Mechanical Behaviour of Clay Liners

Marco Favaretti, Researcher, Nicola Moraci, PhD., Paolo Previatello, Associate Professor
Instituto di Costruzioni Marittime e di Geotecnica - Faculty of Engineering, University of Padova, Italy

Abstract. In addition to water content, degree of saturation, effective pressure and stress history, the hydraulic and mechanical behaviour of clay also depends on its mineralogy and the chemical compounds of the pore-fluid. This fact is fundamentally due to ion exchange and mineral dissolution phenomena occurring between clayey minerals and cations existing within the fluid. This paper deals with a laboratory investigation on the behaviour of four different sand-bentonite mixtures, alternatively permeated with water and leachate. A sodium bentonite and a calcium bentonite were used in order to emphasize the effects caused by different clay minerals. Fixed-wall permeability and oedometric compression tests were carried out; the influences of void ratio, vertical effective pressure and permeant on hydraulic conductivity, compressibility and consolidation coefficients of clayey samples were investigated.

Introduction

The design of urban solid waste disposals needs to solve many important geotechnical problems. If any impervious natural layers do not exist on the site, barriers to fluids must be made in order to avoid pollution of the surrounding environment. The impermeabilization of bottom and sides of a disposal area must be made using economical materials which are able to maintain their impervious function for at least some decades. Impervious liners can consist of compacted clays, clay-sand mixtures, geosynthetic-clays and geomembranes.

When clay-sand mixtures are exploited, bentonitic clays with a significant montmorillonitic component are preferable. The cost of the mixture usually depends on the amount of the bentonite used. Even relatively low percentages of bentonite (≈9%) in a compacted clay-sand mixture allow permeability coefficients lower than 1E-10 m/s to be obtained [1]. The effectiveness of the clay lining in preventing movement of leachate depends on its ability to maintain a very low permeability while in contact with contaminated fluids. Permeant passing through fluid barriers is generally chemically different from test fluids (water) used in laboratory investigations. So an unexpected increase in liner permeability could also increase the seepage of the leachate through the clay liner. Furthermore an increase in compressibility of the clayey layers involved could cause unexpected settlements of the site.

The hydraulic and mechanical behaviour of clayey soils strongly depends on their fabric and on the chemical nature of the permeant fluid. For example, large differences in hydraulic conductivity were noticed using water or leachate as permeant for the same clay [2]. This behaviour seems to be related to ionic exchange and mineral dissolution processes, occurring between clay minerals and chemical compounds of fluid. As a consequence of these physico-chemical interactions the double layer thickness of clay particles could increase or decrease. Increases in hydraulic conductivity may occur because the soil structure is more flocculated and the soil porosity is higher; otherwise if the thickness of the double layer decreases the hydraulic conductivity should increase.

Dissolution of soil minerals can occur under adverse pH conditions, with caustics tending to degrade the silica tetrahedra and acidic permeants causing dissolution of the octahedral layer [3].

Results of laboratory investigations are presented in this paper in order to emphasize in particular the influences of bentonite-sand content and fluid chemistry on the mechanical and hydraulic behaviour of clayey soils.

Investigation and results

Laboratory tests are generally used to determine the hydraulic conductivity of clays. They may be performed with either fixed- or flexible-wall permeameters using constant or falling head methods. A detailed analysis of the advantages and disadvantages of fixed- and flexible-wall permeameters are reported in [4,5].

Three natural soils were considered: a sodium bentonite (K7), a calcium bentonite (C), and a uniform sand. Their main index properties and chemical compounds are briefly summarized in Table 1 and 2 respectively. Laboratory tests were carried out on sand-bentonite mixtures, prepared in the laboratory with different weight percentages, in order to test samples having varying plasticity. Their Atterberg limits are reported in Table 3.

The natural soils were first completely dried, then mixed using different amounts of natural soils, wetted with water (the resulting moisture content was 1.5 times its liquid limit) and were finally homogenized by hand. After mixing the slurry was covered and allowed to hydrate for a week.

The fully hydrated sand-bentonite-water slurry was placed in oedometric consolidation cell, 70 mm wide and 20 mm high. Care was taken to remove as much air as possible, placing small amounts of soil at a time and vibrating it. Conventional step-loaded oedometer tests with permeability determination were carried out.

TABLE 1. The main index properties of soils

SOIL	W_l (%)	W_p (%)	I_p (%)	G_s	A
K7 - Bentonite	346	56	290	2.70	4.3
C - Bentonite	115	83	32	2.75	1.3
Adige River Sand	-	-	-	2.71	-

TABLE 2. Chemical compounds of bentonites

Chemical Compounds	Bentonite C (%)	Bentonite K7 (%)
SiO_2	48.40	58.88
Al_2O_3	21.05	14.62
Fe_2O_3	11.50	5.56
TiO_2	3.20	=
MnO	0.11	=
CaO	3.45	1.09
MgO	0.45	1.36
Na_2O	0.31	0.96
K_2O	0.40	1.20
H_2O	11.80	16.02
Montmorillonite (%)	70	80 to 90

TABLE 3. Index properties of test mixtures

MIXTURE	W_l (%)	I_p (%)
A (C 50% - S 50%)	58	20
B (K7 30% - S 70%)	61	37
C (K7 20% - S 80%)	46	21
D (K7 10% - S 90%)	31	14

After a 24 hour consolidation stage for each step, a head difference of 3 m was applied between the two ends of the sample, connecting its bottom to a 5 mm diameter burette filled with the test permeant. Two pore volumes of permeant fluid were filtered through the sample before starting each test. A falling head permeability test was then performed. During the test, readings of the decrease in hydraulic head over time were recorded for a period of 24 to 48 hours. Vertical permeability was calculated from the following falling head equation:

$$k = \beta \cdot \ln(\frac{h_0}{h_t}) \cdot \frac{a \cdot L}{A \cdot \Delta t} \qquad (1)$$

where β is a correction factor, depending on the dynamic viscosity and temperature of the test fluid, a and A are the cross-sectional areas of the burette and sample respectively, L is sample height, h_0 and h_t excesses of hydraulic head at time t=0 and after time Δt respectively. It was assumed that flow is governed by Darcy law, the steady state condition is established immediately, the permeability is uniform and remains constant throughout the test.

Two pore fluids were used as permeants to investigate their effects on the hydraulic conductivity of mixtures: (1) deaired distilled water; (2) a suitable caustic leachate, whose main chemical compounds (Table 4) are similar to those produced by an urban waste disposal. A rotational viscometer was used to evaluate permeant viscosity with temperature ranging from 10 and 30°C and shear rate from 0 to 300 s^{-1}. Scattering was less than 6%. Natural water was used both for mixing and initial permeation.

The compressibility curves e-log σ' derived from tests with different mixtures and permeants are plotted in Fig.1 a, b. The calcium bentonite-sand mixture (A) shows a compressibility independent on permeant in the range of investigated pressure, whereas mixtures containing sodium bentonite (mixtures B, C, D) show a certain dependence on the permeant. The compressibility index C_c ranges from 0.17 to 0.82 for samples filtered by water, and from 0.39 to 1.35 for samples filtered by leachate (Table 5).

This is due to the higher colloidal activity of sodium bentonite in comparison with calcium bentonite. Furthermore the calcium cation proves very capable of holding montmorillonite sheets together, therefore calcium bentonite generally presents a slight tendency to swelling.

TABLE 4. The main chemical compounds of the leachate (pH=10)

LEACHATE	MOL/LITRE
NH^{4+}	5.0×10^{-2}
Na^+	8.8×10^{-2}
K^+	2.2×10^{-2}
Ca^{2+}	6.1×10^{-3}
Mg^{2+}	1.0×10^{-3}
Cd^{2+}	1.0×10^{-6}
Cl^-	1.0×10^{-7}
NO^{3-}	2.0×10^{-3}

TABLE 5. Compression index of the test samples

MIXTURE	C_c - water	C_c - leachate
A	0.42	0.39
B	0.82	1.35
C	0.49	0.60
D	0.17	0.39

Vertical coefficient of consolidation c_v remains nearly constant with varying vertical pressure (Fig.2 a,b). However the c_v-coefficients of samples filtered by leachate is always higher than those of samples filtered by water. The permeability test results are plotted in Fig.3 a, b, c, d. The hydraulic conductivity strongly depends on void ratio, decreasing when it also decreases. The relation between hydraulic conductivity and void ratio on a

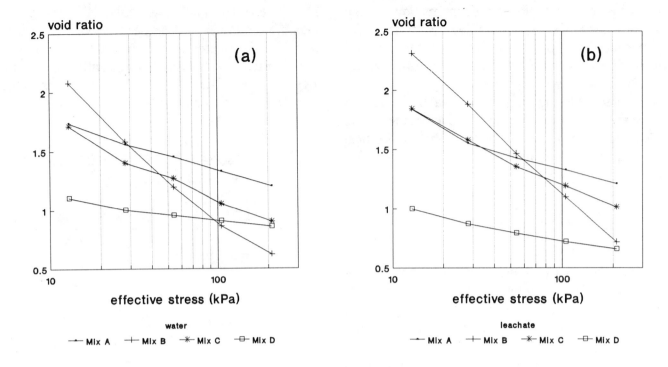

Fig. 1 a,b. Void ratio vs. effective vertical stress curves

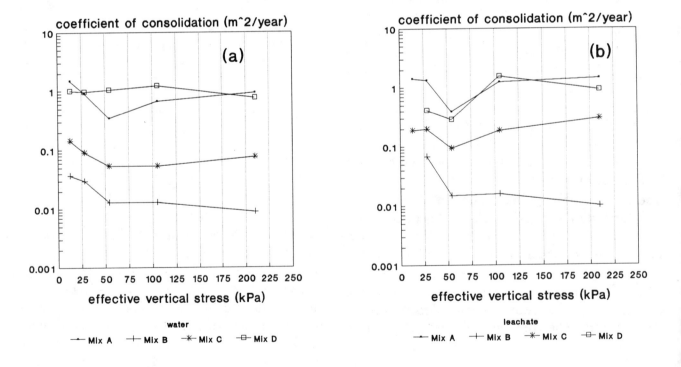

Fig. 2 a,b. Coefficient of consolidation vs. effective vertical stress curves

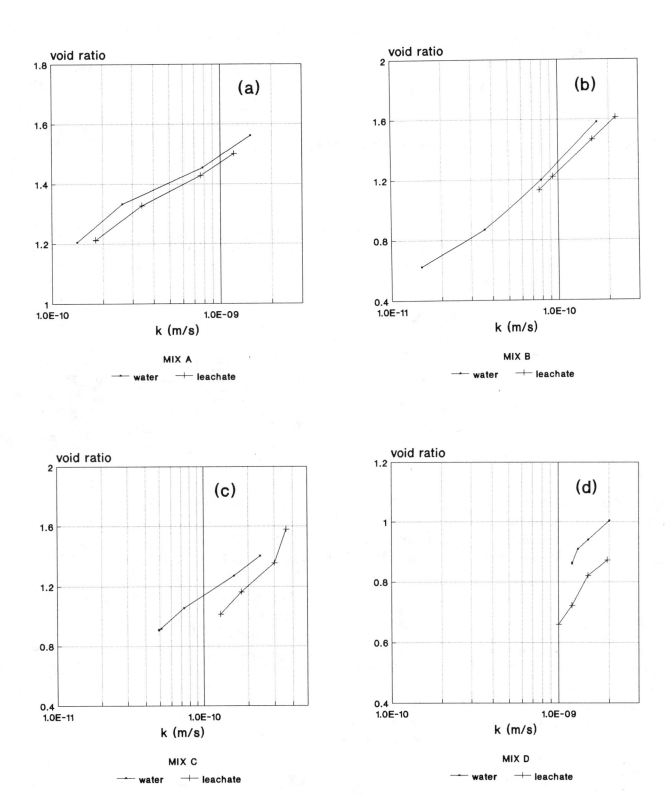

Fig. 3 a,b,c,d. Void ratio vs. hydraulic conductivity curves

semi-logarithmic diagram is nearly linear: its slope can be determined by the following expression:

$$C_k = \frac{\Delta e}{\Delta \ln k} \qquad (2)$$

The mixture A shows the same slope for both permeants. The sodium bentonite-sand mixtures show steeper slopes when they were filtered by leachate (Table 6).

TABLE 6. Slope of e-log σ' plots

MIXTURE	C_k water	C_k leachate
A	0.33	0.33
B	0.90	1.10
C	0.71	0.90
D	0.56	0.76

The hydraulic conductivity of samples filtered by water is always less than that determined with leachate, whereas it seemed not to be dependent on the applied gradient, according to Darcy's law.

Conclusions

Test results showed that sodium bentonite-sand mixtures, saturated and filtered by caustic leachate, present chemical activity which influences the soil structure and causes high compressibility and low hydraulic conductivity. Calcium bentonite-sand mixture showed low compressibility and high permeability, because of its capacity to produce stable structure.

Further laboratory tests, performed with oedometric and triaxial equipments, using acid and neutral leachates are carrying out on sand-bentonite mixtures in order to evaluate the effects of permeant on the behaviour of clayey soils.

Acknowledgment

The authors acknowledge the assistance provided by Francesca Dalla Giustina and Antonio Gobbato in setting up the laboratory tests in this study.

References

[1] DANIEL D.E. (1987). "Earthern Liners for Land Disposal Facilieties". Geot.Pract.Waste Disposal, ASCE, Spec.Publ., n.13, pp.21-39.

[2] ALTHER G., EVANS J.C., FANG H.Y., WITMER K. (1985). "Influence of Inorganic Permeants upon the Permeability of Bentonite". STP 874, ASTM, pp.64-74.

[3] LENTZ R.W., HORST W.D., UPPOT J.O. (1985). "The Permeability of Clay to Acidic and Caustic Permeants". STP 874, ASTM, pp.127-139.

[4] DANIEL D.E., ANDERSON D.C., BOYNTON S.S. (1985). "Fixed-Wall versus Flexible-Wall Permeameters". STP 874, ASTM, pp.107-123.

[5] FAVARETTI M., MORACI N. (1991). "Hydraulic Conductivity Tests on Coesive Soils". Sardinia '91, 3rd Int.Landfill Symp., S.Margherita di Pula, Italy, pp. 699-708.

A New Apparatus for the Measurement of Water-Air Permeabilities

Jean-Marie Fleureau, Said Taibi
Ecole Centrale de Paris, Laboratoire de Mécanique, Chatenay-Malabry, France

Abstract : The paper presents a new apparatus, based on the air pressure technique, for the measurement of water-air polyphasic permeabilities in soils. Three examples of measurements on remolded clayey soils are shown, which point out the important hysteresis between the drying and wetting paths. The comparison of the values of negative pressure for which the water permeability becomes nought, with those of the shrinkage limit highlights the correspondence between the two parameters.

1. Introduction

The study of polyphasic flows in porous media of average permeability is still dominated by the concept of relative permeabilities, first introduced by the petroleum engineers in the '30 [1, 2]. Later, Iffly [3] and others showed that Darcy's law was not valid for gases, even in the case of monophasic flows, while additional problems were encountered in polyphasic flows (end effects, etc.). Schwartzendruber [4], Hadas [5], Nimmo [6] tried to establish the laws describing the flow of water in unsaturated soils. These researches did not lead to very definite conclusions, mainly due to the complexity of the experimental device and procedures. On the other hand, questions arise concerning the validity of Darcy's law for water in the case of clays of very low permeability - e.g. "swelling" clays used as barriers for the storage of nuclear wastes. New approaches have been devised, in which the flow mechanism is associated with a diffusion mechanism that may play the major part [7, 8].

In most cases, however, the polyphasic permeabilities can still be obtained with reasonable accuracy by postulating a flow law and measuring the corresponding coefficients. Two methods are currently used to measure these permeabilities:

- the *unsteady state* method, in which a flow of air or water is imposed on one end of the sample and the change in the state of the material (void ratio, water content) is measured during the advance of the front,

- the *steady state* method, in which both fluids are injected simultaneously at both ends of the sample in such a way that the state of the sample is not modified by the flow.

The first method can be used only if it is possible to measure the local soil parameters at all points (e.g. with a radiographic device [9]), as well as the local pressure gradients. To derive the permeabilities, the measured water content or density profiles must be fitted by means of a computer code, which is often difficult. However, this kind of test can be carried out in the usual triaxial cells.

On the other hand, in the steady state method, the permeabilities are derived directly from the measurements, at least for materials of low deformability, as all the parameters of the sample remain constant during the test. Conversely, the experimental device is complicated by the fact that the flows of water and air must be separated, outside the sample, by semi-permeable membranes.

In fact, in the case of highly plastic or highly compacted materials, in which large strains may develop during drying or wetting, the use of this technique becomes much more complex, due to the fact that a constant negative pressure does not lead to a constant degree of saturation: if the deformation of the sample is prevented during the wetting phase, a swelling pressure will develop and modify the state of the soil; on the other hand, during the drying phase, the deformation of the sample may result in the formation of a void between the sample and the wall of the cell, which will greatly perturb the measurements.

2. Description of the ECP Permeameter

The permeameter developed at the Ecole Centrale Paris is based on the air pressure (or axis translation) technique, in which the control of the negative pressure - or suction - in the sample is achieved by imposing both positive air and water pressures. If u_w and u_a are, respectively, the water and the air pressures, the negative pressure will be equal to $(u_a - u_w)$. The equivalence of this approach with several others methods (osmosis, psychrometry, tensiometry) was validated by a large number of mechanical tests [10]. The experimental device consists of a cylindrical mold, with 35 mm diameter and 50 mm high, which can be fitted either with measurement ends (Fig. 1) or with compaction ends. The measurements can be carried out on undisturbed samples or on remolded samples compacted inside the mold. In the first case, the samples are previously cut at the exact diameter of the cell, at a slightly larger length (depending on the soil studied). The slight compression stress imposed by the fastening of the permeation ends is sufficient to ensure a suitable lateral stress and a good contact between the sample and the wall of the cell. To obtain compacted samples, the compaction of wet powder, at a given water content, is made directly inside the cell using custom-made compaction ends fitted on each end of the mold. The maximum compaction stress is approximately 20 MPa.

TABLE 1 : Characteristics of the membranes used

Membrane	Millipore GS	Millipore MF	Visking	Millipore Fluoropore
Air (Water) entry pressure (kPa)	387	1760	5000	134
Pore diameter (μm)	0.22	0.10	0.10	0.20
Thickness (μm)	150	105		
Conductivity ($cm^3/min/cm^2$) $\Delta p = 70$ kPa	21	16		4

Once the sample is placed inside the cell, the compaction ends are replaced by the measurement ends. These ends feature cellulosic semi-permeable membranes which are placed immediately against the sample. Several types of membranes are used, depending on the soil tested and the range of negative pressure (Table 1). Behind the membrane, repartition grooves allow the resaturation of the circuits.

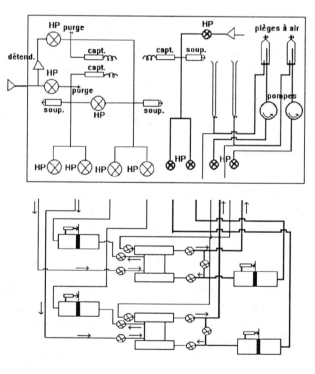

Fig. 2 - Schematic representation of the circuits for the different phases of the test

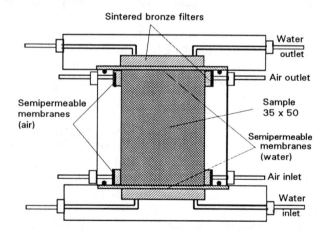

Figure 1- The ECP permeameter

The air inlet and outlet are placed laterally, near the ends of the sample. The air passes through a semi-permeable PTFE (Millipore Fluoropore) membrane (Table 1) and a sintered bronze repartition filter.

Comparisons between the permeabilities of saturated soils measured in the permeameter and measured with other devices showed that the influence of the membranes on the flows of water and air was negligible in the case of clayey materials but that it could reduce noticeably the flow in the case of sands and lead to erroneous results. The presence of a membrane in a circuit can also induce a loss of head and result in a difference between the pressure which is imposed from the outside and the pressure in the sample; this drawback is usually avoided by the very low values of flows used in this kind of measurements.

In the case of materials of low permeability, the measurement of very small quantities of water and air is made by means of specially designed volumeters, the accuracy of which is approximately 1 mm^3. For larger volumes, the flow of air is measured using glass tubes with saphire or metal balls and the flow of water is derived from the volume measured in a burette. Figure 2 shows a global sketch of the different circuits.

3. Materials and methods

The measurements presented in this paper were made on 3 materials: (a) a natural clay used for the construction of La Verne dam, in the south of France; the conditions of compaction are those of the in-situ material, corresponding to the Standard Proctor Optimum water content minus 3% and a density of 95% of the Optimum, (b) an artificial mixture of 3% kaolinite and 97% sand, and (c) a natural loam from Sterrebeek, in Belgium. The two latter materials were compacted in the conditions of the Standard Proctor Optimum. The main characteristics of these materials are shown in Table 2.

After compaction of the samples, a water pressure u_{w0} and an air pressure u_{a0}, homogenous in the sample, are imposed, with $u_{a0} > u_{w0}$. For an initial negative pressure in the sample equal to $-u_{wi}$, the material follows a drying path if $u_{a0}-u_{w0} > -u_{wi}$, or a wetting path in the opposite case. Once the

TABLE 2: Properties of the materials used

Material	M1 mixture	La Verne material	Sterrebeek loam
Granulometry:			
- D_{60} (μm)	59	700	37
- D_{10} (μm)	4.3	80	1
- D_{60}/D_{10}	15	8.8	37
Plasticity:			
- w_L	18	34.5	27
- w_P	3.6	19.2	20
- I_P	14.4	15.3	7
Standard Proctor:			
- w_{OPN} (%)	10	16.5	11.9
- γ_{dOPN}/γ_w	1.72	1.79	1.89
- $(-u_w)_{OPN}$ (kPa)	25	48	

negative pressure equilibrium is reached, after several days or several weeks, depending on the soils, the permeability measurement is made by imposing pressures u_{w1} and u_{a1} at the base of the sample, u_{w2} and u_{a2} at the head, with the following conditions:

$$u_{a1} - u_{w1} = u_{a2} - u_{w2} = u_{a0} - u_{w0} > 0$$

$$u_{a1} > u_{a2} \text{ and } u_{w1} > u_{w2}$$

In these conditions, the flow of the two fluids does not induce a change in the negative pressure of the sample, while maintaining the pressure homogenous in the material. As the sample volume remains constant, all the parameters of the soil (water content, degree of saturation) remain constant during the test. These values are derived from the net balance between the initial quantity of water and the input and output flows. The water and air geometric permeabilities are calculated using the expressions:

$$K_w = \frac{\mu_w Q_w L}{\rho_w (u_{w2} - u_{w1}) S}$$

$$K_a = \frac{2\mu_a Q_a L}{b(u_{a2}^2 - u_{a1}^2)}\left(1 - \frac{\lambda}{\mu_a}\frac{Q_a}{S}\right)$$

where Q_w and Q_a are the mass flows, respectively of water and air, μ_a and μ_w, the kinematic viscosities, ρ_w and ρ_a, the specific densities, L is the length of the sample, S, the area of its cross-section, b, the compressibility ratio and λ, a characteristic length of the porous medium (≈ 10^{-5} m). The latter expression is obtained by assuming a quadratic law for the loss of head [11].

4. Results and Discussion

The results of the tests on the 3 materials are shown on figures 3, 4 and 5. The samples follow a wetting path, from the initial compaction point up to saturation, then a drying path generally leading to complete desaturation. In the case of the M1 mixture, a second wetting-drying cycle was made. The saturated values K_w^{sat} and K_a^{sat} were measured under back-pressures of 200 to 300 kPa and are well correlated with the granulometries of the materials. The figures represent the water and air geometric permeabilities versus the degree of saturation and the corresponding negative pressures. A fifth graph shows the drying-wetting paths followed in the [u_a-u_w ; S_r] coordinate system. As a reference, we have also plotted on the same graph a dashed line representing the drying-wetting paths measured during independent pressure-controlled tests in which the deformation

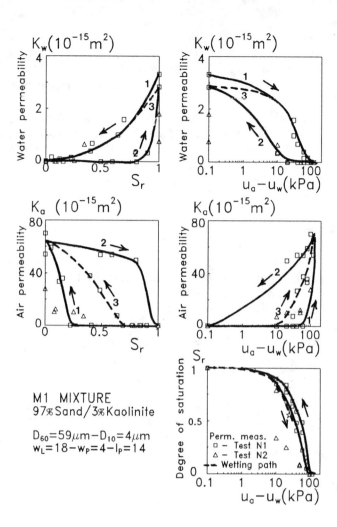

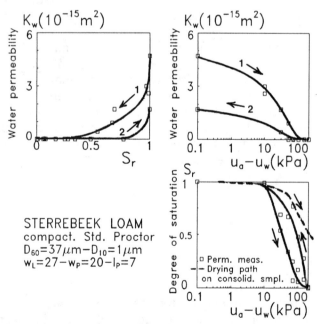

Fig. 3 - Water and air permeabilities of compacted Sterrebeek Loam on drying-wetting path

Fig. 4 - Water and air permeabilities of M1 mixture during drying-wetting cycles

of the samples was free [12, 13]. The difference between the points corresponding to the permeability measurements and the dashed line can be attributed to the change in the boundary conditions, but it must be noted that this difference remains very limited in all the cases. The values of permeability measured in the conditions of constant volume should not therefore be very different from the values measured in conditions of free deformation.

The curves, for the 3 materials, present several common characteristics:
- in the [S_r ; K_w] or [S_r ; K_a] coordinate systems, a rapid change in the water or air permeabilities when $S_r \to 1$ on the wetting path, while the change is much more progressive on the drying path. The degrees of saturation corresponding to a discontinuous water phase ($K_w = 0$) range from 70

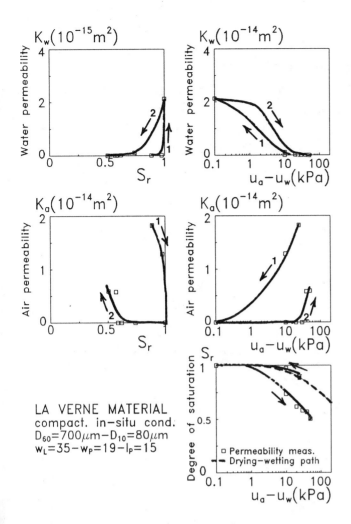

Fig. 5 - Water and air permeabilities of La Verne dam material on wetting-drying path

The permeability measurements can be related to the behavior of the materials on drying-wetting paths. If the deformation of the sample is free, the drying path first follows a normally consolidated or overconsolidated path in the $[u_a-u_w ; e]$ coordinate system, depending on the initial effective consolidation stress of the soil. When the negative pressure is increased above the "shrinkage limit negative pressure", the plastic deformation stops and the void ratio becomes nearly constant. This second phase, which is usually associated with a sharp decrease in the degree of saturation, corresponds to an increase in the normal forces due to the presence of water menisci between the soil particles [13]. It is therefore associated with the passage of water from a continuous to a discontinuous state and should result in the water permeability becoming nought. The value of the shrinkage limit negative pressure depends both on the mode of preparation of the sample and on the path followed.

For the 3 materials studied, the drying-wetting paths were plotted on figure 6. In the case of Sterrebeek loam, the measurements were made on a drying path, on samples consolidated under a 200 kPa stress. The other results were obtained on compacted samples, on a wetting path for the M1 mixture and on a wetting-drying path for La Verne material. The comparison of the values of the shrinkage limit negative pressure was made with the values derived from the water permeabilities

to 90% on wetting paths, and from 20 to 50% on drying paths.

- there is a good correlation between the increase in water permeability and the decrease in air permeability, or the reverse.

- the shape of the air permeability curves appears very different on drying and wetting paths: the permeability regularly increases on the drying path, while, on the wetting path, it remains close to its saturated value in a large range of degrees of saturation before dropping sharply between 80% and 100%.

- when plotted versus the negative pressure, the water and air permeabilities are nearly constant as long as the negative pressure remains below the desaturation pressure: $K_w \approx K_w^{sat}$ and $K_a \approx 0$.

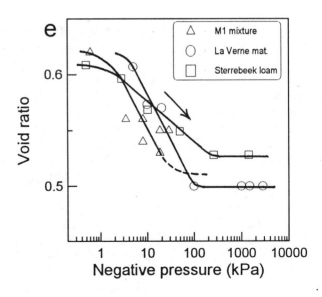

Fig. 6 - Drying paths on the 3 materials studied

TABLE 3 : Comparison of the values of shrinkage limit negative pressure with the values of zero water permeability

	$(u_a-u_w)_{SL}$ (kPa)	$(u_a-u_w)_{K_W=0}$ (kPa)
Sterrebeek loam	150	100
M1 mixture	50	70
La Verne mat.	70	40
Hostun sand	≈10	20

measured during the first drying phase on the 3 materials. Figure 7 shows the water relative permeability curves for the 3 materials, along with the values measured on Hostun sand (D_{60} = 311µm; D_{10} = 170 µm; D_{60}/D_{10} = 1.8).

In spite of the difference in the conditions of the drying and permeability tests, there is a general agreement between the two sets of values, as shown in Table 3. The shape of the relative permeability curves also appears to be related with the granulometry of the soil: on the drying path, the water relative permeability curves are filed according to the D_{10} of the soil, with the sharpest decrease occuring for the sand (D_{10} = 170µm) and much more progressive decreases for the finest soils (D_{10} = 1 and 4 µm).

Conclusion

The permeameter developed at the Ecole Centrale de Paris appears to be well suited to the measurements on clayey soils of low deformability; among other things, it provides useful information on the permeability of soils to gas. This subject, which has received little attention in geotechnical engineering so far, plays a major part in many environmental problems, noticeably in the storage of nuclear wastes. The role of the D_{10} of the soil is very important, both in the case of monophasic and relative permeabilities. The results also highlight the fact that permeability measurements and drying-wetting paths are closely related to each other and that useful information can be derived from that correspondence. This observation confirms the usefulness of representing the permeability changes both as a function of the negative pressure or suction, and of the degree of saturation.

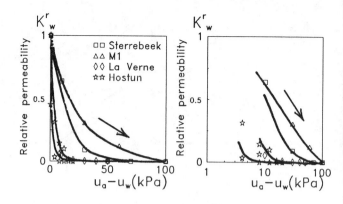

Fig. 7 - Relative permeabilities of the 3 materials studied and of Hostun sand on drying path

References

1. HASSLER, G.L., RICE, R.R. & LEEMAN, E.M. 1936. Trans. AIME, **98**: 116-126.
2. WYCKOFF, R.D. & BOTSET, H.G. 1936. Physics, **7**: 325-332.
3. IFFLY, R. 1956. Revue de l'Institut du Pétrole, **XI** (6): 757-795; (9): 975-1018.
4. SCHWARTZENDRUBER, D. 1963. Soil Sci. Soc. Amer. Proc. **27**: 491-495.
5. HADAS, A. 1964. Israël J. Agric. Res. **14** (4): 159-168.
6. NIMMO, J.R. , RUBIN, J. & HAMMERMEISTER, D.P., 1987. Water Ressources Res. **23** (1): 124-134.
7. PUSCH, R. & HOKMARK, H. 1990. Eng. Geology, **28**: 379-389.
8. ROBINET, J.C. AL-MUKHTAR, M., RHATTAS, M., PLAS, F. & LEBON, P. 1992. Rev. Franç. Géotechnique, **61**: 31-43.
9. ANGULO, R., GAUDET, J.P., THONY, J.L. & VAUCLIN, M. 1993. Rev. Franç. Géotechnique, **62**: 49-57.
10. BIAREZ, J., FLEUREAU, J.M., ZERHOUNI, M.I., SOEPANDJI, B.S., 1988. Rev. Franç. Géotechnique, **41**: 63-71.
11. CHAUMET, P. 1965. in Cours de Production de l'ENSPM, Tome 3: 133-159, Technip ed., Paris.
12. FLEUREAU, J.M., KHEIRBEK-SAOUD, S., SOEMITRO, R. & TAIBI, S. 1993. Can. Geotechn. J. , **30** (2): 287-296.
13. FLEUREAU, J.M. & KHEIRBEK-SAOUD, S. 1992. Rev. Franç. Géotechnique, **59**: 57-64.

Surface Sampling to Detect Unacceptable Human Health Risk

Mark Gemperline
U.S. Bureau of Reclamation, Denver, Colorado, USA

Abstract

An equation is derived to estimate the minimum number of samples required to detect surface hot spot contamination which presents an unacceptable human health risk. It combines common statistical analysis and simplified human health risk analysis. Multiple contaminants having varying degrees of toxicity are considered. The number of samples required is shown to be dependent on a preselected contaminant concentration which may represent the detection limit of an analytical technique.

The equation focuses the user on the problem variables resulting in a better understanding of the uncertainties involved in identifying conditions representing unacceptable human health risk. It can be used directly to calculate the required surface sampling density for a prescribed level of acceptable human health risk or used with an existing sampling plan to help estimate the human health risk associated with a hot spot which may be left undetected.

Introduction

Identifying conditions which present an unacceptable risk to human health is paramount to environmental site investigations. The required number of surface soil samples collected is frequently determined by arbitrarily selecting the "hot spot" size which must be found and applying basic statistical concepts to assure detection with adequate confidence.

The method of selecting hot spot size varies from site to site and often is selected to represent the reasonable size of a contaminant spill. It is based on historical records or observations. This approach does not attempt to assure detection of a condition which represents a human health risk.

An equation is derived to estimate the minimum number of samples required to detect surface hot spot contamination which may present an unacceptable human health risk. It combines common statistical analysis and simplified human health risk analysis. A smallest hot spot which potentially poses an unacceptable threat to human health is hypothesized. Its size depends on a simplified model for contaminant distribution within the hot spot, site specific variables which describe potential for human ingestion and dermal contact, and the toxicity of the contaminant involved.

The equation is restricted to surface soils which may represent direct ingestion or dermal contact hazards and does not address exposure pathways associated with subsurface soil, air, or water contamination. The derivation considers only a single contaminant, however it can be used with multiple contaminants by normalizing contaminant concentrations with respect to relative toxicity. This procedure is discussed.

The number of samples required is shown to be dependent on a preselected contaminant concentration which may represent the detection limit of an analytical technique; the minimum acceptable average concentration of an exposure unit which would result in a threat to human health; and the maximum expected contaminant concentration.

Spacial variation of sample density and the use of successive stages having decreasing sampling density and contaminant detection limit are discussed.

Many site and exposure conditions are simplified for mathematical convenience. These simplifications are discussed so the reader may develop an understanding of the uncertainties involved in the calculation.

Simplified Risk Assessment

An exposure unit is defined herein as the contiguous area containing hot spots of contamination to which a person may be exposed. No exposure is expected from outside this area. A person exposed to a contaminant has a mass M, ingests soil at approximately SI mass-units/day and experiences DE mass-units/day soil dermal exposure. A long exposure period is assumed, consequently the average exposure unit concentration represents the average human exposure concentration. SIA and DEA are terms which represent the fraction of the ingested or dermal contacted contaminant which is absorbed into the body respectively.

The analysis period is the time considered significant to the analysis. For example, a lifetime of 70 years may be considered appropriate for assessing the risk associated with carcinogens. The variable, t_f, is the fraction of the analysis period which a person actually spends within the exposure unit.

Maximum allowable daily dose of contaminant per unit body mass, Rfd*, are published Rfd values for non-carcinogens or calculated by dividing the acceptable carcinogenic risk, R, by a slope factor, SF, for carcinogens. Rfd and slope factors values are found in the USEPA IRIS database as well as other databases and publications. Hazard index is defined as the ratio of the average daily adsorbed mass of contaminant during the analysis period to Rfd* [1].

The average exposure unit surface soil concentration which would result in a hazard index of 1, or a carcinogenic risk R, is approximated by the equation:

$$C_{index} = \frac{Rfd^* \times M}{(SI \times SIA + DE \times DEA) \times t_f} \quad (1)$$

A hot spot, or cumulative sum of hot spots, which causes the average exposure unit concentration to be greater than C_{index} must be located.

Hot Spot Size

Presume for a moment that there is a single hot spot in the exposure unit which causes the average concentration of the exposure unit to be unacceptable. This is subsequently referred to as a significant hot spot. The size of the smallest hot spot which could result in this condition is sought.

It is assumed that the average hot spot contaminant concentration, C_{ave}, is 1/3 of the maximum exposure unit contaminant concentration, C_{max}. This is equivalent to assuming that a single circular hot spot exists which has C_{max} at its center with linearly decreasing concentration toward the perimeter as indicated by figure 1. C_{max} must be estimated based on site history, scoping information and expected fate and transport of site contaminants.

The contribution to the average exposure unit concentration from a single hot spot is:

$$C = C_{ave} \times \frac{A_{hot\,spot}}{A_{unit}} \quad (2)$$

where:
$A_{hot\,spot}$ = area of the hot spot
A_{unit} = area of the exposure unit
C = average exposure unit concentration.

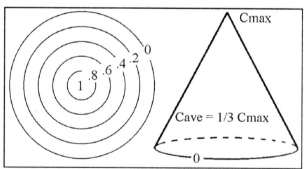

Figure 1. Conceptual Hot Spot Model. Contours of Concentration Normalized with respect to C_{max}.

The smallest hot spot that would cause the average exposure unit concentration to be greater than or equal to C_{index} is sought. The area of this hot spot is obtained by setting C equal to C_{index} in Eq. 2 and rearranging.

$$A_{hotspot} = A_{unit} \times \frac{C_{index}}{C_{ave}} \quad (3)$$

Minimum Number of Samples Required

The number of randomly located surface soil samples is sought which assures that at least one sample will be obtained from within a significant hot spot at a detectable concentration, C_d, or higher. Figure 1 shows the concentration contours within the model hot spot with C_{max} equal to 1. Assume that one of these contours represents the detection limit concentration above which it is desired to detect the hot spot. This value may be equal to, or greater than, the detection limit of a chemical analysis procedure. The circle representing this contour surrounds the detectable area, A1. It is calculated by:

$$A1 = A_{hotspot} \times (1 - \frac{C_d}{C_{max}})^2 \quad (4)$$

The ratio of this area to the total exposure unit area is the probability that a single sampling event at the site will be selected within the detectable region of the hot spot. The probability of having no successes in N trials, α, is given by:

$$\alpha = (1 - \frac{A1}{A_{unit}})^N \quad (5)$$

Rearranging yields:

$$N = \frac{\log(\alpha)}{\log(1 - \frac{A1}{A_{unit}})} \quad (6)$$

The number of randomly selected samples, N, required to assure with 1-α confidence that at least one sample will be from within the detectable area of a hot spot having unacceptable risk is given by Eq. 6.

Although the analysis assumes a single circular hot spot as the worst case, it also applies if it is desired to assume there are multiple hot spots that are smaller in size, however demonstrate the same cumulative concentration distribution as the model hot spot.

Equations 3, 4, and 6 are combined to yield an equation which gives the general solution for N.

$$N = \frac{\log(\alpha)}{\log(1 - 3\frac{C_{index}}{C_{max}}(1 - z\frac{C_{index}}{C_{max}})^2)} \quad (7)$$

where:

$$z = \frac{C_d}{C_{index}}$$

Equation 7 is valid when N decreases as the ratio C_{index}/C_{max} approaches 1 and the detectable area of the hot spot is less than or equal to the exposure unit area. It is shown in the appendix that the first condition is satisfied for values of z between 0 and 4/9. The second condition is satisfied when

$$\frac{A1}{A_{unit}} = 3\frac{C_{index}}{C_{max}}(1 - z\frac{C_{index}}{C_{max}})^2 < 1 \quad (8)$$

If $A1/A_{unit}$ is greater than 1 then, by definition, any sample taken should detect contamination.

Values of z greater than 4/9 may be used, however, a staged sampling plan is required.

Staged Sampling Plan

It is often cost effective to use screening technologies to locate hot spots when the hot spot size which must be found is small and the number of sampling locations large. Screening technologies often have higher detection limits than standard laboratory tests. Consequently z may be greater than $4/9$. Equation 7 is still applicable, however, it must be realized that a significant hot spot, one that poses an unacceptable risk, may be missed. Such a hot spot would be much larger than the hypothesized smallest significant hot spot however, would have a smaller detectable area and a lower maximum concentration. A second sampling stage, using a lower detection concentration, is required to find this hot spot if it exists.

It is shown in the appendix that the number of samples required in the second sampling stage can be calculated by Eq. 7 letting C_{max2} equal $3C_{d1}$. The numeric term added to the subscript refers to the sample stage. A hot spot having a maximum concentration greater than $3C_{d1}$ and less then C_{max1} will be sampled in the first stage with no less than $1-\alpha$ confidence.

If z_2 is greater than $4/9$ then a third stage is required. Stages are added until z is less than $4/9$. Two stages are expected to be sufficient in most cases. Subsequent stages will always require substantially fewer samples.

Decision Units

Often it is desired to sample sub-regions within an exposure unit at different sampling densities. This results from the desire to make independent decisions for these sub-regions, consequently, they are identified as decision units.

Equation 7 can be used to determine the number of samples required in each decision unit. However, C_{index} must be calculated appropriately. C_{indexI} is calculated using Eq. 1 for the Ith decision unit using fractional $Rfd*_I$'s. $Rfd*_I$'s must sum to the $Rfd*$ appropriate for the entire exposure unit. The $Rfd*_I$ values are selected to represent the portion of the total human exposure to contamination which is permitted from a specific decision unit. The portion of total human exposure which may be tolerated in a small decision unit is expected to be greater than the portion tolerated in a larger exposure unit. The addition of decision units always results in more sampling required in the exposure unit.

Normalized Contaminant Concentration

The concept of normalized concentration is introduced to simplify calculations when multiple contaminants are involved. The idea is to convert real contaminant concentration to the equivalent concentration of a user selected standard contaminant so that it reflects an equivalent toxicity or carcinogenic risk. Consequently, calculations are made with a user selected standard contaminant and its corresponding toxicity or carcinogenicity instead of multiple contaminants with varying toxicity.

The following equation is used to calculate Hazard Index for multiple contaminants [1].

$$HI = \frac{M_1}{Rfd_1} + \frac{M_2}{Rfd_2} + \ldots + \frac{M_i}{Rfd_i}$$

where: Rfd_i = reference dose for contaminant i, and M_i = daily mass of absorbed contaminant i.

This can be rewritten as:

$$HI = \frac{K}{Rfd_1}(C_1 + C_2\frac{Rfd_1}{Rfd_2} + \ldots + C_i\frac{Rfd_1}{Rfd_i})$$

where K is a constant associated with exposure and contaminant adsorption. For simplicity K is assumed to be the same for all contaminants. If desired, K can be related to the absorption characteristics of each contaminant resulting in a K_i value associate with each term in parentheses. C_i is the average concentration of contaminant i in the exposure unit. The expression in parentheses is the sum of the normalized concentration of all contaminants with respect to contaminant 1. The Rfd ratios are normalization factors. If all contaminant concentrations are normalized by multiplication by the appropriate normalization factor then they may be treated as the standard contaminant.

The calculation of normalized concentrations for carcinogens is the same as for non-carcinogens except that the Rfd values are replaced with the value of the risk divided by the appropriate slope factor.

Average Exposure Unit Concentration

The presented approach attempts to assure that at least one sampling event occurs in the detectable region of the hot spot. Consequently it is unreasonable to assume that the average of all sample concentrations would guarantee adequate representation of exposure unit average concentration. To assure adequate representation of a hot spot in the determination of average concentration it is recommended that a composite sampling program be conducted in addition to a hot spot detection sampling program.

Composite specimen collection density should guarantee several specimens are taken within a significant hot spot. This will increase the certainty with which the composite concentration, or the average concentration of a group of composites, represents the average exposure unit concentration. The required number of composite sample specimens, M, can be related to N from Eq. 7 by:

$$N = \frac{\log \sum \left(\frac{M!}{(i!(M-1)!)} (p)^i (1-(p))^{M-i} \right)}{\log(1-(p))} \quad (9)$$

where the summation is with respect to i; i is the desired number of specimens to be collected from within the hot spot minus one; p equals $3C_{index}/C_{max}$. The confidence, $(1-\alpha)$, used to obtain N from Eq. 7 is equal to the confidence used to obtain M from Eq. 9. A short computer program which solves Eq. 9 is presented in a previous paper [2].

Example

The following hypothetical situation exemplifies the use of equations 1 and 2.

An old transformer storage site in the midst of a national park has potential PCBs hot spot contamination. The site is approximately 25 acres and it has been 15 years since transformers were stored there. Historical record review, site inspection, and cursory analyses estimating the extent of the PCBs volatilization and leaching are conducted. It is concluded that the maximum expected value of PCBs, C_{max}, that could exist at this time is approximately 4000 mg/kg. The following values are used with Eq. 1 to approximate C_{index} as 10 mg/kg. An average site concentration greater than this is expected to represent an unacceptable risk.

M ≈ 16 kg
SI ≈ 200 mg/day
DE ≈ 10000 mg/day
SIA ≈ 0.3
DEA ≈ 0.1
t_f ≈ 0.02
Rfd* ≈ 1.25×10^{-5} mg/kg/day

Rfd* was calculated as the acceptable risk divided by the slope factor. These are estimated as:

Acceptable risk = 1×10^{-4}
Slope factor ≈ 8 day/(mg/kg)

A three stage program is used to assure that adequate information is obtained for risk assessment. Stage 1 will detect small hot spots having high PCBs concentrations whereas stage 2 will find large hot spots having low concentrations. A third stage will consist of a composite sampling program to be used to determine the average site concentration. It is desired to have a minimum of 5 specimens (i=4) obtained from the smallest significant hot spot. Ninety-five percent confidence is desired.

A PCBs surface soil screening using an analytical method with a 50 mg/kg detection limit, C_{d1}, is used for stage 1. A laboratory analytical method with a detection limit of 1 mg/kg, C_d, is used for stages 2 and 3.

Equation 7 is used to calculate that 408 sample locations are required to be screened in stage 1. Stage 2 sampling requirements are calculated using Eq. 7 by setting $C_{max} = 3C_{d1} = 150$ mg/kg and $C_d = C_{d2}$. Stage 2 requires 14 samples be tested using the laboratory analytical procedure.

Equation 9 is used to calculate the stage 3 composite samples specimen requirement. The number of composite specimens required, M, is determined by trial and error. Composite samples should incorporate 1234 specimens.

It is decided to perform stage 1 screening on a 50 ft. grid having a randomly selected origin. This will result in approximately 440 stage 1 point samples. Grid sampling was selected instead of random sampling for simplicity. Samples will be obtained at 14 randomly selected locations to determine PCBs concentration by the laboratory method. This will accommodate stage 2 data needs.

Twenty-five composite samples will be collected each consisting of fifty specimens, i.e. two randomly selected specimens from each of the sites 25 subdivided acres. This amounts to for 1250 stage 3 specimens. The selection of 25 composites samples was arbitrary for this example.

The average of the composite sample specimens is expected to provide a reasonable estimate of the site average concentration even for the condition of a smallest single significant hot spot. If no significant hot spots exist on the site then the calculated mean is expected to be less than C_{index}. The variance of the composites will provide an estimate of the accuracy with which the mean may be estimated. It is considered unreasonable for composite sampling to indicate a mean concentration greater than C_{index} without point samples identifying significant hot spots. If this happens, the conceptual model is probably in error and the sampling program needs would require reevaluation.

The need to characterize individual hot spots for risk assessment purposes will be determined after the proposed sampling and analyses are complete.

Conclusion

Equations are presented which permit the calculation of the number of surface samples required to assure detection of a significant hot spot i.e., a surface hot spot which may pose a threat to human health. The minimum number of samples required for hot spot detection is also related to the number of composite specimen sampling locations required to assure adequate representation of a significant hot spot in the calculation of mean exposure unit concentration. These equations may be modified and adjusted to meet the specific needs of the user.

It is expected that the numerous simplifying assumptions used in equation development will provide a basis for discussion and future improvement. Understanding how variables influence sampling density will assist the user in developing an adequate sampling program.

Equations 7 and 9 may be solved to estimate minimum sampling requirements or used to help understand the representation of a proposed sampling plan.

The equation is expected to be used in a trial-and-error process in which variables and solutions are optimized for site specific needs.

References

[1] United States Environmental Protection Agency (USEPA). 1989. Risk Assessment Guidance for Superfund, Volume 1, Human Health Evaluation Manual (Part A) Interim Final. Office of Emergency and Remedial Response, Washington D.C. EPA/540/1-89/002.

[2] Gemperline, Mark C. 1993. Surface Sampling to Detect Hot Spots Presenting Unacceptable Human Health Risk, Proceedings of the Superfund XIV Conference, November 30- December 2, 1993, Washington, DC.

Appendix: Equation 7 Limitations

Equation 7 is derived for a hot spot scenario. A hot spot detectable area, $A1$, greater than the exposure unit area A_{unit} no longer constitutes a hot spot scenario and invalidates Eq. 7. When a hot spot has a detectable area, $A1$, equal to the total area, A_{unit}, then the probability of detection, $A1/A_{unit}$, becomes 100 percent. The value $A1/A_{unit}$ is given by equation 8. Observe that Eq. 6 is undefined for $A1/A_{unit}$ greater than 1. Equation 7 variables, z and C_{index}/C_{max}, which result in $A1/A_{unit}$ greater than or equal to 1 are graphically illustrated on figure 2.

Intuitively, an upper limit to the acceptable magnitude of the ratio C_d/C_{max} is expected. It must be assured that the detectable hot spot area for all significant hot spots having a maximum concentration less than C_{max} is larger than

is greater for significant hot spots having lower maximum concentrations. This condition is satisfied when the detection concentration is less than, or equal to, the average concentration of the largest significant hot spot that does not have a detectable area greater than the exposure unit area. Mathematically, this is deduced by noting that the derivative of Eq. 8 with respect to the ratio C_d/C_{max} equals zero when $C_d/C_{max} = 1/3$. This is the condition in which $A1/A_{unit}$ reaches a maximum value. The following sequence shows this analysis.

Equation 8 is rewritten as:

$$f(w, z) = 3 \frac{w}{z} (1-w)^2$$

where $f(w,z) = A1/A_{unit}$, $w = C_d/C_{max}$ and $z = C_d/C_{index}$.

The derivative of $f(w,z)$ with respect to w is given by:

$$\frac{df}{dw} = \frac{3}{z}(1-w)^2 - 6 \frac{w}{z}(1-w)$$

solving for $df/dw = 0$ yields:

$$\frac{df}{dw} = 0 \quad when \quad w = \frac{1}{3}$$

The region in which C_d/C_{max} is greater than 1/3 is indicated on figure 2.

The intersection of the lines marking the boundary of the regions $C_d/C_{max} > 1/3$ and $A1/A_{unit} > 1$, occurs at $z=4/9$ and provides the maximum value of z for which equation 7 may be applied with one contingency, i.e., $A1/A_{unit} < 1$. It is assured that any significant hot spot having a maximum concentration less than C_{max} will have a larger detectable area than the significant hot spot containing C_{max}. Consequently the number of samples required to assure detection of a significant hot spot containing C_{max} will assure with greater confidence the detection of all other possible significant hot spots.

If Eq. 7 is used with z greater than 4/9 then there is inadequate assurance that all possible significant hot spots will be found. Significant hot spots having an average concentration, C_{ave}, below the detection concentration, C_d, experience a shrinking detectable area as the significant hot spot maximum concentration approaches Cd. Consequently, the probability of finding hot spots becomes increasingly lower.

Significant hot spots which are missed by using C_d/C_{index} greater than 4/9 can be found with a second sampling effort. Equation 7 may be used to calculate the number of required samples knowing that the maximum concentration of a significant hot spot which may have been missed in the previous sampling effort is not greater than three times the detection concentration used in previous sampling. A significantly fewer number of samples are required for the second

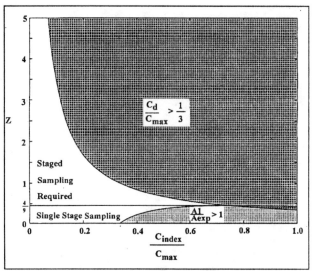

Figure 2. Variable combinations affecting use of Eq. 7.

sampling effort. The second effort uses a lower detection concentration with a lower C_{max}. If C_d/C_{index} for the second sampling set is less than 4/9 then it is assured with $1-\alpha$ confidence that all significant hot spots in the exposure unit having a maximum concentration less than C_{max} have been found. If z is greater than 4/9 then a third sampling effort, designed by the same procedure as the second, is required.

Investigations and Measurements Associated with Environmental Improvement of Inner Bay Areas by Covering Polluted Soft Mud with a Thin Sand Layer

Michio Gomyoh, Research Engineer, Hideo Hanzawa, Director,
Technical Research Institute, Toa Corporation, Yokohama, Japan,

Tej Pradhan Associate Professor
Yokohama National University, Yokohama, Japan

ABSTRACT   The sand capping method where the bottom mud is covered with a thin sand layer is noted as a solution of the environmental improvement of polluted inner bays and lakes. Proper evaluation of the shear strength of the bottom mud and the thickness of the thin sand layer are important subjects in the sand capping method both for design and quality control. In this paper, the shear strength of typical bottom muds found in Japan measured by vane shear and cone penetration tests is presented together with the procedure for measuring the thickness of the thin sand layer. Significant local variation of the shear strength of the bottom mud is demonstrated. The piezo cone penetration test is found to be a useful tool for measuring both the shear strength of bottom mud and the thickness of the thin sand layer.

INTRODUCTION

Enclosed water areas throughout the world have been utilized extensively, resulting in a lot of environmental problems such as water pollution and ecosystem destruction. For example, red and blue tides have quite often taken place in the inner bays of Japan. As the reason for such water pollution, the effect of polluted bottom mud as well as effluent and primary production of marine organisms has been pointed out because of its release of nutrient and oxygen consumption. Quantities of effluent and release from bottom mud in the main inner bay areas of Japan are reported as summarized in Fig.1 (Horie,1987). A series of calculations with the use of a simplified model based on site investigation in Tokyo Bay also reported that the effect of oxygen consumption on overlying water is significant (Sasaki et al.,1993).

As one of the methods for improvement of polluted inner bay water, the sand capping method where the bottom mud is covered with a thin sand layer can be pointed out because of

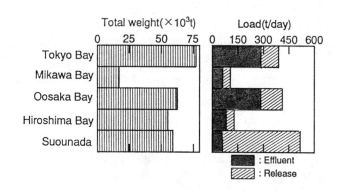

Fig.1 COD effluent and release from bottom mud (after Horie, 1987)

its improvement of bottom mud, decrease of nutrient release and oxygen consumption, and restoration of benthos. Release experiments, evaluation of water quality and ecosystem, and site investigation on actual sand capping works have proved the validity pointed out above (Ohtuki and Gomyoh, 1991, Horie and Horiguchi,1989). On the other hand, it is firstly required to evaluate the shear strength of very soft bottom mud in order to determine the construction method to place the sand on this soft material. In addition, it is also an important subject in quality control to find out a procedure to measure the thickness of thin sand layers ranging from 20 to 50cm.

This paper mainly describes the two subjects pointed out above to be clarified for the sand capping method based on a series of field and laboratory tests.

CLASSIFICATION AND PHYSICAL PROPERTIES OF BOTTOM MUD

Some physical tests were carried out on representative bottom muds found at four sites located at the inner bays and the lake of Japan. Water content, w_N, liquid and plastic limits, w_L and w_P, liquidity index, I_L and ignition loss from the tests are presented in Fig.2. The values of w_N of the upper 30cm at A and C bays reach 2 to 3 times of w_L and from the values of I_L, it is suggested that the bottom mud can be classified into 1) very soft and sensitive mud with I_L from 2 to 5 and 2) soft mud with I_L ranging from 1 to 2. Ignition loss indicates that the bottom mud is classifiable as an organic clay.

SHEAR STRENGTH OF BOTTOM MUD

The shear strength of bottom mud with $w_N = (2\sim3)w_L$ can not be measured by the conventional method such as unconfined compression test. Noticing that vane shear test, VST, and cone penetration test, CPT, are much more convenient for this purpose, shear strength measurements with VST and CPT were conducted on the bottom muds as presented in Fig.2 (Gomyoh and Mimura,1992). VST were

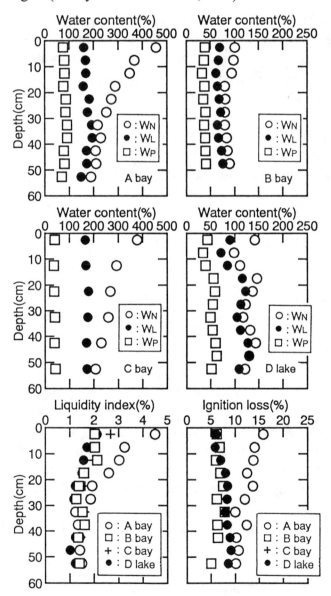

Fig. 2 Physical properties of bottom mud

done on two samples obtained by diver and stationary piston sampler from a pontoon, while CPT with measurement of pore water pressure was made from the pontoon. The vane shear strength, $\tau_{f(v)}$ and point resistance evaluated by $q_T - \sigma_{vo}$, where q_T is corrected cone resistance and σ_{vo} is the total overburden stress, are plotted versus depth in Fig.3. The following comments can be made:- 1) $\tau_{f(v)}$ values significantly varies regionally, 2) $\tau_{f(v)}$ and $q_T - \sigma_{vo}$ values increase with increasing depth showing very low values in the top 20cm, and 3) there is no significant difference in $\tau_{f(v)}$ values on samples by diver and stationary piston sampler. It has been shown that there is a unique correlation between $\tau_{f(v)}$ and $q_T - \sigma_{vo}$. The values of $q_T - \sigma_{vo}$ are plotted versus $\tau_{f(v)}$ measured at D lake in Fig.4. Though there are significant scatters, $\tau_{f(v)} = (q_T - \sigma_{vo})/(6 \sim 10)$ was obtained. Because of easy and faster operations of CPT when compared with VST, and continuous measurement of q_T more than 10 times that of $\tau_{f(v)}$, CPT is recommended for the measurement of the shear strength of bottom mud after $\tau_{f(v)}$ versus $(q_T - \sigma_{vo})$ correlation is obtained.

METHODS OF MEASURING THICKNESS OF THIN SAND LAYER

In order to determine the thickness of a sand layer only several tens of centimeters thick, both survey of the bed before and after sand capping, and sampling method have been used. The survey method always gives a thinner thickness because of the settlement of sand into bottom mud which is inevitable, and the sampling method also gives thinner one because of the volume contraction of sand induced through sampler penetration into loose sand. In this section of the paper, a new procedure for determining the thickness of thin sand layer placed on bottom mud using piezo-CPT, which is an important subject in quality control for the sand capping method, is introduced based on a series of laboratory and field tests results.

Laboratory Tests Result

Since the value of q_T is affected by the compressibility of ground located 20 to 30cm below the end of cone, it is difficult to find out the boundary between sand and mud from q_T value alone. In order to overcome this difficulty, it will be a solution to investigate the difference of generated excess pore water pressure, Δu, behaviour between sand and clay as discussed in detail by the authors (Gomyoh et al.,1993).

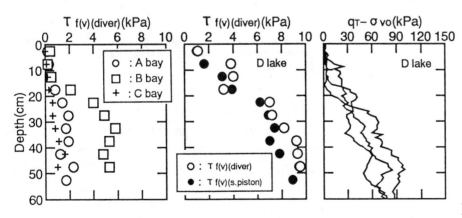

Fig. 3 Shear strength of bottom mud

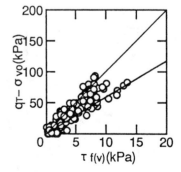

Fig. 4 Relationship between $\tau_{f(v)}$ and $q_T - \sigma_{vo}$

A series of CPT with piezometer installed at the end of the cone was carried out as indicated in Fig.5 where sand layers of 20cm and 30cm thick, and a clay layer of 20cm thick were artificially prepared in an acryl pipe with diameter of 50cm. A typical test result is presented in Fig.6 where $q_T - \sigma_{vo}$ and the normalized Δu over $q_T - \sigma_{vo}$ are plotted versus depth. The following comments can be made: --1) $q_T - \sigma_{vo}$ value increases with depth but turns to decrease in the sand at a depth of 23cm, 7cm above the boundary, which demonstrates that q_T value dose not indicate the boundary correctly as pointed out above, and 2) $\Delta u/(q_T - \sigma_{vo})$ value clearly changes its behaviour at the boundary. The laboratory tests were also carried out on muds from A Bay and D lake and the same results as mentioned above were obtained (Gomyoh et al.,1993).

Measurements in Actual Sand Capping Works

In order to prove the efficiency of the above measuring method for determining the thickness of sand, Ts, field measurements of Ts were made in actual sand capping works in C bay and D lake, where dredging of surface mud to a depth of 20cm was done for D Lake. Outlines of the works and investigation methods of Ts are summarized in Table 1. The values of $q_T - \sigma_{vo}$ and $\Delta u/(q_T - \sigma_{vo})$ measured at D lake are presented in Fig.7 together with location of sand determined from ground level survey and settlement measurement, which is suggested to give the real Ts. A sudden increase of $\Delta u/(q_T - \sigma_{vo})$ shows quite good agreement with the boundary between sand and clay, while it is difficult to determine the boundary depth from $q_T - \sigma_{vo}$. The values of Ts determined by $\Delta u/(q_T - \sigma_{vo})$ are plotted versus Ts from ground level survey and settlement measurement for D lake in Fig.8(a) and it can be said that both Ts show good agreement with each other. In Fig.8(b), Ts from sampling are plotted versus Ts from CPT. Ts values from sampling are about 90% of Ts from CPT on average.

SETTLEMENT MEASUREMENT

Proper evaluation of the settlement of the bottom mud induced by sand capping is also an interesting subject. Settlement plates were installed on the surface of bed at C Bay and D Lake, and the settlement values were 19cm

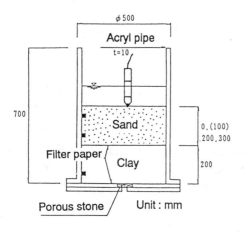

Fig. 5 Experimental apparatus

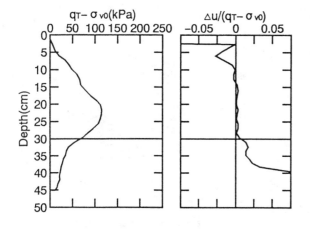

Fig. 6 A typical result by piezo-CPT

TABLE 1--Outlines of sand capping works and investigation method of Ts

	C Bay	D Lake
water depth	3 m	1.5 m
sand capping area	19,190 m²	22,000 m²
dredging of bottom mud		20 cm thickness
thickness of sand	30 cm	20 cm
sand capping method	dry method with backhoe	special hydraulic method
field survey for Ts 　ground level survey 　settlement measurement 　CPT 　sampling	 5 point 12 point 12 point	staff survey 13 point 24 point 24 point

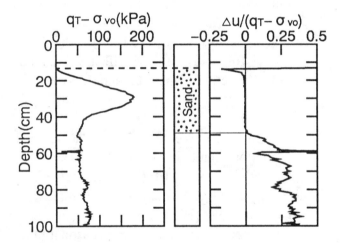

Fig. 7 Sand thickness determined by CPT

and 2.7cm on average, which is consistent with the shear strength difference in the two areas as indicated in Fig.3.

Settlement was evaluated with eq.1.

$$S = P(1-\nu^2) \cdot Is \cdot B/E \qquad (1)$$

where P:stress induced by sand
　　ν :Poisson's ratio
　　Is:settlement coefficient
　　B:width of sand placed
　　E:elasticity modulus (E=100$\tau_{f(v)}$)

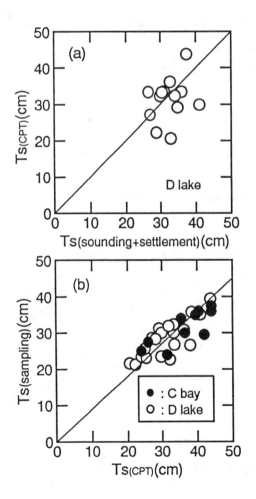

Fig. 8 Sand thickness measured by different methods

The values used for calculation and the values S are summarized in Table 2, where the width of sand placed was assumed to be 1.0m with consideration on the careful construction work. The calculated S values were 16cm for C Bay and 1cm for D Lake, respectively.

TABLE 2--Settlement calculations

	C Bay	D Lake
P (kPa)	3.8	3.1
ν	0.5	0.5
Is	3.70	3.70
B (m)	1.0	1.0
$\tau_{f(v)}$ (kPa)	0.2~0.5	2.5~5.0
S_{cal} (cm)	16	1.0

CONCLUDING REMARKS

Evaluation of the shear strength of bottom mud and the thickness of thin sand layers for the sand capping method have been presented. The following conclusions can be made from the study.

1) The shear strength of bottom mud significantly differs regionally.
2) VST and CPT are effective tools for evaluating the shear strength of bottom mud.
3) Thickness of thin sand layers can be obtained with reasonable accuracy with piezo-CPT from the difference in $\Delta u/(q_T - \sigma_{vo})$ between sand and clay.
4) The sampling method which has been used for measuring Ts gives thinner Ts than actual one, which is quite natural because loose sand contracts during sampling.
5) Settlement evaluated by the ordinary method is in a reasonable range when compared with the measured value.

REFERENCES

Gomyoh, M. and Mimura, N. (1992): "Shear strength and rheological properties of undisturbed soft mud", Proc. of Coastal Eng., Vol.39, pp. 501 - 505 (in Japanese).

Gomyoh, M., Fukasawa, K. and Pradhan Tej (1993): "Measuring method of thin sand layer placed on very soft mud (No.1)", Proc., 48th Annual Conf., JSCE, pp. 1484 - 1485 (in Japanese).

Horie, T. (1987): "Nutrient cycle model and its application to the prediction of seawater and sediment improvement", Rep. of The Port & Harbor Res. Inst., Vol.26, No.4, pp. 57 - 123 (in Japanese).

Horie, T. and Horiguchi, T. (1989): "Predictive model for analyzing the effect of seabed improvement on marine fauna", Proc. of Coastal Eng., Vol.36, pp. 859 - 863 (in Japanese).

Ohtuki, T. and Gomyoh, M. (1991): "Properties of bottom mud in Tokyo Bay", Hedoro, No. 50, pp. 53 - 61 (in Japanese).

Sasaki, J., Isobe, M., Watanabe, A. and Gomyoh, M. (1993): "Poor oxygen phenomenon and model for seasonal variations of dissolved oxygen in Tokyo Bay", Proc. of Coastal Eng. in Japan, Vol.40, pp. 1051 - 1055 (in Japanese).

Geotechnical Aspects of Environmental Violations in Cryolitic Zone

V.I. Grebenets
Research Institute of Bases and Underground Structures, Norilsk, Russia

A.B. Lolaev
Norilsk Industrial Institute, Norilsk, Russia

D.B. Fedoseev
Research Institute of Bases and Underground Structures, Norilsk, Russia

V.A. Savchenko
Committee of Environmental Protection "Norilskecology", Norilsk, Russia

Abstract: Mining of minerals and metallurgical industries are major industrial enterprises that have brought economic development to northern regions of Russia. One of the largest areas is Norilsk industrial region (Northern Siberia). However, a high level of industrialisation significantly affects the geocryological conditions of these regions. In-situ study results indicate tendencies of permafrost degradation of the top geological horizons which are characterised by increasing of the active layer depth, permafrost temperatures and technogenic salt concentrations. Most of technogenic violations relate to the urban territories, industrial zones, places of mining and metallurgical waste landfills creating irregular encircled zones of natural landscapes with different levels of ecological disturbance. The special geotechnical environmental measures directed in permafrost controlling and stabilizing of technogenic complexes are very important in the conditions of the industrial influence increasing and global climatic heating.

Introduction

In Russia rich deposits of minerals are located in permafrost regions and severe climate. Their intensive exploitation in the last 50-70 years led to the formation of the vast zones with disturbed environmental balance. Norilsk industrial region is one of the biggest zones of exploitation and processing of non-ferrous metals ores. The technological influence of the last decades led to significant changes in the frozen soils, as well as to the development of the dangerous cryogenic phenomena and thus to the necessity of the working out of special geotechnical measures for the stabilisation of the environmental situation. Most of the disturbance of the technological origin relate to the previously exploitated construction sites, industrial sites, places of storage of waste materials of coal and metallurgical industries around which are located concentric natural complexes with various degrees of technological disturbance. Concentricity of spreading of the nature-technological complexes is disturbed due to the relief, intensity of the technological influence, location of large pollution objects, difference in the frozen geological structure, wind activity.

The polar cities are seats of the concentrated technologicalimpact on environment. Besides the mechanical impact an important factor is heat impact, different kinds of which are realized through temperature field. Not only natural, first of all climatic geocryologic factors and processes, but also technological systems and various elements of economic activities can greatly change the existing heat state of naturally developed, long-term frozen soils, their engineering and construction properties. In mountain regions of cryolitic zone under technogenesis dangerous glacial and cryogenic processes take place (Grebenets and Fedoseev, 1992).

It is typical that technogenic changes of the geocryological factors start before the exploitation of the territory and construction works on it: degradation is due to acid rains, snowfalls, refuse, drive in tundra with flora disturbance in summer and snow packing in winter as well as temporary construction objects - all this promote emergence of degradation phenomena. These phenomtna are developing at the stage of construction or exploitation of the deposits. The used technologies of foundation building with wet processes while drilling, concreting of foundations and other underground constructions bring into the depth of the frozen soils additionalheat and very often contaminants which store in the seasonly melted active layer or on the surface penetrate into the permafrost.

For example, completion of several foundations on piles 30 - 45 m in depth in Talnah (Northern Siberia) on the territory with high temperature frozen soils (0°C up to -0,8°C) due to heat with the wet technology drilling of holes led to melting of permafrost and destroying of the developed natural- and technological complexes on the territory twice as large as the construction site.

It is evident that the contemporary building works with the developed infrastructure in permafrost regions with the cyclic type of weather lead to the degradation of the frozen soils. The dynamics of the general temperature field can be seen from the deep layers temperature data.

In 1959 in the center of Norilsk 200 m - hole D-1284 (K-203) was bored and regular temperature observations were carried out up to 1995 (Fig. 1). The temperature changes of soils in the period of 30 years on the depth of 10...60 m was 0.5 ... 1.0°C higher showing the degradation of the frozen depth within the precincts of the city.

Changes of the natural environment is very actively continuing on the built-up territory under the influence of many factors, which can be conventionally divided into three groups, such as general, local and specific factors.

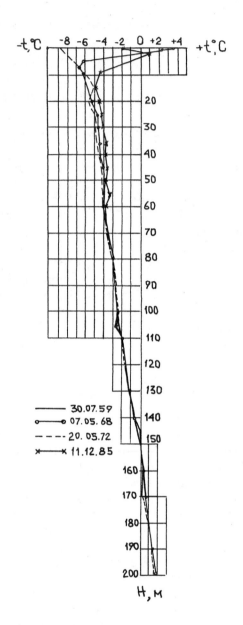

Fig. 1. Changes of permafrost temperature in the central part of Norilsk (Northern Siberia)

Factors of environmental influence

Of the general factors an important one is the influence of the economic activities in the exploitation of the territory on the radiation balance. It is manifested in the weaking of the intensity of the total radiation, decrease of the length of the sun-shine, increase of the absorption of the short-wave radiation and reduction of the effective earth radiation. For example, Norilsk is one of the most unfavourable cities of the country: 2.4 mln. tonnes of sulpher dioxide, dozens of tonnes of dust, chlorine, nitrogen, exhaust gases of the automobiles are polluted into the atmosphere every year. This considerably decreases air transparency of the atmosphere and heat penetration in summer. In summer period contaminants penetrate into the seasonly melting layer and ground temperature becomes higher due to the increase of the heat conductivity of the soil on the "foot" of the seasonly melting layer and in its depth. Experiments were conducted to study the rate of penetration of sulphate salt into the soil during the freezing and thawing processes. Aqueous sodium sulphate salt solutions were added to the surface of a laboratory prepared sample before the beginning of the thawing processes. The time of thawing was kept constant. Three cycles of repeated thawing and freezing were conducted after each addition of salt. A total of ten salt additions were made and studied. This simulate about ten years of salt accumulation in the field. The test results indicate that the depth of active layer increases with each addition of salt. The increase in the depth of the active layer is about 8.4 per cent for ten additions of salt. Simultaneously the temperature of the frozen section of the soil specimen also increases with each addition of salt.

On the built areas considerable changes in turbulent heat circulation take place and this occur under the influence of the specifity of the architectural designs of blocks of houses and separate buildings as well as density and the amount of storeys of the buildings. Turbulent mixing in the near-Earth atmosphere layers is one more factor barring radiation cooling of the surface of the soil and thus stimulating degradation of permafrost. Aerodynamics laws and tendencies of the moving of air masses, with which to a great extent are snowstorage and snow-dislocation connected, depend on the type of upbuilding and gaps in it and on the amount of storeys of the build: in the passages of the overwinded perimeter upbuilding 1.5 m wide along the streets, coaxial with wind direction the velocity of the air flows 1.1 ... 1.3 times exceeds that outside the city; and at the entrance areas of the city facing the overwinded side 1.5 ... 2.0 times, and it makes these areas a kind of wind-tracking pipe-sockets, and the nearby streets - aerodynamic canals. It is natural that in snow-storage areas the increase of the temperature of the frozen foundations is more considerable.

Significant aspect is the shadowout of the surface caused by the neighbouring location of the buildings. According to our natural observations over the thermometrical holes made by the perimeter of the nine-storied house at the distance of 1.5 ... 1.8 m from it, it was found that for eight years of exploitation the temperature of the long-term permafrost at the depth of 8 - 10 m at the northern side of the building become 0.2 ... 0.4°C lower, but the temperature remained the sameat the other sides of the building. Ventilated undergrounds, collectors for engineering communications, technological pourings as well as the conditions of snow storage at the built-up areas change the ground conditions and lead to the beginning of permafrost degradation. Changes of geothermal permafrost regime on the construction site with thick artificial ground layer (thickness of the artificial ground layer is about 8 ... 10 metres) in Norilsk are shown in Fig.2. Active layer depth have increased under engineering collector, at the places of snow accumulation and at the southern slope. At the same time the active layer depth have decreased under the road and under building with cold first floor (cold underground), while the

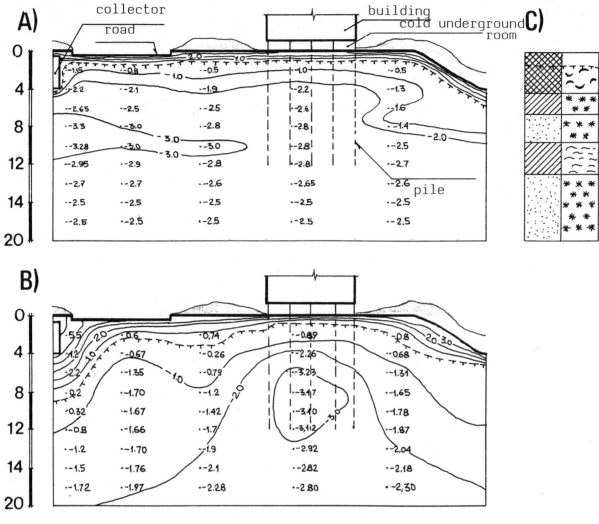

Fig.2. Changes of permafrost thermal regime under technogenic influences.

A) initial temperature   B) situation after 10 years
C) geocryological conditions under the building

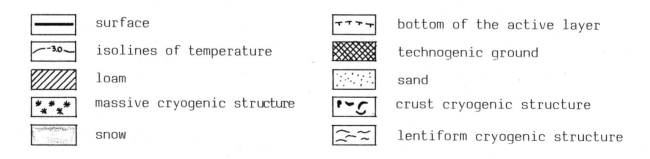

bearing capacity of the frozen grounds increased. The Extreme North natural conditions have predetermined the building construction and exploitation with permafrost preservation. Usually buildings have cold ventilated undergrounds as a geotechnical method of monitoring of the thermoregime of soils. During building exploitation the undesireable phenomena occur: watering of the underground rooms, filling with snow of the ventilation paths, heating from the communications and others. For example, as a result of the disbalance in the exploitation of the underground rooms the deformation affected about 35 per cent of all buildings in Dudinka (Northern Siberia). The system of underground engineering communications with the use of special underground two-storied collectors utilizers for joining of nets serves as means of directed regulation of the heat impact upon the frost situation. It is considered that due to the natural air convection in winter inside the underground canals freezing of the soils occurs. Our natural long-term observations (Grebenets, 1992) showed that inside the canals positive temperature is fixed practically all the year round (+10 ... 14°C for the upper tier; +5 ... 10°C - for the lower one). The existing natural ventilation is uneffective, for example air circulation between the tiers does not exceed 10 per cent. As a result of this phenomenon, even in winter, grounds around collectors don't freeze. Taking into account changeability of the grounds differences in structure and iciness and also the considerable (up to 10 ... 30°C) variability of the heat emission over the cutting and in length and height, uneven melting takes place which causes breaks of the collectors and penetration of the sewage waters into the grounds, which in turn still more activates the process of melting. When the grounds are very reach in ice the thermokarst processes occur. As a rule collectors are interconnected and present a powerful grating-type technological system bored into the ground at the depth of 4 ... 6 m from the surface. The sizes of melting zones around the collectors are 5 ... 8 m in the middle of the streets and places with regular snow cleaning - 3 ... 4 m. Being a basis for erosion the collectors stimulate the development of the underground water paths which are fed by the surface waters from rainfalls and by the sewage waters from engineering communications.

The average annual temperature of the grounds containing the collectors and within the limits of melting zone is from 0.1°C up to +5°C.

Artificial technogenic grounds from poorly sorted coal and metallirgical waste materials are widely used in northern cities. Overcoming the moss flora they destroy its thermoisolation function and this actually stimulates the increase of heat flow into the soils. Besides, significant coefficients of the filtration of the poured materials (10 ... 50 m/day) provide penetration of the surface waters, about 10 - 15 per cent of the heating effect being infiltrated by the rainfalls and high water and the rest-by bringing heat into the soils with the break-down wastes from the communications. For example, constructed in 1978-80 nine-storied houses at the northeast side of Norilsk, the foundations of which contained powerful technological pourings, in some years underwent considerable deformations based on the loss of the carrying ability of soils while melting and weaking. The area snowing and poor sorting of the artificially layed materials blocked the convension of the outer air inside the pouring in winter at the same time permanent escapes from the engineering nets led to the heating of the frozen layers, the temperature of the soils under the available pouring changes from +2°C to +2.5°C. Our calculator data showed that technological wastes being absent and regular snow removed from close to building territories, pourings have the cooling effect, the temperature of the soils lowering up to -2.5 ... 4.5°C.

Redistribution of snow within the limits of the construction area causes significant differences between this snow and that in natural conditions (the regime of snow storage is meant). In small polar towns and villages with low density of upbuilding (less than 10 per cent) snow storage in 80 per cent of cases is happening on the shadowy (as to the main winter winds) sides of buildings, the frozen soils temperature rises at the snow-drift areas and the carrying ability of the soils reduces, that causes deformations of buildings as a result of irregular setting of the foundations.

In large cities of the cryolitic zone which have good snow removing services the blowing paths of the underground rooms and surrounding territory of the building are regularly cleaned as a rule. But snow storage in big (from 1.5-2 up to 5-7 m) waste heaps causes the emergence of the local heating zones. Significant differences in the snow storage for various blocks of houses of the cities lead to the overwatering of some collectors and this stimulates thermokarst processes. In our observations we found that the temperature of the soils on the "foot" of the heat-turns at the areas of the permanent snow storage is 1 ... 2°C higher than at those periodically cleaned out and 2 ... 3°C high than in the soils under the roads regularly cleaned out from snow.

Conclusions

While exploitating the polar regions the disbalance of the environment should be as little as possible. Concerning our task this means that the average integral temperature of the surface should not exceed the temperature of the frozen soils at the depth of the zero annual amplitude of the season oscillation of temperature (it should be lower or equal).

There are two ways to receive the required meaning of the average integral temperature: 1) to control the balance of the components of the territory under construction (the extensive factors); 2) to control the surface temperature within each component (the intensive factors).

The task of changing the balance of elements of the territory under construction is solved by changing the density of upbuilding and by the conformity of the areas to the actual preparation of the surface (irrigation, pleanting- asphalting and concreting). In the conditions of the polar cities the increase of the density of upbuilding decreases the temperature of soil due to cooling-out of its surface in the ventilated underground rooms of the buildings - on the one hand, and to increase of the temperature due to heat effect on the second hand. Generally if there are no wreeks in geotechnical systems functioning the cooling effect increases with the increase of the density of upbuilding.

The means of control of the temperature of the soil surface within the limits of the territory under construction are: 1) some elements of geotechnical systems providing their normal functioning (underground rooms, collectors, pourings); 2) additional devices for cooling of soils (devices of natural and artificial cold); 3) snow-fencing of the territory under construction and arrangement of snow removing service. Calculations and natural observations in the normally ventilated underground rooms show that in the conditions of the Far North it is possible to reach the temperature of the soil surface in the underground room equal to 5.5°C. For this it is necessary: 1) to provide normal functioning of the underground room all the year round making the required number of blowings out and their location (in the windy side of the building) above the snowstorage level; 2) if necessary there can be made additional ventilation mines provided with to enforce the air extraction; 3) usage of hard waterproof coverings for the surfaces of the underground rooms breaking draining of waters from communication; 4) arrangement of asphalt road sides around the building for the removal of the rainfall and high water.

To lower the heat load from the underground heat radiating collectors for communications it is necessary to provide freezing - through of the melting soil area forming around them. For this purpose it is necessary to supply extracting mines of the collectors by the air-convective coolers for reinforcement of the air circulation inside the layers in winter; in the hard frozen soils - to make stimulatively ventilated cavities beneath the bottom of the collector, to equip parallel to it vapour - and liquid thermodevice with horizontal pipebends. It is recommended to make the upper overlap of the canal over the daytime surface and not to store snow here. It is necessary to make antifiltration of clay both around the canal and along the washer across the trenches. Constant meltings around collectors being liquidated removed it is possible to lower the temperature of the surface within the limits of this element of upbuilding from +3.0°C up to +0.1°C.

Making technological pourings it is advisable to use the material the thermophysical properties of which stimulate the cooling effect. It is important to observe the principle: the coefficient of heat conductivity multiplied by the humidity meaning of the latter should be less than the analogous parameter of the soils laying below pouring. It is necessary to make antifiltration horizontal screens, for example of clay, polyethylene films protecting from watering of pouring and lower laying soils. In the region of cyclon type of weather the principal way of control is monitoring of snow storage. For example in the cities of North-West Siberia by snow cleaning it is possible to reduce the average annual temperature of the surface of the soil by 2 ... 3°C. Special architectural and design methods as well as creation of snow-fencing constructions near up building reduce snow slopes by 1.5-2 times. Alongside with the mechanized snow removing from the highways and in-block areas it is necessary to carry out snow beyond the city boundaries.

The above mentioned measures promote significant cooling effect and decrease of the average temperature of the open territories as well as of inblock ones up to +3.5°C. Local freezing of permafrost soils is very important for the thermal regime control. During last decade we carried out the investigation and full utilization of the various ways of artificial lowering of soil temperature both in the foundations of civil and engineering objects and of hydrotechnical constructions. The investigation of the air-convection coolers, liquid seasonly cooling devices and (in crashing situations) mobile cooling stations showed that with their help it is possible not only to freeze local, but also to provide transferring of soils into the hard frozen state, i.e. to reduce their temperature up to -2 ... 5°C. Especially effective were vapour-liquid thermal devices (Grebenets,1990). In the course of investigation the design of the thermal devices was improved developed for the deep (10 ... 15 m) soil freezing as well as ones with the horizontal pipebends. As a rule the temperature of the frozen rocks is 3 ... 6°C lower that in the natural conditions before the construction of the building. Geotechnical aspects of exploitation of the environment are of the first-hand significance for the present-day research of the polar geotechnology. Accomplishment of the two-sided task of preservation of the environment and reasonable exploitation of the permafrost zone will allow in the XXIst centure with profit for mankind to reach new limits in the civilized creation of the harmonic world.

References

Grebenets V.I., Fedoseev D.B. 1992. Industrial influence on glacial process in mountains of circumpolar regions. In Annals of Glaciology, 16, 212-214.

Grebenets V.I. 1992. Underground space utilization for communications construction in permafrost. In "Icusees-92" 5th International Conference on Underground Space and Earth Sheltered Structures. Editor L.L.Boyer. Delft University of Technology. Delft University Press. 864-867.

Grebenets V.I. 1990. Antifiltration curtains construction with the natural cold utilization. In Proceedings of the 6th International EREG Congress, Amsterdam, 6 - 10 August, 1990 . A.A.Balkema (Rotterdam/Brookfield), 1285-1287 .

Use of Geomembranes as Vertical Barrier Liners for Containment on the North Slope of Alaska

Paul G. Hansen, P.E.
ARCO Alaska, Inc., Anchorage, Alaska, USA

G. Robert Crotty
CH2M-Hill, Anchorage, Alaska, USA

Drilling activities associated with development of Alaska's North Slope oil fields since the late 1960s have generated drilling wastes, mostly drilling muds and cuttings, which typically have been stored in above-grade reserve pits formed by constructing gravel berms on the tundra, adjacent to the gravel drill pads. The tundra and underlying thick, continuous permafrost served as a pit floor, and lining materials were not used. During the windswept winter months, extensive snow drifts accumulated in the pits. After the snow melted the following June, the pits were dewatered as necessary to prevent overtopping. It was believed that permafrost would form in the berms, and during summer a frozen, impermeable core would remain. In the mid-1980s, however, it was found the berms thawed and additional measures were needed.

As one response, North Slope operators during the 1984-1991 period installed about 21 miles of "vertical barrier liners" (VBLs) to provide containment around many reserve pits and other facilities. A VBL consists of placing a geomembrane vertically in a narrow, 8-10' deep trench excavated around a facility and then backfilling the trench. The geomembrane forms a curtain extending down through the gravel fill, tundra mat, and native soil, and into the permafrost. The VBL relies on permafrost at the base of the geomembrane to provide a seal, preventing the escape of any pit fluids through the gravel berm and onto the tundra during summer. Vertical containment below the pit floor is provided year-round by some 1,800 feet of permafrost. North Slope VBL installations are best suited 1) to provide containment at existing unlined pits with waste and 2) in other applications, to confine contamination or potential contamination in gravel pads. This paper addresses VBL use on the North Slope.

Introduction

Alaska's North Slope, also known as the Arctic Coastal Plain, is characterized by low hummocky relief, with elevations averaging less than 30 feet at Prudhoe Bay. The typical "wet tundra" terrain is dotted with thaw lakes (covering up to 60-80 percent of the land mass), wetlands, and rivers draining northerly to the Beaufort Sea.

The Arctic climate at Prudhoe Bay is severe. Annual temperatures average about 10°F, with less than 10 inches of precipitation, generally in the form of snow. Winters on the coastal plain are long, dark, and cold, with air temperatures remaining below freezing for about 9 months of the year. During January and February, temperatures average about -20°F but often drop to -40 to -50°F. Summer temperatures range from below freezing to 70°F and average about 40°F.

These extremes have created and preserved permafrost on the North Slope for the last 350,000 years. The permanently frozen ground starts just below the tundra surface and extends to a depth of approximately 1,800 feet in the Prudhoe Bay area. Ground ice is common. The permafrost is overlain by a shallow "active layer" (seasonal frost) that begins thawing in late May and starts refreezing in late September. Depending on several factors, especially vegetative cover, the active layer thickness can range from 6 inches to 5 feet.

Since the late 1960s, North Slope oil fields have been developed by placing facilities on gravel pads designed to keep the ice-rich permafrost from thawing and settling during the brief summer. Exploration activities and oil field development have generated drilling wastes, mostly drilling muds and cuttings, which have typically been stored in above-grade reserve pits at drill sites. The reserve pits were formed by constructing gravel berms, typically about 5 feet high, on the tundra next to the gravel drill pads. The tundra and underlying impermeable permafrost served as a pit floor, and lining materials were not used. During the windswept winter months, extensive snow drifts accumulated in the pits. After the snow melted the following June, the pits were dewatered as necessary to prevent overtopping. It was believed that permafrost would form in the berms, and during summer a frozen, impermeable core would remain. In the mid-1980s, however, it was found the berms thawed and additional measures were needed.

The Alaska Department of Environmental Conservation enacted new regulations in 1987 with stricter requirements for reserve pits, including 1) containment, 2) fluid management, or 3) closure. During the 1984-1991 period, the operators of the Prudhoe Bay oil field and Kuparuk oil field (30 miles west of Prudhoe Bay), as well as a local service company, installed approximately 110,000 linear feet (21 miles) of impermeable geomembrane liners at 44 drill sites and other facilities. In addition, annual fluid management programs, consisting of snow removal in April-May and dewatering in June-July, were started at all reserve pits and other containment facilities.

The discussion that follows is based largely upon the authors' experience with vertical barrier liner installations in the Prudhoe Bay Unit - Eastern Operating Area, operated by ARCO Alaska, Inc.

Concept

A "vertical barrier liner" (VBL), also called a "curtain liner" or "perimeter liner", consists of placing a geomembrane vertically in a narrow, 8-10' deep trench excavated around a pit or remediation site, and then backfilling the trench with the excavated material (Figure 1). The geomembrane forms a curtain around the facility, extending down through the gravel berm or pad, tundra mat, and native soil, and into the permafrost. The VBL relies on permafrost at the base of the geomembrane to provide a seal, preventing the escape of any pit fluids through the gravel fill and onto the tundra during summer. Vertical containment below the pit floor is provided year-round by some 1,800 feet of permafrost. North Slope VBL installations are best suited 1) to contain waste in existing unlined pits and 2) in other applications, to confine contamination or potential contamination in gravel pads.

Two VBL configurations have been used: a "single VBL", a single vertical curtain; and a "double VBL", consisting of a liner folded at the middle into a U-shaped envelope. Geofabric has been used at some double-VBL installations for increased puncture protection.

Design Considerations

Liner Selection. A liner material must be chemically resistant and be able to perform in the severe Arctic environment. It is usually installed in winter, when the ground is frozen and trenches do not slough. The liner must be capable of being installed without physical damage in weather as cold as -45°F. During installation, the liner may be significantly stressed as it is unfolded and placed in the trench. At low temperatures, field seaming is difficult and fabric may become stiff and hard to handle. Once installed, the liner must perform satisfactorily for 20 to 30 years in ground temperatures reaching -25°F. At the top of the liner, the annual temperature will range from about -25 to 55°F; at the bottom, about 10 feet down, the temperature will range from about 0 to 25°F.

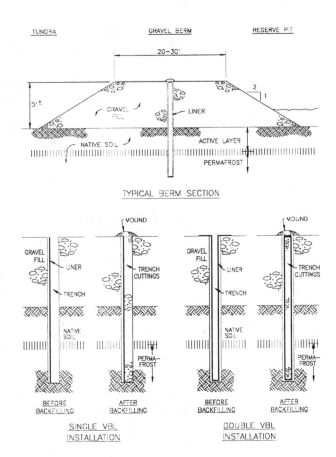

FIGURE 1-- Typical North Slope VBL installation in gravel berms. Some double VBLs include geofabric on the inside and outside of the U-shaped envelope for extra protection.

In 1986 some 23 commercial liners were laboratory tested by BP Exploration (Alaska) Inc., operator of the Prudhoe Bay Unit - Western Operating Area, for cold crack-flexibility; freeze-thaw, to simulate annual temperature swings; and hydrocarbon resistance. Cold-chamber temperatures reached -35, -50, and -65°F.

Based on the 1986 testing, vendor data, and project experience, three liner materials have been used for nearly all North Slope VBL installations: reinforced polyvinyl chloride (PVC), reinforced polyurethane, and high-density polyethylene (HDPE). All have chemical resistance suitable for specific applications, and all have been installed in temperatures well below zero. The flexible PVC and polyurethane liners were prefabricated to the required width and shipped folded on pallets. Two HDPE thicknesses have been used: relatively stiff 80-mil and more flexible 40-mil. The HDPE was supplied in standard 15' and 23' wide rolls.

Liner Configuration. The decision to install a single or double VBL depended on material selection, cost and availability, and technical and installation preferences. Assuming the same material, the double VBL offers more protection; should one side be damaged, the other will provide backup. The double VBL, however, requires about twice as much liner material.

Of the two North Slope VBL configurations, the double VBL was used first, starting in 1984. Reinforced PVC, reinforced polyurethane and 40-mil HDPE have been used successfully in this configuration. The material is sufficiently flexible for conforming to the trench bottom and sides under the weight of backfill. Double VBLs, mostly PVC, total about 85% of the installed North Slope footage. About 40% of the double-VBL footage included nonwoven geofabric for additional protection.

The single VBL was first used in 1990. Robust 80-mil HDPE was used to eliminate panel fabrication, reduce material costs, and simplify installation. A 23' wide roll could be cut in half and installed in a 10' trench, with an allowance for settlement. The thicker material is less susceptible to damage, and it may be as effective as a double VBL of thinner material.

Installation Window. North Slope VBLs have been installed during nearly all months, but mostly during the November to April period when the active layer is frozen and trenches do not slough or fill with water.

In the fall, the active layer starts to freeze in late September and is completely frozen by early November. Working conditions from November to February can be miserable due to darkness, extreme cold, wind and blowing snow, resulting in low productivity. While March is acceptable, the April-May period was found to be the best time because of abundant daylight and daytime temperatures generally above zero. Thawing begins in late May and large projects should not be attempted at this time.

Because trenching in frozen ground is slow, costly, and hard on equipment, summer liner installation was attempted at various Prudhoe Bay drill sites during July, August and September 1987; in addition, the September work included 57 test trenches. During the work, some trenches stayed open until the liner was placed and backfilled; however, others started sloughing and collapsing almost immediately because ground water perched on the permafrost flowed into the trench, causing the saturated gravel to slough. Gains in trenching rates were lost in cleaning out sloughed material. The field study concluded that summer trenching in gravel pads and berms greater than 4 feet thick is too risky.

Liner Routing. For the various installations, liner routes were selected that offered the fewest above- and below-ground interferences while providing sufficient space for workers and equipment. Interferences included buried pipes, electrical and telephone cables, utilidors, above-ground pipelines, and structures. Buried utilities were not always located accurately on drawings, and scrap material was encountered in at least one pad. Narrow berms had to be wide enough for equipment to work, especially at turns. The routes bypassed producing wells at a safe distance to avoid thawed ground around the well.

Where buried interferences were unavoidable, the liner penetration was booted or otherwise sealed to maintain integrity. At some interferences, such as electrical cables, trenching had to be performed with jack hammers and shovels. Trenching was not possible under wide pipe racks, which are up to 20 feet wide. One solution was to bury polystyrene insulation boards just below the surface in order to raise the permafrost into the gravel. The boards overlapped the liner on each side of the pipe rack.

Liner Depth. The VBLs were installed to the estimated depth of maximum thaw, plus a safety factor for unseasonably warm summers and winters. Thaw depths depend upon several factors, including gravel thickness, moisture content of gravel and native soil, air temperatures, and snow cover, and can vary each year.

During summer, thawing progresses rather quickly in the unsaturated gravel and slows once it encounters saturated gravel, the tundra mat, saturated native soil, and any ground ice. Thawing reaches its maximum depth in late September or October. Based on the 1987 test trenches, for pads up to 8 feet thick, thawing generally reached as much as 1-1/2 feet below the tundra surface. For pads greater than 9 feet thick, thawing was limited to about 9 feet and did not extend into the tundra. In gravel berms it was expected to be slightly deeper due to additional warming through side slopes.

While pads and berms are nominally 5 feet thick and generally level, they may range in thickness from 4 to 10 feet due to topography and the tundra's micro-relief. For example, at locations with patterned (polygonal) ground, the gravel fill may be 1-2' thicker at the troughs. The best design approach, it turned out, was to specify a fixed liner depth. A 10' depth is now recommended for most North Slope VBL designs, considering the limit of many trenchers is 10 to 11 feet. While this is usually deeper than required, a larger safety factor is provided. In addition, the incremental cost was found to be minor because trenching in frozen fine-grained, ice-rich native soil is easier than in frozen gravel fill. In one installation requiring a deeper depth, the gravel surface was scalped to lower the trencher, and insulation boards were installed to reduce the thaw depth.

Installation

Trenching. Before trenching began, all underground utilities, such as electrical cables and pipes, were field located, often with the trenching foreman and operations personnel present. A metal detector was found to be useful in checking the route for any buried metal that could damage the trencher and cause delays and cost increases. Trenching was usually not started until all parties agreed to the route, and route changes were documented.

The typical trenching crew included operators, laborers, and an inspector. Nearly all trenching was accomplished using a Vermeer T800B or the larger Vermeer T1450 (Figure 2). Trenches were typically about 1' wide and 8-10' deep. Turns were handled by starting a new trench 90 degrees to the old or by a series of slots. As trenching proceeded, cuttings were windrowed along the trench. Any cuttings spilled on the tundra were cleaned up.

The inspector typically measured the trench depth and pad thickness every 50 or 100 feet with a survey rod or tape measure and checked the trench for sharp protrusions, snow, and cuttings.

A backhoe with narrow bucket was often used to remove any remaining cuttings or drifted snow. Trenches left open longer than usual were covered with plywood for protection and safety. The plywood edges were often sealed with cuttings to keep out drifted snow.

Trenching was stopped short of any buried utility. The obstruction was then uncovered using a backhoe, jack-hammers, and shovels. Occasionally, the trench was enlarged to about 5 feet wide to allow workers to chip or thaw frozen gravel from a pipe or cable to install a boot.

Liner Placement. Before liner was placed, it was visually inspected for defects. For a double VBL shipped folded on a pallet, the liner panel (normally 22-24' wide and 200-400' long) was kept on the pallet or refolded accordion-style on a long painter's pick (scaffold plank). At the work site, the panel was deployed from a loader with forks as it backed up, centered over the trench (Figure 3). Workers lifted the edges and pushed the center area into the trench with a padded pole. The installed envelope was held open and secured at the top using piled cuttings or 2x4 wood runners and blocking until backfill was placed. If a geofabric was specified, the procedure was to unroll the outer geofabric over the trench, deploy the liner, and then unroll the inner geofabric. The 3-layer sandwich was then pushed carefully into the trench.

For 40- and 80-mil HDPE shipped on a roll and used in either VBL configuration, the roll was suspended with a spreader bar from a loader centered over the trench. A single-VBL roll cut in half was about 11 feet wide, a double-VBL roll about 23 feet wide. As the loader backed up, workers unrolled the liner and pushed it into the trench (Figure 4). The top of liner was secured as required.

Liner Seaming. Liner panels were best seamed together by thermal fusion to form a continuous vertical barrier, using a heat gun, wedge welder, or extrusion welder. A double-wedge welder was best for HDPE, and the resulting air channel was pressure-tested; however, extrusion welding was necessary in areas with poor access. Field seams were best made by a qualified installer using equipment and procedures approved by the liner manufacturer, with a rigorous QA/QC testing program.

Backfilling. After liner placement, the trench was usually backfilled soon thereafter to minimize any snow drifting. A small bulldozer was found to be best for backfilling, although a small loader was also used (Figure 5). Cuttings were dropped from no higher than 4 feet to avoid liner damage. Frozen cuttings could not be compacted to conven-tional densities. Those that thawed the following summer settled from their own weight and any wheel loads.

At some later installations, an initial lift of backfill, about 5 feet thick, was placed and saturated with water to help consolidate the loose, frozen cuttings; press the liner against the trench sides and bottom; and facilitate an impermeable zone at the base upon freezeback.

Next, a second lift was placed to within 6 inches of the top of trench. Extra liner was folded over for future retrieval should the backfill settle excessively and drag the liner with it. Other practices included stringing bright survey flagging along the trench centerline and burying a steel plate at each change in direction to help workers find the liner in the future.

Final backfill was mounded over the trench to help offset settlement the following summer. The loader or bulldozer made numerous passes to achieve some compaction, and additional backfill was mounded. Lath was placed to help

FIGURE 2-- A Vermeer T1450 excavating a 10' trench around a Prudhoe Bay reserve pit for a double-VBL installation, March 1989. Note survey rod to check depths.

surveyors as-build the alignment.

Personnel & Equipment Requirements. A typical construction crew and equipment spread included a foreman, laborers, installers, operators, inspector, QA/QC personnel, trencher, small bulldozer, loader, water truck, and backhoe. Having a liner manufacturer's or vendor's representative present was an option.

The installation work was usually orderly, with liner placement, seaming and backfilling immediately following the trenching. An efficient contractor was able to install at least 500 linear feet of liner per 10-hour shift. However, obstacles to a smooth operation included drifted snow, very cold weather, wind, darkness, equipment breakdowns, and trenching interferences, some of which could be formidable.

Monitoring

At a number of installations, thermistor strings were installed along the alignment to monitor ground temperatures and demonstrate that the liner was keyed into permafrost (Figure 6). The strings were installed in plastic pipe within the trench during backfilling, or in a borehole adjacent to the liner. The thermistors, commonly placed at 2-3' intervals down to a few feet below the liner, are usually read at least during late September and October when the active layer is completely thawed.

If the VBL's effectiveness needs to be checked, a piezometer (perforated plastic pipe) can be installed on each side of the liner. During summer a difference in the two water levels should be obvious, once any snowmelt seepage into the pad has dissipated.

Maintenance Work

Maintaining installation integrity is important. After trench backfill settled the first summer, additional gravel was often placed to restore grade. If settlement was significant, the folded liner could be carefully uncovered and pulled up.

Pits with VBLs, like all pits, were dewatered each summer as part of the fluid management program. This averted a high water table and a soft pad condition inward of the VBL.

Typical Installations

Term Well "A" Reserve Pit. A double VBL was installed around an existing 180' x 470' reserve pit in February 1989. About 1,400 linear feet of polyester-reinforced PVC liner (Seaman 8218 LTA in 22.5' x 350' panels) was installed between two layers of nonwoven geofabric (Amoco 4551 and Supac 6NP). The pad thickness

FIGURE 3-- Folded reinforced-polyester liner being deployed from a loader with forks for a double-VBL installation at a Prudhoe Bay reserve pit, February 1989. Note light-colored geofabric layer already rolled out.

FIGURE 4-- 40-mil HDPE being placed as a double VBL at a Deadhorse service company pad, December 1991. A padded 2x4 was used to push and place the liner into the trench. Photo was taken at 10:30 am in the winter darkness, with a full moon and an air temperature of -42°F.

FIGURE 5-- A loader backfilling a double VBL at a Prudhoe Bay reserve pit, July 1987.

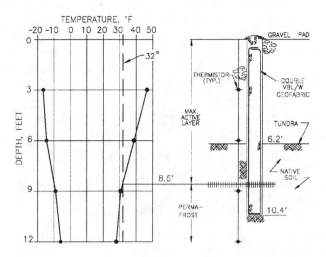

FIGURE 6-- Minimum/maximum ground temperatures for a double-VBL installation at a Prudhoe Bay DS-2 reserve pit, based upon monthly readings from May 1989 to May 1993. The maximum thaw depth was 8.6 feet.

averaged 4 feet. The contractor used a Vermeer T-1450 trencher, Case 450C bulldozer, and Caterpillar 966C loader.

The installation work was uneventful and took about 5 days. Trenching averaged 115 feet/hour, including standby time. During construction, temperatures were relatively warm, ranging from 3 to 31°F. Four thermistor strings were installed to monitor ground temperatures.

Drill Site 16 - Reserve Pit 3. A single VBL was installed around the south end of an existing reserve pit. The reserve pit was subdivided in September 1990 by bulldozing drilling waste aside and installing a gravel berm. The single VBL was installed the following November around the 200' x 900' subdivided area. About 2,300 linear feet of 80-mil National Seal HDPE in 14.8' x 850' rolls was installed in a 10' trench. The pad thickness averaged 6 feet. The contractor used a Vermeer T800B trencher, loader, and bulldozer over 5 days. During construction, temperatures ranged from -29 to 13°F. High winds one day lifted sections of the liner from the open trench. Seams were made with a double-wedge welder. After backfill was placed, liner material protruding above the pad was trimmed.

Deadhorse Pad. A double VBL of 40-mil HDPE was installed in December 1991 around an oil field service company's 12-acre pad in the Deadhorse area near Prudhoe Bay to contain and prevent any pad contamination from reaching the tundra. About 3,000 linear feet of Columbia Geosystems HDPE in 23' x 750' rolls was installed in a 10-11' trench. The pad thickness averaged about 4 feet. The contractor used a Vermeer T1450 trencher, bulldozer, loader, and water truck over 11 days. Working conditions were very cold, from -14 to -44°F, and dark, with only an hour of twilight at midday.

Inadvertent construction metal, such as a grouser bar and bolts, was encountered during trenching at several locations. After the first encounter, the contractor used a metal detector to check the route and make changes. The trencher, proceeding cautiously, still hit metal several more times, and repairs took from 3 hours to 4 days. At obstructions, the contractor used jackhammers and propane burners to remove frozen material prior to installing boots.

The project had several innovations. During backfilling, the initial 5' lift was saturated with water to help consolidate the backfill, fill voids, and press the liner against the trench walls and bottom. For locating the VBL in the future, red flagging was buried along the alignment, with a steel plate at each change in direction. Eight thermistor strings were installed to monitor ground temperatures.

Conclusions

By installing about 21 miles of vertical barrier liners on Alaska's North Slope, oil field operators have developed a method of containment, acceptable to regulatory agencies, that takes advantage of the region's extreme cold and thick permafrost to prevent pit fluids from seeping to the tundra; vertical containment is provided by the permafrost.

Acknowledgments

Cooperation and assistance from numerous individuals at ARCO Alaska, Inc., BP Exploration (Alaska) Inc., and CH2M-HILL are gratefully acknowledged. This paper is presented with permission of the management of ARCO Alaska, Inc. The techniques and conclusions are those of the authors and are not necessarily shared by Prudhoe Bay Unit Working Interest Owners.

Visualisation du déplacement de polluants liquides dans un milieu poreux

Musatapha Hellou, Mohamed El Yazidi
Laboratoire de Géomécanique, Thermique et Matériaux Institut National des Sciences Appliquées, Rennes, France

RESUME

Dans le cadre d'un programme de modélisation du transport de polluants lourds en écoulement dans un sol moyennement perméable, nous visualisons d'abord le déplacement de ces polluants dans un milieu poreux modèle. Nous pouvons ainsi analyser l'influence du régime d'écoulement et de quelques propriétés physiques des fluides sur ce processus de transport. Le milieu poreux est simulé par un assemblage de cylindres en verre.

1. INTRODUCTION

Les écoulements dans les milieux poreux suscitent depuis de nombreuses années l'intérêt des chercheurs. Un tel intérêt n'est pas lié seulement à des considérations théoriques, mais concerne plusieurs secteurs comme, par exemple, l'hydraulique souterraine, le génie pétrolier, l'industrie céramique, le génie atomique, la pollution du sol,...

Les milieux poreux naturels ou artificiels présentent une grande diversité et une structure assez complexe, ce qui rend difficile la connaissance de leurs propriétés. Pour comprendre les phénomènes microscopiques qui ont lieu dans ces milieux, certains auteurs ont fait appel à des modèles macroscopiques. Par exemple, MAALOUF [1] a considéré un réseau de cylindres pour l'étude de la recirculation d'un fluide qui peut se développer entre les pores, ou une coque poreuse pour établir les conditions aux limites à la surface du milieu poreux lors de l'infiltration d'un liquide dans ce dernier. ARIBERT [2] a étudié dans un canal Hele Shaw l'instabilité de l'interface lors du déplacement d'un fluide visqueux par un autre dans un milieu poreux. Actuellement, une équipe de chercheurs, ARNAUD et al. [3], mènent des travaux sur le transport de polluants (gazole routier) dans un aquifère alluvial. La spécifité de leur projet réside dans la conception d'un aquifère expérimental contrôlé, dans le but de vérifier l'efficacité des techniques de reconnaissance et des procédés de décontamination in-situ.

Notre étude consiste également à apporter une contribution à ces problèmes liés à l'environnement en nous intéressant aux mécanismes de transport de polluants, plus denses que le fluide porteur, dans un milieu poreux modélisé par un assemblage homogène de cylindres. Nous nous proposons d'illustrer, par des visualisations expérimentales, l'influence de la nature de l'écoulement du liquide continu sur le déplacement de gouttes dispersées dans ce dernier.

2. DESCRIPTION DU MODELE DE MILIEU POREUX ET DES FLUIDES EMPLOYES

Le milieu poreux considéré est un assemblage en quinconce de cylindres en verre de diamètre 1cm espacés de 1cm également suivant une ligne horizontale. Cet assemblage est disposé dans une cuve transparente en plexiglass, de section rectangulaire, de dimensions 20cm pour la longueur, 4cm pour la largeur et 20cm pour la hauteur. La figure 1 montre une coupe de ce milieu. Les liquides employés sont : l'huile de silicone comme fluide porteur, l'eau, l'huile de glycérine ou leur mélange, plus denses que l'huile de silicone, comme phase dispersée. Nous avons retenu deux huiles de silicone de viscosité cinématique 100 et 1000 centi-Stokes à 25°C. Elles nous permettent de réaliser des expériences avec des mobilités λ différentes (la mobilité est le rapport entre la perméabilité géométrique et la viscosité dynamique). Pour la phase dispersée, le mélange de l'eau et de la glycérine permet de varier leurs densités et leurs viscosités. Le tableau 1 donne les valeurs de quelques paramètres, de ces fluides, qui nous intéressent ici.

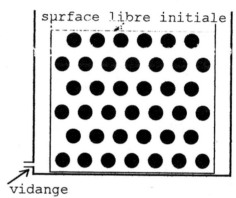

FIGURE 1 :
Modèle du milieu poreux employé

TABLEAU 1 :
Propriétés des fluides employés

	tension interfaciale dynes/cm
eau - silicone 1000	7.5
glycérine - silicone 1000	22
eau + glycérine - silicone 1000	14

	ρ g/cm³	ν (25°C) centi-Stokes
silicone 47V100	0.96	100
silicone 47V1000	0.97	1000
eau	0.99	1
glycérine	1.26	1520
eau(70%)+ glycérine	1.07	3.05

ρ : masse volumique
ν : viscosité cinématique

3. REALISATION DES EXPERIENCES

Dans toutes les expériences qui vont suivre, nous avons considéré le même écoulement du fluide porteur. Celui ci est réalisé par vidange lente à partir d'un orifice situé en bas à gauche de la cuve. La photographie de la figure 2 montre la structure de cet écoulement. La visualisation est obtenue en photographiant des fines particules de rilsan en suspension dans le fluide et éclairées par un nappe laser mince d'épaisseur 1mm. Les gouttes dont on veut suivre le déplacement sont introduites lentement, dans le milieu, pour ne pas leur donner une vitesse initiale importante. Nous avons choisi de les injecter par l'intermédiaire d'une seringue afin de les positionner dans le plan lumineux. Leur volume peut aller jusqu'à 1cm³. Elles sont initialement à une distance suffisante de la surface libre pour apprécier l'influence de celle-ci lors de l'écoulement.

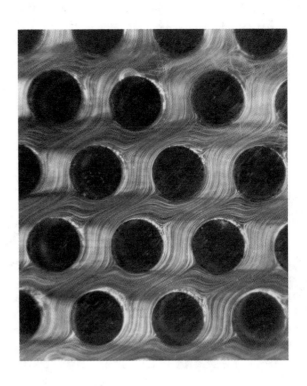

FIGURE 2 Structure globale de l'écoulement du fluide porteur

4. PRESENTATION DES RESULTATS

Nous présentons qualitativement les résultats d'observation du déplacement de gouttes de glycérine, d'eau, et de leur mélange dans chacun des fluides porteurs sélectionnés.

4.1. Le fluide porteur est l'huile de silicone de viscosité cinématique 1000 centi-Stokes

4.1.1. Gouttes de glycérine

Dans ce cas, les gouttes injectées dans le fluide, même immobile, se déplacent facilement par gravité, quelque soit leur volume. Nous avons observé cependant des déformations importantes des grosses gouttes pour s'infiltrer dans le milieu.

4.1.2. Gouttes d'eau

Contrairement à la glycérine, les gouttes introduites dans le milieu s'accrochent toutes à un ou plusieurs cylindres selon leur volume. Par ailleurs, lorsque le fluide continu est en écoulement, les gouttes ne se déplacent que lorsqu'elles sont affleurées par la surface libre. Elles restent pratiquement sur cette surface pendant son mouvement. Cependant, elles restent accrochées aux cylindres lorsque le niveau de la surface libre remonte.

4.1.3. Gouttes de mélange d'eau et de glycérine

Nous avons réalisé des expériences avec des mélanges à différentes proportions. Un des fluides pour lesquels nous avons constaté des phénomènes différents des précédents est le mélange à 70% d'eau et 30% de glycérine. La photographie de la figure 3 montre que deux gouttes 1 et 2 de ce mélange, initialement en contact avec les cylindres, ne subissent aucune déformation ni déplacement lorsque la surface libre est loin. sur cette photo, la surface libre n'est pas visible puisqu'elle est en dehors du champ visualisé. Elle se trouve à la distance 0.3h de la goutte 1, h est le niveau de cette surface libre par rapport au fond de la cuve.

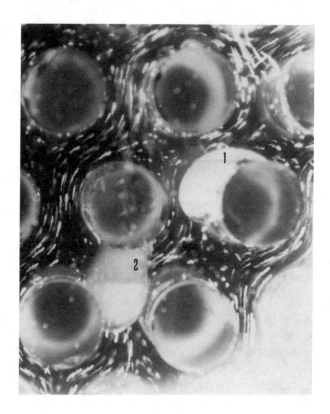

FIGURE 3 : Gouttes de mélange eau-glycérine immobiles dans la silicone 1000

Par contre, sur la photographie de la figure 4, on note un déplacement de la goutte 1 lorsque la surface libre s'en rapproche (cette surface libre est visible sur la photo, elle est notée par L1). En même temps, la goutte 2 s'est scindée en deux parties 3 et 4. La goutte 3, très petite reste en contact avec le cylindre d'origine, alors que la goutte 4 plus volumineuse s'est déplacée pour se mettre de nouveau en contact avec un cylindre de la rangée inférieure.

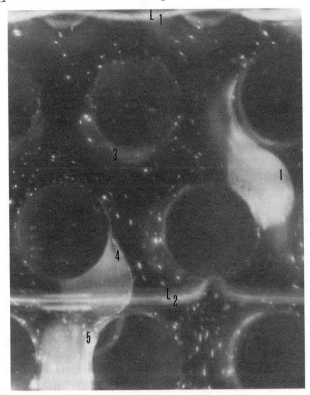

FIGURE 4 : les gouttes précédentes sont en mouvement

Sur cette même photographie a été réalisée une autre prise de vue qui montre l'influence de la surface libre sur la goutte 4 ; la nouvelle position de cette surface libre est notée par L2 sur cette photo. On observe clairement la déformation de la goutte 4 (voir sa trace numérotée par 5).

4.2. Le fluide porteur est l'huile de silicone de viscosité cinématique 100 centi-Stokes

Nous avons visualisé seulement le déplacement du mélange eau-glycérine. Par ailleurs, dans cette expérience, nous avons introduit des gouttes de dimensions plus petites que dans le cas précédent. En effet, le fluide porteur étant dix fois moins visqueux, il offre moins de résistance au déplacement des gouttes.

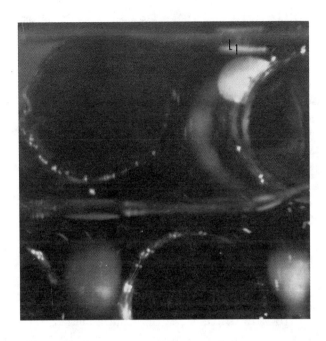

FIGURE 5 : Gouttes de mélange eau-glycérine dans la silicone 100

La photographie de la figure 5 montre que l'influence de la surface libre ne se manifeste que lorsqu'elle est très proche des gouttes mises en contact avec les cylindres au moment de leur injection. Il en résulte un mouvement des gouttes quasiment continu avec des petites déformations lorsqu'elles passent entre les cylindres.

5. CONCLUSION

A partir d'observations directes sur le milieu, nous avons constaté que des polluants plus denses que le fluide porteur peuvent rester accrochés aux éléments du squelette solide lorsque ce fluide porteur est en écoulement lent mais à charge constante. Cependant, les visualisations présentées indiquent que lorsqu'il y a abaissement de la nappe, ces polluants se mettent en mouvement descendant quand la surface libre s'en rapproche. Les premières gouttes influencées sont celles qui se trouvent à l'amont des éléments solides. Par ailleurs, selon la viscosité du fluide porteur, qui simule avec l'indice des vides la mobilité du milieu, les gouttes mises en mouvement, se déplacent de façon quasi-continue ou progressivement jusqu'au substratum imperméable. Les expériences réalisées montrent également que la nature des fluides (viscosité, densité, tension interfaciale) et le volume des gouttes interviennent dans les mécanismes de ce processus de transport. Une analyse plus approfondie sur l'influence de ces paramètres est en cours. La perspective d'étudier l'effet de la nature du squelette solide (en étudiant, toujours du point de vue mécanique, l'écoulement autour d'obstacles en différents matériaux) est également envisagée.

REFERENCES

[1] MAALOUF, A. 1987 Etude numérique et expérimentale de l'écoulement plan autour et au travers d'un cylindre, à faibles nombres de Reynolds ; modélisation d'un milieu poreux par des treillis de cylindres. Thèse de Doctorat ès-Sciences Physiques, Poitiers, France.

[2] ARIBERT, J. M. 1970 Problèmes de déplacement d'un fluide visqueux par un autre. Thèse de Doctorat ès-Sciences Physiques, Toulouse, France.

[3] ARNAUD, Cl. et al. 1993 Un bassin expérimental de grandes dimensions pour l'étude de la décontamination de sites alluviaux pollués. Colloque international "Environnement et Géotechnique", E.N.P.C., Paris, avril 1993, pp. 335-342.

Migration Characteristics and Remediation Efficiency at Organic Compounds Contamination Sites of Volcanic Ash Soil Layer in Japan

Satoshi Imamura, Masanori Shimomura, Toru Sueoka
Technology Research Center, Taisei Corporation, Yokohama, Japan

Abstract :

The infiltration tests of TCE and benzene on unsaturated volcanic ash soil, which is peculiar local soil in Japan, were performed in the laboratory and relative permeability was clarified at various degree of saturation. Furthermore, air permeability was also measured at various degree of saturation. These measured relative permeability are compared with empirical relationships derived by Parker et. al. This relationship is fairly effective for evaluation of relative permeability of volcanic ash soil. However, the theoretical absolute air permeability shows large difference from the experimental results. A field study of soil vapor extraction technique, which was performed at typical TCE contamination sites of volcanic ash soil layer, is introduced. Furthermore numerical simulations are carried out to simulate the underground gas flow by finite element code of multiphase flow and multispecies transport(MOTRANS). In this simulation, air pressure distribution shows a fairly good agreement with observed data, but flow rate cannot be achieved a good agreement with observed data. This paper describes the results of these laboratory and field studies, and compares them with the numerical results.

1. Introduction

Recently, soil and groundwater contamination problems caused by organic solvents such as trichloroethylene (TCE) or tetrachloroethylene (PCE) are getting to be large social problems in Japan. Volcanic ash soil in Japan is a peculiar local soil, and the characteristics are very interesting. Thus it is important to investigate various characteristics of this soil such as multiphase permeability or mass transport parameters. The laboratory infiltration tests of TCE and benzene in unsaturated volcanic ash soil are performed at various degrees of saturation, also the laboratory air permeability tests of TCE are performed.

Soil vapor extraction technique is one of the most effective site remediation measures for soil contaminated by volatile organic compounds. A field study of this technique, which was performed at typical TCE contamination sites of volcanic ash soil layer in Japan, is introduced. Furthermore numerical simulations are carried out to simulate the underground gas flow by finite element code of multiphase flow and multispecies transport(MOTRANS). This paper describes the results of these laboratory and field studies, and compares them with the numerical results.

2. Permeability of Volcanic Ash Soil on Multiphase Flow

Permeant Conductivity

Organic solvents are expected to flow in unsaturated, or saturated soils as two phase (nonaqueous phase liquid or "NAPL" and water) or three phase (air, water and NAPL) porous media systems. However, these theoretical approaches were performed by many researchers, but enough experimental data on multiphase flow in Japanese typical soils have not been collected yet.

Consequently, infiltration tests of TCE and benzen in unsaturated volcanic ash soil were performed [2], as well as the air permeability tests[3]. Infiltration tests were carried out with 10cm diameter mould for permeability measurements as shown in Fig.1. Specimens were adjusted to the same dry density (ρ_d =1.06g/cm^3), and different degrees of saturation. Fig.2 shows the relationships between permeant conductivity and the degrees of water saturation. Permeant conductivity is decreasing according to the increase in degree of saturation. These tendencies are almost similar to their physical properties such as viscosity and surface tension.

Parker et al.[3] proposed the relationships between phase permeabilities, saturations and pressures. This model is described by the three phase extension of the van Genuchten model which takes into account effects of NAPL entrapment on NAPL relative permeability. According to their model, the relative permeability relations are given by

$$k_{rw} = \overline{S_w}^{1/2} \left[1 - \left(1-\overline{S_w}^{1/m}\right)^m \right]^2 \quad (1)$$

$$k_{ro} = \left(\overline{S_t} - \overline{\overline{S_w}}\right)^{1/2} \left[\left(1- \overline{\overline{S_w}}^{1/m}\right) - \left(1-\overline{S_t}^{1/m}\right)^m \right]^2 \quad (2)$$

$$k_{ra} = \left(1 - \overline{S_t}\right)^{1/2} \left[1-\overline{S_w}^{1/m} \right]^{2m} \quad (3)$$

where k_{rp} is the relative permeability of phase p (that is water phase (w), organic liquid phase (o), and gas phase (a)), $\overline{S_w}$ =(S_w-S_m)/(1-S_m) is the "effective" water saturation, S_w is the "irreducible" water saturation, m is porous medium parameter (m=1-1/n), $\overline{\overline{S_w}}$ is the apparent water saturation, S_t is the effective total liquid saturation.

Fig.3 shows the comparison results between experimental relative permeabilities and analytical ones which

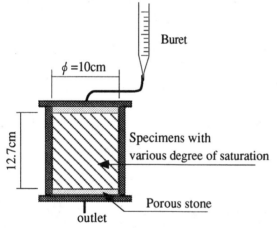

Fig.1 An Overview of Infiltration Test Apparatus

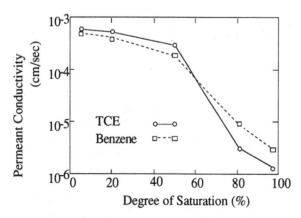

Fig.2 Permeant Conductivity of volcanic ash soil

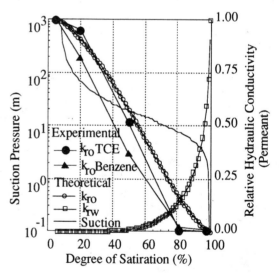

Fig.3 The relationships between relative permeability, degree of saturation and suction pressure of volcanic ash soil

are calculated using Parker's model under two phase states (water-NAPL). Typical relationships between pressure and saturation degree of volcanic ash soil are used in this calculation. These relationships are also shown in Fig.3. The model is fairly effective to evaluate the relative permeability of volcanic ash soil under two phase states. Furthermore, absolute permeant conductivities (Sw=0%) are almost similar to the hydraulic conductivity (Sw=100%). Since relative viscosities of TCE and benzen are approximately 0.6, absolute permeant conductivities can be explained using only viscosity.

Air Conductivity

Air permeability tests were carried out using a triaxial test apparatus, as shown in Fig.4. It is important for measurements of air permeability, to control the generation of air-path between specimen and container, accordingly triaxial cell was adopted to reduce the possibility of air leak. Fig.5 shows the test results. Air permeability is decreasing according to the increase of initial water saturation, however a sharp change is not observed, unlike the test results of infiltration tests. Moreover, the saturation dependency of air permeability increases with the rise of confining pressure. Fig.6 shows the comparison between experimental relative air permeability and analytical ones. The model is not fairly effective to evaluate the relative permeability comparing to the permeant permeability, and the difference between experimental value and analytical value is large (about 2 to 3 times) at the high degree of saturation. However, since air permeability in field often varies a value 2 to 3 times as much, the model can be used for the prediction of air permeability. But, the absolute permeability cannot be explained only using the difference in viscosity. Absolute air permeability is almost 1000 times as large as water permeability,

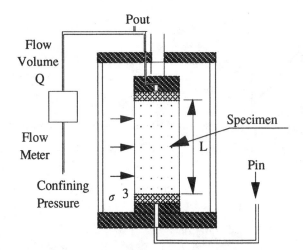

Fig.4 An Overview of Air Permeability Test Apparatus

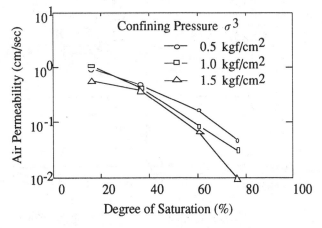

Fig.5 Air Permeability of volcanic ash soil

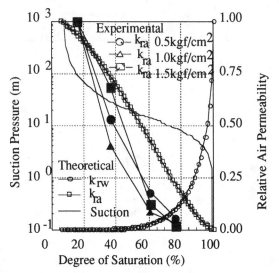

Fig.6 The relationships between relative permeability, degree of saturation and suction pressure of volcanic ash soil

though the viscosity of air is 56 times as large as water viscosity, Volcanic ash soil has well developed frame structures including micro-pore structures, and this micro-pore structures seems to dominate the difference of permeability, that is, air can percolate through the tiny pores which water can not infiltrate through.

3. Field tests for Soil Vapor Extraction Technique

In order to predict the cleanup efficiency of soil Vapor Extraction, it is important to evaluate the flow of soil gas in the field. It is also important to evaluate the many parameters about mass transport. It is really difficult to predict where and how soil gas flows. To understand the flow of soil gas, a series of field experiments were performed at typical TCE contamination sites of volcanic ash soil layer. A top layer of about 12m thick is volcanic ash soil, and underlaid by a gravel layer. Groundwater level is approximately below ground surface 16m in the gravel layer. The location of extraction well and monitoring wells in a field test is indicated as Fig.7, and a sectional view is illustrated in Fig.8.

Experiments are carried out with following procedures.
1) Experiments are conducted by constant volume extraction. The extraction gas volume rate was approximately 140L/min.
2) At first, experiments are performed at the dry condition of the ground surface.
3) Extraction operations are continued to reach the steady state condition, after which monitoring is maintained for 24 hours.
4) Water are sprayed at the ground

Table-1 Soil Parameter

	Depth GL-(m)	Hydraulic Conductivity (cm/sec)	Porosity	Unsaturated Parameter	
				α (m^{-1})	n
Volcanic Ash Soil	0~12	4.5×10^{-3}	0.76	0.10	2.00
Gravel	12~16	2.1×10^{-2}	0.30	5.22	5.69

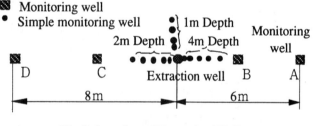

Fig.7 Location of Extraction Well and Monitoring Wells

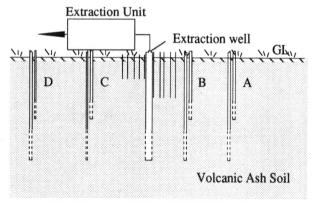

Fig.8 Sectional View of Field Test

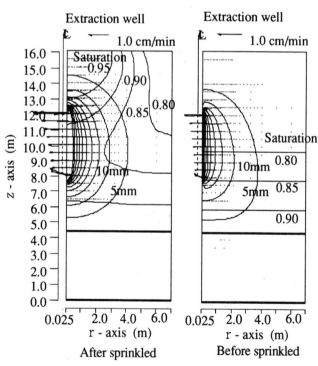

Fig.9 Calculation Results (Steady State)

surface within a 2.5m radius from the extraction well, and experiments are continued for 24 hours at a state where no air intrudes to the underground from the wet surface.

To verify the model described above, verification analysis are carried out. A two dimensional finite element code, MOTRANS [4] is used for verification. Analysis is carried out for the region which is radially expanded 7m, and vertically 16m (ground water level). The determination of effective radius is very difficult, because this parameter is largely affected by permeability and the degree of saturation. In the verification analysis, effective radius is determined by the decrease trend of monitoring pressure. Soil parameters are shown in Table-1. Hydraulic conductivity is determined by a single hole permeability test, however unsaturated parameters of van Genuchten [5] are not observed by experiments. These parameters are determined by the typical existing data [6] and the water content distribution of the site. The calculated relationships shown in Fig.4 are used for verification. Absolute permeability at 0% degree of water saturation is determined from saturated hydraulic conductivity and is considered only the difference of viscosity.

Calculation results at the steady condition are shown in Fig.9. Both the contour on the degree of saturation and flow vector are shown in Fig.9. The change of saturation before and after sprinkled are well simulated qualitatively. Moreover, it is clear that air pressure distribution is equally developed with upper and lower directions after sprinkled. Before sprinkled, air pressure distribution is poorly developed with upper direction because of the supply of fresh air from the ground surface. Furthermore, the flow velocity after sprinkled is larger than that before sprinkled because of higher pressure in extraction well.

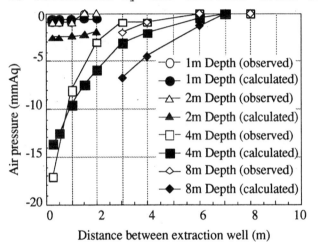

Fig.10 Comparison between experimental and analytical results (before sprinkled)

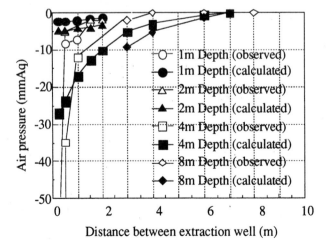

Fig.11 Comparison between experimental and analytical results (after sprinkled)

Fig.10 and Fig.11 show the comparison results between experiment and calculation before and after sprinkled, respectively. As a result of comparison, air pressure are largely affected by sprinkled water, especially at the shallower depth. At the depth of 8m, there is no change on air pressures. The largest difference between experimental results and calculation results is the trend of pressure drop near extraction well. The experimental pressure drop is larger than that of calculation. Especially, in the case of after sprinkled, this pressure drop is fairly large. The cause of this pressure drop is inferred as follows. The degree of water saturation rises up due to the supply of sprinkled water, and the negative pressure of extraction well increases due to the decrease of air permeability. Pore water is gathered to the extraction well for high negative pressure and high flow velocity, as a result, air permeability will be decreased more and more. However, the model assumes the Darcy's flow, and model cannot simulate the phenomena which air flow moves together with water due to the high velocity.

In this verification, air pressure distribution shows a fairly good agreement with observed data, but flow rate cannot be achieved a good agreement with observed data. This difference seems to be caused by the error of absolute air permeability as prescribed in laboratory test. In the laboratory tests, large difference is observed between calculated air permeability and measured air permeability. If this order of difference in absolute air permeability is considered in the field test results, the calculated flow rates shows good agreement with experimental results. A Main mineral of volcanic ash soil is allophane, and this mineral shows large changes of aggregate structures when drying. This phenomena can explain the difference of absolute air permeability.

4.Conclusion and Discussions

The infiltration tests of TCE and benzene on unsaturated volcanic ash soil were performed in the laboratory and relative permeability was clarified at various degree of saturation. Furthermore, air permeability was also measured at various degree of saturation. These measured relative permeability are compared with empirical relationships derived by Parker et. al. This relationship is fairly effective for evaluation of relative permeability of volcanic ash soil. However, the theoretical absolute air permeability shows large difference from the experimental results. This phenomena is caused by the characteristics of volcanic ash soil, which shows a change of aggregate structures corresponding to the change of saturation degree. Accordingly, the determination of absolute air permeability is to be considered in the case of prediction analysis. The change of soil structures is expected to occur, when the NAPL infiltrates into the volcanic ash soil, however, this effects is not so large.

A field test or soil vapor extraction technique is introduced. This test was performed at typical TCE contamination sites of volcanic ash soil. Numerical simulation was carried out to simulate the underground gas flow by finite element models of multiphase flow and multicomponent transport (MOTRANS). The numerical results indicated a fairly good agreement for the pressure distribution. However, calculated flow rates cannot represent well the observed data. This difference is due to the similar reason as for the laboratory results.

In this paper, verification results on mass transport cannot referred due to limit in space. Constitutive laws of mass transport should be verified at field scale level to make the good prediction of soil vapor extraction.

<References>

[1] Imamura,S.,Y,Fukazawa,M. : A study on the Interaction between Hazardous Organic Chemicals and Soil or Groundwater, Taisei Technical Research Report, Vol.24, pp.177-184, 1991 (in Japan).
[2] Imamura,S., Sueoka,T., Nagura,K. : Study on Air Permeability of Volcanic Ash Clay, Proceedings of 28th annual conference of JSSMFE, pp.1953-1954, 1992 (in Japan).
[3] Parker,J.C : "Multiphase flow and transport in porous media", Reviews of Geophysics.27, pp.311-328, 1989.
[4] Environmental Systems & Technologies Inc.:MOTRANS A Finite Element Model for Multiphase Organic Chemical Flow and Multispecies Transport, Program Documentation ver.1.1, pp.1-67,1991.
[5] van Genuchten,M.Th. : A closed form equation for predicting the hydraulic conductivity of unsaturated soils, Soil Sci.Soc.Am.J., Vol44, pp.892-898, 1980.
[6] Nishigaki,M., Kusumi,K. : A Study on Seepage Characteristics of Unsaturated Soil, Symposium on the Properties of Unsaturated Soils, JSSMFE, pp.179-186,1987.

A Weibull-based Reliability Analysis of Waste Containment Systems

Hilary I. Inyang
Geoenvironmental Design and Research (GDR), Inc., Fairfax, VA, USA

Abstract

Waste containment systems degrade as time progresses. This situation stems from environmental conditions which stress various components of any containment system. The generating mechanisms for these stresses are thermal processes, biologic processes, geostatic and geodynamic loads, sunlight, hydraulic processes, and physico-chemical interactions between waste constituents and barrier materials. For a given set of stress conditions, the effectiveness of the entire containment system decays at a rate that depends on the conservatism of its design, quality control during construction, and the frequency of its maintenance. This paper focuses on a conceptual and generalized approach to indexing the potential degradation pattern of waste containment systems. This work is the initial phase of an ongoing effort to develop mathematical models for predicting the performance of waste containment systems using reliability concepts.

Introduction

The effectiveness of a waste containment system is defined in terms of its ability to adequately prevent the generation and migration of waste constituents into the surrounding environment. Systems such as landfills, slurry walls, grout curtains and concrete covers comprise one or a combination of physical barriers. Often, these barriers are made up of geomaterials and/or polymeric materials. Ideally, it is desirable to develop and validate a composite degradation model that comprises submodels, each of which treats the decay of effectiveness of a specific barrier component of the system with time. Towards this end, the durability of various barrier materials under accelerated testing conditions in the laboratory have been vigorously studied within the past fifteen years [1-4]. Unfortunately, the results of laboratory tests alone are insufficient for use in predicting the long term effectiveness of barriers that comprise the tested materials. The effects of long term physico-chemical processes may not be adequately captured in accelerated tests. In addition, in order to extrapolate laboratory test results beneficially to field situations, such data need to be fed into a scheme that accounts for time and spatial scale differences. This paper focuses on the development of such a scheme.

Interactions among relevant factors

The development of numerical relationships between the effectiveness or reliability of multi-component systems and service duration requires the development of a conceptual scheme for environmental processes which may intensify or wane as time progresses. These processes are controlled by several categories of factors among which are system design configuration, site and waste characteristics, intensity of human/animal activity, occurrence of transient events (for example, earthquakes), and hydrological conditions. The variety and numerosity of significant factors imply that any useful attempt at verifying containment system performance models must involve extensive factorial experiments. Obviously, such an undertaking would consume resources excessively.

Furthermore, factor sensitivity analyses implicit to such an approach would not be complete because some factors cannot be fully controlled. Examples of uncontrollable factors are site hydrogeology and some geohazards (for example, earthquakes).

The relatively short experience with modern waste containment systems such as landfills, slurry walls, grout curtains, radioactive waste repositories and confined disposal facilities (CDFs) is such that data is scanty on their long term performance. Notwithstanding the paucity of long-term performance data, numerical assessments of potential performance are necessary for both design and regulatory purposes. Available predictive tools are largely based on the tacit assumption that waste containment systems will exist at the same degree of structural integrity over an extended time period. Also, these tools are usually useful for evaluating a few elements of the overall issue of effectiveness.

Most waste containment systems comprise many components, each of which is susceptible to a different degree, to environmental stresses mentioned earlier. For example, the polymeric materials of a geomembrane layer are more susceptible to ultraviolet radiation than clay liner materials. For shallow-depth and surficial containment systems (for example, landfills), the burrowing activities of rodents may introduce defects into the system in the long term. The uncertainties associated with the occurrence of the phenomena discussed above, and the physical response of composite containment systems to potential synergistic stresses preclude the development of precise, verifiable mathematical models for predicting their performance. It may be difficult to attain a level of accuracy that supersedes that of simple rating schemes exemplified by [5-6].

The conceptual scheme

In this analysis, the effectiveness of a multi-component waste containment system is assumed to vary with time under various scenarios as presented in Figure 1. When built initially, the containment system has an effectiveness Eto, which depends on the conservatism of its design and the adequacy of construction quality control. As time passes, environmental stresses cause a continuous decrease in effectiveness in a pattern described by curve 1. This curve has a convex geometry, implying an increase in the rate of degradation with time. This pattern is typical of constructed facilities and has been recorded for highway pavement structures. A regulatory time frame t_r, corresponding to a minimum acceptable effectiveness E_{tr}, may be specified by appropriate regulatory agencies. In the United States, t_r may correspond to 30-year operational time that precedes the post-closure period for hazardous waste landfills, or 10,000 years of service for radioactive waste repositories. At time t_m, maintenance activities may be implemented at the facility such that an instantaneous gain in effectiveness results as follows.

$$\Delta E_m = E_{tm} - E_2 \qquad (1)$$

ΔE_m = Gain in system effectiveness due to maintenance activity.

E_{tm} = Effectiveness immediately after maintenance at time t_m.

E_2 = Effectiveness without maintenance at time t_m.

Subsequent to the implementation of maintenance activities, the degradation course would follow curve 3.

In addition to continuous degradation processes, transient events such as earthquakes and floods may occur near the site of a waste containment facility. As discussed by Inyang [7-8] and Wright et al. [9], seismic activity poses a threat to the stability of both underground and aboveground containment systems. Ironically, the longer the design life of a facility, the greater the probability of occurrence of a potentially damaging transient event. This is due to the fact that higher impact events have longer recurrence intervals. In Figure 1, the occurrence of a damaging transient event at a time t_g, causes a decrease in system effectiveness as follows

$$\Delta E_g = E_1 - E_{tg} \qquad (2)$$

ΔE_g = Degradation of the systems due to the deformation caused by a transient event

E_1 = Effectiveness of the system at time t_g, without the occurrence of a transient event.

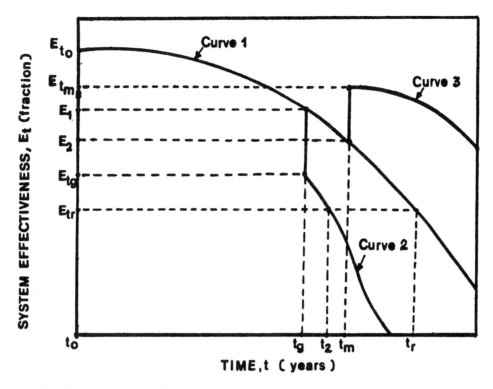

FIGURE 1--Conceptual long term degradation patterns of waste containment systems.

E_{tg} = Effectiveness of the system immediately after the transient event at time t_g.

From a practical standpoint, one of the undesireable effects of the transient event is a subsequent increase in the degradation rate of the facility such that the specified minimum effectiveness E_{tr}, is reached sooner. This implies the closure or remediation of the facility at time t_2, instead of time t_r.

Application of reliability concepts

Reliability can be used instead of the general term "effectiveness" in describing the potential performance of a containment system for wastes. In this case, the reliability of the system is defined as the probability that it will satisfy performance specifications at any future time. The probability that the system will not satisfy performance specifications at a time is the unreliability of the system F.

$$F_t = \int_0^t f(t)dt \quad (3)$$

F_t = Probability of failure from time 0 to t.

$f(t)$ = Probability density function of failures.

$$R_t = 1 - F_t \quad (4)$$

R_t = Reliability of the system at a given time.

The reliability of a containment system can also be described in terms of its failure rate λ.

$$R_t = \exp[-\int_0^t \lambda(t)dt] \quad (5)$$

For a constant failure rate case, equation (5) reduces to equation (6).

$$R_t = \exp[\lambda t] \quad (6)$$

TABLE 1--Reliabilities of a hypothetical waste containment repository at the end of various service time intervals

Time (t-t₀) years	Shape Parameter b	Scale Parameter n, years	$-[(t-t_o)/n]^b$	Reliability R_t (%)
1000	2	9000	-0.0123	98.77
3000	2	9000	-0.1111	89.48
5000	2	9000	-0.3086	73.44
7000	2	9000	-0.6049	54.61
9000	2	9000	-1.0000	36.79
1000	3	9000	-0.0013	99.87
3000	3	9000	-0.0370	96.37
5000	3	9000	-01715	84.24
7000	3	9000	-0.4705	62.47
9000	3	9000	-1.0000	36.79
1000	4	9000	-0.0001	99.99
3000	4	9000	-0.0123	98.77
5000	4	9000	-0.0952	90.91
7000	4	9000	-0.3659	69.36
9000	4	9000	-1.0000	36.79

Waste containment systems are expected to degrade at a rate that generally increases with time. Conceptually, this corresponds to the non-constant failure rate case. The geometry of the degradation-time curve can take many specific forms, depending on the factors discussed earlier. However, the largely convex shape fits the expected degradation pattern. The Weibull distribution provides the range of curve geometries that fits this problem. R_t can be described using the three-parameter Weibull distribution as follows.

$$R_t = \exp[-((t-t_o)/n)b] \quad (7)$$

    t = Time after initiation of degradation.

    t_o = Location parameter at beginning of degradation.

    b = Shape parameter.

    n = Scale parameter.

Equation (7) fits curve 1 of Figure 1 or segments of similar curves without an instantaneous decay in reliability. For increasing failure rate, b>1. The shape parameter value for containment systems is likely to have an approximate range of 2 to 5. This parameter depends on the design conservatism of the containment system relative to the intensity and frequency of occurrence of degradation events. The shape parameter b, is directly proportional to the conservatism of the design of the system. The location parameter t_o, corresponds to the initial time at which R_t = 1. The scale factor n, is essentially a normalization factor for the time interval considered. In order to select an appropriate value for n, the failure function F_t, needs to be plotted against time on a log-log paper. The scale parameter n, is the time duration that corresponds to $F_t = (1-e^{-1}) = 0.632$.

Simple numerical example

Radioactive waste containment systems are intended to serve as effective barriers for 10,000 years. If the degradation pattern of a repository is expected to follow the Weibull distribution with b and n values of 3, and 9000 years respectively, the composite

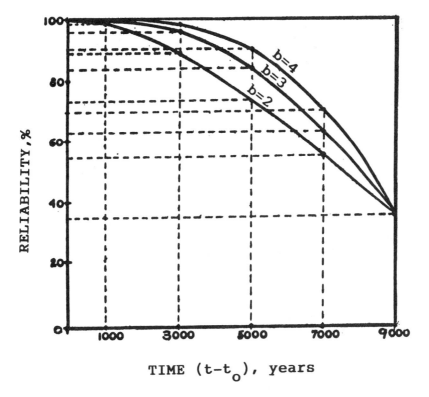

FIGURE 2--Reliabilities of a waste containment system with known Weibull shape and scale parameters.

reliability of the system at the 7000-year mark can be computed as follows.

$$R_{7000} = \exp[-(\tfrac{7000}{9000})^3]$$

$$R_{7000} = \exp[-0.4705] = 0.6247$$

The result of computations for the same problem but with different time intervals and shape parameters are shown in Table 1. They are also illustrated in Figure 2. It should be noted that in this general example, the instantaneous decrease in reliability which can result from the damage of the repository by a transient event is not covered.

Model calibration and conclusion

A logical extension of this analysis is the development of composite models for estimating the probability of failure of containment systems for specific scenarios and time. Following the approach adopted herein, the challenge is to estimate the magnitudes of parameters b and n for such specific situations (without raw, time-distributed data on facility damage). Furthermore, the failure state of each category of multi-component containment system needs to be specified. The failure of a component does not necessarily imply that the entire system has failed.

The system-specific degradation model for waste containment systems that comprise layers of geomaterials must cover the rate of occurrence of physical and physico-chemical phenomena such as cation exchange, barrier material dissolution, and flow channel development due to freeze-thaw action (if applicable), dessication cracking, interface failures and boring activities of rodents. This research is going in that direction. Due to the fact that most of the current configurations of waste containment systems have been in existence for less than three decades, time-distributed failure data are almost non-existent. Many new containment systems are instrumented, hence such data will become more available in the future for use in calibrating degradation models. Nevertheless, it is generally known that all constructed facilities degrade as time progresses. Stochastic schemes will be incorporated into this approach to address the potential of damage of each system component by one or a combination of transient events of specified minimum magnitudes. The utility of this approach is that it can provide the approximate numerical bounds for expected degradation of composite waste containment systems.

References

1. Kanninen, M.F. 1992. Assuring the durability of HDPE geomembranes. ASTM Standardization News, pp. 44-49.
2. Mitchell,J.K. and Madsen, F.T.1987. Chemical effects on clay hydraulic conductivity. Proc. of ASCE Geotechnical Engineering Division, Speciality Conference, Ann Arbor, Michigan, pp.87-116.
3. Daniel, D.E.1984. Predicting hydraulic conductivity of clay liners. Journal of Geotechnical Engineering Division, ASCE, Vol. 110, No.2, pp.285-300.
4. Pierce,J.J. Sollfors,G. and Peel, T.A. 1987. Effects of selected inorganic leachates on clay permeability. Journal of Geotechnical Engineering Division, ASCE, Vol. 13, No. 8, pp. 915-919.
5. Koerner,R.M. and Daniel,D.E. 1992. Better Cover-ups. Civil Engineering, pp.55-57.
6. Inyang,H.I. and Tomassoni, G.1993. Indexing of long term effectiveness of waste containment systems for a regulatory impact analysis. Technical Resource Document, Office of Solid Waste, U.S. Environmental Protection Agency, Washington D.C., 29 pages.
7. Inyang,H.I.1992. Aspects of landfill design for stability in seismic zones. Journal of Environmental Systems, Vol.21, No.3, pp.223-235.
8. Inyang,H.I. 1991. Hazardous waste facilities in seismic zones. AAAS/U.S. EPA Environmental Science and Engineering Fellowship Report, American Association for the Advancement of Science, Washington, D.C., 79 pages.
9. Wright,F.G., Inyang,H.I. and Myers, V.B.1993. Risk reduction through regulatory control of waste disposal facility siting. Journal of Environmental Systems, Vol.22, No.1, pp.27-35.

Application of Cone Penetration Test for a Background Sampling Program

Hank H. Jong, Nestor O. Acedera, Todd F. Battey, Roger R. Batra
Jacobs Engineering Group, Pasadena, California USA

Abstract

Cone penetration testing (CPT) was used during a basewide background sampling program for the remedial investigation/feasibility study (RI/FS) at Vandenberg Air Force Base (AFB), California. The RI/FS includes approximately 50 sites throughout the 98,400-acre military facility. The primary CPT application in the sampling program was to collect soil samples to establish background metal concentrations of different lithologic units underlying potentially contaminated sites. Metal contamination can then be identified and risk assessment can be performed after background metal concentrations are established.

Three main geologic units were identified for the RI/FS based on the differences in depositional source materials and environment. These geologic units include alluvial soils, windblown sands, and weathered bedrock. Each geologic unit may include several subunits depending on the geologic age and the environment of deposition.

To establish the background metal concentrations in the subsurface lithologic units, 30 uncontaminated locations throughout Vandenberg AFB were selected after review of available site use records and aerial photographs. The background sampling program was designed using CPT to distinguish and identify geologic units that showed a distinct pattern of cone tip resistance and friction ratio. After the penetration test, accurate sampling at depth intervals of identified geologic units was achieved using the push cone technology. CPT was also used to identify and sample significantly different soil types within the same geologic units.

The background sampling program was also designed to maximize the advantages of CPT. Investigation-derived wastes were greatly reduced by this drilling technology, which generates minimal cuttings. By using CPT, the sampling program was completed in an extremely tight schedule with cost savings (compared with traditional hollow-stem auger drilling). The main limitation of the CPT was sampling at depth in highly resistant lithology.

Introduction

Cone penetration testing (CPT) was developed in the Netherlands to measure in situ soil parameters which were then used in the foundation design of civil structures. CPT has been commercially available as a geotechnical exploration tool in the United States since 1970s. To satisfy the needs of the environmental industry, CPT was recently utilized to accurately delineate field contamination and obtain subsurface parameters to estimate contaminant transport [1]. ASTM D3441 [2] is a standard procedure developed for electric CPT.

This paper presents discussions and results of an application of CPT for a basewide background sampling program implemented for the remedial investigation/feasibility study (RI/FS) at Vandenberg AFB, California. The background sampling program was designed and successfully completed by using CPT.

General Description of Vandenberg AFB

Vandenberg AFB is located on California's south-central coast, approximately 275 miles south of San Franclsco and 140 miles northwest of Los Angeles (Figure 1). Vandenberg AFB is the third largest U.S. Air Force installation, occupying more than 98,400 acres along approximately 35 miles of the Santa Barbara County coastline.

FIGURE 1 — Location of Vandenberg AFB

Previous investigation [3] has identified about 50 sites at Vandenberg AFB where hazardous materials may have been released. An RI/FS is being conducted at the identified sites to determine the characteristics and extent of potential contamination.

Geologic Conditions

The principal geologic units in the Vandenberg AFB area consist of dune sand, alluvium, and bedrock. Surface soil (0 to 6 inches deep) was sampled as an additional unit because it may contain different background metal concentrations due to natural and artificial processes.

The dune sand unit consists of Quaternary age windblown sand derived from beaches located to the west. These deposits range from medium dense to dense poorly graded sand to silty sand. Generally, the dune sand unit consists of a homogeneous sandy deposit varying from a few feet to a few hundred feet thick.

The alluvial deposits, which are much more heterogeneous than dune sand, were derived locally or from inland areas located to the east. For the background sampling program, the alluvial geological unit includes colluvial and pluvial (lake) deposits in addition to stream-deposited alluvium. These Quaternary alluvial deposits include clay, silt, sand, and minor gravel. Sand layers in the alluvial deposits generally coexist as thin (less than 1 foot) interlayers with clayey and silty materials.

The predominant bedrock formations are the Miocene Monterey and Sisquoc Formations, which both consist of laterally extensive, fine-grained, marine sedimentary deposits [4, 5]. The stratigraphically higher Sisquoc Formation is a white to cream-white diatomaceous claystone and clayey diatomite. The older Monterey Formation consists of white-weathering, thin-bedded, siliceous shale with interbeds of chert [4].

Generally, the fresh bedrock materials do not allow significant contaminant transport due to their relatively low hydraulic conductivity. Contaminants, if present, generally are transported through the fractures (joints or faults) or through the more permeable weathered zone. The weathered bedrock materials as defined in this background study represent the zone on the top of bedrock where sampling can be achieved by hollow-stem auger drilling or CPT. CPT defined a thinner weathered bedrock than the drilling. Refusal drilling and sampling occurred in fresh bedrock.

In general, the weathered bedrock materials are gravelly, heterogeneous, and less than 5 feet thick. Cone tip resistance and sleeve friction of the weathered bedrock materials are higher than those of the alluvial soils.

Geographic Conditions

Vandenberg AFB consists of upland terraces and mountains divided by two major drainages: the Santa Ynez River and San Antonio Creek. The upland areas comprise three physiographic regions referred to as San Antonio Terrace, Burton Mesa, and Lompoc Terrace (Figure 2). A schematic cross section of Vandenberg AFB is illustrated in Figure 3. Each of these three upland terraces represents a relatively distinct geological setting. These three physiographic regions are described from south to north as follows.

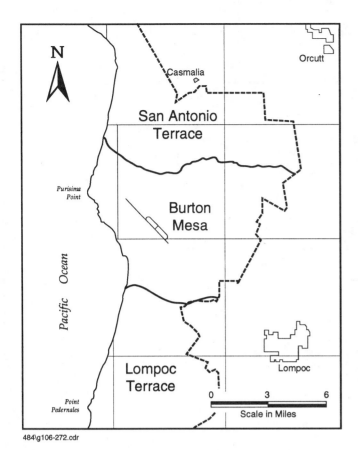

FIGURE 2 — Physiographic Regions Present at Vandenberg AFB.

Lompoc Terrace contains a thick, homogeneous dune sand sequence that overlies bedrock. In addition to the previously mentioned bedrock formations, the mountains located along the south edge of Lompoc Terrace contain volcanic rocks, which provide a distinctive sedimentary source.

To the north of Lompoc Terrace lies Burton Mesa, which contains relatively shallow bedrock overlain by a thin veneer of unconsolidated deposits. In addition to dune sand, the unconsolidated deposits on Burton Mesa contain widespread alluvial deposits.

The bedrock on San Antonio Terrace is moderately deep and the unconsolidated deposits contain both dune sand and alluvium. San Antonio Terrace is immediately south of the Point Sal ophiolite suite, an ultramafic assemblage of igneous rocks that contains high concentrations of metals.

Background Sampling Program

Based on the results of previous investigations at Vandenberg AFB [3], several inorganic contaminants (metals) might have been released as results of prior military operations at the base. The inorganic constituents may exist naturally in soil materials and groundwater. The primary objective of the background sampling program was to establish the baseline concentrations of naturally occurring inorganic constituents in similar soil, and groundwater units underlying Vandenberg AFB. The background program involved sampling surface soil, subsurface soil and groundwater from areas that do not appear to have been impacted by prior military activities or by other localized areas of contamination. Background threshold values could therefore be established by statistically analyzing the chemistry of collected background samples.

The background concentrations were compared to samples taken at environmental sites. If the concentrations detected in environmental samples were greater than the baseline threshold values, the samples were suspected of being contaminated and required further investigation.

CPT was used in the background sampling program for soil material only. Background groundwater sampling at Vandenberg AFB is therefore not discussed in this paper.

Due to the geological differences between the three physiographic regions and their distinctive geological units, the soil sampling program was designed to ensure that an adequate number of soil samples was collected from each geological unit within each physiographic region. A total of 30 boreholes (10 boreholes for each physiographic region) was drilled for the background study. The background sampling program included the following:

- The borehole locations were selected based on their uncontaminated nature.

- The boreholes were drilled to refusal or 100 feet below ground surface, whichever came first.

- Geologic units encountered in each borehole were identified based on CPT logs.

- A sample was collected from each geologic unit encountered in each borehole.

- Weathered bedrock samples were not collected from each borehole because bedrock material did not vary geographically. A total of 10 weathered bedrock samples were collected.

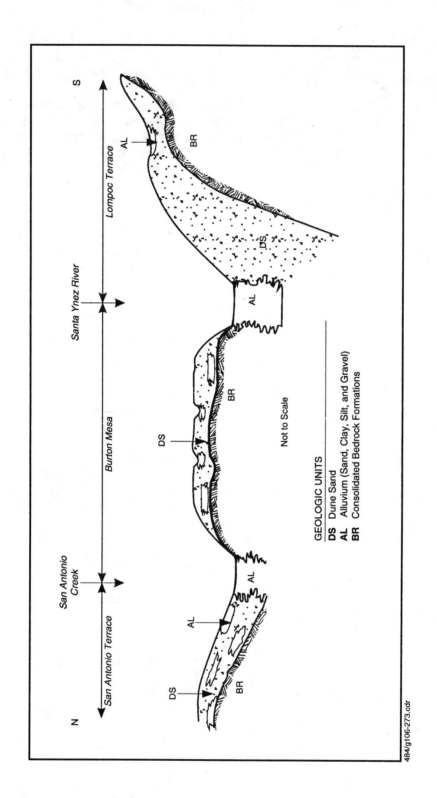

FIGURE 3 — A Schematic Cross Section of Vandenberg AFB.

- An equal distribution of samples were collected from different lithologies (clayey or silty soils identified by CPT) of the alluvial soils.

Most of the boreholes were located in remote areas of Vandenberg AFB. To access the borehole locations, drill and CPT rigs equipped with all-terrain capability were selected to perform background sampling. A CME-750 hollow-stem auger rig was used to drill and install 6 monitoring wells and to drill 7 boreholes for the background sampling. An all-terrain CPT rig with a maximum thrust of 15 tons was used to complete sampling at 23 borehole locations.

Applications of CPT

The background sampling program was successfully completed and the characteristics of the CPT application are described as follows.

Identification of Geologic Units

Drilling conducted by using a hollow-stem rig generally collects samples at every 5 feet. Such drilling can only identify lithological differences and cannot clearly distinguish geologic units. CPT was therefore used to provide a continuous subsurface logging. Encountered geologic units and lithologies were identified from the CPT logs. Sampling depth intervals were immediately determined from the interpretation of CPT logs. Soil samples at determined depth intervals were then collected by CPT equipped with a soil sampler at the tip. The soil sampler was opened at pre-determined depths, and soil samples were then collected while advancing the sampler. The validity of soil classification based on the CPT log was verified by a visual identification of collected soil samples.

Figures 4(a) and 4(b) provide typical CPT logs of the background sampling program. Figure 4(a) shows the sequence of dune sand, alluvial soils, and weathered bedrock, from top to bottom. Figure 4(b) shows a dune sand layer overlying weathered bedrock. Residual gravelly soils may have existed at the top of weathered bedrock materials which represent the unconformity surface (Figure 4b). Background samples were collected from each geologic unit encountered, as shown on Figures 4(a) and (b).

In general, dune sand and alluvial soils can be easily differentiated by CPT. Difficulty sometimes occurs in identifying weathered bedrock, especially at locations where sandy weathered bedrock underlies dune sand. Thick dune sand would cause an early refusal which was not caused by the existence of bedrock materials. The existence of gravelly soils in alluvium could also be misinterpreted as bedrock materials when refusal occurred. It was therefore very important to visually inspect the collected soil samples, especially the weathered bedrock samples. The visual inspection emphasized soil classification and other sedimentary features which could be used to identify weathered bedrock.

Difficult Sampling

Because the diameter of the soil sampler (2.2 inches) was larger than the electric cone (1.7 inches) used in this study, the depths of CPT were reduced when a soil sampler was installed at the cone tip. To obtain a sample at the necessary depth, a pre-sampling hole was sometimes dug by pushing a dummy cone to the top of sampling depth. Prepunching increased the success of the background sampling; however, the existence of thick dune sand deposits or gravelly layers would still cause great difficulty in sampling. Refusal also occurred at locations where only a thin weathered zone existed above fresh bedrock. Sampling attempts by CPT would fail in such conditions; a drill rig was required to collect critical samples. Since a weathered bedrock sample was not required from each borehole, a sufficient number of weathered bedrock samples was collected without the aid of drill rig in the background sampling for Vandenberg AFB.

Analytical Testing

Collected background soil samples were analyzed for metals (aluminum, antimony, arsenic, barium, beryllium, boron, cadmium, calcium, chromium, cobalt, copper, iron, total lead, magnesium, manganese, mercury, molybdenum, nickel, potassium, selenium, silver, sodium, thallium, vanadium, and zinc), semi-volatile organics, pH, soil conductivity, and common anions (chloride, nitrates, nitrites, sulfates, and fluorides). CPT collects samples of about 1.4 inches in diameter and 18 inches long in three sections of 6 inches long. In general, the top section was only partially filled; however, collected soil samples were sufficient for all required analyses.

Comparison with Drill Rigs

Using CPT for the background sampling reduced costs, minimized cuttings, and increased daily production as follows:

- The production rate of CPT was about 60 feet per day

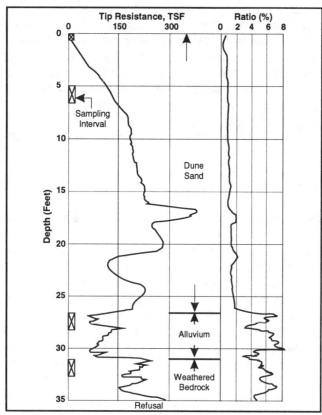

FIGURE 4 — Typical CPT Logs
(a) CPT No.: BG-JB-8

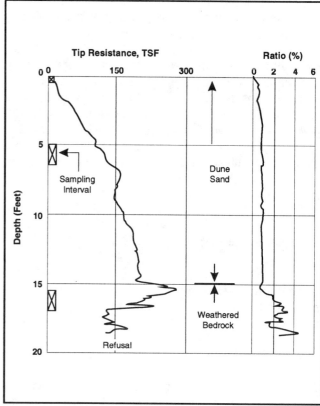

FIGURE 4—Typical CPT Logs.
(b) CPT No.: BG-JB-14

(compared to about 50 feet per day for the drill rig). A lot of time was used to move CPT rigs and hollow stem rigs between boreholes.

- Waste (cutting) minimization is apparent in CPT technology.

- Unit cost of CPT was about $47 per foot (compared to about $55 per foot for the drill rig). The cost includes strategic sampling by CPT and sampling every 5 feet by the drill rig. Mobilization had increased the unit costs of CPT and drilling rig.

Conclusions

CPT was used in a background sampling program for the RI/FS at Vandenberg AFB. Geologic units were identified from the CPT logs and soil samplings were conducted based on interpretation of the CPT logs. When compared to the hollow-stem auger rig, CPT also demonstrated its advantages in waste minimization, cost savings, and higher production rates.

The main problem with sampling by CPT is refusal before reaching the necessary depth. A drill rig is recommended to be included in a sampling program with CPT to collect critical samples.

Acknowledgments

The authors wish to express their gratitude to U.S. Air Force Lt. Col. George Lane for permission of presenting this paper. Lauri Roché, John Lawrence, and Deborah Paulshus of Jacobs Engineering are also appreciated for their efforts of putting this paper in its final shape.

References

[1] Hekma, L.K., K.K. Muraleetharan, and G.F. Boehm. 1990. "Application of the Electric Cone Penetration Test to Environmental Groundwater Investigations," Air & Waste Management Association, 83rd Annual Meeting and Exhibition. Pittsburgh, Pennsylvania. 21p.

[2] ASTM D3441. 1992. American Society for Testing and Materials, Vol. 04.08.

[3] Jacobs Engineering Group Inc. 1993. "Installation Restoration Program, Vandenberg AFB, RI/FS Work Plan." Submitted to Headquarters Air Force Space Command, Colorado.

[4] Dibblee, Thomas, Jr. 1950. *Geology of the Southwestern Santa Barbara County, California*. California Division of Mines, Bulletin 150.

[5] Dibblee, T., Jr. 1988. *Geologic Map of the Lompoc and Surf Quadrangles, Santa Barbara County, California*. Dibblee Geological Foundation. Las Vegas, Nevada. 1:24,000.

Potential Utilization of Waste Rock Powder

Masashi Kamon, Professor
Disaster Prevention Research Institute, Kyoto University, Kyoto, Japan,

Takeshi Katsumi, Research Associate
Disaster Prevention Research Institute, Kyoto University, Kyoto, Japan

SYNOPSIS: A large volume of waste rock powder (WRP) is generated annually as a by-product in crusher plants because a large amount of rubble is produced for use as substitute for natural gravel. These waste materials have been only reclaimed in either exhausted diggings or disposal sites. The objective of this study is to demonstrate the advantage of using the solidified WRP as a construction material. A method, of which WRP and Carbonated-Aluminate Salts (CAS) is filled in the site and then solidified by watering, is proposed as a novel approach. It is different from traditional soil stabilization techniques, and the WRP-CAS mixtures can be utilized not only as subbase or embankment, but permeable subgrade of road and back filling for retaining wall. In particular, a new method called the "Bagged WRP Method," where non-woven fabric bags are filled with the dry mixture of WRP and CAS and are solidified by soaking, is developed and evaluated. Because of the combined characteristics of WRP, CAS and non-woven fabric bags, the Bagged WRP Method may be applied to sunk-levee materials or seafloor ground improvement.

Introduction

Material flow in civil engineering is just entering a new phase in Japan. A large amount of resources are used for construction, some of them are stocked as building or other structures, and large quantities have been disposed of as waste materials. Consequently, the high quality domestic resources are quickly exhausted, while the limited waste disposal sites are expected to be filled in the near future not only with industrial wastes but also with construction wastes. In the field of environmental geotechnology, many tasks have internationally been proposed in international forums which contribute to the preservation of our environment from natural and man-made disasters [1, 2, 3, 4]; The utilization of industrial waste for construction has been promoted from the environmental geotechnical aspect, based on practices of material use in construction and waste management [4, 5, 6].

About half of the industrial wastes are reused due to recent developments in waste management technology. For example, coal ash and iron slag can be utilized for cement manufacturing or road construction, according to the Japanese Industrial Standard (JIS) and Asphalt Pavement Outline. It is waste materials from construction work that should be reused in civil engineering. Surplus soil and waste slurry from foundation works and waste concrete from building reconstruction have been targeted for reclamation [7, 8]. **Waste rock powder** (WRP), whose utilization has received less attention, is generated in large quantities from crusher plants, and, to establish sound material flow in civil engineering, a proper method for WRP utilization is required.

Due to environmental constraints put on the use of natural gravel, alternative sources or methods are required to fulfill the role played by natural gravel. One of the alternative sources can be rubble produced in crusher plants nationwide. Japan produces about 500 Tg of rubble and 10 Tg of WRP as a by-product annually. Although WRP is non-hazardous in nature, under the waste management law it is considered an "industrial waste." Therefore, its utilization has been limited to fill material in exhausted diggings or it is simply disposed of. Taking into account the scarcity of the disposal sites, a suitable method to solidify and utilize the WRP in large quantity is needed.

The objective of this study is to demonstrate the effectiveness of a solidification method that can be employed to use the WRP effectively as a construction material. WRP, as well as surplus soil and discharged waste slurry, originates from the ground, but it is easier to handle because many of the WRP are in dry condition. The characteristics of WRP depend on the mother rock; WRP of limestone is used as the raw materials of cement due to the chemical composition, and the silty WRP of sandstone is reused as a filling material in exhausted diggings. The WRP of sandstone, with a large specific surface area and a large amount of amorphous materials, increases the effect of lime stabilization in soils with a low proportion of fine particles or amorphous materials is present [9]. In this study, a method whereby WRP and Carbonated-Aluminate Salts (CAS) are filled in the site and then solidified by watering is proposed as a novel approach which is different from traditional soil stabilization techniques. The WRP-CAS mixtures can be utilized not only as subbase or embankment, but permeable subgrade of road and back filling for retaining wall. To heighten the supplemental value of the waste utilization, a new method called the "Bagged WRP Method," where non-woven fabric bags are filled with the dry mixture of WRP and CAS and solidified by soaking, is proposed.

Properties of Materials

Waste Rock Powder (WRP)

The characteristics of WRP depend on the types of machinery for crush and collection as well as the properties of the mother stone. WRP is generated from the crusher for rock crush and sand production, in dry and wet conditions. The WRP used in this study is the by-product generated through rock crush under dry conditions. Its properties are shown in **Table 1**. The mother stone is liparite, and the main minerals investigated through X-ray diffraction analysis are quartz, feldspar and amesite. The WRP are composed of particles equivalent to the size of silt grains, and judging from the uniformity coefficient and coefficient of curvature, the WRP is poor in particle size distribution. Moreover, the optimum moisture content is near the liquid limit. Therefore, because the WRP is very difficult to manage by means of compaction, the noncompacted method is recommended for WRP solidification. Because the permeability of WRP is in the range between 10^{-3} and 10^{-5} cm/s as shown in **Fig. 1**, WRP is considered to be a permeable material, and this method can promote its utilization as a well-drained material if the stabilized mixture is as permeable as untreated WRP.

Table 1--Properties of WRP.

Particle density	(g/cm³)	2.65
Liquid limit	(%)	14.3
Plastic limit	(%)	NP
Optimum moisture content	(%)	13.0
Maximum dry density	(g/cm³)	1.97
Particle size distribution		
Sand fraction	(%)	16.8
Silt fraction	(%)	70.4
Clay fraction	(%)	12.8
D_{60}	(mm)	0.031
D_{30}	(mm)	0.009
D_{10}	(mm)	0.004
Uniformity coefficient U_c		7.75
Coefficient of curvature U_c'		0.65
Ignition loss	(%)	1.35
Chemical compositions	(%)	
SiO_2		65 - 75
Al_2O_3		10 - 15
Fe_2O_3		3 - 4
Na_2O		2 - 3
K_2O		1 - 2
Main minerals		Quartz Feldspar Amesite

Basic Characteristics of WRP-CAS Mixture

Expansion Characteristics

CAS, used as hardening materials in this study, leads to rapid formation of ettringite and other hydrates and consequently not only harden but also expand with watering. Therefore, it is important to investigate the expansion characteristics of WRP-CAS mixtures. **Fig. 2** illustrates expansive pressure versus watering time of the mixtures. In this experiment, the consolidation mold filled with the dry mixture was soaked while the pedestal was stationary and the expansive pressure was measured continuously. It is considered that expansive pressure depends on CAS content and the dry density of the mixture, and in this case these two mixtures have similar dry densities, 1.15 g/cm³ and 1.18 g/cm³. The expansive pressure-time curves of the two different mixtures each have one peak and show a similar trend. When WRP:CAS = 10:2, the expansive pressure maximum is only about 8 kPa in a day, while the WRP:CAS = 7:3 mixture has a maximum expansive pressure of 70 kPa, after 4 days of curing.

The expansive characteristics of the stabilized mixture has some advantages; the mixture is stabilized at a low density and shrinkage is prevented when used as a filling material. To prevent harm to the surrounding structure, it is necessary and possible to control expansion by adjusting the CAS content and density.

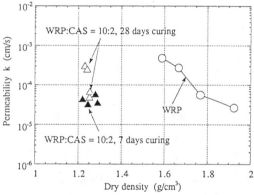

Fig. 1. Permeability of WRP and WRP-CAS mixtures.

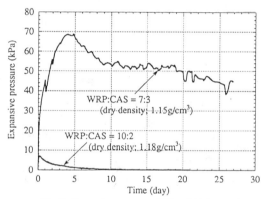

Fig. 2. Expansive pressure of WRP-CAS in watering.

Carbonated-Aluminate Salts (CAS)

One kind of **Carbonated-Aluminate Salts (CAS)**, a newly developed cement-based stabilizer, is used as the admixture. It has been previously shown that CAS are effective as hardening materials for soft clays or waste materials [5, 8, 10]. The effects of the CAS materials are as follows: (1) formation of ettringite and other reaction products and the crystallization of excess pore water in the soils; (2) activation of the pozzolanic reaction in the long term; (3) control of the pH value of the hardened mixtures for reactive acceleration and environmental impact; and (4) activation of hydration through accelerating dissolution from soil/waste materials. CAS used in this study has the composition of Ordinary Portland Cement (OPC): $Ca(OH)_2 : Al_2(SO_4)_3 = 50 : 20 : 30$ (dry weight basis), and hydrated reaction of it leads to immediate formation of ettringite and hardening and expansion of CAS as soon as watering.

Non-Woven Fabric (NWF)

It is proposed that non-woven fabric bags (NWF) are used in the Bagged WRP Method suggested by authors. Nowadays, NWF are widely utilized as geosynthetic materials for purposes such as earth reinforcement or at waste disposal sites because they have reinforcement ability, filtering and separating ability, fluid transmission ability and chemical stability. In the Bagged WRP Method, it is expected that NWF can mainly function as separator, protector, filter and in fluid transmission.

Strength and Density Characteristics

Specimens for unconfined compressive strength tests were prepared basically according to Practice for Making and Curing Noncompacted Stabilized Soil Specimens (JSF T 821-1990). The dry mixture of WRP, which is finer than 2 mm, and stabilizers such as CAS and OPC were filled in the cylindrical mold (10 cm height and 5 cm or 5.6 cm diameter), aiming at some levels of dry density in the range 1.2-1.4 g/cm³. The mold filled with the mixture was soaked while the exposed upper side of the mold was covered by filter paper for water transmission and loaded at 10-20 kPa to prevent harmful expansion.

Fig. 3 shows the strengths of the WRP mixtures which were stabilized by CAS and OPC and cured for 7 days. There is a clear correlation between strength and density of the mixtures. The WRP-OPC mixtures have higher strength and larger density than the WRP-CAS, because these stabilizers have different reaction mechanisms. OPC, when used as a stabilizer, dissolves by means of watering and hardens after a minimum of 1 day of curing, so the mixture shrinks and the density increases to 1.4-1.6 g/cm³, higher than the objective density of 1.2-1.4 g/cm³. The WRP-CAS mixtures harden rapidly and tend to expand by the immediate formation of ettringite as soon as the dry mixtures are soaked. Therefore, the desired low density, 1.2-1.4 g/cm³ were kept as they were. The observations of the mixtures in early time frames prove these phenomena; after only 1 hour of watering, WRP-OPC mixtures from a slurry, but the WRP-CAS has already hardened.

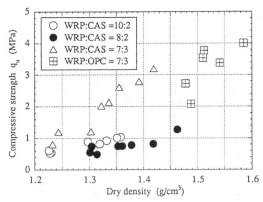

Fig. 3. Strengths of WRP mixtures cured for 7days.

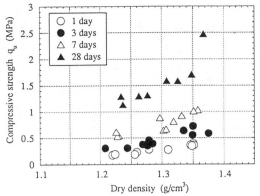

Fig. 4. Strengths of WRP-CAS mixtures (WRP:CAS = 10:2).

The increase of CAS content as well as dry density results in the strength increase. The 70% WRP-30% CAS mixtures exhibit higher strength than 1 MPa, and can be utilized as a subbase material. On the other hand, the strengths of 80% WRP-20% CAS and 100% WRP-20% CAS are in the range between 500 kPa and 1 MPa and these mixtures are considered to be available for use as a subgrade or embankment.

The relationship between strength and curing time of the WRP-CAS mixture are illustrated in **Fig. 4**. As is the case with general stabilized soil, curing causes the strength to increase, and the strength depends on the effect of curing time more than the dry density. Although the WRP-CAS mixtures can not reach the stress of 1 MPa for subbase purpose, the specimens cured for 1 day have higher strength than 100 kPa, and this stabilization method not only makes waste treatment possible but also the utilization of WRP as embankment or backfilling.

Fig. 5 shows the relation between the dry density (γ_d) and water content (w) of the mixtures. The strong correlation between these indexes is independent of the types and content of stabilizer and curing time, and presents the following regressive equation.

$$\gamma_d = 1.886 - 0.0164w \quad (R = 0.995) \qquad \text{(Eq. 1)}$$

It is easy to measure water content of specimens sampled in the site and the unconfined compressive strength can be estimated from the mix proportion, curing time and dry density from the equation based on water content.

<u>Permeability</u>

Specimens for permeability tests were prepared by the same method as the ones for the unconfined compressive strength test. Falling head permeability test was carried out according to Test Method for Permeability of Saturated Soils (JSF T 311-1990).

The results of permeability test are shown in **Fig. 1**. WRP-CAS mixtures with a density of 1.2-1.3 g/cm³ exhibit about 10^{-4} cm/s of permeability, therefore the mixtures can be utilized as well drained materials. Although the permeability depends on the raw materials and the stabilization method, the general soils stabilized by hardening materials such as cement or lime exhibit permeability below 10^{-6} cm/sec and some of them reach 10^{-8} cm/s for the cut-off of water. Nowadays, it is said that the development of permeable ground materials for construction is required for the water circulation from environmental geotechnical viewpoint. Permeable pavement system, group-grained coal ash utilization, and so on have been developed as permeable materials owing to the high degree of technology. The method proposed in this study has the advantage of realizing stabilized permeable materials by a simple process. A combination of the properties of WRP and the reaction mechanisms of CAS leads to the permeable characteristics of the mixture, that is, WRP which itself is permeable, is immediately hardened by CAS on watering and CAS causes the mixture to have a low density due to the expansion characteristics in the early stage.

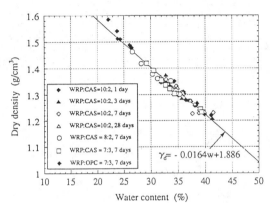

Fig. 5. Relationship between dry density and water content of WRP mixtures.

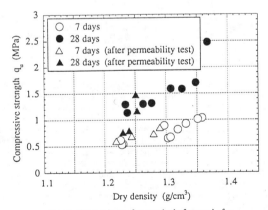

Fig. 6. Comparison of strengths before and after permeability test (WRP:CAS = 10:2).

The compressive strength tests were carried out on the specimens after the permeable test, as shown in **Fig. 6**. The strengths of specimens after permeability testing are as high as those which were not subjected to permeability testing. Therefore, the permeability history has little effect on the properties of the WRP-CAS mixture. As the permeable period was about 1 hour in this experiment, it is important to evaluate the durability of WRP mixtures under more severe conditions such as cyclic drying-wetting or sulfate attack.

Applicability of WRP-CAS mixture

These experimental results show the effective utilization of WRP-CAS mixtures. A method, in which WRP and CAS is filled in the site and then solidified by watering is a novel approach to soil stabilization. With the proper selection of strength and density of the WRP and CAS, mixtures can be utilized for various purposes e.g., mixtures with a high density can be utilized as subbase course, and those with a low density can be used as not only subgrade and embankment material but well drained material such as permeable subgrade of road and back filling for retaining wall, due to the characteristics of strength, shrinkage and permeability.

Applicability of Bagged WRP Method

Basic Concept of Bagged WRP Method

Supplemental value is important to promote waste utilization. To ensure that WRP utilization possesses supplemental value, we propose a new method where the NWF bags are filled with a dry mixture of WRP and CAS which are then solidified by soaking. This method is called the "Bagged WRP Method." Permeability of WRP, expansion of CAS, reactive mechanism of CAS (and WRP), and function as separator, protector, filter and fluid transmission of NWF are combined effectively in the Bagged WRP Method. **Fig. 7** illustrates the relationship of material characteristics used in this method.

Table 2 -- Curing condition of Bagged WRP Method

Mix proportions	WRP:CAS = 7:3, 8:2
Curing water	fresh water(tap water), sea water(artificial)
Curing temperature	14 - 20 $^{\circ}$C

Table 3 -- Characteristics of non-woven fabric

Type		SP	VN-160	VN-300
Mass	(g/m^2)	80.0	160.0	300.0
Thickness	(mm)	0.5	1.5	3.0
Tensile strength	(kN/m)			
lengthwise		4.0	9.0	19.0
widthwise		3.0	7.0	15.4
Permeability	(cm/s)	0.35	0.25	0.15
Open area size	(mm)	0.05	0.07	0.05
Applicability		unsuitable	unsuitable	suitable

Experimental Results

The WRP-CAS mixtures can be hardened effectively by the expansion of CAS and the restriction of NWF. The mixtures filled in NWF bags became hardened immediately on watering; that is, the mixture salvaged after watering for 2-3 minutes behaved as one hardened block rather than a flexible clod. **Fig. 8** and **Fig. 9** show the strengths of the WRP-CAS mixtures sampled from NWF bags. There is the strong correlation between the dry density and the compressive strength as well as the samples from laboratory tests, and the regression plots are considered to be in the range of the ones of laboratory test samples. However, the Bagged WRP Method samples have lower strength and density than the laboratory samples, because the former mixtures are

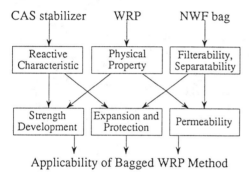

Fig. 7. Relationship of Material Characteristics used in Bagged WRP Method.

Experimental Procedure

To evaluate the applicability of the Bagged WRP Method, NWF of 120 cm length and 50 cm width was doubled, and the two layers were stitched with a sewing machine. Formed NWF bags of 50 cm x 60 cm were filled with 30 kg of premixed dry WRP-CAS, and the opened rest side was stitched so that the mixture may be shut in the bag. The bags filled with the dry mixture were laid flat and soaked in 1 m of either fresh or sea water in a closed cylindrical tank 1.5 m diameter. Five and four bags were soaked in fresh and sea water, respectively. Mixtures hardened by watering were salvaged and cut into the rectangular parallelepiped samples of 5 x 5 x 10 cm for the unconfined compressive strength test. **Table 2** shows the curing conditions in these experiments.

The selection of NWF must be based on the characteristics of NWF such as filterability, tensile strength and resistance. Three types of NWF, as shown in **Table 3**, were tested but only one NWF was selected for the experiments due to its superior characteristics. When using 'SP,' the inner mixture issued from the open area of NWF and the curing water became a little muddy, and this was a problem when the method was applied at the test site. Both 'VN-160' and 'VN-300' functioned effectively as separator and protector and the curing water had been kept clean, but 'VN-160' bag was stretched when carrying the bag filled with the mixture. 'VN-300' is most applicable because of filterability and tensile characteristics.

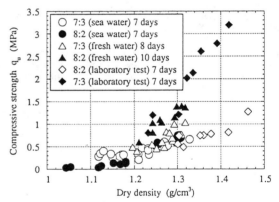

Fig. 8. Strengths of WRP-CAS mixture by Bagged WRP Method and laboratory test.

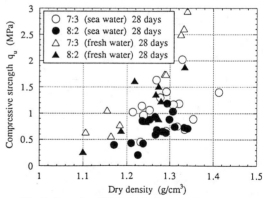

Fig. 9. Strengths of WRP-CAS mixture by Bagged WRP Method.

free to expand as long as NWF bags stretch while the later are restricted by the rigid mold. The strengths of WRP:CAS = 8:2 mixtures cured in sea water are much lower (about 100 kPa) than any other mixtures because of the looseness of filling in the bag. Therefore, it is important that the CAS expand and the NWF provide resistance to compensate for incomplete stuffing. The other mixtures have at least 200 kPa stress and usually have higher strength than 500 kPa after 28 days, which exhibits the sufficient hardening effect.

Fig. 10 illustrates the density and water content of mixtures by laboratory test and Bagged WRP Method. The regression plot is expressed by,

$$\gamma_d = 1.860 - 0.0163w \quad (R = 0.884) \,. \qquad \text{(Eq. 2)}$$

This expression is almost the same as the one for the laboratory tests, therefore the basic properties of mixtures by Bagged WRP Method are same as the later ones, and the estimation of strength is possible by the process stated above.

The hardening reaction occurs gradually from the surface to the center of the mixtures because of water seepage. **Fig. 11** illustrates the relationship between density and water content from three parts separated in bag, that is, the upper and the lower are near NWF and the center is far from NWF. The center has higher dry density than the upper and lower because the outer mixtures react immediately on watering, expand and compact the inner mixtures. It is considered that the decrease in permeability of the outer mixture due to hardening might prevent water from reaching the inner mixture, and size effect must be evaluated to utilize this method in the site.

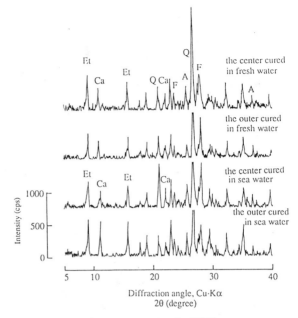

Fig. 12. X-ray diffraction patterns of WRP-CAS mixtures (cured for 28 days, WRP:CAS = 8:2).

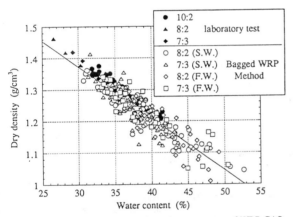

Fig. 10. Relationship between dry density and water content of WRP-CAS mixture by laboratory test and Bagged WRP Method.

According to the results of X-ray diffraction analysis on the mixtures (**Fig. 12**), it is considered that Ettringite ($3CaO \cdot Al_2O_3 \cdot 3CaSO_4 \cdot 32H_2O$) and Calcium Aluminate Hydrate ($4CaO \cdot Al_2O_3 \cdot 13H_2O$) contribute to the strength development because of the presence of Al and SO_4 in CAS. These hydrates react with the large amount of pore water, expand and harden. The curing condition and sampled position hardly affected X-ray diffraction patterns.

The mixtures soaked and cured in sea water have much lower strength than those cured in fresh water, as shown in **Fig. 8** and **Fig. 9**. There are several possible reasons for this phenomenon. Because these tests were carried out in different seasons, a 2-5 °C difference in curing temperature might lead to the deference in strengths, and of course, the variety of WRP properties can not be ignored. The main reason for this difference is the composition of sea water; SO_4^{2-} of 0.2% and Cl^- of 1.5% might affect the hardening reaction, and Mg are considered to behave the ion exchange function of Ca. The durability of hardened materials in sea water is also important; it is well known that $MgSO_4$ in

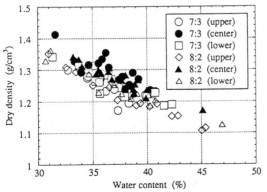

Fig. 11. Relationship between dry density and water content (cured in sea water for 28 days).

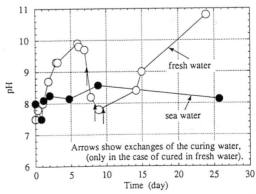

Fig. 13. The pH value in tanks where the mixtures were cured.

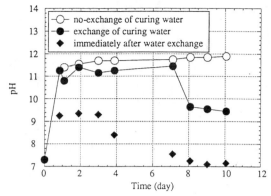

Fig. 14. The pH value by model tests (cured in fresh water).

sea water can destroy the cement bonds and produce the expansive materials such as Ettringite. However, the durability in sea water is not discussed experimentally in detail in this study, because the main phenomenon, the formation of Ettringite is advantageous for the development of strength of WRP stabilization using CAS. It is suspected that the composition of sea water might prevent the increase in pH and consequently inactivate the hardening reaction, as shown in **Fig. 13** which illustrates the pH values of curing water.

The low pH value is acceptable from the environmental viewpoint. The pH value of sea water where the WRP-CAS bags were cured is lower than 9.0, satisfying the environmental standards, but in the case of curing in fresh water, the pH increased due to curing in spite of water exchanges. **Fig. 14** shows the experimental results, where the 100 g dry mixtures filled in small bags were soaked in 1000 cm^3 fresh water. While the pH reached 12 without water exchange, the curing water exchange reduces the pH of curing water to below 10 after 7 days. The Bagged WRP Method is applicable in the open area such as the sea but not in closed areas.

Application of Bagged WRP Method

The Bagged WRP Method is applicable to sunk-levee materials as a substitute for rubble or concrete block, or seafloor ground improvement, considered through these experimental studies. **Fig. 15** illustrates an application of the Bagged WRP Method to tidal flat construction. In Japan, about 40% of tidal flat areas have disappeared due to reclamation or dredging works in the last decade. To minimize the environmental damage and sustainable development, man-made tidal flats must be constructed, for example, as the substitute of tidal flats which will disappear if development continues on its present course. The tidal flat functions as not only in the preservation of ecosystems, purification of sea environment and fish production (fish farming), but also as resort areas and waste utilization (reclamation) areas. The construction of these man-made tidal flat uses the sunk-levee to puncture and hold soil materials of tidal flat, and the ground improvement for sunk-levee stability is necessary because the materials usually used for sunk-levee, such as rubble or concrete block, are very heavy (density of about 2.6 g/cm^3). The Bagged WRP Method has some advantages when it is applied to the sunk-levee. The materials made by this method are light weight (density of about 1.5-1.7 g/cm^3), therefore, even the soft seafloor ground does not required improvement. Due to the deformation of inner mixtures until they harden, the bags can adhere to each other. Of course, one of the most attractive merits is the waste utilization as the substitute of natural rubble. A construction of sunk-levee applying the Bagged WRP Method has been carried out in a harbor area.

Conclusions

Two methods for potential utilization of Waste Rock Powder (WRP) as construction materials are proposed. These methods depend on the characteristics of WRP itself, Carbonated-Aluminate Salts (CAS) and non-woven fabric (NWF). The WRP-CAS mixture has high solidified strength and permeability, therefore the dry WRP-CAS mixture can be filled in the site and then solidified by watering. This method can be applied not only as a subbase or embankment but as well-drained materials such as permeable subgrade of road and back-filling for retaining wall. The "Bagged WRP Method" whereby the NWF bags are filled with the dry WRP-CAS mixture and solidified by soaking is also proposed. While we evaluate the application of WRP to the method, it is considered that coal ash, volcanic ash and other materials as well as WRP can be used as a stuffing material. The method can be applied to sunk-levee materials or seafloor ground improvement based on the future research on utilization systems, design methods, and environmental impact.

Acknowledgment

The authors sincerely acknowledge Mr. S. Oyama, graduate student of Kyoto University, Mr. M. Matsushima of Matsushima-Saiseki Co. Ltd., and Mr. Y. Higuchi of Toyobo Co. Ltd., for their cooperation and support. The financial support by The Asahi Glass Foundation is gratefully acknowledged.

References

[1] Moh, Z.C.(ed.) (1977). Geotechnical Engineering and Environmental Control, *Proc. Specialty Session, 9th ICSMFE*, 480p.
[2] Sembenelli, P. and Ueshita, K. (1982). Environmental Geotechnics, *Proc. 10th ICSMFE*, Vol.4, pp.335-394.
[3] Morgenstern, N.R. (1985). Geotechnical Aspects of Environmental Control, *Proc. 11th ICSMFE*, Vol.1, pp.155-185.
[4] Kamon, M. (1992). Definition of Environmental Geotechnology, *Proc. 12th ICSMFE*, Vol.5, pp.3126-3130.
[5] Kamon, M., Nontananandh, S. and Tomohisa, S. (1991). Environmental Geotechnology for Potential Waste Utilization, *Proc. 9th ARC*, pp.397-400.
[6] Kamon, M. and Nontananandh, S. (1991). Combining Industrial Wastes with Lime for Soil Stabilization, *Jour. Geotech. Eng. Div., ASCE*, Vol.117, No.1, pp.1-17.
[7] Kamon, M., Tomohisa, S., Tsubouchi, K. and Nontananandh, S. (1992). Reutilization of Waste Concrete Powder by Cement Hardening, *Soil Improvement, CJMR Vol.9, Elsevier Appl. Sci.*, pp.39-53.
[8] Kamon, M. and Katsumi, T. (1994). Utilization of Waste Slurry from Construction Works, *Proc. 13th ICSMFE*, Vol.4, pp.1613-1616.
[9] Nishida, K., Sasaki, S. and Kuboi, Y. (1992). Utilization of Waste Rock Powder for the Lime Stabilization of Residual Soil, *Soil Improvement, CJMR Vol.9, Elsevier Appl. Sci.*, pp.55-70.
[10] Kamon, M., Nontananandh, S. and Katsumi, T. (1993). Utilization of Stainless-Steel Slag by Cement Hardening, *Soils and Foundations, JSSMFE*, Vol.33, No.3, pp.118-129.

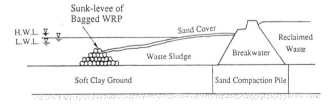

Fig. 15. Application of Bagged WRP Method to Tidal Flat Construction.

Calcium Bentonite and Fly Ash in Slurry Wall Construction Materials

Raj P. Khera, Professor
*Department of Civil and Environmental Engineering, New Jersey Institute of Technology,
University Heights, Newark, New Jersey, USA*

The majority of test data indicate that the hydraulic conductivity of soil-sodium bentonite mixtures increases several fold in the presence of many organic and inorganic waste chemicals. In the present investigation the selected clay materials consisted of calcium bentonite from the US and from Greece. Different proportions of cement, slag, and fly ash were used for preparing bentonite-cement mixtures. With water and organic chemicals (aniline 5000 ppm, phenol 5000 ppm, concentrated trichloroethane, and leachate from a municipal landfill) the soil mix showed a decrease in hydraulic conductivity with time. When a fly ash from Mercer generating station replaced part of the slag, no discernible change in hydraulic conductivity was recorded. The results of consolidated undrained triaxial compression tests indicate that the strength with or without the fly ash was essentially the same. After reaching the peak strength the specimens continued to deform without sudden failure. The results clearly show that calcium bentonite-cement-slag mixtures provide stable chemical barriers.

At waste disposal sites encapsulation rather than removal is becoming more significant as a means of remediation. Underground vertical barriers are constructed using slurry wall technique to prevent the lateral migration of liquid pollutants from entering ground water. The complex interaction of the system of soil and rock containment, fluid barrier, and encompassed fluids requires that the data be consistent, suitable, and valid. The major concern for the design and installation of such impoundments has been over their integrity and ability to fulfill established requirements and satisfy the regulations on a long term basis. Though the proper functioning of barriers may depend upon various factors, the maintaining of low hydraulic conductivity is considered the primary performance criterion.

With the increasing use of slurry walls, many investigators have directed their efforts toward determining the durability of the materials constituting the impermeable barriers. Sodium bentonite, though small in proportion to the other constituents, plays the primary role toward attaining the design hydraulic conductivity. A large majority of test data indicate that the hydraulic conductivity of soil-sodium bentonite mixtures increases several fold in the presence of many organic and inorganic waste chemicals.

For a "contaminant resistant" bentonite, Ryan (1987), reported a hydraulic conductivity increase of over five and a half times with leachate containing hydrocarbons, phenol, acetone, and other organic compounds having individual concentrations of less than 75 ppm. In search for a better material to replace soil-bentonite backfill, Evans et al. (1987) examined plastic concrete consisting of cement, sodium bentonite, and aggregate. The plastic concrete showed a hydraulic conductivity comparable to that of soil-bentonite, its strength was higher and it appeared to offer greater resistance to contaminated pore fluids. The addition of fly ash and bottom ash improved workability but an equal replacement of cement with fly ash resulted in an increase in hydraulic conductivity. Edil et al. (1987) reported a decrease in permeability for fly ash sand mixture with Class C type self cementing fly ash from coal sources in the western united States. Because of the disposal problems associated with the large amount of fly ash being generated by the power industry, considerable research has been devoted to utilizing fly ash in seepage cut off applications. Hermanns et al. (1987) reported a lower hydraulic conductivity for a calcium bentonite-cement mix than for a sodium bentonite-cement mix. When permeated with phenol, the hydraulic conductivity was about 3×10^{-10} m/sec for the calcium bentonite mix after 25 days but for the sodium bentonite mix it was 1×10^{-8} m/sec. After two months of permeation with a reconstituted leachate, the hydraulic conductivity of the mix containing sodium bentonite was four orders of magnitude greater than that containing calcium bentonite. In a recent study Khera (1990) selected calcium bentonite as an alternate to sodium bentonite. Of the four different types of calcium bentonites tested, those from American Colloids and Greece showed the best potential as alternate materials to sodium bentonite.

In this study slag and two Class F fly ashes from eastern United States were used to determine their effect on hydraulic conductivity and undrained strength of cement calcium bentonite mixes. Water, organic chemicals, and a leachate from a closed municipal landfill served as permeants.

Materials

American calcium bentonite (indicated by 'A' in specimen name) was provided by American Colloids and Greek calcium bentonite (indicated by 'G' in specimen name) was provided by IKO Industriekohle of Germany. The proportion of various materials is shown in Table 1.

The diameter of the specimen for hydraulic conductivity tests was about 70 mm and height ranged between 21 mm and 61 mm. For triaxial tests the corresponding dimensions were about 35 mm and 80 mm.

TABLE 1--Mix proportions

Ca. Bent %	Cement %	Slag %	Fly ash %	Name
12	15			AC
12	7.5	7.5		ACL
12	12		4H	A12C4H
12	6	6	4M	A12CL4M
18	5	15		18A5CL
18	5	10	5M	18A5S
15	15			15GC
15	7.5	7.5		15GCL

Hfly ash from Hudson generating station, Mfly ash from Mercer generating station,

Calcium bentonite was mixed with water and left overnight for hydration. Cement, fly ash, and slag were added to the hydrated bentonite as needed and mixed in a mechanical blender. The mixture was poured into the appropriate cylindrical molds which were placed in plastic zip-log bags with wet paper towels and sealed carefully. The specimens were allowed to cure for seven days before testing.

The liquid limits for American and Greek calcium bentonites were 100 and 108, and the plastic indices were 45 and 46, respectively.

Hydraulic Conductivity

A flexible wall permeameter was used for all hydraulic conductivity measurements (ASTM D 5084-90). Complete saturation was assured through proper deairing of the system, application of back pressure, and ensuring a pore pressure parameter B value close to one. Effective stress prior to permeation for test specimens was 8 psi (55 kPa). Both inflow and outflow were measured. The maximum hydraulic gradient was maintained at 30 or below.

TABLE 2-- Hydraulic conductivity test data for 12% calcium bentonite from American Colloid

Specimen	Liquid	Hydraulic conductivity m/s
AC-8W	water	1.80×10^{-8}
ACL-3W	water	3.00×10^{-9}
A12CL4M	water	3.00×10^{-9}
A12C4H	water	1.30×10^{-7}
AC-8P	phenol	3.20×10^{-8}
ACL-2P	phenol	1.00×10^{-8}
A12C4H-8P	phenol	1.03×10^{-7}
15GC-1	water	1.49×10^{-8}
15GCL-1	water	5.68×10^{-11}
15GCL-1A	aniline	5.39×10^{-12}

Effect of Slag on Hydraulic Conductivity

From Table 1 a comparison of specimens AC-8W and ACL-3W, the later with half of the cement being replace with slag, shows almost an order of magnitude drop in hydraulic conductivity. Similarly, a comparison of specimens 15GC-1W and 15GCL-1 shows more than two orders of magnitude drop in hydraulic conductivity. One of the major reasons why cement bentonite is not used in containment structures is the high hydraulic conductivity that such mixtures yield. However, it is quite clear that the addition of slag reduces the hydraulic conductivity considerably yielding values well within the limit of 10^{-9} m/s specified by many of the regulating agencies. Further to be noted is the fact that for specimen 15GCL-1A, where aniline (5000 ppm) was used as permeant the magnitude of hydraulic conductivity dropped by an other order of magnitude.

Effect of Fly Ash on Hydraulic Conductivity

As seen in Table 2, specimen AC-8W has a hydraulic conductivity of 1.8×10^{-8} m/s whereas specimen A12C4H has a hydraulic conductivity of 1.3×10^{-7} m/s, or an order of magnitude greater. This increase in hydraulic conductivity is attributed to the fly ash from Hudson generating station, since one quarter of cement was replaced by the fly ash. For a sand-sodium bentonite mix, Khera et al. (1987) reported an increase in hydraulic conductivity of one and a half orders of magnitude when fly ash from Mercer generating station was added. Specimen A12CL4M, which also contains Mercer fly ash, does not exhibit any increase in hydraulic conductivity. The lack of increase in hydraulic conductivity is attributed to the presence of slag in the specimen.

Hydraulic conductivity data for specimens containing fly ash from Mercer and those without fly ash are presented in Fig. 1 and Fig. 2. In all instances the hydraulic conductivity for specimens which contained fly ash was somewhat lower or about the same as that for specimens containing no fly ash.

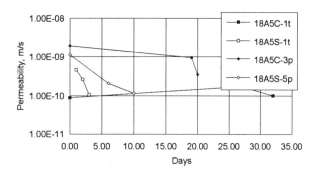

Fig. 1. Hydraulic conductivity of specimens with and without fly ash using concentrated TCE (t) and 5000 ppm phenol (p) as permeants

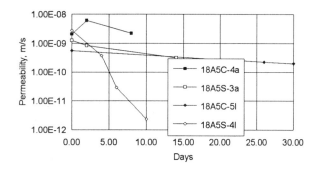

Fig. 2. Hydraulic conductivity of specimens with and without fly ash using 5000 ppm aniline (a) leachate (l) as permeants

Triaxial Test

All triaxial specimens were subjected to an effective hydrostatic stress of 17 psi (117 kPa) for a period of at least 48 hours before shearing. The results of triaxial compression tests for specimens without fly ash are shown in Fig. 3.

Note that $q = \dfrac{\sigma'_1 - \sigma'_3}{2}$ and u = pore water pressure. The maximum shear strength was reached at a relatively low strain of 0.27% to 0.55%. After reaching the peak shear the specimen continued to deform without sudden failure. The maximum shear strength ranged between 76 psi (524 kPa) and 90 psi (620.6 kPa).

For sodium bentonite-cement-fly ash mix containing 15% cement and 10% to 30% fly ash Evans et at. (1987) reported strain at failure to be 0.2% to 1.1% and the shear strength ranged from 106 kPa to 358 kPa. The lower shear strength observed for the present tests is attributed to the absence of coarse aggregate and lower cement contents.

As seen from Fig. 3, pore pressure built up very rapidly and reached its maximum value at strains of 0.31% to 0.1 %. The magnitude of pore water pressure at failure was 22 psi (151.7 kPa) and 15 psi (103.4 kPa). Eventually the pore pressure became negative reaching minimum values of - 2.39 psi (-16.5 kPa) and - 6.36 psi (- 44.1 kPa) at a strains of more than 20%.

The test results for specimen with fly ash are shown in Fig. 4. The strain at failure was 0.50% and the maximum shear strength was 73.6 psi (507.5 kPa).

The pore pressure built up very rapidly. However, its maximum value was only 4.32 psi (29.8 kPa) at a strain of 0.35%. At failure its magnitude was 4.21 psi (29.0 kPa). Note that this value of pore pressure at failure is considerably lower than that for specimens without fly ash. Eventually the pore pressure became negative reaching a minimum value of -27.3 psi (-188.2 kPa). This too is much lower when compared to - 2.39 psi (-16.5 kPa) and - 6.36 psi (- 44.1 kPa) for sample without fly ash.

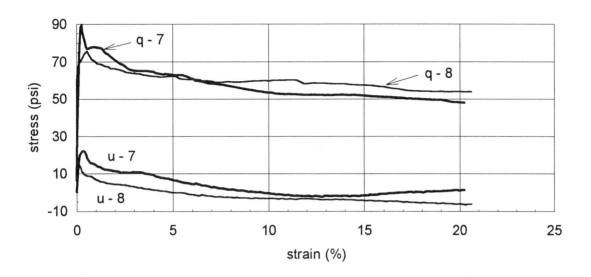

Fig. 3 Undrained shear and pore water pressure versus strain for samples without fly ash (1 psi = 6.895 kPa)

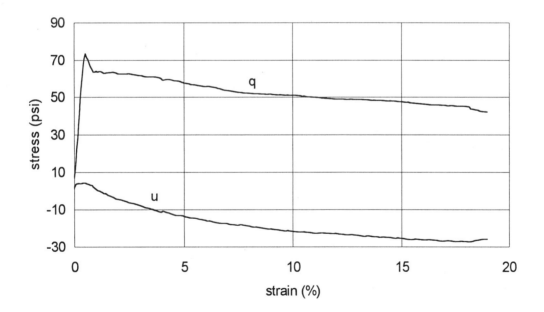

Fig. 4 Undrained shear and pore water pressure versus strain for samples with fly ash (1 psi = 6.895 kPa)

Conclusions

Based on the present investigation the following conclusions are drawn:

A mixture of calcium bentonite from American Colloids, cement, and slag yield hydraulic conductivity values equal to or less than 10^{-9} m/s.

With various concentrated and dilute organic chemicals the hydraulic conductivity also showed a decrease with time. There was no evidence of increase in hydraulic conductivity with time in any of the cases.

Partial replacement of slag with fly ash further reduced the hydraulic conductivity, though only slightly.

There was not much difference in shear strength or strain at failure when slag was partially replaced by fly ash.

Specimens with fly ash showed considerably lower pore water pressures at failure when compared with those without fly ash. In addition the minimum pore pressure recorded was lower for specimens with fly ash.

References

Edil, T.B., Berthouex, P. M., and Vesperman, K. D., "Fly Ash as Potential Waste Liner," Geotechnical Practice for Waste Disposal '87, Geot. Special Pub. No. 13, ASCE, June, 1987, pp. 447-459.

Evans, J. C., Stahl, E. D., and Droof, E. "Plastic Concrete Cut-Off Walls." Geotechnical Practice for Waste Disposal '87, Geot. Special Pub.No. 13, Woods, R. D., Ed., June 1987, pp. 462-472.

Hermanns, R., Meseck, H., and Reuter, E., "Sind Dichtwandmassen beständig gegenüber den Sickerwässern aus Altlasten?" Mitteilung des Instituts für Grundbau und Bodenmechanik, TU Braunschweig, Heft Nr. 23, Ed. Meseck H., Dichtwände und Dichtsohlen, June 1987, Braunschweig, Federal Republic Germany, pp. 113-154.

Khera, Raj P., *New Materials for Slurry Wall Containment Structures*, Hazardous Substance Management Research Center, A National Science Foundation Industry / University Cooperative Center and a New Jersey Commission on Science and Technology Advanced Technology Center, April 1990.

Khera, R. P., Wu, Y. H., and Umer, M. K. "Durability of Slurry Cut-Off Walls around the Hazardous Waste Sites." Proceedings: 2nd Int. Conf. on New Frontiers for Hazardous Waste Management, EPA/600/9-87/018F, Aug. 1987, pp. 433-440.

Ryan, C. R., "Vertical Barriers in Soil for Pollution Containment," Geotechnical Practice for Waste disposal '87, Proc Spec Conf Geot Special Publ. No. 13, Woods, R. D., Ed., June 1987, pp. 182-204.

Acknowledgement

This research was funded by Hazardous Substance Management Center, A National Science Foundation/Industry/University Cooperative Center, through various grants since 1985. The writer gratefully acknowledges the continual financial support of the Center. The materials were provided by American Colloids, Arlington Heights, IL, IKO Industriekohle GH & Co, Marl, Germany, Blue Circle Atlantic of Atlanta, GA, Lehigh Cement Co., and PSE&G.

Site Characterization and Remediation of Hydrocarbon Contaminated Area

Suresh R. Kikkeri, Edward P. Hagarty, P.E., DEE
Woodward-Clyde, Gaithersburg, Maryland, USA

Jimmy L. Wilcher
U.S. Army Corps of Engineers, Baltimore, Maryland, USA

John Bowman
MDW Directorate of Public Works, Arlington, Virginia, USA

Abstract

Construction of new buildings is proposed at two different Military Installation areas at Fort Myer in Arlington, Virginia, where USTs containing gasoline, diesel and heating oil were located. Precision testing performed on these USTs in September 1991 indicated that the system as a whole was leaking. Many of the tanks were removed for permanent closure. A soil gas survey was conducted. Initial Abatement was also carried out immediately. A detailed Site Characterization Study was performed. The study included the investigation of horizontal and vertical extent of subsurface contamination, geology and hydrogeology of the area. Based on initial sampling and analyses, the soil and groundwater were found to be contaminated with petroleum hydrocarbons. Before constructing any buildings in these contaminated areas, remediation must be conducted in accordance with the State, Federal and other applicable regulations. Due to the scheduled construction, a fast track project was conducted to characterize the contamination and begin construction. Site Characterization investigative techniques including Hydropunch™, field Screening using immunoassay tests, and conducting a pilot scale test for dual phase extraction. In order to conduct the pilot test, more monitoring wells had to be installed. This paper discusses the different aspects of site characterization in the vicinity of removed USTs.

Introduction

Characterization of a hazardous waste site involves gathering and analyzing data to describe the processes controlling the transport of wastes from the site. It provides the understanding to predict future site behavior based on past site behavior. It can encompass the characterization of the waste itself as well as that of various transport pathways such as air, surface water, biota, and groundwater [1].

When Cameron Station - a U.S. Army Military installation in Alexandria, Virginia, closes in 1995 under the Base Realignment and Closure (BRAC) initiative, the Military District of Washington (MDW) logistics activities and other operations are scheduled to be moved to the proposed logistics Warehouse and Administrative Building at Fort Myer, another U.S. Army Military installation located in Arlington, Virginia. The proposed Logistics Complex includes the construction of a Vehicle Maintenance Building and Vehicle Storage area;

and a Warehouse and Administration Building.

In order to clear the areas where the construction is proposed, underground storage tanks (USTs) that were not in service were removed. During the time of removal, it was noticed that petroleum products remained in some USTs; some USTs had holes; free product was noticed in the excavation pits; and soil stains were observed. These visual observations triggered an immediate follow up investigation to examine the extent of contamination and subsequent remediation. Ten USTs were identified at the two areas proposed for construction. The site investigations for four of these USTs (Tanks 12-15) are discussed. The regulatory aspects of the project are discussed else where [2]. Further field investigations and immediate corrective measures carried out at the site included:

- Initial abatement and site check
- Site characterization
- Free product recovery

Based on the investigations, the following reports were prepared:

- Site Characterization Report
- Risk Assessment Report
- Corrective Action Plan
- Site Clearance Report.

Field Investigation

In order to comply with the BRAC schedule for the construction of the proposed Logistics Complex, the field investigations were conducted in an accelerated schedule to expedite the remedial actions at the site. Upon completion of the field investigations, a Risk Assessment was performed, the results of which were used in preparing the Site Characterization Report (SCR), Corrective Action Plan (CAP) and Site Clearance Report (SCLR). An SCR for USTs 12-15 was prepared in January 1993 [3] and an Addendum to the SCR was prepared in August 1993 [4]. Sampling locations are shown in Figure 1.

Dual Phase Vapor Extraction: The field work included a dual phase vapor extraction (DPVE) pilot test. The main objective of this pilot test was to evaluate the response of both the saturated and unsaturated soil zones for possible remediation of the petroleum contamination. Four observation wells and one deep well were installed for the pilot test. A vacuum of 20 inches of mercury was applied to extraction well OW-1 by the pilot test system. This system was capable of creating a vacuum of 29 inches of mercury and an actual air flow of 100 cubic feet per minute (ACFM). The average soil vapor extraction rate was 21 standard cubic feet per minute (SCFM), and the resulting groundwater extraction rate was estimated to be approximately 0.1 gallon per minute (GPM). The relatively high vacuum of 20-inches of mercury that was maintained in the well indicated that the soil in the upper water-bearing zone was not very permeable to air. The low rate of groundwater extraction was consistent with this observation, as the shallow soil formation was shown to yield little water. The influence of pilot testing at OW-1 was not symmetric. Upper water-bearing zone heterogeneities, groundwater gradients, or man-made preferential pathways (e.g. utilities) were likely responsible for the non-symmetric response of the upper water-bearing zone water levels.

The laboratory chemical analysis for both groundwater and soil gas samples that were collected during the pilot test show that gasoline-related compounds were effectively extracted during the test. The pilot test also indicated that perched groundwater existed at the site. This information triggered further investigation.

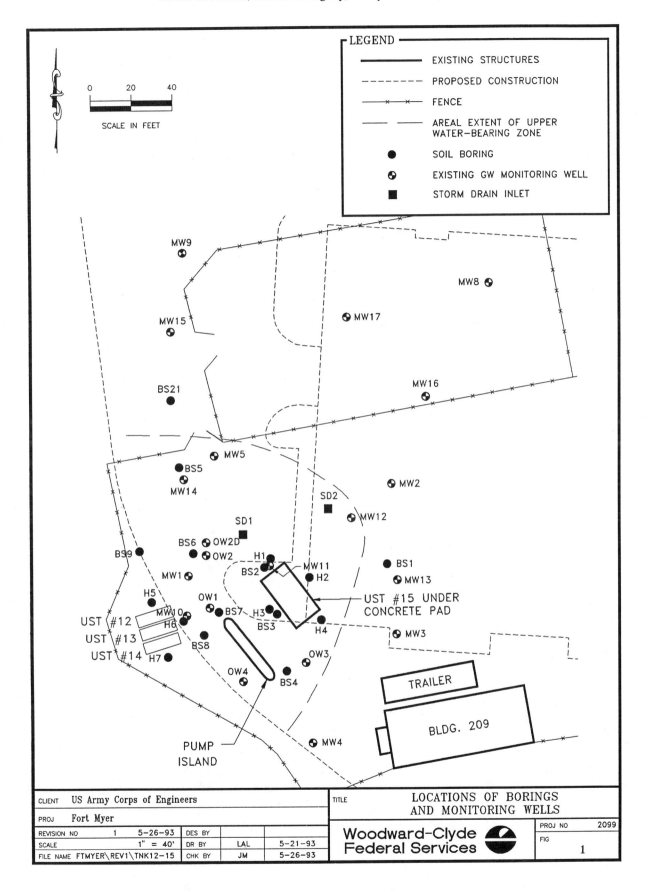

Hydropunch™ Sampling: Hydropunch™ is advantageous in site assessment and hydrocarbon detection, since groundwater samples can be collected without constructing a well. This tool is constructed of stainless steel and is equipped with a driving point. Drilling using hollow-stem augers was performed to drill a borehole just above the desired depth. The Hydropunch™ tool is then driven into the aquifer. When the desired depth is reached, the tool is withdrawn approximately four feet, leaving the point in the ground and exposing the screen so that groundwater and floating product can enter. A Teflon® bailer is lowered through the hollow interiors of the drive casing of the Hydropunch™ to collect the sample. Hydropunch™ and soil samples were collected in the vicinity of observation wells. Because of the relative ease with which groundwater samples can be obtained using the Hydropunch™, this system has the potential for collecting large amounts of data quickly and inexpensively. Groundwater samples thus collected were field screened for TPH using an immunoassay [5] technique.

Immunoassay: Groundwater samples were collected and analyzed for TPH using an immunoassay technique according to SW-846 Method 4030. This semi-quantitative field screening method was used to provide rapid on-site results to assist in determining subsequent sampling locations. The method employs enzyme-linked immunosorbent assay to detect components of petroleum products. Samples are prepared and analyzed at the same time as calibration standards, and the absorbances are recorded using a spectrophotometer. Samples were first run for the 165 ppb and 1650 ppb levels of the test. Some samples were then diluted and rerun to bring them within the calibration range of the test. Selected samples were then analyzed in the laboratory for BTEX, TPH, metals, methyl-tertiary butyl ether (MTBE) and polycyclic aromatic hydrocarbons (PAHs).

Findings of the Field Investigation

Materials encountered at the site include recent fill material overlying Quaternary alluvium. The study indicated the presence of an extensive clayey to silty sand fill material having an average thickness of approximately 2 meters (6 feet) and a maximum thickness of 3.7 meters (11 feet).

Hydrogeological investigations revealed that the perched groundwater was found mostly in possible sand stringers in silt or clay. Field observations suggest that the water levels in the perched zone are influenced by runoff from storms. Water washes down the nearby slopes, infiltrating at the edge of the asphalt, and percolating into the subsurface. Immediately below the asphalt is up to 3.7 meters (11 feet) of fill material (silty sand, gravel, and debris) overlying mostly a silty clay. Water is then trapped by the clay, forming a limited perched zone.

Risk Analysis

In order to define areas of concern, a risk assessment was prepared in accordance with the U.S. Environmental Protection Agency's (EPA) guidance document [6]. Potential health effects were assessed for both a reasonable maximum exposure (RME) and an average exposure. Carcinogenic risks and non-carcinogenic health hazards were assessed by applying generally recognized risk assessment methodology. The resultant estimates for RME exposures for both the scenarios indicate that carcinogenic risks are within U.S. EPA's acceptable risk range (1×10^{-4} to 1×10^{-6}) and non-carcinogenic Hazard Indices are below the "acceptable" threshold value of 1.0. Estimates for average exposures are lower, and also within acceptable levels.

In order to be conservative, the chemical-specific preliminary remedial goals (PRGs) were set at the lower cancer risk level end of the range (1 x 10^{-4}) and a non-carcinogenic Hazard Index of 1.0. PRGs were developed for BTEX and PAHs in groundwater and soil. PRGs were derived from both the construction worker and Logistics Complex worker scenarios, applying RME assumptions. For the construction workers, exposure through inhalation and dermal contact with groundwater were considered; incidental ingestion of soil and dermal contact with soil were found to be minor contributors to overall risks and hazards. A PRG was not developed for lead in soil (lead concentrations were found to be less than 50 μg/kg).

Conclusion

The data generated were compared with the PRGs (Figure 2). The areal extent of contamination was estimated to be 150 square meters (1,400 square feet). The depth of concern was identified to be three meters (nine feet). Area of groundwater contamination was found to be localized near OW-1.

The field techniques used were valuable tools to assist in the determination of the extent of contamination. The DPVE pilot test indicated that a perched zone exists above the deeper aquifer and that remediation using DPVE was feasible. The Hydropunch™ sampling method allowed for rapid collection of groundwater samples without constructing monitor wells. The immunoassay field screening assisted in locating subsequent sampling locations without the expense or delay of 24-hour turn-around-time from an off-site laboratory.

References

1. USEPA. 1991. "Seminar Publication, Site Characterization for Subsurface Remediation." EPA/625/4-91/026

2. Kikkeri, S.R., Hagarty, E.P., and Wilcher, J.L. 1993. "Regulatory Aspects of Hydrocarbon Contaminated Soil and Groundwater Remediation at a Building Construction Site." Presented at the Eighth Annual Conference on Contaminated Soil, University of Massachusetts, Amherst, MA.

3. Engineering Technologies Associates, Inc. 1993. "Site Characterization Report, Tanks 12-15, Fort Myer, Arlington, Virginia."

4. Woodward-Clyde Federal Services. 1993. "Draft Final Addendum, Site Characterization Report, Tanks 12-15 Area, Fort Myer, Arlington, Virginia."

5. Tracy, A., Cline-Thomas, T., and Mills, W. 1993. "The Application of Immunoassay-Based Field Methods to Delineate Hydrocarbon Contamination: A Case Study." Presented at the Eighth Annual Conference on Contaminated Soil, University of Massachusetts, Amherst, MA.

6. USEPA. 1991. Role of the Baseline Risk Assessment in Superfund Remedy Selection Decisions. OSWER Directive 9255.0-30, April 22, 1991.

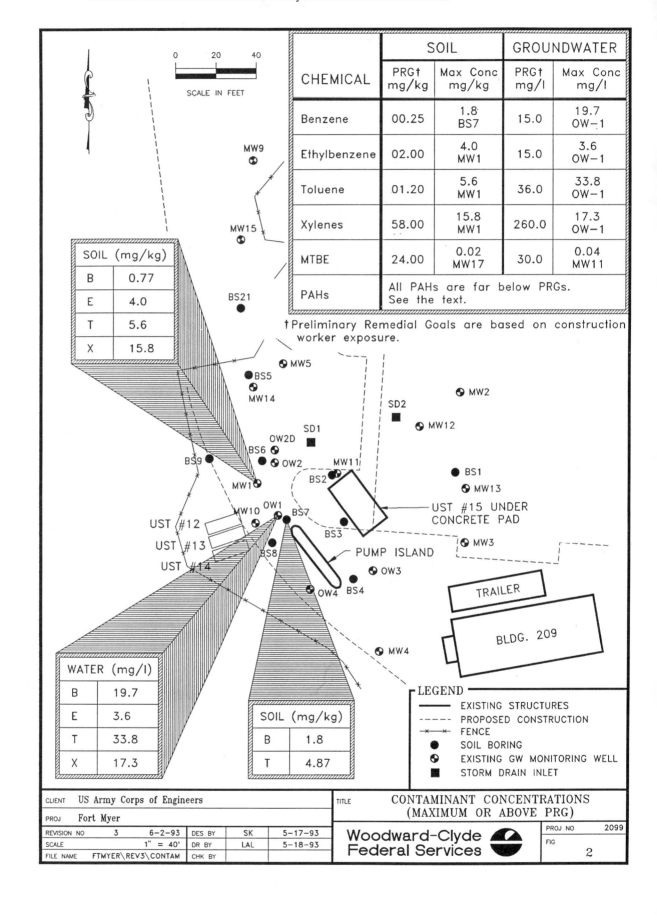

Remediation of a Chemical Landfill at the University of Saskatchewan, Saskatoon, SK

Paul Kozicki, P.Eng., President
GE Ground Engineering Ltd., Regina, Saskatchewan, Canada

Abstract

A chemical landfill had been operated from 1978 to 1986 by the University of Saskatchewan, north of Saskatoon, Saskatchewan. An investigative program determined that contaminants were emanating from the chemical landfill and that they were reaching the Forestry Farm Aquifer and were being transported to, and probably beyond the boundaries of the University property.

The variety of solid and liquid chemicals encountered was extremely wide ranging and including solvents, chlorinated liquids, acids, boxes, pesticides, scintillation phials, crystalline picric acids, gas cylinders, dismantled research equipment, and a large volume of low concentration hospital waste. The retrieval of the liquids was hazardous and characterization of the substances was complex and difficult. All retrieved liquid chemicals were transported off site to Ontario, or British Columbia for disposal at a commercial waste management facility. The retrieval of the liquid and solid chemicals and site remediation was completed in 1990.

A temporary confinement facility was constructed for storing contaminated soils and low grade solid wastes. It is believed that the confinement facility is a state-of-the-art facility which will provide total confinement of all materials stored in it. Any water which might penetrate the top seal will be intercepted and not be able to pass through the stored material and escape to the environment.

The damaged environment has been remediated. This will allow a proposed future residential sub-division to be developed in the area surrounding the chemical landfill site.

Introduction

This paper describes the remediation activities associated with a Chemical Landfill site which was operated by the University of Saskatchewan north of Saskatoon, SK. The size occupied approximately eight (8) hectares of property. In 1978 the University was granted a license to dispose of solid chemical waste in a restricted area on the property.

This area was approximately 40mx100m and was secured by a chain link fence and gate which was kept locked at all times.

The chemical waste which consisted of both solids and liquids, was disposed of within the fenced area in a series of trenches or pits approximately four (4) metres deep. The trenches were filled to within a metre of the top and the waste was covered with soil. Solid chemical waste was sometimes buried in cardboard containers or plastic bags and the liquid chemicals were buried in their original containers or in sealed plastic or steel pails. The liquid chemical containers were generally labelled to show the contents.

The stratigraphic profile at the chemical landfill site consists of approximately four (4) metres of ablation till overlying approximately five (5) metres of pervious sand. The pervious sand is underlain by an undetermined depth of glacial till.(1)

The natural groundwater level is approximately halfway up the sand layer and is known as the Forestry Farm Aquifer. The general ground surface dips gently from east to west towards Petursson's Ravine and the South Saskatchewan River, and the stratigraphic units described above conform roughly with this dip. The flow in the aquifer is in the direction of the dip.

Records were not kept of the chemicals placed in the landfill prior to 1983, but after that, an inventory was kept. The chemicals in the landfill consisted of a wide variety of liquids and solids from the various laboratory facilities at the University. The University stopped placing chemicals in the landfill towards the end of 1986 and began disposal of the chemicals at a commercial facility in Ontario.

In 1981 the Province of Saskatchewan, Department of the Environment, informed the University that leachate from the chemical landfill was contaminating the water in the Forestry Farm Aquifer. On August 21, 1986 the Saskatchewan Department of the Environment issued a Minister's Order to the University instructing it to secure the chemical landfill site and to effect its closure as soon as possible. In addition, the University was directed to carry out a comprehensive study to determine the environmental impact of the facility and if necessary to effect whatever remedial measures were required to mitigate this impact.

In the Spring of 1987, an investigation program was initiated which included the installation of 10 water sampling wells and piezometers. These wells were designed to recover 23 independent samples at different elevations in the aquifer. In 1988 the investigation program was extended to include the adjoining property and Petursson's Ravine. A total of 29 additional sampling wells and piezometers at 27 locations were installed to sample the water in the aquifer. The surface water in the springs in the ravine was also sampled.

As a result of the investigation program, the Consultants recommended to the University that:

---

(1) GE Ground Engineering Ltd. Report, 1987

- All chemicals in the landfill should be retrieved;

- All retrieved liquid chemicals should be transported off site for disposal at a suitable commercial facility;

- All solid chemicals should be disposed of as deemed appropriate;

- All contaminated soil and low grade solid waste should be placed into a confinement facility constructed on the site.

Hydrological studies had determined that once the source of the contaminants was removed, the residual contaminants in the aquifer would be flushed out in approximately three (3) years. Since the levels of contaminants in the aquifer were relatively low, it was recommended that no attempt be made to clean up the residual contaminants.

The University accepted the recommendations and in May, 1988 instructed the Consultants to prepare the necessary contract documents and designs. The contract for retrieval and disposal of the chemicals was awarded during May, 1989 and work commenced on site during the first week of June, 1989.

The variety of solid and liquid chemicals encountered was extremely wide ranging and included solvents, chlorinated liquids, acids, bases, pesticides, scintillation phials, crystalline picric acid, syringes, gas cylinders, dismantled research equipment and a large volume of low concentration hospital waste.

The Contractor had estimated that the retrieval would be completed by October 31, 1989. However, the deteriorated conditions of the containers and chemicals in the landfill, and the wide variety of substances encountered, made retrieval painstaking and hazardous.

The retrieval was completed in mid December of 1989 and all the retrieved materials were either removed from the site or stored in a secure, fully lined confinement facility constructed on the site. The site was then closed for the winter. Construction of the cover seal over the confinement facility and final reclamation of the site was completed in August, 1990.

Monitoring of the water quality in the Forestry Farm Aquifer will continue for several years to determine the effects of the remedial measures.

Investigations & Monitoring of Groundwater Quality

The investigation of groundwater contamination at the Saskatoon Landfill started in the early 1980's with the installation of four (4) stand-pipe piezometers and analysis of the groundwater for inorganic constituents. The results of that program were inconclusive.

In the Spring of 1987, a total of 10 monitoring wells and piezometers were installed on the University property to determine whether or not the Forestry Farm Aquifer underlying the landfill site was being contaminated. As well, enough data was gathered to characterize the direction and rate of groundwater flow and to isolate, if possible, the source of any contamination. The study was complicated by the presence of a decommissioned general purpose landfill located immediately upgradient of the chemical landfill and an older decommissioned municipal type landfill located further south and east, but also generally upgradient.

The investigation was performed in stages, including the installation of additional piezometers and sampling wells. Wells were also installed to determine whether or not there were any strong vertical gradients. The purpose of this was to establish the flow regime coming from the chemical landfill to Petursson's Ravine. Four (4) additional wells at three (3) locations were installed during the fall of 1991 to provide additional groundwater monitoring down gradient of the general landfill.

Water samples were taken in the spring and fall of each year since 1987, except for 1989 when water samples were taken only in the spring.

Analyses was carried out for three (3) groups of compounds: 1) volatile organics, principally halogented solvents, using a purge and trap apparatus coupled to a gas chromatograph mass spectrometer (GC/MS); 2) semi-volatile organics, using liquid extraction and a GC with a Fl and EC detector for routine screening. Specific compound identification was carried out by using a GC/MS; and 3) conventional inorganic constituents, the major ions normally found in groundwater, a suite of metals, nutrients and alpha and beta emitters. Following the first two samplings, the number of compounds being analyzed was reduced.

The wells sampled in the Spring of 1988 indicated the presence of a narrow contaminant plume which extended all the way to Petursson's Ravine. The concentrations of the solvents at the ravine were much less than at the landfill boundary, and generally less than at the property boundary. However, tetrachloroethylene was found at the ravine indicating either wide temporal variations or very discrete and differentiated contaminate flows within the aquifer. In general, there were concentrations of solvents in the plume all the way to the ravine that would result in the initiation of some action by most regulatory authorities.

Development of Contract Documents

A large number of factors had to be taken into consideration in deciding how the retrieval and disposal of the chemicals should be undertaken.

The consultants believed that it would be possible to prepare a document which would provide the University with protection from exploitation, but at the same time to provide fair compensation to the Contractor for the many unforeseen conditions which were anticipated.

The contract was set up in such a way that the University would be protected against exploitation in the event that the time spent on retrieval was escalated either by more time being required to complete the work, or by dilatoriness on the part of the Contractor.

Security and Safety

In view of the hazardous nature of the Work, the Contractor implemented strict security measures and control over the personnel entering or leaving the general site. Security guards were present 24 hours per day, 7 days per week.

Access to the Chemical Landfill security fence area could only be gained by passing through a decontamination station where a strict protocol was observed. Donning of personal protective equipment consisted of stripping to one's undergarments and then putting on cotton overalls, work socks, three (3) pairs of latex gloves, a disposable hooded coverall, steel toed rubber boots and having all cuffs and openings sealed with tape. Full face respirators, hard hats and PVC gauntlets (strong glove with covering for the wrist) were also worn at all times. All personnel were required to disrobe in an approved sequence and shower after leaving the restricted area. All Contractors, University, and Consultants personnel engaged on the site were given training in the security protocol and safety measures, and were required to comply with them.

Throughout the work the Contractor monitored air quality in the vicinity of the site.

Retrieval, Sorting and Disposal of Chemicals

According to the information available from the University, the chemicals had been placed into three (3) trenches, each approximately three (3) to four (4) metres wide by four (4) metres deep, which ran parallel to the largest dimension of the fenced area. The most recent trench had not been completely filled and the containers of the most recently placed liquid chemicals were visible.

The retrieval method adopted was to carefully remove the soil cover over the trench using a Gradall or hydraulic backhoe until the chemicals were exposed. The soil recovered from this operation was stockpiled and subsequently placed in the confinement facility. The soil removal was carried out in short lengths to ensure that the least possible surface area of chemicals was exposed at any time.

The waste containers were generally intermingled with soil and it was necessary for the technicians to use archaeological methods employing simple tablespoons, garden trowels and other small hand tools to uncover the chemical containers intact.

Wherever possible, each object or chemical container was identified as it was retrieved and sorted into plastic bins located in the bottom of the trench. These bins were subsequently transferred to the main sorting area where the contents were re-sorted and characterized.

The liquid chemicals were analyzed using a winnowing procedure and "Haz-Cat" analysis kit then bulked into categories. Liquids which did not fit into categories were treated as "specials" and placed into lab packs.

The solid chemicals included a wide variety of objects and packages. In this category were numerous gas cylinders, plumbing fittings, cardboard containers of paper, filters, pieces of structural steel, lumps of concrete, old steel drums, etc. These objects were sorted, cleaned and disposed of as appropriate. Much of this material such as cardboard containers, plastic pails, pieces of plywood, etc. was shredded on site and mixed in with the soil being placed into the confinement facility.

Shock-sensitive materials such as crystalline picric acid (19 jars) and peroxyacid were periodically detonated on site by the RCMP bomb squad soon after they were recovered.

As the retrieval progressed towards the end of the first trench and into the second trench, an ever increasing amount of low concentration hospital waste was encountered. This material, which had not been included in the inventory records made available by the University prior to commencement of the Work, consisted of wipes and clean-up plastic garbage bags containing pipettes, test tubes, syringes with needles attached, scalpels, absorbent pads, hospital and laboratory specimens, histrology samples and other medical laboratory waste. Many of the plastic bags had been ruptured and the contents spilled into the soil.

Retrieval of this material was particularly hazardous and painstaking and there were many instances where the needles punctured the gloves worn by the technicians which resulted in medical treatment being required. In addition, this material did not fall into any of the disposal methods contemplated under the Contract and special arrangements had to be made to sort it into various categories and package it for transportation and disposal at the University Incinerator.

After sorting and packaging of the liquid chemicals had been completed, the drums and lab packs were collected by a commercial carrier and transported to Ontario or British Columbia for disposal at a commercial waste management facility.

Contaminated Soils

Soil excavated from the landfill site and shredded materials were blended and placed into a confinement facility in layers and compacted to achieve a dense homogeneous fill. The stored material was subjected to a drilling and sampling program to determine its physical state and chemical content before the final cover seal was constructed. Analysis of the soil showed general levels of contamination ranging from 100 parts per billion to 100 parts per million, with the odd pocket of near clean soil.

Generally, the soil in the bottom portion of the confinement facility was found to be the least contaminated and the highest levels of contamination were in the upper two (2) metres. This was consistent with the order in which the soil was removed from the Chemical Landfill site.

Confinement Facility

It was estimated that the retrieval would generate approximately 10 000 cubic metres of contaminated soil, shredded containers and solid chemical waste. It was anticipated that the level of contamination should be relatively low and that the material would probably be acceptable for disposal in a municipal landfill. However, the

Saskatchewan Ministry of Environment expressed reservations about disposing of it in the municipal landfill and it was decided to place it into a secure, fully lined confinement facility constructed on the site and keep it there until such time as it could be decided what should be done with it.

The confinement facility is located immediately to the south of the chemical landfill site and is approximately 60 m by 60 m in overall size.

The bottom of the facility is a square basin 36 m by 36 m which is three (3) metres deep below the natural ground surface and has 1 on 4 side slopes. The bottom of the facility is located approximately two (2) metres above the natural groundwater elevation. The basin is lined with a compacted till seal which is 600 mm thick and has a design objective coefficient of permeability not greater than 10^{-9} m/s.

The lined floor area is graded to a slope of two (2) percent towards a low point at the middle of the west side. The lined floor area is covered with 300 mm of fine granular filter material and a system of 100 mm diameter slotted agricultural drain pipes which are designed to collect any liquid in the filter layer and deliver it to the low point on the west side. The drain pipes are surrounded with coarse granular filter material. The side slopes are also covered with a similar layer of fine filter material, but do not have any slotted drain pipes installed.

A sealed pipe connects the slotted drainage pipe system to a seepage collection manhole adjacent to the west side of the facility.

The purpose of the drainage system is to intercept and collect any seepage which may be discharged from the waste material stored in the facility.

The earthfill waste material was compacted into the facility to form a dense homogeneous mass and, since it is predominantly till, is very impervious.

The top of the compacted waste material rises about two (2) metres above the natural ground elevation and is shaped to form a truncated pyramid with side slopes of 1 on 3. The top of the truncated surface is graded towards the west to prevent ponding of water on the surface.

The entire top surface of the compacted waste is covered with a layer of filter material, together with a system of slotted drainage pipes similar to that over the floor of the facility. This system is connected to a sealed pipe which conveys any seepage which may occur through the cover seal, to the sump adjacent to the west side of the facility.

The entire top surface of the filter blanket is covered with a top seal constructed of a 40 mil thick HDPE geomembrane sandwiched between two (2) layers of compacted glacial till, each 300 mm thick. Finally, the surface is covered with approximately 300 mm of topsoil and seeded with grass.

It is believed that the confinement facility is a state-of-the-art facility which will provide total confinement of all materials stored in it. Any water which might penetrate the top seal will be intercepted and not be able to pass through the stored material and escape to the environment.

SUMMARY

The Chemical Landfill Remediation Project was initiated to clean up the existing Chemical Landfill and decrease the off-site movement of chemical contaminants from the site. With the removal of the chemicals, the concentration of the contaminants in the aquifer will start to decrease as the aquifer is flushed. The estimated travel time for the water in the aquifer to flow from the landfill to the ravine is in the order of three (3) years. Decreasing concentrations at the ravine will therefore not be seen for some three (3) years after they are established in the wells adjacent to the now excavated chemical landfill site. Monitoring of the water quality in the Forestry Farm Aquifer will continue for seven (7) years to determine the effects of the remedial measures.

Contrôle Électrocinétique des Écoulements Souterrains

G. Lefebvre, F. Burnotte
Département de génie civil, Université de Sherbrooke, Sherbrooke, Québec, Canada

P. Rosenberg
Lupien, Rosenberg et associés, Montréal, Québec, Canada

SOMMAIRE

La possibilité de contrôler les écoulements souterrains dans un sable silteux au moyen d'électrodes implantées dans le sol est examinée à partir d'une série d'essais longue durée réalisés en laboratoire. Les résultats montrent qu'une telle barrière à l'écoulement pourrait être efficace pendant plusieurs années et à des coûts relativement faibles.

INTRODUCTION

Bien que l'application de l'électrocinétique soit bien connue pour la consolidation des argiles et considérée comme offrant un bon potentiel pour la décontamination des sols, la possibilité d'utiliser une barrière électrocinétique pour contrôler l'advection d'eau souterraine contaminée a été jusqu'ici peu examinée (1, 2). Casagrande, dès 1949, a utilisé l'électrocinétique pour contrôler les écoulements afin d'améliorer la stabilité d'excavation dans des milieux granulaires fins (3). Le contrôle des écoulements au moyen d'une barrière électrocinétique consiste à provoquer un débit électro-osmotique au moyen d'électrodes implantées dans le sol pour contrer l'effet d'un gradient hydraulique. Le débit électro-osmotique est habituellement relié au potentiel de surface des particules et à la double couche diffuse qui en résulte. C'est le mouvement des ions dans cette double couche qui entraînerait par viscosité l'eau libre des pores. Les potentiels de surface sont considérés significatifs pour les argiles ou encore pour les oxydes de métaux hydratés (4). Le débit électro-osmotique pourrait donc être faible dans les matériaux granulaires fins, tels les silts, où le potentiel de surface est relativement faible. Il est de plus connu que le débit induit par un potentiel électrique tend à s'arrêter avec le temps (5). L'objectif de cet article est d'examiner l'efficacité d'une barrière électrocinétique dans un sable silteux au moyen d'essais longue durée conduits en cellule de laboratoire en examinant la longévité d'un tel système et la puissance nécessaire.

Description des sols testés et programme expérimental

Les quatre sols testés ont été préparés à partir d'un sable silteux naturel de la région de Coaticook (Québec, Canada), soit le sol à l'état naturel, la partie fine seulement (inférieure à 80 µm), le sol lavé (lavage à l'eau du robinet pour enlever les particules argileuses) et le sol lavé à l'acide (lavage au HCl pour enlever les ions de fer et de calcium). Le sol naturel est constitué de 58% de sable, 39% de silt et 3% d'argile (< 2µm). La fraction argileuse est donnée au tableau 2 pour chacun des sols préparés. Les cations du sol à l'état naturel sont présentés au tableau 1. Le fer est présent sous forme d'oxydes de fer hydratés, donnant une couleur brune au sol. La grande quantité de calcium, généralement sous forme de carbonate de calcium, est responsable du pH basique initial du sol (pH = 8,4) et de sa grande capacité à tamponner les acides. Après le lavage à l'acide, le pH a diminué à 5 et la concentration en ions fer a diminué jusqu'à 2700 ppm (diminution de 82%), et à 400 ppm pour les ions calcium (diminution de 92%).

TABLEAU 1 Concentration des cations du silt de Coaticook naturel

Ion	Fe	Ca	Mg	Al	P	Ti	Cu	S
Concentration (ppm)	15 000	15 000	2 500	900	250	120	50	50

TABLEAU 2 Résumé des résultats

Essai No	Type de sol	%<2µ	ω_i (%)	pH_i du sol	Durée totale h	Temps approx h	i_h cm/cm	i_e V/cm	k_h cm/s .10-5	k_e cm2/V.s 10-5	ρ kΩ/cm	Energie totale kW.h/m³
I	Naturel	3	16.0	8.4	300	20	0.2	0.8	5.2	1.4	28	
						250	0.2	0.4	5.2	1.9	28	3
II	Naturel	3	11.1	8.4	4341	500	0.5	0.4	1.4	1.5	33	
						3500	0.3	0.7	1.7	0.7	18	121
IV	<80µm	7	20.0	8.4	3950	500	0.3	0.4	2.3	1.4	23	
						3500	0.2	0.4	1.2	0.3	12	62
V	lavé	1	12.5	8.4	460	40	0.1	0.2	2.3	0.5	32	
						400	0.2	0.9	2.3	0.5	38	4
VI	lavé à l'acide	2	15.0	5.0	1201	8	0.1	0.5	3.4	0.9	44	
							100	0.2	0.6	3.4	1.4	95	
							1200	0.2	0.2	3.4	1.1	100	5

Les 5 essais (numérotés I, II, IV, V et VI) sont effectués dans des cellules cylindriques de plexiglass, d'une longueur d'environ 33 cm et d'un diamètre de 6,3 cm (Fig. 1). Les électrodes sont en acier ordinaire. Les potentiels électriques sont mesurés en 5 points de lecture dans l'échantillon (v_1 à v_5), ainsi qu'aux électrodes et à une référence dans chaque réservoir. Les cellules des essais II et IV sont munies de quatre piézomètres (p_1 à p_4) mesurant la charge hydraulique locale. Un schéma du montage est donné à la figure 1. Les sols ont été compactés dans le cylindre par couche de 2 cm à des teneurs en eau généralement supérieures à l'optimum Proctor afin d'améliorer la saturation. Un essai proctor standard sur le sol naturel a montré une densité sèche maximale de 1.91 à une teneur en eau de 10%.

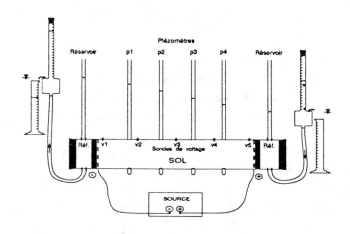

Figure 1 Schéma du montage

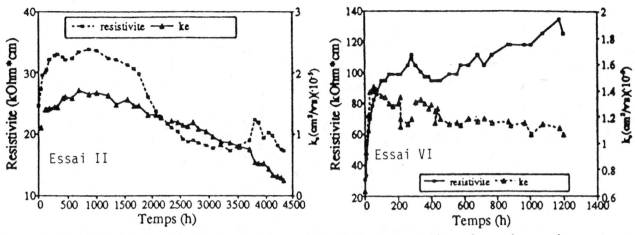

Figure 2. Evolution de la résistivité et de la perméabilité électro-osmotique durant les essais

Les essais consistent essentiellement à provoquer un écoulement sous un potentiel hydraulique et ensuite à contrer cet écoulement en appliquant un potentiel électrique. Après montage de l'échantillon, une charge hydraulique est imposée durant quelques jours pour déterminer la perméabilité hydraulique initiale (k_h). Le potentiel électrique est ensuite appliqué et ajusté jusqu'à arrêt de l'écoulement. Par le principe de superposition des débits hydrauliques et électro-osmotiques, la perméabilité électro-osmotique (k_e) est calculée. Les gradients appliqués sont ajustés en cours d'essai afin de maintenir un débit nul. La source électrique est débranchée à quelques reprises durant l'essai pour effectuer des mesures de perméabilité hydraulique.

Présentation et discussion des résultats

Les caractéristiques des essais sont résumées au tableau 2. Seuls les résultats des essais II et VI sont présentés en détail (Fig.2 et 3). La perméabilité hydraulique dans les cinq essais a varié entre 1.2 et 5.2 x 10^{-5} cm/sec. Les différences de perméabilité entre les différents échantillons reflètent probablement surtout le degré de saturation des échantillons notamment en ce qui concerne les essais 1 et 2, réalisés tous deux sur le sol naturel mais préparé à des teneurs en eau différentes. Au cours des premières cinq cents heures, la perméabilité électro-osmotique, k_e, est voisine de 1.5 x 10^{-5} cm^2/V sec. pour tous les essais sauf pour l'essai V réalisé sur le sol lavé où elle n'est que de 0.5 x 10^{-5} cm^2/V sec. Des essais électrocinétiques semblables actuellement en cours sur des argiles de l'Est du Canada montrent des k_e de l'ordre de 3 x 10^{-5} cm^2/V sec. Il apparaît donc qu'un bon débit électro-osmotique puisse être obtenu dans les sols granulaires fins à la condition que la fraction argileuse soit supérieure à 2 ou 3%. La présence ou non d'une fraction sableuse d'environ 60% n'influence pas le k_e (essais 1, 2 et 5). Les ions fer et calcium ne semblent pas jouer ici un rôle important dans le débit électro-osmotique car une forte réduction de ces derniers par un lavage préalable à l'acide (essai VI) n'a pas influencé sensiblement la perméabilité électro-osmotique. Il n'est donc pas évident que le débit électro-osmotique dans ces sols granulaires puisse s'expliquer par un modèle basé sur la double couche.

Évolution en fonction du temps

Le sol de l'essai VI a une résistivité électrique près de trois fois plus grande que le sol de l'essai II (Tab. 2, Fig.2) malgré une plus faible teneur en eau de ce dernier. On peut penser que cette augmentation de la résistivité est reliée à la forte diminution des ions fer et calcium dans le sol préalablement lavé à l'acide. La figure 3 présente pour les deux essais les variations locales de résistivité pour les deux essais et les gradients hydrauliques locaux mesurés uniquement dans l'essai 2. Pour l'essai II, la figure 3 montre la migration d'un front résistant qui progresse de la cathode vers l'anode. Les variations de résistivité locales sont dues principalement à une variation de la conductivité de l'eau des pores, puisque la partie du courant transmise par les grains solides est négligeable (5). La position de ce front résistant est interprétée comme le lieu où les ions OH^- en provenance de la cathode rencontrent les ions H^+ en provenance de l'anode formant ainsi de l'eau et donc un fluide à faible conductance. La migration de ce front résistant se manifeste donc à l'avancée de la zone basique de la cathode vers l'anode. A la fin de l'essai II, soit après plus de 4000 heures, le front résistant et la zone basique ont complètement traversé l'échantillon pour atteindre la zone de l'anode. La traversée de la zone basique est confirmée par les pH du sol qui sont, après l'essai, plus élevés que le pH initial (Tab. 3). Cette traversée de la zone basique durant l'essai est expliquée par la présence des carbonates qui tamponnent l'acidité en provenance de l'anode. Le débit électro-osmotique et donc l'efficacité du système a chuté rapidement dès que le front résistant a atteint la région de l'anode. L'essai VI, bien qu'arrêté après 1200 heures permet de visualiser un comportement bien différent. La zone la plus résistante demeure localisée dans la région de la cathode alors qu'une résistance très faible est observée dans la région de l'anode. On peut donc en déduire que c'est plutôt la région acide qui progresse durant cet essai de l'anode vers la cathode. Cette progression d'un front acide est confirmée par des pH après essai qui sont plus faibles que le pH initial sauf au voisinage de la cathode (Tab. 3). Dans ce type de sol, avec une faible capacité pour neutraliser l'acidité, on ne peut donc envisager une forte diminution de l'efficacité due à la traversée d'un front basique de la cathode vers l'anode. L'acidification progressive du sol pourrait cependant diminuer les potentiels zéta (6) et résulter en un arrêt ou même un renversement de

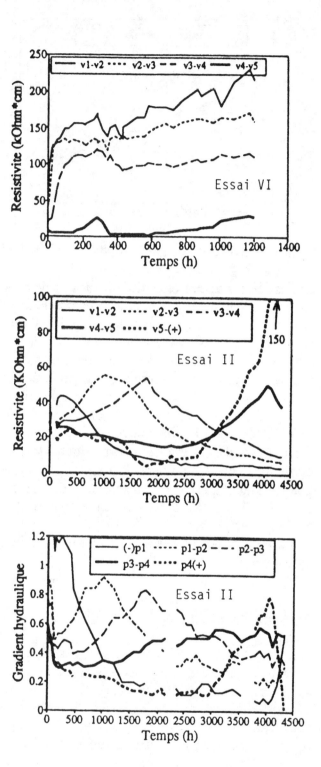

Figure 3. Evolution dans le temps des résistivités locales et des gradients hydrauliques locaux

l'écoulement électro-osmotique (5). Les résistances locales présentées à la figure 3 expriment la perte de potentiel entre deux points et donc les gradients électriques locaux. La mesure des charges hydrauliques locales durant l'essai II montre que les gradients hydrauliques locaux montrent une bonne proportionnalité avec les gradients électriques locaux, semblant confirmer que le débit électro-osmotique est bien dû au gradient de voltage et non à l'intensité du courant (Fig. 3).

TABLEAU 3 pH du sol à la fin des essais II et VI

Position dans l'échantillon	V1	V2	V3	V4	V5	pH$_0$
Essai II	10.8	10.5	10.5	9.9	8.5	8.4
Essai VI	8.9	4.1	4.1	4.1	3.8	5.0

Discussion générale

Les essais présentés ici considèrent uniquement le contrôle des écoulements. La migration des contaminants qui se retrouvent plutôt sous forme moléculaire dans l'eau interstitielle, comme bien des composés organiques, sera contrôlée si les écoulements souterrains sont arrêtés. De même, la migration des contaminants sous forme de cations tels les métaux lourds sera renversée. Par contre, la migration des contaminants sous forme d'anions tels les nitrates et les phosphates, sera accélérée à l'intérieur du champ électrique, même si les débits sont arrêtés (2).

Les gradients de voltage qu'il faut imposer pour contrôler un écoulement souterrain dû à un gradient hydraulique donné peuvent facilement être évalués si les perméabilités hydrauliques et électro-osmotiques sont connues en équilibrant les débits:

$$Q_h = k_h i_h A = Q_c = k_e i_e A \quad (1)$$

et

$$i_e = (k_h/k_e) i_h \quad (2)$$

L'équation (2) est relativement bien vérifiée par les résultats obtenus durant ce programme d'essais. Si on suppose, par exemple, un gradient hydraulique sur le terrain de 0.01 et une perméabilité hydraulique de 1×10^{-4} cm/sec., il faudra appliquer un gradient électrique de 0.1 V/cm, soit un potentiel de 10 V pour un espacement de 2 m entre les électrodes. La puissance nécessaire, P = VI, et donc les coûts peuvent être évalués si la résistivité du sol est connue. En effet, V = RI et en tirant le voltage appliqué de l'équation (2), on obtient la puissance par unité de volume entre les électrodes.

$$P = i_e^2/\rho \quad (3)$$

L'énergie consommée, soit la puissance multipliée par le temps est donc proportionnelle à la distance entre les électrodes et inversement proportionnelle à la résistivité du sol. Pour l'exemple où un gradient électrique de 0.1 V/cm est appliqué, la puissance nécessaire pour opérer une barrière électrocinétique avec des électrodes espacées de 2 m et déployées sur une longueur de 100 m et une profondeur de 10 m serait donc de 1 KW, soit une consommation d'énergie de l'ordre de 750 kWh par mois d'opération. Il faut noter qu'une résistivité de 20,000 Ω cm a été considérée dans cet exemple alors que la résistivité des sols testés dans le programme d'essais a varié généralement entre 20,000 et 100,000 Ω cm.

La période pendant laquelle une barrière électrocinétique demeure efficace apparaît comme proportionnelle à l'espacement entre les électrodes car contrôlée par la traversée d'un front basique ou acide. La période nécessaire à cette traversée sera d'autant plus longue que la distance est grande entre les électrodes. En laboratoire, deux essais (II et IV) ont duré environ 4000 heures et ont montré une bonne efficacité jusqu'à 3600 heures ou 150 jr pour un espacement de 33 cm entre les électrodes. Si un espacement de 2 m entre les électrodes est considéré sur le terrain, on peut donc envisager une période d'efficacité de l'ordre de 2 à 3 ans.

La traversée d'un front basique ou d'un front acide dans le sol est relié à la migration d'ions OH$^-$ et H$^+$ produits par électrolyse aux électrodes. La production de ces ions par électrolyse est proportionnelle à l'intensité du courant dans le système, lui-même proportionnel au gradient de voltage appliqué. A des fins d'expérimentation en laboratoire, des gradients hydrauliques relativement élevés ont été contrôlés par des gradients électriques généralement de l'ordre de 0.4 V/cm. Le gradient électrique appliqué sur le terrain dans l'exemple développé plus haut est de 0.1 V/cm, soit quatre fois plus faible que dans les essais de laboratoire. La production d'ions OH$^-$ et H$^+$ par électrolyse et la vitesse de progression des fronts acide ou basique devrait donc normalement être diminuée d'autant. En ajustant l'espacement entre les électrodes et à cause de la puissance électrique relativement faible nécessaire au contrôle des écoulements, une barrière électrocinétique implantée sur le terrain pourrait donc demeurer efficace pendant plusieurs années.

CONCLUSION

Les essais réalisés montrent que des bons débits électro-osmotiques sont obtenus dans des sables silteux généralement considérés comme ayant un potentiel de surface négligeable. Une fraction argileuse de l'ordre de 2 à 3 % semble cependant nécessaire pour obtenir de bons débits électro-osmotiques dans les matériaux granulaires. La période d'efficacité d'une barrière électrocinétique apparaît reliée à la migration d'un front acide ou basique selon la nature du sol et serait donc proportionnelle à l'espacement entre les rangées d'électrodes et inversement proportionnelle à l'intensité du courant circulant dans le système. Basé sur les résultats d'essais obtenus, il apparaît qu'une barrière électrocinétique implantée sur le terrain dans des sables silteux pourrait demeurer efficace pendant plusieurs années et à des coûts d'énergie relativement faibles.

REMERCIEMENTS

L'étude a été supporté par les subventions de dépenses courantes du CRSNG et par un programme de développement PARI-M supporté conjointement par le Conseil National de Recherches du Canada et la firme Lupien, Rosenberg et associés de Montréal.

RÉFÉRENCES

(1) Mitchell, K.J. (1991) "Conduction phenomena: From theory to geotechnical practice", Géotechnique, vol. 41, No 3, 299-340.

2) Yeung, A.T.C. (1990) Electrokinetic Barriers to Contaminant Transport Through Compacted Clay", Thèse de doctorat, University of California, Berkeley, E.-U. 260 p.

(3) Casagrande, L. (1983) "Stabilization of soils by means of electro-osmosis - State of the art", Journal of the Boston Society of Civil Engineers (ASCE), Vol. 69, No 2, p. 255-302.

(4) Dzombak, D.A., Morel, F.M.M. (1987) "Adsorption of inorganic pollutants in aquatic systems", Journal of hydraulic engineering (ASCE), Vol. 113, No 7, p. 430-475.

(5) Shapiro, A. (1990) Electro-osmotic Purging of Contaminants from Saturated Soils", Thèse de doctorat, Mass. Inst. of Tech., E-U. 136 p.

(6) Shaw, D.J. (1980) "Introduction to colloid and surface chemistry", Boston, Ms E-U, 3e édition, Butterworths, 273 p.

Laboratory Determination of Diffusion and Adsorption Coefficients of Inorganic Chemicals for Unsaturated Soil

P.C. Lim, S.L. Barbour, D.G. Fredlund
Department Of Civil Engineering, University of Saskatchewan, Saskatoon, Saskatchewan, Canada

Abstract A technique is proposed in this paper for determining the diffusion and adsorption coefficients for inorganic chemicals in unsaturated soils. The technique integrates the concept of single reservoir diffusion testing with the axis-translation principle for the control of soil suction. The approach adopted in this study provides a fundamental basis for understanding how decreasing water content alters the pathway, and the open area available for the diffusion of solutes under unsaturated conditions. The main objectives are to demonstrate the feasibility of using this technique to determine both the diffusion and distribution coefficients from a single test, and to establish the functional relationships for the diffusion and distribution coefficients for an unsaturated soil. The test results showed that the diffusion coefficient decreases non-linearly with decreasing water content. Results from mass balance calculations indicate the mass adsorbed per mass of dry soil drops below the equilibrium isotherm as the water content decreases. This seems to suggest that either the distribution coefficient varies with water content or that a fraction of the soil solids may be inactive as the water content decreases.

Introduction

The processes that govern the transport of inorganic chemicals through saturated soils are equally valid for unsaturated soils; however, the nature of the transport parameters may be different. Transport parameters such as hydraulic conductivity for advection, diffusion coefficient for molecular diffusion and adsorption coefficient for adsorption-exchange phenomenon are constant for saturated soil. However, in unsaturated soils, there may need to be functional relationships between these parameters and the water content of the soil. Characterization of these parameters for unsaturated soil is of fundamental importance to the prediction of contaminant migration.

In this paper, a technique for determining the diffusion coefficient and the adsorption coefficient for inorganic chemicals in an unsaturated soil is described. This approach allows the stress state of the soil and the pore water to be controlled throughout the test. The main objectives of this paper are to demonstrate the feasibility of using this new technique to determine both the diffusion and adsorption coefficients of an unsaturated soil in a single test, and to establish the functional relationships for the diffusion coefficient and the adsorption coefficient as the soil desaturates.

Functional relationships for the diffusion coefficient of an unsaturated soil have emerged primarily through the soil science literature. It is known that the diffusion coefficient varies with the water content, but the form of the functionals as reported by different researchers seems to differ. Results showing the relationship between the diffusion coefficient normalised with respect to the diffusion coefficient in a free solution and the volumetric water content of various geologic materials are as shown in Fig. 1.

Data on the adsorption coefficient for unsaturated soils are scarce, and a fundamental understanding of whether the adsorption characteristics change with water content is lacking. The paper by Brown (1953) is the only reference in the literature known to the authors. His results showed that the amount of cation exchange in soil decreases with decreasing water content. The test durations however, were only 4 days and it may be questionable as to whether equilibrium condition had been attained. The computation of the adsorption coefficient from the apparent diffusion coefficient deduced from transient diffusion test is also complicated by its dependency on the solution concentration. The effect of desaturation on the adsorption phenomenon is still unknown at present although, it appears that most researchers have been using the adsorption coefficient obtained by batch-type test, regardless of the degree of saturation of the soil.

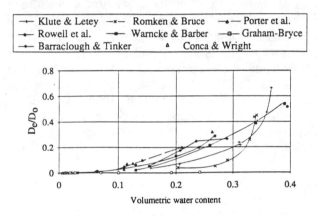

Fig. 1 Functional Relationship between D_e/D_o and volumetric water content (D_e is the diffusion coefficient of soil and D_o is the diffusion coefficient in free solution).

Many researchers attribute the decrease in the rate of diffusion as the water content decreases, to the increased tortuosity of the pathway for diffusion. Mobility of ions through water absorbed on soil surfaces has been reported to be very small (Kemper, 1960 and Porter et al., 1960). The presence of discontinuous water filled pores or water films, and the small mobility of ionic species in the thin water films could have serious implications on the adsorption characteristics of an unsaturated soil. The role of water content on the pathway for diffusion and on the adsorption characteristics are essential information that needs to be verified.

TEST EQUIPMENT

A new diffusion testing technique for unsaturated soil has been developed at the University of Saskatchewan. A cross-section of the apparatus is shown in Fig. 2. This equipment integrates the concept of single reservoir diffusion testing with the axis-translation principle for the control of soil suction. The single reservoir method is similar to the modified column test with a decreasing source concentration as used by Rowe et al. (1988). The diffusion cell was made of plastics. The main features of this equipment are (i) zero advection, (ii) diffusion and adsorption coefficients can be determined independently and, (iii) control of the stress states in the soil and pore water throughout the test.

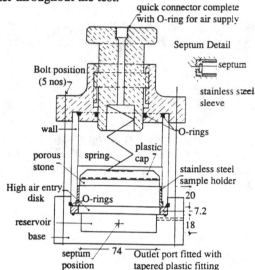

All dimensions are in millimeters.

Fig. 2 A Cross-Section of the Diffusion Cell for Unsaturated Soil Testing.

Material

A natural aeolian sand, called Beaver Creek sand was used for this study. It is an olive brown, oxidized, calcareous, fine to medium sand which is poorly sorted. Sand was selected for this study because of the low silt and clay content, and the low water retention characteristics would provide some insights to the role of thin water films and the evidence of discontinuous water-filled pores or water films. The fines content (i.e., % passing No. 200 US standard sieve) of this clean sand was only 1%. The dominant mineralogy of this sand was quartz mineral and 38% of the fines were of smectite mineral. The cation exchange capacity of the sand was relatively small at 1.45 meq/100g of soil. The major exchangeable cations were calcium and magnesium at 0.86 meq/100g and 0.5 meq/100g, respectively.

Procedure

Laboratory tests simulating one-dimensional diffusion were conducted on the sand at various suctions. The suctions corresponded to a range of water contents where the degree of saturation varied from 100% to residual saturation. Deaired, deionised, distilled water was used throughout the test from preparation stages to sampling.

Prior to sample preparation, the high-flow air entry disk was first saturated, soaked and flushed at least 3 times with water until the level of electrical conductivity was undetectable. The soil was then placed in a slurry form and preloaded under a constant spring load to ensure relatively consistent void ratios for all specimens. The soil was allowed to consolidate under the preload for at least 24 hours before application of a matric suction. At the end of consolidation, the air pressure was applied in order to drain the soil water to the desired matric suction. The applied suction was left to equilibrate for at least 48 hours.

After moisture equilibrium was attained, the reservoir was spiked with a potassium chloride solution by injecting a stock solution into the reservoir below the specimen. The reservoir was stirred continuously during the injection with a magnetic stir bar at slow motion for 2 minutes. A sample was taken immediately to measure the initial reservoir concentration. Decreases in the reservoir concentration with time were monitored by sampling a volume of 0.6 to 0.7 ml of the reservoir solution at various time intervals. The tests were curtailed when chemical equilibrium was attained as indicated by zero or very small change in the reservoir concentration. Stirring of the solution in the reservoir was only carried out for about 2 minutes before and during sampling. A slow motion was used to ensure uniform concentration. Sampling was made with a 1 ml gas-tight syringe and at each sampling, the sampled volume was replaced with deaired, deionised and distilled water.

Upon completion of the test, the final weight of the cell was determined. The solution was then drained from the reservoir before dismantling the diffusion cell to prevent rewetting of the soil specimen. The soil samples were extruded and sectioned into slices for moisture content and mass balance determination. Barium chloride at 0.1N was used for extracting potassium ions by soaking the air-dried soil in solution for at least 24 hours. This procedure was repeated three times. An Atomic Adsorption Spectrometer in the emission mode was used for the analysis of K^+ extracted by Barium Chloride solution. Flame photometry was used for the analysis of the change in K^+ in the reservoir.

Diffusion tests at three different levels of suction were conducted. The suction values were 2.7 kPa, 3.1 kPa and 3.8 kPa.

Presentation of Test Results

The physical properties of the soil specimens tested were obtained at the end of the diffusion test. The volumetric water content profile was fairly uniform across the soil specimen for cases at 2.7 kPa and 3.1 kPa. The volumetric water content and degree of saturation at 2.7 kPa and 3.1 kPa were 0.308 and 0.205, and 81% and 52%, respectively. For the case at 3.8 kPa, the volumetric water content was found to vary with depth. The values of the volumetric water content and the degree of saturation ranged from 0.081 to 0.034 and 21% to 9%, respectively. The void ratio and the dry density of the soil are fairly uniform across the soil profile with an average value of 0.63 and 1650 kg/m^3, respectively. The amount of moisture loss over the test duration varied from 0.2g to 0.36g of water. The test durations were 26, 30 and 48 days at 2.7 kPa, 3.1 kPa and 3.8 kPa, respectively. The concentration of K^+ and Cl^- in the soil at the start of the test was only 1.5 mg/l. This amount was assumed to be negligible.

The adsorption equilibrium isotherm for K^+ is presented in Fig. 3. The adsorption isotherm is nonlinear. There are two aspects of the batch test that are different from a conventional batch test. First, these data were obtained using a diffusion-type batch test where the soil was first saturated with deionised distilled water and then spiked with K^+ at various concentration. The soil sample was left to equilibrate by diffusion over a 5 day period. Second, the presentation of the partitioning relationship differs from conventional form. The mass adsorbed by the soil solids is being normalised with respect to the surface area of the soil solids rather than with the dry mass of the soil.

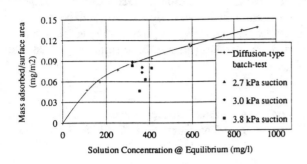

Fig. 3   Partitioning Relationship of K^+ between the solid and the solution phases.

Computation of the surface area for the coarse fraction and the quartz fracton of the fines was based on distribution of particle sizes as described by Bear (1972). For the smectite and illite minerals of the fine fraction, the surface area was obtained by assuming a specific surface of 500 m^2/g and 2.8 m^2/g, respectively. This approach gives a better representation of the interfacial phenomenon and in normalising the adsorption coefficient with respect to the surface area, the isotherm is made independent of the soil type of the same mineralogy.

Results of the mass of K^+ that was adsorbed by the soil at various suctions are also plotted on the same figure. The mass adsorbed was deduced from the total mass released by Barium Chloride after deducting the mass of K^+ in solution and the background K^+ present in the soil. This mass of K^+ was checked against that obtained by simple mass accounting of the initial mass, the mass removed during sampling, and the mass in solution assuming chemical equilibrium at end of test. The difference in computed mass was between 1 to 2%.

Variations in the concentration of K^+ in the reservoir versus time for the soil samples at three different suctions are plotted in Figure 4. In addition, to the experimental data points, a theoretical curve determined using the analytical solution "POLLUTE" (Rowe et al. 1983) was also shown in the same figure. Non-linear adsorption isotherm based on Freundlich equation as given Eq. 1 was used in the numerical simulation. The constant terms K_f and m were 7 and 0.5, respectively. For cases which fall below the equilibrium isotherm, the dry density, ρ_d, associated with the mass adsorption term, $\rho_d K_d$, in the continuity equation was modified to account for the reduction in the mass adsorbed by the soil. It was assumed that the reduction in the adsorption characteristics is due to physical inaccessibility to the adsorption sites. Modification was made by reducing the actual dry density by the ratio of S_e to S_s, where S_e is the actual mass absorbed by the soil per unit surface area, and S_s is the potential mass absorbed by the soil per unit surface area under fully saturated conditions, at the same equilibrium solution concentration. The diffusion coefficient was backcalculated by fitting various values to match the experimental data. The goodness of the fit was measured by the residual sum of squares.

$$Q = K_f C^m \quad (1)$$

where Q = mass adsorbed per unit mass of dry soil (M/M)
C = solution concentration at equilibrium (M/L^3)
K_f = partition coefficient (L^3/M)
m = constant

All simulations were based on a diffusion and adsorption coefficient of 0.75 x 10^{-5} cm^2/s and 0.08 ml/g, respectively, for the high air entry disk. These coefficients were determined from a separate diffusion test on the high air entry disk. The tests on the high air entry disk were carried out in duplicates. No diffusion tests were performed on the porous stone over the specimen but simulations showed that the concentration versus time profile were not sensitive to the parameters of the stone.

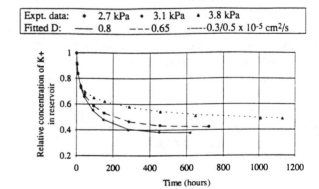

Fig. 4   Concentration-time profile of K^+ in the reservoir for various stages of suctions (D_0 = 1.8 x 10^{-5} cm^2/s at 19.6°C).

The concentration versus depth profile of K^+ in the soil as obtained from the simulation was found to be uniform at the end of the test except for the case at 3.8 kPa. Results obtained from the numerical simulation showed that the solution concentration at the top of the specimen for this case was only 80% of that in the reservoir at the end of test. Back-analysis of the Cl⁻ ions in the soil at the end of test showed that the Cl⁻ profile was uniform across the soil profile for all cases except for the case at 3.8 kPa. This further validates the attainment of chemical equilibrium in the soil. Results of Cl⁻ was however, not presented because of the additional mass loss due to chemical reaction between Cl⁻ and aluminium foil.

Interpretation of Test Results

The results of the concentration versus time profiles for the three cases were reasonably good and consistent. The data showed that the reservoir concentration tends toward equilibrium at large times. The maximum moisture loss encountered throughout the test duration was only 0.36g. Steps have been taken to eliminate moisture losses by minimizing fluctuations in temperature and relative humidity as the plastic diffusion cell responds quite significantly to changes of these factors.

The results on the mass absorbed by the soil were normalised with respect to the total surface area of all the soil solids. For the case at a matric suction of 2.7 kPa as shown in Fig. 3, the test data fall along the equilibrium isotherm line. This shows that all the soil solids are actively participating in the adsorption phenomenon even at a degree of saturation of 80%. As matric suction increases, or as the degree of saturation decreases, the amount of mass adsorbed by the soil drops below the equilibrium isotherm. This seems to suggest either that the adsorption coefficient is dependent on water content or that a fraction of the soil solids is inactive. The latter hypothesis offers a more logical explanation of the reduction in the adsorption characteristics as observed in the experimental data.

The fraction of the active mass of soil solids participating in the adsorption phenomenon is estimated by expressing the actual mass absorbed per total surface area as a ratio to that given by the equilibrium isotherm at the same concentration. The functional relationship for the adsorption phenomenon given by the ratio of active mass of soil participating in the adsorption phenomenon to the total mass of soil is plotted in Fig. 5, as a function of the degree of saturation. The results indicate that there is a gradual but small decrease in the mass of soil participating in adsorption as the degree of saturation decreases from 80% to 20%. At near residual saturation ($S < 20\%$), it decrease rapidly.

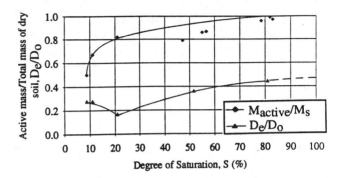

Fig. 5 Effect of water content on the magnitude of active mass of soil solids participating in the adsorption phenomenon, and the normalised diffusion coefficient.

Results of the normalised diffusion coefficient, D_e/D_o, are also plotted as functions of the degree of saturation in Fig. 5. The functional relationship for the diffusion coefficient is slightly nonlinear. The form of the functional relationship is quite similar to that given by Baraclough and Tinker (1982). One anomaly in the data is the increase in the diffusion coefficient at low water contents. The explanation for this may be attributed to the fact that at higher matric suctions the water is confined to small pores congregated at isolated spots where the pathways are less tortuous.

Concluding Remarks

The results obtained using this new technique for diffusion testing of unsaturated soil provided a reasonable basis for the interpretation of the adsorption phenomenon and the diffusion mechanism. The functional relationships for adsorption as reflected by the ratio of active mass to the total dry mass of soil is relatively linear between 80% saturation to near residual saturation. The functional relationships for the normalized diffusion coefficient on the hand is slightly nonlinear. Beyond a 20% degree of saturation, the fraction of soil mass participating in adsorption decreases rapidly but on the other hand, the normalised diffusion coefficient increases. Further studies will have to be carried to verify these results and the anomaly in the diffusion coefficient at low water content.

The key factor affecting adsorption and diffusion phenonmena in unsaturated soil is closely related to distribution of water in the soil at various suctions. The importance of the presence of water is seen in its role in providing physical access to ionic species along continuous water filled pores or water films to the vicinity of the solid surfaces for adsorption to occur.

References

Barraclough, P.B. and Tinker, P.B. (1982). The determination of ionic diffusion cofficients in field soils. I. Diffusion coefficients in sieved soils in relation to water content and bulk density. Journal of Soil Science, 32, 225-236.

Bear, J. (1972). Dynamics of Fluids in porous media. New York, America Elsevier Pub. Co., chap 2, 50-51.

Brown, D.A. (1953). Cation exchange in soils through the moisture range, saturation to the wilting percentage. Proceedings. Soil Science Society of America, 17, 92-96.

Graham-Bryce, I.J. (1963). Effect of moisture content and soil type on self diffusion of ^{86}Rb in soils. Journal of Agricultural Science, 60, 239-244.

Klute, A. and Letey, J. (1958). The dependence of ionic diffusion on the moisture content of nonadsorbing porous media. Soil Science Society America. Proceedings, 22, 213-215.

Porter, L.K., Kemper, W.D., Jackson, R.D. and Stewart, B.A. (1960). Chloride diffusion in soils as influenced by moisture content. Proceedings. Soil Science Society of America, 24, 460-463.

Romkens, M.J.M. and Bruce, R.R. (1964). Nitrate diffusivity in relation to moisture content of nonadsorbing porous media. Soil Science, 98, 332-337.

Rowe, R.K., Caers, C.J. and Barone, F. (1988). Laboratory determination of diffusion and adsorption coefficients of contaminants using undisturbed clayey soils. Canadian Geotechnical Journal. 25, 108-118.

Rowe, R.K. and Booker, J.R. (1983). Program POLLUTE - 1D Pollutant Migration Analysis Program. ASCDA, Faculty of Engineering Science, The University of Western Ontario, London, Ont.

Rowell, D.L., Martin, M.W. and Nye, P.H. (1967). The measurement and mechanism of ion diffusion in soils. III. The effect of moisture content and soil solution concentration on the self diffusion of ions in soils. Journal of Soil Science, 18, 204-222.

Warncke, D.D. and Barber, S.A. (1972). Diffusion of zinc in soil: I. The influence of soil moisture. Soil Science Society America. Proceedings, 36, 39-42.

Additional Lagooning on Stage I of the Gale Common Ash Disposal Scheme: Investigation, Design and Preparatory Works

P.L. Martin, G. Vaciago, N.R. Somers, H.T. Burbidge
Rendel Palmer & Tritton, UK

Stage I of the Gale Common Ash Disposal Scheme is now approaching completion and a landscaped capping over the completed pfa lagoons is required to comply with the conditions of planning consent. In order to make cost savings it was proposed to form most of the capping by additional lagooning instead of dry filling as originally intended. This requires construction of containment bunds over previously lagooned ash, and investigations were necessary to characterise this material with respect to consideration of drainage, stability and potential for liquefaction. Preparatory work has been carried out, including the installation of drainage measures, and the displacement of cenospheres.

Introduction

The Gale Common Ash Disposal Scheme in North Yorkshire is the largest of its type in the UK and provides for the disposal of 49 million m^3 of pulverised fuel ash (pfa) from two 2000 MW coal fired power stations some 5 and 7m distant, as well as 16 million m^3 of spoil from a nearby colliery. Some of the 1 million tonnes of pfa pumped to site annually as a slurry is processed in a vacuum filtration plant to produce conditioned pfa for embankment construction, the remainder is discharged into lagoons.

Stage I of the scheme started in 1967 and is now nearly complete. It covers an area of over 160 hectares and consists of an embankment (crest length 2.2km) containing two lagoons about 45m deep, separated by a narrow bund. The form of construction of the embankment is shown in Fig. 1. The lagoons were raised alternately in 3.3m lifts and the supernatant water discharged via outfall towers into a culvert below the embankment (Fig. 2).

The site was a flat area of marshy ground and is underlain by several metres of lacustrine and alluvial clays, sands and gravel, over bedrock. Coal has been mined from a 1.5m thick seam at a depth of about 700m. The overall scheme is described further by Haws et al[1] and Dennis et al[2] and the geology of the site and geotechnical characteristics of the foundation soils are given by Taylor et al[3].

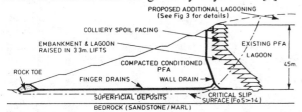

Fig. 1 : Cross section through Stage I embankment

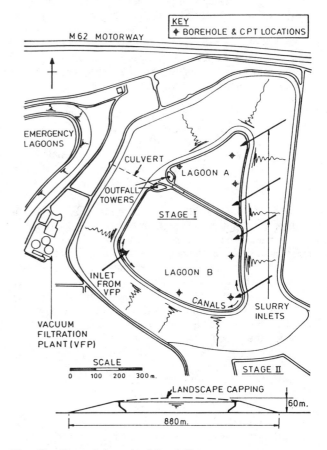

Fig. 2 : General layout of Stage I

Additional Lagooning Concept

A 15m high landscape capping is required over the completed lagoons, to comply with the conditions of

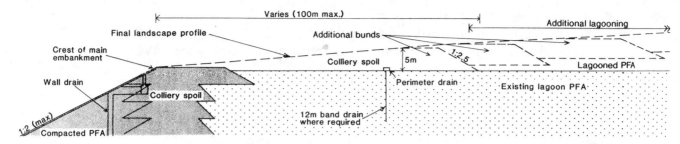

Fig. 3 : Cross section through Additional Lagooning bunds

planning consent. It was originally envisaged that this capping would be constructed by placing and compacting conditioned pfa and colliery spoil over the settled ash. However, it was subsequently determined that cost savings could be made by forming a large portion of the capping by additional lagooning of pfa within bunds built over the lagooned ash. In practice the additional lagooning as described would involve the formation of a slurry reservoir approximately 50m above the surrounding ground level. Instability of the containment bunds could result in flow failure of the ash slurry and therefore potentially serious damage to life and property. Possible failure mechanisms to be designed against included uplift of the outermost slope of the additional bunds, slope instability and liquefaction of the lagooned ash below the bunds. A cross section showing the additional lagooning scheme is given in Fig 3.

As the additional lagoons progress upwards and inwards to conform to the landscape profile, the outfall towers will become isolated from the main lagoon area. To overcome this problem a new outfall tower will be constructed on top of the division bund between the existing lagoons, with a piped discharge to one of the original towers.

Cenospheres (silt sized ash consisting of hollow glass balls with particle density less than unity, colloquially referred to as "floaters") have accumulated to form a layer over the existing lagoons up to about two metres in thickness. Their unsuitable geotechnical characteristics made it undesirable to include them in the foundations of the additional bunds. They also have a significant commercial value eg. as a lightweight filler in plastic, and the scheme is arranged to displace them towards the centre of the lagoons, allowing time for their recovery.

Investigation of Lagoons

Unlike the design of the main embankments, the additional lagooning required the characterisation of the lagooned ash, particularly in terms of its vertical (k_v) and horizontal (k_h) coefficients of permeability and its susceptibility to liquefaction.

Fieldwork comprising piezocone penetration testing (CPT) including dissipation tests, cable percussion boring down to 40m depth, conventional and thin walled piston sampling, Standard Penetration Tests (SPT), gamma - gamma

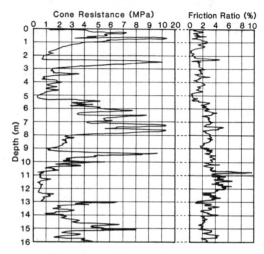

Fig. 4 : Typical CPT profile

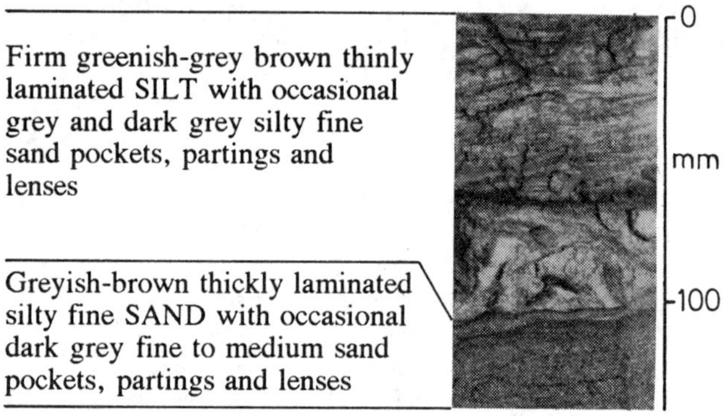

Firm greenish-grey brown thinly laminated SILT with occasional grey and dark grey silty fine sand pockets, partings and lenses

Greyish-brown thickly laminated silty fine SAND with occasional dark grey fine to medium sand pockets, partings and lenses

Fig. 5 : Typical split undisturbed pfa sample showing interlaminations and cross bedding in upper layer

geophysical borehole logging, constant and falling head permeability tests, and installation of piezometers was carried out by conventional equipment from causeways of colliery spoil. Exploratory holes were positioned to investigate the variation of the pfa between the inlet pipes and outfall towers across each lagoon (Fig. 2). Laboratory testing provided information on the index properties, density, permeability, consolidation and shear strength of the pfa. The investigation was carried out in 1991.

The presence of interlaminated to interbedded fabric of the lagooned pfa was identified by CPT profiles (Fig. 4) and confirmed by logging split undisturbed samples (Fig. 5).

TABLE 1 -- Summary of geotechnical properties from laboratory testing of lagoon pfa

TEST[1]	w (%)	ATTERBERG LIMITS		ρ (Mg/m³)	$ρ_s$ (Mg/m³)	CONSOLIDATION VALUES FROM OEDOMETER			SHEAR STRENGTH[3]						
		w_L (%)	w_P (%)			c_v[2] (m²/yr)	m_v[2] (m²/MN)	C_c	c_u (kPa)	c′ (kPa)			φ′ (°)		
										loose	medium	dense	loose	medium	dense
AVE	33	28	NP	1.57	2.28	10	0.09	0.07	72	0	0	2	27	29	33.5
RANGE	17 to 58	21 to 37	NP	1.43 to 1.80	2.05 to 2.58	3.0 to 23.8	0.033 to 0.185	0.04 to 0.105	19 to 138	-	-	0 to 6	-	27.5 to 30	33 to 34
NO. OF TESTS	29	9		5	10	3			5	1	3	3	1	3	3

Notes: (1) All tests carried out in accordance with British Standard 1377 (1990). (2) at $p_o′$ + 100kPa. (3) Initial test density : Loose ρ < 1.55Mg/m³; Medium ρ = 1.55-1.59Mg/m³; Dense ρ = 1.60-1.70Mg/m³. (4) For particle size distribution see Fig. 7. (5) For permeability see Table 3.

TABLE 2 -- Design parameters for lagoon pfa

ρ (Mg/m³)	c′ (kPa)	φ′ (°)	k_v 10^{-6} (m/s)	k_h/k_v	c_v (m²/yr)	C_c
1.68	0	22	1.5	10	>1000	0.07

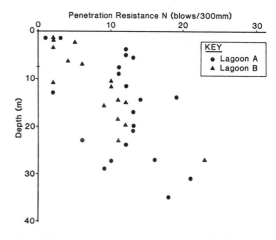

Fig. 6 : Plot of uncorrected Standard Penetration Test results with depth in lagoon pfa

Peaks in the profiles generally corresponded to coarse grained horizons as identified in borehole logging and by particle size distribution analyses. These sedimentary features are consistent with the deltaic type of depositional environment of the lagoons. Superimposed on this fabric is a stratification resulting from operating the lagoons in 3.3m lifts. Geophysical logging did not prove as useful in determining the stratification of the lagooned pfa as had been expected.

Piezometer monitoring indicated downward seepage in both lagoons. In Lagoon B, which was operational at the time of the investigation, the ratio of piezometric head to hydrostatic head was 0.7.

Geotechnical properties and design parameters for the pfa are summarised in Tables 1 and 2. The SPT's show that the pfa is loose to medium dense with a maximum observed N value of 23 (Fig. 6).

Classification test results show the pfa to be predominantly non-plastic, silt sized material with very little clay content and a varying proportion of sand sized material (Fig. 7). Samples containing higher proportions of sand-sized material gave higher bulk density (ρ) and particle density ($ρ_s$) values. The distribution of voids ratio with depth (Fig. 8) is broadly consistent with the compressibility of the ash determined from oedometer tests.

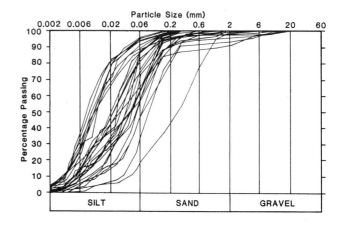

Fig. 7 : Plot of particle size distribution analyses for lagoon pfa

Various methods were used to determine k_v and k_h (Table 3). Using engineering judgement based on the results of the most reliable methods, for example CPT dissipation tests and borehole tests with casing flush at base, and the observed stratification it was concluded that anisotropic conditions exist in the lagoons, with k_h about 10 times k_v.

From in situ tests the coefficient of consolidation (c_v) was found to be large (>1000m²/yr) and long term consolidation in response to applied load is unlikely to occur. Determinations of c_v from oedometer tests is considered unrepresentative of the mass properties of lagooned pfa due to scale, sample disturbance and edge effects associated with the test. The value of compression index C_c adopted in design was 0.07.

Strength testing confirmed the value of zero effective cohesion c' assumed for lagoon and compacted pfa. The results indicate effective friction ϕ' to increase with density, varying between 27° for loose to medium dense and 34° for dense pfa. The value of $\phi' = 22°$ assumed in design relates to the need to guard against the possibility that saturated ash sedimented in a very loose state may show significant brittleness at high mobilised shear stress, resulting in sudden shedding of load and progressive failure. The adoption of $\phi' = 22°$ in limit equilibrium stability analyses, provides an additional factor of safety of 1.4 to the strength of saturated lagoon pfa.

Geotechnical Design

The geotechnical design of the additional lagooning scheme was governed by the need to ensure the stability of both the main embankments and the additional bunds at all times. In accordance with previous practice on the project, a minimum factor of safety of 1.4 was required against overall shear failure of the main embankment under static conditions assuming steady state seepage at all stages of construction, whilst a minimum of 1.0 was required under earthquake conditions using a seismic coefficient of 0.038g in quasi static analysis and assuming, conservatively, that

TABLE 3 -- Permeability test results for lagoon pfa

Method Type	Details	Remarks	Range 10^{-6}(m/s)	Average 10^{-6}(m/s)
Direct Borehole	Casing flush with base	Largely k_v	10 to 0.2 (6 tests)	2
Direct Lab	Undisturbed sample	k_v	17 to 0.04 (6 tests)	3
Indirect Lab	c_v & m_v from oedometer	k_v	(unreliable)	(unreliable)
Direct Borehole	Casing retracted from base	Variable k_h:k_v ratio	30 to 0.03 (6 tests)	7
Direct CPT	Dissipation test	Largely k_h	40 to 1 (20 tests)	9
Direct Piezo	Through filter zone	k_h	1 to 0.04 (8 tests)	0.2
Direct Lab	Remoulded sample	Soil fabric destroyed	7 to 0.1 (3 tests)	3
Indirect Lab	PSD results & Hazen's formula	Soil fabric destroyed	16 to 0.09 (30 tests)	0.9

the lagooned ash had no strength. Construction of containment bunds over the existing lagoons, however, required a more detailed analysis of the liquefaction potential of lagooned pfa, based on a return period of 1000 years with an associated peak ground acceleration of 0.075g [4] and an upper bound earthquake magnitude of 5.5 [5]. Measures to ensure stability of the bunds both for earthquake and static conditions were designed accordingly.

Analyses were carried out using a commercial finite difference program to evaluate the distribution of pore pressures due to seepage of water from the additional lagoons, for use in the stability calculations and drainage design. Different drain layouts were investigated assuming that flow was confined to the pfa lagoons, with the foundation clay and colliery spoil in the embankments providing nominally impervious boundary conditions. These assumptions yield the maximum pore pressures in the pfa lagoons for the given condition being investigated, making evaluations of stability, uplift and liquefaction potential based on these pressures conservative. The design of bund geometry and drainage measures ensures a factor of safety of 1.2 against uplift. Drainage measures, generally consisting of a simple perimeter toe drain, included 12m deep vertical band drains in some areas to counter anisotropy and to reduce uplift pressures where it was not possible to provide a sufficiently thick cover of colliery spoil (Fig. 9).

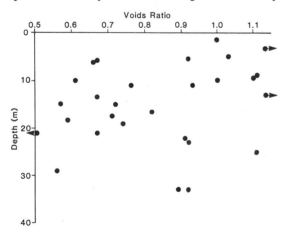

Fig. 8 : Plot of voids ratio with depth

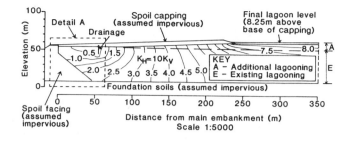

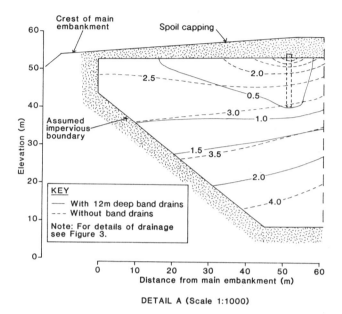

Fig. 9 : Contours of head in metres replotted from computer output (datum is base of capping)

The overall stability of the main embankment, which is controlled by the thin layer of lacustrine clay just below original ground level, resulting in strongly non-circular critical slip surfaces (Fig. 1), was found not to be a problem because conditions pertaining to the additional lagooning were only slightly more onerous than those due to the "dry capping" for which the scheme had been designed.

Preparatory Works

The construction of the bunds has been started by the "canal method". The essential principle is to build a base within the lagoon on which the new bund is built. Continuous control of the incoming slurry is maintained by canalising it progressively along the perimeter of the lagoon, working in both directions from the slurry discharge points (Fig. 10a). This leaves a firm exposed bed of relatively coarse settled ash on which colliery spoil can be placed (Fig. 10b). A second canal is then formed along the new inner face

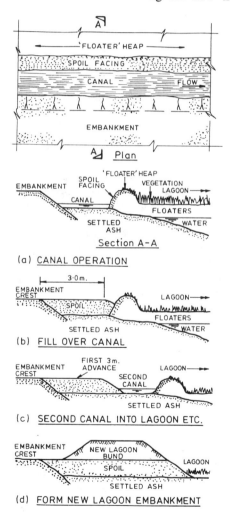

Fig. 10 : Canal method of construction

The design of bund geometry and drainage measures also derived from the requirement to prevent liquefaction of the underlying ash, as discussed above. Susceptibility to liquefaction at any depth is a function of the average cyclic shear stress induced by the applied ground motion and the effective overburden at that depth. Quantitative analyses based on SPTs [6] and the results of dynamic tests on similar ash [7] were used to determine minimum bund geometry and drainage measures. It was concluded that a 5m thickness of colliery spoil gave a high degree of confidence that liquefaction would not take place below the proposed bunds (Fig 3).

The factor of safety against instability of the inner face of the additional bunds before impounding was found to be low, but acceptable in light of previous experience of bund construction on lagooned ash, the assumption of $\phi' = 22°$, and because such a potential failure has limited consequences. A generous crest width was provided in the design of these bunds to allow for possible localised failure during earthquake.

until the desired location for forming the new bund is reached (Fig. 10c). Each canal advances the base approximately 3 metres into the lagoon. The required drainage measures can then be installed by trenching for pipework consisting of slotted 100mm pipe in a 1m x 1m sand filled trench or by push-in mandrel for band drains, and a new bund constructed (Fig. 10d). The canalisation and bund construction sequence is then repeated. Exercising due caution the lagoons are to be raised in 1.65m lifts ie. half the height adopted previously.

Trials of the method showed that it was difficult to remove floaters by the scour effect of the canal and use of an excavator proved necessary. This difficulty is exacerbated by a greater than expected depth of floaters in some areas, probably caused by uneven filling of the lagoons during their long history and accumulation at the lagoon edge as the floaters are dragged in the predominant wind direction. However the commercial value of the floaters is now being exploited and their removal from the lagoon should assist the canal method, particularly for the construction of higher lagoons. In the final lagoon it is proposed to bury any floaters remaining after the extraction process.

Summary and Conclusions

On the basis of the results of the investigation, design calculations confirmed that additional lagooning was feasible, practical and provided a cost-effective alternative to dry capping. The investigation revealed the lagoon pfa to be highly interstratified and thinly bedded, with k_h to k_v of the order of 10. It was determined that a cover of several metres of colliery spoil over the lagoons would provide adequate security against liquefaction of the lagoon pfa. A simple perimeter collector drain was found to be sufficient to deal with seepage pressures and prevent excessive uplift of the outer slope of the additional bunds, except where the geometry of the final landscaping profile provided insufficient cover. Here the installation of vertical band drains was found to be necessary. Preliminary works have confirmed the feasibility of constructing a firm base for the additional bunds by the canalisation method.

Acknowledgments

The authors are grateful to National Power Plc for permission to publish this paper.

References

1. Haws E.T., Martin P.L. and Orange-Bromehead R.A.N. Gale Common ash disposal scheme - concept, design and construction. Proc. 6th British Dam Society Conf. on The Embankment Dam, 1990. Thomas Telford 1991.

2. Dennis J.A., Hillier D.J. and Moggridge H.T. Gale Common ash disposal scheme - planning, environment, operation and restoration. Proc. 6th British Dam Society Conf. on the Embankment Dam, 1990. Thomas Telford 1991.

3. Taylor R.K., Barton R., Mitchell J.E. and Cobb A.E. The engineering geology of Devensian Deposits underlying PFA lagoons at Gale Common, Yorkshire. Quarterly Journal of Engineering Geology, 1976 Vol. 9, 195-216.

4. Irving J. Earthquake hazard in Britain. Earthquake Engineering in Britain. Thomas Telford, 1985.

5. Ambraseys N.N. and Jackson J.A. Long term seismicity of Britain. Earthquake Engineering in Britain. Thomas Telford, 1985.

6. Seed H.B. and Idriss I.M. Ground motion and soil liquefaction during earthquakes. EERI, Berkley, 1982.

7. Haws E.T., Pedley W.A. and Nisbet R.M. Brotherton Ings Ash Lagoons. Proc. 12th Cong. International Commission on Large Dams, Mexico, 1976.

Contamination Assessment and Remediation Approaches in North America

Godwin N. Nnadi
Bromwell & Carrier, Inc., Lakeland, Florida, USA

Abstract:

The North American Free Trade Agreement (NAFTA) between Canada, the United States and Mexico, if ratified aims to improve trade between these countries. Controversy over its potential environmental effects has dogged it because of the recognition of the dependence of the region's economy on the environment. The approach to investigation of environmental contamination in these countries is similar, but the assessment and enforcement do vary significantly. Detail discussion of the relevant legislation in each country is presented. Comparisons are made of their respective approaches to soil and groundwater contamination assessment and remediation processes. The paper also expresses views on how under NAFTA, these approaches may progress and the role of engineers in the environmental industry.

Introduction:

On February 5, 1991, the Prime Minister of Canada and the Presidents of the United States and Mexico announced their intent to pursue a comprehensive and trade liberalizing North American Free Trade Agreement (NAFTA). NAFTA is aimed at further integrating the economies of Canada, United States and Mexico. This agreement will link the divergent economies of the two developed northern countries and Mexico, a developing country. The negotiations present major risks and opportunities for the international environment, the outcome of which may greatly influence regional environmental regulations and enforcements as well as regional trade agreement.

Mexico is one of the world's fastest growing markets, with infrastructure and industrial modernization facilitating an increasing demand for imported goods, services, capital and technology. Mexico is by far Canada's and U.S's largest trading partner in Latin America. By some measures of world development, Mexico is well placed and ranks high in world economies. However, by another yardstick, Mexico is in crisis. Its environmental problems are serious and widespread. Mexico's increasing infrastructure and industrial modernization have taken a profound toll on the environment. Much of the rapid industrialization Mexico experienced in the last decade is as a result of maquiladora industries located along the border with United States. The maquiladora program is a trade agreement between United States and Mexico which allows duty free importation of component parts to factories based in Mexico. These factories in turn export manufactured goods to markets in U.S. and Canada. The maquiladora program and the Mexican economy contributed to a tremendous increase in population along the border because of the employment opportunities. A feature story this year in the Toronto Star described Matamoros, an industrial border town as a futuristic nightmare [1]. Lewis, S.J. et al [2] documented that severe toxic discharges occurred at more than one-third of the 23 industrial sites sampled along the border.

Opponents of NAFTA, are concerned that economic development and the expansion of markets resulting from NAFTA could worsen Mexico's critical deficits in water supplies, sewage treatment and solid waste disposal and add to environmental problems in the area. However, proponents of the treaty counter that accelerating economic growth will provide Mexico with resources to strengthen its notoriously lax enforcement of environmental laws.

The approaches to assessment and remediation environmental contamination in these countries can and do vary significantly. This paper provides an overview of the concepts employed in contamination assessment in these countries and the relevant legislation and regulations in each country. Comparisons are made of the approach to assessment and remediation that would be adopted for contaminated land. This paper also expresses views on how under NAFTA, these approaches may progress and role of engineers in the environmental industry.

Government Polices

The theoretical framework and operational practice of the Environmental Impact Assessment (EIA) have existed for decades. United States EIA came into effect in 1969 in the US National Environmental Policy Act (NEPA). Canada's Environmental Assessment and Review Process (ERAP) came into effect in 1974. Mexico's initial fray into environmental policy making was in 1971 with the enactment of the Federal Law for the Prevention and Control of Environmental Contamination [3].

United States

Early post-World War II environmental legislations left standard setting and enforcement of contamination remediation to the States. However, most states did not have the expertise to set and enforce the standards [4]. In response to the heavy contamination of many rivers; Lake Erie pollution; and raw sewage spewing into San Francisco Bay, the U.S. Environmental Protection Agency was established in 1970 and given the powers of standard-setting and permitting powers for air and water emission. The states, however, have since taken over many of the standard-setting and enforcement tasks.

In 1976, Congress enacted the Resource Conservation and Recovery Act (RCRA) to control disposal of hazardous waste. Included in RCRA is the regulation of solid waste disposal facilities, underground storage tanks and medical waste tracking. With RCRA, EPA adopted regulations that identified hazardous wastes and imposed "cradle-to-grave" controls.

In 1980, the outcry that followed the Love Canal landfill pollution incident led to the introduction of the Comprehensive Environmental Response, Compensation and Liability Act (CERCLA). CERCLA addresses the immediate removal of hazardous substances posing an imminent danger to public health or welfare and provides for the long-term remediation of contaminated sites in the absence of imminent danger. CERCLA also created the Superfund program (a clean-up fund for sites) as amended by the Superfund Amendments and Reauthorization Act of 1986 (SARA). It provides EPA with the necessary resources to identify, evaluate and remediate hazardous waste sites in the U.S. and empowers EPA to recover the costs of the clean-up. Although, Superfund is federally based, individual states have the right, and more often than not choose to, introduce their own legislation to supplement the above.

Due to the 1985 disaster in Bhopal, India, and the release of toxic chemical in Institute, West Virginia shortly thereafter, Congress added the Emergency Planning and Community Right-to-Know Act (EPCRA) to the Superfund Amendment and Reauthorization Act of 1982 (SARA). The EPCRA sets requirements for federal, state and local governments and industry regarding emergency planning and community right-to-know reporting on hazardous and toxic chemicals.

Canada

The Canadian approach to contamination is driven by legislation from three tiers of government: federal, provincial and municipal. In 1970, the Canadian Department of the Environment was formed with a duty to preserve and enhance the quality of the natural environment including water, air and soil. The centrepiece of Canada's environmental legislation is the 1988 Canadian Environmental Protection Act (CEPA). It provides for the protection of the environment and the life and health of Canadians from the effects of toxic substances through a life-cycle approach to management - from development and manufacture through transportation, distribution, and use to final disposal. CEPA includes mechanisms to promote federal-provincial cooperation, strong public accountability and modes of fines and penalties for environmental offenses. It also provides broad powers within which provincial government should act, whilst also providing detailed jurisdiction on certain areas of activity.

Part I of the Act gives the Minister of the Environment responsibility for establishing (i) environmental quality objectives, (ii) limits on the amounts of substances that may be released to the environment, and (iii) environmental codes of practice and formulation of monitoring systems and pollution control. Part II of the Act identifies control procedures for toxic substances and hazardous wastes.

In 1991, Canada released a 'Green Plan' for a health environment as federal government's response to Canadians' concerns and ideas for cleaning up and protecting their environment. The Green Plan sets targets and schedules for government actions on the environment. It also commits $3 billion Canadian in new fund for five years over government annual spending of $1.3 billion on environment [5].

Mexico

Mexico passed its first environmental law in 1971, establishing a Subsecretariat of Environmental Improvement under the Secretariat of Health. But this agency did not accomplish much due to lack of funding. In 1982, Mexico passed the Federal Law on Environmental Protection and created Secretariat of Urban Development and Ecology (SEDUE) with responsibilities similar to U.S. E-PA. SEDUE was organized in three subsecretariats (Housing, Urban development and Environment) with a budget less than one percent of U.S.EPA budget for environmental protection.

In 1988, the comprehensive General Law of Ecological Equilibrium and Environmental Protection was enacted to cover both environmental protection and conservation of natural resources. SEDUE was given the responsibility of implementing the Law. SEDUE was also given considerable powers to shut down plants, but as an agency it was plagued by lack of funds [6]. The general law, however, stated general objectives, covering the full spectrum of environmental issues rather than specific criteria for each of the different media as in U.S. and Canada environmental policies.

In 1992, the Secretariat of Social Development (SEDESOL) officially replaced SEDUE. Its broad mandate includes environmental policy formulation and enforcement, urban planning and national solidarity programs. By the end of 1992, Mexico's federal environmental agency had promulgated four regulations and about 81 standards. These regulations and norms are applicable to environmental impact assessment; hazardous wastes; prevention and control of air pollution; vehicle emission in Mexico City and contamination of the sea. However, the areas of land disposal of hazardous waste; leaking underground storage tanks and cleanup of abandoned hazardous waste sites are yet to be regulated. The general law and the regulations establish SEDESOL's concurrent jurisdiction with the States and Municipalities in specific environmental protection matters of local interest. As in Canada and the U.S; Mexico State laws and municipal ordinances enacted pursuant to the general law must be at least as stringent as the applicable federal regulations or standards.

Mexican controls on waste management tend to be more stringent for new sources than the existing sources. For new sources government authorization including an environmental impact assessment will be required. New facilities must use the "best available technology" while existing facilities are called on to strengthen pollution controls and to recycle [7].

Approach To Remediation

The approach to contamination remediation depends on the nature, distribution and origin of the contamination. Industrial land can be contaminated by:

1. Leakages or spillage from pipes or tanks;
2. Storage of raw materials or products;
3. On-site disposal of wastes; and
4. Deposition of airborne contamination such as fallout from chimney discharges.

The origin will affect its location and distribution on the site. This may take the form of localized concentrated 'hot spot' or of lower levels spread more over a large area. Contamination at a fairly substantial depth do occur as a result of waste disposal, leakage from underground storage tanks, or even by percolation due to rainfall.

Once an area is contaminated, some form of remedial works will be required. These will of course depend upon many factors such as: risk to human or ecological receptors; end use of the site; and site value, but the approach to remediation will be within these categories:

A. Removal and/or replacement of contaminated materials from the site for off-site treatment and disposal;
B. Containment or isolation by superimposing cover and providing, if necessary, in ground barriers to contain migration;
C. Retention of contaminated material on site after treatment to reduce the level of contamination; e.g. in situ process such as vitrification or enhanced biological treatment or by chemical process such as stabilization/ solidification to reduce the toxic or hazardous nature of the chemical in the soil;
D. Treatment to remove or neutralize contamination either on-site or in-situ involving physicochemical or thermal processes such as soil washing, soil leaching, solvent extraction, in-situ vitrification, in-situ vacuum extraction, etc.

The choice of which method or combination of methods to be adopted depends on site-specific requirements and the nature of contamination.

US Remediation Approach

For the past two decades, the U.S. has begun a massive effect to clean up its land and water resources. The Superfund program and the 1976 RCRA are the two pillars of the clean up program. The Superfund program involves the remediation of damages resulting from past hazardous waste disposal sites not currently actively managed. Such sites include landfills, mining areas, manufacturing facilities and illegal hazardous waste dumps. The RCRA program is directed toward current solid and hazardous waste management practices. It is estimated that by the year 2000, U.S. land pollution control will cost over $46 billion [8].

The US approach to investigation and remediation is designed specifically to identify contaminated media that presents a risk to human or ecological receptors. The need for remediation is based on a formalized risk assessment criteria. The rationale is to clean up a site so that it can be used for a range of purposes in the future and will have no impact on groundwater. Although, Superfund is risk based, most states programs and RCRA have at least proposed standards. The EPA has been proactive in identifying past polluters, even if what was carried out in the past was lawful and the best practice at that time. The US has adopted the philosophy that treatment of the problem is the best way forward. Therefore remediation by removal and/or replacement is not widely practised. Among the numerous technologies available to remediation include incineration, solidification, capping with soil, pump and treatment, etc.

Canada Remediation Approach

The Canadian Council of Ministers of the Environment (CCME) has produced an interim remediation criteria for contaminated site which is under constant review. The criteria is based on three end users: agricultural, residential/ parkland, and commercial/industrial.

The criteria is to encourage the lowest level of contamination practicable, with due consideration to both the desired end use and technological limitations. The criteria is not considered absolute but are interpreted on a site specific basis and if the contamination concentrations excess the guidelines, it is an indication that further investigation and/or remedial action should be considered.

The interpretation of the federal guidelines at provincial level vary clearly. For example, the Ontario Environmental Act has no limits for contamination but rather allows the assessment of contamination on a case-by-case basis. Clean-up is considered where a high risk of off-site contaminant migration is considered likely or harm to human health is expected. The responsibility of site clean-up is that of the owner and therefore environmental liability is an important consideration for potential buyers.

Although, Canada has the 'polluter pays' principle, CCME initiated the National Contaminated Sites Remediation Program, a $250 million program to clean up high risk 'orphan' contaminated sites which may pose imminent threat to human health and the environment. The former Expo site in Vancouver is among estimated 50 abandoned contaminated sites in Canada to be cleaned up under this program.

Mexico Remediation Approach

In spite of the advances made in recent years in environmental laws and regulation, the existence of these laws and regulations does not necessarily guarantee their compliance or their effectiveness. Mexico's funding and personnel constraints preclude any comprehensive, effective program to monitor and enforce industries compliance with legal requirement.

Environmental impact and risk assessments are not required for existing industrial facilities, but plants must apply for air, water and hazardous permits [9]. Very little study has been carried out on the quality of groundwater or the extent of polluted land. This lack of information about polluters, pollutants and risk to human and ecological receptors is a major hinderance to improving environmental protection.

Mexico, so far is practising a sectored and reactive approach to managing its environment, attacking problems one by one only after they reach critical stage. For example, the environmental policy in Mexico does not require the installation of double liners underneath landfills, yet landfills receive hazardous wastes. In 1983 and 1986, two landfills in Mexico City Metropolitan Area (MCMA) were shut down due to subsurface leachate containing a complex mixture of contaminant discovered by the Departamento del Distrito Federal (DDF) [10]. In 1992, an explosion erupted from gasoline released into sewer in Guadalajara. After the explosion, investigations of Petroleos Mexicanos (PEMEX) gasoline stations identified gasoline leaks in over 80 per cent of the 253 PEMEX stations in MCMA. The difference between Mexico and US or Canadian environmental laws include the followings:

1. Mexican environmental laws have not provided mechanisms to promote strong public accountability and public participation in environmental issues.
2. SEDESOL has not yet promulgated treatment-oriented land disposal restrictions equivalent to those under the American RCRA or Canadian CCME.
3. SEDESOL has not addressed the issue of leaking underground storage tanks.
4. Mexican environmental law lacks the 'cradle-to-grave' regulation of hazardous wastes required under both US and Canadian environmental laws.
5. Mexico's standard for environmental impact statements fail to require examination of alternative actions and to mandate public participation.
6. Mexico lacks an equivalent to the US Superfund program or Canada's Contaminated Site program.

Implication of NAFTA

NAFTA is the first trade agreement to recognize explicitly the inextricable relationship between the environment, trade and industry [11]. It recognizes the principles of sustainable development and environmental protection.

The three countries have proposed a North American initiative for the environment. Although, only an idea, such initiative could result in an agency to enforce environmental laws and finance environmental projects. The agency could establish continent-wide guidelines for managing natural resources in accordance with the social and environmental needs. It could establish uniform standards on hazardous disposal and clean up.

The environmental agency could help Mexico expand its environmental legislation and help educate Mexican public on environmental issues and train environmental workers. It could establish incentives for pollution prevention and set clear penalties.

Under NAFTA, service providers including engineers are not required to establish an office or take-up residence as a condition of providing service. However, the countries have undertaken to ensure that future professional licensing and certificate restrictions are based on objective criteria such as professional competence. Citizenship or permanent residency requirement are to be removed two years after NAFTA's implementation [13]. NAFTA, therefore encourages mutual recognition of licenses and certificates. It is expected that under NAFTA, engineers and scientists in the environmental industry will have more opportunities to develop the scientific and engineering infrastructure which Mexico needs to combat its numerous environmental problems.

Conclusion

Although the North American countries have comprehensive environmental laws, their approaches to contamination remediation and the techniques applied to abandoned hazardous sites are substantially different. U.S. and Canada are proactive while Mexico is reactive due to lack of fund and the infrastructure needed to control environmental contamination.

It is anticipated that there will be a degree of synergy in the environmental protection between the three North American countries after the implementation of NAFTA. If NAFTA is ratified by the three countries, engineers and scientists in U.S. and Canada will have opportunities to help Mexico control and combat its environmental problems.

References

1. Diebel, L. (1993) Mexico's futuristic Nightmare, Toronto Star, March 13, pp. D1-D2.
2. Lewis, S.J. et al. (1991) Border Trouble: Rivers in Peril, National Toxic Campaign Fund, Boston, Mass.
3. Mumme, S.P. (1991) Clearing the Air: Environmental Reform In Mexico, Environment, Vol.33, pp. 9-10.
4. Kneese, A.V. and Schultze, C.L. (1975) Pollution, Prices and Public Policy, Brooking Institution, Washington D.C.
5. Government of Canada (1990) Canada's Green Plan, Ministry of Supply and Services, Ottawa.
6. Jones, R.S. (1991) Learning from Experience, Business Mexico, October, p. 26.
7. U.S. Environmental Protection Agency (1991) Mexican Environmental Laws, Regulations and Standards: Preliminary Report of EPA Findings, Office of Enforcement, EPA, 2p.
8. Carlin, A. et al. (1992) Environmental Investment: The Cost of Cleaning Up. Environment, Vol.34, No.2, pp. 12-44.
9. Congress of the United States, Office of Technology Assessment. (1992) U.S.- Mexico Trade: Pulling together or pulling apart?, ITE-545, U.S. Government Printing Office, Washington, D.C.
10. Mazari, M and Mackay, D.M (1993) Potential for groundwater contamination in Mexico City, Environ. Sci. Tech. Vol.27. No.5. pp.794-802.
11. Clark, L. (1992) NAFTA, House of Commons Debate, Dec.1, Ottawa.
12. Sanchez, R. (1993) NAFTA and the Environment, Tech. Review, Apr. p.30.
13. Thorburn, C.C. and Murphy, K.M. (1993) NAFTA: Implications for Engineers, Engineering Dimension, Vol.14, No.2, pp. 30-33.

Containment System of an Industrial Wastewater Lagoon, including Surcharge Loading of Sediments for Cap Support

Paul M. Przygocki, P.E., Site Manager, L. Alan Johnston, P.E., Project Engineer, Walter Kosinski, P.E., Principal in Charge
GZA GeoEnvironmental, Livonia, MI, USA

This paper presents a case study of the design and implementation of a containment system for sediments within a hazardous wastewater lagoon. The containment system consists of a soil/bentonite slurry wall keyed into an underlying glacial till strata, a groundwater extraction system to maintain an inward hydraulic gradient, and a geocomposite cap to reduce infiltration of precipitation. Significant cost savings were realized by controlling settlement of the cap through preconsolidation of the extremely soft sediments with a surcharge load, rather than by in-situ solidification. This paper discusses the field and laboratory testing which was completed on the lagoon sediments, to model and monitor their behavior, as well as the difficulties experienced in placing and constructing the fill.

Background

The hazardous wastewater surface impoundment is part of a large industrial plant located in central Indiana, in the United States of America (USA). The impoundment had been in service since 1942 and occupies approximately 8 acres, consisting of two adjacent basins connected by a concrete spillway. The two basins functioned as a single integrated water retention, treatment, and solid particle removal unit. One basin was a concrete lined impoundment which was used for wastewater treatment. This basin measured about 350 feet by 70 feet. The second basin consisted of an unlined lagoon which was used as a wastewater retention and particulate settlement. This lagoon, measuring about 400 feet by 650 feet, had formerly been used as a sand and gravel borrow source, and therefore the sediments extended fairly deep. Figure 1 shows the general lagoon sediment thickness contours at the time it was removed from service.

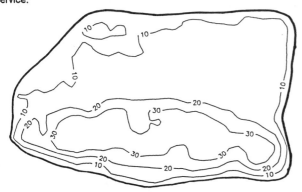

Figure 1 Lagoon sediment initial thickness contours (feet).

The surface impoundment received runoff from the plant's storm sewers, boiler blow-down water, ash quenching water from the power house, water softener rinse water, and non-contact cooling water. Prior to 1988 the impoundment also received effluent water from the plant's on-site wastewater treatment system.

The sludge and sediment that resulted from the wastewater treatment process was listed as RCRA F006 and F009 wastes. The USA Code of Federal Regulations, 40 CFR 261.31, defines these wastes as: "Wastewater treatment sludge from electroplating operations..." and "Spent stripping and cleaning bath solutions from electroplating operations, where cyanides are used in the process...", respectively. Subsequently the sediments in the lagoon became a listed hazardous waste under the same waste codes, under 40 CFR 261.3 (b). Due to changes in regulatory status, the U.S. EPA and the Indiana Department of Environmental Management ordered the surface impoundment to be removed from service and closed.

In April of 1992 the Indiana Department of Environmental Management approved the surface impoundment Closure Plan, which called for: the installation of a soil/bentonite slurry cutoff wall around the entire impoundment and keyed into the underlying fine-grained glacial till layer; in-situ solidification of the lagoon sediments by mixing with cement/fly-ash grout; installation of a groundwater extraction system to create and maintain an inward hydraulic gradient; the construction of a RCRA composite cap over the entire impoundment; and appropriate monitoring controls.

Geologic Setting

The surface impoundment unit lies in a glacial outwash deposit of sand and gravel extending approximately 115 feet deep. The uppermost bedrock is the New Albany Shale. In the vicinity of the impoundment the glacial outwash is divided into an upper and lower unit by a relatively continuous fine grain deposit of glacial till at a depth of approximately 55 feet. This till contains a significant portion of silt and clay. Grain size testing on the till indicated that silt quantities ranged from 19 to 43%, clay from 20 to 25%, sands from 36 to 47%, and gravels from 4 to 7%, by weight. Plastic limits of the fine grained materials ranged from 8 to 10% while liquid limits ranged from 14 to 22%. This material is classified as CL and CL-ML under the Unified Soil Classification System. The vertical hydraulic conductivity of the till deposit ranged from 6×10^{-7} to 2×10^{-8} cm/sec.

The glacial outwash deposits are part of a continuous, unconfined aquifer which is tapped by production wells located at the plant and at some surrounding industries. The general groundwater flow direction in the aquifer is southerly, however pumping by the plant and surrounding industries has created a local groundwater depression. Published literature estimates of the hydraulic conductivity of the outwash sands and gravels to be about 40 to 415 feet/day (i.e. 0.014 to 0.15 cm/sec)[3].

Lagoon Sediment Characteristics

During the development of the Closure Plan, analytical testing was completed on the lagoon sediments to evaluate the chemical constituents and treatability. This testing indicated that while hazardous constituents from the wastewater sludge were present in the lagoon sediments, they did not leach from the sediments at an appreciable rate, nor were they detected in the groundwater monitoring wells adjacent to the lagoon. This allowed for the disposal of water removed from the sediments at the plant's existing wastewater treatment plant.

Due to the extremely soft nature of the saturated lagoon sediments, the collection of deep undisturbed samples was difficult. The crust that had formed on the higher areas of sediment after the surface water had been pumped off, was only approximately 6 inches thick and was not sufficiently stiff to safely support a mechanical drill rig. Therefore, the sediment sampling and analysis program was completed in two phases. Phase I involved the collection of shallow (up to 15 feet) samples by manual means, working off the top of the crust and around the edges of the lagoon. Phase II included sampling and testing throughout the entire depth of sediments using a drill rig, working from on top of the initial sand drainage layer placed over the sediments.

Classification testing on the lagoon sediments included moisture content, dry density, liquid and plastic limits, particle size analysis, void ratio, and specific gravity. The moisture content testing indicated values as high as 200% for sediments less than 7 feet deep, and decreasing levels with depth to approximately 80% at a depth of 30 feet. Figure 2 contains the results of the dry density testing relative to the depth of the sample. Figure 2 illustrates that the shallowest sediments (directly below the desiccated crust) were at a dry density less than 30 pounds per cubic foot (pcf), and the density increased gradually with depth, to a maximum of approximately 50 pcf at a depth of 30 feet.

The average liquid and plastic limits of sediment samples that displayed plastic behavior were 57% and 42% respectively. All but two samples tested for particle size distribution had 99% or more passing the #200 sieve. Two samples had 60% and 89% passing the #200 sieve. The sediment void ratio ranged from 1 to 4, with an average of approximately 2.5. A general decrease in void ratio with depth was apparent although there was a large distribution in the results. The specific gravity of sediments varied widely from 1.3 to 2.6. This is indicative of the wide range of sources of the sediments, including lime from the wastewater treatment to fly ash, and cinders from boiler blow down water.

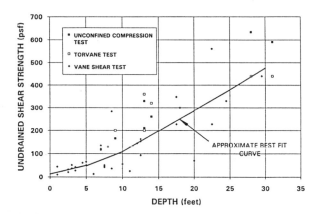

Figure 3   Sediment undrained shear strength versus depth.

Sediment shear strength testing included torvane and unconfined compression testing on undisturbed samples, and in-situ vane shear testing. The unconfined compression testing was completed in Phase II, after the placement of the initial sand lift. Figure 3 presents sediment shear strength data versus depth. An increase in shear strength with depth is apparent, although again the data is widely distributed.

One-dimensional consolidation tests were completed on four sediment samples. Two tests were run on samples during Phase I (HB-7 and HB-

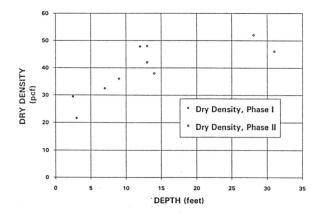

Figure 2   Sediment dry density variation with depth.

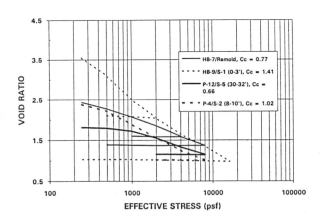

Figure 4   Sediment e versus log P, Consolidation curves.

9), and two tests were run on samples collected in Phase II (P-12 and P-4). The void ratio (e) versus log of the effective stress (log p) curves from these tests are shown on Figure 4. The compression index values for the sediments tested, ranged from 0.66 to 1.41. The deeper samples from a depth of 8 to 10 feet and 30 to 32 feet, displayed a preconsolidation pressure of approximately 300 and 900 pounds per square foot, respectively, indicating that the sediments were generally normally consolidated under the stress of the saturated sediments themselves.

The coefficient of consolidation (Cv) values for each test and stress level are shown on Figure 5. The curves illustrate a general decrease in Cv with increasing stress level. The test run on the sample from a depth of 30 to 32 feet had slightly higher Cv values than the other tests. The Coefficient of Secondary Consolidation values from the consolidation testing, ranged from 0.001 to 0.04, with the lower value also coming from the deepest sample.

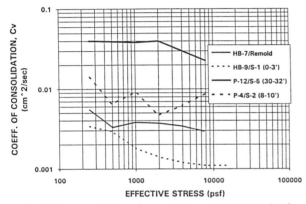

Figure 5 Sediment coefficient of consolidation versus stress level.

Solidification Feasibility and Cost

The Closure Plan originally called for in-situ solidification of the sediments by mixing with cement fly-ash grout. Mixing was to be accomplished using a modified caisson drilling rig, where grout is pumped through a hollow kelly bar and exits via ejection ports on the trailing edges of mixing blades.

In-situ solidification was originally chosen as the most practical means of stabilizing the sediments, to allow for the construction and support of the cap, due to the high sediment moisture content and low densities observed during the development of the Closure Plan.

The direct costs of the in-situ solidification were estimated at $56/cyd, based on an estimated sediment volume of 80,000 cyds. Therefore, the in-situ solidification of the sediments was estimated to cost $4,5000,000.

Consolidation Feasibility and Cost

During development of the Closure Plan, chemical testing was performed on the sediments and leachate, but little geotechnical testing was performed on the sediments. Therefore, the preliminary testing program (Phase I) was conducted to evaluate the physical properties of the sediments to assess the feasibility of consolidation. This program indicated that although the sediments exhibited very low shear strengths (i.e., less than 40 psf near the surface) at their existing moisture content, the material would gain strengths to about 100 psf or more, once the moisture content reached approximately 100%. Additionally, the one dimensional consolidation tests indicated that upon loading, the sediments would consolidate at a reasonable rate and appeared to have consolidation characteristics similar to naturally occurring fine-grained sediments (i.e. silt and clay).

Based on the promising results of the preliminary sediment testing, the possibility of using consolidation appeared feasible for support of the cap, provided that the economics could justify abandoning in-situ solidification. Subsequent cost estimations indicated the total price for surcharge loading, including mobilization charges, would be on the order of $29/cyd of sediment or about $2,300,000. This represented a potential cost savings of approximately 55% over in-situ solidification. Therefore, consolidation of the sediments was an attractive alternative.

The primary remaining concern was the constructability of the initial fill layer over the soft sediments. To limit the amount of sediment displacement and movement during fill placement, a biaxial geogrid was specified to be placed directly over the sediment. Also, dewatering of the sediments was initiated shortly after completion of the slurry wall, in an attempt to initiate the consolidation of the sediments and increase the in-situ sediment shear strength. The dewatering was conducted using a large diameter steel sump, lowered into the sediments. The dewatering caused a noticeable amount of consolidation to occur, and a "rapid drawdown" type failure in the higher sediments, which had the beneficial effect of flattening the sediment surface. These observations provided further evidence that consolidation of the sediments was a viable alternative.

Consolidation Design Considerations

The evaluation of the feasibility and cost for the construction of the preload necessary for the consolidation of the lagoon sediments required analysis of the consolidation and stability behaviors of the sediment/fill system. The following sections present a discussion of the analyses performed.

Predictions of Settlement

An initial goal of the preliminary sampling and testing program was to develop data necessary for the estimation of the amount of sediment settlement which could be expected. In order to model the sediment behavior, one-dimensional consolidation tests were performed on undisturbed and remolded samples. A remolded sample was evaluated because it was expected that the near surface sediments would become highly disturbed during the initial fill layer placement.

The consolidation tests and the sediment thicknesses provided the data necessary for an estimation of fill settlement. Expected settlements were evaluated for a series of sediment thickness/groups (i.e. 0-5 feet, 5-15 feet, 15-25 feet, and 25-35 feet). These were represented by an average sediment thickness in each group (i.e. 5 feet, 10 feet, 20 feet, and 30 feet, respectively). An average top of final cap elevation was assumed for calculation of the final overburden stress, and a conservative Compression Index (i.e. 1.41) was used to evaluate settlement. The expected primary consolidation settlements were then estimated using Terzaghi one-dimensional consolidation theory[2]. Figure 6 illustrates the expected settlement for each sediment thickness/group. The anticipated required fill volumes were estimated using the respective thickness/group surface areas and the corresponding predicted settlements.

Secondary settlement was estimated using the more conservative parameters determined during testing. For the 25 to 35 foot sediment thickness group, 0.8 feet of secondary settlement was estimated to occur after 50 years. Even though the composite cap would be capable of sustaining some differential settlement, it was decided to design the surcharge to remove as much of the anticipated secondary settlement expected to occur in 50 years as possible, during the anticipated 6 month period the surcharge would be in place.

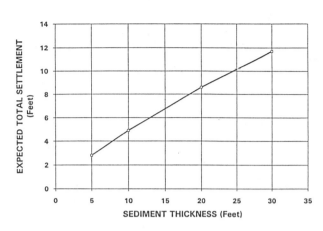

Figure 6   Estimated sediment settlement versus sediment thickness.

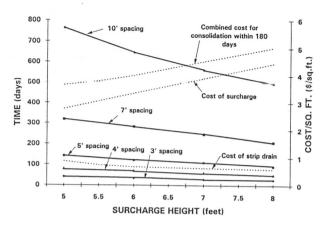

Figure 7   Time required for complete primary consolidation versus surcharge height, for various strip drain spacings, and cost curves for combinations of strip drain spacings and surcharge heights for consolidation within 180 days.

Accelerating Consolidation

To evaluate the feasibility of consolidation of the lagoon sediments, it was also necessary to estimate the time required to complete primary consolidation. As reported earlier the coefficient of consolidation, Cv, of the sediment ranged from 0.001 to 0.04 cm²/sec. Since the testing indicated the average was closer to the lower end, Cv was taken to be 0.002 cm²/sec or approximately 0.2 ft²/day. The pore water drainage path was assumed to be vertical only from the top surface, because static groundwater levels in the lagoon had been observed to be several feet higher than the surrounding unconfined aquifer. Also, the initial excess pore pressure distribution was assumed to be constant with depth. Using Terzaghi's consolidation theory, and a percent consolidation required (Ureq%) of 98%, Tv=1.8, the time required for consolidation at a sediment thickness of 20 feet was estimated to be 10 years. This was obviously an overly conservative estimate, but it demonstrated that acceleration would be required to complete primary consolidation within a reasonable period of time, such as 6 months.

To accelerate the primary consolidation of the sediments, a combination of vertical strip drains and additional surcharge was proposed. An analysis was then undertaken to evaluate the required vertical drain spacing and surcharge loads at various sediment thicknesses. For the purpose of analyzing consolidation by radial drainage, Cvr, was taken to be the same as Cv. Then, using time factors for radial drainage, Tr [1], the time required for primary consolidation was determined. The contribution to consolidation from vertical drainage out of the top of the sediments was also evaluated, but found to have a small effect on rate, for strip drain spacings less than 6 to 7 feet and primarily only during the early portion of consolidation. Vertical drainage was therefore ignored in the analysis. The results of this analysis for a sediment thickness of 20 feet are graphically displayed in Figure 7. This figure shows the amount of time required for complete primary consolidation on the left axis, for various strip drain spacings and surcharge height combinations. On the right axis the cost per square foot of surface is shown for the combinations of surcharge heights and strip drain spacings which will complete primary consolidation in 180 days. The total combined cost is also shown.

It is apparent from Figure 7 that vertical strip drains are much more efficient and cost effective, at accelerating consolidation than is additional surcharge. The actual optimum cost would be at a surcharge height less than is shown on the graph, but a minimum of 5 feet of surcharge was required for sufficient effective stress for removal of significant secondary consolidation. Therefore, at a 20 foot sediment thickness, a combination of 5 feet of surcharge and a strip drain spacing of approximately 5.5 feet was called for. The total cost of this combination was approximately $3.70 per square foot.

Preload Stability

One of the initial questions regarding the use of a surcharge to consolidate the lagoon sediments was whether or not the fill could be placed over the very soft material. After removing the surface water from the lagoon, an investigation of the near surface sediments revealed that a thin desiccated crust (6 to 12 inches) had formed over the higher portions. Beneath this crust and in lower areas, the surficial sediments had little to no shear strength. Deeper probing revealed that the shear strength did increase with depth, but the sediments were too soft to allow for access of heavier equipment necessary for the collection of deeper undisturbed samples.

A preliminary stability analysis was undertaken to evaluate the feasibility of placing the fill, based only on the limited initial data. The sediment strength parameters used in this analysis were a cohesion, c, of 20 psf at the surface and linearly increasing by 1 psf per foot of depth. The shear strength friction angle was conservatively assumed to be zero because of a lack of adequate data and uncertainty regarding pore pressure distribution and dissipation with loading. Stability analyses were then completed using "PCSTABL5" and the Modified Janube Method. Analysis of a three foot sand fill layer placed directly over the sediments indicated a factor of safety less than 1. Therefore, a biaxial geogrid was proposed to help distribute the fill load and "bridge" softer areas of sediment. A discussion of the difficulties placing the initial fill layer will be presented in the next section.

Additional analysis indicated that shallower lifts placed over the initial 3 foot lift had factors of safety greater than one. Also, as previously reported, vane shear testing indicated that the sediments developed higher shear strengths with depth. This was primarily the result of what was a significant increase in effective stress, from just the first lift of fill, over the normally consolidated, saturated, sediments.

Preload Construction

Construction operations for the consolidation of sediments, began in May 1993, with the installation of Tensar BX-1200™ geogrid and the placement of a sand drainage layer (approximately 3 feet in thickness). Manufacturers specifications called for the geogrid to be installed in a shingle like manner, with 3 to 5 foot seam overlap. Seams were to be mechanically fastened on 5 foot centers using plastic zip strip ties. The geogrid was to be anchored around the perimeter of the lagoon in a 10 foot wide bench, cut to the elevation of the sediment surface.

Drainage layer fill was to be placed using a low contact pressure bull dozer, approximately equivalent to a Caterpillar D-4 dozer in gross weight. The fill sequence was such that the dozers were to cover the lapped seams first with about 2 to 3 feet of sand drainage material, then cover the center of the sheets. The dozers were to move fill to the leading edge in a gradual manner, such that a large mound of fill would not be advanced across previously placed fill and only a thin leading edge would be advanced over the sediments.

Several days into this operation it became apparent that the near surface sediments were too weak in certain areas of the lagoon to support this approach. The advancing fill tended to push down the sediments near the leading edge, forming a "mud wave". Additionally, as the leading edge advanced, the height of the wave tended to increase, making it more difficult to control.

A number of alternative placement techniques were attempted in and effort to control the "mud wave" problem. These included reducing the initial layer thickness, using lighter equipment, and installing a series of "stab" wells and pumping to dewater the sediments. While these methods proved effective in many of the lagoon areas, the "mud wave" problem was not completely eliminated. An effective placement method was developed utilizing a track mounted backhoe to place fill in small piles, ahead of the leading edge, allowing it to sit overnight, then typically the dozers were able to directly access the area and level out the fill the following morning.

While this method was the most successful in the difficult areas, it proved to be much slower than anticipated. The drainage layer placement took 67 workdays to complete when it was anticipated to take only 45. Additionally, while covering the sediments, occasional localized stability failures occurred. During these failures the underlying geogrid tore and a large area of fill would drop by 2 to 5 feet. Also, the most fluid sediments would be forced up through cracks in the fill and flow over the leading edge of fill. Typically, once these failures occurred, stresses were relieved the area would stabilize.

Wick drain installation was commenced once a sufficient area of the lagoon was covered with the sand drainage layer. Amerdrain™ wick drains were installed using a 15 ton mandrel attached to a Komatsu 300 backhoe. Drain spacing in the shallow areas (i.e., less than 20 feet of sediment) was on a 5 foot rectangular grid. In areas where sediments were thicker, a 5 foot triangular spacing was used to reduce the drainage path length slightly.

In the Closure Plan the average sediment thickness was estimated to be 10.6 feet. Based on this thickness it was estimated that approximately 100,000 feet of wick drain would be required, provided a 5 foot rectangular spacing was used and drains were not installed in areas where the sediment thickness was less than 5 feet. During the drain installation, it became apparent that the depth of sediment had been underestimated. Sediment thickness averaged 16.2 feet indicating that the sediment volume had been underestimated by 52%. Consequently, a total of 197,000 feet of wick drains were installed in the lagoon sediments.

Shortly after wick drains were installed in an area of sediment, a dramatic drop in pore pressures was observed and a corresponding increase in stability was apparent. This drop in pore pressure can be seen on Figure 8, where elevation heads decreased by 5 to 8 feet after 30 days, during placement of additional fill. This response was typical at all the piezometer locations, following installation of the drainage wicks.

Prior to completing wick drain installation, a gravity drainage system was

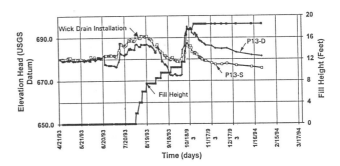

Figure 8 Time dependant pore pressure data from piezometer cluster P-13 (S-shallow and D-deep).

installed in the initial drainage layer to remove water which would become trapped when the general fill was placed. The system consisted of 2 sumps constructed of galvanized steel with finger drains extending radially out in four directions. The sumps were situated in the areas of the lagoon expected to experience the most settlement. The sumps were equipped with submersible pumps that fed into the plants wastewater treatment system.

Once the gravity drainage system was installed, placement of the general fill commenced. The first lift was placed using a Caterpillar D-4 wide-tracked dozer. Subsequent lifts were placed using heavier equipment including Caterpillar D-8 dozers, 977 loaders, and rubber tired scrapers, with no stability problems.

Instrumentation and Monitoring

Field monitoring was considered to be critical for the safe and successful completion of the installation of the fill. A series of pneumatic piezometers and settlement monuments were installed in order to monitor the behavior of the sediments. In deeper areas of sediment, clusters of three piezometers were installed at the top, middle and bottom. Data from this instrumentation was plotted on a daily basis. Figure 8 presents the pore pressure data from one piezometer cluster, designated P-13.

Settlement monuments were installed on top of the initial drainage layer, after the wick drain installation was complete. Figure 9 presents the data for settlement monument SM-6 (the field settlement curve was adjusted by 2 feet, which was the amount of settlement estimated to have occurred prior to the installation of the settlement monument). The curve shows a rapid increase in settlement immediately following installation of the wick drains. On the plot is also a curve for the theoretical settlement which would have occurred since the fill had first been placed at that location assuming radial drainage.

Conclusions

The project presented many interesting geotechnical challenges, some of which were easily addressed and others created significant problems. The limited laboratory testing program proved to be adequate for a reasonable estimation of the consolidation behavior of the lagoon sediments. The greatest difficulty and obstacle in completing the project, was the construction of the initial drainage layer over the very soft lagoon sediments. The early dewatering of the sediments proved to be invaluable in allowing for the subsequent placement of fill. The biaxial geogrid was critical in allowing the

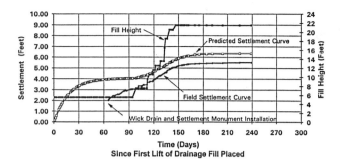

Figure 9 Time dependant settlement at monument SM-6, theoretical and settlement assuming radial drainage, and height of fill at SM-6.

placement of the initial sand layer, which could not have been done without some form of reinforcement. Evaluation of stability accurately predicted the stability of the sediment/fill system, which made it obvious at an early stage that reinforcement would be necessary. Also, the stability after the initial sand layer was placed was accurately predicted, as ease of later construction demonstrated.

The utilization of wick drains for the acceleration of consolidation proved to be cost effective and successful. Wick drain installation also had a dramatic effect in increasing the stability of the sediments.

This project demonstrated that deep, very soft, saturated, normally consolidated, wastewater lagoon sediments, can be preloaded and consolidated for the subsequent support of a cap. This construction method gives engineers and owners a cost effective alternative for the closure of a wastewater lagoon.

Acknowledgments

The authors would like to extend special thanks to: Eric Mendel, client's representative; Martin Battistoni and Thomas Uher, Rust Remedial Services, Inc.; Steve Schleede, Jim Berry, Steve Euker and Jim Thomson, Schleede Hampton Associates; John Cole, John Cole Contracting; and Bill West, GZA GeoEnvironmental, for their assistance and cooperation during this project.

References

1. Das, M. Braja, *Advance Soil Mechanics*, McGraw-Hill Book Company, 1983.

2. Holtz, Robert D. and Kovacs, William D., *An Introduction to Geotechnical Engineering*, Prentice Hall, 1991.

3. Pettijohn, Robert A., *Nature and Extent of Groundwater Quality Changes Resulting From Solid Waste Disposal, Marion County Indiana*, U.S. Geological Survey, 1977.

The Statistical Analysis of Leachate and Water Quality Monitoring Data at a Large Landfill Site

P.B. Rainbow, R.G. Clark, J.H. Donovan
C L Associates, Birmingham, UK

Abstract

Environmental monitoring has been carried out over a six year period for a large landfill used for the disposal of household, industrial and commercial waste. An outline of the geological and hydrogeological setting of the site is presented together with details of the monitoring programme and a statistical database which has been set up to determine trends in leachate, surface water and groundwater quality and any anomalies in the data.

Introduction

The disposal of waste is an activity which in the current climate of increasing environmental concern is one which is attracting more attention particularly in respect of its potential to create pollution. This concern is most specifically reflected in the UK in the controls imposed through the site licensing requirements of the Control of Pollution Act 1974 (soon to be superseded by the requirements of the Environmental Protection Act 1990). These provisions permit the Waste Regulation Authorities (WRAs), and through them the National Rivers Authority (NRA), to exercise control over the methods utilised in the practice of landfilling wastes and in particular those aspects of the process which have potential for causing pollution.

In addition to the controls over the physical practices adopted in the landfilling process, the regulatory bodies also impose requirements for the monitoring of leachate within landfills and the quality of the waters in the surrounding surface and groundwater environments. There is thus a legal requirement on landfill operators to carry out a certain amount of environmental monitoring. In addition, landfill operators may have further reasons to carry out this type of monitoring in order to :

- Improve the effectiveness of their operations;
- Identify potential problems at an early stage; and
- Reduce environmental liabilities.

Dependent on the circumstances this may well lead to quite substantial monitoring programmes which, in turn, generate a substantial volume of data requiring analysis and interpretation. A case in point is the Mucking Landfill Site which is owned and operated by Cory Environmental Limited, a wholly owned subsidiary of the Ocean Group plc. Environmental monitoring data has been obtained for this site over a six year period and C L Associates have been assigned to analyse the information and provide an ongoing appraisal of the results.

Site Description

The Mucking Landfill has an area of approximately 240 hectares and is located on the north bank of the River Thames (see Fig 1). The site consists of former marsh and farmland, part of which has been subsequently used for mineral extraction. In the eastern parts of the site, landfilling has taken place directly on to the saltmarsh which is naturally sealed by clay, silt and alluvial deposits. In the western part of the site, where gravel extraction has taken place, the excavations have been sealed with a clay liner before landfilling.

The geology of the site is relatively simple with a straightforward succession of Estuarine Alluvium over Flood Plain Gravels over Lower London Tertiary Deposits over Upper Chalk (Ref 2). The estuarine alluvium mainly consists of clay which is of relatively low permeability and has a thickness varying from

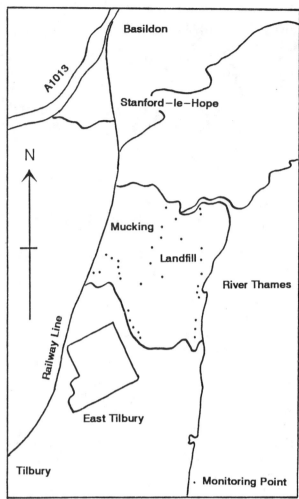

FIG.1 -- Site Plan

2 to 9 m across the site. The Flood Plain Gravels constitute an aggregate deposit of economic significance varying widely in thickness from 1 to 13 m. The Lower London Tertiary strata underlying the site include the Thanet Sands overlain by the Woolwich and Reading Beds. This section of the succession is predominantly silty clay at depth, the overall thickness of the deposit within the site varying between 15 and 29 m. The whole site is underlain by what is believed to be an extensive depth of Chalk although the relevant geological maps do not indicate any outcrop in the general vicinity of the site.

Both the Flood Plain Gravels and the underlying Lower London Tertiary deposits carry groundwater and it is deduced that they are in hydraulic continuity with each other. However, the major aquifer in the area is the Upper Chalk which is actively abstracted as a potable water supply at the nearby Linford pumping station. The balance of current hydrogeological data fortuitously gives rise to the conclusion that the Chalk is not in hydraulic continuity with the overlying Tertiary strata. It is speculated that this is due to the low permeability of the formerly eroded surface of the Chalk and the predominantly clayey nature of the overlying Lower London Tertiary deposits encountered at depth.

Current Operations

The site is licensed under the Control of Pollution Act 1974 for the disposal of household, industrial and commercial solid wastes as well as a limited amount of liquid and semi-solid industrial waste. Currently the site accepts approximately 1,000,000 tonnes of waste per year, half of which is delivered by river barges and half by vehicles.

When the site was first used for the disposal of waste (in the 1950s) landfilling was carried out directly into dewatered aggregate workings (in the north western part of the site). Hence despite the fact that this part of the site has been restored there is a high potential for any leachate generated to contaminate near surface groundwater. Subsequently the eastern part of the site was landfilled on top of the Estuarine Alluvium (typically composed of clays with permeabilities less than 1×10^{-10} m/s) which possibly isolates any leachate from the groundwater system.

More recently (since 1980) landfilling has been carried out in the central part of the site in specifically designed cells (lined with a 2m clay seal) on top of the Lower London Tertiary Deposits. These operations are carried out in accordance with current best practice and should promote a regime of waste (and accompanying leachate) containment isolated from the groundwater environment.

Monitoring Programme

The disposition of monitoring points around the landfill obviously reflects the need at various times to monitor the different facets of the operation. It is, however, a less than an ideal configuration for the sites existing features. The locations of the monitoring points are shown in Fig 1 and comprise 11 leachate points, 11 groundwater points (2 in the Chalk, 5 in the Lower London Tertiary and 4 in the Flood Plain Gravels) and 11 surface water points.

The sampling regime at the site (ie frequency of

	Type of Monitoring Station	
Determinand	Leachate	Clean Waters
Ammonia	Quarterly	Monthly
Chloride	Quarterly	Monthly
Conductivity	Monthly	Monthly
Dissolved Oxygen	–	Monthly
Leachate / water level	Monthly	Monthly
pH	Monthly	Monthly
Redox Potential	Monthly	Monthly
Temperature	Monthly	Monthly
Alkalinity	Quarterly	Quarterly
BOD	Quarterly	Quarterly
Cadmium	Yearly	Quarterly
Calcium	Quarterly	Quarterly
Chromium	Yearly	Quarterly
COD	Quarterly	–
Copper	Yearly	Quarterly
Iron	Yearly	Quarterly
Lead	Yearly	Quarterly
Magnesium	Quarterly	Quarterly
Manganese	Yearly	Quarterly
Mercury	Yearly	Quarterly
Nickel	Yearly	Quarterly
Potassium	Quarterly	Quarterly
Sodium	Quarterly	Quarterly
Sulphate	Quarterly	Quarterly
TOC	Quarterly	Monthly
TON	Quarterly	Quarterly
Zinc	Yearly	Quarterly

TABLE 1 Environmental Monitoring Programme

monitoring and the constituents analysed for) has been subject to variation over time and, as the site has progressed, the regime has been reviewed and refined. Essentially both leachate and "clean" waters (ie groundwater and surface water) are sampled monthly. The frequency of sampling and the suite of determinands analysed for both leachate and clean waters are presented as Table 1.

The method used for obtaining water samples includes measures to :

- Ensure the correct labelling of samples;

- Ensure that the samples collected are uncontaminated;

- Minimise any change in sample chemistry prior to analysis; and

- Collect representative samples using a groundwater pumping system and well designed bailers.

A series of on site testing was also devised in order to obtain results from samples as soon as they are collected.

Database

Monitoring has been carried out monthly over the period from 1988 to 1993 inclusive. The total number of determinands is currently 27, the total number of monitoring points is 52 and hence the theoretical number of data items for the whole monitoring period is approximately 100,000. In practice the actual number of data items is less than this because not all samples are collected on every occasion (as some monitoring points are inclined to dry up during the summer months) and not all samples are subject to the complete suite of analytical tests every month. In addition, over a period of time a proportion of the monitoring points change as site conditions change, new landfilling areas become active and old sampling locations become inoperable (for instance due to damage, silting up or permanently drying out).

A computer database has been compiled in order to record and analyse this considerable quantity of data. The purpose being to :

- Establish a permanent and readily accessible record of the monitoring data;

- Use statistical analyses to assess the data for trends and potential problems; and

- Develop recommendations for improvements and modifications in the methodology for leachate and "clean" water monitoring.

In fact, rather than manipulate the data in a standard database package, a three dimensional spreadsheet (Lotus 123 v 3.4) has been utilised to facilitate data computation and manipulation and ease production of graphical output. This approach has facilitated the examination of data sets both as time series and as groupings of other data subsets (such as groundwater by type of geological succession).

The results are progressively added to the database and then critically reviewed for any apparent trend, discrepancy or seasonal variation in the monthly results. The year as a whole is also compared with previous years to look for longer term trends. Typical examples of the two forms of table produced for each monitoring station are reproduced as Tables 2 and 3.

TABLE 2 Determinand Concentrations at MW11

DETERMINAND	1992											
	Jan	Feb	Mar	Apr	May	June	July	Aug	Sept	Oct	Nov	Dec
ALKALINITY	150	190	250	200	200	200	190	220	240	160	100	150
AMMONIA	0.1	1.5	0.1	0.1	0.1	0.1	0.3	3.4	0.1	0.1	0.1	0.1
BOD												
BROMIDE	0.1	1.0	0.1	1.0	1.0	3.3	1.0	1.0	1.0	1.0	1.0	1.0
CHLORIDE	85	70	40	90	64	65	70	50	53	185	250	160
CHROMIUM	0.01	0.01	0.01	0.01	0.01	0.05	0.01	0.05	0.05	0.05	0.05	0.05
COPPER	0.005	0.005	0.005	0.005	0.005	0.005	0.005	0.005	0.005	0.011	0.005	0.005
IRON	0.62	1.63	0.02	0.42	0.26	0.02	0.34	0.19	0.69	0.21	0.44	0.27
MANGANESE	0.01	0.708	0.618	0.624	0.653	0.63	0.634	0.112	0.01	0.05	0.01	0.010
NICKEL	0.01	0.01	0.01	0.02	0.01	0.01	0.01	0.01	0.01	0.03	0.01	0.02
NITRATE	62	55.3	2.7	4.8	71.7	164.6	26	48	62	7.9	9	17.80
pH	6.8	8.0	7.3	6.8	8.1	8.1	7.7	6.9	7.2	7.1	7.5	7.1
POTASSIUM	5	8	6	7	6	5	5	5	7	12	10	9
SULPHATE	164	175	45	15	189	451	68	155	169	275	1857	233
TOC	7	9	7	4	2	7	6	5	6	17	7	7
ZINC	0.005	0.005	0.005	0.005	0.005	0.005	0.005	0.005	0.005	0.009	0.005	0.005

TABLE 3 Annual Mean Determinand Concentrations at MW11

DETERMINAND	1989	1990	1991	1992
ALKALINITY	161 ± 10	169 ± 16	176 ± 79	188 ± 42
AMMONIA	0.7 ± 0.9	0.1 ± 0.0	0.8 ± 0.7	0.2 ± 0.4
BOD	1.3 ± 0.4	1.3 ± 0.5	1.1 ± 0.1	
BROMIDE	1.0 ± 0.0	0.8 ± 0.3	1.0 ± 0.1	0.8 ± 0.4
CHLORIDE	42 ± 1	46 ± 11	83 ± 45	99 ± 65
CHROMIUM	0.017 ± 0.012	0.026 ± 0.021	0.011 ± 0.004	0.030 ± 0.021
COPPER	0.006 ± 0.000	0.009 ± 0.002	0.005 ± 0.002	0.005 ± 0.000
IRON	0.26 ± 0.15	0.45 ± 0.45	0.93 ± 1.08	0.32 ± 0.22
MANGANESE	0.009 ± 0.005	0.013 ± 0.006	0.030 ± 0.025	0.339 ± 0.321
NICKEL	0.04 ± 0.00	0.04 ± 0.00	0.02 ± 0.01	0.01 ± 0.00
NITRATE	37.0 ± 24.9	60.2 ± 5.4	33.3 ± 31.7	33.4 ± 26.6
pH (pH units)	6.9 ± 0.2	7.3 ± 0.4	7.2 ± 0.5	7.4 ± 0.5
POTASSIUM	5 ± 1	5 ± 1	6 ± 1	7 ± 2
SULPHATE	152 ± 6	158 ± 11	145 ± 105	176 ± 120
TOC	8 ± 5	10 ± 6	9 ± 7	6 ± 2
ZINC	0.006 ± 0.000	0.008 ± 0.002	0.014 ± 0.020	0.005 ± 0.000

Key: Mean ± Standard Deviation
All values in mg/l unless otherwise stated

In its raw form the data obtained frequently contains anomalous readings which make it extremely difficult to detect trends etc. Hence, it is necessary to use some method to screen out these anomalous readings and thus reduce any unrealistic variations in the data. Anomalous readings or "outliers" have therefore been identified by means of the ASTM E 178-68 one sided T test (Ref 1) using the critical values at the 1% significance level.

Once identified outliers are scrutinised to determine whether they are valid data reflecting unusual events or, possibly more likely, whether there are errors in the sampling procedure or problems arising during laboratory analysis. The mean and standard deviation for each data subset (a data subset consists of the data for one monitoring location, for one year and for one determinand) have been calculated following the exclusion of anomalous outliers.

In order to obtain an overall view of the annual performance, the results of all monitoring stations of similar type (for example all monitoring boreholes in the Chalk) are aggregated and mean and standard deviations are calculated (see Table 3).

In addition to the tabular presentation, graphs are produced of some data subsets to display the information and illustrate trends.

Results

The interpretation of the monitoring results has proved to be less than straightforward for a variety of reasons. One major factor is the extent of variation in the results obtained for a single determinand at a single monitoring point (see Fig 2 for a typical example). Apart from the surface water monitoring points (which may be considered to be sampling a relatively swiftly changing environment) the other monitoring points are essentially sampling relatively large liquid bodies which one would expect to only be subject to relatively gradual change. The question therefore arises as to the extent to which the variation can be attributed to a natural "noise"

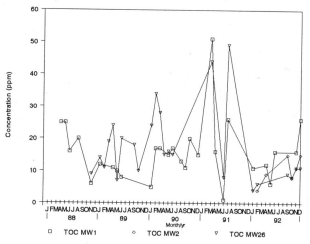

FIG 2 TOC Concentrations at MW1, MW2 and MW26

that would normally be expected for these environments (arising from normal localised variations in the "population"), and the extent to which it is caused by other factors (such as errors in the sampling procedures and method of analysis, etc.).

Despite these difficulties some conclusions have been drawn with a fair degree of confidence. For instance the difference between the groundwater chemistry in the Lower London Tertiary and the groundwater chemistry in the Upper Chalk is quite

DETERMINAND	Monitoring Station Category							
	Leachate	Sump Leachate	Surface Waters	Gobions Stream	Mucking Creek	Flood Plain Gravel Boreholes	Lower London Tertiary Boreholes	Chalk Boreholes
	Mean ± Std	Mean ± Std	Mean ± Std	Mean ± Std	Mean ± Std	Mean ± Std	Mean ± Std	Mean ± Std
ALKALINITY	3892 ± 1437	4714 ± 2369	260 ± 701	153 ± 58	264 ± 63	1315 ± 1444	899 ± 1261	533 ± 679
AMMONIA	485 ± 142	489 ± 208	14 ± 83	0.4 ± 0.6	15 ± 178	158 ± 199	105 ± 172	14 ± 38
BROMIDE	61.9 ± 50	41.2 ± 27.4	2.6 ± 4.1	1.4 ± 1	1.8 ± 3	19.6 ± 34	6.8 ± 10.3	3.9 ± 8.7
CHLORIDE	2710 ± 1472	2819 ± 1663	319 ± 72	270 ± 270	370 ± 875	1050 ± 1010	763 ± 870	321 ± 491
CHROMIUM	0.049 ± 0.035	0.12 ± 0.053	0.028 ± 0.02	0.033 ± 0.02	0.029 ± 0.02	0.028 ± 0.02	0.03 ± 0.013	0.032 ± 0.02
COPPER	0.03 ± 0.063	0.0086 ± 0.008	0.0074 ± 0.011	0.005 ± 0	0.008 ± 0.006	0.0101 ± 0.016	0.0077 ± 0.0081	0.0059 ± 0.0028
IRON	5.44 ± 4.43	4.2 ± 4.47	0.29 ± 0.35	0.34 ± 0.45	0.22 ± 0.26	1.73 ± 4.74	0.95 ± 1.84	0.85 ± 2.1
MANGANESE	0.64 ± 0.53	0.1 ± 0.04	0.42 ± 0.31	0.24 ± 0.3	0.29 ± 0.31	1.06 ± 1.43	0.46 ± 0.33	1.42 ± 2.48
NICKEL	0.086 ± 0.089	0.14 ± 0.12	0.019 ± 0.01	0.015 ± 0.008	0.016 ± 0.005	0.047 ± 0.043	0.023 ± 0.018	0.02 ± 0.018
NITRATE	9.29 ± 30.1	1 ± 0	2 ± 1.75	3.9 ± 4.2	34.4 ± 18.9	16.2 ± 34.7	32.9 ± 47.8	7.1 ± 18
pH	7.6 ± 1	7.6 ± 0.25	7.7 ± 0.5	7.6 ± 0.5	7.7 ± 0.3	7.7 ± 0.8	7.7 ± 0.7	7.5 ± 0.5
POTASSIUM	546.1 ± 201.3	619.3 ± 407.7	37 ± 74.7	14.9 ± 9.9	33.1 ± 14.7	190.8 ± 220	137.3 ± 188.9	33.6 ± 51.7
SULPHATE	187.1 ± 259.3	100.1 ± 96	363.8 ± 94.9	210.3 ± 123.7	232.4 ± 122.5	248.8 ± 423.1	162.4 ± 123.2	196.2 ± 275.6
TOC	375.9 ± 136.6	612.9 ± 390.3	23.4 ± 39.8	11 ± 5.3	21.2 ± 6	91.7 ± 102.9	76.4 ± 115.3	23.3 ± 35.4
ZINC	0.067 ± 0.114	0.038 ± 0.043	0.006 ± 0.004	0.005 ± 0.001	0.029 ± 0.046	0.008 ± 0.010	0.026 ± 0.077	0.008 ± 0.013

KEY: Std – Standard Deviation
All values in mg/l except pH

TABLE 4 Mean Concentrations For Determinand By Monitoring Station Category

pronounced and it has been possible to deduce that in the case of two adjacent boreholes into these strata, samples obtained on several occasions had been interchanged and wrongly labelled.

Similarly in the case of another borehole it was clear that the values obtained indicated that the integrity of the borehole had been compromised. On another occasion a large proportion of the samples analysed (from all sources - leachate, surface waters and groundwaters) showed elevated levels of manganese and this was identified as contamination arising from laboratory analytical procedures.

These observations have given rise to specific recommendations for changes and improvements to the monitoring procedures carried out at the site which should prevent the recurrence of these anomalies in the future. In addition, a more general review of the procedures was undertaken to try to address any systematic errors which may be influencing the results obtained. The object of the review was to examine in detail site practices from sample collection through to analysis and compare the observations with best practice. A number of areas for improvement were identified following the review and recommendations made which should ultimately feedback into providing more accurate data.

As with any interpretation of site data, there can be influence from a host of site specific factors. In the case of Mucking the adjacent estuary can influence sodium and chloride levels in the groundwater. Other landfill facilities in the vicinity can influence both groundwater and surface waters. The pattern of groundwater flows in the area, which would normally be expected to exhibit very gentle gradients toward the estuary, are artificially modified by the de-watering operations to facilitate mineral extraction both on site and nearby. Activities on the adjacent farmland also mean that both surface and groundwaters can be contaminated with agricultural chemicals (particularly nitrates).

Notwithstanding these influences the results for surface waters show relatively little change as they pass the site (ie Mucking Creek to the north and

Gobions stream to the south). Groundwater in the Upper Chalk generally appears to be distinct in chemistry from the groundwater in the Lower London Tertiaries etc. It is hoped that as the improvements to the sampling procedures (referred to above) yield more consistent results these will serve to sharpen the observed differences and make it easier to identify the effects of the other extraneous influences.

Discussion

In as much as the monitoring programme at Mucking fulfils the legal obligations placed on the operator and indeed that the regulatory bodies which receive this information are happy with its content, the monitoring programme at Mucking demonstrably achieves one of its prime aims, in particular. However it also serves a number of other purposes, in particular:

- As a mechanism for measuring the effect the landfill operations have on the surrounding environment;

- To provide early warning of pollution to trigger early remedial procedures (and reduce liabilities);

- To monitor conditions within the landfill; and

- To provide better information on the way landfills behave.

In terms of measuring the actual effect of the landfill on the surrounding water environment then it must be recognised that the monitoring results obtained are at best only a proxy for any real change. The landfill's interface with the environment is both large and concealed and the actual monitoring points are only small "pinpricks" in the potential surface of interaction. Hence, an assumption has to be made that the samples are representative of the wider liquid bodies in which they are situated. Also the very act of taking the samples for analysis will engender some change in the sample (echoes of the Heisenberg's Uncertainty Principle). There appears to be some evidence to support the view that samples from an anaerobic environment brought into, and analysed in, an aerobic environment are liable to lead to significantly different values obtained for the different determinands. The ability of the programme to measure real changes in the environment is therefore a mute point in absolute terms.

However, it is probably of greater importance to consider the ability of the monitoring programme to provide warning of pollution effects in the environment external to the site. In spite of the "noise" in the readings obtained at Mucking, the extent and frequency of the programme should ensure that any major polluting episodes will be identified. Smaller scale leakages over a long time scale again may be discernable as a trend within the monitoring data although the ability to identify such occurrences will depend very much on their scale compared to the amplitude of the background variation or "noise".

The monitoring programme should therefore supply some comfort to the operator that major problems can be identified at an early stage thus providing an opportunity to implement remedial action in a shortened timescale. Similarly the monitoring of leachate chemistry within the landfill should give the operator some guidance on the processes going on within the landfill mass. The programme also helps to contribute to the overall learning process as data from Mucking is collected on a consistent basis and these results together with information from comparable sites are passed to the regulatory bodies. A wealth of information is being acquired requiring cogent analysis prior to being fed back to the waste disposal industry in the form of improved monitoring techniques. To date the data has shown no evidence of significant pollution at Mucking. The only events have been a significant but temporary increase in many determinands considered to be good indicators of the presence of leachate at a borehole to the west of the site. Similarly elevated levels of some determinands have been recorded at several of the boreholes installed through the base of the landfill.

Concluding Remarks

Overall it is clear that the extensive monitoring programme instituted at Mucking reflects current good practice and achieves its principal aims. With the introduction of improved sampling procedures over a period of time it is expected that the use of a computer database will facilitate an easier identification of trends within the data and thus become a more valuable tool in the early detection of any pollution problems and changes in the state of the waste in the landfill.

Acknowledgements

The authors wish to thank Cory Environmental and C L Associates for permission to publish this paper.

References

1) ASTM : 1968 : Recommended Practice for Dealing with Outlying Observations (E178-68), American Society for Testing and Materials.

2) Cory Environmental : 1991 : Planning Application - Environmental Statement.

Composite Liners as Barriers: Critical Considerations

R.Kerry Rowe, M.J. Fraser
Geotechnical Research Centre, University of Western Ontario, London, Ontario, Canada

Abstract

The finite service life of engineered components of composite liner systems is a critical consideration in the design of such systems. Four different barriers incorporating composite liners are examined with respect to service life, leakage through the geomembrane, and the hydraulic conductivity of the geosynthetic clay liner.

Introduction

Composite liners consisting of geomembranes over compacted clay and geomembranes over geosynthetic clay liners (GCL) are gaining wide acceptance in the design of barrier systems for waste disposal. This paper examines a number of key considerations with respect to the design of these systems, with particular emphasis on the finite service life of the engineered systems and the effective hydraulic conductivity of the geomembrane and GCL.

The primary goal of barrier systems in landfills is to minimize the migration of contaminants. The effectiveness of a barrier design can be assessed by examining the impact of the landfill on an underlying aquifer. For the purposes of this paper the migration of chloride and dichloromethane will be examined. Initial source concentrations of 1500 mg/L for Chloride, and 1500 µg/L for dichloromethane are assumed. In addition the mass of the chloride is assumed to represent 0.2% of the total mass of waste, which is has a density of 600 kg/m^3. It is also assumed that the mass of dichloromethane is in direct proportion to the initial source concentration.

The service life of the engineered systems is expected to be finite, due to chemical and biological clogging of the leachate collection systems and ageing (eg. due to chain scission) of the geomembrane. In this analysis the service lives of the engineered systems are assumed to be 50 years for the primary leachate collection system, 125 years for the primary geomembrane, 175 years for the secondary geomembrane, and 200 years for the secondary leachate collection system unless otherwise specified.

Prior to failure of the primary leachate collection system the leachate mound is taken to be 0.3 m above the primary liner, after failure the mound is assumed to build at a rate of 0.25 m/a up to it's full height of 11 m above the primary liner (where the maximum height of the mound is controlled by the thickness of waste in the example being considered).

All the analyses reported herein were performed using a finite layer contaminant transport model [Ref. 5] as implemented in the computer program POLLUTE v.5 [Ref. 6].

Barrier Designs

Four different composite liners are considered for a hypothetical landfill excavated into a relatively permeable silt till which extends 3 m below the top of the primary composite liner, and overlies a 1 m thick aquifer. The silt till is assumed to have a hydraulic conductivity of 1×10^{-7} m/s, a porosity of 0.25, an effective diffusion coefficient of 0.015 m^2/a for chloride and dichloromethane, with the product of soil density, ρ, and dichloromethane partitioning coefficient, K_d, given by $\rho K_d = 2$.

The underlying aquifer is assumed to have a porosity of 0.3, a hydrostatic head of 1 m above the aquifer, and a horizontal flow at the up-gradient edge of the landfill of 20 m/a. Upon failure of the leachate collection systems the mounding of the leachate will cause an increase in the downward Darcy velocity with a resulting increase in the horizontal flow in the aquifer.

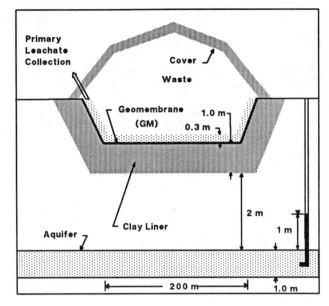

Figure 1. Design 1: Single Liner - Geomembrane & Clay.

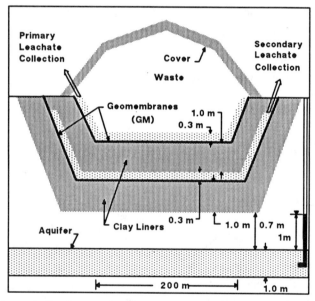

Figure 2. Design 2: Double Liner - Geomembrane & Clay.

The first barrier design incorporates a primary leachate collection system and composite primary liner consisting of a 2 mm (80 mil) geomembrane and 1 m of compacted clay (Figure 1). In this and subsequent designs the geomembrane is assumed to have an effective hydraulic conductivity of 10^{-14} m/s which has been backfigured based on consideration of the likely leakage through a "well constructed" composite liner, with some holes (using information provided by Giroud and Bonaparte [Ref. 1]), and an effective diffusion coefficient of 3×10^{-5} m^2/a, unless otherwise stated.

The compacted clay for this and the second barrier design is 1 m thick and has a hydraulic conductivity of 2×10^{-11} m/s, a porosity of 0.35, an effective diffusion of 0.019 m^2/a. The sorption of dichloromethane is controlled by $\rho K_d = 2$. Leachate collection systems in these designs consists of a granular layer normally 0.3 m thick with a porosity of 0.3. In this design there is 2 m of silt till, below the liner.

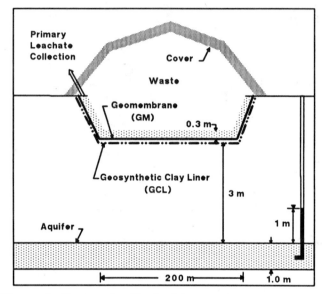

Figure 3. Design 3: Single Liner - Geomembrane & Geosynthetic Clay Liner.

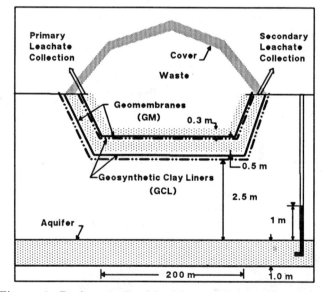

Figure 4. Design 4: Double Liner - Geomembrane & Geosynthetic Clay Liner.

In the second barrier design, primary and secondary leachate collection systems and primary and secondary composite liners are utilized (Figure 2). Both the primary and secondary liners consist of a 2 mm (80 mil) geomembrane over 1 m of compacted clay. The remaining silt till is only 0.7 m thick, if the base of the landfill is maintained at approximately the same elevation as the first design.

The third and fourth barrier designs (Figures 3 and 4) are similar to the first and second designs respectively, except in these designs the composite liners consist of a 2 mm (80 mil) geomembrane over a geosynthetic clay liner (GCL). The thickness of silt till is 3 m below the engineering for the third design and 2.5 m for the fourth design in order to keep the base of the landfill at the same level above the aquifer as in the first and second designs. In these designs the GCL is assumed to have a hydraulic conductivity of 4×10^{-12} m/s, a porosity of 0.75, an effective diffusion coefficient of 0.0047 m^2/a, and the sorption of dichloromethane is given by $\rho K_d = 2$. In the fourth design the secondary leachate collection system was assumed to be 0.5 m thick to allow for a granular - geosynthetic cushioning layer above and below a coarse stone collection layer.

In the analysis that follows the infiltration through the cover is taken to be 0.15 m/a based on experience in Southern Ontario, the waste thickness is 12.5 m, and the landfill length is 200 m in the direction of groundwater flow. The effect of the mass of contaminant was modelled as described by Rowe [Ref. 4]. Due to space limitations, only one hydrogeologic system is considered here. However, as noted by Rowe [Ref. 7], the impact of a given landfill will depend on the interaction between the engineered barrier system and the hydrogeology. Thus care should be taken not to generalize the numerical results beyond the level discussed in the paper.

Service Life of Geomembrane

The service life of the geomembranes is assumed to be finite, due to ageing (eg. chain scission caused by chemical attack). To illustrate the effects of the service life of the geomembrane on the migration of contaminants, a range of service lives of the primary geomembranes were examined for the four designs. Once the primary geomembrane ceases to be effective there will be a significant increase in contaminant contact with the secondary leachate collection system and secondary geomembrane for designs 2 and 4.

This increased contact is then expected to accelerate degradation of these systems. Thus, for this paper the secondary geomembrane and secondary collection leachate system are assumed to fail at 50 years and 75 years respectively after failure of the primary geomembrane. Space does not permit an examination of the effect of this assumption which will be discussed in another paper.

If the service lives of the geomembrane are assumed to be effectively infinite (ie. it's hydraulic containment characteristics are maintained for the entire contaminating lifespan of the landfill) then the impact on the aquifer would be controlled by diffusion of contaminants through the barrier system, even with failure of the primary leachate collection system. For this case the calculated peak increase in chloride concentration would be 14, 10, 12, and 8 mg/L and the peak dichloromethane concentration would be 4, 3, 3, and 2 μg/L for the first, second, third and fourth designs respectively.

For comparison purposes, Figures 5 and 6 show the calculated impact on the aquifer for chloride and dichloromethane, assuming that the service life of the geomembranes is finite and that the service life of the primary geomembrane is between 100 and 150 years. It can be seen that the peak concentration of the contaminants decreases with increasing service life, with the decrease being most noticeable for the barrier designs having primary barriers only (i.e. the first and third designs).

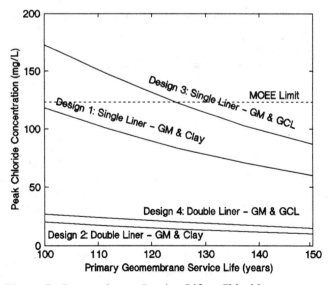

Figure 5. Geomembrane Service Life - Chloride.

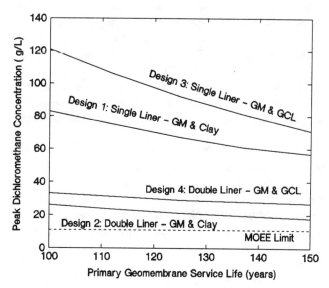

Figure 6. Geomembrane Service Life - Dichloromethane.

The designs with a secondary system (ie. 2 and 4) result in substantially reduced impact compared to those with only a single composite liner (ie. 1 and 3). Neglecting biodegradation of dichloromethane, it is seen that for designs 1 and 3 the calculated impacts are quite significant compared to a drinking water objective of 50 µg/L, even with a substantial service life of 150 years for the primary geomembrane.

In the Province of Ontario, Canada, the Ministry of Environment and Energy's 'Reasonable Use' Policy [Ref. 2] would limit increases in the concentration of contaminants in the aquifer to a maximum of 125 mg/L for chloride and 12 µg/L for dichloromethane, assuming a negligible background concentration. Under this policy the first, second, and fourth barrier designs would be acceptable for chloride for all the service lives examined, however the third design would require a geomembrane service life greater than 125 years to be acceptable for the conditions assumed. None of these designs would be acceptable for dichloromethane with service lives of the primary geomembrane of 150 years or less, although the second design is close to being acceptable if the service life of the primary geomembrane is 150 years.

Leakage through the Geomembrane

The leakage through a geomembrane which forms part of a composite liner system depends primarily on the applied leachate head, the number of 'holes' in the geomembrane and the nature of the contact between the geomembrane and the underlying clay liner [Ref. 1]. In order to allow an 'intuitive' comparison with traditional geotechnical barrier materials (eg. clay liners) it is convenient to backfigure a effective hydraulic conductivity of the geomembrane (considering the factors discussed above) as a measure of the quality of the installed geomembrane and hence to examine the effect of reasonable variation in workmanship in terms of this effective hydraulic conductivity as illustrated in Figures 7 and 8.

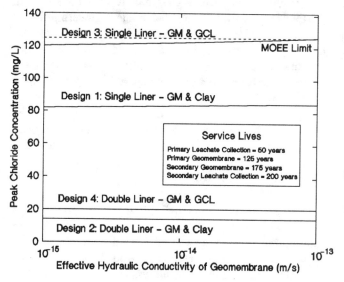

Figure 7. Effective Geomembrane Hydraulic Conductivity - Chloride.

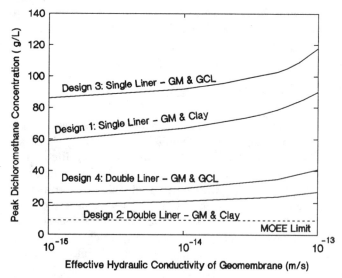

Figure 8. Effective Geomembrane Hydraulic Conductivity - Dichloromethane.

Over the range of effective hydraulic conductivities examined, the peak chloride concentration does not vary appreciably for the four designs (Figure 7). This insensitivity is due to the contaminant migration process being predominantly diffusive while the geomembranes are intact, and then being dominated by advection upon failure of the geomembranes. Due to the effect of sorption, the peak concentration of dichloromethane is more sensitive to the leakage through the geomembrane and the results show a moderate decrease in peak concentration with decreasing effective geomembrane hydraulic conductivity (Figure 8), with the effect being greatest for the single composite liner systems (ie. designs 1 and 3).

The effectiveness of the geomembrane(s) can be assessed by comparison of the peak impacts given in Figures 7 and 8 with those that would be predicted for designs 1, 2, 3, and 4 assuming no geomembrane: namely 199, 40, 349, and 192 mg/L for chloride and 164, 37, 350, and 98 µg/L for dichloromethane.

When examined with respect to the MOEE's 'Reasonable Use' Policy, all of the designs would be acceptable for chloride over the range of workmanship (ie. effective geomembrane hydraulic conductivity) considered. None of the designs would be acceptable for dichloromethane, however the second design comes close to being acceptable for excellent workmanship (ie. an effective geomembrane hydraulic conductivity of 10^{-15} m/s).

Hydraulic Conductivity of GCL

A range of hydraulic conductivities of the geosynthetic clay liner were examined to illustrate the sensitivity of the peak aquifer contaminant concentration to this parameter. Only the third and fourth designs incorporate GCLs and would be sensitive to changes in the hydraulic conductivity of the GCL, in the figures that follow the peak aquifer concentrations of the first and second designs are plotted as constants for reference purposes only.

Figures 9 and 10 show the calculated peak concentration of chloride and dichloromethane in the aquifer, for a range of GCL hydraulic conductivities. The third and fourth designs are quite sensitive to the hydraulic conductivity, since after the geomembrane fails the migration process is predominantly advective with the rate being controlled by the hydraulic conductivity of the GCL. The third design is the most sensitive since there is no secondary leachate collection system to remove contaminant after failure of the geomembrane, whereas in the fourth design the majority of the contaminant is removed by the secondary leachate collection system prior to the failure of the secondary geomembrane.

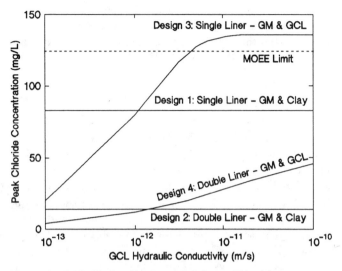

Figure 9. GCL Hydraulic Conductivity - Chloride.

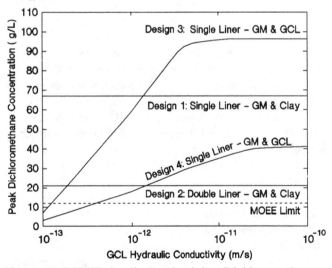

Figure 10. GCL Hydraulic Conductivity - Dichloromethane.

From these figures one can assess the hydraulic conductivity of the GCL at which it outperforms a 1 m thick compacted clay liner with a hydraulic conductivity of 2×10^{-10} m/s. It is seen that for the cases examined this typically occurs for a GCL hydraulic conductivity of between about 10^{-12} and 10^{-13} m/s.

Based on the maximum allowable increase in chloride concentration permitted by the MOEE's 'Reasonable Use' Policy, the fourth design may be acceptable at all the GCL hydraulic conductivities examined, and the third design may be acceptable for GCL hydraulic conductivities less than 4×10^{-12} m/s. For dichloromethane the fourth design would require a GCL hydraulic conductivity less than 4×10^{-13} m/s, and the third design would require a GCL hydraulic conductivity less than 1.3×10^{-13} m/s.

Discussion and Conclusions

Some regulators (eg. see Ref. 3) require that when assessing the potential impact of a landfill on an underlying aquifer, the service life of the engineered systems must be considered. When composite liners are utilized the service life of the geomembranes must also be considered, in addition to the service life of the leachate collection systems. The service life of the geomembrane for barrier systems involving only primary liners has a much greater effect than for systems involving both primary and secondary systems.

Although the geomembranes and geosynthetic clay liners are engineered components, there is still uncertainty regarding their hydraulic conductivity and effective diffusion coefficients. Since the hydraulic conductivity of a well installed geomembrane is very low, the migration process is primarily diffusive and relatively insensitive to the effective hydraulic conductivity of the geomembrane over the range examined. However, after the geomembrane fails the contaminant transport is controlled by advection and for a composite liner incorporating a GCL, the hydraulic conductivity of the GCL becomes the key factor controlling impact (similarly, of course, with a compacted clay liner it would be the hydraulic conductivity of the clay that would control impact).

This paper demonstrates that even with relatively long service lives (more than a hundred years), consideration of the finite service life of components of the engineered systems can have a profound effect on the estimated impact for a modest sized landfill (approximately 12.5 m average waste thickness); the effect could be expected to be greater for a larger landfill. It is concluded that reasonable uncertainty regarding the service life of engineered systems should be considered when evaluating the potential impact and the health and safety risks associated with proposed waste disposal facilities.

Acknowledgements

This paper forms part of a general programme of research into contaminant migration being conducted at the Geotechnical Research Centre, University of Western Ontario, and is made possible by the award of Research Grant A1007, to the senior author, from the Natural Sciences and Engineering Research Council of Canada. [The computations were performed on an R6000 workstation donated to the Environmental Hazard Programme at the University of Western Ontario by IBM Canada.]

References

[1] GIROUD, J.P. and BONAPARTE, R. (1989) "Leakage through liners constructed with geomembranes - Part II, Composite Liners". Geotextiles and Geomembranes, Vol. 8, pp. 71-111.

[2] MOEE (1993a) "Incorporation of the reasonable use concept into MOE groundwater management activities". Ministry of the Environment and Energy, Ontario, Policy 15-08.

[3] MOEE (1993b) "Engineered facilities at landfills that receive municipal and non-hazardous wastes". Ministry of the Environment and Energy, Ontario, Policy 14-15.

[4] ROWE, R.K. (1991) "Contaminant impact assessment and the contaminating lifespan of landfills". Canadian Journal of Civil Engineering, Vol. 18, pp. 244-253.

[5] ROWE, R.K. and BOOKER, J.R (1987) "An efficient analysis of pollutant migration through soil". Numerical methods for transient and coupled systems. Edited by R.W. Lewis, E. Hinton, P. Bettess and B.A. Schrefler. John Wiley & Sons Ltd., New York, N.Y. Chap. 2, pp. 13-42.

[6] ROWE, R.K. and BOOKER, J.R. (1990) "Program POLLUTE v.5 - 1D pollutant migration through a non-homogenous soil: User's manual". Geotechnical Research Centre, University of Western Ontario Research Report GEOP-1-90, London, Canada.

[7] ROWE, R.K. (1992) "Integration of hydrogeology and engineering in the design of waste management sites". Proc. of the International Association of Hydrogeologists Conference on "Modern Trends in Hydrogeology", Hamilton, pp. 7-21.

Sedimentation sur un Fond Permeable

Martin Sanchez, Alain Grovel
Laboratoire de Mécanique et de Géomécanique de Nantes, Université de Nantes, France

ABSTRACT : This paper concerns the study of slurry ponds provided with permeable bottom in order to reduce sedimentation time. A mathematical model of sedimentation on permeable bottom has been carried out on two hypothesis : deposited material is incompressible and there is no effective stresses in slurry. In considering these hypothesis Been and Tan *et al*, show sedimentation velocity on impermeable bottom depends only on concentration, in agreement with Kynch's theory. Then, the mathematical model presented here, is applicable to all the materials concerned by this theory. Sedimentation velocity on permeable bottom is the addition of sedimentation velocity on impermeable bottom and relative velocity of water in the incompressible deposited layer, which depends on hydraulic charge between surface and bottom of this layer, thus, sedimentation time is linked to water drainage gestion. In order to determine sedimentation time, a general methodology is presented ; finally, three simple cases of water drainage gestion are examined and applied to Loire estuary sediment.

RESUME : La sédimentation des solides en suspension est un processus intervenant dans la conception des dépôts confinés. Ces dépôts sont couramment munis d'un fond perméable afin de réduire le temps de sédimentation. Dans ce travail, l'influence des fonds perméables sur la sédimentation est étudiée en considérant les deux hypothèses suivantes : le matériau déposé est incompressible et la suspension est dépourvue de contraintes effectives. Sur ces deux hypothèses, les travaux de Been et Tan *et al*, montrent que la vitesse de sédimentation sur fond imperméable dépend uniquement de la concentration ; la théorie de sédimentation de Kynch est alors valable, si bien que ce développement concerne les mêmes matériaux englobés par cette théorie, notamment les déchets industriels et minéraux, les matériaux issus de dragage, les vases et même les boues argileuses lorsque les contraintes effectives sont faibles. Dans ce travail, on montre que lorsque l'on passe d'un fond imperméable à un fond perméable, le module de la vitesse de sédimentation est augmenté d'une valeur égale à la vitesse relative de la phase fluide au fond (vitesse moyenne sur une unité de surface), laquelle dépend de la charge hydraulique entre la surface et le fond du dépôt incompressible. La charge hydraulique et la vitesse de sédimentation sont toutes les deux liées à la gestion de l'eau de drainage. Ceci entraîne une multiplicité des cas possibles qui gardent cependant, une similitude entre eux. Une méthodologie servant à déterminer le temps de sédimentation est détaillée. Trois cas simples de gestion de l'eau de drainage sont abordés et les résultats sont comparés par rapport au processus de sédimentation sur fond imperméable. Enfin, ayant à la base une vase de l'estuaire de la Loire, un exemple illustre cette méthodologie.

Théorie de Sédimentation de Kynch.

L'évolution de la concentration C à l'intérieur d'une suspension est régie par l'équation de continuité de la phase solide :

$$\frac{\partial C}{\partial t} + \frac{\partial (v_s C)}{\partial z} = 0 \qquad (1)$$

où v_s est la vitesse moyenne des particules solides, z la coordonnée verticale et t le temps.

Si la vitesse de la phase solide dépend uniquement de la concentration, on peut montrer (Courant and Hilbert, 1961) qu'en coordonnées spatio-temporelles (z-t), les caractéristiques associées à chaque concentration sont des lignes de pente W constante donnée par :

$$W = \frac{d Q_s}{dC} \qquad (2)$$

où Q_s est le débit solide défini comme étant le produit de la vitesse de la phase solide par la concentration.

Le long de chaque ligne caractéristique la concentration est constante. Cette propriété des caractéristiques de l'équation (1) pour $v_s(C)$ est à la base de la théorie de sédimentation développée par Kynch (1952).

Dans la théorie de Kynch on considère également que le matériau déposé est incompressible (le processus de consolidation est négligé par rapport à celui de sédimentation). Le comportement incompressible du dépôt veut dire que la concentration ne peut pas dépasser une concentration finie C_F. Cette condition implique que pour une concentration C_F le débit solide est nul et par conséquent une discontinuité apparaît pour C_F dans la loi de variation de $Q_s(C)$.

La discontinuité dans la loi de variation du débit solide entraîne une discontinuité dans le profil de concentration à la frontière entre la couche incompressible et la suspension en cours de sédimentation.

Du côté inférieur de la discontinuité, on observe une concentration C_F et du côté supérieur, une concentration C_d telle que l'équation implicite suivante est satisfaite :

$$C_F W(C_d) = C_d [W(C_d) - v_s(C_d)] \qquad (3)$$

La pente de la ligne de la surface de la couche incompressible W^- dans un espace z-t est celle de la ligne caractéristique pour $C=C_d$.

Lorsque la concentration calculée à partir de l'équation (3) est inférieure à la concentration initiale C_i, la concentration du côté supérieur de la discontinuité est C_i et non celle calculée par (3). Dans ce cas, la pente de la ligne de la surface du dépôt W^- dans un espace z-t est donnée par :

$$W^- = -\frac{C_i}{C_F - C_i} v_s(C_i) \qquad (4)$$

Sédimentation sur un Fond Imperméable.

Dans les suspensions en cours de sédimentation sur un fond imperméable, on observe deux mouvements opposés : la phase fluide s'écoule vers la surface avec une vitesse v_f et la phase solide se dépose. La loi de Darcy modifiée par Scheidegger permet de relier la différence entre ces deux vitesses avec le gradient vertical de surpression hydrostatique. Cette équation s'écrit :

$$\left(1 - \frac{C}{\rho_s}\right)(v_f - v_s) = -\frac{k}{\rho_o g}\frac{\partial u}{\partial z} \qquad (5)$$

où k est le coefficient de perméabilité de la suspension, u la surpression hydrostatique, ρ_s la masse volumique des particules solides, ρ_o la masse volumique de l'eau et g l'accélération de la pesanteur.

Si les deux phases élémentaires de la suspension, l'une solide et l'autre fluide, sont incompressibles, l'équation de conservation de la matière s'écrit :

$$\frac{\partial}{\partial z}\left[\frac{C}{\rho_s}v_s + \left(1 - \frac{C}{\rho_s}\right)v_f\right] = 0 \qquad (6)$$

En intégrant l'équation (6) pour un fond imperméable et en combinant avec (5), on obtient l'équation qui donne la vitesse de la phase solide en fonction du gradient vertical de surpression hydrostatique :

$$v_s = \frac{k}{\rho_o g}\frac{\partial u}{\partial z} \qquad (7)$$

Lors de la sédimentation, il est habituel de négliger les contraintes effectives ; la condition d'équilibre s'écrit alors :

$$\frac{\partial u}{\partial z} = -\left(1 - \frac{\rho_o}{\rho_s}\right)g C \qquad (8)$$

Lorsque les contraintes effectives apparaissent graduellement avec l'augmentation de la concentration, elles peuvent être négligées si les pressions sont faibles (Sanchez et Grovel, 1993). On peut alors utiliser l'équation suivante pour évaluer la vitesse de la phase solide sur un fond imperméable (Been, 1980 ; Tan et al, 1990).

$$v_{so} = -\frac{k}{\rho_o}\left(1 - \frac{\rho_o}{\rho_s}\right)C \qquad (9)$$

Si l'on accepte que le coefficient de perméabilité dépend uniquement de la concentration, alors la vitesse de sédimentation v_{so} selon l'équation (9) n'est liée qu'à la concentration. Ainsi, l'équation (9) justifie formellement la validité de la théorie de sédimentation de Kynch pour un fond imperméable et pour des matériaux ne présentant pas des contraintes effectives.

Sédimentation sur un Fond Perméable.

Des lits de sable sont souvent utilisés afin de drainer par le fond les suspensions en cours de sédimentation. En surface de ces lits on place une membrane filtrante afin d'empêcher que les particules solides de la suspension colmatent le drain de sable.

Pendant le processus de sédimentation, la couche incompressible du fond joue le rôle d'un filtre. Dans ce qui suit, les pertes de charge hydraulique liées à l'écoulement au travers des membranes filtrantes sont négligées devant celles résultant de l'écoulement au travers de la couche incompressible.

L'application de la loi de Darcy à la couche incompressible du fond donne :

$$\left(1 - \frac{C_F}{\rho_s}\right)v_{fF} = -k_F\frac{\Delta H}{z^-} \qquad (10)$$

où v_{fF} est la vitesse de la phase fluide au fond, k_F le coefficient de perméabilité du matériau incompressible, ΔH la différence de charge hydraulique entre la surface et le fond de la couche incompressible du fond et z^- l'épaisseur de cette couche.

Pour un fond imperméable les conditions aux limites sont données par l'équation (10) et par $v_s = 0$. L'intégration de l'équation (6) pour ces conditions aux limites, combinée avec les équations (5) et (8), permet de formuler l'équation donnant la vitesse de sédimentation sur un fond perméable :

$$v_s = -\frac{k}{\rho_o}\left(1 - \frac{\rho_o}{\rho_s}\right)C - k_F\frac{\Delta H}{z^-} \qquad (11)$$

Si l'on compare les équations (9) et (11), on peut observer que la pente des lignes caractéristiques selon (2), est la même pour un drainage exclusivement superficiel et pour un fond perméable. Cependant, selon ce qui est dit au début de cet article la discontinuité pour une concentration C_F est accentuée dans la loi de variation de $Q_s(C)$. Ceci a deux conséquences principales : d'un côté la variation de l'épaisseur de la couche incompressible devient plus importante lorsque l'on passe d'un drainage purement superficiel à un drainage mixte ; d'un autre côté, la discontinuité dans le profil vertical de concentration est accentuée.

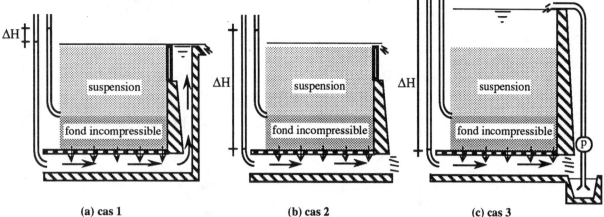

(a) cas 1   **(b) cas 2**   **(c) cas 3**
Fig. 1.- Diagramme illustrant les différentes gestions de l'eau de drainage ici traitées.

L'évolution de la concentration à l'intérieur d'une suspension drainée par la surface et par le fond dépend bien entendu des conditions initiales, mais aussi des conditions de frontière résultant de la gestion de l'eau drainée.

Par la suite sont détaillés trois cas d'utilité pratique pour la conception des dépôts sur des drains plans horizontaux. Les trois cas considèrent une concentration initiale constante.

Cas Type de Gestion de l'Eau de Drainage.

CAS 1. Eau superficielle drainée et pression dans le drain égale à la pression neutre (Fig 1.(a)).

La charge hydraulique ΔH entre la surface et le fond de la couche incompressible est donnée en fonction de z^- par :

$$\Delta H = \frac{1}{\rho_o}\left(1 - \frac{\rho_o}{\rho_s}\right)(C_i H_i - C_F z^-) \quad (12)$$

Première phase de sédimentation.

Il existe toujours dans ce cas, une première phase de sédimentation généralement très courte, ayant pour particularité de présenter un passage direct dans le profil de concentration, de la concentration initiale de la suspension à la concentration finale de la couche incompressible.

Pendant cette première phase, l'évolution de l'épaisseur z^- de la couche incompressible du fond est gouvernée, d'après les équations (4), (11) et (12), par l'équation suivante :

$$z^- W^- = z^- \frac{dz^-}{dt} = \alpha z^- + \beta \quad (13)$$

avec :

$$\alpha = \frac{C_i}{C_F - C_i}\left[\frac{k_i}{\rho_o}\left(1 - \frac{\rho_o}{\rho_s}\right)C_i - \frac{k_F}{\rho_o}\left(1 - \frac{\rho_o}{\rho_s}\right)C_F\right] \quad (14)'$$

$$\beta = \frac{C_i}{C_F - C_i}\frac{k_F}{\rho_o}\left(1 - \frac{\rho_o}{\rho_s}\right)C_i H_i \quad (14)''$$

La solution de l'équation (13) permet de déterminer explicitement le temps correspondant à chaque épaisseur z^- de la couche incompressible :

$$t = \frac{z^-}{\alpha} - \frac{\beta}{\alpha^2}\ln\left(\frac{\alpha z^-}{\beta} + 1\right) \quad (15)$$

Pendant cette phase de sédimentation, l'épaisseur totale de la suspension H, est déterminée selon une condition de conservation de la masse par :

$$H = H_i - z^-\left(\frac{C_F}{C_i} - 1\right) \quad (16)$$

Deuxième phase de sédimentation.

La deuxième phase de sédimentation commence lorsque l'épaisseur de la couche incompressible atteint une valeur critique z^-_c, telle que W^- donné par l'équation (13) est égal à $W(C_i)$. Cette valeur critique de z^-_c est donnée par :

$$z^-_c = \frac{\beta}{W(C_i) - \alpha} \quad (17)$$

A partir de l'instant où la condition (17) est satisfaite, la discontinuité dans le profil vertical de concentrations est graduellement allégée. On trouve à chaque instant, du côté supérieur de cette discontinuité, une concentration C_d telle que l'égalité (3) est satisfaite.

Deux points sont à remarquer pendant cette phase de sédimentation :

a) à chaque instant, la pente W^- de la ligne superficielle de la couche incompressible coïncide avec la pente $W(C_d)$ de la ligne caractéristique de C_d.

b) A chaque instant, prend naissance une ligne caractéristique de concentration C_d, en surface de la couche incompressible.

CAS 2. Eau superficielle drainée et pression nulle dans le drain (Fig 1.(b)).

Pour les mêmes conditions initiales, cette gestion de l'eau de drainage entraîne une augmentation par rapport au **cas 1**, de la charge hydraulique entre la surface et le fond de la couche incompressible. Dans ce cas, la charge hydraulique ΔH est donnée en fonction de z^- par :

$$\Delta H = \frac{1}{\rho_o}\left(1 - \frac{\rho_o}{\rho_s}\right)(C_i H_i - C_F z^-) + \left(\frac{C_i}{\overline{C}} H_i - \frac{C_F}{\overline{C}} z^-\right) \quad (18)'$$

Selon l'équation (18)', la charge hydraulique dépend de la concentration moyenne de la suspension $\overline{C}$. Au cours du processus de sédimentation, cette concentration varie graduellement de C_i à C_F. Afin de simplifier, par la suite dans ce **cas 2**, on considère que la concentration moyenne de la suspension est toujours égale à C_i, de manière que l'expression pour la charge hydraulique ΔH, devient :

$$\Delta H = \frac{1}{\rho_o}\left(1 - \frac{\rho_o}{\rho_s}\right)(C_i H_i - C_F z^-) + \left(H_i - \frac{C_F}{C_i} z^-\right) \quad (18)''$$

De même que pour le **cas 1** de gestion de l'eau de drainage, dans ce cas on assiste à deux phases de sédimentation.

Première phase de sédimentation.

Au cours de cette phase de sédimentation, l'hypothèse introduite ci-dessus est rigoureusement satisfaite, si bien que la première phase de sédimentation est analogue à celle décrite pour le **cas 1**. Egalement, l'évolution de l'épaisseur de la couche incompressible est régie par l'équation (13). Mais, dans ce **cas 2**, α et β sont donnés par :

$$\alpha = \frac{C_i}{C_F - C_i}\left[\left(1 - \frac{\rho_o}{\rho_s}\right)\left(\frac{k_i}{\rho_o} C_i - \frac{k_F}{\rho_o} C_F\right) - k_F \frac{C_F}{C_i}\right] \quad (19)'$$

$$\beta = \frac{C_i}{C_F - C_i}\left[\frac{k_F}{\rho_o}\left(1 - \frac{\rho_o}{\rho_s}\right) C_i H_i + k_F H_i\right] \quad (19)''$$

Alors, les équation (15) et (16) sont valables. Enfin, il faut noter, que par rapport au **cas 1**, la durée de la première phase de sédimentation est plus longue pour des conditions initiales identiques.

Deuxième phase de sédimentation.

Pendant cette phase de sédimentation, l'équation (18)'' est complètement valable dans deux cas extrêmes : en début et en fin du processus de sédimentation. Pour la période intermédiaire, l'utilisation de l'équation (18)'', revient à considérer que la suspension est recouverte d'une couche superficielle d'eau, d'épaisseur variable de l'ordre de quelques centimètres. En bref, l'équation (18)'' permet le calcul approché du temps de sédimentation pour le cas de drainage de l'eau superficielle et une pression nulle dans le drain. Il faut toutefois noter que le temps réel de sédimentation est légèrement supérieur à celui calculé par cette démarche. Si l'on accepte comme valable l'équation (18)'', cette phase de sédimentation a les propriétés qui ont été décrites lors de l'étude du **cas 1** de gestion de l'eau de drainage.

CAS 3. Tirant d'eau constant et pression nulle dans le drain (Fig 1.(c)).

Des trois stratégies de gestion de l'eau de drainage présentées, celle-ci conduit à la plus rapide sédimentation, pour des conditions initiales identiques. Dans ce cas, la charge hydraulique entre la surface et le fond de la couche incompressible en fonction de z^- est :

$$\Delta H = \frac{1}{\rho_o}\left(1 - \frac{\rho_o}{\rho_s}\right)(C_i H_i - C_F z^-) + (H_i + H_c - z^-) \quad (20)$$

Dans ce **cas 3**, on assiste toujours à une première phase de sédimentation ayant les propriétés décrites lors de l'étude du **cas 1**. La deuxième phase de sédimentation peut se présenter ou non, selon les conditions initiales et le tirant d'eau $H_i + H_c$ maintenu constant.

Première phase de sédimentation.

Pendant cette phase, la variation de l'épaisseur de la couche incompressible est régie par l'équation (13) avec α et β donnés par :

$$\alpha = \frac{C_i}{C_F - C_i}\left[\left(1 - \frac{\rho_o}{\rho_s}\right)\left(\frac{k_i}{\rho_o} C_i - \frac{k_F}{\rho_o} C_F\right) - k_F\right] \quad (21)'$$

$$\beta = \frac{C_i}{C_F - C_i}\left[\frac{k_F}{\rho_o}\left(1 - \frac{\rho_o}{\rho_s}\right) C_i H_i + k_F (H_i + H_c)\right] \quad (21)''$$

Pour une épaisseur z^-, le temps correspondant est donné par l'équation (15) et l'épaisseur totale par l'équation (16).

Deuxième phase de sédimentation.

Lorsqu'elle existe, cette phase a les mêmes propriétés du **cas 1** et **2**, lesquelles ont été décrites lors de l'étude du **cas 1** de gestion de l'eau de drainage.

Exemples d'Application à l'Etude de la Sédimentation d'une Vase de la Loire.

Les exemples présentés correspondent à une loi empirique de variation du coefficient de perméabilité donnée par :

$$k = A_1 \exp\left(-A_2 \frac{C}{\rho_s}\right) \quad (22)$$

La validité de cette loi est démontrée (Sanchez, 1992) pour un grand nombre de matériaux étudiés pour la plupart au Laboratoire Central d'Hydraulique de France (Migniot, 1989).

Le matériau étudié par la suite est une vase de l'estuaire de la Loire. Les valeurs des paramètres utilisés ont été déterminés à partir de cinq courbes de tassement (Sanchez et Grovel, 1993). Ces valeurs sont : $A_1 = 0,005$ m/s ; $A_2 = 110$; $C_F = 320$ kg/m^3 ; $\rho_s = 2550$ kg/m^3 et $\rho_o = 1000$ kg/m^3. Chacun des exemples satisfait les conditions initiales suivantes : $C_i = 100$ kg/m^3 et $H_i = 4,00$ m. Enfin, pour le **cas 3** de gestion de l'eau de drainage, une charge complémentaire H_c de 2,00 m est considérée.

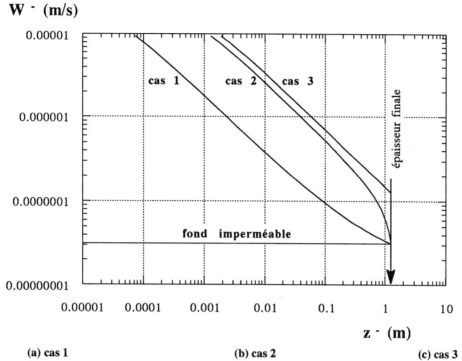

(a) cas 1   (b) cas 2   (c) cas 3

Fig. 2.- Variation de W⁻ en fonction de z⁻ pour les exemples ici présentés considérant d'une part, un fond imperméable et d'autre part, 3 cas de gestion de l'eau de drainage pour la sédimentation sur fond perméable.

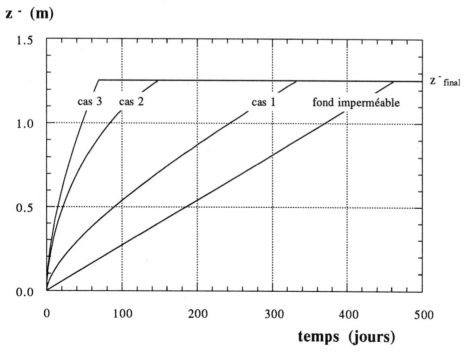

Fig. 3.- Evolution de l'épaisseur z⁻ de la couche incompressi... lu fond pour une vase de la Loire, pour les conditions initiales et les cas de drainage ...illés dans le texte.

Exemple A.- Sédimentation sur un fond imperméable.

Selon la théorie de Kynch, la pente des lignes caractéristiques est constante et donnée en fonction de la vitesse de sédimentation sur un fond imperméable par :

$$W = \frac{k}{\rho_o} \left(1 - \frac{\rho_o}{\rho_s}\right) C \left(A_2 \frac{C}{\rho_s} - 2\right) \quad (23)$$

La ligne superficielle du fond incompressible a une pente telle que l'équation implicite (3) est satisfaite. Pour les paramètres précédents, W^- vaut $3{,}133 \times 10^{-8}$ m/s, la concentration C_d ayant une valeur de $292{,}45$ kg/m^3. La pente W^- étant constante, le temps total de sédimentation T_T est donné en fonction de l'épaisseur finale du dépôt z^-_{final} par :

$$T_T = \frac{z^-_{final}}{W^-} \quad \text{avec :} \quad z^-_{final} = H_i \frac{C_i}{C_F} \quad (24)$$

Alors, pour un fond imperméable le temps total de sédimentation vaut $39{,}9 \times 10^6$ s, soit $461{,}78$ jours.

Exemple B.- Sédimentation sur un fond perméable.

La méthodologie décrite par la suite est valable pour les trois cas de gestion de l'eau de drainage étudiés dans le chapitre précédent.

Première phase de sédimentation.

Lorsque l'épaisseur z^- de la couche incompressible du fond est inférieure à z^-_c, (équation (17)), le temps correspondant à chaque épaisseur z^- est donné explicitement par l'équation (15) avec α et β selon le cas de gestion de l'eau de drainage. La valeur de z^-_c et son temps associé, sont les conditions initiales pour la phase suivante de sédimentation.

Deuxième phase de sédimentation.

Pour chaque épaisseur z^- de la couche incompressible, la vitesse de sédimentation est donnée par l'équation (11) avec ΔH selon le cas de gestion de l'eau de drainage. La concentration C_d de la suspension à la frontière avec la couche incompressible satisfait l'équation implicite (3) ; $W(C_d)$ étant donné par l'équation (23). Ainsi, pour chaque cas de gestion de l'eau de drainage, on a une loi $W^-(z^-) = W(C_d(z^-))$, si bien que l'évolution de l'épaisseur z^- de la couche incompressible du fond est régie par l'équation suivante :

$$\frac{dz^-}{dt} = W^-(z^-) \quad (25)$$

La figure 2 montre pour l'ensemble de cas de gestion de l'eau de drainage ici étudiés, la variation de W^- en fonction de z^-, pour des grandeurs de z^- correspondant à la deuxième phase de sédimentation. Finalement, la figure 3 montre l'évolution de l'épaisseur z^- de la couche incompressible du fond pour un fond imperméable et pour les trois cas de gestion de l'eau de drainage traités dans cet exemple de sédimentation sur fond perméable. Ces courbes sont déterminées en intégrant l'équation (25) par la méthode de Runge-Kutta.

En résumé, les temps de sédimentation pour chaque cas ici étudié, sont :

Sédimentation sur fond imperméable : $T_T = 462$ jours
Sédimentation sur fond perméable, **cas 1** : $T_T = 332$ jours
Sédimentation sur fond perméable, **cas 2** : $T_T > 147$ jours
Sédimentation sur fond perméable, **cas 3** : $T_T = 68$ jours

Conclusions.

Cette étude, fondée sur la méthode des caractéristiques, est une contribution originale à la simulation de la sédimentation sur un fond perméable. Un atout de cette méthodologie est que les comparaisons vis-à-vis de la théorie de sédimentation de Kynch sont directes, car cette théorie, largement utilisée, est fondée sur la même méthode.

Dans cette étude, on montre que la meilleure gestion possible de l'eau de drainage est celle qui permet d'obtenir la plus grande différence de charge hydraulique ΔH, entre la surface et le fond de la couche incompressible du fond. On montre donc, l'intérêt de concevoir des lits de drainage de manière que la pression interstitielle à l'intérieur de ceux-ci soit la plus faible possible. On montre également, l'inconvénient de la pratique courante de drainer l'eau superficielle à la suspension au fur et à mesure que la suspension se tasse. Une meilleure stratégie serait de conserver une colonne d'eau importante sur la suspension, laquelle ne serait drainée que lorsque le processus de sédimentation serait achevé.

Enfin, dans cette étude seulement trois cas de gestion de l'eau de drainage sont abordés ; cependant, cette méthodologie peut être appliquée à l'étude de la plupart des variantes possibles de cette gestion.

Référénces Bibliographiques.

[1] Been, K. (1980). Stress strain behaviour of a cohesive soil deposited under water, Thesis presented to the University of Oxford, at Oxford, England.
[2] Courant, R. and D. Hilbert (1961). Methods of mathematical physics, Vol. 2, Interscience, New York.
[3] Kynch, G.F. (1952). "A theory of sedimentation", Faraday Society Transactions, Vol. 48, pp. 166-176.
[4] Migniot, C. (1989). "Tassement et rhéologie des vases", La Houille Blanche, No. 1 et 2, pp. 11-29 et 95-111.
[5] Sanchez, M. (1992). Application du modèle iso-concentration de tassement dans la simulation des vases étudiées au LCHF, Rapport CFL-92-06, LMG-Université de Nantes.
[6] Sanchez, M. et A. Grovel (1993). "Modélisation du tassement sous poids propre des couches de vase molle et saturée, sur un fond imperméable", La Houille Blanche, No. 1, pp. 29-34.
[7] Tan, T.S., K.Y. Yong, E.C. Leong and S.L. Lee (1990). "Sedimentation of clayey slurry", Journal of Geotechnical Engineering, Vol. 116, No. 6, June pp. 885-898.

Waste Oil Hardening Treatment by Industrial Waste Materials

Kohei Sawa
Department of Civil Engineering, Akashi College of Technology, Japan

Masashi Kamon
Disaster Prevention Research Institute, Kyoto University, Japan

Seishi Tomohisa, Nagahide Naito
Department of Civil Engineering, Akashi College of Technology, Japan

SYNOPSIS : In order to establish a non-hazardous, large-scale and economical method of waste oil treatment, the authors have developed a solidification technique of waste oil absorbed into industrial waste (paper sullage incinerated ash and blast furnace slag) through reaction with cement or lime. The mechanical properties of the solidified waste oil are examined by an unconfined compression test using cylindrical specimens (5 cm in diameter and 10 cm in height). Hardening reaction products of materials were measured by SEM observation. Amount of dissolution of oil was also measured. The potential use of the waste oil solidified with some industrial waste for construction materials is made clear in this study.

1. INTRODUCTION

Due to industrial development, waste oil has greatly increased from machine shops, automobile manufacturing plants and so on. The oil often running into the ocean becomes a serious worldwide problem. The outflow of crude oil during the Gulf War in 1991 and from a stranded tanker in Alaska in 1986 are typical examples of oil spills that have polluted the ocean environment and affected animals and plants (Felzmann, et al., 1992). Through oil leaks from underground storage tanks subsoil and groundwater are contaminated in various parts of the countries (Piotrowski, et al., 1989). Most waste oil can be refined and reused as reclaimed oil or fuel. The waste oil that cannot be refined is generally treated by incineration. However, waste oil with many kinds of mixtures are nonflammable even at high temperature. From the view point of environmental preservation, incineration presents a number of problems because combustion gas contains many kinds of harmful materials such as dust and oxidized sulfur substances.
Industrial wastes have the same problems as waste oil. As they have greatly increased and rarely contain heavy metals and harmful materials, the treatment and effective reuse of such material has to be considered urgently.
The purpose of this study is the development of a treatment method for waste oil using the technique of soil improvement. Some waste oil-absorbing industrial wastes are solidified by hardening agents, and research into the potential use of such wastes with construction materials is presently being done (Sawa, et al., 1992, 1993).

2. SAMPLES

Samples used in this study are oil treated, industrial wastes and hardening agent.

2-1 Oil treated

The following three kinds of oil were used.

(1) Waste oil : This is fuel oil which was recovered from a storage freighter stranded off the Tango Peninsula in Japan in 1990 and contains much sea water. As the volatile matter in waste oil had evaporated before recovery, the waste oil residue remaining was of a viscous consistency.
(2) Crude oil : Arabian blended oil which is low in viscosity and contains little water.
(3) Machine oil : This is lubricating oil or mixing oil. TABLE 1 shows the properties of crude and machine oil used.

TABLE 1 Properties of oil

Kind of oil	Crude oil	Machine oil
Density (15℃)	0.858	0.872
Sulfur (%)	1.41	0.7
Water (%)	0.1	< 0.1
Viscosity (cst)	8.833 (30℃)	9.99 (40℃)

2-2 Industrial wastes

The following four industrial wastes have been selected as oil absorbing material. Specific gravity (Gs) and chemical components of these wastes are shown in TABLE 2. Their grain size distribution curves are presented in Fig.1 and their photographs by a scanning electron microscope (SEM) are shown in Photo 1.
(1) Coal fly ash (CF ash) : This is fine dust in combustion gas of coal which is gathered by an electric dust extractor. Its grains are spherical and their surface is glassy. The grain size of the coal fly ash used is mainly 30-50 μm and the distribution is uniform (uniformity coefficient is 1.7). As the amount of Ca and SO_3 is less than that of normal CF ash, its hardening activity is low.

TABLE 2 Specific gravity (Gs) and chemical component of industrial wastes

Wastes	CF ash	Fine P	PS ash	Slag
Gs	2.20	2.65	2.40	2.87
SiO_2	49.3	74.5	43.6	33.2
Al_2O_3	22.0	19.7	25.3	14.5
CaO	5.4	0	12.7	41.7
TiO_2	1.2	–	1.5	1.4
K_2O	1.2	–	0.4	–
Na_2O	2.0	–	2.9	–
Fe_2O_3	4.4	1.8	1.0	–
MgO	1.1	–	11.0	5.0
SO_3	0.7	0.3	–	2.3

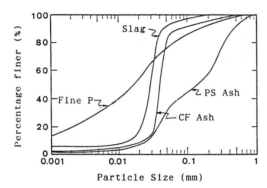

Fig. 1 Grain size distribution curve of industrial wastes

(a) CF ash   (b) Fine P

(c) PS ash   (d) Slag

Photo 1 SEM photographs of industrial wastes

(2) Fine particles of crushed stone (Fine P) : This is a powder which is washed away at the time of production of crushed stone. This waste has a uniform distribution of grain size and has a great number of fine grains among the four types of waste. This waste contains about 75% stable SiO_2 and its hardening activity is the lowest.

(3) Paper sullage incinerated ash (PS ash) : This ash is produced by incineration of paper sullage which is carried out as one part of the paper manufacturing process. It contains comparatively large porous grains (1.0-0.01mm) which have a rough surface. The hardening activity of PS ash is better because it contains the largest amount of Al and Ca.

(4) Blast furnace slag (Slag): This is a by-product which is produced by quick cooling of fused slag from a blast furnace. It has sharp-edged, amorphous and uniformly sized grains under 40 µm in diameter. The high hardening activity (potential hydraulic property) is due to much CaO and SO_3 which are not stable because they solidify without crystallization.

2-3 Hardening agent
The hardening agent used in this study is carbonic aluminate salt (CAS) which consists of cement (80%), $Al_2(SO_4)_3$ (9%), Na_2CO_3 (6%) and CaO (5%). CAS was added in dosages of 3, 6, 9 and 12 % by weight of the industrial wastes which were allowed to absorb oil.

3. METHOD OF EXPERIMENT
The following three series of experiment were conducted.

3-1 Test on oil absorbing capacity of industrial wastes
(1) Compaction test : The industrial wastes which were allowed to absorb various amounts of waste oil were compacted in the mold of 5 cm in diameter and 10cm in height by a rammer of 1.2 kg (compaction energy is 0.55 J/cm^3). The oil absorbing capacity is defined by optimum oil content which corresponds to maximum density.
(2) Absorption test : A piece of blotting paper (2cm $\times$ 10cm) was put on the side of the specimen which was compacted at optimum oil content. After 1 minute the amount of oil absorbed into the blotting paper was measured. The amount of oil on the surface of the specimen was obtained easily.

3-2 Strength test
Specimens were molded in the mold of 5 cm in diameter and 10 cm in height by the following two methods ; (1) compaction method at the optimum oil content with the energy of 0.55 J/cm^3 ; (2) non-compaction method at such state as soft as liquid limit. Method (1) was applied to all four types of industrial waste and (2) to the PS ash alone.
Specimens molded were sealed with resin film and put in a room kept at a constant temperature (20 °C) and humidity (over 95 %) for 1, 3, 7, 28 and 91 days of aging. An unconfined compression test was then carried out.

3-3 Leaching test on dissolved oil
The amount of dissolved oil from treated wastes (PS ash), which absorbed 10% and 30% of machine oil by weight, was examined by the method of Notification No.3 of the Environment Agency in Japan. Specimens were of two types ; one the powder of treated wastes, the other a 1cm molded cubic block. 10g of treated waste was put into 100 ml of distilled water and was well shaken continuously for 2 hours. The amount of oil dissolved into water was measured.

4. OIL ABSORBING CAPACITY OF INDUSTRIAL WASTES

Fig.2 shows the compaction curves of four industrial wastes that absorbed various amounts of waste oil. Optimum oil content of coal fly ash (CF ash), fine particles of crushed stone (Fine P) and blast furnace slag (Slag) is the same at about 25 %, while that of paper sullage incineration ash (PS ash) is about 65 %, or 2.5 times that of the other wastes.

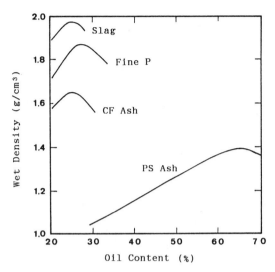

Fig.2 Compaction curves of industrial wastes absorbing waste oil

TABLE 3 shows strength and amount of oil on the surface of the specimen, which was mixed with 3% by weight of the hardening agent (CAS) into industrial waste at optimum oil content and given 7 days of aging. The oil on the surface of the specimen is residual oil which is not absorbed and solidified into waste. As a lesser amount of oil on the surface corresponds to greater oil absorbing capacity of waste, it is clear that the oil absorbing capacity of PS ash is greater than that of the other wastes. As shown in Photo 1, the large, rough and porous grains in PS ash are considered to provide greater oil absorbing capacity.

The strength of the Slag is more than 10.0 MPa at only 3 % by weight of hardening agent and can be considered as lean concrete. PS ash (1.0 MPa), Fine P (0.7 MPa) and CF ash (0.2 MPa) can be used for embankments and back filling.

TABLE 3 Unconfined compression strength and amount of oil on surface of treated wastes

Waste	Strength (MPa)	Amount of oil on surface (g)
CF ash	0.2	0.6
Fine P	0.7	2.2
PS ash	1.0	0.2
Slag	16.0	0.3

5. STRENGTH OF WASTE OIL-ABSORBING SLAG

Fig.3 shows the relation between strength and aging days of waste oil-absorbing Slag at optimum oil content (26%). At the early aging of 1, 3 days, the strength of the specimen untreated with the agent is extremely low, but with only 3% by weight of the agent added it grows in strength to over 10.0 MPa. It is believed that the high strength at the early aging period is caused by the activity of the potential hydraulic property which is the particular hydraulic property of Slag appearing only when it is stimulated by the alkalinity of the hardening agent. The strength developing rate of untreated Slag is the greatest of the five specimens used in the experiment. And with the exception of the specimen treated with 6% of the agent, it can be said that the smaller the amount of agent present, the greater the strength and developing rate of the treated Slag. Consequently, it is believed that the hardening agent in the Slag is effective in promoting high strength in the short term (at early aging period), but does not contribute to sustained strength development over the long term (at long aging period).

Fig.4 shows the results of strength of the Slag which absorbs a mixture of waste oil and water at various ratios ; (waste oil : water) is (30% : 0%), (20% : 8.5%), (10% : 17%) and (0% : 25.5%). Each of the ratios is about optimum oil content and the same consistency. It is clear that the lesser the amount of waste oil and the greater the water content, the greater the strength. This tendency is more evident at long aging periods.

Photo 2 shows the SEM photographs of the 28 days aging specimens of Fig.4. In the specimen with only water, a number of reaction products such

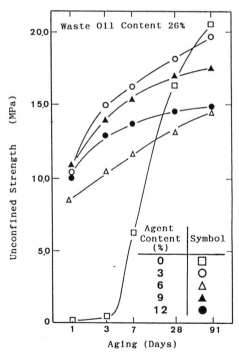

Fig.3 The relation between strength and aging days of Slag absorbing waste oil

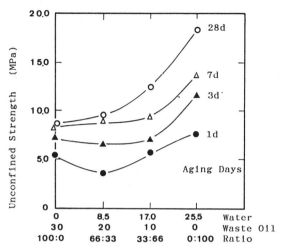

Fig.4 Strength of Slag absorbing waste oil and water

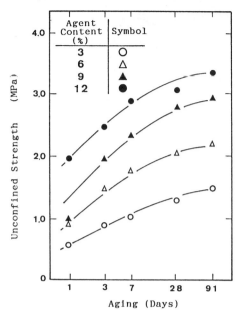

Fig.5 Relation between strength and aging days (waste oil, optimum oil content)

(a) (0 : 25.5)

(b) (10 : 17)

(c) (30 : 0)

(Waste Oil : Water) (%)

Photo 2 SEM photographs of Slag absorbing waste oil and water

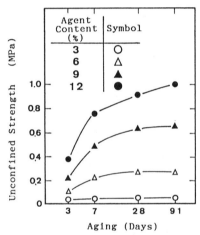

Fig.6 Relation between strength and aging days (waste oil, liquid limit)

as ettringite, $Ca(OH)_2$, $CaCO_3$ and calcium silicic hydrate (CSH) are present, which contribute to greater strength. As the amount of waste oil is increased in the specimen, the Slag becomes covered with oil and the reaction products difficult to find. It means that the long term hydrate reaction cannot progress because of the shortage of water.

6. STRENGTH OF WASTE OIL-ABSORBING PS ASH

Fig.5 shows the relation between unconfined strength and aging days of the waste oil-absorbing PS ash at optimum oil content (65%). It is found that the strength grows considerably after only one day aging and increases further with longer aging period. And the strength with each day of aging increases 0.4-0.8 MPa with every additional 3 % by weight inclusion of the hardening agent. It shows clearly that PS ash has greater hardening activity and the hardening agent works effectively in accordance with the amount of agent mixed in and the number of days aged. As the strength at 7 days of aging is satisfactory for the level of lower subbase course (1.0 MPa) with 3 % by weight inclusion of the hardening agent and the level of upper subbase course (3.0 MPa) with 12% by weight inclusion of the hardening agent, the waste oil-absorbing PS ash is suitable for use as road materials.

The strength of the specimen at liquid limit (oil content is 150 % of waste oil) is shown in

Fig.6. Though the strength grows as the amount of hardening agent increases and the aging periods are long, the strength of all specimens is low (1.0 MPa even with 12% of the agent and 91 days of aging) in comparison with those at the optimum oil content.

7. EFFECT OF THE QUANTITY OF WATER IN OIL

Fig.7 shows the effect of the quantity of water in crude oil absorbed in PS ash at optimum oil content. The mix ratio of oil: water and the mass of the water portion of the oil: water mixture per 100g of PS ash are shown under the horizontal axis in Fig.7. It is seen that the strength at the ratio of (75:25) is lower than that of the other ratios. Compared to the case of only waste oil absorbed in PS ash, the strength of 12 % by weight inclusion of the agent and 91 days of aging is 3.4 MPa (Fig.5). This means that the waste oil used contained some quantities of sea water when it was recovered and consequently hydration occurred even with the PS ash that absorbs only waste oil. As shown in Fig.7, the strength at the ratio of (25:75) is the greatest at under 7 days of aging and the strength at the ratio of (0:100) is the greatest at long aging period. It is clear that both oil and water contribute to the strength at early aging period and that considerably more water is required at long aging period.

Fig.8 shows the relation between the oil: water ratio and the strength of PS ash treated by 9% of hardening agent. As is evident from Fig.8, the strength of the waste oil-absorbing PS ash is greater than that of the crude oil-absorbing PS ash at the ratios of (100:0) and (75:25). This may be explained from the sea water contained in waste oil as mentioned in Fig.7. It can also be seen from Fig.8 that maximum strength is achieved at the ratios of (50:50) or (25:75) and that the strength of the crude oil-absorbing PS ash at these ratios is greater than that of the waste oil-absorbing PS ash. By comparison with Fig.7, it is clear that the strength at the ratio of (0:100) in soft state (contains much oil and water) is less than that at optimum oil content and that the strength at the ratios of (25:75) or (50:50) is not much changed. Therefore, it becomes evident that a small amount of oil in PS ash proves effective for promoting strength and is required in order to sustain strength in the soft state. The residue from evaporation in waste oil may obstruct the growth of strength.

Photo 3 is the SEM photographs of treated PS ash absorbing various ratios of waste oil and water in soft state (12 % of hardening agent and 91 days of aging). PS ash containing only water is covered with many thin needle-like ettringite and CSH reaction products (Photo 3(a)). The reason for a lowering of strength in soft state is considered as follows : as the reaction products are fine, the stiffness of the specimen with large voids remains lower. In the PS ash that absorbed 25% of waste oil, some thick rod-like ettringite and many board-like CAH reaction products can be observed (Photo 3(b)). A good stiffness of grain structure with these products gives higher strength even in soft state. When the amount of waste oil is increased by 50-100 %, ettringite becomes larger crystals but poor in quantity just as with other reaction products (Photo 3(c),(d)). In addition, as the waste oil covers the surface of PS ash and the void of

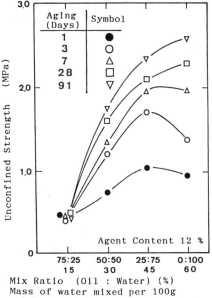

Fig.7 The effect of water in crude oil (Optimum oil content)

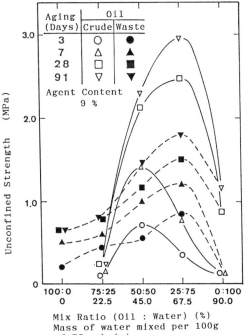

Fig.8 Relation between the oil:water ratio and strength

specimen is large, strength cannot grow significantly greater.

8. DISSOLUTION OF OIL ABSORBED IN PS ASH

Fig.9 shows the results of the leaching test on dissolved oil from treated PS ash. The amount of dissolved oil from the powder type is less than that of the block type (a 1cm cube), because the machine oil used evaporated during

(a) (0 : 100) (b) (25 : 75)
(c) (50 : 50) (d) (100 : 0)
(Waste Oil : Water) (%)

Photo 3 SEM photographs of treated PS ash

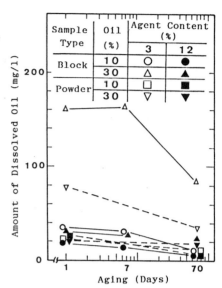

Fig.9 Results of the test on dissolved oil

the aging process. The amount of dissolved oil is decreased as the amount of hardening agency is increased and the aging periods are long.
According to the "law on the treatment and cleaning of industrial wastes" in Japan, the allowable level of oil in industrial wastes to be disposed of in the sea is under 100 mg/l. As shown in Fig.9, the amount of dissolved oil from all treated PS ash samples with 12% of hardening agent is under 40 mg/l. Therefore it will not damage the environment. In the case where 3% of hardening agent, the amount of dissolved oil in 10% of machine oil-absorbing PS ash conforms to the standard. If absorbed oil increases (30% of PS ash) and the hardening agent is poor (3%), about 70 days of aging is required before the amount of dissolved oil reaches the allowable level.

9. CONCLUSIONS
The results obtained in this paper are summarized as follows ;
(1) The oil absorbing capacity of paper sullage incineration ash (PS ash) is greater than that of the other industrial wastes.
(2) It is with blast furnace slag that the treated strength is the greatest among the four industrial wastes that absorbed waste oil. The 28 days aging strength of waste oil-absorbing slag is more than 10.0 MPa at only 3 % of hardening agent and it can be considered as lean concrete.
(3) The strength of slag is caused by the activity of the potential hydraulic property stimulated by the alkalinity of the hardening agents.
(4) In the slag containing only water, many reaction products like ettringite have been observed and contributes to greater strength. As the quantity of waste oil is increased, it covers the surface of slag grains and obstructs hydration. Then the strength is decreased.
(5) The strength of waste oil-absorbing PS ash at optimum oil content increases in proportion to the amount of hardening agent introduced and the number of days aged. It is therefore suitable for use as road materials.
(6) The strength of PS ash absorbing only crude oil is almost zero. But in the case of waste oil, the strength is greater, because the waste oil contains some quantities of sea water.
(7) Both oil and water contribute to the strength of PS ash at early aging period and considerably more water is required at long aging period. The residue from evaporation in waste oil may obstruct the growth of strength.
(8) A small amount of oil in PS ash is effective in promoting strength and sustains strength in soft state. A good stiffness of grain structure with some thick ettringite, which is produced by a small amount of oil, gives greater strength.
(9) The amount of dissolved oil from treated PS ash with 12% of hardening agent is within the allowable level of Japanese standard.
Consequently, the technique described in this paper proved most effective with waste oil treatment.

REFERENCES
Felzmann, H.P, Struck, H. (1992) : Remediation of oil spills, Erdoel Kohle-Erdgas Petrochemie, vereinigt mit Brennstoff-Chemie Vol.45, No.10, pp.371-375.
Piotrowski, D.A, Yost, K.W (1989): Intercept trench technology for remediating waste oil contaminated soil and groundwater; A case study, Proc. 45th Industrial Waste Conf., pp.65-74.
Sawa, K., Kamon, M., Tomohisa, S., Naito, N. (1992) : Studies on the waste oil treatment by the industrial waste materials, Memoirs of the Akashi College of Technology, No.34, pp.35-45, (in Japanese).
Sawa, K., Kamon, M., Tomohisa, S., Naito, N. (1993) : Studies on the waste oil treatment by the paper sullage ash, Memoirs of the Akashi College of Technology, No.35, pp.27-35, (in Japanese).

Using Glacial Till as Liner Material for a Waste Disposal

Markus Sjöholm, Tapio Strandberg
National Board of Waters and the Environment, Water and Environment Research Institute, Helsinki, Finland

About 60% of the soil covering the land area in Finland is glacial till and as the clay formations with a low permeability value are situated only around the coast, the use of clay as liner material is a costly alternative. Many international standards require the permeability of the subbase of waste disposals to be less than $k \leq 10^{-9}$ m/s. As the natural k-value of glacial till on average does not fulfill the required $k \leq 10^{-9}$ m/s, research has been done how the k-value could be decreased with additives like bentonite. Adding bentonite to the glacial till, the permeability of the till can be reduced to approximately $k = 10^{-9} - 10^{-11}$ m/s. The amount of bentonite needed is considerably less than the amount needed in sand-bentonite mixes, due to the higher percentage of fines in till compared to sand. The compactibility of till-bentonite is also considerably better than that of a sand-bentonite mixture. Bentonite content in the till-bentonite mixtures investigated range between 1 - 11 %. The laboratory testing has been mainly measurements of the permeability in triaxial cells. Four different till-bentonite mixtures have been investigated. The swelling pressure of till-bentonite mixes has also been investigated. The swelling pressure as well as the permeability tests indicate that the permeability tests should be continued for at least the time at which the bentonite is still swelling. When using high hydraulic gradients during the permeability tests, the samples showed signs of internal erosion.

Introduction

The contamination of the ground water due landfill seepage has been acknowledged as a problem as well in Finland as in other countries. In previous years the design value for the permeability for subbases for landfills in Finland has been $k < 10^{-7}$ m/s. Many international standards require the permeability of the subbase of waste disposals to be less than $k \leq 10^{-9}$ m/s. As the clay formations in Finland, which satisfy this condition, are situated only around the coast and the majority of the potential new landfill areas are situated on till formations in inner parts of Finland, the use of clay as liner material is a costly alternative.

About 60 % of the soil covering the land area in Finland is till [Okko, 1964]. A typical finnish glacial till contains approximately 20-40 % fines. Clayey till containing a large amount of fines covers only a few percents of the land area in Finland. The natural permeability of glacial till usually is fairly low, the k-value range being between $k = 10^{-5} - 10^{-8}$ m/s. The till has frequently been used in dam construction as core material. The maximum dry densities measured at three dam construction sites were on average $\gamma_d = 19.23$ kN/m^3, relative density $D_r = 93$ % of the modified Proctor density. The permeability at the dam construction sites varied between $k = 3.5 \cdot 10^{-8}$ m/s [Loukola, 1985].

As glacial till formations seldom are used for groundwater acquisition and the bearing capacity of glacial till areas in general are good a demand exists for developing methods to decrease the permeability of the till to satisfy the requirements.

Laboratory Investigations

The hydraulic conductivities of four different glacial tills have been investigated in the laboratory. The selected tills were chosen on a basis of availability, i.e. the investigated soils should represent typical finnish glacial tills. The grain size distribution of the samples have been defined by dry sieving. The fines for samples no. 1, 3 and 4 have been analyzed in the SediGraph 5100 particle size analyzer. The grain size distribution of the different samples are presented in figure 1. Samples no. 1 and 2 are poorly graded sandy tills with a low percentage of fines. Sample no. 1 contains 8% fines and sample no.2 12 % fines. Sample no.3 is a

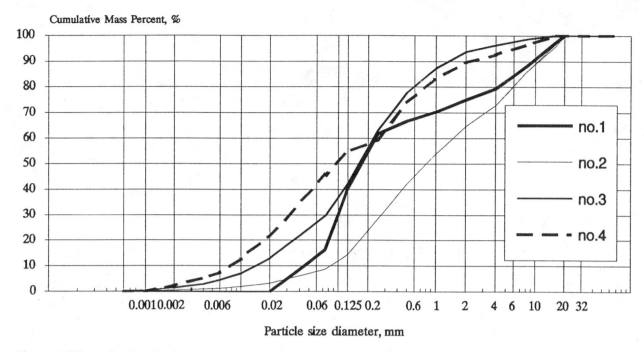

Figure 1. The grain size distribution of the investigated glacial tills.

sandy till, which contains 27 % fines. Sample no.4 contains 40 % fines.

The permeability tests were carried out in a flexible wall permeameter. Back-pressure was not used in the tests. Before loading the permeameters the samples were homogenized and compacted in a Proctor cylinder to Standard Proctor compactness. The natural hydraulic conductivity of the glacial tills was reduced by adding bentonite clays. The bentonites used in the testing program were a Wyoming sodium bentonite and a sodium activated calcium bentonite from Greece. Later on the bentonites are referred to as Na (Wyoming bentonite) and Ca (Greek bentonite).

Before compaction stones with a diameter $\phi > 20$ mm were removed. After compaction the sample was extruded from the mould and 20 mm from both ends was removed so that the compaction would be as even as possible. The lowest permeabilities for liner materials is achieved when the moulding water content is 1 - 2 % higher than optimum [Mitchell et al, 1965].

The mixtures have been compacted at w = 10 % water content, which is 1-2 % bigger than the optimum water content defined for the glacial tills without bentonite. Due to this procedure the lowest theoretical permeabilities have not necessarily been achieved. The bentonite content is expressed as bentonite at natural water content (w = 12%) per dry weight of glacial till, B% = B/T·100. For till no.1 + 5 % sodium bentonite the effect of moulding water content was investigated.

The permeability of the samples were measured for 14 - 45 days. The amount of percolation water has been measured by weighing the water collection bottle. The evaporation from the collection bottles was measured and added to the weight. The confining pressure at the bottom of the sample, the influent end, was maintained at $\sigma_3' < 20$ kPa. The hydraulic gradient during the tests was for experimental reasons varied between i = 8 - 236. The large hydraulic gradient was used to find out the possibilities for short time analysis and to investigate the susceptibility for internal erosion. Using large gradients the confining pressure at the top of the sample, the effluent end, was extensively larger than at the bottom of the sample. The confining pressure at the effluent end while the gradient was i = 236 was $\sigma_3' = 150$ kPa. A large confining pressure closes pores in the sample and consolidates the sample [Bryant and Bodocsi, 1986]. This leads to a lower permeability. However, large gradients might also lead to internal erosion in the sample.

According to the ASTM standard for flexible wall permeameters [ASTM, D5084 - 90] back pressure should be used, to saturate the sample and to prevent the air in the solution to dissolve during the pressure drop. Back pressure has not been used in this investigation.

The required time for the bentonite to swell in a till-bentonite mixture was defined by a swelling test in a Proctor mould. A till-bentonite mixture consisting of till no.1 and 11 % sodium bentonite was compacted into the Proctor mould in the Standard Proctor procedure. The Proctor mould was placed in a triaxial cell with drained ends, so that water needed for swelling was available for the till-bentonite mixture. The swelling pressure was continuously measured by computer.

Results of the Laboratory Investigations

The purpose of the laboratory investigations was to:

- define the effect of moulding water content upon till-bentonite permeabilities,
- define the time required for the test to reach its minimum permeability,
- define the swelling pressure caused by the bentonite,
- define the bentonite content in till-bentonite mixtures to satisfy the requirements for the subbase of a landfill,
- examine the effect of using large gradients in the permeability tests.

The permeability for till-bentonite mixtures as a function of the moulding water content was investigated for till no.2 (fig. 1) with a 5 % sodium bentonite content. The optimum water content of the mixture was w = 9.0 %. The permeability tests were carried out at water contents of w = 5.5 %, w = 7.6 %, w = 12.0 % and 13.5 %. The results proved that the permeability decreases when compacted wet of optimum. However, the laboratory work showed that the water content cannot be increased without limits, due to loss of internal stability in the sample. The results are shown in figure 3.

The time required for the bentonite to swell and for the mixture to reach the long term permeability, which it has in the final liner construction, was defined by a swelling test in Proctor mould.

The maximum swelling pressure was achieved after 150 h, i.e. 6.2 days after the swelling test was started. The development of the swelling pressure as a function of time is shown in figure 3.

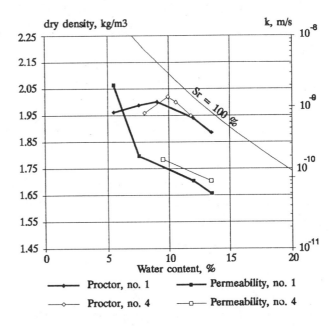

Figure 2. The optimum water content (Standard Proctor) and the effect of the moulding water content on the permeability for material no. 1 + 5 % sodium bentonite and material no.4 + 4 % sodium bentonite .

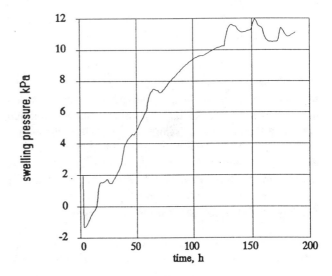

Figure 3. The swelling pressure of a till-bentonite mixture as a function of time. B/T = 11 %.

TABLE 1. The Permeability of the Examined Glacial Tills Depending on the Bentonite Type and Content (figure 1.).

Ballast + type of bentonite	Bentonite content, Bentonite (natural water content) / dry till							
	0 %	1 %	3 %	4 %	5 %	7 %	9 %	11 %
no.1 + Na	$1.1 \cdot 10^{-5}$				$8 \cdot 10^{-11}$	$5 \cdot 10^{-11}$	$3 \cdot 10^{-11}$	$2 \cdot 10^{-11}$
no.2 + Na	$1.2 \cdot 10^{-5}$				$2.5 \cdot 10^{-10}$		$8 \cdot 10^{-10}$	
no.2 + Ca	$1.2 \cdot 10^{-5}$	$4 \cdot 10^{-7}$	$8 \cdot 10^{-9}$		$1.5 \cdot 10^{-9}$	$1.5 \cdot 10^{-10}$		
no.3 + Ca	$2.5 \cdot 10^{-7}$	$1 \cdot 10^{-8}$	$8 \cdot 10^{-10}$	$4 \cdot 10^{-11}$	$3 \cdot 10^{-11}$			
no.4 + Ca	$2 \cdot 10^{-7}$				$1.8 \cdot 10^{-10}$			

Some of the tests showed large permeabilities in the beginning of the test, which can be explained by the fact that the bentonite at this moment has not swollen to its full capacity and therefore does not fill the pores effectively. The rapid decrease in permeability during the first week of the tests and the time required for stabilizing the swelling pressure (fig 3.) supports this. The hydraulic gradient at this stage should be kept $i < 10$ to prevent piping and internal erosion. The permeability value reported in the results is obtained by drawing a line for the stabilized measurements to the permeability axis. This procedure yields a conservative value for the permeability, but it also takes into account the possible decrease in permeability, which the lack of back pressure might produce. To find out the internal erosion in the samples the gradient was increased up to $i = 236$ after 25 days. It seems that the internal erosion in the well graded samples is less than in the poorly graded with a fairly large bentonite content. This tendency can easily be noticed in figure 2. When the gradient was returned to $i = 18$ the permeability was considerably higher than before using the large gradient of $i = 236$. This strongly indicates internal erosion.

A permeability factor less than $k \leq 1 \cdot 10^{-9}$ m/s for glacial tills can according to the test be reached with bentonite contents of 3 – 6 %, depending on the grain size distribution of the till, its natural permeability and type of bentonite used. The permeability factor of the examined samples are presented in table 1 and in figure 4. In figure 4 the permeability of the till samples are presented as a function of the bentonite content.

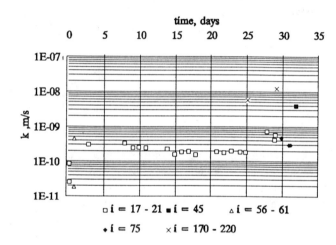

Figure 4. The permeability as a function of time and gradient. Glacial till no.2 (fig. 1) + 5 % sodium bentonite.

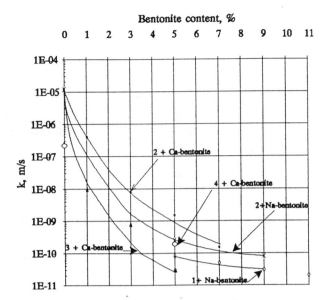

Figure 5. The permeability of glacial till as a function of the bentonite content.

Conclusion and Summary

According to laboratory results even small amounts of bentonite decrease the permeability significantly. Liners with a permeability less than $k < 10^{-9}$ m/s can be constructed with a bentonite content of 3 – 6 %. The natural permeability of the till does not necessarily affect the amount of bentonite needed for achieving the requirements.

The minimum permeability of till-bentonite mixtures is achieved at a water content several percentage unit greater than optimum.

The swelling pressure of a till-bentonite mixture with a bentonite content of 11% was 11 kPa. This equals a protective layer of about 0.7 m loosely compacted soil, which also works as protection against the frost action. If the protective layer is too thin the layer will swell and lose its bearing capacity.

The time needed for the till-bentonite mixture to swell was 150 h, i.e. 6.2 days. Therefore to get reliable permeability results the test should be continued for at least 14 days.

Internal erosion increases the permeability significantly especially for poorly graded tills with high bentonite content. Therefore if the construction is expected to be subject to large hydraulic gradients, the risk for internal erosion and piping has to be evaluated.

References

ASTM D 5084 – 90. Standard Test Method for Measurement of Hydraulic Conductivity of Saturated Porous Materials Using Flexible Wall Permeameter

Bryant, J. and Bodocsi, A. 1986. Precision and reliability of laboratory permeability measurements. EPA Contract No. 68-03-3210-03.

Loukola, E. 1985. Pehmeikölle perustettavan Taasian maapadon jännitykset ja muodonmuutokset. (The strain and deformation in soft clay caused by Taasia earth dam.) National Board of Waters. Helsinki.

Mitchell, J.K., Hooper, D.R., and Campanella, R.G., Permeability of Compacted Clay, Journal of the Soil Mechanics and Foundation Division, ASCE, Vol. 91, No. SM 4, 1965, p. 41-65.

Okko, V. 1964. Maaperä. Rankama (ed.) Suomen Geologia. (Subsoil. Finnish Geology)

Surfactant Enhanced Electrokinetic Remediation of NAPLs in Soils

Mohd R. Taha, Yalcin B. Acar
Dept. of Civil and Environmental Engineering, Louisiana State University, Baton Rouge, Louisiana, USA

Robert J. Gale
Dept. of Chemistry, Louisiana State University, Baton Rouge, Louisiana, USA

Mark E. Zappi
US Army Waterways Experiment Station, Vicksburg, Mississippi, USA

Abstract

Surfactant enhanced electrokinetic remediation of soils is a novel process that utilizes aqueous solutions to create a mobile phase, i.e. micelles, in which the contaminants partition. Preliminary results using hexachlorobutadiene as the contaminant and sodium dodecylsulphate as the surfactant indicated that it is possible to transport the resulting micelles under an electrical field. This technique is promising in remediating non-polar and poorly soluble organic contaminants such as Non Aqueous Phase Liquids (NAPLs) from soils. This paper discusses some of the basic aspects of the surfactant technology together with the desirable characteristics of surfactants in electrokinetic remediation of NAPLs in soils. The requirements for selecting a successful surfactant system are also discussed.

Introduction

Surface active agents or surfactants are an interesting class of materials. Their dual characteristics, hydrophilic (water-loving) and hydrophobic (water hating), make them quite useful. Surfactants play an important role in many processes ranging from the very mundane (soaps for washing clothes and dishes) to the very sophisticated (microelectronics). In civil engineering, additives which give adhesion to wet aggregate in bitumen has been in use for many years. Additives are also used in concrete and mortar as plasticizers, in foaming of concrete or gypsum, in foaming of urea-formaldehyde resins for cavity-wall insulations, etc. These additives are indeed surfactants [1].

Electrokinetic (EK) remediation is one of the decontamination methods that have gained much interest in recent years. The method uses DC currents in the order of mA/cm^2 of processing and its cost efficiency in removing inorganic species from fine grained species have been demonstrated by bench scale tests [2,3,4] and limited field studies [5]. Most data on removal of organic chemicals by EK remediation is associated with miscible species such as phenol [6], acetic acid [7] or BTEX compounds [8] below their solubility limit in which these species become polar. Electrical migration is demonstrated to be the most significant transport process in EK remediation [9]. Electromigration of species requires mobile ions of either negative or positive charge in their transport to respective electrodes upon the application of the current. Therefore, it is hypothesized that non-polar species will not be transported effectively and experiments with hexachlorobutadiene (HCB) in kaolinite with unenhanced EK remediation has shown to be unsuccessful at concentrations of 10 µg/g to 1000 µg/g to emphasize this postulate. However, recent studies by the LSU group have demonstrated that hexachlorobutadiene moved only when a surfactant was used in EK remediation[9, 10]. The demand to remediate fine grained soils contaminated with non-polar organics prompts the need to assess the feasibility and efficiency of using surfactant enhanced (micellar) EK remediation. This paper presents the various fundamental aspects of surfactant enhanced EK remediation of NAPLs in soils.

Micelle Formation

An amphihilic (hydrophobic-hydrophilic) molecule of surfactant consists of a polar hydrophilic head group and a hydrophobic tail (Figure 1). It can be grouped into anionic, nonionic, cationic and zwitterionic (double ionic or amphoteric). These molecules will orient themselves in such a way as to place the water insoluble hydrophobic group (non-polar) in a hydrophobic environment. The water soluble hydrophilic part (polar) is exposed to the polar environment, that is the aqueous phase. Therefore, when the surfactant molecules are exposed to non-polar particles, as in NAPLs, the result is the formation of an adsorbed layer at the interface and the agglomeration of surfactant molecules into micelles above the critical micelle concentration (CMC). Figure 2 shows this action in which a group of organic molecules are captured by the hydrophobic tail and result in the formation of micelles. Below the CMC, the number of surfactant molecules is insufficient to form micelles. Surfactants are present in two possible states in aqueous solutions at concentrations above the CMC: as an unaggregated monomer or as constituents of micelles into which the organic pollutant will dissolve or solubilize. As the total concentration of surfactant is increased beyond the CMC, almost all of the added surfactant goes to increase the micelle concentration [11].

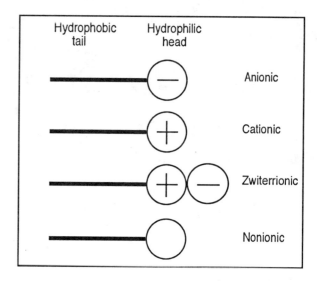

Figure 1 Physical representation of a surfactant molecule and its type

Ionization and Solubilization

The mass transport of contaminants in EK processes involves: (1) ion migration; (2) advection due to electric current (electro-osmosis); (3) advection due to hydraulic gradient ; and (4) diffusion. Ion migration plays a marked role in this transport process [9]. Therefore, it is expected that only ionic species will migrate to the respective electrodes and be transported under the application of an electric current.

NAPLs are non-polar species and ionic migration will certainly be an unavailable transport process. Although there are three other transport processes, it has been shown that there is very little or no movement of species in experiments conducted in HCB spiked Georgia kaolinite [10].

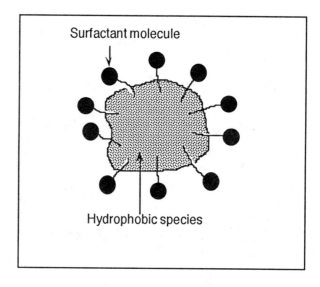

Figure 2 Surfactant molecules encapsulating a hydrophobic globule thus forming a micelle

Therefore, unenhanced EK processes are not expected to be effective in transporting the contaminants. When ionic surfactants are mixed with the non-polar NAPLs, they form charged micelles. These micelles are ionic entities and they should move to the respective electrodes under an electric field to be collected for treatment and/or disposal.

Another well known property of surfactants is their ability to solubilize an otherwise insoluble species in water. Insolubility results from very high interfacial tension (IFT) that retain the insoluble contaminants within the soil matrix. The application of surfactants will decrease the contaminant/water IFT and hence it will increase the solubility of the insoluble compounds.

Many organics, especially NAPLs are either insoluble or have very low solubilities in water. Solubilization by surfactants make it possible for NAPLs to go into solution and the momentum provided by the moving water is expected to move the contaminants towards the electrodes when the driving mechanism is hydraulic gradient or electro-osmotic advection. However, the main purpose of solubilization by surfactants is to prevent precipitation. A known problem associated with EK remediation of metallic species is the precipitation of the contaminants around the cathode. This may be explained by the high concentrations of the contaminants encountered in the vicinity of the cathode as they migrate from anode to the cathode. This results in their ion product (the product of the concentrations of the dissolved ions each raised to its proper power) exceeding the solubility product, K_{sp}, and hence producing a supersaturated solution which eventually will cause precipitation. Precipitation blocks ions from reaching their designated electrodes and surfactants are expected to increase solubility to avoid such precipitation.

Selection of Surfactants

EK is a relatively new technology. There is a wealth of information on the selection of surfactants for many conventional processes such as separation, floatation, dispersion etc. However, the selection criteria for surfactant enhanced EK is not as evident and such criteria need to be developed. In selection of a surfactant for EK remediation, some important desirable characteristics would be: (1) high solubilization capacity for the organic pollutant; (2) ability to form charged micelles to facilitate flow of ions under electrical field; (3) low adsorption capacity with the clay minerals; (4) low monomer concentration, so that little surfactant is wasted; (5) low toxicity; (6) minimal phase separation problems; and (7) cost efficiency.

One desirable characteristic of the surfactant to be used in EK remediation is the need to use a long hydrocarbon chain, since this would result in high solubilization and low monomer concentration. While the resulting large micelle is considered to be an advantage in processes such as membrane separation, larger size would be a disadvantage in EK remediation because migration of the micelles towards the electrode could be impeded due to the small pore space in soils, especially clays. Anionic surfactants are restricted to hydrocarbon chain lengths of about 12 carbons or less if the process is to be applied at room temperature. The Kraft temperature (the temperature at which micelles are formed at specific concentration) is above room temperature for longer hydrophobic groups, so the surfactant would precipitate to make EK ineffective. With cationic surfactants the positively charged micelles will be absorbed onto the negatively charged clay particles hindering their removal. Furthermore, cationic surfactants are more toxic than anionics due to their higher molecular weight. Nonionic surfactants form large micelles, and can have high solubilization capacities and have low monomer concentration in micellar solutions. It has also been found that stable macroemulsions are possible when nonionic surfactants are stirred with some of the organic pollutant [11]. However, solubilization capacities of nonionics are not very high compared to anionics and cationics because of their high molecular weight. Also ion migration, which is the main transport process in EK is not expected to be predominant for uncharged species as a result of non-ionic surfactant systems.

The cost of chemicals involved is another issue that may render surfactant enhanced EK remediation unattractive. The primary features of EK are cost-effective electrical production and transport of species in remediation. The use of surfactant for purely solubilization purposes is not economical especially if metallic species are involved. In this case regular acid solutions such as hydrochloric acid may be used. However, hydrochloric acid is relatively more expensive than the acid produced by the electrical currents. Further discussion on the use of acids in enhancing EK processes is beyond the scope of this presentation.

Therefore, the formulation of an effective surfactant system for EK remediation generally necessitates one or more surfactant species, cosurfactants, inorganic salts and/or bases, and polymers. The primary surfactants, cosurfactants, and inorganic salts should be blended to optimize the phase behavior. Other inorganic salts and bases must be added to reduce surfactant adsorption on the minerals. Polymer inclusions must be considered to improve performance [12].

Tests Results

Some test results from earlier experiments conducted at LSU are discussed in this section. In this study, the hydrocarbon is 1,3-hexachlorobutadiene (HCB), whose solubility in water is 5 ppm and it has also low volatility which complicates its removal from soils by pump-and-treat and vapor extraction technologies, respectively [13]. Two surfactants used were cetyltrimethylammonium chloride (CTAC), a cationic surfactant and sodium dodecylsulphate (SDS), an anionic surfactant. Cylindrical kaolinite samples measuring 10.5 cm in diameter and 11.5 cm in length were used in experiments [10].

As expected, the EK removal of HCB using CTAC system was ineffective. High adsorption of the positively charged micelles on to the negatively charged clay surfaces may have retarded the transport. Further investigation is necessary to delineate the reasons.

The transport of HCB using SDS system is shown in Figure 3. The horizontal axis represents the normalized distance from the anode with 0 depicting the anode position and 1.0 depicting the cathode position. The vertical axis represents the final concentration of HCB with the horizontal line along 1000 µg/g indicating the initial concentration at the start of the tests. At 8 mM of SDS the system is ineffective in moving HCB. The CMC of SDS is around 8 mM and at this concentration, micelles are beginning to form and therefore little transport took place. At 20 mM the results indicated that the hydrocarbon is partially transported by the micelles. The hydrocarbon reduced by about 75% at around the anode region in sections 1 to 3 and increase in concentration of around 100% to 150% in regions 4 and 5 . In the remaining sections, the hydrocarbon concentrations remain the same as the electro-osmotic flow ceases. At the cathode, diffusion together with transport towards the anode is expected to result in a drop of concentration. Test 1 differs from test 2 only in terms of the processing time which are 300 hours and 400 hours respectively. These preliminary tests showed that while surfactant enhanced EK remediation seems feasible, more work needs to be carried out to find the optimum system that would efficiently transport hydrocarbons from fine grained deposits.

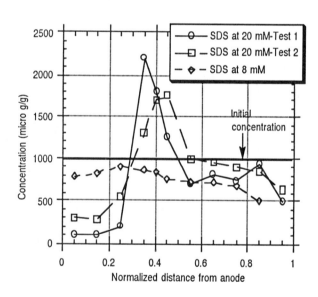

Figure 3 HCB removal using SDS in the anode compartment [10]

One reason that might explain the discontinuance of flow in the experiments is clogging as the positively charged micelles flow from the anode to cathode and the increase in concentration may lead to precipitation. Another possible reason is that the transport of anionic micelles towards the anode may overcome the electro-osmotic flow. The application of the anionic surfactant at the cathode and use of co-surfactants for greater solubilization was recommended. Further studies are in progress.

Conclusions

The basic aspects of surfactant enhanced electrokinetic remediation are described in this paper. The most important process in the proposed technique is the capture of the non-polar organic globules by the surfactant molecule which result in the formation of micelles to ionize and solubilize the contaminant. Some desirable characteristics of surfactants for this process have also been discussed. Test results indicate that the process is capable of transporting hexachlorobutadiene in kaolinite when an anionic surfactant, sodium dodecylsulphate, is used above the critical micelle concentration. However, the complete transport across the specimen was not possible within the specified testing time. It is essential to investigate the fundamentals and the efficiency of the technique as it may provide a technically feasible and cost effective solution to recover non -polar, non-volatile hydrocarbons from fine grained deposits.

References

[1] Karsa, D.R. (1987). " Industrial applications of surfactants-An overview," *Industrial Applications of Surfactants*, Karsa, D.R. (eds), Special Publication No. 59, The Royal Society of Chemistry, London, pp. 1-23.

[2] Acar, Y.B. (1992). "Electrokinetic soil processing: A review of the state of the art," *Geotechnical Special Publication No 30*, ASCE, pp. 1420-1432.

[3] Acar, Y.B. and Alshawabkeh, A.N. (1994). "Modelling conduction phenomena in soils under an electric current," *XIII International Conference on Soil Mechanics and Foundation Engineering*, New Delhi, India, January 1994 (in press).

[4] Acar Y.B., Alshawabkeh, A.N. and Gale, R.J. (1993). "Fundamentals of extracting species from soils by electrokinetics," *Waste Management*, Vol. 13, Pergamon Press, London, pp. 141-151.

[5] Lageman, R., Pool, W., and Seffinga, G. (1989). " Electro-reclamation:Theory and practice, *Chemistry and Industry* , Vol 10, Society of Chemical Industry, London, pp. 585-590.

[6] Acar, Y.B., Li, H., and Gale, R.J. (1992). "Phenol removal from kaolinite by electrokinetics," *Journal of Geotechnical Engineering*, ASCE, V. 118, No. 11, pp. 1837-1852.

[7] Shapiro, A.P., and Probstein, R. F. (1993). "Removal of contaminants saturated clay by electroosmosis," *Environmental Science and Technology*, Vol. 27, No. 2, pp. 283-291.

[8] Bruel, C.J., Segall, B.A., and Walsh, H.T. (1990). "Electroosmotic removal of gasoline hydrocarbons and TCE from clay," *Journal of Environmental Engineering*, ASCE, V. 118 (1), pp 84-100.

[9] Acar Y.B., and Alshawabkeh, A.N. (1993). "Principles of Electrokinetic Remediation," *Environmental Science and Technology*, Vol. 27, No. 12, December (in Press).

[10] Tran, N.K., and Gale, R.G. (1992). "Micellar electrokinetic separation of 1,3 hexachlorobutadiene from kaolinite," *Project Report:Dow-University Grant 1992*, Dept. of Chem., Louisiana State University, Baton Rouge.

[11] Scamehorn, J.F., and Harwell, J.H. (1988). "Surfactant-based treatment of aqueous process streams," in *Surfactants in Chemical/Process Engineering* (Wasan, Ginn, and Shah, eds.), Surfactant Science Series Vol. 28, Marcel Dekker Inc, New York, pp. 77-125.

[12] Schenewerk, P.A., and Wolcott, J.M. (1992). "The utilization of surfactant technology for remediation of the PPI site", in *Evaluation of Current Remedy and Research into Advanced Technologies for Remediation of the Petro Processor Site*, Year 02 Report, College of Engineering, Louisiana State University, Baton Rouge, LA.

[13] Acar, Y.B., Taha, M.R., and Constant, W.D. (1993). "Remedial measures and evaluation of alternatives at a site contaminated with Non-Aqueous Phase Liquids," *International Conference on The Environment and Geotechnics*, ENPC, Paris, April, V.1, pp. 319-326.

Flow Behavior During Electrokinetic Soil Decontamination

S. Thevanayagam, Asst. Prof., J. Wang, Graduate Student
Polytechnic University, New York, USA

Abstract: The dynamics during the electrokinetic process in clayey soils continuously alter the soil conditions and the electrokinetic flow efficiency decreases with time. Through simple experiments, it is shown that, among other factors, even in open electrode configuration, the non-uniform conductivity distribution, and the internal negative pore pressures and hydraulic gradients induced due to this non-uniformity significantly affect the electrokinetic flow behavior.

Introduction

Since its inception as a way of dewatering to improve shear strength of fine-grained soils by Casagrande, electro-osmosis (E-O) has been applied in several geotechnical projects (e.g. Casagrande 1983). Past geotechnical applications have been primarily designed based on empirical relationships based on soil parameters determined from short-term laboratory electro-osmotic tests. The applications were justified, as such cases were often dictated not by technological merits but by geotechnical site conditions which did not suit for more traditional ground improvement technologies. There was not any critical demand to mobilize the full potential of electro-osmotic dewatering capacity.

However, with the ever increasing demand for a fail-safe *insitu* decontamination technology, geo-environmental engineers have turned to electrokinetic soil processing (e.g. Mitchell 1986). The idea was that instead of a closed electrode system utilized in consolidation applications, use of an open electrode system with water available at the anode would help clean the soil due to E-O driven flow (Segall et al 1980). The aim of the ensuing research in the past was to first study the feasibility of this technology as an effective means (e.g. EPA 1986, Leggeman 1989, Hamed et al 1991, Pamukcu and Wittle 1992, Acar 1992, Segall and Bruell 1992, Probstein and Hicks 1993) to decontaminate soils using an open electrode system.

A careful examination of the available data to date indicates: (a) The flow rate and efficiency, k_i, (defined as flow volume per ampere hour) and the contaminant removal rate rapidly falls down within a relatively short time after application of an electrical gradient forcing a longer duration of soil processing, and (b) While the flow rate decreases with time, the power required to maintain the current flow through the soil increases. The combined effect is increase in cost and power consumption. More importantly the tail-end portion of the contaminant removal may practically prove to be a difficult task and may consume most of the total power. In some cases, the removal of the tail-end portion may become impractical.

The past research helped in identifying the

changes in soil conditions such as pH, concentration distributions of contaminants, soil and pore fluid conductivities, atterberg limits, and water content distributions, that take place during the electrokinetic process (e.g. Lockhart 1983, Acar et al 1988, Hamed et al 1991, Segall et al 1992, Pamukcu and Wittle 1992, Mitchell and Yeung 1991). Researchers have recognized that the dynamics during the processing continuously alters the soil conditions and therefore invalidates the use of soil parameters such as laboratory measured short-term performance and electro-osmotic permeability data to reliably design a decontamination system as well as predict feasibility in a given situation. At present, reliable analytical evaluation of the feasibility of electrokinetic soil processing and cost-effectively designing such a system are still difficult tasks without a test pad study. Even then the choices are few that could be utilized to improve the process to obtain a desired level of performance. Recently attempts have also been made to improve the flow efficiency by controlling the pH of pore fluid at the electrodes (Acar 1992 and co-workers) and/or using different types of fluids at the anode (Shapiro and Probstein 1993).

This paper presents some recent results from laboratory and analytical studies, that shed some light on major causes of loss of flow efficiency with time and presents a simplified mechanistic understanding of the overall flow process under electrokinetic process. It is shown that the nonuniform changes in conductivity coupled with internal changes in pore pressures during the process significantly affect the flow behavior. These results can potentially help identify remedial measures to increase or maintain the flow efficiency at acceptable limits.

Background

First it is considered that the *redistribution* of *ion* concentrations is the key to the changes in electro-kinetic flow efficiency. This relies on the premises that the cations are the primary charge carrying ionic conductors that predominantly control the E-O process. Higher the electric conductivity (σ), lower is the resistance to flow of cations (smaller is the drag force exerted on the fluid) and lower is the flow volume that is dragged by the charges.

Based on available data in the literature, during the E-O process there are, among many, three important processes that take place simultaneously and significantly affect the overall flow rate during electrokinetic process. (a) The cations near the cathode are removed by deposition of the cation at the cathode. An equal amount of cations are produced at the anode by anode reaction (H^+ in the case of graphite electrode), (b) The total cation production near the anode is dependent on the amount of cations removed at the cathode as well as excess OH^- production at the cathode. Due to OH^- production at cathode, an equal amount of excess cations (H^+) are produced at the anode in addition to what is produced in (a), and (c) The OH^- and the existing cation in the soil may react and precipitate. These processes have the following implications:

(i) Due to process (a) the cations near the anode become predominantly H^+ whereas the changes in cation type near the cathode is not appreciable. Since H^+ ions are more mobile than any other cation σ in the anode region become larger than the initial σ_o whereas σ near the cathode remains only slightly affected, and

(ii) Due to process (b) the cation concentration near anode region is even further increased due to excess H^+ ions, whereas, near the cathode, there is an increase in OH^- by an equal amount. Due to process (c) the cation concentration may even be reduced near the cathode. σ near the anode will be increased whereas it remains nearly same or reduced near the cathode. Indeed such variations in σ have been observed in several experiments (Hamed et al 1991, Segall and Bruell 1992).

The combined effects of (i) and (ii) are as follows: (1) The double layer thickness of the soil near the anode shrink whereas it is enlarged near the cathode, and (2) Due to high σ near the anode the voltage gradient near the anode decreases from initial value whereas it increases near the cathode region.

Due to (1) and (2), for a constant current (I), the flow rate, solely due to E-O, near the anode will decrease whereas it will increase or nearly remain the same near the cathode. Since the soil elements are connected in series from anode to cathode, the above consequences in the anode and cathode regions begin to set off a chain interaction between the two regions as follows.

Due to decrease in the flow rate in the anode region, there will not be sufficient supply of water towards the cathode to feed the E-O flow capacity in the cathode region. The cathode region will begin to drag additional fluid (in addition to what comes from the anode region) from the soil in the regions away from it. i.e Negative pore pressures are induced in the soil between the anode and cathode.

Even though there is no external hydraulic gradient applied to the soil, the negative pore pressures thus induced create positive and negative hydraulic gradients in the anode and cathode regions, respectively. The flow regime in the two regions is now not only due to electrical gradients (flow due to chemical diffusion is neglected herein for simplicity) but also due to internal hydraulic gradients. These two regions now begin to interact due to newly created hydraulic gradients.

The positive hydraulic gradient induced in the anode region would supplement the weakened electro-osmotic flow in that region. The negative hydraulic gradient in the cathode region will weaken the electro-osmotic flow in that region. The net effect would be that the weakened flow in the anode region is enhanced whereas the net flow rate in the cathode region is retarded. The degree of enhancement in the anode region may not be sufficient to bring the flow rate in that region to the initial rate that prevailed in the initial stages of the process, when there was no differences in the soil properties near the anode and cathode. The overall net effect of all the above interactions is a net decrease in the flow rate through the soil.

As a subsequent consequence the soil between anode and cathode may highly consolidate and develop cracks. Furthermore, the negative pore pressures may reach very high magnitudes leading to cavitation and possibly flow stoppage.

With time, the effects of (a) and (b) will accumulate; the zone of H^+ will expand and migrate towards cathode as well as its concentration will increase leading to further skewed σ distribution and further reduction in flow rate.

In the following sections these effects are exemplified through simple experiments and simplified numerical analyses.

Experiments

Fig.1a shows a conventional setup for 1-D E-O test whereas Fig.1b is a slightly modified setup developed to crudely evaluate the hypothesis described above. Fig.1b depicts a series of 4 soil specimens connected in series from anode to cathode. A geotextile layer separates each specimen. Each end of the soil specimen is connected to a stand pipe filled with tap water. Since they are connected in a series, the current flow through each element is identical regardless of the position. Nearly identical montmorillonite soil specimens were prepared using tap water, Cu^{2+} and Cd^{2+} solutions. A constant voltage

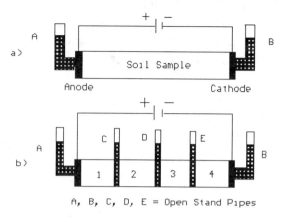

Fig. 1 Schematic E-O Experimental Setup a) Conventional, and b) Modified

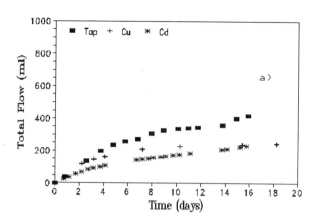

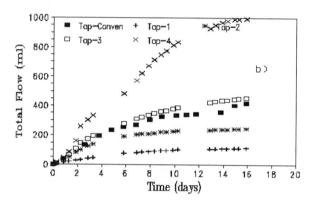

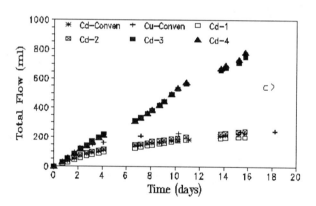

Fig. 2 Flow Vs. Time Data a) Conventional, b) and c) Comparison of Modified with Conventional

gradient of 0.6 v/cm was maintained during the tests. The water inflow or out flow at each end (at stand pipes) of the specimens were measured daily. Therefore the flow rate across each specimen can be calculated for each day. No control was exercised on the pH of the inflow or out flow fluid in any of the tests.

Results: Fig.2a depicts the flow data for to the conventional setup using *tap* water, Cu^{2+}, and Cd^{2+}. Figs.2b-c depict the flow data for each element (denoted by Tap-1,2,3,4; Cd-1,2,3,4) in Fig.1b. Also shown in Fig.2b-c are the flow data corresponding to the conventional tests. In conventional tests the flow rate diminishes rapidly. The results for each of the elements in the modified test are different. The flow pattern for the soil elements near the anode (Tap-1,2; Cd-1,2) resembles the flow pattern for conventional test. The rate is slightly slower than data from conventional setup. Higher flow rates are observed for the elements (Tap-3,4; Cd-3,4) near the cathode. The soil element near the cathode shows the highest flow rate and the soil element near the anode shows the lowest flow rate. This is what was expected according to the hypothesis presented earlier. Even though same current passes through each element, higher H^+ concentrations near the anode leads to high σ and therefore low flow rate, and eventually leads to cessation of flow, whereas the element near the cathode shows a much higher flow rate, because its σ is not altered significantly due to electrokinetic process.

Even though same voltage gradients are setup for both test types, the difference in flow rates between the conventional setup and the modified setup is due to the fact that the interaction due to hydraulic gradients and negative pore pressures between the elements in the modified cell are nullified (or reduced), whereas it is present in the conventional setup.

The above explanation does not imply that there are no internal hydraulic gradients induced inside each small element specimen in the modified setup. In deed negative pore pressures will still be induced in each specimen, but reduced in magnitude, as discussed later. But for practical purposes the overall behavior of each element is compared without considering the induced hydraulic gradients inside each small element. What is implied, however, is only that, if one would make sufficiently small series of soil elements, a similar trend shown in Fig.2b-c would be observed. The experiments are not perfect in many respects, however, they do support the hypothesis presented earlier.

Theoretical Analyses

To further analyze the flow behavior a numerical analysis was performed assuming coupled flow behavior due to electric gradients (dv/dx) and *internal* hydraulic gradients (dh/dx). For simplicity other factors are not incorporated in the analyses. The flow rate q is given by:

$$q(x,t) = -k_e \frac{dV(x,t)}{dx} - k_h \frac{dh(x,t)}{dx} \quad 1$$

where $dv/dx = I(t)/(A \cdot \sigma(x,t))$, k_e = electrokinetic

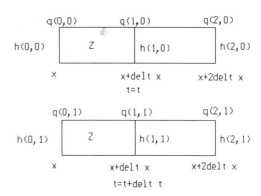

Fig. 3 Typical Soil Element for Numerical Simulation

permeability, k_h = hydraulic permeability. Such coupled approach has been attempted before (including chemical gradients), but assuming that the voltage gradient is constant and does not vary with time (Acar and co-workers, Yeung and Mitchell, 1993). In this paper voltage gradients are assumed to vary due to the *induced* redistribution of σ. The following notations are used for σ, hydraulic head (h), and q: $a_{ij} = a(x+i\delta x, t+j\delta t)$, $a = \sigma, h, q$, where a_{ij} represents σ, h, q at a point at distance $x = x + i\delta x$ from anode at time $t = t + j\delta t$; δx is the thickness of each soil element. Based on mass balance equation, and Eq.1, q, h, and σ for an element (Z, Fig.3) are related by:

$$q_{1,0} + q_{1,1} - q_{0,0} - q_{0,1} = -\alpha (h_{0,1} + h_{1,1} - h_{0,0} - h_{1,0}) \quad 2$$

$$\frac{k_e I}{A} \left(\frac{1}{\sigma_{1,0}} + \frac{1}{\sigma_{1,1}} - \frac{1}{\sigma_{0,0}} - \frac{1}{\sigma_{0,1}} \right) =$$
$$\beta h_{2,0} + \beta h_{2,1} + (\alpha - 2\beta) h_{1,0} \quad 3$$
$$- (\alpha + 2\beta) h_{1,1} + (\beta - \alpha) h_{0,1} + (\alpha + \beta) h_{0,0}$$

where $\alpha = m_v \gamma_w \delta x / \delta t$, $\beta = k_h / \delta x$, and m_v = soil compressibility. The above equations can be solved for in numerical form after substituting for boundary conditions with respect to the initial hydraulic heads at the boundaries of the

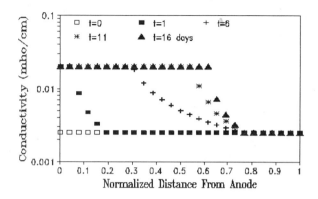

Fig. 4 Typical Conductivity Redistribution

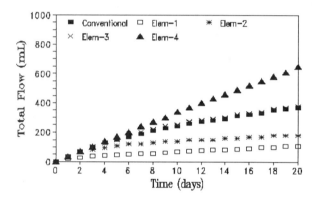

Fig. 5 Comparison of Numerical Results for Flow Rate for Conventional and Modified

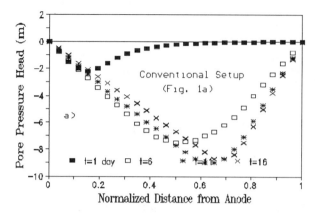

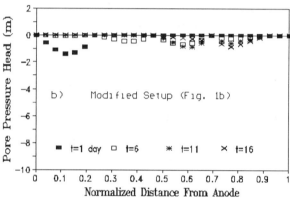

Fig. 6 Internal Induced Pore Pressure Distribution a) Conventional, and b) Modified

soil elements, current (I), and conductivity distribution. For illustration purposes only, a simple σ distribution similar to what has been reported in the literature (Hamed et al 1991) is assumed as shown in Fig.4. Fig.5 shows the numerical results for Flow vs. Time corresponding to conventional and modified setup. Figs.6a-b show numerical results for the internal pore pressure distribution.

Conventional Setup: As anticipated, the flow rate gradually decreases with time (Fig.5). The pore pressures in the soil specimen gradually increase (Fig.6a). With time, the peak magnitude of the negative pore pressure shifts towards the cathode. In deed, such pore pressure distributions have been observed before during actual experiments (Tedashi et al, 1961). It can also be shown that the negative pore pressures can reach high values and cause cavitation depending on current, k_h, k_e and length of specimen, and eventually cause flow cessation.

Modified Setup: A flow pattern (Fig.5) similar to those measured during the experiments (Figs. 2a-c) is observed. The flow rate near the anode decreases faster than the flow near the cathode. The flow during conventional setup falls between the two extremes. Significantly smaller negative pore pressures are observed (Fig.6b). All of these results qualitatively agree with the hypothesis and the observed test data.

Conclusions

The important roles of non-uniform conductivity distribution and the resulting negative pore pressures, even in open electrode configuration, and internal hydraulic gradients induced in the soil on the overall performance of electrokinetic flow are illustrated. While there are many other factors that influence the flow efficiency, the above electro-osmotic and internal hydraulic interaction appears to be significant and can cause a drastic reduction in flow efficiency. While the experiments and the numerical examples are not perfect, they indicate the influence of non-uniform conductivity distribution and the resulting internal hydraulic gradients induced due to the conductivity redistribution on the overall flow behavior. Based on this experiments one could device strategies to improve the flow efficiency during electrokinetic flow process.

Acknowledgements: The authors wish to thank the summer interns, Messrs. E. Nolan and J. Horton, supported by the Youth in Engineering and Science Program, Polytechnic University, for their assistance in the experimental studies.

Appendix I: References

(1). Acar, Y.B. (1992) "Electrokinetic Cleanups", ASCE Magazine, October, 58-60.
(2). EPA (1986)"Workshop on Electro-Kinetic Treatment and Its Application in Environmental - Geotechnical Engineering for Hazarous Waste Site Remediation", Proc. Workshop, Haz. Waste Eng. Res. Lab., Cincinnati, OH.
(3). Hamed, J., Acar, Y.B., and Gale, R. (1991) "PB(II) Removal from Kaolinite by Electrokinetics" ASCE J. Geotech. Eng., 117(11), 241-271.
(4). Leggeman, R. (1989) "Theory and Practice of Electro-Reclamation" NAto Pilot Study, Demonstration and Remedial Action Technologies for Contaminated Land and Groundwater, NATO/CCMS, Copenhagen, Denmark.
(5). Lockhart, N.C. (1983) "Electroosmotic Dewatering of Clays I,II, III" Coll. and Surf. 6, 229-269.
(6). Mitchell, J.K. (1986) "Potential Uses of Electro-Kinetics for Hazardous Waste Site remediation", EPA Workshop 1986, Univ. of Washington, Seattle.
(7). Mitchell, J.K., and Yeung, A. (1991) "Electro-Kietic Flow Barriers in Compacted Clay", TRB 1288, 1-9.
(8). Pamukcu, S. and Wittle, J.K. (1992) "Electrokinetic Removal of Selected Heavy Metals from Soil", Env. Progress, 11(3), 241-250.
(9). Probstein, R.F.and Hicks, ER.E. (1993) "Removal of Contaminants from Soils by Electric Fields", Science (260), 498-503.
(10). Segal, B.A., O'Bannon, C.E., and Matthias, J.A. (1980) "Electro-osmosis Chemistry and Water Quality", ASCE, J. Geotech. Engrg. 106(10), 1143-1147.
(11). Segall, B.A., and Bruell, C.J. (1992) "Electroosmotic contaminant removal processes", ASCE, J. Env. Eng., 118(1), 84-100.
(12). Shapiro, A.P, and Probstein, R.F. (1993) "Removal of Contaminants from Saturated Clay by Electro-Osmosis", Env. Sci. technol., 27, 283-291
(13). Thevanayagam, S. (1993) "Efficiency of Electrokinetic Soil Decontamination", Proc. ASCE/ASME/SES Spec. Conf. Ed. Herakovich, C.T and Duva, M., Meetn 93. Univ. of Virginia, June 6-9.
(14). Yeung, A.E. and Mitchell, J.K. (1993) "Coupled fluid, electrical and chemical flows in soil", Geotechnique, XLIII (1), 121-134.

Quality and Efficiency Control During Soil Air Sampling and Remediation

Dipl.-Ing. Matthias Vogler
Technical University of Darmstadt, Institute of Geotechnics, Darmstadt, Germany

Dr.rer.nat. Helmut Dörr
Trischler und Partner GmbH, Darmstadt, Germany

Prof. Dr.-Ing. Rolf Katzenbach
Technical University of Darmstadt, Institute of Geotechnics, Darmstadt, Germany

Dr.-Ing. Norbert Molitor
Trischler und Partner GmbH, Darmstadt, Germany

0. Abstract

A method to control the quality of soil vapour wells and soil vapour measuring points is based on the use of ^{222}Radon as a tracer for gas-transport and gas mixture processes. The method is also suitable to consider the Efficiency of soil air remediation. This method is comparatively easy to handle especially at difficult site conditions with suitable costs. A relatively exact prediction of duration and success of remediation action is thereby possible.

1. Introduction

Soil vapour contamination with volatile organic hydrocarbons is a well known problem in industrial countries. This contamination is produced by service stations, dry cleaners, metallic industry and others. A common technical solution to protect the environment is soil vapour remediation. As in all sampling and remediation processes quality control is very important. The main question is: From where is this sample gained? The risk to underestimate the real concentration of the contaminant by dilution of soil vapour with atmospheric air depends on the heterogeneity of the soil, sampling equipment, sampling procedure etc.. Quality control is also very important to document the success and the suitability of a remediation technique. Factors determining the remediation technique are design of the plant, costs, target values and remediation goals.

A new method for quality control was developed by Trischler & Partner, Germany, to distinguish between atmospheric and soil air and to determine the efficiency of soil vapour remediation. This method allows to determine the contribution of atmospheric air and soil air, to calculate the needed remediation time, and to estimate the radius of influence of soil venting/ vacuum extraction procedures.

2. Material and Methods

2.1 Material

To measure the concentration in soil air, soil vapour wells and soil vapour measuring points are used.

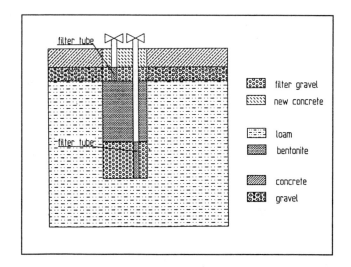

Figure 1 Multi-level measuring point (from [1])

Figure 1 shows a multi-levelling measuring well. The measuring point has a filter pack in the depth where the concentration is required to be measured. The rest of the borehole is sealed with bentonite to prevent breakthrough of atmospheric air or soil air from other depths. This multi-level construction is very important in inhomogeneous soils or with inhomogeneous contamination distribution.

The measuring point is connected with a sampling pump to take a sample of volatile organic hydrocarbon. The air is pumped from the well and different parameters are measured (e.g. methane, carbon dioxide, nitrogen etc., or the ^{222}Radon concentration when measured). When the concentration of these parameters is constant, the soil air sample is taken by enrichment of the volatile organic hydrocarbon on activated charcoal (see Figure 2). The sample can be influenced by air which gets into the system through leakage (not tight fittings, breakthrough of atmospheric air etc.) during sampling, transport or storage. Another possibility to take soil air samples is to pump the soil air into airbags (made of plastic covered aluminium foil). For volatile organic hydrocarbons this method leads to a decreased level of detection, to take air samples for radon analysis this method however is suitable. The ^{222}Radon concentration in the soil air samples is analysed out of the airbags. The results are corrected for radioactive decay. The lower limit of detection (2 σ criteria) by radon analysis is about 0.5 mBq. The amount of soil vapour to determine the Radon concentrations with accuracy of + 5% is about 10 to 100 ml. Direct measurements in the field are also possible. The time needed for one measurement is between 10 and 15 minutes.

2.2 Methods

^{222}Radon is a natural existing radioactive noble gas (half life 3.8 days). In the underground ^{222}Rn is produced through disintegration of ^{226}Radium. In a steady state the concentration distribution of ^{222}Radon in the soil is increasing with depth (see Figure 3) even in inhomogenious and/or anisotropic soils. From a constant concentration in a typical depth of about 2-3 meters depth the concentration decreases to the low concentration in the atmosphere [5].

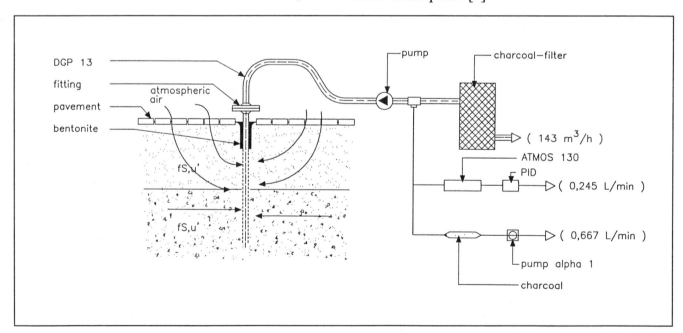

Figure 2 Soil vapour remediation test

At 1 m depth Typical ^{222}Radon concentrations in the soil are in the range of about 1-100 Bq/l. Figure 4 shows a statistical distribution of ^{222}Rn concentrations at 1m depth from the literature [ref. 10 - 17]. The absolute value of Rn in the soil depends on the mineral composition (^{222}Radon content) and the grain size distribution (emanation coefficient [5], figure 5). The production/ emanation of ^{222}Rn in soil air decreases with the grain size. This means, that typically the radon production and Rn concentration in a sand is lower than in a clay. In the atmosphere the ^{222}Radon concentration is about 3-4 orders of magnitude lower than in soil air. Typical outdoor ^{222}Rn concentrations are about 20 Bq/m^3. As inert, noble gas which is only produced in the soil at a well known production and decay rate Radon can be used as a tracer for soilgas transport and gas mixture processes [5].

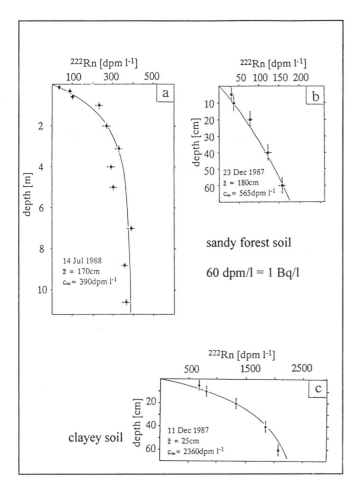

Figure 3 ^{222}Radon concentration in different soils and depth resolution (after [5])

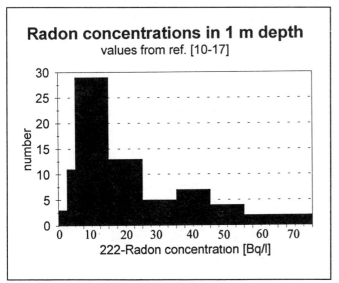

Figure 4 Statistical concentration value distribution

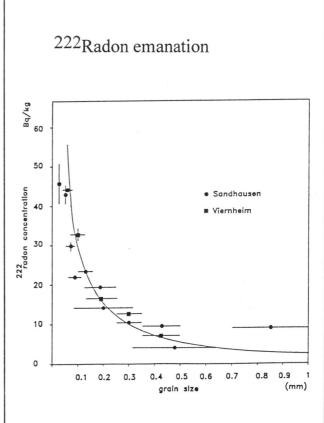

Figure 5 ^{222}Rn concentration over grain size

3. Examples

3.1 Quality control

The Radon concentration in the underground in comparison with the small value in the atmosphere is a good base for quality control. Usually the quality of well installation and soil gas sampling is controlled by measurement of O_2, CO_2 CH_4. The concentration of these parameters, however, do not differ significantly between soil air and atmospheric air. Thus, these parameters may be used for quality control only in special cases which, however must be identified by measurement of these parameters. An alternative method to overcome these principal problems is the "radon-method". The quality of soil air samples is controlled by measuring Radon concentration in soil airs. Figure 6 shows the results of four samplings. The change in concentrations of carbon dioxide, nitrogen, methane and oxygen varies in a small range and is close the atmospheric level. The radon results of RKS 1 - RKS 3 are between 9 and 13 Bq/l. This concentration is in the typical range for soil air between 0.5 and 2.0 meters depth. Generally N_2 is not measured but calculated from the main balance of air composition. Methane in considerable concentration is mainly detected at recent and former landfill sites. In aerobic soils CO_2 is produced by microbial activity (decomposition of organic matter using O_2). Thus, the sum of CO_2 and O_2 concentration must be at the constant atmospheric content of about 21%. The use of O_2 and CO_2 as indicator for soil gas depends on the microbial activity in the sampled soil, which is not predictable before the measurements.

	RKS1	RKS2	RKS3	RKS4
O_2	19.1 %	19.1 %	18.9 %	20.0 %
CO_2	2.4 %	1.6 %	2.2 %	0.0 %
CH_4	0.0 %	0.0 %	0.0 %	0.7 %
Rn [Bq/l]	9.3±0.4	12.3±0.9	9.7±0.5	18.9±0.4
O_2+CO_2	21.5 %	20.7 %	21.1 %	20.0 %

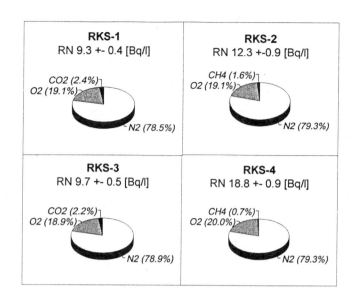

Figure 6 Quality control during soil air sampling

This indicates that the contribution of atmospheric air due to leakage of sealing around the measuring points, pneumatic breakthrough, leakage of samples, equipment can be neglected. If we assume a leakage in the system that causes a 50% contribution of atmospheric air, the resulting Rn-concentration would be about 5 Bq/l, which is easy to distinguish from the measurement 10 Bq/l at standard deviation less than 1 Bq/l. On the other hand, the resulting O_2 and CO_2 concentration would be 20% instead of 19% for O_2 and about 1% instead of 2% for CO_2. In relation to the standards deviation of the individual measurements and the determination of atmospheric O_2 content, these values would be comparable and difficult to interpret. The Rn-concentration in RKS 4 is about 19 Bq/l. This is due to the greater depth of this measuring point. The radon concentration values are about 100 times greater than its concentrations in the atmosphere (atmosphere ~0.02 Bq/l), indicating that in this case a possible admixture of atmospheric air is less than about 1%. Also in this case an admixture of atmospheric air would not be detectable from O_2, CO_2 and CH_4 measurements at a satisfactory precision/accuracy.

3.2 Convective Soil Vapour Remediation

Soil vapour remediation tests are used to evaluate costs, duration and efficiency of soil air remediation. Figure 2 shows schematically the proceeding of a soil vapour remediation test. During the test the Radon concentration was measured directly in the field. The concentration of the contaminant was analysed with a gas-chromatograph on 3 different samples. These samples were taken 30 to 60 minutes, 90 to 120 minutes and 150 to 180 minutes after starting the test. The change in Radon concentration over the extracted soil vapour volume is shown in figure 7.

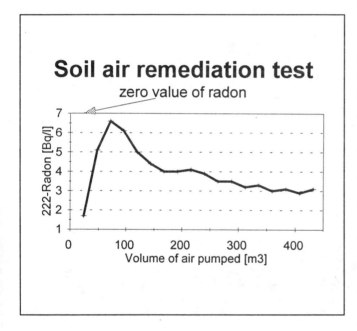

Figure 7 Radon concentration during remediation test

After starting the test the Radon concentration increases nearly up to its initial value, which is measured before the test at a very low volume flux. The Rn-concentration then decreases to a final concentration of about 3 Bq/l. Comparing this value with the initial value of 7 Bq/l shows, that about 40 % of the pumped gas is soil vapour and about 60 % is atmospheric air.

The increase of the radon concentration at the beginning of the test is due to the plushing of the gas-well with its filterpack and the tubes of the sampling equipment. The ratio of soil air pumped is a good indicator for the efficiency of the remediation. The range of influence of the soil vapour remediation is calculated by a radon balance in soil air and pumped air. In this example the total amount of pumped radon until the steady state conditions is 400 m³×4000 Bq/m³(4000 Bq/m³ as average over sampling time) = $1.6×10^6$ Bq. Soil air contains about 7000 Bq/m³ of Radon (initial concentration). The effectively pumped soil air is flux $1.6×10^6$ Bq/7000 Bqm^{-3} ≈ 230 m³, showing that the admixture of atmospheric air is at the average about 57% by volume compared to 60% at the end of the test.

Assuming a cylindrical volume of influence in the soil, the range of the remediation well can be estimated to be about 16 m, what is expected at the site conditions.

4. Conclusions

The "radon-method" is a helpful tool in environmental geotechnics. This method is useful

- for quality control
- to determine the efficiency of remediation action techniques
- to predict the range of soil vapour extraction.

The main advantages of the Radon method are:

- the method is easy to handle
- it's a direct sensitive technique
- it's an efficient (technical and economic) method.
- it's a convincing method

This method was successfully used in a couple of environmental project and proved to be a suitable tool in environmental engineering and geotechnics.

5. References

[1] H. Dörr; R. Gellermann; N. Molitor, Qualitätssicherung und Effektivitätsprüfung bei Bodenluftbeprobung und -sanierung, TerraTech, 2/1993

[2] N. Molitor and P. Ripper, Die Bodenluftspülung als Sanierungsmaßnahme, Schr. Angew. Geol. 9. I-VII, S. 187-198, 1990

[3] B. Toussaint, Kritische Anmerkungen zur Plausibilität der Gehalte leichtflüchtiger Halogenkohlenwasserstoffe in beprobten Umweltmedien, Schr. Angew. Geol. 9. I-VII,S. 93-112, 1990

[4] H.Israel, Die natürliche und künstlische Radioaktivität der Atmosphäre in Kernstrahlung in der Geophysik, Springer Verlag, 1962

[5] H. Dörr and K.O. Münnich, ^{222}Rn flux and soil air concentration profiles in West Germany. Soil ^{222}Rn as tracer for gastransport in the unsaturated soil zone. Tellus 42 B (1) P. 20-28, 1990

[6] I. Levin, P. Bergamaschi, H. Dörr and D. Tapp, Stable Isotopic Signature of Methane from Major Sources in Germany, Chemosphere, 1992

[7] M. Born, H. Dörr and I. Levin, Methane consumption in aerated soils of the temperate zone. Tellus 42 B (1), P. 2-8, 1990

[8] Sogaard-Hansen and A. Damkjaev, Determining ^{222}Rn Diffusion Lengths in Soils and Sediments, Health Physics Vol. 53, No. 5 S. 455-459; 1987

[9] K. K. Turekian, Y. Nozaki, L. K. Benninger, Geochemistry of Atmospheric Radon and Radon Products, Ann. Rev. Earth Planet Sci. Vol 5 P. 227-255, 1977

[10] E. M. Kovach, Diurnal Variations of the Radon-Content of Soil Gas

[11] S. D. Schery, D. H. Gaeddert, M.H. Wilkening, Factors Affecting Exhalation of Radon from a Gravelly Sandy Loam, J. of Geophysical Res. Vol. 89, No. D5, P. 7299-7309, 1984

[12] J. N. Andrews, D. F. Wood, Mechanism of Radon Release in Rock Matrices and Entry into Groundwaters,Trans. Inst. Min. Metall. Sec. B.81, P. 197-209

[13] J. W. Luetzelschwab, K.L. Helweick, K. A. Hurst, Radon Concentrations in Five Pennsylvania Soils, Health Physics, Vol. 56, No. 2 P. 181-188, 1989

[14] G. Hradil, Zur Messung des Emanationsgehaltes der Bodenluft über geol. Strukturlinien, Band 100, Heft 4

[15] H. Bender, Über den Gehalt der Bodenluft an Radiumemanation, Gerl. Beitr. Geophys. 41

[16] H. Israel, Die natürliche und künstliche Radioaktivität der Atmosphäre, Kernspaltung in der Geophysik, Springer Verlag, 1962

[17] High Background Radiation Research Group, China, Health Survey in High Background Radiation Areas in China, Scine, Vol. 209, 1980

Effet de Trois Lixiviats de Decharges d'Ordures Menageres sur le Gonflement de Quatre Bentonites

T. Winiarski
*Laboratoire des Sciences de l'Environnement de l'Ecole Nationale des Travaux Publics de l'Etat,
Vaulx-en-Velin, France*

J. Cavalcante Rocha, G. Didier
Laboratoire de Géotechnique, Villeurbanne, France

RESUME : la réalisation de dispositifs d'étanchéité sous les sites de décharges fait souvent appel aux propriétés gonflantes des bentonites. Deux essais de gonflement sont comparés selon la solution d'hydratation. Dans cette étude, plusieurs solutions sont utilisées : l'eau du réseau et des lixiviats provenant de trois décharges d'ordures ménagères. Les résultats obtenus sur quatre bentonites (deux sodiques activées, une calcique et une Wyoming) sont comparés en prenant en compte les modifications physicochimiques des solutions d'hydratation.

ABSTRACT : swelling properties of bentonites are frequently used as sealing of landfill sites. Two procedures for estimation of the percent swell index values have been compared, versus four different solutions of hydratation. The solution used in this study are tap water and three landfill leachates. The modifications of the physico-chemistry of the solutions, and the swelling results of the four bentonites (two sodium actived ones, one pure calcium and one Wyoming) are compared.

Introduction

Afin de résoudre les problèmes de contamination des eaux souterraines par les lixiviats de décharge [1], les mélanges bentonite-sable sont souvent utilisés comme dispositif d'étanchéité sous les Centres d'Enfouissements Techniques [2]. Ils peuvent donc se trouver en contact avec des lixiviats de décharges. Ce sont les effets de ce contact sol aménagé/lixiviat qui peuvent poser des problèmes [3]. Plus particulièrement les phénomènes de gonflement (les bentonites sont choisies pour leur propriété gonflante influençant ainsi la perméabilité) et de rétention peuvent être affectés par les différents produits véhiculés par le lixiviat [4]. En effet, des études de l'agence américaine de protection de l'environnement (USEPA) [5] ont montré que ce dispositif pouvait être défaillant.

En plus de ce gonflement, il faut cité d'autres phénomènes physiques, tels que l'adsorption de surface et la capillarité qui vont influencer d'une manière directe les coefficients de transfert. Il faut préciser que ce gonflement, pour les montmorillonites sont principalement de deux types : hydratation des sites hydrophiles (mécanisme important pour les montmorillonites sodiques) et adsorption d'eau due à la pression osmotique.

Actuellement, seul le compartiment argile est étudié car c'est le minéral responsable du gonflement, donc influençant d'une manière directe la perméabilité [6]. Il est aussi, grâce à sa structure (surface spécifique importante), un lieu d'échange, de rétention etc...Un mélange argile-sable est donc un lieu réactionnel essentiel [7].

Matériels et Méthodes

Le gonflement a été mesuré selon deux méthodes : soit par sédimentation-floculation de 3 g d'argile dans 100 ml de solution, soit par mesure du gonflement unidirectionnel d'une éprouvette de montmorillonite compactée à une densité sèche de 1.40 [4]. Ces essais permettent de caractériser le potentiel de gonflement de l'argile directement au contact de différentes solutions, mais ils ne font intervenir ni la vitesse de filtration à travers l'argile, ni la masse volumique de l'argile, ni l'échange avec l'eau de saturation de l'argile et la solution étudiée. Les argiles étudiées sont des montmorillonites ; il s'agit plus précisément, de deux bentonites sodiques activées (B_1Na, B_2Na) et de deux argiles naturelles (B_1Ca calcique et une Wyoming notée Wyo). Les argiles présentent, à des degrés différents selon leur nature, une grande affinité pour l'eau. Lorsqu'on part d'un matériau argileux sec, l'adsorption d'eau se fait d'abord sur les surfaces des particules primaires (ensemble de feuillets) et entre les unités morphologiques [7].

TABLEAU 1 : résultats des analyses effectuées sur les lixiviats utilisés.

Paramètres	Eau	Lix1	Lix2	Lix3
pH	7.14	8.26	7.59	7.78
Cond.(μS/cm)	434	4927	7308	15863
DCO(mgO/l)	-	572.5	552.1	2050
TA(méq/l)	0	0.5	0	0
TAC(méq/l)	3.6	11.6	60.2	117.5
Cl^-(mg/l)	30.8	1290	1280	4830
SO_4^{2-}(mg/l)	22.14	7.94	26.57	39.4
NO_3^-(mg/l)	7.52	0	0.72	0.55
PO_4^{3-}(mg/l)	0	0	0	18
Ca^{2+}(mg/l)	123.7	72.9	72.22	107.6
Na^+(mg/l)	8.68	1140	413.8	2160
NH_4^+(mg/l)	0	254.3	544.6	2060
K^+(mg/l)	1.79	300.6	331.8	896

L'adsorption d'eau s'accompagne d'un gonflement. L'importance de celui-ci dépend du type d'argile, des dimensions des unités morphologiques, de la taille et de la valence des cations compensateurs. En fonction du cation prédominant, il y a des différences de potentiel de gonflement et divers états d'arrangement des feuillets. En effet, les montmorillonites sodiques ont un plus fort potentiel de gonflement, tandis que les montmorillonites calciques sont moins gonflantes (même totalement hydratées) et peuvent présenter des structures floculées [8].

Diverses solutions sont utilisées : de l'eau du réseau (notée Eau) qui est souvent utilisée dans les essais, et trois lixiviats prélevés sur trois sites de décharge d'ordures ménagères (solutions complexes très chargées électriquement et organiquement, tableau 1).

Ces trois lixiviats (noté Lix1, Lix2 et Lix3) ont été choisis afin d'étudier les réactions des bentonites mises en contact avec des lixiviats ayant des caractéristiques biophysicochimiques différentes mais provenant d'un même type de déchet, soulignant ainsi la grande variabilité de ces solutions selon le site de prélèvement [9].

Les analyses de surnageants obtenus avec la première méthode permettront d'évaluer les interactions argiles/solutions.

Résultats

La première méthode montre tout d'abord un effet important des solutions (figure 1), en effet, les bentonites au contact des solutions peu chargées (eau du réseau, Lix1) ont le plus gonflé, tandis qu'avec les lixiviats 2 et 3, le volume de sédimentation est moins important. La bentonite B_1Ca présente systématiquement le volume de sédimentation le plus faible. De plus, la bentonite Wyo a le volume le plus important pour les solutions très chargées de type Lix2 et Lix3. En ce qui concerne les bentonites activées, elles semblent bien gonflées dans les solutions peu chargées, tandis qu'en présence de fortes concentrations en ions, leur comportement vis à vis du gonflement est moyen.

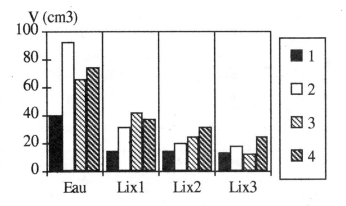

FIGURE 1 : mesures du volume de gonflement V à 20 °C des argiles étudiées après deux semaines de sédimentation. 1 : B1Ca, 2 : B1Na, 3 : B2Na, 4 : Wyo.

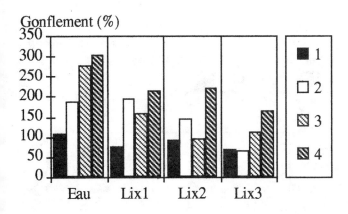

FIGURE 2 : histogramme présentant le taux de gonflement final pour les essais unidirectionnels. 1 : B1Ca, 2 : B1Na, 3 : B2Na, 4 : Wyo.

La seconde méthode semble corroborer une partie de ces résultats (figure 2), en effet, la bentonite calcique B1Ca un faible gonflement, tandis que la Wyo se comporte mieux, et ce quelle que soit la solution. Des comportements intermédiaires sont observés pour les bentonites activées. Les argiles soumises au lixiviat le plus chargé (Lix3) ont un gonflement moins important en comparaison aux autres solutions de saturation, par contre, si on utilise de l'eau du réseau, cette tendance est inversée. Pour les deux autres types d'essais effectués à partir de Lix1 et Lix2, il est plus difficile de tirer un comportement général.

Afin de comparer ces deux types d'essais, il est proposé une série de tests statistiques. Devant le faible nombre d'échantillons (16 pour chaque méthode), il a été choisi les méthodes non paramétriques. Dans le cas présent, la comparaison des données brutes obtenues par les deux méthodes est impossible. Par contre, si on effectue un classement en rang d'une manière décroissante, (de l'échantillon qui a le moins gonflé à celui qui a le plus gonflé) on peut utiliser les tests du coefficient de corrélation de rangs de SPEARMAN et de KENDALL [10].

TABLEAU 2 : coefficients de corrélation obtenus à partir des deux tests effectués sur les deux populations de rangs.

SPEARMAN	KENDALL
$\rho = 0.73$	$\tau = 0.54$
$p < 0.003$	$p < 0.004$

Les résultats obtenus (tableau 2) permettent de conclure qu'il y a dépendance des deux mesures et que leur tendance va dans le même sens à plus de 99%.

Des mesures physico-chimiques sont effectuées sur le surnageant de l'éprouvette, c'est-à-dire sur la solution située au dessus de la couche d'argile sédimentée. Ces mesures permettront de préciser des phénomènes de rétention, ainsi que les effets de ces différents paramètres sur le gonflement. La différence entre ces mesures et les solutions initiales permet d'aboutir à des valeurs négatives ou positives, soulignant respectivement une baisse ou un gain de quantité de matière dans le surnageant. D'une manière générale, le pH a augmenté dans toutes les éprouvettes, semblant exprimer un largage d'ions OH⁻ dans la solution (figure 3). De plus, ce phénomène est plus sensible pour les mélanges eau/bentonites. Les solutions avec les bentonites activées présentent des pH plus importants, tandis que les surnageants des mélanges à partir de bentonites naturelles et en présence de lixiviats possèdent une variation de pH inférieure à 1.

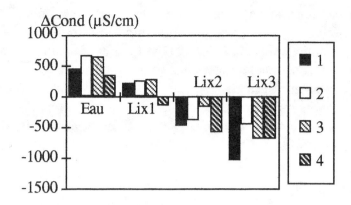

FIGURE 4 : différence entre la conductivité du surnageant et de la solution initiale. 1 : B1Ca, 2 : B1Na, 3 : B2Na, 4 : Wyo.

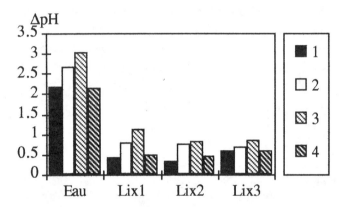

FIGURE 3 : différence entre le pH du surnageant et de la solution initiale. 1 : B1Ca, 2 : B1Na, 3 : B2Na, 4 : Wyo.

En ce qui concerne la conductivité des surnageants (figure 4), quelle que soit l'argile étudiée, une augmentation de la charge électrique du surnageant est observée pour les solutions initialement les moins conductrices (eau du réseau et lixiviat 1), tandis que les solutions les plus chargées au départ ont tendance a être moins conductrices. On observe ce même phénomène avec le titre alcalimétrique complet (T.A.C.) (figure 5). Cette observation met en évidence les relations entre les concentrations en carbonate et bicarbonate (que représente essentiellement le T.A.C.) et la conductivité des diverses solutions.

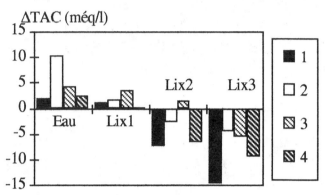

FIGURE 5 : différence entre le TAC du surnageant et de la solution initiale. 1 : B1Ca, 2 : B1Na, 3 : B2Na, 4 : Wyo.

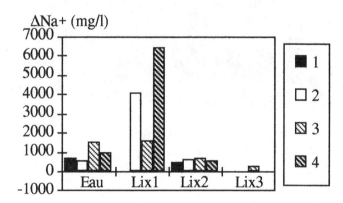

FIGURE 6 : différence entre la concentration en sodium du surnageant et de la solution initiale. 1 : B1Ca, 2 : B1Na, 3 : B2Na, 4 : Wyo.

La mesure de la quantité de sodium dans le surnageant montre que toutes les bentonites sodiques ont tendance à larguer ce cation (figure 6). La bentonite calcique naturelle B1Ca semble montrer un largage de sodium moins important et certains échantillons montrent le phénomène inverse (B1Ca-Lix1 et B1Ca-Lix3). C'est dans le lixiviat le moins chargé (Lix1) que ce phénomène est le mieux observé.

Toutefois, des exceptions sont faites pour B1Na-Eau, B1Ca-Lix2 et Wyo-Lix2 (Figure 7).

La figure 8 indique des valeurs négatives pour les mesures de potassium, cela montre un phénomène d'adsorption, de plus, c'est avec la solution la plus chargée que la variation de potassium est la plus petite et ce, quelle que soit l'argile utilisée.

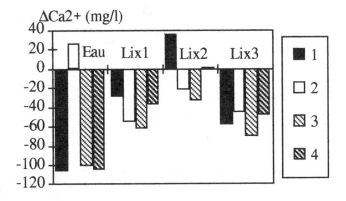

FIGURE 7 : différence entre la concentration en calcium du surnageant et de la solution initiale. 1 : B1Ca, 2 : B1Na, 3 : B2Na, 4 : Wyo.

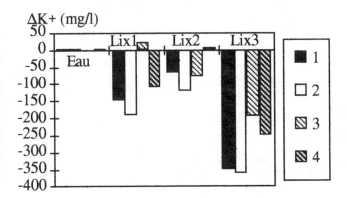

FIGURE 8 : différence entre la concentration en potassium du surnageant et de la solution initiale. 1 : B1Ca, 2 : B1Na, 3 : B2Na, 4 : Wyo.

En ce qui concerne les concentrations de calcium, les analyses montrent une tendance à l'adsorption soulignée par une variation de calcium globalement négatif et cela pour la plupart des bentonites.

Discussion

Si on compare les gonflements obtenus à partir des deux méthodes (Tableau 3), on s'aperçoit que les sept échantillons qui ont le moins gonflé sont essentiellement les bentonites activées soumises aux lixiviats les plus chargés mais aussi tous les échantillons de bentonite calcique. Pour cette dernière, il semble que son aptitude au gonflement soit assez semblable quel que soit le lixiviat utilisé. Tous les échantillons de Wyoming ont gonflé convenablement ainsi que les bentonites activées en contact avec l'eau du réseau ou le lixiviat 1. On observe ainsi le rôle essentiel de la solution, si celle-ci possède une conductivité et un titre alcalimétrique important, il faut s'attendre à un gonflement sensiblement moins grand par rapport à des solutions peu chargées. Il faut aussi noter qu'il semble que les bentonites B1Na et B2Na peuvent très bien se comporter dans l'eau mais réagir insuffisamment dans le cas des lixiviats très chargés.

Quand à l'étude physicochimique des surnageants, on observe, pour les échantillons qui ont le moins gonflé, des concentrations de potassium dans le surnageant généralement plus faibles que dans la solution initiale. De plus, la conductivité qui est, ici, essentiellement expliquée par le T.A.C. semble montrer la même chose, les échantillons qui ont le plus gonflé (Eau et Lix1) ont des surnageants plus chargés que la solution initiale, on observe l'inverse pour les solutions très chargées (Lix2 et Lix3). Ainsi, les deux types d'essais de gonflement utilisés permettent de montrer l'importance de la solution d'hydratation et soulignent les comportements particuliers des bentonites étudiées. D'une manière générale, les lixiviats conduisent à des gonflements plus faibles. Il existe même un gradient de gonflement en fonction de la charge globale de la solution. L'étude des surnageants souligne les mécanismes complexes du gonflement de bentonites en présence de lixiviats de décharge.

TABLEAU 3 : classement des échantillons en fonction des mesures de gonflement obtenues à partir des deux méthodes (du moins gonflant au plus gonflant).

Méthode 1	Méthode 2
B2Na-Lix3	B1Na-Lix3
B1Ca-Lix3	B1Ca-Lix3
B1Ca-Lix1	B1Ca-Lix1
B1Ca-Lix2	B1Ca-Lix2
B1Na-Lix3	B2Na-Lix2
B1Na-Lix2	B1Ca-E
B2Na-Lix2	B2Na-Lix3
Wyo-Lix3	B1Na-Lix2
B1Na-Lix1	B2Na-Lix1
Wyo-Lix2	Wyo-Lix3
Wyo-Lix1	B1Na-E
B1Ca-E	B1Na-Lix1
B2Na-Lix1	Wyo-Lix1
B2Na-E	Wyo-Lix2
Wyo-E	B2Na-E
B1Na-E	Wyo-E

Conclusion

Ces deux types d'essais sont relativement simples et rapides à mettre en œuvre. Ils renseignent sur la sensibilité au gonflement d'un type d'argile, mais aussi sur le rôle prépondérant de la solution d'hydratation dans les mécanismes de rétentions. Ils devraient être mis en œuvre afin de sélectionner un type de géomatériaux pour la réalisation d'une barrière "étanche" résistante aux solutions complexes que sont les lixiviats de décharge. Cependant, il faut être prudent, d'autres paramètres interviennent dans la définition d'une barrière étanche : concentration de bentonite dans le sol, granulométrie de celui-ci, préhydratation avant contact avec le lixiviat, nature des déchets à confiner.

Références

[1] Husain T., Hoda A. and Khan R., 1989. Impact of sanitary landfill on groundwater quality. Water, Air and Soil Pollution, vol 45 : 191-206.

[2] Meisel A., Didier G., Blutsztein M. et Wicker A., 1989. Etanchéification à la bentonite du centre d'enfouissement technique d'Holnon (Aisne). TSM L'eau, 84 éme année, N°11, november 1989 : 601-604.

[3] Chapuis R.P., 1989. Sand-bentonite liners : field control methods. Canadia Geotechnical Journal, vol 27 : 216-223.

[4] Winiarski T., Didier G. & Cavalcante Rocha J., 1993. Effet d'un lixiviat sur le gonflement de bentonites. Géoconfine 1993, Géologie et confinement des déchets toxiques, Symposium International, 8-11 juin 1993, Montpellier, France : 275-280.

[5] Lee G. F. and Jones R. A., 1984. Is hazardous waste disposal in clay vaults safe. Journal of American Water Works Association, september : 66-73.

[6] Fernandez F. and Quickley R.M., 1985. Hydraulic conductivity of natural clays permeated with simple liquid hydrocarbons.

Canadia Geotechnical Journal, vol 22 : 205-214.

[7] Tessier et Pedro, 1976. Les modalités de l'organisation des particules dans les matériaux argileux. Science du Sol 2 : 85-100

[8] Van Olphen H., 1977. An introduction to clay colloid chemistry, second edition. John Wiley & Sons Edition, 320 pages.

[9] Clément B., Delolme C., Winiarski T. and Bouvet Y., 1993. The risks associated with the contamination by landfill leachates of freshwater ecosystems - A review. Sardinia 93, VI international landfill symposium, S. Margherita di Puta, Italy, 11-15 October 1993 : 1155-1166.

[10] Sprent P.,1992. Pratique des statistiques non-paramétriques, édition INRA, collection Techniques et Pratiques, 294 pages.

Liquefaction Characteristics of Coal Fly Ash as a Reclamation Material

Kazuya Yasuhara
Department of Urban and Civil Engineering, Ibaraki University, Hitachi Ibaraki, Japan

Masayuki Hyodo
Department of Civil Engineering, Yamaguchi University, Yamaguchi, Japan

Kazutoshi Hirao
Department of Civil Engineering, Nishinippon Institute of Technology, Fukuoka-ken, Japan

Sumio Horiuchi
Research Institute, Shimizu Construction Co. Ltd., Ecchujima, Taito-ku, Japan

The liquefaction characteristics of Pulverised Fuel ash (PFA) were investigated using cyclic triaxial tests. Through this investigation, the effect of the disposal method for reducing the potential for liquefaction was pursued. Consequently, a slurry disposal method named High Density Ash Slurry (HAS) was recommended for attaining the stable reclamation under cyclic loading. The present investigation also made clear that the mixing procedure of PFA with water was important for producing a stable reclaimed ground of high density. Even if a sufficient lystable ground was not produced, the resistance to liquefaction of PFA with a low density was improved by adding a small amount of cement.

INTRODUCTION

In Japan, three million tons of coal ash (PFA) are disposed of coal power plants, and sea disposal has increased because of insufficient land disposal. Conventional sea disposal is classified into two types:wet and dry disposal. In wet disposal, coal ash is sluiced into the sea, i.e. coal ash is mixed with a large amount of sea water and is hydraulically transported through pipelines to the disposal site. In dry disposal, coal ash conditioned with water is transported with belt conveyors or dump trucks to the shore, and dumped into the sea. According to the information available, grounds reclaimed by the conventional disposal are loose and low in bearing capacity (EPRI : 1979 , Yasuhara et al.: 1988, Sugimoto et al. : 1989). This is due to coal ash's self-hardening property and free sedimentation of coal ash particles.

Slurry disposal method, using underwater placement of coal ash slurry, so called HAS, is a new disposal method to achieve increasing density and strength (Horiuchi et al. :1985, 1993) ; dry density of which is 96% of the maximum dry density, and unconfined compression strength of 200 kPa or more can be obtained for coal ashes containing more than 4% in CaO content. These studies show that the slurry disposal method can be effective for land use. However, the cyclic properties including liquefaction during earthquakes have not been investigated even for dry disposal ground, in spite of their importance. In the present paper, therefore, the cyclic properties of coal ash grounds are discussed based on the results from undrained cyclic triaxial tests.

DISPOSAL METHODS FOR PFA RECLAMATION

Dry, wet and slurry disposal systems are currently be used for PFA in land reclamation in the field. Fig. 1 schematically shows a comparison of these three disposal systems. In the wet disposal system, PFA is mixed with a large amount of water, transported by construction machines, and then thrown into the ash pond. In the dry disposal system (DAS), PFA is mixed with a small amount of water, so that the optimum moisture content is nearly attained.

Generally speaking, both wet and dry disposal systems do not provide sufficient density to enhance strength, because of the loose

	System Process	Appearance
Wet Disposal System (WAS)	① Mixing Fly Ash with a large amount of sea water. ② Pumping the low density FA slurry to the ash pond. ③ Spreading the FA slurry into the ash pond.	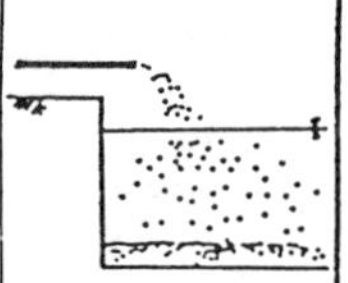
Dry Disposal System (DAS)	① Mixing with sea water at around the optimum water content of Fly Ash. ② Transfering the humidified FA to the disposal site by belt-conveyor or dump trucks. ③ Spread the humidified FA into the ash pond.	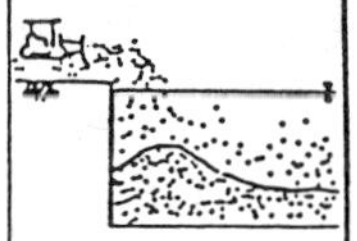
Slurry Disposal System (HAS)	① Mixing with sea water by the double mixing method to make the high density Fly Ash slurry. ② Pumping the high density slurry to the disposal site. ③ Placing the high density slurry onto the bottom of the ash pond.	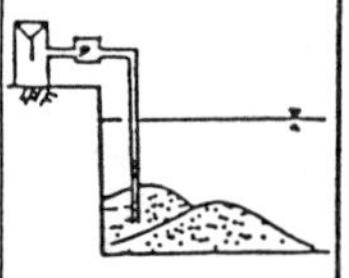

Fig. 1 Diposal methods for PFA reclamation

sedimentation of PFA underwater. In contrast to these wet and dry disposal systems, HAS has the possibility of bestowing a higher density and consequently a higher strength.

The mechanism for the enhanced strength in HAS was originally elucidated by Horiuchi et al. (1985). This method is characterized by the facts that: (1) the density of the PFA increases to up to 95% of the maximum dry density with a double-mixing technique, and (2) the HAS can be placed underwater with less mixing with water than the DAS.

LABORATORY TESTS FOR INVESTIGATION OF THE STRENGTH OF PFA

Sample Preparation

Since it is known from a previous study by Yasuhara et al. (1988) that by the wet disposal system (WAS) for PFA does not produce ground with a high density, PFA grounds produced by DAS and HAS are currently the subject of the investigation of strength characteristics. This was carried out through both plate bearing and cone penetration tests (Yasuhara et al., 1990).

The model PFA ground was produced in a container with volume 1 m^3 and height 1 m. PFA with a content of CaO of 2.0% used for the model tests was produced at a Kyushu Electric Power Co. Ltd power plant.. It was collected with electrical precipitators, and kept dry afterwards. Table 1 shows the chemical and physical index properties. We also carried out laboratory plate loading tests on PFA with the cement and the gypsum added as a hardening material.

Sea water from the port of Kanda in Kitakyushu in Japan was used to prepare the PFA mixtures in the laboratory. Ordinary portland cement was added in a small proportions (2% in weight) to enhance the strength development of PFA. The DAS mixture with sea water was by throwing PFA into the polyacryl tank containing the sea water. In HAS, on the other hand, PFA was slurried by mixing with a given volume of the sea water for 5 minutes using a pan type mixer.

Prior to the model ground preparation in the tank, twelve polyvynyl sampling moulds of 5 cm diameter and 10 cm height and coated by a teflon sheet around the inner wall, were burried at the bottom of the test tank where they would not be influenced by the distribution of load applied onto the plate. After both plate loading and cone penetration tests for investigating the bearing capacity of the model PFA ground were completed, those moulds were taken out and the

Table 1 Conditions of cyclic triaxial tests on PFA

Specific gravity	G_s	2.15
	SiO_2 (%)	60.2
	Al_2O_3 (%)	22.0
	CAO (%)	2.3
Ignition Loss (%)		3.3

samples from them were then used for static and cyclic undrained triaxial tests, and oedometer tests.

Cyclic Triaxial Tests

According to the previous study on liquefaction of PFA carried out by Miki et al. (1985) PFA was easily to liquefy. The main reason for this is that PFA contains fine-grained but inactive particles of silt. This is clear from Fig. 2 which shows that the grain size distribution curve for PFA used in the present study falls in the range for very liquefable to liquefable sand. The present paper focuses on the effects of the disposal method for reclamation on the liquefaction strength. The effect of cement addition was also investigated with reference to the feasibility of liquefaction of PFA disposed by the slurry method.

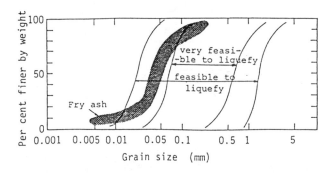

Fig. 2 Grain size distribution curve of PFA

The test conditions used are listed in Table 2. Two preconsolidation periods, 1hr and 24hrs, were used to represent the time for end of primary consolidation and the time for secondary compression, respectively.

The specimen trimmed to 5cm diameter and 10cm height was placed on the lower pedestal in the triaxial cell. Then an isotropic confining pressure of 98 kPa preconsolidation was applied to some specimens for 1hr and other specimens for 24 hr, respectively. A cyclic stress, σ_r, was then applied with a frequency of 0.1 Hz under undrained conditions. Before applying the cyclic stress, we ensured that the B-value in every specimen reached no less than 0.95.

Table 2 Chemical and physical properties of PFA

Test No	T_c*(hr)	ρ_{dc}** (g/cm^3)	e_c***	Stress ratio $\sigma_a/2\sigma_c$
DAS-1	1	1.037	1.205	0.071
DAS-2	1	1.039	1.201	0.092
DAS-3	1	1.031	1.217	0.094
DAS-4	1	1.036	1.207	0.116
DAS-5	24	1.044	1.191	0.100
DAS-6	24	1.050	1.177	0.128
DAS-7	24	1.059	1.159	0.147
DAS-8	24	1.066	1.143	0.123
DAS-9	24	1.047	1.183	0.073
DAS-10	24	1.058	1.161	0.087
DAC-1	1	1.047	1.221	0.198
DAC-2	1	1.033	1.121	0.242
DAC-3	1	1.023	1.273	0.174
DAC-4	1	1.057	1.200	0.180
HAS-1	1	1.134	1.016	0.202
HAS-2	1	1.208	0.892	0.165
HAS-3	1	1.176	0.944	0.194
HAS-4	1	1.165	0.962	0.193
HAS-5	1	1.093	1.091	0.191
HAS-6	24	1.153	0.983	0.191
HAS-7	24	1.125	1.032	0.227
HAS-8	24	1.157	0.975	0.203

* Consolidation period
** Dry density after consolidation
*** Void ratio after consolidation

In general, it is a time consuming task to saturate a specimen of a silty soil such as PFA because it contains fine grained particles. In case of specimens not achieving a satisfactory degree of saturation (B > 0.95), we adopted the procedures given by Koyama and Tatsuoka (1987).

STATIC UNDRAINED BEHAVIOUR OF PFA

To provide information regarding the effects of disposal methods on the static strength of PFA, static undrained triaxial compression and extension tests were carried out under monotonic loading (axial strain rate is 0.1% /min) on the specimens for each of three disposal methods. Each specimen was isotropically consolidated under a confining pressure of 98 kPa, 147 kPa and 198 kPa.

Fig. 3 compares the effective stress paths in p'-q space (p' is mean effective principal stress and q is principal stress difference) obtained from the results of both monotonic triaxial compression and extension tests. Because of the lower density in DAS specimens the effective stress paths descend along the critical state line (CSL) after reaching the maximum value of deviator stress,

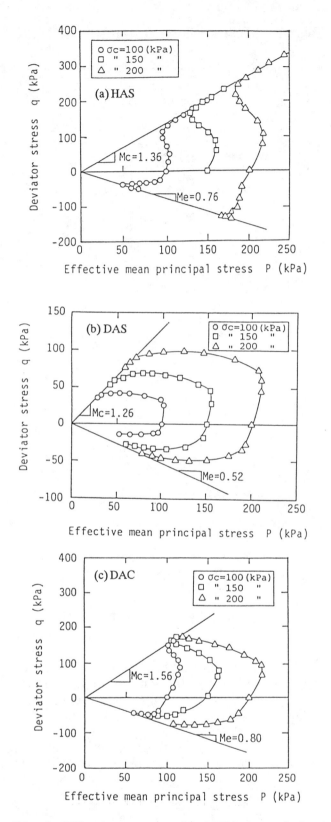

Fig. 3 Effective stress path of PFA in undrained triaxial tests

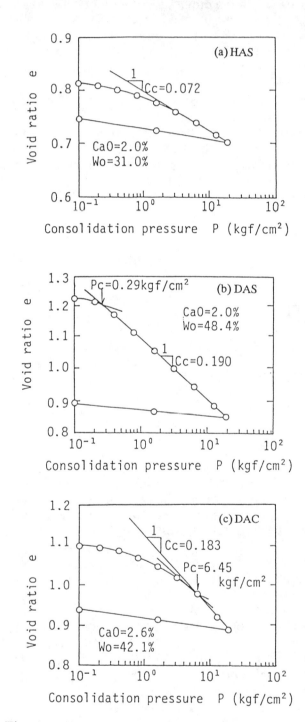

Fig. 4 e - log p' curves of PFA in oedometer tests

similar to the behaviour of loose sands. On the other hand, the stress paths of HAS specimens rise up along the CSL. This is caused by the dilatancy characteristics induced by the higher density produced in HAS. When we compare the results from tests on specimens with DAS and HAS, these dilatancy characteristics are in good accordance with the e - log p' relations from oedometer tests as shown in Fig. 4. Fig. 3 indicates that under the confining pressures used in this study the HAS specimens are in an overconsolidated and DAS specimens in a normally consolidated state. Therefore, it is understandable that during undrained tests both the specimens with DAS and HAS behave as if each were loose sand and dense sand, respectively.

The failure points of DAS with added cement tend to reach a specific point in p'-q space. The same tendency is observed on the extension side. Every specimen exhibited a clear phenomenon of necking at failure. In addition, it can be said that for specimens with all three kinds of disposal method, the critical state parameter, M_C, for the compression side is much larger than the parameter, M_E, for the extension side. The ratio, M_C/M_E, is therefore, larger than 1 for all kinds of PFA, and specifically, is largest for DAS. This fact implies that DAS is provided with the remarkable anisotropy of static strength.

CYCLIC UNDRAINED STRENGTH OF PFA

Influence of Disposal Methods on Liquefaction Strength

Fig. 5 shows typical results from cyclic triaxial tests on an HAS specimen with the time records of axial displacements, axial loads and excess pore pressures. It appears from Fig. 5 that the axial strain increases considerably when the excess pore pressure approaches the initial confining pressure ($\sigma_c = 98$ kPa). At the final stage of undrained cyclic loading, however, as is shown in Fig. 5 the excess pore pressures observed are not completely equal to the confining pressure although axial strains reach 10% or more. A characteristic of Fig. 5 is that the difference between maximum and minimum values of excess pore pressure for each cycle is larger than that one for DAS specimens as is shown in Fig. 6 for comparison. The excess pore pressure in the DAS specimens increases until finally it is equal to the confining pressure.

The deviator stress versus axial strain curves for HAS specimens during undrained cyclic loading are illustrated in Fig. 7. This shows a so called cyclic mobility which is usually observed in comparatively dense sands. Large shear strains are developed on the extension side but small shear strains on the compression side. This implies that HAS is endowed with the strong anisotropy suitable to that normally observed in undisturbed sands. This tendency must originate from the layered sedimented structure of PFA particles due to the HAS disposal method.

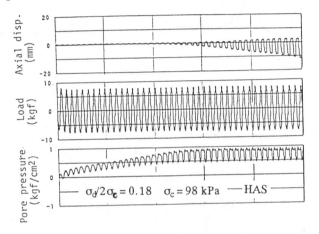

Fig. 5 Time records of HAS during undrained cyclic loading

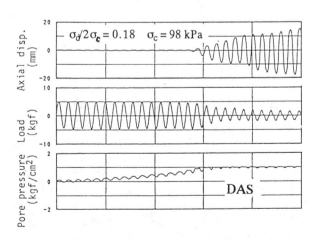

Fig. 6 Time records of DAS during undrained cyclic loading

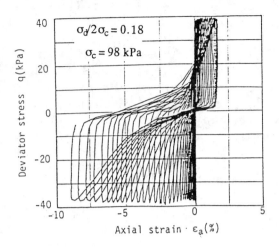

Fig. 7 Stress-strain curves of HAS during undrained cyclic loading

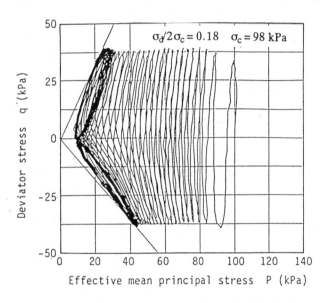

Fig. 8 Effective strss path of HAS during undrained cyclic loading

Fig. 8 illustrates typical effective stress paths using the p'-q representation with a cyclic stress ratio, $\sigma_d/2\sigma_c = 0.18$. Although it does not reach the origin of the p'-q space, it reaches failure shown by a steady loop curve. Against this loop we have drawn the failure envelope passing through the origin as shown in Fig. 8, with the different gradients of failure envelope on the compression and extension sides. This is also an indication of the anisotropy of the HAS specimen. In comparison of these values $M_C = 2.0$ and $M_E = 0.7$ under cyclic loading with the ones under monotonic loading, M_C is larger than M_E, and both M_C and M_E is slightly smaller than the respective values under static undrained monotonic loading as was shown in Fig. 3.

Effect of Cement Addition on Liquefaction Strength

To improve the resistance of DAS to cyclic loading, 1% by cement weight of cement was added. Fig. 9 shows an example of the variation of pore pressures and shear strains with elapsed time during a cyclic triaxial test. In the DAS specimen with cement addition, the excess pore

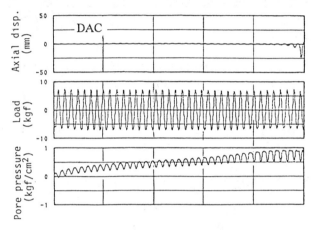

Fig. 9 Cyclic strength curves for three kinds of PFA

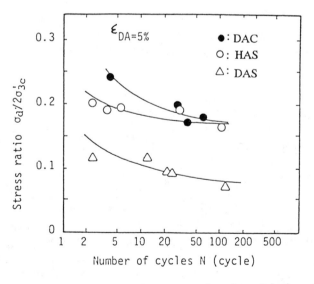

Fig. 10 Cyclic strength curves for three kinds of PFA

pressure increases monotonically with the number of load cycles. The shear strain suddenly increases on the extension side after a certain number of load cycles leading to sudden failure. In addition, shear strain does not develop on the compression side but on the extension side, exhibiting a necking-like failure. The relations for the those PFA disposal methods (HAS, DAS and DAC) between cyclic stress ratio, $\sigma_d/2\sigma_c$, needed for a 5% double amplitude axial strain, ε_{DA}, and the corresponding number of load cycles, N, are combined in Fig. 10. This indicates that DAS exhibits the weekest resistance, among the three kinds of the disposal method, to undrained cyclic loading. As shown in Fig. 10, the cyclic strength of DAC with 1% cement addition is as almost twice that of DAS without cement addition.

Influence of Consolidation Period on Liquefaction Strength

To investigate the effect of preconsolidation period on liquefaction strength, another series of cyclic triaxial tests under the confining pressure, σ_c=98 kPa, were run on PFA. We adopted in this series 24hrs as the preconsolidation period for this series of tests instead of 1hr used in the previous series. Fig. 11 shows the effect of preconsolidation period on the cyclic stress ratio versus number of load cycles relations corresponding to the double amplitude of cyclic axial strain obtained from a series of cyclic triaxial tests on the HAS. It is seen from Fig. 11 that the gradient of R (=$\sigma_d/2\sigma'_{3c}$) versus N curves with the longer period of preconsolidation (tc=24hrs) is steeper than the one with the shorted period (tc=1hr). It is also recognized that the cyclic strength of HAS increases remarkably with the increased period of preconsolidation from 1hr to 24hrs. This is considered to be probably due to densification of slurried PFA particles. The results from cyclic triaxial tests on PFA specimens with the DAS are also plotted in Fig. 11. The same tendency of the preconsolidation effect as in HAS is recognized in DAS.

CONCLUSIONS

1) Although PFA is liquefiable, cyclic strength depends on the disposal method for reclamation. Among the disposal methods currently used in the field, the HAS PFA slurry disposal method exhibits the largest resistance to cyclic loading.

2) The liquefaction potential of PFA is also improved by adding a small amount of cement. This implies that the cyclic strength improvement is governed by both the density and the amount of sodium oxide (CaO).

3) A strong anisotropy of undrained strength in PFA is observed during both monotonic and cyclic loading.

4) Cyclic strength of PFA remarkably increases with the increasing preconsolidation period. This is probably due to densification of fine-grained material contained in PFA during secondary compression.

REFERENCES

1) Coal Ash Disposal Manual (1979) : EPRI, FP-1259.

2) Horiuchi, S. et al. (1985) : Rheological and mechanical properties of high density fly ash slurry, Proc. 7th Intnl. Ash Utilization Symp., pp. 907 - 917.

3) Horiuchi, S., Tamaoki, K. and K. Yasuhara (1993) : Coal fly ash for effective underwater

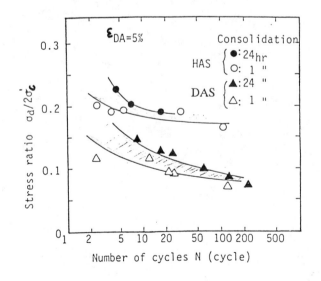

Fig. 11 Effect of consolidation period on cyclic strength curves for PFA

disposal, Soils and Foundations (to be appeared).

4) Koyama, T. and F. Tatsuoka (1985) : A simplified method to saturate undisturbed sandy silt with low permeability, Proc. 42th Annual Conf. JSCE, Vol. III, pp. 638 - 639 (in Japanese).

5) Miki G. et al. (1985) : Liquefaction potential of coarse fly ashes, 20th Japan Conf. SMFE, Vol. 1, pp. 613 - 614 (in Japanese).

6) Sugimoto, T., I. Oguchi and S. Horiuchi (1990) : Effective method for underwater disposal of coal fly ash, Proc. 2nd Intnl. Conf. Environmental Geotechnology, Vol. 1.

7) Yasuhara, K et al. (1988) : Strength and compressibility characteristics of fly ash as a reclamation material, Technical Report of Geotechnical Research Institute, NIT, Vol. 6, pp. 16 - 27 (in Japanese).

A Novel Concept of Design of an Ash Pond

M. Yudhbir, M. Kumar, S. Kumar, A. Kumar
Indian Institute of Technology Kanpur, India

Abstract

A novel ash pond design concept is presented. Induced flow failure of loose saturated fly ash deposit in the sedimentation compartment is utilized to transport it to the storage chamber. To minimize the negative impact of fly ash storage in the ash ponds, clay liners, dual media filter and nuclear waste filter columns are incorporated in design. The outflow water is recycled for making ash slurry. The proposed design is economical, encourages recycling of water and use of fly ash, and is environment friendly.

Introduction

In India around 40×10^6 tonnes of fly ash is produced annually from burning of coal by the thermal power plants. Most of these fly ashes, with the exception of limited quantities produced by burning of lignite, are of category III - low calcium variety which are produced from the burning of anthracite and bituminous coal; contain crystalline minerals like Quartz, Mullite, Hematite, Magnetite, have CaO < 5% and glass content < 50% most of which is of low quality variety (1,5). Furthermore desirable practice of dry storage of fly ash is rarely followed and generally adopted method of storage is to pump fly ash as a slurry in large ash ponds/lagoons.

In this paper the prevailing practice of design and construction of ash ponds is briefly reviewed and details of new design methodology are discussed.

Ash Ponds

Figure 1 shows some typical examples of ash pond designs built over the past one to two decades. Table I gives brief details of these ash ponds. Typically 6 to 9 m high embankments are built with compacted soil (with or without any proper compaction control). The whole pond, consisting of one or two compartments is built right in the beginning and fly ash slurry is pumped through one or more inlets. Outlet structures with filters are built for escape of water from slurry which is usually discharged into an adjoining river or a nallah. The main consideration in the location of outlet structures is their proximity to river or nallah into which the outflow is to be discharged as seen in Figure 1. Only now the issues of recycling of such large quantities of water used for making slurry are being taken up seriously. For example at the HTPP plant in M.P., water recycling plant is being designed

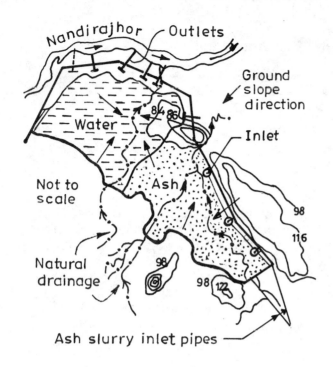

Figure 1 a. NCTPP ash pond

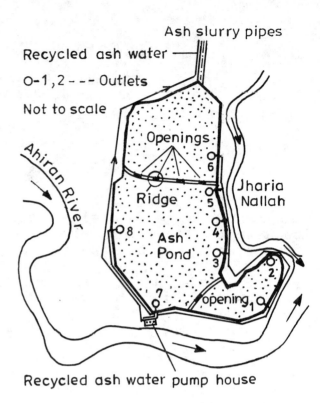

Figure 1 c. HTPP ash pond

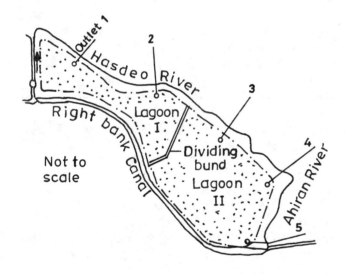

Figure 1 b. KSTPP ash pond

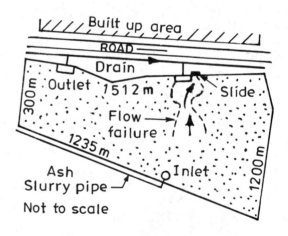

Figure 1 d. PTPP ash pond

as indicated in Figure 1.

With inadequate slurry distribution network (to economize on pipeline costs) and lack of proper grading of the pond bed to direct flow of slurry, most of the ash ponds are not filled uniformly over the whole pond area. Furthermore, except in some cases

(Figure 1.a), the outlet structures are not very effective since the relationship between preferred flow direction of the ungraded pond bed and the location of outlet filters is not considered at the time of design and construction. In fact in many cases the defective location and ineffective performance of outlet structures has led to failure of pond embankments.

Table 1

Pond/Location	Embankment Height m	Pond/lagoon area hecta.	Ash/year ton.$\times 10^6$
PTPP, U.P.	6	100	0.28
HTPP, M.P.	9	125/107*	1.1/1.65*
KSTPP, M.P.	9	304	2.8-3.5
NCTPP, Orissa	9	185/706**	1.64

* Earmarked for 3rd stage to accommodate 1.65×10^6 tonnes of fly ash
** Total area acquired for ash pond.
PKPP - Panki Thermal Power Plant
HTPP - Hasdeo Thermal Power Plant
KSTPP - Korba Super Thermal Power Plant
NCTPP - NALCO (National Aluminium Company Ltd.) Captive Thermal Power Plant

The Failure of Ash Pond Embankment at Panki Plant

At Panki a 4 meter wide section of the embankment next to an ineffective outlet structure, failed and initiated a major flow failure in the ash pond causing deep scour of saturated loose ($\gamma_d = 0.8 t/m^3$) sedimented fly ash extending over a large area behind the embankment (Figure 1.d). Close to the failure site the scour channel extended over a width of 16m and was roughly 2.5 m deep. Fly ash flowed out like a liquid through the 4 m wide orifice created by the small land slide and spread over the area filling the drain channel and the adjoining road.

The clogged outlet structure led to storage of water and rise of water level in the fly ash deposit. Poor design and construction of the embankment caused erosion at its toe due to seepage along the contact between its base and the subsoil. The erosion progressed retrogressively into the embankment and as soon as the erosion channel reached the saturated ash reservoir upstream, it collapsed and the embankment experienced mass failure. Sudden loss of support caused undrained shear of these loose saturated deposits leading to generation of positive pore pressures causing peak failure at small strains followed by sudden loss of strength. The loose fly ash deposits immediately next to the landslide underwent liquefaction and this phenomenon spread backwards retrogressively along the direction from which the water was flowing towards the embankment. The large spread of failure may also be due to the lateral migration of pore pressure since the horizontal permeability of these sedimented ash deposits is 10 to 15 times greater than the vertical permeability (3×10^{-7} m/s). The extent of failed mass was possibly restricted by the adjoining partially saturated ash deposit.

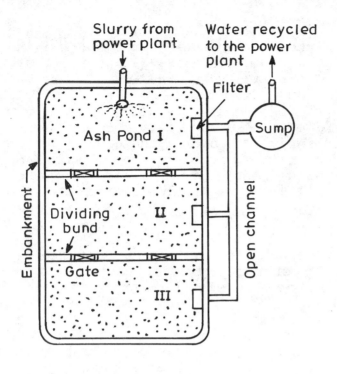

Figure 2. Proposed ash pond layout

While no detailed liquefaction investigation of these loose saturated ash deposits has been conducted, a poor man's cyclic triaxial tests were conducted on saturated samples having in-situ dry density. A triaxial sample consolidated to a pressure of 0.5 kg/cm^2 was loaded monotonically upto 0.33 times the failure stress before starting undrained cyclic loading. After 100 cycles the drainage valve was opened and the sample gave out 3 cc of water indicating development of positive pore pressures during cyclic loading.

Proposed Design Concept of Ash Pond

The proposed design concept is based on the observed liquefaction failure on sudden removal of support and the subsequent spread of failure leading to formation of wide scour channel in areas of saturated loose ash deposit. In a sense the energy of the liquefied mass is proposed to be used to transport ash in a fluid form from the sedimentation chamber to the adjoining storage chambers of the ash pond.

As shown in Figure 2 the ash slurry is discharged into the sedimentation chamber and during initial filling phase the gates in the divide bund are kept closed. The bed of the ash pond is suitably graded to ensure flow of water towards the outlet structure. As the sedimentation chamber fills up, ash is taken out to construct the embankment of the adjoining storage chamber whose bed is kept about 0.5 m lower than that of the sedimentation chamber. When the sedimentation chamber is about to be full, the water outlet is closed so as to convert the ash deposit into a fully saturated ash with water level near the ash surface. The gates are then suddenly opened to induce liquefaction failure and transport of ash as a mud flow into the adjoining storage chamber. The gates are then closed and the slurry is again deposited in the sedimentation chamber with the outlet structures fully open. This process can be repeated to transfer fly ash from the first storage chamber to second storage chamber.

Some of the salient features of the proposed design are: substantial reduction in initial investment in acquisition of borrow areas and construction of ash pond; use of fly ash to construct storage chamber embankment and thus also increasing capacity of

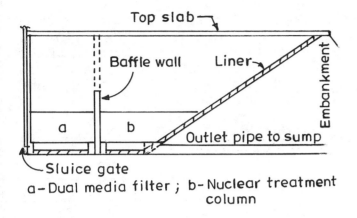

Figure 3. Outlet structure

sedimentation chamber over the designed plant life; recycling water for slurry making; its environmentally friendly nature.

Recycling of Ash Slurry Water

At present at most of the plants, the outflow from the ash pond is generally discharged into natural rivers or specially built drains. Water for slurry making is used in the ratio of 20 to 30:1 (water to ash) and in most cases it is supplied from adjoining rivers or through specially built water supply canals. At Panki about 6×10^6 m^3 of water is used annually for slurry making. The proposed design concept incorporates (Figure 2) the idea of recycling of this water as also the control of slurry water into the ground water through infiltration from the bed of the ash pond.

Design of Liner and Filters

Earthen liner is used as a hydraulic barrier for preventing seepage water and with it leached toxicants in the ground water. In the proposed design a 30 cms thick clay liner compacted at 2% wet of proctor optimum moisture content and having coefficient of permeability of 10^{-7} cm/sec is provided (Figure 3).

As part of the water recycling operation, filters are provided both above the liners and in the outlet structures (Figure 3). A dual media filter, 1.5 m thick is provided with 0.75 m layer of sand, and 0.75 m layer of crushed anthracite, D_{15} for both layers being 0.5 mm. Filtration rate through the media is 20 m^3/m^2/hour. Other design details are given in Kumar et al (3). An other filter for treating nuclear wastes is also provided. Fly ash contains cadmium and in some cases uranium also. This treatment is essential in a scheme envisaging recycling of slurry water so as to prevent concentration of these harmful contaminants.

Design of the nuclear waste filter column (Figure 3) is based on the utilization of microbes such as fungi, bacteria, yeast and algae for removal and recovery of heavy metals and radionuclides. The phenomenon of biosorption appears to offer a potentially efficient economical route to treatment of low level aqueous radioactive wastes. Here Ganoderma lucidum - a non viable, non-edible wood rotting macro fungi is used as the biosorbent. A plan area of 75 m^2 with 1.5 m thickness is provided as the filter column. Details of nuclear waste filter column design are reported by Murleedharan and Venkobachar (4) & Kulshreshta, (2).

Summing Up

The proposed design concept utilizes the energy of an induced flow failure to transport fly ash from the sedimentation chamber. Use of fly ash from the sedimentation chamber to construct embankments of storage chambers will not only be economically attractive, it will also increase the design life of the sedimentation chamber. Recycling of water from the ash pond and incorporation of liners and filters will help mitigate the negative impact on environment arising from the storage of ash in the lagoons.

Reference

1. Das, S.K.(1992). Morphological, Chemical and Mineralogical Characterisation of some Indian Fly ashes. M.Tech. thesis, Civil Engineering Department, IIT Kanpur, India.

2. Kulshreshtra, M. (1992). Bio-sorption of U(IV) from simulated aqueous low level wastes and immobilization of loaded biomass. M. Tech. Thesis, Civil Engineering Department, IIT Kanpur, India.

3. Kumar, M et al (1992). Design of an ash pond for a thermal power station. B.Tech project, Civil Engineering Department, IIT Kanpur, India.

4. Muraleedharan, T.R. and Venkobachar, C (1992). Mechanism of biosorption of copper (II) by Ganoderma Lucidum. Biotechnology and Bioengineering, 35, 320-325.

5. Yudhbir & Honjo, Y (1991). Applications of geotechnical engineering to environmental control. Theme Lecture, IX Asian Regional Conference on Soil Mechanics & Foundation Engineering, Bangkok, Thailand.

Mining

Seepage Patterns in Tailings Dams: Implications for Operation and Remediation

Richard R. Davidson
Woodward-Clyde Consultants, Denver, Colorado, USA

Final manuscript not received in time for inclusion in the *Proceedings*. For additional information, please contact:

> Mr. Richard R. Davidson
> Woodward-Clyde Consultants
> Stanford Place 3, Suite 600
> 4582 S. Ulster Street
> Denver, CO 80237
> USA
> Voice: +303-694-2770
> Fax: +303-694-3946

Environmentally Acceptable Tailings Dams

G.E. Blight
Witwatersrand University, Johannesburg, South Africa

ABSTRACT

When geotechnical engineers consider tailings dams, they usually deal with only the geotechnical aspects of what is a much larger and more embracing environmental problem. The object of this paper is to set out a broad framework for all of the technical aspects of tailings dam design, without going too deeply into the detail of any aspect. In this way it is hoped to provide a framework for understanding the broad principles and practices of tailings management, and in particular, those aspects that directly impact on the environment.

1. INTRODUCTION

Although there many sets of national and provincial guidelines for the design of tailings dams (eg. the Canadian "Pit slope manual" (1), the South African "Chamber of Mines guidelines" (2), the Australian "Code for radioactive wastes" (3), the ICOLD "Manual on tailings dams" (4), and the British Columbia "Health, safety and reclamation code" (5), these usually relate specifically to local conditions and to local mining or environmental legislation. The passages that follow attempt to set out the essential elements for the design of environmentally acceptable tailings dams. An attempt has been made to do this in a neutral way, so that this paper can be applied in any region, with appropriate adjustments for local conditions, such as climate, topography and labour sophistication and supply. The points covered in this survey are not all new developments, but show how the process of tailings dam design is correctly becoming more concerned with the integrated design of a complete system that considers the geotechnology, hydrology, operational and environmental issues associated with tailings disposal, and includes the dam and its appurtenant works.

With regard to labour, in developed economies, the operation of tailings disposal systems may call for a technically-sophisticated, highly-mechanised approach with a small, but skilled labour content. In developing countries which often have high unemployment rates, however, it is more often desirable to use labour-intensive methods that create employment, with little technical sophistication. A prime example of a country where this approach is appropriate is South Africa, where the unemployment rate in 1993 was estimated to exceed 45 per cent of potentially economically active people.

In any case, the object is to design an appropriate system for efficient, safe, economic and environmentally acceptable disposal of tailings or tailings-like industrial wastes.

2. ENVIRONMENTAL CONCERN

The design of every tailings disposal system should be embarked on with an environmental impact study. This study should go through a preliminary phase, before the tailings disposal system is designed, and should consider such matters as alternative sites and proposed end uses of the abandoned and rehabilitated dam, as well as identifying the principal environmental concerns. Cognizance can then be taken of these concerns in the preliminary and final designs, which should each be checked for environmental impact.

A survey shows, however, that most current guidelines to the design and operation of tailings dams do not refer specifically to environmental concerns. Exceptions, and notable developments are the introduction of the EMOS (Environmental Management Overview Strategy) in Queensland, Australia and the similar EMPR (Environmental Management Programme Report) in South Africa (6). The preparation of an EMPR, for example is required for every new and existing prospecting, quarrying and mining project. The EMPR has as its basis, an Environmental Impact Report for the project which should identify and assess the severity of every potential or actual negative impact and describe the design or management measures to be taken to mitigate each impact. The EMPR must give details of how these measures are to be integrated into the day-to-day running of the project, from turning the first sod, through operation, to closure and post-closure maintenance and monitoring. A very important aspect of an EMPR is that it must describe the financial provisions that are necessary to implement the EMPR throughout this extended period. A licence for any project is issued on the basis of a satisfactory EMPR, and satisfactory financial arrangements. The licence can be withdrawn at any stage, should it become apparent that the Environmental Management Plan is not being adhered to.

3. TAILINGS AS A MATERIAL

Tailings consist of the barren residues of ores, left after the commercially-sought minerals have been removed. As the extraction process is never 100 per cent efficient, tailings always contain a small mineral content that may justify re-processing at some time in the future. The extraction process usually requires milling or comminution of the ore to liberate the mineral, Tailings therefore generally consist of sand and silt - sized particles. Because the cost of milling must be offset against the value of the recovered mineral, base-metal tailings are generally not milled as finely as tailings from precious-metal recovery. Figure 1 illustrates the range of particle size distributions presented by tailings arising from various crushed ores.

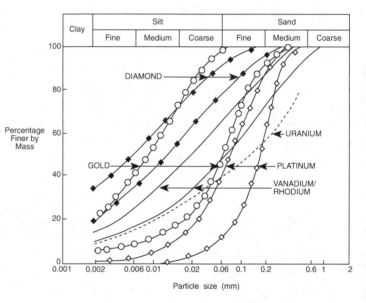

Figure 1: Particle size analyses for tailings from milled ores, diamond, gold, platinum, uranium and vanadium/rhodium tailings.

4. CONSTRUCTION OF TAILINGS DAMS

Mineral extraction usually takes place by wet

processes in which the mineral is separated from the host rock by froth flotation, heavy medium flotation, dissolution by cyanide, washing, etc. As a result, most tailings and many industrial wastes originate as water-borne slurries of solid particles. As hydraulic transportation by pumping or is usually the cheapest method of transporting particulate solids, tailings are usually transported as a slurry and deposited hydraulically

When tailings comprise a wide range of particle sizes, (e.g. the tailings size distribution for diamond tailings shown in Figure 1) it is common practice to cause a separation between the sand, or coarse fraction of the tailings, and the fine fraction, usually termed slimes. The sand is used to build the outer wall or shell of the impoundment, while the slimes is delivered into the body of the impoundment. When the product has a narrow range of particle sizes, and the fine fraction is not clayey, the total product may be used to build the outer wall.

When a tailings slurry is discharged on a tailings dam, it will run from the point of deposition towards the pool (see Figure 2), forming a tailings beach in the process.

The profile of the beach is characteristic for a particular product and slurry water content. If the dimensions (length and elevation) of a tailings beach are non-dimensionalized, then all beaches of the same product, discharged at similar water contents, will adopt the same non-dimensional profile, regardless of the distance and difference in elevation between the point of deposition and the edge of the pool (eg. (7)). The resulting beach contours are similar to those shown in Figure 2. In each case, a section normal to the contours will have the same non-dimensional, or "master profile".

In the process of beach formation, the coarser fraction of the tailings will settle out closer to the point of deposition, while progressively finer material will settle out as the pool is approached.

The result of the gravitational particle sorting is that the permeability of the deposited tailings will be greatest near the point of deposition and will decrease progressively towards the pool.

5. HYDROLOGICAL DESIGN OF TAILINGS DAMS

Good environmental practice requires that any water that falls, or originates outside of the tailings impoundment area should not be allowed to become contaminated by contact with the tailings. A surface water diversion system must be designed and constructed to ensure that the separation of polluted from unpolluted water is maintained.

Similarly, all water falling within the contaminated area must be retained therein. Figure 3 shows how the principle of catchment separation is applied to a ring-dyke dam:

The tailings dam and the area surrounding it is divided into three catchments:

- Catchment 1 consists of the top surface of the impoundment which drains towards the penstock, together with the associated return water reservoir and its catchment.

- Catchment 2 consists of the outer slopes of the ring dyke impoundment together with their erosion catchment paddocks.

- Catchment 3 is the unpolluted water catchment which is separated from catchments 2 and 3 by a storm water diversion drain.

- Catchments 1 and 2 are each designed to be self-contained and to retain any precipitation that falls within them. There will be a draw-off from the return water reservoir, and catchment 2 may be arranged so that it spills into catchment 1.

division wall, so that one half can periodically be drained and de-silted.

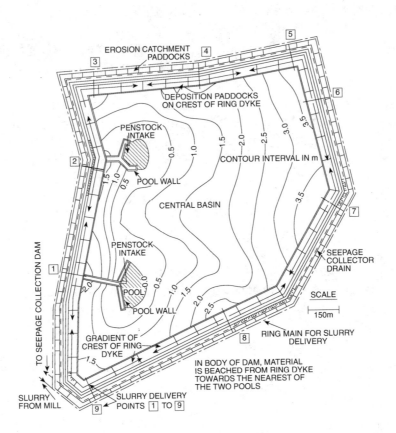

Figure 2: Ring-dyke tailings dam showing slurry delivery points, decant pools and contours of beaches.

Typically, such a system would be designed to retain the maximum precipitation to be expected over a 24 hour period with a frequency of 1 in 100 years and still maintain a freeboard of at least 0.5 m. It should be noted that the return water reservoir is usually designed with a

6. THE WATER BALANCE FOR A TAILINGS DAM

Because the water in a tailings dam system is usually polluted, because of water shortages in arid areas, or for reasons of economy, it is usually necessary to operate the dam and its associated catchments either as a closed system from which water is not released, or as a system from which only controlled water releases at specific times are permitted. It is therefore

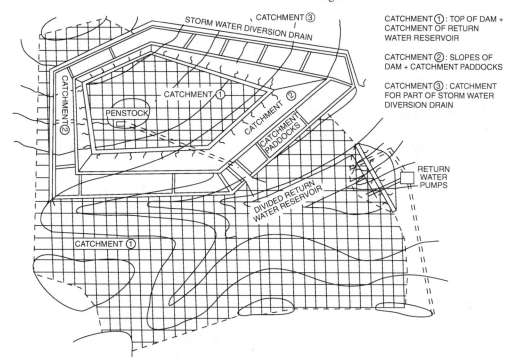

Figure 3: Diagramatic representations of hydrological components of a tailings dam.

necessary to do a water balance calculation for each sub-catchment of each dam to assess whether and/or when it will be necessary to spill water from the system. The water balance is stated as follows:

Water Input - Water Stored - Water Losses = Water surplus

The components of the water balance are shown diagrammatically in Figure 4.

The water input for the tailings impoundment (catchment 1 in Figure 3) consists of:

- water delivered with the tailings,
- precipitation on the surface of the tailings dam,
- any additional water that is disposed of on the dam.

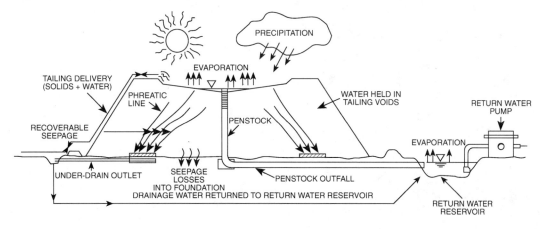

Figure 4 : Diagrammatic representation of the water balance for a tailings dam

The water stored consists of:

- water held in the void space of the tailings,
- water held in the pool around the penstock,
- water held in the return water dam. (This includes seepage into the underdrainage system, which is usually collected and channelled into the return water dam, see Figure 3).

Water losses comprise:

- seepage losses into the foundation of the dam which are irrecoverable.
- evaporation from the beaches and pool of the dam and from the return water dam.

Because the precipitation (as water) and evaporation vary throughout the year, the water balance must be calculated at least on a monthly basis, and preferably at weekly intervals.

If the calculated water surplus is positive for a particular time period, water will accumulate within the system and it may be necessary to allow excess water to spill. If the water surplus is negative, the volume of water within the system will be decreasing, and it may be necessary to make up the water within the system.

After tailings disposal ceases, the water input to the dam reduces to infiltration from natural precipitation. If permanent provision is made for draining the pool of the dam, this infiltration will be quite small, because the surface permeability of the top surfaces of tailings dams is usually not large. It is then possible, even in climates that have an excess of precipitation over potential evaporation, to maintain a permanent water deficit in the decommissioned dam. This will, in the long term, eliminate, or greatly reduce any seepage of polluted water from the dam.

7. TAILINGS DAM STAGE CURVES

Stage curves for a tailings dam consist of a set of relationships between the:

- area of impoundment,

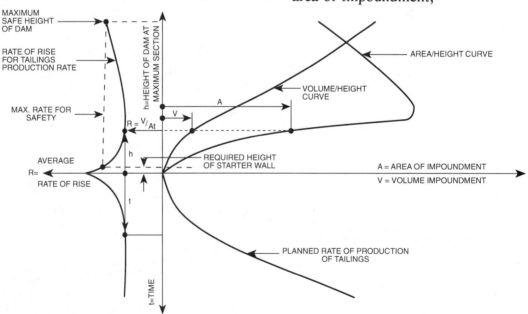

Figure 5 : Typical stage curve for ring-dyke tailings dam

- volume of tailings impounded,
- maximum required height of impoundment,
- rate of rise of outer wall, and
- time.

Stage curves are used both in planning the layout of a dam to meet the basic design requirements of containing the tailings production, as well as in the hydrological design and the stability design of the outer wall.

A typical set of stage curves for a ring-dyke impoundment is shown in Figure 5.

It will be seen from Figure 5 that whereas the volume of tailings impounded increases continuously with height, once the plan area of a ring-dyke dam has been covered with tailings, the top area starts to decrease as the side slopes draw in. With a valley dam, the top area usually increases continuously.

The rate of rise of a dam is usually at a maximum shortly after commissioning the tailings disposal system and declines thereafter. In the case of a valley dam, the rate of rise will usually decrease continuously, and the area covered will increase continuously as the height increases. With a ring dyke dam, the rate of rise will increase again as the top area decreases.

The initial rate of rise will usually exceed the rate at which the deposited tailings can consolidate, and it will be necessary to provide a starter wall to retain the tailings until the rate of rise has reduced to a value for which the tailings wall will be stable.

In the case of both ring dyke and valley dams, a limiting height and rate of rise may be reached at which the dyke or impoundment wall ceases to be stable. This consideration will limit the volume of tailings that the dam can hold.

8. TAILINGS DAM OPERATION AND MANAGEMENT

A prime objective of tailings dam management should be to maximize the tonnage of tailings solids stored within a given volume of impoundment. In other words, the aim should be to minimize the void ratio of the stored tailings, which in turn means that the effective stress, to which the stored tailings are subjected, should be maximized. This in turn will maximize the shear stability of the outer wall of the tailings impoundment for a given configuration.

There is a set of simple rules to follow in order to achieve the above ideal:

- The pool of the tailings dam should be kept as small as possible, the penstock inlet should be kept central in the pool and should be located as far away as possible from the outer wall of the dam. This will keep the phreatic surface in a depressed position and subject the tailings above the phreatic surface to consolidation under capillary stresses.

- The tailings dam impoundment should be designed for as long a life as possible. A long operational life will usually mean a slow rate of rise, and hence dense, fully consolidated stored tailings.

- A tailings beach should, to the greatest extent possible, be allowed to dry out, and, preferably to become sundried between successive depositions.

A most important consequence of operating a dam so as to minimize void ratio, is that the distinction between upstream, centre-line and downstream methods of construction becomes blurred. In the case of a dam with a large water deficit and where extensive sun-drying is possible, the distinction may disappear completely. In a case such as this, an upstream embankment may have an outer wall that is every bit as strong and substantial as the corresponding downstream embankment.

9. CONTROL OF EROSION OF OUTER SLOPES OF TAILINGS DAM

Losses of tailings from the slopes of tailings dams of as much as 500 tons per hectare of slope area per year have been measured, hence erosion can pose a very significant maintenance and environmental problem. Material can be lost by both wind and water erosion. The loss varies seasonally and is dependent on local climatic conditions.

Most of the body of knowledge concerning the mechanics of erosion relates to agricultural fields. Research has shown that erosion losses are principally related to the length and inclination of the ground surface and other factors such as crop cover and intensity of rainfall.

Research in South Africa (8) has shown that for ring-dyke tailings dams:

- erosion losses resulting from rainfall and wind are roughly equal in magnitude;
- most of the erosion occurs from the slopes of a tailings impoundment, and very little from the top surface;
- erosion losses are roughly proportional to slope length; and
- relatively little erosion occurs from slopes flatter than 20° or steeper than 40°.

Erosion losses reach a maximum for slope angles of between 25° and 35°. While vegetational cover on a slope inhibits wind erosion, it does relatively little to prevent erosion by water. The conclusions regarding slope length and angle are shown schematically

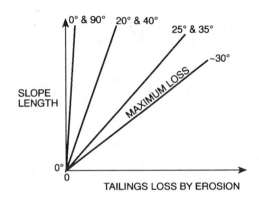

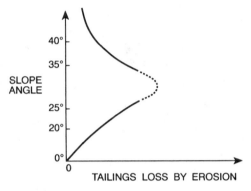

Figure 6 : Effects of slope length and slope angle on rates of erosion from the outer slopes of tailings dams.

in Figure 6. A further rider is that slopes should be protected from water cascading down from either the top of the dam wall, or from slopes higher up. To summarize, to reduce erosional damage to the slopes of a tailings dam:

- slope lengths should be broken by the introduction of berms or steps-back;
- dam crests and the edges of berms should be graded away from the edge, so as to prevent water cascading over the edge. Water must then be drained from the surfaces of berms by means of penstocks and pipes or armoured surface drains.
- slopes should be vegetated as soon as possible. This will reduce wind erosion, but will do little to reduce water erosion.

Finally, because material will inevitably be lost from the slopes of tailings dams, provision should be made to catch the lost material at the toe of the dam by means of silt catchment paddocks or traps, such as those indicated around the perimeter of the dam shown in Figure 2.

10. DESIGN OF THE OUTER SLOPES OF A TAILINGS DAM

The design of the outer slopes of a tailings impoundment depends on a number of factors, most of which have been described earlier in this paper. They include:

- the proposed end use for the completed tailings dam and plans for its rehabilitation;
- an assessment of short and long term erosion losses from the slopes and measures for combatting these;
- the geotechnical characteristics of the deposited tailings, which include characteristics that arise from the methods of deposition and operation.
- features built into the outer slopes, e.g. underdrainage systems, underlining systems, etc.

In the case of coarse, non-plastic tailings, if the dam is correctly operated, the deposited tailings should be:

- completely consolidated, possibly even over consolidated by sundrying, and dense;
- the phreatic surface should be low, and ideally at the level of the base of the dam;
- the outer wall should consist of a substantial wedge of dense material that has either been densified by compaction or by drainage and sun-drying;
- in the case of centre-line or upstream embankments, the transition from outer shell to impounded material should be gradual, again providing a substantial outer wedge of dense, well consolidated and drained material.

11. DESIGN IS AN ONGOING PROCESS

Tailings disposal complexes often have to be designed before the first tailings have been produced, or else when only samples from a pilot scale metallurgical process are available. These samples are often not representative of the tailings that will eventually be produced from the full-scale plant. Also, it is never certain at the design stage how well the extraction process can be controlled, and hence how consistent the tailings properties will be. Furthermore, it is never certain that the operational conditions foreseen for the design will be realized in practice.

Hence the designer and the mine management must very clearly understand that the "final" design on which the preliminary works for the tailings dam will be based, is only provisional, and must be confirmed once tailings are being produced by the prototype plant. Thus the design of a tailings dam is not a once-off process, but must be subject to continuous examination and confirmation during the first year or so of the dam's operation, and to periodic review thereafter.

The objectives of these periodic reviews (which should be carried out at two yearly intervals, at least) are to confirm the following:

- that the proposed method of deposition and operation are feasible in practice and are being realized;
- that the beach angles and profiles achieved on the dam are those foreseen for design;
- that the method of operation is maintaining a small pool that is correctly located around the penstock inlet;
- that the water balance is correct as forecast;
- that the actual stage curves are in agreement with those forecast in design;
- that the properties of the deposited tailings, the position of the phreatic surface, etc. are as

planned;
- that the outer slopes of the dam are being constructed to the correct slope; and
- that the safety of the outer slopes of the dam is as forecast.

The effect of any deviation from the predicted conditions must be assessed and the design, if necessary, modified accordingly.

12. CONCLUSION

This paper has dealt with the overall geotechnical aspects of environmentally acceptable tailings disposal, and has attempted to paint a broad picture, without getting too involved in details. It is hoped that this broad picture will provide a framework that will place in perspective the various detailed aspects that are discussed in the rest of the conference.

It must also be remembered that geotechnology does not cover all aspects of environmentally acceptable tailings disposal. Tailings management is a multi-disciplinary activity that should involve people from a range of disciplines, including mining engineering, geochemistry, geohydrology, microbiology, botany, soil science and agronomy. Each discipline has its role to play in reducing the negative impact of tailings disposal on the environment.

13. REFERENCES

1. Pit and slope manual, Chapter 9, Waste Embankments (1977), Canada Centre for Mineral and Energy Technology, Canmet Report 77-01, Ottawa, Canada.

2. Handbook of guidelines for environmental protection, vol 1 (1979, revised 1983 and 1993), The design, operation and closure of metalliferous and coal residue deposits, Chamber of Mines of South Africa, Johannesburg, South Africa.

3. Code of practice on the management of radioactive wastes from the mining and milling of radioactive ores (1982), Commonwealth of Australia, Department of Home Affairs and Environment, Canberra, Australia.

4. Manual on tailings dams and dumps (1982) International Commission on Large Dams Bulletin No.45. ICOLD, Paris, France.

5. Health, safety and reclamation code for mines in British Columbia (1992) Resource Management Branch, Ministry of Energy, Mines and Petroleum Resources, British Columbia, Canada.

6. Fuggle, R.F. and Rabie, M.A. (1992), Environmental management in South Africa, Juta, Cape Town, South Africa, 823.

7. Blight, G.E. and Bentel, G.M (1983), The behaviour of mine tailings during hydraulic deposition. Jour. S. African Inst. of Mining and Metallurgy, 83(4) 73-86.

8. Blight, G.E. (1989), Erosion losses from the surface of gold-tailings dams. Jour. S. African Inst. of Mining and Metallurgy, 89 (1) 23-29.

Cover Technology for Acidic Tailings: Hydrogeological Properties of Milling Wastes used as Capillary Barrier

M. Aubertin, Professor, R.P. Chapuis, Professor, M. Aachib, Research Assistant, J.F. Ricard, Research Assistant, L. Tremblay, Research Assistant
École Polytechnique, Montréal, Québec, Canada

B. Bussière, Research Associate
Unité de Recherche et de Service en Technologie Minérale (URSTM), Rouyn-Noranda, Québec, Canada

Abstract Mining activities can be the source of various environmental problems. Among these, acid mine drainage (AMD) produced by sulfide tailings (milling wastes) is probably one of the most serious challenges to the mining community, in Canada and elsewhere around the world. AMD induces a lowering of the subsurface water pH, down to values of 2 to 3, which in turn favours solubilization of potentially toxic heavy metals. There are only a few available alternatives to reclaim such tailings ponds, among which the use of covers appears as one of the most practical. Such covers, which aim is to limit the flow of water and/or oxygen (necessary elements for the production of AMD), are traditionally built from fine-grained soils. Because these materials are not always economically available, the authors have proposed the use of fine tailings themselves as capillary material in layered cover systems. In this article, some of the basic mechanisms controlling the efficiency of such cover systems are presented, together with recent hydrogeological laboratory tests results on mine tailings, including consolidation, permeability and capillary characteristic curve.

Résumé L'industrie minière peut être la source de divers problèmes environnementaux. Parmi ceux-ci, le drainage minier acide (DMA) produit par les rejets de concentrateur contenant des sulfures constitue probablement le défi le plus sérieux auquel doit faire face la communauté minière, au Canada et ailleurs dans le monde. Le DMA induit une baisse du pH des eaux souterraines, jusqu'à des valeurs autour de 2 à 3, ce qui favorise la mise en solution de métaux lourds potentiellement toxiques. Il n'existe que quelques solutions applicables afin de restaurer les parcs à résidus miniers générateurs d'eaux acides. L'utilisation de barrières de recouvrement apparaît actuellement comme une des plus pratiques. De telles barrières, qui ont pour but de limiter l'infiltration de l'eau et/ou de l'oxygène (deux éléments nécessaires à la formation du DMA), sont habituellement construites à partir de sols fins. Malheureusement, de tels matériaux ne sont pas toujours économiquement disponibles. Alternativement, les auteurs ont proposé l'utilisation de rejets de concentrateur comme matériau capillaire dans une barrière multicouche. Dans cet article, certains des principaux mécanismes qui contrôlent l'efficacité d'un tel système de recouvrement sont énoncés et quelques résultats d'essais récents sur les propriétés hydrogéologiques de rejets de concentrateur sont présentés, notamment sur la consolidation, la conductivité hydraulique et la courbe caractéristique de capillarité.

1 - Introduction

The mining industry provides an essential contribution to the economy of several provinces across Canada. For instance in Québec, it is estimated that mining activities account for about 40% of the gross interior product for the primary sector, and that it generates more than 10% of the province exports. Nevertheless, mining operations can also be the source of various detrimental effects for the environment. Probably the most serious problem associated with mining activities is the so-called acid mine drainage (AMD) produced by tailings (milling wastes) containing sulfides; AMD is generated when sulfide minerals (essentially iron sulfides) are in contact with water and oxygen (e.g., Ritcey, 1989; SRK, 1989). This acid water may contain high levels of potentially toxic heavy metals, such as lead, cadmium, mercury, and nickel, which constitute a serious hazard for the local ecosystems.

The control of acidic effluents during and after mining operations is often very costly. Although water treatment is an efficient process, used with success by mines for decades, it can become a heavy financial burden on any mining company faced with the prospect of having to control water quality for tens if not hundreds of years.

One possible alternative is to control the production of AMD. This approach is often considered when one wishes to reclaim the land and return it to a productive state. Among the few techniques available for that purpose, the use of covers (or caps) installed over existing tailings ponds is probably the most practical option for the time being. This approach, however, is still somewhat controversial as it is very difficult to predict and control the behavior of such covers for long periods of time when exposed to climatic effects such as freeze-thaw and wetting-drying cycles. Moreover, building such covers can be very expensive, as most estimates are above 100 thousand dollars/hectare. This would mean that reclaiming costs for all mining wastes across Canada would surpass the Can$ 5 billion mark (Itzkovitch and Feasby, 1993). In order to reduce somewhat these costs, the authors have proposed, a few years ago, to use tailings that ideally would be free of acid generating materials, to built such covers. This solution would be advantageous for various reasons, including the fact that such materials are often available close to the site. It is also an interesting solution for mills that treat ores free of sulfides, as the resulting tailings could be used to cover those that are generating AMD. Another application is related to recent projects where mining companies have used a separating technique to produce some tailings without sulfide, as these could be used to build part of a cover system.

In this article, the various components of a multilayer cover system are described, together with their specific functions and the type of materials needed. Because the key for success with such complex cover is the fine grain material layer, which in this project would be built from sulfide free tailings, some of their hydrogeological properties are given, as obtained from recent experimental tests which include consolidation, permeability and capillary characteristic curve.

2 - Cover Systems

When AMD producing tailings are stored on the surface, the bottom of the pond and the retaining dams should be impervious enough so that no contaminated water would flow down to pollute the soil and the natural ground water. By measuring the amount of water in and out of the pond, one can monitor the efficiency of the hydrogeological barriers put in place, and control the quality of the final effluent. Unfortunately, past practices often did not include such protection so that in some cases, it may be difficult to know what portion of the infiltration water is captured and treated.

If a tailings pond is to be reclaimed, it is advisable to try to stop production of AMD. It is often considered that a cover that limits the flow of water and/or oxygen is the most practical approach for that purpose. Such cover is called "humid" when water is used to submerge the tailings, thus reducing oxygen flux to a hopefully negligible level (SRK, 1989). However, the actual efficiency of this technique for old tailings ponds already producing AMD has been questioned by some, following inconclusive experimental results (e.g., Ritcey, 1989, 1991).

Furthermore, such water covers may be difficult to build and maintain over time, as topography and long term stability of the dams become key factors to the success of the project.

One could also make use of geosynthetics (geomembranes) as an impervious layer in a cover, but costs and durability are major concerns in this case.

Because there is a great deal of experience available from the use of so-called "dry" covers built from geological materials, mostly for industrial and municipal wastes (e.g., Oakley, 1987; Oweis and Khera, 1990), it is often considered in reclamation projects for AMD situations. And although such type of cover is not free of potential problems either (e.g., Sutter et al., 1993), it represents one of the most practical solutions presently available to mining companies who wish to close down their operation and reclaim the land on which tailings ponds were installed.

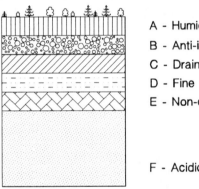

A - Humid layer
B - Anti-intrusion layer
C - Drainage layer
D - Fine grain material layer
E - Non-capillary layer

F - Acidic tailings

Figure 1 - Multilayer cover system
(after Aubertin and Chapuis, 1991)

In order to be efficient to control the production of AMD, it is now generally admitted that such barrier should be made from a multilayer system, each layer having its own specific function (e.g., SRK, 1988; Nicholson et al., 1989, 1991; Collin and Rasmuson, 1990; Aubertin and Chapuis, 1991; Aachib et al., 1993). A schematic representation of such a multilayer cover is presented on Figure 1. Starting from the top, the following layers are encountered: a humid layer to support vegetation (layer A, thickness t ≥ 15 cm), a coarse material layer containing a large portion of cobbles to prevent biological intrusions from roots and animals (layer B, t ≥ 30 cm), a sandy material acting as a drainage layer (layer C, t ≥ 30 cm), a fine grain material layer acting as a capillary retention zone (layer D, t = 50 to 150 cm), a non-capillary layer (layer E, t ≥ 30 cm) to stop capillary rise from the tailings underneath (layer F). Of course, each adjacent layer of the system must satisfy filter criteria (e.g., Lambe and Whittman, 1979) so that no particles migration would affect the integrity of the barrier. In this multilayer structure, the two coarse grain material layers placed above and below the capillary layer play a double role. First, these materials (typically sands) allow the lateral flow of water to the drainage zones built on the site. Second, the grain size contrast with the fine-grained material produces a large difference in suction properties which serves to maintain the middle layer close to saturation (Akindunni et al., 1991; Yanful and Aubé, 1993; Crespo, 1994). Having a saturation ratio close to 100% in this capillary layer is essential to provide an efficient barrier to oxygen transport into the tailings below.

In this complex cover, it should be possible to use various types of mining wastes for the construction of the different layers. The fine fraction of tailings, obtained by natural segregation or by the use of hydrocyclones, could be used to build the capillary layer (layer D on Figure 1). The coarse fraction (sands) of these tailings could then be used in layers C and E, depending on their availability and hydrogeological properties. On the other hand, layer B could include some waste rock from the mine. Finally, the humid layer A should be made from the excavated overburden material, with the original topsoil (stacked and protected) used as the final cover.

Because the efficiency of such a cover system depends on its capacity to reduce water and/or oxygen flow, the most critical component becomes the material used in layer D. This is why most of this ongoing experimental study is aimed at this material, which in this case is simply tailings recovered from various sites located in Québec. In the following section, some laboratory tests results are presented.

3 - Properties of Homogeneous Tailings

This project started with the sampling of over 30 mining sites across the province, most of which being located in the Abitibi region (Aubertin and Bussière, 1991). Preliminary tests, including mineralogical analysis, grain size determination and Atterberg limits, allowed the selection of three sites where the tailings were subsequently sampled again for more detailed studies. The grain size curve of these three tailings (identified as BE, SE and SI) are shown in Figure 2; these curves represent average grain size for hard rock mine tailings. According to the Unified Soil Classification System (Casagrande, 1948), these materials are sandy silts or silty sands with low plasticity. None of the three tailings contain sulfides.

Consolidation: Tailings sampled in bulk were homogenized and then submitted to various tests in the laboratory. Consolidation properties were studied with

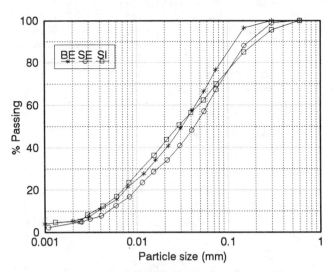

Figure 2 - Sieve analysis results for the 3 tailings

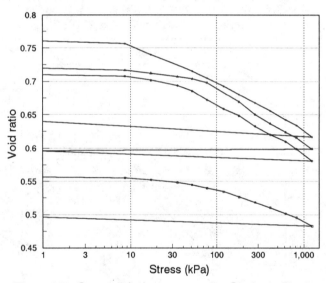

Figure 3 - Consolidation curves for SI site tailings
(after Bussière, 1993)

conventional oedometer apparatus. In this case, the compaction energy for placement of the material was controlled so that the initial void ratio e could be varied from 0.5 to 1.1; the required energy was determined from compaction tests. Figure 3 shows some typical results. For the different tailings, the observed compression index C_c varied from 0.05 to 0.1 and the coefficient of consolidation c_v was between 10^{-3} and 10^0 cm^2/s; more details on these properties are given in Bussière (1993). The consolidation properties of the homogenized tailings are well within the range of what is usually found for similar materials (e.g., Vick, 1983).

Hydraulic conductivity: The hydraulic conductivity k is one of the most important properties of any material used in a cover system. In order to evaluate it, three types of tests have been performed on the homogenized tailings: rigid wall permeameter tests with constant head and falling head conditions, and permeability tests in the oedometer cell with constant total stress and varying water pressure.

These tests were conducted on the three tailings for different void ratios. This allowed an evaluation of the effect of different factors on the k value. Among the existing relationships established to quantify the influence of these factors, it was found that the Kozeny-Carman equation describes fairly well the observed behavior. This equation can be written as follows (Lambe and Whitman, 1979)

$$k = \frac{1}{k_o S^2} \frac{\gamma_w}{\mu} \frac{e^3}{(1+e)} \qquad (1)$$

where k_o is a shape factor, S is the specific surface area (calculated according to Chapuis and Légaré, 1992), γ_w is the unit weight of water and μ is water viscosity.

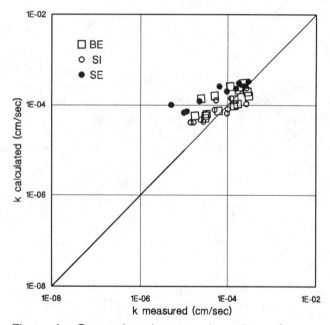

Figure 4 - Comparison between the values of k measured and k calculated using equation (1)

Figure 4 shows a fairly good correlation between the measured and calculated values for the three tailings studied. Other relationships used gave approximately the same precision (Bussière, 1993). The practical appeal of such type of relationship is that it allows an approximate evaluation of the hydraulic conductivity of homogeneous tailings, and its evolution as a function of the void ratio and of the other parameters appearing in Eq.1. Of course, the measured and calculated k values are given for a total saturation (S_r = 100%) and these should be corrected for unsaturated conditions.

As one can see in Figure 4, the hydraulic conductivity of the three tailings is higher than that of most impervious soils. This is due to the fact that the void ratio was generally maintained high to reflect existing conditions, and also because the tailings were fairly coarse. Using Eq.1, it is easy to show that the k value could be reduced by an order of magnitude and more, to values of about 10^{-6} cm/s, simply by using only the fine fraction corresponding to the overflow from hydrocyclones and by bringing the void ratio closer to optimum density conditions. Furthermore, one should not be overly concerned with the hydraulic conductivity value as it is usually recognized that the ability of covers to control AMD production is primarily related to its ability to limit the flow of oxygen. This aspect is discussed below.

Other properties: Among the other properties playing an important role in the efficiency of a cover system, the ability to control oxygen transport appears critical. It should be noted that it is usually considered (at least for preliminary calculations) that oxygen flux is controlled by Fickian type diffusion and that pressure and temperature gradients effects are negligible (Nicholson et al., 1989; Aachib et al., 1993).

In a Fickian flow, the oxygen flux is largely dependent upon the effective diffusion coefficient of oxygen D_e, which in turn depends on grain size, porosity, tortuosity and volume-tric water content. This latter factor is very important, as the diffusion coefficient in water is about 10^4 times lower than in air. Unfortunately, the precise measurement of D_e as a function of the above mentioned parameters is not an easy task, and somewhat conflicting results have been reported in the literature (e.g., Reardon and Moddle, 1985; Collin, 1987; Yanful, 1993). Nevertheless, some preliminary calculations, using various empirical models, have shown that the D_e value becomes close to that of water ($D_w \approx 2.5 \times 10^{-9}$ m²/s at 20°C) when the saturation ratio is about 90% (Aachib et al., 1993). Remembering that water covers are often considered as the most effective mean to control the production of AMD (e.g., SRK, 1989), it appears that a "dry" cover system having a layer with a high water content could prove to be a good alternative for that purpose. Laboratory measurements are underway to evaluate D_e for different tailings as a function of various influence factors.

In order to be able to evaluate the potential for oxygen flux reduction, one has to estimate the actual volumetric water content θ in the system. Such parameter can be obtained from the moisture characteristic curve of the material, which gives the relationship between θ and the negative pore pressure (or suction) ψ (Kovacs, 1981; Fredlund et Rahardjo, 1993). A schematical representation of such curves is given on Figure 5. On this Figure, one can see the typical shape of this curve, which is often expressed in a mathematical form using three parameters: ψ_a, the air entry value, ψ_r, the residual pressure corresponding to the residual water content θ_r. These three parameters vary according to properties of the particulate media, and various equations have been proposed to express these interactions. For instance, the different relationships proposed in the literature for ψ_a (e.g., Bear, 1972; Kovacs, 1981; Nicholson et al., 1991; Aachib et al., 1993) can be reduced to the following general formulation

$$\psi_a = f(D_x, n, Y) \qquad (2)$$

where D_x is the grain size, n is the porosity and Y is a complementary function (often omitted) to take into account specific effects such as adsorption and osmotic pressures.

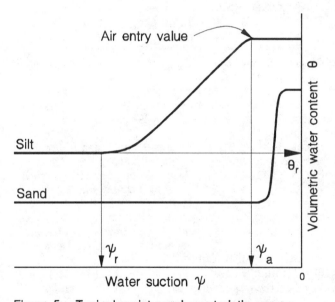

Figure 5 - Typical moisture characteristic curves

Figure 6 shows a typical result for homogenized tailings. The results obtained so far indicate that typical ψ_a values range between 1.5 to 3.5 m (about 15 to 35 kPa) and that ψ_r is in the range of 10 to 100 m (or 100 to 1000 kPa). According to such properties and to preliminary calculations, a fine material layer (placed between two sand layers) having a thickness of about one meter seems to be close to the optimum choice (Aachib et al., 1993).

It should be noted however that our experimental results are somewhat different than those recently published by Marchibroda et al. (1993), but the latter were obtained on processed tailings using the so-called thickened discharge method (e.g., Ritcey, 1989). Other tests are presently underway to evaluate the characteristic curve of the three tailings at different void ratios.

The final aspect of this research program, aimed at studying the properties of tailings used in cover systems, is to conduct numerical calculations to predict the behavior of multilayer barriers. These will be compared to measured performance obtained from laboratory column tests (described by Aachib et al., 1993) in which the fine and coarse layers alternate so that the capillary material can remain close to saturation. Large scale tests in situ are also under consideration for investigating scale effects, durability, mechanical stability, installation methods and other related problems.

4 - Conclusion

In order to control the production of acid mine drainage (AMD), a cover system able to limit the flow of water and/or oxygen appears as one of the few practical alternatives available. Such type of cover is usually built from fine-grained soils. In this paper, the authors present a conceptual approach in which tailings could be used as the construction material in the multilayer system. After describing the component of the cover and the function of the various layers, some of the main hydrogeological properties of the studied tailings are presented. Consolidation characteristics, hydraulic conductivity and a preliminary moisture characteristic curve are presented and briefly discussed. The presented results are part of an ongoing research project on a laboratory study of cover systems built from tailings.

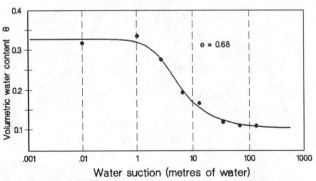

Figure 6 - Moisture characteristic curve for an homogenized tailing; the full line is given as an indication only.

5 - Acknowledgements

A financial support was provided by the MEND program, through the NEDEM-Québec branch managed by the Centre de recherches minérales.

6 - References

Aachib, M., Aubertin, M. and Chapuis, R.P. (1993). Étude en laboratoire de la performance des barrières de recouvrement constituées de rejets miniers pour limiter le DMA - un état de la question. Rapport EPM/RT-93/32, École Polytechnique de Montréal, 180 pages.

Akindunni, F.F., Gillham, R.W. and Nicholson, R.V. (1991). Numerical simulations to investigate moisture-retention characteristics in the design of oxygen-limiting covers for reactive mine tailings. Canadian Geotechnical Journal, Vol. 28, pp. 446-451.

Aubertin, M. and Bussière, B. (1991). Étude préliminaire - Évaluation des barrières sèches construites à partir de résidus miniers alcalins. Report submitted to Centre de recherches minérales. Project C.D.T. P1610, École Polytechnique de Montréal, 46 pages.

Aubertin, M. and Chapuis, R.P. (1991). Considérations hydro-géotechniques pour l'entreposage des résidus miniers dans le nord-ouest du Québec. Proc. 2nd Int. Conf. on the Abatement of Acidic Drainage, Montréal, Vol. 3, pp. 1-22.

Aubertin, M., Chapuis, R.P., Bussière, B. and Aachib, M. (1993). Propriétés des résidus miniers utilisés comme matériau de recouvrement pour limiter le drainage minier acide (DMA). Geoconfine 93, Arnould, Barrès and Côme (eds), Balkema, pp. 299-308.

Bear, J. (1972). Dynamics of Fluids in Porous Media, Dover Pub.

Bussière, B., (1993). Évaluation des propriétés hydrogéologiques de résidus miniers utilisés comme barrières de recouvrement. M.Sc.A. Thesis, Department of Mineral Engineering, École Polytechnique of Montréal, 171 pages.

Casagrande, A. (1948). Classification and identification of soils. Trans. ASCE, Vol. 113, pp. 901-930.

Chapuis, R.P. and Légaré, P.P. (1992). A simple method for determining the surface area of fine aggregates and fillers in bituminous mixture., ASTM STP 1147, pp. 177-186.

Collin, M. (1987). Mathematical modelling of water and oxygen transportation in layered soil covers for deposits of pyritic mine tailings. Doctoral Thesis, Royal Institute of Technology, Sweden.

Collin, M. and Rasmuson, A. (1990). Mathematical modelling of water an oxygen transport in layered soil covers for deposits of pyritic mine tailings, Acid Mine Drainage: Designing for Closure, GAC-MAC Annual Meeting, pp. 311-333.

Crespo, J.R. (1994). M.Sc.A. Thesis, Department of Mineral Engineering, École Polytechnique de Montréal (to be published).

Fredlund, D.G. and Rahardjo, H. (1993). Soil Mechanics for Unsaturated Soils, John Wiley & Sons.

Freeze, R.A. and Cherry, J.A. (1979). Groundwater, Prentice-Hall.

Itzkovitch, I.J. and Feasby, D.G. (1993). Le programme de neutralisation des eaux de drainage dans l'environnement minier. Proc. NEDEM '93, Val d'Or, pp. 1-16.

Kovacs, G. (1981). Seepage Hydraulics, Elsevier Scientific Pub.

Lambe, T.W. and Whitman, R.V. (1979). Soil Mechanics, SI Version, John Wiley & Sons.

Marchibroda, R.L., Wilson, G.W. and Barbour, S.L. (1993). Evaluation of the net infiltrative fluxes across the surface of exposed mine tailings. Proc. 46th Canadian Geotechnical Conference, Saskatoon, pp. 167-175.

Nicholson, R.V., Gillham, R.W., Cherry, J.A. and Reardon, E.J. (1989). Reduction of acid generation in mine tailings through the use of moisture-retaining cover layers as oxygen barriers, Canadian Geotechnical Journal, Vol. 26, No. 1, pp. 1-8.

Nicholson, R.V., Akindunni, F.F., Sydor, R.C. and Gillham, R.W. (1991). Saturated tailings covers above the water table: The physics and criteria for design, Proc. 2nd Int. Conf. on the Abatment of Acidic Drainage, Montreal, Vol. 1, pp. 443-460.

Oakley, R.E. (1987). Design and performance of earth-lined containment systems. Geotechnical Practice for Waste Disposal, ASCE, pp. 117-136.

Oweis, I.S. and Khera, R.P. (1990). Geotechnology of Waste Management, Butterworths.

Reardon, E.J. and Moddle, P.M. (1985). Gas diffusion coefficient measurements on uranium mill tailings: implications to cover layer design. Uranium, Vol. 2, pp. 111-131.

Ritcey, G.M., (1989). Tailings Managements, Problems and Solutions in the Mining Industry. Elsevier.

Ritcey, G.M. (1991). Deep water disposal of pyritic tailings. Proc. 2nd Conf. on the Abatement of Acidic Drainage, Montréal, Vol. 1, pp. 421-442.

SRK (Steffen, Robertson and Kirsten Inc.) (1988). Cover technology for acid mine drainage abatment: Litterature survey, Report No. 64702/1, Norwegian State Pollution Control Authority.

SRK (Steffen, Robertson and Kirsten Inc.) (1989). Draft Acid Rock Drainage Guide, BC AMD Task Force, Vol.I

Sutter, G.W. II, Luxmoore, R.J. and Smith, E.D. (1993). Compacted soil barriers at abandoned landfill sites are likely to fail in the long term. J. Environ. Qual., Vol. 22, pp. 217-226.

Vick, S.G. (1983). Planning, Design, and Analyses of Tailings Dams, John Wiley & Sons.

Yanful, E.K. (1993). Oxygen diffusion through soil covers on sulphidic mine tailings. Journal of Geotechnical Engineering, ASCE, Vol. 119, No. 8, pp. 1207-1228.

Yanful, E.K. and Dubé, B. (1993). Modelling moisture - retaining soil covers. Proc. Joint CSCE-ASCE. National Conference on Environmental Engineering, Montréal, Vol. 1, pp. 273-280.

Numerical Analysis of the Sedimentation and Consolidation of a Neutralized Red Mud

R.F. Azevedo
Catholic University of Rio de Janeiro (PUC-Rio), Brazil

M.C.M. Alves
Federal University of Rio de Janeiro (UFRJ), Brazil

T.M.P. de Campos
Catholic University of Rio de Janeiro (PUC-Rio), Brazil

ABSTRACT: This paper deals with the numerical analysis of the sedimentation and consolidation of a neutralized red mud observed during laboratory column tests described in a companion paper presented to this Conference (de Campos; Alves; Azevedo and Sills, 1993). Initially, the paper describes the equations that are considered to govern each problem separately and both problems together. Subsequently, the numerical solution adopted is explained and, finally, based on the experimental results, a proposed model is calibrated and comparisons between analytical and laboratory results are presented. Good agreement between numerical and experimental results is achieved.

INTRODUCTION

The continuous exploration of mineral resources has been making tailings storage an increasingly important problem, specially because of environmental concerns.

From the geotechnical point-of-view, the storage of these tailings (very soft materials) requires the development of a sedimentation and consolidation theory which may be validated by laboratory sediment column tests.

The deposition of a sediment in a column may be considered as having two stages which occur simultaneously. At the uppermost part of the column, the material is in a state of a dispersion and settles by sedimentation. At the bottom of the column, soil is continuously been formed by sediment accumulation and consolidates under its own weight (Pane and Schiffman, 1985).

An exact solution for the sedimentation of a dispersion must satisfy both equilibrium and continuity equations of the mixture. Kynch (1952) has developed a sedimentation theory supposing a hindered settling in which the basic assumption was that the speed of the falling particles is determined only by the local density (void ratio). This approximate solution satisfies only the continuity of the solid phase of the mixture, ignoring the possibility of effective stress development during the sedimentation process. Kynch's development, however, allows for different modes of settling, including abrupt changes of void ratio that may lead to linear and non-linear settling modes.

Consolidation induced by the self-weight of these very soft soils naturally generates large deformations. In these circumstances, the small strains conventional Terzaghi's consolidation theory is not applicable, enhancing the needs for using the large strains consolidation theory developed by Gibson, England and Hussey (1967). This theory, however, although quite general, does not take into consideration the sedimentation process.

Been (1980) has demonstrated that, during a deposition process, the hindered settling equation can be deduced from the consolidation theory by setting the effective stress equal to zero.

Been and Sills (1981) presented comparisons between analytical and experimental results of column tests similar to the ones utilised in this paper. However, the analytical results disregarded the sedimentation process and used linear constitutive relationships.

Pane and Schiffman (1985) proposed an extension of the classical principle of effective stress and developed a theory that considers the processes of sedimentation and consolidation together. In the following, the basic equations of this theory is shown, the numerical solution adopted is explained and, based on the experimental results, the theory is calibrated and comparisons between analytical and laboratory results are presented.

THEORY OF SEDIMENTATION AND CONSOLIDATION

The extension of the classical principle of effective stress proposed by Pane and Schiffman (1985) is:

$$\sigma = \beta(e)\sigma' + u_w$$

where, σ is the total stress, e is the void ratio, σ' is the effective stress, u_w is the pore-pressure and $\beta(e)$ will be explained later. Using this expression and following the same steps of Gibson's development, Pane and Schiffman (1985) obtained the equation:

$$\frac{d}{de}[\gamma_r k_z]\frac{\partial e}{\partial z} + \frac{\partial}{\partial z}[\frac{k_z}{\gamma_w}\frac{d\sigma'}{de}\beta\frac{\partial e}{\partial z}] + [\frac{k_z}{\gamma_w}\sigma'\frac{d\beta}{de}\frac{\partial e}{\partial z}] + \frac{\partial e}{\partial t} = 0$$

that controls the sedimentation and consolidation processes simultaneously. In this equation, z is the reduced coordinate given by:

$$z(a) = \int_0^a \frac{1}{1+e(\alpha,t)}d\alpha = \int_0^\xi \frac{1}{1+e(\zeta,t)}d\zeta$$

where, a and ξ are the lagrangian and the eulerian coordinates of a given point,

$$k_z = \frac{k}{(1+e)}$$

is the absolute coefficient of permeability (k is the coefficient of permeability) and

$$\gamma_r = (\frac{\gamma_s}{\gamma_w} - 1)$$

is the relative density. γ_s and γ_w are the specific weight of the solids and pore-liquid, respectively.

If function $\beta(e)$ is equal to zero, only sedimentation is occurring and the general equation becomes:

$$V_z(e)\frac{\partial e}{\partial z} + \frac{\partial e}{\partial t} = 0$$

where,

$$V_z(e) = \frac{d}{de}[\gamma_r k_z]$$

that is analogous to Kynch's sedimentation equation.

On the other hand, if β(e) = 1, only consolidation is happening and the general equation becomes:

$$\frac{d}{de}[\gamma_z k_z]\frac{\partial e}{\partial z} + \frac{\partial}{\partial z}[\frac{k_z}{\gamma_w}\frac{d\sigma'}{de}\frac{\partial e}{\partial z}] + \frac{\partial e}{\partial t} = 0$$

that is exactly Gibson's consolidation equation.

If the height of solids, h_z, increases with time, this last equation becomes:

$$-\frac{1}{h_z^2(t)}\frac{\partial}{\partial y}[\frac{k_z}{(1+e)\gamma_w}\frac{\partial \sigma'}{\partial e}\frac{\partial e}{\partial y}] + \frac{1}{h_z(t)}\frac{\partial e}{\partial y}[y\,R_z +$$

$$(1-\frac{\gamma_z}{\gamma_w})\frac{d}{de}\frac{k_z}{(1+e)}] = \frac{\partial e}{\partial t}$$

where,

$$y = \frac{z}{h_z(t)}$$

is the, so-called, non-dimensional reduced coordinate and,

$$R_z = \frac{d h_z(t)}{dt}$$

is the solids deposition rate.

Figure 1 illustrates a typical β(e) function. For e > e_s the material is a dispersion that settles only by sedimentation. For e < e_m the material is already considered as soil and settles due to consolidation. Michaels and Bolger (1962) and Been and Sills (1981) suggested the existence of a transition zone between dispersion and soil. In this region, defined by two limiting void ratio e_s and e_m that probably are soil parameter, 1 > β(e) > 0.

NUMERICAL SOLUTION

The direct experimental determination of function β(e) in the transition region is difficult, if possible. Therefore, a simplified way to solve sedimentation/consolidation problems is to neglect the existence of this transition region, assuming that β(e) is equal to:

β(e) = 0 for e > e_m

and

β(e) = 1 for e < e_m

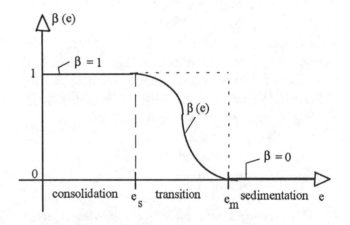

Figure 1 - Typical β(e) Function

where e_m is a limiting void ratio that bounds the sedimentation and the consolidation processes (dotted line, Figure 1). This limiting void ratio can be determine experimentally and seems to be a soil parameter.

Assuming this simplified form for $\beta(e)$ function, the general equation can be solved separately during the sedimentation and the consolidation regions of a settling column.

To solve the sedimentation equation, a computer program called ETSED was developed based on the method of characteristics (Abreu, 1989 and Alves, 1992). The input soil parameters are: the initial void ratio, e_0, the relative density of solids, γ_r, and the relationship k_z x e obtained from experimental results. Besides, the end of sedimentation process must be estimated by means of the limiting void ratio e_m. The output results are the height of the mud-liquid interface, the void ratio distribution and the solids deposition rate, R_z.

The numerical solution of the consolidation problem with accreting bottom layer was found by means of a computer program called ADF, using the finite difference method (Pinto, 1988; Azevedo and Sado, 1990; Alves, 1992). In this program, the input soil parameters are the solids deposition rate, Rz, calculated previously by ETSED, and the constitutive relations σ' x e and k_z x e. Besides that, it is necessary to know the initial conditions, $e_0 = e_m$ = uniform, and the boundary conditions, impervious on the base and pervious on the top (de Campos; Alves; Azevedo and Sills, 1993).

Finally, the coupled (sedimentation/ consolidation) solution was obtained with the addition, for each time step, of these two separate solutions, from the beginning of the test till time T2 (Figure 2). After time T2, only consolidation persists, therefore, only the consolidation equation is solved.

MODEL CALIBRATION AND COMPARISONS

Based on laboratory tests (de Campos; Alves; Azevedo and Sills, 1993) it was seen that the constitutive laws were reasonably well defined for values of void ratio less than approximately 12 and poorly defined for void ratio greater than that. On the other hand, as e_m is defined as the limit between sedimentation and consolidation and, if the settling column tests are long enough to let all consolidation under its own weight happen, e_m can be estimated as the highest void ratio at the end of the settling column tests, which vary between 10.4 and 11.7.

To start calibration, an average value for e_m was obtained. The effective stress x void ratio relationship was directly adjusted from the test results for values of void ratio less than e_m. The experimental permeability x void ratio relationship presented a large scatter for values of void ratio greater e_m and, although more accurate for values of void ratio less than e_m, the best adjustment for this relationship was found by a trial-and-error procedure that best fitted the laboratory and the numerical results.

When calibrating the model it was noticed that the void ratio e_m had little influence on the height of the mud-liquid interface. Therefore, an adequate choice of this parameter could not be done by observing the movement of this interface. On the other hand, different values of e_m led to variations on time T2 and, consequently, on strong variations on the effective stress distribution with height. Therefore, the effective stress measurements were a much sensible (better) way to evaluate e_m than the height of the mud-liquid interface movement.

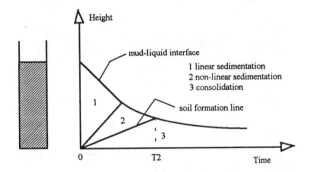

Figure 2 - Typical Mud-Liquid Interface Height versus Time Curve

According to the above described methodology, the parameters that best represented the red mud behavior were:
- for sedimentation:
$$k = (1+e)[-1.486+0.627\ln(e)] \text{ (mm/min)}$$
- void ratio $e_m = 11.3$
- for consolidation:
$$e = 4.885\,(\sigma')^{-0.168} \quad \sigma' \text{ in (kPa)}$$
$$k = 4.18078\text{E-}11(e)^{4.549} \text{ (m/s)}$$

Figure (3) shows typical results of measured and calculated density profiles. It can be seen a reasonable agreement between calculated and measured values specially for large time values.

Typical results for effective stress distribution with column height are shown on Figure (4). It can clearly be noticed that for time values much smaller than T2 the calculated and measured distribution are different, the experimental one showing effective stress in the sedimentation region, which is not take into account by the theory. As the test time approaches the value of T2 and for values larger than T2, a good agreement was clearly obtained between analytical and experimental results.

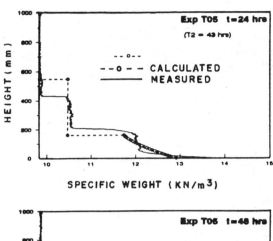

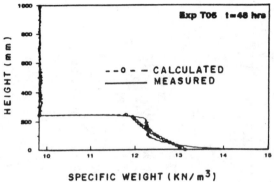

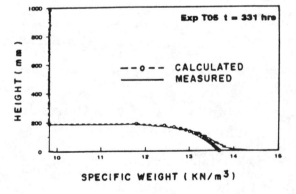

Fig. 3 - Comparisons between Measured and Calculated Density Profiles

CONCLUSIONS

This paper dealt with the numerical analysis of the sedimentation/consolidation behavior of a red mud observed during laboratory

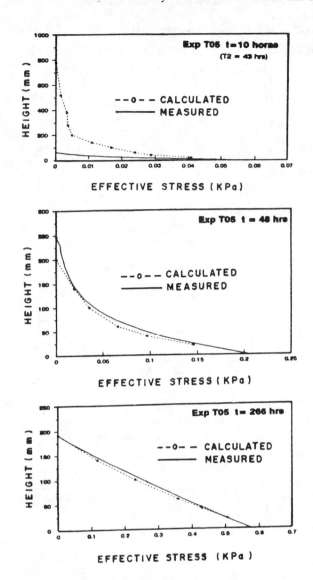

Fig.4 - Comparisons between Calculated and Measured Effective Stress Distribution with Height

the value of the limiting void ratio (e_m).

- On the other hand, the effective stress distribution with height and time was very much influenced by the value of the limiting void ratio and served well to evaluate it.

- The limiting void ratio used was very close to the average value of the highest void ratios at the end of the settling column tests. Therefore, this can be an easy way of estimating e_m.

- After some adjustment of the permeability versus void ratio relationship, displacements, pore-pressures and effective stresses calculated with the proposed model during sedimentation and consolidation of the red mud studied reproduced reasonably well the observed experimental results, specially for time value larger than T2.

ACKNOWLEDGMENTS

The present work was supported by PADCT/FINEP (Brazilian) National Research Program, CAPES (Minister of Education/Brazil) and the Federal University of Pernambuco/Brazil. This support was essential for the research and it is greatly appreciate by the authors.

REFERENCES

1. **Abreu, F. R. S.** - "Numerical Model to Analyse the Sedimentation of Industrial Tailings" (in Portuguese). M.Sc.Thesis, PUC-Rio, Brazil, 1989.
2. **Alves, M. C. M.** - "Sedimentation andConsolidation of a Red-Mud" (in Portuguese). Ph.D. Thesis, PUC-Rio, 1992.
3. **Azevedo, R. F. and Sado, J. S.** - "One-Dimensional Analysis of Tailings

column tests. After describing the equations proposed to govern the problem, the numerical solution and the calibration of the proposed model were presented. Finally, comparisons were made between analytical and laboratory results. The main conclusions are:

- The velocity of the mud-liquid interface showed to be independent of

Consolidation of a Red-Mud" (in Portuguese). Ph.D. Thesis, PUC-Rio, 1992.

3. **Azevedo, R. F. and Sado, J. S.** - "One-Dimensional Analysis of Tailings Dam Reservoir Fillings by means of The Finite Strain Consolidation Theory" (in Portuguese). Proceedings of the IX Brazilian Conference on Soil Mechanics and Foundation Engineering", Salvador, Brazil, 1990.

4. **Been**, K. - "Stress Strain Behavior of a Cohesive Soil Deposited under Water", DPhil. Thesis, University of Oxford, 1980.

5. **Been, K. and Sills, G. C.** - "Self-Weight Consolidation of Soft Soils: An Experimental and Theoretical Study", Geotechnique 31, No. 4, 519-535, 1981.

6. **de Campos, T. M. P.; Alves, M. C. M.; Azevedo, R. F. and Sills, G. C.**- "Laboratory Settling and Consolidation Behavior of a Neutralized Red Mud". First International Congress on Environmental Geotechnics, Edmonton, Alberta, Canada, 1993.

7. **Gibson, R. E.; England, G. L. and Hussey**, M. J. L. - "The Theory of One-Dimensional Consolidation of Saturated Clays. Part I: Finite Non-linear Consolidation of Thin Homogeneous Layers",Geotechnique 17, 261-273, 1967.

8. **Kynch, G. J.** - "A Theory of Sedimentation". Transactions of the Faraday Society, 48, pp. 166-176, 1952.

9. **Michaels, A. S. and Bolger, J. C.** - "Settling Rates and Sediment Volumes of Flocculated Kaolin Suspensions", Journal of Industrial Engineering Chemistry 1, No. 1, 24-33, 1962.

10. **Pane, V.** - "Sedimentation and Consolidation of Clays". Ph.D. Thesis, University of Colorado, Boulder, Colorado, USA, 1985.

11. **Pane, V. and Schiffman, R. L.** - "A Note on Sedimentation and Consolidation", Geotechnique, March, 1985.

12. **Pinto, W. P.** - "The Finite Strain One-Dimensional Theory of Consolidation" (in Portuguese). M.Sc. Thesis, PUC-Rio, Brazil, 1988.

Ground and Surface Water Pollution at a Tin Mine in Transvaal, South Africa

S.E.T. Bullock, F.G. Bell
Department of Geology and Applied Geology, University of Natal, Durban, South Africa

ABSTRACT

In order to ensure that both surface water and groundwater is not adversely affected after the closure of a tin mine in the Transvaal an investigation was undertaken to assess possible sources of pollution. This revealed that acid waters were emanating from pyrite dumps. Analysis of water from the pyrite dumps showed it to have pH values in the order of 2.5 to 4. Soil sampling from trenches dug around these dumps indicated that soils at a depth of 2.2 m had pH values around 4.5. Analyses were carried out on the solid material from other dumps to determine its composition, most of these proved to be non-acid generating.

A number of drillholes were sunk near the mine in the lease area to determine whether the groundwater had been polluted as a result of mining operations. Those near the pyrite dumps contained acidified water of unacceptable quality. Those drillholes to the west of the mine workings and not sunk into the Boschoffsberg Quartzite contained water which was not affected by acid drainage. The water was comparable with samples taken off site.

Domestic refuse was being disposed of in an abandoned opencast pit and it was feared that leachate from the landfill might pollute the groundwater. Analysis of liquid from the base of the landfill indicated that only the chemical oxygen demand was unsatisfactory.

INTRODUCTION

An investigation was undertaken to assess the possibility of pollution occurring to water supplies around a tin mine in the west-central Transvaal some 250 km north-west of Pretoria. The mining activities at the site have included extensive underground mining, limited small scale surface opencasting and mineral processing. With the recession, rates of production have been reduced and a rehabilitation programme has been implemented. This has involved stripping and demolition at defunct operations, and restoration of some of the surface where mining has taken place.

As the area falls within a region of water shortage rehabilitation included an assessment of whether pollution of ground and surface water is occurring and, if so, to locate the sources of pollution and to offer solutions. Hence an evaluation of the surface and subsurface water resources was also carried out.

Mean temperatures in the area reach a maximum in December/January and a minimum in June/July. For example, the maximum mean temperature in January is around 29° to 31°C, the minimum falling to 17°C. In July the maximum mean temperature is around 22°C, the minimum being between 3° and 6°C.

The mean annual precipitation is 620 mm. The rainy season lasts from October to May, with the most rain falling in January. About 50 to 80 rain days may be expected each year. In the period 1963 to 1992, there were six drought seasons indicating a drought expectancy every 5 years. Evaporation exceeds precipitation every month of the year, the potential water loss reaching a maximum during August, September and October.

Topographically the area is characterized by undulating terrain comprising rolling hills and wide valleys. The average elevation is some 1100 m above mean sea level.

As the mine is situated in a semi-arid region, there is little surface run-off, the flow in all streams being seasonal with run-off being limited to short periods of heavy rainfall. The water courses and watersheds in the area are indicated in Fig. 1. The general drainage direction is from the south to the north.

There are two reservoirs in the area with capacities of 312,500 m³ and 210,000 m³ which satisfy the needs of the mine and are used for watering gardens and such like. Surface water, however, is not used for human consumption at the mine, this water being obtained from wells. Because of the non-perennial nature of many of the streams, farmers in the immediate vicinity of the mine do not rely on surface water, again drawing much of their supply from wells.

GEOLOGY AND GEOHYDROLOGY

The tin deposits are hosted by rocks of the Rooiberg Fragment, a triangular shaped portion of the uppermost Transvaal sequence. The fragment is found within the western lobe of the Bushveld Complex and is surrounded by sheet-like granitoid intrusions of the Lebowa Granite Suite. The tin field is typified by two diverse lithological components, firstly, the sedimentary and volcanic rocks of the Pretoria and Rooiberg groups of the Transvaal Sequence which constitute the Fragment and host the various tin deposits, and secondly, the surrounding granitoids of the Bushveld Complex.

The Pretoria Group contains the Leeupoort Formation and the Smelterskop Quartzite Formation. The Boschoffsberg Quartzite Member forms the lower part of the Leeupoort Formation and consists of thick sequences of rudaceous and arenaceous rocks most of which are feldspathic quartzites or arkoses. It is from this member that the majority of the tin is obtained. The succeeding Blaaubank Shale Member is subdivided into two units, namely, the Shaley Arkose, which represents a gradational transition between the arkoses and the shales, and the Main Shale. The Shaley Arkose is made up of well bedded micaceous and feldspathic sandstone layers rhythmically alternating with pink sandstone, grey-green micaceous shale and grey to purple silty shale bands. The Main Shale unit is comprised of shale, siltstone and fine grained quartzite. The sediments of the Smelterskop Formation vary greatly in grain size and generally fine upwards. The Rooiberg Group consists of massive fine grained felsite at the base, succeeded by banded felsite and porphyritic felsite with quartz porphyry occurring at the top of the group.

Dykes and sills of various ages have been intruded and are particularly well developed in the so called dyke belt. They are frequently discontinuous, thin and occupy north-south and east-west trending fractures.

The soils are mostly residual, predominantly comprising sandy loams with lesser quantities of silty loams and silty clay loams. Some surficial transported sandy silty soils are found on the flanks of the local hills. Sandy loams are best developed where they overlie feldspathic sandstones. Silty loams are best developed over shales. The soils range up to several metres in thickness.

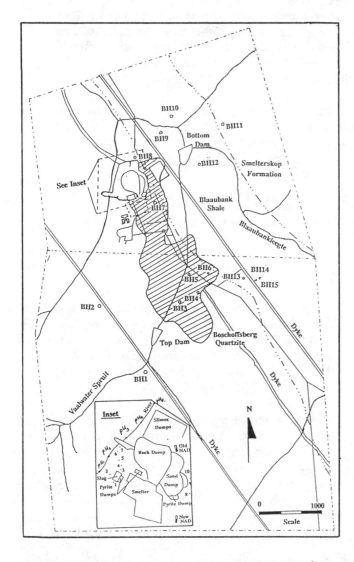

Fig 1. Mine lease area showing surface drainage, geology, mined out area (shaded), locations of drillholes (BH1 etc), pits (1, 2 etc), stream sampling points (pH1 etc) and dumps.

In terms of intergranular flow, the above mentioned rocks can be regarded as impermeable. Groundwater is primarily contained within and flows via discontinuities.

Most of the groundwater in the area occurs in aquifers associated with Boschoffsberg Quartzite Member. These aquifers are weathered and highly fractured, and lie directly beneath the soil surface. The quantity of groundwater stored in these aquifers is limited. Even so groundwater provides the primary source of supply of potable water for both domestic purposes and for stock. Hence pollution of groundwater resources would adversely affect both the mining and agricultural communities in the area.

A survey was undertaken to assess the groundwater resources. This indicated that the occurrence of groundwater is geologically controlled in that those drillholes, put down during the survey, with moderate to high yields were associated with dykes. The intrusion of the dykes caused fractures and joints to develop in the host rocks and these discontinuities may have significant water-holding potential. The yield of the drillholes, when tested, varied from 2 to 5 m^3/h. In addition, a perched water table was intersected by digging a series of trenches in the mine area during the survey. This water table is impounded by a relatively impermeable layer of ferrocrete in the soil. The water associated with the perched water table is of poor quality.

Groundwater movement in the Boshoffsberg Quartzite takes place along three major fracture directions. However, the situation is complicated by the presence of a number of dykes which compartmentalize the groundwater. Hence water moves along fractures until it comes in contact with a dyke. Lateral movement along dykes is thought to be from the south east to the north west. Extensive underground workings in the mine area further complicate groundwater movement, the presence of underground tunnels overriding the aforementioned geological control, and abandoned workings act as groundwater reservoirs. Groundwater movement in the perched water table is topographically controlled.

ASSESSMENT OF POLLUTION POTENTIAL

During the extraction of tin, waste is produced at various stages. This waste material has been disposed of in a series of dumps, that is, the rock dump, sand dump, pyrite dumps, slime dumps and slag dump (Fig. 1). A series of tests were performed on the different dump materials in order to assess their potential to pollute the surface water.

The pyrite dumps possess the greatest potential to pollute water. An estimated 8 125 tonnes (5 000 m^3) of pyrite material has been deposited in three dumps. The pyrite contains around 45% of sulphur and 35.6% of iron. Consequently ground and surface water around the dumps have low pH values due to the formation of sulphuric acid when pyrite breaks down. For instance, the values of pH and electrical conductivity taken from the tributary of the Vaalwater Spruit and some of the pits are as follows:

From stream			From pits		
Location	pH	EC (mS/m)	Location	pH	EC (mS/m)
1	6.5	35	1	4.4	152
2	6.2	2210	2	7.8	120
3	2.8	1228	3	5.0	182
4	4.4	279	4	4.6	360
5	6.7	47	5	2.9	469
			6	7.1	342
			7	3.7	314

The maximum permitted limits of pH and electrical conductivity values are 5.5 to 9.5, and 400 mS/m respectively as laid down by the Water Act of 1956. Hence at a number of locations the quality of the water, both surface and subsurface, falls outside the permitted limits.

The pyrite breakdown reactions involved are as follows:-

$$2FeS_2 + 2H_2O + 7O_2 - 2FeSO_4 + 2H_2SO_4 \quad (1)$$
$$4FeSO_4 + O_2 + 2H_2SO_4 - 2Fe_2(SO_4)_3 + 2H_2O \quad (2)$$
$$Fe_2(SO_4)_3 + 6H_2O - 2Fe(OH)_3 + 3H_2SO_4 \quad (3)$$
$$4Fe^{2+} + O_2 + 4H+ - 4Fe^{3+} + 2H_2O \quad (4)$$
$$FeS_2 + 14Fe^{3+} + 8H_2O - 15Fe^{2+} + 2SO_4^2 + 16H^+ \quad (5)$$

Bacteria are also involved in the breakdown process, notably *Thiobacillus ferroxidants*. *Thiobacillus ferroxidans* converts the ferrous ion of pyrite to the ferric form. The formation of sulphuric acid in the initial oxidation reaction and concomitant decrease in the pH make conditions more favourable for the biotic oxidation of pyrite. For instance, the biotic oxidation of pyrite is four times faster than the abiotic reaction at pH 3.0. The rate of pyritic oxidation is also increased with increasing temperature.

As the surface temperature of the pyrite dumps at the tin mine exceeds 35°C in the summer months, together with the fact that *T.ferroxidans* cultures are present on the dumps, it was calculated that sulphate production rates probably are greater than 3.5 moles/kg tailings per month. Further proof of the very high generation rates of sulphate was obtained from the results of acid-base account tests done on the pyrite material. These gave an acid potential of 1 410 g $CaCO_3$/kg, a neutralizing potential of -40.6 g $CaCO_3$/kg and a net neutralizing potential of -1451.7 g $CaCO_3$/kg. Samples with a negative net neutralization potential and a ratio of neutralizing potential to acid potential of less than 1:1 have a high potential for acid generation. The acid-base account results therefore clearly demonstrate that the pyrite dump material has a high potential to produce acid waters.

Several pits were excavated around the major dumps and soil samples taken. These samples were scanned for a number of elements using an I.P.C. mass spectrometer (Table 1). This revealed that the concentration of metals in the soil was high. Acidic waters associated with the pyrite dumps have leached metals from the dumps and these have accumulated in the soil. Pit 10 (Fig. 1) showed the highest concentration of metals. The pH of the soil sample taken from this pit was 3, and its electrical conductivity was 3 100 mS/m. The denuded patches of vegetation around pit 10 are attributed to both low pH and high metal concentrations in the soil. Aluminium and copper are regarded as major causes of plant toxicity in soils with pH values less than 5. For example, most plants die in the presence of 0.1% total copper in the soil. The soil in pit 10 has a copper concentration of 3%. The analysis of material from pit 2 is given in Table 1 for comparison since this material is more or less unaffected.

Table 1. Analysis of material from pits 10 and 2, and from the sand and slag dumps.

	Pit 10	Pit 2	Sand Dump	Slag Dump
pH	3.0	7.8	8.2	8.1
EC	3 100	120	47	2 000
Mg	47 295	25	130	19 419
Al	30 781	46		8
Ca	147	95	428	3 836
Mn	46 201		6	14
Fe	24 731	3	1	16
Co	3 916			1
Ni	484			
Cu	30 012	1		5
Zn	226			

Most of the sulphides have been extracted from the material which is deposited in the slimes dumps or tailings lagoons, only 0.3% of sulphur being present compared with 6.4% of iron oxide. Material from these dumps was subjected to analysis by X-ray fluorescence, the results of which are shown in Table 2. Although some of the dangerous metals (Mn, Cu, Pb, Cd, Zn) are present in trace amounts, they may be taken into solution and concentrated by acidic water from the pyrite dumps. The acid-base account results of tests done on the slimes material revealed an acid potential of 5.0 g $CaCO_3$/kg, a neutralizing potential of 40 g $CaCO_3$/kg and a net neutralizing potential of 35 g $CaCO_3$/kg. Therefore the slimes material has a potentially acid consuming character.

Table 2. Results from XRF analysis on slimes material

Major elements (> 1%)	Silica Aluminium Sodium Potassium	Tin Iron Calcium Magnesium
Minor elements (1% - 0.1%)	Rubidium Titanium	Zirconium Phosphorous
Trace elements (< 0.1%)	Manganese Sulphur Copper Cadmium	Chromium Barium Lead Strontium

The material in the sand dump is disposed of without sulphide being removed. The composition of this material can be seen in Table 1. The acid-base results for the sand material indicate an acid potential of 30 g $CaCO_3$/kg, a neutralizing potential of 38 $CaCO_3$/kg and a net neutralizing potential of 8 $CaCO_3$/kg. Therefore the material falls within the uncertainty zone between that of acid consumption and acid generation.

The slag dump consists of crushed slag from the smelter and has an estimated volume of 5 000 Mg/m^3. Most of the sulphides have been extracted prior to the material being disposed of, so that 0.35% of sulphur was present, compared with 3.6% of Fe_2O_3, which was present. An analysis of material from the slag dump is given in Table 1. The results of acid-base account tests indicate that it has a potentially acid consuming character.

Both the surface water and groundwater from wells and

drillholes were sampled for chemical analyses. Table 3 lists the results obtained from analyses carried out on surface water in the mine lease area. The standing water was taken from next to a pyrite dump. It has a very low pH value, and a very high sulphate content. The stream water was taken from a stream over 200 m to the west of the same dump. Its pH value is outside the permitted levels. The water in the other surface supplies is not satisfactory for drinking purposes, for example, it either has too much sulphate or too high a chemical oxygen demand (COD). The amount of fluorine in the Bottom reservoir is much too high. The high sulphate concentration and low pH in some of the surface water samples is attributed to the acid drainage which results from the oxidation of sulphide minerals.

The character of the groundwater present in drillholes and wells on and off the site is given in Table 4. It can be seen that many of the drillholes near the pyrite dumps are polluted with sulphate, have low pH values, and the contents of Ca, Mg and Na are unacceptable, as is their electrical conductivity. Those drillholes to the south west satisfactory for drinking purposes, for example, it either has too much sulphate or too high a chemical oxygen demand (COD). The amount of fluorine in the Bottom reservoir is much too high. The high sulphate concentration and low pH in some of the surface water samples is attributed to the acid drainage which results from the oxidation of sulphide minerals.

The character of the groundwater present in drillholes and wells on and off the site is given in Table 4. It can be seen that many of the drillholes near the pyrite dumps are polluted with sulphate, have low pH values, and the contents of Ca, Mg and Na are unacceptable, as is their electrical conductivity. Those drillholes to the south west of the mine workings, and in the Blaaubank Shale and Smelterskop Formation, as well as off the mine lease area contained acceptable groundwater. However, as pointed out above, the main aquifers occur in the Boschoffsberg Quartzite. Many of the drillholes sunk in this member indicated that groundwater was being polluted.

Domestic waste is being disposed of in an abandoned opencast pit in the Boschoffsberg Quartzite. Old mine workings exist beneath the opencast pit and consequently the ground is fractured in places. At present the landfill occupies only a small part of the pit but the possibility of pollution of groundwater could exist, especially as the landfill grows in size. Hence an analysis was made of water taken from the base of the landfill (Table 3). With the exception of the COD, the other factors determined fall within permissible limits.

CONCLUSIONS

A rehabilitation programme at a tin mine in the west-central Transvaal included a survey of the surface and subsurface water resources to determine whether

Table 3. Character of the surface water in the mine lease area and from landfll.

	Standing water	Stream water	Bottom Dam	Vaalwater	Top Dam	Landfill	Some permitted limits
pH	2.4	4.1	7.4	7.1	7.3	7.1	5.5 - 9.5
COD	1 410	38	-	83	-	109	75
EC (mS/m)	865	57.2	65	82	138	275	400
Total Hardness	32 409	219	249	523	663	523	
Total N	38.4	2.5	<0.2	0.2	<0.2	8	
Ca	163	64	82	66	93	66	200
Mg	74	12	32	30	22	27	125
Na	4	11	65	51	97	51	
K	<1	9	8	23	10	118	
SO_4	11 189	236	174	119	571	341	400
Cl	18	6	42	34	42	71	600
F	<0.1	<0.1	1.5	0.7	8.4	0.1	0.8 - 2.4
Fe	3 580	2.4	-	0.2	-	0.2	

All units expressed as milligrams per litre, except pH and where stated.

Table 4. Analysis of water from wells and drillholes on and off site.

Sample	pH	Ca	Mg	Na	SO4	HCO3	K	NO3	Cl	TDS	EC	Hardness	Alk
BH2	7.43	17.1	14.3	364	427	372.1	30.6	3.9	121.0	1167	187	102	310
BH3	6.3	401	289	418	2937	24.4	18.5	11.8	26.7	4114	444	2171	20
BH4	5.9	539	373	202	3275	18.3	23.8	4.9	73.5	4404	421	2866	15
BH7	3.01	493	3376	228	47720	0.0	12.7	29.4	86.9	51946	3130	-	0
BH9	2.9	627	621	102	14580	0.0	6.9	64.8	33.4	16039	2370	4121	0
BH8	2.8	509	561	70	27070	0.0	0.7	19.6	53.5	28284	1680	3579	0
Old NAD	7.54	475	430.8	367	3463	280.6	20.1	4.7	25.4	4933	497	2729	240
New NAD	7.74	153	123.4	131	805.5	353.8	5.2	1.0	62.2	1467	190	601	305
BH1	7.99	79	74.2	54	405.6	85.4	19.1	42.9	63.6	781	129	434	70
BH6	7.73	129	99.5	85	683.4	122.0	10.0	57.5	49.4	1175	182	632	100
BH5	8.15	106	85.7	92	581.2	112.0	8.8	61.4	48.6	1057	165	520	103
BH10	7.2	275	3.1	69	12.0	374.4	9.0	0.2	4.3	-	69	317	375
BH11	7.3	315	3.2	68	15.0	374.3	6.0	0.4	12.6	-	71	334	375
BH12	7.8	76	3.7	138	263.0	296.2	4.0	<0.2	12.3	-	112	452	298
BH13	7.2	112	4.5	54	62.0	263.4	3.0	<0.2	17.4	-	58	226	265
BH14	7.1	213	3.5	20	7.0	308.6	3.0	0.4	10.3	-	51	286	309
BH15	7.5	145	3.3	91	7.0	407.0	9.0	0.2	4.6	-	75	314	409
Visser 1	7.65	22.3	57	71.7	23.0	536.8	2.1	2.6	6.4	435	81	439	440
Visser 2	7.58	55.7	85.5	19.5	4.1	610	3.3	1.1	4.6	478	78	491	500
Sleepwa	8.57	44.9	34.4	19.7	76.5	122	2.0	2.3	31.9	279	51	154	110
Knoppieskraal	7.74	32	23.6	37.6	5.6	189.1	2.4	20.8	18.5	235	45	22	155
Strydom	7.18	62.9	46.4	33.3	3.7	536.8	1.7	2.5	5.1	424	68	349	440
Blockdrift	7.32	76.5	54.0	30.0	2.0	494.1	1.4	0.0	6.8	9	77	405	390
Nieupoort	7.16	74.3	56.9	76.7	15.4	530.7	0.3	13.5	80.8	583	107	420	435
Blaaubank	8.06	12.3	7.9	18.3	18.8	36.6	11.4	63.4	10.2	161	36	23	30

All values expressed in mg/l except pH and where stated.

pollution was taking place in the mining area. The waste material from the mine has been disposed of in a series of dumps, including pyrite dumps, slime dumps, a sand dump and a slag dump. In addition, domestic landfill occupies part of an abandoned opencast pit. The dumps and the landfill were investigated to see if they represented sources of water pollution.

Surface water sources within the mine lease area consist of non-perennial streams and two reservoirs. Most of the groundwater is found in aquifers in the Boschoffsberg Quartzite Member which also is the principal host rock of the tin deposits. The water occurs in and moves via the discontinuities within the rock masses.

The pyrite dumps represent the most serious source of pollution, giving rise to acidic waters which have contaminated the soil around them. The water in nearby drillholes has been acidified, as has water in a tributary of the Vaalwater Spruit, although not to the same extent. The sand dump may also be responsible for some pollution, however, the slim dumps and the slag dump do not appear to be acid generating. It is possible that the acid waters could be collected and treated with neutralizing additives such as limestone, lime or sodium hydroxide. Alternatively such additives could be mixed with the acid generating material. Unfortunately there is no source of such additives available locally. Hence it may prove more cost effective to cover the offending dumps with a clay soil to inhibit access of oxygen and water, thereby restricting the production of acidified waters.

Fortunately the pollution of water resources would not appear to have travelled beyond the mine lease area. This is primarily due to the impermeable nature of the rock types in region and the compartmentalization of groundwater by dykes.

At present the landfill does not represent a serious source of groundwater pollution. However, as the landfill increases in size, its potential for pollution will also increase. As the rocks exposed within the opencast pit are fractured and as some of these fractures are associated with the mine workings beneath, it would be wise to provide a lining to the base and sides of the pit to prevent any future pollution of the groundwater by leachate.

Geotechnical Problems in Construction over a Thick Layer of a Mine Waste

G. Calabresi
University of Rome 'La Sapienza', Italy

V. Pane
University of Perugia, Italy

S. Rampello
University of Rome 'La Sapienza', Italy

O. Bianco
ENEL S.p.A., Construction Division, Italy

Abstract: The construction of a power plant over a thick layer of a clayey mine waste encountered geotechnical problems deriving from its strong heterogeneity and the presence of consolidation settlements still in progress. In these conditions, reducing negative friction on piles and soil stabilisation under shallow foundations were of the utmost importance. Electro-osmosis treatment for soil stabilisation was shown to be successful in increasing the undrained strength by a factor of 2 to 3. Values of the undrained strength on samples cored from the stabilised columns were 5 to 8 times greater than those on natural soil samples; however, difficulties were encountered in obtaining homogeneous soil stabilisation with depth.

Introduction

A new power plant is being built by the Italian State Electricity Agency (ENEL S.p.A.) at Pietrafitta in Central Italy. The site for construction involves a mine waste, 15 to 25 m in thickness, consisting mainly of silty-clay back-fill. Field instrumentation showed excess pore water pressures in the mine waste and in the underlying natural clay layers; these were variable through the area depending on the thickness of the fill and the time elapsed since its laying down. Field monitoring showed surface settlements occurring at an average rate of 1 cm per year, due to the consolidation process of the waste alone [1].
To reduce the influence of the excess pore water pressures on the overall behaviour of the power plant, the ground elevation was lowered, and vertical drains were installed to accelerate the consolidation process [1]. For deep foundations, attention was also focused on reducing the down-drag forces applied on piles by the mine waste.
Finally, soil improvement was carried out to stabilise vertical excavations and to reduce differential settlements under shallow foundations, due to the strong soil heterogeneity.
In this work two different techniques of soil improvement are examined and their effectiveness evaluated for the mine waste. Both of them have been used successfully in the past for soft clay deposits, but less experimental evidence exists for their application in a composite soil such as the one to hand. They are the electro-osmosis, used in full scale field tests, and the stabilised column treatments, extensively used at the site to improve the mechanical behaviour of the clayey waste. Both of them were seen to be effective, the advantage of electro-osmosis being the attainment of a uniform improvement of mechanical properties, while that of the treated columns being a greater, but not uniform, increase in shear strength.

Site Description

The mine waste consists mainly of partially weathered elements of medium to stiff silty-clay. The natural soil was excavated in the Sixties for the exploitation of an underlying layer of brown coal; then it was disposed in the mine basin in two different stages, about 10 years apart. It contains inclusions of silty-sand and lignite. The heterogeneity of the waste is caused by the irregular mixing of the excavated soil, and by the different extent of swelling and weathering processes experienced by the clay elements during the mine exploitation and fill-

TABLE 1: Characteristics of the mine waste

w_L %	I_P %	G_s	<2μm %	γ kN/m³	w_o %	I_L %	c_u kPa
52	26.9	2.72	40.2	18.2	37.6	0.56	53.3

ing stages. Table 1 lists the index properties of the mine waste, together with its in situ liquidity index. Due to the nature of the soil the experimental data are widely scattered; the coefficient of variation (c.o.v.) is about 30% for the moisture content and the limits of Atterberg, and about 51% for I_L. Also listed in table 1 is the mean value of the highly scattered values of the undrained strength c_u, as obtained by standard undrained triaxial tests (TX-UU): c.o.v. is about 47%. In 54 TX-UU tests it was in the range of 10 to 110 kPa, the mode being 30-35 kPa. The scatter of c_u values can be explained partly by the soil heterogeneity, partly by the occurrence of contacts, at the sample scale, between the excavated clay elements. In these conditions failure is likely to occur along the relatively weak boundaries between the elements; therefore the shear strength on these surfaces will be measured, rather than in the intact clay. Evidence of such contacts was not immediate since, due to the swelling and weathering experienced by the element boundaries, a sort of welding occurred between the clay elements, still undisturbed in their inner part, under the soil self weight. The mine waste thus appears as a continuum deposit, in which however an irregular net of weakness surfaces is present. From the data as a whole, the mine waste can be considered as a heterogeneous, slightly overconsolidated, silty-clay deposit. A more detailed description of soil profile and compressibility characteristics is given elsewhere [1].

Electro-osmotic Treatment

Electro-osmosis, that is the flow of water induced by an electric field applied to the soil mass, in absence of hydraulic gradients, has been successfully used in a number of practical applications. Among them are those relative to the modification of shaft shear stresses on driven steel piles (e.g. [2]), and the mechanical and electro-chemical hardening of soft clay deposits (e.g. [3]). At Pietrafitta, these kinds of applications were tested in full scale field tests, as an alternative to other more common techniques.

The effectiveness of electro-osmosis in reducing negative friction on piles was analysed in a full scale test on a large bored pile; analysis of the test results may be found in [4]. The pile, 80 cm in diameter and 45 m in length, was iron jacketed down to a depth of 18 m from ground level. An electric current was applied at a voltage of 60 V between the pile and two electrodes, 15 m in length, driven at 6.5 m from its axis. Strain gauges installed in 8 instrumented sections allowed for the calculation of the shaft shear stresses. The cathodic treatment produced a reduction of the adhesion factor $\alpha = \tau/c_u$ from 0.35-0.4 to 0.1 along the iron jacket, and an increase of α underneath, the overall effect being a transfer of the shaft shear stresses to the lower portion of the pile. However, it was observed that a relative displacement subsequent to the treatment was characterised by a strong increase in α. For the problem at hand, because of the ongoing progress of consolidation, this would have required a periodic application of the treatment to release the down-drag forces on piles. Therefore, preference was given to more conventional bitumen covered jackets.

To assess the effectiveness of electro-osmosis for the stabilisation of the clayey mine waste, 4 full scale field tests were carried out using different electrode patterns and spacing; a description of the test characteristics and an analysis of the test results are given in [5].

In the following, reference is made to field tests C (3x3.5 m²) and D (7x8 m²). Iron pipes 80 mm in diameter were installed in a rectangular pattern; 0.5x0.75 m² for test C and 1x2 m² for test D, the greater spacing being that between anodes and cathodes. Electrodes were 6 and 12 m in length for tests C and D. The applied voltage was kept constant at 60 V during the tests.

A different waste composition characterised fields C and D. In the first a prevalence of brown silt of medium plasticity (I_P = 25 %) was observed. The waste deposit was fairly homogeneous with values of w = 20-30 % and a mean undrained strength of 50.3 kPa (c.o.v. =31 %). In the second field the

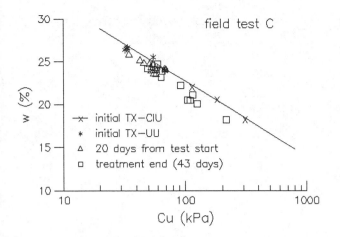

Figure 1: Electro-osmosis effect on w-c_u relationship (after [5])

mine waste consisted mainly of grey clay of higher plasticity (I_P = 30-40 %) with frequent inclusions of brown silt; this produced a pronounced heterogeneity and a high scatter of the soil characteristics within the field test area: w = 22-55 % and c_u = 28 kPa (c.o.v. = 47.4 %).

After the treatment, cone penetration tests showed an increase of the tip resistance q_c by a factor of 2.5-3 in test C while of about 1.25 in test D; on the contrary, the undrained strength, as obtained by triaxial unconsolidated undrained tests (TX-UU), was increased by a factor of 2 in both the tests. However, an increase in the scatter of the measurements was also observed at the end of the treatment [5]. The data in figure 1 show that the measured values of c_u and water content at failure are consistent with the relationship between w and c_u of the untreated soil, as derived from triaxial consolidated undrained tests. This suggests that the hardening of the clayey waste, caused by electro-osmosis, can be mainly ascribed to the effect of consolidation, thus excluding appreciable influences of electro-chemical effects. No changes evident in the index properties after the treatment confirmed this hypothesis [5].

Since soil improvement by electro-osmosis appeared to be highly dependent on the heterogeneity of the clayey waste, choices for design were directed towards the use of stabilised columns under flexible structures.

Stabilised Column Treatment

The stabilised columns were obtained by the mechanical mixing of the clayey mine waste with a stabilising powder agent. In the applied treatment the soil is first thoroughly remoulded in the drilling stage by screwing down the cutting blades; then, in the raising stage under a reverse rotation, the stabilising agent is forced out into the soil by compressed air. A twin head rig allowed the simultaneous installation of two columns 100 cm in diameter and 150 cm apart. The method, first developed in Sweden for stabilisation of soft clayey soils (e.g. [6]), has been modified for application at Pietrafitta. Initially it involved a dry mixing procedure, but, since the clayey waste is stiffer than a soft clay deposit, small quantities of water were added in the first stage of the treatment to improve the remoulding of the soil. In the treated columns the stabilising agent reacts with water in the soil during hardening and develops cemented bonding between the clayey aggregates. Therefore the method differs from an electro-osmosis treatment in which soil stabilisation is mainly attributable to consolidation effects.

At Pietrafitta, Portland or pozzolanic 325 cement was used as a stabilising agent; the cement amount, 18 to 21% of the dry unit weight of the treated soil, was characterised by small values of the coefficient of variation (6-8%). Figure 2 shows the three different zones of the area in which the stabilised columns were installed; table 2 summarises the characteristics and the applications of the treatments.

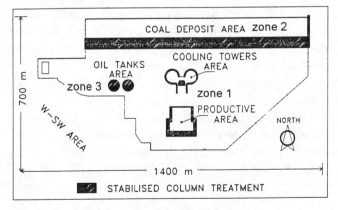

Figure 2: Plan view of the treated zones

TABLE 2: Treatment extent and applications

zone	N_{col}	L_{col}	soil stabilisation for:
1	3464	15-25 m	excavations 5 m deep
2	4490	18-32 m	coal deposit
3	2486	17-23 m	oil tanks

TABLE 3: Test results after treatment

	45 days c_u		90 days c_u	
remoulding stage	1	2	1	2
mean (MPa)	0.68	1.33	1.07	1.54
stand. dev. (MPa)	0.58	0.69	0.81	0.71
c.o.v. (%)	86	52	76	46
tests number	30	41	655	134

Visual inspection of the treated samples and of the excavation faces in zone 1 showed a non homogeneous mixing of the soil with the cement along the columns' depth. Due to the presence of stiffer clay elements or lignite levels, in some zones the cement appeared to concentrate in sub-horizontal thin levels a few centimetres in thickness. It has proved very difficult to avoid these inconveniences because of the nature of the soil involved. To improve the treatment homogeneity a different procedure was attempted in zone 1, where, for a certain number of columns, the remoulding stage was repeated twice before mixing the cement.

The influence of the remoulding procedure on the treatment homogeneity was evaluated on 30 boreholes from zone 1 through the index of Rock Quality Designation (RQD). For the present problem the stabilised portions of the sample were considered as intact material. Small differences in values of RQD were observed between the columns obtained after a single or a double remoulding stage: RQD was 64% for the former while 70% for the latter. Since the values of c.o.v. were lower than 10%, the above evaluations may be considered as reliable. It is worth noting that the above values of RQD should be considered as a lower bound of the portion of the sample actually stabilised, since by definition, stabilised elements shorter than 10 cm are assumed to be fissured.

Treated samples were cored from about 100 columns for laboratory testing. The undrained shear strength was measured in unconfined compression tests on triaxial specimens 38 mm in diameter, sheared under a constant strain rate of 0.26 %/min. Since the shear strength of the stabilised soil gradually increases with time, laboratory tests were carried out after a time of 90 days after the treatment end had elapsed.

To evaluate the strength increase with time, samples from zone 1 were also tested at 45 days after treatment.

Table 3 summarises test results for the single and the double remoulding procedures; the time elapsed since the end of the treatment is also indicated.

The high values of c.o.v. = 46-86 % show the strong heterogeneity of the treatments. However the coefficient of variation is seen to decrease with time and to be lower for columns treated with the double remoulding procedure. The influence of the remoulding procedure on the undrained strength decreases with time since the ratio of the c_u values for columns treated with either two or one stages of remoulding is about 2 and 1.44 at 45 days and 90 days respectively. In fact, the increase in strength with time is greater for columns obtained after a stage of remoulding (57 %) than for those remoulded twice (16 %).

Due to the pronounced scatter observed in the test results, a normal distribution was assumed for c_u and the value corresponding to a lower confidence limit of 16% was used in design. With reference to the undrained strength at 90 days on samples from columns remoulded once before mixing the cement, a value of $c_u = 260$ kPa is obtained. This is about 5 times greater than the undrained strength of the untreated soil if compared with the mean value, while it is 8 times higher when related to the mode value.

Figures 3 and 4 show histograms of the undrained strength at four different levels of depth; data refer to samples cored from zones 2 and 3 only. It is seen that values of $c_u > 1$ MPa were reached in 52 % of samples in zone 2, while 34 % in zone 3.

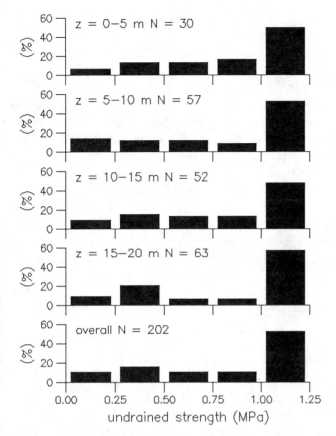

Figure 3: Histograms of c_u in zone 2 (coal deposit)

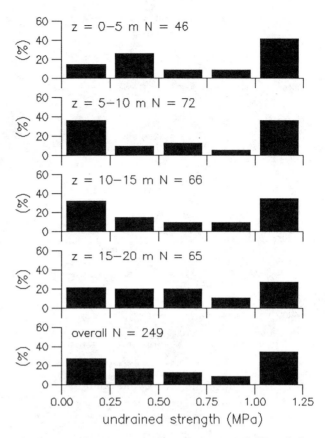

Figure 4: Histograms of c_u in zone 3 (oil tanks)

Also, in zone 3 a percentage of samples as high as 27% showed low values of undrained strength in the range 0-0.25 MPa. The shape of the histograms with depth do not show significant changes apart from a slight prevalence of low values of c_u from 5 to 10 m in zone 3. This range of depths corresponds to the contact between the two portions of the waste which were back-filled 10 years apart. The difference in the effectiveness of the treatment in zones 2 and 3 is probably due to the moisture content in zone 3 (30-60 %) being higher than in zone 2 (25-45 %).

In addition to the laboratory tests, in situ loading tests were carried out on 36 columns of zone 1; 13 obtained after one stage of remoulding, while 23 after two stages. For these tests the ratio K between applied loads and measured displacements was analysed at three different levels of loading; these were in the range 150-250, 350-450 and 600-750 kN. The corresponding values of the secant stiffness are indicated by K_I, K_{II} and K_{III} in table 4. Data from the table allowed evaluation of the influence of the number of the remoulding stages on the overall vertical stiffness of the columns. It can be observed that columns which underwent two stages of remoulding are stiffer for each level of loading. The difference between the treatments increases with the level of loading; the ratio of the K values for columns treated with the double and the single remoulding stages is 1.3 for K_I and 1.9 for K_{III}. However, the scatter of the results is high for both the treatment procedures and it increases with loading.

TABLE 4: Loading tests on treated columns

	K_I	K_{II}	K_{III}	K_I	K_{II}	K_{III}
remoulding stage		1			2	
mean (MN/m)	74	33	13	99	51	24
stand. dev. (MN/m)	55	42	16	69	50	28
c.o.v. (%)	75	125	127	70	98	116
tests number		13			23	

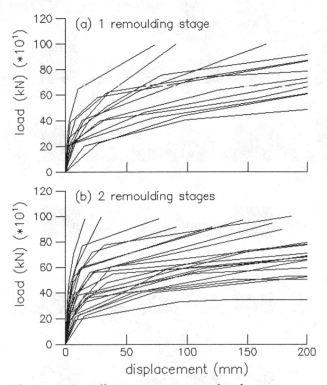

Figure 5: Loading tests on treated columns

Figure 5a-b shows the load-displacement relationships as obtained by the in situ loading tests; the curves have been distinguished according to the number of the remoulding stages carried out before mixing the cement. A non linear behaviour is evident for both the kind of treatments; values of displacement as high as 20 cm show that collapse of some columns was reached at the third, and even occasionally, at the second level of loading. This could be explained as an effect of the squeezing of the untreated portions of soil in the columns. However, the loads applied under working conditions are in the range of the first step of loading, for which the differences in the values of K_I are lower. Histograms of K_I show that for columns obtained after a stage of remoulding, about 50% of the values are in the range 0-50 MN/m and 20% in the ranges 50-100 and 100-150 MN/m; columns remoulded twice before mixing the cement show values of K_I uniformly distributed in the range 0-150 MN/m with a slight prevalence in between 50-100 MN/m. A value of $K_I = 18.3$ MN/m, corresponding to a lower confidence limit of 16% in the hypothesis of normal distribution, was assumed in the design for columns treated with a single stage of remoulding.

Conclusions

The examined mine waste is characterised by a great heterogeneity. Therefore any value describing its physical and mechanical properties is highly scattered, and the statistical evaluation of many test results is necessary. The two kinds of mechanical improvement tested, i.e. electro-osmosis treatment and stabilised columns, were successful in decreasing the mine waste compressibility and increasing its shear strength. The column treatment produced a much greater, but less uniform, shear strength increase. The statistical analysis of many experimental data proved to be the only means to evaluate the effectiveness of the soil treatment. Therefore the cost of field and laboratory investigations to gather the necessary information was justified.

References

1. Rampello S., Pane V., Calabresi G. and Bianco O. (1994): Geotechnical characterization of a clayey mine waste. Proc. of the First Int. Congr. on Environmental Geotechnics, Edmonton.
2. Davis E.H. and Poulos H.G. (1980): The relief of negative skin friction on piles by electro-osmosis. Proc. III Austr. N. Zeal. Conf. on Geomechanics, vol.1:71-77.
3. Bjerrum L., Moum J. and Eide O. (1967): Application of electro-osmosis to a foundation problem in Norwegian quick clay. Géotechnique, 17, n.3:214-235.
4. Rampello S. and Ascoli Marchetti V. (1989): Effetto di un trattamento elettro-osmotico sulla resistenza laterale di un palo di grande diametro. Proc. XVII Convegno Nazionale di Geotecnica, A.G.I., Taormina, vol. II: 263-269.
5. Tamagnini C. and Calabresi G. (1991): Esperienze sul consolidamento elettrosmotico di terreni argillosi teneri. Rivista Italiana di Geotecnica, R.I.G. 25, n.2: 115-136.
6. Broms B. and Boman P. (1977): Stabilization of soil with lime columns. Design Handbook, Dept. of Soil and Rock Mechanics, Royal Inst. of Technology, Stockolm.

Solution Cavern Disposal of Solvay Process Waste

B.C. Davidson, M.B. Dusseault
University of Waterloo, Waterloo, Ontario, Canada

R.J. Demers
General Chemical Canada Ltd., Amherstburg, Ontario, Canada

Abstract

An alternative method for the disposal of non-toxic, non-hazardous Solvay process waste is presently being evaluated by General Chemical Canada Ltd. Research is currently focused on subterranean disposal of waste material in abandoned salt solution caverns in Southwestern Ontario. The geological conditions of the proposed site and material behaviour indicate that long-term environmental security is attainable. Solution cavern disposal appears to be both an environmentally superior and an economically cost effective alternative to conventional surface waste beds.

1.0 Introduction

General Chemical Canada Ltd. (GCC), produces soda ash and calcium chloride at a chemical plant in Amherstburg, Ontario. Presently, GCC is evaluating alternative methods of disposal of inert wastes generated through the Solvay process. The Geomechanics group at the University of Waterloo in co-operation with GCC has initiated a multi-disciplinary research program to evaluate the engineering and scientific aspects of Solvay process waste disposal in abandoned solution caverns. The concept of solution cavern disposal of Solvay process wastes represents an environmentally superior and practical alternative to conventional surface waste beds. Surface disposal facilities currently utilized by GCC have security problems and long-term integrity is not guaranteed. Chloride infiltration of existing monitoring wells and localized dyke seepage are an increasing concern, given the increasing levels of environmental scrutiny. Solution cavern disposal represents a positive engineering approach which addresses the needs for environmental security over geological time. This paper provides an overview of the proposed waste disposal concept and presents some initial quantitative material behaviour results.

2.0 Comparison of Disposal Methods

2.1 Surface Waste Beds

The hydraulic placement of 140,000 - 160,000 tons (dry basis) of voluminous thixotropic Solvay process wastes within an above ground, compacted clay dyke containment area began at **GCC** in 1982 [1]. Currently, 180 acres of productive farmland have been consumed for disposal purposes. At present production rates, the current waste bed will be functionally full by 1998, and an addition 10 acres/year of farmland will be required for Solvay process waste disposal. Although **GCC** is utilizing the most common, economically available waste bed technology, several disadvantages of this method have been identified:

- Conventional technology will continue to consume valuable, productive farmland at a rate of 10 acres/year.
- There is no guarantee that future permits for surface disposal will be granted.
- There are high developmental costs associated with future waste beds.
- Perpetual care of the waste beds is required.
- There may be significantly high close-out costs if the waste beds are to be decommissioned.

2.2 Solution Cavern Disposal

Since 1919, GCC has solution mined about 30×10^6 tons of salt creating a subterranean void of approximately 13×10^6 m^3. With the existing volume of waste generated per ton of soda ash produced being 1.1 m^3, GCC could dispose of Solvay Process wastes into exhausted caverns for 20 years. If GCC's current efforts to reduce waste volumes to 0.8 m^3/ton are achieved, this would equal the volume of salt dissolved per ton of soda ash produced and GCC would have guaranteed long-term subsurface disposal capacity. Waste reduction may also allow current waste facilities to be remediated and reclaimed by reslurrying the solids and injecting the slurry into exhausted brine caverns. The most notable advantages of implementing solution cavern disposal of Solvay process wastes include:

- High quality, productive farmland will not be consumed for disposal purposes.
- There is no perpetual care associated with a cavern.
- Surface subsidence associated with a cavern closing over time will be reduced.
- There are minimal close-out costs associated with cavern disposal.

3.0 Cavern Disposal Technology

3.1 Cavern Development

Solution caverns developed for the purpose of extracting saturated brine for industrial processes are created using conventional oil-well drilling techniques. In most cases, a wellbore is drilled to the base of the salt formation, the well is cased and cemented down to the top of the salt bed, and a centre tube is fixed in the well extending to the bottom of the salt. Fresh water is circulated down through the centre tube and saturated brine passes up the annulus between the centre tube and the well casing. In a bedded salt, such as the Salina Formation, the vertical height of the cavern is limited by the thickness of the salt bed. Therefore, formation dissolution tends to produce flat horizontal cavities which gradually develop outward from the wellbore. Outward propagation and upward growth of the cavity (overall size), can be controlled by applying a pressurized gas cap during cavity dissolution. A generalized representation of a solution cavern in bedded salt is presented in Figure 1.

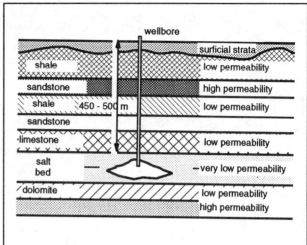

Figure 1 A schematic diagram of a solution cavern in a salt bed

3.2 Waste Placement Technology

Pipeline technology for transporting the slurried waste material to the solution caverns is well established, and it is currently employed to transport Solvay process wastes to existing waste beds. A simplified flow diagram of Solvay Process waste disposal from GCC's plant site to their brine field operations is shown in Figure 2.

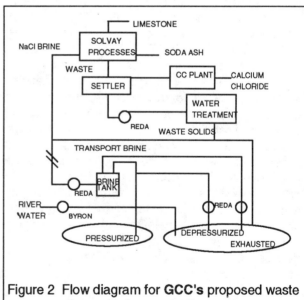

Figure 2 Flow diagram for GCC's proposed waste disposal/brine production system

A generalized procedure for the proposed waste placement technique is as follows. Solvay process wastes are washed and vacuum filtered to remove CaCl$_2$. The filter cake obtained through this process is then reslurried with raw NaCl saturated brine and granular salt and is mechanically pumped to GCC's brine field 2 km east of the plant site. The slurry is introduced to a depressurized, partially brine filled cavern through a moveable tailpipe situated near the bottom of the cavern. As the cavern fills, the tailpipe is gradually withdrawn maintaining a waste/brine interface that is approximately horizontal because the slurry acts as a dense fluid. Because the slurry density in the tailpipe is 1.45-1.55, whereas the brine expelled from the cavern is 1.18, the pressure differential maintains circulation with a minimum amount of mechanical pumping [2]. The waste/brine density difference also allows a stable density current to develop in the cavern, flowing laterally outward with minimal mixing.

The hydraulic behaviour of Solvay process waste placement has been verified through a bench-scale laboratory simulation. The simulation consisted of a lucite model, 45 cm wide having "obstacles" to represent fallen anhydrite blocks typically found at the bottom of a solution cavern, and a moveable tailpipe to inject the wastes. During waste injection, a flat density current developed and flowed laterally outward, filling the entire model including all obstructed areas.

During large scale waste placement, excess brine is displaced and pumped to a brine holding tank where it is mixed with fresh, saturated NaCl brine and pumped back to the plant site for use in the Solvay process. This in effect creates a closed loop where excess displaced brine is continually used and no subsequent disposal or treatment of the brine is required.

Consolidation and a corresponding decrease in porosity of the Solvay process waste in a cavern will be a function of both long-term self-weight compaction and cavern creep closure. he latter is quite slow at the modest salt depths of 450-500 m, thus cavern capacity is limited by total volume, height, and self-weight equilibrium void ratio and consolidation time. Typical intermediate stress void ratios of the waste material range from 1.4 to 1.7, therefore, depending on initial cavern volume, it is feasible that more than one waste injection procedure will be required to fill a cavern to its maximum capacity. Once filled, a disposal cavern may be decommissioned by removing the tailpipe and casing and cementing the entire length of the remaining borehole. Gradual closure will continue to compact the waste, expelling brine into bounding permeable strata.

4.0 Site Stratigraphy

Saturated NaCl brine is currently obtained by **GCC** from wells drilled into the bedded salt-bearing strata of the Salina group, of the middle to late Silurian period. The Salina group is underlain by dolomites, shales and sandstone of the early Silurian period and is overlain by dolomites, shales, limestones and sandstones of the early Devonian period [3]. Figure 3 shows a columnar section of the subdivisions of the salt-bearing and younger strata of Southwestern Ontario. Figure 4 displays the distribution of salt in the subsurface of Southwestern Ontario. Although brine expulsion will occur over geologic time as creep closure of the cavern takes place, the bounding strata of the Salina group provide a natural horizontal permeable channel to decrease the probability of brine interaction with shallow and surface groundwater systems.

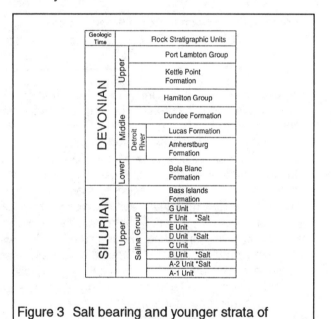

Figure 3 Salt bearing and younger strata of Southwestern Ontario

5.0 Cost Analysis

Although solution cavern disposal of Solvay process waste appears to represent a more environmentally acceptable and secure method of disposal compared to surface waste beds, consideration must be given to the costs associated with implementing this technology. It would not be in the best interest of GCC to implement solution cavern disposal if the cost of this technology continually exceeded that of the surface disposal methods currently employed. A comparison of solution cavern disposal to surface

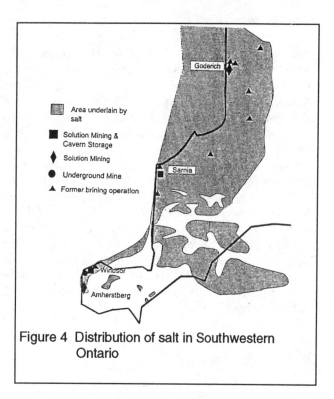

Figure 4 Distribution of salt in Southwestern Ontario

6.0 Material Properties

6.1 Solvay Waste Composition

Solvay process waste is a non-toxic, inert, fine-grained mineral in a saturated brine. The waste is a byproduct of the Solvay process which generates Na_2CO_3 and $CaCl_2$ from a saturated NaCl brine and finely ground limestone ($CaCO_3$); the resulting sodium carbonate com-prises a major constituent of glass and other products. The chemical composition of Solvay process waste is presented in Table 1.

Table 1 Chemical Composition of Solvay Process Waste
40% - 50% $CaCO_3$ calcium carbonate
10% - 20% $Ca(OH)_2$ calcium hydroxide
8% - 15% SiO_2 silicon oxide (quartz)
8% - 15% $CaSO_4$ calcium sulphate (anhydrite)
3% - 4% Al_2O_3 aluminum (III) oxide
3% - 4% Fe_2O_3 iron (III) oxide

6.2 Consolidation behaviour of Solvay waste

Understanding the consolidation behaviour of Solvay process waste under stress conditions approximating those found in solution caverns 450 - 500 m in the subsurface is an important parameter for design, monitoring and implementing the proposed disposal technology.

To study the consolidation behaviour of Solvay Process waste, one-dimensional oedometer testing was performed using a 100 mm x 400 mm brass cell equipped with an internal axial piston. Axial deformation is achieved by applying a gas pressure over the axial piston with changes in height being monitored by an LVDT. Figure 7 shows an e vs. log p plot for a consolidation test performed on Solvay process waste at intermediate stresses. Values of permeability (k), coefficient of consolidation (c_v) and volume compressibility (m_v) were calculated as, k = 4.34×10^{-7} m/sec, c_v = 4.336 m^2/year and m_v = 0.273 m^2/MN. The values obtained for the three parameters indicate that the waste material is of low permeability, medium compressibility and medium

disposal is given in Figure 5. Incremental yearly costs have been indexed for inflation and are projected over a 40 year operating period. Initially, the costs associated with implementing solution cavern disposal exceeds that of surface waste beds. This is because of new equipment acquisition, the set-up of slurry transport facilities, monitoring facilities, etc. Following initial capital expenditures, the principal cost associated with solution cavern disposal is yearly operating and maintenance expenses. In contrast, surface disposal is initially less costly, but as new land acquisitions are made to expand the disposal facility the incremental costs go up dramatically and in the end, far exceed the costs associated with cavern disposal.

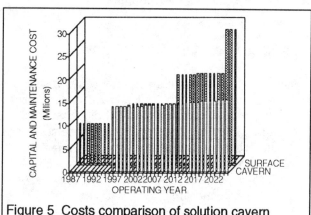

Figure 5 Costs comparison of solution cavern disposal versus surface waste beds

plasticity [4].

Test results support earlier statements which suggest that self-weight compaction in a cavern will be a slow process and several waste injection phases may be required to fill a cavern to its maximum capacity.

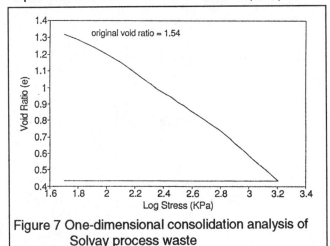

Figure 7 One-dimensional consolidation analysis of Solvay process waste

6.3 Mechanical properties of salt from GCC

Creep closure rates of a salt solution cavern which is filled or partially filled with waste material or is an empty void is a geomechanical issue which must be addressed for several reasons. Determining closure rates will allow computer models to predict the short-term and long-term interaction of a closing cavern and host strata, the influence a closing cavern has on the consolidation properties of the waste material may be better understood, and calcula-tions of long-term volume changes can be made.

Triaxial creep tests on 89 mm diameter intact salt-rock core form GCC's brine field operation have been initiated in the WATSALT Test Facility, University of Waterloo. Equipment design and test procedures have been previously reported by others and will not be discussed here [5].

Figure 8 shows a creep test on natural dirty salt (about 3% solids content, mainly along grain boundaries). Some differences between natural dirty salt and clean salt behaviour are noted. First, the transient creep magnitude in the first cycle is exceptionally large, even after the hydrostatic healing phase [6]. Also, a sudden change in behaviour after 14 days during Stage 2 remains unexplained. At this time, we attribute these anomalies to the effect of clay on the boundaries between halite crystals; rather than dislocation glide and pressure solution, there may be effects of clay sliding and strain-har-dening. This issue continues to be investigated.

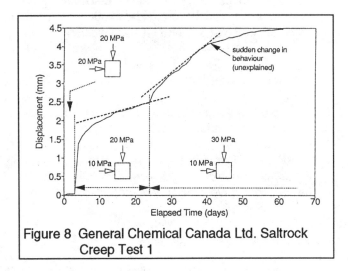

Figure 8 General Chemical Canada Ltd. Saltrock Creep Test 1

7.0 Environmental Considerations

From an engineering standpoint, the technology being proposed for solution cavern disposal of Solvay waste is safe and cost effective. As stated, the waste material is non-toxic and inert, therefore, probability of negative interaction with the biosphere is vanishingly small. We acknowledge that brine will be expelled during cavern creep closure and it is has been suggested that brine contamination of shallow groundwater and surface water systems will be a concern. Low pressure gradients, a density stratified system (1.18 at depth versus 1.00 at surface), horizontal flow regimes in flat-lying strata, and near-surface dilution by precipitation are all indications that brine contamination of shallow waters will never be an issue [2]. Million-year security is reasonable; if brine does reach the surface, rainfall dilution ($7.5 \times 10^5 m^3/km^2/yr$) will drop concentrations close to resolution limits. The benefits of solution cavern disposal clearly out-weigh potentially problematic surface waste beds; furthermore, farmland is preserved and surface subsidence is reduced.

8.0 Conclusions

Solution cavern disposal of Solvay process wastes represents an environmentally sound alternative to current surface waste beds which consume hundreds of acres of high quality farmland. A subterranean void of approximately 13×10^6 m^3 currently exists and

GCC and other companies, may be able to use this storage capacity in geological strata where million year security is not unreasonable. We propose that salt solution cavern solid waste disposal by slurry placement should be con-sidered as well for more noxious industrial and civil terminal wastes.

9.0 Acknowledgements

The co-operation of General Chemical Canada Ltd. is greatly appreciated. The Waterloo Centre for Groundwater Research, Ontario Ministry of Northern Mines and Development and the Ontario University Research Incentive Fund have contributed research funds which have allowed this project to continue.

9.0 References

[1] Demers, R.J., 1993. Amherstburg plant strategic plan, soda ash waste disposal. Internal report of General Chemical Canada Ltd., Amherstburg, Ontario, Canada.

[2] Dusseault, 1993. Solution cavern entombment of granular wastes. International Symposium on Geology and Confinement of Toxic Wastes, Montpellier, France, 8-11 June 1993, pp. 47-54.

[3] Hewitt, D.F., 1962. Salt in Ontario. Ontario Department of Mines Industrial Mineral Report No. 6., Toronto, Ontario, Canada.

[4] Head, K.H., 1982. Manual of soil laboratory testing. Halsted Press, pp. 651-730.

[5] Baleshta, J.R. and Dusseault, M.B., 1988. Triaxial testing of intact salt rocks: pressure Control, pressure systems, cell and frame design. ASTM Special Technical Publication 977, pp. 155-168.

[6] Allemandou, X. and Dusseault, M.B., 1993. Procedures for cyclic creep testing of salt rock, results and discussions. 3rd Con-ference on the Mechanical Behaviour of Salt. Ecole Poly-technique, France.

Laboratory Settling and Consolidation of a Neutralized Red Mud

Tácio M.P. de Campos
DEC/PUC-Rio, Brazil

Maria Cristina M. Alves
UFRJ, Brazil

Roberto F. Azevedo
DEC/PUC-Rio, Brazil

ABSTRACT: This paper presents laboratory results on the sedimentation-consolidation behaviour of a neutralized red mud, discharged in the field with void ratios ranging from 15 to 36. An instrumented sedimentation column and two types of slurry consolidation devices were employed in the experimental study. Characteristics of sedimentation and self-weight consolidation of the mud are discussed. Compressibility and permeability data, for effective stresses varying from circa 0.01 to 300 kPa, are evaluated. The red mud presents a clear change in behaviour after reaching a condition of pure consolidation. The void ratio corresponding to such condition appears to be a material characteristic.

INTRODUCTION

Large volumes of caustic wastes are generated in the industrial process of production of aluminium. Such wastes, know as red muds, usually are disposed as slurry in settling ponds.

The knowledge of the characteristics of sedimentation-consolidation of the mud is required for both the design of the slurry retaining structure, which may be built in steps along the time, and of the environmental reintegration of the resulting pond.

Little information is available on the sedimentation behaviour of red muds. As far as the authors are concerned, published data refers to results of settling column tests in which only the variation of the height of the interface slurry-water is monitored with time (e.g., Pacheco and Melo, 1987). In such case, relations between effective stress and permeability with void ratio, required in analytical studies (e.g., Azevedo, Alves and de Campos, 1994), are not obtained.

Consolidation properties of slurry wastes are usually determined, in the laboratory, through some type of oedometer test. In such tests, however, the initial void ratio of the samples is generally much smaller than those corresponding to field discharge conditions.

This work presents experimental results of a research programme, in which the behaviour of a red mud is studied through a combined set of laboratory sedimentation and consolidation tests. By doing so, it was possible to obtain effective stress and permeability data from void ratios corresponding to discharge conditions to those related to effective stresses of circa 300kPa.

MATERIAL CHARACTERIZATION

The waste under study is from an industrial plant located near the city of Ouro Preto, in the state of Minas Gerais, Brazil. In this plant, the caustic waste is neutralized through the addition of sulfuric acid before being discharged in the settling pond, at solid contents varying from 9% to 19% (void ratio = 34 to 15).

The red mud comprises an well graded silty material, with 25% of fine sand, 47% of silt and 28% of clay fractions in average. For the material previously dried, the liquid and plasticity limits are, respectively, of circa 44% and 12%.

Chemical analysis indicated the occurrence of circa 48% of iron in the mud, what explains the high value of the relative density of its grains, of 3.47 in average. Mineralogical analysis indicated the absence of clay minerals in the slurry, which is in agreement with the use of sulfuric acid in its neutralization process.

EXPERIMENTAL PROGRAMME

Sedimentation and self-weight consolidation characteristics of the red mud were determined using an instrumented sedimentation column developed at the University of Oxford (e.g. Been and Sills, 1980). An X-Ray system is used to monitor changes of the mud density with time, along the height of the column. Profiles of density throughout sedimentation and self-weight consolidation could, therefore, be obtained with no disturbance to the settling process of the slurry. Excess of pore water pressures were also monitored, in ten points along the height of the column, using a high resolution pore pressure measurement technique, refer to Bowden (1988).

Further consolidation characteristics of the mud were determined using two types of oedometer: a restricted flow device (Sills et al, 1986), and a constant rate of displacement (CRD) consolidation cell (e.g., de Campos et al, 1991).

The restricted flow tests were performed on undisturbed samples (ϕ = 100mm; h = 20mm) retrieved from the base of the column device at the end of the sedimentation tests. The CRD tests were carried out on large slurry samples (ϕ = 216.5mm, h = 61.5mm), under rates of displacement ranging from 0.01 to 0.04 mm/min.

The slurry samples were homogenized just before pouring then into the sedimentation column. The column tests lasted from 14 to 290 days. The initial void ratio of the slurry in the column and oedometer tests varied respectively from 15.1 to 36.5 and from 3.3 to 5.0, see Table I.

TESTING RESULTS AND DISCUSSION

Sedimentation–Self-Weight Consolidation

Figure 1 shows typical data from the instrumented sedimentation tests. Part (a) presents selected density profiles obtained at different time intervals after starting the test. Figure 1b shows corresponding excess (over hydrostatic) pore water pressure profiles.

The profiles corresponding to 2h exemplify initial conditions found in the tests. The slurry presented a nearly constant density along the height of the column; there was detected the beginning of formation of a higher density layer at the bottom of the column in the first measurements. This probably indicates some sort of segregation, with the heaviest, iron rich, components of the slurry settling quickly. Small,

TABLE 1 — Initial Slurry Conditions in the Three Testing Series

TEST No.	SEDIMENTATION COLUMN				OEDOMETER					
					RESTRICTED FLOW			C R D		
	H (mm)	H_s (mm)	e_o	% W_s	H_s (mm)	e_o	% W_s	H_s (mm)	e_o	% W_s
1	609.0	33.1	17.4	16.6	3.1	5.0	41.0	10.4	4.9	41.5
2	952.3	25.4	36.5	8.7	3.4	4.9	41.5	12.3	4.0	46.5
3	1202.5	51.8	22.2	13.5	3.8	4.3	47.7	14.3	3.3	51.3
4	1034.9	64.3	15.2	18.6	2.9	4.8	42.0	% W_s = solid content		
5	989.1	26.4	36.5	8.7	3.5	4.6	43.0	e_o = initial void ratio		
6	1031.9	45.5	21.7	13.8	3.7	4.4	44.1	H_s = height of solids		
7	1014.5	63.0	15.1	18.7	H = initial slurry height in the column					

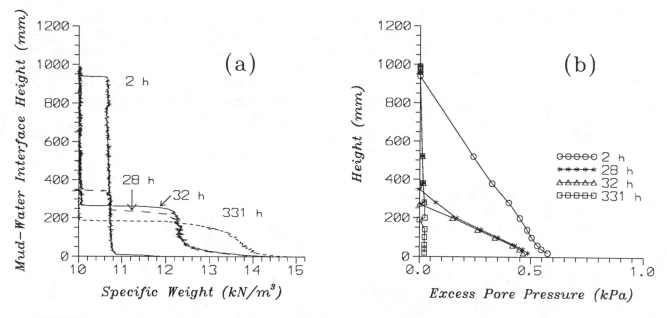

FIGURE 1 – Typical Results Obtained from the Instrumented Column Tests: (a) Specific Weight and (b) Excess Pore Water Pressure Profiles at Different Time Intervals

but detectable, effective stresses were obtained within the first 24 h of each test, independently on the initial solid content of the slurry. Such results suggest that, if a Stokes type of sedimentation occurs as part of the overall settling process of the red mud, it may not be relevant in practice.

The profiles corresponding to 331h in Figure 1 exemplify conditions observed at the end of the tests. An unique soil layer, showing a continuous increase in density with depth, is formed at the bottom of the column. The excess pore water pressures were almost fully dissipated, indicating that primary consolidation of the slurry was practically finished at the end of the tests.

The profiles corresponding to 28h and to 32h in Figure 1 show typical intermediate settling conditions. Consolidation is occurring, as indicated by the decrease in excess of pore water pressures in the profiles in Figure 1b. The 28h profile in Figure 1a shows a condition where three distinct layers occur in the settling column: a top water layer ($\gamma \approx 10.0 kN/m^3$); a bottom layer, with higher densities ($\gamma > 12 kN/m^3$), and a transition sedimentation layer. The sharp density changes observed in this transition layer indicates the occurrence of a discontinuous sedimentation mode in the slurry (Kynch, 1951).

The 32h profile configuration in Figure 1a is considered, in this work, as representative of the beginning of formation of an unique soil layer at the bottom of the column. There are evidences which suggest that a coupled hindered-sedimentation – self-weight consolidation process still occurs in the beginning of the formation of this unique layer (Alves, 1992).

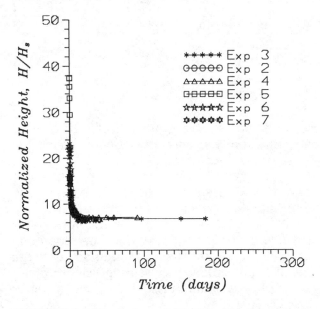

FIGURE 2 – Variation of the Height of the Mud-Water Interface with Time

Figure 2 shows the variation, with time, of the height of the mud-water interface, H, normalized by the height of solids, H_s, for all sedimentation tests. It is apparent in this Figure that the final thickness of the sedimented slurry is independent on its initial void ratio; being a function only of the height of solids.

Density measurements taken at the end of the tests also indicated that the discharge void ratio did not affect the void ratios found at the end of the self-weight consolidation process, see Table 2. The maximum and the minimum void ratios shown in this Table refer to the values measured at the top and bottom of the unique soil layer.

It is also of interest to notice that the initial void ratio of the oedometer samples did not affect the consolidation characteristics of the slurry.

TABLE 2 – Void Ratios in the Unique Soil Layer at the End of the Sedimentation Tests

Test Nº	Void Ratio		
	Max.	Min.	Mean Value
2	11.7	4.7	8.2
3	11.2	4.7	8.0
4	10.4	4.2	7.3
5	11.4	4.0	7.7
6	11.7	3.7	7.7
7	11.3	4.3	7.8
Average	11.3	4.3	7.8

σ' - e - k Relationships

Figure 3 shows log-log plots of effective stress and permeability against void ratio, which include results of the three series of test.

In the sedimentation tests, the compressibility data were obtained directly from integration of the density profiles, combined with the pore water pressure measurement results. Corresponding permeability data were computed through determinations of average settling velocities and excess of pore water pressure gradients, refer to Been (1980).

The results of the restricted flow oedometer tests, which did not provide information on permeability, refer to Alves (1992), were obtained according to Sills et al (1986). The compressibility and permeability data from the CRD tests were computed according to the large strain consolidation theory, as suggested by Znidarcic et al (1986).

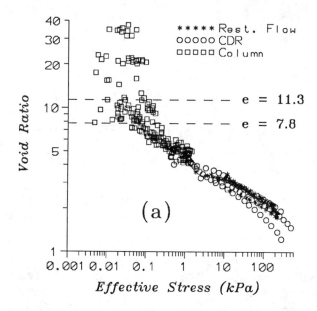

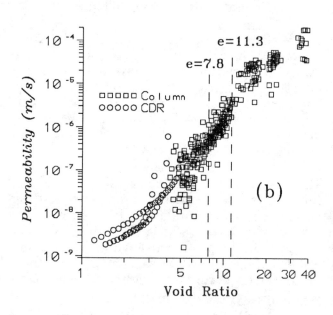

FIGURE 3 – Combined Results of the Sedimentation and Consolidation Tests:
(a) Effective Stress - Void Ratio
(b) Permeability - Void Ratio

As it can be seen in Figure 3, the combined results from the column and oedometer tests showed a fairly good general agreement, both in what refers to the definition of compressibility (Figure 3a) and permeability (Figure 3b) characteristics of the red mud.

This Figure also indicates a clear change in the general slurry sedimentation-consolidation pattern at fairly small effective stresses.

Taking void ratios as reference, it was considered that, at the end of the sedimentation tests, a consolidation-only process would be active in the unique soil layer formed at the bottom of the testing column. Average values of void ratios representing mean condition ($e = 7.8$), and found at the top of such layer ($e = 11.3$, refer to Table 2), are indicated by the dashed lines in Figure 3. Taking into account both permeability and compressibility data, it is apparent that the above higher value of void ratio constitute a better representation of a limit void ratio, below which a consolidation-only process occurs in the mud.

Based on this reasoning, the following equations provided a best fit to the data shown in Figure 3, with σ' in kPa and k in m/s:

a) For e smaller than 11.3, when a consolidation-only process prevails,

$$e = 4.885 * \sigma'^{-0.168} \quad (1)$$

$$k = 4.181 * 10^{-11} * e^{4.549} \quad (2)$$

b) For e greater than 11.3, when both hindered sedimentation and self-weight consolidation occur,

$$e = 14.93 * \sigma'^{-0.114} \quad (3)$$

$$k = (8.04 * \ln e - 2.03) * 10^{-4} \quad (4)$$

CONCLUDING REMARKS

The combined use of the instrumented column and oedometer tests provided means for the definition of a threshold void ratio, below which a consolidation-only process prevails in the slurry.

The limit value experimentally defined, found to be independent on the initial void ratio of the slurry, was confirmed by numerical analysis of the sedimentation – self-weight consolidation of the mud, as reported in a companion paper to this Conference (Azevedo, Alves and de Campos, 1994).

Although not affecting the general findings here reported, large scatter of data was observed, within different ranges of effective stresses or void ratios, in the obtained $\sigma' - e - k$ relationships. Effects related to the absence of clay minerals and to grain density variations in the iron rich slurry; and the existence of fairly small pore pressures at the end and at the beginning of, respectively, the column and the CRD tests, may explain that; the former effects being more relevant to the compressibility and, the latter, to the permeability data. A discussion on that shall be presented elsewhere.

ACKNOWLEDGEMENTS

This work was developed within a research project funded by PADCT / FINEP. The second author was supported by CAPES in the DSc programme developed at PUC-Rio and the University of Oxford. The authors are grateful for all support given by these Institutions, as well as to ALCAN-Brazil, who provided means for sending the red mud to England.

REFERENCES

Alves, M.C.M. (1992) "Comportamento de sedimentação e adensamento de uma lama vermelha" - DSc Thesis, Civil Eng. Dept., PUC-Rio, 213 p.

Azevedo, R.F.; Alves, M.C.M. and de Campos, T.M.P. (1994) "Numerical analysis of sedimentation and consolidation of a neutralized red mud" - 1st International Congress on Environmental Geotechnics - Edmonton, Canada.

Been, K. (1980) "Stress-strain behaviour of a cohesive soil deposited under water" - D.Phil Thesis, University of Oxford.

Been, K. and Sills, G.C. (1986) "Self-weight consolidation of soft soils: an experimental and theoretical study" - Geotechnique 31:4, 519-535.

Bowden, R.K. (1988) " Compression behaviour and shear strength characteristics of a natural silt clay" - D.Phil Thesis, University of Oxford.

De Campos, T.M.P., Villar, L.F., Azevedo, R.F. and Guimarães, J.L. (1991) "Consolidation analysis of a tailings reservoir" - IX Pan. Conf. SMFE, Viña del Mar, Chile, V.3, pp.1021-1033.

Kynch, G.J. (1951) "A theory of sedimentation" Transactions of the Faraday Society, 48, pp. 166-176.

Pacheco, E.B. and Melo, C.E. (1987) "Sedimentação, drenagem e compressibilidade de rejeitos em testes de grandes dimensões" - I Simpósio sobre Barragens de Rejeito e Disposição de Resíduos, ABMS, Rio de Janeiro, V. 1, pp.23-38.

Sills, G.C., Hoare, S.D.L. and Baker, N. (1986) "An experimental assessment of the restricted flow consolidation test" - Consolidation of Soils: Testing and Evaluation, ASTM STP 892, pp. 203-216, R.N. Yong & F.C. Townsend Eds.

Znidarcic, D.; Schiffman, R.L.; Pane, V.; Groce, P.; Ko, H.Y. and Olsen, H.W. (1986) "The theory of one-dimensional consolidation of saturated clays: part V, constant rate of deformation testing and analysis" - Geotechnique 36, nº 2, pp. 227-237.

Engineering and Environmental Considerations in Salt Tailings Management and Disposal Underground

Euler De Souza
Department of Mining Engineering, Queen's University at Kingston, Canada

Abstract: The beneficiation of potash ore produces sodium chloride, a waste material which comprises some 60% of the processed ore and which is disposed of on surface in massive tailings piles. Brine is also produced by the beneficiation process and is stored in extensive surface ponds. Surface disposal of salt tailings has a number of detrimental environmental effects including groundwater, soil and vegetation contamination. A salt tailings assessment program and its application for improved mine safety and environmental control is presented, and methodologies and engineering strategies that may eliminate the risks and consequences of salt tailings contamination are discussed in this paper.

Introduction

With present mining and processing technology, the beneficiation of salt and potash ores results in millions of tons of salt tailings which are normally disposed of on surface. Contamination of surface and ground water by brine and of soil and vegetation by salt waste represents some of the major environmental concerns facing the industry.

This paper discusses the environmental concerns produced by surface salt tailings and presents alternative waste management solutions to salt tailings and brine disposal which offer the potential to minimize the disruption of local and regional ecosystems. Of the various methods of long term salt waste management, the most attractive and practical option is that of underground storage, that is, a process that utilizes salt tailings as backfill to support the underground excavations. Such a system utilizing salt backfill would assist in the control of surface subsidence and potential brine inflows and would have a very positive impact on the long term adverse environmental effects associated with surface salt tailings disposal.

A salt backfill assessment program and its application for improved mine safety and environmental control has been studied. Backfilling techniques which can provide improved ground control, minimize the risk of mine flooding and minimize surface subsidence are described. The results of rock mechanics studies to evaluate the behaviour of salt tailings as backfill and of the effectiveness of salt backfill in terms of local and regional mine stability and ground support are presented. A centrifuge testing program developed to evaluate the mechanism of brine infiltration in tailings piles and ground contamination, the effectiveness of salt backfill as a bulkhead used to isolate areas of potential brine inflow and to evaluate the effect of brine inflow and brine infiltration on long term backfill behaviour is also presented. Applicability of the results to practical underground soft rock mining and salt tailings management is also discussed.

Environmental Concerns and Waste Management Options

The main environmental concerns associated with the surface disposal of salt tailings and with the surface storage of process brine are the possibility of contamination of surface water and groundwater, and air contamination resulting from wind erosion of the tailings pile since wind blown salt may have a negative impact on vegetation and soil. Brine contamination may occur in a number of situations. Brine may escape storage ponds to find its way into fresh groundwater. Brine pond dikes are also susceptible to failure with catastrophic results.

Groundwater contamination may also occur during deep well injection if careful monitoring and control are not established. Precipitation falling on the tailings also results in large volumes of saturated brine which may move into the tailings pile and migrate through the soil and into the underlying groundwater. Subsurface migration of brine may extend several kilometres from the waste management area, and the salinization of fresh groundwater and of agricultural soil would pose a serious risk of loss of fresh water supply and of productive land.

Current tailings containment and maintenance techniques include the use of large brine storage ponds formed by extensive systems of dykes, the establishment of tailings piles and the use of brine injection wells. Of these, gravity driven injection well systems are the most practical for long term tailings management. Alternative waste management solutions sought to eliminate environmental contamination include dry stacking, capping the tailings to isolate the pile from the environment, surface burial of salt waste, potash tails dissolution and deep well disposal, and underground disposal.

Dry stacking, a method of disposing of tailings on surface in a dry state via a system of conveyors, rather than in a slurry form, represents a viable scheme to reduce the amount of brine that is stored on surface, thus minimizing brine seepage and possible contamination of surface water, ground water and soil. Dry stacking allows the construction of more stable tailings piles with a decrease in the brine elevation in the pile, thus reducing tendency for the brine to recharge the local groundwater.

The capping of the tailings with a multi-layered soil system to isolate the pile from the environment and to eliminate potential brine discharge into the groundwater is a potential method as it has been effective and commonly used in land reclamation practices. Preservation of the soil cover would be achieved through the use of vegetation; this would control erosion and soil degradation and would also stabilize the pile. The soil cover acts as a barrier to reduce movement and percolation of precipitation into the tailings pile thus minimizing acidic discharge into the ground. Some dissolution of salt may however occur at the base of the pile. Also, upward migration of salts through the soil cover may occur, leading to contamination of the top vegetation rooting zone and to soil erosion, to eventually result in partial exposure of the tailings. Studies concerning soil characteristics, soil blanket thickness and capillary barriers, vegetation and species growth, climate and precipitation would have to be conducted to verify the applicability of the method.

The burial of salt tailings would require excavation of the surface and stockpiling of soil to provide a burial site for the salt tailings. Additional costs would be associated with the transportation and placement of the tailings, replacement of top soil and revegetation. This is an impractical and economically unfeasible scheme.

Of the various methods of salt waste management and contamination control, the most appealing option is that of underground storage. Practical engineering solutions for the successful management of existing salt tailings accumulated on the surface and for handling and disposing of salt waste which will be produced in the future, using this option, are discussed below.

Two practical options could be used for the elimination of existing tailings piles from the surface: underground disposal in old panels and salt dissolution and injection into deep formations. The placement of tailings (dry stowing or hydraulic paste fill) in the mined out panels underground is one of the most practical methods for disposing of the tailings stored on surface. However, storage of tailings in mined out panels underground may, at present, be uneconomical. Also, because most of the old mined out panels are closed or inaccessible, only a very limited volume of waste would be eliminated from surface. In this case, dissolution and deep well injection is the most applicable method. A number of potash operations use deep well injection to eliminate excess brine stored on surface. However, the economic viability of the proposed method is questionable due to the high costs associated with salt dissolution and pumping practices.

A practical option could be used for the elimination of future waste: the disposal of process waste into the mining area as paste backfill during mining. If the pre-processing of ore is also made underground, incorporating electrostatic separation, it would greatly reduce the amount of waste to be hoisted to surface. With this option, the technology, mining methods and layouts currently utilized would be maintained. The mining operation would have to be adapted to allow the introduction of paste fill during panel mining. Paste fill would maintain the structural integrity of the mine, control surface subsidence and ground closure associated with mining, and prevent the occurrence of brine inflows and mining induced seismicity. Initial capital costs would be required for the establishment of the backfilling program and several mining trials would have to be explored before full production is implemented. Additional costs would also be required for the introduction of machinery for the preparation, transportation and placement of salt paste fill.

Engineering properties of salt backfill

Determination of the flow of brine through salt tailings requires that the saturated and unsaturated properties of the tailings be established. One-dimensional infiltration and seepage models have been developed by a number of researchers, including Philip (1957) and Freeze (1969), and general numerical solutions of multidimensional unsaturated flow have been developed by Freeze (1971), Papagianakis et. al. (1984), Lam et. al. (1987) and others. For any of the above models, the basic relationships required for flow analysis are fluid content versus matric suction and permeability versus matric suction. Wong et. al. (1985, 1987) have specifically evaluated and modelled brine flow through a tailings pile. They have found that brine infiltration rates are strongly dependent on the permeability and degree of anisotropy of the tailings. Ho et. al. (1987) have also evaluated the effect of brine contamination on fine grained soils. Brine contamination was found to have a significant influence on some soil properties (liquid limit, dispersivity and clay mineral double layers) and very little on others (grain size distribution, density, plastic limit and hydraulic conductivity).

An extensive testing program was developed to assist in the feasibility assessment of salt backfilling (De Souza, 1993). This study aimed at evaluating the behaviour of salt tailings as backfill; its suitability to control underground deformations and surface subsidence; and the effectiveness of salt backfill in terms of brine inflow control. The effect of granulometry, water content (in this paper, the term water content represents the relationship between the mass of *brine* and the mass of dry backfill material), density and consolidation pressure on the creep and consolidation behaviour of the waste salt has been studied; salt backfill creep-consolidation tests were performed to evaluate the time dependent behaviour and strength development of the backfill; and a centrifuge testing program was developed to evaluate the effect of brine inflow and brine infiltration on backfill behaviour.

Index Properties of Salt Tailings Backfill

The material constituting the backfill consisted of Halite (NaCl), Sylvite (KCl), and some clay minerals. The salt tailings was shipped in two forms, a dried material and a material with 2% water content. The density of the salt was measured as 2.175 g/cm^3, and the density of the brine was determined as 1.195 g/cm^3. The bulk density of the received dry material was estimated as 1.3 g/cm^3 and the porosity calculated as 40%. The water content at which this material would reach full saturation was determined as 31%. For the 2% material the bulk density averaged 1.05 g/cm^3 with an estimated porosity of 52%. The degree of backfill saturation was estimated as 4%. Grain size distributions indicated that the material size ranged from 1.18 mm to under 0.038 mm. It has an effective grain size (d_{10}) of 0.277 mm and a coefficient of uniformity ($C_u = d_{60}/d_{10}$) of 2.97, indicating a relatively uniform material. The 50% size passing (d_{50}), was estimated as 0.693 mm. Standard dynamic compaction tests indicated that the maximum dry density, approximating 1.67 g/cm^3, is achieved at a water content of approximately 12%. Backfill materials, constructed of salt tailings, should thus be placed at an ideal minimum water content of 12%, with which a high density backfill is produced.

Backfill Creep Consolidation Behaviour

The creep consolidation testing program was designed to simulate the approximate conditions as encountered underground. A consolidation cell, made of steel hydraulic pipe with an effective diameter of 10.16 cm and an effective length of 30.48 cm, and with a base which permitted free brine drainage, was designed and built. Test procedure was to pack samples of tailings salt into the consolidation cell to give maximum preconsolidated densities determined in the standard Proctor tests. Load was applied to the specimens using an electro-hydraulic servo-controlled compression loading frame with a capacity of 890 kN. Free drainage during the test was permitted (to avoid pore water pressure build-up) through the cell filter and the brine outflow was collected and monitored within a drainage system located at the bottom of the cell. A range of consolidation loads were tested (3.45, 6.89, 13.79 and 20.68 MPa) to simulate stress levels encountered underground. Once the test applied load was reached, samples were allowed to stabilize and creep under constant load for 6 hours. Following consolidation, the specimens were extruded from the cell, weighed and tested for water content. The collected brine volume was measured to calculate the final water content of the specimens, density and other related parameters.

Figure 1 presents compression curves for the material, tested under different maximum consolidation stresses. The load-strain relationships were almost identical for the specimens tested, indicating equivalence in backfill consolidation behaviour when under conditions of increasing stress concentrations. A quasi-linear relationship between consolidation strain and applied stress is evident for all levels of stress; the slope of curves at any point is a measure of the compressibility or modulus of deformation of the material. Plots of specimen axial deformation versus time are presented in Figure 2. These curves show the instantaneous, irrecoverable strain which occurred at load application. This initial strain is due to brine expulsion and due to rearrangement of the grains. The instantaneous deformation and the creep rates are a function of the applied stresses, and increase with increasing stress level. The maximum amount of consolidation that the material may undergo in confined, drained conditions is also dependent on the strain related to the volume of expulsed brine. Further strain then continues as grain angularities are broken and the porosity decreases. This is followed by a stage of time dependent creep behaviour, in which strains are also related to plastic deformation of

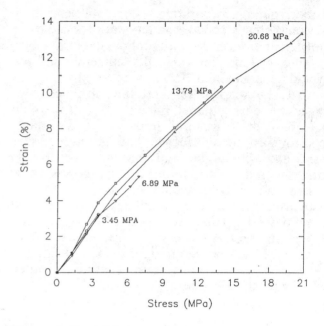

Figure 1. Backfill Compression Curves

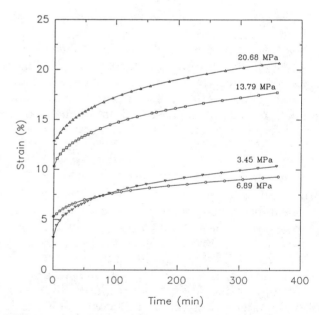

Figure 2. Creep-Consolidation Curves

the grains. The secondary creep rates increased slightly with increase in stress level.

For maximum applied stress, backfill axial strains reached 20.7% of the original column height, indicating its potential for ground subsidence control. Reductions in initial water content during creep reached 49%, producing final water contents of 6%. Increases in density approached 19%, to give equivalent final densities of 1.85 g/cm^3. Test results indicated that the material exhibits poor capability for retaining brine during the early stages of consolidation; approximately 50% of the initial brine volume is drained at full load and after creep. Porosities changed to 48% of the initial specimen values, to produce final values of 14.8%. At these final levels of density and porosity, it is believed that the material would be able to retain a potential brine inflow. A bulkhead constructed with this material should thus provide minimum strain and maximum support, being ideal to seal off mined out areas.

Centrifuge Modelling

The centrifuge testing program was designed to evaluate the behaviour of salt tailings as backfill; the effectiveness of salt backfill as a bulkhead used to isolate areas of potential brine inflow and the potential use of salt backfill to control surface subsidence. A 33 g-tonne, geotechnical centrifuge, designed and built at Queen's University, was used (Mitchell, 1989). It was designed to rotate at speeds up to 350 rpm with 100 kg masses on each end of the main beam (3 m radius). At 330 rpm the centrifuge is exerting an outward acceleration of approximately 330 g on a model, creating stress similitude with a 100 m high prototype.

The variables incorporated into the testing program included applied stress, time and brine inflow volume and pressure. The following tests, summarized in Table 1, have been conducted to verify the impact of each variable on backfill behaviour: Model 1 - the model is loaded to 50 g (0.6 MPa) and then creep consolidated for 12 months; Model 2 - the model is loaded to 50 g (0.6 MPa), 15 ml (1879 litres) of brine is then injected and the backfill creep consolidated for 3 months; Model 3 - the model is loaded to 50 g (0.6 MPa) and creep consolidated for 6 months, 15.1 ml (1891 litres) of brine is then injected and the backfill creep consolidated for 3 months; Model 4 - the model is loaded to 50 g (0.6 MPa) and creep consolidated for 12 months, 15.2 ml (1904 litres) of brine is then injected and the backfill creep consolidated for 3 months; Model 5 - the model is loaded to 50 g (0.6 MPa) and creep consolidated for 12 months, 30.5 ml (3811 litres) of brine is then injected and the backfill creep consolidated for 3

Model	Stress (MPa)	Consolidation Time Prior to Injection (months)	Injected Brine (litres)	Consolidation Time After Injection (months)
1	0.6	12	-	-
2	0.6	0	1879	3
3	0.6	6	1891	3
4	0.6	12	1904	3
5	0.6	12	3811	3
6	1.2	12	3808	3

Table 1. Centrifuge Model Tests

months; Model 6 - the model is loaded to 100 g (1.2 MPa) and creep consolidated for 12 months, 3.8 ml (3808 litres) of brine is then injected and the backfill creep consolidated for 3 months.

In the developed experimental procedure, a column of fill is placed in the centrifuge strongbox and allowed to drain under the influence of the increased gravity field of the centrifuge. After a predetermined length of time has passed, brine is introduced at a point source at the top of the fill column. The brine tracer is allowed to infiltrate into the soil for a set length of time, after which the fill column is removed from the centrifuge and partitioned into horizontal slices for determination of water content and brine concentration.

A schematic of a typical model is shown in Figure 3. A fill column mold, 30 cm in length and 10 cm in diameter and a base plate with a porous filter to permit uniform drainage of brine from the end of the fill column were used. A moveable top plate was used to transfer load to the fill column, to measure fill consolidation and to allow brine addition to the column to simulate inflow. A reference (fixed) plate was used to house a direct current displacement transducer (dcdt) used to continuously monitor the consolidation of the fill column. Fill consolidation was measured by means of a rod which connected the column top plate to the dcdt. A data acquisition system was used for dcdt data collection, the sampling rate was set at 15 seconds. A column made of stainless steel tubing filled with lead shots was built and used to apply surcharge load to the fill. The surcharge column had a centre hollow tube to permit placement of the brine injection tubing and consolidation measuring rod. The lead shot column was placed directly on top of the fill column (top plate). The column weighed 9.8 kgf, producing prototype surcharge stress of 0.612 MPa (50 g model) and 1.224 MPa (100 g model). A brine feed tank assembly, mounted in the centrifuge arm, near the center of rotation (shaft), was used to supply a premeasured quantity of brine to the fill column while in flight. The brine tracer is put in the tank prior to the start of a test. A solenoid valve, controlled by means of a switch in the centrifuge control room, was used to permit brine delivery to the fill column, at a specific time, via a feed plastic tubing (3.175 mm diameter). A brine receiving tank was placed below the fill column to receive and measure the outflow brine during flight. The tank was weighed before and at the completion of a modelling sequence in order to estimate brine drainage volumes.

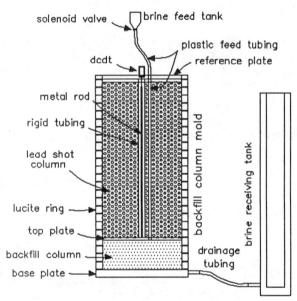

Figure 3. Centrifuge Creep-Consolidation Model

Standard centrifuge modelling laws were used for the prediction of prototype behaviour from experimental data obtained from a scaled model. Figure 4 presents creep curves for models 2 to 6. The salt fill presented a maximum strain of 4.11% at a consolidation stress of 0.6 MPa and 4.62% at 1.2 MPa. These strain values were noted after 12 month creep, with drainage of preparation brine and prior to brine injection. The occurrence of inflow, simulated by the injection of brine, increased the maximum strain to 4.51% and 5.25% at a consolidation stresses of 0.6 and 1.2 MPa, respectively. These results clearly show the influence of stress and brine inflow after consolidation on backfill behaviour. The volume of injected brine did not seem to influence backfill behaviour. By doubling the amount of brine inflow volume (1904 litres in model 4 and 3811 litres in model 5) no substantial difference in backfill strain was noted. Although the applied stresses were much lower than expected field values (1.2 MPa vs 20 MPa), the above results indicate that salt tailings could probably be applied underground as bulkheads to seal off areas of potential inflow.

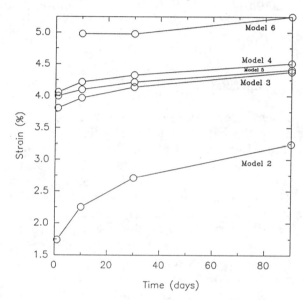

Figure 4. Centrifuge Creep Curves

Figure 5 presents brine drainage curves after injection for models 2 to 5. Models 2, 3 and 4 presented similar drainage volumes as the volume of injected brine was approximately the same. The effect of doubling the volume of brine inflow is represented by model 5, which presented much higher brine drainage volumes. In general, 80% of the initial fill brine volume is expulsed during consolidation thus indicating the need for a drainage system if the fill is used underground. Because the applied stresses and fill consolidation were relatively low, the 'fill bulkhead' did not retain all of the injected (inflow) brine. Approximately 60% of the injected brine (inflow simulation) was expulsed. At typical field stresses it is expected that the fill would be able to seal off panels with brine inflows.

The salt fill presented a final water content of approximately 2% after 12 month creep and prior to brine injection, thus representing a reduction in

85% of the initial water content. After injection and 90 day creep, final water content values approximated 3%. These very low values of water content reflect the rapid brine drainage experienced by the fill column.

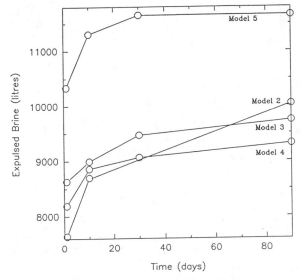

Figure 5. Prototype Brine Volume Expulsion

Conclusions

A study to provide realistic and practical engineering solutions for the successful introduction of salt tailings as backfill to efficiently and safely mine soft rock underground has been presented in this paper. The most important aspect of the study was careful consideration with respect to mine integrity and safety and to environmental control. A salt backfill characterization study indicated that a suitable backfilling material for underground soft rock mines can be used as a support mechanism, to control mine subsidence and to control brine inflow. Test results also indicated that salt tailings could probably be applied underground as bulkheads to seal off areas of potential inflow with the additional advantage of controlling room closure and local back instability.

References

De Souza, E.M. (1993) 'Geotechnical Assessment of Salt Tailings as a Backfill Material', research report submitted to the Saskatchewan Potash Producers Association, pp. 154.

Freeze, R.A. (1969) 'The Mechanism of Natural Groundwater Recharge and Discharge. 1. One-Dimensional Vertical, Unsteady, Unsaturated Flow Above a Recharging or Discharging Groundwater Flow System'. Water Resources research, vol. 7, pp. 153-171.

Freeze, R.A. (1971) 'Three-Dimensional, Transient, Saturated-Unsaturated Flow in a Groundwater Basin'. Water Resources Research, vol. 7, pp. 347-366.

Ho, Y.A. and Pufahl, D.E. (1987) 'The Effects of Brine Contamination on the Properties of Fine Grained Soils', Proceedings of the Geotechnical Practice for Waste Disposal Conference, pp. 547-561.

Mitchell, R.J. (1989) 'Model studies on the Stability of Confined Fills', Canadian Geotechnical Journal, vol. 26, no. 2, pp. 210-216.

Philip, J.R. (1957) 'Numerical Solution of Equations of the Diffusion Type With Diffusivity Conductivity Dependent II'. Australian Journal of Physics, vol. 10, pp. 29-42.

Taylor, R.L. and Brown, C.B. (1967) 'Darcy Flow With a Free Surface'. ASCE Journal of the Hydraulics Division, vol. 93(HY2), pp. 25-33.

Wong, D.K.H., Barbour, S.L. and Fredlung, D.G. (1987) 'Modelling of Flow Through Potash Tailings Piles', Canadian Geotechnical Journal, vol. 25, pp. 292-306.

Wong, D.K.H. and Pufahl, D.E. (1985) 'Studies of the Infiltration and Migration of Brine in Salt Tailings', Proceedings of the Canadian Society for Civil Engineering Annual Conference, Saskatoon, Saskatchewan, vol. 1A, pp. 357-374.

Papagianakis, A. and Fredlung, D.G. (1984) 'A Steady State Model for Flow in Saturated-Unsaturated Soils'. Canadian Geotechnical Journal, vol. 21, pp. 419-430.

Lam, L., Fredlung, D.G. and Barbour, S.L. (1987) 'Transient Seepage Model for Saturated-Unsaturated Soil Systems: A Geotechnical Engineering Approach'. Canadian Geotechnical Journal, vol. 24, pp. 565-580.

Comparison of Fine Tailings Properties and Disposal Methods for the Alumina, Phosphate, and Oil Sands Industries

W.A. Ericson, P.E., C. Winkler III, P.G., M.E. Plaskett, E.I.
BCI, Lakeland, Florida, USA

ABSTRACT

Experience with several fine-grained, mine and industrial tailings has provided the opportunity to compare and contrast their properties and the methods of disposal. Understanding the similarities and differences in material characteristics are important factors in encouraging technology transfers on disposal and reclamation practices between the various industries.

Bauxite is typically shipped to alumina refineries where the refining process generates fine grained, plastic alumina red mud. Red mud disposal methods vary from wet ponds to "dry stacking" techniques, often dictated by climatic conditions, land availability and operator preference.

The phosphate mining industry of Florida and North Carolina has been handling large quantities of highly plastic clays since the 1940s. For several years, the phosphate industry considered the low solids content ponds as a nuisance. However, changes in technology and increased environmental awareness have resulted in disposal and dewatering techniques to speed reclamation.

The oil sands surface mining industry of northern Alberta has grappled with issue of bitumen rich fine tailings since mining began in the 1960s. Even though the fines content in the raw materials and the tailings waste stream is typically less than 15 percent, by dry weight, segregation of the plastic, clayey, bitumen fines occurs resulting in large volumes of slowly consolidating, gel-like deposits. Techniques to improve disposal characteristics continue to be studied by the oil sands operators.

Introduction

The extraction and processing of minerals results not only in useful products, but also large volumes of residues and wastes. While industries such as phosphate, alumina, and oil sands try to constantly improve process efficiencies, reduce costs, and reduce or eliminate wastes; they still have to deal with millions of tons of semi-solid wastes or by products.

In the case of the phosphate industry, highly plastic clay wastes are produced during the beneficiation process. The clays have traditionally been deposited in above ground settling areas, often each area being more than 100 hectares in size. Some operators are currently using methods to thicken the clays and then mix them with sand tailings or gypsum prior to disposal. Though considered inert, the phosphatic waste clays demonstrate poor consolidation properties.

The alumina and oil sands industries have wrestled with many of the same problems as phosphate mines. Although material properties and volumes often differ greatly, there is still the issue of what to do with the fine grained, plastic tailings. Considering that tailings management is generally a liability because no one pays you for your waste products, tailings disposal ponds were used but not managed. Changes in environmental awareness, regulatory requirements to restore waste disposal areas, and growth within the concept of sustainable development [Atkins 1992] have prompted the exchange of waste disposal and reclamation technologies.

As discussed in detail in Carrier et al. 1983, determining the engineering properties and the affects these properties have on material behavior is critical to the understanding and planning of disposal and reclamation activities. The key engineering properties of fine tailings and their relationship to consolidation behavior, as discussed in that paper are still valid. However, improvements in testing and computer analysis methods have helped engineers, mine operators, and regulators make advances in this field.

Waste Material Characterization

The response of fine grained waste materials to changes in stress is a key factor in developing waste disposal and reclamation plans. The basic properties of plasticity (Atterberg limits), compressibility, and permeability are the key consolidation factors that need to be determined. This section briefly describes these factors and updates in test methods used to determine compressibility and permeability of slurried mineral wastes.

Atterberg Limits Correlation

As discussed by Carrier and Beckman 1984, the compressibility and permeability of a plastic material are directly related to its Atterberg limits. Typically, as the value of the plasticity index (PI) increases, the void ratio at any given effective stress also increases, i.e., the solids content decreases. For fine grained materials, such as mineral wastes, as void ratio increases, permeability also increases. Based on this relationship, effective stress ($\bar{\sigma}$) and permeability (k) as a function of the void ratio can be estimated. Consolidation parameters derived from Atterberg limits correlations provide good estimates of consolidation behavior when no additional laboratory data are available.

Restricted Flow Consolidation Test

The Restricted Flow Consolidation (RFC) test apparatus consists of a piston and chamber which allows low percent solids content tailings to be consolidated under very low gradients created by restricted flow of water through the sample. Measurements of the volume of water entering the chamber with time provides continual monitoring of the sample height change and the flow rate of pore water out of the sample.

Seepage Induced Consolidation Test

Drs. Abu-Hejleh and Znidarcic [1992] at the University of Colorado have developed a method of testing soft soils and mineral wastes termed a seepage induced consolidation (SIC) test. The SIC test uses a stainless steel syringe, with a precisely controlled mechanism driving the piston, to produce a variety of flow rates through the sample in a modified triaxial cell. Consolidation of the sample produces a hydraulic head difference across the sample which is measured with a differential pressure transducer. Analysis is completed using an iteration algorithm to determine consolidation parameters. The compressibility and permeability relationships used for the analysis to predict consolidation of various mineral wastes are listed below.

Consolidation Parameters

As defined by the following relationships, consolidation of plastic materials is a function of two physical properties - permeability and compressibility. Compressibility characteristics dictate the _amount_ of consolidation, and the permeability controls the _rate_ at which consolidation takes place.

Compressibility

$$e = A\bar{\sigma}^B$$

$$e = A(\bar{\sigma} + Z)^B \text{ [Abu-Hejleh, 1992]}$$

$$e = A\bar{\sigma}^B + M \text{ (oil sand tails)}$$

Permeability

$$k = \frac{Ee^F}{1+e}$$

Where:

e = Void Ratio
$\bar{\sigma}$ = Effective Stress, (kPa)
k = Permeability (meters per second)

The consolidation parameters A, B, E, F, M and Z are input variables for analyzing finite strain non-linear consolidation of mineral waste slurries, [Carrier et al. 1983].

Consolidation Modeling

For comparison purposes, consolidation modeling for a variety of materials was completed using two finite strain computer programs that predict consolidation during filling and quiescent periods. Typical input parameters for the computer programs include settling area geometry, flow rate, and initial clay solids. Engineering properties of the material include specific gravity and consolidation parameters. Figure 1 shows plots of a hypothetical three-year fill and a subsequent five-year quiescent consolidation period for a variety of materials.

Industry Overview

The three industries discussed in this paper produce millions of tons of mineral waste products each year. Similarities between these industries have allowed them to share disposal and reclamation technology and gain experience, often without having to reinvent the wheel.

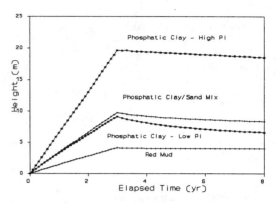

Figure 1 - Combined Filling and Quiescent Consolidation

Phosphatic Clay

The phosphate mining industry of Florida has been handling large quantities of fine grained, highly plastic tailings since the 1940s. Phosphatic clays typically consist of the highly plastic clay minerals smectite and palygorskite/sepiolite [McClellan & Van Kauwenbergh, 1990]. PI values for phosphatic clays range from 70 to 220.

When flotation was introduced (about 1940) to separate sand-sized phosphate from quartz sand, the industry was forced to develop techniques for waste clay disposal. For years it has been standard practice to pump a low solids content clay slurry to above-grade impoundments where the clay is allowed to consolidate. For several decades, the phosphate industry considered these pseudo-stable ponds as a nuisance and little effort was put into reclamation efforts. By 1975, there were about 40,000 hectares of unreclaimed clay impoundments which comprised about 40 percent of the total mined acreage.

However, financial, aesthetic and environmental considerations have resulted in improvements in disposal and dewatering techniques to speed reclamation. First, the industry realized unreclaimed clay settling areas were a tremendous liability that could be converted to an asset if reclaimed. Reclamation also lowered the exterior embankments, returning the areas to near pre-mining topography.

The major problem with reclamation of clay settling areas is dewatering the soft clay. Dewatering usually begins with draining standing water, followed by excavating a drainage ditch along the pond perimeter.

Isolated depressions and the gradual formation of a large bowl shaped clay surface often require using low ground pressure equipment on the clay. Initially, large rubber tired equipment was used to excavate interior ditches to route water to the perimeter. This equipment was used in tandem to allow for towing vehicles mired in the clay. Recently, very low ground pressure tracked equipment (less than 5 kPa) has been developed that allows the clay surface to be worked when the surficial clays are between 20 to 40 percent solids content.

Post-reclamation land use options have also increased in recent years. Agricultural use utilizing macro-beds have allowed alfalfa, corn, and other cash crops to be grown on reclaimed clay settling areas. One operator is experimenting with pecan trees and several areas are being reforested to create wildlife corridors.

Red Mud

Bauxite refineries produce alumina (Al_2O_3) from either raw or washed bauxite. The alumina is principally used as a feed stock for the aluminum reduction industry. Currently there are active bauxite refineries throughout the world, with the largest production capabilities being located in Australia.

Red mud is the precipitate and undigested solids from caustic suspensions of sodium aluminate. It is transported, typically as a slurry to ponds or stacks at solids contents ranging from about 12 percent to 60 percent. The resulting red mud remains caustic (pH values of 11 to 13) and is high in dissolved solids, iron, silica, aluminum, calcium, and sodium. The red mud is often high in clay size minerals with the tailings waste stream containing less than five percent sand, by dry weight. Red mud streams in western Australia are typically sandier.

The engineering properties of the red mud are typically characterized by a low to medium plasticity range (10 to 20 PI). The red mud materials generally show a smaller change in void ratio with increases in effective stress than phosphatic clays and as such also show smaller permeability changes with changes in void ratio.

Disposal of slurried red mud varies widely around the world. Typical red mud disposal discharge rates are on the order of 200,000 to 1,000,000 metric tons per year. The predominant disposal method is the placement of the red mud in at-grade or above grade impoundments. The red mud slurry is transported via pipeline, either by gravity or pumps to the disposal site(s). The red mud ponds are typically 50 to 150 hectares in size with some approaching 1,000 hectares. The depths of these red mud disposal impoundments are generally a few to a few tens of metres.

Sampling of red mud ponds and data from literature sources shows that the red mud typically quickly achieves a solids content value of about 50 to 55 percent even if the initial solids content value is in the teens. Saturated red mud ultimately reaches solids content values of about 65 to 70 percent.

Other disposal methods for red mud include variations of "dry stacking". Dry stacked mud is typically delivered to the disposal area at a higher solids content of 50 to 65 percent.

Dry stacking on a sloping surface has been used at Alcan's operation in central Jamaica. The mud is deposited in thin layers and allowed to dry, [Chandler 1986].

Alternate dry stacking/ponding methods are used at some locations in western Australia and Canada. In Australia, red mud is placed in one metre lifts and then allowed to dry and crack. The drying process is augmented by surface tracking using amphibious equipment. Where this process is used, the mud is thickened and the sand separated from the tailings stream [Gale and Coffey, 1992].

"The alkaline nature of the muds will limit plant growth on the dried, closed impoundments." [USEPA 1990] Some firms have made progress in studying the restoration and revegetation of red mud ponds. Studies are on-going at the Point Comfort, Texas facility to evaluate the feasibility of using dredged spoil material as a cover and growing medium over the red mud [Krishnanohan, et al. 1992]. Other revegetation/restoration activities have been on-going in Australia, where red mud ponds are supporting grasses and trees with minor amounts of soil amendments such as gypsum and sewage sludge [Wong and Ho, 1992; Bell and Ward, 1992]. Most of the inactive red mud ponds support little or no volunteer vegetation and therefore will likely require a planned effort by the operators to close these areas.

Oil Sands Tailings

The oil sands industry of northern Alberta has been surface mining the bitumen rich sands for over twenty-five years. The major constituents of the oil sands are quartz sand, clay, water, and bitumen. Once mined, the oil sands are processed to remove the bitumen which in turn is upgraded to "synthetic" crude oil.

In order to extract the bitumen from the oil sands, the ore is typically treated using a hot water froth flotation process. The waste tailings from the extraction process consist of a mixture of sand, clay, hot water and caustic soda (NaOH), and un-extracted bitumen. The mineral fraction of the tailings is composed mostly of quartz, kaolinite, illite, and trace amounts of montmorillite. The amount of bitumen remaining in the tailings varies, but is generally on the order of 5 to 12 percent. Total solids contents are often on the order of 40 to 55 percent.

The tailings are pumped as a slurry for disposal in large diked ponds. Typically, these dikes are raised with the sand fraction of the tails using upstream construction techniques.

As of 1991, approximately 300 million cubic metres of tailings had been produced. The tailings are presently stored in disposal areas that occupy over 35 square kilometres [MacKinnon and Sethi 1993].

As the tailings are deposited in the disposal areas, approximately 50 percent of the fines settle out with the coarse fraction to form beaches. These beaches are used to raise the impoundment dikes and contain significant amounts of unrecovered bitumen. The fine fraction that does not become entrained in the beach, enters the disposal area at a relatively low solids content and sediments rather quickly (within a few weeks) to about 20 percent solids, where the process of consolidation begins. Although most of the unrecovered bitumen in the tailings stream is entrained in the beaches, a small amount (about 2 to 5 percent) remains with the sludge as it collects in the center of the pond.

The engineering properties of the fine fraction "sludge" are characterized by a plasticity index that typically ranges between 30 to 65. Based on the values of the Atterberg limits, the sludge is classified as a highly plastic clay (CH) according to the Unified Soil Classification System. Because of the high plasticity and the influence of the bitumen on the sludge, the consolidation of the fine tailings is very poor.

Current disposal methods for oil sand process tailings do not allow for reclamation of disposal areas in a reasonable period of time. In order for the industry to reclaim their disposal areas, alternative methods of disposal and/or treatment will likely be required in the future. Research is being conducted on methods to stabilize the sludge that presently exists in the disposal areas and to improve the consolidation characteristics of tailings deposited in the future. Possible alternatives include addition of chemicals to increase the rate of consolidation for the fine tails.

One study being completed in conjunction with the University of Alberta concentrates on creating a tailings stream comprised of a nonsegregating sand/fine mix. This technique has proven successful in the Florida phosphate industry as it increases the total solids content, thus decreasing the volume required for disposal. Preliminary results indicate the use of this alternative disposal method appears to be promising. Existing fine tailings could also be dredged, mixed with sand, and treated with chemicals before being returned to the disposal areas [Scott et al. 1993].

Once the tailings surface has been stabilized with the addition of chemicals and/or sand, the disposal areas could be dewatered with equipment similar to that used in the phosphate and alumina industries.

After a program of alternate disposal has been developed, and the oil sand tailings surface has densified enough to permit low contact pressure vehicles to access the ponds, revegetation could be initiated.

Bibliography

Abu-Hejleh, A., Znidarcic, D., Robertson, A., "Results of Seepage-Induced Consolidation Tests on Phosphatic Clay," Prepared for the Florida Institute of Phosphate Research, October, 1992.

Atkins, P.R., "Global Environmental Issues and How They Might Affect Bauxite Residue Management," an International Bauxite Tailings Workshop, Perth, Western Australia, November 1992.

Bell, David T., and Ward, Samuel C., "Selection of Trees for Rehabilitation of Bauxite Tailings in Western Australia," an International Bauxite Tailings Workshop, Perth, Western Australia, November 1992.

Carrier, W.D., III and J.F. Beckman, "Correlations Between Index Tests and the Properties of Remolded Clays," Geotechnique, V. 34, No. 2, pp 211-228, 1984.

Carrier, W.D., et al. "Design Capacity of Slurried Mineral Waste Ponds," Journal of Geotechnical Engineering, V. 109, No 5, pp 699-716., 1983.

Chandler, J.L., "The Stacking and Solar Drying Process for Disposal of Bauxite Tailing in Jamaica," Proceedings of an International Bauxite Tailings Conference, Kingston, Jamaica, W.I., October 1986.

Gale, A.J., and Coffey, P., "Environmental Management of Mineral Processing Residues in the Bauxite/Alumina Industries," an International Bauxite Tailings Workshop, Perth, Western Australia, November 1992.

Krishnanohan, R., Herbich, J.B., Hossner, L.R., Williams, Fred S., "Geo-Environmental Monitoring Techniques Used for Investigating Options for Closure of a Bauxite Residue Disposal Area," an International Bauxite Tailings Workshop, Perth, Western Australia, November 1992.

MacKinnon, M.and Sethi, "A Comparison of the Physical and Chemical Properties of the Tailings Ponds at the Syncrude and Suncor Oil Sands Plants," in Oil Sands - Our Petroleum Future Conference, April 1993.

Scott, J.D., Liu, Y., Caughill, D.L., "Fine Tails Disposal Utilizing Nonsegregating Mixes," University of Alberta, Department of Civil Engineering, 1993.

USEPA, Report to Congress on Special Wastes from Mineral Processing, Summary and Findings Methods, Analyses Appendices, Chapter 3 EPA 530-SW-90-070C, July 1990.

Wong, J.W., and Ho, G.E., "Viable Technique for Direct Revegetation of Fine Bauxite Refining Residue," an International Bauxite Tailings Workshop, Perth, Western Australia, November 1992.

The Influence of Evaporation on the Consolidation Behaviour of Gold Tailings

Martin Fahey, Associate Professor, Yoshimasa Fujiyasu, Research Student
Department of Civil Engineering, The University of Western Australia

Abstract: This paper examines the factors which influence the rate of consolidation of tailings from the gold mining industry in Western Australia (WA). Most of the mines are in arid areas with very high annually rates of evaporation. This can make an enormous contribution to the rate of consolidation of fine-grained tailings. However, the ground water used in processing is often extremely saline, and, as shown by laboratory evaporation tests, this can severely reduce the rate of evaporation. It is clear that any useful numerical modelling of tailings consolidation must be capable of including the effects of evaporation, but the detrimental effect of high salinity must also be recognised.

Introduction

Gold mining is an important industry in Australia, with the country producing about 200 tonnes in 1992 (about 12% of the world total), of which Western Australia (WA) is responsible for about 140 tonnes. Much of the production in WA comes from large open cut mines with low-grade ores, resulting in the production of very large volumes of tailings. These tailings can have a significant clay content, which, coupled with the faster rates of tailings disposal involved in modern open-pit operations, could lead to excessively long consolidation times if only self-weight consolidation is involved.

An outline map of WA is shown in Fig. 1. Much of the gold-mining industry is located in the area around and north of Kalgoorlie – in the Norseman–Wiluna–Meekatharra triangle on this map. Superimposed on this map are contours (in metres) of annual net pan evaporation. Thus, it can be seen that the gold mines are located in arid or semi-arid areas, where net evaporation rates are of the order of 3 to 4 m per year. (Note than peak rates in January can be twice the annual rates). Under these conditions, any numerical modelling of tailings consolidation must take the beneficial effects of evaporation into account, particularly where formation of a surface crust is a prerequisite for rehabilitation of the tailings disposal areas. This type of modelling shows that, even for deep deposits of fine-grained tailings, the rates of consolidation and strength gain, particularly near the surface, can be dramatically increased if evaporation is included. The contribution of evaporation is particularly important where seepage through the base of the area is prevented, and where the clay

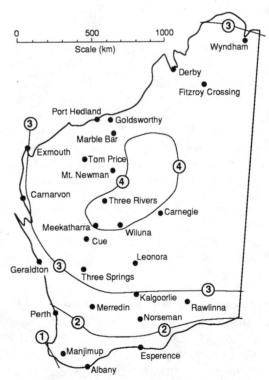

Fig. 1. Contours of net annual pan evaporation rate (m) for WA (data from [1]).

content of the tailings is significant, as is common in the gold mining areas of WA.

In the central part of WA, fresh water is scarce, so that ground water is generally used in the gold refining process. This ground water is extremely saline in many areas, with TDS of over 10% (100,000 ppm) being common. (Concentrations are expressed as weight of salt divided by weight of solute. Thus, a concentration of 10% wt/wt is equivalent to 107 g of salt per litre of solute). After processing, the tailings water can have TDS levels of well over 20% in some areas. This level of salinity can have a dramatic effect on the rate of evaporation from fine-grained tailings.

The paper describes experimental and numerical modelling work which demonstrates the enormous contribution which the high potential rates of evaporation can make to strength and density increase in fine grained tailings. The effect of high salinity is examined, with the results of laboratory evaporation experiments being presented to show comparisons between evaporation from tailings with pure and saline pore water. This work shows that salinity levels must be taken into account if realistic modelling of evaporation under these conditions is to be undertaken.

Consolidation Behaviour of Tailings

The proper management of an area used for disposal of slurried tailings, both during and after the active disposal period, requires an understanding of the processes that control the movement of water in the tailings, the rate and final amount of surface settlement, and the changes in the strength profile through the depth of the tailings. This involves understanding the processes of initial sedimentation, self-weight consolidation under various boundary conditions, and formation of surface "crusting" due to the combined effects of water table lowering in the tailings and surface desiccation due to evaporation.

At the University of Western Australia (UWA), a combination of physical (mainly centrifuge) and numerical modelling techniques has been used to predict the behaviour, and carry out the "what if" studies required in design of tailings disposal areas. The first phase of the work concentrated on consolidation in the absence of evaporation. This work is described in [2], [3] and [4]. The centrifuge facility at UWA is described in [5]. The centrifuge serves two purposes in the tailings consolidation work: tests with simple boundary conditions can be carried out for calibration of the numerical model, and some parametric studies and model verification can be carried out by varying the drainage boundary conditions.

The instrumentation used in the centrifuge tests permits profiles of pore pressure and shear strength to be determined at any stage during the test. Tailings can be added at any stage, and the base drainage conditions and base piezometric head can be varied using solenoid-operated valves. The miniature "Druck" pressure transducers used to measure pore pressures can, with proper de-airing, measure suctions as low as about −90 kPa (gauge pressure). At the end of the test, the sample is dissected to obtain subsamples for determination of the final density and water content profiles, and to check the final location of pore pressure transducers.

Numerical Modelling of Consolidation

The finite difference computer program developed to model the large-strain consolidation behaviour of tailings is described in [2], [3] and [4]. The program uses total pore pressure as the dependent variable. In the updating phase after each increment, fresh tailings can be added at the surface, as would occur in practice during the active life of the disposal area. The boundary conditions can also be changed – for example changing the water table in any base drainage layers, where natural or engineered base drainage is present. Water draining to the top surface can be either decanted, or left to accumulate on the surface. If the base is permeable, with a piezometric head lower than the top surface of the tailings, then eventually only downward drainage occurs. When any water accumulated on the surface has drained back in through the surface, lowering of the saturated front below the surface is prevented until the suction just below the surface reaches the air entry value of the tailings. The suctions so generated produce an increase in effective stress at the surface, and hence an increase in shear strength. Thus, even in the absence of evaporation, a surface "crust" can be developed due to base drainage.

Centrifuge testing was used extensively during the development and verification of the numerical model. Centrifuge testing is also used to provide parameters for any new tailings material. Tests with simple boundary conditions are carried out, during which pore pressure, settlement and strength measurements are taken. At the end of the test, the sample is dissected to obtain the final density profile. Non-linear permeability – void ratio (k–e) relationships and void ratio – effective stress (e–σ') relationships are used, with the coefficients in these relationships being obtained by fitting the numerical results to the observed behaviour.

Having derived the model parameters, some verification is provided by carrying out a second centrifuge test with different drainage conditions. Typically the first test has only one-way drainage (i.e. impermeable base), with base drainage being allowed in the second test, generally with a reduced piezometric head in the base drainage layer. The model calibrated on the first test is then used to predict the behaviour in the second test. If the match is good, the calibration is assumed to be appropriate.

Full verification of this type of approach requires comparison with full-scale behaviour. In a number of projects where tailings were deposited subaqueously (and hence evaporation is not an influencing factor), this methodology has been used with some success to predict the measured full-scale behaviour, and hence it is believed to be reliable under these circumstances.

Effect of Evaporion

With annual evaporation rates of 3–4 m per year being common in the gold-mining areas, it is obviously important to assess the potential contribution of evaporation to consolidation and strength gain of tailings, particularly to the development of a surface "crust" capable of supporting backfill.

The factors controlling the evaporation behaviour have been investigated using both 1g and centrifuge tests, with the emphasis to date being on the 1g tests. In the centrifuge, the effect of evaporation was observed in the early tests. However, evaporation was prevented in most of the subsequent testing, so that other effects could be investigated. It is now planned to continue the work with controlled rates of evaporation. To date, it has been shown that it is possible to achieve a wide range of evaporation rates using rather simple control measures.

The 1g evaporation tests consisted of subjecting slurried samples of clays and silts to artificial drying conditions in the laboratory. Most of these tests so far have been on pure kaolin clay, mixed with water containing various concentrations from 0 to 20% (0 to 230 g/ℓ) of common salt (NaCl). In the most recent series of tests, the samples were prepared at about 100 – 120 % water contents (i.e. about twice the liquid limit) and placed in 200 litre drums. Arrays of infra-red lights and fans were suspended over the drums to produce the artificial drying conditions. In all cases, drums of fresh (tap) water were simultaneously subjected to identical drying conditions to provide the benchmark potential evaporation data. Evaporation loss was determined by weighing the drums. After testing, the strength profile in the drums was obtained using a miniature vane shear device, and the profile of density and water content was then obtained from core samples.

The results obtained from one such series of drying experiments are shown in Figs 2 and 3. The potential rate E_p (measured for the container of water alone) was about 1 cm/day. This corresponds to an annual rate of 3.6 m/year which is typical of the rates in parts of the goldfields (Fig. 1). The evaporation rates shown in Fig. 2 are normalised by dividing by E_p. The plots show how the rates vary with time from the start of the experiment. With fresh water and kaolin, the evaporation rate drops to about 60% of E_p within about 10 days, but the rates from the saline mixtures reduce to 10 to 20 % of E_p. Thus, the effect of the salt is to reduce the evaporation rate from the tailings by a factor of 3 to 6.

Careful sampling from the top surface of the non-saline samples showed the material appeared to be fully saturated practically right to the surface. Fig. 4 shows the void ratio profiles in two of the drums at the end of the tests. In both cases, the data shown were obtained from vertical profiles at different distances from the sides of the drums. The difference in void ratio is reflected in differences in strength profiles, as measured with a small shear vane. This showed that the strength in the 5% sample was less than about 2 kPa, compared to 10–25 kPa in the non-saline sample.

In other smaller-scale tests, it was found that the evaporation rate from non-saline tailings was much closer to E_p than shown in Fig. 2. This may be because the evaporation rate was lower than the 1 cm/day rate used in the tests reported here. Experiments are continuing at lower evaporation rates to determine if the trends evident from these tests are also seen at lower evaporation rates.

Numerical Modelling of Evaporation

The problem of modelling evaporation from the surface of a drying soil is a very complex one. However, a simple approach is to assume that evaporation from tailings is practically the same as from open water, at least while the tailings remain saturated. Thus, evaporation occurs at the maximum potential rate (E_p) dictated by environmental conditions (temperature, humidity, wind speed, radiation etc), and not by the properties of the soil. According to [6], this persists up to some point of transition, when the evaporative demand is no longer being met by the rate at which water can be transported to the evaporation front.

The results in Fig. 2 shows that this assertion is probably not too unreasonable (even if a little non-conservative) for saturated non-saline tailings. However, in the presence of salinity, the evaporation rate is severely reduced right from the start, to less than 20% of E_p.

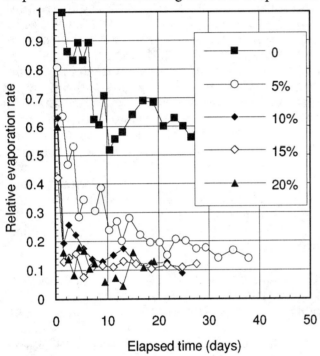

Fig. 2. Effect of salinity (expressed as % wt/wt) on evaporation rate.

The approach taken to modelling evaporation is completely empirical. In this approach, the evaporation rate (a percentage of E_p which depends on salinity level) is imposed as a top boundary condition. If the rate of drainage to the top surface due to self-weight consolidation is faster than E_p, then evaporation only contributes to the removal of this surplus water (leaving less to decant, if that is the option chosen). If not, the hydraulic gradient required to give sufficient drainage to satisfy the evaporation rate is set as a boundary condition. This results in the required amount of water to satisfy the evaporative demand being brought to the surface due to the imposed hydraulic gradient.

This situation is assumed to continue up to the point at which de-saturation begins – i.e. this is taken as the point of transition. For any particular soil, this point can be established approximately using a shrinkage limit test, since the shrinkage limit corresponds to the point of air entry. The simplest version of this test is just a variant of the test for linear shrinkage, with the only addition being that the length and water content are measured periodically during drying. A plot of length versus water content remains sensibly linear up to

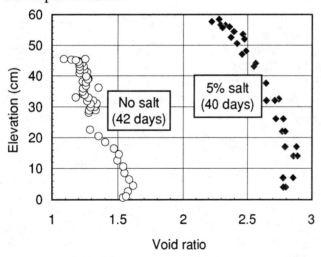

Fig. 3. Final void ratio profiles in laboratory evaporation experiments.

the shrinkage limit, with no further shrinkage beyond this point.

For most clayey soils, the suction at the shrinkage limit can be deduced to be very high (typically more than 1 MPa), which is much higher than can be achieved just by water table lowering in most tailings deposits. With evaporation past this point, much higher suctions are generated. This does not result in significant further shrinkage, but it can give a very significant increase in shear strength, and hence can play a very important part in rehabilitation of the tailings area.

Shear Strength

The consolidation model does not give shear strengths directly. However, the shear strength can be deduced from the effective stress using the Cam Clay model, and this has been incorporated into the program. It is assumed for this calculation that the effective stress is given as total stress less total pore pressure, even when the pore pressure is negative. This clearly holds at least until the suction reaches −100 kPa (gauge), and probably at least until the start of desaturation.

The ability to carry out strength measurement on the centrifuge at any stage of the test allows the model predictions of strength to be verified, at least qualitatively. Similarly, vane shear testing on the samples used in the $1g$ evaporation tests provides shear strength data from these tests.

Application of the Numerical Model

To illustrate the importance of evaporation on the consolidation behaviour, the simple case of a 20 m thick deposit of clayey tailings (with initial void ratio of 3.0) has been modelled. The full deposit is created instantaneously, and the piezometric head in the base drainage layer is maintained at 7 m. In one case, no evaporation is allowed, and in the second, a rate of 1 m/year is imposed. The results are shown in Figs 4 (no evaporation) and 5 (evaporation of 1 m/year) as strength versus elevation and void ratio versus elevation. In Fig. 4, some strength is eventually developed at the surface due to negative pore pressures above the water table, but this is a very slow process. However, even a modest evaporation rate of 1 m/year very quickly results in significant strength gain at the surface. Note that the minimum void ratio (after 6.4 years) is about 0.8, which is equivalent to a water content of about 30%. This is close to the point at which desaturation might be expected to start for most tailings, and hence the simple model would not be appropriate past this point. However, even at this stage, the model predicts that a strong crust has developed, so that backfill could be added at this stage, if required under the local rehabilitation regulations.

Conclusion

The paper has outlined the approach being taken in modelling the consolidation behaviour of tailings in the gold-mining industry in WA. For consolidation in the absence of evaporation, a combination of centrifuge and numerical modelling techniques is used as a predictive tool in the both the design and operating phases of tailings disposal facilities.

Current research effort is being directed towards the modelling of evaporation. This has shown that evaporation can produce dramatic improvements in the rate of consolidation and strength gain, particularly near the surface of the tailings. However, the potential rate of evaporation can be severely reduced in the presence of the salinity levels common in the gold-mining areas of WA.

The approach used to date to model evaporation has been empirical − assuming that the rate of evaporation from the tailings is directly related to E_p. The effect of salinity on evaporation is expressed as a reduction factor on E_p. The laboratory work has shown that the reduction can be very severe. Though no specific field work on this aspect has yet been carried out, observation of tailings deposits often shows a very thin white salt crust over very soft tailings, with the crust obviously severely inhibiting evaporation. Due to this, the rate of strength increase in these deposits is often very much lower than expected.

Acknowledgements

The work described in this paper was supported by a grant from the Australian Research Council. The second author is supported by an Overseas Postgraduate Research Studentship from the

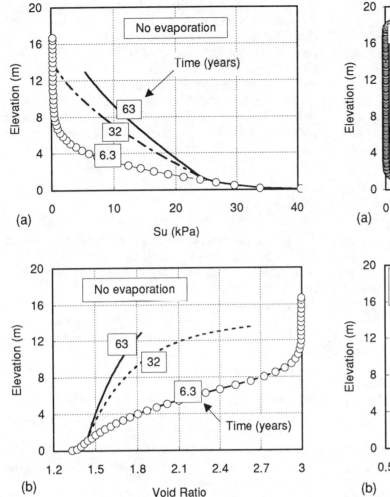
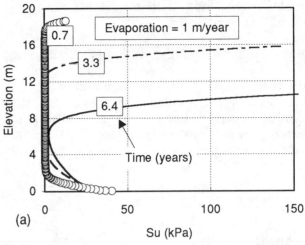
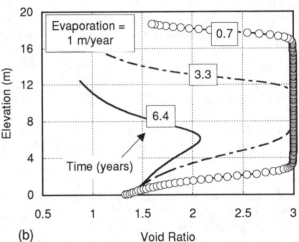

Fig. 4. Consolidation of 20 m thick layer with no evaporation.

Fig. 5. Consolidation of 20 m thick layer with evaporation rate of 1 m/year imposed.

Australian Government, and by a Studentship from the Geomechanics Group, UWA.

References

[1] Luke, G.J., Burke, K.L. and O'Brien, T.M. (1987). *Evaporation data for Western Australia*, Technical Report No. 65, Salinity and Hydrology Branch, WA Department of Agriculture.

[2] Fahey, M. and Toh, S.H. (1992). Physical and numerical modelling of mine tailings. *Australian Geomechanics*, **22**, 17–25.

[3] Fahey, M.F. and Toh, S.H. (1992). A methodology for predicting the consolidation behaviour of mine tailings. *Proc. Western Australian Conf. on Mining Geomechanics*, Kalgoorlie, WA, 445–452, Curtin University of Technology, WA.

[4] Toh, S.H. (1992). *Numerical and centrifuge modelling of mine tailings consolidation*. PhD Thesis, Department of Civil Engineering, The University of Western Australia.

[5] Randolph, M.F., Jewell, R.J., Stone, K.J.L. and Brown, T.A. (1991). Establishing a new centrifuge facility. *Proc. Int. Conf. Centrifuge 1991*, Boulder, Colorado, 3–9, Balkema, Rotterdam.

[6] Swarbrick, G.E. and Fell, R. (1992). Modelling desiccation behaviour of mine tailings. J. Geotech. Engng, ASCE, **118**(4), 540–557.

Use of Dynamic Compaction for the Rehabilitation of a Mining Backfill Area

A.B. Fourie
Department of Civil Engineering, University of the Witwatersrand, Johannesburg, South Africa

B. Wrench
Steffen, Robertson and Kirsten, Johannesburg, South Africa

Abstract

The use of the dynamic compaction technique for the compaction of rockfill into a 20m deep excavation, in order to prepare the ground surface for construction of settlement-sensitive structures, is described. Three techniques were used to determine the quality of compaction, viz Menard pressuremeter, horizontal plate loading tests and laboratory oedometer tests. The results of these tests were used in a probabilistic analysis of likely settlements. The results of three years of post-compaction monitoring have shown settlements that are well within the stipulated requirements.

Introduction

Open cast mining is frequently used to recover shallow ore bodies. On completion of mining, the excavation is usually backfilled using the overburden materials, and the surface re-contoured. This is done to reinstate the original contours as closely as possible, or to ensure that a surface profile is created that allows adequate control of surface water runoff. The backfill is invariably loosely placed, and large surface settlements typically occur. Field measurements have shown that these settlements may continue for many years, although the magnitudes generally decrease with the logarithm of time.

The increasing demand for land for development often leads to the requirement of construction of services or buildings over undermined or backfilled land. In the case of backfilling of open cast workings the magnitude of post-compaction settlements, and the time over which they occur may place severe constraints on the nature of the possible land use developments.

This paper describes the use of the Dynamic Compaction technique for the compaction of open cast backfill to ensure that subsequent surface settlements were within prescribed limits for the construction of settlement sensitive structures. Details are given of a case history in which 20m of backfill were compacted using the technique, prior to development of the site.

Evidence of Problems Encountered With Building on Backfilled Open Cast Workings

Prior to discussing a specific case history, it is useful to review some of the published information concerning problems that have been encountered with settlement of buildings on backfilled open cast workings. In general, the two predominant effects appear to be time-dependant compression (creep), and inundation-induced settlement, caused by re-establishment of the water table once mining activities have ceased.

Some of the most comprehensive literature on this topic relates to the settlement characteristics of loosely placed rockfill used in embankment dams. This data indicates settlements of about 0.01% of the embankment height per year, continuing for many years (or even decades). As discussed by Terzaghi [1], very large pressures are developed at the contact points in the rockfill, and the time dependant settlements are a result of crushing of the rock at these contact points. This phenomenon is exacerbated by softening of the rock at the contact points that results when the fill becomes inundated. In general, he found that settlements were largest when only large rocks were used in the fill, and reduced as fine material was introduced. The relevance of these conclusions is that mining backfill frequently comprises of rockfill with little or no fines, and will thus be potentially susceptible to post placement settlements.

Consolidation settlement as a consequence of the self-weight of the backfill usually occurs during, or soon after, placement of the fill because of the highly permeable nature of the fill. The major concern with construction on backfilled areas is thus the additional consolidation settlements due to the load of the surface structures, and particularly inundation induced settlements. The importance of inundation induced settlements was described by Penman [2], who performed laboratory tests on oolitic rock and boulder clay backfill recovered from open cast iron ore mining at Corby in the U.K. Samples were set up in a 1m diameter oedometer at their in-situ moisture content, and loaded to a vertical stress of 120 kN/m^2. After only three days, total compression of 9.5% had occurred, and on saturation a further 10.9% compression occurred. Further evidence of the magnitude of the problem was provided by Reed et al [3], who presented surface settlement measurements from three backfilled open cast mining sites in the northern United Kingdom. Surface settlements of up to 500mm as a consequence of the re-establishment of the water table were measured. Re-establishment of the water table may take many years; for the above sites it was found that the rate of rise varied from about 1m per year, up to 4m per year. Resulting settlements only became noticeable some years after the water table began to rise. Once wetting-induced settlements start to occur, however, they may continue for decades, at a rate that decreases linearly with the logarithm of time, (Charles et al [4]).

Smythe-Osborn [5] reported the performance of a lightly loaded steel framed factory that was built on the surface of a backfilled coal pit approximately 18 years after the pit had been backfilled. Distortion of the structure was only noticed about six years after construction, and continued for a number of years thereafter. Eventually some 360mm of settlement occurred, causing extensive damage to the factory, requiring costly remedial work.

In redeveloping backfilled land it is therefore usually necessary to either first consolidate the backfill (eg by surcharging), or to design the structures to accommodate the expected movements. It is usually not possible to use conventional systems such as piles to support structures, because of the thickness (typically in excess of 20m) and the nature of the backfill material. It is extremely difficult to design structures to accommodate movements of the magnitude noted in the preceding discussion, and the preferable solution is to improve the backfill quality prior to re-development of the land.

Burford [6] described a trial in which a 7m thick surcharge was placed on 24m of backfill to accelerate settlements. Surface settlements of about 300mm were measured, with an interesting finding that at a depth of 10m the settlements were only about 10mm. Subsequent tests showed that settlements of less than 7mm could be expected under lightly loaded structures founded on the surface. The use of a surcharge may thus be a relatively cost-effective way of improving the compression characteristics of backfilled land, particularly if the work is carried out at the same time as the workings are backfilled. The major disadvantages of the method (aside from cost), are the need for rehandling the surcharge material, and the time required for the water table to re-establish in order to maximise the benefit of the surcharge approach.

An alternative approach, and one that was followed in this paper, is to compact the backfill during its initial placement. The blocky nature of the backfill, and the considerable thickness of material to be placed usually makes conventional methods of compaction untenable because they require placing and compacting of the fill in layers using heavy equipment. As discussed by Lukas [7], the Dynamic

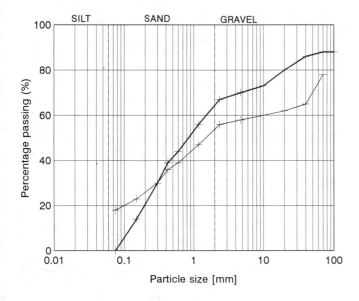

Figure 1. Particle size distributions of backfill material.

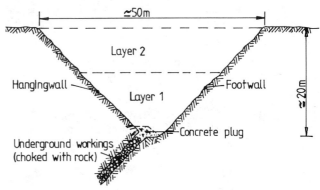

Figure 2. Cross-section through backfilled pit.

Compaction technique avoids many of the problems caused by the blocky nature of the backfill, and as shown below, is ideally suited to the compaction of open cast backfill.

Application of Dynamic Compaction Technique: A Case History

The site described in this paper was at the Vogelstruisfontein Gold Mine, which is approximately 30km from Johannesburg in South Africa. A shallow gold-bearing reef had been mined by excavating an open pit approximately 20m deep, with side slopes of approximately 50° to the horizontal. The soil and wasterock removed from the pit had been stockpiled, and were available for use as backfill. Typical particle size distributions for the fill material are shown in Figure 1. The surface rights at the site belonged to a third party who planned to construct a new factory on the site. It was therefore agreed that the mining company would reinstate the site to a condition such that the proposed development could proceed shortly after mining was completed, and such that excessive structural limitations would not be required. To this end it was agreed that surface settlements would not exceed 100mm over the period from 6 months to 60 months after completion of mining.

After evaluating the potential of using conventional compaction techniques, it was decided to proceed with rehabilitation of the site using the Dynamic Compaction technique. A dyke was found to cut across the mining area, and a ventilation shaft associated with underground mining operations was located within this dyke. Accordingly it was decided not to mine the dyke, and two separate pits, denoted the East and West pits were excavated. A section through the East pit is shown in Figure 2; it can be seen that underground mining from the base of the excavated pit had resulted in adits that had to be plugged prior to backfilling of the pit. Concrete plugs were used to cover these adits, and in addition a large concrete cap was built over the vertical ventilation shaft.

It was decided to place and compact the backfill in two layers, the first being between 8 and 12m thick, and the second about 8m thick to the surface. Field trials were undertaken to determine the optimum amount of dynamic compaction for the site in question. Trials were carried out using a 15t mass falling through a height of 20m, and the effectiveness of the approach was evaluated by measuring the penetration of the mass into the fill with each blow. The amount of penetration decreased with the number of blows, and after 12 blows additional settlements were negligible. Total settlements of nearly 2m were achieved after 12 blows.

In order to achieve the required post-backfilling performance, the contractor proposed the following Dynamic Compaction procedure based on the results of the above trial:

- Primary compaction on an 8m grid, using 12 blows of a 15t mass falling through 20m per imprint site.

- Secondary compaction on intermediate sites, using the same compactive effort as that for the primary phase.

- A third and final 'ironing' phase on a 2m grid, comprising single blows of the mass falling through 5m, in order to compact the upper 2m of backfill.

The primary form of quality control adopted was the use of Menard pressuremeter tests, carried out after the second phase noted above had been completed. Provision for a third phase of primary compaction was made, in case the pressuremeter tests showed unacceptable material variability. The stringent requirements for post-compaction settlements led to a decision to undertake additional quality control testing, aside from the pressuremeter tests noted above, because of concerns that the pressuremeter tests could not effectively measure the bulk stiffness of large blocks of rockfill. Horizontal in-situ plate loading tests were carried out in trial pits, and laboratory tests were carried out on representative samples of the fill material. These latter tests consisted of both conventional tests to determine the compression index of the fill, as well as collapse potential tests, (where a 'collapse potential' test consists of loading a soil specimen in a conventional oedometer to a prescribed load, at its natural moisture content, and then inundating the specimen with water. The resulting strain is termed the 'collapse potential' - see Jennings and Knight [8] for more details).

Results from Quality Assurance Tests

i) Pressuremeter tests: Pressuremeter tests were carried out in both layers and in both the East and West pits. Difficulties were indeed experienced with the size and number of boulders within the backfill, and in many locations it was not possible to advance the pressuremeter beyond depths of 3m. Twenty one tests were completed in Layer 1, and twenty nine in Layer 2. In one particular location, the test results were not satisfactory, and a third round of primary compaction was carried out before the final 'ironing' phase over the entire site. Fifteen additional pressuremeter tests were carried out at this location after the additional compaction was completed. The results of all the tests that were considered to have passed the quality control criterion (and thus required no further compaction) are summarised in Table 1.

TABLE 1--Results of pressuremeter tests.

LAYER	MENARD MODULUS (MPa)		
	Mean value	Standard Deviation	Coef. of Variation
1	30.1	11.8	0.39
2	21.6	10.1	0.47

ii) Plate load tests: Horizontal plate load tests were carried out in 750mm diameter trialholes that were drilled into the backfill. Nine holes were drilled into the compacted backfill in layers 1 and 2. Ten plate load tests were carried out in layer 1, and fourteen tests in layer 2. The tests were carried out by preparing two parallel surfaces on opposing faces of the trialhole, setting up the testing equipment horizontally across the hole, and jacking two rigid 200mm diameter plates into the fill. It was found that tests could only be performed in areas locally free of large boulders. The tests were interpreted in accordance with the rectangular hyperbola curve fitting method, as discussed by Wrench [9]. The results of the tests are summarised in Table 2.

TABLE 2--Results of plate load tests.

LAYER	DRAINED MODULUS E (MPa)		
	Mean value	Standard Deviation	Coef. of Variation
1	11.7	4.5	0.38
2	27.2	7.0	0.26

TABLE 3--Oedometer test results.

Depth (m)	Dry Density (kg/m^3)	Degree of Saturation (%)	Constrained Modulus D (MPa)
2.0	1942	54.7	11
4.0	1930	54.5	28
6.0	1809	54.2	20
8.0	2001	83.3	14
9.0	2010	61.3	14

iii) Laboratory Oedometer Test Results: Oedometer tests were carried out on five representative samples of the backfill that were recovered from a trialhole in layer 2 in the Western pit. The results of these tests are given in Table 3. The large variability in the degree of saturation is a consequence of the work carried out during the wet summer months; in any event, it seems to have relatively little effect on the dry density. The constrained modulus was calculated from the normal consolidation phase of the oedometer tests, using the equation discussed by Coumoulos [10]:

$$D = \frac{[1+e_0]\sigma_v'}{0.434 C_c} \quad (1)$$

where e_0 is the initial void ratio, σ_v' is the vertical effective stress, and C_c is the compression index.

The constrained modulus is related to the modulus of deformation E according to:

$$D = \frac{1}{m_v} = \frac{E(1-\nu)}{(1+\nu)(1-2\nu)} \quad (2)$$

where m_v is the coefficient of volume compressibility, and ν is Poisson's ratio.

Estimation of the Deformation Modulus of the Backfill

According to Ménard [11], results obtained from pressuremeter tests may be converted to a modulus of deformation (E), by multiplying by a factor α. The approach used in this project was to use an α value of 0.5 if the limit pressure was greater than 3MPa (usually indicating the presence of large rock fragments), and otherwise to use a value of 0.3. The laboratory constrained modulus values were converted to modulus of deformation values using eq.2, with a Poisson's ratio of 0.2.

Table 4 summarises the average values of the modulus of deformation, E, obtained from the various field and laboratory tests, as well as a back-calculated value from observed settlements. The observed surface settlements were used to calculate an E value by assuming an average bulk unit weight of 18kN/m^3 for the 20m thick layer.

TABLE 4--Comparison of deformation modulus values for layer 1

Modulus of deformation, E (MPa)			
Menard	Plate tests	Oedometer	Back-analysed
15	11.7	15.3	7.2

Although fairly crude assumptions were made in carrying out the above calculations, the agreement amongst the various values is extremely good. The back-analysed data results in the lowest modulus, which is due to this calculation being done for the entire thickness of fill, and not just the upper layer, (since the observed surface settlement cannot be ascribed to layer 2 alone).

Additional laboratory tests carried out on representative samples showed negligible collapse potential (<0.4%).

Predicted Post-construction Settlements

A probabilistic settlement analysis was carried out in which two parameters were varied, i.e., the thickness of layer 2 of the backfill, and the modulus of deformation. The point estimate method proposed by Rosenbleuth [12] was used with various combinations of these two parameters. A Beta distribution was fitted to the resulting settlement predictions, thus allowing the determination of the probability of the surface settlement being greater than a specified amount. The results of this analysis showed that the probability of the settlement exceeding the limiting value of 100mm over the period from 6 to 60 months after compaction was 12%. This was judged an acceptable risk in view of the conservative estimates of water level rise that were made in the analysis. The compaction work was completed in February 1990. Precise levelling onto settlement beacons located over the surface of the site have to date (3½ years later) shown surface settlements of less than 5mm.

Conclusions

This paper has illustrated the usefulness of the Dynamic Compaction technique for compacting open cast backfill in order to achieve very small post-compaction settlements. Three different techniques of quality assurance were used, and all three techniques gave similar results, which indicated that adequate stiffness of the fill had been achieved. Furthermore, the improved stiffness was achieved throughout the thickness of the 10m compacted layers. Measured post-compaction settlements have to date been well within requirements. Comparison of these results with other methods of compaction and rehabilitation reported in the literature indicate that this method is ideal for the type of application discussed in this paper.

References

1. Terzaghi,K. (1964) Mission dam - An earth and rockfill dam on a highly compressible foundation. Geotechnique, 14, pp 14-50.

2. Penman,A., and Godwin,E. (1974) Settlement of experimental houses on land left by opencast mining at Corby. BGS Conference on Settlement of Structures, Cambridge, pp53-61.

3. Reed,S., Hughes,D.,B. and Singh,R. (1987) Backfill settlement of restored strip mine sites - case histories. Int. Jnl of Mining and Geological Engineering, Vol.5, pp161-169.

4. Charles,J.A., Hughes,D.B. and Burnford,D. (1984) The effect of a rise of water table on the settlement of backfill at Horsley restored opencast coal mining site, 1973-1983. Proc. 3rd Int. Conf. on Ground Movements and Structures, Cardiff, U.K., Ed. J.D.Geddes, pp423-442.

5. Smythe-Osborn,K. and Mizon,D. (1984) Settlement of a factory on opencast backfill. Proc. 3rd Conference on Large Ground Movements and Structures, UWIST, Cardiff, Ed.J.D.Geddes.

6. Burford,D. (1991) Surcharging a deep opencast backfill for housing development. Ground Engineering, Vol.24, No.7, pp36-39.

7. Lukas,R.G. (1980) Densification of loose deposits by pounding. Jnl. Geotech. Eng. Div, ASCE, Vol.106, No.GT4, pp435-444

8. Jennings,J. and Knight,K.(1975) A guide to construction on or with materials exhibiting additional settlement due to 'collapse' of grain structure. Proc. 6th Reg. Conf. for Africa on Soil Mech. and Foundation Eng., Durban, Vol.1, pp99-105.

9. Wrench,B.P. (1984) Plate load tests for the measurement of modulus and bearing capacity of gravels. The Civil Engineer in South Africa, Sept.1984, pp429-437.

10. Coumoulos,D.G. (1979) Determination of settlement design parameters from conventional oedometer tests. Proc Conf on Design Parameters in Geot. Engineering, Brighton, U.K. Vol.1, pp111-118.

11. Ménard L F (1975) The Interpretation of Pressuremeter Test Results. Sol Soils, 26, pp7-43.

12. Rosenbleuth E (1975) Point Estimates for Probability Moments. Proc. Nat. Acad. Sci. USA, 72, No.10.

Recycling Derelict Land

Prof. Dr.-Ing. Dieter D. Genske
Delft University of Technology, The Netherlands

Prof. Dr. Jean Thein
Universität Bonn, Geologisches Institut, Bonn, Germany

Abstract: This paper discusses strategies to recycle derelict land such as abandoned industrial plants or coal mining sites. Procedures to work out a cost effective remediation plan and means to harmonize the remediation activities are pointed out. The use of computers to visualize the remediation steps are demonstrated. Finally, a case study dealing with a land recycling project in the German Coal Mining District is presented.

1 Introduction

The Ruhr Coal District in Germany, being one of the oldest and largest mining and industrial regions in Europe, has been in a process of constant change over the last 40 years. Since the end of the second world war the coal industry has been declining. This is partly due to the exploitation of the easy accessible coal and lignite resources. A major reason is the fall of the coal prices on the world market and the gradual increase of safety standards for mining in Germany. Besides the short but intensive coal boom during the oil crisis in the 70s it became much cheaper to buy coal from abroad. As a consequence, large areas formerly used by the mining industry and associated industries such as coking plants, chemical factories or steel works are now abandoned.

During the last decade there have been large efforts to establish new industries on derelict sites. The remediation of the former mining sites in the Ruhr-District and the re-establishment of alternative industries have now become both a challenge for city planners and a prestigious attribute for ambitious politicians. It has become the declared goal of the German government to convert the Ruhr District into the greenest industrial region in the world. One important piece in this campaign is to let the 'Internationale Bauaustellung' - an international city planning fair - take place in the Ruhr District, a plan which was settled in 1989. The first project will be presented to the public in 1994.

Although there are a number of governmental programs, initiatives of the European Community and activities from private investors there is one crucial problem to be faced if derelict sites are to be recycled: a high percentage of the reclaimable land must be considered as seriously contaminated. A broad variety of contaminants has leached into the ground, thus polluting the soil and groundwater.

The key to an effective land re-utilization plan is a harmonized management of ground investigation, risk assessment and clean-up strategies, as described for example in BRÜGGEMANN et al. (1991), GENSKE & NOLL (1994). This paper reports on experiences with major land recycling projects made in the German Ruhr District.

2 Remediation strategies

There are three major steps to remediate a contaminated site: the risk assessment, the remediation strategy and the realization of the remediation measures chosen. The first step - the risk assessment - involves the documentation of the status quo, i.e. the present situation on the site, a historical research on the former utilization, i.e. the reconstruction of the generations of fabrication plants established on the site, the layout and realization of a field campaign to investigate the local contamination which is expected to occur in the vicinity of former factories, and finally based on this evaluation the risk assessment report. In this report the history of the site is discussed, contaminated sectors are identified and possible environmental hazards due to these contaminations are pointed out.

The second step - the remediation strategy - starts with a feasibility study, the main purpose of which is the discussion of possible remediation plans such as the excavation and dumping of contaminated soil, the confinement of polluted sectors by means of vertical and horizontal barriers, hydraulic measures to clean the contaminated ground water, etc.

Criteria to analyze certain remediation strategies are the remediation costs, the realization time, the technical feasibility, the compatibility with governmental regulations and standards, and the public acceptance. The feasibility study will be presented for discussion to all parties involved including the governmental agencies and the public. Based on this discussion it will be decided which remediation strategy will be realized.

The last step will be the realization of the remediation plan, i.e. the clearing of the site, removing or confining of contaminated sectors and revitalizing the site.

Harmonizing all three steps has to be considered an interdisciplinary and rather complex task. Geologists and financing experts, chemists and governmental officials, city planers and environmentalists, politicians, the public, they all have to cooperate closely in order to avoid the loss of time and money. With this in mind the elaboration of appropriate presentation material is essential. This presentation material must be understandable, to the point, and most of all financiable. In the next chapter it will be demonstrated how suitable presentation material can be generated by means of PC's and digital tools.

3 Remediation maps on the PC

There is quite a variety of maps to be prepared when dealing with a remediation project. Before starting with the first map which would be the reference map defining the status quo on the site one has to decide whether a regular CAD application or a geographic information system shall be employed. Most CAD software has the advantage of being user friendly and inexpensive whereas a geographic information system has the advantage to integrate spreadsheets and reports into the map. Furthermore, it has to be decided whether a 2D or a 3D presentation of the data is desired. The presentation may be as sophisticated as depicted in figure 1 which gives a snapshot from a video where digital 3D images where animated and taped on a demo to visualize the history, the contamination and the remediation strategy for a derelict mining site. However, for the production of this kind of presentation material a powerful workstation and a variety of soft- and hardware tools are needed.

For most remediation projects a suitable CAD application is sufficient to generate all the necessary maps which are:
1. A status quo worksheet depicting the present topographic situation on the site. The status quo worksheet is the basic reference map upon which all the additional information from the following steps will be plotted and harmonized.
2. A map and possibly some vertical sections giving information on the geological and the hydrogeological situation.
3. A historical evaluation or "multitemporal" analysis of the site depicting all generations of former use. Since the industrialization of some European regions such as the Ruhr District has begun already in the last century the prepara-

tion of such maps is a fairly complicated and time consuming affair.

4 A map dealing with the field campaign to investigate the ground conditions, the inhomogenities, and the contamination of the ground. In order to lay out an optimal and cost effective field campaign the information from the historical analysis are used. Since the areas prone to pollution are known from the historic information costly site investigations (drillings, pits, etc.) can be concentrated on certain locations.
5 A map giving the contaminated sectors and the type of contamination.
6 A map depicting restrictions for the future use of the site due to the contamination and the inhomogenities in the ground.
7 A remediation map with the remediation measures chosen and the future utilization of the site.
8 Additional maps dealing with construction details.

As mentioned already the first map to be drawn is the status quo worksheet. This map can easily be generated by applying a digitizing tableau, working with a scanned topographic map section,

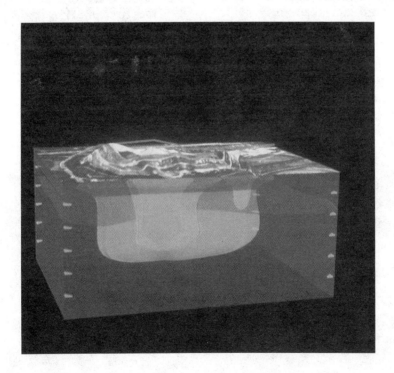

Fig. 1: Snapshot from a digital video demonstrating the migration of contaminants underneath a derelict mining site. See also the case study at the end of this paper. (Bernd Girod & Ervin Keve, Kunsthochschule für Medien, Köln, Germany, and Dieter Genske and Christof Olk, DMT-IWB, Essen, Germany. © Kunsthochschule für Medien, Köln, Germany).

importing an image in form of a data file from another CAD application, or importing information from a geographic information system.

More complicated is the multitemporal analysis. In order to prepare an appropriate map historic documents and building permissions are scanned and integrated into the status quo worksheet. Since the scans have to be adjusted in scale and rotated to match the present topography a powerful CAD program and sufficient hardware have to be provided. The precision of the matching is verified by applying the zoom function of the software.

Especially interesting is the digital processing of aerial photographs. In Germany, the earliest coverages date from the 1920s. They are supplemented by stereoscopic photography from allied reconnaissance and mapping sorties during World War II and after that by air covers taken at regular intervals of 2-3 years since the late 1950s. As a result, there is an aerial photographic documentation of most sites comprising 15 to 25 and sometimes even more covers, which are available for multitemporal analyses and mapping.

The destruction of the production plants and facilities during World War II has certainly contributed to a major part of the contamination of the sites. Furthermore, some of the bombing craters were quickly filled with all kinds of material just at hand. Nobody paid special attention to possible environmental hazard which might arise later from these fillings. Today we know that they rather often are serious sources of pollution. It is therefore fairly important to map all bomb impact craters and adjust the investigation program accordingly (DODT et al. 1993).

4 Brauck Park: a case study

Brauck Park is located at the former mine and coking plant 'Graf Moltke 3/4' in the northwestern part of the Ruhr-District close to the city of Essen and covers an area of about 230.000 m². In 1873 the first shaft was excavated, three more followed within the next 30 years (fig. 2). A coking plant was built in 1903/04 followed by benzene and ammonia factories. Within the next 50 years

Fig.2: Historical view of the Graf Moltke coal mining site from 1912.

additional industries were established, turning the site into a multiuse industrial complex (fig. 3a). The coal production went up to 1 million tons per year in the 60s until the mine was closed down in 1971.

In the following years the Graf Moltke Coal Mine became a typical industrial wasteland with all its negative attributes. The positive point, however, was the infrastructure of the surroundings and the immediate neighborhood of the site to the A2 Autobahn, one of the most frequented highways and an important east-west traffic connection which especially in the light of the German reunification has become a rather important aspect. It was therefore decided to remediate the site as a project of the 'Internationale Bauausstellung' and establish a high quality industrial park (fig 3c). The European Community provided the project with appropriate funding (European Fund for Regional Development EFRE) so that only about 50% of the remediation costs had to be supplied by the owners of the former mining site.

As to the geological situation the subsoil of the Brauck site can be divided into three parts: an upper filling with foundation fragments (2 to 9 meters), a middle stratum of quaternary sediments (about 10 meter below surface) and a fractured cretaceous marl as bedrock. The groundwater table is located at a depth of about 5 meters.

A serious hydrocarbon contamination has been detected in the vicinity of the former coking plant (fig 3b). Because of the lack of landfill space available for toxic wastes it was required by the environmental protection agency to minimize all excavation. A special surface covering technique was chosen for the heavily contaminated sectors. The cover system had to meet three tasks: 1) it had to be waterproof to prevent the infiltration of precipitation into the contaminated ground; 2) it had to be gas proof to stop the migration of toxic gas to the surface; 3) it had to be stiff enough to allow the construction of streets and buildings.

The third aspect accounts for the fairly inhomogeneous ground conditions. As mentioned already, massive fragments of the former foundations of the dismantled buildings have remained in the subground next to loose fillings, thus causing severe structural problems as to possible differential settlements of future structures. To satisfy all three objectives, a reinforced geotextile sandwich system was designed. Basically it is composed of three elements: a lower reinforced support layer, a drain and seal system, and an upper layer of reinforcement elements (grits) to account for the vehicular and structural loads. This cover and seal technique is explained in detail in GENSKE et al. (1993).

5 Conclusion

Land recycling is an important environmental issue today. This paper discusses a number of key issues and presents a case study from the German Ruhr District. Dealing with land recycling projects is an interdisciplinary task which calls for innovative ideas. Land recycling also appears to be a growing market in both industrial and developing countries. Geotechnical engineers need to speak up and formulate their ideas now.

6 References

BRÜGGEMANN, J., SCHRODT, D. & J. THEIN 1991: Reaktivierung von Industriebrachen. - Energie, 12/91.

DODT, J., D. D. GENSKE, T. KAPPERNAGEL, P. NOLL 1993: Digital Evaluation of Contaminated Sites. *International Conference on Digital Image Processing: Techniques and Application in Civil Engineering*, February 28 - March 5, 1993, Kona, Hawaii.

GENSKE, D.D., H. KLAPPERICH, P. NOLL & B. THAMM 1993: Surface Confinement Techniques of Derelict Industrial Sites. – *GEOCONFINE 93 International Symposium on the Geology and Confinement of Toxic Wastes*, 8-11 June 1993, Montpellier, France.

GENSKE D. & P. NOLL 1994: Managing Land Recycling - Examples from the German Coal Mining District, *3rd. International Conference on Environmental Issues and Waste Management in Energy and Mineral Production*, August 29th - September 1st 1994, Perth, Australia.

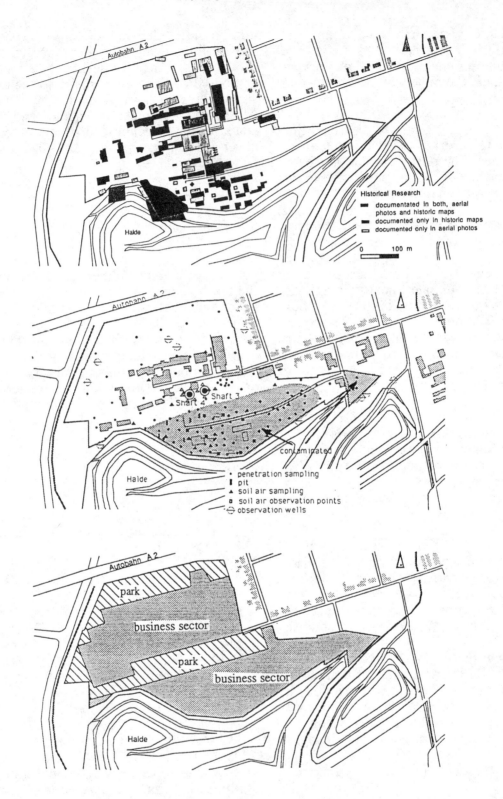

Fig. 3: The Graf Moltke Remediation Plan: 3a) the multitemporal analysis (upper figure); 3b) the risk assessment map with the field investigations and the contaminated sectors (middle figure); 3c) the future utilization of the site (lower figure).

Determination of Tailings Properties for Seismic Response Analysis of a Tailings Dam

T. Germanov, Dr. Eng., V. Kostov, Eng.
Department of Geotechnics, University of Architecture & Civil Engineering, Sofia, Bulgaria

ABSTRACT: The results from laboratory and in situ study for assessing the stress-strain characteristics of tailings of the "Benkovski" spigoted downstream tailings dam are presented. The basic tailings properties (density and water content) were determined by testing undisturbed samples taken from different depths of the dam. The change of the density with depth was assessed by compression tests. Nonlinear stress-strain properties of the tailings (parameters of the initial tangent modulus, failure ratio, Poisson's ratio) for static stress analysis were estimated using static triaxial tests. The dynamic properties (strain-dependent shear modulus and damping) were investigated by means of a shear apparatus with a wide range of cyclic loading, specimen size and accuracy of measuring. The results obtained will be used for static and dynamic response analysis of the tailings dam considered.

1. Introduction.

The seismic stability evaluation of an earth dam requires the determination of the real physical and mechanical characteristics of the embankment materials. A pseudo dynamic approach for seismic stability analysis was applied recently in Bulgaria. This approach, however does not take into account the real shear strength characteristics and damping properties of the dam materials. The tailings dam response under static and dynamic excitation is quite different from the response of earth and rockfill dams. On the other hand, at present, there are not sufficient investigations of the tailings properties. The difficulties are mainly due to the specific behaviour of the tailings materials which as a fine grained medium, form a stress-strain state different from that of the natural soil deposits.

Bearing in mind that many dams in Bulgaria, built before the strong earthquakes on the Balkan Peninsula (Skopje - 1963, Vranca - 1977, Strajitza-1986) were designed without taking into account the seismic impact, a new design and monitoring pattern of the behaviour of the existing dams has been initiated.

2. Description of the Tailings Dam under Consideration.

The "Benkovski" tailings dam is part of the Elatzite Copper Mining Plant and is located about 70 km east of Sofia on an extremely large territory. The area of the tailings dam at the final stage is to be 268 ha, and that of the catchment area - 330 ha [1]. The tailings dam is being constructed using the downstream method, by hydrocycloning. The operation of the copper plant was started in 1981. The tailings dam was designed to reach an ultimate height of 145 m, and at the end of 1992 the height reached 95 m. The principal cross section, with annual increase of the height is given in Fig.1.

Bearing in mind the natural conditions of the tailings dam, (the ground water table and the compaction of cycloned sands during the operation), the experimental study for assessing the static and dynamic properties of the tailings was carried out on samples with two densities and water contents, corresponding to the stress-strain states at a depth of 10.0 m (air-dry, W = 10%) and 40.0 m (saturated).

3. Basic Tailings Properties.

The physical and mechanical properties of the tailings deposit were determined by using undisturbed and disturbed samples taken from different depths and different sites of the dam. The ground water table, during this study was approximately 30.00 m below the top of the tailings dam.

Undisturbed samples were taken from three depths (0.5, 6.0, and 8.0 m) by hand drill. The density and the water content were determined from these samples. The other properties (grain size distribution, compressibility, permeability and shear strength) were established using disturbed samples.

The compression tests were carried out by means of samples with initial density corresponding to the density at the dam surface.

The shear strength tests were carried out on prepared samples with 50 mm diameter and 100 mm height under the triaxial drained consolidated conditions.

The results obtained are given in Table 1 and Fig.2.

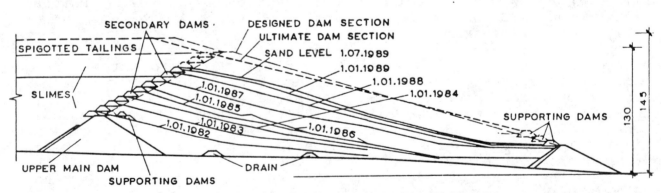

Fig.1.-Principal cross section of the "Benkovski" downstream tailings dam.

TABLE 1.- Average tailings characteristics

Characteristics	Values
1. Dry density, ρ_d, g/cm^3	1.38
2. Water content	
-Air-dry, %	10.0
-Saturated, %	35.3
3. Void ratio, e	0.952
4. Grain size distribution, %	
0.50 - 0.25 mm	11
0.25 - 0.10 mm	19
0.10 - 0.01 mm	37
< 0.01 mm	3
D_{50} mm	15
5. Coefficient of permeability, k, m/s	8.94×10^{-5}
6. Angle of friction, φ	
-Air-dry	31.5°
-Saturated	35.0°
7. Cohesion, c, kPa	
-Air dry	44
-Saturated	30

4. Tailings Properties for Static Stress Analysis.

For a complete comprehensive analysis of the response of a tailings dam to an earthquake, it is necessary to perform first a static stress analysis to determine the stress distribution throughout the dam before the earthquake. The determination of the nonlinear characteristics of the tailings is performed applying the method developed by Duncan and Chang [2]. This study involves the estimation of the experimental relations of the relative axial strain versus the stress deviator.

Poisson's ratio (ν) was calculated using test data from measuring the volumetric and axial strains. No big differences were found in the Poisson's ratio values for different stress-strain states.

The representative experimental stress-strain curves, for constant confining pressure, are presented in Fig.3.

Using of the hyperbolic stress - strain curves (Fig.3) and their linear transformation (not given here), the failure ratio is calculated by the expression [3]:

$$R_f = (\sigma_1 - \sigma_3)_f / (\sigma_1 - \sigma_3)_{ult}, \quad (1)$$

where R_f = failure ratio; $(\sigma_1 - \sigma_3)_f$ = stress difference at failure; $(\sigma_1 - \sigma_3)_{ult}$ = asymptotic value of stress difference; σ_1 and σ_3 = major and minor principal stresses.

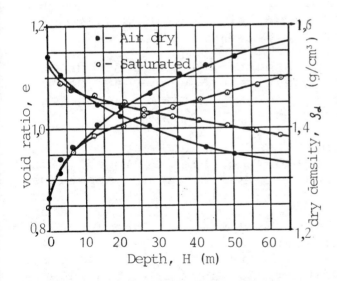

Fig.2.- Void ratio and dry density versus depth.

TABLE 2.- Summary nonlinear stress strain parameters

Type of Tailings	ν	R_f	K	n
Air-dry	0.380	0.960	20.8	0.477
Saturated	0.336	0.843	11.7	0.652

The values of the initial tangent modulus E_i, can be calculated from the linear transformation curves $\varepsilon_a = f[\varepsilon_a/(\sigma_1 - \sigma_3)]$ by the relationship [2,5],

$$E_i = K p_a (\sigma_3 / p_a)^n \qquad (2)$$

where:
K = modulus number (K = E_i for σ_3 = 1 atm); n = modulus exponent determining the rate of change of E_i with σ_3, n = $\log(E_i/K)$ for σ_3 = 10 atm; p_a = atmospheric pressure in the same units as σ_3.

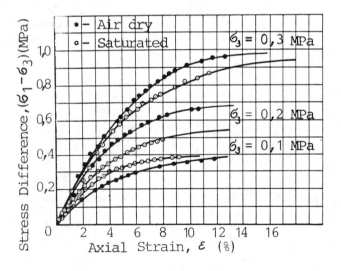

Fig.3.- Experimental stress-strain curves.

The values of the nonlinear tailings characteristics for the two states investigated are given in Table 2.

5. Tailings Properties for Dynamic Stress Analysis.

The dynamic analysis to evaluate the seismic tailings dam response involves the determination of the history of the varying stresses during an earthquake. To perform such an analysis, the dynamic properties of the tailings should be estimated. This procedure involves determination of the shear modulus at very small levels of strain, the strain-depending modulus and damping characteristics, and also the behaviour of the materials of the tailings dam under cyclic stress.

The dynamic properties are usually obtained from field tests, laboratory tests, or from the knowledge of the properties of similar soils.

The resonant column test device and the torsion shear test device are more appropriate among the laboratory equipment for determining the dynamic soil properties. However, our laboratory has been equipped with triaxial dynamic shear apparatus with a wide range of characteristics: axial loading up to 100 kN; confining pressure in the cell up to 1000 pa; samples sizes (diameter/height) 50/100mm, 70/140mm and 100/200mm; maximum piston amplitude - 50mm; frequency range from 1×10^{-8} to 10 Hz; accuracy of stress measuring 0.5 kPa, and strain measuring 0.001mm. This apparatus allows the performance of large scale testing, and the determination of dynamic tailings properties with accuracy sufficient for the design.

Tailings sand specimens with 100mm diameter and 200mm height were prepared by placing air-dry tailings sand into the forming mould, or by pouring saturated tailings sand in the mould and then tapping it to obtain the desired density.

Tailings sand specimens were tested under an isotropic consolidation and drained scheme, and under a stress-strain state corresponding to a depth of 10.0m (air-dry, W = 10%) and 40.0m (saturated). When the consolidation process was completed a cyclic deviator (σ_{dp}) with different strain amplitude (ε) was applied, and after 5 - 6 cycles unloading and reloading, the stress-strain hysteresis loops were registered. During the cyclic loading the volumetric and axial strain were measured, and after that Poisson's ratio was again calculated.

Applying the equations of the theory of elasticity, the $\sigma = f(\varepsilon)$ hysteresis loops were transformed into the $\tau = f(\gamma)$ hysteresis loops (τ - shear stress, γ - shear strain).

If the tailings properties are assumed not to change with the progression of cycles, then the stress-strain curve stays unchanged for a constant amplitude of the unload-reload cycles. Where sufficient number of $\tau = f(\gamma)$ hysteresis loops are available, the shear characteristics of tailings for steady-state cyclic loading can be represented by the secant shear modulus, G, which is defined by

$$G = \tau_a / \gamma_a \quad (3)$$

where τ_a and γ_a denote the amplitude of the shear stress and the shear strain.

$$\tau_a = \sigma_{dp} / 2 \quad (4)$$

The damping characteristics of tailings are presented by the damping ratio, D, which is defined by

$$D = \Delta W / 4\pi W \quad (5)$$

where ΔW is the damping energy, i.e. the area within the hysteresis loop, and W is the equivalent strain energy defined by

$$W = \gamma_a \tau_a / 2 \quad (6)$$

The results from the computation of the dynamic shear modulus and the damping ratio for different shear strain amplitudes are presented in Fig.4.

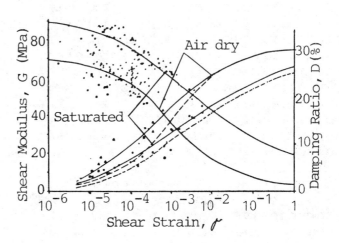

Fig.4.- Shear modulus and damping ratio versus shear strain.

The dash lines on **Fig.4.** give the relations of the damping ratio versus shear strain computed by the empirical formula [4]

$$D = D_o(1 - G/G_o) \qquad (7)$$

where D_o is the asymptotic damping ratio for the large strain ($\gamma = 1$) and G_o is the initial tangent shear modulus for $\gamma = 10^{-6}$. The values of G_o are computed directly from the plots of Fig.4.

G_o can be calculated by empirical empirical formula depending on void ratio and mean effective confining pressure [4]. The values of D_o can by accepted [3] 33% (for air-dry) or 28% (for saturated) reduced by $lg(N)$ (N - number of cycles for large strain). The corrected relations $D = f(\gamma)$ are represented with a full line on the Fig.4.

6. Conclusions.

A complete set of data for the basic standard tailings characteristics, nonlinear and dynamic properties are estimated. A comparison of the values of characteristics shows some essential differences between the two investigated states - air dry (above water table) and saturated (below water table).

The results obtained in this study will be applied for a comprehensive analysis of the stresses and strains induced in the tailings dam by earthquakes and for determining the resuling deformations applying the Finite Element Method.

References

1. Abadjiev,C.B.,and A.A.Caradimov, "Tailings dam of copper mining plant "Elatzite" after eight years of operation", Proc., 6th Conference The Embankment Dam, The British Dam Society, University of Nottingham, 12 - September 1990, pp. 47-50.
2. Duncan,J.M.,and Chin-Yung Chang, "Nonlinear Analysis of Stress and Strain in Soil", Proc. ASCE, Vol.7513, SM5, 1970, pp. 1629 - 1653.
3. Hardin, B.O. and Drnevich,V.P., "Shear Modulus and Damping of Soils: Design Equations and Curves", Proc.ASCE, Vol.98, SM7, 1972, pp. 667-692.
4. Iwasaki, T. and Tatsuoka, F., "Dynamic Soil Properties with Emphasis on Comparison of Laboratory Tests and Field Measurements", Proc. VI World Conference on Earthquake Engineering, New Delhi, January, 1977.
5. Seerf, N., Seed, H.B ., Makdisi, F.I., Chang, C.-Y., "Earthquake Induced Deformations of Earth Dams", Report No EERC-76-4, University of California Berkely, September 1976.

An Assessment of the Co-mingled Waste Disposal Method for ARD Control

J. Hadjigeorgiou
Université Laval, Quebec City, Quebec, Canada

R. Poulin
The University of British Columbia, Vancouver, B.C. Canada

K. Aref
CSIR, Johannesburg, South Africa

T. Boyd
The University of British Columbia, Vancouver, B.C., Canada

Abstract: Co-mingling is the process of joint disposal of waste rock and mine tailings. Current work aims to assess the suitability of this method in controlling acid mine drainage. A laboratory research programme employing column testing on co-mingled waste rock and mill tailings is described. Preliminary results suggest that by modifying the ratio of fines to coarse material, and consequently the overall material permeability, the generation of acidic water is reduced. This work is a prelude to a full scale site investigation.

Introduction

Co-mingling is the process of co-disposing mine waste rock with tailings. This paper reports on current research aiming to assess the suitability of co-mingling as a viable alternative for controlling acid rock drainage (ARD).

The complete elimination of acid generating waste is not at the present time a realistically attainable goal for reactive waste facilities. Consequently the focus is on developing a successful control strategy of the ARD process by employing prevention and/or abatement techniques. Prevention techniques are designed based on the assumption that ARD will be a by-product of mining. Abatement measures are implemented at facilities where ARD generation was not anticipated or where the designed control measures have proven ineffective [1].

Conventional waste disposal methods differentiate between dry and wet waste. Dry waste is primarily comprised of mined rock below the economical cut-off grade, but also includes unsaturated material recovered during overburden removal. Dry waste is dumped at pre-defined sites using either truck and shovel or conveyor systems. As a result of the predominance of coarse material, in dry waste, the resulting waste piles stand at angles of repose ranging from 35-45°.

Wet waste, or tailings, is the fined grained by-product of the mineral processing mill. Tailings are generally discharged as a slurry to an impoundment (tailings pond). Subsequently, discharged tailings, tend to dewater as a result of gravity drainage and, in arid climates, due to evaporation. These processes are, however, slow. Consequently, tailings ponds must be designed to hold both the solids and process water. Furthermore, it has proven difficult to fully reclaim such impoundment facilities. This can severely restrict future land use options [2].

Two approaches, both employing a variation of the co-disposal method, have recently been proposed as alternatives to the traditional methods of waste disposal. At the Jeebropilla coal mine, in south east Queensland, Australia, combined pumping is employed where tailings and coarse material are mixed together to produce a heterogeneous material [2]. The other alternative, investigated in this study, is the co-mingled waste disposal method where waste is inter-layered with dewatered tailings. Co-mingling was employed at the American Girl mine in southern California in the United States [3].

This paper describes the preliminary results of a laboratory investigation in assessing the effectiveness and applicability of the co-mingling waste disposal method. This is a prelude to a full scale field investigation that will consider both the practical and economic considerations.

Co-mingling method

Co-mingling requires mechanical dewatering of the tailings and some means of transportation and mixing with waste rocks. The blending process of tailings and waste rocks can take place prior to the waste disposal, as at the Jeebropilla mine. Alternatively waste rock can be placed with tailings introduced at various disposal stages. At the American Girl mine, tailings were pressure filtered to reduce their moisture content to 20% and subsequently placed on the waste piles on a weekly basis where they were promptly covered.

Co-mingling is a possible means for controlling ARD and mitigating its impact. Water and oxygen are necessary components in the reaction that produces low pH. Compacting or introducing low-permeability material is intended to reduce the bulk permeability of the waste rock dump, and consequently to mitigate ARD generation [4]. It should be noted, however, that co-mingling is a new process that has not yet been implemented in areas with high precipitation. Furthermore, the resulting economic and technical implications have yet to be fully investigated. For co-mingling to become a viable long term waste disposal alternative that can provide adequate ARD control, these concerns will have to be further addressed.

Design of the experiment

A two phase program was developed to evaluate co-mingling. In the first phase, a laboratory experiment was designed to address whether co-mingling is a viable method of waste disposal. In particular the influence on ARD generating potential of selected waste/tailings ratios was examined. Consideration was given to the impact of cementing agents to the tailings. The second phase of this project will examine in situ considerations, including pile stability, and the development of an appropriate placement methodology.

The design of the laboratory experiment was based on the prevailing conditions at the Gibraltar copper mine which is located 160 km south of Prince George in central British Columbia. This is an open pit operation where the recovered ore is concentrated using conventional crushing, grinding and flotation processes. The generated tailings are discharged to an impoundment area where approximately 80% of the water is recycled as process water.

The Gibraltar deposit lies entirely within a quartz diorite pluton. The rock is equigranular with an average grain size of 2-4 mm. It is composed of quartz (25-30%), a mixture of albite-epidote-zoisite-muscovite (50-55%), and chlorite (20%). Opaque minerals present are pyrite and chalcopyrite with minor amounts of magnetite, bornite and molybdenite. Oxidized phases of these minerals are present as secondary mineralization [5].

Material characterization

The waste material used in the laboratory columns was exposed to natural weathering processes at the mine site for approximately five years. It had already displayed signs of oxidation. This weathering in effect, reduced the time necessary to initiate a reaction in the columns. In order to respect scaling of maximum rock size to the column diameter (15 cm), the waste was crushed to -38 mm with a jaw crusher and was subsequently riffled, respecting standard sampling procedures. Fine material generated during crushing and passing a #6 sieve, was discarded. The crushed waste material was further analyzed to determine its shape characteristics. Defining "a" as the largest, "b" the intermediate and "c" as the smallest fragment dimension, then

the elongation ratio, $q = b/a$
the flatness ratio, $p = c/b$
the shape factor, $F = p/q$

and the sphericity is determined by:

$$\psi = \frac{12.8 \, (p^2 q)^{1/3}}{1 + p(1 + q) + 6[1 + p^2(1 + q^2)]^{1/2}}$$

The waste rock was found to display average values of $q = 0.658$, $p = 0.611$, $F = 0.994$ (i.e. equidimensional) and a sphericity of 1.252.

The shape of granular waste influences not only its density but also the workability of the system, its internal friction and its permeability [6].

The particle size distributions for both the processed waste and mine tailings are presented in Figure 1. Sieve analysis was conducted on the waste material placed in the columns. In order to complete the information on the fine tailings, sieve testing was complemented by micro-cyclone tests. Based on the results presented in Figure 1, for the crushed waste rock the coefficient of uniformity, C_u, was equal to 2.60 and the coefficient of curvature, C_c was 1.86. For the tailings the corresponding values were $C_u = 16$ and $C_c = 1.78$.

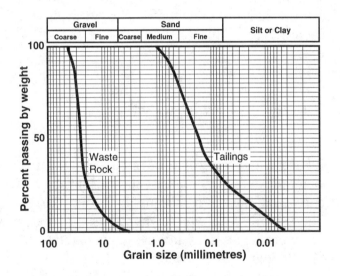

FIGURE 1. Particle size distribution curves.

Null pycnometer tests resulted in specific gravity values of 2.82 for the waste rock and 2.79 for the tailings. Variable head permeability tests on the tailings, for different void ratios, yielded the results presented in Table 1. The obtained hydraulic conductivity values are well in agreement with values predicted in the literature [7].

TABLE 1. Tailings hydraulic conductivity.

void ratio e	hydraulic conductivity K (cm/s)
0.83	2.14×10^{-3}
0.73	9.34×10^{-4}
0.60	5.57×10^{-4}

The frictional properties of the tailings was determined by direct-shear tests. The tailings had an angle of friction of 34.5° for an initial void ratio of 0.903 and 42.0^0 for a void ratio of 0.680. For the waste material the angle of repose was measured as 44°. Past experience relating the angle of friction to the angle of repose suggest that the angle of friction is usually 2 to 3° lower.

Description of the laboratory set-up

Figure 2 is a photograph of the erected four clear plexiglass columns, 180 cm long and 15 cm in diameter. Columns 1, 2, and 3 were filled with waste which was interspersed with dewatered layers of tailings compacted using the standard Proctor hammer. Column 1 was prepared to a 7:1 waste to tailings weight ratio, column 2 at 5:1 and column 3 at 7:1. The tailings in column 1 were enriched by the addition of 3%, by weight, of cement. Column 4 was the control column containing only waste rock.

Prior applications of the co-mingled method were in arid conditions. Consequently it was of interest to investigate the applicability of the method in areas of higher precipitation. The rainfall data for the Gibraltar mine were analyzed over a 20 year period, with the results

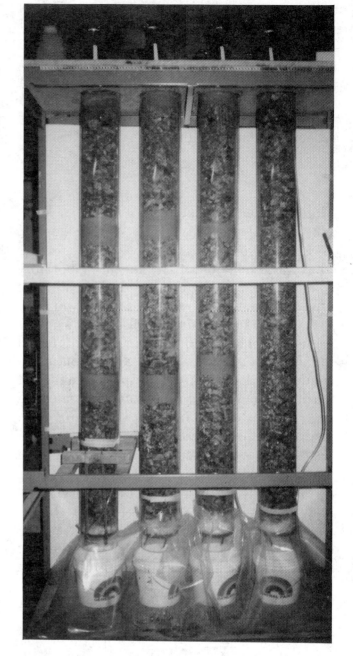

FIGURE 2. Test columns.

given in Figure 3. The precipitation data were grouped in three month intervals. In the laboratory set-up each time 3 month period was represented by a week.

Based on these assumptions the following column inflow rates were arrived at:

week #1: 0.185 ml/min
week #2: 0.112 ml/min
week #3: 0.272 ml/min
week #4: 0.180 ml/min

Consequently a column running for 4 weeks is equivalent to field exposure of one year's duration. A Masterflex low rpm multi-channel peristaltic pump was employed to control water flow in the columns. Distilled water is pumped at the predetermined rates at the top of each column where it is dispersed by a layer of glass wool.

Eh (mV H°)
conductivity (mS/cm)
alkalinity/acidity
Cu, Mo, Fe (ppb)
Sulphate (g/l) and pH.

Preliminary results suggest that co-mingling provides a mitigating effect on the generated ARD. The weekly monitored fluctuation in pH, sulphate and copper concentrations are presented in Figures 4 to 6.

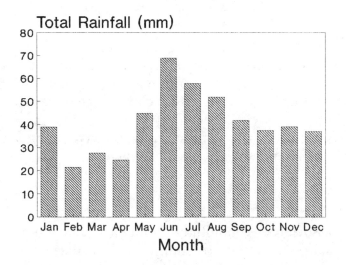

FIGURE 3. Average monthly rainfall at the Gibraltar site.

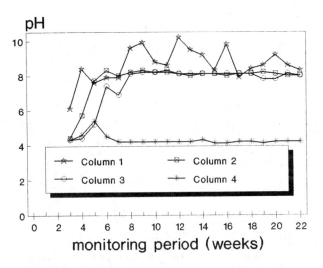

FIGURE 4. Fluctuations of pH levels.

Monitoring results

The experiment is currently in operation with weekly monitoring of column leachate being recorded. The monitoring program has been designed to evaluate the effect of co-mingling on water quality of column effluents. It comprised of the following measurements:

volume of water added (ml)
volume of leachate (ml)

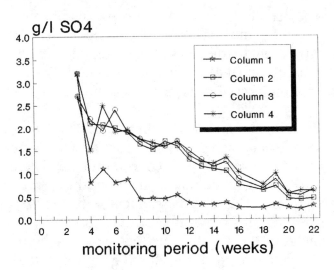

FIGURE 5. Sulphate content fluctuations.

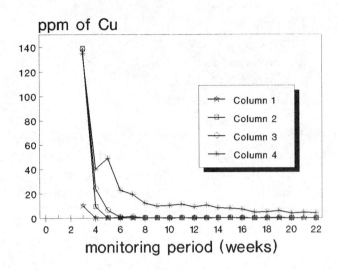

FIGURE 6. Copper content fluctuations.

The addition of cement in column 1 significantly influences the properties of the collected effluent. The pH level in the collected leachate from columns 2 and 3 is lower than that recorded in column 1, Figure 4. This is still considerably higher than the level recorded in the control column (no tailings). The pH fluctuation in column 1 can be linked to the water inflow, with pH increasing proportionally to the amount of water introduced in the column.

The sulphate concentration is significantly lower for the effluents of column 1. This is attributed to the presence of calcium in the cement mixture which contributes to the precipitation of calcium as gypsum.

In the columns where tailings and waste were co-mingled the high pH levels are also reflected in the recorded copper concentrations. These are a magnitude lower than in the control column. This is in agreement with previous observations of decreased Cu solubility at higher pH levels, [8]. The recorded decrease in sulphate and copper is due to high water flows which result in flushing of the rock surface.

Conclusions

Preliminary laboratory results suggest that co-mingling influences the ARD potential of the investigated waste and tailings. The resulting chemical reactions generated by the introduction of dewatered tailings in waste can significantly improve the quality of the collected water effluents. Different waste/tailings ratios have been investigated, and the possibility of adding cementing agents to the tailings was addressed. The present experimental and monitoring format is projected to be in operation over a period of two years.

The next phase of this investigation will involve the construction of a field scale test plot. This will permit the further evaluation of co-mingling as an ARD control measure, as well as the geotechnical stability of such a structure. The optimum design of co-mingled waste piles will be defined after due considerations of the associated operating costs. The long term benefits of reducing or eliminating wet tailings deposits and the associated rehabilitation problems can counter some of the construction costs. Furthermore, co-mingling can facilitate rehabilitation of mine waste piles for future land use as well as reduce the ARD potential of waste disposal areas.

Acknowledgments

The authors would like to acknowledge the financial support of the MEND program. The collaboration of R.J. Patterson of Gibraltar Mines is acknowledged as well as the technical input of Dr. R.W. Lawrence and Ms. D. Lister of The University of British Columbia.

References

[1] Lawrence R.W., Poling G.W., Ritcey G.M. and Marchant P.B. (1989). Assessment of predictive methods for the determination of AMD potential in mine tailings and waste rock. Proceedings of the International Symposium on tailings and effluent management, Halifax, pp. 317-331.

[2] Williams D.J. and Kuganathan V. (1992). Co-disposal of fine and coarse grained coal mine washery wastes by combined pumping. International Journal of Environmental Issues in Minerals and Energy Industry, 1 (2): 53-58.

[3] Aref K. (1993). Report on site visit to the American Girl mine, California, (unpublished).

[4] Sengupta M. (1993). Environmental impacts of mining: monitoring, restoration, and control. Lewis Publishers, Boca Raton, 494 p.

[5] Drummond A.D., Tennant S.J. and Young R.J. (1973). The interrelationship of regional metamorphism, hydrothermal alteration and mineralization at the Gibraltar Mines copper deposit in B.C., CIM Bulletin 66(730): 48-55.

[6] Winterkorn H.F. and Fang H.Y. (1975). Soil technology and engineering properties of soils. Foundation Engineering Handbook, pp. 67-120.

[7] Freeze R.A. and Cherry J.A. (1979). Groundwater, Prentice-Hall, N.J., 603 p.

[8] Pourbaix M. (1966). Atlas of electrochemical equilibria in aqueous solutions. Pergamon Press, 644 p.

A Review of Mine Tailings Management Practises for Closure

Peter C. Lighthall, Harvey N. McLeod, Serena Domvile
Klohn-Crippen Consultants Ltd., Richmond, B.C. Canada

This paper presents a review of emerging technologies for design and management of mine tailings disposal systems for closure. The purpose of the paper will be to outline the major considerations in closure design and to provide a review of closure technologies and their applicability to specific applications.

The paper provides examples where closure has gone wrong. These examples show where inadequate consideration of: a) geotechnical aspects, or b) water management, or c) waste chemistry, have led to "failures" of tailings structures or negative environmental impacts following closure. The review of case examples demonstrates how high closure costs, or costly failures, could have been avoided through alternative design. An overview is presented of the objectives of closure design.

A comprehensive design for closure requires integration of a number of key technical areas. Innovations in closure design and tailings management are described in the areas of site characterization, tailings management, water management, groundwater monitoring and control, chemical stability and risk management. Key areas identified requiring further research or practical development include ARD prediction and modelling, tailings dam liquefaction assessment and cyanide in groundwater.

1. Introduction

Current practice in design of mine tailings disposal systems requires consideration of the long term physical and chemical stability of tailings deposits. As well, legislation in most jurisdictions is requiring that closure plans be prepared for presently operating tailings systems. Based on the lessons of older mining operations and on research in the past decade, a number of new technologies for closure are emerging. This paper presents:

- lessons from case histories where closure has gone wrong
- major considerations in closure design
- emerging technologies in closure design

2. Examples of Closure Failures or Problems

Successful closure of tailings deposits requires full consideration of potential long-term problems and implementation of appropriate remedial measures. The best lessons can be learned from review of cases where closure has gone wrong. The following examples show where inadequate consideration of: a) geotechnical aspects, b) water management, or c) waste chemistry, have led to failures of tailings structures, negative environmental impacts or high ongoing maintenance costs.

Geotechnical failure of permafrost foundation - At the Clinton Creek Asbestos Mine, Yukon Territory, dry tailings material was placed by a stacker conveyor over a slope adjacent to the mill. Thawing of the permafrost foundation led to massive sliding of the tailings pile, which dammed a creek and underwent considerable erosion. Expensive drainage control measures were only partially successful. The problem could have been entirely avoided through adequate geotechnical investigations and alternative siting.

Seismic tailings dam failure - A number of tailings dams have failed during large earthquakes. A notable example of seismic failure was at the El Cobre Nuevo Tailings Dam, which failed during a 1965 earthquake in Chile. A volume of 2 400 000 m^3 of liquefied copper tailings flowed 10 km down a valley and buried 300 people in the mine camp (Cohen and Moenne, 1991). The tailings dam was inadequately designed to resist strong earthquake motions in a well known zone of seismic activity. Many currently operating tailings dams may have inadequate seismic resistance and should be assessed for safety for both continuing operation and for closure.

Overtopping failure due to long term degradation of diversion channel - At the abandoned Matachewan Gold Mine in northern Ontario, a breach of a tailings dike occurred in 1990, resulting in release of 195 000 m^3 of tailings. The tailings inundated private properties downstream, damaged a 1 km section of highway, and contaminated water systems in a river up to 100 km downstream. Investigation showed that the pond overtopped when a diversion channel became blocked by natural processes over the 35 year period following mine closure, backing up water levels in the pond. Expensive remedial measures are required to rehabilitate the site. The failure could have been prevented by ongoing monitoring of water levels and maintenance of the channel.

Acid drainage - Numerous examples exist of tailings deposits with ongoing heavy-metal-bearing acid drainage resulting from oxidation of sulphides. A well documented example is the Heath Steele Mine in Ontario, where Noranda Minerals will continue to operate a drainage collection and lime treatment plant for many years in the future. Current understanding of acid rock drainage provides ample measures for prediction and prevention so such ongoing liabilities can be avoided.

Oil sands sludges - An undesirable component of Athabaska Oil sands bitumen extraction plants is the fines component of the tailings, which forms a low density "sludge" (referred to as mature fine tails). The sludges form vast pools of material which represent future reclamation challenges. Major research efforts are underway to identify alternative extraction methods to reduce sludge formation or tailings management methods to encapsulate the sludges.

3. Objectives of Closure Design

Mine closure planning should be carried out with the following overall objectives:

- To identify post closure land use which will provide either enhanced value or, at worst, have a neutral impact.
- To eliminate or minimize ongoing care and maintenance of a site following closure.
- To characterize waste with regard to its potential reactivity, leachability or undesirable components and to develop mitigating strategies.
- To develop a sound water management plan for closure.
- To maintain high water and air quality standard.
- To provide long-term geotechnical stability.

4. Closure Technologies

Comprehensive design for closure requires integration of a number of key technical areas. Innovative new technologies or innovative applications of existing technologies are being developed to respond to increasingly stringent requirements or more complex problems. Some of the new technologies in use are presented in the following sections.

4.1 Site Characterization

Assessment of existing tailings impoundment for closure most often focuses on the two key aspects of seismic stability and acid drainage potential. Conventional geotechnical investigation methods of drilling, sampling and laboratory testing are often too expensive and time consuming for large sites. Alternative technologies can provide much more rapid characterization. These include:

- Cone penetrometer testing (CPT) is ideally suited to in situ site investigation for seismic stability assessments of existing hydraulic fill tailings dams. Recent work (Plewes et al., 1992) enables accurate correlation of fines content and standard penetration test (SPT) values for liquefaction assessments. Groundwater pressures, phreatic levels and permeabilities can be determined in situ. CPT testing is superior to conventional drilling in that it provides continuous profiling, is repeatable and is generally considerably less expensive.
- The resistivity cone is a tool being developed as a supplement to the CPT apparatus (Campanella et al, 1993). The resistivity cone provides a continuous profile of soil resistivity and is an excellent tool for tracing development of acidic porewaters or other chemical constituents in tailings deposits.

4.2 Tailings Management

Probably the most significant change brought about by modern closure planning is the variety of tailings handling practices being employed to provide long-term environmental protection. Some of these include:

Tailings Dewatering - Several mines, including Placer Dome Inc.'s La Coipa Gold Mine in Chile, are using belt filtration to dewater tailings to less than 20% moisture content. Tailings are placed in a stack by conveyor. The system provides the dual benefits of minimizing entry of cyanide to the groundwater and also desaturating the material to prevent earthquake liquefaction.

Freezing - New mining projects in permafrost regions of northern Canada are planning to place tailings in layers which will freeze each winter and remain permanently frozen. By so doing, reactive tailings will not oxidize nor will oxidation products enter the water system.

Solidification - Chemical or cement stabilization is being considered for some particular tailings products, such as chemically unstable autoclave tailings.

Accelerated Consolidation - Numerous tailings systems are employing methods to speed consolidation of fine tailings. The purposes of enhanced consolidation techniques are: a) to maximize use of storage capacity by increasing tailings density, and b) to reduce post-operational settlements so the surfaces can be quickly reclaimed to their planned eventual use. Techniques for accelerated consolidation include:

- densification through freeze-thaw or evaporation of tailings placed in thin layers
- provision of under-drainage
- wick drains.

Covers - Current practice for sulphide tailings is to provide covers to exclude oxygen for long-term prevention of acid generation. The most widely accepted practice is a water cover. Extensive study, under the Mine Environment Neutral Drainage (MEND) program, of sites where reactive tailings have been placed on lake bottoms has shown essentially no lasting impacts on aquatic ecosystems. Ocean disposal is also very effective but permitting is difficult because of negative perception by the public and environmental groups. Water covers are also being applied to conventional tailings ponds where impervious dams and a natural inflow will provide long-term flooding of tailings surfaces. Pervious soil covers, consisting of non reactive tailings or other materials covering reactive tailings, can also be effective. They are particularly effective if the phreatic surface is within the non reactive cover, thus excluding oxygen from the reactive zone. Impervious soil or composite soil/geomembrane liners require a high degree of long-term integrity to be effective. With their high capital cost, they should be considered only in special situations.

Separation of tailings streams - Several mines with potentially reactive tailings are planning installation of supplementary flotation circuits to separate the sulphide component of their tailings stream. The sulphides can then be placed separately in central locations in the tailings impoundments which will remain permanently flooded upon closure. In one case, the sulphides may be marketable so the earlier, reactive tailings mass can be covered with a non reactive cap for the last several years of operation before closure.

4.3 Water Management

Surface water control is an aspect of closure design requiring not innovation, but in general an extra degree of conservatism. Experiences such as the unprecedented 1993 Mississippi River floods show that we must design for the long return period, unimaginable flood event to have long-term security against overtopping or flood erosion. Where failure of a structure could result in loss of life or significant environmental damage, probable maximum precipitation (PMP) events should be used for closure design.

The overtopping failure at Matachewan showed that ongoing monitoring and maintenance of water handling facilities for closure may be necessary. Where the integrity of structures depends on long term performance of water handling facilities, a mechanism such as a monitoring and maintenance fund should be established.

4.4 Groundwater Monitoring and Control

It is imperative for mining operators to clearly establish their impacts on groundwater resources. Once contaminated by heavy metals or other contaminants, it is difficult to definitively clean up aquifers. Thus, thorough background surveys should be undertaken, either to establish background for new projects or to establish present levels of impacts for operating mines. Groundwater monitoring should usually continue for some period beyond closure. Some recent developments in groundwater characterization and monitoring are:

- Geophysical methods have found valuable application in tracing groundwater contamination in reactive tailings sites (King and Pesowski, 1993). Airborne electromagnetic (EM) surveys are excellent screening tools for identifying conductive plumes. On a site scale hand held EM profiling equipment has been used very successfully not only to trace plumes but also to prepare depth profiles of acidic groundwater. Conventional resistivity surveys and radar profiling have also been successfully applied to groundwater detection.
- Push-in technology is also being developed for groundwater sampling. Patented samplers allow retrieval of groundwater samples from a push-in cone apparatus.
- For convenience of sampling, many monitoring networks are being installed with dedicated sampling pumps in monitoring wells. Packers may be installed above and below sampling points to greatly reduce pumping volumes and sampling times. Lysimeters (suction samplers) have been developed for sampling in unsaturated wastes.

Increasingly tailings ponds are being constructed with geomembrane, soil, or composite soil/geomembrane liners to reduce groundwater contamination. More complex systems are also being developed with drainage layers directly above the liners to reduce seepage head on the liners.

4.5 Chemical Stability

Long-term chemical stability of tailings deposits is becoming a critical aspect of closure. An example is the large porphyry copper-molybdenum mines in British Columbia. Low residual levels of molybdenum are being leached into tailings drainage water in several of these mines, making the water unacceptable for discharge to the environment. Considerable effort is being expended on identifying viable water treatment technologies.

A second example of long-term chemical stability concerns is the long lag time in some cases (up to several decades) prior to development of acid drainage. The imprecision of predictive methods is requiring more conservatism in interpretation of ARD testing results.

Acid drainage can present not only water pollution problems but can also lead to tailings dam instability. Cases have been noted where precipitates from acid drainage have clogged tailings dam toe drains, resulting in raising of phreatic surfaces. In response, designs have been developed which keep drains permanently submerged to avoid formation of precipitates.

4.6 Risk Management

Mine operators are often faced with extremely high costs of remedial measures for closure. The techniques of risk assessment have proven valuable in developing rational priorities for expenditure. Probabilities, impacts and the costs of various failure scenarios are compared to determine where, and if, expenditures should be made. It has been found through risk assessment studies that remedial measures are not always cost effective and owners may be able to show that the modest costs of ongoing monitoring and maintenance provide security equal to expensive capital remediation programs.

5. Further Research and Practice Needs

A number of mine closure issues remain for which further understanding is required. At present, we are forced to either design with excessive conservatism, or apply uncertain technologies, because of our complete lack of understanding. Some of these key areas include:

<u>Acid Rock Drainage Prediction and Modelling</u> - The key closure concern for both new and currently operating mining projects is assessment of long term acid drainage potential. To assess long term performance, researchers and practitioners are beginning to develop site-specific predictive models of acid generation. These models must encompass waste chemistry, oxidation rates, transport of oxidation products and mitigative measures. Development and calibration of reliable mathematical ARD models will be a key area of future development.

<u>Tailings Dam Liquefaction Assessment</u> - Large numbers of older tailings dams throughout North America are constructed by simple upstream spigotting methods, with no compaction or internal drainage measures. Many of these structures may be susceptible to liquefaction failures in the event of a large earthquake. Considerable work needs to be done to characterize these older structures and to develop realistic assessments of earthquake potential to assess the risks of failure.

<u>Cyanide in Groundwater</u> - A large emphasis is given in mine environmental design to containment of cyanide wastes. However, it is also well recognized that cyanide degrades rapidly in the environment and does not present a long term hazard. Less understood is the long term fate of cyanide in groundwater. A better understanding of cyanide degradation in groundwater would allow more rational assessment of design objectives for long term containment.

6. Conclusions

Tailings design and management for closure is a rapidly evolving field. Past failures have shown the many problems which can develop with abandoned tailings impoundments. As we are now approaching closure of a number of the very large tonnage mining operations which developed in the latter part of this century, more economical techniques are required for closure assessment of these large scale operations. As well, more sophisticated understanding is required to procure permits for new operations. This paper has summarized some of the new technologies for both characterization of older deposits and for design and management of new operations. As well, key areas of further research and practice needs for closure are identified.

7. References

Campanella, R.G., M.P. Davies and T.J. Boyd (1993). "Characterizing Contaminated Soil in Groundwater Systems with In situ Testing"; Proceedings, ASCE/CSCE Joint Environmental Conference, Montreal, July.

Cohen, M. and G. Moenne (1991). "Tailings Deposits at the El Soldado Mine Compañía Minera Dispauada de Las Condes"; Special Volume, Seismic Design, Abandonment and Rehabilitation of Tailings Dams, IX Panamerican Conference on Soil Mechanics and Foundation Engineering.

King, A and M. Pesowski (1993). "Environmental Applications of Surface and Airborne Geophysics in Mining"; CIM Bulletin, Vol. 86, No. 966, January, pp. 58-67.

Plewes, H.D., M.P. Davies and M.G. Jefferies (1992). "CPT Based Screening Procedure for Evaluating Liquefaction Susceptibility"; Proceedings, 45th Canadian Geotechnical Conference, Canadian Geotechnical Society, Toronto.

Oil-Sands Fine Tailings Reclamation Involving Sand-Fines Mixtures

Robert C. Lo
Klohn-Crippen Consultants Ltd., Richmond, B.C. Canada

Edward R.F. Lord, Julian M. Coward
Syncrude Canada Ltd.

ABSTRACT

The two oil sands plants (Syncrude and Suncor) located near Fort McMurray, Alberta, Canada produce over 90 million barrels of oil each year, enough to supply over 16% of Canada's petroleum needs. During the production of oil from the oil sands, large amounts of slowly densifying fine tailings (tails) in a fluid state are produced. Currently, the two plants have accumulated over 300 M (million) m³ of fine tails in their tailings settling basins. One method to reduce the volume occupied by fine tails is to increase the amount of fine tails entrapped in the tailings sand deposits.

A laboratory test program was conducted to determine the geotechnical properties of various sand and fine tails mixtures, with fines (<22 μm) contents ranging from 5% by weight (current beach deposits) up to 20%. Laboratory observations were made on aspects of sedimentation, consolidation, permeability and shear strength. The results indicate that with 5 to 10% fines, the deposits will sediment and consolidate very rapidly. At higher fines contents, the sedimentation and consolidation rates are reduced. Similarly, test data indicate that the shearing strength decreases with increasing fines content. However, even at 20 % fines content, the shear strength characteristics of the mixture are still governed more by the sand structure than by the fines matrix.

Two large scale field tests were conducted to evaluate techniques of generating sand deposits with increased fines contents. Tailings slurries were enriched with fines and the fines contents of the resulting deposits were measured. The results from these tests have been encouraging, and this technique has the potential to reduce the volume of fluid fine tails; providing alternative non-fluid ways to reduce the large volume of fine tails which are generated from the oil sands plants.

Introduction

The Athabasca oil sands deposits cover an area in Alberta of over 30,000 km² and consist of bituminous quartz sands containing up to 18% by weight of bitumen. This is one of the largest oil reserves in the world, with almost 1000 billion barrels of oil; over four times the amount in Saudi Arabia [1]. The recoverable reserves from the Athabasca oil sands are sufficient to supply Canada's current petroleum needs for several hundred years. At present there are two large oil sand surface mining operations in the Athabasca region:

- Suncor, which has been operating since 1973, at a current synthetic oil production rate of about 70,000 barrels per day;
- Syncrude, which has been operating since 1978, at a current production rate of about 193,000 barrels per day.

Synthetic crude oil from these two oil sands plants represents about 16% of the total oil production in Canada, and over 1% of the North American needs.

During the production of bitumen from the oil sands, large quantities of tailings are generated from the process [2]. The tailings slurry consisting of about 55% by weight solids in water, is pumped to the tailings settling basin. The coarse solids, which are predominantly a fine sand, readily settle out to form the beaches and dykes around the settling basin. The finer solids flow into the basin, where they settle in a few months to form a fine tails containing about 20% solids by weight. Over a period of a few years, the fine tails dewater further to about 30% solids. The water released from the fine tails is recycled to the plant, and reused in the process. To date, the two plants have accumulated over 300 M m³ of fine tails in the settling basins. Various reclamation techniques have been proposed, including methods of fine tails disposal as viable water capped lakes, and as dry disposal if the fine tails are mixed with overburden or sand [3]. This paper describes studies undertaken to assess the feasibility of increasing the fines content of the sand deposits, thereby reducing the volume of fluid fine tails.

Laboratory Testing

A laboratory testing program was performed in 1988 by Klohn Leonoff (presently Klohn-Crippen Consultants Ltd.), on Syncrude mixtures of tailings sand and fine tails. The purpose of the work was to determine the depositional characteristics of various sand and fine tails mixtures, to evaluate their geotechnical properties and to identify alternative dry fine tails disposal schemes based on an improved understanding of the materials. The focus of the laboratory testing program was to evaluate sand and fine tails mixtures deposited from pumpable slurries.

Samples of tailings sand, fine tails and water were collected from the Syncrude settling basin. Average index properties of the sand and fine tails are shown in Table 1.

The amount of fines captured within the sand-fine tails matrix of the deposited materials with minimal segregation is an important factor controlling the mixture characteristic. Four kinds of laboratory tests were selected as screening tests to identify promising sand-fine tails mixtures for further study:

- single cylinder sedimentation column tests for preliminary screening;
- double cylinder sedimentation column tests for evaluating segregation potential,
- flow cone tests (ASTM Designation C939) for assessing pumpability
- bearing capacity tests for studying load bearing characteristics.

Sand and fine tails mixtures with 5% to 20% fines (<22μm sizes) and 45% to 90% by weight solids content were evaluated utilizing 18 single cylinder and five double cylinder sedimentation column tests to screen these mixtures for least segregation potential. As expected, the sedimentation rate of the mixture reduces with increasing fines content. The double cylinder apparatus, developed specifically for this project, is approximately 1 m high with an inside diameter of approximately 90 mm. The inner cylinder consists of a series of detachable stacked rings of fixed height held together by the outer cylinder. The mixtures were left in the double cylinder for approximately ten days. After the completion of each test, the inner cylinder was disassembled and the material in each of the rings retrieved for index property testing. Profiles of dry density, water, solids and fines content provided definitive proof of the non-segregating nature of mixtures with 15% to 20% fines and 65% to 72% solids content (1.1 to 1.3 g/cc dry density) in quiescent sedimentation conditions. These results are consistent with the findings reported by Pollock [4] using large-scale (10 m high and 1 m diameter) self-weight consolidation models. Sixteen flow-cone tests were carried out which indicated mixtures to be pumpable up to 72% by weight solids content. Simple bearing capacity tests were performed which showed non-segregating mixtures to have a surface load bearing capacity much higher than the fine tails alone.

Additional laboratory testing was carried out to determine the geotechnical properties of sand and fine tails mixtures with 5% to 20% fines and initial dry density of 1.1 to 1.5 g/cc. Four conventional oedometer primary consolidation tests and three oedometer secondary consolidation tests were carried out to determine the compression index, coefficient of consolidation and rate of secondary compression for these mixtures. The results are summarized in Table 2. They indicate that the consolidation rate of the mixture also reduces with increasing fines content.

Direct permeability measurements were carried out utilizing four oedometer specimens and eight triaxial specimens. All data showed the trend of decreasing permeability with both the increase of fines content and the decrease of void ratio. However, significant scatter exists among test data at similar fines content and initial dry density. The reasons for these differences may be related to the testing apparatus, the methods of specimen preparation, the degree of saturation and the migration of fines during the tests. Similarly, large differences exist among Syncrude data obtained by others [4]. The bitumen content was an additional variable among these earlier tests.

Ten consolidated-undrained triaxial compression shear tests with pore pressure measurements were carried out to determine the effective and total shear strength parameters and stress-strain behaviour of the sand-fine tails mixtures. All test specimens with fines content ranging from 5% to 20% were prepared at an initial dry density of about 1.5 g/cc, which is close to the lower density limit achievable by preparing the triaxial test specimen at the wet state. This density is also close to the upper density limit achievable in the sedimentation tests described earlier. Test data indicated that, in general, the shearing strength increases with decreasing fines content. However, even at 20% fines content, the shearing strength characteristics of the mixture are still governed more by the sand structure than by the fines matrix. The range of the effective strength parameters are $c' = 0$, $\phi'_{peak} = 16°$ to $22°$, and $\phi'_{ultimate} = 32°$ to $40°$. Range of the total strength parameters are $c = 0$, $\phi_{peak} = 12°$ to $19°$. The values of these strength parameters reflect the influence of pore pressures measured at the peak and ultimate states. Significant differences exist in the stress-strain curves of these consolidated-undrained tests. At low confining pressures, the stress-strain curve of the mixtures with 5% fines does not have a peak, while all other stress-strain curves show a peak at low strain values.

Field Testing

In 1990, Syncrude conducted a large scale field test to evaluate methods to increase the fines content of the tailings sand deposits [5]. In one test, fine tails at a solids content of 38% by weight, were added to the tailings slurry to replace tailings line make up water which is normally added (spiking). The spiked tailings slurry had a fine tails concentration of about 20% by volume. A large deposit (13000 m³) was formed (Test 3) and samples taken for analysis of fines content. Similarly samples were taken from two deposits formed in the normal manner (Tests 1 & 8). Typical tailings particle size distribution is shown in Figure 1 which would classify it as a uniform fine sand. About 10% is classified as silt/clay size. The fine tails consists predominantly of clay/silt sized particles.

The fines content was evaluated from measurements of particle size distribution and using sieve and Microtrac, a laser light scattering technique [6]. The relative amounts of solids: <88µm, <44µm, <22µm, <11µm, and <5.5µm, in the deposits are shown in Figure 2. As can be seen in Figure 2, the deposit (Test 3) contains more fines than in the base cases (Tests 1 & 8). Similar trends were obtained for fines content determined by the Methylene Blue Test [7]. In these test deposits, the fines content was about double that of normal tailings deposits. This test proved that the fines content of the tailings deposits could be increased by spiking.

A second test conducted in 1991 by OSLO (Other Six Lease Owners) on the Syncrude site and a similar result was achieved [8]. An offtake from a tailings line provided feed to a hydrocyclone. The cyclone underflow product was discharged into a pump box where a predetermined quantity of fine tails at a solids content of 27.5% by weight, were added. The blended slurry was pumped to a test area where a number of deposits were formed. The deposits were sampled and the particle size distributions determined by sieve and hydrometer tests. The relative amounts of solids: <176µm, <88µm, <44µm, <22µm, <11µm, <5.5µm, and <1.3µm in one of the deposits (Test 6) (396 m3) are shown in Figure 3, together with corresponding values for the Base Case (Test 1). Similar to the 1990 test results, the fines content in the test deposit was about double that of the normal tailings deposits.

Conclusions

The laboratory test results indicate that the sedimentation, consolidation and shear strength characteristics of a sand-fines mixture are influenced by its fines contents (<22 µm). However, even at 20 % fines content, the shearing strength characteristics of the mixture are still governed more by the sand structure than by the fines matrix. Field testing indicates that the fines content of the coarse tails deposits can be increased relatively easily, up to about 8%, by enriching (spiking) the tailings slurry with fine tails. Where appropriate, Syncrude is committed to adopting this approach to reducing the volume of fluid fine tails.

Acknowledgments

The authors would like to thank their colleagues at Syncrude in particular G. Cuddy, W. Shaw and W. Mimura, and at Klohn-Crippen in particular P. Lighthall and S. Rice for their helpful advice and contributions during these studies and the preparation of this paper. Permission of Syncrude Canada Ltd to publish this paper is also gratefully acknowledged.

References

[1] Outrim, C.P. and Evans, R.G. Alberta's Oil Sands Reserves and their Evaluation. 28th Meeting of the CIM, Calgary, May, 1977.

[2] List, B. R. & Buckles, J. M., The evolution of Tailings Management Practices at Syncrude Canada Ltd. Proceedings of the International Congress on Mine Design, Kingston, August 1993. pp 83 - 90

[3] Lord, E. R. F., MacKinnon, M. D., Fair, A. E. & Friesen, B. C., Disposal of Fine Tails Produced at he Syncrude oil Sands Mine. Proceedings 1993 Joint CSCE-ASCE National Conference on Environmental Engineering, Montreal, July, 1993. pp. 203-212.

[4] Pollock, G. W., Large Strain Consolidation of Oil Sand Tailings Sludge, M. Sc. Thesis, University of Alberta, Edmonton, Alberta, 1988.

[5] Cuddy, G., Lahaie, R. L., & Mimura, D. W., Spiking Tailings as Method to Increase Fines Retention in Beach Deposits. Proceedings Fine Tailings Symposium. Oil Sands - Our Petroleum Future Conference, Edmonton, April, 1993. Paper F21 pp. 1 - 21.

[6] Muly, E. C., Frock, N. H. & Weiss, E. L. The Application of Fourier Imaging Systems to Fine Particles. Journal of Powder and Bulk Solids Technology, 2, 1978 3.

[7] ASTM Standard Designation C 837 - 81 (Reapproved 1988) Standard Test Method for Methylene Blue Index of Clay pp. 275 - 276

[8] Shaw, W., Livingstone, B. & Beck, J. Disposal of Fine Tailings utilizing Hydrocyclone Techniques. Proceedings Fine Tailings Symposium. Oil Sands - Our Petroleum Future Conference, Edmonton, April, 1993. Paper F19 pp. 1 - 20.

TABLE 1 --Index Properties of Tailings Sand and Fine Tails

Material	Water Content $\frac{W_w}{W_s}$ (%)	Fines Content $\frac{W_f}{W_s}$ (%)	Solids Content $\frac{W_s}{W_w+W_s}$ (%)	Bitumen Content $\frac{W_b}{W_s}$ (%)	Liquid Limit	Plastic Limit
Tailings Sand	12.1	4.7	89.2	-	-	-
Fine Tails	164	87	37.7	2.61	50	18

Notes: W_w = weight of water $\qquad$ W_s = weight of solids
$\qquad\quad$ W_b = weight of bitumen $\qquad$ W_f = weight of fines (<22μm)

TABLE 2 -- Summary of Consolidation Test Data for Sand - Fine Tails Mixtures

Initial Dry Density g/cc (See Note 2)	Fines Content (%<22μm)	Compression Index C_c	Rate of Secondary Compression $C\alpha$	Coefficient of Consolidation C_v (10^{-4} cm^2/sec)
1.1	10-20	0.05-1.7	0.0005-0.003	0.5-900
1.5	5-20	0.02-0.09	0.0002-0.0005	(See Note 1)

Note 1: Materials with 5% to 10% fines consolidate too rapidly to allow accurate determination of C_v. The coefficient of consolidation, C_v, ranges from 1 to 180 x 10^{-4} cm^2/sec for the material with 20% fines.

Note 2: Typical dry density values of coarse tailings deposits [2]
        Cell construction   1.66 g/cc
        Beach above water   1.58 g/cc
        Beach below water   1.51 g/cc

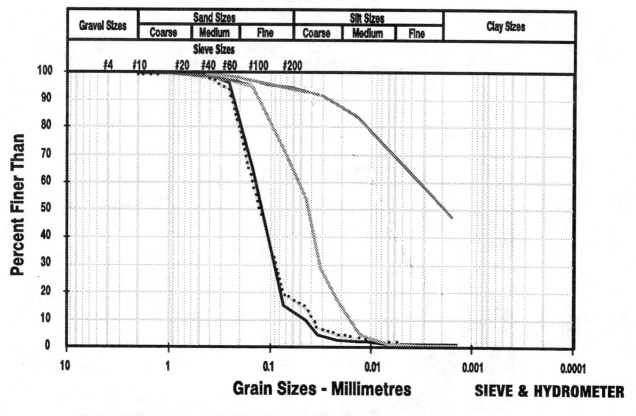

FIGURE 1 Typical Tailings Particle Size Distribution - (Syncrude 1990 Field Test)

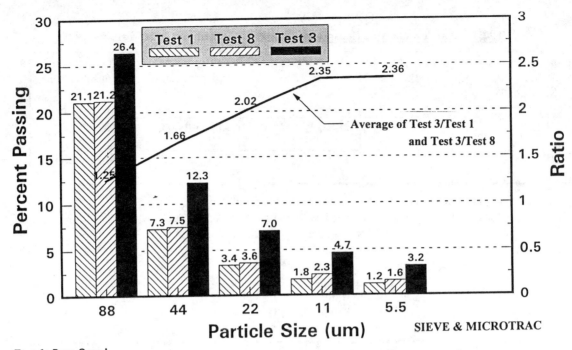

Test 1: Base Case I
Test 8: Base Case II
Test 3: Spiking

Note: Test data from 1990 field test
Test 1 (19 samples); Test 8 (6 samples); Test 3 (10 samples)

FIGURE 2 Comparison of the Particle Sizes in the Tailings Sand Deposits - Syncrude Field Test

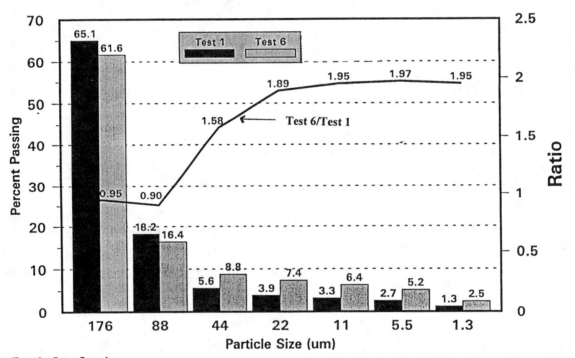

Test 1 : Base Case I
Test 6 : OSLO DBM Spiking

FIGURE 3 Comparison of the Particle Sizes in the Tailings Sand Deposits - OSLO Field Test

Consolidation and Pollutant Transport Associated with Slurried Mineral Waste Disposal

L.J. Potter, C. Savvidou
University of Cambridge, Cambridge, U.K.

R.E. Gibson
Golder Associates

Abstract

The objective of any modelling exercise is one of simulation and prediction. In the particular case of subsurface contaminant transport, modelling provides valuable information about the movement, spread and growth of pollutant plumes and indicates the effectiveness of various containment strategies.

This paper presents analytical and experimental modelling of problems related to the disposal of slurried mineral wastes. The scheme under consideration involves the instantaneous dumping of a compressible waste over a permeable stratum. The waste may be covered by a cap layer of soil. Consolidation of the waste causes a discharge of a finite volume of polluted water; some is discharged into the pores of the underlying soil, and some upwards into the capping layer. The design of the cap layer in practice involves choice of the material and of the layer thickness such that there is minimal risk of cracking to the capping material or of unacceptable pollutant discharge into any overlying water.

The present work outlines analytical-numerical solutions of consolidation and pollutant transport under one dimensional, small strain conditions. The analytical data is compared to centrifuge model test data. A small diameter centrifuge was used as a tool for preliminary experimental modelling of the consolidation, and associated contaminant flow from a slurried waste overlying a permeable stratum. The results highlight the principal mechanisms involved in mineral waste disposal problems.

1. Introduction

The disposal of slurried mineral wastes presents major technical and environmental problems that challenge the ingenuity of the engineering profession. Several industrial wastes which come under this category have been analysed by Carrier et al [2].

Slurried mineral wastes are commonly deposited hydraulically in containment ponds. Design problems associated with these ponds include storage capacity, embankment stability, migration of contaminated pore fluid from the waste and final land use. To date, industry has had the use of computer codes for the prediction of consolidation rates and final surface levels. However, present commercial predictions are not able to anticipate the contaminant flow from the waste layer into the overlying and underlying layers that is associated with the consolidation of the waste. They are also not able to predict damage to capping layers due to large pore pressures if the capping is relatively impermeable, or contamination of the lake layer if the capping is relatively permeable. Hence the prediction of consolidation and the associated contaminant flow from slurried mineral wastes presents a particularly challenging problem. It requires new developments in both experimental and mathematical modelling.

2. General Problem

The consolidation and associated contaminant transport from a slurried mineral waste layer, overlying a permeable stratum, is considered. The problem is idealised as follows.

An extensive lake is filled with still, clean water. It is underlain by a thick, uniform deposit of incompressible silty sand. It is in turn underlain by a layer of coarse sandy gravel, but not so coarse that the silty sand enters it. There is slow but persistent seepage in the sandy gravel, which is, nevertheless, considerably faster than the flow of pore fluid into it from the overlying silty sand. The sandy gravel rests on an impermeable base. At some instant, a waste layer is instantaneously dumped through the water to come to rest over the silty sand. Figures 1(a) and 1(b) show that the waste layer may be covered by a capping layer.

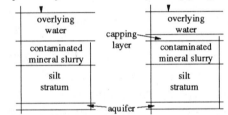

FIGURE 1 -- Mineral waste disposal options

It is assumed that the waste is fully saturated, and that before placement it is in a slurry state. The term 'waste' covers such a vast range of materials, that it is necessary to state here that it is envisaged that the waste is a soil-like, particulate medium.

The finite difference technique is used to produce a numerical model. As a first attempt it is assumed that strains are small, and that soil properties are constant during consolidation. A comparison between two different types of capping layer is made. The experimental simulation of the problem is achieved by centrifuge testing. Comparisons between experimental and numerical predictions are presented [1].

---

[1] All parameters are defined at the end of the paper.

3. Numerical Modelling

3.1 Governing Equations

The flow of pore fluid in the soil is assumed to be governed by Darcy's law:

$$v' = \frac{ki}{\eta} \quad (1)$$

The one dimensional, infinitesimal strain, consolidation equation applied in the waste, has the form [7];

$$\frac{\partial u}{\partial t} = C_v \frac{\partial^2 u}{\partial z^2} \quad (2)$$

where the excess pore pressure u, is defined as;

$$u = p - \gamma_w x \quad (3)$$

The transport of contaminants in rigid porous media can be described using the classical advection-dispersion equation, which has the form;

$$D_{hl} \frac{\partial^2 c}{\partial z^2} - v' \frac{\partial c}{\partial z} = \frac{\partial c}{\partial t} \quad (4)$$

in which the first term represents the dispersive component of contaminant movement and the second term the advective component [1]. In order to yield a conservative prediction, diagenetic reactions such as sorption, bio-transformation and radio active decay are assumed to be negligible.

3.2 Coupled Consolidation and Contaminant Transport

During the process of consolidation of a contaminated compressible medium, transient pore fluid velocities will be induced as the generated excess pore pressures dissipate. This implies that the migration of contaminants through such media will be affected by the consolidation process. For a numerical simulation of such a scenario the consolidation theory and the contaminant transport theory must be used in parallel.

3.3 Non-Dimensional Representation

To enable comparisons between the numerical and the experimental simulations to be made, the variables can be non-dimensionalised. Thus the infinitesimal strain consolidation equation becomes

$$\frac{\partial U}{\partial T_c} = \frac{\partial^2 U}{\partial Z^2} \quad (5)$$

and the one-dimensional advection-dispersion equation becomes

$$\frac{\partial^2 C}{\partial Z^2} - V'\frac{\partial C}{\partial Z} = \frac{\partial C}{\partial T_{ct}} \quad (6)$$

The non-dimensional groups for consolidation are;

$$U = \frac{(p - \gamma_w x)}{p_o}, \quad T_c = \frac{C_v t}{L^2}, \quad Z = \frac{z}{L},$$

where p and u are related by Eq. 3, and p_o is defined in section 3.4. For contaminant transport the non-dimensional groups are;

$$C = \frac{c}{c_o}, \quad V' = \frac{v'L}{D_{hl}}, \quad T_{ct} = \frac{D_{hl} t}{L^2}, \quad Z = \frac{z}{L}$$

The continuity equations that have been used here can be found in reference [5].

3.4 Boundary Conditions

Consider a waste layer of depth L_w, underlain by a layer of depth L_u. The waste may be overlain by a capping layer of depth L_o. The total depth of the three layers is L as shown in figure 2. Thus;

$$P(0,T) = \frac{p_{base}}{p_o}$$
$$P(1,T) = 1$$
$$C(1,T) = 0$$

where p_o, the pressure on the upper boundary, will increase as the waste layer consolidates, and p_{base}, the pressure at the base of the silty sand, will be constant with time.

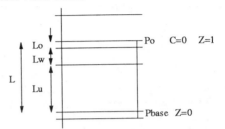

FIGURE 2 -- Mathematical notation showing boundary conditions

3.5 Initial conditions

The pore fluid in the waste is contaminated. The concentration of the pore fluid in the overlying and the underlying layers is assumed to be that of clean water. The pore pressures in the waste layer are generated by self weight, the weight of the capping layer and the weight of the overlying water. Since both the underlying layer and the capping layer are incompressible, and relatively more permeable than the waste layer, steady state flow is achieved rapidly and hence the pore pressure profiles are linear in these layers.

3.6 Finite difference formulation

As a first step a piece-wise, explicit, iterative finite difference technique has been used to numerically solve the problem of coupled consolidation and contaminant transport from a mineral waste layer.

Typical material properties were taken from Hellawell [3], and from Carrier et al [2], and are shown in table 1.

TABLE 1-- Material properties for the computer simulation.

parameter	waste	silt stratum & silt capping	sand capping
η	0.9	0.4	0.45
$D_{hl}\ (m^2/s)$	3.14×10^{-10}	1.5×10^{-9}	1.0×10^{-5}
$k\ (m/s)$	1.0×10^{-6}	1.0×10^{-5}	1.0×10^{-3}
$C_v\ (m^2/s)$	2.5×10^{-5}	N/A	N/A

3.7 Results

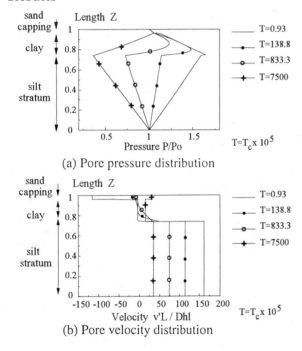

(a) Pore pressure distribution

(b) Pore velocity distribution

FIGURE 3 -- Output from the computer simulation for the situation with a sand capping

Figure 3(a) shows typical isochrones of total pore pressure as the consolidation process continues. It shows the initial creation of excess pore pressures due to the surcharge of the capping, and the self weight of the clay layer. With time these excess pressures dissipate, to a final equilibrium condition as consolidation ceases. Figure 3(b) shows the distribution of pore velocities. The apparent discontinuities on the boundaries are due to a difference in porosity of the layers. In particular it should be noted that initially the velocity in the capping layer and in the top of the waste layer is negative (i.e. upwards). This will have obvious effects on the movement of pollutants from the waste layer, as can be seen in figures 4(a) and 4(b).

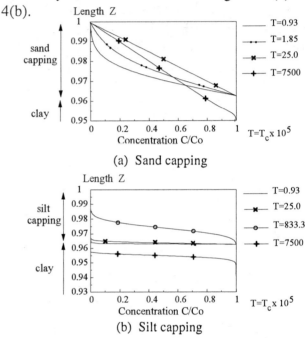

(a) Sand capping

(b) Silt capping

FIGURE 4 -- Comparison of concentration profiles at the capping-waste interface for two types of capping layer

Figures 4(a) and 4(b) show the concentration profiles in the capping and in the top of the waste. They show the comparison in profiles for a sand capping and a silt capping. All inputs were the same for the two scenarios, except for the properties of the capping layer.

It can be clearly seen in figure 4(a) that in the period when velocities are negative, the pollutant plume reaches the lake. However figure 4(b) shows that for a less permeable capping the velocities become positive before the pollutant plume reaches the overlying water, hence the less permeable layer acts as a barrier to pollutants.

4. Physical Modelling

4.1 Centrifuge Modelling

The mechanical behaviour of a prototype soil mass of depth l, under the earth's gravity, g, can be replicated in a small scale model of l/n experiencing a centrifugal force of ng. If the product of depth and acceleration is the same in the model and the prototype, then the stress distribution throughout the model will be identical to that throughout the prototype [5,6]. Movement of a pollutant in water bearing media is heavily dependent on the permeability. Since the permeability of a soil mass is related to the effective stress within the soil, then the achievement of identical stress states at homologous points in the model and prototype is essential. The centrifuge has been proved to be a valuable tool for the study of pollution transport problems [5,4].

The relevant scaling laws for centrifuge modelling of consolidation and associated contaminant transport processes, pertinent to the problem under investigation are ;

$$t_p = n^2 t_m \quad (7)$$
$$v'_p = \frac{1}{n} v'_m \quad (8)$$
$$c_p = c_m \quad (9)$$
$$u_p = u_m \quad (10)$$

Where m and p denote the model and prototype respectively, and n is the scaling factor.

4.2 Centrifuge Test

The Mistral is a small, commercially available, table-top centrifuge, which has been adapted for the purpose of soil mechanics modelling. It was used for the present study to provide preliminary data for future large scale centrifuge tests. A package has been used in which the boundary conditions are strictly controlled. At the top of the sample a constant head of water is applied using a drip and overflow system, and at the base there is free draining into a cavity, with breather pipes to prevent the build up of air pressure.

For this study speswhite kaolin clay, contaminated with a 0.1M sodium chloride solution and with an initial water content of 300% was used to simulate the waste layer. The underlying layer consisted of a stratum of 180 grade silica flour. The capping layer was not simulated in the centrifuge tests.

A miniature 4-pin resistivity probe was used in the underlying silt layer to detect the progress of the pollutant plume. A DRUCK PDCR81 pore pressure transducer was used in the waste layer to monitor the dissipation of excess pore pressures as consolidation occurred. A multiplexer was used to energise each probe individually, thereby avoiding interference between probes.

Before the test, the silt layer was consolidated in the centrifuge. The contaminated clay slurry was then poured over the silt at $1g$, and immediately the package was taken up to $14g$. The readings of the two probes and the temperature were recorded throughout the test. At the end of the test the position of each probe was determined by careful excavation.

5. Comparisons of the Numerical and Experimental Models

The finite difference computer program was executed assuming no capping layer, and using the parameters presented in table 1. The concentration and pressure profiles from the computer simulation were compared to the Mistral test results.

Figure 5 shows good correlation between the numerical and experimental data for the concentration at a particular point in the silt layer. The numerical curve is more square than the experimental curve, indicating a discrepancy in the hydrodynamic dispersion coefficients in the two models. Both models show the contaminant front passing over a period of 1.5 hours which is equivalent to 12 days 6 hours in prototype time.

The waste layer experienced consolidation to less than half its original thickness. It was therefore difficult to determine the exact location of the pore pressure probe in the waste layer. As a result the pore pressures for the two models differed by a factor of ten. No comparison can be made in the early stages since data acquisition commenced after the large initial pore pressures had dissipated. Pressures were shown to be negative in the waste, near the silt-waste interface by both models. The disparity between the experimental and numerical simulations may be accounted for by the assumption in the numerical analysis that strains are small (see section 6).

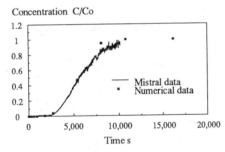

FIGURE 5 -- Comparisons of concentration at a point in the underlying silt for the numerical and experimental predictions

6. Conclusions

Numerical and physical modelling of contaminant transport from a consolidating waste layer has been undertaken. A number of interesting observations have been made.

Large excess pressures are generated as soon as the waste is placed, these pressures dissipate rapidly causing pore fluid flow into the capping layer. The capping layer may act as a barrier to pollutants depending on its permeability. Good agreement was obtained between the numerical and experimental models for the concentration of contaminant in the silt layer.

At present, large scale centrifuge model tests and more complex, large strain, analyses are being undertaken to complement the current findings.

Notation

C	c	concentration
C_v		coefficient of consolidation
D_{hl}		dispersion coefficient
i		hydraulic gradient
k		permeability
L		depth of soil layer
P	p	total pore pressure
T	t	time
U	u	excess pore pressure
V'	v'	pore fluid velocity
x		Cartesian co-ordinate positive downwards
z		Cartesian co-ordinate positive upwards
η		effective porosity
γ_w		unit weight of the pore fluid

References

[1] Bear J. (1972) 'Dynamics of Fluids in Porous Media.' pub. American Elsevier

[2] Carrier W.D., Bromwell L.G., Somogyi F. (1983) 'Design capacity of slurried mineral waste ponds.' Journal of Geotechnical Engineering, A.S.C.E., Vol. 109 No. 5

[3] Hellawell E.E. (1991) 'Modelling of clean-up operations in contaminated land.' Report for a Certificate of Postgraduate study, University of Cambridge, England

[4] Hensley P.J. (1989) 'Accelerated physical modelling of transport processes in soil.' PhD dissertation, University of Cambridge, England

[5] Hensley P.J. Savvidou C. (1993) 'Modelling of coupled heat and contaminant transport in groundwater.' International Journal of Numerical and Analytical Methods in Geomechanics, Vol. 17, No. 7, 493-527

[6] Schofield A.N. (1980) 'Cambridge geotechnical centrifuge operations.' Geotechnique 30:3:227-268

[7] Terzaghi K. (1943) 'Theoretical Soil Mechanics.' 5th ed. John Wiley, New York, N.Y.

Geotechnical Characterization of a Clayey Mine Waste

S. Rampello
University of Rome "La Sapienza"

V. Pane
University of Perugia

G. Calabresi
University of Rome "La Sapienza"

O. Bianco
ENEL S.p.A. - Construction Division

Abstract: A power plant is currently under constrution over a clayey waste area at Pietrafitta, Central Italy. The waste and the natural soils below it are undergoing a slow subsidence and exhibit high excess pore pressures. Attention is focused on the extensive site investigation and monitoring so far carried out, and on the interpretation and future trends of the observed phenomena.

Introduction

A new power plant is currently under construction in the proximity of the village of Pietrafitta, in central Italy. The main structures and facilities of the plant are to be founded in a waste area, approximately 1400 x 700 m in extent, gently sloping in the northern direction towards a 2 million m^3 water reservoir. A schematic plan-view of the plant indicating some of the major plant areas is shown in Fig.1; the reservoir southern edge, not shown in the figure, is located approximately 600 m north of the Coal Deposit Area. The waste, 15 to 25 m in thickness, is constituted by a silty clay back-fill which is experiencing a slow subsidence due to self-weight consolidation; this poses various problems for proper design, construction and maintainance of the power plant.

This paper briefly describes the site investigation and monitoring carried out to date for a geotechnical characterization of the site, and the interpretation and possible characterization of the consolidation process. Other aspects of the study, such as the artificial improvement of the mechanical characteristics of the waste, are examined elsewhere [1].

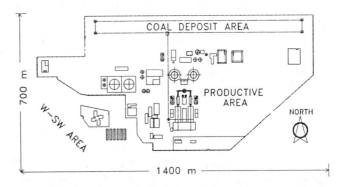

Figure 1 - Schematic lay-out of the power plant

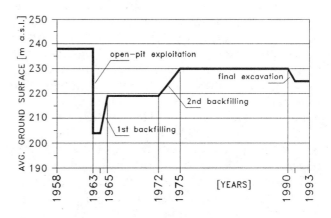

Figure 2 - Recent history of the site

TABLE 1 - Field instrumentation

Instrument	Type	Characteristics	total	non-functioning
piezometers	multipoint (Westbay)	packers and pressure transducers	14	0
	ceramic transducer	pushed in place by CPT rods	49	21
	zero control	strain gage	102	3
	Maihak	vibrating wire	38	1
	Geonor	vibrating wire	1	0
	modified Casagrande	manual or remote reading	37	2
	stand pipe		34	1
settlem. gauges	multipoint (BRS)	magnetic	27	1
	two-point (AP)	top and bottom of waste	37	0
	extensometer	rod-type, 3 point	8	0
inclinometers		accelerometric transducer	42	5

Recent History of the Site and Soil Profile

For a proper understanding of the observed phenomena it is necessary to gain some information about the history of the site and the origin of the waste. The area has in fact been subjected to a rather complicated recent history, which is qualitatively depicted in Fig.2 and outlined as follows.

The exploitation of the site started in the early sixties, when the natural clay present at the site was open-pit excavated down to depths of 30-35 m for the extraction of an underlying layer of lignite (brown coal). Once the exploitation activities were concluded, about 15 m of the excavated clay were back-filled in 1964-1965, while the remaining 5-10 m were back-filled ten years later. As shown in Fig.2, in 1975 the resulting ground surface was considerably below the original one. In 1986 the Italian State Electricity Board (ENEL S.p.A.) planned the construction of a new power plant over the existing waste; therefore the whole area had to be levelled and graded to accomodate plant machinery and facilities. The first earth-moving works began in 1990-1991, when in the major part of the site the upper 5-10 m of the waste were re-excavated down to the present ground surface (elevation 225 m). At the same time, a comprehensive site investigation and monitoring programme commenced.

The above excavation/back-filling activities led to the present soil profile outlined below, in order of increasing depth :
- clayey waste (Soil S1), medium to stiff, with lignite fragments, bedded with rare lenses of sandy silt and ash (thickness: 15-25 m);
- lignite (Soil S2), constituting the unexploited lowermost portion of the original lignite stratum (thickness: 3-4 m);
- lacustrine peaty clay (Soil S3) and blue silty clay (Soil S3'), stiff, bedded with thin lenses of sand (thickness: 25-30 m);
- stratified silty sands (Soil S4) and sandy silts (Soil S5) of lacustrine origin, occasionally bedded with layers of silty clay (thickness: > 60 m).

Site Investigation and Monitoring

The site investigation carried out from 1990 to date is extensive. It includes more than 200 borings, 200 Cone Penetration Tests (CPT's), laboratory tests on undisturbed samples, and several full-scale (prototype) tests. In addition, the site is currently instrumented and continuously monitored by means of piezometers, shallow and deep settlement gauges, inclinometers, and topographic surveys. Table 1 provides a partial indication of the extent of the field instrumentation; it is noted that 7 different types of piezometers and pressure transducers, as well as 3 different types of settlement gauges, have been installed at the site. The amount of

TABLE 2 - Index and consolidation properties of fine-grained soils

Soils	γ (kN/m^3)	e_0	w (%)	w_L (%)	I_P (%)	I_L	C_c	C_s	OCR	c_v (m^2/sec)	K (m/sec)
S1	18.2	1.05	37.6	52.0	26.9	0.46	0.36	0.10	0.9	2.7 10^{-8}	6.0 10^{-11}
S2	14.5	1.99	78.0	95.0	42.0	0.78	0.83	0.22	3.6	1.2 10^{-7}	8.3 10^{-11}
S3	17.2	1.25	47.0	78.0	44.0	0.29	0.61	0.18	1.2	2.2 10^{-8}	2.1 10^{-11}
S3'	19.8	0.70	26.1	57.0	33.4	0.10	0.30	0.11	1.0	2.5 10^{-8}	1.4 10^{-11}

γ = bulk unit weight; e_0 = in-situ void ratio;   w = water content; w_L = liquid limit;
I_P = plasticity index I_L = liquidity index; C_c = compression index; C_s = swelling index;
OCR = overconsolidation ratio c_v = vertical coefficient of consolidation;   K = vertical coefficient of permeability

instruments which have failed is also indicated in the same Table; this provides an indication of the reliability and durability of each instrument.

Characterization of the Clayey Waste

The processes and activities which have generated the clayey waste provide an insight to the mechanical and consolidation properties of this material. Throughout the open-pit exploitation of the site the natural clay was excavated in the form of blocks, a few decimeters in size, which were progressively laid down on the ground surface for a considerable time before back-filling. Although the documentation regarding the exploitation phases are rather scarce, it is quite likely that during this time the clayey blocks were subjected to considerable water absorbtion, swelling and superficial softening, and that such softening increased with decreasing block-size and increasing exposure time. Upon back-filling, the inner portions of the larger blocks retained the consistency of the parent natural clay, while the softened outer portions provided a partial welding among the blocks under the overburden pressure. Voids among the blocks are partially filled by sand, silt and ash in an irregular fashion. Thus the resulting waste deposit is highly heterogeneous on a macroscopic scale, and characterized by an irregular net of discontinuities, spaced a few decimeters apart. Table 2 indicates the mean values of the index and consolidation properties resulting from laboratory tests on the clayey waste and on the underlying natural clays; the listed values of OCR indicate the normally consolidated state of such soil types.

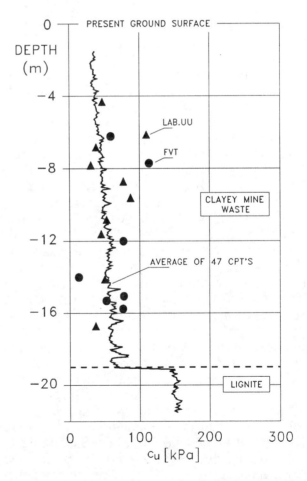

Figure 3 - Profile of undrained shear strength in the Productive Area

An indication about the strength characteristics of the waste can be inferred from Fig.3, which reports the variation of undrained shear strength c_u with depth in the Productive Area as determined from laboratory triaxial tests (lab UU), CPT's and field vane tests (FVT's). It is noted that the laboratory and vane c_u values in the clayey waste exhibit a wide scatter and do not show a definite trend with depth, reflecting the origin of this material. In particular, the lowest values of c_u may be attributed to existing contacts, at the sample scale, between the excavated blocks. On the other hand, the values of c_u from CPT's show a moderate increase of the undrained strength with depth.

Consolidation Settlements

Although the values of c_u shown in Table 2 and Fig.3 are typical of a slightly overconsolidated deposit, the field instrumentation unambiguously reveals that the waste is undergoing a slow consolidation process, and that high excess pore pressures exist, at least in the upper 50 m of the subsoil. A typical profile of piezometric head (H) down to a depth of 135 m below ground is shown in Fig.4; the field data refer to 37 measuring points along 5 instrumented boreholes in the Cooling Towers Area, approximately 5000 m² in extent. The measured values of H at the bottom of the waste are about 10 m higher (H=235 m) than present ground, and decrease moving upwards within the waste until reaching a value equal to the ground surface (H=225 m). High gradients of H also exist in the natural clays underlying the lignite layer. The data clearly show that such gradients disappear rather abruptly in the sandy-silty soils encountered below elevation 170 m; it is interesting to note that in such coarse-grained soils the average piezometric head measured throughout the site is equal to 207 m, and that this elevation corresponds to the water level of the northern water-resevoir mentioned above. Figure 4 also shows that the maximum values of H are close to the theoretical values corresponding to rapid deposition of material in undrained conditions (H_{max} = 235 m). This suggests that ongoing consolidation of the waste is proceeding at an extremely slow rate.

The extensive data so far collected give rise to a number of interpretations and questions. The first question regards the long-term (equilibrium) distribution of the piezometric head at the site, H_f. This is of considerable importance since it determines the magnitude of the present excess pore pressures and of the future settlements. Three possibile H_f profiles are shown by the dashed lines in Fig.4. Profile labeled as c) corresponds to hydrostatic equilibrium conditions below a groundwater table located at 207 m; this situation, however, is unrealistic since it leads to unexplainingly high values of excess pore pressures in the proximity of the free-draining ground surface. Profiles labeled as a) and b), which are deemed more realistic, correspond to a downward steady-state flow within the fine grained soils (Soils S1 through S3'), fed by rainfall and superficial water.

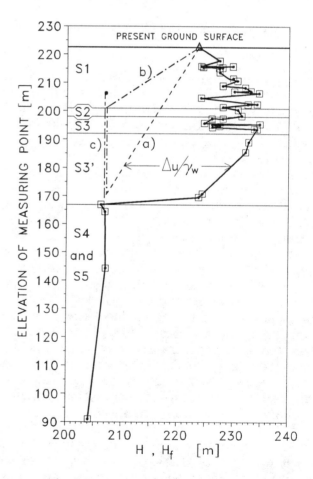

Figure 4 - Present and long-term profiles of piezometric head in the Cooling Towers Area

TABLE 3 - Progress of consolidation settlements

Area	Layers	w_f (cm)	U_{avg} (%)	t_{70} (years)	$\Delta w/\Delta t$ (cm/year) predicted	$\Delta w/\Delta t$ (cm/year) measured	R = meas/pred
Productive Area	S1	30	34	35	0.62	0.32	0.52
	S3-S3'	21	46	48	0.25	0.07	0.28
Chimney	S1	64	25	40	1.45	0.93	0.64
	S3-S3'	31	39	56	0.38	0.23	0.59
Cooling Towers	S1	67	17	43	1.66	1.92	1.16
	S3-S3'	44	16	76	0.74	0.40	0.54
Coal Deposit	S1	71	48	25	1.41	1.48	1.05
	S3-S3'	33	42	52	0.40	0.36	0.90
Oil Tanks	S1	40	38	133	0.23	0.32	1.39
	S3-S3'	17	44	50	0.20	0.15	0.75

The second question regards the high values of H observed in the natural clays (S3 and S3'); in fact, Fig.2 makes it clear that for such materials the overall loading history of the site has led to a ground surface and to present overburden stresses which are sensibly lower than the ones existing prior to the site exploitation, and that the duration of the unloading/excavation phases has been much shorter than the reloading phases. These observations suggest that the values of the piezometric head presently existing in the natural clays cannot simply be attributed to mechanical loading processes, and that the transient downward seepage observed in these materials is a direct consequence of the ongoing consolidation of the overlying waste.

Determination of the present excess pore pressures from the piezometric profiles $[\Delta u = \gamma_w (H-H_f)]$ allows the final consolidation settlements in the different areas of the power plant to be evaluated. Considering the one-dimensionality of the problem, and assuming the equilibrium piezometric profile denoted by a) in Fig.4, the surface settlement (w_f) due to Δu dissipation may be calculated as:

$$w_f = \Sigma \ \Delta z/(1+ e_0) \ C_c \log(1+ \Delta u/\sigma'_v) \quad (1)$$

where the summation is extended to both the clayey waste and the underlying natural clays. The predicted values of w_f are listed in Table 3 for several important areas of the power plant; they indicate that the final consolidation settlements are likely to range between 50 and 100 cm, and that the contribution to these settlements deriving from the natural clays (S3 and S3') is relevant.

As usual when dealing with macro-structured deposits such as the one under consideration, the calculations of the progress of settlements with time are considerably more uncertain. As a first approximation, such calculations have been performed assuming mean laboratory values of the coefficient of consolidation (c_v) for the clayey waste and for the natural clays, and considering the lignite layer as a draining boundary. A representation of the progress of consolidation is provided in Table 3, which shows the estimated values of the present average degree of consolidation (U_{avg}), the time necessary to reach 70 % consolidation (t_{70}), and the average settlement rate in the next 10 years ($\Delta w/\Delta t$). The long time necessary to reach a significant degree of consolidation is noted from the Table; the calculated values of t_{70} range between 40 and 80 years for most areas of the plant, and the values of t_{90} are about twice as long.

It is pointed out that the calculations reported in Table 3 were made in 1991. At that time, the only information regarding surface settlements was provided by topographic surveys of limited reliability. In the last two years, however, good quality settlement data has become available from

TABLE 4 - Vertical drains of the Trial Areas

Trial Area n.	area extent (m^2)	drain type	d (cm)	spacing (m)	n_{piez}	t_{obs} (days)	U_l (%)	c_h (m^2/sec)
1	6 x 6	sand	25	3.0	3	218	62-87	3.5 10^{-7}
2	6 x 6.5	jetted	20	2.5	1	144	85	4.4 10^{-7}
3	15 x 15	wick	5	1.5	4	160	10-22	3.3 10^{-8}

d = drain diameter; n_{piez} = number of piezometers within the waste; t_{obs} = observational period

the two-point (AP) settlement gauges extensively installed throughout the site in 1990. This provides a meaningful comparison between measured and predicted rates of settlement for the different areas of the power plant; from Table 3 it is noted that their ratios (R) generally range between 0.5 and 1.4. Such agreement is deemed to be surprisingly good, if one recalls the interpretations and approximations invoked in the analysis. It is hoped that the future field data will allow the analysis to be refined.

Vertical Drains

As described elsewhere [1], several solutions have been employed to limit the amount of consolidation settlements and/or to hasten the progress of consolidation. It is worth mentioning the comprehensive field tests carried out in 1990-1991 to ascertain the effectiveness of vertical drains in reducing, within reasonable times, the high excess pore pressures observed in the clayey waste. In particular, three Trial Areas (Trial Areas 1 - 3) of limited extent were selected to investigate the efficiency of various types of vertical drains, namely, mechanically perforated drains (sand drains), hydraulically perforated drains (jetted drains) and prefabricated drains (wick-drains). Table 4 shows the characteristics of the Trial Areas and of the drains, the observational period, and the number of pore pressure transducers installed for monitoring of the waste. For all cases, the distance between the transducer and drain was equal to the radius of the equivalent cylinder of influence of the drain, i.e., equal to the maximum horizontal drainage path.

The local degree of consolidation [$U_l = U_l(t)$] achieved at the transducer measuring point has been estimated as:

$$U_l = \Delta u / \Delta u_o = (H_o - H) / (H_o - H_f) \qquad (2)$$

where $H_o(z)$ and $H_f(z)$ are the initial and final values of the piezometric head, and $H(z,t)$ is the current value at the generic time t. This allows a back-calculation of the horizontal coefficient of consolidation of the clayey waste, c_h. The calculations have been carried out using Barron's [2] free-strain solution and are summarized in Table 4. In particular the Table shows, for each Trial Area, the range in the values of U_l measured by the various transucers at the end of the observational period, and the values of c_h corresponding to the lowest measured value of U_l. It is noted that for the mechanically and hydraulically perforated drains (Trial Areas 1 and 2) the field values of c_h are about one order of magnitude greater than the average value of c_v measured in the laboratory (c_v=2.8 10^{-8} m^2/sec); this confirms the significant macrostructure of the waste deposit. On the other hand, for the prefabricated drains (Trial Area 3) the field and laboratory values of the coefficient of consolidation are comparable; this apparently contradictory result may be a consequence of the driving installation procedure for this type of drain and the resulting smear effect, and of the small diameter of the drain.

Following the promising results obtained from the

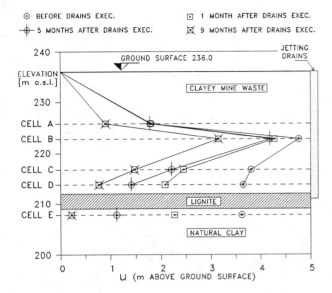

Figure 5 - Pore pressure isochrones after drain installation in the W-SW Area

field tests, sand and jetting drains have been subsequently installed in some critical areas of the plant. Figure 5 shows the evolution of excess pore pressure isochrones recently measured by a multipoint piezometer located in the W-SW area. A direct comparison with adjacent untreated areas reveals the beneficial effect of the jetted drains, and preliminary calculations substantially confirm the field values of c_h obtained from the Trial Areas.

Conclusions

The mechanical behaviour of the thick deposit of clayey waste is very complex and its characterization was a difficult task. Due to the heterogeneity and the macrostructure of the man-made deposit, the prediction of the consolidation process had a low reliability. The great effort in instrumenting and monitoring the area with an exceptionally high number of measuring points allowed an understanding of the ongoing consolidation process and the design the proper procedures for controlling its effects on the plant facilities.

Acknowledgements

We wish to thank Mr. P. Gigli, Mr. L. Samorì and Mr. F. Pirrone of ISMES S.p.A. for the valuable help in collecting some of the field data and preparing plots and drawings.

References

[1] Calabresi G., Pane V., Rampello S., Bianco O. 1994. *Geotechnical problems in construction over a thick layer of mine waste.* First Int. Congress on Environmental Geotechnics, Edmonton, Canada, July 1994.

[2] Barron R.A. 1948. *Consolidation of fine-grained soils by drain wells.* Trans. ASCE, vol.113, pp.718-754.

Computer Modeling to Define the Extent of Groundwater Contamination at a Coal Refuse Disposal Facility

Krishna R. Reddy, Assistant Professor
Department of Civil Engineering, Mechanics & Metallurgy, University of Illinois, Chicago, Illinois, USA

Jeffrey C. Schuh, Vice President
Patrick Engineering Inc., Glen Ellyn, Illinois, USA

ABSTRACT

This paper presents the selection and construction of groundwater flow and contaminant transport models to assist in defining the extent of groundwater contamination so that a Groundwater Management Zone (GMZ) could be established at an existing coal mine facility in southern Illinois. The models will ultimately be used to determine an efficient extraction well system for the control and treatment of contaminated groundwater. The mine has been in operation since 1977 and has utilized groundwater and surface water in coal processing operations. The refuse resulting from the coal processing is disposed on site by constructing perimeter embankments with coarse refuse and then depositing fine refuse into the impoundment formed by the embankments. The uppermost aquifer has been impacted by leachate from the refuse piles and a GMZ needed to be established for groundwater control and treatment. Computer modeling of groundwater conditions was performed to assist in defining the extent of contamination and was used to determine locations for long term monitoring of groundwater conditions.

Introduction

A coal mine, located in southern Illinois, has been in operation since 1977. The raw coal is processed on-site and then transported by trains to utility companies. Refuse (waste) generated from coal processing operations is disposed on-site at two areas known as Refuse Disposal Area No.1 (RDA-1) and Refuse Disposal Area No.2 (RDA-2). Refuse consists of two types: coarse refuse and fine refuse (slurry). The coarse refuse is transported in off road haul trucks while the fine refuse is conveyed through pipes which are directed to the disposal areas. The refuse is disposed by constructing perimeter embankments with coarse refuse and then depositing fine refuse into the impoundment formed by the embankments. The embankments are approximately 60 feet high. Drainage ditches and holding ponds along the periphery of the refuse disposal areas control stormwater runoff. The mine facility includes two artificial lakes, known as Fresh Water Lake (FWL) and Recirculation Lake (RCL). Clarified water from the refuse disposal areas and water from RDA perimeter drainage ditches and holding ponds is stored in the RCL. The remaining site drainage is directed into the FWL. Controlled discharges have been permitted from the FWL and RCL into the adjacent Grassy Branch. Figure 1 shows the locations of the refuse disposal areas, lakes, holding ponds, Grassy Branch and Sugar Creek.

Water quality in the refuse impoundments, holding ponds, FWL, RCL and Grassy Branch has been monitored on a regular basis. Groundwater quality has been also monitored using a network of monitoring wells. The monitoring well data indicated that the groundwater within the uppermost aquifer has been impacted by the refuse disposal areas and other on-site sources (including the FWL and RCL). The groundwater quality (specifically concentrations of chloride, sulfate, total dissolved solids, iron and manganese) in some parts of the site exceed the regulatory maximum allowable values. In an attempt to reduce the degradation of groundwater, a remedial program consisting of three pumping wells was implemented in 1981. Figure 1 shows the locations of all monitoring wells at the project site as well as the pumping well locations.

In spite of the continued operation of the pumping wells, degradation of groundwater was observed in some of the monitoring wells. Because of increased groundwater contamination, the Illinois Environmental Protection Agency required that the extent of contamination be determined, a Groundwater Management Zone (GMZ) be defined, and a plan be developed and implemented to control and ultimately remediate the contaminated groundwater.

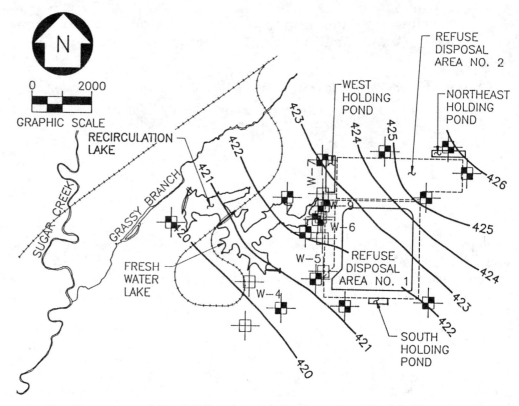

Figure 1. Site Features and Static Piezometric Surface Contour Map

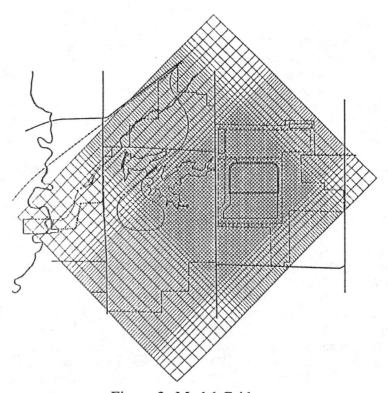

Figure 2. Model Grid

This paper briefly presents the site conditions including site hydrogeology and groundwater quality at the mine site. Details on modeling methodology, model calibration, and model results are then presented. The usefulness of the modeling to predict the plume location and the benefits of modeling in the determination of a GMZ are then discussed.

Site Conditions

Site Geology. The site geology was defined using information from over 125 borings drilled at the site and published literature from Illinois State Geologic Survey. The site geology has been generalized into six geologic units denoted by Units A through F as shown in Table 1.

TABLE 1-- Summary of geologic units

Unit	Thickness (ft.)	Description
A	12-22	Silty clay to clayey silt
B	0-35	Sand with fine gravel
C	0-48	Silty clay
D	0-10	Sand or fine gravel
E	10-50	Shale
F	125-225	Sandstone

Site Hydrogeology. A limited number of borings drilled at the site were converted into water supply wells, monitoring wells or remediation wells. These wells were used to measure water levels and to perform aquifer tests. Based on this data, three aquifers are located within the project area. They are: (1) Pearl Sand Aquifer or Unit B, (2) Lower Sand Aquifer or Unit D, and (3) Trivoli Sandstone Aquifer or Unit F. The other units (Units A, C and E) are low permeable units and act as aquitards or acquicludes.

The Pearl Sand aquifer is the uppermost aquifer that has been affected by the mine operations. The aquifer exists under confined conditions, yields moderate to large amounts of water, and it is considered to be the major aquifer at the site. The other two aquifers have not been impacted by refuse leachate because Units C and E serve as hydraulic barriers to the downward flow of groundwater. Therefore, only groundwater flow within Units A and B were analyzed for the determination of the GMZ.

Based on laboratory triaxial tests, the vertical hydraulic conductivity of Unit A averaged 0.032 ft/day. Based on pump tests, the transmissivity of the Pearl Sand aquifer (Unit B) ranged from 5760 ft^2/day to 15700 ft^2/day. The storativity of this unit ranged from 1.1×10^{-4} to 6.7×10^{-2}. Water level measurements were utilized to interpret the flow conditions within the Pearl Sand. Groundwater flow is generally in the northeast to southwest direction as shown in Figure 1.

Groundwater Quality. Water quality sampling and testing has been performed at monitoring wells, remediation wells and surface water monitoring locations since the mine opened. The parameters tested include chloride, sulfate, total dissolved solids (TDS), manganese, iron, iron bacteria, pH, dissolved oxygen, ammonia, nitrate (as N), total alkalinity, total acidity and total hardness. Water quality data suggested that the refuse disposal areas as well as the on-site lakes have impacted the water quality within the Pearl Sand aquifer.

Modeling Methodology

Modeling was performed in an effort to simulate the leaching from the refuse disposal areas and leakage from the on-site lakes into the Pearl Sand aquifer. Calibrated models would be used to select new monitoring well locations for the physical determination of the GMZ. Models would also be used to evaluate groundwater collection and treatment options.

Models Used. The computer models used for this study were: (1) Modular Three-Dimensional Finite Difference Groundwater Flow Model, known as MODFLOW, developed at the United States Geologic Survey (USGS) by McDonald and Harbaugh in 1988, and (2) Modular Three-Dimensional Contaminant Transport Model, known as MT3D, developed with funding from the United States Environmental Protection Agency (USEPA) by Papadopulos & Associates in 1990. MODFLOW simulates groundwater flow conditions, and MT3D simulates the contaminant transport (using the flow conditions determined by MODFLOW). Complete details on the selected models are available in the respective documentation manuals [1,2].

Model Grid. Both the MODFLOW and MT3D models are based on the finite difference method and require a grid system. The size and orientation of the grid system used for this study is shown in Figure 2. The grid was developed to encompass the entire mine facility. The grid spacing ranged from 125 feet near the refuse disposal areas and lakes to 500 feet at the model boundaries. The selected variable grid spacing provides higher resolution at the potential source areas (refuse disposal areas and the lakes), enabling the model to simulate the conditions accurately in the areas of concern.

The orientation of the model grid was selected to simulate groundwater conditions with minimal effect from boundary conditions. The model size was limited to the project site to avoid excessive extrapolation of hydrogeologic data.

Model Layers. The purpose of this study was to model contaminant migration and dispersion from the refuse disposal areas, lakes and ponds through Unit A and into Unit B. As such, the physical and hydrologic properties of Units A and B were needed for modeling of the Pearl Sand aquifer. The thicknesses of Units A and B are variable at the site, and interpreted thickness contour maps were used in the models.

Only one layer representing the Pearl Sand aquifer (Unit B) was used in the model. The thickness and hydraulic conductivity of Unit A was incorporated as a conductance term. This representation was used because: (i) leakage through Unit A can be represented by general-head boundaries with known conductance values, and (ii) consideration of one layer instead of two layers is computationally efficient. This representation also allowed the incorporation of leachate seepage through the refuse disposal areas.

Because of the dynamics of the groundwater regime in areas where pumping wells are located, the Pearl Sand aquifer (Unit B) may locally change from confined to unconfined conditions. Also, the transmissivity of the aquifer is variable because of the variable thickness of the aquifer at the site. Based on these considerations, the aquifer was specified as Layer-type 3 [1].

Model Boundaries. Review of the available geologic and water level data indicates that the Pearl Sand (Unit B) aquifer is not in direct hydraulic communication with either creek located to the west and to the northwest of the site. No other hydrogeologic boundaries are located near the model boundaries. For the purpose of modeling, artificial (or distant) boundaries were utilized.

Based on groundwater level measurements in the existing wells, groundwater flow within the Pearl Sand aquifer occurs generally from northeast to southwest (Figure 1). While not shown on Figure 1, the groundwater flow is altered in the vicinity of the on-site pumping wells due to pumping operations.

For the groundwater flow model (MODFLOW), the northeast and southwest boundaries were set as specified head values to allow discharge from the model towards discharge zones to the southwest. The other two boundaries of the model were assumed to be no-flow boundaries based on the flow pattern (Figure 1). Based on historical water level measurements (1991-1993) in the monitoring wells, piezometric levels for the Pearl Sand fluctuate seasonally, with the change in water levels on the order of 2 to 3 feet. This fluctuation in water levels was accounted for in the model by specifying time-dependent constant head boundaries. Plots of water levels versus time for the upgradient wells were used in setting the northeast constant head boundary.

Historical water quality data were used to establish initial chemical concentrations in the Pearl Sand aquifer for the contaminant transport model (MT3D). The northeast and southwest boundaries were set as constant concentration boundaries at background groundwater concentrations. The other two boundaries (no-flow boundaries) were assumed to be no-mass flux boundaries.

Sources. The following sources were incorporated into the models: Refuse Disposal Area No.1, Refuse Disposal Area No.2, Fresh Water Lake, Recirculation Lake, West Holding Pond and Northeast Holding Pond (Figure 1). These sources were modeled as general-head boundary conditions. Conductance values for each model cell within the source areas were calculated using the following equation and were then input to the model.

$$C = K \frac{LB}{T} \qquad (1)$$

where C=conductance (ft^2/day), K=hydraulic conductivity of Unit A (ft/day), L and B=dimensions of model cell (feet), and T=thickness of Unit A (feet). In the refuse disposal areas, the thickness and hydraulic conductivity of both the coarse refuse and Unit A were used to calculate equivalent conductance values. The general-head and water quality for each source were specified based on measured data.

Sinks. Existing water supply and mitigation pumping wells at the site were included as sinks in the modeling. The wells considered in modeling were: W-4, W-5, W-6, W-7 and W-9. The locations of these wells are shown in Figure 1. The measured pumping rate for each well was averaged for each quarter and this average rate was input to the model.

Recharge. Recharge was specified in the model to simulate percolation from precipitation. The recharge rate was calculated using the recorded precipitation and temperature data, and estimated runoff and evapotranspiration rates.

Evapotranspiration. Evapotranspiration losses were modeled using the procedures described in the MODFLOW manual by specifying the maximum evapotranspiration rate and the extinction depth.

Contaminant Constituents. Chlorides, sulfates, total dissolved solids (TDS), total iron and manganese are the primary indicators of groundwater contamination at the site. The initial concentrations of chloride, sulfate, TDS, total iron and total manganese in the Pearl Sand aquifer (Unit B) used to establish the initial chemical concentrations were based on the fourth quarter 1991 groundwater quality data obtained for the existing monitoring wells within the study area.

Model Calibration

Because of continued waste disposal and other operational activities, transient hydrogeochemical conditions exist at the project site. Prior to calibration of the models for transient conditions, steady state conditions needed to be defined. Consequently, a steady state analysis was performed for the first quarter and the results were then used to calibrate the models for transient conditions for subsequent quarters.

The model calibration was initiated with time zero being the beginning of the first quarter of 1992. Since this time, complete data on mining activities as well as groundwater quality and pumping well operations was available for model construction. The models were calibrated using the data for the first quarter 1992 through the fourth quarter 1992.

A good match between the measured concentrations and predicted concentrations was observed at several monitoring well locations. However, poor predictions were made at a few locations. This may be attributed to several simplifications and assumptions of the site hydrogeochemical conditions. The assumptions included the following:

1. The Pearl Sand aquifer is assumed to be isotropic and homogeneous.

2. The vertical distribution of contaminant concentrations within the aquifer is neglected. The concentrations at any location represent the averaged conditions over the aquifer thickness.

3. The quarterly averaged conditions are simulated, therefore, the short term effects of intermittent pumping and waste disposal operations are not accounted for.

4. No attenuation (other than dispersion) of chemical parameters is assumed to occur within the aquifer.

5. Source concentrations are assumed to represent the breakthrough concentrations into the aquifer.

6. The Pearl Sand aquifer, a continuum geologic medium, has been represented by discrete cells, and the models have been simplified to represent the averaged conditions within each model cell.

Results and Discussion

The models as calibrated show that groundwater within the Pearl Sand aquifer is being impacted by seepage from all the on-site sources. The initial contamination observed at the site appears to have occurred from RDA-1 and the on-site lakes. Breakthrough concentrations from RDA-2 and the holding ponds appear to be starting to impact groundwater. RDA-1 will remain a potential source of contamination due to the large quantity of waste deposited in this area.

The models calibrated in this study were used to predict the extent of groundwater contamination in the Pearl Sand aquifer for the third quarter 1993. The predicted concentrations of chloride and TDS are shown in Figure 3.

The developed models represented the best estimate of contaminant plume movement within the aquifer. Because of the limited information available, it was not expected that the models would predict the exact location of the plume fringe. The sole purpose of the initial modeling was to provide guidance on the selection of new monitoring well locations and to provide a model which could be refined as more data is collected.

The use of computer models to predict the contaminant plume location proved to be beneficial in that monitoring well locations could be selected based on simulated site conditions rather than by judgement only. As computer modeling was ultimately needed to evaluate remediation options, construction of the models early in the groundwater evaluation allowed for the continued calibration of the models as additional data was collected. This approach to modeling hydrogeological conditions resulted in a model which accurately predicted the plume fringe and reduced the number of monitoring wells needed to define the GMZ.

 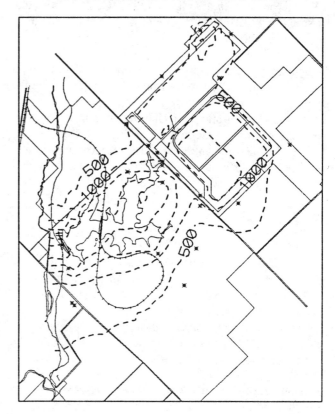

Figure 3. Model Predictions

Summary

The available hydrogeochemical data has been used to construct a groundwater flow model (MODFLOW) and a contaminant transport model (MT3D) for a coal mine facility to determine the extent of contamination and to ultimately evaluate remediation options. The models were calibrated using historical groundwater quality data obtained for the last quarter of 1991 and all four quarters of 1992. The models were then used to predict the extent of groundwater contamination and to optimize the number of monitoring locations for long term monitoring of groundwater conditions.

References

1. McDonald, M.G., and Harbaugh, A.W. (1988), "A Modular Three-Dimensional Finite Difference Ground-Water Flow Model," USGS TWRI, Chapter 6-A1, pp.586.

2. Zheng, C. (1992), "MT3D: A Three-Dimensional Transport Model," Version 1.8, S.S. Papadopulos & Associates, Inc., Bethesda, Maryland.

Water Resisting Barriers at Coal Mining Activity Sites

Yakov M. Reznik
Commonwealth of Pennsylvania, Department of Environmental Resources, Pennsylvania, USA

ABSTRACT   Efficiency of low-permeable water resisting barriers in many cases must be increased. The present "know-how" experience demonstrates that in reality, the defects of constructed and/or installed liners create certain technical problems. Although the design and construction of these structures became an undesirable investment, environmental protection considerations continue to dominate the further development of improved water resisting barriers. The selection of a certain type of low-permeable water resisting barriers (liners, covers, slurry walls, etc.) depends not only on project requirements, but also site conditions, and technical qualities of the proposed barrier material. The cost-effective calculations as well as the relative impact of the barrier on the environment must be seriously considered prior to making a decision. The availability of an ideal material for the construction of a water resisting structure is questionable. Unavoidable defects of those structures must be considered and included in all final evaluations of proposed water protecting systems. In addition, utilization of industrial by-products as well as recycling of some used construction materials is a very important theoretical and practical aspect of the environmental protection strategy. The paper presents examples of liners constructed at coal mining activity sites, analyzes errors made during construction stages, recommends measures required to avoid similar construction mistakes in the future, and discusses environmental advantages of different engineering decisions in minimizing groundwater contamination.

INTRODUCTION

The common recognition of the necessity to prevent groundwater contamination is undisputable. The general understanding of the above mentioned problem opens the door for a friendly discussion of existing regulatory requirements and their role in the fight for the clean environment.

Pennsylvania Rules and Regulations incorporate special provisions for the safe disposal of different waste materials. Although the coal mining industry is not associated with generation/accumulation of hazardous wastes, contamination of the Commonwealth waters is attributed in many cases to the past or present coal mining activities. Undesirable contamination sources have been observed not only at the abandoned deep/surface mine sites, inactive coal refuse disposal areas, and unreclaimed processing facilities, but also in the vicinity of active mines, coal preparation plants, house coal yards and coal distribution centers.

Coal refuse is a by-product of coal mining processes and comprises rocks and minerals separated from coal. Improper disposal of this material and neglecting the construction of water resisting barriers which divert uncontaminated water away from disposal sites causes aquifer degradation and jeopardizes biological equilibrium of surrounding areas.

Chapters 86, 87, 89, and 90 of Pennsylvania Code (Title 25. Environmental Resources) contain general rules for the design of coal mine waste disposal sites. They outline major requirements and special procedures for minimizing surface and groundwater contamination. Rules and Regulations contain specific requirements for technical evaluations of chosen sites which must be based on laboratory/field soil and rock test results. Accepted rules increase costs of design and construction procedures which ensure a compliance of proposed and/or existing coal mining activity sites. A familiarity with specific site geomorphologic, geologic and hydrogeologic conditions at selected areas as well as an evaluation of on-site soil/rock properties allows one to consider a wide range of design alterternatives which leads to inexpensive and enviromentally safe engineering solutions. This paper contains several case studies resulting from cooperative efforts of regulatory agency and coal opeperators who achieve necessary production goals without jeopardizing established standards of groundwater quality.

CASE STUDIES

Site No. 1

A company had submitted a permit application for a coal preparation and loading facility site located along a river bank. The area in question was developed in the beginning of this century by an uncontrolled disposal of industrial wastes accompanying the unprecedented growth of of the American steel industry. Due to the high permeability of the disposed materials (predominantly metallurgical slag), no surface water accumulation has been observed at the site. According to geologic investigations which were conducted later, numerous discontinuities of different sizes existed within the fill. The approval of handling and processing of potentially toxic materials at this site without a construction of a liner system became impossible. Although the applicant considered the recompaction of on-site material as an economically effective solution for the problem, the granular character of disposed materials did not allow implementation of that idea. The second suggested alternative was to construct clay liners at coal storage areas. The business nature of this company which has been involved in continuous relocation of large volumes of stored mineral products/materials using heavy equipment made it impossible to guarantee the purity of processed and transported products, which mainly was coal. Simple calculations performed by the Applicant indicated that a 2-3% coal weight loss accepted by all his customers could be avoided by constructing concrete or asphalt pads. The interested party agreed to finance the construcion of those pads. This action not only eliminated undesirable infiltration of contaminated surface water but also prolonged the life of expensive loading equipment.

Site No. 2

A coal processing/coal loading facility site, that is located on the top of an abandoned spoil pile, was used at one time as a huge drying bed for fine coal discards. Creators of this gigantic drying bed decided to change their business affiliation. As a result of that move, the spoil pile being exposed to the atmospheric precipitation became a source of ground water contamination. Owners of the adjacent to the spoil pile area, which also had been used as a drying bed, had closed their mining business due to economic reasons and left a layer of coal fines on the ground.

Acid producing materials had been observed within chaotically disposed coal mine gob and overburden rock fragments at numerous spoil pile exposures. A drilling of new bore holes required by Pennsylvania Regulations for more precise identification of site geologic conditions could not reveal new useful lithologic information. The existence of the disturbed surface area covered with a layer of coal fines increased the probability of ground water contamination at the site.

At the meeting between the applicant and representatives of the Department of Environmental Resources, the following agreements were reached:

1. The applicant would approach the owners of the property adjacent to the spoil pile with a proposal to remove coal fines from that area using the applicant's equipment and work force.

2. The applicant would regrade and reclaim drying bed areas adjacent to the site in question as well as parts of the proposed permit area which would not be used.

3. With the elimination of potentially acidic recharge zones surrounding the proposed permit area as well as liner construction above the operation zones no geologic/hydrogeologic bore holes or spoil sampling and testing would be required.

No liner material had been selected during that meeting. The unusual solution of that problem had been found when the Pennsylvania Department of Transportation had started the reconstruction of an old highway. Large concrete blocks deteriorated along their edges had created hazardous problems for moving vehicles. The repair of voids developed between those blocks was economically ineffective and blocks had to be removed prior to the highway repavement. High costs of long distance transportation and expenses for the disposal of removed re-inforced blocks had been avoided by the "recycling" of that material: the applicant had offered to use those blocks as elements of a gigantic liner covering the entire spoil pile. The spoil pile regrading had been suggested prior to the installation of concrete blocks for the proper collection and treatment of surface runoff. All spaces between blocks had been filled with a compacted clayey material to prevent surface water infiltration. The coal operator's "know-how" experience, personal energy of coal company owners and the involvement of the applicant's work force had been major ingredients necessary for the construction of that liner which was completed in 1987. Areas adjacent to the proposed permit boundary had been successfully reclaimed.

Site No. 3

Two diked ponds were proposed to collect and hold acidic water in the vicinity of a relatively new coal preparation plant. Three soil samples had been recovered at the borrow area. Based on laboratory test results, suitability of this soil for a pond liner material was approved by the regulatory agency. To ensure proper placement and compaction, it was decided to construct a test fill in the vicinity of the proposed ponds. To verify the earth work quality, nuclear densometer tests were performed by an independent geotechnical laboratory on the test fill site. As it has been reported by the nuclear densometer operator, all coal operator efforts to achieve maximum dry density of compacted material were unsuccessful. Comparison of natural water contents of a clay soil delivered to the site and laboratory compaction test results conducted by the regulatory agency indicated that in-situ water contents exceeded optimum water contents by 7-8%. The coal operator who diligently followed regulatory requirements had not been warned by his geotechnical consultant that 74% of maximum dry density achieved at the experimental pad site had to be expected in accordance with laboratory compaction test results. If the clay soil delivered to the site had been air dried prior to the placement and compaction, the coal operator would not have experienced economical losses as well as technical and emotional inconveniences associated with inadequate geotechnical services. The regulatory agency representative helped the operator overcome all technological difficulties and the proposed ponds were successfully completed.

Site No. 4

Gradual degradation of groundwater was observed in the vicinity of a coal refuse disposal site. Field inspections revealed the following:

1. Coal refuse delivered to the site contained a large volume of combustive material, pieces of old metal cables, corroded pipes, and used machine parts

2. Although the refuse pile in question had been designed for the disposal of deep mine rock, pond settlings from adjacent sedimentation ponds were also spotted at the site.

3. Coal refuse delivered to the site was not spread and compacted immediately after delivery that increased the probability of surface water infiltration in the previously disposed refuse.

4. Field density tests demonstrated that coal refuse had not been compacted properly. The regulatory agency representative was witnessing an attempt to compact a 3.0 to 4.0 foot thick layer of refuse with a light bulldozer.

A problem developed at the site in question was discussed with the coal company employee and the regulatory agency permitting and compliance personnel during several office and field meetings. As a result of those meetings, the following site remediation strategy had been developed:

1. The coal company personnel had been instructed to remove all "foreign" material from the pile. It had been emphasized that only material indicated in the permit application was approved for disposal at this area. No material from sedimentation or treatment ponds had been permitted to be disposed at this site.

2. Several bag samples representing "old" and "fresh" coal refuse were sent to the laboratory to update the information relating to physical and mechanical properties of that material.

3. It had been proposed to examine coal refuse property changes during disposal and compaction operations within the test fill area to develop a productive compaction technique utilizing the on-site equipment. The achieved degree of compaction had to be verified by sand cone and nuclear densometer tests.

4. The reclamation of pile outslopes and benches, construction of clay and/or top soil cover(s) would be initiated immediately after achievement of designed grades and elevations.

The coal company has developed a very effective field compaction technique utilizing hauling trucks and other available rubber tire equipment. The regulatory agency personnel has reported the significant improvement of groundwater quality at that site.

DISCUSSION

Design and construction of water resisting barriers is commonly recognized as a necessary and very costly component of any disposal site project. Existing theories describe and predict in general terms negative events associated with permeation of pure liquids or solutions through naturally deposited or artificially placed soils. How stringent must the design requirements be for coal mining activity sites which in some cases represent less hazardous conditions than existing polluted areas adjacent to those sites? What is the degree of engineering accuracy that is allowed to substitute a costly testing program with a mathematical modeling to obtain a problem solution benefiting a coal operator and satisfying permit application reviewers? The rule of a thumb generally used in the review process is as follows: a specific discharge through the soil or fill immediately below a coal refuse pile or an impoundment containing acidic water or fine coal refuse must not exceed $5*10^{-5}$ or $5*10^{-7}$ cm/s, respectively. How realistic are these numbers? What is the minimum number of soil/fill laboratory or field test results required to calculate those magic keys that open technical reviewers' hearts?

It is commonly known that infiltration rates calculated utilizing large scale field permeability tests are 100 to 1000 times greater than the values calculated using laboratory test results. Laboratory specimens, on the other hand, characterize the finest fraction of a material suggested for a water resisting barrier. Uniformity of compacted laboratory specimens can be achieved easier and controlled more accurately than a degree of material compaction in the field. What additional criteria have to be considered by engineers and soil scientists for a successful search for correction coefficients negating unrealistic assurances obtained with the help of laboratory data? How will collapsibility of compacted matarial affect correction coefficient values? Assuming that laboratory permeability coefficients can be used with safety factors ranging between 100 and 1000, we have to agree with a fact that liners can be accepted as a last resort and are not a safe solution. This discovery lead us to a necessity to consider a prevention technique which was illustrated in cases 1 and 2. Comparison of water flow velocity along properly constructed hard surfaces similar to ones described in those examples even for the highest possible infiltration rates shows that less than 0.01% of atmospheric precipitation has a chance to infiltrate the space below water resisting barriers. If we agree that the quantity of transported contaminants is proportional to the total volume of water entering aquifers, we conclude that the above mentioned illustrations represent very conservative engineering methods utilizing inexpensive construction techniques which minimize the probability of aquifer contamination.

Incorrect implementation of design ideas can nullify the best engineering efforts to prevent groundwater degradation. The success of earth work procedures depends on a provided construction quality control. Problems which were faced by the coal operator during pond liner construction at Site No. 3 could be avoided if his geotechnical consultant had sent a qualified person to supervise earth work operations. The availabilty of high tech geophysical equipment such as nuclear densometer requires a higher degree of knowledge from the personnel operating this equipment. Case No. 4 confirms this statement.

To accelerate the review process of permit applications, many regulatory agencies have developed manuals or instructions delineating a sequence of narrative conttents. The intent of manuals or instructions is to describe the minimal scope of information required from the applicant. Manuals (instructions) leave a broad field for environmental consultants to introduce new ideas associated with the prevention of any undesirable problem. A permit application may not be construed as a standard project design. It may include any innovation minimizing any type of negative impact on the environment. It may include "safeguards" in proposed projects. Coal refuse piles designs may be presented as "modular" projects comprising independent structural "blocks" or "cells". When an independently constructed unit reaches its design dimensions it will be temporarily or permanently covered/reclaimed. Therefore, a sudden interruption of coal refuse disposal operations will affect only a small portion of the permitted area and that, in turn, may have a minimal negative impact on the environment. The achievement of a minimal nature equilibrium disruption under any circumstances is a basic requirement of our dynamically growing society.

CONCLUSIONS

1. Assesments of the effectiveness of water resisting barriers based only on rigid technical requirements in some cases are unreasonably conservative. Realistic ecologic analyses and economic considerations must preview any proposed industrial project. A scope and modifications of laboratory and/or field tests selected for the design may not be defined only on the basis of standard technical requirements included in regulations. The type and the number of proposed tests must be approved with respect to the complexity and anticipated hydrogeologic consequencies at a particular site.

2. Specific features of each industrial project do not exclude a detailed examination of all geomorphologic, geologic, hydrogeologic, and engineering information accumulated for the site in question and adjacent areas. Utilization of sophisticated mathematical modeling in day-by-day work is always admirable. However, incorporating the most advanced computers in the design process must co-exist with a traditional "brain storming" process which explores all economically and technically effective ideas.

3. The availability of high tech laboratory and field testing equipment can not substitute knowledge, dedication, and experience of the personnel operating that equipment. A field job was always the most important component of industrial development. Only the most knowledgeable employee must be involved in the sampling and testing processes.

4. Industrial growth is always accompanied by environmental changes. No progress can be slowed down by conservative attitudes and nonscientific decisions. Economic effectiveness and industrial progress are dynamic events. The best environmental laws always reflect the oldest fundamental scientific ideas. The best environmental laws are those where fairness to the public and industry was verified many times.

ACKNOWLEDGEMENTS

The Author is grateful to Ms. L. Kirwan and Messrs G. Camus, J. Kernic, J. Koricich, J. Pari, and W. Plassio for the editorial help and technical suggestions.

Estimation de la Stabilite des Terrils Interieurs sur la Base de la Theorie du Risque

Prof.Dr.Ing. Pechka Stoeva, Dr. Ing. Paoulin Zlatanov
Université de Mine et Geologie, Sofia, Bulgarie

RÉSUMÉ: L'exploitation des lignites dans les mines à ciel ouvert en Bulgarie est liée à l'excavation de grands volumes de couverture. Une partie des masses argileuses est mise en terrils intérieurs. Les calculs de la stabilité ont été effectués sur 4 modèles de déformation avec 4 groupes de paramètres. On a estimé 4 stratégies liées à la théorie du risque par lesquelles on peut obtenir le schéma de déformation utilisable et le coefficient de sécurité optimum.

L'exploitation des lignites dans les grandes mines à ciel ouvert en Bulgarie-"Maritza-Est" est liée à l'excavation de grands volumes de couverture (variétés des argiles de Pliocène). Pendant l'excavation et le transport, les argiles se modifient en une masse déstructurée. Une partie de cette masse est mise en terril dans les tranchées des mines à ciel ouvert sous la forme de terrils intérieurs. La hauteur des terrils est de 50 à 65m. La technologie de la construction du terril dépend des machines minières utilisées. La construction des machines permet la mise en terril de la masse molle par quelques horizons. Le nombre des horizons et leurs hauteurs se déterminent à la base de la stabilité de chaque couche.

Les argiles de Pliocène à l'état structuré (du bord) et déstructuré (du terril) se caractérisent par: une faible résistance; une grande possibilité de déformation et une tendance du fluage et de glissement du terrain. La mise en terril se fait sur le substratum constitué d'argile organique, plastique et dans certaines zones, fissurée et mouillée. Au cours de travaux d'exploitation on a été obtenu les miroirs tectoniques horizontaux dans le substratum.

Les conditions défavorables dans les matériaux argileux non-consolidés dictent des déformations de long-terme et provoquent des déplacements et des glissements du terrain dans les terrils. Ce phénomène a créé des difficultés lors du transport des matériaux argileux, de la technologie de mise en terril, de l'organisation de travail et de l'exploitation des lignites (en effet le premier gradin des lignites se situe sous la masse du terril glissant).

L'exploitation des mines à ciel ouvert, pendant 30 ans, s'accompagne de glissements du terrain sur les bords et dans les terrils qui sont un facteur complexe du risque de la rupture par rapport au processus technologique, à la sécurité des machines et des appareils et à l'état psychologique des équipes de mineurs.

L'étude du comportement d'ouvrages a été faite à partir des observations géodésiques de longue durée sur le terrain. Les résultats obtenus donnent une possibilité de distinguer 4 (quatre) schémas fondamentaux réels de déformation de pente générale des terrils. Le mécanisme de déformation se détermine par la spécificité du massif de terril et le substratum de mise en terril:

- une surface cylindrique (circulaire) qui coupe le massif du terril (presque homogène) jousqu'au pied de la pente (Fig.1 - schéma D_1);
- une surface cylindrique (circulaire) qui coupe le massif du terril et l'argile organique fissurée du substratum dans le fond de la mine (Fig.2 - schéma D_2);
- un prisme de pression active, un déplacement horisontal du bloc principal sur le contact mouillé terril - substratum (Fig.3 - schéma D_3);
- un prisme de pression active, un déplacement horisontal du bloc principal sur le miroir tectonique dans le substratum (Fig.4 - schéma D_4).

Fig.2. Schéma circulaire sous le pied de pente: 1-surface potentielle; α-angle général du terril; H-hauter du terril.

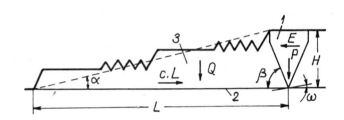

Fig.1. Schéma circulaire classique: 1-surface potentielle; α-angle général du terril; H-hauter du terril.

Fig.3. Schéma par un prisme de pression active et contact mouillé: 1-prisme de pression active; 2-contact mouillé terril-substratum; 3-bloc principal du déplacement horizontal; c.L-résistance du substratum; Q-poids du bloc principal; E-force active; α-angle général du terril; β-inclinaison du prisme actif.

Dans sa totalité le massif du terril est constitué par un mélange polyphasique de l'argile avec une structure et une cohésion dérangée. Les caractéristiques physiques et mécaniques du matériau de terril sont toujours difficiles à évaluer; cela tient tant à son hétérogénité qu'à sa granulométrie. Ce fait explique les différents paramètres de calcul de différents auteurs et instituts, obtenus en laboratoire et sur le terrain.

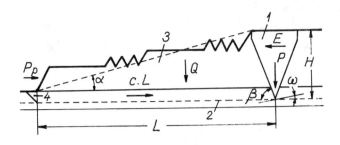

Fig.4. Schéma par un prisme de pression active et contact de miroir tectonique: 1-prisme de pression active; 2-miroir tectonique dans le substratum; 3-bloc principal du déplacement horizontal; c.L-résistance du substratum; Q-poids du bloc principal; E-force active; α-angle général du terril; β-inclinaison du prisme actif.

Karatcholov [1] a obtenu les paramètres P_1 par la réalisation d'un essai triaxial non consolidé et non drainé (masse volumique - ρ=1,80 g/cm³, angle de frottement - φ=3,5°, cohésion - c=0,34 x 10⁵ Pa). Guéorguiev [2] a déterminé les paramètres P_2 par le calcul inverse sur la base du glissement de terrain avec l'utilisation du schéma 4 (D_4 - Fig.4), (ρ=1,70 g/cm³, φ=8°, c=0,30x10⁵ Pa). Institut "Minproject" [2] a obtenu les paramètres P_3 aussi par le calcul inverse (ρ=1,75 g/cm3, φ=6°, c=0,28x10⁵ Pa). Todorova et Stoeva [3] proposent les paramètres P_4 sur la base des essais de cisaillemenet non consolidé et non drainé (schéma de Skempton [5]), (ρ=1,82 g/cm3, φ=11°, c=0,25x10⁵ Pa).

L'existence de 4 schémas (modèles) differents de déformation et 4 groupes de paramètres de calcul crée une indétermination qui est liée au risque de rupture pat rapport à la stabilité du terril.

La méthode de variation utilisée jusqu'à maintenant pour le choix de coefficient de sécurité (F) a habituellement un caractère intuitif, qui se base sur l'expérience des spécialistes suivies d'observations sur le terrain et d'essais en laboratoire.

Pour le calcul du coefficient de sécurité (F) on utilise les méthodes suivantes:
- la méthode de Fellenius en rupture circulaire;
- la méthode du prisme de la pression active. L'expression de F est sous la forme [2,6]:

$$F = \frac{c.L + Qtg\varphi \cos\omega + P_p + Q\sin(-\omega)}{E + Q\sin(+\omega)} \quad (1)$$

où: P est masse du prisme actif, kg
Q - masse du bloc central (entre les deux prismes), kg
E - force glissante du prisme actif le long de la surface du glissement, kg
P_p - force passive, kg
L - longuer de base du bloc central, m
ω - angle d'inclinaison de la surface du glissement: (+) - inclinaison vers la mine, (-) - vers le massif, x°
c - cohésion de la base du bloc central, kPa
φ - angle de frottement interne sur la surface du glissement du bloc central, x°

L'estimation d'une situation défavorable peut se produire par la théorie du risque qui se base sur la méthode des jeux et de solution statistique [4]. La stratégie des jeux est développée pour les deux éléments fondamentaux: modèle réel de déformation (D_i) et variation de paramètres (physiques et de rupture) de calcul (P_j). Le changement de D_i et P_j dans différentes combinaisons dicte des valeurs variables correspondant au coefficient de sécurité pour les mêmes profils technologiques (tabl. 1).

Tableau 1

D_i/P_j	P_1	P_2	P_3	P_4	α_i
D_1	0,95	1,10	1,15	1,30	0,95
D_2	1,17	1,23	1,25	1,35	1,17
D_3	1,22	1,50	1,27	1,40	1,22
D_4	1,15	1,26	1,20	1,23	1,15
β_j	1,22	1,50	1,27	1,40	

Les données obtenues pour F, représentées dans le tableau 1 ont des valeurs proches qui ne peuvent pas employer le choix de critère d'estimation de la stabilité. D'après cela on utilise le risque comme une caractéristique supplémentaire.
Les valeurs qui donnent satisfaction à la condition de la stabilité de pente générale s'estime par la matrice du risque (tabl. 2), constituée sur la base de la définition suivante [4]:

$$r_{ij} = \beta_i - F_{ij} \qquad (2)$$

où: r_{ij} est le risque pour le couple D_i et P_j,

$$\beta_i = \max F_{ij} \qquad (3)$$

F_{ij} - valeur concrète de F pour le couple de combinaison de D_i et P_j (i= 1,2,3,4 et j=1,2,3,4).

Tableau 2

D_i/P_j	P_1	P_2	P_3	P_4	r_j
D_1	0,27	0,40	0,12	0,10	0,40
D_2	0,05	0,27	0,02	0,05	0,27
D_3	0,00	0,00	0,00	0,00	0,00
D_4	0,07	0,24	0,07	0,17	0,24

Pendant la comparaison de F pour les différentes stratégies (tabl. 1) on constate que le coefficient de sécurité a des valeurs équivalentes comme par exemple pour les cas suivants:

I cas
F=1,15 pour D_4 et P_1
F=1,15 pour D_1 et P_3
II cas
F=1,22 pour D_3 et P_1
F=1,23 pour D_2 et P_2
F=1,23 pour D_4 et P_4

Ces caractéristiques équivalentes ne correspondent pas au sens simple, au point de vue de l'application pratique, bien que les cas représentés dans le tableau 1 ont F>1. L'analyse qui a été faite à l'aide du tableau 2 du risque montre:
- pour le premier cas le risque est: $r_{4,1}$=0,07 et $r_{1,3}$=0,12;
- pour le deuxième cas le risque est: $r_{3,1}$=0,00, $r_{2,2}$=0,27 et $r_{4,4}$=0,17.

Les différences entre les valeurs du risque s'expliquent avec le fait que les 4 schémas et les 4 paramètres utilisés ont différentes efficacités. La solution pour D_4-P_1 (I cas) est meilleure de D_1-P_3, bien que ce risque est plus petit. La solution pour D_3-P_1 est préférable à celle pour D_2-P_2 parce que le risque est minimal ($r_{3,1}$=0,00).
A l'indétermination dans l'information complexe pour obtenir une solution favorable on utilise:
- le critère de Wald, selon lequel la meilleure stratégie donne satisfaction à la condition:

$$\alpha = \max_i \ \min_j \ F_{ij} \qquad (4)$$

c'est-à-dire α exprime la valeur maximale des coefficients minimaux de sécurité F_{ij}. Pour notre cas la condition a été respectée par F=1,22 pour le couple D_3-P_1 (tableau 1, colonne $α_j$);

- le critère de Sevidj, selon lequel la meilleure stratégie donne satisfaction à la condition:

$$S = \min_i \max_j r_{ij} \quad (5)$$

c'est-à-dire S exprime la valeur minimale des risques maximaux;

La condition a été respectée pour les stratégies D_3 - (P_1, P_2, P_3, P_4).

L'existence de 4 schémas de déformation et de 4 groupes de paramètres de calcul complique le choix du coefficient de sécurité optimum. La résolution du problème se base sur les résultats obtenus par l'utilisation de la matrice du risque. Dans ce cas le schéma optimum correspond au modèle D_3. Le même modèle se confirme par le critère de Sevidj, valable pour tous les paramètres (P_j) de calcul. Avec le critère de Wald on rend concret le même schéma de déformation (D_3) et pour les paramètres de calcul P_1. Le couple D_3-P_1 est une condition sur laquelle on peut compter pour l'estimation de la stabilité.

CONCLUSION

1. Sur la base de la configuration des pentes du terril, des paramètres du calcul des matériaux et des schémas de stabilité on a déterminé le shéma utilisable D_3 avec le coefficient de stabilité (F_j) de 1,22 a 1,50, mais correspondant au risque égal à zéro.

2. D'un point de vue pratique on a trouvée une possibilité technologique de la construction des pentes des terrils dans les mines à ciel ouvert "Troyanovo 1,2,3" correspondant au schéma (modèle) D_3. Cela se réalise par l'excavation des zones avec un miroir tectonique dans le substratum du terril.

3. La gestion technologique atteinte au processus de mise en terril est liée à la diminution considérable du risque de rupture par le glissement du terrain et en même temps est liée à l'augmantation du volume de la masse non-productive; à l'amélioration et la recultivation du sol de la surface du terril; à la diminution des terrains ruraux détruits pendant l'exploitation minière et à la création d'une sécurité des hommes et des machines.

REFERENCES

1. Karatcholov P. 1990. Etude de la stabilité des pentes du terril en Maritza-Est. Thèse,pp.32-37. (bul.)
2. Gueorguiev G. 1981. Manuel méthodologique de calcul de stabilité des pentes. Sofia,pp.160-190. (bul.)
3. Todorova M., P. Stoeva. 1973. Propriétés physiques, mécaniques et rhéologiques des argiles du bassin minier "Maritza - Est". Travaux de Minproject, v. XII, Sofia, p. 64-95. (bul.)
4. Ventzel E.S. Etude d'operation. 1988. Naouka, Moscou, pp. 195-207. (rus.)
5. Skempton, A.W. 1964. Long-term stability of clay slopes. Géotechnique, 14, No 2 pp. 75-101
6. Zlatanov, P., P.Stoeva 1990. Application de la technique d'ordinateur au calcul de stabilité des bords des mines à ciel ouvert construites dans un massif à plusieurs couches. Proc. 6th Inter. Ass.of Engin Geology, Amsterdam, pp 2341-2345

Large Scale Consolidation Testing of Oil Sand Fine Tails

Nagula N. Suthaker, J.Don Scott
University of Alberta, Department of Civil Engineering, Edmonton, Alberta, Canada

Abstract

A self weight consolidation test in a 10 m high standpipe has been conducted for more than 10 years. The objective of the experiment was to determine whether the consolidation properties measured in large strain slurry consolidation tests could predict the large scale, long term consolidation behavior of fine tails. The rate of consolidation in the ten metre standpipe, as shown by the rate of settlement of the water-fine tails interface, was in good agreement with the predicted values based on the slurry consolidometer tests and the finite strain consolidation theory. The predicted settlement was more influenced by the measured permeability than by the compressibility. Consolidation was experienced throughout the depth of the standpipe with a higher solids content increase near the top surface of the fine tails. The consolidation equipment and testing methods are described. Two metre standpipe tests coupled with field measurements in the tailings pond indicated that sedimentation was fairly rapid and could be considered to be complete when the solids content reaches 15%. Therefore the long term settlement behavior is self weight consolidation not sedimentation.

Introduction

In Northern Alberta, oil sand deposits are mined and processed to recover heavy oil from two oil sand plants, Syncrude Canada Ltd. and Suncor Inc. Syncrude and Suncor, respectively, produce about 480,000 tonnes per day and 170,000 tonnes per day of tailings which varies in solids content from 40% to 60%. The tailings streams are composed of 75% to 90% sand and 10% to 25% fines by weight with a small amount of heavy oil which bypasses the extraction plant. The disposal of the tailings stream is accomplished by allowing the sand to settle out to construct dykes and beaches which accumulate at the rate of 225,000 m^3 per day. Much of the fines are carried into the pond as a fine tails stream around 8% to 10% solids. When the stream flow slows, sedimentation of the fine tails particles begins.

Sedimentation is the process in which the solids fall through an ambient fluid where stress is not transferred from one particle to another. At a solids content of around 15%, the oil sand fine tails solids begin to form a matrix such that stress can be transferred from one particle to another. After reaching this solids content, any increase in fine tails density arises from the process of self weight consolidation. The fine tails forms a deposit at about 30% solids after 1 to 2 years of consolidation. This fine tails deposit is accumulating at a rate of 45,000 m^3 per day. Due to the slow process of self weight consolidation the solids content increases slowly above 30% solids and the fine tails must be stored in large tailings ponds which are a major environmental problem.

The rate and amount of consolidation of fine tails can be measured in the laboratory by large strain slurry consolidation tests. The results from these tests can be extrapolated to predict the long term behavior of the fine tails in the tailings ponds using a finite strain consolidation theory. The extrapolation of short term, laboratory tests to long term field behavior, however, requires verification so that the extrapolated results can be used with confidence for long term planning and design of tailings ponds. A large scale self weight consolidation test in a ten metre high standpipe has been conducted for more than ten years to provide this verification. The objective of this study is to review the results of the testing programs and compare the consolidation test performance and the 10 m standpipe performance.

Two Metre Standpipe Tests

Self-weight standpipe tests from 180 cm to 200 cm in height have been conducted to determine the completion of sedimentation and subsequent progress of consolidation of the fine tails. The two metre height was selected in order to obtain a measurable effective stress. The standpipes (Figure 1) were continually monitored for settlement of the fine tails-water interface. Pore pressure changes were measured at 20 cm height intervals by small diameter manometers. Samples (5 ml) were obtained from another set of ports and the density and solids content were determined.

Figure 2 [1] shows different stages of a 180 cm standpipe sedimentation and consolidation test on a fine tails sample (initial solids content of 10% and density of 1.05 g/cm^3). The initial pore pressure at the base of the column was equal to the total mass of the column of fine tails showing that no effective stress existed in the slurry at the beginning of the test. During sedimentation of the fine tails, a density increase has started from the bottom of the tube, where the the fine tails particles had developed effective stress between them. As the water-fine tails interface settled, the effective stress front travelled upwards. After 2.5 days, the full column had developed an initial interparticle stress. No significant consolidation had taken place at this time. Consolidation proceeded downward from the fine tail surface not upwards from the bottom as consolidation theory predicts. This phenomenon was also observed in the ten metre standpipe and an explanation is provided later in this article. Consolidation was practically

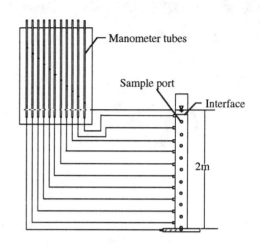

Figure 1. Two Metre Standpipe

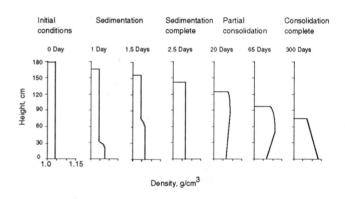

Figure 2. Stages of Sedimentation and Consolidation [1]

complete after 300 days. The interface continued to settle slowly after this time due to secondary consolidation.

Such laboratory tests coupled with field measurements in the tailings pond showed that sedimentation of the fine tails is fairly rapid and could be considered to be complete when the solids content had increased to approximately 15%. Further increases in density of the fine tails has to take place by consolidation. Imai [2] pointed out that the solid content at which sedimentation is complete is not unique for a material but depends on the initial solids content of the slurry. Similar results were found with the fine tails, but the range of solids content at the end of sedimentation was between 10% and 15% for fines-water mixtures with initial solids contents of 5% to 8%.

Slurry Consolidometer Tests

The slurry consolidometers were designed for samples 20 cm in diameter and 30 cm in height. With this apparatus, strains up to 80% of the initial sample height may be measured. A typical experimental arrangement for a step loading consolidation test is shown in Figure 3.

Because the finite strain consolidation theory used requires a knowledge of the variation of the coefficient of consolidation of the material during a test, the slurry consolidometers were designed so that void ratio - effective stress and void ratio - permeability relationships could be measured for each load increment. After consolidation under a load increment was complete, a permeability test was conducted on the sample. During the permeability tests, the hydraulic gradient was kept small so that seepage induced consolidation was minimized. Figure 3 shows a method of overcoming the tendency for seepage induced consolidation. When consolidation under a load increment was complete, the loading ram was locked in place and an upward flow constant head permeability test was performed. This procedure eliminates the risk of having seepage induced consolidation.

A step loading consolidation test was performed with a 29% initial solids content (water content of 245%) fine tails. After each increment as described earlier constant head permeability tests were also performed. Figure 4 shows the relationship between void ratio and effective stress. The variation of permeability with void ratio is depicted in Figure 5 [3]. When modelling the consolidation of fine tails, it is necessary to use the consolidation parameters in

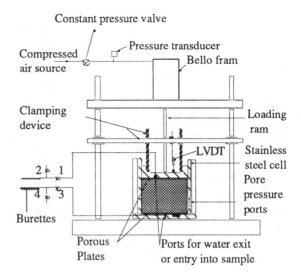

Figure 3. Slurry Consolidometer

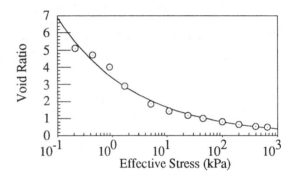

Figure 4. Compressibility of Oil Sands Fine Tails

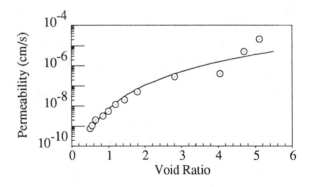

Figure 5. Permeability of Oil Sands Fine Tails

the form of power law constants for input to the finite strain computer program used.

The relationship for compressibility and permeability are

$$e = 28.71 \cdot \sigma'^{(-0.3097)} \quad (1)$$

where σ' is in pascals.

$$k = 7.425 \times 10^{-11} \cdot e^{(3.847)} \quad (2)$$

where k is in m/s.

These relationships described the laboratory data well (Figure 4 and Figure 5), for both permeability and compressibility.

Ten Metre Standpipe Test

A self weight consolidation test in a 10 m high by 1 m diameter standpipe on 31% solids content fine tails was commenced in 1982 to determine whether the material properties measured in large strain slurry consolidation tests could predict the consolidation behavior of fine tails when used in the finite strain consolidation theory.

The ten metre standpipe (Figure 6) is similar to a two metre standpipe except that pore pressure and sample ports are located at 1 m intervals. Additional ports at 0.5 m intervals are available to allow closer measurements, but were not used except at the top and bottom of the standpipe where rapid changes occurred.

To obtain density and solids content measurements a sample probe was inserted to a desired location through a sample port which has an O-ring seal. Specimens were selected at different distances from the port to reduce disturbance at any single location. The ball valve remained closed until the probe tube passed the seal. The valve was then opened and the tube was inserted to the location to collect the sample. For high solids content fine tails which did not flow easily, a vacuum was applied to the sample bottle. The main measurements on the samples were density and solids content measurements. Extraction tests and hydrometer tests were carried out to measure bitumen content and settling of coarse sand grains in the mix.

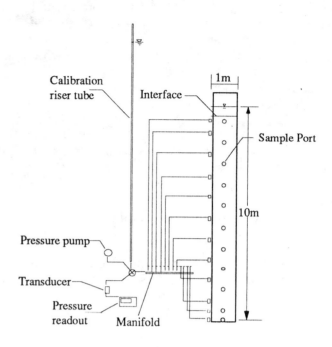

Figure 6. Ten Metre Standpipe

Temperatures of the fine tails were measured in the standpipe using a probe similar in diameter to the sampling probe. The variation in temperature in the standpipe has been constant throughout the test, in summer or winter, with an increase of about 3°C from the bottom to the top of the standpipe. This gradient is not considered sufficient to have any significant effect on the consolidation behavior.

The consolidation of fine tails in the self weight consolidation test is characterized by the settlement of the water-fine tails interface. Figure 7 shows the test results and the predicted values based on the test results from slurry consolidation tests and a finite strain consolidation numerical model. The numerical model consists of a finite difference computer program [3] based on the the general non-linear equations developed by Gibson, England and Hussey [4]. The compressibility and permeability were expressed as power functions [5] in order to define the consolidation parameters over a large range in void ratio. The predicted settlement was found to be more sensitive to permeability than to the compressibility.

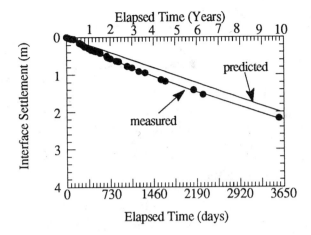

Figure 7. Comparison of Measured and Predicted Consolidation

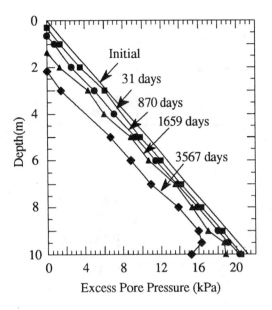

Figure 8. Variation of Pore Pressure with Time

Pore pressures were measured with a pressure transducer which was calibrated against a column of water as shown in Figure 6. Figure 8 shows the variations of excess pore pressure with time along the height of the standpipe. Figure 9 shows the measured and predicted excess pore pressures after 10 years and a very good agreement is noticeable. Larger sand grains appeared to be settling through the fine tails and accumulating at the bottom of the standpipe. A small layer of segregated sand could have created a reduction in the pore pressure at the bottom of the standpipe.

The measured solids content variations with time and depth are shown in Figure 10. Figure 11 shows measured and predicted solids contents after 10 years. Similar to the predicted excess pore pressure profile, the increase in predicted solids content began from the bottom, which is to be expected from normal consolidation behavior. However the data shows an increase in solids content along the entire depth of the fine tails with the greatest increase near the top of the fine tails. Therefore, the measured data indicate that consolidation is occurring at all depths with the highest solids content near the top. This might be due to free gas in the voids or to creep and thixotropy occurring in the standpipe which was not considered in the theory. Due to the high compressibility and high void ratio of the fine tails, they exhibit very high creep rates. The fine tails also display a large increase in

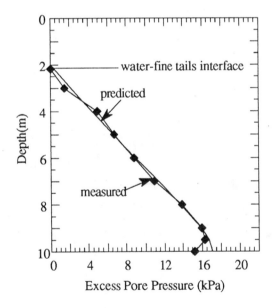

Figure 9. Predicted and Measured Pore Pressure Profiles after 10 years

thixotropic strength which is probably due to the organic bitumen and chemical additives. The pore pressure reduction found at the bottom of the standpipe after 10 years from sand accumulation explains the solids content increase found at the bottom of the standpipe.

Conclusions

A slurry consolidometer test has been performed to determine the void ratio-effective stress and the void ratio-permeability relationships for the fine tails. The laboratory data for permeability and compressibility were well described by power laws. Sedimentation in the two metre standpipes was fairly rapid and was complete in a short period of time, indicating that is not a governing factor of the long term behavior of the fine tails. The ten metre standpipe consolidation results are generally in good agreement with the predicted values from the slurry consolidometer test results employing the finite strain consolidation theory. Such agreement will allow the laboratory consolidation test results to be extrapolated with confidence to predict the consolidation behavior of oil sand fine tails.

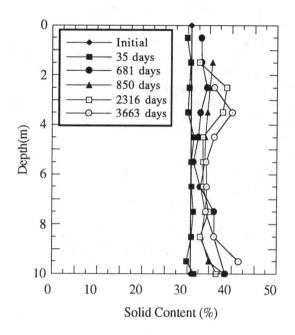

Figure 10. Variation of Solids Content with Time

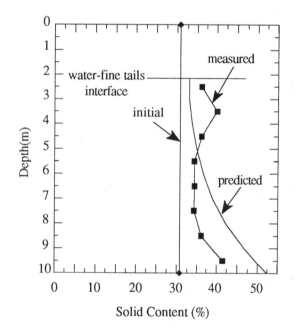

Figure 11. Predicted and Measured Solids Content Profiles after 10 years

References

[1] Scott J.D., Dusseault M.B. and Carrier III W.D., 1986. Large-Scale Self Weight Consolidation Testing, Consolidation of Soils: Testing and Evaluation, ASTM STP 892, Philadelphia, 1985, pp. 500-515.

[2] Imai G., 1981. Experimental Studies on Sedimentation Mechanism and Sediment Formation of Clay Materials, Soils and Foundations, Vol. 21, No. 1, March, pp. 7-20.

[3] Pollock G.W., 1988. Consolidation of Oil Sand Tailing Sludge, MSc. Thesis, Department of Civil Engineering, University of Alberta, Edmonton, Canada, 276 p.

[4] Gibson R.E., England G.L. and Hussey M.J.L., 1967. The Theory of One Dimensional Consolidation of Saturated Clays, I, Finite Non-Linear Consolidation of Thin Homogeneous Layers, Geotechnique, Vol. 17, pp. 261-273.

[5] Somogyi F., 1980. Large Strain Consolidation of Fine-Grained Slurries, presented at Canadian Society for Civil Engineering Annual Conference, Winnipeg.

The Use of Small Scale Experiments to Predict Desiccation of Tailings

Gareth E. Swarbrick
School of Civil Engineering, University of New South Wales, Australia

Abstract: This paper gives a summary of an established deposition model for which parameters obtained by small scale tests may be used. A procedure using small scale laboratory tests is given and the method by which the required parameters are determined is described. A comparison is then given between parameters obtained by small scale tests and those obtained by larger lysimeter based tests.

Introduction

The disposal of slurried mineral wastes such as mine tailings and dredged material generally involves significant costs. In addition to the costs due to the design and construction of a storage facility there are costs as a result of maintenance and rehabilitation.

Current emphasis for the prediction of the deposited properties is focussed upon large strain saturated consolidation models [8,4,14,15]. These models are well suited to subaqueous deposition methods but not to subaerial deposition, particularly where the tailings are allowed to desiccate in thin layers after deposition.

Sub-aerial deposition techniques which employ solar desiccation have several associated benefits including [12,13].:
- increased storage capacity
- reduced permeability and seepage
- reduced compressibility
- increased shear strength
- facilitates reclamation
- minimal cost

There are several published cases which demonstrate such benefits [2,12,5].

Due to the potential benefits of solar desiccation, there is a demand for accurate methods for predicting the deposited properties of wastes under these conditions. Until recently such methods were based upon trial and error method which does not allow for efficient design and planning of the storage facility. New research has shown that the desiccation of mine tailings can be predicted with reasonable accuracy [17,19,18].

The current model

Laboratory and field scale studies of deposited mine tailings undertaken at the School of Civil Engineering, University of New South Wales have enabled the development of a semi-analytical model PRED. This model can accurately predict the post depositional behaviour of mine tailings undergoing sedimentation and desiccation induced consolidation.

The model may be summarised as follows. The adopted approach is to predict the *average* properties for a given layer of tailings. This approach is only suited to relatively thin layers (0 - 2 m) commonly used in subaerial techniques. More details concerning the model method may be found in [17,19].

Sedimentation

A column settling test is performed to determine the height of settled solids over time. The interface height is differentiated to calculate the interface velocity over time. The velocity data is then fitted to the Eq. (1) using regression techniques [20].

$$v = v_o 10^{k_o \frac{V_s}{H}} \quad (1)$$

H is the interface height, V_s is the total volume of solids per unit area within the column and v_o and k_o are fitting parameters. The relationship is nonlinear so PRED uses this relationship to calculate sedimentation using an appropriate time step (one minute).

Once the final settled height is reached the stress distribution through the column is a function of effective stress only. By assuming a common effective stress relationship of the form [3]

$$e = A\sigma^B + e_f \quad (2)$$

the relationship between settled height, H_{sett}, the unit weight of water, γ_w, the soil particle relative density, G_s and the effective stress parameters may be found by integration of Eq. (2) over the column length, ie

$$H_{sett} = V_s \left[\frac{A}{B+1} (\gamma_w [G_s - 1] V_s)^B + e_f + 1 \right] \quad (3)$$

Generally the minimum void ratio, e_f, may be found by air drying a sample. The other parameters are found from two or more column tests by regression techniques. These tests should be significantly different in height in order to obtain a good correlation for the parameters.

Desiccation

Desiccation is modelled as a two stage evaporative process [9]. One the rate of bleed from the waste surface (ie v) exceeds the potential evaporation rate, stage one evaporation begins to occur. As the first stage of evaporation is equal to the potential evaporation the important consideration is to predict when stage one evaporation will end. This is achieved by assuming the quasi-analytical relationship [7]

$$\theta = \frac{1}{a} \ln \left(1 + \frac{a e_p H^1}{2 D_o} \right) \quad (4)$$

Where θ is the volumetric water content, e_p is the potential evaporation rate, H^1 is the height at which stage one evaporation commences. The fitted parameters a and D_o describe the diffusivity of the soil as an exponential relationship [6]. They are best obtained by fitting Eq. 4 to laboratory experiments. This requires at least four different desiccation experiments in order to achieve enough data with which to regress.

The second stage of evaporation is characterised by a falling rate of actual evaporation. A modified sorptivity relationship is used to model this stage of the form:

$$E = b_s V_w^1 \sqrt{t - t_o} + E_o \quad (5)$$

where E is the cumulative evaporation, b_s is the normalised sorptivity, V_w^1 is the total volume of water per unit area to be evaporated at the commencement of stage one and E_o and t_o are fitting parameters. Unlike most sorptivity relationships [1,10,16,11], this approach uses two fitting parameters to describe this stage. They are dependent upon the potential evaporation rate and the time at which stage two evaporation commences as given by Eqs. 6 and 7.

$$E_o = e_p t_1 - \frac{(b_s V_w^1)^2}{2 e_p} \quad (6)$$

$$t_o = t_1 - \left(\frac{b_s V_w^1}{2 e_p} \right)^2 \quad (7)$$

Substituting Eqs. 6 and 7 into 5 gives a quadratic equation in E. Using this equation regression techniques may be used to determine b_s, e_p and t_1 from the desiccation data.

As well as predicting sedimentation and desiccation, PRED uses empirical relationships to cater for settlement during desiccation, rewetting due to rainfall and continuous deposition.

Current limitations

Because PRED relies upon empirically derived constants, the method by which these parameters are obtained is very important. The sedimentation characteristics are currently obtained by analysing both large diameter (100 mm) settling columns 2 m in height. The drying parameters are found by laboratory scale tests on $0.5 \times 0.5 \times 0.5$ m samples in specially built lysimeters [17,19]. These tests typically run for 3-4 months before the drying cycle is complete and the data may be analysed.

questionable. If the procedures outlined above are adopted, desiccation predictions can proceed with some degree of verified accuracy.

Small scale experimentation

To date the prediction of the deposited behaviour of subaerially deposited tailings has been successfully predicted using large scale tests. The method of parameter determination is as discussed previously. The important differences for small scale tests are summarised below.

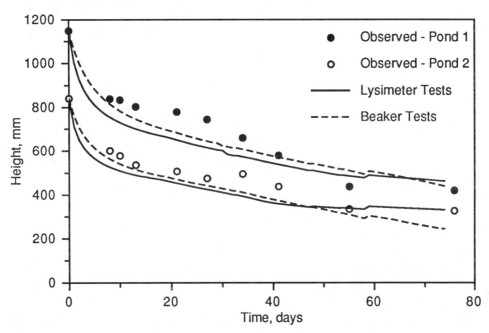

Figure 1: Observed vs predicted height for Riverside tailings

It has been recognised that for feasibility scale design, the current requirements in terms of cost and time required for parameter determination are unsuitable. Such recognition has promoted the need for smaller scale testing in order to predict model parameters with suitable accuracy for feasibility design.

In addition to the reasons given above, there is also a need to examine the accuracy of small scale desiccation experiments in general. Currently, estimates of the anticipated effects of desiccation are achieved through the use of small, simple drying tests. The ability to extrapolate this data to field scale with any degree of accuracy is

Sedimentation

The sedimentation parameters may be successfully attained using short (0.5 m) columns. The use of such columns reduces the settling time dramatically. Two tests are recommended and the resulting parameters v_o and k_o from Eq. 1 should be averaged.

As only one height is used, the determination of parameters A and B using Eq. 3 is difficult. It is suggested that published data be used to define B [3] thus allowing A to be determined from Eq. 3.

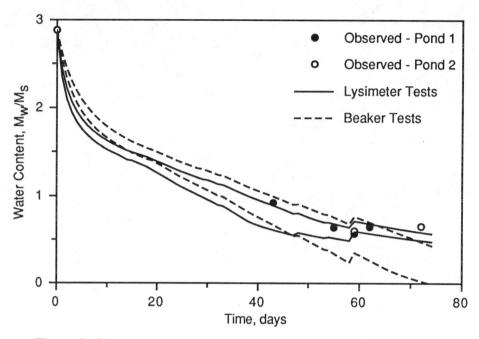

Figure 2: Observed vs predicted water content for Riverside tailings

Desiccation

The determination of the desiccation parameters a, D_o and b_s has been shown to be very sensitive to the laboratory procedure used. The laboratory lysimeters are exposed to radiation induced evaporation using four 500 Watt linear halogen light sources. The radiation sources are fluctuated diurnally to simulate sunlight. Initial experiments using diurnal fluctuations on small beaker samples resulted in unrealistic desiccation parameters. This was attributed to the prolonged duration of stage one evaporation as a result of diurnal fluctuations.

Subsequent analysis has shown that small beaker samples of 100 mm and 200 mm settled height under constant evaporative conditions are able to produce material parameters with acceptable accuracy.

Parameter comparison

A comparison between parameters obtained by the small scale method and those obtained by the original lysimeter tests is shown in Figs. 1 and 2. Also included are the observed data for two experimental ponds. In general the comparison between the original parameters (lysimeter tests) and the small scale parameters (beaker tests) is favourable.

Some of the observations of the small scale parameters are
- settlement is slightly underestimated
- the duration of stage one evaporation is overestimated (as shown by the extended linear loss of water in Figures 1 and 2)

These observations are expected due to the reduced scale of the tests. A summary of the parameters is given in Table 1.

TABLE 1: Lysimeter vs beaker test parameters

Parameter	Original	Small scale
v_o	24	4.5
k_o	-13.4	-10.9
A	1.94	1.485
B	-0.162	-0.143
D_o	18	1054
a	11	47.6
b_s	0.093	0.075

Conclusions

The results show that small scale testing can produce drying parameters of sufficient accuracy for feasibility scale tests. The comparison between lysimeter scale and beaker scale tests shows that sedimentation is generally under predicted while desiccation is over predicted. It must be remembered that thus far the technique has been tested for high clay content tailings. More research is required to verify this conclusion for several other materials.

Although small scale testing has been proved successful, it is important to consider how the observed behaviour may be interpreted for large scale prediction. In this instance the drying parameters have been derived in accordance with an established tailings deposition model. The extrapolation of such data simply by purely empirical means is questionable.

References

1 Black, T.A., Gardner, W.R. and Thurtell, G.W. (1969). "The prediction of evaporation, drainage and soil water storage for a bare soil," Soil Science Society of America Proc. **33**:655-660.
2 Blight, G.E. and Stephen, O.K.H. (1979). "Geotechnics of gold mining waste disposal" in Current Geotechnical Practice in Mine Waste Disposal, ACSE, Michigan, pp. 1-52.
3 Carrier, W.D. and Beckman, J.F. (1984). "Correlations between index tests and the properties of remoulded clays," Géotechnique **34**(2):211-228.
4 Carrier, W.D., Bromwell, L.G. and Somogyi, F. (1983). "Design capacity of slurried mineral waste ponds," Journal of Geotechnical Engineering, ASCE **109**(5):699-716.
5 Cooper, D. (1988). "Tailings management in Australia" in *Hydraulic Fill Structures*, Ed(s): D.J. van Zyl and S.G. Vick, American Society of Civil Engineers, Colorado State University, pp. 130-141.
6 Gardner, W.R. (1959). "Solutions of the flow equation for drying soils and other porous media," Soil Science Society of America Proc. **23**(3):183-187.
7 Gardner, W.R. and Hillel, D.I. (1962). "The relation of external evaporative conditions to the drying of soils," Journal of Geophysical Research **67**(11):4319-4325.
8 Gibson, R.E., Schiffman, R.L. and Cargill, K.W. (1981). "The theory of one-dimensional consolidation of saturated clays, II. Finite nonlinear consolidation of thick homogeneous layers," Canadian Geotechnical Journal **18**(2):280-293.
9 Hillel, D.I. (1979). *Fundamentals of soil physics*, Academic Press, New York.
10 Idso, S.B., Reginato, R.J. and Jackson, R.D. (1979). "Calculation of evaporation during the three stages of soil drying," Water Resources Research **15**(2):487-488.
11 Jalota, S.K., Prihar, S.S. and Gill, K.S. (1988). "Modified square root of time relation to predict evaporation trends from bare soil," Australian Journal of Soil Research **26**:281-288.
12 Knight, R.B. and Haile, J.P. (1983). "Sub-aerial tailings deposition," *Proceedings* 7th Pan-American Soil Mechanics Conference, Vol. 2, Vancouver, Canada, Canadian Geotechnical Society, pp. 627-639.
13 Lighthall, P.C. (1987). "Innovative tailings disposal methods in Canada," International Journal of Surface Mining **1**:7-12.
14 Schiffman, R.L., Pane, V. and Gibson, R.E. (1984). "The theory of one-dimensional consolidation of saturated clays, IV. An overview of nonlinear finite strain sedimentation and consolidation" in *Sedimentation/Consolidation Models - Prediction and Validation*, Ed(s): R.N. Yong and F.C. Townsend, ASCE, San Francisco, California, pp. 1-29.
15 Schiffman, R.L., Vick, S.G. and Gibson, R.E. (1988). "Behaviour and properties of hydraulic fills" in *Hydraulic Fill Structures*, Ed(s): D.J. van Zyl and S.G. Vick, American Society of Civil Engineers, Colorado State University, pp. 166-202.
16 Sparrow, G.J. (1981). "Calculation of maximum ponding depth to dewater a clay tailing"No: MCC 290 CSIRO, Division of Mineral Chemistry, February 1981.
17 Swarbrick, G.E. (1992). "Transient unsaturated consolidation in desiccating mine tailings," PhD thesis presented to the School of Civil Engineering, UNSW, 1992.
18 Swarbrick, G.E. (1993). "An approximate method for the design of tailings dams using sub-aerial deposition," Geotechnical Management of Waste and Contamination, Ed(s): R. Fell, T. Phillips and C. Gerrard, Sydney, Australia, 22-23 March, A.A. Balkema, pp. 463-472.
19 Swarbrick, G.E. and Fell, R. (1992). "Modelling the desiccating behaviour of mine tailings," Geotechnical Engineering Div. Jour., ASCE **118**(4):540-557.
20 Thomas, D.G. (1964). "Turbulent disruption of flocs in small particle size distributions," Journal of American Institute of Chemical Engineering **10**(4):517-523.

Approach to Closure Designs for Mine Sites with Severe AMD

M.B. Szymanski, H.L. MacPhie
GEOCON, Division of SNC♦Lavalin Environment Inc., Mississauga, Ontario, Canada

Abstract

The approach to closure designs for mine sites with severe acid mine drainage (AMD) is discussed from the practitioner's perspective. Designing for mine closure is a relatively new discipline and there is a need to exchange the experience gained from actual mine closure design projects. The paper focuses on those aspects of the approach to closure designs which are characteristic for mine sites with severe AMD. The need for a site specific approach to closure designs and for investigation of all plausible closure design options is discussed and emphasized throughout the paper.

Introduction

It has been stated on many occasions that, from a closure cost and environmental impact perspective, existing mines with actual or potential severe AMD represent the most significant problem facing the mining industry in Canada. On the other hand, designing for mine closure is a relatively new discipline which has evolved over the last 10 to 15 years only. Hence, there is a strong interest within the mining community in planning and designing for mine closure as demonstrated by the successes of the Mine Environment Neutral Drainage (MEND) program and Montreal's Conference on the Abatement of Acidic Drainage (Reference 1). The purpose of the present paper is to convey some of the Authors experience gained over the last several years on more than ten major closure design projects carried out for existing or new mines located in New Brunswick, Quebec, Ontario and British Columbia. The type of mines included underground and/or open pit gold, base metals and graphite mines.

The two fundamental conclusions that have emerged from these projects are seemingly contradictory: on one hand, each project site requires a very specific approach to closure design and, on the other hand, the experience and concepts developed from one project can be almost always applied to another project. This apparent contradiction results from the fact that each mine site is characterized by specific features (such as distribution and properties of mine wastes, characteristics of the receivers, climatic, hydrologic, geologic and geomorphologic conditions, available materials and other resources, mining objectives and legislative requirements, etc.) which have to be accounted for in the designs and yet the design experience from a number of projects will usually allow for a quick identification and evaluation of closure options that could be potentially applied to a new site. Looking closely at the above conclusions, it follows that designing for mine closure is consistent with other engineering disciplines.

As with any new discipline, there may exist a

tendency to carry out studies or investigations, in support of closure designs, which are not always well justified. For instance, installation of some monitoring wells and carrying out a number of kinetic tests may be sometimes considered just because the same has been done at other mine sites. One of the objectives of this paper is to emphasize the need for a site specific approach to all the aspects of designing for mine closure.

The intent of this paper is to discuss a general approach to designing for closure of mines with severe AMD rather than to describe case histories. References to actual projects are made to emphasize and illustrate certain aspects of the discussion. Selecting a rational approach to designing for closure is, in the Authors opinion, fundamental to a successful completion of any major mine closure project.

Severity of AMD

Determination of the severity of AMD at a mine site is a good example of the necessity to approach each project on a site specific basis. The acid generation potential, as determined by acid base accounting testing and confirmed by kinetic tests, can be quite a misleading factor in establishing the severity of AMD. For instance, two mine waste materials from different sites can have the same properties in terms of net neutralization potential and acid generation rates, however, the severity of AMD at each site can be completely different in terms of environmental impacts. In this context, the severity of AMD depends on many other factors such as size, geometry and gradation of the waste, hydrologic, climatic and subsurface conditions, metal leaching rates, distance to the receivers, attenuation/assimilation capacity between the wastes and the receivers, sensitivity of the receivers, public perception, etc.. Furthermore, the severity of AMD must also be judged in terms of time. Even if AMD is presently fully developed at a site and is not judged severe from the perspective of water quality, it can become severe in the future as the attenuation capacity of the downstream environment is depleted/reduced and/or the AMD products accumulate.

Sensitivity of the receiving environment is always a very important factor in defining the severity of AMD. However, even in this regard, determination of AMD severity may present difficulties. For instance, the question of which receiver is to be considered as the reference and at what location is often not clear. It is quite obvious that in any receiving system there is a point sufficiently downstream at which AMD will not be considered severe, regardless of the specifics of the mine waste.

Based on the above discussion, we will assume that AMD is severe if it causes, or if it will cause at some time in future, an unacceptable environmental impact on the downstream receivers or terrestrial resources at a location which can be selected and justified based on environmental design criteria.

Design Criteria

It is convenient to divide the criteria on the base of which designs for mine closure are developed into three groups: General, Engineering and Environmental. The General Criteria would include those which are very site specific and usually quite obvious, for example, if it is evident that a portion of the mine site will require long term collection and treatment of AMD, then the availability of a treatment plant must be taken into consideration in developing the closure designs for other components of the site. Examples of Engineering and Environmental Criteria are given in the two following (italic) paragraphs, respectively. These are excerpts from the design criteria developed in conjunction with a closure design project completed recently. It is emphasized that these criteria were developed for a specific mine site and would not necessarily be appropriate for other sites.

Engineering Criterion - Example

The minimum freeboard requirements under extreme hydrologic events are to be:

major structures: *0.3 m under PMP event*
   1.0 m under 1:100 year event

minor structures: *0.0 m under PMP event*
   0.5 m under 1:100 year event

where the terms "major" and "minor" refer to the consequences of a failure of a site structure in terms of public safety and environmental or property damage, and PMP denotes Probable Maximum Precipitation.

Environmental Criterion - Example

To minimize potential for a long term environmental impact on water quality in the downstream receivers, the water retention structures are to be designed to ensure that the collected contaminated runoff is not allowed to discharge to the environment under a hydrologic event with a return period of up to 1:100 years. This 1:100 criterion may refer to one of the two following hydrologic events: ▸ *rainfall event selected as the 24-hour 1:100 year rainfall, or* ▸ *a long duration runoff sequence, defined from continuous simulations of rainfall and/or snowmelt, which generate the largest runoff occurring with a return period of 100 years. For each of the designed structures, the more stringent of these two events is to be selected as the governing criterion.*

The above paragraphs illustrate the importance of separating the design criteria established for a mine closure project into engineering and environmental criteria. Spillway design details, for instance, can be different when based on engineering criteria (where the integrity of a water retention structure has to be assured) or on environmental criteria (where a discharge of contaminated runoff has to be restricted). The design intervals and the probabilities of occurrence, which determine the selection of design return periods, may also be different depending on whether environmental or engineering criteria are considered.

The Authors believe that establishing one set of uncompromising engineering and environmental design criteria for all mine sites, or for all parts of a single mine site, is not feasible. In other words, engineering and environmental criteria applied to mine closure designs should be set up on a site specific basis. Consider, for example, the design issue of controlling the discharge of contaminated runoff to receivers as a surface or subsurface flow. On a mine closure project in Ontario, a ditch blasted in rock and a grouted bedrock barrier were designed and constructed to reduce the discharge of contaminated surface and subsurface flows originating at a waste rock pile. The results of seepage analysis showed that the groundwater discharge would be reduced by 50 % (minimum estimate). On a New Brunswick project, a substantial flow of contaminated groundwater under an existing water retention dam has been observed for years. An option to provide a grouted bedrock cutoff was briefly considered also at this site. However, as it turned out, the preferred design option for this site is to permanently lower the water level upstream of the dam to reduce the hydraulic gradient to zero so that no seepage would bypass the dam under normal operating conditions reducing, therefore, the existing seepage rate by practically 100%. The zero gradient approach to seepage reduction was not possible at the Ontario project site and an environmental criterion that would require close to

100% reduction in the contaminated groundwater flow rate at that site could not be practically met.

Nevertheless, a meticulous analysis of the site conditions and a design effort can often substantially reduce the environmental impacts and reduce the closure costs as well. On another project in Ontario, a cutoff trench had been proposed to reduce the flow of contaminated groundwater through a confined aquifer overlain by 10 metres of soft clay. Some contaminated seepage would still bypass this cutoff. Owing to the existence of the soft clay deposit, the expected construction cost of the cutoff was very high, estimated at about $3,500,000. Subsequently, we carried out a detailed groundwater flow study which showed that installation of gravity relief wells, at a fraction of the above cost, would allow for a complete interception of the contaminated groundwater flow.

Even if a considered closure option satisfies all the design criteria, other plausible rehabilitation options should still be developed and compared, at least at a feasibility design level. This is because of typically high rehabilitation costs that are associated with the closure of mine sites with severe AMD. The investigated design options should then be judged in terms of closure option evaluation criteria.

Closure Option Evaluation Criteria

There usually exists a number of potential closure design options that can be applied to a mine site with severe AMD. For the record, the Authors believe that a "zero impact" closure design option is not possible for the majority of the existing sites with severe AMD although, in some cases, achieving a situation in which the long term environmental impacts can not be measured is feasible (on a project in Ontario, it has been determined that a major tailings disposal facility is impacting water quality in the downstream receiver, however, the resulting contaminant levels are below the detection limits). The investigated closure options should be evaluated in terms of, in order of priority:

- chance of success
- closure cost vs environmental benefits

In other words, no matter how great the environmental benefits would be, there has to be a reasonable chance of success for the closure design to be selected. Although closure designs can be developed so that any mine site can be closed out with the long term environmental impacts below measurable limits, at the majority of existing mine sites with severe AMD the associated closure costs would be enormous and the mine owners (and the public as a whole) could not afford such an expense. Nevertheless, achieving "extra" (i.e., beyond a reasonable minimum required) environmental benefits at an acceptable cost is in many cases possible. For example, at a mine site in Ontario, a contaminated runoff interception dam was designed and constructed in conjunction with mine closure so that any short or long term runoff event, including that resulting from a PMP event, would be intercepted. Although the design criterion called for the interception of the runoff so that no spill would occur, on the average, more often than once in hundred years, by raising the dam height by about 0.6 metres it was possible to exceed the design criterion very significantly, at an acceptable additional cost.

Using well established engineering principles in designing for closure obviously improves the chance of success. As it turned out, this is the approach that has been taken in conjunction with most of the mine closure design projects in which the Authors were involved. Such an approach may be somewhat conservative, however, the designs, when carried out based on newer creeds, did not often produce satisfactory results in terms of the chance of success. This was perhaps most visible in considering the use of dry soil covers to reduce infiltration and/or oxygen flux (to reduce AMD to acceptable levels). It has been concluded that the available techniques to predict the performance of dry covers present a significant problem. For example, modelling of infiltration into dry soil covers to be placed over waste rock piles was carried out in conjunction with two projects using the well known, and sometimes highly recommended, HELP computer model. It has been determined, however, that this model is not well suited for the Canadian conditions and, in addition, it includes a heat generation subroutine developed for municipal landfills rather than for mine wastes. Even after appropriate modifications to that subroutine were made, the modelling results were not sufficiently convincing, in terms of the chance of success, to select a soil cover as the preferred closure option. Moreover, the most obvious (and frequently emphasized) weakness with regard to the chance of success of dry soil covers is their long term performance as related to drying, cracking, long term erosion resistance, root penetration, freeze-thaw effects etc..

Determination of the net neutralization potential, acid generation and metal leaching rates may also present a problem with regard to the chance of success since obtaining definite and accurate field parameters based on laboratory test results is often very difficult, if possible at all. This may be particularly applicable to borderline cases. Based on this consideration, it was decided to recommend the removal of a 'borderline' waste rock pile at a mine site in Ontario after the acid base accounting tests indicated that the net neutralization potential was slightly on the negative side.

The use of a plastic liner over a mine waste to prevent oxygen entry and inhibit acid generation would ease the design evaluation task significantly, however, the high cost of installation and uncertainty with regard to limited longevity of plastic liners are the factors which, in many cases, may render such an option impractical. Consider, for instance, a medium size tailings deposit generating severe AMD and located in northern Ontario (350 mm runoff depth), with a footprint size of 100 ha and a total watershed area of 150 ha, which is well suited for plastic liner installation. The cost of plastic liner installation including bedding (where required), protective soil cover, anchoring, etc. would be in the order of $20M. On the other hand, a collect and treat approach would involve the annual cost of treatment (common AMD characteristics) in the order of $0.3M/yr, the capital cost of providing runoff collection and treatment facilities in the order of $3.0M and the cost of tailings surface stabilization in the order of $2.0M. Depending on the longevity of the liner and the type of rehabilitation measures applied after liner deterioration, the total cost of the site rehabilitation, at 3.5% of the real rate of return, would be (millions of today's dollars):

	Longevity = 50 yrs	Longevity = 200 yrs
A	$20+$3.6=$23.6	$20+$0.02=$20
B	$20+$2.4=$22.4	$20+$0.01=$20

where Case **A** refers to the situation in which the liner would be replaced after 50 or 200 years, and Case **B** refers to the situation in which a runoff collection and treatment operation would be realized after these time periods. The $20M component represents the cost of the plastic liner installation upon mine closure and the remaining components represent the costs that would be incurred after 50 or 200 years including capital and operating costs, as applicable (two simplifying assumptions have been made here: ▸ the costs of plastic liner installation and replacement are the same and ▸ the tailings surface will have to be stabilized after the liner deteriorates in Case **B**).

Case **A** shows that $3.6M would be saved in terms of liabilities if an assurance can be gained that the liner will last for 200 years rather than 50 years. Comparison of Cases **A** and **B** for the longevity of 50 years illustrates that $1.2M would be saved by simply declaring that after the liner deteriorates it would not be replaced but rather a collection and treatment operation would commence.

Consider also the same example assuming that a collection and treatment operation is implemented upon mine closure, i.e., no plastic liner is provided. Computations show that the total closure cost, including the capital and the long term operating costs, would be $13.6M. Therefore, savings of at least $6.4M could be achieved without the necessity to prove the longevity of the plastic liner. Similar discussion could apply to a dry soil cover and to other rehabilitation options.

The above discussion has focused on two issues inherent to the design for closure of mine sites with severe AMD: *longevity* of rehabilitation measures and *perpetual* collection and treatment of AMD. Both of these issues are relatively new to the design engineers.

How Long is Long ?

Is it possible to design rehabilitation measures that will last forever, or more realistically for, say, 1000 years ? This question pertaining to the longevity of closure measures appears in conjunction with the design of runoff interception and collection systems, dry or wet covers, tailings and water retention structures, spillways, etc. In the Authors opinion, the answer is yes, however, some monitoring and maintenance will in practical terms be always required (unless an "unlimited" budget to close a mine is available). Therefore, a complete "walk-away" closure scenario, which would not require any monitoring or maintenance over the next 1000 years, is not realistic in conjunction with designing for closure of most of the existing mine sites with severe AMD (there are obviously exceptions, for instance, if all acid generating wastes from a mine site can be economically disposed of in an open pit which would remain naturally flooded, no long term monitoring or maintenance would be, in all likelihood, required).

With regard to perpetual collection and treatment it is considered that the term "perpetual", even if applied to a period of say 300 or 500 years, is misleading. This is because it is unrealistic to expect that a collection and treatment operation will be executed in "perpetuity" (i.e., until sulphides in the portion of a mine waste exposed to water and oxygen are depleted or become inactive). Taking into account the technological progress made over the past 300 years (between 1694 and 1994 this progress has continued in an exponential form), it would be unreasonable to assume that the expensive and unproductive AMD collection and treatment operations, including wastage of the sludge and other resources, will continue in the year 2294 (300 years into the future). There is a very high probability that these operations will only continue for a relatively short time when considered in the above context. An use for AMD will, in all likelihood, be found if cost benefits can be achieved. A possibility to use AMD for coagulation of municipal wastewater, for example, was investigated and reported on recently in Reference 2. On the other hand, AMD is already used for treatment of effluent from an acid plant at a property in Quebec.

It follows that the concept of a long time, as applied to the two extreme mine closure scenarios (complete "walk-away" and "perpetual" collection and treatment), has to be carefully considered in designing for closure, particularly if one looks also at other possible closure scenarios in more detail.

Closure Scenarios

The Authors experience in mine closure designs indicates that the following seven general closure scenarios applicable to mine properties with actual or potential severe AMD can been identified:

(i)   immediate walk-away
(ii) walk-away after transition period(s)
(iii) perpetual collection and treatment
(iv) delayed perpetual collection and treatment
(v)   perpetual maintenance
(vi) perpetual maintenance after transition period(s)
(vii) passive treatment

Scenarios (i), (iii) and (vii) correspond to the walk-away, active care and passive care design categories identified, for example, in Reference 3. The "extended" classification of closure scenarios presented above refers specifically to AMD sites and it accounts for four additional situations which the Authors frequently considered. Firstly, Scenario (ii) accounts for the fact that in many cases a "walk-away" situation would be possible only after a transition period during which collection and treatment of AMD would still be required (e.g., after flushing out accumulated AMD products from a reactive material deposit in conjunction with the construction of a soil or water cover). Should such a transition period be relatively long (say, more than 10 years), both economical and environmental consequences can be very significant. Secondly, scenario (iv) accounts for the fact that in some cases the commencement of an AMD treatment operation may by delayed (until, for example, an open pit fills up with contaminated runoff). This may have significant consequences with regard to closure costs. Thirdly, Scenario (v) accounts for the situation in which closure of a mine site with severe AMD would involve long term operating and/or maintenance requirements, however, no treatment of AMD would be necessary. Finally, Scenario (vi) accounts for the situation in which perpetual maintenance would commence after the end of a transition period involving treatment of AMD.

Except for Scenario (vii), all of the above scenarios have been selected as preferred closure design options for a mine site, or for a portion of a mine site, on our past or current projects. If, however, allowing for assimilative capacity of the downstream environment is classified as "passive treatment", then Scenario (vii) will be automatically included in closure designs for the majority of mine sites with severe AMD.

Generic Project Approach

An emphasis has been made in this paper on the need for a site specific approach to designing for mine closure. Nevertheless, it seems that a generic approach that could be applied to most of mine closure projects can be defined and this is outlined in a point form below.

- Review and Analysis of Existing Information
- Site Mapping
- Identification of Closure Concerns
- Analysis of Mine Development Stage vs Closure Requirements
- Evaluation of Current Environmental Conditions
- Establishing Project Criteria
- Identification of Viable Closure Scenarios and Design Options
- Feasibility Evaluation of Identified Design Options
- Development and Execution of Support Investigation and/or Monitoring Programs
- Selection of Preferred Closure Design Option
- Engineering Analyses and Designs
- Modelling of Long Term Environmental Impacts

Albeit all these closure design tasks are equally important, it is emphasized that the need to identify and consider all plausible design options for a mine site, or even for a portion of a mine site, is particularly essential. On one of our recent mine closure projects, as many as ten design options were considered for a mine site component before the Authors felt satisfied that all plausible solutions were investigated. Although this is an extreme example, two or three design options for a mine site, or for a site rehabilitation area, can often be identified such that only after some preliminary design considerations one of the options can be declared preferable. As stated before, the typically high cost of the rehabilitation of a mine site with severe AMD and the potential for long lasting environmental damage demand that no stone be left unturned.

Acknowledgements

In the course of carrying out mine closure design projects, the Authors benefited very substantially from discussions with numerous mining people. In particular, the Authors benefitted from input to many projects received from Messrs. V. Coffin, R. Siwik and W. Sencza of Noranda Minerals Inc.

References

1. Second International Conference on the Abatement of Acidic Drainage, Montreal, September 16, 17 and 18, 1991.

2. S.R. Rao, R. Gehr, M. Riendeau, D. Lu and J.A. Finch (1991), "Potential of Acid Mine Drainage as a Coagulant", Second International Conference on the Abatement of Acidic Drainage, Montreal, Paper No. 9.8.

3. Ontario Ministry of Northern Development and Mines (1992), "Rehabilitation of Mines - Guidelines for Proponents", Version 1.2

Municipal

Remedial Works in a Highly Industrialized Region

Hans L. Jessberger
Ruhr-University Bochum, Germany

Abstract: Site assessment, risk analysis and development of remediation concepts for contaminated land and abandoned landfills are discussed. Two remediation projects show the application of these aspects. Design alternatives are described.

1. Introduction

The Ruhr Industrial Area is heavily populated with 5.5 M inhabitants and highly industrialized for more than 150 years. There is a strict demand to reactivate the areas of closed factories rather than to settle new industries in the remaining green zones. Further, abandoned landfills can be the source of heavy polution of air, soil and water. In investigations it has to be sorted out, what is the best remedial measure for the individual site, based on site assessment, investigation, considering alternative solutions, leading to decisions which are based on the real situation, specially considering the effectively possible contaminant emission.

In this paper two examples of remediation projects are presented. One project is the remediation of an abandoned landfill with mainly industrial waste including construction debris and some highly toxic waste. The investigations, the design and the execution took place about 6 years ago. The landfill has been contained by a capping system including active hydraulic and pneumatic measures. The second project is the location of a large steel mill, which has been shut down for a few years. The task is the reactivation of the contaminated area. The new utilization of the location has been chosen according to the remediation concept which consists mainly of providing containment or immobilisation. The design of the remediation concept is in the final state and the execution will follow soon.

Both projects are characterized by the fact that the decision for the remediation concept and the design of the remediation has been based on a careful risk analysis, undertaken in a qualitative manner following extensive field and laboratory investigations. In both cases containment including additional measures has been chosen and confirmed by the authorities.

2. Basic Considerations on Remediation of Contaminated Land and Abandoned Landfills

2.1 Site Investigation

For the site investigation of contaminated landfill and abandoned landfills experience is available and the standard investigation procedure is as follows.

a) Information on the types of products, the production steps, the input material, the location of factories, buildings, possible accidents etc. are taken from the files or if possible from former employees or adjacent residents. Extremely useful is the multi-temporal air-photo interpretation which provides detailed information on the historical development of the investigated site.

b) Based on these results an appropriate sampling strategy has to be chosen in order to recognize the irregular spatial distribution of the contamination in heterogenic soil conditions. It should be intended to avoid as much as possible uncertainties with respect to the extent and the type of contaminations. Guidelines for the investigation of soil contaminations based on statistical or non-statistical methods are reviewed by Bosman (1993).

c) Ground water, soil and air sampling has to be conducted in such a way, that for a contaminated site or an abandoned landfill the initial site conditions or range of aggressive substances present can be reliably determined, effects on the environment can be recognized and remedial methods be introduced in good time. Additional contamination of soil or ground water by using contaminated sampling equipment previously used in contaminated areas must be avoided, see R 1-2 (ETC 8, 1993).

d) A test strategy for the chemical analysis of the ground water, soil and air samples has to be developed, based on the known or the expected contamination. The investigation should start with screening tests, looking for combined parameters like BTX or PAH. Field measurement techniques may be helpful.

Finally it is necessary to have the results of step by step and contaminant specific analyses. In general the analyses contain some parameters like

- COD, BOD_5
- TOC, hydrocarbons
- phenol (total)
- AOX, CN^-, S^{2-}

PAH = polyaromated hydrocarbons
BTX = benzene, toluene, xylenes
COD = chemical oxygen demand
BOD_5 = biological oxygen demand
TOC = total organic carbon
AOX = adsorbable organic halogen compounds

The following metal ions may be relevant:

- Al, As, Pb, B, Cr, Cd, Cu, Ni, Hg, Zn.

2.2 Risk Assessment

In the risk assessment the results of the different investigations are evaluated with respect to the need and to the kind of remediation. This evaluation has to consider the geogenic contamination and other background data. On the other hand the existing or the future utilization of the contaminated site has to be taken into account. In some countries, e.g. The Netherlands, or in some cities, e.g. Hamburg, there are values available showing the tolerable content of distinct contaminations in soil or ground water. But besides such limiting, yielding or transfer values in the risk assessment the following aspects have to be considered:

a) Contaminated ground water
- specific hydro-chemical properties like stability or degradation of the contaminants
- specific hydro-mechanical properties of the aquifer like direction and velocity of contaminant transport
- site specific aspects like cover layers, which can influence the contaminant migration into the ground

b) Contaminated soil
- not only contaminant concentration in the soil but primarily the mobility of the contaminants and their elution capacity and transfer behaviour depending on geological and hydrogeological site conditions
- existing or future utilization of the contaminated site
- transfer behaviour into plants depending on
  · type of contaminants and binding form on the soil particles
  · chemical and physical soil properties
  · type of plants
  · duration of vegetation period
  · depth of rooting soil
  · precipitation and climate during vegetation period

c) Health risk
- regulations, e.g. related to contaminant content in plants and food
- acceptable daily intake (ADI) depending on e.g. age of persons etc.
- consideration of the multi-steps impact path (e.g. soil-plant-animal-human; soil air-cover-exterior-human; ground water-contaminant migration-drinking water extraction-human).

In summarizing it should be mentioned that risk assessment must be based on sound site investigations and results of field and laboratory tests, on possible impact on people and on goods needing protection and also on the existing or future utilization of the site.

2.3 Remediation Concept

Different methods of remediation for contaminated sites and abandoned landfills are available. The Council of Experts on Environmental Problems (SRU, 1990) defines as remedial measures:

- Restriction in utilization
  · this restriction can lead to the change of the existing utilization of the site in the direction of avoiding sensible use like playground etc.
  · within the remediation concept it is possible to define the use of the site after remediation in that way, that it is compatible to the remediation method, e.g. industrial park rather than playground
- Providing containment by cutting off migration pathways by
  · capping systems
  · vertical cut-off walls
  · basal lining systems
  · active pneumatic and hydraulic measures
  · immobilization
- Decontamination of contaminated soil by
  · thermal treatment
  · microbiological treatment
  · soil washing
- Replacement of contaminated soil or waste; this method should not be used.

According to SRU (1990) providing containment can be considered as being equivalent to decontamination since any risk for humans and the environment are eliminated, for instance by encapsulation. Containment is in principle a temporary remedial action and must allow for clean-up activities to be undertaken in the future; an effective monitoring system will show if and when additional remediation is necessary.

The selection of an appropriate remediation concept should follow this methodology:

- Characterization of the initial situation
  - protection objectives
  - trigger values
  - individual planning intentions
- Pre-evaluation
  - technical applicability test
  - plausibility test/comprehensive test
- Site specific remedial scenarios
  - modular development of relevant partial processes
- Determination of the remediation objectives
  - determination of the site specific importance of main and secondary objectives
- Detailed evaluation
  - determination of the scenarios with respect to the degree of realization
  - classification of the existing alternative concepts concerning effectiveness and residual risks
- Economic consideration
  - qualified cost estimate
  - consideration of cost effectiveness
- Proposal of remedial concept with choice of the most suitable site specific remediation concept.

For the detailed evaluation of the remediation scenarios the system of objectives shown in Table 1 can be helpful. After economic considerations a remedial concept is proposed by the consultant to the responsible authorities. This proposal contains the scope of the remediation and the measures to be taken in individual cases.

Table 1: Site specific selection of remediation concepts

main objectives	secondary objectives
reduction of existing risk	soil specific applicability
	pollution specific applicability
	good site specific adaptability
	reliable and simple monitoring- and control possibilities
availability	stage of development
	short duration of the remedial action
	minimal requirements on infra-structure
	minimal limitations of use/applicability
minimization of new environmental impacts	minimal impact by emission
	minimal energy and resource consumption
	preservation of soil specific characteristics
	minimal consequences of groundwater lowering
	minor residual products
high safety standards	safe soil excavation
	safe soil treatment
	safe additional measures

3. Remediation of the Abandoned Industrial Waste Landfill "Richter" in Bochum

Outside of Bochum city, but close to residential areas, sand for construction purpose has been borrowed. About 20 pits have been refilled in the early seventies by different materials without sealing measures. During a general site assessment campaign for all these places in 1986, which are used as farmland, in one of these landfills toxic industrial waste was found. Based on the results of an intensive site investigation, a remediation concept has been developed and the remedial work of building a capping system including hydraulic and pneumatic measures has been executed.

3.1 Site assessment

From geological maps it was known that in this area the pleistocene meander of the Ruhr-river in the carbonian rocks stratum has been filled with sandy and silty melt-water sediments, where the sand pits have been installed. With 20 borings the geological and hydrogeological situation has been investigated with these results (see Fig. 1):

- The soil formation around the former sand pit consists mainly of silt with $k < 1 \cdot 10^{-9}$ m/s
- There is one small "channel" of fine to medium sand with $k = 1 \cdot 10^{-5}$ to $1 \cdot 10^{-6}$ m/s crossing the pit in a depth of 12 to 15 m below surface
- Ground water not influenced by the landfill lays about 25 m below surface within a sand/gravel aquifer stratum
- In the landfill itself the leachate liquid reaches a level of about 7 m below surface, gradually decreasing in the vicinity of the landfill to the uninfluenced ground water level of about 18 m below.

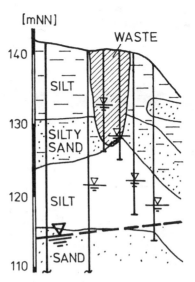

Fig. 1: Soil and ground water conditions of landfill "Richter"

The surface and the inground contour of the landfill has been investigated by multi-temporal air-photo interpretation and has been reconfirmed by dynamic soundings with the result, that the geometric situation of the landfill with the volume of about 160000 m³ is well known.

The waste material filled into the former pit could be recognized from the files of the landfill operator indicating large amounts of slag and construction debris but also some smaller portions of high volatile halogen compounds. This anticipated waste material has been reconfirmed by the results of the chemical analysis. Only one sample of leachate liquid contained a very small content of polychlorated dibenzodioxine and dibenzofurane, whereas in the borings outside of the landfill nothing of these toxic substances has been found.

The results of the soil-air analysis are:
- Varying amounts of methane in the soil air

- In some measuring points high to medium volatile hydrocarbons
- The content of high volatile halogenated hydrocarbons in the soil air has been insignificant.

3.2 Remediation concept

Prior to the remediation the landfill has been covered by a layer of silty soil and a restoration layer for farmland use. The cover is permeable and the leachate could reach the above mentioned level within the landfill well on top of the original ground water level.

In the course of developing the remediation concept some alternative proposals have been discussed. The first proposal of an on site- or off site-decontamination could not be realized due to the fact, that no suitable decontamination technology was available at that time.

The next proposal was encapsulation with a vertical cut-off wall surrounding the landfill together with a capping system. The benefit of this proposal was very limited because of the following reasons:
- The mobility of the toxic substances in the leachate is extremely low and these substances will not move outwards if they are not forced e.g. by infiltrating surface water. Therefore the cut-off walls are not needed if the capping system is effective.
- The contaminants that have already migrated into the surrounding soil and ground water can not be brought back.
- Because of the other abandoned landfills nearby with undoubtedly existing pollution it does not make sense to keep just one spot clean.

The final design of the remediation concept contains an encapsulation in combination with hydraulic and pneumatic measures (see Fig. 2):

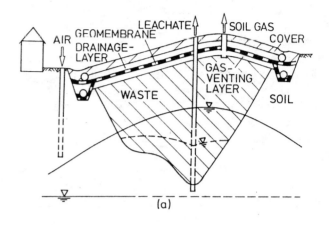

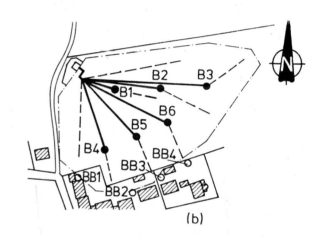

Fig. 2: Cross-section (a) and ground plan (b) for remediation of landfill "Richter", not to scale
B   Leachate extraction well
BB Ventilation well
--- open   --- closed gas venting system

- The capping system has to prevent surface water penetration into the landfill from the top; details of the capping system are given in Fig. 3.
- The whole landfill is surrounded by a drainage trench preventing surface water from penetrating into the landfill from the side (see Fig. 3).

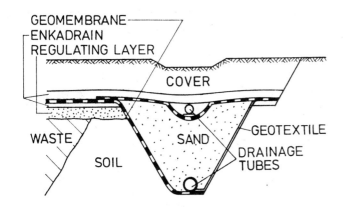

Fig. 3: Drainage trench including capping system

- The leachate liquid and the landfill air are extracted and treated.
- A number of vertical wells are arranged in front of the housing area providing soil air movement away from the residential area and also supporting microbiological degradation of organic compounds in the landfill.

The extracted leachate and the landfill gas is pumped to the service station in the northwest corner of the area, where the leachate is cleaned by activated carbon and the gas is incinerated. The clean surface water is collected at two locations and delivered into the community rain water collection system.

3.3 Soil-air extraction

In the service station three air blowers with a capacity of 300 m³/h each were installed, two of them in continuous operation and one as stand by unit. With this configuration a partial vacuum of 6.9 mbar was maintained in the soil-air ventilation system.

During a test operation in fall 1990 a methane content in the soil-air of about 2 % was measured. This content decreased to about 0.2 % within one month and then remained unchanged over a period of about 1.5 years. According to Fig. 4 the CO_2 content of about 1.5 % and the oxygen content of 14 to 18 % also remained unchanged during this period. No hydrogen sulfide has been detected during this time. These results showed that anaerobic atmosphere continues in small portions of the landfill, where methane is produced. However in general an aerobic atmosphere is predominant due to the relative high contents of CO_2 and oxygen.

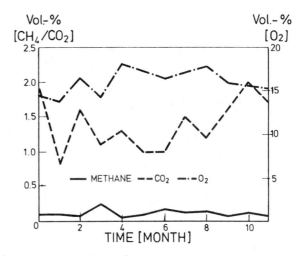

Fig. 4: Soil-air components during one year period

4. Remediation of an Area of the Abandoned Steel Mill "Henrichshütte" in Hattingen

The steel mill "Henrichshütte" in Hattingen has been milling steel for 150 years. Fifteen years ago, the steel production had to be reduced drastically. Therefore a large portion (45 ha) of the area of the factory now has to be remediated. The new utilization of the area should be a landscape structure and an industrial park.

The area to be treated is subdivided into two portions. One part is the so-called "Alter Ruhrarm" which means that 35 years ago the bed of the Ruhr-river has been filled to reclaim new land for the factory. The second part is the location of the former coking plant. Fig. 5a shows a photo of "Alter Ruhrarm", taken in 1970. In Fig. 5b the ground plan of the reactivated location with new factories and parks is shown.

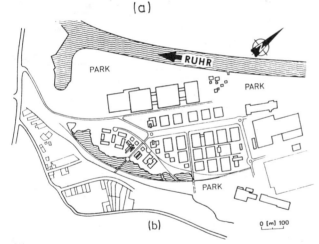

Fig. 5: Section of the abandoned steel mill "Henrichshütte"
(a) Photo   (b) Future utilization

4.1 Site Assessment

Based on a multi-temporal air photo interpretation the reconnaissance has been conducted. More than 100 soundings, about 150 test pits - due to the coarse grained refill material - and 16 borings for ground water measurement have been executed. The mean distance between the points of investigation has been 40 m. About 220 samples have been investigated in a chemical laboratory according to a special test program adopted to the contaminant inventory of heavy metals and coking plant specific contaminants.

The results of the field investigations show that on top of the carbonian rock stratum there is a gravel layer of maximum 10 m thickness sedimented by the Ruhr-river. In certain areas there is a silty loam cover on top of the gravel. The area under consideration is covered with waste material from the steel mill like blast-furnace slag, construction debris etc. up to a thickness of 4 to 5 m. Fig. 6 is a schematic drawing of this situation. The ground water is about 5 m below the surface.

Some results of the chemical analyses are presented in Table 2 and Table 3. The relatively high heavy metal contamination especially with lead and chrome contamination is shown. The contamination is not concentrated in distinct zones in the ground but it is diffusively distributed over the area of investigation. Further it is characterized by the filling of the Ruhr-river arm with waste, leading to the fact that the contaminant content is very non-uniformly distributed.

A big question has been whether or not the heavy metals are soluble. Elutriation tests indicated that a very small portion of the heavy metals is soluble in water presumably due to the high pH-value (pH > 7) and the silicate bonding form of the heavy metals. Further the low solubility of the heavy metals in water is also indicated by the very small heavy metal content in the ground water (see Table 2).

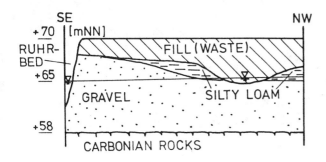

Fig. 6: Soil and ground water conditions of "Henrichshütte"

Table 2: Heavy metals in soil and ground water

parameter	soil (mg/kg)	ground water (mg/l)
lead	30,9- 38.500	< 0,001- 0,078
chrome	9,3- 4.900	0,002- 0,026
cadmium	< 0,3- 68,3	0,0001- 0,0003
mercury	0,02- 1.190	< 0,0001- 0,0002
copper	18,2- 768	0,008- 0,02
nickel	1,5- 2.030	0,004- 0,017
zinc	8,6- 75.000	< 0,01- 0,47

At the location of the abandoned coking plant not only heavy metals were found but also coking plant specific organic and inorganic contaminations of the soil and the ground water (see Table 3).

The chemical analysis of the soil air in this zone did not show significant contamination. Therefore the emission path "soil-air" is not decisive for the remediation concept.

Table 3: Hydrocarbons in soil and ground water underneath the former coking plant

parameter	soil (mg/kg)	ground water (mg/l)
BTX total	n.n.- 813**	n.n.- 0,0076
phenol	n.n.- 296**	n.n.
PAH (EPA)	n.n.- 141.430**	0,00207- 0,1394
PAH (TrinkwV*)	n.n.- 32.300	n.n.- 0,01

* drinking-water regulation
** with tar in phase

4.2 Remediation Concept "Alter Ruhrarm"

Due to the geological conditions and the results of the chemical analyses the main emission paths are:
- Wind blown dust
- Direct contact to contaminants
- Contaminant transfer via soil-plant/animal-human
- Ground water contamination with organics underneath the abandoned coking plant.

After a consideration of several remediation alternatives the following concept for this zone of "Alter Ruhrarm" with about 1 M m³ of fill material has been chosen:
- Cutting off the above mentioned emission paths by
  · covering the surface with clean, low permeable soil not less than 0.5 m thick
  · sealing the surface by buildings, roads etc.
  · controlling the ground water quality continuously

- Restriction of the utilization of the area
  · industrial utilization
  · sensible utilization only in connection with special containment measures.

One specific problem has been the volume of contaminated form sand that has been filled together with other waste material in the former Ruhr-bed. As this material could not remain in the area of industrial utilization, it will be encapsulated in a landscape structure with no buildings etc. Fig. 7 shows schematically this structure which is placed between the natural slope of the shore of the former Ruhr-river and an artificial wall. The basal sealing system consists of 0.5 m of mineral sealing. The capping system has been designed according to the regulations for a landfill capping system. For the water front structure a tied back sheet pile wall has been chosen.

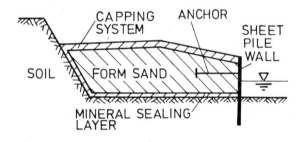

Fig. 7: Encapsulation of contaminated form sand

4.3 Remediation concepts for the abandoned coking plant

At the location of the former coking plant there is contamination randomly distributed consisting of heavy metals and organics like BTX, phenol and PAH. The decontamination by on site/off site-treatment techniques is not considered to be an exclusive solution, but a combination of different techniques should be evaluated including containment, hydraulic measures, decontamination and immobilisation. The following three remediation concepts are under discussion based on the authority's decision to use this area as green and exhibition park.

- The first idea is the choice of a capping system together with vertical cut-off walls as encapsulating elements. In addition there should be an active dewatering system, which would produce an inverted hydraulic gradient (see Fig. 8a). The lowered water level prevents washing out of the organic contaminants.

- The second proposal combines decontamination and containment. The mobile contaminants like fluid coal tar and contaminated ground water would be decontaminated off site. The slightly contaminated soil should be placed well above the ground water level (see Fig. 8b). In the zone of changing ground water level inert soil should be installed. A capping system prevents infiltration of surface water.

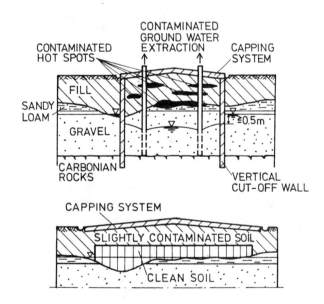

Fig. 8: Remediation concepts for the abandoned coking plant

- The consultants favoured proposal is the immobilisation of the contaminated soil by bentonite admixture. The bentonite admixture provides close encapsulation of the contaminants, reduction of hydraulic conductivity and of elutriation behaviour in connection with high adsorption capacity of heavy metals. The volume of 6000 m^3 contaminated soil would be excavated, mixed with bentonite, refilled and compacted. The restoration layer would be the covering layer.

5. Conclusions

After basic considerations on site assessment, risk analysis and development of remediation concepts, two remediation projects are described. Based on site investigations the requirements and the realistically available remediation techniques are defined. Alternative remediation concepts are discussed. The selected solutions are to provide containment, as well as additional measures.

Literature

Bosman, R. (1993): Sampling Strategies and the Role of Geostatistics in the Investigation of Soil Contamination. Arendt, Annokkée, Bosman & van den Brink (eds.), Contaminated Soil '93, Kluwer Academic Publishers, pp. 587-597

ETC 8 (1993): GLR-Recommendations "Geotechnics of Landfill Design and Remedial Works". Ernst & Sohn, Berlin, 2nd edition

SRU (1990): Rat der Sachverständigen für Umweltfragen. Sondergutachten Altlasten. Metzler-Poeschel, Stuttgart

Municipal Solid Waste Landfilling

Robert M. Quigley
Geotechnical Research Centre, The University of Western Ontario, London, Canada

SYNOPSIS: The short and long term performance of clayey barriers (the cheapest way to encapsulate waste) is the subject of this paper. Municipal solid waste leachate varies from a moderately saline, slightly organic, slightly acid liquid when fresh to a non-threatening liquid once aged and diluted. Biological activity within the waste is responsible for extensive carbonate and sulphide dumping which tends to clog drainage systems. Concurrent advection and diffusion play major roles in salt and organic transfer through clay barriers. Typical salt fluxes are presented for barriers of differing thickness to illustrate the great importance of diffusion as a transfer process.

Field compaction of natural clays wet of Standard Proctor optimum is advocated as the best way to obtain a "soft" pliable barrier which will self heal under the impact of rapid migration of chemicals before and during increases in stress levels from the waste loading. Provided the barrier mix is designed to self heal, even clay mineral incompatibility problems associated with c-axis contraction (vermiculites and smectites) should not adversely affect long term performance.

A design option available to reduce saline fluxes through barriers is to incorporate thin, very low permeability layers (bentonite or HDPE membranes) into thicker barriers of lower quality soil.

INTRODUCTION

Standard international practice is to encapsulate municipal solid waste (MSW) within a barrier system designed to inhibit migration of polluting chemicals to local groundwater supplies. An integral part of the design is normally a leachate collection system that prevents development of a leachate mound and simultaneously removes large amounts of salt from the landfill thus greatly reducing long term impact on ground water by transferring the salt load to streams after treatment. There are many, many design schemes varying from multilayer composites to hydraulic traps (inward flow against outward diffusion) to very simple rural schemes of direct dumping into clay pits with no drainage system. It is not the purpose of this paper to review design schemes but take a look at some patterns of MSW behaviour. The patterns will include:

1) a brief look at leachate and how it changes with time;

2) the behaviour of good barriers with time, properly compacted wet of Standard Proctor optimum;

3) a look at actual diffusion profiles at two sites; and

4) the size of diffusion fluxes through clay barriers as a function of their thickness.

MSW LEACHATE

Curbside solid waste consists of miscellaneous rubbish (~ 50%), wet plant matter (~ 14%), food wastes (~ 12%) and non-combustibles (~ 24%) (Ham et al, 1978). Recycling apparently will have little net effect on these overall percentages (Ham, personal communication) even though the total waste volume might decrease by up to 50% in some very efficient programs. The resulting elemental composition should remain fairly stable therefore at ~ 21% water, 28% carbon, 4% hydrogen, 22% oxygen, 0.5% nitrogen and 25% non-combustibles.

Leachate is produced during groundwater percolation through waste as it biologically degrades, aerobically at the very top and increasingly anaerobically towards the bottom. The result is production of CO_2 gas at the top, CO_2 and methane at mid-depth and mostly methane at the base of a thick (30 m +) landfill. Simultaneously, the leachate becomes increasingly saline towards the base to levels of about 5 to 10 g/L for a leachate extraction system and the top open for rainfall infiltration. Typical components of leachate include:

Chloride	~ 4 to 6 g/L
Sodium	~ 3 to 5 g/L
Potassium	~ 0.3 to 1 g/L
Magnesium	~ 200+ mg/L
Calcium	~ 300+ mg/L
Ammonium	~ 1 g/L
Heavy metals	~ 0.2 to 0.4 g/L

Whereas most of the salts are extracted by leaching from the waste, NH_4^+ is probably biologically produced. Of the above, K^+ and NH_4^+ represent some danger to any vermiculites and smectites in the clay barrier since both may fix into holes in the clay structure, causing "c" axis contraction and potential decreases in CEC and clay mineral volume (Quigley, 1989, 1993).

When stored in the laboratory, fresh new leachate rapidly changes composition even in an anaerobic state as illustrated on Figure 1.

The pH rapidly rises from slightly acid in most but not all cases as the Eh drops to -250 to -300 mv. Reduction in organic food supply results in a rapid drop in the bacterial count as the leachate becomes dormant. During this period both calcite and amorphous iron sulphide are dumped from solution forming a slimy solid mass as illustrated by Quigley et al (1989) and shown by the decreases in Ca^{++} and Fe^{++} on Figure 1.

These dumping processes apparently occur in the field in a similar manner (King et al, 1993, Brune et al, 1991) often resulting in serious obstructions of drainage layers. Much more needs to be done to understand and rectify these drainage layer clogging problems because they can create serious leachate mounding and high hydraulic heads resulting in much higher advective salt fluxes than originally designed.

The organics present in leachate are originally dominated by short chain carboxylic acids similar to acetic acid. With time, however, other organics appear in the leachate as discussed by Barone et al (1993). These typically may include benzene, ethyl benzene, toluene and xylenes, all of them probably dumped with the waste. A further volatile organic causing some concern is dichloromethane which has shown up at Keele Valley in Toronto at levels of 1-3 ppm (Rowe, 1993, personal communication). This fairly toxic liquid is also fairly volatile and its source remains obscure (biogenic or dumped waste). Since it is a powerful solvent, an industrial source seems more a probability than a biological source within the waste pile.

CLAYEY BARRIER PLACEMENT

It is this author's opinion that natural inactive clayey barriers (excluding bentonites) should be placed wet of Standard Proctor optimum by a process of excessive kneading compaction. This corresponds roughly to the plastic limit and produces a clay soft enough to self heal by consolidation as the waste load is applied and as the barrier comes under chemical stress.

Some authors advocate compaction anywhere along but wet of the line of optima, however, a clay compacted wet of Modified Proctor is a very dry, stiff, unyielding material. On the other hand, clays compacted wet of Standard Proctor are somewhat thixotropic (Mitchell et al, 1965) and may show small increases in hydraulic conductivity (k-values). Such clays may also display an apparent preconsolidation pressure, σ_p', of 100 to 150 kPa. Deep landfills applying stresses exceeding this σ_p' cause significant consolidation and major reduction in k-value. At Keele Valley these reductions amounted to 1.5 orders of magnitude, greatly improving the barrier compared to its as-compacted state. Figure 2 adapted from Leroueil et al (1992) and King, Quigley et al (1993) demonstrates the recommended zone for compaction and the expected improvement in behaviour on addition of effective stress loadings and improvement by exchange of Ca and Mg on most natural clays by Na^+ from typical MSW leachates.

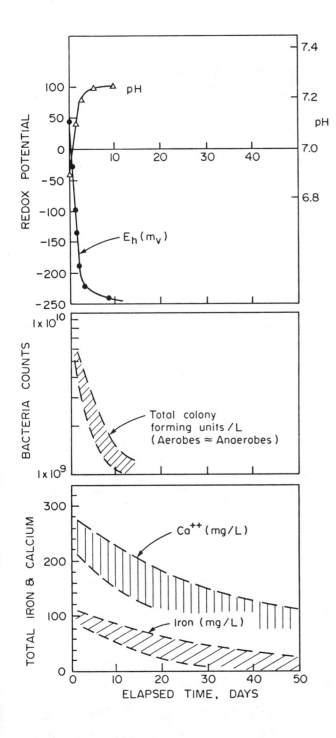

Figure 1. Chemical changes shown by leachate stored anaerobically in the laboratory.

CONTAMINANT MIGRATION

For this presentation, our landfill will be

assumed to be clay lined and with a net zero hydraulic head across the barrier. Under this scenario, diffusion is the only process of migration. This is in contrast to a downward flow scenario where both advection and diffusion fluxes would have to be added together, and the upward flow scenario or hydraulic trap, where the advective flow blocks downwards diffusion. Extensive studies at the Confederation Road Landfill on deep clay near Sarnia, Canada and on the Keele Valley Landfill, Toronto, Canada will be used as examples.

Transient migration by diffusion only follows Fick's second law and produces a pattern of migration illustrated on Figure 3 (Quigley et al, 1987).

As discussed at some length by Quigley et al (1987) diffusion is initially a very rapid process (Figure 3b) slowing down dramatically as the gradient dc/dx decreases (i.e. increasing distance from the source). A thin bentonite geosynthetic clay liner would be traversed in a matter of hours, a 30 cm sand/bentonite barrier in a matter of months, and a 1.2 m barrier in a matter of 5 years or so as illustrated on Figure 3a. The corresponding advective breakthrough times might be up to 100 years or more for the 1.2 m barrier at low gradients compared to 5 years for diffusion.

Actual diffusion profiles for common Na^+ and Cl^- are presented on Figures 4 and 5 for comparison to the theoretical plot on Figure 3. At the Confederation Road Landfill in Sarnia (Figure 4) Cl^- has migrated ~ 2 m, exactly as predicted for a D_{Cl} of 6.5×10^{-6} cm²/s over 12 years. Similarly at Keele Valley (Figure 5) Cl^- has migrated 90 cm in 4.25 years to yield a diffusion coefficient of 6.5×10^{-6} cm²/s.

The distance of migration from the base of the waste at Keele, which is the top of the sand

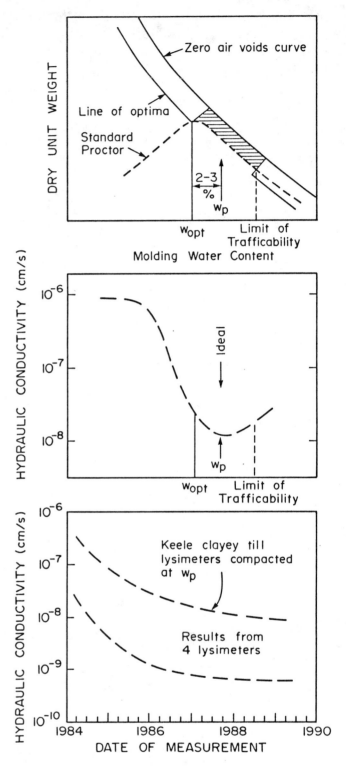

Figure 2. Compaction, permeability and field performance as a function of molding water content (Adapted from Leroueil et al, 1992 and King et al, 1993).

cushion corresponds closely to the theoretical diffusion plots on Figure 5 as well as to apparent conductivity plots in the clay barrier presented by King et al (1993).

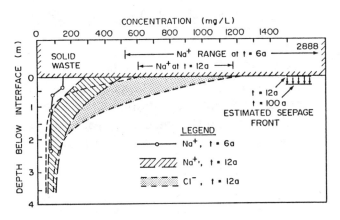

Figure 4. Diffusion profile for Na and Cℓ at t = 6 and 12 years, Confederation Road Landfill. (Adapted from Quigley and Rowe, 1985; Crooks and Quigley, 1984; and Goodall and Quigley, 1977).

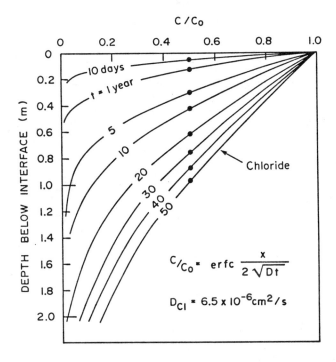

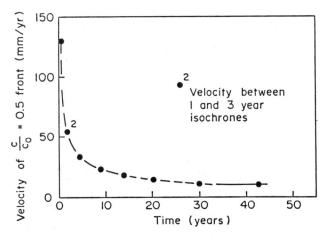

Figure 3. Time rate of migration by diffusion: (a) relative concentration-depth-time plots: (b) velocity of migration of $c/c_0 = 0.5$ front (After Quigley et al, 1987).

In order to provide a complete pattern of diffusion, the cation profiles are presented on Figure 6 as measured on the same samples as chloride in Figure 5. Na^+ has obviously migrated nearly as far as chloride, again starting at the base of the waste at the top of the sand cushion. K^+ is greatly retarded by the vermiculitic clay barrier, migrating only 3 or 4 cm into the clay compared to 35 or 40 cm for Na^+. Finally Ca^{++} shows a desorption halo or concentration hump, so is migrating downward probably paired with $Cℓ^-$. These patterns of behaviour are identical to those observed during laboratory compatibility testing and early field results from the Keele lysimeters (King et al, 1993).

A final plot of liquid hydrocarbon diffusion is presented for Keele Valley after Barone et al, 1993. This work, also done at Western, showed clearly that organics migrate by diffusion in exactly the same way as inorganics and within clayey barriers may be subject to little biodegradation. At the time the plots on Figure 7 were prepared (t = 4.25 years) there seemed to be no dichloromethane yet present in the leachate.

In conclusion, an overwhelming body of field data now confirms that diffusion is often the most important mechanism of leachate species migration in well designed sites with very low k-values and little or no advective flow.

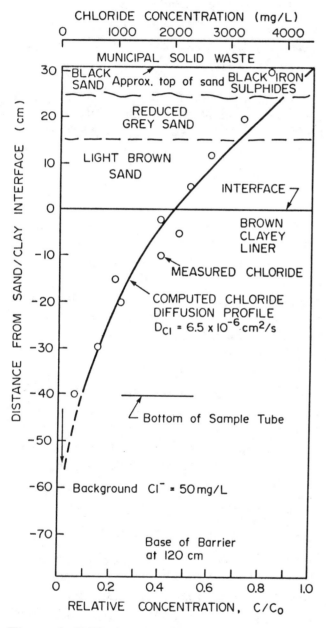

Figure 5. Diffusion profile for chloride and the cations at Keele Valley at 4.25 years (From King, Quigley et al, 1993).

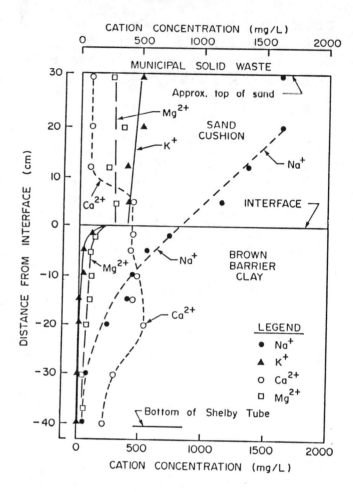

Figure 6. Porewater cation profiles accompanying the chloride profiles on Figure 5 (From Quigley et al, 1990).

SIZE OF CHEMICAL FLUXES THROUGH BARRIERS OF VARIABLE THICKNESS

The magnitude of a diffusion only chemical flux, F, is calculated using Fick's first law, namely, $F = D.n.dc/dx.A$, where D is the relevant diffusion coefficient, n the porosity, dc/dx the chemical gradient across the barrier and A the area of the landfill.

Two figures are presented to illustrate the magnitude of flux as calculated for a 1.2 m thick barrier and a 2 cm thick bentonite geosynthetic clay barrier. Figure 8 for a 1.2 m barrier covered by 30 cm of sand and a linear concentration profile yields a salt diffusion flux of 2×10^7 kg/a/100 hectare site for zero advective velocity. This would correspond to some 600 dump trucks of salt. The same barrier would yield a similar advective flux of 600 trucks of salt for a barrier $k = 10^{-8}$ cm/s and a gradient near unity. A weakening of the k specification to 10^{-7} cm/s increases the advective salt flux by a factor of 10 to 6000 trucks, at which point diffusion can be essentially forgotten as a critical process.

A similar exercise presented on Figure 9 for a 2 cm thick GCL (bentonite) yields a diffusion only flux of some 60,000 trucks/a from our 100 ha site for a leachate concentration of only 5 g/L. Compare this with a very low advective flux of 100 trucks/a assuming that a low k-value of 10^{-10} cm/s can be achieved by the barrier.

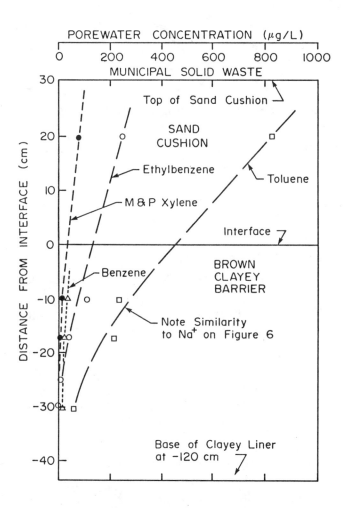

Figure 7. Diffusion profiles for volatile liquid hydrocarbons measured in the barrier soils at Keele (Adapted from Barone et al, 1993).

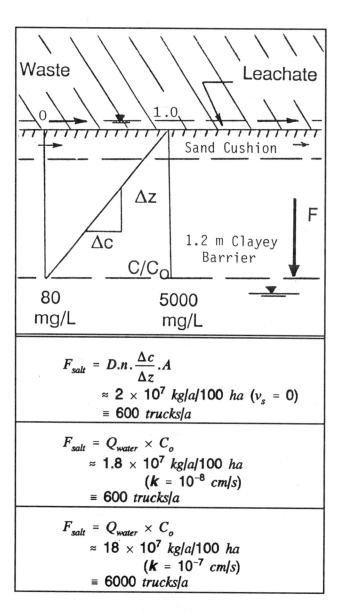

Figure 8. Comparison of diffusion only flux with advection only fluxes for k-values of 10^{-8} cm/s and 10^{-7} cm/s. Clayey barrier 1.2 m thick, sand cushion 30 cm thick, gradient ≈ 1.

This calculation clearly shows that this barrier should transmit copious quantities of salt to the groundwater environment unless backed up by plastic membranes or liner composites designed to increase the diffusion distance.

A final drawing showing one way to obtain a relatively cheap barrier using mostly inferior soil is presented as Figure 10. A thin layer of impervious material is placed at the base of the barrier. This could be a thin local high quality clay, a geosynthetic clay liner, a HDPE membrane or possibly a combination of two of them would form the hydraulic barrier. Above them a thick inferior clay layer would be used as a diffusion barrier in the same way the sand cushion operated at Keele.

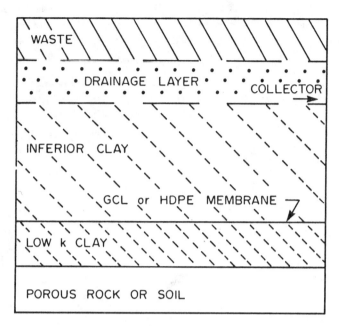

Figure 10. Clayey liner option using high quality clay, GCL or membrane near the base overlain with inferior clayey soil to operate as a diffusion barrier.

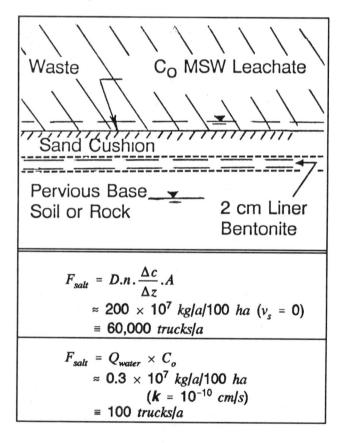

Figure 9. Comparison of a diffusion only flux with an advection only flux through a geosynthetic clay liner (bentonite) with $k = 10^{-10}$ cm/s and a gradient of ~ 3.

CONCLUSIONS

This paper has reviewed the performance of clayey barriers with respect to municipal solid waste disposal. A review of leachate itself followed by barrier performance in contact with leachate at two low advection sites has demonstrated the great importance of diffusion as a migration process. This was further demonstrated by flux calculations for a thin (2 cm) barrier and a thicker (1.2 m) clayey barrier.

At sites where relatively high k-values of 10^{-6} or 10^{-7} cm/s are tolerated, diffusion may be a relatively unimportant migration process since it is swamped by advection. At sites where barrier k-values of 10^{-8} cm/s are mandated or where advection is absent, diffusion is dominant and very important.

REFERENCES

Barone, F.S. Costa, J.M.A., King, K.S., Edelenbos, M. and Quigley, R.M., 1993. Chemical and mineralogical assessment of in situ clay liner - Keele Valley Landfill, Maple, Ontario. Proc. Joint CSCE-ASCE Nat'l. Conf. on Environmental Engineering, Montreal, July 1993, Vol. 2, pp. 1563-1572.

Brune, M., Ramke, H.G., Collins, H.J. and Hanert, H.H., 1991. Incrustation processes in drainage systems of sanitary landfills. Proc. Sardinia 91, Third Int'l. Landfill Symposium. S. Margherita di Pula, Cagliari, Italy, Oct. 1991.

Ham, R.K. in Lockland & Assoc., 1978. Final Report to EPA. Contract #68-03-2536. Recovery, processing and utilization of gas from sanitary landfills, U.S. EPA.

King, K.S., Quigley, R.M., Fernandez, F., Reades, D.W. and Bacopoulos, A., 1993. Hydraulic conductivity and diffusion monitoring of the Keele Valley Landfill liner, Maple, Ontario. Canadian Geotechnical Journal, Vol. 30, No. 1, pp. 124-134.

Leroueil, S., LeBihan, J.P. and Bouchard, R., 1992. Remarks on the design of clay liners used in lagoons as hydraulic barriers. Proc. Canadian Geotechnical Journal, Vol. 29, pp. 512-515.

Mitchell, J.K., Hooper, D.R. and Campanella, R.G., 1965. Permeability of compacted clay. Journal of Geotechnical Engineering Division, ASCE, Vol. 91, SM4, pp. 41-65.

Quigley, R.M., 1993. Clay minerals against contaminant migration. Geotechnical News, Vol. 11, No. 4, pp. 44-46.

Quigley, R.M., 1989. Effects of waste on soil behaviour. Proc. 12th International Conference on Soil Mechanics and Foundation Engineering, Rio de Janeiro, August, Vol. 5, pp. 3135-3138.

Quigley, R.M., Yanful, E.K. and Fernandez, F., 1990. Biological factors influencing laboratory and field diffusion. In: Microbiology in Civil Engineering, (Ed.) P. Howsam, pp. 261-273. (Proc. Fed'n. of European Microbiological Soc. Symp., Cranfield Inst. of Technology, U.K., Sept. 1990.)

Quigley, R.M., Yanful, E.K. and Fernandez, F., 1987. Ion transfer by diffusion through clayey barriers. In: Geotechnical Practice for Waste Disposal '87. ASCE, Geot. Spec. Publ. No. 13, pp. 137-158.

Geotechnical Properties of a Clay Contaminated with an Organic Chemical

A. Al-Tabbaa
Lecturer in Soil Mechanics, The University of Birmingham, UK

S. Walsh
Research Assistant, Oxford University, UK

ABSTRACT Some of the geotechnical properties of kaolin clay contaminated with poly(ethylene glycol) were investigated as the concentration of this organic chemical was increased. The results suggest that the contaminant's solution concentrations between 25 and 300g/100ml can be divided into three zones of dilute, moderate and concentrated solutions with respect to their effects on the clay's geotechnical properties. Strengthening as well as weakening of the properties was observed. It was possible to explain most of these effects in terms of soil-contaminant interactions.

Introduction

Natural clays and clay fills have been extensively used to form the lining of landfill sites to act as barriers for the containment of contaminants. An understanding of the behaviour of materials used for these liners is essential in the design of such barriers. Some types of contaminants have been shown to change the properties of their host soils and this behaviour has been shown to be dependent on the concentration of the contaminant solution (Fernandez and Quigley, 1988; Anderson et al., 1985). This paper concentrates on the effect of a particular contaminant, as its concentration is increased, on some of the geotechnical properties of natural and compacted kaolin clay. The chosen contaminant is a soluble organic polymer in the form of poly(ethylene glycol) which has wide applications in industry as a solvent and soil conditioner and aggregate stabilizer. The research work presented here forms an initial part of a research programme at Birmingham University investigating the effects of various types of organic contaminant solutions on clay structure hence its geotechnical properties.

Clay and contaminant

The clay used in all the tests presented here is speswhite kaolin clay, the physical and mechanical properties of which are well documented. Kaolin has 80% of particles smaller than 2µm. The contaminant poly(ethylene glycol), see Figure 1, has an average molecular weight of 2000 and a viscosity of 1.022 and will be referred to as PEG in the rest of the paper. PEG is a neutral polar organic chemical which does not exhibit a net charge. However, it has an asymmetric distribution of electron density resulting in an appreciable dipole moment, which is an indicator of its polar characteristics. The variable in this study is the concentration of the PEG solution in the soil. Eight different concentrations were used ranging from 25 to 300 g/100ml of water which gives molarity concentrations of 0.1M and 1.2M respectively.

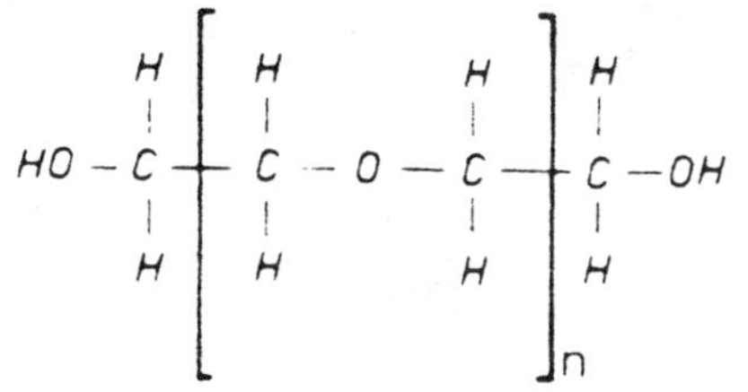

Figure 1. Structure of PEG, $[HO(C_2H_4O)_nH]$.

The bonding between PEG and the clay particles occurs through hydrogen bonding between the OH functional group of the polymer and the oxygens of the clay surface. When clays are permeated with an organic chemical, changes in interlayer spacing could occur. This has been related to the dielectric constant of the organic chemical. The dielectric constant of a medium is defined as ε in the following equation: $F=(QQ')/(\varepsilon r^2)$, where F is the force of attraction between two charges Q and Q' separated by a distance r in a uniform medium. The interlayer spacing between the clay particles is proportional to the dielectric constant of the pore fluid. The dielectric constants of water and PEG are 79.5 and 37.7 respectively. Hence, if ε decreases the interlayer spacing decreases and the double layer around the clay particles shrinks. For organic chemicals with dielectric constants lower than that of water, such as PEG, the individual clay particles will contract as a result of thinner interlayer spacing and this provides the opportunity for the clay particles to orientate themselves into a flocculating structure. This effect applies to polymers when applied in moderate concentration. However, as the concentration of the polymer solution is increased it has the opposite i.e. a dispersive effect. This action is ascribable to the formation of a coat or layer of the polymer molecules around the solid particles. By this means, the particles are sterically prevented from entering each other's attraction sphere and is known as steric stabilisation. This effect is shown on Figure 2.

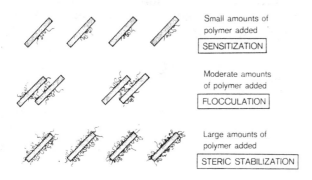

Figure 2. The effect of various concentrations of PEG on clay structure (Theng, 1979).

Testing, results and discussion

Soil samples of the eight different PEG solution concentrations were tested for their geotechnical properties using the liquid and plastic limits, compaction, unconfined compression strength and permeability. In addition, a set of uncontaminated samples was also tested for comparison. The contaminant was introduced to the soil using two methods. These are explained in the relevant section below. Deaired water was used in all the tests. The moisture content in all the tests was calculated taking into account (subtracting) the weight of the solid PEG particles in the dry soil.

The liquid and plastic limits (LL & PL)

The cone penetration test was used to obtain the LL and the 3mm thread method for the PL (BS1377). Figure 3 shows a plot of both limits against the concentration of the PEG solution. The LL and PL for the uncontaminated soil were measured at 69 and 36 respectively. The figure shows that up to a PEG solution concentration of 100g/100ml both limits were lower than the uncontaminated values and reduced to minimum values of 64 and 30 respectively which are 93% and 83% of the uncontaminated values. For concentrations higher than 100g/100ml both limits gradually increased up to values of 145% and 117% at 300g/100ml concentration of PEG. It has been reported that as a clay becomes more flocculated, its LL, PL and plasticity index decrease (Lambe, 1958). When clays are permeated with concentrated organic liquids, with dielectric constants lower than that of water, the LL, PL and the plasticity index increase (Foreman and Daniel, 1986). Both effects are observed in Figure 3 and this can be explained in terms of Figure 2. This suggests that a PEG solution concentration of 100g/100ml could be the dividing line between moderate and concentrated solutions i.e. between a flocculated and dispersed clay structure.

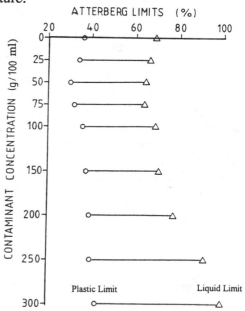

Figure 3. The variation of Atterberg limits with PEG concentration.

The decrease in the LL was small compared to the subsequent increase while the opposite was observed in the PL. Since the increase in the LL was greater than that in the PL, the plasticity of the soil increased as the concentration of PEG solution increased. This is shown on Figure 4.

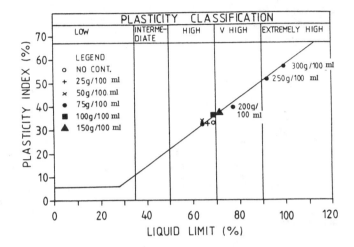

Figure 4. Plasticity index vs. liquid limit for different PEG concentrations.

Compaction tests

The samples used for the compaction tests were formed by mixing the dry soil with the PEG solution. Samples of all eight concentrations were tested for their compaction behaviour using the Dietert apparatus (Head, 1980). Moisture contents ranging between 10% and 60% were used with five to six different moisture contents for each contamination batch. The soil was compacted in a 38mm diameter mould using 10 blows on both ends of the sample and the resultant dry density calculated. Two samples from each concentration batch were tested and the average values of the dry density reported. The results were within 10% of the average value. A plot of dry density against moisture content is shown on Figure 5 which shows a gradual decrease in maximum dry density as the concentration of PEG solution increases. This does not show different effects for the different concentration levels.

It has been reported (Head, 1980) that as the clay's plasticity increases, the clay becomes more difficult to compact. This behaviour can be seen in Figure 5 where as the concentration of PEG solution increased, the maximum dry density at the optimum moisture content reduced. The maximum dry density for the uncontaminated samples was found to be, for these sets of tests, 1.42Mg/m^3 and this value reduced to 1.24Mg/m^3 for the 300g/100ml concentration samples which is a reduction to 87%. The value of the optimum moisture content increased from around 25 up to 40 as the concentration of PEG increased. Figure 5 shows the line of zero air voids showing that there is less air in the sample at the maximum density as the PEG concentration increases.

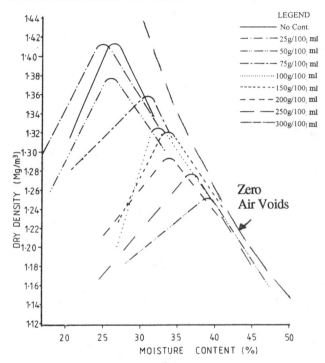

Figure 5. Dry density vs. moisture content for different PEG concentrations.

Unconfined compression strength (UCS)

The 38mm diameter soil samples compacted in the Dietert apparatus were trimmed to 76mm and then tested in the unconfined compression apparatus for their UCS (BS1377). Again two compacted samples from each concentration batch were tested and the average result reported. The results were within 30% of the average value. The shape of the UCS curves is very similar to the compaction curves shown in Figure 5. The maximum value of the UCS corresponded reasonably well with the optimum moisture content in the compaction tests. The average maximum value of the UCS from each concentration is plotted against PEG solution concentration on Figure 6. This figure shows a reduction in strength for dilute solutions. Al-Tabbaa (1993) reported a similar effect for PEG concentration of 38g/100ml. For 'moderate' concentrations, 50-100g/100ml, there is an increase in UCS above the uncontaminated value (flocculated structure). For 'concentrated' solutions, 150-200g/100ml, the strength was lower than the uncontaminated value (dispersed structure). For concentration above 200g/100ml the increase in strength must be due to a different effect.

It has been well documented that soil conditioning polymers such as PEG increase the mechanical strength of soils (Lambe and Michaels, 1954). It has, however, been suggested (Theng, 1979) that as the concentration of the polymer solution increases, the percentage of polymer molecules being adsorbed onto the adsorbing sites changes and hence there is a threshold at which interparticle bonding efficiency decreases and the strength decreases. This threshold could be at 50g/100ml. Comparing the UCS results with the Atterberg limits in Figure 3 shows some correlation but not in all the results.

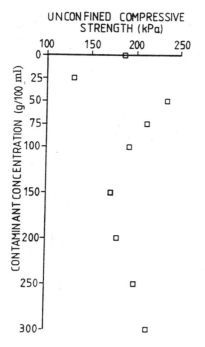

Figure 6. Maximum UCS against PEG concentration.

One factor which might have contributed to the scatter and inconsistency of the results is the time lapse between mixing and compacting and between compacting and the UCS tests. The time gaps between tests should have been either maintained constant or recorded but neither was the case. This time gap varied between 1 and 5 days. It is hence not possible to come to any conclusion regarding its effect on the results. This is the subject of a current study.

Permeability

For the purpose of the permeability tests, the clay was consolidated in two 150mm diameter hydraulic Rowe cells (Rowe and Barden, 1966). Uncontaminated clay was consolidated from a slurry at a moisture content of 120% (approximately twice the liquid limit). The samples were consolidated under equal vertical strain conditions, using a rigid platen, in increments up to 250kPa and then unloaded. Both the vertical (k_v) and horizontal (k_h) permeabilties were measured. Drainage during consolidation in the vertical permeability Rowe cell was vertical to porous discs at the top and base of the sample. To facilitate measurements of the horizontal permeability, a Rowe cell was modified (Al-Tabbaa and Wood, 1987) so as to enable the installation of a central porous column (50mm diameter porous metal column). Drainage during consolidation in this cell took place towards inner and outer (1.5mm diameter porous plastic sheet) drainage boundaries. It was not possible to use any back pressure in those tests. Up to this pressure no contaminant solution was introduced. At 250kPa the permeability of the clay to water was measured. Then two PEG concentration solutions were introduced one at a time in the permeability tests. In addition the permeability to one solution concentration was compared under different hydraulic gradients. Because each permeability test took about a week to carry and due to lack of time, only fourteen tests were carried out.

The permeability values were measured using the flow pump technique. The flow pump used was a Watson Marlow 503S/RL peristaltic flow pump capable of operating at 0.4 to 10rpm with a selection of manifold tubing giving a range of flow from 0.0008ml/rev to 0.22ml/rev. The vertical permeability was measured by applying a head difference between the top and bottom of the sample. For the horizontal permeability, a head difference was applied between the centre and the edge of the sample. The flow arrangement in the two cells is shown on Figure 7. The permeant solution at a given concentration was forced to flow through the sample at a constant flow rate (Δq). At the inflow, a connection was made to a pore pressure transducer to measure the pore fluid pressure (Δu) developed at the inflow position. This was connected to a four channel data logging system, connected to a printer via a printer-interface unit. This enabled the readings to be recorded continuously. The pore fluid pressure at the outflow end was taken as zero. The flow was allowed to continue until the pore water pressure reached equilibrium. Given the thickness of the sample, which was around 2cm, it took two days for the solution in the sample to be fully replaced by the inflow solution. After that the flow was monitored for 1-4 days. In all the tests the height of the sample during the flow pump tests was monitored and the pore water pressure difference across the sample was maintained at 10-15% of the applied vertical stress. Monitoring of the height of the sample showed that there was no seepage induced consolidation taking place. From those tests the permeability was calculated according to Darcy's law using the following two equations:

$$k_v = [h \, \Delta q \, \gamma_w] / [\pi \, r^2 \, \Delta u]$$

$$k_h = [\Delta q \, \gamma_w \, \ln(r_1/r_2)] / [2 \, \pi \, h \, \Delta u]$$

where h and r are the current height and radius of the sample, r_1 and r_2 are the outer and inner radii of the sample respectively and γ_w is the unit weight of the pore water.

The vertical permeability of kaolin to water at 250kPa vertical stress was measured to be 2.6×10^{-6}mm/s. This value agrees with Al-Tabbaa and Wood, 1987. When a solution of 25g/100ml concentration of PEG was introduced using the same method and at the same hydraulic gradient, the vertical permeability value reduced to 2.36×10^{-6}mm/s which is a decrease to 91% of the water permeability value. When this test was repeated with a hydraulic gradient 80% of that used above, the permeability value increased to 5×10^{-6}mm/s. This means that as the hydraulic gradient was reduced the permeability value increased and it increased beyond the water permeability value. When a solution of 100g/100ml concentration was used, the permeability increased to 15.3×10^{-6}mm/s which is 6 times greater than the permeability to water. This value was less fluctuating with change in the hydraulic gradient.

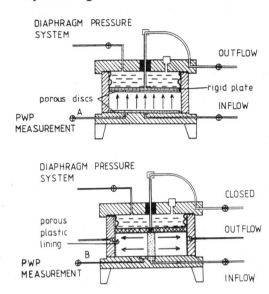

Figure 7. Vertical and Horizontal flow arrangement for the flow pump tests.

Similar results were observed with the horizontal permeability measurements. The permeability of the sample to water at 250kPa vertical pressure was measured to be 8.3×10^{-6}mm/s. A similar result was reported by Al-Tabbaa and Wood (1987). With a PEG concentration of 25g/100ml at the same hydraulic gradient, the permeability reduced to 5.3×10^{-6}mm/s which is a reduction to 64%. Work carried out by Davies (1993) using concentration of PEG of 35g/100ml showed a reduction in permeability down to 40% at a number of vertical stress levels up to 250kPa.

When a higher and a lower hydraulic gradients of 125% and 50% of the initial hydraulic gradients were applied to the same sample, they both increased the permeability to around 9.8×10^{-6}mm/s which is an increase to 120%. With 100g/100ml concentration solution the permeability increased to 1.54×10^{-4}mm/s which is 18 times greater than the permeability to water. Based on the results above, no conclusions can be reached regarding the effect of varying the hydraulic gradient on the hydraulic conductivity except that the permeability value seem to be sensitive to hydraulic gradient variations.

Published results on the effect of dilute and concentrated solutions of organic contaminants (Fernandez & Quigley, 1988) suggest that for dilute solutions, the dielectric constant of the solution will be similar to that of water and its effect of shrinking the double layer is negligible. In this situation it is the viscosity of the solution which has the dominating effect on the changes in the permeability. Since the viscosity of the PEG solution is higher than that of water, the permeability should decrease at low concentrations which was observed in some of the tests using 25g/100ml solution concentration. As the concentration of the solution is increased it is expected that the effect of viscosity on the flow is superseded by the effect of the dielectric constant which shrinks the double layer causing flocculation and an increase in permeability which was observed for the 100g/100ml solution concentrations. For higher solution concentrations, a decrease in the permeability would be anticipated because of the expected dispersed structure of the clay.

Conclusions

This paper presents the results of a preliminary investigation into the effect of poly(ethylene glycol), as its solution concentration is increased, on the geotechnical properties of kaolin clay. This work forms an initial part of a research programme at Birmingham University into the effect of contaminants on the geotechnical properties and structure of clays. These tests show that the effect of the PEG solution concentration on the soil's geotechnical properties is not due to simple relationships. The results do however suggest that the tested concentrations of 25-300g/100ml can be classified as dilute, moderate and concentrated in terms of their effect on the clay's geotechnical properties, some strenghening and some weakening the properties. It was possible to explain most of the observed behaviour in terms of soil-contaminant interaction. A more extensive experimental programme is required for more specific conclusions to be made.

Acknowledgements

The experimental work was carried out by the second author as part of an M.Sc. degree course at the University of Birmingham funded by the SERC who are gratefully acknowledged. The authors wish to express their thanks to David Lynock for his technical assistance.

References

Anderson, D. C., Brown, K. W. and Thomas, J. C. (1985) Conductivity of compacted clay soils to water and organic liquids, Waste Management and Research, Vol. 3, pp 339-349.

Al-Tabbaa, A. and Wood, D. M. (1987) Some measurement of the permeability of kaolin, Geotechnique 37, No. 4, pp 499-503.

Al-Tabbaa, A. (1993) Some observations on the behaviour of an engineered contaminated cohesive fill, International Conference on Engineered Fills, Thomas Telford, London.

British Standards Institution, Methods of tests for soils for civil engineering purposes: BS1377: 1990, BSI, London.

Davies, K. A. (1993) The effects of some contaminants on the permeability of kaolin clay, Undergraduate Research Project, The University of Birmingham.

Fernandez, F. and Quigley, R. M. (1988) Viscosity and dielectric constant control on the hydraulic conductivity of clayey soils permeated with simple liquid hydrocarbons, Can. Geotech. J. 25, pp 582-589.

Foreman, D. E. and Daniel, D. E. (1986) Pemeation of compacted clay with organic chemicals, J. of Geotechnical Engineering, ASCE, No. 7, pp 669-681.

Head, K. H. (1980) Manual of soil laboratory testing, Volumes 1 and 3, Pentech Press, London.

Lambe, T. W. and Michaels, A. S. (1954) Altering soil properties by chemicals, Chemical and Engineering News: 32, pp 488-492.

Lambe, T. E. (1958) The engineering behaviour of compacted clay, J. Soil Mech. Fdns. Div., ASCE 84, No. SM2.

Rowe, P. W. and Barden, L. (1966) A new consolidation cell, Geotechnique16, pp162-170.

Theng, B. K. G. (1979) Formation and properties of clay-polymer complexes, Elsevier.

Walsh, S. (1993) Some mechanical properties of a contaminated soil, M.Sc. Thesis, The University of Birmingham.

Stability of Wastes on Lining Systems Analysis and Design

Olivier Artières, Philippe Delmas
Bidim Geosynthetics S.A., France

Michel Correnoz
France-Déchets, France

Odile Oberti
FD Conseil, France

ABSTRACT : Multi-layers structures used in lining systems require a specific study of the interaction between their different components, specially for mechanical characteristics. A numerical analysis of stability problems due to low friction interfaces inside the system is presented and the influence of several parameters is discussed. Reinforcement methods with geotextiles offers interesting solutions, as it was designed in the case of the french Torcy landfill.

1 Introduction

The developpement during the last 15 years of the problems related to the environment has had the consequence to involve a larger panel of specialists. This evolution was particularly marked in the case of waste disposal projects. The design of such sites includes their geological and hydrogeological aspects, but also uses civil engineering techniques (soil mechanic, geotechnic, hydraulic, chemistry, etc...), due to a higher level of specifications and more precise administrative requirements.

For exemple, lining and drainage systems are now usually specified as a multi-layered structure including geosynthetics and natural materials, instead of or in complement to the existing natural clayey layer.

All the possible interactions between these different materials, which can affect their own functions and their durability, must be kept in mind. It is particularly the case regarding the problems of stability due to slipping between two layers.

This paper analyses the influence of geometrical parameters of the landfill on this type of failure and gives methods able to improve the stability of wastes in this specific case.

a) Failure Through the Waste Pile

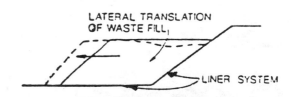

b) Failure By Sliding Along the Landfill Liner System

Figure 1 : Types of failures in landfills analysed in this paper.

2. Stability over Lining System

Some types of failures in waste diposal sites were listed by Mitchell and Mitchell [1].
One of them, the global failure along the lining and the drainage systems (figure 1-b) was observed on several sites, landfills or not.
It is very important to be aware of this specific aspect, because lining and drainage system often introduces low friction interfaces due to the supperposition of geosynthetics layers.
A simple view of this problem is shown in figure 2 : a water bassin lined with a geomembrane and covered with a nonwowen needle-punched geotextile and a soil protective layer. Friction angle between soil and geotextile is at least 3 times higher than friction angle between geomembrane and geotextile. When the geotextile is not specifically designed to resist to high tensile strength due to the weight of the soil and to the difference of friction angles, it usually fails during soil installation. Such a design mistake can be easily rectified in this case. The consequences in a waste disposal site already filled with millions of m³ of wastes are more critical.
Furthermore, the stability study is much more difficult because of a higher number of layers. For instance, in case of landfills, failures can occur even if the lining slope is small, depending on the waste pile geometry on it. Let us note that the use of a geomembrane with a rough texture on its both surfaces doesn't solve the problem, and is even dangerous, because the failure zone moves from the geotextile to the geomembrane, and this is not desirable.

3. Parametric stability analysis

A numerical study was undertaken to evaluate the influence of several parameters (bottom slope, height of wastes, type of friction interface, water level in the waste) on the stability of the wastes over the lining system.

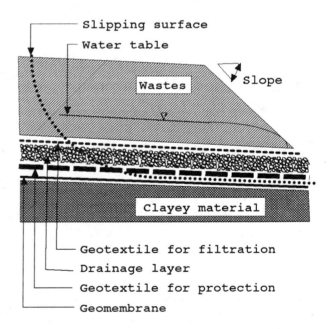

Figure 3 : Typical section of the studyed lining system

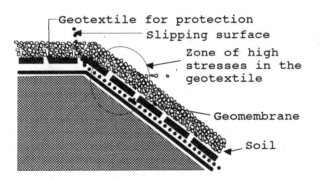

Figure 2 : Failure of sloped lining system covered with soil.

TABLE 1 :

Geotechnical characteristics of each material and their interface

Material	$\gamma^{(1)}$ (kN/m³)	$Cu^{(2)}$ kPa	$\varphi'^{(3)}$ (°)
clay	19	150	-
clay/gmb$^{(5)}$	-	50	-
gmb/gtx$^{(4)}$	-	-	8
gtx/gravel	-	-	30
waste	10	-	19 $^{(6)}$

Notes
(1) γ : Unit weight of soil
(2) Cu : Cohesion measured under undrained conditions
(3) φ' : Friction angle
(4) gtx : Geotextile
(5) gmb : Geomembrane

The computed profile is described on figure 3. It is a lining system composed from the bottom to the top by a clayey layer, a HDPE geomembrane, a nonwowen needle-punched geotextile for protection, a granular drainage layer, and the wastes.
The characteristics of the materials are given in table 1. The internal friction angle of the waste (19°) can be found small comparing to the usual values for municipal waste. In fact, if the mean is arround 35°, the variation is higher than for well known granular material. It was therefore judged better to apply an additionnal safety factor (equal to 2) on the internal friction angle of the wastes $^{(6)}$.
The numerical methods are the Cartage and Etage computer programs developped by the french Laboratoire Central des Ponts et Chaussées (LCPC). These methods have been already presented in former papers [2]. Both use an isostatic limit-state equilibrium method, named method of perturbations. The normal stress along the failure surface is written as follow :

$$\sigma_n = \sigma_o \cdot (\lambda + \mu \cdot \tan \alpha)$$

with :
σ_n : normal stress along the surface
σ_o : value of the normal stress near the solution (for instance, σ Fellenius)
λ and μ : variables computed to verify the equilibrium relationships
α : angle between σ_n and the vertical.

This method presents the advantages to verify all the equations of the static equilibrium, and has been used with success for 20 years by specialists of landslide.
The Etage program has been developped to evaluate the stability of multi-layer lining systems. It gives the global safety factors for the computed failure surface and the value of the minimum reinforcement force which must be applied to stabilise the structure.
The Cartage program solves the problem of stability of a soil structure and verifies a reinforcement system with geotextiles.

The first step of the study was the evaluation of the influence of the height of wastes and the level of water over the geomembrane (see Fig 3) on the stability of the system, when the other parameters are fixed: lining slope (6%) and waste front slope (1/1).
The length of the failure surface along the lining system is chosen equal to the height of the wastes (this surface is not necessarily

the most pessimistic one), and the safety factor is FS = 1,3 (short term stability corresponding to the exploitation phase).

The conclusions are the following [3]: a) the most critical interface corresponds to the lower friction angle (here, the interface between geomembrane and geotextile); b) the efficiency of the drainage over the geomembrane is of great importance on the stability;

Another set of calculations was carried out to evaluate this time the influence of the external slope of the embankment. In this case, the water level equals to zero.

As expected, the results indicate that a lower slope of wastes increases the stability of the embankment (table 3). But its influence is rather low : for 10 m height of wastes, the external slope without reinforcement must be so flat that it is practically and economically not possible.

The use of specific methods to reinforce the body of waste is quite justified. For instance, reinforcement with geotextiles, either near the lining system, or inside the body of waste is a solution which allows on one hand an increase of the global stability during a period corresponding to the improvement of the geotechnical characteristics of the wastes (these characteristics tend to increase when domestic wastes are consolidating), and on the other hand, a closer design of the landfill to the specific constraints of the site : better adjustment to the local topography (variable slopes of the embankment are possible), increase of the stored waste for a given land area, etc...

TABLE 2 :

Influence of the height of wastes and of the water level on the stability of the waste embankment

Height of wastes (H)	Water level over the lining system	Safety factor without reinfor-cement	Reinfor-cement force to maintain FS=1,3
5 m	0 m	1,35	Stable
	0,3 m	1,31	Stable
	1 m (H/5)	1,16	11 kN/m
10 m	0 m	1,20	27 kN/m
	0,3 m	1,18	34 kN/m
	2 m (H/5)	1,01	80 kN/m
15 m	0 m	1,14	95 kN/m
	3 m	1,12	109 kN/m
	3 m (H/5)	0,95	208 kN/m

TABLE 3 :

Influence of the slope on the stability of the waste embankment

Height of wastes (H)	Slope of the embankment	Reinforcement force to maintain FS=1,2
5 m	45° (1/1)	9 kN/m
10 m	34° (3/2)	100 kN/m
15 m	27° (2/1)	210 kN/m

4. The Torcy landfill case

The Torcy landfill is a french municipal waste disposal owned by the waste management company France-Déchets. The studied zone covers an area of 2,5 ha.

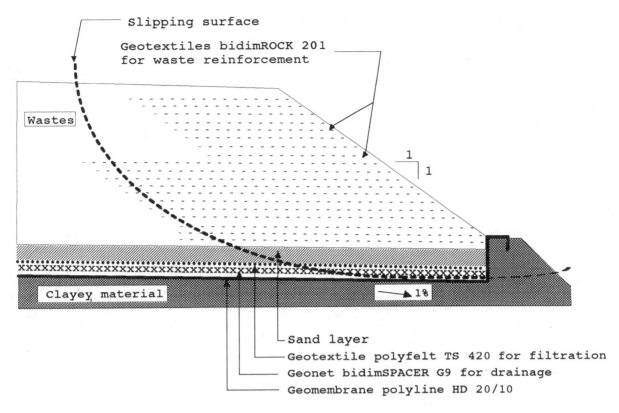

Figure 3 :
Cross section of the Torcy landfill.

According to the local administrative specifications, the lining system consists of the association of an 2 mm thickness HDPE geomembrane in conctact with a clayey layer. The drainage structure consists of a HDPE geonet covered with a geotextile for filtration, polypropylene needle-punched nonwowen, and 40 cm of sand. This sand layer has only a protection function regarding to the traffic of the engines for wastes installation (figure 3).
Due to local topography, the maximum height of wastes is planed to be 40 m with an external slope of 45°. The slope of the lining system is 1 % in this direction.

The internal stability of the wastes and the stability of the pile over the lining system was examined using the numerical methods previously presented. The geotechnical characteristics of the material are the same as those given in table 1. In the Torcy case, the sliding surface passes through the waste and follow the lining system between the geomembrane and the geonet. The two materials have the lower interface friction angle of all the structure. The global tensile force which is necessary to obtain a safety factor of 1.2 is about 2800 kN for 1 m width of the embankment in the short term case.

The reinforcement structure designed with the CARTAGE program, by the authors, consists of 31 layers of high modulus geotextile. The space between the layers is 1 m with a length between 25 and 70 m. The geotextile is the bidim-

ROCK 201, a bi-function product for reinforcement applications. It is the association of a needle-punched nonwowen polypropylene geotextile, knitted with polypropylene yarns. The yarns give a tensile strength of 200 kN/m with a modulus of about 1400 kN/m, when the nonwowen placed above the yarns protects them against mechanical damage due to puncturing by the waste or by the traffic. This particular point allows to lower the safety factor on the product according to the french standardization on geotextiles for soil reinforcement.
Polypropylen offers the best chemical compatibility geotextile-leachate, particularly at high pH values. Partial safety factors tacke into account durability and possible degradation of the product.
The internal stability of the wastes (figure 1-a) was also studied, but it doesn't require in this case additionnal reinforcement than that defined previously.

It must be noticed that the assumptions on the wastes shear characteristics were chosen on the safe side and induce a high level of reinforcement. More precise evaluation will certainly allow a reduction of the reinforcement costs, and be more economic, in particular for municipal wastes.

4. Conclusion

New containment sites for waste storage are increasingly designed with multi-layers systems for lining and drainage. The stability of wastes over such structures must be carrefully checked, because of low friction angles which can appear between these materials. Numerical methods, like the Etage program, give an useful help to simulate these complex interactions and to find the best design.

In case of instability, the reinforcement of the wastes with geotextiles provides a satisfactory way to reach the required safety factor : low cost comparing to the usual reinforcement techniques and ease of installation directly by the owner of the landfill. They also offer the possibility to model external embankments in such a way to be well integrated in the landscape.

5. References

[1] Mitchell J.K. and Mitchell R.A. (1991). Stability on landfills. PR XV CGT, Turin, 20-22 nov. 91, 47 p.

[2] Soyez B., Delmas Ph.,Herr Ch., Berche J.C. (1990). Computer evaluation of the stability of composite liners. Proc. of the 4th Int. Conf. on Geotextiles, Geomembranes and related Products, La Haye, G; Den Hoedt (Ed.), Balkema, pp 517-522.

[3] Delmas Ph., Soyez B., Berche J.C. and Artières O. (1993). Stability of wastes on lining systems : Models and analysis. Proc. of Int. Symp. Geoconfine 93. Arnould, Barres & Côme (eds), Balkema, Rotterdam. pp 505-510.

Compared Performances of Bottom and Slope Lining Systems of Municipal Landfills

C. Bernhard
CEMAGREF-France

J.P. Gourc
IRIGM, France

I. Le Tellier
CGEA-ONYX, France

Y. Matichard
LRPC Nancy, France

ABSTRACT

A comprehensive real scale experimental research program has been set up by one of the most important waste management companies in FRANCE (CGEA-Onyx) and French governmental laboratories (CEMAGREF, LCPC, University of Grenoble) with the support of the French Environmental Agency. The behaviour of four different landfill bottom and slope liner systems (1 : compacted clay, 2 : HDPE geomembrane, 3 : bentonite membrane, 4 : composite clay/HDPE geomembrane) are compared mainly on the basis of leakage rate control. The leachate flow and hydraulic head in each cell, the temperature of the geosynthetics, the settlement of the waste and the displacement of the geosynthetics and their protection on the slopes are also monitored. Each cell has a 2500.sqm surface area, a depth of 5.5 m and is filled with municipal waste which is compacted under usual service conditions. Monitoring data will be collected for a minimum period of 3 years (starting January 1994). The paper describes the design and installation procedures and the quality assurance plan. The first monitoring results are discussed.

1 Introduction

A research and test program on the performance of a number of synthetic and mineral landfill lining systems has been undertaken since May 1991 at the municipal landfill of Montreuil-sur-Barse (France).This operation involves a waste management company (CGEA - ONYX) through the CREED (Research and Experimental Centre on Environment and Waste) and public laboratories (CEMAGREF, LCPC, IRIGM) with the support of the Ministry of the Environment.

The main objective of the study is to assess how the main lining systems commonly used in landfills (compacted clay, HDPE geomembrane, bentonite geocomposite, geomembrane combined with compacted clay) when linked to the same drainage system, perform. The leakage rate under the tested lining systems is measured by means of a double liner system (geogrid associated with a HDPE geomembrane).

This study allows also the comparison of the technical difficulties associated with their installation and the cost of each lining system.

2 Design of the cells

2.1 Site conditions

The site, a former clay quarry which had supplied a brick factory, was converted to a landfill in 1986. The artificially excavated area has a maximum depth of 20 meters. It covers some 14 ha. About 5 ha has been set aside for experiments in the construction and operation of landfill sites. The cells used for the lining project cover a 1.5 ha area.

The local geology is composed of Gault clay as well as that of the Armance. The clays go down 10 to 15 meters. Underneath the clay layer, there is a non productive and untapped aquifer over a large area around the site.

The climate is oceanic. Annual average rainfall reaches up to 700 mm. Average temperature is 10°C.

200 to 300 tons per day of domestic and commercial waste from the city of Troyes (160 000 people), and a few nearby districts, are being processed at the Montreuil-sur-Barse municipal landfill.

2.2 The four lining systems

All the experimental cells have a similar geometrical structure. They have been excavated and the dikes have been compacted with the same material (Gault clay). Drawn on a plan, each compartment is a 50 m x 50 m square. The waste height can rise up to 5.5 meter. The walls at each cell are slopped in part 1/2 and 1/1. The stability study of the geosynthetics on the slopes imposed an intermediate step trench, one meter wide, placed approximately 1.5 meter from the bottom.

The various lining systems under study represent recently developed concepts used in France and elsewhere. The choice criteria for these systems have already be provided by Le Tellier and al., 1993. Their structure is given by Figure 1:

All lining systems include two parts :
- the lining system itself which can vary depending on the cells
- the overlaying layer for leachate drainage, identical for all the cells (30 cm non calcareous gravel with 20/40 mm particle size).

2.3 The leakage detection system

Figure 1 also shows the structure of the leakage detection system underneath the various tested lining systems. It is achieved by a 4 mm thick HDPE geonet (transmissivity of the order of 1.10^{-4} m²/s under 200 kPa lateral compression pressure). This system is only in use at the bottom of the cells up to an intermediate step trench 1.5 meters from the bottom. The leakage can be collected in a sump located at the lowest point of each cell. A PVC pipe (200 mm diameter) running along the slope penetrates the primary lining system at its top and allows a submersible pump to be periodically lowered into the sump to measure the amount of leakage (Fig 2).

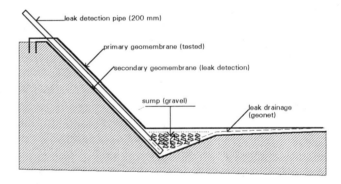

Fig 2 Detail of the leak collection and pumping system (cell GO)

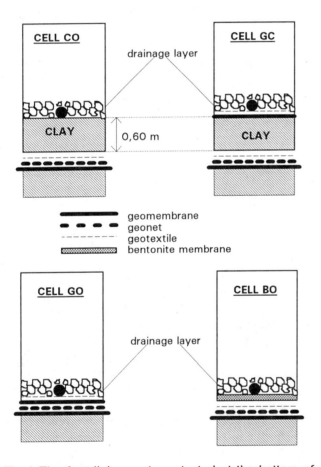

Fig 1 The four lining systems tested at the bottom of the cells

A micro-irrigation system (supplying 4 litres/hour each 30 cm) placed entirely around the geonet is used to introduce clear water in the leak detection system, providing the possibility to test its efficiency (especially before the cells are filled with waste). Another utilisation consists in flushing, and so detecting, small leaks of leachate as a tracer in a larger amount of clear water. Since the micro-irrigation pipe can be supplied by parts, it should even be possible to locate roughly the source of the leaks.

2.4 Leachate removal system

Figure 3 shows the principle of the leachate removal system by means of a low volume sump and a manhole. A submersible pump runs automatically when the leachate level becomes too high. This system is better than evacuating the leachate by gravity because there is no perforation of the membranes by any pipe.

2.5 Protection of the geosynthetics on the slopes

The protection of the geomembranes on the slopes against puncturing is achieved by a soil layer pushed in place by steps as the cells are filled with waste.
The bentonite membrane needs a protection just after its installation. On the 1/1 slope, tri-dimensional geotextile cells filled with soil have been used. On the 2/1 slope, a simple layer (30 cm thick) of gravel and of a silty sand has been tested. The displacements of the geosynthetics and their cover are monitored in the cells GO and BO. Fig. 4 is a picture of cell GO after completion.

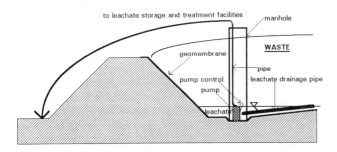

Fig 3 Principle of the leachate removal system

Fig 4 General view of cell GS after completion

3 Lessons drawn from the work progress

Precise specifications have been set up for the clay compaction and for the installation of the geosynthetic materials, especially for the geomembranes. The work was carried out by two teams belonging to the same company (compaction team; geosynthetic installation team). As a general observation, the first obstacle was due to the complexity of the work to be carried out on a relatively small scale (access for heavy earthmoving machines). This leads to emphasise the need for a general manager for the organisation and operation of the site as well as quality control. This person should be able to co-ordinate the various teams at work and to record all the relevant information.

The following remarks can be given concerning the difficulties encountered during the material installation:

-clay : the compaction of a clay layer over the leak detection system may be a concern since the geonet is placed immediately below the compacted clay with only a geotextile as an intermediate barrier. The geotextile can be designed as a filter but even with this precaution, a test pad in real field conditions has been carried out. This showed that the compaction machinery was not affecting the geonet and the geomembrane below.
The use of double weld machines for HDPE seams on a clay layer is difficult when the clay is too dry (dust) or too wet (mud). A geotextile locally installed under the seams can provide a clean surface (but the transmissivity of this geotextile must be as low as possible in order not to spread the leakage coming from a perforation of the geomembrane over a large area)

-geomembranes : the welding of HDPE seams is the main concern. A large amount of extrusion weldings was carried out on sensitive areas (three fold overlaps; corners) with difficulties both for the achievement of the seams by the installation team and for their control. The quality and control ability of the double weld seams are much more better (control achieved by 2 bar air pressure in the central canal; 10% drop of pressure was allowed in 10 mn). However, the seaming parameters have to be checked as often as necessary. Otherwise, a great amount of seams can be of poor performance.

Another point is the unloading and the handling of the geomembranes on the site. The machinery must allow for the rolls to be unloaded without being twisted lengthways (hoist fitted with independent arms). The geomembrane panels must also be ballasted before seaming to withstand any weather conditions.

4 First results

Only a few results are available since filling the cells with waste in January 1994. The results presented below refer to the period prior to filling.

4.1 Leak detection

Tests of the leak detection system consisting in leak measurements without waste in the cells have been carried out in a first stage. These led to the conclusion that leaks existed in both geomembranes in the cells GO and GC. In the case of GO, the leakage rate was more than 1500 litre/day when the cell was filled with 1 meter of clear water. An electrical leak sensor system was used for the localisation of the defects, mainly due to puncturing of the geomembrane by the drainage gravel under the load of the machinery. The defects were repaired and tested again.

The clay liner in CO was tested with 1 meter of clear water in the same way and its permeability has been evaluated for several months. Fig. 5 indicates the variations of the leakage rate through the clay liner (for the 900 m² of surface). The average value is 36 litre/day. Assuming that the flow is steady and unidirectional, with a hydraulic gradient of 2, we obtain an average permeability coefficient $k = 2.3 \cdot 10^{-10}$ m/s. This value is close to the results of laboratory tests conducted on undisturbed samples taken from the compacted clay liner on trial pads.

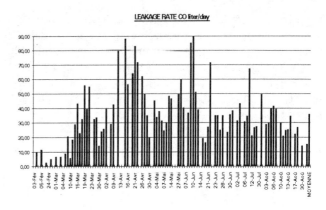

Fig. 5 - Leakage rate trough the clay liner

4.2 Temperature of the geomembrane on the slopes

The temperature of the geomembranes exposed to solar radiation has been measured and values up to 75°C have been observed during the summer. Pierson and al., 1993 have used this measurements for the validation of a theoretical model predicting the wrinkles in the HDPE geomembrane.

4.3 Displacement of the geosynthetics on the slopes

The displacement of the geosynthetics and their protection are measured on the slopes of GO and BO. In the future, the settlement of the waste will also be monitored.

Before filling the cells with waste, there are only negligible displacements of the HDPE geomembranes on the slopes.

In the case of BO, three types of cover over the bentonite geoproduct are tested. The silty sand cover was unstable on the 1/2 slope gradient whereas the gravel layer only induced a few centimetres of displacement. On the 1/1 gradient slope, the tri-dimensional geotextile cells filled with soil (silty sand) remained stable after an immediate displacement of approximately 40 cm. However, in response to a rainy period in September 1993, some of the cells failed, but were rapidly repaired.

5 Conclusion

The success of a work like the construction of the experimental landfill cells at Montreuil-sur-Barse is, at first, the result of a clear design which takes the installation constraints into account. But the full result cannot be achieved without a quality and control plan (installation, control and repair specifications). Despite this control during the installation of the geomembranes, leaks appeared just after completion of the cells. This emphasises the need for a general test of such cells, after the drainage gravel layer has been installed over the geomembranes. The experimental cells will now be filled by the waste and new results provided by the monitoring systems should be available in the next years, allowing for a better understanding of the behaviour of geosynthetics and clay liners in a landfill.

References

Pierson P., Pelte T., GOURC J.P., 1993. Behaviour of geomembranes exposed to solar radiation. Proceedings Sardinia 93, fourth international Landfill Symposium, S. Margherita di Pula, Cagliari, Italy; 11-15 October 1993, CISA, Cagliari, 350-356.

Le Tellier I., Bernhard C., Gourc J.P., Matichard Y., 1993. Real scale experimentation on the bottom and slope lining of municipal landfill. Proceedings Sardinia 93, fourth international Landfill Symposium, S. Margherita di Pula, Cagliari, Italy; 11-15 October 1993, CISA, Cagliari, 137-145.

A Geotechnical Investigation of Soil-Tire Mixtures for Engineering Applications

Brent A. Black, Abdul Shakoor
Department of Geology and the Water Resources Research Institute, Kent State University, Kent, Ohio, USA

Abstract

Soil-tire mixtures containing 0%, 10%, 20%, 30%, 40%, and 50% shredded tire, by weight, were tested for moisture-density relationships, permeability, shear strength, unconfined compressive strength, and consolidation behavior. The results show that the density and compressive strength of soil-tire mixtures decrease, and permeability increases, with increasing shredded tire content for all three soil types (sand, silt, and clay) and all three shredded tire sizes (4mm-7mm, 1mm-4mm, and <1mm) used. The addition of all three shredded tire sizes improves the friction angle for silt and the cohesion values for sand and clay. The compression index values for silt and clay decrease with the addition of 4mm-7mm size shredded tire but increase when the two smaller size ranges are used. Bulk chemistry and leachate analyses indicate that all three tire sizes used in this study exceed the maximum concentration levels (MCLs) for chromium, iron, and zinc. Based on the test data, soil-tire mixtures can be used where improvement in the drainage characteristics of low permeability soils is desired or where a lightweight fill material is needed.

Introduction

In the United States 279 million scrap tires are generated each year [5]. Of these, nearly 85% are landfilled, stockpiled, or illegally dumped. This constitutes about 1.2% of the nation's municipal solid waste stream [5]. Currently, 2.5-3 billion scrap tires are stockpiled across the United States. Stockpiled tires represent a waste of resources, an aesthetic nuisance, and a public health hazard because they serve as an ideal breeding ground for rats and mosquitos. Mosquito borne diseases, associated with stockpiling of scrap tires, cost approximately $5.5 million per year and the annual cost of extinguishing scrap-tire fires exceeds $2 million [5]. Scrap-tire fires frequently have devastating effects on air quality, ground-water quality, and public health due to liquid and gaseous emissions from the burning tires.

The best way to reduce the environmental and health hazards associated with scrap tires is to minimize, and ultimately eliminate, stockpiling. A large scale potential use of scrap tires can be in soil stabilization. Shredded scrap tires can be mixed with different soils to improve their engineering properties for specific applications. To date, only a few studies have been conducted to evaluate the potential engineering applications of shredded scrap tires [1, 4, 6, 9]. Additional research is needed to supplement the results of these preliminary studies.

Objectives of the Study

The present study was undertaken to investigate both the engineering properties and chemical characteristics of soil-tire mixtures. The specific objectives were:

1. To test the engineering properties of soil-tire mixtures containing 0%, 10%, 20%, 30%, 40%, and 50% shredded tire by weight so that the optimum amount of shredded tire material needed for improving the desired soil properties could be determined.

2. To investigate the effect of the size of shredded tire material on the engineering properties of interest.

3. To evaluate the suitability of tire-stabilized soil for use as: (a) daily cover material for sanitary landfills; (b) general purpose fill material; (c) highway embankment material; (d) lightweight material for reconstruction of failed slopes; and (e) material that would create hydraulic barriers to ground water flow.

Research Methods

Sample Collection and Preparation: Three different size ranges of shredded tire (4mm-7mm, 1mm-4mm, and <1mm) and three different types of soil (sand, silt, and clay) were used in this study. Samples of shredded tire were obtained from R.B.W. Industries, Evans City, Pennsylvania. The tread rubber was stripped from the tire by the supplier before shredding, resulting in steel-free pieces of shredded tire. Bulk samples of sand, silt, and clay were collected from local borrow areas. The soil samples were oven dried at 105°C for 24 hours, cooled to room temperature, pulverized to pass through #10 sieve (2 mm), and stored in air-tight plastic bags for later use.

Laboratory Investigations: All three soil types were tested in the lab to determine their pertinent engineering properties including grain size distribution, Atterberg limits, compaction characteristics, permeability, shear strength parameters, unconfined compressive strength, and consolidation behavior. Soil-tire mixtures, containing 0%, 10%, 20%, 30%, 40%, and 50% shredded tire by weight, were tested for compaction characteristics, permeability, shear strength, unconfined compressive strength, and compressibility. These mixtures were prepared using all three soil types and each of the three size ranges of shredded tire. The permeability, shear strength, and unconfined compressive strength tests were performed on samples compacted to at least 95% of the maximum dry density and within ±2% of the optimum water content, except for sand-tire mixtures which were compacted at 5% water content. The consolidation tests were performed on compacted samples of silt and clay, and silt-tire and clay-tire mixtures containing 30% shredded tire by weight. All engineering tests were performed according to the methods specified by the American Society for Testing and Materials (ASTM) [2].

Samples of shredded tire were analyzed for bulk chemistry and leachate composition, using an inductively coupled plasma (ICP) spectrophotometer. Loss on Ignition (LOI) tests were performed to quantify the organic content of shredded tires. The purpose of chemical analyses was to determine if the leachate generated from the shredded tires exceeded the tolerance limits of trace metals set for hazardous wastes by the U.S. Environmental Protection Agency [11].

Engineering Properties of Soils, Shredded Tire, and Soil-tire Mixtures

The engineering properties of the soils used in the study are shown in Table 1. Based on their grain size distribution and Atterberg limits, the soils were classified as a well-graded sand (SW), non-plastic silt (ML), and low plasticity clay (CL), respectively. The relatively high friction angle value for clay may be attributed to its somewhat sandy nature. Table 2 shows the dry density and shear strength parameters for the three size ranges of the shredded tire. The variation of engineering properties with increasing tire content for clay-tire mixtures is shown in Figures 1-6. The trends for sand-tire and silt-tire mixtures were found to be quite similar and, therefore, clay-tire mixtures are considered to typify the results for all three soil types.

TABLE 1--Engineering properties of the soils used.

Physical Properties	Soils Used		
	Sand	Silt	Clay
Liquid Limit (LL)	-	21.5	32.7
Plastic Limit (PL)	-	21.5	23.5
Plasticity Index (PI)	-	0	9.2
Maximum Dry Density (pcf)	115.5	107.4	101.1
Optimum Water Content (%)	6.6	12.2	20.0
Permeability (cm/sec)	9.08×10^{-5}	2.55×10^{-7}	2.44×10^{-8}
Friction Angle (°)	41	29	35
Cohesion (psf)	0	504	0
Shear Strength (psf)	869	1058	727
Unconfined Compressive Strength (psf)	-	3851	5213
Soil Classification (USCS)	SW	ML	CL

Note: 1 psf = 0.048 kN/m^2 ; 1 pcf = 0.016 Mg/m^3

TABLE 2--Dry density and strength properties of shredded tire.

Physical Properties	Shredded Tire Size		
	4mm-7mm	1mm-4mm	< 1mm
Dry Density (pcf)	33	33	33
Friction Angle (°)	27	31	30
Cohesion (psf)	130	70	100
Shear Strength (psf)*	707	671	677

* The values listed are for a normal stress of 1000 psf.
1 psf = kN/m^2 ; 1 pcf = 0.016 Mg/m^3

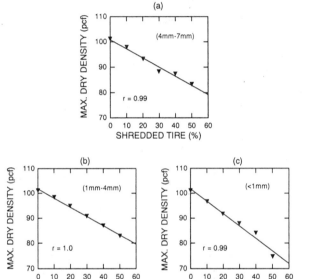

Figure 1: Maximum dry density versus shredded tire content for clay-tire mixtures.

Compaction tests on soil-tire mixtures revealed that the maximum dry density decreased with increasing tire content for all three soil types and all three tire sizes. An example of this trend is provided in Figure 1 which also shows that the largest tire size (4mm-7mm) results in the smallest decrease in density for a given proportion of shredded tire.

Permeability of soil-tire mixtures was found to increase with increasing shredded tire content for all three soil types and all three tire sizes. The maximum increase in permeability for all three soil types occurred with the addition of 4mm-7mm shredded tire size. Figure 2 shows a plot of permeability vs shredded tire content for clay-tire mixtures.

The strength characteristics of soil-tire mixtures were determined using the direct shear and the unconfined compression tests. Figures 3 and 4 show the variation of friction and cohesion, respectively, with respect to shredded tire content and shredded tire size. The effect of tire content on friction angle values for clay is unclear because of the scattered nature of the data points (Figure 3). Cohesion, however, shows a significant increase with increasing tire content, the maximum increase being 560 psf (26.8 kN/m^2) at 30% tire content for both 4mm-7mm and <1mm size ranges. It should be noted that the initial cohesion of clay was zero because of the consolidated-drained conditions under which the clay was tested.

Compared to clay, silt showed an overall increase in friction angle, and a decrease in cohesion, with increasing tire content. Because of these inconsistent results, it was decided to relate shear strength, rather than individual strength parameters, with the shredded tire content. The shear strength for soil-tire mixtures was computed using an assumed normal stress value of 1000 psf (47.9 kN/m^2). A plot of shear strength vs tire

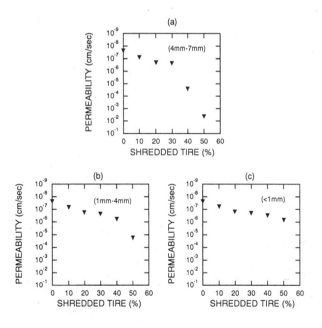

Figure 2: Permeability versus shredded tire content for clay-tire mixtures.

content for the clay-tire mixtures is shown in Figure 5. The addition of all three shredded tire sizes improves the shear strength of clay-tire mixtures. The maximum increase of 1310 psf (62.7 kN/m^2) occurs when 40% shredded tire in the 4mm-7mm size range is added. The results for sand-tire and silt-tire mixtures were similar, but less consistent.

A plot of unconfined compressive strength versus shredded tire content for clay-tire mixtures (Figure 6) shows that the unconfined compressive strength decreases with increasing tire content. Silt-tire mixtures showed similar trends. The test was not performed for the sand-tire mixtures.

The results of consolidation test are summarized in Table 3. The compression index (C_c) values for both silt and clay decrease with the addition of 4mm-7mm shredded tire but increase when 1mm-4mm and <1mm sizes are added.

Chemical Characteristics of Shredded Tire

A major concern regarding the use of shredded scrap tires is the potential for toxic substances to leach from the material upon interaction with water, thus posing a threat to surface and ground water resources. In order to address this concern, the bulk composition and leachate characteristics of shredded tire were analyzed for selected trace metals. The results of leachate analysis were then compared with the results of two others studies, one conducted by the Minnesota Pollution Control Agency (MPCA) and the other by the Rubber Manufacturers Association (RMA).

TABLE 3--Compression index values for silt, clay, silt-tire, and clay-tire mixtures at 30% shredded tire content.

Shredded Tire Size	Compression Index (C_c)	
	Silt-Tire Mixture	Clay-Tire Mixture
	0.080 (silt only)	0.117 (clay only)
4mm-7mm	0.076	0.080
1mm-4mm	0.131	0.163
<1mm	0.169	0.158

LOI and Bulk Chemical Analyses: The loss on ignition (LOI), determined for 20 samples of shredded tire following ASTM method D2974, ranged from 2.0% to 5.7%, with a mean of 2.9%. The LOI represents the loss of petroleum hydrocarbons or polynuclear aromatic hydrocarbons. The bulk chemical analysis was also

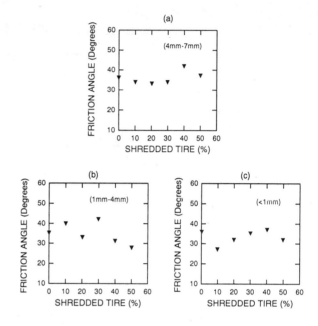

Figure 3: Friction angle versus shredded tire content for clay-tire mixtures.

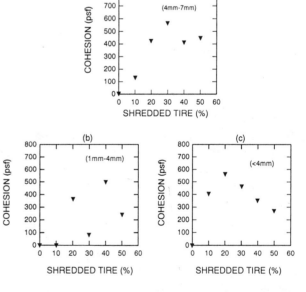

Figure 4. Cohesion versus shredded tire content for clay-tire mixtures.

normalized to present the results in mg/kg units. The results of bulk chemical analysis are presented in Table 4. The bulk chemistry gives some indication of the quantity of metallic constituents that would leach out under a "worst case" scenario, i.e. if the shredded tires were to completely degrade and all inorganic components were to be released. Table 4 shows that concentrations of zinc and iron are high for all three shredded tire sizes.

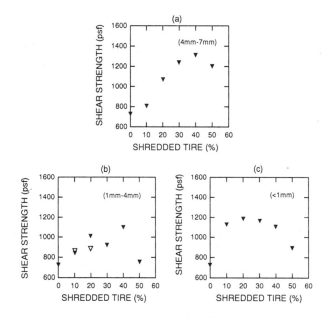

Figure 5: Shear strength versus shredded tire content for clay-tire mixtures. Open triangles represent repetitions.

TABLE 4--Results of bulk chemical analyses.

Sample	Tire Size	Normalized Concentration (mg/kg)				
		Fe	Cd	Pb	Zn	Co
13	4mm-7mm	360.0	3.72	42.9	17610	1.63
14	"	961.0	4.05	114.0	13360	2.97
17	"	473.0	0.38	9.5	14680	1.15
20	"	692.0	2.01	44.8	15310	4.46
50	"	657.0	1.91	66.9	14370	7.38
97	"	40.3	0.01	2.3	1730	0.20
Mean	"	530.0	2.03	46.7	12840	2.97
15	"	1790.0	5.92	149.0	8460	4.34
18	"	1720.0	1.26	88.8	8060	1.16
21	"	1520.0	3.82	94.2	9680	6.92
51	"	1470.0	5.61	300.0	7490	7.48
54	"	1430.0	2.42	123.0	7240	3.96
56	"	1550.0	3.11	221.0	7520	8.82
Mean	"	1580.0	8.07	155.0	7750	8.59
16	<1mm	3360.0	5.11	108.0	13970	4.30
19	"	2300.0	3.10	76.6	13970	6.40
22	"	840.0	2.08	54.6	13510	2.08
32	"	1750.0	3.48	105.0	13930	1.19
52	"	2890.0	2.99	95.4	11960	5.89
55	"	2840.0	4.09	98.5	11890	96.10
57	"	3020.0	4.03	96.3	11700	9.62
Mean	"	2430.0	3.55	90.7	12960	17.90

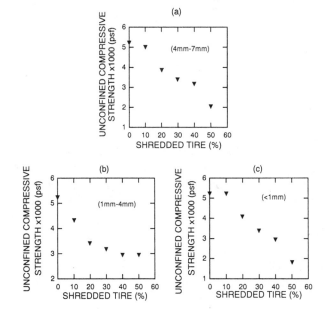

Figure 6: Unconfined compressive strength versus shredded tire content for clay-tire mixtures.

performed on 20 samples of shredded tire using an inductively coupled plasma (ICP) spectrophotometer. The analysis was confined to the following suspected metals: iron, cadmium, lead, zinc, and cobalt. The measured concentrations of these metals were

Leachate Analysis: All three sizes of shredded tire were subjected to leachate extraction tests in accordance with the EPA method SW-846-1310 [10]. The method involved placing 50g of shredded tire material in 300 ml of deionized distilled water (DIDW). The solution was agitated for a period of 24 hours, while maintaining the pH of DIDW at 5 ± 0.2 by the manual addition of 0.5N acetic acid. The extract solution was then analyzed, using ICP techniques, for barium, iron, cadmium, chromium, lead, and zinc. The results of leachate analysis, presented in Table 5, show that the chromium,

TABLE 5--Results of leachate analyses.

Samples	Leachate Concentration (mg/l)							
	Size	pH	Ba	Fe	Cd	Cr	Pb	Zn
70	4mm-7mm	≅5.0	0.04	0.44	ND	0.63	ND	6.65
71	1mm-4mm	≅5.0	0.11	0.61	ND	1.66	ND	17.70
72	<1mm	≅5.0	0.31	15.10	0.02	3.28	0.66	36.50
MPCA-1	25.4mm	3.5	0.21	500.00	0.13	0.24	0.05	23.50
MPCA-2	25.4mm	5.0	0.62	23.30	<0.01	<0.01	0.05	17.50
RMA	<10mm	≅5.0	0.07	-	-	ND	0.02	-
MCLs	-	-	2.00	0.30	0.005	0.10	0.015	5.00

Notes: -: Sample not tested, not applicable.
ND: Not Detected.
Size: Size of shredded scrap tire.
MPCA: Minnesota Pollution Control Agency (1990).
RMA: Rubber Manufacturing Association (1990).
MCLs: Maximum Concentration Levels (EPA CFR 57)

iron, and zinc concentrations for all three shredded tire sizes exceed the maximum concentration levels (MCLs) as established in EPA CFR 57 [11]. The smallest shredded tire size (<1mm) exceeds all MCLs. Due to the presence of chromium at levels exceeding those specified in EPA CFR 57 [11], the shredded scrap tire tested in this study can be categorized as a hazardous material. It should be noted, however, that the above results represent the worst case scenario since the tests were conducted on 100% tire material. When tire is mixed with soil and placed in a compacted state, the concentration of the leachate is expected to be much lower than the results in Table 5.

Table 5 also compares the results of the present study with those obtained by the Minnesota Pollution Control Agency [7] and the Rubber Manufacturer's Association [8]. The results are quite similar. The MPCA and RMA studies used the same procedure for leachate extraction as that used in this study. A close examination of Table 5 reveals that metals are more likely to leach from smaller sizes of shredded tire and under acidic conditions (low pH).

Engineering Applications of Tire-Stabilized Soils

Based on the engineering properties of soil-tire mixtures discussed previously, tire-stabilized soils have the potential for the following applications:

Landfill Cover and Liner Material: The permeability requirements for cover and liner materials for sanitary landfills are 1×10^{-5} cm/sec and 1×10^{-7} cm/sec, respectively, or less [10]. Although the addition of shredded scrap tire increased the permeability of each soil tested, the permeability values for some soil-tire mixtures (silt and clay with smaller percentages of shredded tire content) were still quite low. Based on the permeability values obtained in this study, both silt-tire and clay-tire mixtures can be used as a daily cover material for sanitary landfills as long as the mixtures contain less than 30% shredded tire by weight. Clay with 10% shredded tire content also meets the permeability requirements of a landfill liner. Therefore, a substantial amount of shredded tire can be used in such applications.

Football Fields and Playgrounds: Football fields and playgrounds require permeable and nondeformable soils that allow easy growth of grass. Since the addition of shredded tire increases both the permeability and the shear strength, without significantly increasing the compressibility, it is believed that soil-tire mixtures are well suited for construction of football fields and other playgrounds, especially if the original soils are clayey. In fact, this is already being practiced in states like Colorado, Michigan, and Pennsylvania where such applications consume an average of 12,000 tires to treat a single football field [3].

Lightweight Fill Material: The requirements of a lightweight fill material include low density, high shear strength, and good drainage characteristics. Since tire-stabilized soils fulfill these requirements, they can be used to construct highway embankments on soft ground, such as peat and clay. Soil-tire mixtures can also be used to reconstruct already failed, or potentially unstable, slopes.

Conclusions

The conclusions of this study can be summarized as follows:

1. The dry density and compressive strength of soil-tire mixtures decrease, and the permeability increases, with increasing shredded tire content for all three soil-types and all three shredded tire sizes.

The addition of all three shredded tire sizes improves the shear strength of clay but only the 4mm-7mm size material improves the shear strength of sand. The shear strength of silt improves slightly with the addition of shredded tire of <1mm size. The compression index values for silt and clay decrease with the addition of 30% shredded tire in the 4mm-7mm size range, but increase slightly with the addition of the other two sizes.

2. All three shredded scrap tire sizes used in this study exceed the Maximum Concentration Levels for chromium, iron, and zinc. The metals were found to leach in the highest concentrations from the smaller tire sizes (1mm-4mm and <1mm) under acidic conditions. Based on these leachate characteristics, it is suggested that tire-stabilized soils be used in places where the water table is deep and where extreme variations in pH are not expected. Also, only 4mm-7mm or larger sizes of shredded tire should be used.

3. The optimum amount of shredded tire material needed to improve soil properties depends upon the intended engineering application. Clay with the addition of 10% shredded tire in the 4mm-7mm size range is suitable as a landfill liner. Both silt and clay with the addition of 10%-30% shredded tire meet the requirements of daily cover material. The addition of 30% shredded tire to clay appears to be the optimum amount of shredded tire material needed to improve its shear strength and drainage characteristics, and lower its density, for use as a lightweight fill. For sand and silt, 10-20% shredded-tire content appears to optimize the desired engineering properties.

References

1. Ahmed, I., 1992, Laboratory Study on Properties of Rubber Soils: Report No. FHWA/IN/JHRP-91/3, School of Civil Engineering, Purdue University, West Lafayette, Indiana, 145 p.
2. American Society for Testing and Materials, 1990, Soil and Rock, Building Stones, Geotextiles: Annual Book of ASTM Standards, Vol. 4.08; Philadelphia, Pennsylvania, 1092 p.
3. BioCycle, 1990, New Sports Fields from Old Tires: July, Vol. 37, No. 7, J.G. Press, Inc., Emmaus, Pennsylvania, p. 84.
4. Bosscher, P.J., Edil, T.B., Eldin, N., 1992, Construction and Performance of a Shredded Waste-Tire Embankments: Department of Civil and Environmental Engineering, University of Wisconsin-Madison, Wisconsin, 17 p.
5. House Bill S2462, 1990, Tire Recycling Incentive Act of 1990: 101st U.S. Congress, Washington, D.C.
6. Lamb, R., 1992, Using Shredded Tires as Lightweight Fill Material for Road Subgrades: Materials and Research Laboratory, Minnesota Department of Transportation, Maplewood, Minnesota, 30 p.
7. Minnesota Pollution Control Agency, 1990, Waste Tires in Subgrade Road Beds: Minnesota Pollution Control Agency, St. Paul, Minnesota, 34 p.
8. Rubber Manufacturers Association (RMA), 1990, RMA Status Report: June 19, Vol. X, No. 2, Page 2: RMA, Washington, D.C.
9. Upton, R.J. and Machan, G.M., 1993, Use of Shredded Tires for Lightweight Fill: Oregon Department of Transportation, Salem, Oregon.
10. U.S. Environmental Protection Agency, 1990, Federal Register CFR 40, No. 61, Washington, D.C., pp. 11798-11877.
11. U.S. Environmental Protection Agency, 1992, Federal Register CFR 57, Washington, D.C., p. 60848.

Landfill Gas Protection Measures for Industrial Developments

R.G. Clark, BSc., MSc., DIC, MICE, Eur Ing.
C L Associates, Birmingham, UK

Abstract

The types of gas protection measures for industrial buildings are reviewed and the criteria for selecting a particular type are discussed together with a brief summary of currently available guidelines in the UK. Examples are given of the various components of a system including membranes, vented granular blankets, undercrofts, passive and active venting, instrumentation and control systems and precautionary gas detection inside the building.

Introduction

This paper seeks to provide a broad outline of the factors which should be taken into account in designing and installing gas protection measures and provides examples of some of these features.

Although consideration is primarily given to the requirements for landfill gas, the details are in many respects also applicable to providing protection against other types of gas. Domestic dwellings should not be built on or adjacent to landfills and are therefore not considered.

The recommendations and design details given in this paper assume that a thorough site investigation has been carried out and that as a result the nature of the ground materials, the types and concentrations of gas and the potential emission rates from the ground are known. The ground should be investigated for gas for any development within 250 metres of a landfill site (Ref. 5).

Although reference is made in this paper to landfills and landfill gas, it should be remembered that other fills containing biodegradable materials and natural organic strata can also produce the same problems.

Guidance Documents

There are presently three recognized published documents in the UK which give outline guidance in respect of the criteria for deciding whether a building on a particular site requires protection measures. These are DoE "Waste Management Paper 27, Landfill Gas" (Ref 5), the UK Building Regulations "Approved Document C" (Ref 6) and the Building Research Establishment publication entitled "Construction of New Buildings on Gas Contaminated Land" (Ref 1). These documents refer to each other and credit each other with certain recommendations as follows :-

i) No further protection is required if the methane concentration in the ground is less than 1% by volume and a ventilated suspended floor slab is provided.

ii) There is a need to consider possible protection measures if carbon dioxide concentrations exceed 1.5% by volume.

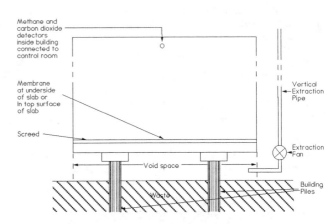

FIG. 1 SCHEMATIC SKETCH SHOWING VENTED UNDERCROFT SYSTEM

iii) Specific design measures are required if carbon dioxide concentrations exceed 5% by volume.

iv) The amount of gas in the ground and its pressure relative to atmosphere also need to be considered. High concentrations of gas have less of an impact when the quantity of gas is low.

v) The concentration of gas found in the ground does not necessarily relate to the concentration of gas which would be found in a building (that does not have protection measures).

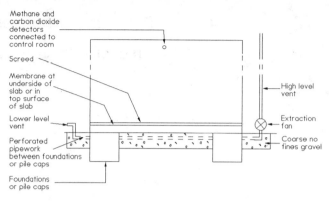

FIG. 2   SCHEMATIC SKETCH SHOWING VENTED GRANULAR BLANKET SYSTEM

Although providing some guidance, these documents are not sufficiently comprehensive to enable the designer to decide on the type and extent of measures that are required for certain gas concentrations. The problem facing the designer is therefore to decide how extensive the control measures must be in order to ensure safety but at the same time be reasonable in terms of the commercial viability of the project.

Types of Protection Measures

For some projects it may be economically viable to remove all gas producing materials from the ground and to either transport them off site or to place them in an area of the site where they cannot have an adverse effect on the development. Alternatively it may be appropriate to construct an in ground barrier, such as a diaphragm wall, grout curtain or cement bentonite cut off (with or without a geomembrane) between the source of the gas and the proposed structures. There are, however, a considerable number of projects where gas protection measures incorporated into the building itself are the best option.

The production of landfill gas from biodegradable materials can continue for 40 years (Ref 2) and in some cases considerably longer. Whatever measures are chosen, they must therefore be able to protect the building for at least this order of time.

Protection measures can either be simply a gas proof membrane or a vented undercroft (sub slab void) and membrane or a vented granular blanket and membrane. Venting can either be active or passive. Clark et al (Ref 3) have reviewed the various types of protection measures (see Figs 1 and 2).

In the majority of cases where gas protection measures are required the provision of only a membrane will not be sufficient. The situations where the author has recommended this form of protection have been where concentrations of carbon dioxide in the ground are in the range 1.5 to 5% and where methane concentrations are considerably less than 1% by volume.

Gas Proof Membranes

In order to fulfil its required function, a gas proof membrane must have both low permeability to gases and sufficient strength characteristics to prevent tearing or puncturing during handling, laying and subsequent construction activities. Secondary criteria influencing the choice of membrane are cost and ease of handling. Factors that determine the permeability and strength of the membrane are the thickness and type of material used for its manufacture.

Perhaps the most commonly used membrane is 2.0 mm thick High Density Polyethylene (HDPE). However, a 1.5 mm Very Low Density Polyethylene (VLDPE) membrane has comparable puncture and tear resistance and although its gas permeability is higher it is still very low and acceptable. The VLDPE membrane has the advantage of being more easy to handle in confined spaces and to lay over irregular shapes. Either type is preferable to the use of PVC which does not possess the required criteria and, despite being inexpensive, should not be used as a gas proof membrane. Both HDPE and VLDPE are usually used below the floor slab and in the case of a vented undercroft the membrane requires some form of additional support or fixing to the underside of the slab.

Gas resistant aluminium or bitumastic membranes can be used above the floor slab but will require a screed or boards laid over the top to ensure protection. This form of membrane is therefore not suitable where high skid forces have to be resisted by the floor eg. a building with vehicle access or where stacking equipment is operating.

Durability is also important in that the material comprising the membrane must not degrade and must be resistant to corrosion from any of the substances with which it could possibly come into contact. Some membranes degrade under UV light.

The structure may be piled and therefore the membrane will have to be sealed around each one of the piles unless the building design is such that the membrane can pass across the top of all the pile caps. For some of the projects on which the author has been involved this has meant sealing the membrane around in excess of 1000 piles. If precast piles or otherwise smooth shafted piles are used then this may not be a problem. But if irregularly shafted piles such as Continuous Flight Auger (CFA) piles are used then providing an effective seal may not be possible. In this situation it may be necessary to break down the pile shafts to below the level of the membrane and to recast the piles back up to final level using smooth formwork. Alternatively, some type of smooth permanent former has to be used at the correct level when the piles are first cast. Once a smooth shaft has been achieved the seal can be effected by either using preformed "top hats" or fabricated "skirts" of membrane (see Fig 5) which are first fixed to the piles with tightened alloy bands and mastic or neoprene strips and then extrusion welded to the main membrane.

Membranes must be continuous across any wall cavities and must be effectively sealed to any damp proof course. In effect a gas proof barrier must extend in both directions from one outside wall face of the building to the opposite outside wall face.

The overall permeability of a membrane can be very seriously influenced by the number of undetected punctures, tears or poor welds. Construction Quality Assurance of the work is therefore vital to ensuring the integrity of the membrane. This should be in respect of inspections, destructive and non destructive testing of welds (and other forms of joints) and general strict control of operations. The design of the membrane configuration must also take into account the consequences of any differential settlement between the structure and the ground. This will particularly be the case if the development is directly on top of a landfill.

Vented Granular Blankets

Due to air resistance through the gravel it is questionable as to whether naturally vented granular blankets actually work. The author is not aware of any evidence that confirms they can be effective. Without effective air flow the system may just be a reservoir within which gas can accumulate and therefore granular blankets, unless under a small floor area, should always be actively vented.

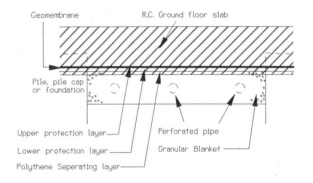

FIG. 3 TYPICAL SECTION

A typical cross section through a vented granular blanket (usually not less than 400 mm deep) is shown in Fig 3. The protection to the geomembrane should be particularly noted. Obviously the geomembrane must be laid on a sufficiently smooth surface that will not cause it to be damaged. Work of this type is also being carried out in a construction site environment which means that the membrane is vulnerable to damage from equipment, plant and falling objects. It is therefore necessary to provide protection on top of the membrane prior to steel fixing or other related activities. The protection above and below the membrane can be small aggregate concrete, sand/cement screed, fibre boards or similar.

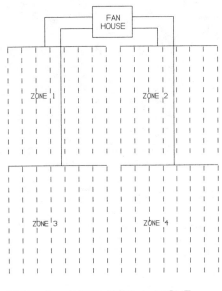

FIG. 4 PIPEWORK LAYOUT

The rate of generation and concentration of methane and/or carbon dioxide within the ground together with the floor area of the proposed building will determine the spacing of the pipes. The stone should be single size (say 40 mm), rounded and have no fines to ensure a high permeability thus allowing gas to move freely. It should not break down under load or compaction and should be only lightly compacted. Suitable materials for the pipework are HDPE or polypropylene.

Fig 4 shows a possible layout. In this case the area has been divided into zones in order to ensure that preferential extraction does not take place from one side of the building leaving the other side unprotected, which might be the case if the pipes run the full length. One project involved eleven such zones (Ref 4). The horizontal pipes are not connected to inlet pipes in order to ensure that air flow takes place through the blanket and not just through the pipes. Condensation within venting pipes is always a problem and therefore condensate traps must be provided.

A satisfactory air flow into the granular blanket must be ensured. Just providing a stone filled trench at the intake side of the blanket may not be sufficient as this could become blocked by sediment from surface water, leaves, snow or similar. As a precaution it is therefore advisable to also provide short intake stacks at intervals along the trench or some other means of ensuring positive air flow.

Vented Undercrofts

The choice of an undercroft or granular blanket appears to be subjective. Whilst not insurmountable by any degree, the problems associated with supporting a slab during construction without the aid of the ground lead to increased costs and increased construction time. Nevertheless, a number of undercroft type systems have been successfully constructed using permanent soffit shuttering.

The means of providing ventilation for an undercroft should result in a minimum of 1500 mm^2 per metre run of wll or 500 mm^2 per square metre of floor area, whichever gives the greater area of opening (Ref 1). The configuration of the void should be such as to allow free flow of air and there should be plenty of air gaps in any sleeper walls in order to ensure that there will be no pockets of still air where gas could accumulate.

Services

The membrane must be integral with the damp proof course which should also have low gas permeability and all services should ideally enter the building above the membrane. Every effort should be made to design the building so that toilets and kitchens etc are located adjacent to outside walls so that drains can pass directly through the walls and any manholes, rodding eyes etc can be positioned outside the building. The drains themselves must be gas tight and the design must ensure that any subsequent ground settlement does not open joints in the drain pipes or pipe connections. Where services passing through the membrane cannot be avoided then these should be suitably sealed or puddle flanged.

Provision must also be made for service trenches and service ducts to be vented or sealed before the services enter the building (see Fig 5). The annulus within a duct must not act as a conduit which will allow gases to enter the structure.

Passive Venting

A passive venting system utilising natural air movements may be suitable in certain

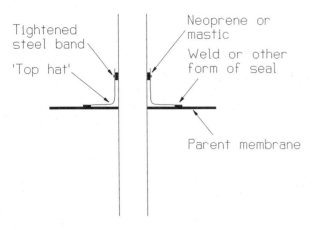

FIG. 5 TYPICAL TOP HAT DETAIL

circumstances for small floor area buildings. In these situations the undercroft or the perforated pipes in a granular blanket are connected to low level stacks at one side of the building and high level stacks with rotating cowls at the other side in order to maximise the use of natural air movements.

Active Venting

In a mechanically vented system motor driven fans can be used, with a greater fan capacity often being required for venting an undercroft compared to a granular blanket. A general rule of thumb for a mechanically vented system is that the fans must have the capacity to change the air within the void or pipes and granular blanket one to two times per hour. This should be considered in the light of the actual gas concentrations and emission rates for a particular site. A standby fan will also be required as a precaution against mechanical failure. The fans can either be triggered by flow sensors in the extraction pipe or by gas sensors placed beneath the floor slab. The fans and the motors that drive the fans must be intrinsically safe and must be capable of working in an explosive atmosphere.

Some projects may require more than one duty fan. This could either be because the general arrangement of the building dictates that each zone has to be kept separate and therefore a fan is required for each zone or because with all zones connected to the same manifold, more than one fan is necessary in order to provide the required number of air changes per hour.

Fig 6 shows a schematic arrangement for a two fan system. In this type of configuration it may be necessary to run only one fan continuously in order to provide a balanced suction pressure to each zone, the balance being controlled by non closing electrically operated valves on each of the inlet pipes leading to the main manifold. In the event of predetermined acceptable gas levels being exceeded, the control valve relative to the affected zone opens and the second fan starts up and runs until the level is reduced. Should the extraction rate still be insufficient then additional fans (if present) can be started up and run until acceptable levels are achieved. In order to equalise the wear on the fans, inbuilt timers allow the lead fan to run for a predetermined period and then become a standby as the second fan takes the lead and so on. In the event of a motor/fan failure then the next available fan is automatically started (see below).

Instrumentation and Control Systems

Some configurations may be sufficiently complex to warrant the use of an electronic control system in order to record data and to operate the various functions. The instruments linked to the control system may include the detection of methane, carbon dioxide and vacuum in the delivery pipes, flow rate and temperature in the outlet stack, running time of the motors and any motor failure (see Fig 6). In addition visual and audible alarms can be provided in order to indicate a motor failure, a no flow condition in the stack, whether gas levels in the delivery pipes have exceeded a certain level and whether gas concentrations are continuing to rise even with all fans running.

Electronic data loggers or chart recorders can be used to record information from the various instruments thus providing a permanent record and enabling an assessment to be subsequently made of any events or malfunctions. On some systems a flame arrestor has been fitted in the outlet stack. In addition a thermocouple can be installed in the stack which in the unlikely event of a fire will automatically shut the plant down and close all valves. The plant can be set to automatically restart when the thermocouple registers cold again.
The fans, motors, instruments and control system

should preferably be located in a separate Fan House adjacent to the main structure. In some cases it has been possible to construct a self contained Fan Room within the main building, in which case the room should have structural walls between it and the remainder of the facility in order to ensure that the extraction system does not itself become the means whereby gas enters the building.

Gas Detection Inside the Building

No gas protection measures can be regarded as infallible and therefore permanent methane and carbon dioxide sensors need to be installed within the building in order to demonstrate that the measures are effective (see Figs 1 and 2). A criteria that has been adopted by the author is that all rooms on the ground floor (including switch rooms, cleaners cupboards and small kitchens etc) need to be protected by sensors. On other floors only rooms that have toilets or wash basins or similar facilities need protection by sensors. This is because on these upper floors the most likely means of gas entry is through the drainage system.

All sensors inside the building can be connected to a central control panel which has visual and audible alarms to show that a sensor has been triggered. Some fire brigades prefer this panel to be located at the main point of entry to the building. Gas detection equipment should also be provided in the Fan House or Fan Room in case of leakage from pipework or the fans themselves. These gas detectors should be linked to the alarm system.

Concluding Remarks

Gas protection measures can satisfactorily be designed and installed to enable industrial buildings to be constructed on or adjacent to landfills. However, it is necessary to ensure that the appropriate degree of sofistication is incorporated into these measures consistent with the gas concentrations and rate of generation appertaining to a particular site. Some form of protection may also need to be considered for adjacent car parking and hardstanding areas.

References

1) BRE : 1991 : Construction of New Buildings on Gas - Contaminated Land, BR 212, Building Research Establishment.

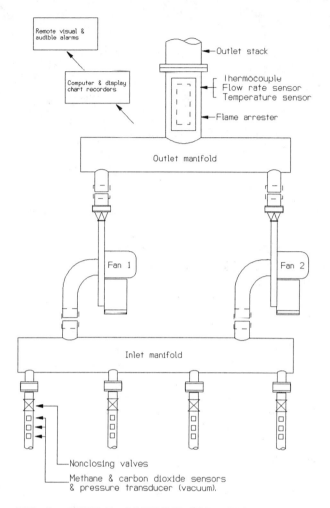

FIG. 6   TYPICAL CONFIGURATION FOR TWO FANS

2) CLARK R G : 1991 : The Hazards from Old Landfills. Symposium on the Containment of Pollution and Redevelopment of Closed Landfill Sites, Leamington Spa.

3) CLARK R G and WARBY I S : 1991 : Gas Protection Measures for Buildings on or Adjacent to Landfill Sites. The Planning and Engineering of Landfills, Midland Geotechnical Society Conference, Birmingham.

4) CLARK R G and DAVIES G : 1992 : Gas Protection Measures for an Industrial Development. Second International Conference on Construction on Polluted and Marginal Land, Brunel University, London.

5) DoE : 1991 : Waste Management Paper 27, Landfill Gas, Department of the Environment, HMSO.

6) DoE : 1991 : The Building Regulations, Approved Document C, Department of the Environment.

Remblais Alleges en Dechets de Matieres Plastiques

C. Coulet, M. Boudissa, L. Curtil
Laboratoire Génie Civil Habitat Environnement, Villeurbanne Cedex, France

Résumé : Les déchets de matières plastiques en mélange, fortement compactés pour former des blocs, sont employés pour la construction de remblais allégés sur des sites compressibles ou à l'arrière de soutènements. Des essais de laboratoire ont montré que ce milieu orthotrope est viscoélastique non-linéaire. Deux chantiers d'application confirment la validité de ce procédé Plastbloc. Pour le premier, 600 m³ de blocs ont été disposés au-dessus d'une canalisation enterrée, afin de permettre la construction d'une route, sans augmentation de contrainte sur la canalisation. Le deuxième concerne un élargissement de virage dans un site montagneux ; 500 m³ de blocs ont été mis en remblai derrière le rideau de palplanches afin de minimiser les poussées.

Introduction

Les normes, de plus en plus draconiennes, relatives à l'élimination des déchets, ainsi que la pénurie, dans certaines régions, de granulats et de bons matériaux naturels, contribuent à l'émergence de procédés innovants favorisant la préservation de l'environnement. Les travaux publics, qui chaque année mettent en oeuvre de gros volumes de matériaux constituent une filière très intéressante et économiquement rentable pour la valorisation de certains déchets en tant que matériaux de substitution.
Ainsi les déchets de matières plastiques, dont le gisement s'est considérablement accru ces dernières années (2 Mt par an en France), posent de réels problèmes à l'environnement. Leur recyclage, tout particulièrement les plastiques en mélange, reste marginal et leur élimination par mise en décharge est de plus en plus onéreuse. Quant à l'incinération, qui permet une valorisation énergétique, elle émet cependant des rejets gazeux nocifs dans l'atmosphère.
Ces déchets plastiques, fortement compactés et ligaturés pour former des blocs parallélépipédiques d'un volume voisin de 1 m³ et de densité variant de 0.3 à 0.6 selon les utilisations envisagées, peuvent servir de matériaux de construction pour des remblais allégés établis sur des sites compressibles ou à l'arrière de soutènements. Ce nouveau procédé, appelé Plastbloc, est donc simplement constitué d'un empilement de blocs en quinconce mis en corps de remblai [1]. Des essais de laboratoire, sur la

compressibilité et le fluage des blocs, ont permis d'appréhender le comportement de ces matériaux. La validité du procédé a été vérifiée par deux applications de chantier sur des sites différents, suivies par notre laboratoire avec la collaboration de J. Perrin du bureau d'études Ingeval.

Etude du Comportement du Bloc

Formé d'un empilement de couches de matières plastiques fortement comprimées, le bloc est assimilé à un milieu orthotrope de révolution. De nombreux essais de laboratoire sur des échantillons de taille réduite, compressibilité et fluage, ont abouti à une modélisation du comportement viscoélastique non-linéaire de ce matériau [2]. L'exploitation des mesures, effectuées sur des blocs réels et sur un remblai expérimental de 4 m de hauteur [3], a permis de vérifier la validité du procédé et de confirmer que le comportement du matériau est très influencé par la densité initiale du bloc qui, sous charge, est soumis à une déformation instantanée importante tandis que le tassement au cours du temps reste acceptable (figure 1). Toutefois cette densité initiale, fonction de la nature des déchets plastiques constitutifs des blocs, ne peut être un critère absolu du comportement. Le mode de fabrication des blocs, tout particulièrement le gonflement plus ou moins maîtrisé avant ligaturage, semble aussi un facteur influençant le comportement [4].

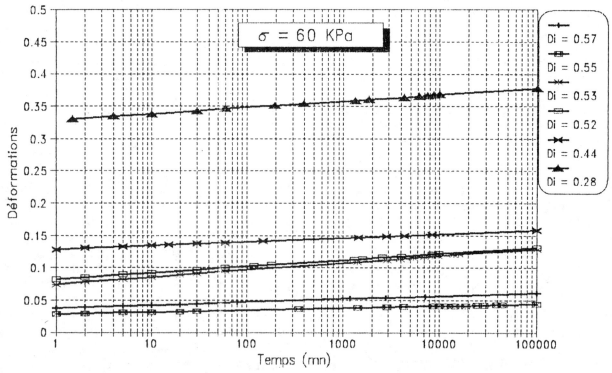

Figure 1 : Déformation au cours du temps du bloc pour différentes densités initiales Di sous contrainte de 60 KPa.

Remblai Plastbloc sur Conduite enterrée

L'aménagement d'une nouvelle voirie, en Haute-Savoie, imposait l'élévation d'un remblai sur un secteur traversé par une canalisation d'eaux usées enterrée. La solution de base lancée à l'appel d'offre consistait à construire un pont au droit de cette conduite. Proposé en variante, le procédé Plastbloc a été retenu pour deux raisons. D'une part il permettait, avec sa faible densité de 0.5, de franchir en remblai la canalisation sans en accroître la contrainte initiale. D'autre part le coût des travaux était très nettement inférieur à l'estimation établie pour un ouvrage béton armé. Pour respecter la valeur initiale de la contrainte au-dessus de la canalisation, évaluée à 72 kPa, il a été procédé à une excavation du sol en place, comblée par le remblai Plastbloc (figure 2). Ainsi 600 m^3 de blocs, disposés en hauteur sur 4 rangées, ont été mis en oeuvre à la pelle hydraulique (figures 3 et 4). L'ensemble des blocs a été enveloppé par un film polyéthylène, afin de minimiser les circulations d'eau à l'intérieur du massif Plastbloc. Un géotextile de protection a été déposé à l'interface des blocs et des couches de chaussées d'une épaisseur de 1.30 m.

Deux repères de tassement ont été posés respectivement en fond de fouille et sur la dernière rangée de blocs afin d'en suivre le comportement au cours du temps. Après deux mois de mise en circulation, intervenue début août 1993, les tassements mesurés sont millimétriques, donc négligeables.

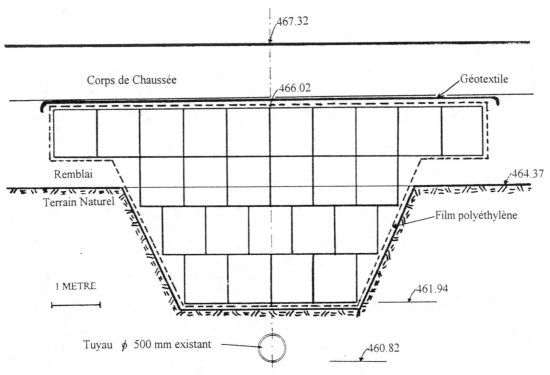

Figure 2 : Coupe longitudinale du remblai.

Figure 3 : Les premiers blocs

Figure 4 : Rangée supérieure

Remblai derrière Palplanches

Pour l'élargissement d'un lacet d'une route de montagne en Savoie, dans un site sensible soumis à des glissements de terrain, il a été décidé d'établir un rideau de palplanches ancrées par deux lignes de tirants. Afin de minimiser les efforts derrière le rideau, le remplissage a été effectué sur 5,40 m de hauteur avec un remblai Plastbloc (figure 5). Compte tenu de la faible densité (0.5) et de l'absence de déformation latérale des blocs sous sollicitation verticale, on peut estimer une réduction de poussée de 20% à 30% par rapport à un remblai normal. Cependant pour augmenter les frottements aux interfaces entre deux rangées de blocs et éviter des reports d'efforts du terrain naturel sur le rideau de palplanches, un lit de gravier concassé a été répandu au dessus de chaque couche de blocs. 500 m^3 de blocs ont été ainsi posés avec une grue et arrangés avec une mini-pelle descendue dans la fouille (figure 6). Deux repères de tassement permettent un suivi du comportement dans le temps. Le repère 1 est posé au-dessus de la troisième rangée de blocs et le repère 2 sur la cinquième et dernière rangée. Les courbes de la figure 7 confirment la forte compressibilité instantanée du matériau et l'admissibilité des tassements au cours du temps : 1 cm après 2 mois de mise en service.

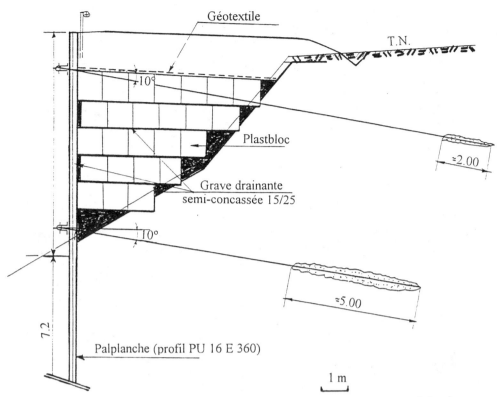

Figure 5 : Coupe transversale du remblai

Figure 6 : Pose de la première rangée

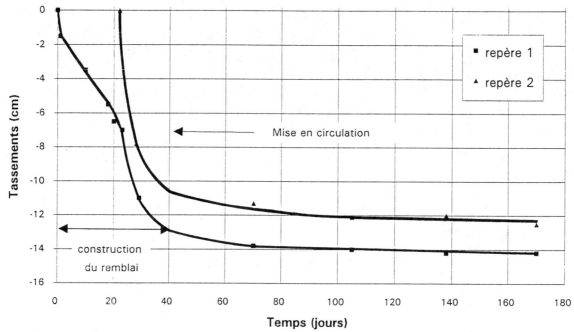

Figure 7 : Tassement mesuré au repère n°1.

Conclusion

Les deux premières applications de Plastbloc démontrent la validité de ce nouveau procédé de remblai allégé. Tant sur le plan géotechnique, puisqu'il permet de résoudre des problèmes de remblais impossibles à construire avec des sols classiques, que sur le plan économique car il est parfaitement compétitif avec les procédés concurrents.

De plus, gros consommateur de déchets de matières plastiques en mélange (550 tonnes valorisés pour les deux chantiers), Plastbloc permettrait, si cette technique était développée et étendue à d'autres applications (remblais sur site compressible, de protection contre les chutes de blocs, d'amortissement de vibrations), de résoudre le difficile et onéreux problème posé par l'élimination de ces déchets.

Références :

[1] Daudon D., Remblais allégés en déchets de matières plastiques étude du procédé Plastbloc. Thèse de Doctorat, Université Claude Bernard Lyon1, France, 1992.

[2] El-Ghoche H., Expérimentation et modélisation de blocs compressibles de matières plastiques pour des remblais allégés. Thèse de Doctorat, Université Claude Bernard Lyon1, France, 1992.

[3] Daudon D., Coulet C., El-Ghoche H., Low-weight plastic waste embankments, Geocoast 91, Yokohama, Japon 1991.

[4] El-Ghoche H., Coulet C., Daudon D., Plastic waste for low-weight embankments. ASCE, Conference on Grouting Soil Improvement an Geosynthetics, New Orleans, USA, 1992.

Site Investigation for Rehabilitation of a Municipal Landfill

**Francisco Casanova de O. e Castro, Mauricio Ehrlich,
Maria Claudia Barbosa, Marcio S.S. de Almeida**
COPPE-UFRJ, Graduate School of Engineering, Federal University of Rio de Janeiro, Brazil

Abstract

Site investigation of a large municipal landfill resting directly upon soft clay on shore of Guanabara Bay is reported here. The landfill site has been investigated with the aim to subsidize informations for a pollution control project in the area. A salty solution in a high and homogeneous concentration constitute the clay chemical environment. Free leachate and clay pore water salinity values are also of the same order. The diffusion process is very complex because, in general, the ions pore water concentrations are higher than in the leachate. The great difference between the permeability of the waste and the clay result that most of the leachate generated is directed to the landfill periphery either by internal or surficial flow. This contamination process should be considered the most important in the area.

Introduction

The Gramacho Landfill is located on the shore of Guanabara Bay, near to the city of Rio de Janeiro. It has been in operation since 1978, and presently receives about 5000 tons/day of all kinds of urban wastes. The Gramacho landfill covers an area of 1.2 Km^2. Presently its average height is 7.7 m, thus being possibly the largest landfill in South America. The average annual leachate production of the landfill is estimated about 1000 m^3 daily [1].

Gramacho site has been investigated with the aim to subsidize informations for a pollution control project in the area [2] described in a companion paper [1].

Site investigation and laboratory tests to define parameters and mechanism of contamination process in the area have been performed, including permeability and infiltration tests on the waste [2]. Pore water of the clay specimens under the landfill was squeezed out for chemical analyses. Clay specimens nearby and distant from the landfill but of the same deposit area, and free leachate have also been collected and submitted to tests. It is the purpose of the paper to present and discuss these results. This is part of a large programme of depollution of Guanabara Bay.

Site description

The Gramacho landfill is bordered North by Sarapui canal, West by Iguaçu river, East by a Sarapui derivation, and South by Guanabara Bay. A very complex hydrologic regimen that periodically submergs the landfill and reverses the water rivers flow in the area occurs due to the tidal oscillations.

The landfill rests upon a soft low permeability clay stratum 5 to 18 m thick on the estuary of the Iguaçu and Sarapui rivers, and has already been significantly compressed by the landfill since 1978. This deposit is 3 km distant from

Sarapui soft clay reference site that have been extensively studied by means of laboratory and in situ test and field trials [3-4].

The overall region is called Fluminense Plains and is densely occupied by small and big industries and poor population. The region lacks in sewage collection and treatment system. Most of the sewage is launched straight into the rivers. Thus, the soil samples collected outside the landfill do not represent the natural condition but may be used as reference soil samples for analysis of contamination impact of the landfill itself.

The stratum is composed by a thiomorphic organic-salty clay developed from fluvial-marine sediments corresponding to the last sea retreat about 3000 years ago. Kaolinite is the major clay mineral, illite and also montmorillonite occur in less extent. The clay overlays sandy and sandy-clayey material also of marine and fluvial origin with different textures and thickness [5].

Site investigations

Twenty one boreholes were distributed over the whole area in order to define the soil landfill contact surface and to characterize the foundation layers. Clay specimens 3 km inland (Sarapui site) and nearby (SP-A) from the landfill but of the same deposit area and also free leachate were collected and submitted to tests. Atterberg limits (W_L = 135% to 155% and W_P = 35% to 65%) are of the same order of those determined at Sarapui site. The natural water content (range from 145% to 250%) is greater than at Sarapui site.

Pore water of the clay specimens was squeezed out at 100 kPa pressure for chemical analyses. Testing of turbidity (UTF), organic matter (OM), chemical and biochemical oxygen demand (COD and BOD), specific electrical conductance (SEC), pH, redox potential (Eh), alkalinity, sulfates, chlorites and heavy metals have been performed. Tables 1 and Figs. 1 to 4 show the results of chemical analyses.

Permeability test performed on the waste under variable gradient show a permeability coefficient equal to 10^{-3} cm/s.

TABLE 1 - Chemical analyses results

test	clay pore water		Sarapui	free eachate
	Landfill			
	max	ave		
TF	19.0	5.0	4.2	>1000
M (%)	5.8	3.6	5.0	5-7
OD(mg/l)	4406	2157	3610	7000
OD(mg/l)	194	146	-	580
EC(ms/cm)	38	23	14	40

Analysis of the results

A salty solution of chlorides, sulfates and sodium in a high and homogeneous concentration constitute the chemical environment of the soft organic clay stratum. The pore water environment is slightly alkaline and, as expect, reducing (Eh= 150-200 mv). Test results do not show any clear variation horizontally and with depth. In addition, results are of the same order of magnitude, no matter if the sample origin is from the landfill

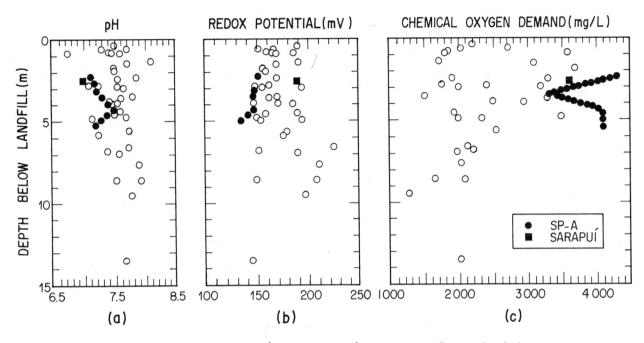

Figure 1 – pH, Redox Potential and Chemical Oxygen Demand of the porewater versus Depth Below the Landfill.

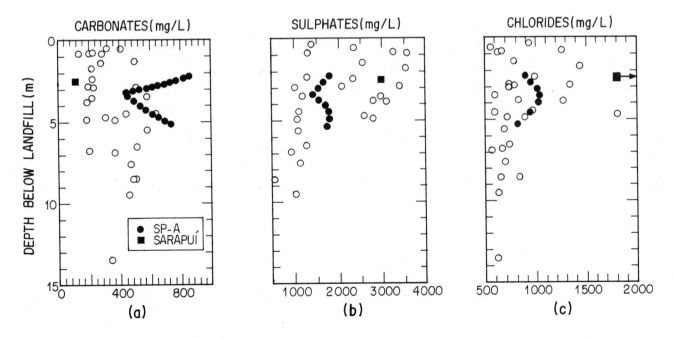

Figure 2 – Chemical Analyses of the porewater versus Depth Below Landfill.

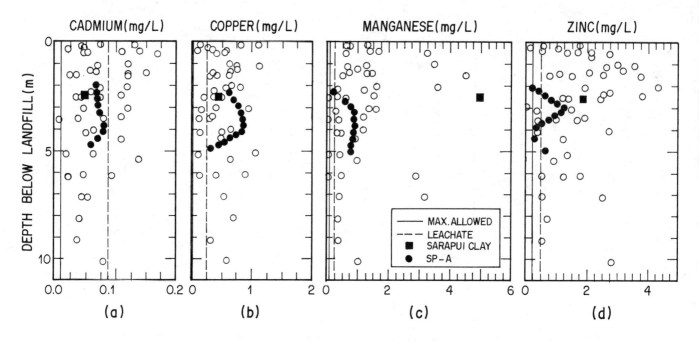

Figure 3 – Heavy Metals in porewater versus Depth Below Landfill.

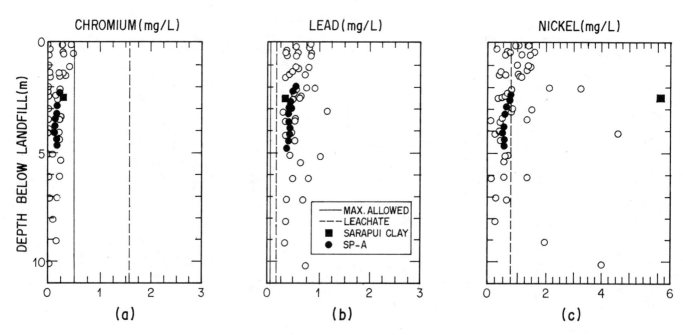

Figure 4 – Heavy Metals in the porewater versus Depth Below Landfill.

foundation or not (Sarapui site and SP-A, see Figs. 1 and 2).

Leachate and clay pore water salinity values are also of the same order. However, the amount of solids in suspension in the leachate is much higher than in the clay pore water, the turbidity tests of the clay pore water show values less than 1% of the leachate results.

The chemical oxygen demand (COD) values reflect the total amount of organic matter present in the leachate and in the clay pore water. COD is usually taken as measurement of the organic contamination. Fig. 1c shows that high COD values were measured in the clay pore water. Surprisingly, Sarapui site's samples, collected faraway from the landfill, also shows high COD values. Nevertheless, the soil is naturally rich in organic matter. Thus, the results on its own can be considered useless to assure the soil contamination.

The biochemical oxygen demand (BOD) value is dependant only on the amount of biodegradable organic matter and can be associated to the bacteriologic activity and, therefore, to the most significant contamination. Analyses of the clay pore water show low BOD values (lower than 200 mg/l).

It is not clear which transport mechanism most contribute to the advance of contaminants through the landfill foundation nor if the soft clay could be considered a natural barrier in any sense. Advection is not a significant contaminants migration process. Indeed, the hydraulic conductivity of landfill and soft clay are 10^{-3} cm/s and 10^{-8} cm/s, respectively. The great difference between the permeability of the waste and the clay induce that most of the leachate generated is directed to the landfill periphery either by internal or surficial flow.

The diffusion process is very complex in the area because the ions pore water concentrations are higher than in the leachate, except for the chromium. The diffusion may occur in both directions. The soil samples collected outside the landfill may not represent the natural condition but may be used as reference soil samples for analysis of contamination impact of landfill itself. Heavy metals concentration in the clay pore water give similar results no matter if the sample origin is from the landfill foundation or not (Sarapui site and SP-A, see Figs. 3 and 4). There is no clear variation between results, horizontally and with depth. In any case, the heavy metals concentration in the clay pore water and in the leachate are above the recommended Brazilian standards [6].

Conclusions

The consequence of the great difference between hydraulic conductivities of the waste and the clay is that most of the leachate generated is directed to the landfill periphery either by internal or surficial flow. This contamination process should be considered the most important in the area.

It is not clear which transport mechanism most contribute to the advance of contaminants through the landfill foundation nor if the soft clay can be considered a natural barrier in any sense. Advection is not expected to be an important contaminations migration process.

Chemical diffusion, ions adsorption by organic matter and clay particles and ions complexation by free organic matter in the pore water may be the most probable transport and interaction mechanism through the landfill foundation. Nevertheless, the diffusion process is very complex because the ions pore water concentrations are higher than in the leachate, except for the chromium. The diffusion may occur from the waste to the clay and vice-versa.

Acknowledgments

These studies were done in collaborative programme between COPPE and COMLURB (Rio de Janeiro Municipal Waste Disposal Company). The authors thank the support of COPPE and COMLURB's staff.

Bibliography

1. Ehrlich, M., Almeida, M.S.S. and Barbosa, M.C., Pollution Control of Gramacho Municipal Landfill, First Int. Congress on Environmental Geotechnics, Edmonton, Canada, July 1994.

2. Coppetec, Design of Gramacho Landfill Rehabilitation System. COPPE/UFRJ report to COMLURB, November 1992 (in Portuguese).

3. Ortigão, J.A.R., Lacerda, W.A. and Werneck, M.L.G., Embankment failure on clay near Rio de Janeiro, ASCE, Journal of Geotechnical Eng. Division, 109(11), p 1460-1479, 1983.

4. Almeida, M.S.S. and Fereira, C.A.M., Field, in situ and laboratory consolidation parameters of a very soft clay, Predictive Soil Mechanics, Wroth Symp., T. Telford, 1993.

5. Amador, E.S., Cenozoic Sedimentary Unity of Guanabara Bay, Brazilian Academy of Science Report, Vol. 52, No 4, p 743-761, 1981 (in Portuguese).

6. Brazilian National Environmental Commission, Resolution No 20, July 1980 (in Portuguese).

Shear Resistance Tests on Solid Municipal Wastes

O. Del Greco, C. Oggeri
Dipartimento Georisorse e Territorio, Politecnico di Torino, Italy

ABSTRACT: Waste disposal involves the resolving of various geotechnical problems during the design progress of a landfill. Among these problems is that of analyzing the stability conditions of complex structures also made up of waste materials. A reliable knowledge of the geotechnical parameters of wastes is required for a correct design. This consideration, together with the lack of research so far carried out on this subject, induced the authors to develop an experimental study with the purpose of better knowing the shear behaviour of urban wastes.
In the paper the results of some direct shear tests carried out on municipal waste bales and interposed materials are presented.

INTRODUCTION.

The lifetime of a landfill involves several phases: design, construction, filling and maintenance. During all these phases safety criteria must be assumed in order to avoid contamination problems in the surrounding areas. For this purpose a landfill can be considered as an envelope which is isolated by a lining structure in which the waste is placed. The leachate drainage systems and biogas outlet tubes constitute the only openings through the lining, with the function of controlling fluid products of the wastes.
In order to obtain isolation, the lining structure should be suitable for avoiding transport phenomena from the body of the landfill to the surrounding environment and be able to accept large displacements (in the order of tens of centimeters) without interrupting the continuity of the lining. The whole structure, and in particular the cover system, must prevent rainwater inlet and leachate outlet, especially during the maintenance phase.
The above mentioned aspects involve the solution of related geotechnical problems which deal with stability analyses of the landfill body and of the hosting geological formation, and the deformability of the whole stucture under loading actions.

GEOTECHNICAL PROPERTIES OF WASTE MATERIALS.

Some potential failure modes, involving geological formation, the waste mass or the isolation structures are here summarized: a) instability of the natural sidewall and base of the landfill site; b) failure through the waste pile; c) failure through the foundation soil, the waste fill and the liner; d) pullout of the liner system components from the anchor trenches and sliding along the slope towards the base area; e) sliding of waste pile along the composite liner system; f) sliding within caps and cover interfaces situated on the slope sides of the waste pile; g) excessive or differential settlement in the foundation soil, in the liner system or in the waste pile.

From this schematic description it appears that a correct geotechnical characterization of both waste and materials in contact is necessary. "Waste" is a generic term for various types of materials of different origin and composition: consequently, its mechanical characterization presents several difficulties.

The mechanical properties of waste are described on the basis of various geotechnical methods, as waste can be divided into two groups, one of which is analogous with soils. Construction material debris, excavation rock muck, ash and ore mill tailings can be listed in this group, meanwhile municipal waste and other similar industrial wastes being part of the second group.

The mechanical properties, such as strength and compressibility, of this second group of waste, show a wide range of variability, which presents a heterogeneity of the components, wide size distribution, a variable moisture content and a time dependent behaviour due to the decomposition of the organic part. It is thus necessary to develop appropriate testing procedures, as standard soil mechanics tests are inapplicable.

The evaluation of geotechnical waste properties has been, till now, carried out using indirect methods due to the difficulties of sampling. For example, in order to determine shear strength, laboratory tests have been performed on reconstituted samples or on compacted bales of refuse (1). Alternatevely in situ testing (by means of penetrometer, vane shear and plate loading tests) or back calculations from field tests have been developed for various design purposes. Observation of failure modes in waste landfill could allow one to obtain useful values of parameters by means of the relationships which are typical of soil mechanics, even though these methods would involve some inaccuracy of the results.

Some strength properties for municipal waste based on the results of these various methods are reported in the references (2),(3),(4). The bibliographic data has a certain range of values, so there is an uncertainty in assuming the proper values of shear strength parameters.

SHEAR TEST PROCEDURES.

A characterization of municipal solid wastes using a direct shear test has been studied because of the previously mentioned reasons. Municipal solid wastes are loose materials but are not granular, their deformability is high - both in the sample and in the landfill disposal - and the mechanical yielding in the waste pile does not produce an appreciable change in the waste structure. These facts lead us to perform tests which have some aspects in common with typical geotechnical field and laboratory testing methods, while recognizing the above mentioned differences between the wastes and soils.

A direct shear test on bales of municipal solid waste has been carried out in order to obtain the shear resistance parameters of the surface contact between the bales. This type of test does not reproduce the real behaviour of a waste pile in a landfill, as the two structures are different in size, shape and constitution, but it provides an initial approach to more careful procedures.

The equipment for this test (figures 1 and 2) is made up of a steel frame anchored to a concrete base to oppose the vertical force, while the horizontal reaction is provided by a concrete wall. Loads have been applied using manual hydraulic jacks equipped with manometers. The bales were 40*50*60 cm with a weight of about 50 kg, in order to be representative of the waste pile, and were made up using a hay baler. The range of density values changed from 400 to 600 kg/m3 after compaction during testing. The testing procedure was as follows: application of vertical loads onto the two superimposed bales; completion of compaction and settlement of the bales; application of horizontal load; completion of lateral deformation of bales; relative sliding (fig. 3).

During the test, large relative displacements occurred, so that some distancing wooden elements were used to enable the jacks to be reset.

The vertical pressures were varied from 10 kPa to 60 kPa, reaching a maximum value of 100 kPa. This choice of values corresponds to a

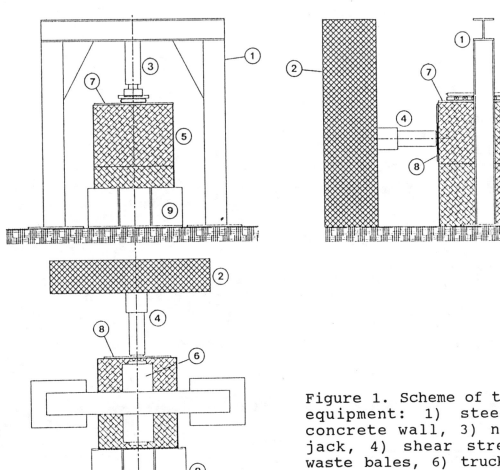

Figure 1. Scheme of the shear test equipment: 1) steel frame, 2) concrete wall, 3) normal stress jack, 4) shear stress jack, 5) waste bales, 6) truck, 7-8) steel plates, 9) lower bale stop.

Figure 2. Shear test on waste bales at direct contact.

waste pile of about 15 m in height, after compaction during the filling operations. The upper value of normal stress was also conditioned by the need to keep the bales intact. Higher pressures could be reached with a confining action on the bale, using improved equipment.

SHEAR TEST RESULTS.

The measured values of normal and shear stresses of the tests performed on the bales in direct contact are reported in figure 4. One can observe both that the dispersion of the experimental values is limited, and that the waste compaction influences the shear resistance parameters, as these are higher while the compaction level rises. Some aspects which are similar to the behaviour of classical geotechnical testing results can be observed: for example the anisotropic structure of the waste bales (due to the baling procedure) produces a dilatancy effect under low vertical stresses when overlapping of elements along the contact surface occurs during shearing; the apparent cohesion is revealed under higher vertical stresses.

At low normal stress levels the friction angle has a higher value, while at high normal stress levels the asperities at the contact surface are levelled and the sliding occurs in "residual" friction conditions, unless large scale waviness is present at the contact surfaces. The relative low values of friction angle could be explained by the amount of plastic materials in the tested bales.

The shear tests have also been performed with the interposition of different materials (such as high density polyethylene geomembrane, geotextiles, clay and permeable soil) between the waste bales. These arrangements furnished the values of the shear

Figure 3. Relative sliding of the bales at the end of a shear test with soil interposed between the waste bales.

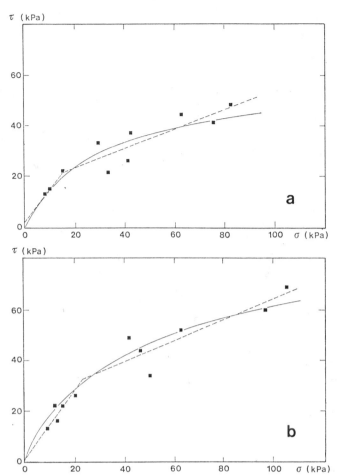

Figure 4. Diagrams σ-τ of shear tests on waste bales at direct contact: a) low density bales, b) high density bales.

parameters listed in Table 1. The relationships of the normal and shear measured stresses are also reported in figure 5.

In particular, one can notice the linear relationship in the HDPE geomembrane interposition case (only frictional behaviour). The dispersion is due to the waviness effects formed after the bale settlement under vertical loading (analog as to a smooth discontinuity in rock).

Evidence of a linear relationship has also been found in the coupling of clay with the HDPE geomembrane (see also (5)). In this case, sliding movement took place along the contact surface between the clay and the geomembrane.

Another kind of test analyses the presence of a sandy gravel layer adopted in the landfill to protect geomembrane from damages. The interposition of granular material determines a sliding movement analogous to the case of a rock joint with a filling material, as shear occurs mainly in the interposed soil layer. The soil characteristics influence the measured shear parameters. Finally, the coupling of geomembrane and geotextile showed a very low friction parameter

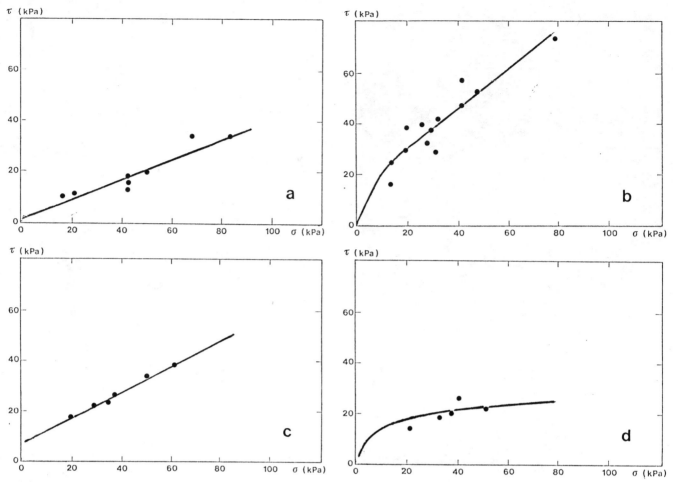

Figure 5. Diagrams σ-τ of shear tests with different slide contacts: a) waste-HDPE geomembrane, b) waste-sandy gravel soil, c) clay-geomembrane, d) HDPE geomembrane-geotextile.

value, with a marked non linear behaviour and the occurrence of dilatancy effect. The effect of waviness disappears at high vertical loads and the friction resistance becomes lower.

DISCUSSION OF RESULTS.

The mechanical properties of the materials involved in a waste landfill are fundamental in order to analyse the stability conditions and the deformations of these structures.
The tests that have been performed represents a direct procedure to determine the mechanical properties of the materials composing a waste landfill.
A preliminary set of values has been obtained for these direct shear tests. Starting from a conventional soil mechanics testing procedure, it would be possible to obtain more detailed results both in the sense of resistance parameters and deformability rates. This knowledge would be useful for a correct geotechnical analysis of landfill structures; other aspects, such as decomposition phenomena are also required for long term stability analysis.
The loose structure of the waste determines the high deformability

TABLE 1 -- Shear parameters of waste bales and liners.

Test types	Test layouts	Shear parameters c (kPa)	φ
Waste-waste coupling ($\gamma = 0.5$ kg/dm3)		16	21°
Waste-waste coupling ($\gamma = 0.7$ kg/dm3)		24	22°
HDPE geomembrane-waste coupling		0.0	17°
Sandy gravel soil - waste coupling		15	38°
HDPE geomembrane - clay coupling		8	26°
HDPE geomembrane - geotextile coupling		0.0	14°

feature of these materials. As a consequence of this waste feature, failure in the waste landfill often occurs on several sliding surfaces which are difficult to identify.

It is therefore not possible to distinguish "non disturbed" waste samples from remolded samples, until the waste pile has been subjected to aging phenomena which cause a particular structure.

The obtained shear resistance values have to be interpreted taking into consideration the fact that the direction of lateral loading was perpendicular to the "foliation" of the waste bales and that there was a low velocity in the shearing movement. However the measured values allow one to recognize some typical behaviour, which can be usefully employed as guidelines in stability analyses.

CONCLUSIONS AND DEVELOPMENTS.

A series of problems are now being studied in order to develop a better testing procedure:

a) the sample must be representative not only in the sense of relative dimension to the size of the constituing elements, but also in the sense of the real structure of a waste pile. In situ testing would overcome this problem but it is still possible to obtain a compromise in laboratory tests obtaining useful results.

b) A new testing equipment could permit a more flexible procedure, with a larger range of applied loads, and continuous shear displacements. A shear box could allow one to obtain these features.

c) The paths followed in the authors test were related to simple shearing along pre existing discontinuities in the waste. Further development of the research would also permit testing the shear in the body of the waste, with continuous samples.

d) Deformability and anisotropy in waste require large scale cells in order to predict reliable settlement rates, which so far have been based mainly on site investigations and monitoring.

It would be of considerable interest to develop further studies in this field, and in particular on industrial waste tailing, whose aspect is similar to soils even though the mechanical behaviour can be different, and on waste characteristics in old sanitary landfills.

REFERENCES

1) Fang H.Y., Slutter R.G. and Koerner R.M. (1977) "Load bearing capacity of compacted waste disposal materials". Proc. 9th ICSMFE, Tokyo.

2) Singh S. and Murphy B. (1990) "Evaluation of the stability of sanitary landfills". Geotechnics of Waste Fills, ASTM STP 1070, Landva and Knowles ed.

3) Mitchell J.K. and Mitchell R.A. (1991) "Stability of Landfills". Proc. XV Conf. Geotecnica, Torino.

4) Jessberger H.L. and Kockel R. (1991) "Mechanical properties of waste materials". Proc. XV Conf. Geotecnica, Torino.

5) Seed R.B. and Boulanger R.W. (1991) "Smooth HDPE-clay liner interface shear strengths: compaction effects". J. of Geotech. Eng., vol.117 n.4.

A New Piezometer for Sanitary Landfills

Eduardo C. do Val, Director & Partner, Luiz Antoniutti Neto, Engineer
Vector Projetos Integrados S/C Ltda., São Paulo, Brazil

Abstract: The main problems affecting the performance of Casagrande piezometers in sanitary landfills are presented. These problems are related to the presence of gas and scum in the piezometers, causing false detection of the true water level and correct detection of a false water level. An alternative design to circumvent those problems - the 'Vector' piezometer- is proposed and its qualities and limitations are discussed.

Introduction

In any man-made landfill with saturated wastes, the knowledge of the pore pressure distribution within the mass is very important if stability and other geotechnical analyses are to be carried out with a reasonable degree of accuracy. For this purpose, several types of piezometers have been developed so that the field pore pressure conditions can be directly assessed.

In sanitary landfills, the common practice has been to use standpipe (Casagrande) piezometers for their low cost and ruggedness. Sometimes, however, such reliability is not confirmed. At the Bandeirantes Municipal Sanitary Landfill in São Paulo, SP, Brazil, where an average of 5,000 to 6,000 t of waste is poured in every day within a hilly area of about 700,000 m², 34 Casagrande piezometers were installed and have been in use for many years, showing pore pressure variations that puzzled the engineers involved in the job. A recent investigation detected that the performance of piezometers is abnormal, and possibly not the pore pressures within the waste mass.

Problems Associated with the Casagrande Piezometer in a Municipal Sanitary Landfill

The basic problem found with the use of the Casagrande type of piezometer at the Bandeirantes Landfill was that its readings were not consistent. At a given location, the head in the piezometer would rise several decimeters for many consecutive days without any apparent reason such as rain or an increase in fill height. Typically, other piezometers in the same vicinity would not confirm such trend and sometimes would even detect a drop in the piezometric head. After a few days, that "problematic" piezometer would often show a head drop on the order of meters, turning an alarming situation into an optimistic one.

Although the authors had been involved in the project for only a few months when the problem of inconsistent readings was detected, similar behavior had been reported in the past by the staff which was previously in charge of the field instrumentation. The former staff also complained that the presence of gases in the piezometers often hampered the reading operations.

All the piezometers installed at Bandeirantes used a ⌀25 mm diameter PVC pipe in which an electric dipmeter was inserted to measure the water (leachate) level. Sometimes, when the dipmeter was being lowered inside the pipe, the piezometer suddenly became "alive", bursting out gas and spills of leachate.

Erratic readings were frequently detected: one day there would be a several meter rise in piezometric head, the next day the readings would return to its original value, without any incident.

After several reports on these observations the difficulties encountered, two types of errors were identified: false detection of the true water level and correct detection of a false water level.

Correct detection of a false water level, associated with the steady rise and burst phenomenon, was probably caused by gas bubbles that would build up inside the piezometer stand pipe, forming a column of gas under a few millimeters to many centimeters of leachate. Even a small gas pressure would be enough to raise the liquid level - and hence the readings - many decimeters. Figure 1 shows this situation. With this model it is easy to explain the sudden burst of an otherwise "dead" instrument: when the dipmeter is lowered, it punctures the liquid membrane on the top of the gas column, blowing out the gas bubble. When this happens, the turbulence forms a somewhat thick scum that easily sticks to the internal pipe walls and stays there for long periods.

The first type of error, associated with the erratic readings, was attributed to the presence of the scum deposited by the gas bubbles. When the dipmeter is lowered into the pipe, the scum closes the dipmeter circuit, falsely indicating that the liquid level has been reached. Fortunately, in most cases this type of error was easy to notice. The sticky nature of the scum would keep the circuit closed - and the instrument beeping - even after the probe was taken out of the pipe.

Remedial Actions: the 'Vector' Piezometer

In order to overcome the shortcomings associated with the presence of gas in the Casagrande piezometer, some modifications were introduced in the original design, rendering a new instrument which has been referred to as the "Vector" piezometer.

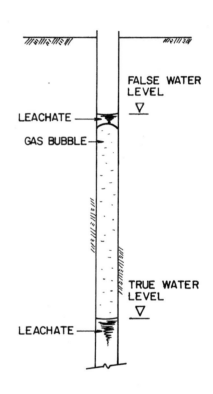

FIG. 1 - GAS BUBBLE BUILD UP EFFECT

Description

The basic difference between the conventional Casagrande piezometer and the Vector piezometer not one, but two concentric pipes are inserted into a borehole. As shown in Fig. 2, these pipes are placed several meters below the depth at which the pore pressure is to be

Landfill there was no agreement on how much this pressure would be, so, a 6 m extension was specified, supposedly far on the safe side. It was estimated that the gas pressure would vary between 5 and 20 kPa. Since the leachate unit weight is about 10 kN/m³, the height of leachate column that the gas pressure could push would be about 2 m.

The zone in which pore pressures are measured is referred to as the "chamber". Along the chamber length (500 mm), the outer pipe is slitted and protected against clogging by a filter of granular material. The outer pipe is plugged at the tip and, along the rest of its length, it is surrounded by impermeable bentonite paste. The inner pipe is perforated only at the tip and its top is open. The annular space between the two pipes is sealed at the top by appropriate couplings. A valve is placed laterally in the outer pipe just below the top sealing, so that a pressure gage can easily be attached.

Working Principle

The waste fluids (leachate and gas) at the measuring depth are allowed inside the outer pipe through the filter and slitted pipe chamber. The heavier leachate goes down all the way to the tip of the outer pipe and then up inside the inner pipe until its height above the chamber level balances the pore pressure at that level, as long as the valve at the top of the outer pipe is open. The lighter gas will tend to go up through the outer pipe and escape to the air through the open valve.

The dipmeter is inserted into the inner pipe and, after stabilization of readings, the valve is closed and the gas pressure is allowed to build up. A pressure gage is fastened to the valve which is then opened, so that readings of gas pressure can be directly made on the gage. At the same time, the dipmeter is again inserted into the inner pipe, giving a new reading of the pressure head.

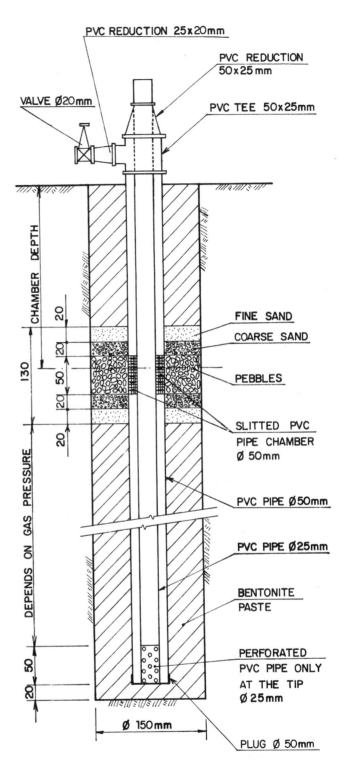

FIG. 2 - "VECTOR" PIEZOMETER

The difference between this reading and the initial (stabilized) reading, multiplied by the unit weight of the leachate, should equal the gas pressure directly measured by the gage.

Installation Details

As in any standpipe piezometer, a good chamber seal is essential to the good performance of the instrument. In the Casagrande piezometer, where the chamber is placed at the tip of the pipe, only the portion above the chamber needs sealing. In the installation of the Vector piezometer, attention should be paid to ensure proper sealing both above and below the chamber. At the Bandeirantes Landfill it was specified that the bentonite be placed first in the borehole, at the liquid limit. Then the piezometer pipe would be pushed into the bentonite paste, ensuring good envelopment. Care should be taken to place a correct amount of bentonite paste into the borehole, so that the slitted part of the outer pipe would neither sink below nor stay far above the surface of the bentonite paste. Small amounts of bentonite can be added into the borehole to reach the desired level after the pipe is pushed in without much danger of clogging to the slitted openings. However, it is virtually impossible to remove excess bentonite and ensure clean slitted openings.

The granular filter is typically made with quartz sand and pebbles. At the Bandeirantes Landfill, the leachate chemical action on crushed granite and gneiss used for drainage practically decomposed those hard rocks within a time span of no longer than a year. Therefore, if quartz sand and pebbles are not available, synthetic (glass marbles, for instance) or other appropriate material should be used.

Multiple Chamber Piezometers

The major cost in installing conventional standpipe piezometers lies in the borehole drilling. In the Vector piezometer, although more than twice the length of pipe is used, boring still represents the major installation cost, for it must be remembered that this kind of piezometer is applied in waste landfills where boreholes are particularly difficult (and expensive) to drill. Therefore, the use of multiple chamber piezometers so that pore pressure can be measured at different elevations on the same borehole becomes economically attractive. At the Bandeirantes Landfill due to the placing of the waste in the landfill in horizontal layers intercalated by clayey covers, the water table becomes "perched". Therefore, a number of piezometers are needed at different elevations in the same profile, thus making multiple chamber piezometers highly desirable and cost effective.

At the Bandeirantes Landfill, all the Casagrande piezometers had four chambers. In many of them, however, sealing problems were detected between chambers. Hence, a limit of two chambers per piezometer was specified for the Vector piezometers.

Comments and Conclusions

A modified version of the traditional Casagrande piezometer has been developed to overcome the problems associated with the presence of gas when measuring pore pressures in sanitary landfills.

Similar to the Casagrande piezometer, this new instrument - the Vector piezometer - also needs a relatively large quantity of flow in order to yield the "true" readings. Therefore, the Vector piezometer also has a relatively long time lag response when compared to constant volume piezometers such as pneumatic ones.

Besides the low cost, the ruggedness and the reliability that it preserved from the traditional standpipe, the Vector piezometer has the ability of separating the gas pressure from the liquid pressure when measuring the pore pressure (disregarding the gas dissolved in the liquid). One of the greatest virtues of the Vector piezometer, however, is that, by comparing pressure gage and dipmeter readings, proper functioning of the instrument can immediately be assessed.

References

Casagrande, A. (1958). *Piezometers for Pore Pressure Measurements in Clay.* Mimeographed, Harvard University.

Pollution Control of Gramacho Municipal Landfill

M. Ehrlich, M.S.S. Almeida, M.C. Barbosa
COPPE-UFRJ, Graduate School, Federal University of Rio de Janeiro, Brazil

Abstract

Stability and pollution control studies of a large municipal landfill resting directly on a soft clay at the shore of Guanabara Bay are reported here. The present embankment height is about 8 m. Stage construction stability analysis have shown that the embankment could be taken to a full height of 32 m (further six years of use) at an acceptable factor of safety. At present all generated leachate goes straight to Guanabara Bay. It is proposed that a peripheral system of leachate collection should be installed to direct this leachate to treatment units.

Introduction

The Gramacho Municipal Landfill, about 14 years old at present, receives daily 5000 tons of waste (20000 m^3 uncompacted), 78% of which comes from Rio de Janeiro city and 22% from four local counties. However it has not being functioning strictly as a sanitary landfill, as waste is not contained within cells covered with low permeability compacted soil. Consequently, about 1000 m^3 of leachate are generated daily. As the landfill is located on the shore of Guanabara Bay and rests directly upon a 5 to 18 m thick soft organic clay stratum, virtually all generated leachate goes straight into Guanabara Bay. The total area occupied by Gramacho Landfill is 1.2 km^2 and the average height is about 7.7 m, thus being apparently the largest of this type in the world.

Due to lack of available large areas, the Gramacho landfill shall still be used for up to late nineties. Consequently, COPPE was hired in 1991-1992 [1] to propose a system of leachate control and treatment and to define maximum landfill heights. The present paper describes these studies which have been based on a thorough site investigation [2] to define geoenvironmental parameters and mechanisms of contamination process in the area. All these studies are part of a large government programme of pollution control of Guanabara Bay.

Site description

The Gramacho landfill, Fig. 1, is located at Duque de Caxias County, about 20 km from downtown Rio de Janeiro. The landfill is bordered South by Guanabara Bay, North by Sarapuí channel and East by Iguaçu river. The hydrological regime is fairly complex in the area due to the presence of rivers and tidal oscillation, causing reversion of flow in small rivers and drainage channels. The overall region is called Fluminense Plains and is well known by its large poor population and infra-structure problems, the most important of which being lack of local sanitation. Consequently, sewage is launched straight into the rivers which causes greater pollution to Guanabara Bay, than that caused by Gramacho landfill itself.

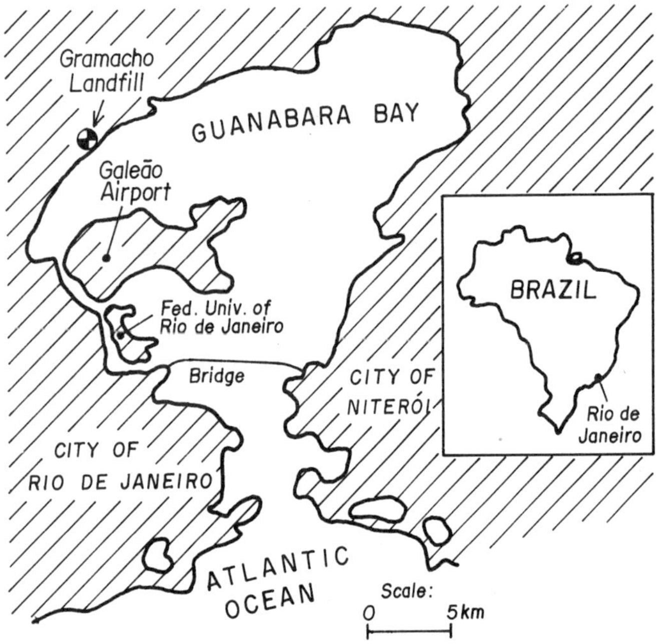

The soft clay deposit under Gramacho landfill is 3 km distant from the Sarapuí soft clay reference site, which has been comprehensively studied by means of laboratory and in situ tests and field trials [3]. Twenty one boreholes were distributed over the whole of Gramacho landfill in order to define the soil-landfill contact and to characterize the foundation layers. A longitudinal cross section is shown in Fig. 2. Atterberg limits ($w_L = 145\%$; $w_p = 57\%$) of organic clay at Gramacho landfill are of the same order of those determined at Sarapuí site. However the water content under Gramacho landfill is greater than at Sarapuí site. Consequently the Gramacho clay is even softer than Sarapuí clay, a finding confirmed by piezocone tests performed in both areas. As the local bay shore area was previously flat, it can be concluded that landfill settlements (from 1978 till 1991) caused by consolidation and local failure range from 0.8 to 4.5 m.

Fig. 1 - Location of Gramacho landfill

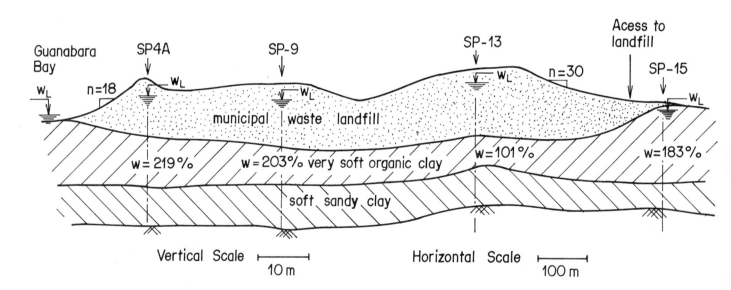

Fig. 2 - Typical Gramacho landfill cross section

Municipal waste strength parameters

About 80% of Gramacho Municipal Landfill are domiciliar and public waste. Organic matter (34%), paper (27%) fine agregate (15%) and plastic (13%) are the main components of Gramacho landfill. As a result municipal waste shows high compressibility. As far as strength is concerned the municipal waste behaves similarly to a reinforced soil as the fibrous elements and plastic interlocked act like reinforcements. Stress-strain curves of milled waste [4] have shown that failure cannot be observed even at high sample compression and that waste hardens with deformation. For this reason failure with shear planes through the waste material have only been reported when weak zones (e.g., the soft clay in the present case) are present. In addition the mechanical properties of the waste are time dependent due to biochemical degradation processes.

Notwithstanding the above difficulties the engineer has to decide which values of c_r and ϕ_r to use in stability analyses. The strength values adopted in the present stability analyses have been based on back-analysis of a 9 m high vertical cut observed in the top region of Gramacho landfill which produced a range of possible c_r and ϕ_r pairs. The pair $c_r = 20.4$ kPa and $\phi_r = 20°$ is just within the recommended region [5] of laboratory data and field (back-analysed) data. Other back-analysed data fall outside that region.

Stability analysis of landfill

The remaining amount of waste allowed to be placed at a landfill is controlled by the maximum embankment height and side slopes. In other words long term safety has to be proved on the basis of appropriate stability analysis. A wedge type stability, analysis, Fig. 3, has been performed for Gramacho landfill, for which the factor of safety can be computed by equation

$$F_s = (P_p + S_{u2}L)/P_a \qquad (1)$$

where $L = n\,h$; n and h are, embankment slope and height respectively, P_a and P_p are active and passive earth pressure forces, S_{u2} is the representative undrained strength under landfill slope, assumed to be the average between S_{u1}, the unconsolidated clay undrained strength, and S_{u3}, the consolidated undrained strength. S_{u1} and S_{u3} have been estimated by means of piezocone tests. The active force P_a inside the landfill is given by the well known Rankine equation computed with waste strength parameters c_r and ϕ_r defined above, and also including the fluid pressure, (computed with the height of leachate h_w inside embankment - assumed equal to 0.5h). Thus P_a equation is quite cumbersome and is not presented below. P_p is given by equation

$$P_p = 0.5\,\gamma_c\,d^2 + 2\,S_{u1}.d \qquad (2)$$

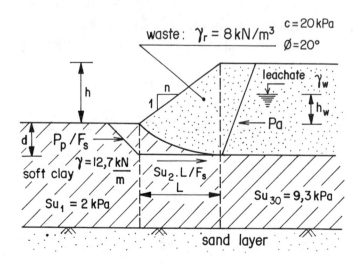

Fig. 3 - Wedge type stability analysis of Gramacho landfill

where $\gamma_c = 12.7$ kN/m³ is the soft clay bulk weight and d is the depth of landfill penetration into soft clay, as show in Fig. 3.

Stability analysis assumed a gain in S_{u3}, hence a stage construction type of analysis. The value of S_{u2} was again computed by the average between S_{u1} and S_{u3}, altough the former, measured outside the embankment toe was assumed to remain constant. The gain in S_{u3} was computed using Terzaghi theory with a 9m thick double drained clay layer and $c_v = 1.6$ m²/year. Consolidation times Δt (see Table 1) for each landfill height were computed assuming 5000 tons of waste placed daily at $\gamma_r = 8$kN/m³. It is also assumed 80% of waste per m³ of embankment. Computations were as follows: $t = \Sigma \Delta t$; $T_v = f(t)$; $U = f(T_v)$; $\Delta \sigma'_v = U.\Delta \sigma_v$; $\Delta S_{u3} = 0.25.\Delta \sigma'_v$; $S_{u3} = S_{u30} + \Delta S_{u3}$; $S_{u2} = 0.5(S_{u1} + S_{u3})$.

The maximum allowed landfill height (restricted by the presence of Rio's International Airport nearby) was h=30m, which is reached in six years as shown in Table 1. Variation of the factor of safety with embankment height is shown in Fig. 4. It is seen that, at h=32 m, an average slope n=10 can be adopted. This corresponds to a factor of safety equal to 1.21. However if leachate level within the embankment is lowered ($h_w = 0$) the total active thrust P_a decreases, hence F_s increases (see

TABLE 1 S_u gain in stage construction

Landfill Ellev m	Waste Volume x10⁶ m³	Δt years	U %	ΔS_{u3} kPa
7.74	-	-	-	-
11.8	3.0	1.56	30	2.4
16.9	3.1	1.59	32	3.2
21.9	2.4	1.26	28	2.8
27.0	1.8	0.96	19	1.9
32.0	1.3	0.96	17	1.7

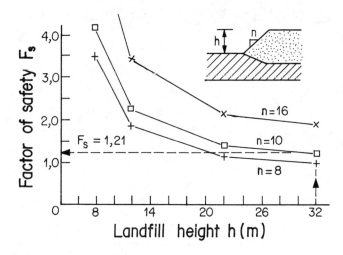

Fig. 4 - Variation of factor of safety with landfill heigth and slope

Eq.1). Assuming that n=16 is typical of the present situation, for further heightening n=8 can be adopted. This results in n=10 average slope.

Leachate collection system

In order to design systems for leachate collection, leachate volume has to be estimated. The estimate of the leachate volume used the water balance method[6]. The relevant input variables are: the preciptation P, the infiltration I, the surface runoff R_o and the input water from surrounding surface runoff SR (for final design conditions this can be assumed zero). Values of I have to discount the actual evapotranpiration AET, the change in moisture storage in soil cover ΔS_o and in waste ΔS_r, and add W_d the water from decomposition of solid waste. Based on infiltration tests performed locally (the soil cover is rather small and fairly permeable) and on pluviometric data, it is estimated that about 40 mm can infiltrate daily. The rain in excess to that will flow surficially. Computations of leachate volume using P values measured monthly by the local weather bureau and assuming a landfill area equal to 1.2 km² resulted in 820 and 1234 m³/day, respectively for 1987 and 1988.

As the soft clay permeability is about $k = 10^{-8}$ cm/s and the waste permeability is about $k = 10^{-3}$ cm/s, virtually all leachate seeps to the side slopes, thus reaching Guanabara Bay. Therefore peripheral drainage lines, should be built internally to the side slopes to intercept and collect this leachate and direct it to a treatment plant, as shown in Fig. 5. These leachate drainage pipes, which have a gradient of 1:10000 to promote drainage to a collection point, should be strong enough to accomodate further settlements and to be capable of inspection and maintanance. Regarding the last feature, a maintenance road runs along the drainage line, Fig. 6, giving access to maintenance shafts.

It is also proposed that a 0.6 m thick compacted soil liner be built around the landfill toe in order to avoid the contribution of tidal oscillation for leachate generation (surface run-on). This compacted liner is protected by a 0.4 m thick layer of sand and rip-rap. A surficial peripheral drainage line has also been planned to collect the run-off water. The external liner and the run-off drainage system are shown in Fig. 6.

Because of landfill settlement, a bottom leachate reservoir will continue to exist if leachate is not pumped upwards. Modern sanitary landfill operation recommends that all leachate should be drained (a maximum hydraulic head of 0.3 m is recommended) in order to decrease migration of pollutants (by advection, diffusion or sorption) as well as degradation dependent settlements. However at the time COPPE was hired (1991) the local authority (COMLURB) accepted that this bottom reservoir should be tollerated. This is becausse funding was not be available to built and operate a leachate pumping system. It is however planned that when the landfill reaches its final height a final cover

Fig. 5 - Leachate collection system

system will be installed together with the leachate pumping system.

Leachate treatment system

Biodegradability tests in laboratory [1] were not effective in reducing both Chemical Oxygen Demand (COD) and Biologic Oxygen Demand (BOD). The explanation for these uncommon features may be that the leachate generated at Gramacho Municipal Landfill presents both higher COD/BOD ratio and amount of total dissolved solids as compared to literature data. On the other hand a two-stage sequential treatment, physico-chemical, followed by aerobic, was shown to be effective. The location of these two treatment units is shown in Fig. 5.

It is recognized that laboratory conditions differ from field ones. This is because leachate composition is not only quite variable along the year but is also affected by storage periods.

Considering these important aspects, a pilot plant should be implemented initially and tested for about three months. This will allow a better basis for subsequent planning of the treatment system.

Conclusions

Stability and pollution control studies of Gramacho Municipal landfill have been reported here. Gramacho landfill is about 8m high and rests directly on a very soft impermeable clay at the shore of Guanabara Bay. Stage construction stability analyses considering a gain in undrained strength with time have shown that the embankment could still be used for further six years reaching 32 m at an acceptable factor of safety.

As far as pollution is concerned, at present all generated leachate seeps to the side slopes going straight into Guanabara Bay. It is then proposed that a peripheral system of leachate collection should be installed to direct this leachate to treatment units.

Acknowledgements

The studies reported here have been undertaken through a collaborative program between COPPE and COMLURB (Rio de Janeiro Waste Disposal Company). Stability analyses had the support of G. Leal and J.E.M, Medina from IESA Engineering.

References

1. COPPETEC, Design of Gramacho Landfill Rehabilitation System, Report to COMLURB, November 1992 (in portuguese)

2. Castro, F.J.C., Ehrlich, M., Barbosa, M.C. and Almeida, M.S.S., Site investigation at Gramacho Landfill, First Int. Congress on Environmental Geotechnics, Edmonton, Canada, July 1994

3. Ortigão, J.A.R., Lacerda, W.A. and Werneck, M.L.G., Embankment failure on clay near Rio de Janeiro, ASCE, Journal of Geotechnical Eng. Division, 109(11), p 1460-1479, 1983

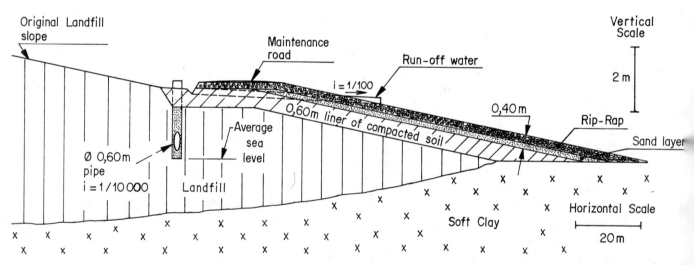

Fig. 6 - Cross section of leachate collection system

4. Jessberger, H.L. and Kockel, R. Determination and assessment of mechanical properties of waste materials, Int. Symp. of Geotechnics Related to the Environment, Bolton, England, 1993.
5. Sanchez-Alciturri, J.M., Palma,J., Sagaseta, C. and Canizal, J., Mechanical properties of waste in a sanitary landfill. Int. Symp. of Geotechnics Related to the Environment, Bolton, England, 1993.
6. EPA, Environmental Protection Act SW-168, 1975

Aerobic Biological Clogging of Soil/Geotextile/Geonet Filter in Leachate Collection Systems

Richard B. Erickson, Echol E. Cook, Braja M. Das
Southern Illinois University at Carbondale, Carbondale, Illinois, USA

Abstract The results of a number of long-term laboratory falling head hydraulic conductivity tests are presented. These tests were conducted to evaluate the long-term aerobic biological clogging of soil/geotextile/geonet filters used in primary leachate collection systems of municipal landfills. The hydraulic conductivity of the filter appears to be primarily controlled by the smallest grain size of the soil.

Introduction

In municipal landfills, bottom liners and leachate collection and removal systems prevent migration of contaminated liquid from the waste cell into the adjacent soil and ground water. The primary leachate collection system generally consists of a granular drainage layer underlain by a layer of geotextile and then a layer of geonet.

The leachate seepage from a landfill site may continue for a period of about 30 years until its generation is stopped. Hence, it is important that the drainage capability of the granular soil/geotextile/geonet filter system remain in order for a long period. Clogging of the filter system will reduce its hydraulic conductivity and, hence, diminish the effectiveness of the leachate removal system. The clogging which will result in the reduction of the hydraulic conductivity of the filter system may be due to the following three major factors: (a) silt or clay-size particles carried by the leachate being deposited in the void spaces of the granular soil layer and geotextile; (b) densification of the granular soil layers, and movement of the finer granular particles into the openings of the geotextile (tuning); and (c) bacterial growth in the void spaces of the granular filter and the openings between the fibers of the geotextile.

Experimental studies leading to the evaluation of the first two factors have received considerable attention in the past. The purpose of this paper is to present some recent long-term laboratory falling head hydraulic conductivity test results to evaluate the effect of biological clogging on the hydraulic conductivity of granular soil/geotextile/geonet filter systems. For the present tests, a soluble synthetic leachate containing compounds and minerals typically found in municipal landfills was used. Research to date indicates a high degree of potential clogging by biomass given the proper conditions for cell growth within the soil/geotextile filter systems. The results of these studies were widely different, varying from low effects on geotextiles to considerable potential for clogging due to biological activity [2,3,4].

Laboratory Tests

In order to conduct the long-term biological clogging tests a totally soluble synthetic leachate representative of typical organic and inorganic components found in municipal leachate discharge was used. The details of the synthetic leachate are given in Table 1. A sand collected from the field was used for granular filter. The grain-size distribution of the sand collected from the field is shown in Fig. 1. The sand was sieved in the laboratory to separate and collect plus 40, 20-40, and 10-20 U.S. sieve fractions. Geotextile samples taken from the same lot were used for all hydraulic conductivity tests. The geotextile had a thickness (ASTM D-1777) of 0.18 cm (70 mils) and its weight was 33.9 g/m^2 (4.2 oz/yd^2).

TABLE 1--Synthetic leachate constituents

Item	Description
Temperature (°C)	27
pH	3.9
COD (mg/l)	24,900
Organic acids (mg/l)	
Acetic	2,000
Propionic	1,600
Butyric	2,300
Isobutyric	400
Valeric	500
Isovaleric	200
Inorganic components (mg/l)	
Alkalinity	0
Chloride	74.2
Hardness	44.5
Nitrogen (total)	2.0
Potassium	76.0

Falling head hydraulic conductivity tests were conducted in Plexiglas tubes having internal diameters of 102 mm consistent with ASTM test

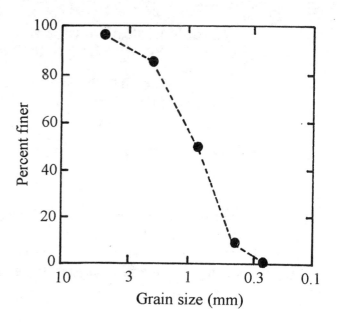

Fig. 1. Grain-size distribution of the sand collected from the field

designations D-5101 and D-4491. The tests were conducted over a five-month period. A schematic diagram of a test column is shown in Fig. 2. The filter system consisted of 305-mm thick sand layer(s) underlain by a layer of geotextile and then a layer of geonet. The 305-mm thickness of sand is the minimum recommended by EPA [1]. The initial placement of relative density of compaction of the sand layers in the Plexiglas columns varied between 65% to 70% with an average value of 67%.

Two series of hydraulic conductivity tests were conducted. In Series A, the sand/geotextile/geonet columns were seeded with a concentration of cells acclimated to the soluble synthetic municipal landfill leachate. The biomass seeded into each column was activated sludge bacteria acclimated to the synthetic leachate over a two-week period utilizing a fill and draw process. Over the acclimation period, the cells were exposed to increased concentrations of the synthetic leachate until full strength leachate was utilized. The columns were fed weekly with the soluble

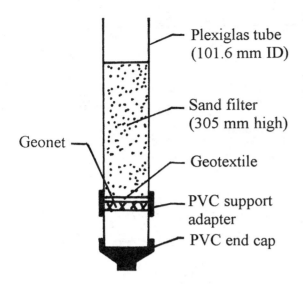

Fig. 2. Column configuration of sand/geotextile/geonet system used for the hydraulic conductivity tests

synthetic leachate and were maintained under continuous moist conditions.

Hydraulic conductivity tests were conducted just prior to the weekly feeding. In Series B, blank continuous column tests were run in conjunction with the biological continuous column tests (Series A) to distinguish between the reduction of hydraulic conductivity due to filter tuning of the composite columns and that due to biological clogging. Only tap water was passed into the blank test columns which had not been seeded with acclimated cells. The volume of water fed weekly into the blank columns was equivalent to the volume of leachate passed into the biological columns over the five-month test duration. From the blank column tests, clogging due to tuning of the composite filters was then differentiated from clogging due to biological activity. Table 2 gives the details of the hydraulic conductivity tests. The test columns were left open to the air throughout the testing. Ambient air temperatures varied from 25°C to 28°C over the five-month test duration and moist conditions were maintained continually.

TABLE 2--Details of hydraulic conductivity tests

Test No.	Type of sand column
A-1	Unsieved
B-1	Sand (Fig. 1)
A-2	Plus 40 U.S. sieve
B-2	Sand
A-3	20-40 U.S. sieve
B-3	Sand
A-4	10-20 U.S. sieve
B-4	Sand
A-5	102 mm 10-20 sand (top layer)
B-5	102 mm 20-40 sand (middle layer)
	102 mm Plus 40 sand (bottom layer)

Note: Test Series A--columns seeded with acclimated cell; Test Series B--blank column test

Laboratory Test Results

Figure 3 shows the plots of $k_t/k_{t=0}$ (where k_t = hydraulic conductivity at time t days, and $k_{t=0}$ = hydraulic conductivity at time $t = 0$ days) for all tests given in Table 2. From the plots shown in Fig. 3, it can be seen that the hydraulic conductivity for all Series B tests, which are due to filter tuning only, stabilized after about 30 days. However for seeded column tests (Series A), the hydraulic conductivity continued to decrease even after 155 days. Thus the percentage of decrease at any time t due to biological clogging can be expressed as $(k_t/k_{t=0})_{Series\ A} - (k_t/k_{t=0})_{Series\ B}$. The percentages of clogging due to tuning and biological clogging at $t = 155$ days are given in Table 3. Based on the results shown in this table and Fig. 3, it appears that the 10-20 sieve sand filter had the least biological clogging.

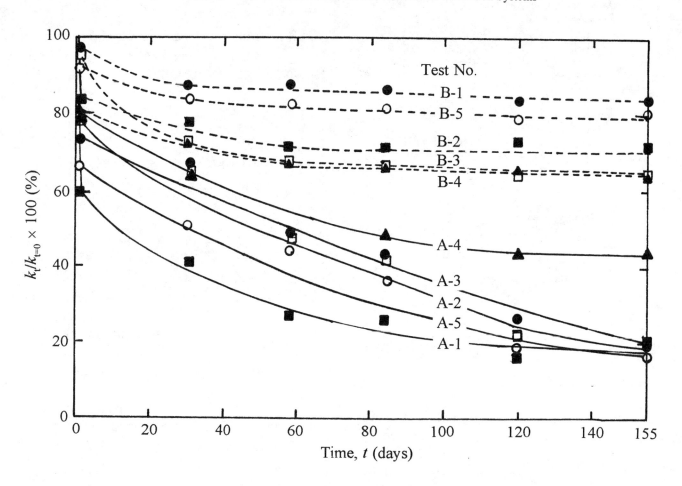

Fig. 3. Variation of $k_t/k_{t=0}$ for all hydraulic conductivity tests

TABLE 3--Clogging due to tuning and biological clogging

Filter	Clogging-tuning (%)	Clogging-biological (%)
Unsieved sand	27.9	51.2
Plus 40 sand	33.9	46.1
20-40 sand	15.7	64.2
10-20 sand	35.2	18.6
Layered sand	19.4	62.1

Observations of all the test columns after five months showed that the majority of all visible biological growth and clogging occurred within or adjacent to the aerobic regions open to the atmosphere, specifically, within the upper 90 mm of the granular filters as well as encompassing the underlying geotextile. In the upper 20 mm of the 90 mm portion of all sand columns, an extensive soil/biofilm formation was observed. The most cohesive filter cake included sand columns with grain size less than No. 20 sieve. The 10-20 size sand filters obtained the lowest level of soil/biomass agglomeration.

Upon completion of testing, representative geotextile specimens from each of the five continuous biological filter columns were analyzed utilizing scanning electron micrographs (SEM). Two specimens from each sample were analyzed as follows:

1. The first group of specimens was prepared utilizing a series of chemical fixatives and buffer rinses. They were then dehydrated in alcohol, critical point dried, and finally coated with a film of gold plating prior to scanning. These preparations were considered standard to preserve the biological cells and formations as well as to decrease static electricity buildup on the samples.

2. The second group of specimens was oven dried at 60°C prior to analysis. Although the limited preparation of the second group of specimens could possibly result in burning of the specimen surface during electron scanning, it was predicted that the specimens prepared in the first group may possibly wash away a portion of the actual biofilm buildup on and within the polyester fibers of the geotextile.

Figure 4 shows scanning electron micrographs with a magnification factor of 1700 for the geotextile specimen used in Test No. A-4 (10-20 U.S. sieve sand column). Based on the scanning electron micrographs, it appeared that the geotextiles used in Tests No. A-2, A-3, and A-5 had the greatest biofilm growth and bridging between the fibers. The geotextile underlying the 10-20 sieve soil (Test A-4) revealed the least bridging between the geotextile strands. No sand grains were found within the geotextiles photographed.

Based on the present tests and general observations, it appears that bacteria and associated slime growth are the predominant (if not sole) clogging mechanisms.

Conclusions

The results of a number of falling head hydraulic conductivity tests on a granular soil/geotextile/geonet filter system have been presented to evaluate the effect of aerobic biological clogging in leachate collection systems. Based on these test results the following conclusions can be drawn.

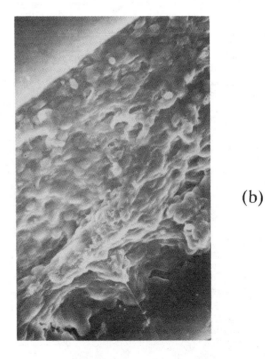

Fig. 4. Scanning electron micrographs for the geotextile specimens used in Test A-4: (a) fixed film preparation; (b) oven dried preparation

1. The 10-20 sieve sand/geotextile/geonet composite filter showed the least biological clogging. The hydraulic conductivity reached practically a constant value after about two months.

2. The composite permeability in the filter system appears to be regulated by the smallest grain size present. Therefore, selective placement and gradation of soil covers above the geotextile may derive no benefit in regards to minimizing biological growth and ultimate clogging.

3. The biological growth and associated clogging of the geotextile appears to be greatly dependent on (a) the spacing and ultimate bridging distance of the biofilm growth between the geotextile fibers, and (b) the flow velocity of leachate through the geotextiles and resulting biomass shearing, controlled exclusively by the overlying soil cover. The velocity at which cells are sheared from growth bridging the individual fibers appeared to be in the range of 0.38 to 0.55 cm/sec.

References

1. Environmental Protection Agency (1989). Requirements for Hazardous Waste Landfill Design, Construction, and Closure, Publication No. EPA-625/4-89-022, Cincinnati, OH, U.S.A.

2. Koerner, R.M. (1990). Geosynthetics, 2nd Edition, Prentice Hall.

3. Koerner, G.R. and Koerner, R.M. (1989). "Biological clogging in leachate collection systems," in Durability and Aging of Geosynthetics, Ed. R.M. Koerner, Elsevier Applied Science, pp. 260-277.

4. Koerner, G.R. and Koerner, R.M. (1990). "Biological activity and potential remediation involving geotextile landfill leachate filters," Geosynthetic Testing for Waste Containment Application, STP No. 1081, ASTM., pp. 313-334.

Numerical Analysis of Effectiveness of Imperfect Underground Barriers

F. Federico
University of Rome "Tor Vergata", Italy

Abstract - Steady-state seepage flow through imperfect underground barriers is investigated by a numerical method. Results indicate that barrier effectiveness very much depends on its depth and integrity. The effectiveness may well be less than the corresponding design values if the barrier is not completely keyed into an underlying less permeable base formation or the barrier is affected by constructive defects, especially if these occur within permeable horizons of the foundation soils.

1. Introduction

Underground barriers are frequently used in Geotechnical Engineering for different purposes, such as the reduction of neutral pressures in foundation soils of dams (Casagrande, 1961) and river levees, the abatement of leakages from reservoirs, the storage of fluids by underground dams (Kamon, Aoki, 1987), the protection of aquifers from pollution (A.S.T.M., 1984; Rowe, 1988). For all these cases, the control of seepage flow plays a key role.

General design rules for evaluating the effectiveness of the mentioned geotechnical structures are not yet available. However, for simple geometric schemes and boundary conditions, some analytical solutions have been proposed in a closed form.

Although useful to explain the role of geometric and physico-mechanical parameters on seepage flow, oversimplified hypotheses strongly limit the applicability of closed form solutions to practical cases, that are generally characterized by heterogeneity of foundation soils, anisotropy of the coefficients of permeability and by complex geometric schemes and boundary conditions. To overcome these difficulties and to throw light on some peculiarities frequently arising in geotechnical design of containments, the effectiveness of a barrier, partially penetrating in heterogeneous soils, has been investigated in the present paper by a numerical method.

In the analysis, steady state conditions, i.e., long term conditions after the installation of the barrier, have been considered, starting with a determination of the effectiveness.

The effectiveness of the barrier has firstly been determined for different depths of the barrier.

The effect of a localized increase of the permeability of the barrier (geotechnical "defect") on the seepage flow was finally examined and is reported.

2. Excerpt of literature review

The response of ground-barrier systems has so far been analysed under some symplifying assumptions regarding to foundation soils, seepage flow, depth and thickness of the barrier, distribution and anisotropy of coefficients of permeability both of the barrier and of the surrounding soil. The relation between seepage and deformation processes has generally been neglected.

Telling et al. (1978) considered partial or fully penetrating imperfect cut-off walls of finite thickness under *confined* steady seepage conditions. The flow problem of a perfect (i.e. impervious) partially penetrating barrier of neglegible thickness has been solved by Polubarinova - Kochina (1962).

Few studies have been devoted to the case of *unconfined* seepage flow. The steady state problem has been considered by Polubarinova-Kochina (1962) in the case of a watertight barrier partially penetrating in a homogeneous half-space, when the thickness of the barrier is negligible. Permeable barriers fully penetrating in a homogeneous layer have been investigated by Outmans (Bear, 1972) and Sato (1982).

The transient free-surface seepage flow through fully penetrating permeable barriers with finite thickness has been dealt with by Sato (1982) and by Federico (1990). The solutions have been obtained by the method of successive steady states.

Experimental data are very rare. Some are reported by Sato (1982) and are compared with theoretical results by the same author and by Federico (1990). For both authors, numerical and experimental results agree satisfactorily.

Finally, the consequence of a defect on the seepage process and on the effectiveness of a thin, impervious cut-off structures has been analysed theoretically by Mc Lean and Krizek (1971).

3. Flow through unsaturated soils

The previously mentioned analyses always neglect the variability of the coefficient of permeability k when water seeps through unsaturated soils.

The coefficient of permeability is an indirect measure of the space available for water to flow. This space varies only with void ratio for a saturated flow and with void ratio and degree of saturation (or volumetric water content) for an unsaturated soil.

Simplified, the coefficient of permeability may be considered only as a function of the volumetric water content of the soil (Fredlund, 1981).

Under unsaturated conditions, only some parts of the pores are filled with water and only these parts convey water. Hence, a decrease of the volumetric water content reduces the size and the number of elementary flow channels, as well as the continuity of pore network, thereby decreasing the capacity to conduct water through the soil.

Since the volumetric water content is a function of pore-water pressure and the coefficient of permeability depends on the volumetric water content, the permeability coefficient, in turn, may be expressed as a function of pore-water pressure. If so, it is possible to analyse problems involving both unsaturated and saturated soils and it is possible to extend the flow to the whole physical domain. As a consequence, the flow domain is no more longer bounded by an unknown free surface.

4. Statement of the problem

Whichever method is employed to make a barrier, either by inserting structural elements into the ground or by modifying the foundation soil or rock properties within an assigned volume of the ground, in practice, the barrier is not watertight owing to the intrinsic, hydraulic conductivity of the composing material, as well as to possible constructive defects. Moreover, it frequently is not completely keyed in an underlying impervious soil. Therefore, the barrier is "imperfect".

The role of location and thickness of a permeable barrier inserted in a homogeneous ground mass on the effectiveness of the barrier has already been analysed, e.g. by Federico (1990).

In the current paper, attention is focused on the role of *depth* and *integrity* of a barrier inserted in *heterogeneous* soils.

To this purpose, a partially penetrating vertical barrier of 120 cm thickness and maximum value of the coefficient of permeability $k_M = 1E-10$ m/s, that penetrates through a river levee ($k_M = 1E-5$ m/s) into the underlying foundation soils, has been considered (fig. 1). The upper layer is composed of silt and sandy silt with permeability coefficient $k_M = 1E-6$ m/s; the lower layer is composed by silt and clayey silt soils with $k_M = 1E-7$ m/s.

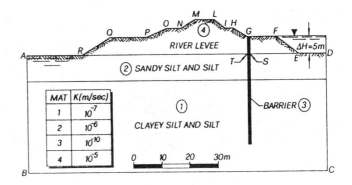

Fig.1 - Reference scheme.

The assumed relationships k(u) between the coefficients of permeability k of the materials vs the pore-water pressure u are reported in fig. 2.

As far as boundary conditions are considered, the physical, two-dimensional domain is delimited by two vertical planes and by a lower horizontal boundary; the vertical line AB and the contours FED and AR (fig.1) are equipotential lines; the horizontal base line BC, line CD and the boundary of the river levee FGHILMNOPQR are impervious and therefore can be assumed as streamlines.

The water level in the river has been fixed equal to $\triangle H = 5$ m.

In absence of sources and by assuming that an incompressible fluid flows both through saturated and unsaturated soils following the Darcy's law, the continuity equation reduces to the well known Laplace's equation.

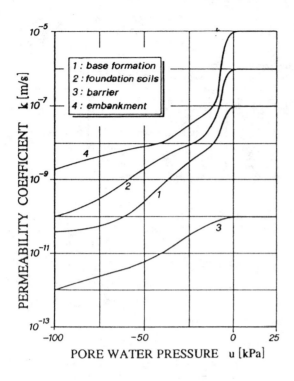

Fig.2 - Coefficients of materials permeability vs pore-water pressure.

5. Numerical analysis and results

Solution of the governing equation has been carried out by a numerical method, through a finite element scheme, taking into account the progressive modification of permeability of each element, according to the assigned permeability vs pore-water pressure law (Papagianakis, Fredlund, 1984).

The flow domain has been firstly subdivided in isoparametric elements, eight nodes quadrilateral and six nodes triangular elements. Due to the presence of secondary nodes, head distribution within each element is non-linear.

The thickness of each element is considered to be constant over the entire element (two-dimensional seepage flow).

The finite element equations have been solved through a Gauss elimination technique, following repeated substitutions in the iterative process, allowing for variation of the coefficient of permeability of the elements.

Calculations have been performed in double precision. The measure of convergence is defined as the Euclidean norm of the pressure head vector.

The iterative process continues until the results satisfy the convergence criteria; this last one, in turn, consists of a difference between two values of the norm, corresponding to two consecutive steps in the calculations, less than 1%.

Seepage discharge has been computed in correspondence of some vertical sections.

Effectiveness of barriers is defined (Telling et al., 1978) as $E_H = \triangle h / \triangle H$, $\triangle h$ being the difference between the piezometric heads, measured at the level $z=H_v$, at the contact soil-barrier (points T,S, fig.1), while $\triangle H$ is the difference of piezometric heads H_m, H_v at the boundaries FED and AB, respectively ($\triangle H = H_m - H_v$).

In an alternative, but complementary way, the effectiveness may be evaluated through the ratio $E_Q = (Q_o - Q)/Q_o$, Q_o being the original discharge flow in absence of the barrier, and Q the discharge flow, reduced by the presence of the barrier.

The first definition seems more useful since E_H may be easily determined by measuring the piezometric heads.

Numerical results, synthetically reported in fig.3, clearly put into evidence the dependence of the effectivenesses E_H, E_Q upon the depth of the barrier.

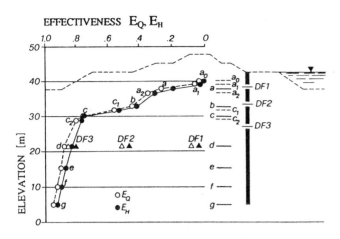

Fig.3 - Effectiveness of partially penetrating barrier vs depth of the barrier.

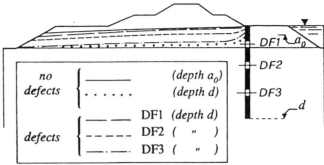

Fig.4 - Zero pore-water pressure line positions for partially penetrating barriers with or without defects.

Howewer, the rate of increase in effectiveness with depth is reduced when the barrier penetrates into the less permeable layer. As a matter of fact, with the base of the barrier at the interface between the two foundation layers, just about 8m below the natural ground level, E_H and E_Q are both equal to approximately 0.8. To increase this value of about 10%, it is necessary to deepen the barrier another 10 meters.

6. The role of defects

In order to enlighten the role of defects on the effectiveness of the barrier, the numerical analysis was used with the bottom of the barrier in a position labelled in fig.3 as d (fig.1).

The barrier is in this case affected by a constructive defect, modelled by increasing, in a zone of 1m height, the permeability coefficient of the barrier up to that of the surrounding soil.

Three sub-cases have been considered, depending upon the position of the defect: within the embankment (DF1) and within the upper (DF2) and the lower (DF3) foundation soil layers, respectively (figures 1 and 3).

The results of the numerical analysis, reported in fig.3, show that the role of the defect in reducing the effectiveness of the barrier can hardly be overemphasized. The effect is more evident as higher is the permeability of the soil surrounding the defect. In the case of position DF1, the barrier is, in practice, not effective at all.

Analogous effects can be outlined concerning the line characterized by values of the neutral pressure equal to zero. This line is located in proximity of the base of the levee in the case of absence of defect (fig.4); it progressively raises with the position of the defect from the case DF3, to the cases DF2 and DF1. In the case DF1 the barrier is equivalent to a very short barrier (bottom fixed at level a_0) without defect.

In presence of a defect, the zone of seepage flow domain where positive neutral pressures act is larger than the corresponding zone of the ideal case (barrier without defect, bottom at position d); the seepage flow exhibits, moreover, a downstream exit point.

One can conclude that the presence of a defect reduces the effectiveness of the barrier and thus the stability of the river levee. This is due to the increase of neutral pressures in the levee body and due to the presence of a downstream exit point, exposing the levee to the risk of piping and erosion phenomena.

7. Concluding remarks

The most important application of underground barriers in Geotechnical Engineering concerns the control of seepage flow.

Barriers are never *perfect*, that is watertight to seepage flow. In case of fully penetrating barriers in homogeneous soils, the degree of imperfection may be defined by the ratio k_b/k_{fs} between the permeabilities of the barrier (k_b) and of the surrounding soil (k_{fs}), respectively.

For partially penetrating barriers, the ratio between the distance from the bottom of the barrier to the impervious base to the total difference of piezometric levels, must be considered.

In the paper, reference has been made to a river levee resting on heterogeneous fine grained soils, made of a horizontal upper layer of sandy silt and silt overlying a horizontal layer of clayey silt and silt. The role of the depth of a partially penetrating imperfect barrier intercepting the embankment and the foundation soils has been investigated.

The consequences of a constructive defect in the barrier has been examined and the conclusion was drawn that even small defects could definitely impair the effectiveness of the barrier, especially when defects occur within pervious horizons of the foundation soils.

References

A.S.T.M. (1984) - Hydraulic barriers in soil and rock. Special Technical Publication, n. 874.

BEAR J. (1972) - Dynamics of fluids in porous media. American Elsevier, New York.

CASAGRANDE A. (1961) - Control of seepage through foundations and abutments. Geotechnique, Vol. 11, n. 3, pp. 161-181.

FEDERICO F. (1990) - Underground barrier for seepage control. 1th Italian Conf. on "Protection and control of groundwater" (in italian), C.N.R., Modena, Italy.

FREDLUND D.G. (1981) - Seepage in unsaturated soils. Panel discussion: Groundwater and Seepage Problems. 10th I.C.S.M.F.E., Stockholm, vol. IV, pp. 629-641.

KAMON M., AOKI K. (1987) - Groundwater control by Tsunegami underground dam. IX E.C.S.M.F.E., Vol. 1, pp. 175-178, Dublin.

MC LEAN F.G., KRIZEK R.J. (1971) - Seepage characteristics of imperfect cut-offs. J. of Soil Mech. and Found. Div., ASCE, n. SM1, January.

PAPAGIANAKIS A. T., FREDLUND D. G. (1984) - A steady state model for flow in saturated - unsaturated soils. Canad. Geotech. J., vol. 21, pp.419-430.

POLUBARINOVA-KOCHINA P.M. (1962) - Theory of groundwater movement. Princeton Univ. Press.

ROWE R. K. (1988) - Contaminant migration through groundwater. The role of modelling in the design of barriers. Canad. Geotech. J., vol. 25, pp.778-798.

SATO K. (1982) - Hydro-dynamic behaviour of groundwater in confined and unconfined layers with cut-off wall. Soils and Found., 22, n. 1, pp. 14-22.

TELLING R.M., MENZIES B.K., COULTHARD J.M. (1978) - A design method for assessing the effectiveness of partially penetrating cut-off walls. Ground Engineering, November.

Design and Performances of a Clay Liner for Containment and Impermeabilization of an Aeration Waste Water Plant in Ghana

Angelo Lucio Garassino, Massimo Giambastiani
Garassino s.r.l., Milano, Italy

Bruno Salesi
ABB - SAE-SADELMI S.p.A., Milano, Italy

ABSTRACT: The construction of a waste water plant in Ghana met local soil conditions that suggested to use a clay liner to reduce the percolation from the bottom of the ponds. The clay liner was built with quarry material from the nearby sites. The results of in situ tests in the preliminary soil investigation showed that the soil in its in situ state was unsuitable for such a use. By a careful examination of laboratory data and from an additional soil investigation it was proved that the material coming from the Layer B soil could be used for the clay liner if properly compacted. The paramount importance of the working procedure in order to obtain satisfactory results is enhanced coupled to control measurements by field infiltration tests to get a very good quality control.

PLANT AND SITE DESCRIPTION

In Ghana, according to the Volta River Authority development program, it was planned to build three large reservoirs in the Akosombo region for the aeration of sewerage wastes before treatement and use for agricultural purposes, see Figure 1.

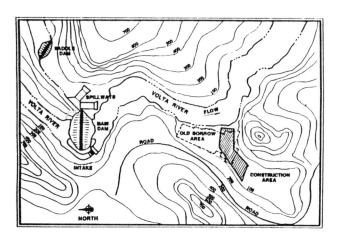

FIGURE 1 - Location plan

Two of these basins were built first, the third one (Maturation Pond B) is to be completed later. The characteristics of the constructed ponds are:

- total surface = 116000 m^2;

- maximum elevation of the earth dikes = 6.50 m;

- total length of the earth dikes = 2175 m.

The basin bottoms were to be at different elevations, see Figures 2 and 3.
The design elevations of the bottoms of the ponds coupled to the site morphology led to two quite different conditions with the bottom of the ponds partly on impervious material (Layer B) and partly on a pervious layer overlying impervious soils (Layer A). Where the thickness of the pervious layer was limited it was removed, where the thickness increased it was left in place and a seal with a clay liner was planned. The possibility of using the sandy silty clay, locally detected and

quarried in the construction area, was examined.

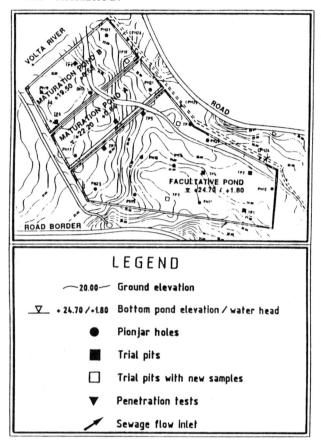

FIGURE 2 - Water treatment plan

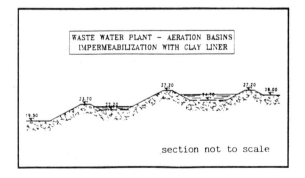

FIGURE 3 - Section of waste water plant

A procedure for overlying the natural graded soil with this clayey material was put forward and in situ tests were performed in a field trial in order to check the suitability of: selected thickness, clay material characteristics and the compaction procedure adopted.

A very limited amount of water percolating through the clay liner can be tolerated. The area on which the sewage water treatment was to be built was a relatively flat area sloping gently toward the river with few meters difference in level by hills on three sides and not far from the Volta River bank on the remaining side.

SOIL PROPERTIES

Soil properties were investigated limited to the upper few meters by boreholes, field and laboratory tests in two phases which led to the soil characterization herebelow described:

LAYER A: silty clay clayey sand

LAYER B: silty sandy clay

LAYER C: gravelly clay (with patches of decomposed phyllite, overlying the bedrock), see Figure 4.

In the first phase Layer B was particularly tested in order to find its suitability as base clay liner, both in its natural and recompacted state.

In situ permeability tests by falling head or rising head method showed $K = 1$ to 6×10^{-6} m/s and laboratory permeability tests gave $K = 2 \times 10^{-7}$ m/s.

By a deep examination of the sampled material it followed that the investigated material was mainly very fine material (see Figure 4) and, as a consequence, that the Layer B material properly compacted could furnish impervious clay liner with permeability far lower than that obtained by in situ tests.

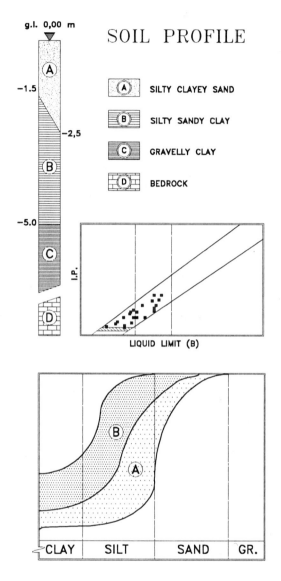

FIGURE 4 - Soil profile and soil properties

By adopting a correlation between permeability and results of sieve analysis (Hazen) as equation 1 shows:

$$K_m = C_1 \times D_{10}^2 \quad (m/s) \quad (1)$$

C_1 = empirical factor 1×10^{-5} to 1.5×10^{-5};

D_{10} = diameter of the 10% passing (mm)

the obtained results gave $1.6 \times 10^{-8} \leq K \leq 2.4 \times 10^{-8}$ m/s on the basis of a conservative evaluation of D_{10} thus allowing Layer B to be considered as very low permeability soil. Then additional samples were taken and laboratory tests on remoulded and recompacted soil, according to ASTM D 698 Standard Proctor, were performed in a permeameter reproducing the actual final condition of 2 m waterhead pressure at the bottom of the pond.

The results indicated permeability values $1.6 \times 10^{-10} \leq K \leq 2.68 \times 10^{-10}$ m/s thus confirming that Layer B material, properly compacted, should have been suitable as bottom liner to limit the percolation from the sewage pond.

LINER THICKNESS

The design of the liner thickness followed the Client's requirements and specifications.
The calculations were carried out according to the seepage model shown in Figure 5 as Equation 2 shows: seepage rate per unit area:

$$q = K \frac{h_1 + L}{L} \quad (m \times s^{-1}) \quad (2)$$

A minimum thickness of 0.4 m was considered suitable, if properly constructed, thus allowing a vertical downward flow rate $q = 1.28 \times 10^{-9}$ m/s per unit area, that was considered acceptable.
The large scatter between laboratory and in situ tests needed a full scale test to get reliable data on the effectiveness of the adopted procedure under the actual field conditions including scale effects.

compaction nonuniformity and microfractures.

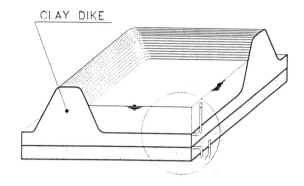

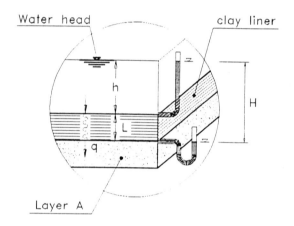

FIGURE 5 - Seepage model through liner

On this field trial in situ infiltrometer tests were performed following Daniel (1984, 1989) and Fernuik and Haug (1990).
The calculations of the infiltration rate I and of the hydraulic conductivity K have been executed according to the relationships (D.E. Daniel, 1986):

$$I = \frac{Q}{At} \quad \text{and} \quad K = \frac{I}{i} = \frac{I}{\dfrac{H + L + y_f}{L}}$$

I = rate of infiltration (cm/s)
Q = quantity of flow (m^3)
A = area of the ring (m^2)
t = elapsed time (s)
K = hydraulic conductivity (m/s)

i = hydraulic gradient
H = water head (m)
L = liner thickness cm)
y_f = suction head (m).

According to Daniel (1989), the suction head value y_f has been neglected, leading to a slightly overestimation of the hydraulic conductivity K at the beginning of the test before the total liner saturation.

In order to carry out a representative test (Figure 6) the following procedure was followed:

- the test area was selected where the more pervious layer, Layer A, was still present;

- the natural topsoil was removed and the surface was graded and compacted;

- then two layers of material coming from Layer B were spread and compacted by wetting and drying it to reach the optimum water content.
The thickness of the spread clay layers was 0.30 m to 0.35 m before compaction for a total thickness 0.4 m after compaction; the material was compacted by a sheeps-foot roller and then the finished surface was completed with some passes of a smooth wheeled roller;

- a check was made on the stockpiled material coming from Layer B collected in different quarry areas in order to be sure of the uniformity of the material used and that the full scale test was representative; the results are presented in Figure 7.
Stockpile L5 was rejected and the field tests were performed with both material from L8, considered the most suitable, and from stockpile L4 the less suitable one;

- then the infiltrometer was placed on the surface and its cutting bottom was pushed to the top of Layer A (figure 6);

- then the infiltrometer was filled with water to the operative level 1.8

m above the top of the clay liner and the test was carried out.

Permeability values $K = 5 \times 10^{-9}$ m/s for stockpile L8, to $K = 7.5 \times 10^{-9}$ m/s for stockpile L4, were found, not very far from the ones coming from laboratory tests on compacted samples. Seepage unit quantities $q = 4.1 \times 10^{-8}$ (m × s^{-1}) could still be acceptable.

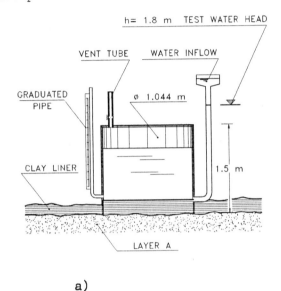

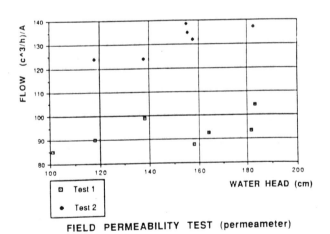

FIGURE 6 - Permeability tests:
a) field measuring device;
b) in situ results on 0.4 m thick clay liner

FIGURE 7 - Stockpiled material soil properties

Finally, it was specified that the thickness of the clay liner be 0.4 m for those zones in which it overlays Layer B and 0.6 m for those zones where the liner lays on Layer A (more sandy, with higher permeability). A program of control tests was sheduled during the works both on stockpiled materials and on the finished liner, in addition some more permeameter tests were planned in order to indicate the final suitability of the liner.

COMPARISON BETWEEN AS BUILT AND DESIGN

The design prescriptions were not followed over the full extension of the lining with regard to compaction and control tests. Neither the prescribed permeability values were performed nor the suggestion to avoid excessive drying of the liner was followed. Pond A, characterized by unfavourable natural subsoil conditions, was initially graded, compacted and coated with the clay liner; a proper maintenance of a sufficient minimum water head on the constructed sections or wetting was not carried out. In such a way dried portions of the bottom pond surface were present locally. Subsequently the main Facultative Pond, was constructed and immediately filled. Although interesting partially the same natural subsoil, it did not show any failure. When it was decided to fill the lower Pond A, previusly constructed and partially dried during the very hot season, the poor results both from laboratoy and in situ infiltrometers tests were ignored, and the risk that mud cracks will reduce the impermeability of the clay liner was underestimated.

After a certain time had elapsed from the beginning of ponding an amount of water, outcoming from the ponds, was detected, much higher than was to be expected, thus indicating the opening of some failure points. After the ponds had been emptied an accurate survey was carried out, samples were taken and in situ permeability tests were carried out. The results showed clearly that some sinkholes were present in the clay liner. The tests showed that close to the damaged zones the clay was either poorly compacted, thus being much more pervious than expected, or fissured as a consequence of extreme drying due to sun exposure. The mud cracks were in the upper 0.1 m portion of the clay liner and were found filled with sand, the thickness of intact lining being still higher than 0.4 m. Close to the sinkholes area the inspection trenches excavated showed the presence of sandy seams a few centimetres thick and some meters wide in both horizontal directions included into the liner. In situ density tests gave results far lower than the design values and the tests carried out with the permeameter showed infiltration rates far higher than those obtained in the field trial. The area interested by the presence of the sinkholes and mud cracks was, once again, investigated to detect the quality of the material utilized as clay liner. The utilized material, that once controlled was accepted according to the initially stated specifications, was properly recompacted to obtain a 0.60 m thick clay liner. The material that was detected of unsuitable quality was replaced.

CONCLUSIONS

As was proved by the field evidence the permeability of a clay liner is far more important than the thickness. The obtained permeability values are primarily governed by the type of material selected and secondly by the procedure adopted for the construction of the clay liner. The spreading and compacting and testing procedure is of paramount importance for the effectiveness of the lining.

Differences in the hydraulic conductivity of the order of 100 times can be achieved between proper and unsafe procedures, thus affecting the success of the design: i.e. that the clay liner design cannot leave aside the field aspects of the work.

REFERENCE

DANIEL D.E. (1984) - Predicting Hydraulic Conductivity of Clay Liners - ASCE - Geotechnical Engineering Division - Vol. 110, n° 2 pp. 285.

DANIEL D.E. (1989) - In situ Hydraulic Conductivity Tests for Compacted Clay - ASCE - Geotechnical Engineering Division - Vol. 115, n° 9, pp. 1205 ÷ 1226.

DAY S.R., DANIEL D.E. (1985) - Hydraulic Conductivity of Two Prototype Clay Liners - ASCE - Geotechnical Engineering Division - Vol. 111, n° 8.

FERNUIK N., HAUG M. (1990) - Evaluation of in Situ Permeability Testing Methods - ASCE - Geotechnical Engineering Division - Vol. 116, n° 2, pp. 297 ÷ 311.

MUNDELL J.A., BAILEY B. (1985) - The design and testing of Compacted Clay Barrier layer to Limit percolation Through landfill Covers - Hydraulic Barriers in soil and rock, ASTM STP 874, pp. 246 ÷ 262.

A Study on the Characteristics of Marine Clay and Admixed Liners

S.W. Hong, Senior Research Fellow,
Geot. Engng. Div. Korea Institute of Construction Technology, Seoul, Korea

Y.S. Jang, Assistant Professor
Dept. Of Civil Engineering, Dongguk University, Seoul, Korea

H.I. Jeong, Senior Researcher
Geot. Engng. Div. Korea Institute of Construction Technology, Seoul, Korea

SYNOPSIS : The applicability of the Kimpo marine clay, fly ash and granitic materials as the liners of waste landfills is investigated by examining their compaction and hydraulic conductivity characteristics and the compatibility with representative chemicals within landfills. The amounts of bentonite needed to make hydraulic conductivity of the admixed liners below 1×10^{-7} cm/sec are 18, 30, 10% by weights for Seochon anthracite fly ash, Samchonpo bituminous fly ash and weathered granitic soil, respectively. The hydraulic conductivities of all liner materials tested are increased by permeating pure chemicals. The rate of the hydraulic conductivity increase is lowest for Seochon fly ash admixed liner and highest for natural marine clay liner. The influence of chemical solutions on the liner materials is least for acetic acid and most for aniline.

INTRODUCTION

The environmental regulations in many countries are being strengthened significantly, while the amount of wastes that need to be disposed of by landfill increases. Most regulations require the installation of clay liner materials together with polymeric membranes. However, some mountainous areas lack clay materials at the landfill site and clay has to be transported from remote areas. In this case utilization of admixture liners may be a good alternative that can meet the stringent regulations for constructing waste landfills.

In this paper, the compaction and the hydraulic conductivity characteristics of admixed liners produced by mixing bentonite with flyash or weathered granitic soils are examined. Compatibility tests are performed using chemicals that can represent the leachate within landfills.

SELECTION OF THE LINER MATERIALS

Two types of fly ash, one produced from anthracite coal called Seochon fly ash and other produced from bituminous coal called Samchonpo fly ash, and weathered granitic soil are used as the major substances of liner materials and sodium bentonite is used as an additive material. Using fly ash as a major substance in admixed liners has two advantages; (1) by reducing the volume of fly ash deposited in the pond, the life span of the ash pond for a coal fired power plant can be extended significantly, and (2) it would be an inexpensive alternative to an earth liner where clay is scarce. The performance of the weathered granitic soil for admixed liners is examined, since it is the most abundant material available in Korea.

Many waste landfills in Korea are located at seashores and formed with marine clay foundations since the sites can be easily obtained and many of them are near to urban areas. Hence, marine clay sampled from a waste landfill site in the western part of Korea is included in the chemical compatibility tests.

ENGINEERING PROPERTIES OF THE LINER MATERIALS

Basic Properties

The basic properties of the major substances and the additive are shown in Table 1 and the grain size distribution curves are also shown in Figure 1.

From Table 1. it can be seen that the specific gravity of the bentonite is significantly lower than that of the other materials. The coefficient of uniformity(C_u) shows that weathered granitic soil and bentonite are well graded and the other materials are relatively

Table 1. The basic engineering properties of the liner materials

Material types		Specific gravity	Grain size distribution				Soil classification
			D_{10} (mm)	D_{60} (mm)	Cu	Percent of passing No.200 sieve	
Fly ash	Seochon*	2.30	0.003	0.02	6.6	93.0	ML
	Samchonpo**	2.14	0.024	0.053	2.2	80.0	ML
Weathered granitic soil		2.54	0.1	1.0	10.0	5.0	MC
Bentonite		1.70	0.0008	0.04	50.0	76.1	CH
Marine clay		2.70	0.0083	0.037	4.5	90.0	CL

* Anthracite fly ash.
** Bituminous fly ash.

uniform. Most materials except the weathered granitic soil are composed of fine materials (silt and clay) which is indicated by the percent passing the No.200 sieve.

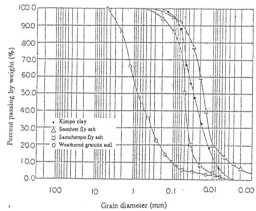

Figure 1. Grain size distribution of major substances of the liner

Fly ash and weathered granitic soil can be classified as silt and silty clay, respectively, and the other materials are classified as clay. The grain size distribution curves show that the grain size of the weathered granitic soil is significantly larger than that of the other materials. The Samchonpo fly ash procduced from bituminous coal has more uniform distribution compared with the other materials.

Maximum Dry Density and Optimum Water Content

When liner materials are installed in landfills, they are often compacted to reduce the hydraulic conductivity below the value specified in the regulatory guide and to have enough strength to support the weight of wastes. Hence, it is important to test the maximum dry density and optimum water content of compacted liner materials.

Compaction tests are performed by using the standard proctor method (KSF 2312, A-1 type) and by adding bentonite until the hydraulic conductivities of the compacted materials become less than 1×10^{-7} cm/sec. The variations of the properties with respect to the percent of bentonite by weight of the sample are shown in Figure. 2.

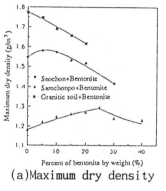

(a) Maximum dry density

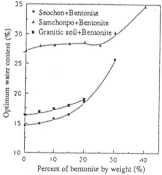

(b) Optimum water content

Figure 2. Variations of maximum dry density and optimum water content with respect to the pecent of bentonite by weight

The maximum dry density is highest for the weathered granitic soil which has well graded grain size distribution and lowest for Samchonpo fly ash which has uniform grain size distribution. It can be recognized that the well graded soil can have compact grain structures and the density is high. The peak of γ_{dmax} is reached by adding 5% bentonite by weight for Seochon fly ash, contrary to 25% for Samchonpo fly ash. This means that the structure of grains for Samchonpo ash has more voids than that of Seochon ash and bentonite has to fill up the voids until the peak value of γ_{dmax} is reached. The γ_{dmax} of weathered granitic soil is reduced continuously as the amount of bentonite increases. This means that the structure of weathered granitic soil is already compact and the bentonite, which has smaller specific gravity than the other soils, can make the density of admixed materials lower than the density of the soil without bentonite. The downward curve of Seochon fly ash with the percent of bentonite greater than 5% is resulted by the same reason as the case of the weathered granitic soil.

The optimum water content of admixed materials increases as the percent of bentonite increases (Figure 2(b)) and the magnitude of optimum water content for Samchonpo fly ash is higher than that of the other materials. It is interesting to find that the dramatic increase of W_{opt} for Samchonpo fly ash at 25% bentonite is consistent with the peak of γ_{dmax} for Samchonpo fly ash in Figure 2(a). This is the point at which the amount of bentonite influences both on the density and on the W_{opt} of admixed Samchonpo fly ash. The increase of optimum water content with the increase in the percent of bentonite seems to be caused by the increased hydrophilic tendency of the admixed materials with the addition of bentonite.

HYDRURIC CONDUCTIVITY TEST OF ADMIXED MATERIALS

Hydraulic conductivity of the liner is a really important property that can control the amount of seepage and leachate from the landfills. Hydraulic conductivity tests were performed using water and chemical solutions as permeants and by using the rigid wall hydraulic conductivity tester. The specimens for the hydraulic conductivity test are produced at 95% of γ_{dmax} and at 3% moist side of optimum water content. The water content of the compacted samples was selected based on the fact that liner materials compacted at moist side of the optimum water content have dispersed soil structures and can have the lowest hydraulic conductivity at a water content between +2 to +4% of optimum water content (Mitchell, 1976).

The hydraulic conductivity tests described in this paper have two stages:
(1) By percolating water through liner materials, the minimum amount of bentonite needed to make the hydraulic conductivity of the liner below 1×10^{-7} cm/sec, which is the required minimum value of K for the liners, is estimated. The minimum amount of bentonite is important for specifying the mixing ratio of bentonite and major materials for producing admixed liners in the field.
(2) By using the chemical solutions which can represent the substances of the leachates in the wastefills, the compatibility of the admixed liners with the chemical solutions is analyzed.

Preparation of the Experiment

Different types of hydraulic conductivity test are used depending on the objectives of the test. The variable hydraulic conductivity test with a hydraulic gradient of about 5 was used for measuring the hydraulic conductivity of the admixed materials with water. When the chemical solutions were used as permeants, the pressurized constant head hydraulic conductivity tester using nitrogen gas (Figure 4.) was used.

The dimensions of the specimens were 10 cm diameter and 8 cm height. Gas pressures varied from 0.3 kg/cm^2 for saturation of the samples to a maximum of 2.4 kg/cm^2 for the experiment with chemical solutions. Gas pressures were increased step by step to prevent fractures between the sample and the mould or the holes within the sample. The hydraulic gradient that corresponds to the maximum gas pressure of 2.4 kg/cm^2 was 300.

Permeants

The permeants used in the experiments are water and the chemical solutions. Four types of organic chemical solutions : acetic acid,

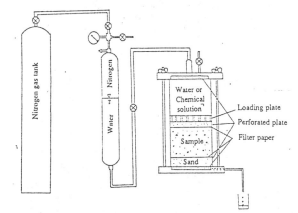

Figure 3. Schematic diagram of the pressurized hydraulic conductivity tester used for chemical compatibility test

methanol, heptane, aniline, which may represent the various types of chemical substances in the leachates of waste fill are used. Acetic acid is the acidic solution which can be found frequently in the leachates at the actively decomposing stage of a municipal waste fill. Methanol is the hydrophilic and heptane is the hydrophobic organic material. Aniline is the basic solution which belongs to the amine group.

Hydraulic conductivity of the Liner Materials

The hydraulic conductivity of the admixed liner materials with respect to the percent of bentonite is shown in Figure 4. According to Figure 4, the hydraulic conductivity of the liner materials is reduced continuously as the mixing ratio of the bentonite increases. The cause of the hydraulic conductivity decrease is that the bentonite can swell up to 5 to 10 times its original volume with absorption of water. The percent of bentonite needed to reduce the hydraulic conductivity of the liner materials below 1×10^{-7} cm/sec is 18% for Seochon anthracite fly ash, 30% for Samchonpo bituminous fly ash and 10% for weathered granitic soil. It can be recognized that the weathered granitic soil is most efficient for producing liner material and Samchonpo fly ash produced from the bituminous coal is the least efficient.

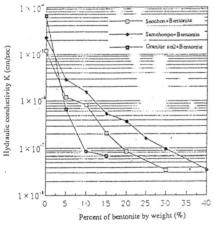

Figure 4. Variation of hydraulic conductivity with different percent of bentonite by weight

It is interesting to find out that the fly ashes which have different geological origins may have significantly different hydraulic conductivity characteristics. At least from the results of these experiments, it can be recognized that Samchonpo fly ash is not recommendable for admixed liner materials, because Na-bentonite is quite expensive to be used as liner materials with large mixing ratios.

The cause of the significant difference in the hydraulic conductivity of the two types of fly ashes is that the grain size of Seochon ash is 3.5 times smaller than that of the Samchonpo ash (see Figure 1) and Seochon ash forms smaller voids compared with the Samchonpo ash. The smaller mixing ratio for the bentonite with weathered granitic soil for achieving the required hydraulic conductivity comes from the gradation, although the grain size of the weathered granitic soil is quite large.

Compatibility to Chemical Solutions

The chemical compatibility of the admixed liner materials and the marine clay to the chemical solutions are examined. The admixed materials used in these tests are Seochon fly ash and weathered granitic soil. Samchonpo fly ash is discarded in this experiment because the amount of bentonite needed to reach the required hydraulic conductivity is too large to be used as a liner materials. Instead, marine clay is used in this experiment because many wastefills in Korea are constructed at seashores and the compatibility of the marine clay to the leachate solution is important for long term behavior of these landfills. Also, this gives a good opportunity to compare the performance of the admixed materials to that of the natural marine clay liner.

In this test, of pure and 75% concentration and chemicals are used and the pressure is increased step by step up to 2.4 kg/cm^2 by percolating the water as the permeant in the initial stage. When the pressure reaches to 2.4 kg/cm^2, chemical solutions are percolated through the liner materials until a significant change in the hydraulic conductivity is obtained. The results are shown in Figure 5. Figure 5 shows the variation of the hydraulic conductivity of the liner materials with respect to the use of various pure chemicals. The magnitude of K increase for marine clay is more significant compared with other materials except the case using the acetic acid as the permeant.

In the case of acetic acid (Figure 5(a)), the hydraulic conductivity decreases in all materials at the initial stage and increases subsequently to higher values. The magnitude of increase was highest for admixed weathered grantic soil and lowest for admixed Seochon fly ash. The mechanism for the initial decrease is considered to be the result of the desolution of the soil constituents by the acid and of the precipitation and clogging of the pores by the disolved constituents, as pH decreases. The hydraulic conductivity increase of the next stage seems to be caused by the progressive soil piping that eventually cleared the initially clogged pores.

In the case of using methanol and heptane as the permeant(Figure 5(b) and (c)), the initial decrease in hydraulic conductivity is not observed and the hydraulic conductivities increase continuously by more than one order of magnitude to some constant values.

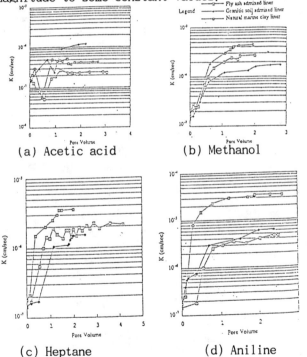

Figure 5. Variation of the hydraulic conductivity for various chemical solutions

The magnitutede of K increase was highest for marine clay and lowest for weathered granitic soil. The cause of K increase for the two permeants is that they have lower dielectric constants than water and can decrease the interlayer spacing of the clay minerals when these organic permeants replace the pore water in the specimens. The structural changes caused by the replacement of pore water were visible as cracks and holes on the surface of the specimens.

The magnitude of K increase with methanol may be greater than that of heptane because methanol is a miscible solution and can replace the pore water completely. Contrary, heptane which is a neutral nonpolar solution, has limited ability to penetrate interlayer spaces of the clay minerals. The effects of this difference is shown in Figure 5.(b) and (c), For all materials, the increase of K developed by permeating methanol is a little greater than that developed by permeating heptane.

It is interesting to find that the hydraulic conductivity of all these materials is most sensitive to aniline and least sensitive to acetic acid, which shows that pH of solutions has a significant influence on the hydraulic conductivity of liner materials. The hydraulic conductivity by percolating aniline becomes to one or two orders of magnitude higher compared with the hydraulic conductivity at the beginning of permeant exchange, i.e. the time at pore volume of permeated materials equals to zero. It can be recognized that pure organic base causes extensive structural changes in the specimens of liner materials and the massive structure of the liner materials can be altered by aniline into an aggregated structure which is manifested by the visible hole and cracks on the surface of the specimens.

The hydraulic conductivity of three liner materials is tested using 75% methanol and heptane with the concentation equivalent to minimum solubility(53 g/l). When the permeant is methanol, the hydraulic conductivity of Seochon fly ash admixed material and of the Kimpo marine clay increases a little and converges to constant values (Figure 6(a)). The hydraulic conductivity of the weathered granitic material decreases initially and converges to a certain value after the increase of a minor amount. The hydraulic conductivity after permeating the solutions of several pore volumes is 2.2 times greater than the initial hydraulic conductivity for Seochon fly ash and Kimpo clay, and is almost the same for the weathered granitic soil.

When the permeant is heptane, the hydraulic conductivity decreases initially and increases significantly after permeation of the solution of 1.5 pore volumes (Figure 6(b)). This change of hydraulic conductivity is caused by the non-polar characteristics of the heptane

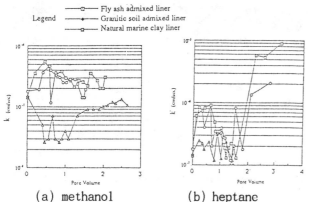

(a) methanol   (b) heptane

Figure 6. Variation of hydraulic conductivity to the permeation of a solution of 75% concentration

solution, which makes water and heptans into two separate solutions even when they are mixed together. The solution permeated at the initial stage has concentration of 53g/l of heptane and at the later stage is almost the same as the pure solution.

It can be recognized from the results of the tests that the liner materials in this chemical compatibility test are stable to the solutions of quite high chemical concentrations although the results should be tested more by various concentrations and time periods.

SUMMARY AND CONCLUSIONS

The appropriateness of the marine clay and the admixed liner materials using fly ash and the weathered granitic soil is evaluated by performing the compaction test and the hydraulic conductivity test with water and chemical solutions. The following conclusions are made from the experiments :

1)The difference of the properties of the compacted liner materials, e.g. maximum dry density and optimum water content, with the addition of bentonite was quite different. This difference seems to come mainly from the difference in specific gravity and grain size distributions of the liner materials.

2)The percentage of bentonite by weight needed to make the hydraulic conductivity below 1×10^{-7} cm/sec was 18% for Seochon anthracite fly ash, 30% for Samchonpo bituminous fly ash and 10% for weathered grantic soil. It was recognized that the amount of bentonite would be quite different even between the fly ashes if they have different geologic histories.

3)The hydraulic conductivities of all liner materials including marine clay are sensitive to the percolation of pure chemical solutions and the aniline, an organic base solution in amine group, has the highest ability for changing the hydraulic conductivity of the liner materials.

4)The increase of hydraulic conductivity is not signficant with the solution of concentration 75%. This results should be tested more by various concentrations and time periods.

REFERENCES

Acar, Y. B., A. Hamidon, S. D. Field, and L. Scott, "The Effect of Organic Fluids on Hydraulic Conductivity of Compacted Kaolinite", Hydraulic Barriers in Soil and Rock, ASTM STP 874, pp. 171-187, 1985.

Bowders, J. J. "The Influence of Various Concentrations of Organic Liquids on the Hydraulic Conductivity of Compacted Clay", Geotechnical Engineering Dissertation GT85-2, The University of Texas at Austin, 218 pp, 1985.

Goldman L. J., L. I. Greenfield, A. S. Damle, G. L. Kingsbury, "Design, Construction, and Evaluation of Clay Liners for Waste Management Facilities", US EPA, EPA Contract No. 68-01-7310, 1987.

Behavior of Bordo Poniente Landfill in Texcoco Zone, Near Mexico City

Rosario Iturbe, Ana E. Silva
Instituto de Ingeniería, UNAM, México

Abstract

This study illustrates the migration of contaminants under Bordo Poniente landfill which is the biggest in Mexico City, through microfissures produced as a consequence of the clay characteristics and to the garbage weight.

Introduction

The landfill denominated Bordo Poniente, is located in the Federal Zone of the Texcoco Lake area, NE of Mexico City. The landfill first phase (phase I), of 50 hectares approximately, started its operations in 1983, and phase II, of 60 hectares, in 1987. Phase III presently in operation has an approximate extension of 120 hectares.

The site was initially considered adequate due to the low permeability of the soil (average $k = 5.5 \times 10^{-9}$ m/s). However, due to the high compressibility conditions of the subsoil and to the garbage weight, phases I and II suffered considerable settlements originating fissures under the cells in addition to those previously existing due to the periodic desiccation of the soil.

Phase IV of the landfill will be constructed at the same site, in a 400 hectare area, which is considered enough to bear the solid wastes of Mexico City for a ten year period.

Because of the possible vertical migration of leachate through the soil microfisures, it was decided to perform a study to evaluate the site's present conditions and make the necessary changes to avoid such migration, since the Valley of Mexico aquifer system is the main source of water of the almost 20 million inhabitants of Mexico City and its surrounding area.

Description of the site

The clay of the Texcoco lake zone has a high compressibility, more than 400 % moisture and a permeability of approximately 5.5×10^{-9} m/s. The clay thickness of the site is between 60 and 70m with intermediate strata of sandy silt from 5 to 10 m, known as "hard layers". Abundant remains of rotten organic matter are detected in the clays, corresponding to plants developed during the different times of the lake (ref 1).

The lacustrine deposits in the Texcoco lake zone, contain high natural concentrations of sodium chloride (NaCl) and sodium bicarbonate (Na_2CO_3), reason why they are an important source for industries such as Sosa Texcoco who has worked in the site for more than 30 years extracting the brine from the subsoil, which is the raw material for the sodium bicarbonate. The brine is obtained from the water contained in the shallow sandy silt strata.

The Texcoco lake zone comprises independent permeable strata: two shallow, which are the ones exploited by Sosa Texcoco, and a deep aquifer containing high quality water, which is an extension of the Valley of Mexico's main aquifer.

Objectives

The objectives of this study were the following:

a) To evaluate the present contamination status under the landfill, of phases I to IV, for which soil samples were obtained (0 - 40 m depth), leachate (phases I and II), wastewater from the channels surrounding the area and from their sediments, as well as water from the acuitard.

b) To determine the leachate and the channels influence on the contaminants migration through the soil.

Sampling and analysis

In order to evaluate the susceptibility of the soil to the leachate migration, soil sampling were performed in eight sites at different depths, close to the landfill zone as well as at distant places from it, so as to establish comparisons (ref 2). The sites were selected considering the location of phases I, II and III of the landfill with respect to the flow direction of the shallow acuitard (ref 3). Sites 1 and 2 are out of the landfill zone; sites 3 to 7 are located at the edges of the three stages of the landfill, and site 8 is exactly located under the landfill (phase I) (Fig 1).

Also, samples of the leachate generated in the first two phases of the landfill were obtained. Samples from the upper acuitard were obtained in order to complement the information.

Since the landfill zone is surrounded by erodible channels which evacuate the wastewaters from great part of Mexico City, it is feasible that these waters migrate through the subsoil an also influence its contamination, therefore, samples of these waters were obtained from eight sites. On the other hand, and in order to estimate the type of contaminants tending to deposit on the soil, sediment samples from a regulation lake, which also contain wastewaters, were analyzed.

Each sample was determined the following parameters: pH, oxidation-reduction potential, electric conductivity, temperature, organic and ammoniacal nitrogen, sulphates, some total metals (As, B, Cr, Fe, Ni, Pb and Zn), MBAS, total organic carbon, BOD_5. Also, organic compounds: halogenated volatile organic compounds, extractable organic compounds in a basic and acid media, were determined (ref 2).

Results of the chemical analysis of the soil

Figure 2 presents the contaminants distribution in each sampled site, grouped according to their chemical characteristics.

Firstly, the metals group is shown as well as the sum of metals indicating contamination (As, Pb, Zn, Cr, Cd, Hg). The organic group corresponds to the sum of the identified organic compounds. The extractable organic compounds group by HPLC, is the sum of the aromatic polynuclear hydrocarbons (such as naphthalene, anthracene, fluorene, pirene, etc.). The total organic halogenated group is the sum of the halogenated hydrocarbons (such a trichlorethylene, vinyl chloride, tetrachloroethylene, etc).

Due to their location, sites 1 and 2 are considered out of the landfill's influence, and show very low concentrations of contaminants, which could be considered as natural of the place, or due to the influence of the wastewater channels that occasionally overflow and flood the area where these places are located.

A migration of metals is clearly observed in site 5, which increases accordingly to the depth. The organic compounds present a similar behavior. This site corresponds to the place where the debris produced by the 1985 earthquake were deposited, and is a clear indication of vertical migration, with organic compounds accumulation (toluene and octilphtalate) at the layer of greater permeability (15 m).

Site 7 presents the maximum value of organic compounds found in the soil samples, excepting site 8 which is located within the same landfill. Site 7 corresponds to the youngest phase of the landfill (phase III); it is distant from the wastewater channels; shows a concentration profile of toluene, decane and arsenic whose maximum corresponds to the permeable layer at a 15 m depth.

Site 8 presents the maximum concentrations of organic compounds and heavy metals between 15 and 25 m. Instead, the HPLC and the organic halogens are similar to the remaining sites. The results confirm the vertical migration of the landfill towards the soil.

Only for the metals, a direct association was found between the leachate contaminants and those found in the subsoil. This can be explained, since during the time in which some compounds migrate through the soil, the garbage suffers various decomposition processes, and it is said that the leachate is "getting older".

Results of the leachate analysis

The results show a slightly basic leachate, with an oxidation-reduction potential slightly reducer (between -0.5 and 0 volts), with a high ammoniacal nitrogen content (2 244 mg/l) and Kjeldhal total (2 460 mg/l), high electric conductivity (EC). The metals content is in general low, even though Cr may have a concentration of 5.3 mg/l in a phase I well.

According to its oxidation status, the metals contained in the leachate tend to form precipitates, however, the possible change in pH and oxidation-reduction potential produced by soil conditions, may disturb this status and favor or restrict their migration.

Results of the chemical analysis of water

a) Wastewater
The wastewater arriving in the Texcoco lake area are domestical and industrial sewage type (ref 2).

The chemical analysis results show that the metal values are homogeneous and smaller than those of the norm for potable water. The MBAS values are homogeneous in the eight sites (average: 13.7 mg/l). The BOD_5 concentrations are characteristic of wastewaters with low organic content.

b) Phreatic water
The groundwater samples obtained from the sites located in phases I and II of the Bordo Poniente landfill indicate that the alkalinity value is very high (50 000 to 70 000 mg/l as $CaCO_3$) due to the easy solubilization of the soil carbonates, whose concentrations in the water are from 34 000 to 45 000 mg/l. Also, the high concentrations of chlorides (27 000 to 40 000 mg/l), and sodium (35 000 to 50 000 mg/l) are confirmed (ref 4). These characteristics are not due to the landfill, but they are proper of the type of soil due to their formation from the sediments of a brackish lake.

The ammoniacal nitrogen concentration in the groundwater is of approximately 100 mg/l; neither nitrates nor nitrites were detected; the phosphates concentration varies between 20 and 200 mg/l.

c) Sediments
The results of the average concentrations of metals in the sampled sediments from the regulation lake (ref 2), present concentrations similar to those of the soil, excepting for Pb, whose concentrations in sediments (77.34 mg/Kg) are much greater than those in the soil (<10 mg/Kg). Cd and Hg were only detected in the sediments, and were not found in the wastewater samples nor in the soil samples. Great part of the organic compounds detected in sediments belong to the alkane group. It is possible that some elements of the wastewater may have suffered transformations in the sediment, for example, the octilphtalate is found in the water, and the dioctilphtalate in the sediment. On the other hand, no benzene derivatives nor the acids of large chain (more than 10 carbons) were detected in the sediments, but were indeed found in the wastewater channels.

A clear relationship is presented among the different alkane (hexadecane, heptadecane, octadecane) which are usually utilized in products for lubricants and oils used for the industry. The concentration of dibutilphtalate in leachate is very large and it is not observed in the soil, however, there is a high concentration of dioctilphtalate, which also exists in the leachate. The greater concentration in the sediments, corresponds to the dioctilphtalate, while the octilphtalate was detected in the wastewater.

Contaminants migration

According to the results shown (Fig 2) and the corresponding tables (ref 2) it is possible to conclude that there exists a significant influence of the solid wastes deposits on the concentration of the compounds found in the soil. This influence is greater in the soil which is exactly located under the landfill (site 8), than in the sites contiguous to the cells, which implicates that there is a preferably vertical migration, at least down to the hard layer (40 m).

It is outstanding the fact that the greater concentrations of contaminants in the soil of site 8, consisting of organic compounds as well as metals, were found at a 15 m depth, which corresponds to the sandy stratum located in the sampled interval in this area. This same phenomenon was observed in sites 3 to 7, for different metals and compounds.

The above can be explained considering that the vertical migration occurs through the fissures formed in the clayey material. The contaminants travel through these microfissures and upon arriving in the zone of greater permeability (approximately 15m) the contaminants concentrate in this stratum where they can move laterally, to later on continue vertically downwards through the fissures; this implicates that many fissures are not necessarily found in an aleatory drilling of the soil and therefore, the contaminants concentration may be low, instead, the permeable stratum will always present a similar content of contaminants, that will, in general, be greater than the one found in the fissures and microfissures. However, the lateral flow has not been sufficiently high so that the contaminants displacement be greater in the horizontal direction than in the vertical one.

It is considered that the iron in site 5 has had a migration rate of approximately 5 m/year and that the toluene and octilphtalate have had a similar rate, with the difference of an accumulation in the stratum at 15 m.

In site 7 the migration rate is much greater, at least for the toluene and decane, since the aging of the garbage deposit at the time of the soil sampling, was of approximately 6 months. The greater influence of the leachate is presented in the first 10 meter depth.

It was possible to observe the wastewater influence in sites 1 and 2, in particular due to its contribution of iron, toluene, 1,2 dimethylbenzene, 1,3 dimethylbenzene and octilphtalate; however, the influence magnitude of these channels is lower than that of the leachate of the landfill.

As far as the halogenated organic compounds are concerned, in site 8 which crosses the landfill phase I, a concentration profile gradually decreases accordingly with depth. This is also observed, although in a lower proportion, in sites 3 and 4 close to phases I and II, respectively. Once again, the advance of these contaminants stands out in sites 5 and 7; in the first one, they are found at 37 m depth, and in site 7, an accumulation in the layer at 15 m is observed, although the advance is registered down to 30 m.

Conclusions

According to the results presented in the study, it is evident that there exists a migration of contaminants generated by the landfill as well as by the wastewater, which travel through the fissures of the clayey soil.

If the presence of the organic compounds is not clearly appreciated in the soil as it is in the leachate, it does not mean that there exists no influence of the landfill, since it is feasible that the organic compounds migration be so fast that the sampling in the "old" phases do not permit to find the same types of contaminants.

There is information available on the existence of fissures in the zone of the clayey stratum (0 to 15 m) and of microfissures down to the hard layer(40 m), therefore, if these would not exist in the underlying strata, a low risk could be contemplated for the leachate compounds and wastewater to arrive in the deep aquifer, considering the low permeability of the soil. However, there is a possibility that part of the contaminants upon arriving in the greater permeability stratum, may migrate laterally and reach materials with different characteristics.

Due to the above, it is highly recommended the impermeabilization of the new cells (phase IV). In order to decide the best option, three methods are presently under testing in experimental cells in-situ: synthetic membranes (geotextile); clay and bentonite (ref 5).

Acknowledgementes

The authors thank to Departamento del Distrito Federal that sponsored the study, specially to Ing Jorge Sanchez and Ing Felipe López that adviced and reviewed thr project

References

1. Lozano G.S. (1989), Geofísica Internacional. Vol 28, pp335-362, México.

2. Iturbe A.R., Silva A.E (1993) Institute of Engineering, UNAM. Technical Report.

3. CAVM (1986), Soil Mechanics Bulletin, No 10, SARH, México.

4. GIISA (1993) Technical Report of groundwater characteristics in the Bordo Poniente zone.

5. PROCESA (1993), Technical Report of experimental cells at Bordo Poniente landfill.

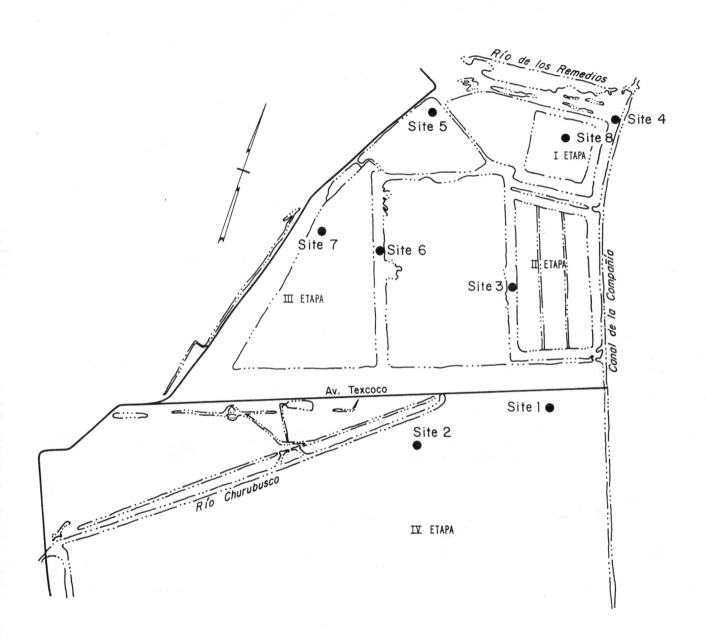

Fig. 1 Bordo Poniente landfill zone

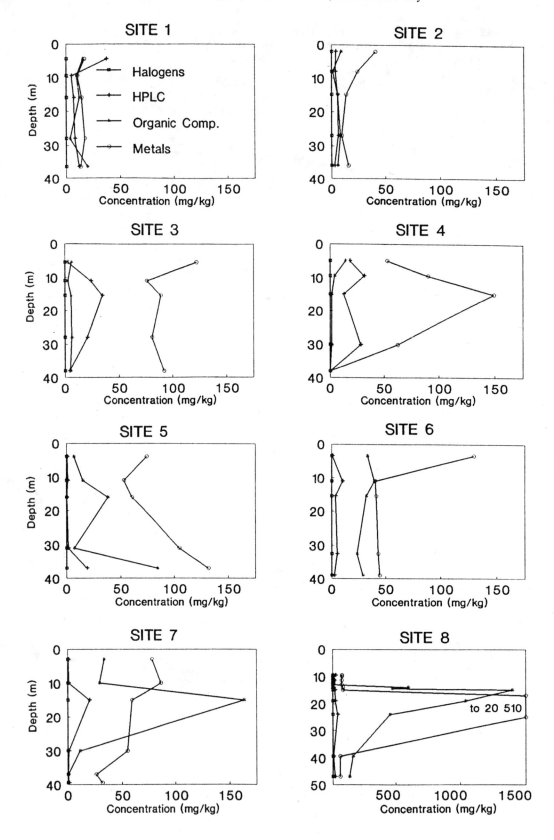

Fig. 2 Concentration of contaminants in soil samples at Bordo Poniente

Numerical Modelling of Contaminant Migration in Two Municipal Landfill Sites

Y.S. Jang, Assistant Professor
Department of Civil Engineering, Dongguk University, Seoul, Korea

K.Y. Lee, Assistant Project Manager
Sunkyung Engineering & Construction Ltd., Seoul, Korea

S.S. Kim, Professor
Department of Civil Engineering, Chungang University, Seoul, Korea

G.S. Lee, Staff Engineer
Sunkyung Engineering & Construction Ltd., Seoul, Korea

SYNOPSIS: Contaminant migration in two municipal landfills is studied using the finite element numerical program of contaminant transport. For Nanji landfill in which waste are disposed over the last 2 decades and closed at 1992, model parameters were validated using the in-situ concentration data and the possible behaviour of transport next 30 years is predicted. For Kimpo landfill which is constructed as an alternative site of Nanji landfill, contaminant movement through the foundation of pheripenal embankment is analyzed and the reinforcement of barrier systems is suggested.

Nanji Landfill

Nanji landfill located at the southern metropolitan area of Seoul beside the Han River is the site in which all of the municipal wastes in Seoul has been disposed in last two decades (Figure 1). The area of the landfill is about 3km^2 and maximum disposal height is 70m, when the landfill is closed at the end of 1992. The operation of this landfill has begun at 1978 without any barrier systems and caused the contamination of the surrounded area including Han River.

The soil profile of this site is composed by the upper disposed waste layer, intermediate river transported sand and gravel layer and the weathered bed rock. The thickness of the upper waste layer is 10.5~13.1m near the Han River and 1.1~1.6m at Sang-Am Dong, northern part of the landfill. Intermediate layer can be divided

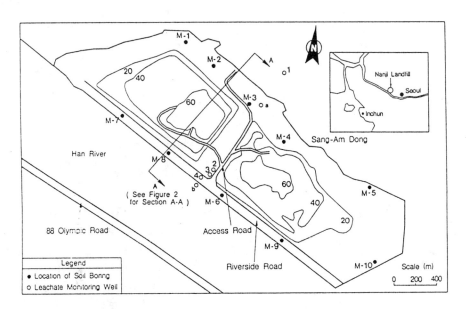

Figure 1. Plan View of Nanji Landfill with the Location of Soil Boring and Leachate Monitoring Well

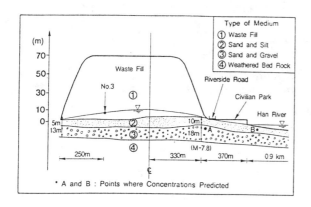

Figure 2. Typical Section of Nanji Waste Fill

further into the upper sand and silt layer and the lower sand and gravel layer whose thickness are about 8.6 and 10m, respectively. The typical landfill section passing boring between M-7 and M-8 (section A-A in Figure 1) is drawn in Figure 2.

Hydrogeological Data

The hydrogeological data at the subsurface area of the landfill is summarized in Table 1 based on the literature which has studied this area. It shows that the hydraulic conductivity (K) obtained from pumping test is more than an order of magnitude greater than the ones obtained from other methods. Considering the limitations of slug test and laboratory hydraulic conductivity test which have small soil volume and a difficulty to repeat the in-situ soil density in the laboratory, the K obtained from pumping test may be closer to the site K data and the K used in the analyses was 1.1×10^{-4} m/sec.

The longitudinal dispersivity test was performed using injection and monitoring well whose distance is 7m apart. Since the dispersivity depends on the scale of the experiment, the representative dispersivity value should be greater than 5m. Concentration of non-reactive chloride (Cl^-) component which can be a useful tracer for identifying the extent and direction of pollution was measured by Choi(1992) and is shown in Table 2.

Table 1. The Hydrogeological Properties of the Landfill

Description	Soil	Measurement type	Measured value	Reference
Hydraulic Conductivity (m/sec)	Surface sand/ Silt layer	slug test	4.0×10^{-5}	Choi(1992)
	Surface sand/ Silt layer	slug test	1.03×10^{-5}	Hans Eng. (1992)
	Deep sand/ Gravel layer	pumping test	5.5×10^{-4}	Hans Eng. (1992)
	Surface sand/ Silt layer	Laboratory test	4.6×10^{-6}	Saegil Eng. (1992)
	Deep sand/ Gravel layer	Laboratory test	2.1×10^{-5}	Saegil Eng. (1992)
Longitudinal Dispersivity (m)		tracer test	> 5m	Hans Eng. (1992)

Table 2. The Concentration of Chloride(Cl^-) in the Landfill (after Choi, 1992)

Location*	1	a	2	3	4	b
Concentration (mg/l)	125	140.2	1406.1	1397.1	1250.6	26.6

* Measurement location is shown in Figure 1.
* Location a, b : surface water ; 1,2,3,4 : monitoring well

According to this measurement, concentration of Cl^- to the Han River (location 2,3,4) was much higher than the one to Sang-Am Dong (location a and 1). The degree of contamination would be much greater at Han River than to the opposite direction.

Validation of Model Parameters

Since the disposal of waste in this site is uncontrolled throughout the operation period, site specific data were scarce and many of them had to be inferred from the available data (e.g. groundwater levels in the landfill and variation of source concentration). For the groundwater level, the previous groundwater level (=15m) measured in 1985 when the maximum waste height is 25m in the landfill is used, and the water levels inside the landfill is predicted (15~25m) based on the assumption that curve of water level increment would have similar shape as that of waste

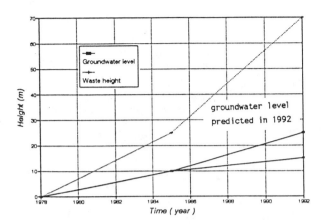

Figure 3. A predicted change of groundwater level based on the change of waste height through the period of waste disposal

The concentration contours computed by the model at the year 1992 is presented in Figure 4. The average concentration computed using source concentration 2400 mg/l and $C/C_0=0.62$ was 1488 mg/l which is quite closer to the concentrations monitored at the well 2 and 3 (see Table 2 and Figure 1).

Prediction of Contaminant Migration

Several analyses were performed to investigate the status of contaminant distribution during the period of 30 years after the landfill closure. The emphasis of the analyses are given to:
(1) comparing the variation of source concentration with the concentration at two detection points A and B at various times. The point A is located near the toe of the landfill and the point B is near the riverside (see Figure 2).
(2) finding out the influence of the uncertainty of groundwater level in the landfill and the magnitude of dispersivity on the concentrations at the location B at the riverside.

The results of the first analysis are shown in Figure 5. The variation of concentration at two points have basically same trend with that of source concentration but the time of peak concentrations is lagged by 10 years for the point A and 17 for point B.

height (Figure 3).
Variation of source concentration was assumed to increase linearly until the waste landfill is closed in 1992 and would degrade to zero during the next 30 years. The degradation curve was assumed to be first order.
Model validation was performed using the data in Table 2 and the computer model, CTRAN/W, which is a finite element analysis program of contaminant transport developed by Geo-slope International Ltd. (1991). Contaminant type used in the analysis is chloride which is non-reactive and has no decaying action. The measured concentration of chlorides in the landfill (source area) in 1992 is about 2400 mg/l. This value was used as the source concentraion.

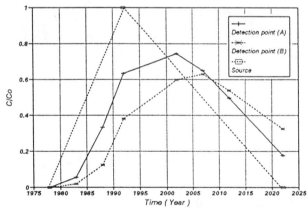

Figure 5. Relative concentration versus time at the toe of the landfill (point A) and at riverside (Point B)

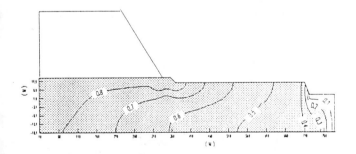

Figure 4. Contours of relative concentration (C/C_0) at 1992 for Nanji Landfill

The influence of the groundwater level and longitudinal dispersivity on the concentration of location A and B at the year 1993 is shown in Figure 6. Concentrations on both points increase as the magnitude of two parameters increase. The rate of concentration increase with respect to the increase of groundwater level, while the rate of concentration increase converges to a certain value as the magnitude of dispersivity increases.

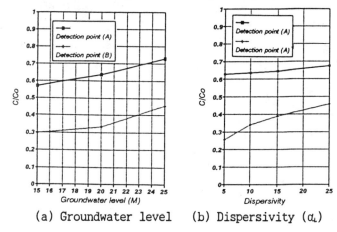

(a) Groundwater level   (b) Dispersivity (α_L)

Figure 6. Effects of hydraulic heads and dispersivity on the relative concentration for Nanji Landfill

Kimpo Metropolitan-Area Landfill

Kimpo Landfill is located approximately 40km west of Seoul and has an area of 20.8km^2 (Figure 7). This site is constructed on a seashore as an alternative site of Nanji Landfill with minimum barrier systems and waste disposal has started at the end of 1992. The planned period of waste disposal is about 25 years.

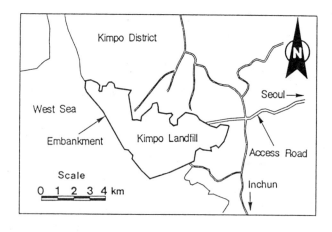

Figure 7. Geographical Location of Kimpo Landfill

The ground underlying the landfill is the marine sediment composed of silts, silty clays and sandy gravel. The layer has average thickness 12m and soil types are varying significantly from site to site. The layers below the marine sediment are weathered residual granite and bed rock.

Barrier system of this site consists of pheriperal embankment, leachate and gas collection pipes, and rain water collection pipes (Figure 8.a). Any special barrier materials were not built at the bottom of the landfill. The embankment has height 5m and HDPE geomembrane sheets is constructed at the upstream side of the embankment to prevent the movement of leachate through the body of the embankment (Figure 8.b).

The leachate collection pipe was made as bundle type and installed at the gradient 0.4~0.6% with the separation distance of 400m (Figure 8.c).

Table 3 shows the physical and mechanical properties of the marine clay subsurface layer. The subsurface layer is generally weak and silt is a dominant material. Hydraulic conductivity of the subsurface layer is not sufficient to satisfy the required K value (i.e. 1×10^{-7} cm/sec).

Table 3. The Physical and Engineering Characteristics of Kimpo Marine Clay

Description	Content	Remark
Fines Content	Silt : 60~80% Clay : 10~20%	
Liquid limit (w_L)	LL=1.59Z+32.49	increases with depth Z (meter)
Plasticity index (I_P)	PI=1.41Z+10.77	
Void ratio (e_0)	e_0=0.057Z+0.88	
Unit weight (KN/m^3)	17.9 arg	
Water content (%)	28~42	below G.W.L.
Undrained Shear Strength (S_u) (KN/m^2)	18~36	UU Triaxial Test
Compression Index (C_c)	0.20~0.24	
Hydraulic Conductivity (cm/sec)	1.80×10^{-5} 3.76×10^{-6}	upper layer lower layer

Prediction of Contaminant Migration

As stated above, the subsurface of this landfill is weak and may have leachate moundings in the landfill due to malfunction of leachate collection facility, which may be caused by the differential settlement of clay layer, by the impact of waste

load at the time of disposal, and by high temperature formed within the waste, etc.. The hydraulic conductivity of the subsurface layer is also heterogeneous between $10^{-5} \sim 10^{-6}$ cm/sec.
Hence, in this analysis, the influence of water mounding on the distribution of concentrations on the section of given geometry (i.e. Part C in Figure 8.a) and the concentration at the two points a and b near the toe of the embankment.
The source of contaminant is assumed to be continuous and is given to the surface of the liner in the landfill. Adsorptions and decaying characteristics are not considered. Longitudinal and transverse dispersivity of the medium are assumed 0.7m and 0.2m.
The distributions of concentrations for various levels of the groundwater level in the landfill after 30 years of waste disposal are shown in Figure 9. It can be recognized that the upper clayey silt layer which is about 5 times more pervious than the lower layer becomes the channel of contaminant migration. Contaminants are spreaded and penetrated into the lower layer by the influence of HDPE at the upstream of the embankment. Penetration of the contaminants near the stream was caused by the impervious geotextiles installed at the side slope of the stream.
Comparison of Figure 9.a, 9.b and 9.c shows the faster movements of contaminants into the lower layer as the level of groundwater increases.
Figure 10 shows the change of concentration at the two points a and b near the stream with respect to time.

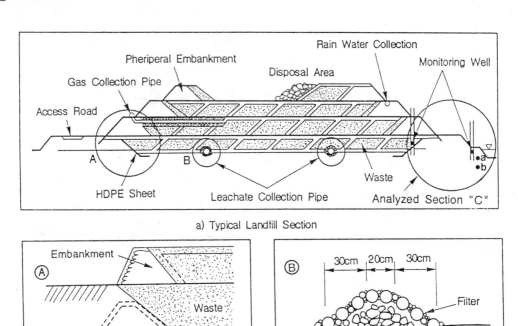

a) Typical Landfill Section

b) Embankment with Geomembraine Sheet (Detail "A")   c) Section of Leachate Collection Pipe (Detail "B")

Figure 8. Typical Section of Kimpo Landfill with Detailed Configuration of Embankment and Leachate Collection pipe

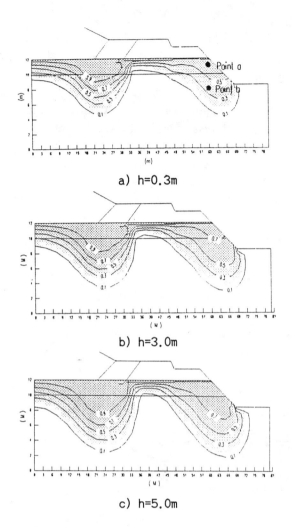

a) h=0.3m

b) h=3.0m

c) h=5.0m

Figure 9. Contours of the relative concentration for different hydraulic heads after 30 years

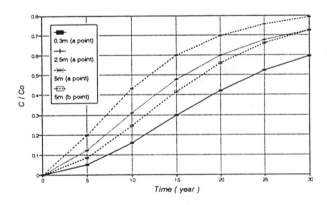

Figure 10. Relative concentration versus time for different hydraulic heads and Location a and b

Concentrations gradually increases from quite early time (e.g. 5 years) and goes up to $0.6 \sim 0.8 C_0$ after 30 years. This means that the subsurface medium is not able to cut off contaminants efficiently and vertical wall type barrier systems are needed at the bottom of the embankment to sufficient depth into the lower clay layer.

Conclusions

Following conclusions are obtained from the analyses of contaminant transport in two landfills.

For Nanji landfill, the contamination of Han River would reach the maximum 17 years after the closure of the landfill in 1992 and the concentration of chloride near the Han River at this time would be 1488mg/l.

The increase of two parameters, the level of leachate mounding in the landfill and the magnitude of dispersivity, results in the increase of concentrations of contaminant near Han River. However, the rate of concentration increase becomes higher for the rise of the leachate mounding level, while it tends to converge a certain concentration for the increase of the magnitude of dispersivity.

For Kimpo landfill, the upper pervious medium which has hydraulic conductivity, 1.8×10^{-5} cm/sec becomes the path of contaminant migration and the points near downstream of the embankment shows contaminant from early period of waste disposal. Installation of vertical barrier systems at the bottom of the embankment to a sufficient depth into the lower clay layer is recommended.

References

1. Choi, Sei-Young (1992), *"A study of Groundwater Pollution by Landfill Leachate,"* Ph.D. Thesis, Seoul National University.
2. Geo-Slope International Ltd. (1991), Finite Element Contaminant Transport Analysis (CTRAN/W), Ver. 2, Calgary, Alberta, Canada.
3. Hans Engineering Co. (1992), *"The Hydrogeological Investigation of Nanji Landfill,"* Engineering Report.
4. Saegil Engineering (1992), *"The Geotechnical Investigation of Nanji Landfill,"* Engineering Report.

Impact of Geo-Engineering Mapping on Waste Management in Greater Dhaka City, Bangladesh

Mir Fazlul Karim, Mohammed Jamal Haider
Geological Survey of Bangladesh, Segun Bagicha, Dhaka-1000, Bangladesh

Abstract

The expansion of Dhaka city is taking place rapidly. A large number of engineering projects are in progress. Consequently, there is an increase in the generation of waste. The quality, quantity and distribution of municipal and industrial wastes have been studied. The result indicates that the wastes are adversely affecting the environment.

The effect on the environment due to interaction amongst the geological aspects, existing engineering structures and waste disposal has been studied. A geo-engineering model to identify environmental hazards is proposed in this paper. It reveals that a detailed geo-engineering mapping and geotechnical database may help for safe urban planning and waste management.

Introduction

Greater Dhaka city has developed on relatively elevated and north south elongated Quaternary Deposits. It is surrounded by depositional plains. About 70% of the expanding city lies below the highest flood level of 1988. In the last three decades a population influx has caused the city to expand rapidly. Infra-structural development has taken place including an expansion of industrial and residential area. This has led to an increase in waste production. Figure 1 shows the existing distribution of urban settlement, industrial area and other engineering structures.

Wastes and waste disposal

Industrial waste: There are about 250 industrial units within the study area which includes tanneries, textiles, pharmaceutical, steel and iron, electronic and electroplating units etc. These type of industries pose high magnitude of environmental degradation to land quality, surface and ground water, socio-economic, health and aesthetic aspects. Ahmed (1993) estimated a total polluting load and effluent generating from industries of Greater Dhaka City as 49,000 kg/day and 31,000 m^3/day respectively. 50% of this waste is generated in densely populated areas.

The study of Ahmed (1993) and Karim (1992) indicates that most of the industries do not have waste treatment plants. A survey in August, 1993 showed that the water bodies of Tejgaon and Hazaribag industrial area to be highly polluted by thick cover of chemical foam, polythene bags, organic decomposed materials and paper intermixed with water hyacinth. Table 1 shows the typical effluent quality of major industries. It may be noted that this type of untreated effluent has 10 to 100 times more polluting effect than treated effluent (Ahmed 1993). It is estimated that about 5.9×10^5 m³ of untreated effluent is discharged in the river Burhiganga per day.

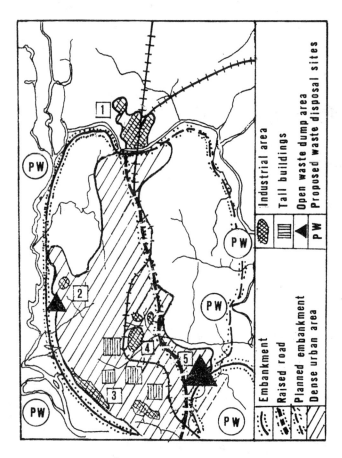

Figure 1. Urban Settlement and Engineering Structures of Greater Dhaka City. Tongi-1, Mirpur-2, Hazaribag-3, Tejgaon-4, Jatrabari-5.

Municipal Waste: Municipal waste generation in Dhaka City is about 0.50 kg/capita/day, which is relatively low. But with increasing population the total amount of waste is becoming high. Kazi (1988) estimated the municipal waste to be 700-1,100 tons/day. This has increased to 2,000-3,000 tons/day (Islam, 1992). The municipal wastes are composed of 33.5% organic matter, 33.8% dust and fine soil, 18.8% vegetables matter, 10.5% stone and brick chips and 6.2% rags and papers. These wastes are collected and disposed off at Jatrabari and Mirpur waste dumps. The disposal sites consists of open dumps and are located near densely populated areas (Fig. 1).

The leachets entering to the system contain 13,000-30,000 mg/l of total solids, 1,300-5,000 mg/l of chloride, 0-200 mg/l of nitrate, 0-15 mg/l of phosphate, 5,000-15,000 mg/l of BOD(Biochemical Oxygen Demand) and 5000-17000 mg/l of COD (Chemical Oxygen Demand) (Islam, 1992). The total BOD load entering the system are about 90,000 tons/day. Beside this about 200 million liter/day waste water is disposed off from the domestic and commercial sources. Only 20% of the city is served by controlled sewerage system the remaining use septic tanks and sanitary pits.

From the above, the dependence of waste removal on the natural drainage system is clear. To fully comprehend the waste removal aspect, it is necessary to understand the geo-morphological and geotechnical characteri-stics of the area.

Geo-engineering mapping

The area is divided into three geomorpho-logical units (Fig. 2).

1. Central high area : This unit is an elongated table-land above flood level. Average eleva-tion is 7 m above mean sea level (AMSL).

2. Complex of high and low areas: This unit consists of narrow strips of benches or foot slopes, rounded to elongated erosional saddles of unit 1, narrow and shallow gullies and valleys. Most of the area lies below normal flood level.

TABLE 1 -- Typical Effluent Quality of Major Polluting Industries (Values in mg/l).

Industries	pH	BOD	Suspended Solid	Toxic Substance
Tanneries	7.5-11.0	450-3000	2000-8000	Chromium 1.5-10.5 Ammonia 110
Textiles & Dying	9.8-11.8	500-1000	100-2000	Chromium 12.5

TABLE 2 -- Geotechnical Land-Use Classification System.

Characteristics of the units	GLU-I	GLU-II	GLU-III
Geotechnical Limitations	Low	Moderate	High
Suitability for Development	High	Moderate	Low
Engineering cost for Development	Normal	High	Very High
Intensity of Site Investigation Required	Normal	Intensive	Very Intensive
Geological Constraints	Swelling Soil	Slope Failures, Flood	Flood, Soft Clay, Silt, Liquefaction
Typical Geomorphological Characteristics	Elevated Land	Dissected, Undulated	Low Flood-plain
Site Suitability for Waste Disposal	Not Suitable	Small Scale with Synthetic Lining	Suitable by Trenching/ Protective Wall
Susceptibility to Earthquake	Low	Moderate	High

3. Complex of low area : It consists of the surrounding floodplain which is annually inundated. The unit is flat and the elevation is less than 2 m AMSL.

(GLU) are shown in Figure 3 and the Table 2 shows the geotechnical characteristics of the GLU classes.

Figure 2. Engineering Geomorphology of Greater Dhaka City (modified from Karim et al., 1993).

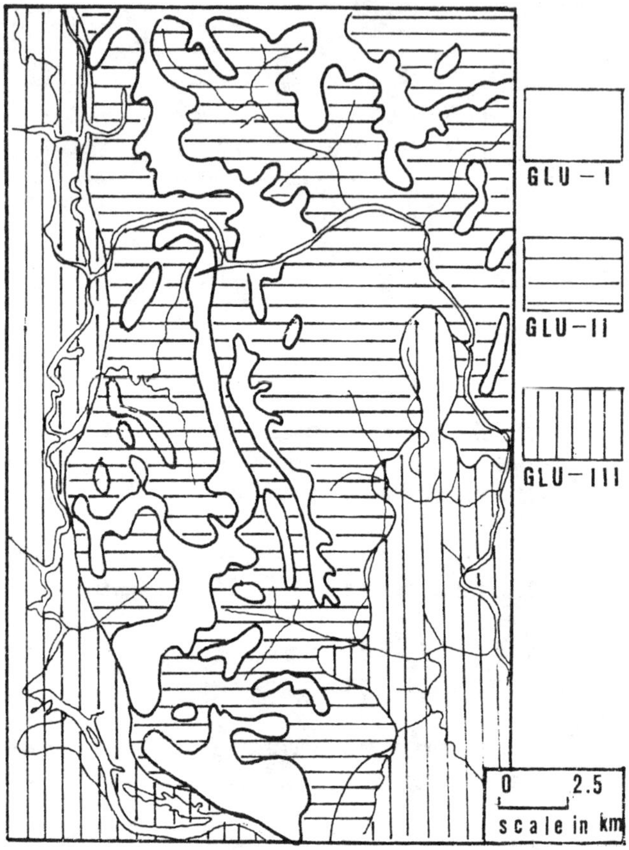

Figure 3. Geotechnical Land-Use Map of Greater Dhaka City.

The drainage pattern of the area is dendritic to trellis. Some streams are shallow while others are incised. The drainage system can be divided into two major water sheds flanked by the central high area.

Based on the geomorpological characteristics, a Geotechnical Land-Use Map has been prepared. The identified Geotechnical Land-Use Units

Urban and geological environment

The GLU-I and part of GLU-II are the source areas of polluted industrial and municipal wastes. The untreated wastes and effluent are either dumped or disposed off through the streams and storm system of GLU-II and III.

Previously, GLU-I and II were drained by

innumerable intricate streams. But as the city expanded, most of these streams have been filled up or blocked for infrastructural development. This ill planned development of engineering structures has increased polluted water logging to an alarming degree. On the other hand, an embankment (Fig. 1) has been built on GLU-III to protect the city from floods. But as there is no adequate water discharge facility, the embankment traps the storm water and industrial effluent. Internal floods caused by rain distribute the wastes in the area, particularly to the populated unit of GLU-II. At places, the stagnant polluted water in depressions comes in direct contact with ground water.

The present field observations and previous studies of Karim (1992), Islam (1992) and Khondaker and Chowdhury (1992) show that at places there are significant settlement of the embankment due to the presence of highly compressible organic clay-silt in the subsurface. There are also slope failures, cracking, erosion and siltation at the regulator sites. At some places, the local people have cut across the embankment to let out the logged polluted water.

The master plan of Greater Dhaka City involves numerous municipal development projects in the GLU-III area. These include storm water drainage system, Greater Dhaka flood protection embankment and walls, cross-roads, bridges, culverts, regulators and buildings. It is evident that the GLU characteristics (Table 2) and maps (Fig. 2 & 3) have to be taken into consideration or else this project is liable to suffer from hazards. Like the other structures, the storm water drainage system will also be susceptible.

Geo-engineering model

A three-dimensional model has been prepared to understand the surface and subsurface geological conditions. The proposed area for waste dumps can be later used for Garbage Gas Generation (Fig. 4). The model also shows that the dumps on the east of the area will provide minimum hazard as these will be on thick clayey soil. Only flood protection cell will be adequate. But dumps

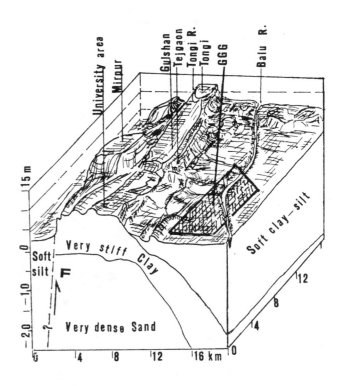

Figure 4. Geo-engineering Model for Dhaka City. (F-Fault, GGG-Proposed location for Garbage Gas Generation).

on the western side will lie on relatively permeable silty sediments. The aquifer (dense sand bed) has vertical contact with the silt bed. The ground water will have to be protected from seepage of leachets by synthetic lining.

Generally, geotechnical data are not used for waste management. For overall waste management, a geotechnical database is necessary (Fig. 5). The terrain- and waste-data are incorporated in the geotechnical database.

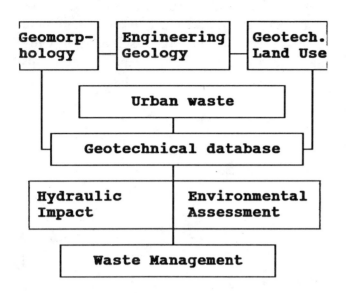

Figure 5. Schematic Model for Urban Waste Management.

This database then can be used to simulate hydraulic and environ-mental sub-models which will lead to proper planning and management.

Conclusion

Development of an effective waste management system depends highly on the geotechnical characteristics of the area. The following points are important for the Greater Dhaka area.

Geotechnical database should be used to deve-lop waste management system.

Infrastructures should be planned so as not to disrupt the natural drainage system.

Dumping ground should be in GLU-III area, away from thickly populated areas for land cost and environmental reasons.

References

Ahmed, M.F. (1993); Sector Review: Industry; Environmental Management Training Project, (EKB-07); Training of Trainers Course Manual, Vol.I, Version I, Department of Environment, Bangladesh. pp.1-18.

Islam, A.K.M. (1992); Impact of Greater Dhaka City Flood protection on the Environment With Specific Reference to Domestic Waste Disposal. M.Sc. Thesis, Dept. of Civil Engg., Bangladesh University of Engineering and Technology (BUET), Dhaka. pp.1-169.

Karim, M.A. (1992); Impact of Greater Dhaka City Flood protection on the Environment With Specific Reference to Industrial Waste Disposal. M.Sc. Thesis, Dept. of Civil Engg., BUET Dhaka. pp.1-150.

Karim, M.F., Haider, M.J., Chowdhury, M.A. and Kabir, S. (1993); Engineering Geomorphology for Ground Improvement in Part of Dhaka City and Tongi Area. First Bangladesh-Japan Joint Seminar on Ground Improvement, organized by Bangladesh Society for Geotechnical Engineering, Dhaka, Bangladesh. pp.185-194.

Kazi, N.M. (1988); Characterization, Treatment and Disposal of Sewage of Dhaka City. M.Sc. Thesis, Dept. of Civil Engg., BUET Dhaka. pp.1-154.

Khondaker, M.A.M. and Chowdhury, J.U. (1992); Effect of Dhaka City Flood Protection Embankment upon Surface Water Drainage Systems and Environment. IFCDR, Bangladesh University of Engineering and Technology, Dhaka. pp-74.

Non-Intrusive Rayleigh Wave Investigations at Solid Waste Landfills

Edward Kavazanjian, Jr., Michael S. Snow, Neven Matasovic
GeoSyntec Consultants, Huntington Beach, California, USA

Chaim J. Poran
GEI Consultants, Inc., Raleigh, North Carolina, USA

Takenori Satoh
Vibration Instruments Company, Ltd., Tokyo, Japan

Abstract

Non-intrusive Rayleigh wave investigation using the Spectral Analysis of Surface Waves (SASW) technique is a powerful tool for characterization of solid waste disposal sites. Applications of SASW surveys for site characterization at solid waste landfills include development of shear wave velocity profiles for the waste mass, evaluation of cover thickness, location (depth) of the waste/native ground interface, and identication of the presence of underground obstructions. Measurements at 10 solid waste landfill facilities in southern California have clearly demonstrated the potential applications of SASW.

Introduction

Spectral Analysis of Surface Waves (SASW) is a non-intrusive geophysical technique used primarily for evaluating subsurface shear wave velocity profiles. SASW is a particularly attractive method of investigation for landfill engineering. The non-intrusive nature of SASW eliminates many of the health and safety concerns typically associated with conventional borings for geoenvironmental investigations. SASW results are representative of the average properties of a relatively large mass of material, mitigating the potential for misleading results due to the non-homogeneity of many solid waste deposits. Furthermore, SASW can be a very cost-effective method of investigation. The ease and rapidity of field measurements and automated algorithms for data processing and inversion allow for evaluation of subsurface conditions at a relatively large number of points at a fraction of the cost of conventional intrusive exploration techniques.

A schematic representation of SASW testing is presented in Figure 1.

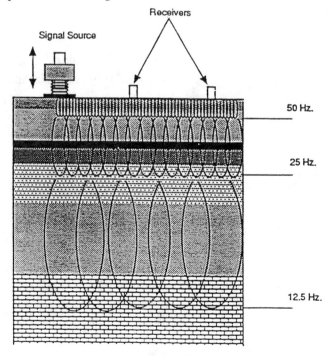

Figure 1. SASW Test Set Up

Excitation at the ground surface is used to generate surface, or Rayleigh, waves of various frequency. By spectral analysis of the ground surface response (velocity or acceleration) at two points a known distance apart, the Rayleigh wave velocity can be obtained at discrete frequencies. Usually, an inversion process is used to determine the velocity profile. At sites where wave velocity increases gradually with depth, the velocity profile may be determined directly from the field data. By assuming a value for Poisson's ratio, Rayleigh wave velocity may be relateded to average shear wave velocity. The depth over which reliable measurements can be made depends upon the energy and frequency content of the source excitation and the consistency of the subgrade material. Measurements are not affected by the depth to the water table.

The concept of measuring the velocity of Rayleigh waves of different frequencies to determine the profile of shear wave velocity with depth was first proposed by Jones [1], in Great Britain, for pavement surveys and by Ballard [2], at the Waterways Experiment Station in Vicksburg, Mississippi, for geotechnical analyses. These investigators used impact loading as the source excitation and developed an analysis based upon the assumption of a uniform, homogenous layer. Stokoe and Nazarian [3], at the University of Texas, Austin, extended the spectral analysis to consider multi-layered media. These investigators also used a surface impact as the source excitation, and thus reliable measurements were typically limited to maximum depths on the order of 10 meters by the relatively low energy content of the excitation at greater wave lengths.

Satoh and his co-workers [4,5], in Japan, developed an electro-magnetic controlled vibrator for use as the source excitation. Using a large (2000 kg) mass, Controlled Source Spectral Analysis of Surface Waves (CSSASW) equipment capable of penetrating over 100 meters below ground surface has recently been developed.

Southern California Landfill Investigation

Using a grant from the National Science Foundation's Small Grants for Exploratory Research (SGER) program, supplemented by internal research funds from GeoSyntec Consultants, CSSASW testing was performed at eight municipal solid waste landfills and 2 industrial solid waste landfills in southern California. Testing was performed using two vibrators: a small-mass (42 kg) source, capable of penetrating to depths of 10 to 15 meters, and an intermediate-mass (350 kg) source, capable of penetrating to depths of 40 to 50 meters.

Test sites encompassed the range of climatic conditions representative of southern California, from coastal sites in Ventura and Orange Counties to desert sites in San Bernardino and Riverside Counties. The equipment required for CSSAW testing, shown in Figure 2, was loaded on a flat bed truck mounted with a small crane to handle the 350-kg source. A total of 43 measurements were made at 26 locations (two or three locations per landfill) by a two man crew in a three and one-half day period at sites up to 160 km apart, demonstrating the speed and simplicity of field measurements using the CSSASW system.

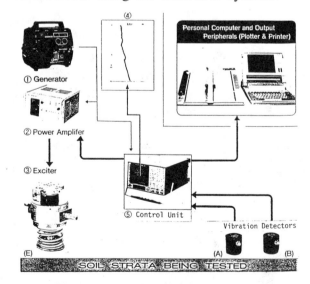

Figure 2. CSSAW Equipment

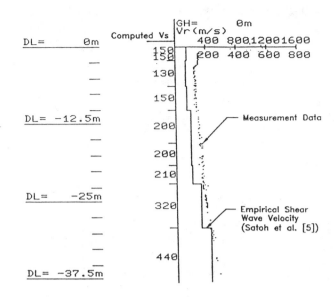

Figure 5. 350-kg Mass Results, Coastal Site

Shear Wave Velocity of Municipal Solid Waste

Shear wave velocity measurements were obtained at eight southern California municipal solid waste landfills. Results from the Coastal site, presented in Figures 3 through 5, were typical. Shear wave velocities at the Coastal site appeared to increase relatively linearly from about 130 m/s near the surface to 200 m/s at 20 meters (Natural ground is at about 22 meters depth). Waste was last placed at the Coastal site in 1989. At the Bailard site, another unit of the same Ventura County landfill, waste was placed through 1991, at least two years later than at the Coastal site. The near surface wave velocities at Bailard were approximately 20 to 30 m/s less than at Coastal.

At the City of Los Angeles Lopez Canyon landfill, over 50 meters of waste has been placed since 1989 in Area AB+. Shear wave velocities in this "young" waste deposit were the lowest of any municipal waste site, varying from 90 m/s near the surface to 150 m/s at a depth of 20 meters below the landfill deck. In Area B at Lopez Canyon, the last lift of waste was placed in 1991 and most waste was placed much earlier than that. The shear wave velocities of the "old" waste in Area B were almost double the shear wave velocities in Area AB+, varying from 170 m/s near the surface to 290 m/s at a depth of 20 meters and 335 m/s at a depth of 30 meters.

At Los Angeles Site Number 1, a commercial municipal solid waste facility, some of the highest wave velocities were recorded. A shear wave velocity of 550 m/s was calculated at a depth of 30 meters. This site has accepted a significant amount of inert waste in the past, including construction debris, and puts a lot of effort into waste compaction to maximize capacity.

The Orange County Olinda-Olinda Alpha Landfill is another site where the velocity of young and old waste could be compared. The Olinda waste unit has been relatively inactive for several years, being used primarily for wet weather operations. The deck at Olinda has been used to stockpile fill and is well compacted by vehicular traffic. Most waste received at the site in the past few years has been placed in the Olinda Alpha unit. Shear wave velocities obtained at Olinda Alpha varied from 100 m/s near the surface to 180 m/s at a depth of 20 meters. Shear wave velocities in the older waste of the Olinda unit were typically 50 percent higher than in the younger waste at the Olinda Alpha unit.

Shear wave velocities obtained at the Gothard landfill, a closed Orange County landfill in a coastal environment, at the East and West units of the Cajon landfill in San Bernardino County, and at Los Angeles Site Number 2 provided a set of measurements relatively consistent with those discussed above. The only major discrepancy in the trend of the measurements was at the Highgrove Landfill, a desert site in Riverside County, where shear wave velocities in waste over 10 years old were at the lower end of the range of measured values for the municipal solid

The primary objective of the landfill investigation was to collect information on the shear wave velocity of municipal solid waste for use in seismic response analysis of landfills. Recently promulgated federal regulations on design of municipal solid waste landfill facilities have made seismic design an important consideration. The lack of information on dynamic modulus and damping of municipal solid waste introduces significant uncertainties into the design process. By assuming a value for the density of the waste mass, small-strain dynamic moduli for use in seismic response analyses can be computed from shear wave velocity.

Test sites were chosen to provide shear wave velocity information for both young and old waste at municipal and commercial facilities in wet and dry climates. Other facets of the landfill investigation included evaluation of the ability to establish the depth to native soil and to identify the location of obstructions. The investigation also demonstrated the ability of CSSASW testing to evaluate the thickness of the landfill soil cover.

Measurements using the 42-kg source employed a pseudo-random signal with a pre-determined frequency content designed to maximize the resolution of the test [5]. Results from the 42-kg source test at the Coastal Landfill in Ventura, typical of results at most sites, are shown in Figure 3 as a plot of phase velocity versus frequency. Figure 3 is sometimes referred to as the dispersion curve. The data in Figure 3 is plotted as "x"s in Figure 4, a plot of phase velocity versus representative depth (assumed equal to half the wave length). Shear wave velocity profiles and their respective Rayleigh wave modes were computed using the inversion process described by Poran et al. [6]. The first three computed Rayleigh wave modes are shown in Figures 3 and 4. Measurements using the 350-kg source were made by sweeping through a range of frequencies with a pure sinusoidal excitation. The difference in arrival time at the two transducers was used to calculate Rayleigh wave velocity directly and generate a plot of phase velocity versus depth in real time.

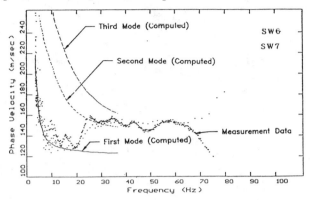

Figure 3. Dispersion Curve, Coastal Site

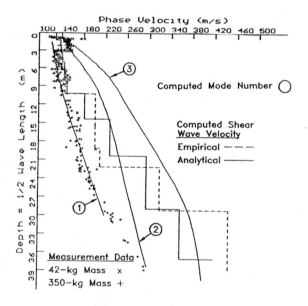

Figure 4. Processed Data, Coastal Site

Results from the 350-kg source, shown as "+"s in Figure 4, were consistent with the results of the 42-kg source. However, the 350-kg source penetrated to much greater depths than the 42-kg source. The layered profile for the 350-kg mass in Figure 5 was computed using an empirical algorithm developed by Arbeleaz [7]. The empirically computed profile in Figure 4 was based upon the layered profile in Figure 5. The analytical profile in Figure 4 was developed using the method of Poran et al. [6].

waste sites. Shear wave velocities obtained at Highgrove varied from 95 m/s near the surface to 165 m/s at a depth of 20 meters.

Cover Thickness and Waste-Ground Interface

The cover soil layer at the municipal solid waste landfill sites was clearly discernable as a velocity contrast in both the raw and processed data. In the raw data, a change in slope of the data is indicative of a velocity contrast. Accurate resolution of cover thickness to fractions of a meter requires sampling at high frequencies and expansion of the scale of data plot.

The waste-natural ground transition was also usually discernable as a velocity contrast. At several sites, the waste-natural ground interface was discernable from both the velocity contrast and the uniformity of the data. Data points within the waste deposit tended to be scattered, while the data within the natural ground was relatively uniform. The Coastal site data in Figure 5 and the data in Figure 6, results from the 42-kg source at a shallow industrial waste pit at Point Magu in Ventura County, illustrate this point.

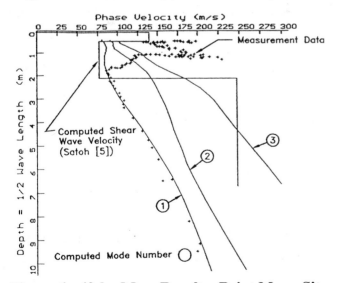

Figure 6. 42-kg Mass Results, Point Magu Site

Identification of Subsurface Obstructions

CSSASW testing was performed at one location where attempts to install a ground water observation well were aborted due to the presence of a subsurface obstruction. While neither the nature nor, from a single test, the size of the obstruction could be identified from the CSSASW test, the erratic signal clearly indicated the presence of an irregular object in the subsurface. One potential application of the 42-kg mass device could be to establish locations for borings or wells "clear" of near surface obstructions.

Summary and Conclusions

A 350-kg source CSSASW system was used to obtain shear wave velocity at depths of up to 40 meters at eight municipal solid waste landfills and two industrial waste landfills in southern California. Sites were selected to provide a range of different "age" profiles and to encompass different types of waste, from municipal solid waste to inert industrial waste and construction debris.

Shear wave velocity profiles developed at eight municipal solid waste sites demonstrated a marked difference between young and old waste. Shear wave velocities obtained near the surface in young waste were as low as 90 m/s. In older waste units, the shear wave velocity near the ground surface was sometimes as high as 170 m/s. Twenty meters below the surface, shear wave velocities in young waste were typically on the order of 140 to 170 m/s, while in older waste they were as high as 290 to 350 m/s. Shear wave velocities as high as 550 m/s were obtained at one site at a depth of 20 meters. In all cases, shear wave velocity in municipal solid waste appeared to increase systematically with depth.

The shear wave velocity profiles developed using the CSSAW technique in this study are invaluable to engineers evaluating the seismic response of solid waste landfills, a topic of greatly increased significance with the promulgation of new federal minimum design standards for municipal solid waste landfill facilities (RCRA Subtitle D).

Results of the southern California landfill study also clearly demonstrated the ability of CSSASW to determine cover thickness and, within the depth limitation of the particular vibrator in use, the depth of the waste-native ground interface. At one location, the CSSASW technique was successful in verifying the presence of a subsurface obstruction encountered during installation of a ground water monitoring well.

CSSASW is an ideal technique for obtaining shear wave velocity profiles at solid waste landfills. This technique eliminates many of the technical and health and safety issues associated with the borings required for conventional intrusive cross-hole and down-hole measurements. Furthermore, CSSASW facilitates rapid evaluation of the subsurface conditions at a relatively large number of points at a fraction of the cost of conventional intrusive exploration techniques.

Acknowledgement

The work described in this paper was funded by the National Science Foundation, Biological and Critical Systems Division, Earthquake Hazard Mitigation Program under the Small Grants For Exploratory Research Program, Grant No. BCS-9312744. The authors are grateful to the Foundation and to Program Director Dr. Clifford J. Astill for their support. The authors are also indebted to Dr. J.P. Giroud of GeoSyntec Consultants for his invaluable review comments on the manuscript.

References

1. Jones, R. (1962) "Surface Wave Technique for Measuring Elastic Properties and Thickness of Roads: Theoretical Development," *British Journal of Applied Physics*, Vol. 13, pp. 21-29

2. Ballard, R.F., Jr. (1964) "Determination of Soil Shear Moduli at Depth by In Situ Vibratory Techniques," Miscellaneous Paper No. 4-691, U.S. Army Waterways Experiment Station, Vicksburg, MS, USA

3. Stokoe, K.H., II, and S. Nazarian (1985) "Use of Rayleigh Waves in Liquefaction Studies," *Proceedings of a Session on Measurement and Use of Shear Wave Velocity for Evaluating Soil Dynamic Properties*, ASCE National Convention, Denver, CO, USA, May, pp. 1-17

4. Satoh, T., C.J. Poran, K. Yamagata, and J.A. Rodriguez (1991) "Soil Profiling by Spectral Analysis of Surface Waves," *Proceedings of the Second International Conference on Recent Advances in Geotechnical Earthquake Engineering and Soil Dynamics*, St. Louis, MO, USA, March, Vol. II, pp. 1429-1434

5. Satoh, T. (1989) "On the Controlled Source Spectral Rayleigh Wave Excitation and Measurement System," VIC Ltd., Tokyo, Japan

6. Poran, C.J., J.A. Rodriguez, and T. Satoh (1993) "Controlled-Source Surface Wave Measurements for Shear Wave Velocity Profiling of Soils," *Proceedings of a National Science Foundation Workshop on Geophysical Techniques for Site and Material Characterization*, Georgia Institute of Technology, Atlanta, GA, USA, June, pp. 231-245

7. Arbeleaz, M.C. (1992) "Controlled Source Spectral Analysis of Surface Waves for Soil Profiling," MS Thesis, University of North Carolina, Charlotte, NC, USA

Design and Construction of Soil-Bentonite Liners and Two Case Histories

Paul Kozicki, P.Eng., Steven Harty, E.I.T., John P. Kozicki, E.I.T.
GE Ground Engineering Ltd., Regina, Saskatchewan, Canada

Abstract

Low permeability soil liners can be constructed from natural clays, well graded soils with only a small clay content, or sandy soils blended with a small quantity of imported clay such as bentonite. The permeability of such materials properly placed with good quality control, generally lies in the range of 10^{-9} to 10^{-11} m/s, and can be designed to meet minimum prescriptive standards set by regulatory agencies. Important benefits resulting from the use of soil-bentonite liners include, the slow rate of penetration of a saturated wetting front, the decrease in permeability of the liner as it consolidates under the applied load of effluent and attenuation of potential contaminants in the waste waters resulting in clean fluid emanating from the soil, ie. dirty water in/clean water out. Synthetic liners can be attributed a finite permeability as a result of minute perforations and seaming deficiencies and possess none of the above properties. Synthetic liners are also, by their very nature, relatively thin and even a small head build-up above the liner will result in a large hydraulic gradient across it.

Introduction

Facilities to store special wastes can require the construction of some form of liner for environmental protection. These liners are designed to isolate the special wastes from the surrounding environment. In many instances, designers elect to use synthetic membranes to construct the liner because it is a quick and simple solution. However, in many locations natural materials exist which can be readily used to construct relatively economical, efficient and effective liners. The range of materials suitable for soil liners varies widely and a significant clay content is not necessarily required. A deficiency in silt and clay sized particles can often be overcome with the addition of small quantities of bentonite.

This paper presents procedures for calculation of the bentonite application rate and moulding moisture content on the basis of standard laboratory test procedures and basic soil mechanics. The principal design objectives are to achieve a system whereby seepage discharge to the environment will be reduced to negligible proportion during operation of the facility. Also, case histories are presented for two projects:

.1 An underseal for a Special Waste Storage Facility and Expansion constructed at Key Lake, Saskatchewan, 1988 and 1992.

.2 An underseal for a Settling Pond constructed at McArthur River, Saskatchewan, 1993.

SPREADING BENTONITE WITH SPREADER TRUCK

Soil-Bentonite Liner Technology

Sodium bentonite is a natural clay mineral with the same chemical composition as other clays, but with a unique molecular structure that permits it to absorb many times its weight of water. In doing so, bentonite swells enormously - up to 15 times its dry weight when fully wetted.

PULVI-MIXER EQUIPMENT

Bentonite's affinity for water and its swelling capability are its strengths in stopping the passage of water (Figure 1). Under confined conditions, as in soil-bentonite liners, the wetted, swelled bentonite particles will be forced against each other and into voids between soil

Figure 1

PUG MILL MIXING EQUIPMENT

particles to form a barrier against further passage of water and fluids. If kept in a moist state, bentonite never sets or hardens, and maintains its "water bound" condition forever.

Bentonite works best for most horizontal liner applications when applied uniformly and in relatively small quantities (3 to 8 percent by weight of dry soil). Thorough mixing is essential to ensure a uniform distribution of bentonite through the soil. Mixing can take place in situ using pulvi-type mixing equipment or in a stockpile using pug mill type equipment. The water conditioned soil bentonite mixture is compacted to form a durable low permeability liner.

Suggested Method for Determining the Bentonite Application Rate and Moulding Moisture Content (1)

The bentonite application rate should be sufficient to coat all soil particles with bentonite and fill the pore spaces between the soil particles at <u>maximum density</u>. The required bentonite addition rate is strongly dependent on the density and grain size distribution of the material being modified. It is recommended that the material being modified have a minimum silt and clay content of 10%. At the correct moulding moisture content, the soil-bentonite mixture can be easily compacted with about 4 to 5 passes using a 10 tonne vibratory steel drum roller. Also, the compacted soil-bentonite liner has very high shear strength and is not easily damaged by construction equipment.

On the basis of standard laboratory test procedures and basic soil mechanics formula, specific procedures were developed to calculate a bentonite application rate and moulding moisture content. The procedure was developed from initial test work undertaken for constructing the underseal for the Special Waste Storage Facility at the Key Lake Uranium Project in Northern Saskatchewan. The procedures developed were as follows:

.1 Use basic soil mechanics formulae to determine the porosity (n) of the base material at the maximum dry density (γ_d max) determined by ASTM D-1577. (Modified Proctor Test).

.2 Calculate the weight of the bentonite required to fill the voids using Eq. 1, assuming an empirical swell factor (S) of 4.0 and a dry weight of bentonite of 1.0 t/m^3:

$$\text{wt of bentonite} = \frac{n}{S} \ (t/m^3) \quad (1)$$

.3 Calculate the bentonite application rate for the designed liner thickness (t) in mm, using Eq. 2.

$$\text{application rate} = \frac{t \ n}{S} \ (kg/m^2) \quad (2)$$

.4 Use Eq. 3 to calculate the moulding moisture content on the basis of that required to achieve 90% saturation at the maximum density determined by ASTM D-1557, such that:

$$\text{moulding moisture content} = \frac{90 \ n}{\gamma_d \ max} \ (\%) \quad (3)$$

---

(1) Construction of Underseals for the Key Lake Project, Sk. by J.P. Haile, P. Eng., Knight & Piesold Ltd. paper presented at the First Canadian Engineering Technology Seminar on the Use of Bentonite for Civil Engineering Applications, March 1985, Regina, Saskatchewan.

After performing the above calculations to determine the amount of bentonite that is required to fill the voids, a moisture density test should be performed on the soil-bentonite mixture to determine the moulding moisture content for compaction. Depending on the gradation of the insitu soils, the calculated moulding moisture content may have to be adjusted. In the above formula, the calculation for the moulding moisture content is on the basis of that required to achieve 90% saturation. This amount of moisture may present construction problems insofar as compacting the soil-bentonite mixture to the maximum density. It is not necessary for the soil-bentonite mixture to be compacted at 90% saturation since the bentonite will take on additional moisture as the wetting front penetrates the liner. The bentonite, which is an active type clay mineral, will gradually expand, or swell, as the wetting front advances through the liner, decreasing the hydraulic conductivity of the liner further. The amount of swelling or expansion is dependent on the water quality of the permeant and the quality of the bentonite. The quality of the sodium bentonite material is generally dictated by the smectite mineral content, however, other parameters usually specified are Plasticity Index, Modified Free Swell Index, Specific Surface, Cation Exchange Capacity and the Ratio of Na:Ca:Mg.

Construction of effective soil-bentonite liners requires thorough mixing of the bentonite with native soil, correct moisture conditioning and compaction. A poor understanding of these procedures has contributed a large extent to previous failures. Due consideration must be given to controlling such factors as desiccation, erosion, osmotic consolidation and subgrade preparation.

In controlling seepage from lined storage facilities, soil-bentonite liners provide some distinct advantages. Firstly, although initial wetting under unsaturated conditions may be fairly rapid, it takes a significant period of time for the saturated wetting front of permeant to travel through the liner. Secondly, the permeability of the soil will generally decrease with increased loading or confining pressure, further reducing the rate of infiltration. Thirdly, there may be a significant reduction in the concentration of waste constituents in the leachate as a result of dispersion, diffusion and absorption within the soil matrix.

Deilman Special Waste Pad and Expansion, Key Lake, Sk.

The Deilman Special Waste Pad was constructed in 1988 and covered an area of 12 526 square metres. In 1992, the existing pad was expanded to cover an additional 42 000 square metres. The native overburden materials in the area consist of extensive deposits of glacial till and outwash sands which have coefficients of permeability in the order of 3×10^{-6} and 1×10^{-4} m/s, respectively. Locally available till material was screened over a 100 mm mesh and imported into the waste pad area for construction of the liner. The completed underseal thickness was designed to be 200 mm with a maximum hydraulic conductivity of 1×10^{-9} m/s.

The complete storage area was covered with 200 to 300 mm of imported, screened till material. The screened till material had a silt and clay content of 15%. The bentonite was then applied at the designed application rate by means of a spreader truck. The application rate was measured by placing a 500 mm square pan in the path of the spreader truck and then weighing the bentonite in the pan after the truck had passed over. The bentonite and till materials were then mixed to a depth of 200 mm using a Bomag Model MPH 100 Pulvimixer. A second pass was then made with the mixer, however, this time water was added through the mixer to bring the moisture content of the mixture up to the desired level. The moisture conditioned mixture was then compacted using a vibratory steel drum roller.

Screened till material, sufficient to construct a 200 mm thick underseal on the embankment slopes, was imported and placed on top of the finished floor liner. The screened till material was then modified with bentonite, moisture conditioned, pushed onto the slope and compacted. Once the entire underseal was complete, a piping collection and drainage system comprised of perforated piping was installed on top of the underseal. Surface infiltration is intercepted and conveyed to a sump, through the piping system, where it is treated and disposed of. Upon completion of the underseal and piping system, a one metre thick drainage blanket of outwash sand was placed on top for protection.

A bentonite application rate of 20 kg/m^2 was used to construct the 200 mm thick underseal for all floor areas and the expansion area embankments. The application rate was reduced to 16 kg/m^2 for the underseal on the 1988 embankments. A field moisture content of 10.0% (optimum) $\pm$ 1.5% was specified for the 1988 underseal. Generally, a moisture content at or above the optimum value results in a material that is spongy and difficult to compact. The resulting underseal is soft and weak, and could be easily damaged by construction equipment.

Moisture density tests were carried out after the 1988 underseal project was completed. A total of 13 test results were obtained from the soil-bentonite liner constructed on the floor of the special waste pad. The moisture content ranged from 8.3% to 11.3%, with an average value of 9.8%. The average relative compaction ranged from 97% to 103%, with an average value of 100.7% of Standard Proctor Density. Hydraulic conductivity values of 8.2×10^{-11} m/s were reached after 67,000 minutes (46 days). All permeability tests were performed at the University of Saskatchewan, Civil Engineering Soils Laboratory, using triaxial type permeameters.

A total of 16 moisture density tests were performed on the underseals for the 1992 expansion of the special waste pad. The moisture content ranged from 7.8% to 12.6%, with an average value of 10.0%. The average relative compaction ranged from 95% to 102%, with an average value of 99.5% of the Standard Proctor Density. Hydraulic conductivity values of 2.0×10^{-10} m/s were reached after 67,500 minutes (47 days). The permeability tests were performed in the laboratory using a solid wall type permeameter.

X-Ray diffraction analyses were performed on two (2) separate samples of the bentonite material used for the expansion underseal. The smectite mineral content, determined by a semiquantitative analyses of the constituent minerals, ranged from 70% to 82%.

Settling Pond, McArthur River, Sk.

The settling pond was constructed in 1993 and covered an area of approximately 3 100 square metres. The completed liner thickness was designed to be 250 mm with a maximum hydraulic conductivity of 1.0×10^{-9} m/s.

The predominant native overburden materials in the McArthur River area consist of a clean, uniform sand material. Initially, it was proposed to construct the underseal using the native sand material, however, laboratory grain size analysis tests indicated that the native sand material had only about 1.5% silt and clay size particles. The native sand was therefore considered unsuitable material to use for constructing a soil-bentonite liner.

Locally available till material had to be located and imported to construct the liner. It was screened over a 100 mm mesh to remove oversize rock. The screened material had a silt and clay content of 21%. Gradation analyses indicated that mixing equal portions of the uniform native sand with imported screened till material would produce a fairly well graded material. However, when this was attempted in the field, the resulting mixture was not homogeneous and contained pockets of clean sand which was considered unacceptable. Since the bentonite was already blended with the sand/till mixture, it was decided to leave this material in the bottom and use it as the working floor for mixing the till bentonite material required for the embankment slopes.

Screened till material, sufficient to construct a 250 mm thick liner, was imported and spread over the floor area. Dry bentonite was spread at the designed application rate by laying 40 kg bags at precalculated spacing. The bags were cut open and the bentonite dumped out and spread evenly over the application area. The area was then mixed to a depth of 250 mm using a Bomag Model MPH 100 Pulvimixer. It was then mixed a second time with water being added through the mixer to bring the moisture content of the mixture to the desired level.

This method of adding water through the mixer is very effective and provides a very uniform moisture content. Upon achieving the desired moulding moisture content, the material was pushed onto the embankment slopes, spread and compacted using a vibratory steel drum roller. Once all the embankment slope underseals were completed, the floor area was reconstructed by importing sufficient screened till material to construct a 250 mm thick liner. Bentonite was spread and mixed with the screened till material and the mixture was moisture conditioned and compacted. The completed underseal was finally covered with a 150 mm thick layer of sand for protection.

Bentonite application rates of 25 kg/m^2 for the floor and 23 kg/m^2 for the slopes were used to construct the 250 mm thick underseal. A total of 16 moisture density test results were obtained from the completed soil-bentonite liner. The moisture content ranged from 5.5% to 8.0%, with an average value of 6.6%. The optimum moisture content obtained from the Standard Proctor Test was 9.6%. The average compacted density ranged from 98.0% to 101.1%, with an average value of 99.3% of maximum Standard Proctor density.

Hydraulic conductivity values for the underseal material sampled from the floor and slope areas, reached 1.43×10^{-10} m/s and 1.79×10^{-10} m/s respectively, after approximately 40 000 minutes (28 days).

A semiquantitative analysis of the constituent minerals was performed on a sample of the bentonite material used and a smectite mineral content of 70% was obtained.

Summary

Soil-bentonite liners constructed at Key Lake and McArthur River in 1988, 1992 and 1993 were located in areas where no natural materials existed that could be used to construct low permeability liners.

Locally available till material was imported and modified with bentonite. The resulting mixture was moisture conditioned and compacted to construct a low permeability underseal. The resulting hydraulic conductivity values of the completed underseals meet all required specifications. Also, the completed underseals exhibit very high shear strengths and cannot be easily damaged by construction equipment.

Geotechnical Characteristics of Refuse Ash Dumped in Phoenix Project in the Amagasaki Offshore Site

M. Miyake, M. Wada, A. Maruyama
Technical Research Institute, Toyo Construction Co. Ltd., Japan

H. Iwatani
Service Center for Waterfront Environment in Osaka Bay, Japan

Abstract This paper describes the following contents: 1) a construction outline, including the location of an ash disposal pond, the soil profile of the existing alluvial clay stratum, and a new dumping method for filling uniformly and softly; 2) prior to the commencement of refuse ash dumping, the density distributions at depths from +2.0 to −9.0 m in elevation at the site predicted with centrifuge model tests; 3) strength and deformation characteristics of refuse ash obtained from CIU triaxial tests and CID triaxial tests on standard and large size specimens, large size consolidation and permeability tests and standard size cyclic triaxial tests for liquefaction, under undrained conditions.

Introduction

Based on the Osaka Bay Phoenix Plan, the Service Center for Waterfront Environment in Osaka Bay, organized by the Ministry of Construction, the Ministry of Transport, the Ministry of Welfare and the prefectural governments of Wakayama, Osaka and Hyogo in Japan, is constructing two final disposal ponds in the Amagasaki(113 ha) and Izumiotsu(203 ha) Offshore Sites which will receive refuse ash from 149 cities in the Kansai area. In the reclamation and utilization of the reclaimed land, it is necessary to select reclamation methods which will not disturb the existing ground under sea, to analyze the bearing capacity as a subsoil for construction roads, to estimate for possible accommodation of volume of ashes including self-compression, and to examine the dumped materials including various waste items such as garbage and sewage mud ashes, cinders of metals, timber and plastics, metal and wood chips, and pieces of broken glass. The reclaimed ground profile can be expected to be inhomogeneous. However, the geotechnical characteristics of refuse ash have been not been sufficiently. This paper focuses on the geotechnical characteristics of refuse ash obtained from triaxial tests, consolidation tests, permeability tests and cyclic triaxial tests.

Construction Outline

Disposal location and existing ground condition

The final disposal pond in the Amagasaki Offshore Site is located at the Amagasaki port area in the interior of Osaka Bay, and lies on the alluvial clay layer in the mouth area of the Mukogawa, Kanzaki, and Yodo rivers, as illustrated in Fig.1. In the typical existing ground, the alluvial clay layer, of approximately 20 m in thickness, has undrained shear strengths (Cu in tf/m²; depth Z in

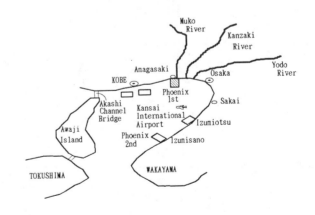

Fig.1 Site location

m) of 0.175Z to 0.25Z with water contents of 80% to 120% to a depth of 15 m. The water depth is 10 m.

Reclamation system

Several reclamation systems were examined, including the thin layer dumping method with floating conveyers, the end-on system with bulldozers, the direct dumping reclamation system with bottom door type hopper barges, and a thin layer reclamation system placed by dump trucks on a movable floating pier(hereafter referred to as the 'Floating Pier System'). The Floating Pier System was considered the most feasible, for the following reasons: the use of thin layers for reclamation facilities, the creation of uniform thickness ground, and prevents failure of the existing clay layer; the facilities require no maintenance and can receive ash continuously; and the construction cost is low.

The refuse ash is transported by hopper barges from the shipment bases to the landing berths of the disposal pond, transferred by excavator from the barges to dump trucks, and is dumped in the sea by the trucks from a floating pier. The reclamation procedure consists of two stages of reclamations by the Floating Pier Method, one stage of end-on reclamation by bulldozers in water, the intermediate covering of the refuse ash and a final stage of end-on reclamation by bulldozers on land, as illustrated in Figs.2(a) and 2(b). The slope is one to four to maintain stability.

Strength and Deformation Characteristics of Refuse Ash

Physical properties of refuse ash

The general garbage ash and the sewage mud ash are mixed at a standard ratio of 1 to 0.2 at the shipment bases, and then dumped into the disposal pond. In the disposal pond, the mixture ratio is from 1 : 0.2 to 1 : 0.6. The physical properties of the refuse ash are highly variable with the differences in the refuse, the efficiency of the individual incinerators and the incineration temperature.

TABLE 1 -- Physical properties

	general garbage ash				sewage mud ash
gravel(%)	50	59	31	0	3
sand (%)	39	36	40	61	44
under silt (%)	11	5	29	39	53
maximum size (mm)	19	19	25	2	19
uniformity coefficient : Uc	43	21	67	–	13
coefficient of curvature : Uc'	3.1	1.2	0.5	–	0.18
ash density (g/cm^3)	2.68	2.51	2.55	2.60	2.92

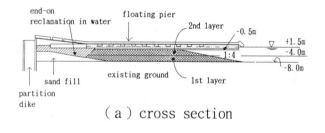

(a) cross section

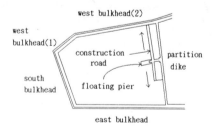

(b) ground plan

Fig.2 Reclamation system by parallel movement

The general garbage ash and the sewage mud ash samples used in this series of tests were collected randomly at the Amagasaki Shipment Base. The physical properties and grain size distributions are shown in Table 1 and Fig.3, respectively. The general garbage ash and the sewage mud ash are classified sandy soil and fine-grained soil respectively by the Japanese Unified Soil Classification.

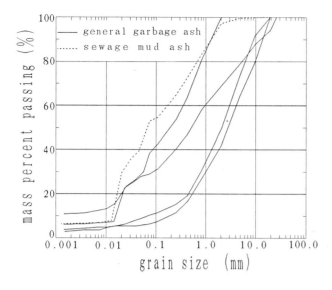

Fig.3 Grain size distribution

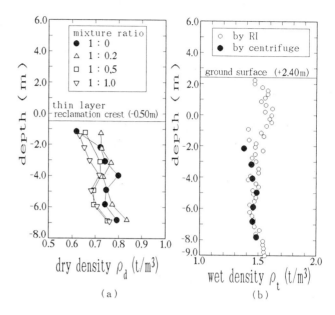

Fig.4 Density distributions

Prediction with centrifuge model tests on the density distribution with depth

Prior to reclamation, centrifuge model tests were conducted to estimate the distribution of the refuse ash with the depth. The centrifuge used in this test is owned by Toyo Construction Technical Research Institute (Miyake et al.,1988). The direct dumping by dump trucks was modeled with a sand hopper system at a reduced scale 1:45, under a gravitational acceleration of 45g, in a container measuring 55 cm long by 15 cm wide. The refuse ash was dumped into 20 cm of water (9.0 m in prototype scale) from a height above the waterline of 16.7 cm (7.5 m in prototype scale). The refuse ash used in the tests is a mixture of all general garbage ash and sewage mud ash passing a 4.76 mm sieve, mixed at ratios of 1:0.2, 1:0.5 and 1:1.0. The distributions of the dry density of the ash mixture with the depth obtained from the centrifuge model tests are shown in Fig.4(a).

The distributions with depth are slightly dispersed because the refuse mixtures separated into coarse-grain and fine-grain sizes during sedimentation. In every case, the dry density increments with depth are small, and the difference is approximately 0.15 t/m^3. Because the difference of the dry density by the mixture ratio is little, the initial dry density of specimens in mechanical tests were determined from this results. In-situ density tests with a radio isotope were conducted at landed sites to verify the prediction based on centrifuge model tests. Because the prediction agrees well with in-situ test results, as shown in Fig.4(b), the centrifuge model test is a useful method to simulate the ground reclaimed with refuse ash.

Compression and permeability tests of refuse ash

The oedometer tests on the general garbage ash and the ash mixture were conducted using specimens of 10 cm diameter and 7 cm height (Test code : C1), and specimens of 30 cm diameter and 30 cm height (Test code : C2). The ashes used in C1 and C2 are the fractions passed through 9.52 mm and 50.8 mm sieves, respectively. In C1, the coefficients of permeability were obtained directly from falling head permeability test at each load stage during compression tests. A back pressure of 1.0 kgf/cm^2 was applied to raise the degree of saturation during the tests. The

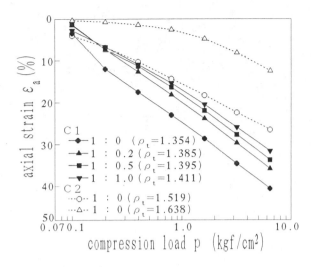

Fig.5 Relationships between ε_a and $\log_{10}P$

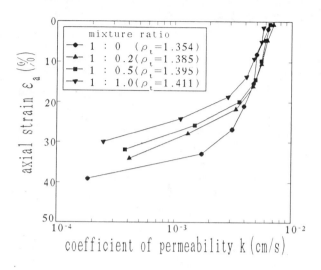

Fig.7 Relationships between ε_a and $\log_{10}k$

relationships between ε_a (axial strain of the refuse ash) and $\log_{10}P$ (applied load) obtained from the compression tests are shown in Fig.5. The compression index tends to increase as the mixture ratio increases. The compression index in C2 is 0.434 and is approximately 2/3 of C1. The relationship between the initial dry density and the compression index is shown in Fig.6. The inclination in C2 with the increase of the initial dry density is less than that in C1. From this result, we have concluded that the compressive characteristics of the refuse ash scarcely affect the ground compression because large grain size impurities support most of load. Accordingly, the surcharge-dependent ground settlement is significant, and the strength increment resulting from embankment preloading can not be expected.

The coefficients of permeability (k) tend to decrease similarly independent of the mixture ratio in the range of ε_a less than 20%, as shown in Fig.7. The k corresponding to the equivalent ε_a tends to decrease as the mixture ratio increases.

Shear strength and deformation characteristics of general garbage ash

The relationships between the deviator stress(q) and volume strain (ε_v) obtained from CIU triaxial tests under the effective lateral pressure of 1.0 kgf/cm² are shown in Fig.8. The result obtained from a large size triaxial test (diameter 30 cm, height 60 cm) conducted to study the influence of impurities is shown in Fig.9. The standard size specimen, except for the large size triaxial test, has a diameter of 5 cm and a height of 10 cm. Only in the case of a dry density of 1.111 t/m³ is the peak deviator stress confirmed clearly. The volume strains at a dry density greater

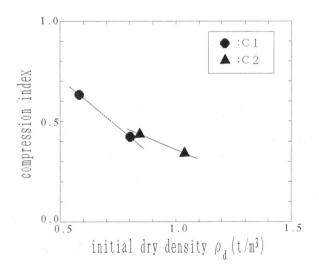

Fig.6 Change of the compression index with respect to the initial dry density

than 0.891 t/m³ tend to decrease initially and reach at the trough, and then swell. For a dry density less than 0.891 t/m³, the volume strains decrease with increasing axial strains and show large volume changes, in the range of 8.30% to 10.15%, at the axial strain of 15%. The relationships between the deviator stress(q) and the volume strain(ε_v), and the axial strain(ε_a) obtained from the large size test at the dry density of 1.022 shows likely the tendency of the result from the standard size test at the dry density of 0.891 t/m³. The relationships between the stress ratio (q/p') and the strain increment ratio ($-d\varepsilon_v/dr$) under the effective lateral pressure of 1.0 kgf/cm² are shown in Fig.9. Here p is the effective mean principal stress, $d\varepsilon_v$ the volume strain increment and dr the shear strain increment. Although the dilatancy characteristic in the standard size tests is similarly independent of the difference in the dry density, the dilatancy in the large size test is larger than in the standard size tests. Apparently, this depends on the large test's dry density, in spite of its large void ratio, because of the inclusion of impurities with a high density relative to the ash, such as metal fragments. The angle of internal friction(ϕ_{do}) obtained from CID tests
increases with increasing dry density, as shown in Fig.10. From the results, it can be considered

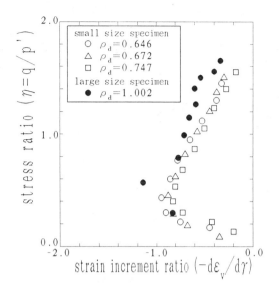

Fig.9 Dilatancy characteristics

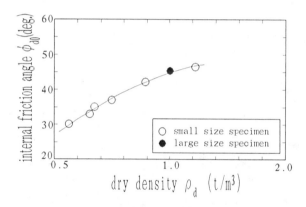

Fig.10 Relationships between the dry density and the internal friction angle

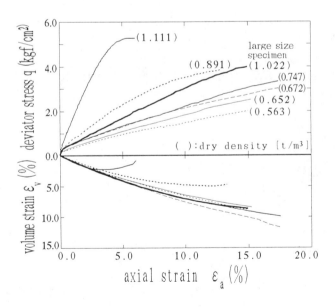

Fig.8 Relationship between stress and strain in CID test

that the deformation and strength characteristics of the refuse ash-reclaimed land can be estimated from the results obtained from the standard size triaxial tests without these dense impurities. However, because the dry density as an index expressing the conditions of the specimen is seriously affected by impurities, it is necessary to find another index by accumulating the data obtained from the large size triaxial test.

CIU tests were conducted on specimens with the mixture ratios of 1 : 0, 1 : 0.2 and 1 : 0.5. As shown in Fig.11, the angle of internal friction correlated linearly with the dry density, and was independent of the mixture ratio.

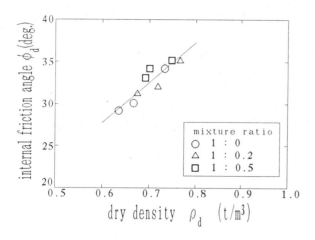

Fig.11 Influence of the mixture ratio on the internal friction angle

Strength and deformation characteristics under cyclic loading

The loosely deposited refuse ash ground is prone to liquefaction during earthquakes. The cyclic triaxial tests with standard size specimens were conducted under both consolidated drained and consolidated undrained conditions, in order to study the ground behaviour in an earthquake. A cyclic load was applied by our stress control system with a frequency of 0.01 Hz. The relationships between the cyclic stress ratio ($\sigma_d/2\sigma'_c$) and the volume strain at the 20th load cycle are shown in Fig.12. The volume strains begin increasing sufficiently from the cyclic stress ratio of approximately 0.25, and reached 6% at the cyclic stress ratio of 0.4.

The relationship between the dry density and the liquefaction strength ratio (R_{L20}), and the cyclic stress ratio at which the strain amplitudes reach 10% at loading cycle number 20, is shown in Fig.13. The relationship between the static strength ratio ($\sigma'_{1max}/2\sigma'_c$) obtained from CIU tests and the dry density is also shown in Fig.13. For dry densities less than 0.717 t/m³, R_{L20} is constant. The increment of the R_{L20} is slight in comparison with the increment of $\sigma'_{1max}/2\sigma'_c$ with the dry density. Because the R_{L20} has an unique

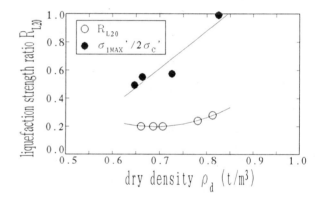

Fig.13 Relationships between the liquefaction strength ratio and the dry density

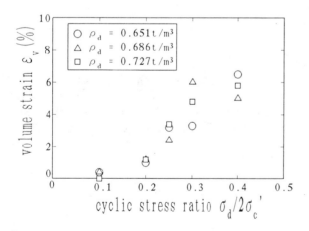

Fig.12 Relationships between the cyclic stress ratio and the volume strain

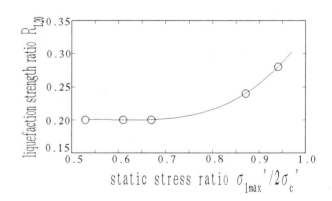

Fig.14 Relationship between the liquefaction strength ratio and the static strength ratio

correlation with $\sigma'_{1max}/2\sigma'_c$ independent of the dry density, as shown in Fig.14, it is suggested that the liquefaction strength of the refuse ash can be estimated from the CIU test. From the aforementioned results, the refuse ash ground shows a large volume change for cyclic shear stress, and has little liquefaction strength. Accordingly, it is necessary to perform countermeasures for liquefaction during earthquakes.

Conclusion

Currently, many people in various fields are discussing the reduction of waste volume as a social problem. Although garbage volume can be reduced to approximately 30 times by incineration, effective ash disposal is a lingering problem. Disposal methods which utilize the ash constructively remain issues of the future. The waterfront redevelopment in the Osaka Bay Phoenix Project utilizes environmentally responsible refuse ash disposal to create up a urban spaces for the 21st century.

Reference

M.Miyake, H.Akamoto & H.Aboshi(1988). Filling and quiescent consolidation including sedimentation of dredged marine clays, Centrifuge 88, pp.163-170.

Assessment of Hydrogeological and Contaminant Transport Properties of a Fractured Clay in Eastern Ontario

Vince O'Shaughnessy, Graduate Student, Vinod K. Garga, Professor
Department of Civil Engineering, University of Ottawa, Ottawa, Ontario, Canada

Abstract

Physical conditions and visual inspection of fractures of the Leda subsoil was investigated with the use of boreholes and test pits. Piezometers were installed using the conventional auger method and the double-Shelby method developed at the University of Waterloo. Hydrogeological aspects of the upper twelve metres were analyzed using water level fluctuations, hydraulic head and hydraulic conductivity profiles. The geochemistry analysis was done by measuring the major ions and tritium concentration distribution in groundwater samples obtained from piezometers. Laboratory tests were performed on five clay samples to determine the effective diffusion coefficients and the retardation factors of the major ions. Results from water level monitoring, maximum seasonal variations and hydraulic head profiles revealed a highly hydraulically active fractured zone exist at all studied sites. However, the extent of this fractured zone varies in depth from site to site ranging from 3.2 to 6.0 m. Hydraulic conductivity values in the upper active fractured zone range from 1.85×10^{-8} to 2.05×10^{-5} m/s. In contrast, the hydraulic conductivity values measured in the deepest piezometers range from 8.20×10^{-10} to 1.40×10^{-9} m/s. This represents a difference of 1 to 4 orders of magnitude. The geochemical analysis indicated the presence of three hydrochemical facies. Analytical profiles were fitted to the observed field data by using laboratory measured effective diffusion coefficients and retardation factors with advective parameters measured during field investigations. The analysis indicated that Champlain Sea clay deposits of 12.0 m or less may not protect any underlying aquifer from surficial or buried waste, in the long term.

Introduction

Because of their low hydraulic conductivity, many natural clayey till deposits are considered a safe place for waste burial. The protection of the underlying aquifers from contamination is dependent on the nature of these clays. The amount of aquifer protection will depend on thickness, advective and diffusive properties of the clay deposit and most important, if present, the depth of penetration of hydraulically active fractures. These fractures provide an effective pathway for contaminant migration [1,2,3].

The depth to which hydraulically active fractures penetrate are site specific and vary substantially, as revealed by literature review. Researchers have reported hydraulically active fractures to depths of 8 to 10 metres below ground surface for glaciolacustrine clay deposits in Canada [4,5,6,7].

New waste disposal sites will be required in the Ottawa region, and some studies have been commissioned to examine potential sites. It is sometimes assumed that due to the presence of Champlain Sea clay deposits, there would be little risk of contamination of the groundwater. The objective of this paper is to present some the results of an investigation into the hydraulic, geochemical and solute or contaminant transport properties of four fractured Leda clay (Champlain Sea clay) deposits in Eastern Ontario. A particular important aspect of the study was to establish the depth to which the fractured network is hydraulically active. These four sites are known as NRC, Fallowfield, Renfrew and Casselman.

Methodology

Physical conditions and depth of fractures of the Leda subsoil was investigated with the use of boreholes. In addition, visual inspection of the fracture network was possible at NRC and Fallowfield sites by excavated test pits. Piezometers were installed using the conventional auger method and the double-Shelby method developed at the University of Waterloo. Hydrogeological aspects of the upper 12 metres were analyzed using water level fluctuations and variation profiles, hydraulic head and conductivity profiles. The geochemical analysis was made by measuring the major ions and tritium concentration distribution in groundwater samples obtained from piezometers.

To determine the contaminant migration characteristics of Leda clay two different laboratory diffusion tests were performed on five Leda clay samples. One test involved the placement of a single salt solution (NaCl) in contact with a fully saturated undisturbed clay sample, thus allowing downward chemical migration throughout the sample by diffusion only. The second test involved the placement of deionized water in contact with the undisturbed saturated clay, thus allowing a upward chemical migration by diffusion only. During the testing period which ranged from 15 to 16 days, samples were taken from the source solution and analyzed for chemical species of interest (i.e. Cl^-, Na^+, K^+, Mg^{2+} and Ca^{2+}). At the termination of the test, the soil samples were sectioned to determine the porewater and adsorbed concentrations as a function of depth. From plotted "adsorption isotherms", retardation factors were evaluated and the diffusion coefficients were determined by "back-figuring", a finite differences procedure.

Results and Discussion

Test pits revealed a complex system of fractures, roots, root holes, paleoroots, pores and sand lenses. If these geological features are well interconnected and hydraulically active, they define the active groundwater zone. The many small sand lenses observed at Fallowfield, highly influenced the hydraulic properties of the upper clay. Well defined fractures below a depth of 1.1 m probably controlled the hydraulic characteristics of the clay deposit at NRC. These fractures extend beyond the base of the test pit at 4.0 m. Fracture spacing steadily increased with increasing depth. The measured fracture spacing varied from one fracture every 40 mm at 1.5 m depth to one fracture every 400 mm at 3.5 m.

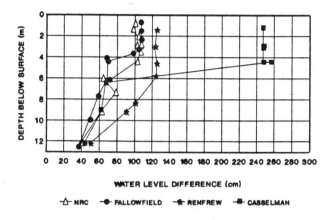

Figure 1. Maximum water level variation.

Results from water level monitoring, maximum seasonal variations and hydraulic head profiles revealed a highly hydraulically active zone at all four sites. The profiles of maximum water level variation with depth for all four sites (Fig.1) were near-vertical in the upper weathered and highly fractured zone. The presence of an upper depth over which the maximum variation of water level remains essentially constant (Fig.1) is indicative of a hydraulically active zone where the fracture network is well connected and open. The active zone defined by hydrogeological evidence extends to 4.4 metres at NRC and Casselman, to 3.2 metres at Fallowfield, and to 6.0 metres at Renfrew. The variation in maximum seasonal piezometric heads were significantly reduced just below the active zone and continued to decrease with depth. This could indicate a reduction in the contribution of groundwater flow by open fractures, or that the influence of fractures was reduced due to the increase in fracture spacing with depth, and an associated decrease in fracture connectivity.

The hydraulic conductivity profiles for the four sites are shown in Fig. 2. In the upper fractured and active zone, hydraulic conductivity values ranged from 1.85×10^{-8} to 2.05×10^{-5} m/s for measurements in piezometers installed using the double-Shelby method. The hydraulic conductivity values in the upper active zone were highly scattered, indicating three orders of magnitude difference in k. These high values were associated with the direct fracture intersection or with more hydraulic conductive sand lenses as observed at Fallowfield. The hydraulic conductivities reported in the upper active and fractured zone were approximately 1 to 4 orders higher in magnitude than those evaluated for clays beneath this zone. The hydraulic conductivity measured in the deepest piezometers ranged from 8.2×10^{-10} to 1.4×10^{-9} m/s. Reported hydraulic conductivity values ranging from 10^{-11} to 10^{-10} m/s [8], which were associated with unfractured, massive Leda clays. These observations for Eastern Champlain Sea clay deposits support earlier findings [9,10], which indicate that the hydraulic conductivity in the upper near surface zone is controlled by a fracture network rather than the intergranular pore spaces.

only moderately dominant at Fallowfield and Renfrew. All SO_4^{2-} profiles gradually decreased from maximum concentration levels in the upper active zone to values less than 0.23 M/m^3 below 8 m depth. These major ion distribution trends are typical of Champlain Sea clay deposits [11].

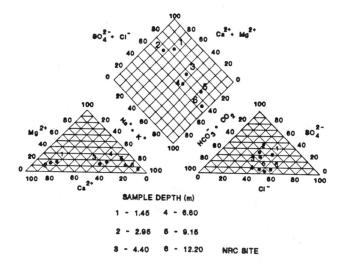

Figure 3. Groundwater trilinear diagram for the NRC site

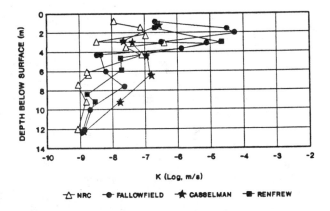

Figure 2. Hydraulic conductivity profiles

The hydraulic conductivity profiles suggest that a highly conductive zone extends to a depth of 4.4 m at NRC, 3.7 m at Fallowfield, 6.0 m at Renfrew and 6.4 m at Casselman.

The chemical analyses indicate that sodium is the predominant cation in solution for Leda clays, while chloride is the more dominant anion in solution. A prominent feature of the major ion profiles was the gradual increase in concentration of Na^+, K^+, Cl^-, HCO_3^- with depth, except at the Renfrew site. Calcium and magnesium were dominant in the upper weathered sections but decreased significantly in concentration below this zone. Carbonates were variable between sites and were dominant at NRC and Casselman and were

The geochemical analysis was best represented by trilinear Piper diagrams (Fig. 3). Three hydrochemical facies were observed at the NRC, Fallowfield and Casselman sites while only one facie could be used to represent groundwater at Renfrew. This was probably due to surface salt contamination.

These hydrochemical facies were established for groundwater samples with respect to vertical depth only and were defined as follows:

1) a shallow "active" facie which represents groundwater which has undergone considerable chemical change from its original composition due to fresh water infiltration;

2) a deep "inactive" facie which represents deep groundwaters which did not interact with the near surface groundwaters and were completely different in chemical composition. This hydrochemical facie may represent the initial groundwater chemical composition or a geochemical change due only to an upward diffusion from the clay bedrock interface to the upper hydraulic active zone [10], and;

3) Intermediate "transition" facie which represents a geochemical transition between the surface and deep groundwaters. This transition zone may have been influenced by fracture flow; or, it is the result of an upward or downward advective and diffusive transport mode or both.

Based upon trilinear Piper diagrams, the highly active groundwater zone extends to 3.0 m, 3.0 m and 4.4 m for the NRC, Fallowfield and Casselman sites, respectively. The transition zone ranges from at least 6.1 m at Fallowfield, to 9.0 m at NRC and to 6.4 m at Casselman. In the case of Renfrew, the presence of relatively young salts at ground surface (possible contamination from a road salt stockpile)and the subsequent downward infiltration indicate that the active zone here ranges between 6.0 and 8.0 m and with a transition zone that extends well below 12.0 m.

The interpretation of the geochemical analysis and chemical trends were limited due to the fact that major ion concentrations below a depth of 12 metres remain unknown for each site.

TABLE 1. A summary of the depth at which the "hydraulic active zone" is defined by different investigative techniques

Site	Test Pit (m)	Hydro-geo-logi-cal (m)	k (m)	Geo-chemical (m)	Tritium, concentration value less than 1 T.U. (m)
NRC	>4.0	4.4	4.4	3.0	6.1
Fallowfield	3.5	3.2	3.7	3.0	7.6
Renfrew	—	6.0	6.0	9.0	—
Casselman	—	4.4	6.4	4.4	12.2

A summary of the depths at which the "active zone" (fracture network is hydraulically interconnected) is defined by the different investigative techniques is shown in Table 1. The minimum tritium value criteria is clearly conservative. The estimate of the hydraulic active zone using the other criteria gave similar results. One of the limitations of the use of Piper plots in the investigation is that the geochemical tests were not carried out for the entire clay depth, hence the criteria for the active zone became more subjective. Criteria based on the water level fluctuation appears to be a simple and practical way to establish the hydraulically active zone. This test has the advantage that the criteria does not depend on results obtained from analysis of data which is necessary to evaluate in situ k values.

Table 2 provides a summary of the estimates of the retardation factors (R) and effective diffusion coefficients (D_e) obtained in this study.

The calculated R and D_e values for Eastern Ontario Leda clay fall within the reported spread and indicate the same trends [13,14], with Ca^{2+} and K^+ as the highest retained cation, while Na^+ is the most mobile cation and chloride is the most mobile ion in solution.

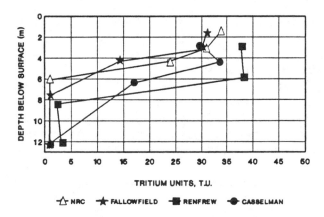

Figure 4. Tritium concentration profiles.

Groundwater with a tritium concentration level below 1.0 T.U. indicates that groundwater recharge has occurred prior to the nuclear tests in 1953 [12]. Tritium concentration levels below 1 T.U. (Fig. 4) occurred at depths of 6.1 m, 7.6 m and 12.2 m at the NRC, Fallowfield, and Casselman sites, respectively. A tritium concentration of 3.0 T.U. was measured for the Renfrew site at 12.2 metres which represents the maximum depth investigated.

In order to evaluate contaminant migration characteristic of fractured Leda clays, theoretical profiles were generated from analytical solutions [2]. Theoretical profiles for sodium for the Renfrew site are presented in Fig. 5. Migration of sodium from a salt stockpile is known to have occurred at the Renfrew site for at least

TABLE 2. Effective Diffusion Coefficients (D_e) and Retardation Factors (R)

Source Solution	Species	Effective Diffusion Coefficient (D_e) (x 10^{-6} cm^2/s)	Retardation Factor (R)
Deionized Water	Cl$^-$	6.5	1
	Na$^+$	3.4	1.5 - 1.6*
	K$^+$	6.0	17.2 - 22.8
	Mg^{2+}	3.8	12.4 - 4.9^a
	Ca^{2+}	5.8	33 - 17^a
NaCl	Na$^+$	7.0	2.2#
	Cl$^-$	6.5	1
Deionized Water	Cl$^-$	6.0	1
	Na$^+$	4.0	2.0 - 2.1*
	K$^+$	7.5	36 - 48
	Mg^{2+}	4.6	4.8 - 8.1
	Ca^{2+}	6.3	18 - 32
NaCl	Na$^+$	7.0	2.7#
	Cl$^-$	6.0	1
NaCl	Na$^+$	6.8	3.6#
	Cl$^-$	5.8	1

\# Retardation Factor, $R = 1 + \rho K_d/\theta$

* Retardation Factor, $R = 1 + [(\rho/\theta)(\beta\gamma \exp(-\gamma C))]$, where R is a function of solute concentration (C). The range given is the difference between lowest and highest solute concentrations.

^{a} Showed greater adsorption capacity at lower solute concentrations.

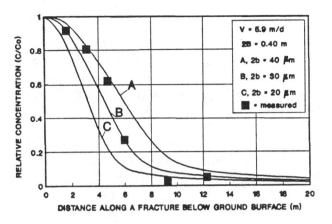

Figure 5. Theoretical profiles for sodium for the Renfrew site after 30 years.

30 years. Analytical profiles were fitted to the observed field data by using laboratory measured effective diffusion coefficients and retardation factors with advective parameters measured during field investigations. Simulated sodium concentration curves in Fig. 5 were generated using a fracture flow velocity (v) of 5.9 m/d, a fracture spacing (2B) of 0.4 m, and varying the fracture aperture (2b) from 20 to 40 µm. The analyses indicated that the effectiveness of fractured Leda clay to attenuate contaminants is significantly reduced when dealing with low and non reactive solutes. After only 30 years, sodium has migrated well beyond a depth of 12.0 metres.

Conclusions

1) Hydrogeological and geochemical investigations revealed the existence of a highly hydraulically active zone at all four sites. The extent of this active zone is site specific. This zone ranges from 3 to 6 metres in depth.

2) Observations of water level fluctuations in piezometers appear to provide a convenient method to assess the depth of the hydraulically active zone.

3) The effective diffusion coefficients for Na$^+$, K$^+$, Mg^{2+}, Ca^{2+}, and Cl$^-$ were evaluated from laboratory experiments and range from 3.4 to 6.5 x 10^{-6} cm^2/s, 6.0 to 7.5 x 10^{-6} cm^2/s, 3.8 to 4.6 x 10^{-6} cm^2/s, 5.8 to 6.25 x 10^{-6} cm^2/s and 6.5 to 7.0 x 10^{-6} cm^2/s, respectively.

4) The retardation factors for Na$^+$, K$^+$, Mg^{2+}, and Ca^{2+} range from 1.5 to 3.6, 17 to 48, 4.8 to 12.4, and 17 to 33, respectively. Based upon calculated retardation factors, chloride is the most mobile ion in solution while calcium and potassium are highly attenuated by Leda clay.

5) Contaminant migration analysis indicated that Champlain Sea clay deposits of 12.0 m or less may not protect any underlying aquifer from surficial or buried waste, in the long term.

References

[1] HARRISON, E.A., SUDICKY, E.A., and CHERRY, J.A., 1992. Numerical analysis of solute migration through fractured clayey deposits into underlying aquifers. Water Resources Research, Vol. 28, No. 2, pp. 515-526.

[2] SUDICKY, A.E., and FRIND, E.O., 1982. Contaminant transport in fractured porous media: Analytical solutions for a system of parallel fractures. Water Resources Research, Vol. 18, pp. 1634-1642.

[3] GARGA, V.,K., and O'SHAUGHNESSY, V., 1994. The hydrogeological and contaminant transport properties of fractured Leda clay in Eastern Ontario. Part 2: Contaminant transport. (Submitted to the Canadian Geotechnical Journal).

[4] GRISAK, G.E., and CHERRY, J.A., 1975. Hydrologic characteristics and response of fractured till and clay confining a shallow aquifer. Canadian Geotechnical Journal, Vol. 12, pp. 23-43.

[5] DAY, M.J., 1977. Analysis of movement and hydrochemistry of groundwater in fractured clay and till deposits of the Winnipeg area, Manitoba. M.Sc. Thesis, Department of Earth Sciences, University of Waterloo, Waterloo, Ontario.

[6] D'ASTOUS, A.Y., RULAND, W.W., BRUCE, J.R.G., CHERRY, J.A., and GILLHAM, R.W., 1989. Fracture effects in the shallow groundwater zone in weathered Sarnia area clay. Canadian Geotechnical Journal, Vol. 26, No. 1, pp. 43-56.

[7] O'SHAUGHNESSY, V., and GARGA, V.K., 1994. The hydrogeological and contaminant transport properties of fractured Leda clay in Eastern Ontario. Part 1: Hydrogeological properties. (Accepted for publication in the Canadian Geotechnical Journal)

[8] TAVENAS, F., JEAN, P., LEBLOND, P., and LEROUEIL, S., 1983. The permeability of natural soft clays, II, Permeability characteristics. Canadian Geotechnical Journal, Vol. 20, pp. 645-660.

[9] LAFLEUR, J., GEROUX, F., and HUOT, M., 1987. Field permeability of the Champlain Sea crust. Canadian Geotechnical Journal, Vol. 24, pp. 581-589.

[10] DESAULNIERS, D.E. and CHERRY, J.A., 1989. Origin and movement of groundwater and major ions in a thick deposit of Champlain Sea clay near Montreal. Canadian Geotechnical Journal, vol. 26, pp. 80-89.

[11] TORRANCE, J.K., 1988. Mineralogy, pore-water chemistry, and geotechnical behaviour of Champlain Sea and related sediments. In: The Late Quaternary Development of the Champlain Sea Basin, N.R. Gadd (Editor). Geological Association of Canada, Special Paper, No. 35, pp. 259-275.

[12] ROBERTSON, W.D., and CHERRY, J.A., 1989. Tritium as an indicator of recharge and dispersion in a groundwater system in Central Ontario. Water Resources Research, Vol. 25, No. 6, pp. 1097-1109.

[13] ROWE, R.K., CAERS, C.J., and BARONE, F., 1988. Laboratory determination of diffusion and distribution coefficients of contaminants using undisturbed clayey soil. Canadian Geotechnical Journal, Vol. 25, pp. 108-118.

[14] BARONE, F.S., YANFUL, E.K., QUIGLEY, R.M., and ROWE, R.K., 1989. Effect of multiple contaminant migration on diffusion and adsorption of some domestic waste contaminants in a natural clayey soil. Canadian Geotechnical Journal, Vol. 26, pp. 189-198.

Evaluation of Soil Bentonite Slurry Wall by Piezocone

**Issa S. Oweis, President, Edward M. Zamiskie, Principal Engineer,
M. Golam Kabir, Staff Engineer**
Converse Consultants East, Parsippany, New Jersey, USA

ABSTRACT

Hydraulic performance of a soil bentonite (SB) slurry wall surrounding an active landfill is being investigated by an in situ device called a "piezocone". The SB wall is nominally three feet thick and generally extends vertically from ground surface a minimum of three feet into clay or bedrock. The on-going state-of-the-art investigation consists of in situ, field, and laboratory testing of the SB wall. The paper discusses the usefulness of the piezocone in verifying the overall competency of the wall and bottom key into low permeability material. The paper also presents a comparison of the present in situ hydraulic conductivity of the wall with the hydraulic conductivity of the wall at the time of its installation.

INTRODUCTION

The application of a soil bentonite (SB) slurry wall (cutoff wall) to impede the lateral subsurface migration of hazardous materials requires an understanding of the hydraulic conductivity characteristics of the wall. Due to the widespread use of SB slurry walls for containment of hazardous materials and inherent high risk involved with failure, it is important that the overall integrity of the constructed wall and the fundamental aspects of the hydraulic conductivity behavior of slurry wall as a function of time be investigated. It is imperative that the degradation of hydraulic conductivity with time due to various factors as well as the effects of micro cracks on the macro-permeability behavior of SB slurry wall be incorporated into the performance analysis of the wall.

As a part of the hydraulic performance evaluation of a SB wall, the in situ testing device "piezocone" is being used to investigate the integrity of the SB slurry wall at an active solid waste landfill. The landfill is surrounded by an approximately 15,800-foot long and 10 to 60-foot deep SB slurry wall which was designed and built as a containment barrier. The construction of the SB slurry wall was performed in two phases. The SB wall is nominally three ft. thick and was designed to extend vertically from ground surface to a minimum of three ft. into clay or bedrock. The laboratory measured hydraulic conductivity of the wall was less than 1×10^{-7} cm/sec at the time of its installation.

The paper examines the key considerations in assessing the integrity of the SB slurry wall. The following aspects of the investigation are presented in this paper:

a) comparison of the present in situ hydraulic conductivity of the wall with the hydraulic conductivity of the wall at the time of its installation;

b) verification of the overall competency of the wall, i.e., absence of large windows or weak layers within the wall, squeezing or other lateral movement of the wall, or other defects; and

c) verification of the keying of the wall sufficiently into material of low conductivity.

The on-going state-of-the-art investigation consists of in situ, field and laboratory testing of the SB wall. Currently the evaluation of piezocone profiling and dissipation test data along the wall for the purpose of identification of apparent wall anomalies or defects is in progress. Based on these findings, borings have been performed within the SB wall at selected locations to verify the piezocone profiling and to obtain samples for identification and laboratory testing, including permeability and grain size. The laboratory

---

[1] President, Principal Engineer and Staff Engineer respectively, Converse Consultants East, 3 Century Drive, Parsippany, NJ 07054.

tests include permeability tests on good quality "undisturbed" SB samples simulating long-term conditions to assess possible degradation by high salinity water and site leachate (a high ion-content permeant). In addition, several pumping tests have been conducted at selected locations to produce a large hydraulic gradient across the wall to assess the effectiveness (gross hydraulic conductivity) of the SB slurry wall. On the basis of this study, hydraulic performance of the wall, and if incompetent areas are identified, the nature of defects as well as possible remedial actions, will be evaluated.

SITE GEOLOGY

The underlying bedrock at the site consists of reddish-brown shale with interbedded siltstone and occasional layers of sandstone. A white to gray clay, 0 to 10 ft. thick, overlies the bedrock. Generally, the SB wall is keyed into the clay layer where it exists in sufficient thickness, or into weathered bedrock where the clay layer is absent or thin. Above the clay or bedrock is a sand stratum, approximately 20 ft. thick, which is generally a clean, medium to coarse-grained gray to white sand containing seams of gravel and clay. An approximately 15-ft. thick soft marsh deposit overlies this sand and is present at or near the surface.

INVESTIGATION OF PIEZOCONE TESTING & EVALUATION

The piezocone consists of a stainless steel cylinder 35.8 mm in diameter, with a 60° cone and a base area of 10.0 sq. cm. The cone measures both tip resistance (Q_c) and sleeve friction (F_s). A porous element to measure pore pressure (U_t), 5.0 mm thick, is located 5.7 mm behind the base of the cone. The rate of cone penetration is typically 20.0 mm/sec. The advantages of the piezocone in investigating the SB wall are minimizing disturbance and exposure to contaminated materials, eliminating risk of hydraulic fracturing, high production rate and lower cost than standard borings.

Piezocone testing has been performed at approximately 400-ft. intervals along the centerline of the slurry wall penetrating full design depth of the wall or until refusal was met. A typical piezocone profile is shown in Fig. 1. In general, the piezocone profiles along the slurry wall show similar results. The piezocone data measured during profiling were used to classify the soil according to soil behavior type classification chart proposed by Robertson (1990). This classification compares well with the borings done for confirmation. The upper 5 to 10 ft. consists of fill-type materials with high cone tip and friction sleeve resistances, and negative pore pressures. Below that, in the SB wall, cone tip and friction sleeve resistances drop to the range of 0-10 tsf and 0-0.2 tsf respectively, and the pore pressures increase gradually with depth. Occasionally, the cone tip and friction sleeve resistances increase with a corresponding drop in pore pressure due to presence of thin sand layers. At or near the design depth of the wall, the cone tip and friction sleeve resistances suddenly increase with or without a drop in pore pressure due to presence of very dense sand or bedrock.

A review of the piezocone tests data revealed a competent slurry wall except for two zones along the wall where further investigation are required. In these zones, most of the piezocone testings failed to reach the design (as-built) depth of the wall. In the first zone, nine piezocone tests were conducted along the wall. As-built drawings of the wall show that the depth of wall varies from 50 to 56 ft. from ground surface, and is keyed into clay material. In this area, most of the piezocone tests were terminated between 30 to 40 ft. below ground surface due to the presence of dense silty sand or sand with silt. Dissipation tests at these depths exhibit quick dissipation of generated excess pore pressure, i.e., the presence of highly permeable material. In the second zone, four piezocone tests were conducted in the slurry wall. The piezocone profiling stopped short of the reported design depth of wall by 6 to 10 ft. because of dense material. As-built drawings of the wall show that the wall is keyed into clay/residual soil in this area. The piezocone testing data show that the material encountered at the bottom was dense silty sand or sand with silt. In addition, dissipation tests were performed at the bottom, and similar behavior was observed.

Test borings were performed in these zones within the SB slurry wall, and disturbed samples were collected for the purpose of characterization. Based on the grain size distribution curves of silty sand or sand with silt encountered below SB wall, the average values of coefficient of uniformity, coefficient of curvature and D_{10} size of particles are 3.45, 1.65 and 0.13 mm respectively. According to NAVFAC DM 7.1 (1982), based on D_{10} size of particles, the hydraulic conductivity of silty sand/sand with silt varies between 0.001 to 0.01 cm/sec. The chracteristics of the SB wall are as follows:

Brown clayey sand with gravel (SC)
LL = 29; PL = 17
Sp. Gravity = 2.745
Natural Water Content = 18.4 %
Unit Dry Density = 111.5 pcf.

It is possible to explain failure of some of the piezocone tests to reach the design depth of the wall by horizontal deviations of the piezocone sounding. However, it is highly unlikely that such is the case for a relatively shallow depth of penetration. On the other hand, a deviation of only 3° from vertical could lead the piezocone out of the wall within 30 ft., and a 5° deviation requires only 17 ft. for the piezocone to be out of the wall. The sandy material at the wall base could also be explained by the partial collapse or movement of the wall, accumulation of sediments in the bottom during construction, or inadequate depth of excavation along these zones. In fact, at some locations borings done at very close spacing perpendicular to the wall indicate that some outward bulging of the wall (in the soft marsh soils) has occured. For the purpose of verification of wall integrity, test borings as well as pumping tests have been conducted in these zones. A typical setup of the pumping test is shown in Fig. 2. The test well (PW-15) consisted of a 4-in. diameter Schedule 40 PVC pipe and No. 20 slotted screen assembly installed in a 10-in. diameter hole. The sand layer, which is from 40 to 50 ft., is fully screened. Three 1-1/2 in. porous tube piezometers (P-15ID, P-15IR, and P-150D) were also installed as shown on Fig. 2. Piezometers P-15ID and P-15IR are located inside the landfill installed at depths of 50 and 66 ft. below ground surface respectfully. The piezometer P-15ID and P-15IR are installed in sand layer and bedrock respectively. The piezometer P-150D is located outside of the landfill as shown in Fig. 2, installed in the sand layer at 48 ft. depth. The responses of the piezometers were recorded after the test well has been pumped at a constant rate of 26 gpm for 180 minutes. The drawdowns of the test well, installed piezometers, and existing piezometers are also in Fig. 2. Based on the result, it is concluded that the SB wall at this zone is ineffective.

The low tip resistance encountered during piezocone profiling in the slurry wall and lack of significant increase with depth are an indication of the low confining pressures maintained in the wall. It is believed that large arching forces may be developed within a narrow deep trench, preventing the soil bentonite mix from experiencing any increased confining pressures with depth. Frictional resistance along the interface between soil bentonite mix and the trench walls opposes subsidence and thus supports part of the weight of the soil bentonite slurry (Schneebeli 1964). As a result, settlement in slurry trenches is severely limited by the effects of arching (Engemoen and Hensley 1986). It has also been observed by Engemoen and Hensley (1986) that the lack of a significant variation in tip resistance between soil bentonite slurries of different ages strongly suggests that any further increases in confining pressures beyond 2 months due to drainage are negligible. Low confining pressures could create a greater potential for hydraulic fracturing across the slurry wall.

The rate of pore pressure dissipation during piezocone tests is theoretically related to the coefficient of consolidation. Also, the permeability is a function of both coefficient of consolidation and soil compressibility. In the case of the piezocone, it is generally recognized that both vertical and horizontal or radial drainage occur if the porous element is located at or on the tip. On the other hand, horizontal or radial drainage dominates the flow if the element is located behind the tip or along the shaft of the probe. Theoretical solutions (Torstensson 1975, Baligh and Levadoux 1986) show that dissipation is slightly faster for a spherical cavity than for a cylindrical cavity. It has been shown by various researchers (Kabir and Lutenegger 1990, Robertson et al. 1992) that the dissipation data can be used for reliable estimates of the coefficient of consolidation as well as permeability of the soil being tested.

Several dissipation tests during profiling were performed to estimate the permeability of the SB slurry wall. A typical dissipation curve is shown in Fig. 3. In general, the empirically estimated values of permeability based on Robertson, et al. (1992) fall in the range of 1×10^{-9} to 1×10^{-7} cm/sec, which suggest a competent wall with respect to hydraulic conductivity. From limited laboratory constant head permeability tests with water, the hydraulic conductivity of the SB wall is around 4.0×10^{-8} cm/sec at 20°C. Also, these values are in the range of the hydraulic conductivity of the wall measured at the time of its installation. In addition to piezocone dissipation and pumping tests, undisturbed samples will be obtained to perform laboratory permeability tests with leachate for further verification of the hydraulic conductivity of the wall, including the simulated long-term effect of exposure to site leachate.

In general, the following factors should be considered for evaluating the hydraulic conductivity of the slurry wall:

1. Properties of backfill materials of the wall, and presence of cracks, windows or pockets within the wall as well as the integrity of the base key. The permeability of a completed SB slurry wall is a function of both the filter cake that forms on the walls and the permeability of the backfill (soil gradation and the quantity of bentonite) placed in the trench.

2. Permeation of a bentonite filter cake or SB backfill by industrial pollutants generally leads to an increased permeability. Concentrated organic fluids can result in significant increases in laboratory permeability; permeability increases due to permeation with aqueous solutions of organic fluids would be expected to be considerably less. Where permeability increases would be expected due to high concentrations of contaminants, the permeability increases would approach, as an upper limit, the permeability of the base soil of the backfill.

3. An increase in stress leads to consolidation of the backfill, which will decrease its void ratio. This decrease, in turn, results in a lower hydraulic conductivity.

4. Effects of hydraulic gradient on the wall.

5. The temperature of the permeant governs its viscosity, which in turns, affects hydraulic conductivity. In addition, the rate of chemical reaction between the permeant and certain backfill constituents may double for every 10° C increase in temperature (Strum and Morgan 1981).

In addition, the uncertainties involved in estimating hydraulic conductivity in situ or in laboratory should be considered through statistical means. Laboratory measurements of "k" in structured fine-grained soils can underpredict in situ values by several orders of magnitude (Rowe and Burden 1966, Rowe 1972, Grisak 1975, Grisak and Cherry 1975). On the other hand, experience with relatively unstructured clays shows that under the best conditions of sampling and testing coupled with superior engineering analysis and judgement, field values of "k" can be predicted within a factor of 2 and 3 but also can be in error by a factor of 10 (Bishop and Al-Dhahir 1970).

CONCLUSIONS

Piezocone testing is an excellent means of exploring in situ conditions of a SB wall because of it's effectiveness in identifying apparent anomalies (non-uniformity and discontinuity in the wall) or defects (cracks, windows or pockets). The parameters measured during piezocone profiling can be used to classify the behavior of materials encountered during penetration. In addition, dissipation of generated excess pore pressures at a specified depth is used to determine the hydraulic conductivity of the material being tested. Use of piezocone eliminates other methods for preliminary investigation for the competency of the SB wall. It is good practice to verify 10 % to 20 % of the piezocone sounding results by standard test borings. This is especially useful when there are natural occuring soils in the area which may have similar properties as the SB material.

REFERENCES

Baligh, M.M., and Levadoux, J.N. (1986), "Consolidation after Undrained Piezocone Penetration II: Interpretation", J. Geot. Engg., ASCE, 112, 727-745.

Bishop, A.W., and Al-Dhahir, Z. A. (1970), "Some Comparisons between Laboratory Tests, In Situ Tests and Full Scale Performance, with Special Reference to Permeability and Coefficient of Consolidation", Proc. Conf. In Situ Investigations in Soils and Rocks, May 13-15, 251-264.

Engemoen, W.O., and Hensley, P.J. (1986), "ECPT Investigation of a Slurry Trench Cutoff Wall", Proc. Use of In-Situ Tests, Virginia, 514-528.

Grisak, G.E. (1975), "The Fracture Porosity of Glacial Till", Can. J. Earth Sci., 12(3), Mar., 513-515.

Grisak, G. E., and Cherry, J. A. (1975), "Hydrologic Characteristics and Response of Fractured Till and Clay Confining a Shallow Aquifer", Can. Geot. J., 12(1), Jan., 23-43.

Kabir, M.G., and Lutenegger, A.J. (1990), "In Situ Estimation of the Coefficient of Consolidation of Clays", Can. Geot. J., 27(1), Feb., 58-67.

NAVFAC, (1982), Soil Mechanics Design Manual 7-1, Dept of the Navy, Naval Facilities Engineering Command, May.

Robertson, P.K. (1990), "Soil Classification using the Cone Penetration Test", Can. Geot. J., 27(1), Feb., 151-158.

Robertson, P.K., Sully, J.P., Woeller, D.J., Lunne, T., Powell, J.J.M., and Gillespie, D.G. (1992), "Estimating Coefficient of Consolidation from Piezocone Tests", Can. Geot. J., 29, 539-550.

Rowe, P.W. (1972), "The Relevance of Soil fabric to Site Investigation Practice", 12th Rankine Lecture, Geotechnique, 22 (2), 195-300.

Rowe, P.W., and Burden, L. (1966), "A New Consolidation Cell", Geotechnique, 16(2), 162-170.

Schneebeli, G. (1964), " Le Stabilite des tranchees profondes forees en presence de boue", Houille Blanche, 19(7), 815-820.

Strum, S., and Morgan, J. (1981), Aquatic Chemistry, 2nd Ed., Reaction Rates of Elementary Process, 357.

Torstensson, B.A. (1975), "Pore Pressure Sounding Instrument", Proc. ASCE Specialty Conf. on In Situ Measurement of Soil Properties, Raleigh, NC, 2, 48-54.

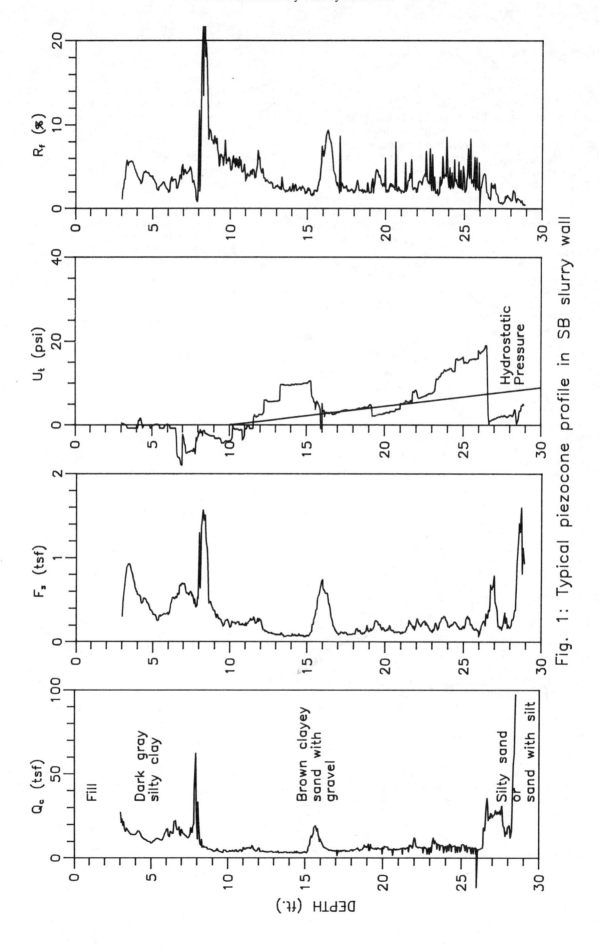

Fig. 1: Typical piezocone profile in SB slurry wall

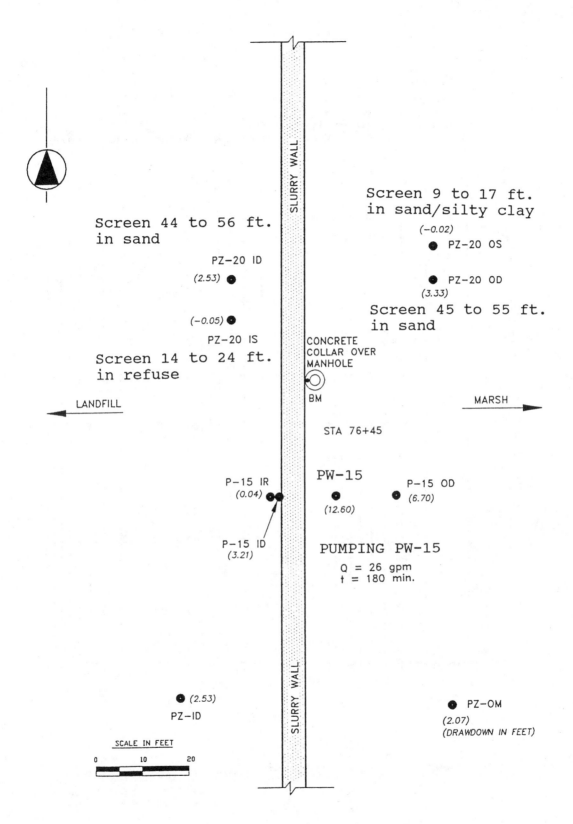

Fig. 2: Distribution of drawdown

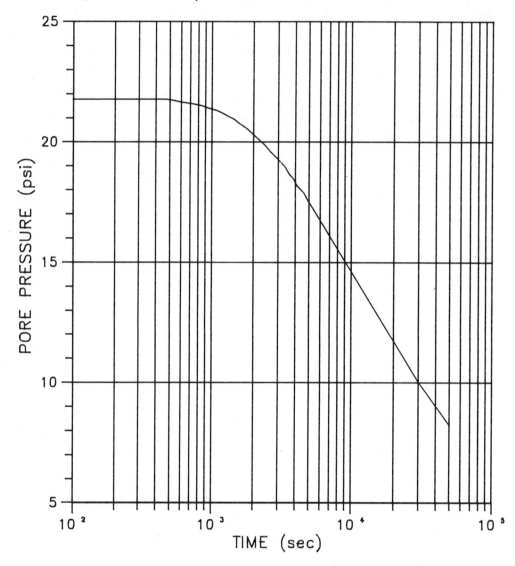

Fig. 3: Typical dissipation curve in SB slurry wall

Deep Subsurface Wastewater Injection in Ioannina, Greece

D.K. Papademetriou
Project Studies and Mining Development Corp. SA (GEMEE SA), Athens, Greece

G.A. Parisopoulos
National Agricultural Research Foundation, Athens, Greece

G.T. Dounias
Edafos Ltd., Athens, Greece

ABSTRACT: A case study for the underground injection of treated municipal wastewater in the town of Ioannina, Greece is presented. The construction, testing and evaluation of an exploratory borehole are described. The injection interval is from 1000 to 1515 m within Triassic Breccia. Cautious injection of the effluent is not expected to cause any environmental problems in the area. The method was used for the first time in Greece, and demonstrated its potential in cases where the effluent quality does not meet the standards for surface disposal, as well as where temporary storage of treated wastewater is required.

1. Introduction

The town of Ioannina is situated in NW Greece (Fig. 1) with a population of around 100,000. It is an old town built on the shore of Pamvotis lake which is overflowing to Kalamas River. The sanitation system consisted up to 1980's of individual pits, while there were numerous illegal connections to the drainage system of the town. The waste from the pits was collected and disposed to swallow-holes not far from the town.

The construction of the central sewage system and waste treatment plant was completed in 1991. The plant has an average capacity of 900 m^3/h and provides secondary treatment. The effluent was planned to be discharged into Kalamas River which has a good water quality and many communities are built along its banks. The area of the river's estuary has a potential for touristic development.

Despite numerous environmental reports and articles which favoured this disposal policy, as the only reasonable solution, the communities concerned strongly objected to the disposal of effluent into the river and forced the cessation of plant operation. After one year of discussions the local authorities agreed to upgrade the effluent quality by adding a tertiary treatment stage. As a temporary short term arrangement it was agreed to inject the effluent into a deep well which was thought to be capable of accepting the present discharge of 200 m^3/h. Deep well injection is a widely used method for the disposal of a variety of wastes [1], [2]. This is the first time it was applied in Greece.

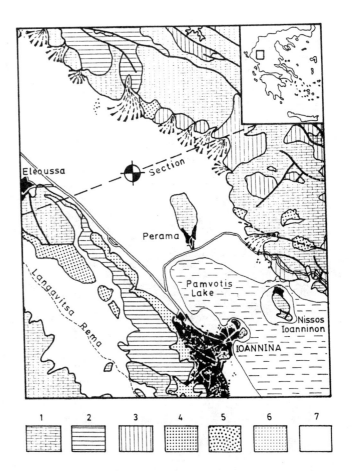

1. Upper Senonian Limestone 2. Cretaceous Limestone
3. Eocene Limestone 4. Pliocene Clays
5. Recent Talus 6. Quaternary Silicate Talus
7. Quaternary Alluvia

Fig 1. Geological map (1:100.000)

2. Well Planning and Construction

Geological information for the area came from shallow (up to 300 m) water wells and surface maps [3]. The injection interval was planned to be within the Pantokrator Limestone (Middle Lias) or the Triassic breccia. These two formations exhibit a high degree of primary and secondary porosity.

In order to locate the best position for the well an electrical geophysical survey was performed according to the Schlumberger method. Measurements were carried out in four profiles in the vicinity of the plant. There were no great differences among these profiles and so it was decided to place the well inside the compound of the treatment plant. The elevation of the well site is 470 m above sea level.

According to the geophysical survey a non-permeable formation consisting of Shales-with-Posidonias (upper Lias) would be encountered at a depth of about 1000 m. Its thickness was expected to be around 50 m. The Pantokrator Limestone was expected to be found under this formation, extending well below 1500 m, the target depth for the well.

The rig used for the drilling was a 150 ton Cooper LTO 550, equipped with two mud pumps (one duplex, one triplex), 3 mud tanks of 110 m^3 capacity, 5000 psi Blow Out Preventers and two 250 KVA generators.

There were no special problems during drilling. The mud was kept at a specific density of 1.10 to suppress an artesian aquifer encountered at about 400 m. The pressure of this aquifer was 5 bar at the wellhead. The cement used for the casing was Class A and was placed at a specific density of 1.75.

The Shales-with-Posidonias was encountered at 600 m and persisted down to 970 m. The Pantokrator Limestone was found between 970 and 1140 m. Underlying this formation was Triassic Breccia within which drilling was completed at 1515 m. A geological cross section through the well as presented by Karakitsios (1993), is depicted in Figure 2. The location of this cross section is indicated on Figure 1.

The well was constructed following closely the initial design as follows:

0 to 350 m: Drill with 17-1/2", case with 13-3/8"
350 to 1000 m: Drill with 12-1/4", case with 9-5/8" liner
1000 to 1515 m: Drill with 8-1/2", set 6-5/8" slotted screens.

Figure 3 provides details of the construction and the geology of the well.

3. Testing and Evaluation [5]

After the completion of drilling and before setting the slotted screen the following logs were performed:

1. Gamma Ray
2. Spontaneous Potential
3. Resistivity
4. Density
5. Neutron Porosity
6. Sonic
7. Dipmeter
8. Temperature
9. Hole geometry

Injection and production tests were also performed. It was found that the well could produce about 4 m^3/h with artesian flow, whereas the injection capacity was 8 m^3/h at 40 bar. In order to improve the characteristics of the formation it was decided to use hydrochloric acid.

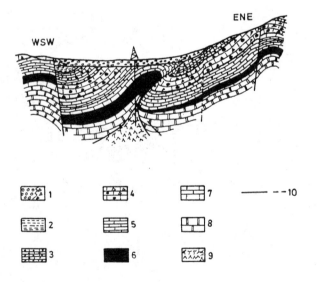

1. Neogene and Quaternary formations 2. Flysch
3. Eocene-Paleocene Limestone 4. Senonian Limestone
5. Vigla Limestone 6. Shales-with-Posidonias
7. Sinies Limestone 8. Pantokrator Limestone
9. Triassic Breccia 10. Tectonic Contact

Fig. 2. Geological section through the well (Karakitsios, [4])

Acidizing took place in five stages at intervals of 2-3 days to allow for the complete removal of reaction products and the evaluation of the results.

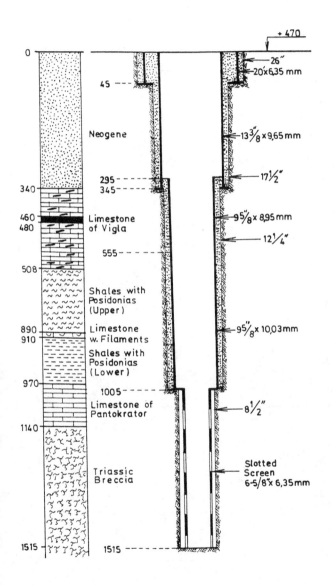

Fig. 3. Geological profile and well construction details

In the first and second stages, 15 m^3 of 15% hydrochloric acid were simply placed within the annulus and left there for about one hour to destroy the mud cake and clean the area around the wellbore. After the second stage the formation could accept 30 m^3/h at 40 bar.

In the third and fourth stage, 30 m^3 of 18% hydrochloric acid containing 70 g/l of $CaCl_2$ to act as a retarder were injected into the formation followed by several hundred m^3 of water. After the 4th stage the injection capacity was 95 m^3/h at 40 bar.

In the fifth stage, 60m^3 of 18% hydrochloric acid containing 70 g/l of $CaCl_2$ were placed into the well followed by several thousand m^3 of water in a 36-hour injection test at a constant rate of 120 m^3/h. (During the first 12 hours the injection rate was 135 m^3/h, but it was decreased because of operational problems). Towards the end of this test the pressure at the wellhead was stabilised at 50 bar (Table 1).

TABLE 1-- Time versus wellhead pressure and injection rate for the 36 hour injection test

Time (hours)	Wellhead Pressure (bar)	Injection Rate (m3/h)
0.0	12	135
2.0	35	135
4.0	40	135
6.0	42	135
8.0	42	135
10.0	43	135
12.0	44	135
14.0	45	122
16.0	46	122
18.0	47	122
20.0	48	122
22.0	48	122
24.0	49	122
26.0	49	122
28.0	50	122
30.0	50	122
32.0	50	122
34.0	50	122
36.0	50	122

Figure 4 gives the injection rate versus wellhead pressure.

Flowmeter tests measured the variation of the injection rate within the injection interval. Table 2 presents this variation by showing the percentages of the total injection or production corresponding to the different formations.

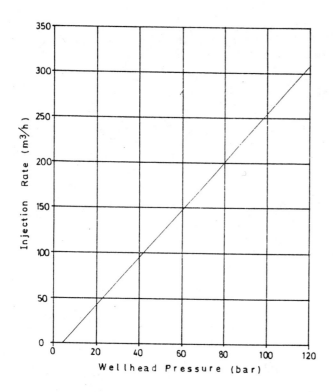

Fig. 4. Injection rate versus wellhead pressure

TABLE 2-- Per cent distribution of injected and produced water along the injection interval.

Interval (m)	Injection (%)	Injection (%/m)	Production (%)	Production (%/m)
998-1045	0	0	0	0
1045-1073	20	0.71	11	0.39
1073-1151	4	0.05	2	0.03
1151-1274	22	0.18	22	0.18
1274-1450	39	0.22	39	0.22
1450-1515	15	0.21	26	0.37

The average porosity of the injection interval was 15%, the average permeability 2×10^{-7} m/sec and the rock compressibility 3×10^{-5} kpa.

After acidizing the artesian flow was 30 m^3/h to be reduced to 28 m^3/h in three months. During this period samples of the water were collected and analysed. Table 3 shows the analyses of the formation water (3 months after acidizing) and the plant effluent (average of several samples over a period of 3 months).

TABLE 3-- Chemical characteristics of formation water and plant effluent (pH is reported in dimensionless units, conductivity in µMHOS/cm and the rest in ppm).

Chemical Characteristic	Formation Water	Plant Effluent
pH	7	7.4
Conductivity	496	1017
Alkalinity p	0	0
Alkalinity m	140	207
Total Hardness	265	341
Ca	88	126
Mg	11.2	7.7
K	1.2	22.4
Na	10.2	72.9
Fe (Total)	0.03	0.04
Mn	0.13	0.06
Cu	0.1	0.03
Zn	0.02	0.03
Cl	60	86.4
SO_4	35.4	74.2
HCO_3	171	253
P as PO_4	0.03	10.8
SiO_2	9	17.4
N as NO_3	0.1	32.3
N as NO_2	0.05	1.75
N as NH_4	0.1	2.21
Residue 110 °C	350	710

The total quantity of chlorine in the outflowing water for these three months is 27,230 kg as can be calculated from Fig. 5, in which the chlorine concentration is plotted against time. The injected chlorine during the acidizing stages 3, 4 and 5 was 26,000 kg. Taking into account that the steady state chlorine concentration is about 60 mg/l it appears that all the injected chlorine came back to the surface. This suggests that there is no migration of the injected fluids and therefore the possibility of the effluent contaminating remote aquifers is very small.

4. Injection Implementation

The deep well injection of the plant effluent was suggested as a temporary solution and will be replaced in a few years by tertiary treatment and disposal of effluent into Kalamas River.

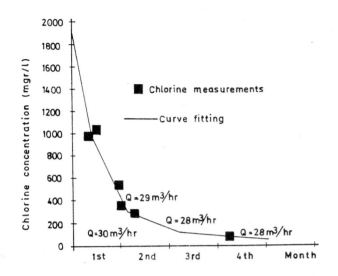

Fig. 5. Chlorine concentration in well flow after acidizing

In this area of the country there is a great demand for irrigation water from May till October. It was therefore suggested - and the local authorities accepted it - that the effluent will be injected only during the winter months. The rest of the year it will be used for irrigation together with the water produced by the well. Care should be exercised, though, in controlling the quality of the water used for irrigation.

The legislation in Greece regarding wastewater injection is practically non-existent. However, the conditions for the Ioannina well are rather favourable for temporary storage on account of the following reasons:

1. The injected water is non-toxic and has a low concentration of pollutants.

2. The injection zone is isolated from the upper aquifers by a thick impermeable layer of shale.

3. Lack of underground movement of aquifer water that could lead to uncontrollable migration of injected effluent.

Among the disadvantages of the method are:

1. Complicated tectonics, abundant faults, many permeable formations, lack of precise geological data.

2. The formation water is of very good quality.

5. Conclusions

1. The characteristics of the formation, the well construction and the mass balance of Cl^{--} suggest a safe temporary disposal of the treated effluent.

2. Acidizing significantly improved injection rates.

3. Temporary storage of effluent during winter and reuse for irrigation during summer is an environmentally sound approach.

6. Acknowledgements

The owner of the deep well is the Municipal Corporation for Water and Sanitation of Ioannina. Their help and permission for publishing the results is gratefully acknowledged.

References

1. Kerr R.S., "An Introduction to the Technology of Subsurface Wastewater Injection". U.S. Environmental Protection Agency, 1977.

2. Smith, M.E., "Solid Waste Disposal: Deepwell Injection", Chemical Engineering, April 9, 1979.

3. Insitut de Geologie et Recherches du Sous-sol - Institut Francais du Petrole., "Etude Geologique de l' Epire", (Grece nord - occidentale), Editor Technip, Paris, 1966.

4. Karakitsios V., "Lithological Study of the Cuttings of a 1515 m Well near the Town of Ioannina.", GEMEE's Internal Report, Athens, 1993.

5. Mehnert E., Gendron R., Brower R., "Investigation of Deep-Well Injection of Industrial Wastes" EPA, 1990

Reemploi de Pneumatiques Usages en Remblais Alleges

J. Perrin
INGEVAL, Veyssilieu, France

J.P. Khizardjian
CETE LYON, France

C. Coulet
IUT A Génie Civil, Lyon, France

Résumé : La recherche de matériaux légers permet d'optimiser les techniques et les coûts de construction en sites compressibles. Le nouveau procédé breveté Pneurésil, qui répond à ces exigences, consiste à empiler des pneus poids lourds en colonne sans les remplir de sol. Des essais de chargement en laboratoire et un remblai expérimental ont permis de valider ce procédé et de tester en vraie grandeur la mise en oeuvre des pneumatiques. Une application en chantier réel est présentée. Il s'agit d'un ouvrage routier franchissant un canal, construit en zone marécageuse compressible saturée. Le procédé Pneurésil a été adopté pour remblayer à l'arrière des culées constituées d'un rideau de palplanches. Ainsi le remblai d'accès à l'ouvrage (densité de l'ordre de 0.4) ne provoque aucune augmentation de contrainte sur le sol support compressible.

Introduction

La construction en site compressible s'avère d'autant plus délicate et onéreuse que les contraintes sont plus élevées, tant verticalement qu'horizontalement. La recherche de matériaux légers permet donc d'optimiser les techniques et les coûts. Le nouveau procédé Pneurésil, qui consiste à empiler des pneus poids lourds en colonnes sans les remplir de sol, répond à ces exigences avec une densité de l'ordre de 0.4. Ainsi les Travaux Publics constituent un débouché important, pour la multitude de pneumatiques retirés chaque année de la circulation (20 000 000 par an en France), qui serviront de matériaux de susbtitution.

Des essais de compressibilité et de fluage ainsi qu'un remblai expérimental ont montré la validité du procédé, qui a été appliqué en Haute-Savoie (France) pour remblayer à l'arrière de culées d'un ouvrage construit en site marécageux [1].

Expérimentations

Les premières expérimentations ont porté sur le chargement d'une colonne constituée de pneus poids lourds, simplement empilés et sans liaison entre eux. Les pneumatiques testés sous charges permanentes étaient usagés et destinés normalement à être mis en décharge. Ces tests ont permis de mesurer la déformabilité de la colonne et de noter que le gradient de déformation verticale diminuait très rapidement avec la charge (figure 1). Ainsi dans le cas d'une application en remblai routier, il est prévisible que la majeure partie du tassement est acquise sous charge permanente (les couches de chaussée) et que la surcharge due à la circulation n'a qu'une faible influence. De plus la déformation horizontale des pneumatiques, sous charge, reste très faible : constatation intéressante dans le cas d'utilisation à l'arrière de soutènements et de culées.

Pour valider le procédé, un remblai expérimental a été édifié sur un site marécageux à Sévrier (Haute-Savoie, France), (figure 2). L'empilement des pneus, en colonnes, d'une hauteur de 2 m, a été effectué sur une couche de 0.50 m de grave sableuse rapportée sur le terrain naturel. Les pneumatiques utilisés ont un diamètre de 1.10 m pour un poids de l'ordre de 60 kg (figure 3).

Les flancs de blocage sont en remblais traditionnels. Afin que le sol ne pénètre pas à l'intérieur des colonnes de pneus, une double structure (géotextile et treillis-soudés) est disposée avant la construction des couches de chaussée (figure 4). Deux types de chaussée, établis sur une couche de forme de 1 m, ont été testés : une chaussée rigide en béton armé de 0.30 m et une chaussée souple de 0.50 m d'épaisseur. Les tassements mesurés sous charge permanente et sous circulation ont permis de conclure que, sous chaussée rigide, le procédé convient parfaitement avec des tassements insignifiants. Par contre, sous chaussée souple traditionnelle, il est nécessaire de différer l'exécution de la couche de roulement définitive pour minimiser ensuite les tassements sous circulation (figure 5).

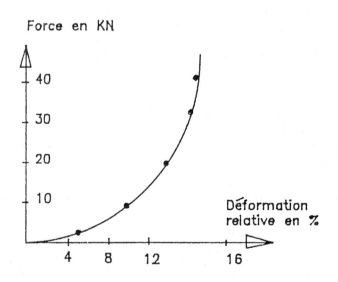

Figure 1 : Essai de Chargement d'une colonne de 4 pneus

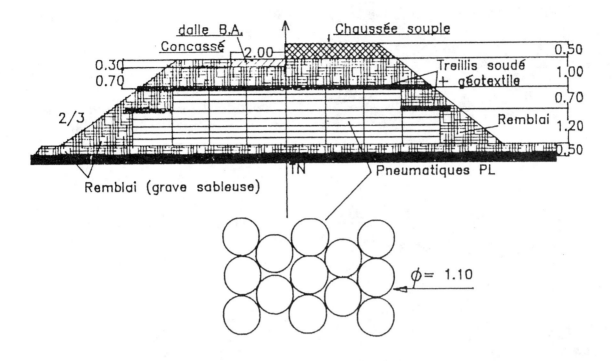

Figure 2 : Profil en travers

Figure 3 : Vue générale du remblai expérimental

Figure 4 : Pose du treillis soudé et du géotextile

Chantier d'Application

Le procédé Pneurésil a été choisi pour construire les remblais d'accès à l'ouvrage de franchissement du canal de Calvi (Haute-Savoie, France) [2]. Etabli en site marécageux compressible saturé, cet ouvrage est fondé sur rideaux de plaplanches encastrés dans le sol porteur. Afin de ne provoquer aucune augmentation de contraintes sur la couche de sol intermédiaire compressible et de minimiser les efforts horizontaux à l'encontre du rideau soutenant les culées, une telle méthode de construction par substitution du sol en place s'est avérée indispensable (figure 6). En outre la solution Pneurésil a permis de tenir des délais très courts et a conduit à une limitation de l'emprise foncière. Préalablement à la mise en oeuvre, les pneumatiques ont été disposés par 5 en colonnes précomprimées et ligaturées (figure 7). Une telle disposition a anihilé le tassement instantané très important évalué à 0.50 m pour une colonne de 10 pneus. Dans ces conditions le massif Pneurésil, sous contrainte permanente de 20 KPa, présente un poids volumique de 3,5 KN/m^3, soit environ 6 fois plus faible que celui d'un remblai ordinaire.

A l'interface entre le massif Pneurésil et la dalle de transition de l'ouvrage a été disposé un treillis soudé recouvert d'un géotextile (figure 8). Ce chantier, constituant une première mondiale, a permis le réemploi de 2000 pneumatiques usagés poids lourds.

Depuis la mise en circulation de l'ouvrage, en octobre 1989, aucune déformation préjudiciable n'a été observée au niveau de la chaussée.

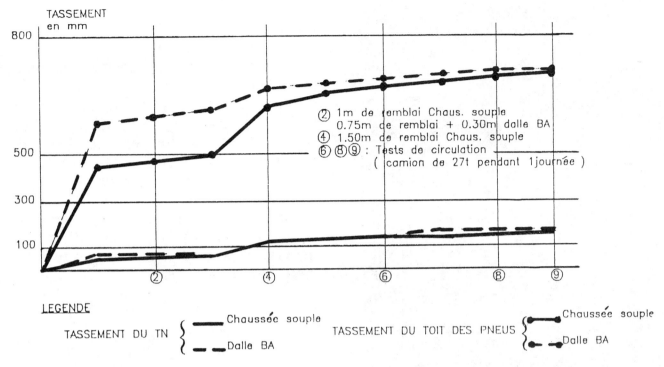

Figure 5 : Tassements mesurés sur le remblai expérimental

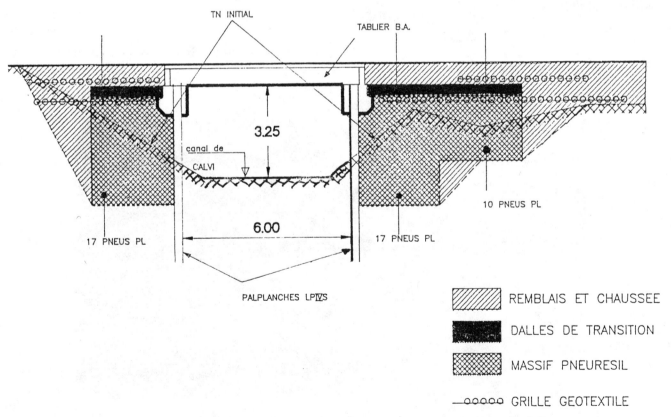

Figure 6 : Coupe du franchissement du Canal de Calvi

Figure 7 : Pose de pneus

Conclusion

Ce chantier d'application en remblais légers démontre le bon comportement du massif Pneurésil et la fiabilité de la méthode utilisée. Ce procédé peut également être adopté en remblai de remplissage au dessus des conduites enterrées (réduction notable de l'Effet MARSTON) et pour la constitution de bassin de rétention hydraulique enterré. Ainsi les travaux publics constituent un secteur important de valorisation des pneumatiques usagés.

Références :

[1] Marchal J., Remblais allégés et remblais renforcés. Symposium International, Tiaret, Algérie, 1989.
[2] Marchal J., Perrin J., Rauber S., Utilisation des déchets pour l'édification de remblais sur sol compressible. Rapport CETE LYON, 1990

Figure 8 : Pose du treillis soudé et du géotextile

Mechanical Properties of Stabilized Light Soil Using Expanded Polystyrene

Tej B.S. Pradhan, Associate Professor, Goro Imai, Professor
Yokohama National University, Yokohama, Japan

Masami Uchiyama, Chief Researcher
Dainippon Ink and Chemicals, Inc., Tokyo, Japan

ABSTRACT: A series of consistent laboratory tests were performed in order to investigate the mechanical properties of a new light geotechnical material called Stabilized Light Soil (SLS). The new material SLS is fabricated by mixing sand and Expanded Polystyrene (EPS; either beads or crush) with some amount of water and some amount of cement additive as stabilizer. Unconfined compression tests were carried out on specimens fabricated under different conditions as (a) unit weight, (b) percentage of cement additive and (c) curing time. It was found that the unconfined compressive strength q_u and stiffness E_{50} of this new material is significantly affected by the density, the cement additive and the shape of EPS. It is also shown that the mechanical properties of this new material SLS is quite similar to other geotechnical materials.

INTRODUCTION

Recently as one of the municipal wastes, the treatment of boxes made of Expanded Polystyrene beads used as buffer material for the transportation of sensitive equipments like computers etc., is becoming a great environmental problem for its treatment. Also a great problem is arising on the treatment of excavated soil (or residual soil) produced from the ground improvement site specially using replacement methods. The main objective of the present study is to reuse the above mentioned waste materials and to manufacture a new light geotechnical material. The main objective of using such light weight construction material is to reduce (a) the settlement of soft subsoil, (b) the earth pressure behind retaining structures and (c) driving moment in slope stability calculations. So far different kinds of light material such as EPS blocks, air-entrained mortar, slags and hollow pipes have been used. However, these methods are quite costly and the treatment of the left soil is always a great problem. As has been pointed out by Shimazu(1989), it will be economical reasonably to fabricate light weight material using the insitu spoil soil. One of the countermeasures is to mix the insitu soil with EPS (either beads or crush) and add some amount of cement as stabilizer. Here this new light material is called STABILIZED LIGHT SOIL (SLS). The second purpose of this SLS is the reuse of the waste EPS scrap. For that reason crushed as well as beads of EPS have been used to fabricate the SLS in the laboratory. The main objectives are to investigate (1) the strength and stiffness of the new material SLS fabricated under different conditions and (2) the possibility of effective reuse of waste EPS crush.

In this paper, the results obtained from a series of laboratory unconfined compression tests on SLS specimens fabricated under different conditions are discussed and a strength envelope has been proposed.

FABRICATION OF SPECIMEN

Kimizu sand (γ_s=27.3kN/m³; D_{50}=0.19mm; U_c=1.75; fine contents <5%), EPS beads (γ=0.2kN/m³; diameter ≈1mm), EPS crush (γ=0.25kN/m³; size≤8mm), ordinary Portland cement were used for the preparation of specimen. In order to know the adequate mixing ratio of sand and EPS, preliminary compaction tests were performed following JIS1210 (based on low compaction energy). The mixing ratio for each specimen density and the specimen conditions are shown in Table-1.

Table-1 Specimen Conditions

MIXING RATIO OF DRY SAND AND EPS	
DENSITY	RATIO BY WEIGHT
0.9Mg/m³	1000:28
1.1Mg/m³	1000:20
1.3Mg/m³	1000:14.6
CEMENT ADDITIVE BY WEIGHT	
4%, 6%, 10% RELATIVE TO DRY SAND	
WATER ADDITIVE BY WEIGHT	
10% RELATIVE TO DRY SAND	
CURING TIME	7, 14, 28 Days

A fixed quantity of each additive is measured and mixed well with sand. The mixture is then poured into a plastic mold (diameter 5cm and height 12 cm with collar) in three layers and each layer was tamped to get the prescribed density. The mold was then wrapped by a thin plastic sheet to prevent the evaporation of water and cured in a room with constant temperature (23°C) and humidity (60%). After a specified curing time, the specimen (diameter 5cm; height 10cm) is taken out from the mold, dimensions measured, weighed and subjected to unconfined compression test. The compressive axial strain speed was kept constant at 1%/min.

TEST RESULTS AND DISCUSSIONS

For each identical condition (mixing ratio, curing time) three tests were performed in order to check the reproducibility of the specimen. Fig.1 shows a typical test results of SLS with either EPS beads or crush for three specimens fabricated under identical conditions. It can be seen from the figure that the reproducibility of the fabricated specimen is quite satisfactory. The fluctuation in the strength is utmost 5%. In all the test results hereafter, the average value of the data obtained from three specimens are used.

Figs 2(a), (b) show SLS specimens (density: 1.1Mg/m³, cement additive: 6% and curing time: 14days) at failure fabricated using EPS beads and EPS crush respectively. SLS with using EPS crush seemed more ductile as compared with that using EPS beads.

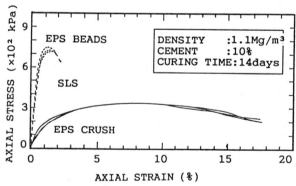

Fig.1 Typical Stress Strain relation of SLS

Effect of Cement Additive

Unconfined compressive strength q_u

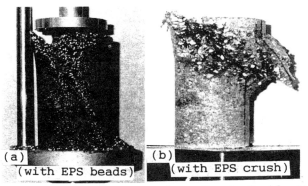

Fig.2 Specimens of SLS at failure

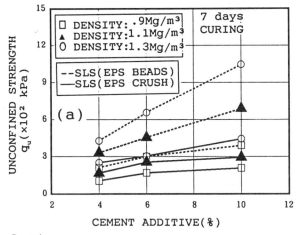

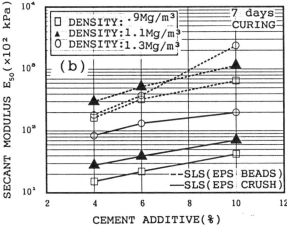

Fig.3 Effect of Cement additive on SLS

against percentage of cement additive is shown in Fig.3(a) for both SLS specimens using either EPS beads or crush with curing time of 7 days. It can be seen that a q_u value of greater than 100kPa can easily be achieved just by the addition of 4% of cement. q_u increases with the increase in the percentage of cement additive but the tendency seems greater for the specimen using EPS beads. For the identical condition, q_u of SLS with EPS crush is about 40-60% of that for SLS with EPS beads.

Stiffness in terms of secant modulus E_{50} against percentage of cement additive is shown in Fig.3(b). E_{50} is significantly affected by the quantity of cement additive. However, E_{50} for SLS (EPS beads) is about 10 times as compared with that for SLS (EPS crush). The reason may lie on the fact that when a EPS bead is crushed the rigidity of the material itself is drastically decreased.

Effect of Density

q_u against density of a specimen is shown in Fig.4(a) for both SLS specimens using either EPS beads or crush with curing time of 7 days. It can be seen that q_u increases proportionally with the increase in density. For the higher density since the amount of EPS present is small, the strength of the SLS seems to be governed by the sand.

E_{50} versus density is shown in Fig.4(b). The increase in E_{50} is predominant at the higher density for the SLS with EPS crush as compared with that for SLS with EPS beads. This figure shows that a similar E_{50} can be achieved for a SLS using EPS crush by increasing the density of the mixture. Similar tendency was observed for other curing times.

Effect of Curing Time

q_u against curing time is shown in Fig.5(a). At the lower cement content, q_u seems almost independent of the curing time.

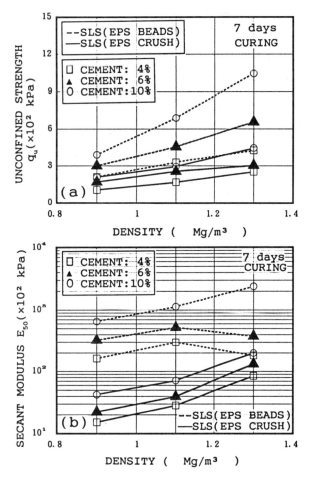

Fig.4 Effect of density on SLS

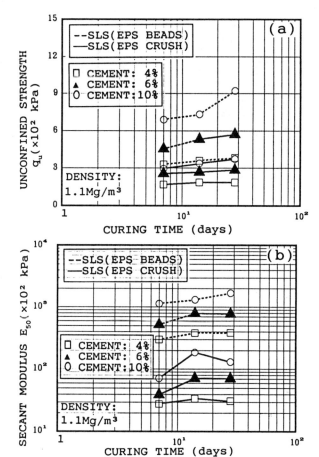

Fig.5 Effect of curing time on SLS

But at the higher cement content q_u seemed to increase somewhat with the curing time. However, more than 80% of the strength has been exhibited at 7 days of curing time. E_{50} against curing time is shown in Fig.5(b). E_{50} seems to be almost independent of the curing time for the range used in the present study. It can be concluded that q_u and E_{50} exhibited at 7days of curing can be sufficient to be used for designing purposes.

Strength Envelope

In order to realize the overall effect due to density and quantity of cement additive on the unconfined compressive strength, a three dimensional representation is shown in Figs 6(a) and (b). Nine averaged data points for specimens with curing time of 7 days are presented for SLS with EPS beads {Fig.6(a)} and crush {Fig.6(b)}. A curved surface connecting those points can be considered as a strength envelope for each type of SLS. It is quite obvious that the strength envelope is rising up towards the density axis implying that the effect on q_u due to density is more predominant than that due to cement content. This tendency is quite similar but the degree seems different for SLS with EPS beads and EPS crush. q_u can be estimated for different

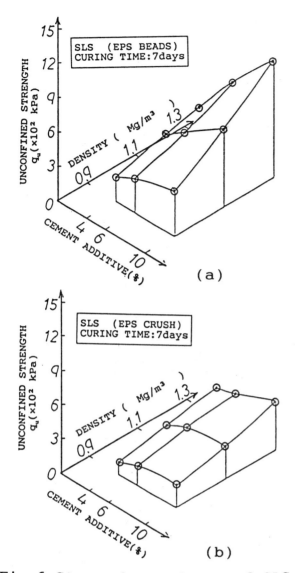

Fig.6 Strength envelopes of SLS

mixing conditions from this strength envelope.

Correlation Between Strength and Deformation Properties

In order to investigate the deformation behavior of the new material SLS, axial strain at failure (ϵ_f) is plotted against the unconfined compressive strength (q_u) in Fig.7 for all the specimens under different conditions. Following points can be noticed from the figure: (1) for the SLS specimens with EPS beads, strain at failure is about 3% (fracture mode is brittle) and is almost independent of the mixing ratio, (2) for the SLS specimens with EPS crush, ϵ_f largely depends on the unconfined compressive strength. When q_u is small, the fracture mode is rather ductile. As a general tendency, the deformation mode largely depend on the density rather than the cement content and is predominant for the SLS with EPS crush.

Stiffness E_{50} is plotted against q_u in Fig.8 for all the specimens. E_{50} increases proportionally with q_u in both logarithmic scale, whereas the tendency seems a somewhat different for the two types of SLS especially at lower value of q_u. In the same figure, data for Kaoline clay, Tokyo Bay clay and cement

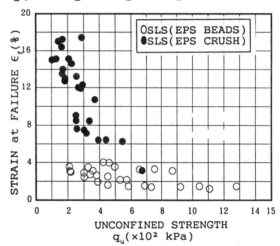

Fig.7 q_u versus strain at failure

treated sand are shown (cited from Tatsuoka et.al., 1991). It is obvious from the figure that the data for different kinds of materials lie in a certain range showing similar properties. It can be said that the new material SLS is not an unusual material showing peculiar behavior but can be confidently used as a new light geotechnical material.

Estimation of Required Strength for SLS with EPS Crush

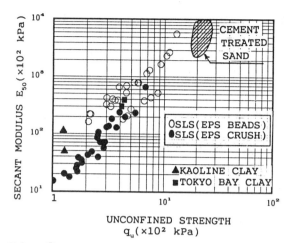

Fig.8 q_u versus secant modulus

A method of estimating a required design strength or stiffness for SLS with EPS crush for a specified density has been proposed in Fig. 9. For example, the strength q_u of 420 kPa that can be achieved from SLS with EPS beads in 7 days can also be achieved from SLS with EPS crush by increasing the cement additive either to (a) 6.5% in 7 days curing, or (b) 7.9% in 14 days curing, or (c) 9.5% in 28 days curing. An optimum value of cement additive can be chosen so as to get a required strength from SLS with EPS crush.

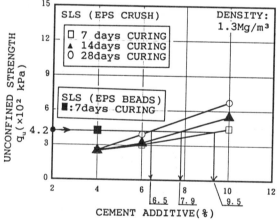

Fig.9 Estimation of q_u for SLS with EPS crush

CONCLUSIONS

A method of reusing environmental wastes like expanded polystyrene materials, spoil soil as a new light geotechnical material has been proposed. Mechanical behavior of the new material called Stabilized Light Soil (SLS) has been investigated in the laboratory.
Followings are the findings from the present study:
(a) The effect of density on the behavior is predominantly higher as compared with that due to cement additive or curing time.
(b) Strength and stiffness of the SLS with EPS crush is 40-60% less than that of the SLS with EPS beads fabricated under the identical conditions.
(c) Strength envelopes with respect to specimen density and cement quantity for this new material are presented.
(d) SLS with EPS crush or beads can be used as a new light geotechnical material and can secure a sufficient strength and stiffness to be used in the field.

REFERENCES

Shimazu, A. (1989). A prospect of lightweight fill construction method. Tsuchi to Kiso, 37-2, (in japanese).

Yamada, S., Nagasaka, Y., Nishida, N. and Shirai, T. (1989). Light soil mixture with small pieces of expanded polystyrol and sand. Tsuchi to Kiso, 37-2, (in Japanese).

Tatsuoka, F. and Shibuya, S. (1991). Deformation characteristics of soils and rocks from field and laboratory tests. Keynote lecture in 9th Asian regional Conference on SMFE, Bangkok.

Pradhan, T., Imai, G., Hamano, M. and Nagasaka, Y. (1993). Failure criterion of a new light geotechnical material SLS. Proc. 3rd Int. Offshore and polar Engg. Conf, Vol.1, Singapore.

Environmental Impact Assessment of Greater Dhaka City Flood Protection Structures

M.H. Rahman, B.Sc.Engg., M.Sc.Eng.(BUET), Ph.D. (UK), MIEB, MISSMFE
Department of Civil Engineering, Bangladesh University of Engineering & Technology (BUET), Dhaka-1000, Bangladesh

Abstract : This study focuses on the environmental degradation of Greater Dhaka City (GDC) areas due to pollution created by domestic wastes, industrial wastes and to some extent agricultural wastes. These wastes are not handled in an organized way. Only 15% of the total population of GDC are being served by water and sewage authority. Rest of the population either have their own sanitation facilities or nothing at all. The surface drains indented to carry storm water, also carry human excreta, organic solid wastes and waste water from industrial and agricultural sources. These environmental problems will be amplified with the construction of Greater Dhaka City flood protection structures (GDCFPS). In this study, an attempt has been made to evaluate the impact of the projects with the context of major salient environmental issues.

Introduction

GDC the capital of Bangladesh is situated in the central part of the country and is surrounded by the rivers and canals (khals) on all four sides (Figure 1). The area of GDC is 265 sq. km. The population of the city was 4.8 million in 1990 [1]. The climatic condition of this area is tropical monsoon type. The temperature varies from 21^oC to 35^oC in summer and 10^oC to 30^oC in winter. Monsoons start in July and stay to October. This period accounts about 90% of total rainfall and the average annual rainfall is about 2000 mm.

Although, Bangladesh is subjected to perpetual floods every year, she suffered two of the most devastating floods on record in 1987 and 1988. Vast areas of the country including GDC were flooded to an unprecedented degree with a flood level, 1.5 m higher than the normal level for a period of about four weeks. About 77% of GDC area were submerged to depths ranging between 0.3 m to 4.5 m and about 60% of city population were directly affected by these floods. City life was totally disrupted during this period, causing enormous suffering of the dwellers. In this circumstances, the Government of Bangladesh decided to construct flood protection structures to surround the GDC area in collaboration with other international support agencies. The most of the construction work of the phase I (Figure 1) has already been completed. The construction of phase I work was started so hurriedly that it lacks the assessment of impact of these structures on GDC environment. Therefore, this study focuses on the impacts assessment of GDCFPS for major environmental issues.

Wastes disposal in GDC

In GDC, the total generated solid wastes is about 2050 Mg/day to 3000 Mg/day. About 50% of these

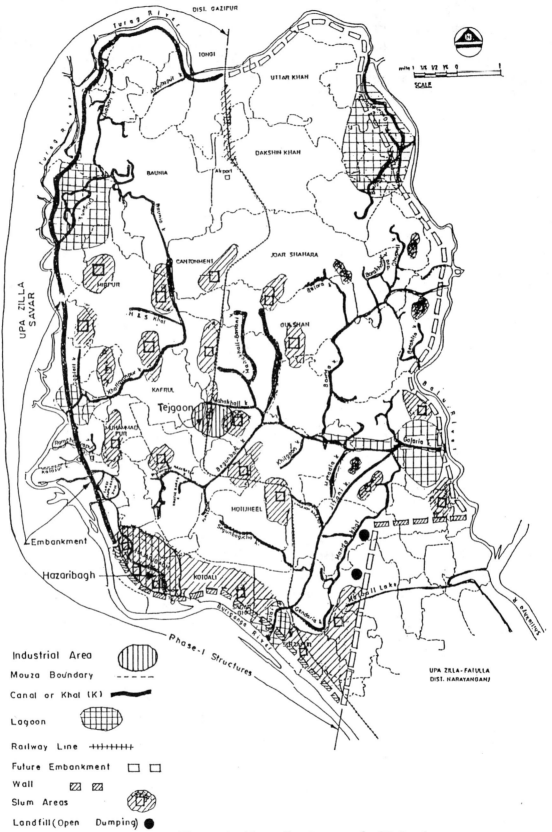

Figure 1: Key Features of GDC Area.

solid wastes is being collected by Dhaka City Corporation. Therefore, the remaining 50% of solid waste is commonly discarded into street, drainage ditches, khals and open spaces. Again, the collected solid wastes are also being dumped in open spaces. Exposed dumping of organic solid wastes, which contain 80% to 85% food wastes [2] will produce strong leachate. The characteristics of leachate collected from the bottom of the bins shows that it contains 5000 mg/l to 15000 mg/l of BOD_5 and 3000 mg/l to 14000 mg/l of suspended solids [2]. The overflow of these leachates and other organic wastes will ultimately cause ground and surface water pollution. Since the solid wastes in Bangladesh mostly contain organic wastes, the low-cost technology for their safe disposal would be production of bio-gas and compost. Bio-gas generated can be used as energy and sludge from bio-gas plant, be composted, and also be used as fertilize and soil conditioner.

There exists a large number (about 1125) of slums in GDC area (Figure 1). About 30% of total GDC population lives in these slum areas [3]. Dhaka Water and Sewage Authority (DWASA) can not provide drinking water to about 45% total GDC population. Therefore, the most of the slum dwellers do not have any definite source of water for their domestic uses. Most of them rely on open water bodies for their daily use. Hence, environmental degradation mainly due to water pollution is the major cause of diseases in the slum settlements. According to the report of the Task Force [4], at any time 30% to 40% of slum dwellers suffer from water or air-borne diseases. So the present practice of water supply and waste disposal system require special attention to improve the situation.

The present status of GDC sanitation practice as presented in Table 1 [2] clearly indicates that 30% of city's population have an unsanitary system consisting of kutcha latrines or open defecation directly into the living environment.

Table 1: Sanitary Coverage of GDC in 1990

Sanitary Practice	% Coverage
Septic tank	40
Kutcha latrines or open defecation	30
Pourflush latrines	15
Sewage system	15

Where sanitary facilities are not available in GDC, the septic tank effluent and sludge, bucket latrine wastes and decomposed residential, commercial and industrial wastes are being discharged into storm drains or open water bodies without regard to the effluent quality and their detrimental effects on the living environment. The existing sewage system (15%) of GDC is currently operating at its maximum capacity. The present sanitation practice can be improved by providing comprehensive sanitation policies. Forty percent of the population have a septic tank system that can be modified by connecting the septic tank to a small-bore sewage system. This system will be less expensive compared to a conventional sewage system. As thirty percent of the population of GDC are slum dwellers and in most cases they do not have proper sanitation practice, awareness regarding environmental sanitation has to be created among them by providing low-cost latrines. Then they will use latrines instead of open defecation into the living environment.

About 554 different types of industry are situated within GDC. Among these, 160 tannery industries are concentrated in a particular area (Hazaribagh) on the bank of the River Buriganga, 166 textile industries, with some other industries, are concentrated in the central part of GDC (known as Tejgaon industrial area) (Figure 1) and the rest of the industries are scattered in different areas of GDC. These are small to medium-scale enterprises. A huge amount of water is therefore required for different industrial uses. Only a portion of that water is incorporated in their product and lost by evaporation; the rest finds its way into open water courses as waste water. The typical effluent qualities of selected industries are presented in Table 2 [2].

Table 2: Typical Effluent Quality of Selected Industries

Industry	Tanneries	Textile or Dyeing
pH	8.6-10.4	6.8-11.8
BOD_5 (mg/l)	660-2800	180-1000
Suspended solids (mg/l)	1300-1500	100-5000
Toxic substance (mg/l)	Chromium = 0.05-10.5 Ammonia = 100-135	Chromium = 9.6

Therefore, to improve the waste water quality of these industries, a clearly defined effluent quality standard is required so that the different industries will employ suitable treatment methods for the safe disposal of their wastes into natural bodies of water.

At present about 45% of total GDC area (265 sq. km.) is being used for agricultural purposes. After the completion of GDCFPS, there will be reduction of agricultural landuse due to switch over of agricultural land to other profitable landuse. However, the overall agricultural activity will increase due to use of more agricultural land throughout the whole year instead of dry seasons, and due to investment of more money and labour in agricultural land for the safe return (as there will be reduction of high flood damage with the construction of GDCFPS). But, the indiscriminate use of chemical fertilizers and pesticides/insecticides could cause ground and surface water pollution. To minimize these problems, the general public must made aware of this situation, so that they will use different fertilizer or pesticides and only under the supervision of trained personnel.

Impact of the project

After the completion of the GDCFPS, the existing surface run-off towards the river system will be cut-off. Hence, the waste water from different waste disposal system will tend to store in five lagoons (Figure 1) and other low-lying areas (mostly in agricultural land). Most of the time of the year except dry seasons (November to February) some pool of water remains at the low-lying area of the city. The water and waste water quality of presently formed two lagoons, different khals and surface wash from slum areas are presented in Table 3. From Table 3, it is apparent that the lagoon water is polluted but not in an alarming degree. However, it has to be noted that GDCFPS is not yet functioning. From the waste water quality data presented in Table 3 for different khals, it is clear that the dissolved oxygen at different khals is zero.

Table 3: Typical Water and Waste Water Quality of Lagoons, Khals and Surface Wash from Slum Areas

Collection sources	Lagoons	Khals	Surface wash from slum areas
pH	6.7-7.1	6.0-6.7	6.4-6.7
Suspended solids (mg/l)	150-300	650-2100	800-3000
Dissolved oxygen (mg/l)	-	Nil	Nil
BOD_5 (mg/l)	40-150	220-460	810-1610
COD (mg/l)	60-150	810-1610	850-1900

After the completion of this GDCFPS, these khals will discharge into the five lagoons and it can be inferred that the quality of lagoon water will be very low. At present waste water from drainage khals and low-lying areas are being discharged at many points and thus utilize the self assimilating capacity of the river. After the completion of the project, waste water will be discharged at a few selected locations. Under these conditions, the river's self purification capacity may fail due to discharge of huge pollution loads at selected points. Then the river system might become a discrete pool of water with a number of anaerobic zones. Thus the river resources will be unusable and aquatic life will be destroyed.

Although Bangladesh is a rural country, according to the World Bank [5] the total urban population is quite high and its increasing rate is quite significant, 6.2% compared to 5.6% by the least developed countries. The annual population growth rate of Bangladesh is less than 3% but that of GDC area is about 8% to 10%. Again, it is anticipated that after the construction of GDCFPS, the area will become a flood free zone. This will encourage in-migration mainly slum pollution. Hence, it will create additional load to the waste disposal as well as water supply system. To improve this situation, a physical planning and landuse control legislation is required. Introduction of good transport facilities and shifting of labour intensive industries outside GDC area will also minimize these problem.

At present in GDC, a number of organizations are working at different sectors with their organizational capacities. DWASA is looking after for water supply and waste water disposal system. Dhaka City corporation is dealing with solid waste management system. Rajdhani Unnayan Katripakha is responsible for housing development. Department of Environmental is responsible for monitoring purpose. Therefore, proper co-ordination is essential among different organizations.

Conclusion and recommendations

From the present study, it is evident that before the construction of the GDCFPS, the waste water of different drainage system are highly polluted (Table 3). Then the waste water from these drainage system discharges into the surrounding rivers at many points thus utilizing the self assimilating capacity of the rivers. But after the construction of GDCFPS, there will be the accumulation of polluted water from domestic, industrial and agricultural sources inside the protective structures and will severely deteriorate the surface and ground water quality. Again, the discharge of these waste water from low-lying areas or lagoons will be at selected location of the rivers. Then, the river's self purification capacity may fail and the river system might become discrete pool of water with anaerobic zones. Therefore, proper steps have to be taken to minimize these problems by improving sewage system, solid waste management system and through slum development. On a critical examination of the existing waste management systems [2], the following recommendations are made to improve the situation:

1) The general public of GDC have to be made aware of environmental sanitation and health education.
2) Low-cost sanitation facilities have to be provided for slum dwellers for the improvement of their sanitation practices. Small-bore sewage system can be employed for the disposal of septic tank effluent.
3) The development of a strategy for proper solid waste management is required to control surface and ground water pollution. This can be done by using sanitary land filling or anaerobic digestion and composting for the production of bio-gas and compost.
4) Industrial wastes have to be treated before their disposal into water bodies and into the living environment. Again, to implement this, well-defined effluent quality standard and legislation are required.
5) The physical planning and landuse control legislation is urgently needed.
6) To improve the overall wastes management system in GDC, it is essential that the agencies involved in the city's development and in monitoring and controlling the GDC environment should recognize the nature of the problem for the development of pollution control legislation and standards.

With proper steps to overcome the problems, GDCFPS will be a very beneficiary project. It will protect the capital city of 265 sq. km. from flooding and ensuring more than 4.8 million people a flood free life.

Ackwoledgement

The author acknowledges United Nations Center for Regional Development (UNCRD), Japan for providing financial support for conducting this study.

References

[1] JAPAN INTERNATIONAL COOPERATION AGENCY (JICA, 1991) Dhaka Protection Project, Flood Action Project FAP 8A.

[2] RAHMAN, M.H. (1983), West Management in Greater Dhaka City, The International Journal of Environmental Education and Information, Vol-12, Number 2, pp. 129-136.

[3] RAHMAN, M.H. AND ISLAM, A.K.M.N. (1992), Impact of Wastes from Slum Areas on Environment in Greater Dhaka City, Proc. of the Eight International Conference on Solid Waste Management and Secondary Materials, Philadelphia, U.S.A.

[4] TASK FORCE (1991), Report of the Task Force on Social Implication of Urbanization, Planning Commission of Bangladesh.

[5] WORLD BANK (1992), World Development Report, World Bank.

Etudes et Applications du Renforcements des Sols par le Procede Plasterre

J.D. Rakotondramanitra, C. Coulet, R. Azzouz
Laboratoire Génie Civil Habitat Environnement, Villeurbanne Cedex, France

Résumé : Les déchets de matières plastiques d'origine industrielle ou ménagère représentent actuellement une quantité importante de matériaux qui soulèvent quelques problèmes pour l'environnement : leur élimination et leur recyclage sont difficiles et coûteux. Nous présentons dans cet article le procédé Plasterre qui permet de valoriser ces déchets, matériaux récupérables à faible coût, en les utilisant dans le domaine des Travaux Publics. De nombreux essais de laboratoire, un remblai expérimental ainsi que des applications réelles ont permis de démontrer que ce procédé améliore considérablement les caractéristiques mécaniques des sols et permet ainsi le raidissement des talus et la diminution des poussées des terres derrière les soutènements.

Introduction

Les déchets de matières plastiques produits par nos sociétés sont de plus en plus abondants et créent des nuisances vis-à-vis de l'environnement. La mise en décharge est coûteuse car elle doit être contrôlée. Leur incinération provoque des rejets gazeux dans l'atmosphère et devient une source dangereuse de pollution. Leur recyclage qui nécessite des tris et des lavages préliminaires est difficile et onéreux.

Le procédé Plasterre [1] utilise ces déchets sans aucune transformation préalable. La technique consiste, lors de la construction de remblais, à inclure une nappe de déchets de matières plastiques au milieu de chaque couche de sol à compacter. Ils sont simplement répandus pour former chaque nappe.

Les matières plastiques peuvent être en mélange de différentes natures (polyoléfines, PVC, ...) et de différentes formes (films, emballages, lanières, fils,...) et doivent être exemptes de traces de produits contaminants pour le milieu récepteur.

L'étude a consisté d'abord à quantifier l'amélioration des caractéristiques mécaniques des sols ainsi renforcés en adoptant une démarche comparable à celle utilisée pour l'étude de la terre armée [2]. Des essais de laboratoire, suivis d'une étude en vraie grandeur sur un remblai expérimental, ont permis de montrer la pertinence du procédé, qui a donné lieu à la réalisation de deux chantiers d'aménagement routier représentant la mise en oeuvre de 120 tonnes de déchets plastiques.

Etudes en Laboratoire

Des études en laboratoire ont été d'abord menées afin de caractériser les propriétés mécaniques des sols ainsi renforcés. Elles ont principalement consisté à réaliser :

- des essais de cisaillement de grandes dimensions à l'appareil triaxial (diamètre = 20 cm, hauteur = 40 cm) et à la boîte de Casagrande (section = 60 x 60 cm).

- des essais de poussée dans une cuve parallélépipédique de 25 m^3.

La démarche suivie pour chaque type d'appareil a été d'effectuer dans un premier temps des essais avec le sol seul, non renforcé. Ensuite le sol est renforcé par des nappes continues et enfin par des nappes de morceaux mélangés. L'analyse et la comparaison des résultats ont permis de dégager les conclusions suivantes :

Essais de Cisaillement au Triaxial et à la Boîte

- Le renforcement se traduit par l'apparition d'une pseudocohésion directionnelle qui augmente avec le nombre de nappes et la quantité de renforts par nappe [3]. L'angle de frottement interne varie peu (figure 1).

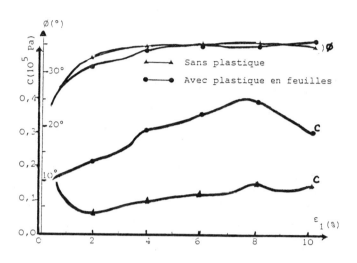

Figure 1 : Evolution de la cohésion C et de l'angle de frottement ϕ interne en fonction de la déformation axiale.

- Une équivalence a été établie entre une nappe d'éléments discontinus et une nappe d'éléments continus et de même nature. Pour les matières plastiques étudiées elle est de 4 masses d'éléments discontinus pour une masse d'éléments continus.

Cette équivalence est nécessaire pour pouvoir appliquer les méthodes de dimensionnement des ouvrages renforcés [4].

Ce résultat a été obtenu exclusivement à partir d'essais triaxiaux. La boîte de cisaillement classique ne nous a pas permis d'approfondir cette notion.

En effet, dans ce cas, la direction du plan de cisaillement est horizontale, et celle des nappes de renfort verticale.

C'est pourquoi, a été conçue la nouvelle boîte de cisaillement de grandes dimensions (longueur = 60 cm; largeur = 40 cm ; hauteur = 30 cm) dont la spécificité réside dans le fait que deux paramètres importants sont plus conformes à la réalité (figure 2a).

Le plan de cisaillement est vertical et la nappe de renfort disposée horizontalement (figure 2b).

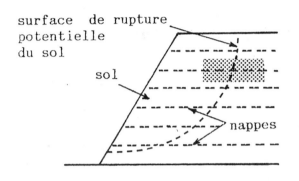

a) Configuration in-situ

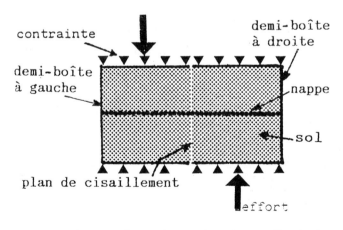

b) Configuration dans la nouvelle boîte de cisaillement

Figure 2 : Directions de la surface de rupture et des nappes de renforts.

Essais en Cuve de 25 m³

Des essais en semi-grandeur ont été réalisés dans une cuve parallélépipédique de 25 m³ (longueur = 5 m ; largeur = 2,5 m ; hauteur = 2 m) [5]. Ils ont permis de démontrer que le renforcement apporte une réduction de la poussée des terres derrière les soutènements et une augmentation de la pente des talus (figure 3).

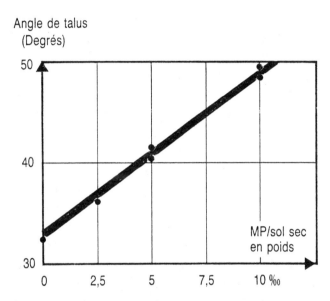

Figure 3 : Variation de l'angle de talus.

Remblai Expérimental

Un remblai expérimental en vraie grandeur de 1500 m³ (longueur = 30 m, hauteur = 4 m, largeur en tête = 4 m, largeur à la base = 12,30 m) a permis de confirmer les résultats précédents. La pente de talus dans la zone renforcée était de 60°. Le sol constituant ce remblai a été choisi de qualité médiocre. Les déchets de matières plastiques utilisés étaient composés essentiellement de films d'emballages industriels en polyéthylène et de bouteilles en PVC. Ils ont été répandus à raison de 4 kg/m². L'espacement des nappes était de 0,40 m.

Construction d'un Merlon de Protection Phonique

Le procédé Plasterre a également été appliqué en 1991 pour la construction d'un merlon anti-bruit le long d'un tronçon d'autoroute dans le département de la Savoie (France). La hauteur est de 1,80 m, la pente de talus 60°, l'emprise au sol de 3,10 m (figure 4). Les nappes de déchets plastiques ont été étalées tous les 0,30 m à raison de 3 kg/m². Le compactage a été réalisé selon la réglementation des terrassements routiers. Un léger toit a été aménagé en crête du merlon pour éviter la stagnation des eaux au sommet. Le merlon est entièrement végétalisé.

Le suivi de ce merlon a montré une stabilité permanente de l'ouvrage.

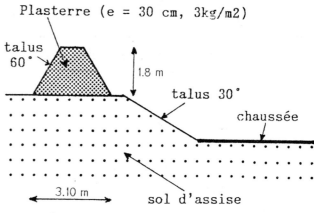

Figure 4 : Coupe transversale

Aménagement d'une Route de Montagne

Le procédé Plasterre a été appliqué en 1987 pour l'aménagement d'une route de montagne : la route départementale n° 81 dans le département du Rhône (France). Les travaux ont consisté à l'édification d'un remblai de 11000 m³, d'une hauteur de 8 m avec une pente de talus de 55° (figure 5). Les nappes de renforts, espacées de 0,40 m, étaient constituées de déchets de matières

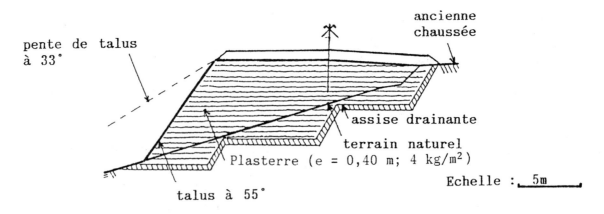

Figure 5 : Coupe transversale de l'aménagement de la route de montagne

plastiques épandus à raison de 4 kg/m² [6]. Elles ont nécessité 110 t de déchets issus de rejets industriels et constitués de films et bandelettes de polyéthylène et de polypropylène.

Le remblai a été équipé de repères topographiques et d'un inclinomètre pour le suivi des déformations verticales et horizontales (figure 6).

Les observations effectuées ont montré un bon comportement de cet ouvrage (figure 7).

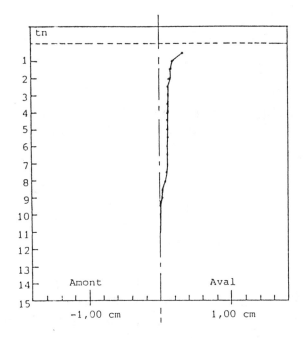

Figure 6 : Mesures inclinométriques (déplacements relatifs)

Figure 7 : Vue générale d'une route de montagne aménagée

Conclusion

L'étude, conduite en collaboration avec le bureau d'études Ingéval (Isère, France), a permis de dégager un certain nombre de points importants qu'on peut tirer du procédé Plasterre :

- sur le plan environnemental : valorisation des déchets de matières plastiques, réduction de certaines pollutions inhérentes à ces déchets.

- sur le plan économique : procédé à faible coût, réduction des emprises foncières grâce au raidissement des talus, diminution des volumes de terre mis en oeuvre.

- sur le plan mécanique : amélioration des propriétés mécaniques des sols, diminution des poussées derrière les soutènements, mise en oeuvre simple.

Références

[1] "Procédé pour armer des matériaux naturels ou non et, notamment, le sol" Procédé PLASTERRE - BREVET n° 88 - 420 048 - 6

[2] Long N.T., Guegan Y., Legeay G. (1972) - "Etude de la terre armée à l'appareil triaxial". Rapport de Recherche du L.C.P.C. n°17, PARIS.

[3] Rakotondramanitra J.D. (1988), Renforcement des sols par nappes de déchets de matières plastiques. Thèse de doctorat, Lyon, France.

[4] Delmas P., Berche J.C., Gourc J.P. (1986) - "Le dimensionnement des ouvrages renforcés par géotextile : Programme CARTAGE" Bulletin de liaison LPC n°142

[5] Coulet C., Rakotondramanitra J.D., Dali-Braham M. (1987).
Expérimentation en vraie grandeur du renforcement des sols par déchets de matières plastiques, 9 th African Regional Conference, 421 - 428, Lagos, Nigéria.

[6] Gielly J., Coulet C., Perrin J., (1989). Renforcement des sols avec le procédé Plasterre, Int. Conf. on soil Mech. and Found. Eng. 1249 - 1250, Rio, Brésil.

Aftercare Research of Closed Sanitary Landfills

Jouko Saarela
National Board of Waters and the Environment, Finland

ABSTRACT

Aftercare of sanitary landfills is in many ways the environmental problem. One of these is the problem of how to cover and rehabilitate old landfill sites so that the leachate is minimun and how they should be vegetated after their use. For this reason The Finnish National Board of Waters and the Environment has started several projects concerning these matters and they include among other things the following points.

1) Inventory of closed sanitary landfills (157).
2) Inventory of geotechical stuctures of covers of closed sanitary landfills.
3) Inventory of landscaping of closed sanitary landfills.
4) There is also purpose to prepare practical directives and code of practice how to cover sanitary landfills after their use.
5) Development of 6 different types of cover structures of sanitary landfills.

In this connection here is discussed mainly nutrient analyses of surface soil samples of a landfill, because it has the important role in the success of the revegetation of the sanitary landfill. The research concerning the vegetatation and landscaping included among other things the following points: Chemical, nutrient metal, gamma, and radon analysis of surface soils of three sanitary landfills. There is also some details about the development of the water balance model by finite element method for final cover design and two types of cover structures developed in this project.

INDRODUCTION

Samples were taken in Sebtember 1990 from the surface layer of the Vuosaari landsfill near Helsinki. Dept of the samples were 10-20 cm. They represent different vegetation types: samples 8, 9 and 10 represent well growing areas and samples 6 and 7 represent poorly growing areas (table 2).

Table 1 shows the physical properties of soil samples, water content, humus content and permeability. All the samples contain clay and silt about 40-60 %, which lower their permeability. Sample 6 contains most clay (35%). Humus content of samples were low and they are below 5%. Water content is low and it varies in different samples.

RESULTS OF SOIL ANALYSES

Chemical, nutrient metal, metane, gamma and radon analyses were made on the samples. The following nutrients were researched: nitrates, phosporus, potassium, calcium, and magesium (table 2). Nutrient analyses show the usable quantity of nutrients for the vegetation.

Table 1. Physical properties of soil samples taken from the surface soil of Vuosaari landfill.

Sample	Type of soil (%)				Humus %	Water content w%	Permeability m/s
	Clay	Silt	Sand	Gravel			
6 (poor growing)	36	30	24	10	2.32	12.1	$10^{-8.79}$
7 (poor growing)	11	33	38	18	4.68	4.3	$10^{-7.29}$
8 (well growing)	13	36	35	16	4.25	9.3	$10^{-7.31}$
9 (well growing)	21	29	30	20	2.82	10.9	$10^{-8.53}$
10 (well growing)	13	26	38	23	3.35	5.2	$10^{-8.95}$

Table 2. The quantity of main nutrients in samples of Vuosaari landfill.

	Point 6 No vegetation	Point 7 Poor growing	Points 8,9,10 Well growing
	mg/l	mg/l	mg/l
nitrate (NO_3^- –N)	4.2	3.5	4.4, 2.0, 6.0
phosphorus	2.2	11.0	2.5, 2.1, 9.0
potassium	200	175	220, 240, 180
calcium	1700	2300	2700, 1900, 3000
magnesium	280	225	590, 325, 195

Table 3. Different recommendations and limits concerning the quantity of nutrients and metals in soils (Assmuth, 1989).

Nutrient or metal	Grass Readily soluble Finnish recommendation for green building	Coniferous forest	Dutch limits for Contaminated soils total content	EPA limits for Contaminated soils total content	Limit values for Causing root damages[x]
nitrate	10–40 mg/l	10–40 mg/l	–	–	–
calcium	1500–3000 mg/l	700–2000 mg/l	–	–	–
potassium	150–350 mg/l	80–150 mg/l	–	–	–
phosporus	20–100 mg/l	5–20 mg/l	–	–	–
magnesium	200–400 mg/l	100–200 mg/l	–	–	–
zink	2–50 mg/l	2–50 mg/l	300 mk/kg	500 mg/l	1.3 mg/l
copper	5–30 mg/l	5–20 mg/l	500 mg/kg	100 mg/l	0.02 mg/l
manganese	10–100 mg/l	10–100 mg/l	–	–	0.05 mg/l
lead	–	–	600 mg/kg	5 mg/l	1.7 mg/l
nickel	–	–	500 mg/kg	–	0.18 mg/l
cadmium	–	–	20 mg/kg	0.5 mg/l	2.1 mg/l
iron	–	–	–	–	9.3 mg/l

[x] prevent the growing 50 %, varies in different precies (Willamson et al. 1982).

Table 4. Content of readily soluble metals, total content of metals and conductivity and pH.

Sample	Content of readily soluble metals						
	Cu mg/kg	Ni mg/kg	Mn mg/kg	Pb mg/kg	Zn mg/kg	Fe mg/kg	Cd mg/kg
P6	6	2	15	8	21	487	0.11
P7	4	2	24	23	21	254	0.10
P8	2	1	27	11	7	422	0.07
P9	6	2	24	9	12	537	0.67
P10	6	3	29	36	29	390	0.13

Sample	Total content of metals						
	Cu mg/kg	Ni mg/kg	Mn mg/kg	Pb mg/kg	Zn mg/kg	Fe %	Cd mg/kg
P6	27	24	223	31	114	1.95	0.33
P7	24	17	201	50	111	1.43	0.31
P8	19	15	173	23	60	1.38	0.16
P9	34	29	238	28	105	2.41	1.16
P10	33	26	241	85	145	2.10	0.37

Sample	pH	Conductivity µS/cm
P6	6.59	218.0
P7	6.76	323.0
P8	6.34	156.7
P9	5.79	134.5
P10	6.67	235.0

It was seen that quantity of main nutrient changes in different sampling points. Sample 6 is taken from the point which is poorly growing area. However in this research point only the quantity of calcium is less than in well growing points or poorly growing point 7. The quantity of calcium and magnesium are in some well growing points higher than in poorly growing points (Figure 1). In the quantity of other nutrients there is not this kind of diffrences in the succees of vegetation.

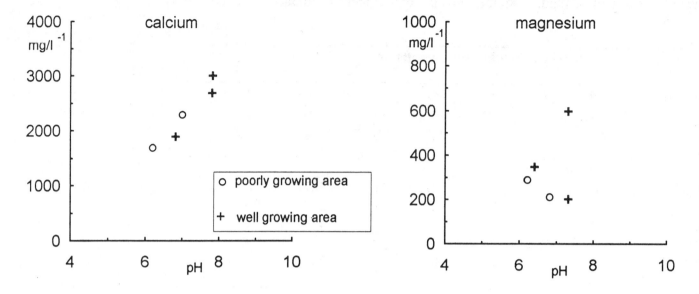

Figure 1. The solublity of calcium and magnesium in the relation to pH in the samples taken from well or poorly growing vegetation areas of Vuosaari landfill.

According to the results the quantity of the nutrient is not dependent on the quantity of fine soils or humus. Only calcium increases when pH increases (figure 2). Only quantity of nitrate and phoshorus is low when one compares the level of main nutrient to the Finnish recommentations for green building (tables 2 and 3). The level of these nutrients has not great differencies between well vegetated and poorly vegetated areas.

Metal content of soil samples were compared to the Finnish recommendations for green building, EPA and Dutch recommendation concerning contaminated soils.

The content of readily soluble metals does not differ considerable from each other (table 4). The level of metals seems to have no meaning in the success of vegetation. Zink, copper, and magnesium do not exceed the Finnish recommendations for green building concerning the quantity of nutrient. Compared to the EPA recommentations concerning contaminated soils there were measured higher values in lead in all sampling points and cadmium in one point. However the total content of metals does not exceed standards used in Holland for contaminated soils.

As a conclusion it can be said that the content of the readily soluble metals are low and they do not have the clear correlation to the succeess of the vegetation. However the small number of the samples makes the estimation difficult for the whole landfill. Figure 2 shows the quantity of cadmium and the relation of the conductuvity to pH in samples taken from well or poorly growing vegetation areas of Vuosaari landfill.

Same researches were made on two other closed landfills. It was seen more clearly on those two landfill than in this landfill that in some research points lack of nutrients or great quantity of metals prevented the success of vegetation. However the lack of nutrients was more decisive than the great quantity of metals in the preventing of the success of vegetatation.

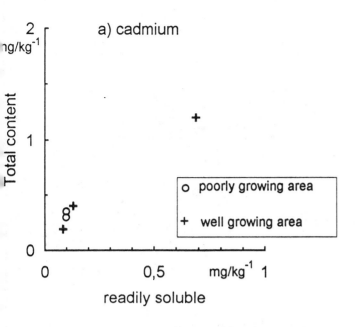 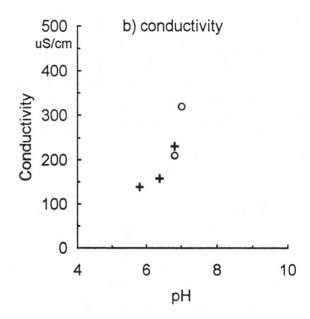

Figure 2. Quantity of cadmium (a) and the relation of conductivity to pH (b) in samples taken from well or poorly growing vegetation areas of Vuosaari landfill.

OTHER RESEARCHES

For the final cover design there was developed the water balance model by finite element method.

In the following there are some facts about the model. The model needs for example the following data about the sanitary landfill:
- permeability of cover
- thickness of cover
- gradient of cover
- vegetation on the cover etc.

In the development and testing of the model the following research results were got. Researches were made mainly on three great landfills near Helsinki.

- weight sounding of the sanitary landfills was very difficult due to the unhomogenous waste material for example building rubble containing concrete pieces.
- water table was 5–10 m from the cover of landfill. The total height of landfills was between 10–20 m
- unit weight was in the depth of 20 m between 17...19 kN m^{-3} and in the surface 12 kN m^{-3}
- from the resistance of rub drilling it was estimated that the angle of friction was in the depth of 20 m approximately 42...45° and near the surface 32...36°
- frost dept was between 40–90 cm. It is less than in natural soils
- conductivity was between 95–2020 μs cm^{-1}. The value is low and wastes contain little salt and ionized compounds
- pH of wastes was 5,16–9,15. The value is high containing alkaline compounds
- bearing capacity of landfills was 10–25 Mpa. The value is low and for example the bearing capacity of sport field is approximately 50 Mpa
- permeability of covers was $10^{-5.49}$–$10^{-8.91}$ m s^{-1}

- metane measurements showed that in the resarch points, where metane comes to the surface there is no vegetation and there is also very high temperature at those points, maximum 46°C

- nutrient analysis of surface soils showed that there was high metal content in some samples but high metal content of soils were not so decisive in the failure of vegetation as for example poor nutrient content of soil.

- there was not higher gamma and radon content in the soil samples taken from the surface of sanitary landfills as in the surrounding.

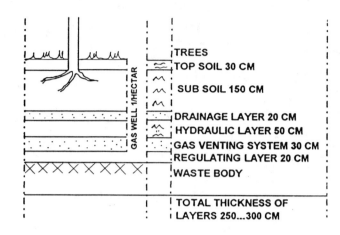

Figure 3. The most superior structure.

REHABILITATION AND TYPES OF THE COVER STRUCTURES

In principle the rehabilitation of the closed landfill can be done in the following ways:

- cover it,

- remove it, for example because it contaminates the groundwaters

- inject or build cut of wall around the landfill to prevent the contamination of groundwaters

However the most common rehabilitation way is to cover the landfill properly and in this connection there is discussed only cover structures because it is the most common way of the rehabilitation.

Figures 3 and 4 show the most superior and simplest types of cover structures.

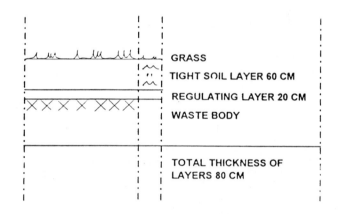

Figure 4. The simplest cover structure.

Figure 3 shows the most superior structure, which comes up to the full reguirements of modern landfill technology. It is used when there is the intention to build on the closed sanitary landfill for example a modern and complex recreation area. It can also be used on the hazardous waste landfills. The total thickness of the cover system varies between 2.5–3.0 m and it can be used for planting trees on the sanitary landfills.

Figure 4 shows the simpler and lighter cover structure. It is used for example on old landfills where environmental risks are already decreasing. The light structure is also sufficient for small landfills which are situated far from the settlement when there is no need for activity purposes for example for recreation. Due to it`s structural simplicity it is not sufficient for preventing leaching water and for controlling gas discharge. The cheapest cover type costs 37 000 US $ and the most expensive one costs 403 000 US $ per hectare. Costs are calculated in the year 1990.

REFERENCES

Assmuth, T., 1989. Riskikaatopaikat ja saastuneet maa-alueet -teollisuusmaiden ympäristönsuojelun painoalue. Vesitalous 4/1989, 20–26 ss.

EPA, Environmental Protection Agency 1980. Idenfication and listing hazardous waste. Federal Register 45 (May 19, 1980), USA.

Suomen Geokemian ATLAS, 1989. The Geochemical Atlas of Finland. GTK, ESPOO.

Williamson, N. A., Johnson, M. S. ja Bradshaw, A. D., 1982. Mine Wastes Reclamation. Mining Journal Books, London, 103 s.

Saarela, J. (1992). Kaatopaikan pinnan loppukäsittely ja sen simulointimalli. Väliraportti. Vesi- ja ympäristöhallitus. (Surface structures of closed sanitary landfills and their simulation model: in Finnish; not published) National Board of Waters and the Environment.

Saarela, J. (1993). Lopetettujen kaatopaikkojen inventoinnin tulokset. (Inventory of closed sanitary landfills: in Finnish; not published) National Board of Waters and the Environment.

Saarela, J. 1991. Landscaping aspects of sanitary landfills in Finland. Third International Landfill Symposium 14 – 18 October 1991. Sardinia -91.

VIATEK Oy, (1991). Käytöstä poistuvien kaatopaikkojen pintarakenteiden kustannukset. Esiselvitys. (Costs of surface structures of closed sanitary landfills: in Finnish).

Sensitive and Simplified Methods for Monitoring of Chemical Species in Ground

Hiroyuki Sakai, Osamu Murata, Hisashi Tarumi
*Soil Mechanics & Foundation Engineering Laboratory,
Railway Technical Research Institute, Tokyo, Japan*

Recently, railways laid in deep ground have been studied and planned to resolve a traffic jam in urban areas in Japan. The construction of a deep tunnel is carried out with special technique in limited space of ground so that the operation is in need of great care taken to avoid contaminating the environment such as soil and groundwater. In conjunction with the construction, construction waste such as excavated soil and disused materials for construction must be kept in strong containers or on the field. When these waste are stored in them, monitoring of contaminating substances originating from wastes is necessary.

Therefore, monitoring of chemical species is also very important in construction/operation of railway. With an increasing interest taken in environmental sciences all over the world and in various fields, sensitive and simple methods for detection of chemical species in environment are required. So we studied new methods for monitoring the chemical species which became the center of interest in environmental sciences, such as heavy metals and undesirable anions in groundwater. In consequence of investigation, chemiluminescence detection was investigated and used to develop new methods for sensitive determination of these species at low concentrations.

INTRODUCTION

Construction of railway yields large quantities of construction waste. They are kept in containers or on the field, but, chemical substances exuded from those wastes contaminate the environment. Especially, in recent years, the regulation of contamination caused by chemical species has been enforced strictly. So monitoring of chemical species such as heavy metals and oxoanions in ground close to waste dump is necessary to keep away the contamination.

On the other hand, Japan Railways (JR line) operate an extended communication network of single and double track lines. Out of the more than 20 thousand kilometer railway track, 16% comprises of bridges and 9% runs through tunnels. In the maintenance of concrete structures, particularly bridges and tunnels, an accurate knowledge of the silicate levels in the water seeping out of these structures is of great importance. Recently, new railway lines have been constructed in urban area. The tracks are laid deep underground so that it is necessary to improve grounds. On that account, grouting underground is often done. For estimation of grouting effect on subway construction, monitoring of the silicate concentration resulting from grouting ground is essential. Besides there are a large number of species contaminating in the railway environment, the monitoring of contaminants is important therefore from a viewpoint of environment protection.

Accordingly, sensitive, low-cost, simple

methods for monitoring the contaminating species in ground or groundwater flowing from waste dump were studied. As a result, new analytical methods for monitoring them were developed, and using these methods low concentrations of heavy metals, oxoanions such as silicate, arsenate and phosphate, these species could be determined sensitively with ease operations. Thus, there has been taken an increased interest in the development of a simple and accurate method for the determination of those species at trace levels.

As typical methods for determination of heavy metals, atomic absorption spectrometry (AAS) and inductively coupled plasma-atomic emission spectrometry (ICP-AES) are employed. When anions are analyzed, high performance liquid chromatography (HPLC), including ion-exchange chromatography (IEC) and ion-exclusion chromatography (ICE), and flow injection (FI) analysis which are combined with fluorescence spectrometry, ultraviolet/visible absorptiometry or conductometry are general by used. But these methods need expensive instrumentation and are required special technique to operate. Specially, for the spectrophotometric determination of silicate, the detection based on a reaction between monomeric silicate and ammonium molybdate to form molybdosilicate, known as the molybdenum blue method(1-3), has been used. However, the formation of exact or similar heteropolymolybdate analogs by phosphate, arsenite/arsenate, etc. reduces the specificity of the method. Recently, a molybdenum blue post-column detection system has been developed for the determination of those oxoanions using flow injection (FI) techniques(4-7). However, kinetically the reaction of silicate was considered too slow for use in chromatographic detection. There is a similar drawback to spectrophotometric determination of arsenite/arsenate, phosphate and other anions.

Chemiluminescence (CL) reaction is usually rapid, the instrumentation used is simple and inexpensive. The dynamic range for analysis is wide with high sensitivity. Thus CL systems are considered to be useful for a post-column chromatographic detection(8). However, only a few non-metal analytes can enhance CL emission(9).

In the inorganic trace analysis, a major problem associated with CL methods using the luminol reaction is interference from other species, particularly transition metal ions(9-13). Therefore, separation of analyte is usually required before analysis (14). An effective separation by chromatographic performance before CL analysis has been successfully demonstrated by several workers using low-capacity ion-exchange columns and various eluents (4, 15-18).

ICE actually involves the chromatographic separation of molecular species rather than ions although an ion-exchange column is generally used(19). Silicate was removed and converted to silicic acid by ICE and the CL detector responded well to it rather than to the eluent. This allowed the development of an FI system based on the direct combination of the ICE separation with the CL detection. One of the advantages of ICE is its ability to remove ionic interferents like transition metal ions. Silicic acid was also separated by the ICE from other weak acids which would otherwise interfere with the silicate determination.

Conductivity detector is widely used for ICE(19). However, conductivity detection signals are generally weak owing to a small degree of ionization of weak acids. A relatively small increase in a high background signal of the eluent is measured and therefore sensitivity of the detection is very poor. In this work, it was demonstrated that when combined with the CL detection, the ICE separation technique is an effective means to improve selectivity and to enhance sensitivity. In addition, the advantages of FI analysis were incorporated into the present method to prove an automated, rapid and reproducible analysis. Using optimized experimental conditions, satisfactory results were obtained for the determination of silicate in tap, river water samples and extracts samples from soil into water by the proposed ICE-CL method.

CL methods have many advantages such as low detection limits, wide dynamic range and simplicity of instrumentation; this technique is low in cost. The CL detections were carried out to determine heavy metals; cobalt(II), chromium(III), copper(II), iron(II/III) and manganese(II), and other species, anions; arsenite/arsenate, phosphate, silicate and sulfide, A luminol CL detection/FI analysis technique coupled with IEC or ICE has been examined for the selective determination of cobalt(II) at pg mL^{-1} levels.

EXPERIMENTAL SECTION

Heavy Metals. A cation-exchange chromatograph with a 100-μL sample loop, equipped with a 50-mm x 4.6-mm i. d. cation separation column, was employed. The column was connected to an FI analyzer. Two pumps of the FI analysis device were used to drive the 1.0 mM luminol-20 mM sodium hydroxide solution and 20 mM hydrogen peroxide solutions though the flow system. In the system, luminol solution is first mixed directly with hydrogen peroxide solution and then with the column effluent from the ion chromatograph, just before entering a quartz flow cell (2.4-mm long, 40μL). The CL was detected by a fluorescence spectrophotometer without wavelength discrimination. Mixing the luminol solution with the hydrogen peroxide solution alone resulted in a relatively high baseline CL signal; the analytical signal was taken as the difference in the observed CL intensities for the analyte and the baseline. Therefore, the flow system was optimized for a stable and constant baseline signal, which is important for obtaining a lower detection limit and better reproducibility. Thorough rapid mixing of the reagent solutions and the column effluent is also important for obtaining a stable baseline and reproducible peak heights(13). This was achieved by the use of mixing tees in the flow system.

A strip-chart recorder was used so that peak height and shape could be observed. The flow-rates were chosen to optimize the CL sensitivity for cobalt(II): luminol, 0.5 mL min^{-1}; hydrogen peroxide, 0.5 mL min^{-1}; barium chloride eluent, 1.0 mL min^{-1}. PTFE tubing (0.5-mm i. d.) was used between all components in the flow system.

Silicate. A cation-exchange chromatograph, with a 100-μL sample loop, was employed. A peristaltic pump was used for driving CL reagent. For separation, a 200-mm x 4.9-mm i. d. cation-exchange column was employed. 4.4 mM perchloric acid was used as an eluent. PTFE tubing (0.5-mm i. d.) was used throughout in the flow system. Samples were introduced into the eluent stream of perchloric acid. The CL reagent solution, mixture of 0.10 mM luminol, 5.0 mM potassium hydroxide and 5.0 mM hydrogen peroxide was uniformly delivered to the system via a pump of the chromatograph device.

The column effluent from the ICE was mixed directly with the CL reagent in a 70-μL coiled flow cell housed in front of the photomultiplier tube of a bio-chemiluminescence monitor. The CL signals produced were recorded on an ordinary strip-chart recorder. The peak height of the CL signal was read against the reagent blank. The constant flow rates for the eluent and reagent solutions were set at 1.0 and 2.0 mL min^{-1}, respectively.

The analyte solution was prepared as follows: To 100 mL of the water sample in a 100-mL PTFE beaker, 0.2 g of sodium hydrogencarbonate was added. After heating in a boiling water bath for 20 min, the resulting solution was cooled to room temperature on standing and then filtered through a filter paper. The final solution was made up to 100 mL with water and was immediately analyzed.

For silicon determination in water samples, parallel measurements by flame atomic absorption spectrometry (AAS) or inductively coupled plasma atomic emission spectrometry (ICP-AES) were also performed. A polarized Zeeman effect atomic absorption spectrometer with a nitrous oxide-acetylene flame at 251.6 nm wavelength and a high-resolution ICP atomic emission spectrometer at 251.611 nm (Si I) were used, respectively.

Arsenite/arsenate. An HPLC pump, with a 100 μL sample loop, was employed. The other HPLC pump was the elution. For separation, a 200-mm x 4.9-mm i. d. cation-exchange column was employed in conjunction with 0.75 mM perchloric acid as an eluent. PTFE tubing (0.5-mm i. d.) was used. Samples were introduced into the stream of perchloric acid. The CL reagent was uniformly delivered to the system via an HPLC pump.

The column effluent from the IEC was mixed directly with the CL reagent in a 70-μL coiled flow cell housed in front of the photomultiplier tube of bio-chemiluminescence monitor. The CL signals produced were recorded on an ordinary strip-chart recorder. The peak height of the CL signal was read against the reagent blank.

The constant flow rates for the eluent and reagent were set 1.0 mL min^{-1}. All the experimental parameters were optimized in a 10 μg mL^{-1} arsenic(III/V) standard solution.

Phosphate. An HPLC pump was employed for the driving reagents. The other HPLC pump with a 100-μL sample loop was used for the elution. For separation, a 250-mm x 4-mm i. d. anion-exchange column was employed in conjunction with 10 mM ammonia-15 mM potassium nitrate solution as an eluent. PTFE tubing (0.5-mm i. d.) was used. Samples were introduced into the stream of the eluent. The CL reagent was uniformly delivered to the system via a pump.

The column effluent from the IEC was mixed with heteropoly acid formation reagent at a tee connection and run into reaction coil, 3-m x 1.0-mm i. d. and then mixed directly with the CL reagent in a 70-μL coiled flow cell housed in front of the photomultiplier tube of a bio-chemiluminescence monitor. The CL signals produced were recorded on an ordinary strip-chart recorder. The peak height of the CL signal was read against the reagent blank.

RESULT AND DISCUSSION

Heavy Metals. Figure 1 illustrates typical response curves obtained by injection of 100-μL dilute cobalt(II) standard under the optimum operating conditions. The minimum detectable concentration for cobalt(II) was 1.0 pg mL^{-1}, where the detection limit is defined as the concentration producing a signal-to-noise ratio of three.

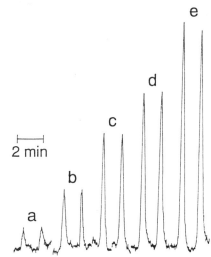

Figure 1. Chemiluminescence-time response for various concentrations of Co(II). Peaks: (a) 1; (b) 10; (c) 20; (d) 30; (e) 40 pg mL^{-1} Co(II).

The magnitude of the baseline noise can be also seen in Fig. 1. The calibration graph was linear from 10 pg mL^{-1} over six orders of magnitude. The relative standard deviations of the CL signals (ten replicate measurements) for 30 pg mL^{-1} and 10 ng mL^{-1} cobalt(II) were 3.8% and 1.3%, respectively. An attempt was made to test the application of this method to the direct determination of cobalt(II) in the primary coolant of a boiling-water reactor without prior preconcentration. However, as there was no standard coolant concentration for cobalt(II) and no practical sample was available, an artificial coolant sample was prepared according to the literature (20), containing cobalt(II) 70, iron(II) 4000, zinc(II) 250, nickel(II) 150 and manganese(II) 50 pg mL^{-1}.

The analysis of this artificial sample by direct calibration gave a value of 67 ± 5 pg mL^{-1} cobalt(II) for triplicate measurements, in good agreement with the content of cobalt(II) in the sample. The results show that the proposed procedure could be employed for the direct analysis of such a coolant sample. The method was also applied to the determination of trace cobalt(II) in commercially available copper(II) standard solution. Samples of 1000 μg mL^{-1} copper(II) standard solutions for AAS purchased from several companies were each diluted 10-fold with water and analyzed in duplicate. The results obtained by direct calibration are given in Table I; a typical chromatogram (Fig. 2) shows that cobalt(II) and copper(II) are resolved from each other, indicating that copper(II) can be tolerated in weight ratios at least up to 1 : 3 x 10^7.

Table I. Determination of Co(II) in Commercially Available 1000 μg mL^{-1} Cu(II) Standards

Sample	Co(II) added / pg mL^{-1}	Co(II) found / pg mL^{-1}	Recovery ,%
A	--	105, 115	--
A	100	207, 222	97, 112
B	--	62, 74	--
C	--	42, 44	--
D	--	41, 43	--
E	--	35, 35	--
F	--	33, 35	--
F	100	129, 133	95, 99

This method could also be applied to the determination of elements other than cobalt(II), such as chromium(III), copper(II), iron(II/III) and manganese(II) in the samples such as river water, groundwater streaming from dumping ground of waste.

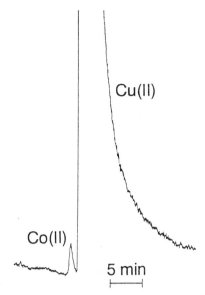

Figure 2. Typical recorder tracing for the analysis of a commercially available Cu(II) standard solution. The samples was prepared by a 10-fold dilution of a 1000 μg mL^{-1} Cu(II) standard.

Silicate. This method was applied to the determinations of silicate in tap and river water samples. Silicates are present in water in different forms, such as monomeric and polymeric silicates. One of these species, probably monomeric silicate ion, is assumed to be eluted in the form of silicic acid and then enhances the CL emission in the present system. Other polymeric silicates should be converted into the monomeric species by digesting the sample with sodium hydrogencarbonate(21). A chromatogram obtained from injection of the river water sample, A, was shown in Figure 3. The results of direct determination of silicate using a calibration curve method are given in Table II.

In order to investigate the possibility of accurate silicate determination by the present method, recovery tests on the tap water sample, A, to which a certain amount of silicate was added were performed. The results indicated good recovery of Si(IV) (Table II). Furthermore, to verify the results, measurements by flame AAS or ICP-AES were performed with

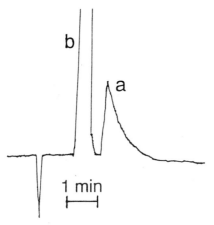

Figure 3. Chromatogram for river water sample A (100-μL) containing Si(IV) (a) and carbonate (b) ions.

Table II. Determination of Silicate in Water Samples

Sample	Si(IV) / μg mL^{-1}	
	this method	other method
(Tap water)		
A (Kokubunji)	12.6, 13.3	13 [a]
A + 10 μg mL^{-1}	23.0, 24.4	--
B (Kudamatsu)	4.6, 5.1	6.4 [b]
C (Higashi–Hiroshima)	5.6, 5.9	5.5 [b]
(River water)		
A (Saijo)	5.1, 5.5	5.1 [b]
B (Kurose)	6.6, 7.0	7.3 [b]

[a] AAS. [b] ICP-AES.

Table III. Determination of Silicate in Extract Samples from Grouting [a] into Water

Sample	Si(IV) / μg g^{-1}
A(close to grouting point)	501 ± 3.80
B	341 ± 0.670
(at a spot of 0.5 m apart from grouting point)	

[a] Collected in Tokyo.

same water samples used in the ICE–CL analysis. These results were consistent with those obtained by present CL method (Table II).

The method was also applied to determination of silicate in extract samples from the grouting into water. The results obtained by direct calibration are given in Table III. It shows the difference of concentrations of Si(IV) in the ground between just the grouting point and a spot a little apart from the grouting point. As a result, the extent of grout in the ground operated grouting.

The analytical results indicated that the proposed ICE–CL method applicable to the determination of silicate in the samples such as tap and river waters, and seeping waters from ground is valid and rapid. This ICE–CL method would also be applicable to the determination of silicate in the environmental samples. Besides, phosphate, hydrogencarbonate and sulfide could be determined by this method (22).

Arsenite/Arsenate. Figure 4 shows a typical chromatogram resulting from injection of a mixture of 5.0 μg mL^{-1} arsenite and 1.0 μg mL^{-1} silicate and post–column reaction with luminol, using the cation–exchange column under the optimum conditions.

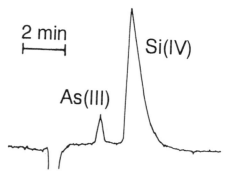

Figure 4. Chromatogram for a synthetic mixture (100 μL) of As(III) and Si(IV): As(III), 5 μg mL^{-1}; Si(IV), 1 μg mL^{-1}.

This ICE-CL method would also be applicable to the determination of arsenite/arsenate, hydrogencarbonate, nitrite, phosphate, sulfide and silicate in environmental and biological samples.

Phosphate. The method was applied to the determination of phosphate in river water samples and the results are given in Table IV. Samples of river

Table IV. Determination of Phosphate in River [a] Water

Sample	P(V) / ng mL^{-1}	
A	277,	286
B	265,	261
C	371,	369

[a] Tamagawa.

water collected at three points in Tamagawa river, were directly injected into the system after filtration with 0.45 μm pore-size membrane filter. An AAS or IEC with suspension equipment was also applied to these samples and the agreement between the two/three methods was satisfactory.

CONCLUSION

The IEC/ICE-CL systems are simple, inexpensive and high by sensitive methods for monitoring environmental samples such as groundwater, river water, water from a drain and water sample extracted from soil. Using these methods, monitoring of important chemical species for environment can be carried out with simple operations.

LITERATURE CITED

(1) Marino, D. F.; Ingle, Jr., J. D. *Anal. Chem.* **1981**, *53*, 294.
(2) Hartkopf, A.; Delumyea, R. *Anal. Lett.* **1974**, *7*, 79.
(3) Neary, M. P.; Seitz, W. R.; Hercules, D. M. *Anal. Lett.* **1974**, *7*, 583.
(4) Boyle, E. A.; Handy, B.; van Geen, A. *Anal. Chem.* **1987**, *59*, 1499.
(5) Montano, L. A.; Ingle, Jr., J. D. *Anal. Chem.* **1979**, *51*, 919, 926.
(6) Gord, J. R.; Gordon, G.; Pacey, G. E. *Anal. Chem.* **1988**, *60*, 2.
(7) Seitz, W. R.; Suydam, W. W.; Hercules, D.M. *Anal. Chem.* **1972**, *44*, 957.
(8) Nau, V.; Nieman, T. A. *Anal. Chem.* **1979**, *51*, 424.
(9) Marino, D. F.; Wolff, F.; Ingle, Jr., J. D. *Anal. Chem.* **1979**, *51*, 2051.
(10) Babko, A. K.; Lukovskaya, N. M. *Zavod.Lab.* **1963**, *29*, 404.
(11) Burguera, J. L.; Burguera, M.;Townshend,A. *Anal. Chim. Acta* **1981**, *127*, 199.
(12) Nordmeyer, F. R.; Hansen, L. D.;Eatough,D. J.; Rallins, D. K.; Lamb, J. D. *Anal. Chem.* **1980**, *52*, 852.
(13) Kobayashi, S.; Imai, K. *Anal. Chem.* **1980**, *52*,1548.
(14) Klopf, L. L.; Nieman, T. A. *Anal. Chem.* **1983**, *55*, 1080.
(15) Sakai, H.; Fujiwara, T.; Yamamoto, M.; Kumamaru, T. *Anal. Chim. Acta*, **1989**, *221*,249.
(16) Jones, P.; Williams, T.; Ebdon, L.; *Anal. Chem. Acta* **1989**, *217*, 157.
(17) Yan, B.; Worsfold, P. J. *Anal. Chim. Acta* **1990**, *236*, 287.
(18) Jones, P.; Williams, T.; Ebdon, L. *Anal. Chim. Acta* **1990**, *237*, 291.
(19) Fritz, J. S. *J. Chromatogr.* **1991**, *546*, 111.
(20) Tanaka, Y.; Mizuniwa, F.; Maekoya, C.; *Anal. Chem.* **1979**, *51*, 2337.
(21) Motomizu, S.; Oshima, M.; Araki, K. *Analyst(London)* **1990**, *115*, 1627.
(22) Sakai, H.; Fujiwara, T.; Kumamaru, T. *Bull. Chem. Soc. Jpn.* **1993**, *66*, 3401.

The Capillary Barrier Effect in Unsaturated Flow Through Soil Barriers

**Charles D. Shackelford, Associate Professor, Chung-Kan Chang, Graduate Student,
Te-Fu Chiu, Graduate Student**
Department of Civil Engineering, Colorado State University, Fort Collins, Colorado, USA

Abstract: The influence of a capillary barrier effect on the migration of water through an initially unsaturated soil barrier is analyzed. The analysis is based on an existing one-dimensional model for flow through a fine-grained soil (e.g., clay liner) overlying a coarse-grained soil (e.g., sand). The results of sample analyses indicate that there is a significant time-lag (> 10 years) before complete transmission of flux into the coarse-grained soil for situations in which the initial degree of saturation of the fine-grained soil is relatively high ($S_{if} \geq 80\%$). An example problem illustrates that the measurement of the saturated hydraulic conductivity of field compacted clay liners with sand underdrains on the basis of saturated flow conditions may be not only incorrect but also unconservative (too low) during the time-lag period. The implication of the analyses is that the capillary barrier effect cannot be ignored even for the case of compacted clay liners for waste disposal with relatively high initial degrees of saturation.

Introduction

A capillary barrier effect results when unsaturated flow occurs through a fine-grained soil overlying a coarse-grain soil. Due to the capillary barrier effect, only a fraction of the flux associated with the wetting front is transmitted into the underlying coarse-grained soil. The remaining fraction of the flux is reflected upward into the overlying fine-grained soil. As a result, the evaluation of soil barriers used for waste containment based on saturated flow conditions may be not only incorrect but also unconservative. Therefore, the purpose of this paper is to illustrate the potential ramifications resulting from the existence of a capillary barrier effect in practice with application to soil liners and covers for waste disposal sites. The illustration will be based on analyses using the theory for one-dimensional unsaturated flow through a capillary barrier as previously developed by Morel-Seytoux [2].

The Capillary Barrier Effect

A capillary barrier effect results when unsaturated flow occurs through a relatively fine-grained soil, such as a clay barrier, overlying a relatively coarse-grained soil, such as a layer of sand. Due to the different moisture-characteristic properties of the two soils, only a fraction of the water flux reaching the interface between the two soil layers is transmitted into the coarse-grained soil. The remaining fraction of the water flux is reflected upward into the overlying fine-grained soil.

For example, consider the moisture-characteristic curves for a fine-grained soil and a coarse-grained soil schematically illustrated in Fig. 1. The normalized volumetric water content, θ^*, shown in Fig. 1 is defined as follows:

$$\theta^* = \frac{\theta - \theta_r}{\theta_s - \theta_r} \qquad (1)$$

where θ is the water content, θ_r is the residual water content, and θ_s is the water content at natural saturation.

In Fig. 1, point (a) represents the capillary pressure head just inside the fine-grained soil side (- side) of the interface at incipient time (i.e, the time at which the wetting front just reaches the

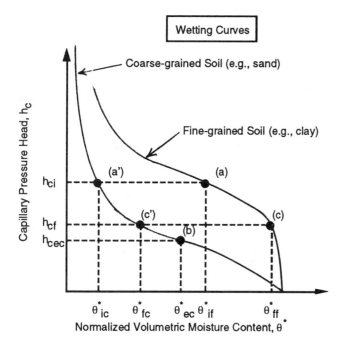

FIGURE 1--Schematic of two moisture-characteristic curves ([2], [3]).

interface). Due to the requirement for continuity across the interface, both the capillary pressure head, h_c, and the flux of water, v, based on Darcy's law must be the same. As a result, the capillary pressure head on the coarse-grained soil side (+ side) of the interface must be at point (a') to satisfy the requirement for continuity of capillary pressure. However, to satisfy the equality of flux requirement, the water content on the coarse-grained soil side of the interface must be at some value θ^*_{ec} which is greater than the incipient value, θ^*_{ic}, due to transmission of water into the coarse-grained soil. Since points (a) and (b) clearly do not represent the same values for the capillary pressure heads (h_{ci} and h_{cec}, respectively), the two interface conditions cannot be satisfied if the water content in the fine-grained soil remains at its incipient value (θ^*_{if}) while the water content in the coarse-grained soil increases. As a result, there must be a simultaneous increase in water content in both soils to values of θ^*_{ff} and θ^*_{fc} corresponding to points (c) and (c'), respectively, in Fig. 1 such that the capillary pressure head at the interface is the same value, h_{cf}. Such an increase in water content on the fine-grained soil side of the interface implies that some of the water which was migrating downward through the fine-grained soil must be reflected back upward into the fine-grained soil and, therefore, only a fraction of the incipient flux is transmitted into the underlying coarse-grained soil.

Problem Geometry

The capillary barrier problem analyzed in this study is described by Morel-Seytoux [3]. The geometry of the problem is illustrated schematically in Fig. 2. In this problem, a supply rate of water, r, at a magnitude which is somewhat less than the saturated hydraulic conductivity of the fine-grained soil, K_f, is steadily and continuously applied to the surface of an unsaturated fine-grained soil. The depth of the upper layer, D, is sufficiently large such that the infiltration flux is essentially gravity driven by the time the wetting front reaches the interface between the two layer of soil (i.e., incipient time), and the water content behind the wetting front is essentially uniform at a value $\theta^*_{if} < 1$. As a result, the infiltration rate, or flux, r, is given by the following form of Darcy's law:

$$r = K_f k_{rwf}\left(\theta^*_{if}\right) \qquad (2)$$

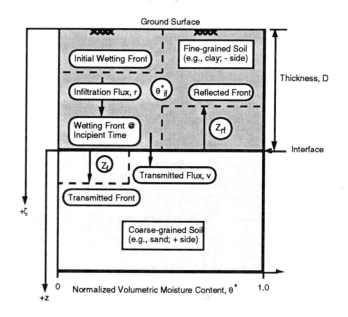

FIGURE 2--Geometry of capillary barrier problem (after [2], [3]).

where the subscript "f" designates a fine-grained soil and k_{rw} is the relative permeability to water, a function of the volumetric moisture content (θ^*), given as follows:

$$k_{rw}(\theta^*) = \frac{K_u}{K} \quad ; \quad 0 < k_{rw} \leq 1 \quad (3)$$

where K_u is the unsaturated hydraulic conductivity. At incipient time (t=0), only a fraction of the flux is transmitted into the underlying coarse-grained soil resulting in a transmitted wetting front, Z_f. The remaining fraction of the flux, r, is reflected and migrates upward into the overlying fine-grained soil layer.

As the reflected front, Z_{rf}, migrates upward into the fine-grained soil, the water velocity (flux) on the fine-grained soil side of the interface (- side) decreases from the incipient value, r, because the capillary drive now opposes gravity whereas prior to incipient time, capillarity and gravity combined to drive the flux r downward. The reflected front also results in a further increase in the volumetric water content in the fine-grained soil to its saturated value (i.e., $\theta^*_{ff} = 1$) and, therefore, a decrease in the capillary suction (i.e., $0 \leq h_c \leq h_{ce}$, where h_{ce} is the entry pressure). The reflected front will continue to propagate upward until the capillary force is reduced to a value which is offset by the gravity force at which time the transmitted flux, v, is equal to the infiltration flux, r (i.e., v/r = 1).

Governing Equations

The theory for analyzing the above problem is presented in detail by Morel-Seytoux [2]. The governing equations are summarized as follows:

$$\frac{dZ_f}{dt} = \frac{v}{\theta_c^+ - \theta_{rc}} \quad (4)$$

$$\frac{dZ_{rf}}{dt} = \frac{r - v}{\theta_f - \theta_{if}} \quad (5)$$

$$v = K_f \left\{ \frac{[1 - f_w(\theta^*_{if})]Z_{rf} + \Delta h_{cc}}{D\mu_r(\theta^*_{if}) + [1 - \mu_r(\theta^*_{if})]Z_{rf}} \right\} \quad (6)$$

and

$$v = K_c k_{rwc}(\theta^{*+}) \quad (7)$$

where the subscript "c" refers to the coarse-grained soil, Δh_{cc} is the difference in capillary pressure head in the coarse-grained soil ($=h_{cc}(\theta^{*+})-h_{cc}(\theta^*_i{}^+)$), and the functions f_w and μ_r depend on the water content and are defined as follows:

$$f_w = \frac{k_{rw}}{k_{rw} + \frac{\mu_w}{\mu_a} k_{ra}} \quad (8)$$

and

$$\mu_r = \frac{f_w}{k_{rw}} \quad (9)$$

where k_{ra} is the relative permeability with respect to air, and μ_w and μ_a are the viscosity for water and air, respectively. These four equations can be solved in closed form for the four unknowns, Z_{rf}, Z_f, v, and θ^{*+} (see [2], [3]).

Results

The unsaturated flow equations were used to analyze flow through a compacted clay barrier overlying a sand layer. Practical examples of this scenario include clay covers underlain by sand drains, clay liners underlain by leachate detection and removal systems, and the use of sand collection basins (known as sand underdrains) beneath compacted clay liners to measure field hydraulic conductivity of the liner. The values for the pertinent parameters in the analyses are given in Table 1. These values are thought to be representative of values in practice. The moisture-characteristic behavior of the soils, including the relative permeability functions, were modeled using the well-known Brooks-Corey [1] functions as follows:

- fine-grained soil:
  $h_{cf} = 40(\theta^*)^{-4}$; $k_{rwf} = (\theta^*)^{11}$;
  $k_{raf} = (1-\theta^*)^2[1-(\theta^*)^9]$

TABLE 1--Values of parameters used in analyses.

Parameter	Values	
	Fine-grained Soil	Coarse-grained Soil
Saturation, S_{if} (%)	80, 85, 90	NA
Thickness, D (m)	1	NA
Residual Moisture, θ_r	0.2	0.1
Saturated Moisture, θ_s	0.5	0.4
Air-Entry Pressure, h_{ce} (m)	0.4	0.02
Hydraulic Conductivity, K (m/s)	10^{-9}, 5×10^{-10}, 10^{-10}	10^{-5}

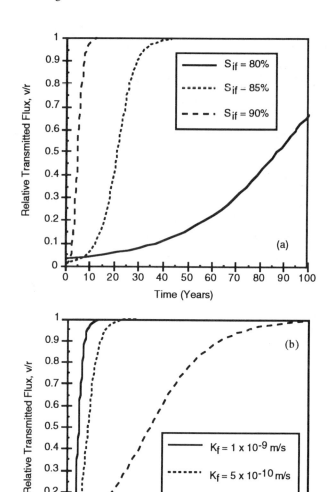

- coarse-grained soil:
  $h_{cc} = 2(\theta^*)^{-1}$; $k_{rwc} = (\theta^*)^5$

where h_{cf} and h_{cc} are in cms.

The results of the analyses are presented in terms of relative transmitted flux, v/r, versus time in Fig. 3. As shown in Fig. 3, the capillary barrier effect results in a time-lag of > 10 years before complete transmission of the infiltrating flux (i.e., v/r =1) into the underlying coarse-grained soil. This time-lag effect increases the lower the initial degree of saturation, S_{if}, of the fine-grained soil (Fig. 3a), and the lower the saturated hydraulic conductivity, K_f, of the fine-grained soil (Fig. 3b). These trends can have important implications with respect to the use of sand underdrains for the measurement of saturated hydraulic conductivity under the assumption of saturated flow conditions.

For example, consider the case where the initial degree of saturation, S_{if}, is 90 percent, and the saturated hydraulic conductivity, K_f, is 10^{-9} m/s. The infiltration flux, r, for this analysis based on Eq. 2 is 1.345×10^{-10} m/s. Five years after placement of the waste, the relative transmitted flux, v/r, is approximately 0.52. Therefore, if saturated flow conditions are assumed to apply, application of Darcy's law assuming a unit hydraulic gradient for the barrier (i = 1) would result in the following estimate for the saturated hydraulic conductivity, K_f:

FIGURE 3--Relative transmitted flux versus time for a 1-m-thick barrier as a function of (a) initial degree of saturation, S_{if}, for a saturated hydraulic conductivity, K_f, of 1×10^{-9} m/s, and (b) K_f for S_{if} of 90%.

$$K_f = \frac{(v/r)r}{i} = \frac{0.52(1.345 \times 10^{-10} \text{ m/s})}{1}$$
$$= 7.0 \times 10^{-11} \text{ m/s}$$

Since the unsaturated flow analysis is based on a saturated hydraulic conductivity of 1×10^{-9} m/s, this result is not only significantly incorrect but also is unconservative (14.3X lower). For this scenario, an accurate estimate of the saturated hydraulic conductivity could not be made using a saturated form of Darcy's law before about 13 years at which time v/r = 1. The results would be more unconservative for lower initial degrees of saturation and/or lower saturated hydraulic conductivities.

Although the above analysis is limited by the assumptions inherent in the governing equations, the potential influence of a capillary barrier effect on the design and evaluation of soil barriers based on the analysis is significant. Thus, more elaborate analyses including laboratory and field investigations are warranted.

Conclusions

The influence of a capillary barrier effect on the migration of water through an initially unsaturated soil barrier is illustrated. The illustration is based on an existing one-dimensional model for flow through a fine-grained soil (e.g., clay liner) overlying a coarse-grained soil (e.g., sand). The results of sample analyses indicate that there is a significant time-lag (>10 years) before complete transmission of flux into the coarse-grained soil for situations in which the initial degree of saturation of the fine-grained soil is relatively high ($\geq$ 80 %). The time-lag increases with decrease in the initial degree of saturation and/or the saturated hydraulic conductivity of the fine-grained soil layer. An example problem based on the results of the analyses shows that the measurement of the saturated hydraulic conductivity of field compacted clay liners during the time-lag period using sand underdrains in accordance with the assumption of saturated flow conditions may be not only incorrect but also unconservative. The example problem indicates that a significant elapsed time (> 10 years) may be required before accurate measurements of the saturated hydraulic conductivity of field compacted clay liners can be made using a saturated form of Darcy's law.

References

[1] Brooks, R. H., Corey, A. T. (1964). Hydraulic Properties of Porous Media. Hydrology Paper No. 3, Colorado State University, Fort Collins, CO.

[2] Morel-Seytoux, H. J. (1992). The Capillary Barrier Effect at the Interface of Two Soil Layers with Some Contrast in Properties. HYDROWAR Report 92.4, Hydrology Days Publications, 57 Shelby Lane, Atherton, CA 94027-3926, 109 pp.

[3] Morel-Seytoux, H. J. (1993). Dynamic Perspective on the Capillary Barrier Effect at the Interface of an Upper Fine Layer with a Lower Coarse Layer. *Proc., Symp. on Engineering Hydrology*, ASCE, C. Y. Kuo (ed.), July 25-30, San Francisco, CA, 467-472.

A Preliminary Survey of Leachate Development in Some Landfills in the Greater Durban Area

A.J. Sillito, F.G. Bell
Department of Geology and Applied Geology, University of Natal, Durban, South Africa

ABSTRACT. A number of waste disposal sites in the greater Durban area were studied with a view to determining leachate generation. Water balance calculations were attempted for one of the sites based on rainfall figures for 1980 to 1990. Evaporation was estimated by the Thornthwaite-Mather method for the same period. This indicated that if liquid waste was being disposed of, then larger quantities could be disposed of during the months April to September inclusive. An attempt was also made to determine the residence times of the liquids passing through the site.

Sampling was carried out at from sumps at four other sites. The pH values and COD's formed the basis of the leachate analysis, and indicated that the sites produced alkaline leachate. However, when these values were compared with rainfall figures, it was seen that both the pH value and COD declined appreciably after periods of high rainfall and then rose during drier periods. The values of COD and pH were generally similar for all sites.

INTRODUCTION

Modern day first world society is generating increasingly larger volumes of waste materials. For economic reasons these materials have to be disposed of in waste disposal sites as close to the source of generation as possible. One of the major problems associated with domestic and much industrial waste disposal is the generation of leachate. Leachate is produced when rainwater, groundwater or codisposed liquids flow through a landfill dissolving the soluble fraction of the waste. This occurs once the absorbent characteristics of the refuse are exceeded. The composition of resultant leachate depends on the materials present in a landfill and the environmental conditions existing at the site. It also varies with time. At many landfill sites leachate has moved into the soil, groundwater, or surface water and this can cause pollution. Ideally leachate production should be kept to a minimum and monitored. Generally the largest contributor to leachate generation is the precipitation which percolates through the waste, reacting with it during the process. In codisposal operations (where both liquids and solids are disposed of together), leachate generation is enhanced by the addition of these liquid wastes.

CASE HISTORY 1

The site occurs within a south facing valley which has steep sides. Drainage ditches were constructed around its perimeter to protect the site against runoff. Drainage to the site which existed prior to the start of operations was redirected around it. Codisposal operations commenced in June 1986 and the site is still operational.

Most of the site is underlain by medium to coarse grained sandstones. They belong to the Natal Group. These are generally hard sandstones except when weathered. Individual beds of sandstone vary in thickness from around 0.3 to 3 m. Lenticular horizons of shale occur within the sandstone but their thickness rarely exceeds 0.5 m. Two major joint sets are present in the sandstone together with secondary random jointing. The joints vary from discontinuous and closed to continuous and widely separated containing clay gouge.

In the northern part of the site a fault throws the sandstones of the Natal Group against the Pietermaritzburg Shale. This consists of a uniform succession of dark bluish-black silty shales and mudstones. Harder indurated horizons are present within the shale and vary up to 0.3 m in thickness.

An appreciable thickness (up to 32 m) of Berea Red Sand overlies these sandstones and shales in the northern and eastern part of the site. Certain horizons within the

Berea Red Sand contain between 25 and 35% clay content. A bed of pebbles and boulders, in a clayey matrix, occurs at the base and is overlain by sands containing a low percentage of clay.

Generally sandy colluvium and underlying clayey residual soil is fairly extensive over the central parts of the site. A stiff residual brown sandy clay overlies the Natal Group sandstones where they are not overlain by the Berea Red Sand and is itself overlain by a thin cover of greyish brown colluvial silty sand.

Within the site boundaries the water table generally lies below the upper surface of the bedrock formations. The exception occurs along the north eastern boundary of the site where the water table occurs in the leached basal Berea Red Sand. An intermittent water table periodically occurs in the residual clayey soils in the valley axis over the southern part of the site where it forms a spring-line in a shallow erosion gully towards the southern boundary of the site. Groundwater flow is confined within the valley defining the site. The main component of groundwater flow therefore is directed down the valley to the river south of the site.

Tests indicated that the permeabilities of the sandstones of the Natal Group vary from approximately 10^{-6} to 10^{-7} m/s. The clay bearing horizons of the Berea Red Sand, have permeability values from 10^{-6} to 10^{-8} m/s.

An estimate of the volume of leachate that will be produced can be obtained from a water balance determination for a landfill, which consists of an evaluation of liquid input and output from the site A water balance calculation for this landfill site was carried out in an attempt to predict the amount of leachate that could possibly be generated. Data was available for quantities of solid and liquid/sludge entry. A compaction density of 650 kg/m^3 was used for the solid waste material. Campbell (1983) showed that such waste material has an absorptive capacity around 0.102 m^3/Mg, which represents 66.2 l/m^3 of liquid. The rainfall figures used were those obtained from Louis Botha airport, the nearest meteorological station. The site covers an area of approximately 5.5 ha. A simplified water balance equation was used to derive the water balance for the landfill and was as follows

$$L_g = P \times A + L_c - L_s$$

where L_g is the amount of leachate generated, P is the percentage percolation from precipitation, A is the area of the landfill, L_c is the quantity of liquid codisposed and L_s is the quantity of liquid retained in storage which primarily represents the absorptive capacity of the waste.

Table 1 gives the amounts of solids, their absorptive capacity, and the quantities of liquids associated with the landfill. The Table indicates that a significantly larger quantity of liquid waste is disposed of into the landfill than the precipitation which ultimately percolates through it. It was assumed that 5% of the rainfall percolated into the landfill. This assumption was made on the basis that during the period 1980 to 1990 inclusive, the average annual rainfall recorded at Louis Botha airport, Durban, was 1051.5 mm. The average annual evapotranspiration calculated for the same period by the Thornthwaite-Mather (1957) method was 1023.6 mm. Rainfall was therefore only slightly higher than evapotranspiration so that this figure is probably a reasonably conservative one to choose. It is evident that the total liquid entry into the landfill appreciably exceeds the absorptive capacity of the solid waste and so a notable volume of liquid is available for the formation of leachate. The average monthly amount of liquid exceeding the absorptive capacity of the solids is 2 285 840 l. However, when this quantity is compared with the average monthly amount of leachate as monitored by the Umgeni Water Board, it is almost three times higher (the approximate figure given by Umgeni Water Board is 1 000 000 litres per month). Unfortunately the figures for codisposed liquid and for codisposed sludge were not distinguished. Obviously only a part of the sludge would be available as liquid and some liquid would be held within voids while some would react with the solid materials present and some would be lost due to evaporation. Consequently this will reduce the amount of liquid available for leachate generation. If 20% of the liquid/sludge is unavailable due to the aforementioned reasons, then the amount of leachate generated per month is 1 013 763 l which compares favourably with the figure quoted by the Umgeni Water Board (Table 1).

After the formation of the landfill, chemical analyses were carried out on samples taken from the sump near the toe of the landfill over a five year period from 1986 to 1991 (Table 2). Table 2 shows some remarkably high values which presumably are associated with the type of waste disposed of. For instance, much of the sludge which goes into the landfill consists of hop waste from breweries. This accounts for the high levels of chemical oxygen demand, oxygen absorption, suspended solids, total dissolved solids, and conductivity.

In order to avoid the problem of pollution, the leachate produced at this site is pumped directly from sump to

TABLE 1. Solids, their absorption capacity, and liquids associated with Site 1

Date	Solids (m³)	Absorptive capacity (l)	Liquid/sludge (l) (a)	Liquid/sludge (l) (b)	Precipitation 5% of total (l)	Total liquid entry (l) (a)	Total liquid entry (l) (b)	Quantity of liquid exceeding absorptive capacity (l) (a)	Quantity of liquid exceeding absorptive capacity (l) (b)
Jan '90	107 640	7 125 768	9 891 000	7 912 800	277 000	10 168 000	8 189 000	3 042 232	1 064 032
Feb	66 130	4 277 806	6 609 000	5 287 200	376 000	6 986 000	5 663 200	2 707 194	1 385 394
Mar	68 647	4 544 431	7 059 000	5 647 200	651 500	7 710 000	6 298 200	3 166 069	1 754 269
April	57 841	3 829 074	5 951 000	4 760 800	108 000	6 059 000	4 868 800	2 229 926	1 039 726
May	70 309	4 694 176	7 355 000	5 884 000	56 000	7 411 000	5 940 000	2 716 824	1 245 824
June	67 095	4 441 689	6 769 000	5 415 200	28 000	6 797 000	5 443 200	2 355 311	1 001 511
July	61 997	4 104 201	6 291 000	5 032 800	3 300	6 294 300	5 036 100	2 190 099	931 899
Aug	63 890	4 229 518	5 854 000	4 683 200	41 650	5 896 350	4 725 550	1 667 132	496 332
Sept	52 288	3 461 466	4 819 000	3 855 200	141 500	4 760 500	3 996 700	1 463 034	499 234
Oct	62 604	4 144 385	5 334 000	4 267 200	338 500	3 672 500	4 605 700	1 528 115	461 315
Nov	60 195	3 984 909	6 612 000	5 289 600	153 000	6 765 000	5 442 600	2 780 091	1 457 691
Dec	50 911	3 370 308	5 047 000	4 037 600	320 000	5 367 000	4 357 600	2 176 692	1 167 292
Jan '91	59 000	3 905 800	5 094 000	4 075 200	505 000	4 580 200	5 599 000	1 693 200	674 400
							Average	2 285 840	1 013 763

Column 'a' takes into consideration the total amounts of liquid/sludge disposed of while in columns 'b' the liquid/sludge values have been reduced by 20%

TABLE 2. Chemical analyses of leachate from sump at Site 1 taken over a 5 year period from 1986 to 1991.

	pH	Suspended solids (mg/l)	Total dissolved solids mg/l	Conductivity (mS/m)	Oxygen absorption (mg/l)	Chemical oxygen demand (mg/l)
Max	8.9	6044	45041	8080	6200	70900
Min	6.6	200	900	450	145	11.2
Mean	7.7	965	17029	1174	649	13546

	Sodium (mg/l)	Potassium (mg/l)	Calcium (mg/l)	Magnesium (mg/l)	Sulphate mg/l	Ammonium (mg/l)
Max	8249	2646	1236	759	2237	3530
Min	80	355	60	70	6.5	92
Mean	1393	613	146	109	837	1093

sewer and conveyed to a sewerage plant for treatment. Consequently the quantity of leachate generated is less significant than the character of the leachate.

OTHER SITES

Four other sites which had ceased operation and were within 1 km of each other were considered. In these cases the COD and pH value of the leachate collected in the sumps associated with the sites were examined and an attempt was made to assess the influence of rainfall on the composition of the leachate produced.

All four areas are underlain by Dwyka tillite above which occurs Berea Red Sand. The latter only covers part of each site, generally occurring on the spurs between the valleys. Unweathered Dwyka tillite is an unstratified dark bluish-grey strong mudstone containing abundant inclusions of older rocks. The matrix is fine grained and consists of small angular fragments of quartz and rock material embedded in a fine base of rock floor. Near the surface the tillite is characterized by well developed joints which are closely spaced with secondary silicification frequently evident. The Dwyka tillite has a coefficient of permeability of 10^{-9} m/s or less, in other words it can be regarded as impermeable. Preferential weathering develops along the joints and gives rise to corestones. The tillite on initial weathering is brown in colour and is still a hard rock. It then grades into a softer yellowish-brown material and when completely weathered is a very soft yellowish brown clay. This clayey residual soil is generally overlain by colluvium, particulary in the valley floors. It consists of an upper sandy layer, up to 2 m thick, and an underlying clayey layer, ranging between 2 m and 3 m in thickness.

Domestic and industrial wastes were disposed of at each site. The volume of liquid waste codisposed was in accordance with a 5% rule. In other words a weight of liquid equivalent to 5% of the dry weight was disposed of in each landfill. This represents 52.5 l per cubic metre of compacted waste with a density of 650 kg/m^3. The wastes were disposed of in terraces up to 2 m in height which then were compacted. At the end of each day the waste in each terrace was covered by 0.1 m of clay.

Two of the landfills (sites 2 and 3) were located in adjacent valleys. As these two sites were filled, the waste overtopped the intervening valley spur so that they merged to form one landfill. This covers an area of 71 560 m^2 and has an approximate volume of 707 500 m^3. Cut-off drains run around the perimeter of the merged site to intercept run-off. These landfills were in operation between October 1983 and June 1987. On completion the landfill at these sites was capped by 0.5 m of clay material (i.e. highly weathered Dwyka tillite) and 0.3 m of topsoil was placed over the capping and grassed. A sump was located immediately downstream of each site from which samples were obtained.

During the 13 months June 1986 to June 1987 inclusive the volume of rainfall which infiltrated into these two sites was 7945.5 m^3, the lowest figure being in July (i.e. 43 m^3) and the highest in December (i.e. 2043 m^3). The average monthly amount of leachate generated was 2 669 000l, the highest volume being in March, 3 422 000 l, and the lowest in October-November, 422 000 l. The average value of COD of the leachate collected in the sump associated with site 2 was 743 mg/l. The highest and lowest values recorded from the sump were 2560 mg/l and 80 mg/l respectively. The average pH values was 8.4, the highest and lowest values being 7.3 and 8.7 respectively.

The highest COD's correspond with periods of low rainfall and vice versa. Obviously a lower volume of rainfall results in a lower volume of flow through a landfill leading to a more concentrated leachate. Heavy rainfall following a dry period initially produces leachate with a high concentration but a gradual reduction occurs after this initial peak as the initial leachate is removed. The lag time between heavy rainfall and the lowest values of COD recorded can be up to 4 months at site 2. In the case of site 2 the highest COD peak corresponded with a period of low rainfall which occurred some 3 months previously. It can be assumed that the residence time of rainfall in the leachate is slightly less than 3 months.

During periods of low rainfall the pH values of the leachate in the sump ranged from 8.2 to 8.9 whereas following periods of heavy rainfall the values dropped to 7.3. Hence the alkalinity of the leachate is reduced appreciably by heavy rainfall. The lag time between rainfall peaks and troughs, and corresponding low and high values of pH is around 4 and 5 months respectively.

At site 3 the average value of COD of the leachate in the sump was 974 mg/l with the highest and lowest values being 2240 mg/l and 421 mg/l respectively. The pH values were alkaline ranging from 7.6 to 9.0 with an average value of 8.6. Again the peaks in COD could be related to periods of low rainfall and the lowest values to high rainfall. For example, in March, 1990, the rainfall was unusually high, 238 mm being recorded. This reduced the COD and alkalinity of the leachate in late June and July.

Sites 4 and 5 are also located adjacent to one another, occupying an area consisting of one large and two small deeply incised valleys which drain in an easterly direction. Again cut-off drains surround the sites to intercept run-off and sumps are located immediately downstream of the sites. The landfill at site 4 occupies an area of 1.62 ha and has an estimated volume of 34 000 m^3. That at site 5 covers 3.1 ha and its estimated volume is 40 000 m^3. Both sites were operational from October, 1983 until June, 1986. On completion they were capped in the same way as sites 2 and 3. During their operational life time the lowest monthly average rainfall was 1.0 mm and the highest was 240.7 mm. This translates into an average input of rainfall into sites 4 and 5 of 203.8 m^3 and 388 m^3 respectively. The leachate generated per month by sites 4 and was respectively 642 351 l and 1 319 326 l.

The maximum, minimum and mean values of COD and pH of leachate recorded from the sumps at sites 4 and 5 are given in Table 3. Similar relationships between periods of high and low rainfall, and low and high values of COD and pH were noted at these sites. The lag times were again around 3 and 4 months.

CONCLUSIONS

The Dwyka tillite and the Pietermaritzburg Shale, in the greater Durban area, are impermeable and therefore landfills located on these rocks are naturally contained. No pollution to groundwater has been recorded from the four sites referred to which are situated on these rocks. The Natal Group sandstone, although having a low primary permeability, is well jointed and bedded and small leachate plumes could develop from landfills located on this rock type. Consequently at the site 1 leachate is conveyed from the sump to a sewer system.

Water balance calculations are complex and reliable figures are difficult to derive because of the number of factors involved. Nonetheless they provide some idea of the amount of leachate that could possibly be generated by a landfill and as such represent a worthwhile exercise.

The four sites where COD and pH values were obtained demonstrated the influence of rainfall on the composition of leachate produced by completed landfills. In other words the COD and pH value and, in turn, the concentration of leachate, is diluted by heavy rainfall and increases are associated with dry periods. However,

Table 3. Maximum, minimum and mean values of rainfall, chemical oxygen demand (COD), mg/l and pH value recorded at sites 4 and 5.

Site 4	Rainfall	COD	pH
Maximum	245.3	960	9.2
Minimum	1.2	14	6.8
Mean	82.3	217.6	7.8
Site 5			
Maximum	245.3	2640	9.3
Minimum	1.2	57	6.6
Mean	82.3	406.1	7.9

these changes do not happen immediately and lag times of 3 to 4 months characterize these landfills.

REFERENCES

Campbell, D. 1983. Understanding water balance in landfill sites. Waste Management, 11, 594-601.

Thornthwaite, C. and Mather, J. 1957. Instructions and tables for computing potential evapotranspiration and water balance. Publications in Climatology, 10, No 3,. Laboratory of Climatology, Drexel Institute of Technology, Centerton, New Jersey.

Replacement Wetland Emphasizes Biodiversity

Robert E. Smiley, RPF, Senior Environmental Scientist
RUST Environment & Infrastructure, Richmond, Virginia, USA

Abstract

The expansion of an existing landfill located in Pennsylvania, impacted areas regulated as waters of the United States. The landfill is owned and operated by a private corporation and accepts solid waste from municipalities in the area. The Pennsylvania Department of Environmental Resources (PADER) has a requirement for mitigation of permitted impacts to aquatic systems. As well, corporate policy of the owner and operator of the landfill stipulates "no net loss" of wetland and "no net loss" of biodiversity. Because of these requirements, an aquatic system was designed to replace the impacted area. PADER regulations require that replacement systems be located adjacent to impacted systems where practicable. Hydrology is a key parameter in the successful creation of an aquatic system, therefore, an extensive investigation was conducted of the hydrologic and geologic conditions in the vicinity of the impacted area. This investigation determined that ground-water extraction wells associated with the existing unlined landfill, a Superfund site, had influenced the lateral flow in the vicinity of the impacted waterway. Drawdown from the wells had effectively drained the tributary and the area adjacent to the tributary with the result that locating a functioning aquatic system adjacent to the impacted site was not possible.

A search was initiated to locate an area nearby in which a suitable replacement system could be created. Such a site was identified along a perennial stream in the same watershed as the impacted area. A detailed analysis was performed to determine the topographic, hydrologic and soils components of the replacement site. With this information, a replacement system was designed to adhere to "no net loss of wetlands" and "no net loss of biodiversity" requirements. To accomplish this goal, the plan was designed to achieve multiple functions for the system. Design functions include water dependent vegetation, floodflow alteration, streambank stabilization, sediment and nutrient removal and/or transformation, and aquatic diversity and abundance.

The replacement wetland is located adjacent to a perennial stream. A meandering stream was created within the wetland to add interest and to increase biodiversity. With the stream, three vegetative habitats were created: Palustrine Forest, Riverine/Emergent, and Riparian Fringe. Hydrology for the system will be dominated by ground-water discharge with a periodic influx of floodwaters from the creek.

Modern Landfill site consists of approximately 432 acres of land located southwest of the Borough of Yorkana in the Townships of Windsor and Lower Windsor, Pennsylvania, USA. The coordinates of the tract are 39°57'36" north latitude and 76°35'29" west longitude. Modern Landfill is located on a hill bounded on the north, east and west by unnamed tributaries of Kreutz Creek. The ridge on the southern boundary of the site is occupied by a 66-acre unlined landfill at an elevation of approximately 700 feet above mean sea level (msl). The 66-acre unlined landfill slopes to the north to a 34-acre double-lined expansion area which has base grade elevations of approximately 510 feet above msl.

The facility is located within the Conestoga Valley Section of the Piedmont Province. Topographically, this province is characterized by well-developed

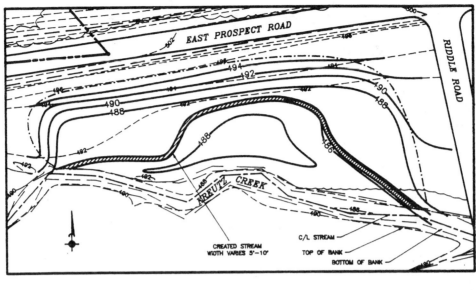

GRADING PLAN
NOT TO SCALE

REPLACEMENT WETLAND
MODERN LANDFILL
YORK, PENNSYLVANIA

northeast-southwest trending valleys and drainage patterns.

The Southwest Expansion area consists of approximately 94 acres of land adjacent to the southwest boundary of the existing landfill operations. This includes a noncontiguous 67-acre disposal area and associated support areas. An unnamed and intermittent tributary lies along the northeast boundary of the expansion area and drains northward towards Kreutz Creek. The drainage was previously identified as a jurisdictional area in a wetland delineation. In this study, the subject drainage was found to meet the criteria for waters of the U.S. (waters). Construction activities for the proposed expansion required that this drainage be filled and relocated. Pennsylvania Department of Environmental resources (PADER) regulations require the replacement of regulated areas. The purpose of this plan was to provide for the replacement of approximately 2,000 linear feet of the intermittent drainageway that was impacted by construction activities.

Before current construction took place, the waterway in question had been severely impacted by the ground-water extraction system. Fifteen ground-water extraction wells are located adjacent to, and throughout the length of, the tributary. The drawdown created by the extraction systems influences the lateral flow component from the landfill towards the tributary and prevents the recharge of the tributary. Similarly, the extraction system induces additional flow from outside the extraction system eastward across the tributary toward the extraction system. Thus, the tributary no longer receives ground-water discharge and the system has been effectively drained.

The primary functions of the waterway appear to have been to collect and filter surface runoff, to contribute habitat for wildlife, and to provide streambank stabilization. This long, narrow riparian forest was surrounded by farm land creating a corridor with an ample amount of "edge effect" preferred by wildlife. This area had its greatest value to upland wildlife as a source of food and cover.

Vegetation along the waterway consisted of a deciduous forest composed predominately of upland species. This plant community appears to have occupied the banks of the waterway and consisted of well defined tree, shrub/sapling, and herbaceous strata.

The tree stratum was dominated by mature tulip tree, black cherry, and red maple. In the understory, shade tolerant white ash had reached sapling size. White ash is a climax species. These conditions indicate that the forest was in an intermediate climax stage. All of the tree species mentioned above are

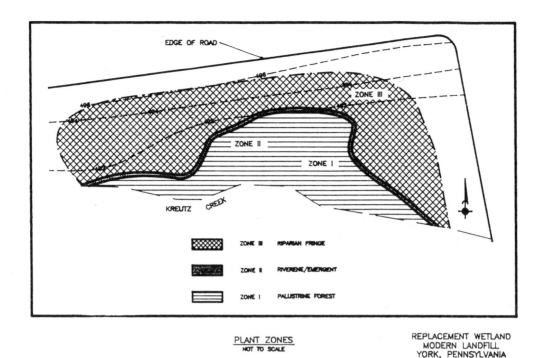

PLANT ZONES
NOT TO SCALE

REPLACEMENT WETLAND
MODERN LANDFILL
YORK, PENNSYLVANIA

typical of deciduous hardwood forests of the Piedmont Plateau. Each species is well suited for the both the moderately well-drained soils and the topographic position that existed along the waterway.

In the understory, along with white ash saplings, spicebush is a dominate species. Spicebush is a common member of the understory of eastern hardwood forests.

On the forest floor, the six herbaceous species that were identified prefer the habitat of a mixed deciduous forests. This stratum is composed entirely of upland species, whereas vegetation in the woody strata is marginally hydrophytic. The likely explanation for this situation is that the deeply rooted woody plants are the result of the more moist conditions at the site prior to the operation of the ground-water extraction system. The shallow rooted, upland herbaceous plants probably invaded the area after the natural water regime was changed.

Ground-water extraction wells will continue to depress the water table at the waterway; therefore, its primary function will continue to be the collection of surface runoff. By relocating the channel and directing flow to Best Management Practices (BMPs), water quality leaving the site will be improved.

Mitigation Plan

Under normal circumstances, a wetland replacement plan should create an area with similar functions and values to those of the impacted wetland. As stated previously, the impacted area was not functioning as a wetland; thus, exact duplication of that environment is not desirable. The purpose of this plan was to incorporate some functions that the impacted area may have possessed prior to dewatering and to create a wetland with enhanced biodiversity. This plan was designed to achieve growth of wetland dependent vegetation, floodflow alteration, streambank stabilization, sediment/nutrient removal/ transformation, and plant diversity/abundance.

The chosen mitigation site is a bottom land meadow situated between a steep hill on the northern side of the site and Kreutz Creek, a small perennial stream, along the western and southern boundaries. Kreutz Creek meanders along the southern boundary of the replacement site for approximately 400 feet. The width of the meadow between the toe of the slope of the hill and the creek averages approximately 170 feet and the meadow is approximately 4 feet in elevation above the creek. Hydrologic studies for the landfill permit indicate that Kreutz Creek and its environs receives ground-water discharge the year around. In placing the wetland system between the hill and the creek, a source of hydrology for the system is assured. A minor portion of Kreutz Creek will be directed into the wetland's channel to augment low flow level in the waterway. In addition, the replacement wetland will receive flooding from over-topping of the creek thereby providing intermittent inundation and flushing of the wetland.

A total of five soil samples were taken to a depth of approximately 4 feet. Based on these samples, hydric soil conditions appear at an average depth of approximately 30 inches. Grades in the mitigation area were adjusted to intercept ground-water flow. Final plan elevations call for a final floor elevation about 36 inches below the existing the existing grade, Thus, the substrate will consist of a silty clay hydric soil and is expected to be well suited to support the hydrophytic vegetation.

A range of final elevations are specified across the profile of the wetland for the purpose of providing variable degrees of wetness, which will support a diverse range of vegetative types. A feature of the plan that will play a key role in the water budget is the establishment of a small meandering waterway that will run the length of the replacement area. The waterway will roughly parallel Kreutz Creek in both plan and profile and will connect to Kreutz Creek via an inlet and outlet channel. The plan calls for the creation of three planting zones:

Zone I, Palustrine Forest, consists of an island between Kreutz Creek and the created stream. The area will receive ground water from both waterways and will be periodically inundated, saturated to the surface during the winter and spring, and saturated at shallow depths during dry periods.

Zone II, Riverine/Emergent, contains the created stream and its environs. In addition to receiving inflow from the existing creek, the created stream will intercept the upper level of the water table and receive ground-water discharge. Water will flow in the deepest part of the stream year around. From the center of the stream the banks will slope gently to higher elevations and provide a range of environments that will support a variety of vegetation.

Zone III, Riparian Fringe, is an ecotone. Surface topography should range from saturated conditions adjacent to Zone II to periodic saturation near the cut slope. As with the created stream, ground water is the primary source of the hydrology. The hydrologic cycle will range over the profile of the zone from permanently saturated and periodically inundated near the stream to widely fluctuating and occasionally inundated in the higher reaches of the zone.

A variety of design features have been incorporated into the wetland mitigation plan in order to maximize the wetland's potential to satisfy the desired goals.

In the following discussions each function of the proposed replacement wetland is examined in relation to the design features that promote the function.

Wetland Dependent Vegetation

Wetlands provide habitat for numerous species of birds, mammals, reptiles, amphibians and fish. The primary goal in providing habitat for wetland dependent fauna is to create an environment which promotes diversity and abundance of these species. A variety of habitat conditions within a wetland is critical to creating species diversity.

Vertical layering and horizonal overlap of habitats of forested or shrub/scrub wetlands support a greater diversity of wildlife species. These functions are developed in the mitigation plan by providing varied bottom contours to create integrated patches of different hydrological regimes and different vegetation classes. This is accomplished by creating the meandering stream channel that bisects the wetland area. The stream will be fed by ground-water discharge and by surface flow from Kreutz Creek. An elongated island is proposed parallel to the creek. The contours will rise from Kreutz Creek to form the island, dip into the permanently fed channel of the created channel, and gradually rise to upland. This configuration allows the establishment of three distinct vegetative zones and corresponding ecotones.

A palustrine forest will be established on the island. A riverine/emergent zone will be established along the creek. The third zone consists of a riparian fringe that abuts an upland created along the cut slope. This arrangement provides a dense and complex layering of vegetative strata, creates a vegetation mosaic pattern, and makes optimal use of edge effect.

Flood Flow Alteration

Flood flow alteration is the process by which peak flows from runoff, surface flow, and precipitation are stored or delayed. Wetlands, as well as upland areas, act to detain flood waters by intercepting sheet flow and flood waters. By lowering flood peaks, wetlands act to decrease flood-related damage. Vegetation slows floodwaters by creating frictional drag in proportion to stem density. Wetlands with dense stands of vegetation are capable of slowing floodwater. Channel roughness, and thus the ability to retain floodwater, increases with increased vegetation density. The vegetation to be installed in the replacement wetland will serve this function well. Sheet flow, rather than channel flow, offers greater frictional resistance. The created wetland has a morphology that allows water to spread out rather than remain in the channel.

Streambank Stabilization

Streambank stabilization is the binding of soil by plant roots and the dissipation of erosive energy caused by currents. Streambank stabilization by plants protects areas from erosion and has the net effect of minimizing the deposition of sediments downstream. The frictional resistance a wetland offers to erosive energy depends on the density and height of the vegetation relative to incoming currents. The persistent emergent and woody vegetation proposed for the replacement wetland will offer the frictional resistance to effectively stabilize the area.

Sediment/Nutrient Retention

Sediment and associated nutrients and contaminants are sometimes carried by runoff or channel flow into wetlands, where they can be removed from the water column by sediment deposition, chemical breakdown, and/or assimilation into plant and animal tissues. Sediments may also be temporarily retained by a wetland before moving further downstream. The principal factor affecting a wetland's ability to trap sediments is the change in the velocity or energy level of incoming water. Decreased water velocity results in sediment deposition. The replacement wetland has the ability to retain water by the following physical and biological factors:

- The gentle gradient in the wetland basin will slow water velocity, and

- the dense wetland vegetation will act to slow water velocity, to force water to flow through a longer course, to retain it longer in the basin, and to discourage re-suspension of bottom sediments.

Plant Diversity/Abundance

A replacement wetland is most valuable if it increases the diversity of the existing plant and animal communities. The ability of a wetland to fulfill this function is dependent upon the type and quality of habitat it provides. Diverse vegetation in the replacement wetland will provide a source of nutrients, protective cover, and temperature moderation by providing shade.

The replacement site plan features some of the dominant species found in the impacted area along with select species found on other undisturbed jurisdictional areas on the property. The majority of the planted canopy species will be red maple and green ash. These species were present on the impacted site, have a wide range of moisture tolerance, are hardier, and have a greater chance of survival than less common species that occur on the property. The relatively fast growth of these species will form a canopy that will allow for the development of shade tolerant shrubs and forbs in the understory. Shrub/scrub and perennial forbs are also proposed to be planted in the various water regimens to create diversity with the replacement wetland.

Immediately following completion, the replacement wetland will assume its functions as a ground-water discharge area and as a flood storage site. The remainder of the functions are dependent upon the establishment and maturing of the vegetation. This will occur gradually, over time as the vegetation, especially the woody species, mature and grow in size. Wildlife values will exist after the first year and increase in value with each subsequent year. Some vertical stratigraphy should be observable within 3 to 5 years. After about 15 years Zone I will begin to take on the appearance of a palustrine forest.

REFERENCES

Created and Natural Wetlands for Controlling Non-point Source Pollution, U.S. Environmental Protection Agency, Published by C.K. Smoley Boca Rotan, Florida, 1993.

Gore, James A., *The Restoration of Rivers and Streams*, Butterworth Publishers, Boston, Massachusetts, 1985.

Hammer, Donald A., *Constructed Wetlands for Wastewater Treatment*, Lewis Publishers, Inc. Chelsa Michigan, 1989.

Kusler, Jon A. and Mary E. Kentula, *Wetland Creation and Restoration, the Status of the Science,* Island Press, Washington D.C., 1990.

Marble, Anne D., *A guide to Wetland Functional Design*, Lewis Publishers, Boca Rotan, Florida, 1992.

Mitsch, William J. and James G. Gamselink, *Wetlands*, published by Van Nostrand Reinhold Company, New York, 1986.

Phase I Remedial Investigation Report, 1990, Golder Associates, Mt. Laurel, New Jersey.

Wetlands and Waters of the U. S. Jurisdictional Boundary Determination, 1988, RMC.

A Method for Prediction of Long Term Settlement of Sanitary Landfill

K.C. Sohn, Principal
Soil Foundation Systems, Inc., California, USA

S. Lee, Assoc. Professor of Civil Engineering
Seoul City University, Seoul, Korea

Abstract: Landfills undergo settlement at significant rate under self weight for several decades. When a landfill site is proposed for development one of the critical tasks confronting the engineer is the prediction of the post-construction settlement of the improvements. This task requires prediction of the magnitude and rate of residual settlement of the landfill under self weight.

The process of landfill settlement is highly complex due to the heterogeneous nature of the landfill and the erratic and continuous changes in the material properties and characteristics with time. However, it is indicated that the highly erratic pattern of settlement is dominant in the first year or two, and thereafter the settlement gradually attains an orderly pattern, characterized by a linear relationship between the settlement rate and the fill age. This relationship is used to model long term settlement behavior of landfill. The model parameters were obtained through analysis of reported long term settlement data. It is shown that the settlement rate is uniquely determined in terms of the height and age of the fill.

Introduction

Sanitary landfill sites are increasingly becoming a potential source for development as more readily developable lands are rapidly depleted in many parts of the world today. One of the critical tasks confronting the engineer in the planning and design of land development and reclamation of a landfill site is the prediction of post-construction settlement of the improvements.

Although landfill sites that are developed for construction purposes are usually old, evidence [1,2,7] indicates that residual settlement of old landfills can still be significant, both in magnitude and rate. Therefore, it is necessary to consider the residual settlement of the landfill under self weight in the prediction of post-construction settlement of the improvements when a landfill site is proposed for development.

Some insight into long term settlement behavior of sanitary landfills has been provided by Yen and Scanlon [8] through analysis of rather extensive settlement platform data from several landfill sites, gathered over a period of nine years following the end of filling. Yen and Scanlon have shown that the settlement rate is linearly related to the logarithm of time, as measured from the mid-point of the construction period, for a given fill height and construction period. This paper examines further details of long term settlement behavior of landfills and presents a method for prediction of long term settlement of landfills under self weight.

Mechanism of Long Term Landfill Settlement

The overall settlement of sanitary landfill involves various mechanisms, ranging from purely mechanical to biochemical in nature [7], and the settlement process is highly complex due to the heterogeneous nature of the landfill and the erratic and continuous changes in the material properties and characteristics with time.

Many of the materials in the landfill, such as dry paper product and garden refuse, are initially resilient to a large extent. These resilient components of the landfill hinder the effectiveness of mechanical compaction for reducing the fill volume during placement of the fill. However, a large compression of the landfill results due to bending, distortion, reorientation and crushing of the materials, as the fill height is increased. Sowers [7] describes this initial phase of the settlement as being completed in less than one month.

In addition to the above process, the materials in the landfill lose their resilient characteristics through adsorption of moisture from rains or other sources in the landfill, and also by decay. With the loss of the resilient character of the materials, the fill is compressed into a more dense packing under its self weight. This natural process is responsible not only for a significant portion of the overall settlement, but also for eliminating much of the complex and erratic pattern of the settlement. Landfills placed at vastly different densities would be brought to relatively similar conditions with respect to wet density through this phase. This phase is generally completed in the first year or two [3], depending on the regional climate and environmental conditions.

Following the early phases discussed above, the fill settlement is gradually transformed into a more orderly pattern characterized by a linear relationship between the settlement rate and the fill age. Therefore, the complex and erratic settlement pattern that dominates the early phases is not directly relevant to long term settlement of landfill. The dominant mechanism controlling the settlement in this phase (secondary compression phase) is the combined effect of biochemical decomposition and mechanical creep of the materials. Citing a case history of an old landfill where the settlement continued with time after the preloading surcharge had been removed, Charles [1] suggested that the ongoing long term settlement was most probably due to biodegradation of the materials. It could also be viewed that the mechanical creep potential of the landfill is enhanced by biodegradation during long term settlement process. Regardless of the precise nature of the possible interaction between the biodegradation mechanism and the mechanical creep mechanism, the settlement process may be characterized by a linear relationship between the settlement rate and the fill age for sanitary landfills older than one to two years. This linear relationship is a convenient basis for modeling the settlement behavior controlled by the combined effect of biochemical decomposition and mechanical creep during the secondary compression phase.

Linear Relationship Between Settlement Rate and Fill Age

Figure 1 presents relationships obtained by Yen and Scanlon [8] between the settlement rate and the fill age, designated as (A), (B), (C), and (D). In Fig. 1, the median age of the fill, t_1, is defined as the time as measured from the mid-point of the construction period, t_c. Fill height, H_f, is the height of the fill at the conclusion of filling. The Yen-Scanlon relationships were determined by the method of least squares applied to groups of rather extensive settlement platform data from three landfill sites, collected over a period of nine years after completion of filling.

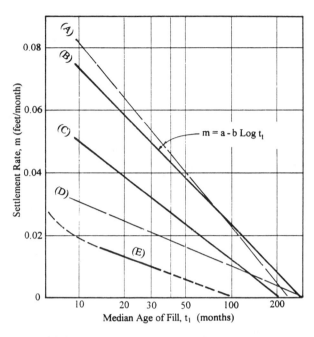

(A) $m = 0.142 - 0.060 \, \text{Log} \, t_1$; $H_f \geq 100$ feet, $70 \leq t_c \leq 82$ months
(B) $m = 0.124 - 0.051 \, \text{Log} \, t_1$, $80 \leq H_f \leq 100$ feet, $70 \leq t_c \leq 82$ months
(C) $m = 0.088 - 0.038 \, \text{Log} \, t_1$, $40 \leq H_f \leq 80$ feet, $70 \leq t_c \leq 82$ months
(D) $m = 0.052 - 0.021 \, \text{Log} \, t_1$, $H_f < 40$ feet, $t_c < 12$ months
(After Yen and Scanlon [8])

(E) $m = 0.037 - 0.0185 \, \text{Log} \, t_1$, $H_f = 5$ feet   (Data from Rao, et al. [5])

Fig. 1 Linear Relationships Between Settlement Rate and Fill Age

Figure 1 also presents a settlement rate vs. fill age relationship determined by using settlement data obtained by Rao, et al. [5] from an experimental cell filled with 5 feet (1.5 meters) of household refuse, designated as (E). A noteworthy observation from (E) is the linearity developing beyond the age of 10 to 12 months.

The linear relationships depicted in Fig. 1 may be represented in the following form:

$$m = a - b \, \text{Log} \, t_1 \qquad (1)$$

where, m = settlement rate; and "a" and "b" = settlement rate parameters.

Determination of Long-Term Settlement Rate

In order to determine settlement rate from Eq. 1, the two settlement rate parameters, "a" and "b", must be known. It was suggested elsewhere [6] that the settlement rate parameters could be uniquely determined in terms of the fill height. This suggestion was originally based on examination of the Yen-Scanlon relationships presented in Fig. 1. For the purpose of this paper, the Yen-Scanlon data were re-analyzed and the settlement rates obtained from the Rao data [5] were also used to determine the settlement rate parameters in the manner described next.

The Yen-Scanlon data set for (B) in Fig. 1 was chosen for analysis because it had less scatter in the data distribution than the other sets. Analysis of the data set, with the exclusion of a few data points that were exceptionally distant from the center of the data distribution, provided the following relationship:

$$m = 0.118 - 0.048 \, \text{Log} \, t_1 \qquad (2)$$

where, m = settlement rate in feet/month and t_1 = median age of the fill in months. From Eq. 2, a = 0.118 and b = 0.048 are obtained.

The settlement rate vs. fill age relationship determined based on the Rao data was chosen as another reference, which is designated as (E) in Fig. 1. It is noted that, beyond the fill age of 10 months, there was no significant scatter of data points that affected determination of the relationship for (E). From this relationship, a = 0.037 and b = 0.0185 are obtained.

By using an average fill height of 90 feet for Eq. 2 and a fill height of 5 feet for (E) in Fig. 1, "a" and "b" points were plotted as shown in Fig. 2 to obtain the a-line and b-line. Other points of "a" and "b" from the Yen-Scanlon relationships shown in Fig. 1 are also plotted in Fig. 2 by using an assumed average fill height of 110 feet for (A), 60 feet for (C) and 30 feet for (D).

It is seen that "a" and "b" from (C) plot exactly on the a-line and b-line, respectively. However, the points from (A) and (D) deviate significantly. This is also apparent from examination of Fig. 1, in that (B), (C) and (E) present a pattern of correlation between them, while (A) and (D) are random. In order to explain possible causes for the deviation exhibited by points from (A) and (D), the pertinent Yen-Scanlon data sets were examined. (A) is a plot for fill height of over 100 feet and construction duration of 70 to 82 months. When compared to (B) and (C), the anomaly of (A) was thought to be the result of the extensive scatter among the data points. (D) is a plot for fill height of less than 40 feet and for a construction period of less than 12 months. The inclusion of a relatively large amount of data points for early stages (less than 12 months) was a direct cause of the anomaly in (D). As discussed earlier, a linear relationship between the settlement rate and fill age is generally attained when the fill is older than 12 months. When the data points for ages less than 12 months are excluded, the settlement rate for (D) is represented by the following equation:

$$m = 0.064 - 0.029 \, \text{Log} \, t_1 \qquad (3)$$

The "a" and "b" points from Eq. 3 are plotted as "p" and "q", respectively, in Fig. 2 for a fill height of 30 feet.

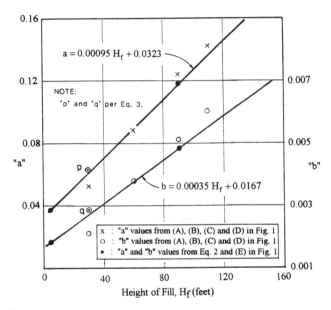

Fig. 2 Determination of Settlement Rate Parameters

From Fig. 2, the following equations may be derived for "a" and "b" in terms of fill height:

$$a = 0.00095 \, H_f + 0.0323 \qquad (4\text{-a})$$

$$b = 0.00035 \, H_f + 0.0167 \qquad (4\text{-b})$$

where, H_f = height of fill, in feet, at the conclusion of filling. The use of H_f and t_1 in the derivation of the above equations follows the adoption by Yen and Scanlon [8] in their data presentation.

From Eq. 1 and Eq. 4, the rate of landfill settlement due to self weight of the landfill is uniquely determined in terms of the fill height, H_f, and fill age, t_1. Settlement rates as determined from Eq. 1 and Eq. 4 are plotted against the median age of fill in Fig. 3 for various fill heights.

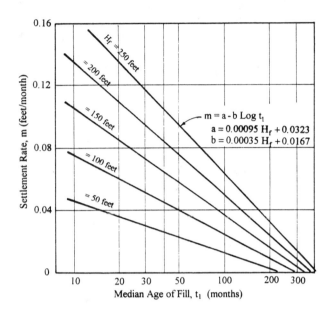

Fig. 3. Settlement Rate Variation with Fill Height and Age

Magnitude of Residual Settlement

Residual settlement of a given landfill under self weight for a time span of interest can be obtained by integrating the settlement rate over that time span. Therefore,

$$S_r = \int m\, dt = \int_{t_1}^{t_1+\Delta t} (a - b\, \text{Log}\, t)\, dt \qquad (5)$$

where, S_r = residual settlement in feet, t_1 = median age of the fill at the beginning of the residual settlement computation period in months, Δt = time span for which the residual settlement is computed, in months, and "a" and "b" are obtained from Eq. 4 in terms of the fill height, H_f, in feet. It should be noted that the time span in the above integration should be limited to:

$$t_1 + \Delta t \leq 10^{(a/b)} \qquad (6)$$

Discussions

Janbu, et. al [4] found for saturated soil deposits, including peat and organic deposits, that settlements continue long after the excess pore pressure has been substantially dissipated in primary consolidation, and have shown the settlement rate (or creep rate) after completion of primary consolidation to vary in accordance with the following equation:

$$m \cong \frac{H}{R\,t} \qquad (7)$$

where, m = settlement rate, t = time since application of vertical loading to the deposits, and R = creep resistance, which ranges from 10 to 100 for peat and organic clays and from 100 to 300 for clays and silts. Edil, et. al [3] indicate that the time-settlement characteristics of sanitary landfill are similar to that of peat or organic deposits. Indeed, it can be noted that Eq. 1 and Eq. 4 together portray generally similar settlement characteristics as portrayed by Eq. 7 in the secondary compression, or creep, range.

Integration of Eq. 7 with respect to time, t, yields the residual settlement, S_r, as:

$$S_r = \frac{H}{R} \ln\left(\frac{t_2}{t_1}\right) \qquad (8)$$

where, $t_2 = t_1 + \Delta t$ in Eq. 5.

From Eq. 1 and Eq. 4, completion of landfill settlement due to self weight would take approximately 18 years for a fill height of 50 feet (15 meters) and more than 24 years for a fill height of 100 feet (30.5 meters). These results substantially agree with reported experiences of long term settlement of landfills in a similar height range.

Residual settlements as calculated from Eq. 5 and Eq. 8 are compared below:

H, feet (m)	t_1, months	S_r, feet (m) Eq. 5	S_r, feet (m) Eq. 8	R
50 (15.3)	60	1.17 (0.36)	64/R	55
	120	0.38 (0.12)	29/R	78
100 (30.5)	60	3.04 (0.93)	158/R	52
	120	1.44 (0.44)	88/R	61

The creep resistance of Janbu, et al, R, was calculated from Eq. 8 so as to realize the same amount of settlement as calculated from Eq. 5. The above results indicate that a landfill of five years old and in the range of 50 to 100 feet in thickness could be considered as the average of peat or organic clay deposits, in terms of the creep resistance.

The proposed settlement rate solution provides an important requirement with regard to the acquisition of data relating to settlement of landfills. Since the settlement rate depends on both the fill height and age, settlement data acquisition should include the thickness and age of the fill.

The proposed solution is useful for correlating the settlement rate data for different fill heights and ages.

Concluding Remarks

On the basis that sanitary landfill older than one to two years attains a linear relationship between the settlement rate and fill age, a model was developed that determines the settlement rate. The model contains two parameters (referred to as settlement rate parameters in the paper) that are determined in terms of the fill height. The relationship between the settlement rate parameter and the fill height was empirically established through analysis of reported settlement data from several landfill sites, ranging from 5 feet to over 100 feet in fill height.

For the range of fill heights cited in the study, both settlement rate parameters, "a" and "b", are linearly related to the fill heights, as shown by Eqs. 4-a and 4-b. Whether these equations would hold for fill heights of much greater magnitude needs to be verified with pertinent field data as they become available.

As is common in the derivation of empirical relationships in general, it would be possible to refine the empirical relationships of Eqs. 4-a and 4-b by supplementing additional quality data, following the methodology discussed in the paper.

Acknowledgment

The subject treated in this paper is a part of a long range research program at Seoul City University, Seoul, Korea, on municipal landfill settlement behaviors. Valuable comments were received from several graduate students during a seminar on the subject at Seoul City University. Special appreciation is due to Patrick Fain of Soil Foundation Systems, Inc. for his careful review of the manuscript and constructive suggestions.

References

1. Charles, J. A., 1991, "The Causes, Magnitudes and Control of Ground Movements in Fills", Proceedings, 4th Int'l Conference on Ground Movements and Structures, Pentech Press, London, pp. 3-28.

2. Druschel, S. J., and Wardell, R. E., 1991, "Impact of Long Term Landfill Deformation", Proceedings, ASCE Geotechnical Engineering Congress 1991, Vol. II, Geotech. Special Publication No. 27, pp. 1268-1279.

3. Edil, T. B., Ranguette, V. J., and Wuellner, W. W., 1990, "Settlment of Municipal Refuse", ASTM Special Technical Publication 1070, pp. 225-239.

4. Janbu, N., Svanø, G., and Christensen, S., 1989, "Backcalculated Creep Rates from Case Records", Proceedings, 12th International Conference on Soil Mechanics and Foundation Engineering, Rio de Janeiro, Vol. 3, pp. 1809-1812.

5. Rao, S. K., Moulton, L. K., and Seals, R. K., 1977, "Settlement of Refuse Landfills", Proceedings, ASCE Geotech. Engrg. Div. Speciality Conference on Geotechnical Practice for Disposal of Solid Waste Materials, pp. 578-598.

6. Sohn, K. C., and Johnson, A. M., 1991, "Factors Affecting Determination of Stability and Settlement of Sanitary Landfills", Proceedings, 5th International Symposium on Solid Waste Management Technology, Seoul, Korea, pp. 207-241.

7. Sowers, G. F., 1973, "Settlement of Waste Disposal Fills", Proceedings, 8th International Conference on Soil Mechanics and Foundation Engineering, Moscow, pp. 207-210.

8. Yen, B. C., and Scanlon, B., 1975, "Sanitary Landfill Settlement Rates", Journal of Geotech. Engrg., ASCE, Vol. 105, No. GT5, pp. 475-487.

Experimental Studies on Prevention of Reddish Soils Contamination of the Sea

H. Uehara, H. Hara
C.E. Dept., Univ. Of the Ryukyus, Okinawa, Japan

1 Introductory Remarks

Because of land reclamations and public works on those smaller islands, the subtropical, lateritic soils (reddish soils; so-called "Kunigami Maaji Soils"[1],[2]) are rapidly weathered and eroded by rainfalls. These reddish soils then flow down to the streams and finally into the sea. Consequently the inland soils are lost causing agricultural and geotechnical problems. The soil contamination of the rivers and the sea has impacted on the inland and sea ecosystems in the Ryukyu Islands.

The field observations of the erosion of reddish soils and flow-out of the soils and the contaminatiom of the sea water have been performed to detrmine effective measures against the disaster, as shown by Pictures 1-3.

Experimental studies to stabilize and protect the slope soils have been carried out for the control and prevention of the erosion and failures Also the investigative studies on the occurrence and the developing mechanisms of the slope erosion and failures by using the big model (at the site) and the small model (in the laboratory) have been executed.

It is quite difficult to take measures against the pollution of streams and the sea caused by public

Picture 2. Flow-out of reddish soils.

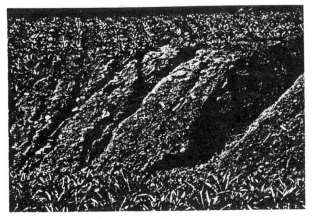

Picture 1. Eroded farm land.

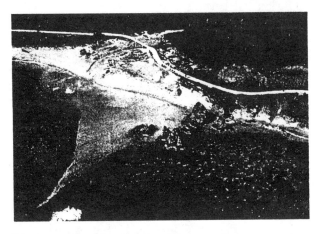

Picture 3. Contaminations of sea water by reddish soils from the river.

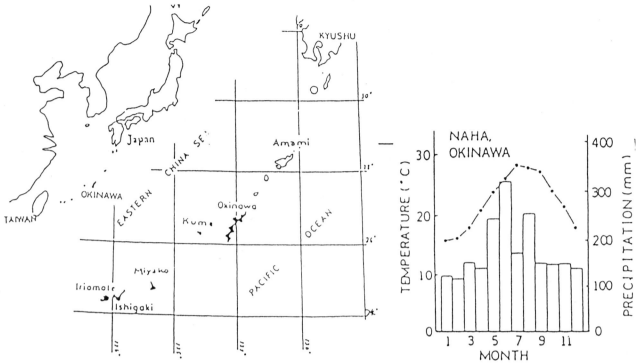

Fig.1a Map of the Ryukyu Islans showing location of Okinawa.

Fig.1b Mean monthly temperature and precipltation.

works and reclamations to step up the regional development. For the geotechnical purposes (environmental geotechnics), some of the results of the investigative studies are reviewed in the present paper to improve the serious condition of the Ryukyu Islands and to decrease the contamination of the sea surrounding those islands as the grobal problems.

2. Location and Kunigami Maaji Soils
a) Location
   The Okinawa prefecture in the Ryukyu Islands is located at the southern part of the Japan Islands as shown on the Fig.1a. There are four major factors that control the climate (Fig.1b) in the Ryukyu Islands;latitude, the warm current, proximity to the large land mass of Asia (China), and the typhoon (hurricane).

b) Properties of Kunigami Maaji Soils
   The distribution of Kunigami Maaji soils in the Ryukyu Islands is shown in

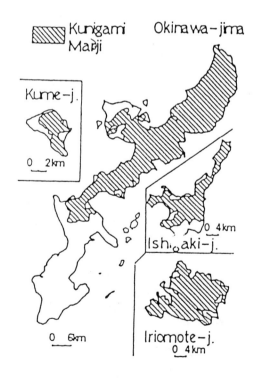

Fig.2 Distribution of Kunigami Maaji Soils.

Table 1. Results of Physical Tests

Samples	Nakijin SOIL	Higashi SOIL
G_s	2.74	2.71
w_L(%)	51.3	29.4
w_P(%)	28.2	20.9
I_P(%)	23.1	8.5
Gravel(%)	3	6
Sand(%)	10	51
Silt(%)	85	35
Clay(%)	2	8
Japan Unified System		
Plasticity chart	C H	S F
Triangular Chart	F	S F

Fig.2. Physical properties of the representative soils of Nakijin area and Higasi area in Okinawa main island are shown in Table 1. and the compaction curves refering to the stabilized soils with cementing material on the Fig.3a,b. Some other geotechnical properties may be refered to the references 1),2),3).

Here, the characteristics of the temperature changes in the reddish soil surface that may effect the weakening and the weathering of the slope soils are presented on the Fig.4.[4]

3. Test Results

Stabilization tests have been carried on using a certain cementing material[5] The following results are presented here as follows.

a). Strength Test
    Unconfined Tests, CBR
b). Slaking Test
c). Erosion Test
    Laboratory Model Test
    Field Model Test

4. Discussion

It is clear from these test results that the reddish soils(slope soils) can be improved by adding the cementing material against the weathering and erosions, as follows,

(a) Mixing moisture content may be decreased to the dry side of optimum moisture content (OMC) of nonstabilized soil to obtain maximum density
(b) Unconfined compressive strength and CBR value may effectively be increased by increasing the mixing

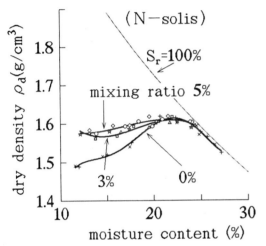

Fig.3(a) compaction curves

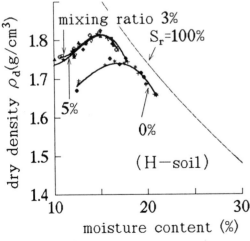

Fig.3(b) compaction curves

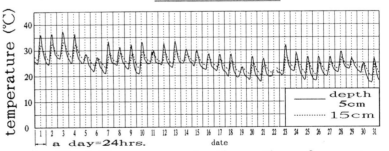

Fig.4 Temprature in the soil surface

ratio of the cementing material and curing time.

(c) From slaking test results, the residuals in water may effectively increased to promote the durability against water (rainfall), and the

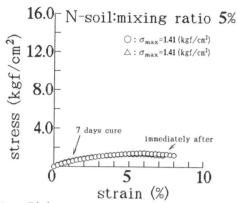

Fig.5(c) stress-strain curves

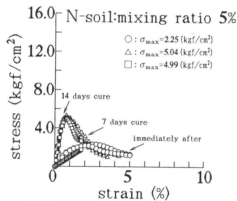

Fig.5(b) stress-strain curves

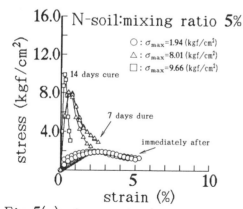

Fig.5(a) stress-strain curves

Fig.5 Unconfined compression test results

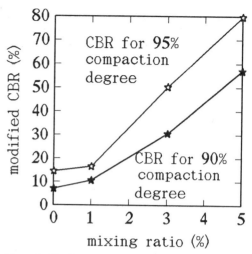

Fig.6 Relationship mixing ratio and modified CBR(N-soil).

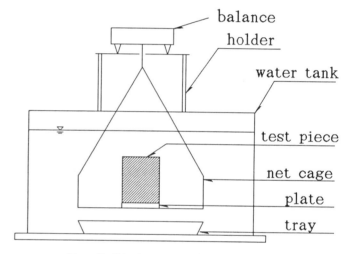

Fig.7 Slaking test

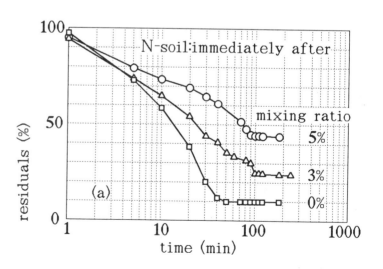

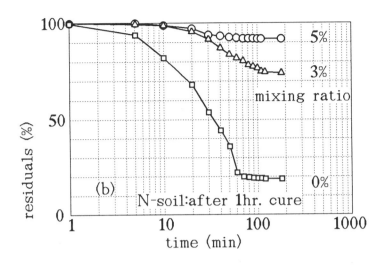

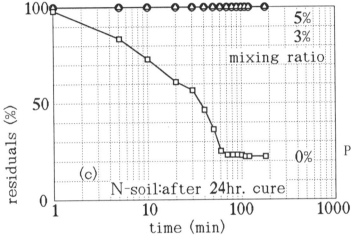

Fig.8 Relationship between residuals and time

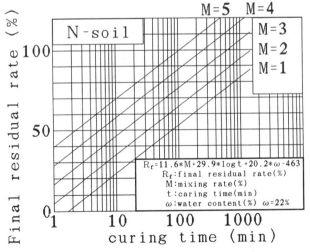

Fig.9 Relationship between final residual rate and curing time

Picture 4. Model test at the Laboratory.

Picture 5. Model test at the site.

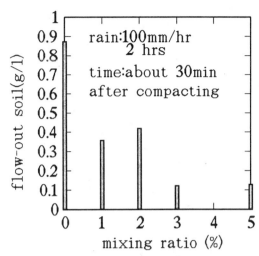

Fig.10 Relationship between flow-out soil and mixing ratio

breakdown of the soil mass can especially be improved by increasing the curing time.
(d) From the model test at the site, with increasing the mixing ratio of the cementing material the erodivity (decreasing flow out of soils) is effectively improved.

5. Conclusive Remarks

There have been great economic loss caused by erosion of the reddish soils due to land reclamations and public works for fisheries, tourism, industies and water supply dams. The government and municipal community have eagerly taking leadership in preventing these problems[6], yet it is very difficult at present.

The investigative and experimental studies must be continued in advances of continued regional developement. Some measures have technically developed here in Okinawa, and those suggested as follows,
1) Planning;
   Control of rainfall drainage
   Control of suface erosion and failures
   Control of project performance in advance
2) Under construction, after completion;
   Control of performance at the site
   Watching and control the performance of erosion control measures
   Temporary erosion control measures at the site,
   Quick planting or soil surface protections (soil stabilization, earth reinforcement)
Also following geotechnical measuring methods may be suggested,
(1) At the site (occurence)
Slope stabilization, Drainage facilities Sand sink tanks, and dams, Planting and grass spreading, Sheeting etc.
(2) In the streams (flow out and down)
Cloudy water filterring, Artificial sedimentation by chemicals and machineries etc., checkdams.

References
1) Uehara, H.(1985) Geotechnical Characteristics of "Kunigami Maaji" Soils on Okinawa Island, JAPAN
Proc. 1st. Int. Conf. on Geomechanics in Tropical Soils, Brasilia, pp. 187-198
2) Uehara, H.(1988) Geotechnical Properties of "Kunigami Maaji" Soils on Okinaw Island, JAPAN
Proc. 2nd. Int. Conf. on Geomechanics in Tropical Soils, Singapore, pp.223-229
3) Uehara,H. & Hara,H(1984) Engineering Properties of "Kunigami Gravelly Soils" 19th Annual Meeting, Japan Society of Soil Mechanics & Foundation Engineering, pp165-166(in Japanese)
4) Uehara,H. & Hara,H.(1992) ,Temprature changes in surface soils and it's effect on weakening and failure of slope soils, 47th Annual Meeting, Japanese Society of Civil Engineering (in Japanese),pp756-757
5) Uehara,H. & Goya,K (1992) Stabilization studies of slope surface of Kunigami Maaji soils, 47th annual Meeting, JSCE(in Japanese),pp992-993

6) Okinawa Prefecture(1993),(1990)
Report of contamination due to Reddish soils and suffering,pp1-204(in Japanese)
Mannual of Prevention Measures against Reddish Soils Flow-Out, pp1-228(in Japanese)

Siting a Municipal Solid Waste Landfill - A Case Study

Mark E. Unruh, Richard P. Kraft
IT Corporation, San Diego, California, USA

Abstract

Siting, design, and construction of new municipal solid waste landfills (MSWLFs) will be a growth industry during the coming years. Federal regulations promulgated in Title 40, Code of Federal Regulations, Part 258 (40 CFR 258), dated October 9, 1991, require the closure of any MSWLFs not meeting minimum siting criteria. The waste streams currently accepted at these facilities will need to be diverted to transfer/recycling centers and/or existing MSWLFs or newly constructed MSWLFs that meet these criteria. Because most existing MSWLFs do not meet these criteria, the demand for new landfills is expected to grow.

A case study of a 1989 landfill siting investigation in Los Angeles County is presented to illustrate the size, scope, and potential problems associated with such projects. The scope of the project is roughly equal to the scope of more recent landfill siting studies performed in the Southern California area and may serve as a rough order-of-magnitude example for similar studies. Some of the problems to overcome during this investigation included a lack of overland access to drilling sites, supplying water to drill sites, a difficult drilling and testing environment, and the site's proximity to faults of problematic age. Solutions to these problems included the use of portable drilling equipment flown by helicopter into areas with no overland access, construction of temporary water supply piping systems, the use of downhole casing hammer drilling methods instead of rotary methods, and careful mapping of faults in consultation with the geologist responsible for the initial U.S. Geological Survey (U.S.G.S.) field mapping to provide a better understanding of on-site faulting. In addition, surface water sampling during the siting investigation provided a unique opportunity to collect background water-quality samples.

The results of this investigation indicated that the site meets the regulatory criteria and could be used to site a landfill.

1.0 Introduction

This paper presents a case study of a landfill siting/feasibility investigation in Blind Canyon in southern California.

The Blind Canyon site is located north of the San Fernando Valley in Los Angeles County (Figure 1). The site was one of six investigated as potential landfill sites. All six sites were evaluated during the environmental impact report (EIR) process, and the most appropriate sites were selected for further investigation.

2.0 Regulatory Requirements for Landfill Siting

The siting/feasibility investigation addressed siting requirements outlined in Title 23, California Code of Regulations (23 CCR) Chapter 3, Subchapter 15, Article 5 (Waste Management Unit Classification on Siting) and the (1989) proposed Federal Rule 40 CFR 257, Subpart B (Criteria for Classification of Solid Waste Disposal Facilities and Practices - Location Restrictions). Although this investigation was completed in 1989, it should substantially fulfill all the requirements for landfill siting in effect today. Since 1982, California has had substantially the same requirements as the newer Subtitle D. This landfill siting study was intended to meet both the California requirements and the anticipated Federal requirements. Table 1 presents State and Federal siting requirements and measures taken in this investigation to address those requirements.

3.0 Field Investigation

This section summarizes the field investigation conducted as part of the Blind Canyon siting study.

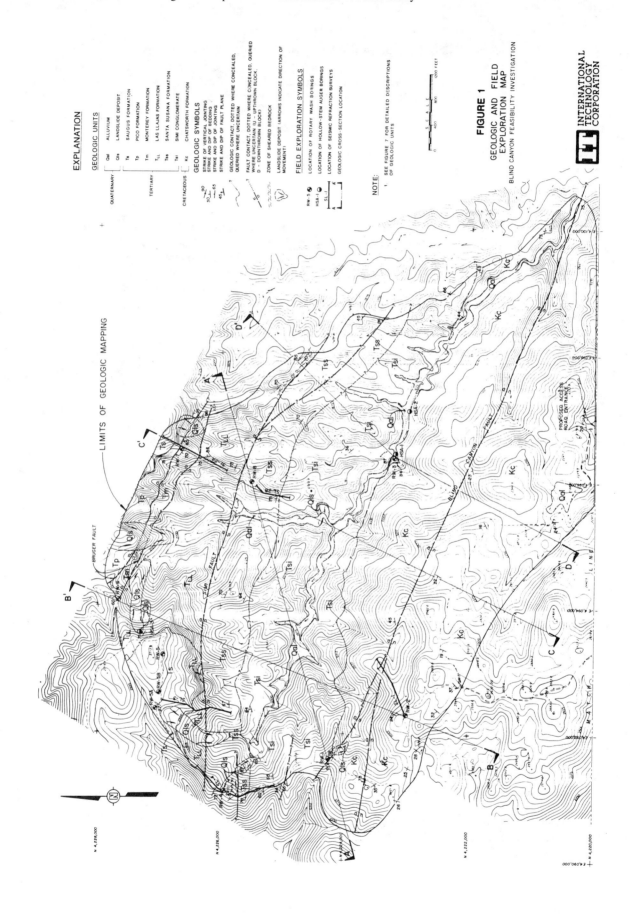

3.1 Geologic Characterization

The geology and groundwater quality of the site and the surrounding vicinity was investigated using a combination of literature review, aerial photograph inspection and surface mapping. Detailed geologic mapping was performed over a 20-day period on a base map of scale 1 inch equals 400 feet (1:4800) and a contour interval of 25 feet. The site geologic map is presented in Figure 1.

3.1.1 Regional Geologic Setting

Blind Canyon is located in the Santa Susana Mountains, which are within the Transverse Ranges province of southern California (Figure 1). The Santa Susana Mountains are situated within the western segment of the Transverse Ranges province. The western segment is composed of clastic marine sedimentary rocks of late Mesozoic and early Cenozoic age and marine, continental, and volcanic formations of Oligocene to Pleistocene age.

The Transverse Ranges have, as a whole, experienced late-Cenozoic structural deformation caused by crustal shortening due to the onset of large-scale displacement on the San Andreas Fault System. The Santa Susana Fault Zone, which is the major structural feature in the site vicinity, is a typical example of the reverse faulting and folding that has facilitated crustal shortening during the late Tertiary and Quaternary.

3.1.2 Stratigraphy

The site vicinity is underlain by a diverse assemblage of sedimentary rocks of Cretaceous to Holocene age. Dominant rock types exposed in the mapping area include friable to well-cemented sandstone, conglomerate, and shale and lesser amounts of coquina, limestone, and diatomaceous shale. A stratigraphic column for the units encountered in the study area is presented in Figure 4. The stratigraphic column provides detailed lithologic descriptions and local formation name designations and ages.

3.1.3 Structure

The Blind Canyon mapping area is underlain by a thick sequence of sedimentary bedrock units that have been tectonically uplifted and tilted so that the strata dip very steeply, with a general east-west strike. Because of the steep dips, the lithologic units form continuous east-west-trending bands across the site. Three high-angle faults cross the site in a general east-west direction, subparallel to the axis of the canyon and the dominant direction of the strike of bedding. These faults are the Blind Canyon Fault, which trends through the southern portion of the mapping area; the Simi Fault, along the northern flank of the canyon; and the Bruger Fault, near the extreme northern boundary of the mapping area. The faults are continuous across the site and represent conspicuous structural features that have played a major role in the development of the canyon.

3.1.4 Recency of Faulting

The Santa Susana Mountains underwent extensive deformation and uplift during Pliocene to late Quaternary time, and many faults in the region show evidence of displacement within the last 11,000 years (Holocene time). The 1971 San Fernando earthquake, centered approximately 20 miles east of the site, demonstrates the continuing faulting activity in the region. Landfill siting requirements prohibit the construction of a landfill within 200 feet of a fault with known Holocene displacement. As evaluated according to surface geomorphic features and identification of oldest units definitely displaced by faulting, none of the faults identified in the mapping area showed evidence of Holocene displacement; however, soil horizons, which would have allowed more accurate age-dating of Holocene fault activity, were not available in the mapping area.

3.2 Geophysical Survey

A seismic-refraction survey was conducted in Blind Canyon and along the access road to determine structure, degree of weathering, and rippability characteristics of the geologic materials beneath the canyon ridges and depth to bedrock and bedrock rippability along the proposed access road. Eight refraction lines, totaling more than 11,000 feet in length, were surveyed. Data were processed and interpreted using the generalized reciprocal method (GRM). The GRM is a seismic-refraction interpretation method designed to allow accurate mapping of undulating surfaces that may have lateral variations in seismic velocity, as would be expected in an area with nearly vertical dipping beds such as Blind Canyon.

The surveys recorded seismic velocities ranging from 800 feet per second (ft/s) for highly weathered bedrock and soil horizons, to more than 11,000 ft/s in fresh, competent sandstone bedrock. A rippability standard developed by Caterpillar Tractor Company was used to develop rippability estimates. The surveys showed, that in most areas of Blind Canyon, rippable materials could be found to depths of about 100 feet. However, in the vicinity of the proposed access road, generally high velocities (over 8,500 ft/s) indicated that these materials were marginally rippable to nonrippable. This

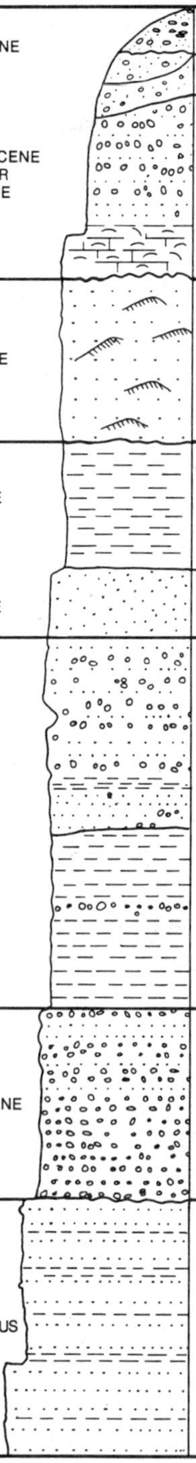

Age		Description
HOLOCENE		Stream Alluvium (Qal) - Loose, brown silty sand with abundant cobbles and boulders.
		Landslide (Qls) - Disrupted mixture of loose soil and various lithologies of rock debris.
LOWER PLEISTOCENE TO UPPER PLIOCENE		Saugus Formation (Ts) - Alternating beds of tan to blue-gray fossiliferous sandstone and coquina; buff to white, poorly cemented, medium to coarse-grained sandstones; and reddish brown, non-fossiliferous conglomerate, with well-rounded pebbles and cobbles. Base of unit is marked by up to 100 feet of gray, well cemented, highly fossiliferous sandstone which is equivalent to the Las Posas sandstone of Dibblee (1987). Basal unit forms locally prominent cliffs. Abundant megafossils including; Pecten sp., Ostrea sp., and Dendraster sp.
PLIOCENE		Pico Formation (Tp) - Buff to light orange, massive to thickly bedded silty sandstone, friable, typically fine to medium grained, locally conglomeratic and fossiliferous, poorly exposed cross-bedding. Forms steep slopes and cliffs on north rim of Blind Canyon.
UPPER MIOCENE		Monterey Formation (Tm) - White to light gray to tan, silty and clayey diatomaceous shale, punky, thinly bedded, highly fractured, fresh shale is dark gray due to high organic content.
MIDDLE MIOCENE		Topanga Formation (Tt) - Thick bedded to massive, gray to brownish sandstone, with some siltstone and conglomerate, buff to orange weathering. This unit was not mapped within the study area, but is part of the local stratigraphic sequence.
MIDDLE EOCENE		Las Llajas Formation (Tll) - Reddish calcareous sandstone with brown conglomerate interbeds. Sandstone is fine to coarse-grained, thin to thickly bedded. Conglomerate consists of well rounded pebbles with some cobbles. Occasional gray shale interbeds which are micaceous and platey. Locally fossiliferous, including abundant Turritella sp.
LOWER EOCENE		Santa Susana Formation (Tss) - Greenish to bluish-gray silty shale. Basal portion of unit marked by alternating thin beds of brown sandstone and gray shale. Conglomerate bed, up to 40 feet thick, found near middle of unit. Overall, this formation crops out as smooth, low slopes which reflect its poor resistance to erosion.
PALEOCENE		Simi Conglomerate (Tsi) - Consists of an upper section of alternating thin to medium beds of sandstone and conglomerate, underlain by massive, reddish brown conglomerate. Conglomerate clasts consist of polished, well rounded, iron stained, pebbles, cobbles, and occasional boulders of various igneous and metamorphic lithologies. Matrix composed of moderately indurated, reddish sandstone.
UPPER CRETACEOUS		Chatsworth Formation (Kc) - Thick bedded to massive, tan to buff to gray arkosic sandstone, fine to coarse-grained, micaceous. Sandstone beds separated by thin interbeds of hard, fissile, greenish-gray micaceous shale. Crops out as resistant, stepped cliffs.

FIGURE 2
STRATIGRAPHIC COLUMN
BLIND CANYON FEASIBILITY INVESTIGATION

PREPARED FOR

COUNTY SANITATION DISTRICTS OF
LOS ANGELES COUNTY

INTERNATIONAL TECHNOLOGY CORPORATION

NOTE:
FORMATION THICKNESS NOT TO SCALE

information was essential in developing economic and constructible access road design and alignment options. As such, the seismic surveys were able to provide information that would be useful in developing future design characteristics. The surveys were also useful in providing subsurface information that could be correlated with the geologic map and the exploratory boring logs. Additionally, the Simi and Blind Canyon Fault traces were identified as lateral seismic velocity variations.

3.3 Exploratory Drilling/Field Hydraulic Conductivity Testing

A drilling program was conducted to evaluate subsurface materials and to collect samples for laboratory testing in selected areas within the landfill perimeter and along the proposed access road alignment. Following drilling and sampling, hydraulic packer testing was conducted within selected intervals of each boring to determine in situ hydraulic conductivity (K) for various rock types. Continuous coring was conducted in bedrock units, to facilitate accurate logging of subsurface materials.

Because several drill sites were not accessible by improved roads, four-wheel-drive, truck-mounted Mobile B-80 and skid-mounted Chicago Pneumatic (CP-8) drill rigs were used to complete the drilling program. The skid-mounted drills and associated equipment were transported to each drilling site by Sikorsky 58-T twin turbine helicopter.

Within the proposed landfill perimeter, eight borings were advanced to an average depth of approximately 100 feet using continuous-coring drilling methods. Along the proposed access road alignment, six borings were cored to an average depth of 20 feet. In addition, three hollow-stem auger borings were advanced to shallow depths in alluvium at the base of the canyon.

A review of regional groundwater data suggested that depth to groundwater at the site could be considerable (greater than 500 feet); as such, groundwater was not expected to be encountered during the drilling program. However, determining in situ hydraulic conductivity characteristics of the various rock types in the unsaturated zone was considered critical to understanding potential leachate migration properties. Because traditional pump tests could not be performed, packer-type hydraulic conductivity testing was performed in each rotary wash test hole. The testing program employed a double-packer system and followed methods outlined in the Ground-Water Manual prepared by the U.S. Department of the Interior (1985). A double packer that seals the selected interval is used, and water is injected into the test interval through the perforated center pipe. By injecting water at constant pressure over a fixed period of time, the quantity of water injected relates directly to the in situ hydraulic conductivity of the interval being tested. As expected, the resulting K values varied considerably according to the rock type and nature of fracturing. K values ranged from 2.1×10^{-2} to 2.5×10^{-7} centimeters per second (cm/s).

3.4 Geotechnical Testing

Alluvial soil samples from the hollow-stem auger borings were analyzed for sieve analysis, Atterberg limits, specific gravity, direct shear, unit weight, unconfined compression, and consolidation.

Laboratory tests were also conducted on core samples to determine hydraulic conductivities using EPA Method 9100. Laboratory K values range from 4.2×10^{-4} to 2.0×10^{-11} cm/s for the 20 samples tested. Laboratory K values are typically two to three orders of magnitude lower than K values determined from the packer testing in equivalent intervals. The discrepancy between field and laboratory K values is related to the fact that groundwater migration is enhanced in Blind Canyon bedrock materials by extensive jointing and fracturing. However, core samples for laboratory analysis must be intact and, thus, represent a less fractured portion of the field tested interval.

3.5 Water Quality Testing

To aid in evaluating the potential impact of a landfill on groundwater and surface water, water-quality analyses should be performed to develop baseline water quality characteristics. This information can be used as background data for comparison with future downgradient water-quality data following landfill construction. Although the considerable depth to groundwater at the site precluded the direct collection and analysis of local groundwater, two surface-water pond samples were collected and analyzed following a storm event. These samples were analyzed for volatile organic compounds (EPA Method 624), semivolatile organic compounds (EPA Method 625), and metals (EPA 6000/7000 series). Samples were tested in the field for pH, temperature, specific conductivity, redox potential, alkalinity, carbon dioxide, and dissolved oxygen. Analytical results indicated that manganese and sulfate concentrations exceed EPA secondary drinking water standards. This background water-quality information will be essential in evaluating potential future water-quality impacts due to the landfill.

4.0 Discussion

The Blind Canyon field program provided the necessary technical and analytical data to determine the suitability of the canyon for consideration for future landfill development. The study also collected information that allowed rough order-of-magnitude construction cost estimates that are also necessary during the preliminary site selection and planning stage.

4.1 Field Program

Important findings of the study are briefly summarized in Table 1.

4.2 Considerations for Siting New Landfills

The Blind Canyon study and recent ongoing studies for the County of San Diego have identified some important considerations for future landfill siting investigations, as follows:

- *Access* - As with Blind Canyon, future landfill sites must be located relatively near major population centers to reduce waste transportation costs by providing for the disposal of waste near generators (the State of California requires localities to be responsible for their own wastes). However, the "not in my backyard" (NIMBY) syndrome virtually ensures that policy makers will require that landfills be surrounded by as few neighbors as possible. As a result, future landfill sites will likely be located in areas with difficult access, similar to the Blind canyon site.

- *Recommended Field Program Options* - The Blind Canyon field program was specifically designed for the conditions that exist at the site. For other sites, the program may be modified to suit the specific data needs and site conditions. For example, at most sites groundwater conditions must be characterized. Therefore, the field program should include installing and sampling of groundwater monitoring wells. The collection and analysis of water samples will allow the determination of local water quality. Additionally, water level measurements and pump tests should be conducted to determine hydraulic flow conditions. Geophysical surveys should be designed specifically for the conditions that exist at the site. For example, at fractured crystalline bedrock sites, Very Low Frequency (VLF) surveys have proved useful at identifying fracture patterns that control groundwater flow migration. The survey thus assists in positioning the groundwater monitoring well locations. When identification of the location and recency of faulting must be demonstrated, trenching is recommended.

- Whenever possible, a siting investigation should be coordinated with the EIR contractor to avoid redundancy and adversarial study conclusions.

The siting/feasibility study indicated that with the proper engineering and construction constraints, Blind Canyon is a feasible location for a Class III landfill. The site satisfies the State and Federal siting criteria. The size and scope of the Blind Canyon siting/feasibility investigation was adequate to address the siting requirements in both the state and federal regulations.

TABLE 1 -- Municipal Solid Waste Landfill Siting Requirements

Item	Federal (RCRA Subtitle D)	State (23 CCR - RWQCB)	Investigation Activities Addressing Requirements and Findings
Geologic Setting	Not addressed.	Landfills shall be sited where soil characteristics, distance from waste to groundwater, and other factors will ensure no impairment of beneficial uses of surface water or groundwater.	Packer and laboratory testing indicated secondary permeability dominates flowpaths in most geologic formations at the site. Fluid migration is largely controlled by fractures rather than by lithology. The seismic-refraction survey determined that materials underlying the site are generally rippable and that portions of the access road are marginally rippable to nonrippable.
Airports	Landfills must be located 10,000 feet away from any airport runway end used by turbojet aircraft or 5,000 feet away from any runway end used by only piston-type aircraft or demonstrated that the WMU does not pose a bird hazard to aircraft.	Demonstrate that the WMU does not pose a bird hazard to aircraft.	Not addressed in this investigation.
Floodplains	Landfills shall be located outside the 100-year floodplain or demonstrate that flow will not be restricted and that washout will not occur.	Landfills shall be designed, constructed, operated, and maintained to prevent inundation or washout due to floods with a 100-year return period.	Literature search indicated the site was not located within the floodplain.
Wetland	Landfills shall not be located in wetlands.	Not addressed.	Surface mapping indicated the site was not in wetlands.
Fault Areas	Landfills shall not be located within 200 feet of a Holocene fault unless it can be demonstrated that setback of less than 200 feet will prevent damage to the structural integrity of the WMU and will be protective of human and environmental health.	Landfills shall not be located on a known Holocene fault.	Geologic mapping identified three faults (Blind Canyon, Simi, and Bruger) trending across the site. No evidence of Holocene displacement was observed along the traces of these faults.
Seismic Impact Zones	Landfills shall not be located within seismic impact zone unless it is demonstrated that all containment structures, including liners, LCRS, and surface water control systems are designed to resist the maximum horizontal acceleration for the site.	The site is situated in a seismically active area. The expected maximum peak horizontal rock acceleration at the site is 0.57g due to a maximum probable earthquake (M=5.7) on the Santa Susana Fault (1 mile from site). A Class III landfill in Blind Canyon could be designed, constructed, and maintained to preclude failure resulting from rapid geologic change.	Literature search, seismicity study.
Unstable Areas	Landfills located within unstable areas must demonstrate that engineering measures have been incorporated into the design to ensure the integrity of the structural components will not be disrupted.	Landfills may be located within areas of potential rapid geologic change if containment structures are designed, constructed, and maintained to preclude failure.	Slope failures due to slumping and landsliding were identified in the steep slopes that make up the northern flank of the site.

Nuclear

Status of the Yucca Mountain Site Characterization Program

Carl P. Gertz
US Department of Energy, Yucca Mountain Project Office, Las Vegas, Nevada, USA

Final manuscript not received in time for inclusion in the *Proceedings*. For additional information, please contact:

        Mr. Carl P. Gertz
        US Department of Energy
        Yucca Mountain Project Office
        P.O. Box 98608
        Las Vegas, NV 98198
        USA
        Voice: +702-794-7964
        Fax: +702-794-5348

UMTRCA Regulations - Evolution of the Technical Basis

Thomas A. Shepherd, Louis L. Miller, Robert L. Medlock
Shepherd Miller, Inc., Fort Collins, Colorado, USA

Introduction And History Of UMTRCA

The Uranium Radiation Control Act of 1978 (the UMTRCA), which amended the Atomic Energy Act of 1954 (the ACT), was enacted by the U.S. Congress in response to concern that uranium mill sites, active and inactive, may be hazardous to the public health and the environment. This concern resulted in large part from instances in Grand Junction, Colorado and Monticello, Utah where uranium tailings had been removed from nearby mill sites and were used as fill around residential building foundations.

Public use of the tailings was possible and problematic for several reasons. First, access to the radioactive tailings was possible because they were uncontrolled and unregulated and were located in or near towns. The potential for misuse of these tailings (as was the case at Grand Junction and Monticello) was therefore high. Second, there was a suspected enhanced cancer risk from exposure to radon gas for the impacted home owners. Third, it was apparent that substantial additional uranium milling would be required to fulfill fuel demand for the nuclear power plants that were forecast to come on line.

These issues, along with an increased public awareness of the environment and an increased fear of cancer and awareness of its potential link to radiation, caused legislative regulatory, and public attention to be focused on uranium tailings and the attendant mill sites.

The primary purpose of UMTRCA was to ensure that every reasonable effort be made to provide for the stabilization and disposal of tailings in a safe and environmentally sound manner in order to prevent misuse, minimize the release of radiation into the environment and prevent or minimize other environmental hazards postulated to be associated with tailings.

UMTRCA directed the U.S. Nuclear Regulatory Commission (NRC) to develop and administer a comprehensive uranium mill tailings regulatory program. The Act established two programs to regulate uranium mill tailings: Title I, required the closure of inactive sites (sites with no identifiable owner which would require funding and cleanup by Federal and State Governments), and Title II, which established a comprehensive regulatory program for the disposal and stabilization of uranium mill tailings at active mill sites. To establish a regulatory home for uranium tailings Section 11e. (2) of the ACT was amended to reverse the definition of byproduct material to include:

> "the tailings or wastes produced by the extraction or concentration of uranium or thorium from any ore processed primarily for its source material content."

UMTRCA also directed the U.S. Environmental Protection Agency (EPA) to establish environmental standards of general application for NRC to use to promulgate regulations for decommissioning and reclaiming milling sites and uranium mill tailings.

As an initial step to promulgating regulations, UMTRCA required NRC to study the issues and the technical and scientific aspects related to uranium tailings management and to prepare a generic environmental impact statement (GEIS) to provide background and support for its regulatory program. Based on the information and conclusions developed in the GEIS, the NRC proposed regulations for the operational and post-operational control of both radioactive and non-radioactive hazards associated with uranium and thorium mill sites and tailings disposal facilities. Defined in this regulatory program was provisions for formal agency review of tailings disposal and decommissioning plans, financial surety to ensure sufficient funds for decommissioning and reclaiming mills and tailings, preoperational and operational monitoring, transfer of title for the land and tailings to effect long-term government control of disposal sites, financial surety to ensure sufficient funds for decommissioning and reclaiming mills and tailings, funding for long-term site surveillance following decommissioning and reclamation, and unrestricted use of the reclaimed site following surety release. Further, the GEIS concluded that reclamation plans must be designed to ensure that no active maintenance will be required to achieve the long-term site integrity sought by the regulations. The regulations took the form of a set of thirteen performance criteria as presented in Appendix A of Chapter 10, Part 40 of the U.S. Code of Federal Regulations (10 CFR 40). These performance criteria in some cases set out very specific design limits or performance standards. In other cases, they provided slightly contradictory guidance. Never the less, the criteria allowed flexibility for both the regulator and the operator to interpret relevant data to determine the final outcome of a design. Notwithstanding this flexibility, the underlying objective of operations and final reclamation are clear; to control the release of radon gas from the reclaimed tailing area to a specific limit, to provide for reclamation and closure to ensure the long-term physical stability of the site, to prevent the release of radioactive material, and to protect ground water quality for very long time periods without reliance on active maintenance or institutional control. These somewhat qualitative objectives, the conditions established within which one could demonstrate acceptable performance, and the lack of definitive understanding of how to demonstrate performance for the very long time periods required, produce a design and regulatory environment that required innovation, imagination and compromise.

UMTRCA thus spawned a new philosophy for site closure which required the integration of a multiplicity of scientific and technical disciplines. This philosophy has impacted the approach now applied to regulate other industry segments. Although this philosophical concept had been developing for some time in the U.S., the enactment of UMTRCA and the development of the GEIS and the Appendix A criteria was the first instance in which it became the clear, driving intent of a regulatory program.

Once in effect, the new regulations placed a significant burden on the operators, engineers, scientists and regulators responsible for compliance. Prior to UMTRCA, reclamation and final closure of a mill site was accomplished using more problem specific, narrowly focused methodologies. The norm had been to do what seemed "right" for a specific issue (e.g., provide stable slopes, establish vegetation) without regard for or the need to investigate other potentially related problems and define how they might impact other control facets of closure plans. Typically, if problems developed later, they were addressed as they arose.

This basis for operating and closure was rejected by UMTRCA and replaced with requirements to find comprehensive engineering designs and scientific solutions that could comply with the demanding requirements of the new law. These requirements called for engineering designs to incorporate reclamation and closure as an integral element of operational design and planning and to effectively analyze long-term performance to ensure that the approved plan would achieve the objectives of protecting the environment and public health.

Out of the UMTRCA regulatory process and its attendant philosophical mandate has evolved predictive engineering and scientific methodologies that are generally better equipped to deal with the complex problems that confront us as we struggle with a wide variety of waste management and human and environmental impact issues. As a technical community we have learned the practical limitations to removing uncertainty from the decision process and have developed an evaluation process that integrates once divergent engineering and scientific disciplines into a comprehensive but flexible set of design tools.

Regulatory Objectives

The NRC regulations intended that site closure plans provide design features that would ensure the minimization or elimination of operational and post-closure risks. Further, this was to be done with no active maintenance and essentially no reliance or institutional control needed after closure of a site. The closure of uranium and thorium mill and tailings sites thus had to be done in a manner that would minimize risk and not burden future generations.

To the policy makers and the environmentally concerned public this was an appropriate and sound goal, especially when considering the depth of concern that was prevalent over the long-lived radioactive tailings constituents. Without entering into debate about the absolute or relative level of risk posed by uranium and thorium tailings, there is no question that it is desirable to design and implement a reclamation and closure plan that will meet stringent performance goals over extended time periods. We believe that the current NRC regulations and guidelines generally achieves these objectives in a fundamentally sound manner. While these objectives can and usually do cause extreme, and sometimes seemingly unnecessary, measures to be used in closure design, the imposition of these objectives on the design process has caused the engineering and scientific community to improve predictive methodologies, integrate natural science and engineering disciplines in evaluating design performance, and become rather innovative in the development of operational and closure plan designs. It is our belief that this evolutionary process has substantially improved industry's overall ability to develop effective closure and reclamation designs, and, in that regard, has contributed in a positive way to the advancement of our technical capabilities. This has resulted in the development and implementation of a comprehensive and inter-disciplinary approach to operational and reclamation design and performance analysis. It has required innovation, adaptation and basic research to provide the engineering and scientific basis needed to achieve the objectives set out by the regulations. In so doing, this process has generally improved our overall capability to effectively and efficiently achieve the broad societal objectives protection of the environmental and public health.

Evolution of Closure Objectives

Initially, the NRC concluded that three meters of cover, maximum 5h:1v slopes and 100,000 years were the appropriate design-life objectives for reclamation plan designs because of the long-lived nature of the perceived radioactive hazard present. This very long-term design life objective, along with the requirement that performance could not rely on active maintenance or institutional control, created two interesting conditions.

First, it was clear that, for perhaps the first time in a comprehensive design sense, natural processes, such as erosion and geomorphic land form development, had to become and integral element of development and analysis of design solutions. Further, because of the very long time frame being considered, extreme and unlikely natural events, such as the Probable Maximum Precipitation event (PMP), were appropriate conditions to consider in design in the closure plan. Second, it was clear that traditional engineering performance evaluation technics were generally inadequate and/or unable to project performance predictions as far into the future as required by the new regulations. In addition, very few soil-based structures could serve as models to demonstrate the long-term effectiveness of engineered designs.

It was immediately clear that new design evaluation and analysis processes were needed and that to facilitate the development of these processes, several evolutionary things had to occur. First, in the GEIS, the NRC evaluated an exhaustive range of potential long-term natural events which could potentially affect the long-term stability of a uranium tailings site. Many of those initially considered (glaciation, pestilence, etc.) were correctly rejected as either not important or so extreme that they could not be reasonably incorporated into a design process. Others, mainly those associated with geomorphic processes including erosion, mass wasting and fluvial processes, have remained an important and appropriate consideration in design development and evaluation.

Second, once it was determined that natural processes should be included in the design development and evaluation process, the means to do this had to be found. Standard engineering design did not contain this capability and neither did the interpretive methodologies traditionally employed by natural scientists. This obvious and serious gap between need and ability was bridged by a generally productive undertaking by all parties involved. This undertaking included government funded applied research, adaptation of standard engineering and scientific methodology to meet the demands for long-term predictions, and innovation by both regulations and the operators and their consultants to provide meaningful analysis of long-term stability and performance.

This evolutionary process has taken time and is not yet complete. The design-life target has been lowered from 100,000 years to a time frame of 200 to 1,000 years. This time frame has allowed more realism to be introduced into design performance analyses but has not removed the necessity to design to accommodate extreme natural events. The design and analysis process related to reclamation surfaces has also evolved into a system that can reliably determine what can be accepted as a final land form.

The design analysis process related to establishing that a reclamation surface achieves the long-term stability performance criteria has thus progressed to the point where little controversy now exists.

This evolution of design and performance analysis has evolved from application of the relatively qualitative judgements possible by the application of classical geomorphic interpretation of land-form stability and the Universal Soil Loss Equation (USLE) to much more highly refined, and at least, semi-quantitative analytical methodologies. The basis for now accepted analytical process is presented in the NRC's Final Staff Technical Position, Design of Erosion Protection Covers for Stabilization of Uranium Mill Tailings Sites (NRC, 1990). One very important tenet of this process is that to be effective, a design must accommodate natural weathering processes without failure of the features which encapsulate the tailings. In addition and corollary to this process, analysis methodologies and design approaches for control of radon emanation have been extensively investigated, reviewed and debated and a final methodology established.

Finalization of the ground water protection element of closure designs is pending. NRC publication of a guidance document (Staff Technical Position) for establishing alternate concentration limits (ACL's) for ground water constituents that cannot be reasonably controlled to achieve background or established regulatory limits. The NRC has written a draft ACL guidance document (NRC, 1992) that is currently being reviewed by the [Nuclear Regulatory] Commission. However the technical and scientific elements of the process, that is geohydrologic and geochemical evaluation and prediction, have been thoroughly addressed and developed to a high level of technical proficiency and reliability. It is fair to state that the uranium industry has been responsible for advancing this general scientific area significantly, particularly where partially saturated flow and contaminant migration geochemistry applied to practical design and performance evaluation. As stated above, final ground water guidance protection acceptance waits for regulatory policy decisions which seem to always lag our scientific and technical understanding of a situation.

Interdisciplinary/Multidisciplinary Decision Approach

One of the most significant contributions that the UMTRAP process has provided are its interdisciplinary or multidisciplinary approaches to regulatory evolution and problem solution development. The National Environmental Policy Act (NEPA) of 1969 preceded UMTRCA by about 10 years. NEPA clearly mandates an interdisciplinary approach to environmental impact analysis which extends to impact mitigation and solutions development. It was clear to the framers of this law that environmental impact analysis required a broadening of the traditional focus of problem analysis. This philosophical approach is embodied in the EIS process.

Significant advancements in mine and tailings closure technology have come as a result of the very practical realization that reliance on a single science or engineering discipline, as was historically the practice, no longer was feasible. That is, it was not possible to support conclusions about the long-term performance of a reclamation design solely on the basis of a single, traditional engineering analysis. The available period of actual performance testing and functional experience with engineered designs was far too short to provide the confidence required by the new regulations. We believe that a large part of this regulatory paradigm shift resulted from the UMTRCA process, wherein the multidisciplinary approach became an accepted element of regulatory decision making and the application of the concepts of such relatively obscure disciplines as geomorphology became an integral element of stability analysis and design justification.

The new regulatory paradigm also required a shift in the mindset of those engineers designing reclamation plans because multiple objectives were not normally integrated in these traditional design approach and in actual fact were sometimes not compatible in terms of objective fulfillment. The most easily recognized example of contradictory closure objectives lies in the design of tailings covers. The tailings cover is required to control radon, be erosionally stable for long periods of time and reduce moisture infiltration to levels necessary to protect long-term ground water quality. Much debate has ensued as to the most appropriate design of a cover system. The universally accepted conclusion that no one design is appropriate for all sites and situations. Further, the objectives, if pursued independently, very often result in contradictory designs. For example, long-term surface stability is most easily achieved by construction of either a very flat cover surface or by covering steeper surfaces with rock. The result of both of these designs with respect to infiltration is to maximize (compared to a steep, vegetated soil surface) the amount of infiltration that can be predicted to occur. The outcome of these designs are not the most appropriate when one considers the requirement to protect groundwater quality for the long-term. This can be a frustrating dilemma. The solutions lie in the integration of many, varied cover designs. And the basis of these designs and the performance analyses used to prove their acceptability have almost always been based on understanding and adequately predicting performance based on application of several different scientific disciplines. If rock cover is needed to provide surface erosional stability, the geomorphology shape of the surface may become less important, but the long-term durability of the rock then becomes an integral element of the design analysis. In addition, since a rock cover acts as a mulch and increases infiltration - especially when compared to a soil/vegetation cover system - the analysis of the plan must also demonstrate that the increased amount of water that might infiltrate will not mobilize contaminants in unacceptable qualities. The resolution of this quandary may rely on integrated understanding of how the chemical environment evolves in the tailing mass once it is covered, the geochemical interaction that will occur beneath the impoundment between the chemistry of the seepage and the chemistry and physical nature of the subsurface geologic materials which lie along the flow path, and the rate (quantity) at which seepage will leave the tailing impoundment over the long-term. In the end, the final resolution is likely a compromise or balance between the competing objectives, optimizing performance within the limits imposed, and using scientific disciplines and procedures to evaluate and analyze the balance sought. With all of the factors considered, the final design might employ a vegetated rock mulch which applied over a highly compacted clay cover with a relatively flat slope would reduce radon emanation, reduce the permeability of the cover and further enhance surface stability and reduce potential infiltration by increasing evapotranspiration.

In current application this multi-disciplinary evaluation and proof is not difficult or overly complex in concept or actual application. However, the evaluation and integration of the combination of disciplines necessary, and the adaptation of technologies to provide a compatible decision framework has not been without its challenges. Happily, this integration has been successfully achieved and has allowed the uranium industry to proceed with reclamation design in a somewhat sensible manner and to pass along this integration to other waste disposal and reclamation problems.

Regulatory Continuity And Growth - The Desired State

It is interesting and instructive, and an element of the regulatory process worthy of commendation, to note in closing, that the evolution of the UMTRCA process has been very open and amenable to change and modification as the state of our understanding of uranium tailings closure technology has developed. Many of the "truths" that characterized the initial regulatory posture have been overturned in a positive sense for new truths. At the head of the list of modified truths is the three meter cover and the "absolute" 5h to 1v and 10h to 1v side slope requirements. It is not likely that a three meter cover will ever be constructed, or, that the NRC would ever require a 5h to 1v slope just because the regulations say 5h to 1v. What has developed is regulatory objectivity that allows licensees to demonstrate, by employing a range of accepted analytical approaches without prescribing specific design requirements, that a design achieves the desired performance of objective. We feel that this has been possible because the regulations are based on performance objectives rather than static, inflexible numeric standards. This approach to regulation imposes a heavy responsibility on both the regulator and the operator to not only to conduct business on the basis of mutual respect and professional competence, but an open exchange and, acceptance and understanding of original ideas and new, innovative approaches. The industry, NRC and the Agreement State personnel are to be commended for nurturing and allowing this environment to prevail. The authors feel that this type of approach would well serve other elements of the regulatory arena.

References

Draft Staff Technical Position. "Alternate Concentration Limits for Title II Uranium Mills. Standard Format, Content Guide and Standard Review Plan For Alternate Concentration Limit Application". U.S. Nuclear Regulatory Commission. December 1992.

"Final Generic Environmental Impact Statement on Uranium Milling", Office of Nuclear Material Safely and Safeguards, U.S. Nuclear Regulatory Commission, NUREG 0706, September 1980.

Final Staff Technical Position. "Design of Erosion Protection Covers for Stabilization of Uranium Mill Tailings Sites" U.S. Nuclear Regulatory Commission. August 1990.

Present Trends in Nuclear Waste Disposal

Vern Rogers
Rogers & Associates Engineering Corporation, Salt Lake City, Utah, USA

Abstract

In the past, low-level radioactive waste (LLW) was often disposed of in barrels and cardboard boxes that were randomly dumped into a shallow earthen trench. Seepage of contaminants from these containers into the ground was considered acceptable. Concern then arose over the actual and potential contamination of potable aquifers, leading to the construction of some commercial disposal facilities in very tight clays. These facilities soon presented problems, however, and a systems approach was therefore developed for siting and licensing facilities. The systems approach and associated regulations optimized the long-term safety of LLW disposal facilities by allowing some release of radionuclides. The most recent trend in LLW disposal is to provide additional engineered barriers in the disposal facility that would prevent the short-term release of contamination. The most common material used in these engineered barriers is concrete. Thus, the latest designs and performance assessments of LLW disposal facilities consider the long-term integrity of the form of the solidified waste, the concrete engineered barriers, the cover materials, and the foundation soils, as well as site geological, hydrological, and meteorological properties. This paper discusses design features of disposal facilities that are presently being considered.

Introduction

To date in the United States, most LLW has been disposed of by shallow land burial. Some shallow land burial facilities were developed at sites that have not contained radionuclides to the extent desired. Operating practices and institutional controls were also inadequate at some facilities. At the present time, only two of the six shallow land burial facilities licensed for commercial operation in the United States are still operating. Two of the closed facilities had contamination migrating through the groundwater. A third site, built in impermeable clay in a humid area, had contamination in surface water that overflowed the shallow trenches and entered nearby creeks and streams. A fourth site, located in an arid region, was closed as part of the compact process.

In part as a result of the experience with shallow land burial, several states and compacts of states have adopted laws that forbid using shallow land burial as a means of LLW disposal. Even where shallow land burial has not been specifically forbidden, there is strong interest in using alternatives to shallow land burial to dispose LLW. Several other countries have constructed alternative disposal facilities that incorporate concrete barriers. Almost all of the new facilities being planned in the United States employ concrete barriers as an important part of the design.

Alternative disposal facility designs seek to overcome the problems experienced in shallow land burial by using materials that prevent or reduce water flow and contaminant transport, improved waste forms that can prevent structural failure of the facility, rigid waste containers made from manufactured materials that do not easily deteriorate under disposal conditions, and multiple barriers to contaminant migration.

In general, a LLW disposal facility must meet the following performance objectives:

- No member of the public shall receive an annual dose exceeding an equivalent of 75 millirem to the thyroid, 25 millirem to the whole body, and 25 millirem to any other organ. Reasonable effort should be made to keep releases to the general environment as low as reasonably achievable.

- Inadvertent intruders must be protected after institutional controls are removed. An inadvertent intruder is a person who might unknowingly occupy a waste disposal site after closure and engage in normal activities such as farming, dwelling construction, or other pursuits through which the person might be exposed to radiation from the waste.

- Operations at the disposal site must keep exposures as low as reasonably achievable.

- Facility siting, operation, and closure should achieve long-term stability of the site and minimize the need for active maintenance after facility closure.

Descriptions of Disposal Technologies

This paper contains descriptions of current technologies for the disposal of LLW. The disposal technologies discussed include:

- Below-Ground Vaults.

- Above-Ground Vaults.

- Modular Concrete Canisters.

- Earth Mounded Concrete Bunkers.

- Augered Holes.

- Mined Cavities.

These technologies cover a wide range of features. The descriptions are of whole designs rather than design components. (An example of a design component is the cover over a shallow land burial trench.) The descriptions serve as a means of comparing advantages and disadvantages of different approaches. They do not, however, represent the only technologies available.

Shallow land burial is also included as a reference design against which to evaluate the other technologies. In this way, the potential benefits of alternatives can be compared to shallow land burial.

LLW disposal designs can be classified according to three functional features [Ba87]:

- Relationship to the natural level of the ground surface, the grade (above or below grade).

- Depth of cover (shallow cover or deep cover).

- Disposal units provide structural stability.

For each of the disposal designs considered here, these functional features are specified and additional descriptive information is given to help characterize the design and provide information about major advantages and disadvantages.

Shallow Land Burial: An illustration of a shallow land burial unit is shown in Figure 1. Shallow land burial of LLW historically has involved placing the waste in unlined earthen trenches within the upper 30 meters of the earth's surface. This has been performed without employing engineered, structurally reinforced enclosures made from manufactured materials for the waste or specially treating the waste itself. The objective of all LLW disposal designs is to contain radionuclides, but shallow land burial is unique in that it relies primarily on the geohydrological and topological features of the site to accomplish these objectives. Other disposal facility designs augment these features with materials that reduce or eliminate waste flow and contaminant transport. In order to be considered an alternative to shallow land burial, a disposal facility design must provide significantly increased contaminant containment for a long period of time after the waste is emplaced.

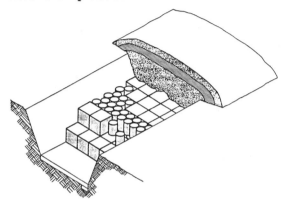

Figure 1. Illustration of shallow land burial.

In terms of the disposal design classification system, shallow land burial means that:

- The waste is disposed of below grade;

- There is a simple, thin earthen cover over the waste; and

- The disposal units do not provide structural stability.

Near the end of 1993, the State of California issued a license for a shallow land burial facility in an arid region of the state. In the design of this facility, the disposed waste is covered with at least 6.4 m of earthen material. A major function of this cover is to minimize percolation of water into the disposal trenches. For this facility, it is predicted that there will be no net percolation through the cover to the waste. While existing commercial LLW disposal facilities use covers of soil excavated from the disposal trench itself, more complex covers are being designed that use different earthen materials, such as clay, sand, gravel, cobble, boulders, riprap (crushed rock), or topsoil. Each of these components of the cover system has a particular function. These include acting as a moisture barrier (clay), a moisture conduit for shedding water that has infiltrated layers above (sand or gravel), an intrusion barrier (cobble or boulders), a growth medium for vegetation (topsoil), and providing resistance to erosion (cobble, boulders, riprap). The top of such covers are graded to encourage precipitation to flow away from the trench and thereby reduce water infiltration into the trench.

Below-Ground Vaults: A below-ground vault disposal unit is illustrated in Figure 2. Disposal of LLW in below-ground vaults involves placing the waste containers in an engineered structure which is then covered with earth. The vault can be above or below natural grade. The structure consists of reinforced concrete floors, walls, and roof. If the vault is constructed below grade, it is covered with an earthen cover in a manner similar to that described above for shallow land burial. If the vault is constructed above the natural grade, the cover would resemble the cover placed over the earth mounded bunker design.

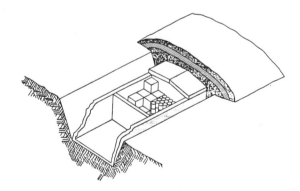

Figure 2. Illustration of below-ground vault disposal.

Generally, the below-ground vault is a disposal design in which:

- The waste is disposed of below ground level;

- There is both an engineered concrete cover and an earthen cover over the waste; and

- The disposal units provide structural stability.

The inclusion of both a concrete vault and an earthen cover enhances the ability of this disposal concept to restrict water infiltration into the waste, to prevent plant or animal intrusion, and to reduce gamma radiation exposure rates at the ground surface. The concrete vault also provides structural stability that is not present in the shallow land burial design. Placement of the vaults below grade makes the facility less likely to attract intruders and less vulnerable to disruptive events and processes that can occur at the land surface. Covering the vault with soil may preserve the integrity of the disposal unit for a longer period than if it were exposed to the atmosphere.

Below-ground vaults should satisfy requirements for minimizing active maintenance after site closure. However, if remedial actions are found necessary, the waste may prove more difficult to remove than from a shallow land burial facility because the massive concrete structure must be breached and removed first.

Below-ground vaults could be appropriate for the disposal of all activities of LLW because the structure itself may provide the required structural integrity and the required protection from inadvertent intrusion. However, in order to meet these goals, the structure would have to endure without substantial degradation for at least 500 years. A general lack of long-term experience with concrete structures makes is difficult to project the performance of below-ground vaults for these time periods with a high degree of confidence. Recently, computer codes have been written to predict the long-term performance of concrete [EPRI88; Shu91]. However, these codes have not been used to satisfy regulatory requirements because it is extremely difficult to validate them. They have mainly been used in design, operations, and performance assessments.

Above-Ground Vaults: Disposal of LLW in above-ground vaults involves placing the waste containers in an engineered structure located above the surface of the ground and that remains uncovered after the facility has been closed. This means that the vaults will also be above the natural grade at the disposal site. The floor, wells, and roof of the structure are made from reinforced concrete. Fill material could be used to fill the spaces among the waste containers inside the structure. The nature of the structure -- with a roof above the waste -- would make placing and compacting the fill difficult. This difficulty could be reduced by placing the roof over each disposal unit or vault after the unit has been filled. In summary, the description of above-ground vaults implies that:

- The waste is disposed of above ground;

- There is an engineered concrete cover over the waste, but there is not an earthen cover; and

- The disposal units provide structural stability.

The above-ground vault concept is envisioned as a large steel-reinforced concrete enclosure with either limited horizontal access (as through a doorway) or vertical access (as through on open top). The structure could be constructed either way, and this distinction is a detail of design that is not important in discriminating between the different designs. An above-ground vault with vertical access is illustrated in Figure 3.

Because of the lack of long-term experience with structures made with modern concrete, it is difficult to determine the time period over which the structural integrity of the above-ground vault would be maintained. When compared with the below-ground vault concept, the above-ground vault appears to have two major disadvantages in the long term. First, the lack of an earthen protective cover could make the above-ground vault more susceptible to environmental factors such as wind and water erosion, freeze-thaw cycles, and other factors, such as acid rain, that could hasten degradation of the structure. Secondly, in the event that the structural integrity of the vault were impaired, there would be no additional protective cover to act as a final barrier to waste migration.

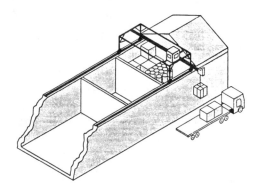

Figure 3. Illustration of above-ground vault disposal.

It should be noted that the above-ground vault described here could be designed to overcome many of the difficulties described in the paragraph above by providing it with an earthen cover. In that case, the design would be described as being above grade but below ground. However, the above-ground vault design described here is more often suggested for LLW disposal than covered above-grade vaults.

Modular Concrete Canisters: A modular concrete canister disposal facility is illustrated in Figure 4. This disposal concept is planned for a LLW disposal facility in an arid region in Texas. Modular concrete canister disposal consists of placing individual waste containers in modular concrete structures (i.e., placing the individual waste containers in reinforced concrete canisters) and placing the concrete canisters in trenches below natural grade. Except for the use of the concrete canisters, the physical details of below-ground modular concrete canister disposal resemble those of below-ground vaults. Each canister can be viewed as a mini-vault. Within the canisters, grout probably would be used as backfill between individual waste containers. The description implies that:

- The waste is disposed of below grade;

- There is a thin earthen cover over the waste; and

- The disposal unit provides structural stability.

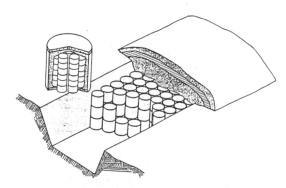

Figure 4. Illustration of modular concrete canister disposal and cutaway of a canister.

As in the case of below-ground vault disposal, the inclusion of both a concrete canister and an earthen cover as barriers enhances the ability of this disposal concept to restrict water infiltration into the waste, to prevent human, plant, or animal intrusion, and to reduce gamma exposure rates at the ground surface. However, water infiltration into the trench containing the canisters is similar to that for shallow land burial. The concrete canisters provide the structural stability that is not present in the shallow land burial concept. Covering the canisters with soil may preserve the integrity of the canisters for a longer time period than if they were exposed to the atmosphere. If a canister were breached, the soil covering would still provide a barrier to isolate the waste from the general environmental and water intrusion.

A major technical difference between disposal of LLW using modular concrete canister disposal and disposal using below-ground vaults is the modular nature of the structure. By using small, self-contained structures, stresses are reduced and failure of a single structure releases smaller amounts of radionuclides. The individual canisters would be easily transportable and, if a long canister life could be assured, remedial action would be much easier since most or all of the non-failed canisters could be removed intact. In most other important technical respects, except for the modular nature of the structure and operating procedures for placement of the waste, modular concrete canister disposal and below-ground vaults would be similar.

Earth-Mounded Concrete Bunkers: The earth mounded concrete bunker is a concept that has already been developed and used for the disposal of LLW in France [NRC83]. An illustration of this disposal design is presented in Figure 5. An earth mounded concrete bunker involves the use of two distinct disposal designs at the same site. It includes both the above grade disposal of waste with an earthen cover (the tumulus), and the below grade disposal of waste in a concrete bunker (the monolith).

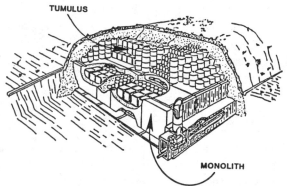

Figure 5. Illustration of earth mounded concrete bunker disposal.

For the tumulus:

- The waste is disposed of above grade;

- There is a thin earthen cover over the waste (typically less than 5 meters); and

- The disposal units do not provide structural stability.

For the monolith:

- The waste is disposed of below grade;

- The extent of the cover depends on whether the monolith is located under a tumulus. If it is, there is a deep, but complex, cover. Since the monolith is separate from the tumulus there would probably be a thin cover over the structure; and

- The disposal units provide structural stability.

A disposal cell in the monolith is prepared by constructing the foundation, floor, and walls which, together with the roof, constitute the boundaries of a monolithic cell. After a layer of waste containers is placed in the cell, all void spaces between waste containers are filled with concrete, which provides additional structural stability. When the cell has been completely filled, the roof is constructed and the cell closed. Because structural stability is provided by both the structure itself and the fill material, and because the structure represents a significant intrusion barrier, the monolith is suitable for disposal of all LLW classes without regard to the inherent structural stability of the waste or the need for additional intruder protection over the disposal cell.

All waste placed in the tumulus is either solidified prior to shipment to the disposal facility, or is supercompacted and grouted within concrete canisters. In either case, structural stability is provided prior to disposal of the waste. The concrete canisters and steel drums (in the case of solidified waste) are stacked on top of completed monoliths or specially constructed above-grade concrete foundation pads in such a way that the sides of the stacks slope inward as depicted in Figure 5. Void spaces among the waste containers are backfilled with earthen material and the tumulus mounded over with an earthen cover.

Because the tumulus places containers above natural grade, it has a higher potential for releasing radionuclides to the air and ground surface and is less able to use features of the site, such as good hydrology, to protect the public from radionuclide migration due to failures in engineered safety features. It may also be considered more vulnerable to intrusion and to other disruptive events and processes at the surface, especially erosion, than shallow land burial trenches. However, because of the earthen cover system provided over the tumulus, the vulnerability to some processes, such as freeze-thaw cycles, is probably not as great as for exposed above-ground vaults. Remedial action involving removal of the waste from the tumulus, if necessary, would probably be easier than for shallow land burial.

The previous discussion of below-ground vaults is generally applicable to disposal in the monolith. However, remedial action for the monolith would be costly and very difficult because of the large quantities of concrete used in the monolith as backfill material.

Augered Holes: Certain relatively high-activity LLW is being disposed of in augered holes at several disposal facilities of the U.S. Department of Energy (DOE). The holes are both unlined or lined with a fiberglass, concrete, or metal liner to improve structural stability and reduce the percolation of water into the disposal unit. For unlined augered holes:

- The waste is disposed of below ground;

- There is a deep cover over the waste (at least 5 meters); and

- The disposal units do not provide structural stability.

For a lined augered hole with a structurally capable hole liner:

- The waste is disposed of below ground;

- There is deep cover over the waste (at least 5 meters); and

- The disposal units provide structural stability.

An augered hole disposal unit with a liner is illustrated in Figure 6.

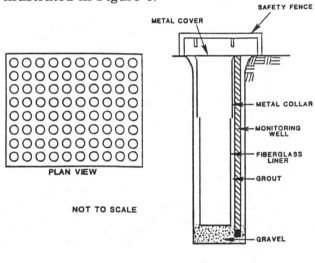

Figure 6. Illustration of augered hole disposal.

Specialized equipment would be used to construct the boreholes. In a humid climate, the bottom of the borehole would be more than 2 meters above the level of the water table. Waste would be placed in the bottom of the hole, leaving room for about 3 meters of below-grade engineered cover. One large engineered cover 3 meters thick would also be placed over the whole facility. This extent of cover material is sufficient to satisfy requirements for human intrusion protection and provides a significant barrier against biological intrusion. There appear to be no difficulties with augered hole disposal with regard to requirements for minimizing active maintenance or vulnerability to disruptive events and processes at the surface. Remedial action involving removal of the waste, if required, would be difficult and costly because of the amount of material that would have to be removed to reach the waste.

Because of the relatively small capacity of each borehole and the associated costs, augered hole disposal would be expensive for the entire spectrum of LLW. It may be possible to dispose of a small fraction of all waste effectively using augered hole disposal at a facility where some other disposal concept is used for the bulk of the waste.

Mined Cavities: An illustration of mined cavity disposal is shown in Figure 7. Mined cavities, whether based on existing or specially-developed mines, have been suggested for the disposal of LLW. Generally, the mined cavity facility is perceived as being a relatively deep facility which is well-isolated from the general environment by its depth or placement. Cavities in salt, coal, granite, or limestone beds have been considered.

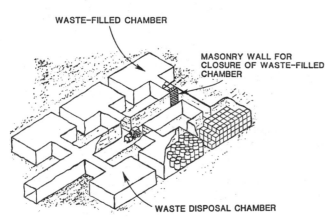

Figure 7. Illustration of mined cavity disposal.

With mined cavities:

- The waste is disposed of below grade;

- There is deep cover over the structure; and

- The disposal units provide structural stability.

A mined cavity would have little vulnerability to disruptive events and processes at the surface because of its remoteness from the surface. The physical setting of the cavity would represent a significant barrier to biological and human intrusion, though with previously existing cavities future intrusion would be invited by the known presence of minerals in the region. Barriers against waste intrusion and radionuclide migration would be largely site-dependent in the case of mined cavity disposal. The potential for water infiltration due to subsidence would be small, though subsidence of land above coal and salt mines is not uncommon. Remedial action once mined cavities are sealed would be difficult and expensive because of their depth and the care with which the cavities would be closed.

Because of the different characteristics of the geologic formations that can be used for mined cavities, it seems unlikely that the efficiencies that flow from standardization in the design, construction, and operation of a mined cavity facility could be achieved. Mined cavity disposal would be unsuitable for disposal of small volumes of waste because the high cost of development makes such disposal very costly on a unit volume basis. Presuming that the mined cavity chosen has sufficient structural capability, mined cavity disposal could be appropriate for unsegregated disposal of all classes of LLW without stabilizing the waste form.

Canisters in Vaults: Another disposal technology being planned for several humid sites in the eastern United States involves placing concrete canisters containing the LLW in either an above grade vault or a below grade vault. This technology provides an additional barrier for containing the waste. An illustration of this technology is shown in Figure 8. This technology combines the benefits of both the concrete canister and vault technologies. However, it is significantly more costly and involves additional operations on the waste.

Summary

In summary, present LLW disposal technologies incorporate extra engineered barriers to further isolate the waste from the environment. Concrete is the material usually used for the additional engineered barriers. Because the long-term integrity of the concrete barriers cannot be assured their main use is to provide additional protection for the first few hundred years after disposal.

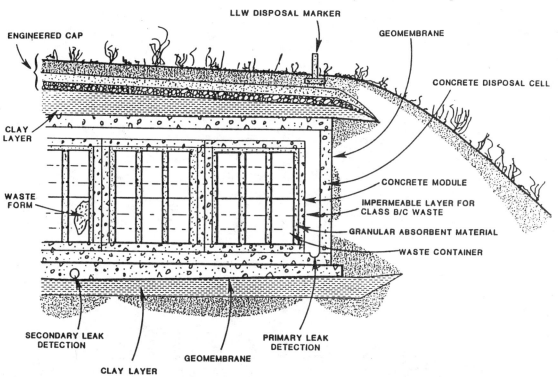

Figure 8. Canisters in an above-grade vault.

References

Ba87   Baird, R.D., et al., "Design and Cost Methodologies for Low-Level Waste Disposal Facilities," prepared by Rogers & Associates Engineering Corporation for Electric Power Research Institute, interim report, RP 2691-1, March 1987.

EPRI88   Electric Power Research Institute, "Performance Assessment for Low-Level Waste Disposal Facilities," report NP-5745SP, prepared by R. Shuman, V.C. Rogers, N. Chau, G.B. Merrell, and V. Rogers, Rogers & Associates Engineering Corporation, April 1988.

NRC83   U.S. Nuclear Regulatory Commission, "Symposium on Low-Level Disposal Facility Design, Construction, and Operating Practices," NUREG/CP-0028, CONF-820911, Vol. 3, March 1983.

Shu91   Shuman, R., V. Rogers, and N. Chau, "Improved Modeling of Engineering Barriers for Low-Level Waste Disposal," in the Proceedings for Waste Management '91, p. 757, Tucson, AZ, 1991.

Analysis of a Compacted Clay Barrier in a Radioactive Waste Disposal Scheme

E.E. Alonso, A. Gens, A. Lloret
Geotechnical Engineering Department, Civil Engineering School. UPC Barcelona, Spain

C.H. Delahaye
Universidad Nacional de San Juan, Argentina

Abstract

The behaviour of swelling clay barriers located between a natural clayey soil and a nuclear canister has been analyzed in some detail by means of a simulation exercise. A coupled hydromechanical model for unsaturated expansive/shrinking materials has been used. The computed results have shown that a complex mechanical phase develops during the transient swelling stage. High tension and shear stress states have been computed and physically explained. The results of no tension and nonlinear elastic type of analysis have been compared. Differences in the water transfer process are slight. However the more realistic no tension calculation leads to a larger outward displacement of the expansive backfill. A longterm open gap between canister and backfill is also predicted by the analysis.

1 Introduction

Expansive clays either compacted 'in situ' or as dense prefabricated blocks have been proposed as a suitable barrier to isolate nuclear wastes from the surrounding geological media. One of the proposed schemes involves the installation of cylindrical canisters in boreholes drilled from underground galleries. The annular space between the canister wall and the natural ground is then filled with the expansive clay barrier. Progressive hydration of the expansive barrier takes place as the natural water content of the ground migrates towards the manufactured clay barrier. The volumetric expansion of the clay will hopefully create a tight and impervious barrier to isolate the canister.

When this scheme is installed in natural clay deposits the barrier hydration process implies a parallel transient drying of the natural clay. Thus, hydration and expansion of the barrier implies a parallel shrinking and drying episode of the natural ground. The transient phase of the whole phenomenon is therefore rather complex and requires a proper understanding of the coupled hydromechanical behaviour of natural and compacted clays, in unsaturated conditions.

This paper provides a fundamental description of the basic phenomena involved and presents the results of a simulation of the field problem outlined above. The soil properties used in the simulation are reasonably consistent with an educated guess of the real properties of an expansive barrier and a natural overconsolidated clay but they do not necessarily reproduce any real scheme or site.

The hydration and drying processes mentioned before imply the progressive saturation of an initially unsaturated active clay and a desaturation of an initially saturated natural clay. Both types of phenomena can be conveniently accomodated within a general framework for the hydromechanical behaviour of unsaturated soils. A model of this kind, described in Alonso et al. (1988), has been used in this paper. In order to introduce the application of this model to describe the barrier problem, a brief presentation of the main ideas and assumptions behind this model will be first made. Thermal effects will not be covered in this paper.

2. Outline of the theoretical framework

2.1. Constitutive relations

Two sets of effective stresses are supposed

to independently control the mechanical behaviour of an unsaturated soil: the 'net' stress or excess of total stress, $\boldsymbol{\sigma}$, over air pressure, p_a, and the water suction, which has an isotropic character ($s = p_a - p_w$), where p_w is the water pressure. In a general way, the mechanical behaviour of an unsaturated soil can be expressed as

$$d\boldsymbol{\sigma}^* = \boldsymbol{D}(d\boldsymbol{\epsilon} - d\boldsymbol{\epsilon}_o) \quad (1)$$

where $\boldsymbol{\sigma}^*$ is the net stress, $\boldsymbol{\epsilon}$ the total strain and $\boldsymbol{\epsilon}_o$ the strain induced by suction changes. Any constitutive law may be used to define $\boldsymbol{D}$. In the work described herein a nonlinear elastic model has been defined through a tangent compressibility coefficient, K_t and a tangent shear modulus, G_t.

Volumetric strains, $(\epsilon_o)_v$, have been specified through the concept of state surface (Matyas and Radakrishna, 1968). The following general expression, proposed by Lloret and Alonso (1985) for a wide variety of unsaturated soils, has been adopted:

$$(\epsilon_o)_v = a_e(\sigma - p_a) + [b_e + c_e(\sigma - p_a)](p_a - p_w) \quad (2)$$

where $(\sigma - p_a)$ is the net mean stress, and a_e, b_e, c_e are constants. K_t is obtained from equation (2) through differentation.

Soil water retention characteristics should also be defined. The concept of state surface is also suitable for these purposes. Experimental results collected by Lloret and Alonso (1985) suggested the following expression for the variation of degree of saturation with suction and confining net stress,

$$S_r = 1 - (1 - exp^{-a'(p_a - p_w)})[b' + c'(\sigma - p_a)] \quad (3)$$

where a', b' and c' are constants.

2.2. Flow properties

Generalized Darcy's laws for water and air flow are used. Alonso et al. (1987) made a review of existing empirical relationships relating the air and water permeabilities to water suction, degree of saturation and soil porosity. In the analysis presented later the following relationships suggested by Lloret and Alonso (1980) were preferred:

$$K_w(e, S_r) = A_e \left(\frac{S_r - S_{ru}}{1 - S_{ru}}\right)^3 10^{\theta e} \quad (4)$$

$$K_a(e, S_r) = B \frac{\gamma_a}{\mu_a} [e(1 - S_r)]^c \quad (5)$$

where A_e, S_{ru}, θ, B and c are constants, γ_a the specific weight of air and μ_a the viscosity of air.

Equation (4) combines the change of water permeability with void ratio e, described in Lambe and Whitman (1964) and the variation with saturation degree proposed by Irmay (1954). Equation (5) has been proposed by Yoshimi and Osterberg (1963).

2.3. Field equations

Three basic sets of equations have to be solved: mechanical equilibrium, continuity of air and continuity of water:

Equilibrium

$$\frac{\partial(\sigma_{ij} - \delta_{ij}p_a)}{\partial x_j} + \frac{\partial p_a}{\partial x_i} + b_i = 0 \quad (6)$$

Air continuity

$$\frac{\partial}{\partial t}[\rho_a n(1 - S_r + HS_r)] + div[\rho_a(\boldsymbol{v}_a + H\boldsymbol{v}_w)] = 0 \quad (7)$$

Water continuity

$$\frac{\partial(\rho_w n S_r)}{\partial t} + div(\rho_w \boldsymbol{v}_w) = 0 \quad (8)$$

where b_i are body forces, ρ_a and ρ_w densities of air and water, n the porosity and H Henry's constant.

The above set of differential equations have been discretized using Galerkin's procedure and a finite element computer program, NOSAT, was written to solve this type of problems. NOSAT has interesting capabilities to solve geotechnical problems dealing with unsaturated soils. Construction and excavation sequences may be specified. In addition to conventional types of boundary conditions, two special types have been implemented for the flow part of the analysis: seepage surface and drain conditions, as described in Lloret and Alonso (1994). It can also handle double porosity materials which are a suitable representation for highly expansive clays and rocks (Gens et al., 1993). The results discussed herein were obtained with the basic formulation outlined before. Concerning mechanical behaviour two types of analysis have been performed: a nonlinear elastic and a no tension version of the preceding case.

3. Modelling the disposal scheme

3.1. Initial and Boundary conditions

The analyzed geometry is schematically represented in Fig. 1. The problem has radial - symmetry and was solved under plane strain conditions. Fig. 1 is a quarter space

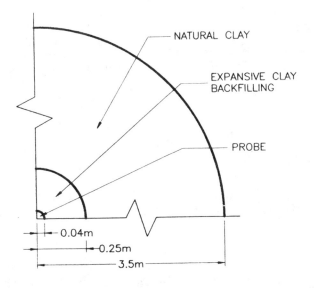

Fig. 1: Definition of the case analyzed.

representation of a cross section through a central rigid cylinder (probe) which acts as a drain, the compacted fill and the natural clay.

Note that the compacted sealing clay fills the annular space between a shaft (ϕ 0.50m) drilled vertically into the natural clay and the steel probe (ϕ 0.08 m). As far as the numerical discretization is concerned the outer boundary was located at a radius of 3.5 m.

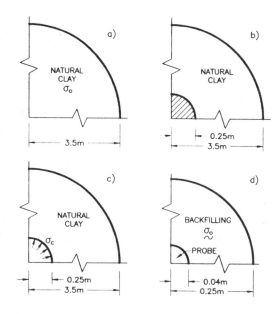

Fig. 2: Simulation of probe installation. a) Initial stress; b) Shaft excavation; c) Compaction of backfill; d) Final state.

Initial stress conditions were obtained through a simulation of the actual installation procedure (see Fig. 2). The vertical shaft (ϕ 0.50 m) is drilled from an underground tunnel located roughly at 250 m depth. It was assumed that the natural clay was subjected to an isotropic initial compressive stress $\sigma_o = \sigma_{ox} = \sigma_{oy} = \sigma_{oz} = 5\ MPa$ (Fig. 2a). The excavation of the shaft was then simulated (the shaded area in Fig. 2b is removed). The installation of the probe and the compacted backfill introduces a compaction boundary stress (σ_c) against the natural clay (Fig. 2c). This stress corresponds to the assumed initial stress conditions in the compacted material (Fig. 2d): $\sigma_{ox} = \sigma_{oy} = 1\ MPa$ and $\sigma_{oz} = 0.25\ MPa$. Finally a rigid contact was assumed at the probe-backfilling interface: Displacements in the radial direction are zero if soil moves towards the center. Otherwise a stress free surface was imposed. The finite element mesh adopted is shown in Fig. 3. It had 135 quadrilateral elements.

The mesh was refined in the backfill material and in the vicinity of the natural clay-backfill interface.

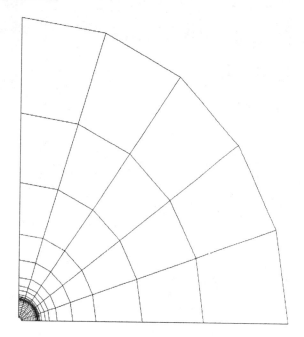

Fig. 3: Mesh used in the analysis.

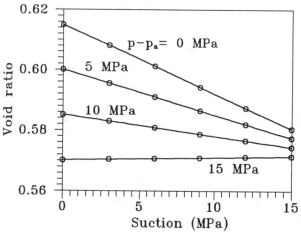

Fig. 4: State surface for void ratio of natural clay.

As far as the initial flow conditions is concerned, the natural clay was fully saturated at a constant pore pressure of 2 MPa. A relatively strong initial suction (15 MPa) was assumed in the compacted backfill.

The outer boundary of the discretized geometry was assumed impervious to air and water. Along the probe-backfill interface a 'seepage face' condition was imposed. (The probe acts as a drain to any incoming water)

3.2. Material properties

A few relevant material properties have been represented in Figs 4, 5, 6, 7 and 8. Fig. 4 shows the state surface for volumetric deformation adopted for the natural clay. Note that an increase in suction (drying) induces a small volumetric shrinkage, which decreases with applied net confining stress. The equivalent state surface for the backfill is plotted in Fig. 5. It is shown that wetting induces a considerable swelling in this material for the whole range of expected confining stresses.

A common state surface for degree of saturation was adopted for the backfill and natural clay (Fig. 6). The variation of water and air permeability with degree of saturation for both materials is plotted in Fig. 7 and Fig. 8.

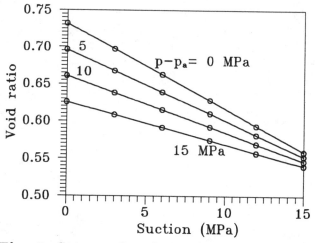

Fig. 5: State surface for void ratio of backfill.

Poisson's ratio was fixed at $\nu = 0.30$ for the natural and compacted clay.

3.3. Results

Computed results have, in general, been presented as the time variation of a selected parameter, property or state variable at a number of locations (identified by their radial coordinate from the center).

The induced flow is conveniently visualized through a plot showing the evolution of degree of saturation at those locations (Fig. 9). The initially unsaturated backfill is progressively wetted as a saturation front driven by the high suction of the expansive fill progresses inwards. This flow induces a transient drying in points of the natural clay located in the vicinity of the fill. Eventually, at times larger than 150 days the systems reaches steady state conditions. The time history of suction (Fig. 10) is consistent with the interpretation above.

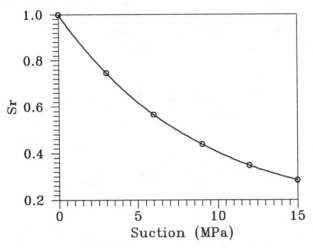

Fig. 6: Retention curve for natural clay and compacted backfill.

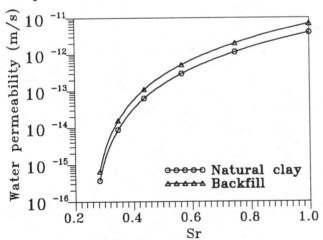

Fig. 7: Relationship between water permeability and degree of saturation.

Fig. 11a shows the computed radial displacements. Maximum displacements at steady state conditions take place at the fill-natural clay interface.

It is interesting to note that the inner boundary experiences a longterm outward displacement once the fill has swelled. A small gap, close to 1 mm in width, was computed.

The interpretation of the wetting process is facilitated if the evolution of stresses is examined. The evolution of the circumferential principal stress is plotted in Fig. 12a. The points in the backfill experience an extension-compression cycle as they are reached by the wetting front. Long term conditions indicate a stable compression in the wetted fill. On the contrary, the natural surrounding clay is pushed away by the expanding inner fill and circumferential stresses become tension stresses.

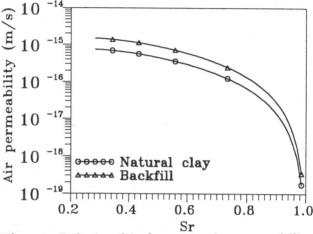

Fig. 8: Relationship between air permeability and degree of saturation.

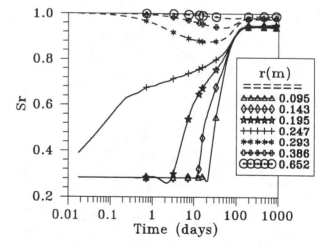

Fig. 9: Evolution of degree of saturation at some selected points.

It was realized that the computed tension could

not possibly take place in the soils involved and a hopefully more realistic no tension analysis was carried out.

Circumferential stresses are now given in Fig. 12b. Equilibrium states under compression stress states could be found after the necessary convergence process. Stresses in the natural clay have decreased to moderate values. Inside the backfill a different distribution of values compressive stresses is computed although the maximum values remain similar to the values computed in the nonlinear elastic analysis. The no tension analysis leads to larger overall displacements if compared with the elastic calculation. This is shown in Fig. 11b which may be compared with Fig. 11a. Note also that an open gap at the probe-fill contact is again computed.

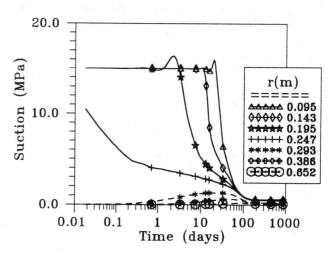

Fig. 10: Time history of suction at some selected points.

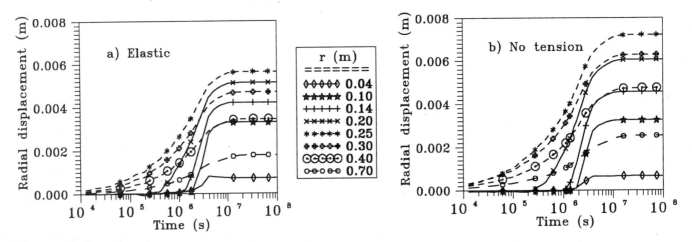

Fig. 11: Computed radial displacements a) Elastic analysis b) No tension analysis.

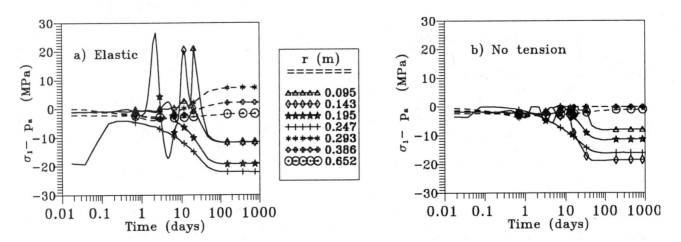

Fig. 12: Computed major principal stresses a) Elastic analysis b) No tension analysis.

A further insight into the mechanical phenomena developing during the backfill wetting is provided by the stress paths followed by the selected points. Fig. 13 shows the variation in net mean stress as suction changes under the no-tension hypothesis. Typically the backfill follows a swelling pressure type of path. Confining net mean stress increases as suction decreases. A loading-unloading cycle is computed at some intermediate stage as a result of the wetting front crossing through the given location. Points located in the natural clay, close to the backfill boundary, experience, however, a drying-wetting episode and, as a result, a transient shrinkage volumetric deformation. More information is provided by the stress paths plotted in the (q,p) plane where q is the deviatoric stress and p the net mean stress (Fig. 14). In Fig. 14 a the stress paths correspond to the nonlinear elastic analysis. Loading-unloading cycles are again clearly identified. Tension states are common as well as strong shear stresses.

It is obvious that these stress paths cross any reasonable strength envelope which could be assigned to both the backfill and the natural clay. Tension mean stresses, in particular, seem to be unreasonable and this was the main argument to perform the no tension analysis.

Stress paths for this second case have been plotted in Fig. 14b. Shear stresses are again important and they probably reach the shear strength in some cases. These results point out the need for improving the analysis reported here, introducing elastoplastic criteria. It should be finally mentioned that the no-tension analysis, if compared with its elastic counterpart induces a minor effect a the flow part of the analysis.

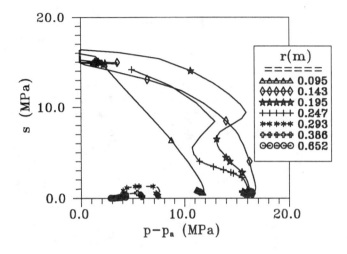

Fig. 13: Stress paths in the (p, s) plane. No tension analysis.

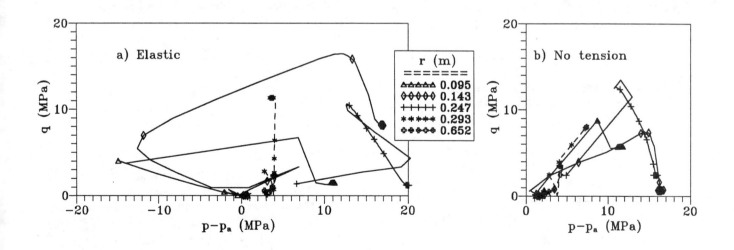

Fig. 14: Stress paths in the (p, q) plane. a) Elastic analysis b) No tension analysis.

4. Discussion and conclusions

At first sigth, the idea of installing a swelling clay around a canister, as a suitable barrier, seems to lead to a simple hydromechanical phenomenon. Examined in more detail the transient hydration process is fairly complex and tends to induce strong shear and tension stresses in both backfill and natural ground. The process can be physically described as follows: the high initial suction of the backfill draws water from the surrounding saturated clay. This flowing water progresses inwards as a sharp wetting front. As a consequence, a thin ring of compacted fill swells, compresses circumferentially and moves outwards. In this way the migrating swelled ring pulls outwards the inner backfill zone and this is the reason for the computed tension states found in the elastic analysis. A given point inside the backfill experiences first a reduction of the initial compression stress and, eventually, reaches a tension state as the wetting front becomes closer. The passage of the wetting front may reverse the sign of the stresses, specially the circumferential ones. Two interesting consequences of this process are the computed net outward displacement of the backfill and the tendency towards failure stress states both in tension and in shear. Failure in tension, specially, may be associated with crack development and a change in the hydraulic conductivity of the soil, which are phenomena not considered in the analysis reported here. As the transient process tends towards steady state conditions an overall compression state develops in the backfill. However, even under steady state conditions a finite gap between the central probe, and the soil has been computed in both elastic and no-tension analyses. The natural surrounding clay, close to the backfill interface, is initially subjected to an advancing drying front whose effect has to be added to the long range effect (overall expansion) of the backfill hydration. Computed stress changes in the natural soil are, however, less marked than in the inner backfill as the plotted stress paths indicate.

The numerical analysis carried out, despite its limitations, has shown the complex nature of the transient hydration process and the strong differences between transient and steady state stress states. The implication of this behaviour of the barrier scheme are not sufficiently understood. Experimental information and a more refined analysis, incorporating plastic and some kind of damage criteria are required to improve our present knowledge of the behaviour of expansive compacted clay barriers.

Acknowledgements

The authors wish to thank the financial support provided by the European Community RADWAS Research Program through Projects FI2W-CT90-33 and CT90-102.

References

1. Alonso, E.E., Batlle, F., Gens, A. and Lloret, A. (1988). "Consolidation analysis of partially saturated soils. Application to earthdam construction. Numerical Methods in Geomechanics". Innsbruck. p. 1303-1308.
2. Alonso, E., Gens, A., and Hight, D.W. (1987). "Special Problem Soils: General Report". Proc. IX E.C.S.M.F.E. Dublin.
3. Gens, A., E.E. Alonso, A. Lloret and F. Batlle (1993). "Prediction of long term swelling of expansive soft rocks". A double-structure approach. Geotechnical Eng. of Hard Soils-Soft Rocks. Anagnostopoulos et al. eds. Balkema. 495-500.
4. Irmay, S. (1954). "On the hydraulic conductivity of unsaturated soils". Trans. Amer. Geoph. Union. 35 p. 463-468.
5. Lambe, T.W. and Whitman, R.V. (1968). "Soil Mechanics". J. Wiley. N. York.
6. Lloret, A. and Alonso, E.E. (1985). "State surfaces for partially saturated soils". Proc. XI I.C.S.M.F.E. San Francisco. Vol. 2 p. 557-562.
7. Lloret, A., and Alonso, E.E. (1994). "Unsaturated flow analysis for the design of a multilayer barrier". XIII ICSMFE, New Delhi. (In press).
8. Matyas, E.C. and H.S. Radhakrishna (1968). "Volume change characteristics of partially saturated soils". Géotechnique, 18: 432-448.
9. Yoshimi, V. and J.O. Osterberg (1963). "Compression of partially saturated cohesive soils. Jnl. Soil Mech. Found. Eng. Div. ASCE 89 (SM4): 1-24.

Granular Halite Creep and Compaction for Backfilling

Maurice B Dusseault, P.Eng, J Carlos Santamarina
U. Of Waterloo, Waterloo, Ontario, Canada

Christine Trentesaux, Patrick Lebon
Agence Nationale pour la Gestion des Dechets Radioactifs, Fontenay-aux-Roses, France

Eduardo E. Alonso
U. Polytech. de Cataluna, Barcelona, Catalonia, Espana

ABSTRACT: Natural halite deposits have potential for use as repositories for solid civil wastes, both radioactive and toxic. These uses are outlined; in particular, transmission vectors for mass transport are important, thus sealing of repositories is vital. Granular halite backfill compaction behaviour is similar to other granular materials, but consolidation is radically different because of solubility and creep. Behaviour is a function of density and moisture content, and all parameters may be considered as time-dependent because of creep-consolidation. Backfilling openings in halite deposits or admixing salt to slurried solid wastes for storage in solution caverns depend upon natural opening creep closure for ultimate sealing. The dominant creep process seems to be dissolved salt transport in the fluid phase.

1 INTRODUCTION

Permanent highly toxic solid civil waste disposal sites require high security and, if radioactive, retrievability. The latter implies that waste recovery remains possible, the former that all biosphere transmission vectors be identified and secured. Because most vectors are fluid transport related, hydrogeological security is paramount. Repositories in salt domes and bedded salt strata are singularly attractive because of salt's low permeability, and because viscoplastic flow heals induced fractures (Allemandou and Dusseault 1993). Natural salt is unjointed, and fluid flow is constrained to anhydrite, limestone, or clay seams, or to transport along grain boundaries with thin water films (Spiers et al. 1989, 1990).

A salt repository may be a mine or a solution cavern. In a mine, waste placement is followed by backfilling, and the shaft as well as the placement horizon must be sealed. Fig. 1 shows the stratigraphy of the Sifto Canada salt mine at Goderich, Ontario, being studied as a site for non-radioactive civil waste storage. Similar studies are being done conducted in the United States, France, and Britain. In Germany, salt mine radioactive waste placement is the option of choice; extensive studies have been conducted for 10 years. Sealing will likely have to be achieved through placement of granular salt backfill.

Salt deposits are being exploited for brine by solution mining throughout SW Ontario, Ohio, Michigan, and western New York (Davidson *et al*, 1994). Similar strata are found throughout the world. In Ontario, approximately 72 caverns exist with a storage capacity of several tens of millions of cubic metres. In France (Boucly 1982), seasonal natural gas storage has shown the feasibility of cavern use for wastes. Flow security is vital once a cavern is filled with waste; wastes should be placed with 30-50% well-graded granular salt to accomplish this in the halite horizon. This gives a denser placement porosity, and the permeability of the placed waste becomes extremely low as pore throats are blocked with solid salt during closure. Economics is a major advantage of solution cavern slurried solid waste placement. Cavern placement operations can probably be achieved for 10-15% of mine repository costs.

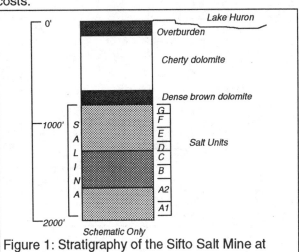

Figure 1: Stratigraphy of the Sifto Salt Mine at Goderich, Ontario

Mine repository sealing with granular salt is advantageous because:

a. It is geochemically compatible;
b. Openings close, reducing permeability with time; and,
c. The material is available locally.

Placement variables include granulometry, moisture content, and compactive effort. In solution cavity slurried wastes placement, behavior of halite mixed with brine-saturated solid waste is also of interest. At long times under stress, granular halite compacts to a rock similar to the surrounding halite, re-establishing a "natural" barrier to flow, yet it can be mined to retrieve wastes if required. To design seals and carry out transmission vector probabilities, quantitative understanding of the creep-compaction behavior of granular halite is necessary; considering options and conditions, many tests are required.

2 DEFORMATION MECHANISMS

Experimental data on the mechanical and hydraulic behavior of salt aggregates is relatively scarce if compared with the available information on rock salt. Fig. 2 shows a typical set of creep data on an intact polycrystalline domal halite of isotropic fabric (n≈ 0.6%; D_{50} ≈ 8mm; >99% NaCl). The rate sensitivity to shear stress (J2) is evident, and a function of the following form is generally used:

$$\dot{\varepsilon}_{ss} = A_1 \left(\frac{J2}{\sigma_o|_1} \right)^{n_1} e^{\frac{-Q_1}{RI}} + A_2 \left(\frac{J2}{\sigma_o|_2} \right)^{n_2} e^{\frac{-Q_2}{RI}}$$

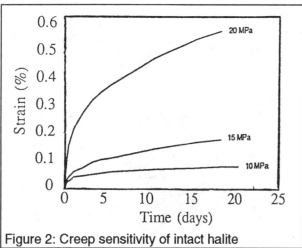

Figure 2: Creep sensitivity of intact halite

This equation supposes coexistence of multiple mechanisms contributing to $\dot{\varepsilon}_{ss}$, each characterized by a Norton power law exponent (n), a thermal activation energy (Q), a constant (A), and a different normalizing stress σ_o. Other stress-level activated mechanisms can be included in hyperbolic or power law form using a Heaviside function to activate the mechanism (Munson et al. 1993). Fig. 3 is a simple rheological model representation of eq. 1 (note that time-deformation parameters are highly non-linear). In eq. 1, n is presumed to be 1.0 for fluid-assisted diffusional transport creep (FADT), typically 3.0 to 5.5 for viscoplastic lattice dislocations such as glide and cross-climb, and perhaps higher (with J1 dependency) for stable micro-cracking (Dusseault and Fordham 1993). For processes such as grain boundary sliding and cataclasis (grain crushing), a broad range of exponents would be expected, depending on stress levels and the presence of moisture.

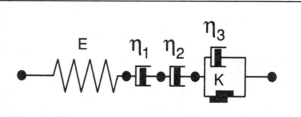

Figure 3: Rheological representation of multiple creep mechanisms

Creep law form and physical processes in granular salt are less clear because of grain size and surface area effects, moisture content effects (unsaturated), and the presence of phases which may interact with mass transport mechanisms. The dominant physical processes must be known for a broad range of stresses, porosities and temperatures to make predictions of long-term crushed halite consolidation.

In a repository, salt may be placed with a high compactive effort, or pressed salt bricks may be used to construct a barrier. High compactive effort causes grain crushing, increasing the surface area, and accelerating creep compaction once the salt is moist and loaded. In the extreme, bricks of compressed salt have porosities as low as 10%, and consist of particles ranging from sand sized to micro-sized particulars. If granular salt is placed dry (or with small amounts of moisture), minimum porosities of about 40% are expected; if placed under brine, higher initial porosities could be achieved with well-graded salt. However, in the long term, extremely slow stress rates are applied to the backfill as the openings close, thus slow creep tests are as vital as compaction tests.

3 TEST PROGRAM DIMENSIONS

Test programs for halite have been initiated; we report some of these results here. Test program features include:

a. Single load step creep consolidation ($\varepsilon_r=0$), Figure 4.
b. Multiple load step creep consolidation ($\varepsilon_r=0$), Figure 9.
c. Spherical creep consolidation ($\sigma_1=\sigma_2=\sigma_3$).
d. Oedometric creep consolidation, Figure 8.
e. Temperatures from 20°C to > 100°C.

Variables of intrinsic state include:

a. Different moisture contents (0% to saturated), Figure 6.
b. Different granulometries and densities, Figure 7.
c. Different co-mixed wastes (fly ash, quartz sand, ...), Figure 10.

Measurements include:

a. Index properties.
b. Electromagnetic and sonic transmission properties.
c. Mechanical stiffness and creep properties.
d. Saturated and unsaturated permeabilities.
e. Semi-quantitative microscopy.

The goal is a relatively complete constitutive understanding of granular halite to allow rational design of repository sealing and cavern placement strategies.

4 EXPERIMENTAL RESULTS

Granular halite may be used dry, moist or west with grain sizes from powder to 8-10 mm with initial porosities of 17% to as high as 47% without cataclasis. Compressed salt bricks created under high stress have particle sizes in the silt to sand range because of cataclasis, and have porosities of 10% or lower.

4.1 Dry granular halite compression

A typical void ratio versus time creep curve at 21°C is shown in Fig. 4 for an oven-dried (≈100°C) 0.54-1.2 mm salt from the Suria Mine (Barcelona, Spain), composition of 97.2% NaCl, 0.9% KCl, traces Mg, $CaSO_4$. The salt was placed at a porosity somewhat higher than 45% at 10 kPa, and loaded to 100 kPa. After 200 hours, a porosity of 44% was reached, and a small amount of continued creep is observed. The void ratio versus creep rate plot of the same test (Fig. 5) shows the initially high deformation rates and the "steps" characteristic of episodic dry grain slip. Based on similar tests (Fordham 1989), if care is taken to eliminate all H_2O that may surface absorb, creep rates attenuate, suggesting that FADT is a key mechanism and serves also to facilitate other mechanisms.

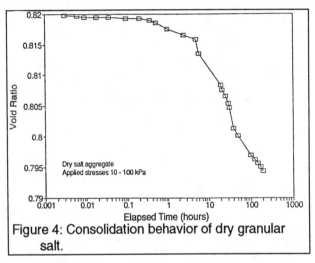

Figure 4: Consolidation behavior of dry granular salt.

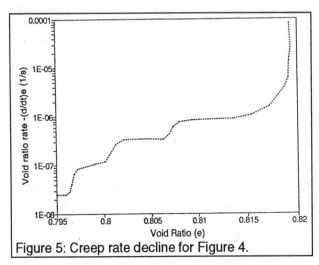

Figure 5: Creep rate decline for Figure 4.

Figure 6 shows data directly comparing moisture effects on compaction at 10 MPa on tailings salt (with 1-1.5% KCl) at various moisture contents. For the oven dried tests, the post-stressing porosity of

34.7 decreased to 31.2 after 48 days; at 5% moisture, the identical material achieved 20% porosity in one day, and 8.5% porosity in 48 days (0.4×10^7 s).

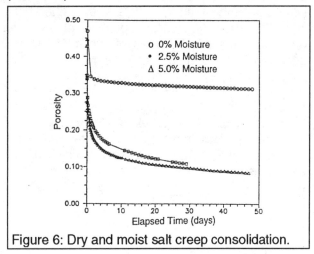

Figure 6: Dry and moist salt creep consolidation.

4.2 Moist granular halite, rapid compaction

For study of mechanical compaction and cataclasis, rapid high stress rate (80 MPa/min) compaction tests from 15 to 200 MPa were carried out on specimens with 0.5-1.4% H_2O. At 15 MPa, the porosity range was 30-37%, at 50 MPa it was 18-26% at 200 MPa it was 1.4%, with a linear e-$\log_{10}\sigma_v$ curve and a Compression Index of 0.48 for the range quoted. Final density (2.12-2.17) and compaction behavior seem totally insensitive to water content over the range tested. Note that at about 200 MPa, complete elimination of non-liquid filled porosity can essentially be achieved; however, this does not necessarily imply high mechanical strength.

The initial porosity achievable through manual densification ("kneading") of a brine-saturated halite tailings with a broad gradation (Fig. 7) was tested. This tailings material from a Saskatchewan potash extraction plan was used "as-received", and densified manually to porosities of 16-20% through rodding and tapping without applying any significant load and without detectable grain crushing. Densified specimens were used for creep-consolidation tests.

On moist specimens (3-4% H_2O by weight) of the same material, high applied stresses lead to cataclasis and rapid crushing, as reported above.

4.3 Creep-consolidation behavior

Long-term tests over large load and temperature ranges are required to identity creep mechanisms. Fig. 8 shows one-dimesional oedometric data for a narrow gradation saturated Suria salt aggregate (grain sizes from 0.50 to 0.54 mm) with e_o = 0.78. Four stages were used, but in none of the early stages had creep terminated before the next load added. (In fact, because of FADT, it is likely that for any stress creep consolidation will continue until the porosity becomes occluded). Assuming that the porosity occlusion (pore throat closure) is reached a 6-9%, it is clear that at the end of the test the specimen has reached an "interconnected" void ratio close to zero. (Note that FADT still is occurring along water films on crystal boundaries, and over tens of years at 10-30 MPa, these materials would approach porosities of 1-2%, similar to intact halite). Plotting the strain rate versus the void radio shows the "bimodal" curves reported by Fordham (1989), and interpreted as grain-sliding dominated, FADT-assisted creep in the period just after load application, then a shift to FADT creep with time.

Figure 9 shows similar results on 34% pure granular salt from the Goderich Mine in Ontario; lower stresses were used, and tests were carried out on 100 mm diameter by 200 mm long specimens in a triaxial cell with $\sigma_a = \sigma_r$. Note the large proportion of instantaneous densification associated with the loading phase of the 4 and 8 MPa increments, not evident in Fig. 8. We attribute this to a reduced (1-D) kinematic freedom for grain boundary slip in the oedometers, where slip along inclined boundaries requires lateral reaction against a rigid container, and is thus impeded. After the initial compaction, the creep behaviors are qualitatively the same

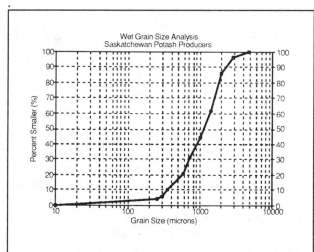

Figure 7: Grain size distribution of halite tailings from Saskatchewan.

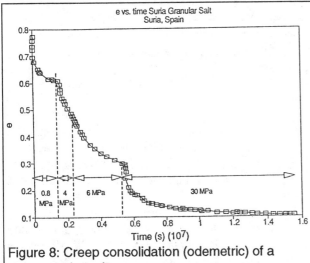

Figure 8: Creep consolidation (odometric) of a granular salt.

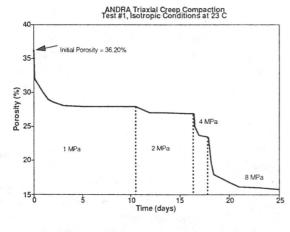

Figure 9: Creep consolidation (triaxial cell) of a granular salt.

To study the effect of stress level, Fordham (1989) carried out many brine-moistened (2.5%) oedometric tests on halite tails (D_{10} = 0.20 mm, D_{50} = 0.54mm, D_{100} = 2mm) between 30 and 39% initial porosity, showing a wide range of slopes on a logϵ-log(σ_V) plot. At high stresses (>5 MPa) and higher porosities, slopes of 3-4 imply grain sliding dominated creep, facilitated by FADT as mentioned above. At low stresses (<1 MPa) and lower porosities, slopes of 1.5 - 2 imply that FADT is the dominant mass-transfer mechanism. Thus, no single mechanism explains the full creep behavior range for stresses, even for isothermal tests outside of the range of cataclastic deformation.

Finally, for the solution cavern placement planning, Figure 10 presents saturated creep consolidation tests results at 500 kPa on three mixes: 25% and 50% quartz sand in salt, and a 25% fly ash in salt mix. The quartz sand and the salt are similar grain sizes, but the fine-grained fly ash tended to fill interstices between salt grains, giving a high initial porosity. The 25% sand mix essentially stopped consolidating after the porosity dropped from 35% to 28%, which took 70 days. The 50% sand mix went from 34 to 30% porosity and stopped after less than 20 days. The fly ash mix dropped from 22% (after initial porosity drop) to about 16%, but in 2 to 3 days, an order of magnitude more rapid for a similar porosity drop, compared to the sand mixes. Evidently, even though the pore throats are blocked by the fly ash particles, the presence of this material accelerates the mass transfer process. This has quite interesting implications for the disposal of fly ash with other wastes in solution caverns.

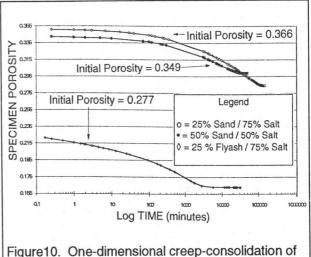

Figure 10. One-dimensional creep-consolidation of salt with additivies.

5 DISCUSSION

Compaction and cataclasis of totally dry granular halite is essentially the same as for dry sands, but cataclastic effects are strong because of the weak grains. In silicates, H_2O is known to reduce fracture strength; we presume the same in halite, and we presume that the same in halite, and we presume that cataclasis stops soon after loading. The low and high stress dry consolidation tests showed small continued creep after initial compaction, long after we would have expected cataclasis to terminate. This may be due to FADT taking place because of minute amounts of H_2O (hygroscopically acquired or from extremely small fluid inclusions). Evidently, the rate-controlling factor is transport rate in this case, rather than precipitation or solution rate. Cataclasis will only be an important mechanism in mined repositories if large dynamic loading is used to pre-compact salt backfill (e.g., cataclasis dominates brick manufacture). Clearly, the greater the placement density, the sooner low closure rates are achieved, other factors being equal. In bricks, creep rates may be even faster than in placed backfill because of the larger surface area, once H_2O is imbibed.

In moist granular salts, we interpret high slope processes at early times after loading as grain boundary slip, abetted by concentrated dislocation zones along crystal boundaries, and aided by FADT. The FADT aid is inferred by the lack of FADT creep in dry salt, and we believe that extremely high rate FADT can act locally to give kinematic compatibility for grain slip and dislocation glide, rather than generating dilatant hardening, as would be evident in quartz sands, for example. As ancillary evidence, the healing of microflaws and its effect on transient salt creep (Allemandou and Dusseault, 1993) is also attributed to FADT processes acting locally within flaws rather than between contacts and free grain surfaces.

At high porosities and low stresses, where grain sliding is impeded by negligible dislocation effects or frictional movement, significant porosity reductions by FADT occur because of the "sharpness" of the contacts (the efficacy of minor contact dissolution in reduction of high porosities is well-known). At low porosities and all stresses, but after long times, FADT is dominant or other processes simply because grain boundary slip and dislocation are greatly impeded by the assemblage density.

At present, it seems as if FADT will be the dominant process in almost the entire range of backfilling conditions because salt will absorb water hygroscopically, and only 1-2% H_2O is needed for FADT to proceed rapidly. Finally, we do note that a comprehensive physics based model (Olivella et al, 1992, 1993), in contrast to the more conventional empirical models (Fordham, 1989), has been developed for FADT processes and is being improved through testing.

6 CONCLUSIONS

Selected compaction and creep tests on granular halite from Spain and Canada have been presented and discussed. Three deformation mechanisms may be identified when dry or moist samples are tested for a wide range of porosities and applied loads. Rapid transient compaction, evidenced in dry to saturated specimens, is cataclasis-dominated. High stress compaction tests show a linear relationship between void ratio and log (vertical stress). The influence of initial water content of the aggregate on the attained porosity for a given applied load is negligible.

If there is sufficient moisture to support fluid-assisted transport of salt in the dissolved state (FADT), two other mechanisms are evidence during oedometric and spherical creep-consolidation. The early-time process with a steep $log\varepsilon$-$log\sigma$ slope is associated with dislocation processes concentrated on the grain boundaries, leading to grain slip, but aided by local FADT nonetheless. The long-term process with a less steep $log\varepsilon$-$log\sigma$ slope is associated with more static grain boundaries and FADT-dominated mass transport from the contacts to the free grain surfaces. The latter process is expected to dominate in repository and cavern closure.

7 REFERENCES

Allemandou, X. and Dusseault, M.B. 1993. Proc. Geot. Eng. of Hard Soils and Soft Rocks, Balkema, Rotterdame, 1581-1590.

Boucly, P. 1982. Rev. Fran. Geot. 18, 49-57.

Davidson, B., Dusseault, M.B. and Demers, R. 1994. 1st Int. Conf. on Waste Geotechniques, Edmonton, in press.

Dusseault, M.B. and Fordham, C.J. 1993. Comprehensive Rock Eng. (ed. Hunson, J.A.), V 3, 119-149, Pergamon Press.

Fordham, C.J. 1989. PhD Thesis, Ear Sci., U. of Waterloo, 181 p.

Munson, D.E., DeVries, K.L. Fossum, A.F. and Callahan, G.D. 1993. Proc. 3d Conf. Mech. Behaviour of Salt, Palaiseau, France, 31-44.

Olivella, S.; A. Gens; E.E. Alonso and J. Carrera (1992). Num. Models in Geomech Pande and Pietruszczak eds. Balkema. 179-189.

Olivella, S.; A. Gens; J. Carrers and E.E. Alonso (1993). Proc. 3d Conf. Mech. Behaviour of Salt, Palaiseau, France, 255-270.

Spiers, C.J., Peach, C.J. Brzesowsky, R.H. Schutjens, P.M.T.M., Liezenberg, J.L. and Zwart, H.J. 1989. Comm. of Euro. Communities Contr. FI1W-0051-NL, Utrecht, The Netherlands.

--- 1990. In Deformation Mechanisms, Rheology and Tectonics, Geol. Soc. Spec. Pub. 54, 215-227.

Scellement de Forage sur les Sites de Stockage de Déchets Radioactifs

M. Gandais, F. Dufournet-Bourgeois, A. Esnault
S.I.F. Entreprise BACHY, France

J.F. Ouvry
B.R.G.M. Ingénierie Géotechnique, France

Résumé : Dans cette communication nous présentons une réflexion sur le rebouchage des forages de reconnaissances autour et au droit d'un site potentiel de stockage. Les matériaux candidats pour le rebouchage doivent présenter les propriétés nécessaires pour assurer un scellement de manière durable. Les matériaux de scellement doivent être adaptés au milieu naturel en prenant en compte des paramètres comme la fracturation du massif, la conductivité hydraulique et l'espace des vides entre les épontes des fractures. Un plot d'essai réalisé dans la région de Limoges, France, est présenté.

Abstract : This paper presents a reflexion on the sealing of exploration boreholes in and around a potential radioactive-waste disposal site. The potential back filling products must have the necessary properties in order to insure a lasting sealing. The sealing products must be adapted to the natural environment conditions, by taking into account parameters such as the local fracturation, the hydraulic conductivity and the void of distribution of fractures. A grouting trial, realized around Limoges, France, is presented.

1 Introduction

L'évacuation en milieu géologique profond vise à immobiliser les déchets radioactifs et à les isoler de l'environnement humain pendant une durée et dans des conditions telles que tout rejet ultérieur éventuel de radionucléide à partir du dépôt n'entraînera pas de risques radiologiques inacceptables même à long terme (AEN/OCDE, 1988).

La réalisation d'installations souterraines nécessite des reconnaissances préliminaires géologique, hydrogéologique, géotechnique, géochimique qui sont réalisées à partir de forages de reconnaissance. Ces investigations in-situ sont effectuées par sondages depuis la surface ou à partir d'ouvrages souterrains (galeries, puits) constituant les laboratoires souterrains de recherche.

Les concepts fondamentaux de sécurité sur les stockages demandent que tous ces forages qui constituent autant de chemins d'écoulements préférentiels pour les fluides et les éventuels radionucléides soient rebouchés ou scellés de façon durable.

Cette communication présente une réflexion sur le rebouchage des forages de reconnaissance dans un milieu géologique cristallin ainsi que l'application qui a été réalisée.

Cette étude est menée dans le cadre d'un projet de recherche soutenu par les Communautés Européennes sur les problèmes de sûreté concernant le stockage de déchets radioactifs.

2. Propriétés Recherchées Pour Les Scellements De Forage

Le scellement des forages doit permettre de reconstituer au mieux les caractéristiques initiales du massif avant perturbation. Pour remplir durablement cette tâche, les matériaux de scellement doivent posséder un certain nombre de propriétés (Meyer, Goodwin, Wright, 1980), (Come, 1984), rapport Mott, Hay & Anderson, 1984) :

- une stabilité physico-chimique vis à vis de la roche encaissante et des circulations des eaux souterraines,

- une déformabilité suffisante pour suivre sans se rompre les déformations ultérieures du massif,

- une très faible perméabilité,

- une grande longévité leur permettant d'assurer ces fonctions pendant une très longue durée.

A ces propriétés peuvent s'ajouter des exigences supplémentaires si le matériau de scellement est destiné à des ouvrages situés dans le champ proche des dépôts :

- stabilité vis à vis de l'augmentation de température,

- propriété de rétention vis à vis des radionucléides.

Une liste de matériaux candidats pour la formulation d'un scellement a été établie en 1984 par Mott, Hay & Anderson Consulting Engineers. Parmi ces matériaux on retiendra les argiles et plus particulièrement les smectiques, les ciments hydrauliques, et les bitumes.

Les argiles sont des matériaux géologiques stables pour des conditions de température et de pression normales. Il existe des sites archéologiques où des ciments ont été utilisés et qui montrent une grande stabilité. Quant aux bitumes, les gisements géologiques suggèrent une longévité importante.

3. Caractérisation Du Milieu Cristallin A Traiter

Pour pouvoir reconstituer les caractéristiques initiales du massif avant perturbation par des forages de reconnaissance, il est nécessaire d'acquérir une connaissance géologique et structurale de celui-ci. Celle-ci passe par un relevé pétrographique et structural avec orientation des carottes de forage complété par des diagraphies géophysiques. Ce relevé doit permettre d'établir une répartition en familles des différentes diaclases et fractures relevées (J.L. Bles, Feuga, 1981), chacune est caractérisée par une orientation et un pendage.

Pour chaque famille on peut définir une morphologie, une extension en continuité, un écartement, une épaisseur et éventuellement la nature du remplissage et le degré d'ouverture libre. Ces paramètres ne sont pas toujours faciles à quantifier mais leur étude doit être systématisée pour les travaux de scellement de forage.

Un massif cristallin fracturé est le plus souvent le siège de circulations d'eau qui s'effectuent par le réseau de fractures et plus particulièrement par celles qui sont suffisamment ouvertes et interconnectées.

Les essais d'eau en forage sont un moyen de déterminer de manière satisfaisante certaines caractéristiques de la fracturation, c'est même le seul moyen de déterminer de manière satisfaisante un paramètre aussi important que l'ouverture hydraulique des fractures (D. Billaux, 1990).

Pour compléter la caractérisation du milieu fissuré, le BRGM a mis au point les méthodes en laboratoire permettant de quantifier l'espace des vides entre deux épontes d'une fracture (S. Gentier, D. Billaux, 1989). Il s'agit d'une technique de moulage, dont le principe est le suivant :

Les vides laissés entre les deux épontes sont remplis au moyen d'une résine silicone colorée, ce moulage est étudié par transmission de lumière, et l'image résultante numérisée par une caméra. Sur l'image en niveau de gris obtenue, les zones les plus sombres correspondent aux vides les plus épais. Les courbes de la fig. 1 montrent pour 3 fractures d'un site expérimental la distribution des vides obtenus. Ces fractures ont été caractérisées hydrauliquement, elles ont une valeur d'absorption unitaire en forage d'environ $1.4 \cdot 10^{-7}$ m^3 s^{-1} m^{-1} sous 4 MPa.

4. Concept De Traitement Des Forages

Durant ces dix dernières années, de nombreuses études en laboratoire et essais in situ de type pilote ont mis en évidence les conditions d'emploi et les insuffisances de tel ou tels matériaux et méthodes de scellement (A. Esnault, J.F. Ouvry).

Les matériaux et les méthodes de scellement de forage doivent être adaptés aux milieux traversés par le forage. Par exemple, pour les portions de forage recoupant un milieu non fracturé peu susceptible d'être le siège de circulations de fluides, l'utilisation de bentonites sèches et compactées comme bouchon peut constituer une solution.

Par contre, dans une portion de forage recoupant un milieu fracturé soumis à des circulations potentielles de fluides, ces bouchons de bentonite sont susceptibles d'être érodés (J.Y. Boisson). Il est donc nécessaire de concevoir un scellement qui traite d'une part la périphérie du forage en colmatant les fractures, et bouche d'autre part le forage proprement dit (fig. 2).

Le matériau à injecter pour traiter la périphérie du forage a pour objectif de combler l'espace des vides entre les épontes des fractures de manière à diminuer les circulations de fluides vers le forage.

Ce scellement doit combiner certains des matériaux candidats énumérés au paragraphe 2 et présenter dans

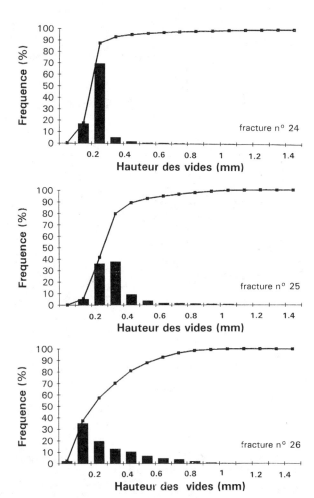

Fig. 1 : Répartition des hauteurs de vides de trois fractures.

son état final les propriétés décrites dans ce même paragraphe.

La granulométrie des produits de base constituant le mélange à injecter doit donc être adaptée pour combler un espace des vides caractérisé par une distribution d'épaisseur.

Le bouchon du forage doit posséder des propriétés permettant de pallier un défaut de colmatage des fractures ; un matériau combinant argiles et ciments avec une cohésion devrait constituer une bonne base pour retarder les phénomènes d'érosion.

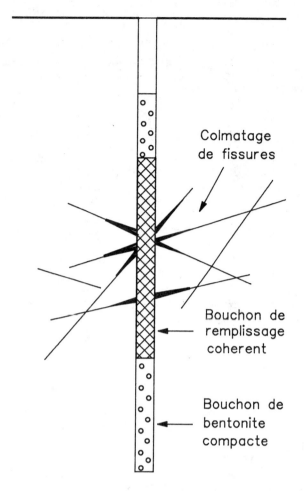

Fig.2 : Scellement de forage adaptatif.

5. Coulis d'Injection Pour Le Colmatage De Fissures En Milieu Cristallin

Le colmatage des fissures en milieu cristallin peut être réalisé par injection d'un coulis constitué par une suspension granulaire.

Les matériaux constitutifs des coulis doivent satisfaire un certain nombre de critères :

- adéquation physico-chimique avec l'encaissant,

- comportement de ces matériaux sous forme de suspensions granulaires,

- relation entre les propriétés du coulis et la géométrie du volume poreux à injecter,

- caractéristiques du coulis à l'état solide.

Les matériaux constitutifs retenus pour cette étude ont été sélectionnés suivant deux critères : la compatibilité physico-chimique avec la roche hôte d'une part, la granularité des constituants en relation avec l'épaisseur des fissures à traiter d'autre part.

Le coulis retenu pour la réalisation du plot d'essai renferme les composants suivants :
- ciment fin (CLK, Spinor A12, Ciments d'Origny, France),
- émulsion de bitume (Jean Lefebvre, France),
- bentonite calcique (FoCa, extraite à Fourges-Cahaignes, France),
- fluidifiant.

Le ciment employé est un CLK de granularité fine (fig. 3). L'émulsion de bitume incorporée au coulis a été sélectionnée parmi plusieurs émulsions mises au point spécialement par l'entreprise Jean Lefebvre. Le diamètre médian des micelles est de 2,7 µm. Elle est stable en milieu alcalin. La bentonite choisie est la bentonite FoCa, sa distribution granulométrique est

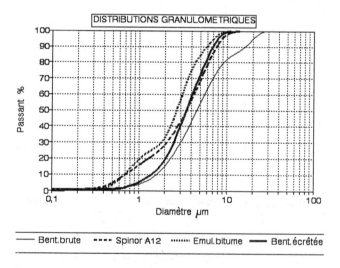

Fig.3 : Courbes granulométriques des matériaux de scellement.

grossière, comparée à celle des deux autres constituants (fig. 3). De plus, l'épaisseur des fissures la plus fréquente se situant aux environs de 0,30 à 0,35 mm (fig. 1), ce matériau a préalablement été écrêté en laboratoire. Le produit obtenu a un diamètre médian de 3,5 µm, le diamètre maximum des particules est de 10 µm (fig. 3) alors que ces deux caractéristiques avant traitement sont respectivement égales à 4,6 µm et 28 µm.

L'incorporation de bitume dans le coulis a pour but d'améliorer ses propriétés après durcissement, notamment la déformation à la rupture, la perméabilité et l'adhérence sur les épontes des fissures. Ce dernier point fera l'objet d'une étude spécifique en laboratoire. La proportion de bitume dans le coulis est relativement élevée, le rapport pondéral bitume sur ciment est égal à 1,6.

A l'état liquide, ce coulis est stable, l'essorage est faible (filtrat à 30 mn mesuré sous 7 bar : 100 ml). Les propriétés rhéologiques ont une grande importance vis à vis de la pénétration, le coulis sélectionné a un seuil d'écoulement faible (0,4 Pa) et une viscosité basse (4,8 mPa.s). Sa densité est 1,17. Après durcissement, ses propriétés déterminées en laboratoire sont satisfaisantes. Les caractéristiques mesurées à la rupture lors de l'essai triaxial consolidé drainé sur des échantillons âgés de 28 jours sous une pression de confinement de 300 kPa sont :

- déviateur à la rupture : 1179,5 kPa,

- déformation axiale à la rupture : 10,8 %.

Le coefficient de perméabilité à 28 jours est égal à $2,9.10^{-10}$ m/s.

6. Plot d'Essai Sur Site

Un plot d'essai a été réalisé dans une mine dans la région de Limoges (France). Une portion de forage traversée par quatre fractures a été isolée avec un obturateur double. La répartition des hauteurs de vide entre les épontes, exprimée en fréquence (Fig. 1) concerne trois d'entre elles représentatives des trois familles structurales du site.

Le forage a préalablement fait l'objet d'une circulation d'eau prolongée sous 6 bar ce qui a permis de nettoyer et de saturer les fissures. Les quatre fractures ont ensuite été injectées avec une pompe à piston en travaillant à pression constante. La pression d'injection est de l'ordre de 20 bar dans un premier temps puis de 40 bar dans un second temps. Le critère d'arrêt de l'injection est un débit faible sous 40 bar. Le volume total injecté est de 14,9 l, compte tenu de l'épaisseur moyenne des fissures on peut estimer que l'auréole d'injection est de l'ordre de 13 à 8 m² par fissure ce qui est très satisfaisant.

7. Conclusion

Dans le cadre de stockage de déchets en milieu géologique, il est nécessaire de reboucher les forages de manière pérenne pour éviter des circulations de fluides vecteurs de risques de propagation de pollution.

Cette étude propose une méthodologie de reconnaissance du milieu rocheux à traiter qui permet ensuite d'adapter le procédé de remplissage des vides en choisissant les matériaux, leur granularité et la technologie de mise en oeuvre.

Les observations faites lors de la réalisation du plot d'essai ont permis de vérifier l'adéquation du couple morphologie de la fracture/Coulis d'injection. Toutefois, ces observations sont insuffisantes pour quantifier l'efficacité de cette injection.

Celle-ci sera vérifiée courant 94 par des mesures de perméabilité en forage et un carottage des fissures injectées.

Cette étude conceptuelle est réalisée sous contrat à frais partagés entre la CCE, le BRGM, SIF BACHY MOTT MAC DONALD dans le cadre du programme de R&D sur la gestion et le stockage des déchets radioactifs.

Références

AEN/OCDE, 1988 - Evacuation des déchets radioactifs dans les déformations géologiques. Recherches et études in situ dans les pays de l'OCDE ; [AEN/OCDE pub., 142 p.]

BLES B., FEUGA B., 1981 - La fracturation des roches ; [Manuels et Méthodes n° 1, Ed. BRGM.]

BILLAUX D., 1990 - Hydrogéologie des milieux fracturés, géométrie, connectivité et comportement hydraulique. [Documents BRGM n° 186.] 277 p.

BOISSON J.Y., 1992 - Colmatage de fissures et de forages en formations géologiques fracturées (granites). [Rapport final Commission des Communautés Européennes, Sciences et techniques nucléaires, EUR 1409 3 FR.]

COME B., 1984 - Aspect d'ingéniérie de colmatage et scellement de dépôts de déchets radioactifs, [Rapport CCE n° EUR 9283.] 23 p.

DUFOURNET-BOURGEOIS F., OUVRY J.F., ROUVREAU L., 1993 - Scellement de forages dans les milieux cristallins, [3rd Progress report of CEC contract n° F12W - CT910072.] 56 p.

ESNAULT A., OUVRY J.F., 1992 - Scellements de forages dans les milieux cristallins, [1er Progress report of CEC contract n° F12W-CT910072.] 41 p.

GANDAIS M., DELMAS F., 1987 - High penetration C3S bentonite-cement grouts for finely fissured and porous rock, [1987, International Conference on foundations and tunnels, Proceedings Vol. 2 pp 29-33, Ingeneering technics press.]

GENTIER S., BILLAUX D., 1989 - Laboratory testing of Voids of a fracture. [Rock mechanics and Rock Engineering 22, pp. 149-157]

LOMBARDI G., 1990 - La perméabilité et l'injectabilité des massifs rocheux fissurés, [Revue Française de Géotechnique 51, pp. 5-19]

MEYER D., GOODWIN R.H., WRIGHT, 1980 - Repository sealing evaluation of material reseach requirements, [Compte-rendu de la réunion AEN sur le colmatage des forages et des puits. Columbus (USA) OCDE pub. pp. 43-63]

MOTT, HAY & ANDERSON, 1984 - The backfilling and sealing of radioactive waste repositories, [Rapport CCE n° EUR 9115 vol. 1 et vol. 2.]

Influence of Nuclear Power Stations on the Groundwater

Jozef Hulla
Slovak Technical University, Bratislava, Slovak Republic

Július Plsko
Ekosur, Jaslovské Bohunice, Slovak Republic

Mohamed Taha
Ain Shams University, Cairo, Egypt

Abstract

One of the dangerous sources of pollution is the nuclear power stations. In Slovak Republic, there are two working nuclear power plants (Jaslovské Bohunice, V-1, V-2) while the third one (Mochovce) is under construction. Leakage of radioactive substances into subsoil arose at the first power station (Jaslovské Bohunice, A-1 - off operate) and consequently contaminated the groundwater in the nearby porous environment.

A system of observation wells is being constructed. It has already enabled us to prove that underground water contains radioactive pollutants such as ^3H, (^{60}Co, ^{90}Sr, ^{241}Am, $^{239,240}Pu$ - locally). It has also enabled us to gain a conception about the extent of contamination and settle the main hydrogeological and migration properties.

Two-dimensional mathematical model taking into account adsorption and decay of radioactive substances is considered the best to evaluate and fit the experimental breakthrough curves. Prediction of radioactive materials transmission is being carried out. A real hydrogeological model based on detailed hydrogeological survey as well as a model of spreading the contamination is being created on the territory.

1. Introduction

The problem of pollutant transport has become increasingly important in recent years, adding a new dimension to groundwater investigations. This problem is much obvious near the nuclear power stations due to the possibility of groundwater pollution by radioactive materials. The interest is focused on how contaminated water moves in various media. Representative parameters are necessary to reproduce and simulate groundwater flow and solute transport. Tracer technology, in combination with solute transport models, is a valuable tool in this area of research.

Previous studies showed that it is not sufficient to solve the problem of solute transport in natural aquifers by laboratory experiments. Due to different scales, the transfer of results from laboratory to the site is difficult and often impossible [2]. Dispersion coefficients measured in the laboratory have been found to differ from those estimated from field studies by an order of magnitude [6]. Therefore, a natural experimental study area in a porous aquifer was created. Revealing of tritium contamination under the site of Nuclear Power Plant (NPP) A-1 in Jaslovské Bohunice, works to secure a complex program for examination and protection of the groundwater have started in 1990 [5].

The defined scope of this work which staged for several years is to, (1) define the actual (NPP A-1) and potential (NPP V-1, V-2) sources of radioactive contamination in the groundwater, (2) computerize hydrogeological model of regime and model of contaminant transport in groundwater, (3) have a routine and emergency monitoring scheme for the quality control of underground water and verification of leakage, (4) evaluate the risk for inhabitants by actual and potential contamination, (5) forecast the great accidents connected with releasing of radioactive materials in groundwater and suggest protective measures.

Because only characteristics of groundwater and pollutant transport were of interest, non-reactive tracer (Natrium Chloride) was selected for the experiment. The experiment was initiated with a pulse release of a known quantity of tracer solution into the aquifer. The tracer plume was monitored in three dimensions periodically during the experiment. In addition, a detailed characteristics of the test site was carried out emphasizing the spatial variability of hydraulic conductivity. Measurements of aquifer porosity, density, and grain size analysis were also performed.

2. Site Description

The study site is located at Jaslovské Bohunice, in the middle west of Slovakia, (Fig. 1).

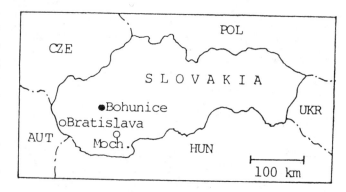

Fig. 1 Location of the test area.

The unconfined aquifer which immediatly underlies the site consists of a covering layer of loess extended to a depth of 20 m from ground surface. The main layer of gravel to gravely sand extends from 20 m to 40 m depth and is followed by an extended impermeable clay layer. Radioactive substances penetrate the unsaturated layer of loess to the layer of saturated gravel in which groundwater flows. The heterogenity of the aquifer is observed in the great variation of the average diameter (d_{50}), from 1.4 to 22 mm, with respect to both depth and position. The uniformity coefficient ($U = d_{60} / d_{10}$) also varies from 2.2 to 100. Fig. 2 illustrates the grain size distribution curves for the region used in the tracer experiment.

The observation boreholes have perforated tubes and filter packing throughout the depth of the gravel layer. Except the monitoring scheme for the groundwater quality control, nine boreholes have been initiated for the tracer experiments in an area of 13*20 m. The set-up of wells on the test site allows the observation of the transport without disturbing the groundwater flow system by the extraction of groundwater.

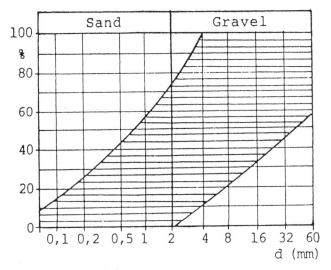

Fig. 2 Grain size distribution curves for the region used in the tracer experiments.

3. Nuclear Wastes Transmitted to the Groundwater

As a result of monitoring the real radiation, the volume activity of tritium (^{3}H) in $Bq.l^{-1}$ is given in Fig. 3. From the observation of the radiation level around the position of NPP we found that:

1- The site of NPP is an actual source of radioactive pollution to the groundwater with some local extremities.

2- The groundwater under the site surface of NPP A-1 are mainly contaminated by ^{3}H, having a volume activity $< 10^5$ $Bq.l^{-1}$. The following radionuclides were found: ^{60}Co in the range of activities $10^{-1} - 10^0$ $Bq.l^{-1}$, ^{90}Sr, $^{234, 238}$U, $^{238, 239, 241}$Pu, and ^{241}Am in the range of $10^{-3} - 10^{-2}$ $Bq.l^{-1}$.

3- Only tritium is transported by groundwater outside the site of NPP. The volume activity outside the site is $< 10^3$ $Bq.l^{-1}$. In the vicinity of the village Žlkovce near the NPP, the main contributor to the tritium activity in the groundwater has been the infiltration from Manivier's canal (leakage from waste pipe of special drainage system of vapour condensates into the Manivier's canal). The release of contaminated water from the bedlayers of NPP A-1 site into the vicinity of this village caused the contamination of groundwater in the range of activities 300 - 500 $Bq.l^{-1}$.

4- No activity of gamma nuclides in groundwater except natural ^{40}K has been found in the interested territory outside the NPP site.

5- It was derived, from the evaluation of radiobiological risks for

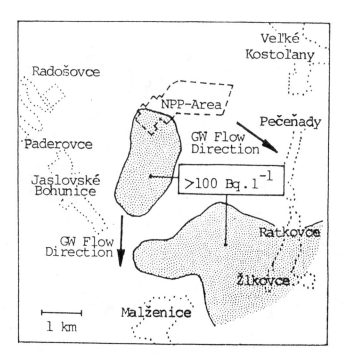

Fig. 3 The present volume activity of ^{3}H in the groundwater.

inhabitants, that the contamination of groundwater water in the place of their potential using fulfill the minimal criteria, even negligible importance of pollution. The value of tritium intake is, on basis of existing approved legislative in Slovakia, limited to 10^5 Bq.l^{-1} [5].

4. Contaminant Transport

After penetration of polluting materials into flowing water through the unsaturated zone, transport and dispersion of these materials take place. The parameters influencing dispersion may be classified into three groups: (1) parameters describing the porous media; (2) characteristics of the fluid; (3) characteristics of displacement. Of course, the properties of the polluting materials are very important.

The choice of a proper model is neither free from ambiguities nor absolutly physically clear. This is especially true in field experiments where many parameters are neither known nor controlled, and the initial and the boundary conditions are difficult to satisfy [3]. A typical problem is the determination of the amount of pollution flowing through an aquifer and the mean residence-time of this pollution in the aquifer. The simplified mathematical model used in our tracer experiments in two dimensional flow of conservative matter [1], is:

$$\frac{\partial c}{\partial t} = D_L \frac{\partial^2 c}{\partial x^2} + D_T \frac{\partial^2 c}{\partial y^2} - v \frac{\partial c}{\partial x} \quad (1)$$

The solution of this equation according to Lenda and Zuber [4], is

$$c(x,y,t) = \frac{c_o V_o}{2hn_{ef}\sqrt{\pi D_L t}} \exp\left[\frac{-(x-vt)^2}{4D_L t}\right]$$

$$* \frac{1}{2\sqrt{\pi D_T t}} \exp\left[\frac{-y^2}{4D_T t}\right] \quad (2)$$

where, D_L and D_T are the longitudinal and transverse coefficients of dispersion respectively, v- the real velocity in x direction (v = x/t_o), c_o - the initial concentration in the injection borehole, V_o - the volume of the solution, c - the concentration at time t, x and y- the coordinates of the observed boreholes, h - the depth of water in the filtered part of the borehole, n_{ef}- the effective porosity, t_o - the time for maximum concentration in the observed borehole.

For longitudinal dispersivity, the next equation is valid:

$$\alpha_L = D_L / v \quad (3)$$

and the longitudinal dispersion coefficient

$$D_L = (\sigma_t v) / 2t_o \quad (4)$$

where σ_t is the standard deviation.

The method illustrated by Lenda and Zuber [4], is used to adjust the mean transit time.

5. Migration Characteristics at the Site

The migration of polluted water in porous media is influenced by the characteristics of fluid, porous medium, and pollutants. The dependence of longitudinal dispersivity on grain size distribution is clear in the following equation, which induced from our laboratory experiments for x = 0.5 - 2.0 m:

$$\alpha_L = 0.001 + 4(d_{50} - 0.00025) + 0.0012 * 10^{0.0874U} \quad (5)$$

The optimum range of Eq. (5) is at $d_{50} = 0.00025 - 0.02$ m, and maximum value of $U = 25$.

In our field experiments, the tracer amount of Natrium Chloride (750 kg NaCl + 5.5 m^3 H$_2$O) was injected during 20 minutes to the injection well. Transport of tracer solution was observed in all observation wells. Breakthrough curves (concentration - time) were drawn and fitted at various depths. Comparison between the theoretical calculated tracer concentration and the experimental values observed for all boreholes has proved the applicability of used equations with the estimated dispersion coefficients. Real velocity distribution, both longitudinal and transverse coefficients of dispersion and consequently the coefficients of dispersivity were gained. The values of longitudinal dispersivity (α_L) in the tracer experiments ranged between 0.63 and 3.0 m. The longitudinal dispersivity was also calculated using grain size distribution curves. The dependence of α_L on the distance (x) was investigated, our results were confirmed by the previously published results. Following equation was derived:

$$\alpha_L = \exp\{-6.908 + [11.513^2 - (\ln x - 11.513)^2]^{0.5}\} \quad (6)$$

6. Prediction of Pollution

The protection of groundwater used for public supply is dependent on an adequated understanding of the fundamentals of geotechnics and groundwater hydrology. In general, the first decision will be to choose the type of structure of pollution and its modelling having in mind that the most dangerous and frequent pollution of an aquifer are miscible with the water of the aquifer, but that immiscible compounds may behave like tracers and also be treated as miscible substances.

Approximate prediction of spreading radioactive materials in groundwater can be achieved using the following equation:

$$c_{max}/c_o = \frac{L}{2\sqrt{\pi \alpha_L R}} \exp\left(-0.693 \frac{xR}{vT}\right) \quad (7)$$

where, c_{max} is the maximum concentration at distance x, R - the retardation factor, T - the half life of the radioactive tracer, L = 5.42/x - initial thickness of pollutant (in our conditions).

Figure 4 illustrates the predicted range of pollutant transport with some comparison between the measured and calculated concentration using equations (6, 7).

From these results, we can distinguish the small influence of T on ^{3}H, and that the maximum distance for ^{60}Co, ^{131}I (tracer), is at x = 1000 m. Also we can notice the great influence of dispersion on all pollutants, for example, the conservative tracer achieved the value $c_{max}/c_o = 10^{-5}$ at x = 1000 m

If it would be possible, due to self purification processes, to expect that the maximum concentration of pollutants would be lower in the sites of groundwater resources used for drinking supply than the permissible concentration (Fig. 4), then no protective works would be necessary.

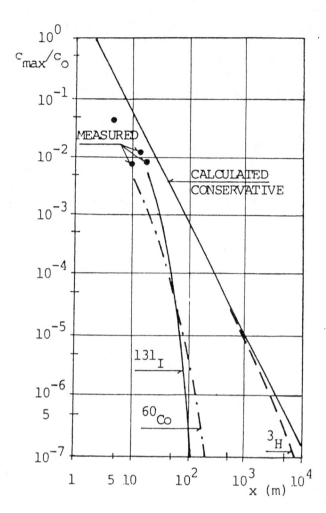

Fig. 4 Predicted range of pollutant transport.

Maps for volume activities in the territory surrounding NPP for tritium concentration after 1, 5, 10, 20, and 30 years have been prepared. These results indicate, that no special suppressive measures for the use of underground water in the villages around the NPP are needed. The radioactive contamination which has been spread by groundwater from the site of NPP A-1 has a very small or even negligible influence from the point of view of its radiobiological risk to the inhabitants of villages. Despite this, rational protective arrangements based on pumping, diluting and sealing of contaminating substances are being prepared. Nevertheless, all villages in the nuclear power station vicinity are supplied by central water-main with water from unmenaced sources.

References

[1] BACHMANT, Y.; J. BEAR,(1964). *The general equation of hydrodynamic dispersion in homogeneous, isotropic porous media*. J. Geophys. Res.69, P 2561.

[2] BOGGS,J.M.; L.W.GELHAR, et.al. (1992). *Field study of dispersion in a heterogeneous aquifer, 1, Overview and site description*. Water Resour. Res. Vol.28, No.12, PP. 3281-3291.

[3] FRIED, J.(1975). *Groundwater pollution*. Elsevier, Amsterdam.

[4] LENDA, A.; ZUBER, A. (1970). *Tracer dispersion in groundwater experiments*. Isotope Hydrology. Int.Atomic Energy Agency, Vienna, PP. 619-641.

[5] PLŠKO,J.; KOSTOLANSKÝ,M.; POLÁK,R. (1993). *Radioactivity pollution and protection of underground waters within the location of nuclear power plants Jaslovské Bohunice*. International Conference on Nuclear Waste Management and Environmental Remediation. Vol.II. American Society of Mechanical Engineers, Prague: 229-232.

[6] TOMPSON,A.F.; W.G.GRAY, (1986) *A second-order approach for the modeling of dispersive transport in porous media, 1, theoretical development*. Water Resour. Res. Vol.22, No.5, PP. 591-599.

Evaluation of Swelling Characteristics of Compacted Bentonite

Hideo Komine, Research Engineer, Nobuhide Ogata, Senior Research Fellow
Central Research Institute of Electric Power Industry, Abiko, Japan

Abstract Compacted bentonites are attracting greater attention as back-filling (buffer) materials for repositories of high-level nuclear waste. The function of compacted bentonite is to create an impermeable zone using swelling characteristics around the canister containing the high-level nuclear waste. For this purpose, a model of the swelling characteristics of compacted bentonite was constructed on basis of the results of laboratory tests the authors had performed. Furthermore, the evaluation formula about the swelling characteristics of compacted bentonites was proposed. The validity of this evaluation was shown by the laboratory tests results on the swelling deformation and swelling pressure of compacted bentonites.

Introduction

Compacted bentonites are attracting greater attention as back-filling (buffer) materials for high-level nuclear waste (SKBF/KBS, 1983). Fig. 1 shows an example of the deposition hole and tunnel (Pusch, R., 1982). This deposition hole is currently planned to be constructed to a depth of several hundred meters, and compacted bentonites are planned to be used as buffers in this hole. The function of the buffer material is to create an impermeable zone around the canister containing the high-level nuclear waste since nuclear waste must be kept separate from the surrounding environment. The buffer material must show swelling properties because cracks in the surrounding rocks may appear and need to be filled up. Compacted bentonites will therefore be used as buffer materials.

However, there are few studies about the swelling characteristics of bentonites, especially compacted bentonites, so it is necessary to clarify the swelling characteristics in detail. We investigated the swelling deformation and swelling pressure of compacted bentonites by various laboratory tests (Komine, H. and Ogata, N., 1992a).

In this study, the swelling mechanism of compacted bentonite is considered to be based on expansive clays such as montmorillonite, and hence a model of the swelling characteristics of compacted bentonites is formulated. Furthermore, the evaluation formula for the swelling characteristics of compacted bentonites is proposed by combining the swelling model and the diffuse double layer theory to estimate the swelling characteristics of clay such as montmorillonite.

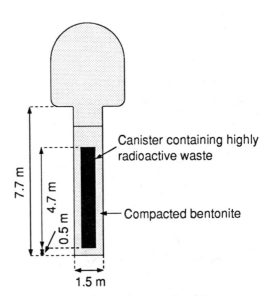

Fig.1 Cross section of tunnel with deposition hole (Pusch, 1982)

Outline of diffuse double layer theory

Soils containing the expansive clay such as montmorillonite expand by absorbing water. A montmorillonite mineral is a 2:1 layer composed of two tetrahedral silica sheets and an octahedral one of aluminum or magnesium (Grim,

R.E., 1953). A montmorillonite particle consists of several layers and interlayer water as shown in Fig. 2. It expands by absorbing water into interlayers. So, the swelling pressure and deformation of clay containing montmorillonite particles are considered to be caused by the repulsive forces occurred between two layers.

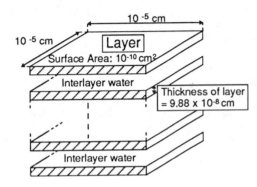

Fig.2 Schematic drawing of a montmorillonite particle

The Gouy-Chapman theory of the diffuse double layer (Chapman, D.L., 1913; Mitchell, J.K., 1976) expresses that the repulsive pressure between two parallel layers in a montmorillonite particle can be calculated from the following equation:

$$P = 2nkT (\cosh u - 1) \quad (erg/cm^3)$$
$$= 2nkT (\cosh u - 1) \times 10^{-4} \quad (kPa) \quad Eq.1$$

$$\text{for } u = 8 \tanh^{-1}\left\{\exp(-\kappa d) \cdot \tanh\left(\frac{z}{4}\right)\right\}$$

$$\kappa = \sqrt{\frac{8\pi n v^2 e'^2}{\varepsilon kT}}$$

$$z = 2 \sinh^{-1}\left(2.9 \times 10^7 \cdot \frac{B}{S} \cdot \sqrt{\frac{\pi}{2\varepsilon nkT}}\right)$$

in which P is the repulsive pressure between two parallel layers, d is the half distance between them (cm), v is the ionic valence, e' is the electronic charge (=4.8×10^{-10} esu), k is the Boltzmann constant (=1.38×10^{-16} erg/K), T is the absolute temperature (K), n is the concentration of cation or anion (number/cm^3), ε is the dielectric constant of pore water, B is the cation exchange capacity (meq/g), S is the specific surface (m^2/g) of clay.

Swelling model of compacted bentonite

Equation 1 evaluates the relationship between the repulsive pressure P and the distance 2d of two layers. Therefore, this equation is considered to be fundamental of swelling characteristics of compacted bentonite.
Bentonite contains not only swelling clay particles such as montmorillonite but also nonswelling particles. Furthermore, it is considered that the clay particles such as montmorillonite expand and fill the voids in the compacted bentonite without increasing the overall volume if the total volume of compacted bentonite is restricted (Komine, H. and Ogata, N., 1992a, 1992b; Pusch, R., 1980b). However, Eq.1 can not evaluate the content of montmorillonite and the filling of the voids. So, Eq.1 can not be used to evaluate the swelling characteristics of compacted bentonite directly, and to generalize Eq.1 is needed.

In this section, the swelling mechanism of compacted bentonite is considered on the basis of laboratory test results (Komine and Ogata, 1992a), and described as following.
(1) The volume of swelling clay such as montmorillonite increases by absorbing water into the interlayers, and the voids in the compacted bentonite is filled by this volume increase.
(2) Provided that compacted bentonite can expand at a constant vertical pressure, the volume of compacted bentonite increases until the swelling pressure of clay particles is equal to vertical pressure (see Fig. 3).
(3) If the total volume of compacted bentonite is restricted, the clay particles such as montmorillonite expand and fill the voids without increasing the overall volume. The voids were filling up completely, the volume of expansive clays can not change in the compacted bentonite, and the resultant pressure can be taken as the swelling pressure of the compacted bentonite (see Fig. 4).

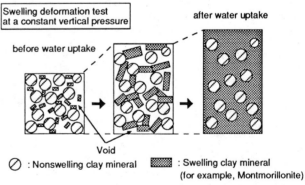

Fig.3 Mechanism on the swelling deformation of the compacted bentonite

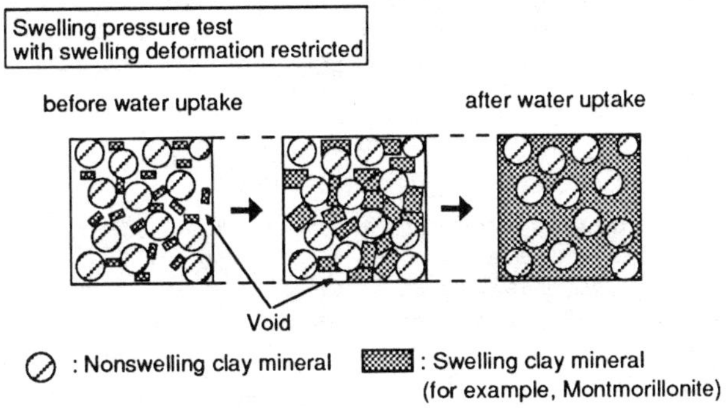

Fig.4 Mechanism on the swelling pressure of the compacted bentonite

Evaluation formula of swelling characteristics of compacted bentonite using diffuse double layer theory

It is widely known that the expansive clay controls the swelling behavior of soils. So, the expansive clays such as montmorillonite are considered to control the swelling behavior of compacted bentonite. In this study, since the swelling clay particles are mainly montmorillonite, let us consider the swelling behavior of compacted bentonite by considering the swelling behavior of montmorillonite.

Fig. 5 schematizes the composition of compacted bentonite, i.e. voids, nonswelling particles and swelling clay particles. Furthermore, the formula evaluating the relationship between the swelling volumetric strain of montmorillonite (ε_{sv}^*) and the swelling strain of compacted bentonite (ε_{smax}) is suggested on the basis of the swelling mechanism described above and the compacted bentonite composition shown in Fig. 5 (Komine, H. and Ogata, N., 1992b).

$$\varepsilon_{sv}^* = \frac{\left(1 + \dfrac{\varepsilon_{smax}}{100}\right) \cdot \dfrac{\rho_s}{\rho_{d0}} - 1}{(C_m/100)} \times 100 \quad (\%) \qquad \text{Eq.2}$$

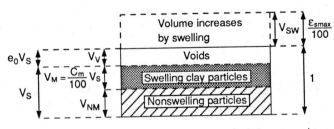

Fig.5 Composition of compacted bentonite

in which C_m (%) is the content of montmorillonite in bentonite by percentage weight, ε_{smax} (%) is the maximum swelling strain of compacted bentonite, ρ_s is density of bentonite, ρ_{d0} is initial dry density of compacted bentonite, and ρ_w is the water density.

ε_{sv}^* can be also evaluated from swelling deformation of montmorillonite particles shown in Fig. 2. The model to simulate the deformation of a montmorillonite particle and the formula evaluating the relationship between ε_{sv}^* and the distance 2d between two parallel layers are proposed. The structure of montmorillonite particle is shown in Fig. 2 as described above. Therefore, the assumption that there is no interlayer water between each layers as shown in Fig. 6(a) is standardized. ε_{sv}^* can be evaluated by Eq.3 on the basis of the thickness increment of a montmorillonite particle (see Fig. 6(b)).

$$\varepsilon_{sv}^* = \frac{(N-1) \times 2d}{N \times t} \times 100 \quad (\%) \qquad \text{Eq.3}$$

in which d is the half distance between two layers (cm), t is the thickness of layer (cm), and N is the number of layers in a montmorillonite particle. Eq.3 can be rewritten as,

$$d = \frac{\varepsilon_{sv}^*}{100} \cdot \frac{t}{2} \cdot \frac{N}{N-1} \qquad \text{Eq.4}$$

The number N of layers in a montmorillonite particle can be obtained theoretically by the specific surface and the montmorillonite content in bentonite. The nonswelling particle in bentonite is much bigger than the layers of montmorillonite. So, the specific surface of the montmorillonite layers is much larger than the nonswelling particle, and the specific surface of bentonite is considered to be almost equal to the specific surface of montmorillonite layers (Nakano, M. et al., 1984). The nitrogen molecules can not penetrate into the interlayer of montmorillonite particle. So, the specific surface of bentonite measured by the BET method (Brunauer, S. et al., 1938) is considered to be the specific surface of the montmorillonite particles. Therefore, equations 5 and 6 should hold by the above mentioned assumptions.

$$N = \frac{S_1}{S_N} \qquad \text{Eq.5}$$

$$S_{BET} = \frac{1 - C_m}{100} \cdot S_{nm} + \frac{C_m}{100} \cdot S_N \qquad \text{Eq.6}$$

S_1 is the specific surface of montmorillonite

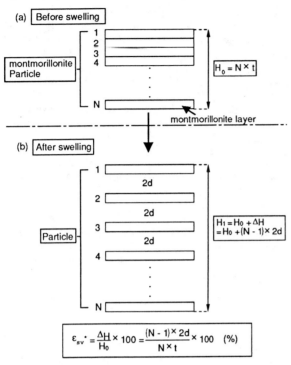

Fig.6 Concept of swelling volumetric strain of a montmorillonite particle

- H_0: Thickness of a montmorillonite particle on condition that there is entirely no interlayer water
- H_1: Thickness of a montmorillonite particle on condition that each distance of interlayers is 2d
- ΔH: Increment of thickness of a montmorillonite particle on condition that each distance of interlayers is 2d
- N: Number of mineral layers in a particle
- t: Thickness of a mineral layer

layers, S_N is the specific surface of a montmorillonite particle, S_{BET} is the specific surface of bentonite measured by BET method, S_{nm} is the specific surface of nonswelling particles. From the above considerations, $S_{nm} \ll S_N$ is considered to be true, and according to Eqs.5 and 6, we obtain:

$$N = \frac{C_m}{100} \cdot \frac{S_1}{S_{BET}} \qquad Eq.7$$

The specific surface S_{BET} of the bentonite used in this study is 46 - 51 m^2/g by BET method. The montmorillonite content calculated from the methylene blue test is approximately 48% (Komine, H. and Ogata, N., 1992b). The specific surface S_1 of montmorillonite layer, which is determined from the crystal structure, is 810 m^2/g (Dyal, R.S. and Hendricks, S.B., 1950). So, the value of N is 7.6 - 8.4 from the result calculated by Eq.7. Nakano also proposed N=8 in case of the bentonite which the montmorillonite content is 65 % (Nakano, M. et al., 1984). Therefore, the number N of montmorillonite layers in a particle is taken as the following in this study.

$$N = 8 \qquad Eq.8$$

The concentration n (number/cm^3) of cation or anion is considered to be dependent on the number of montmorillonite particles per unit volume of pore water. The concentration n is considered to be unchangeable provided that the number of montmorillonite particles per unit volume of pore water increases. So, the concentration n is considered to decrease as the volume of montmorillonite particle increases by water absorption. Therefore, the concentration n must be corrected by Eq.9 considering the volume increase of montmorillonite particles.

$$n = \frac{n_0 \ (mol/l) \times N_A \times 10^{-3}}{1 + \dfrac{\varepsilon_{sv}^*}{100}} \ (number/cm^3) \qquad Eq.9$$

in which N_A is the Avogadro's number (6.023×10^{23}).

The swelling characteristics of compacted bentonites can be evaluated by combining Eqs.1, 2, 4, 8, and 9.

Formula validations

In this section, the validity of the evaluation formula described above is shown from the laboratory tests on the swelling deformation and swelling pressure of compacted bentonites. Please refer to the papers (Komine, H. and Ogata, N., 1992a) written by the authors for details of the laboratory tests. The parameters of the evaluation formula are shown in Table 1. The ionic valence υ, the dielectric constant ε of pore water, and the cation exchange capacity B shown in Table 1 are quoted from Mitchell (1976). Bolt showed that the concentration n_0 of cation or anion in pore water was about 10^{-2} mol/l from the experimental results (Bolt, G.H. and Bruggenwert, M.G.M., 1978). Furthermore, the concentration of water around the compacted bentonite in the swelling pressure test is approximately 0.02 - 0.04 mol/l from the measurement by ICP-atomic emission spectrometer. Therefore, n_0 is considered to be almost equal to the concentration of water around the compacted bentonite, and the value of n_0 is determined as shown in Table 1. The other

Table 1 The parameters in the evaluation formula on swelling characteristics of compacted bentonite

Density of bentonite ρ_s (g/cm^3)	2.79
Montmorillonite content of bentonite C_m (%)	48
Cation exchange capacity (meq/g)	0.76
Ionic valence	1.5
Number of montmorillonite layers in a particle N	8
Absolute temperature T (K)	293
Dielectric constant of pore water ε	80
Concentration of cation or anion in pore water n_0 (mol/l)	0.02 or 0.03 or 0.04

parameters are quoted from Komine and Ogata (1992a).

Fig. 7 shows the relationship between the maximum swelling strain ε_{smax} of compacted bentonite and vertical pressure. In this figure, the maximum swelling pressure measured with the swelling deformation restricted corresponds to the vertical pressure at $\varepsilon_{smax}=0$. The relationship between the maximum swelling strain and vertical pressure calculated from the evaluation formula proposed is illustrated by the curves in Fig. 7.

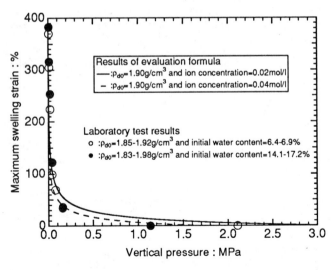

Fig.7 Relationship between the maximum swelling strain and vertical pressure

Fig. 8 shows the relationship between the maximum swelling pressure P_{smax} of compacted bentonite and dry density on condition that the swelling deformation of compacted bentonite is restricted. In this figure, the same relationship calculated from the evaluation formula is represented by curves.

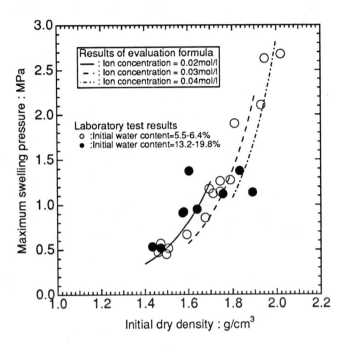

Fig.8 Relationship between the maximum swelling pressure and dry density

The curves calculated from the evaluation formula proposed in this study are in almost agreement with the experimental results as shown in Figs. 7 and 8. The curves evaluated with relatively large value of n_0 is closely agreement with the experimental results of relatively high-dry density. It is generally considered that the concentration n_0 of cation or anion increases as the dry density of compacted bentonite increases, and the result mentioned above is thought to be caused by increasing concentration.

Conclusion

The following conclusions were drawn from this study:

(1) The swelling mechanism of compacted bentonite was considered to be based on the behavior of expansive clay such as montmorillonite. Furthermore, a model of

swelling characteristics of compacted bentonite was constructed, and equations estimating this model were proposed.

(2) The evaluation formula on the swelling characteristics of compacted bentonites was proposed by combining the equations of swelling model and the diffuse double layer theory to estimate the swelling characteristics of the clay particles such as montmorillonite.

(3) The validity of this evaluation formula was shown by the laboratory tests on the swelling deformation and swelling pressure of compacted bentonites.

Acknowledgments

We wish to thank Dr. Kokusho, Dr. Nishi and Dr. Okamoto of CRIEPI, Dr. Olsen, of U.S.G.S. and Mr. Suzuki of KUNIMINE INDUSTRIES CO.,LTD for their guidance. We would also like to thank Mr. Sugawara, and Mr. Tashiro for their help with the experiments.

REFERENCES

Bolt, G.H. and Bruggenwert, M.G.M. (1978). Soil chemistry, a basic elements, Elsevier.

Brunauer, S., Emett, P.H. and Teller, E. (1938). Adsorption of gases in multimolecular layers, J. Amer. Chem. Soc., Vol.60, pp.309-319

Chapman, D.L. (1913). A contribution to the theory of electrocapillarity, Philosophical Magazine, Vol.25, No.6, pp.475-481

Dyal, R.S. and Hendricks, S.B. (1950). Total surface of clays in polar liquids as a characteristic index, Soil Science, Vol.69, pp.421-432

Grim, R.E. (1953). Clay Mineralogy, McGraw-Hill Book Co., New York.

Komine, H. and Ogata, N. (1992a). Swelling characteristics of compacted bentonite, Proc. 7th inter. conf. on expansive soils, pp.216-221

Komine, H. Ogata, N. and Nishi, K. (1992b). Mechanical properties of buffer materials for repositories of high-level nuclear waste (Part 1) - Experimental study on swelling characteristics of compacted bentonite -, CRIEPI report U92039 (in Japanese with English abstract).

Mitchell, J.K. (1976). Fundamentals of soil behavior, John Wiley & Sons, Inc., pp.112-134

Nakano, M., Amemiya, Y., Fijii, K., Ishida, T., and Ishii, A. (1984). Infiltration and expansive pressure in the confined unsaturated clay, Trans. of JSIDRE, Vol.112, pp.55-66. (in Japanese with English abstract).

Pusch, R. (1980b). Swelling pressure of highly compacted bentonite, SKBF/KBS technical report 80-13

Pusch, R. (1982). Mineral-water interactions and their influence on the physical behavior of highly compacted Na bentonite, Jour. Can. Geotech., Vol.19, No.3, pp.381-387

SKBF/KBS (1983). Final storage of spent nuclear fuel -KBS- 3 Barriers

Ground Testings of Radon as a Source of Ecological Danger

**B.I. Kulachkin, V.A. Ilyichev, V.S. Yamschikov, A.I. Radkevich, V.I. Sheinin,
L.R. Stavnitser, I.P. Korenkov, O.G. Polsky, V.F. Kirillov, V.M. Pankov**
NIIOSP, Moscow, Russia

A.P. Van den Berg
Heerenveen, The Netherlands

According to the estimates of UN Scientific Committee for Atomic Radiation Activity, radon, together with daughter products of radioactive decay, is responsible for approximately 3/4 of the annual individual effective equivalent radiation dose received by population from ground radiation sources.

Radon forms in geological structures, accumulates in them and then emanates from the ground everywhere. Its concentration in the air substantively changes in the different parts of the Earth, that is considerably defined by the conditions of its formation and emanation. The ground (soil) as an object has not yet found a proper attention from the scientists engaged in the radon problem. Study of the processes of radon ground accumulation, distribution and emanation will make it possible to receive information for assessment of ecological situation, making up of regional maps and forecasting of radon danger.

It should be noted that at present there are no reliable and mobile means for radon testings in situ. Technology, methods and apparatus for assessment of ground concentrations of radon, as well as for control of radon concentrations at the radioactive wastes disposals, have been developed by the authors of the present paper. Testings can be practically conducted without violation of the ground massif integrity and without possible spontaneous release of radon.

Testing technology for ground (soil) radon concentrations is based on the complex of drilling and sounding on the basis of drilling. This technology makes it possible to widen substantially the field of application of the method and to conduct testings in practice at the depth of about 60 m in any engineering and geological conditions.

The substance of the suggested technology is the use of drilling units to realise the method of static sounding on their basis without changing the standard technology of drilling as well as sounding. Constraints on the method of static sounding at the presence of pebbly deposits, coarse-grained soils and etc. can be lifted by the application of combined sounding (with the use of drilling). Drilling units UGB-50M, UGB-IBC and UGB-2A-2 are the specific examples of this technology. This method makes it possible to conduct testings at any pre-defined depth in the case of sonde stop.

Apparatus for radon testings have been developed. One of the apparatus is PIKA-IO (Russian abbr. for field instrumentation unit) for measurings of static sounding parameters and ground gamma-logging. PIKA-IO is the small, lightweight instrumentation apparatus to be used at any static sounding and drilling unit.

Three parameters are to be measured with the help of this apparatus: ground resistance to sonde cone; ground resistance on the friction coupling; and gamma-level of background. It includes sonde with detectors of the indicated parameters, uncontact detector of the depth of sonde immersion and measuring amplifier, connected with sonde cable. Information received is displayed on the digital board of measuring amplifier.

Another apparatus is called "Radon". Technology and method for ground testing of radon by this unit is similar to PIKA-IO. "Radon" apparatus includes sonde, communication line and recorder. Sonde is equipped with detectors for radon identification and testing of its activity; communication line is a multi-core cable; recorder forms useful signals to display the tested parameters in physical values on digital board. Apparatus operates and controls unattended (according to the programme) as well as by operator's instructions.

As it is known, radon and thoron are chemically inert precious gases. They can be found almost in every rock, most of them (90% and more) are incorporated in rocks, containing their precursors 226_{Ra} and 224_{Ra}. The methods are known for determination of uranium 238_U and 232_{Th} contents according to the gamma-background of the rocks, as most of gamma-radiation of radioactive rows 238_U and 232_{Th} is attributed by daughter products of radon and thoron radioactive decay.

According to this methods the assessment technique for gamma-level of rock background with the use of PIKA-IO unit has been developed, and engineering and geological studies have been conducted in different regions of Russia. In particular, the values of gamma-background from 7 to 2I μR/h have been registered in the studies of upper and middle-quarter deposits of Volgograd in Volga region. Radon contents in the rock could be presented qualitatively along the depth of the tested massif as well as in the plane.

To assess qualitatively and quantitatively all kinds of grounds and and relative contents of radon in them the criterion $\lambda = I/I_o$ has been developed, where:

t- similar criterion of Begeman,
I_o - initial (zero) gamma background,
I- natural gamma background.

A number of local regional maps with quantitive contents of radon in the rocks were drawn up. The new apparatus "Radon" being passed through laboratory and field testings makes it possible to identify and assess radon concentration in the rocks.

Effets de la Thermo-Plasticite des Argiles sur le Comportement des Puits de Stockage

Lyesse Laloui
Laboratoire de Mécanique, Ecole Centrale Paris, France

Hormoz Modaressi
B.R.G.M., France

RESUME

L'analyse du comportement des puits de stockage des déchets radioactifs dans des formations argileuses présente un grand intérêt pour les concepteurs dans l'évaluation de la sûreté de ces ouvrages. Le comportement d'un sol adjacent à un puits, contenant des déchets radioactifs de hautes activités, en conditions non-homogènes et quasi-statiques, est étudié à travers des simulations numériques. Différentes configurations et conditions de drainage sont considérées pour ce puits soumis à des sollicitations thermiques engendrées par ces déchets. L'influence de la variation de la température sur le gradient hydraulique et l'écoulement transitoire de l'eau interstitielle est prise en compte dans ces analyses. Le comportement thermo-mécanique du milieu argileux se caractérise par de fortes non-linéarités. Il est modélisé avec une loi de comportement multimécanisme thermo-plastique développée par les auteurs. Cette loi de comportement tient compte de l'écrouissage thermique et de l'évolution de la surface de charge avec la température. Dans certaines conditions de chargements thermiques, nous pouvons observer une diminution du volume de sol avec l'augmentation de la température. En outre, le modèle prend en compte la variation des modules élastiques sous l'effet thermique, phénomène important dans les conditions presque non drainées (sollicitations rapides, matériaux à faible perméabilité, etc...). Parmi les principaux résultats de ces simulations, l'apport de l'utilisation de cette loi de comportement thermo-plastique par rapport aux approches plus classiques est étudié. En particulier, nos résultats montrent que, dans certains cas, une approche thermo-plastique pour les matériaux argileux prédit une réponse très différente de celle issue des méthodes plus couramment employées.

Introduction

La stabilité mécanique des cavités souterraines durant leur exploitation constitue un élément essentiel de l'étude de la faisabilité technique du stockage de déchets. Elle soulève des difficultés exceptionnelles en géotechnique. En effet, les concepteurs sont appelés à s'assurer que la sécurité de l'ouvrage est garantie avec un niveau exceptionnel pendant une très longue période. Ce type d'ouvrage fait intervenir des phénomènes complexes de nature mécanique, hydraulique et thermique. La prédiction numérique du comportement de ces ouvrages nécessite d'une part, des formulations mathématiques qui tiennent compte des différents couplages entre les phases solide et fluide, en plus de l'effet de la variation de la température, et d'autre part, des lois de comportement appropriées pour les différentes phases.

Formulation Thermo-Hydro-Mécanique

Nous utilisons une formulation mathématique qui permet de suivre l'évolution d'un milieu poreux à travers lequel s'écoule un fluide qui sature l'espace interstitiel [1]. Ce milieu peut être soumis à des chargements Thermique, Hydraulique et Mécanique (THM). L'originalité de ce modèle réside dans la description des deux phases solide et fluide du milieu par une seule fonction d'état définie par les déformations et la température comme paramètres d'état [2].

Dans le cas des argiles, dont la perméabilité est très faible, nous pouvons supposer que le couplage hydro-thermique (effet de la convection) est négligeable. Les déformations du squelette solide sont reliées aux contraintes effectives par la loi rhéologique du matériau. Cette dernière sera présentée dans le paragraphe suivant. Le fluide interstitiel est supposé s'écouler suivant la loi de Darcy généralisée sous l'effet du gradient de la pression interstitielle. Ce gradient peut être généré par la déformation du solide ou par la variation de la température. Nous faisons l'hypothèse qu'aucun transfert de masse ne se fait aux interfaces entre le solide et le fluide, et que la température dans la phase solide est égale à celle de la phase fluide.

En admettant que le principe des contraintes effectives de Terzaghi est applicable aux argiles, l'équilibre global du milieu poreux saturé est donnée par la relation suivante:

$$\mathbb{DIV}\, \sigma' - \operatorname{grad} p + (n\, \rho_f + (1-n)\, \rho_s)\, g = 0$$

avec:

- σ' tenseur de contraintes effectives,
- p pression interstitielle,
- n porosité,
- ρ_s masse volumique du solide,
- ρ_f masse volumique de l'eau,
- g accélération de la pesanteur.

L'équation de la conservation de la masse s'écrit:

$$\operatorname{div} \partial_t \mathbf{u} + \operatorname{div} \mathbb{V} + \left[n\beta_f + (1-n)\beta_s \right] \partial_t p = \left[n\beta'_f + (1-n)\beta'_s \right] \partial_t T$$

où $\mathbb{V}$ est la vitesse relative du fluide par rapport au squelette solide et $\mathbf{u}$ le déplacement du squelette. Cette vitesse est reliée à la pression interstitielle par la loi de Darcy généralisée:

$$\mathbb{V} = -\mathbb{K} \cdot \operatorname{grad}(p + \rho_f\, g.x)$$

avec x le vecteur des coordonnées du point matériel et $\mathbb{K}$ le tenseur de perméabilité du matériau poreux. Ce dernier est fonction de la géométrie des pores et de la viscosité du fluide. Les termes β'_f et β'_s sont les coefficients de dilatation de chacune des phases, fluide et solide. β_f et β_s représentent les compressibilités de chacune des deux phases.

La description du comportement thermique est faite à l'aide de l'équation de conservation de l'énergie. Cette dernière en exprimant un bilan entre les flux entrant dans le volume de contrôle et l'augmentation de l'énergie interne, permet la formulation de la relation suivante:

$$[n\, \rho_f\, c_f + (1-n)\, \rho_s c_s]\, \partial_t T - \operatorname{div}[(n\, \lambda_f + (1-n)\, \lambda_s)\, \operatorname{grad} T] = 0$$

où $\rho_f\, c_f$ est la chaleur spécifique de la phase fluide et $\rho_s c_s$ celle de la phase solide. Dans cette relation, la conduction thermique est supposée être donnée par la loi de Fourier. Le coefficient de proportionnalité entre ce flux et le gradient de la température est la conductivité thermique du milieu poreux. Cette dernière s'exprime à travers les conductivités du fluide λ_f et du solide λ_s.

Cette formulation du comportement thermique correspond à l'hypothèse que la convection thermique peut être négligée dans le cas des argiles où, la faible perméabilité impose une très faible vitesse de filtration. Cette hypothèse aboutit au découplage de l'équation énergétique des deux autres équations de conservation. Ainsi, une fois déterminée par cette dernière équation, la température ne serait qu'un élément de chargement extérieur pour les autres équations de conservation.

Cette description du milieu poreux montre que l'ensemble des équations ci-dessus sont couplées à travers des variables dépendantes que nous ramenons au vecteur déplacement du solide, à la pression du fluide et à la température du milieu.

Loi de Comportement Thermo-Plastique

Les effets du couplage thermo-mécanique dans les argiles sont pris en compte avec un modèle de comportement cyclique multimécanisme thermo-mécanique LTVP [3]. Dans ce modèle, selon les concepts de l'élastoplasticité, il est supposé que l'incrément de la déformation engendré dans le milieu par une sollicitation thermo-mécanique se décompose en parties réversible et irréversible. En notant par $\dot{\varepsilon}^{Te}$ le taux de déformation thermoélastique, nous avons:

$$\dot{\varepsilon}^{Te} = \dot{\varepsilon}^e(p',T) + \frac{\beta'_s(T,p,p_{c0})}{3}\, \dot{T}\, \mathbb{I}$$

où $\dot{T}$ est l'incrément de température, p' la pression moyenne effective. p_{c0} indique l'état de surconsolidation de l'argile. La dépendance du taux de déformation élastique non-linéaire mécanique ($\dot{\varepsilon}^e$) de la pression moyenne effective entraîne sa variation avec la température du matériau. La prise en compte dans la loi de comportement thermo-plastique du lien existant entre le comportement mécanique, la température et les déformations irréversibles d'origine thermique est effectuée quantitativement par une variation continue de certains paramètres mécaniques (relatifs aux comportements isotrope et déviatoire), et qualitativement par une transition vers un comportement plus ductile (dans un cas de chargement). La composante anélastique de la loi de comportement s'écrit ainsi:

$$\dot{\varepsilon}^{Tp} = \frac{1}{H(\sigma,T,\alpha)} (\partial_T f . \dot{T} + \partial_\sigma f : \dot{\sigma}) . \Psi(\sigma,T,\alpha)$$

où H est le module thermoplastique déterminé à partir de l'équation de consistance, f représente la surface de charge et α l'ensemble des variables internes.

La direction de cet incrément de déformation thermoplastique est donné par Ψ qui dépend, entre autre, de la température.

A partir de la connaissance de l'incrément de déformation thermo-mécanique (bilan des incréments de déformations thermoélastique et thermoplastique), il est possible de déduire l'incrément de contrainte généré dans ces conditions, en utilisant la relation qui régit la génération des contraintes à partir des déformations thermoélasto-thermoplastiques:

$$\dot{\sigma}' = \mathbb{D}:(\dot{\varepsilon} - \dot{T}a) - \frac{(\partial_T f - \Phi:\mathbb{D}:a)\dot{T} + \Phi:\mathbb{D}:\dot{\varepsilon}}{\Phi:\mathbb{D}:\Psi - \partial_\alpha f L} \mathbb{D}:\Psi$$

avec $\mathbb{D}$ la matrice d'élasticité et a la matrice des coefficients de dilatation thermique. Φ représente la variation de la surface de charge par rapport à l'état de contraintes et L une fonction constitutive qui dépendra de la température entre autre.

Simulation du Comportement d'un Site de Stockage de Déchets

L'approche numérique est mise en œuvre à partir de la formulation variationnelle des différentes équations présentées précédemment, et leurs discrétisations en espace (par la méthode des éléments finis) et en temps (par un schéma en θ-méthode) [4]. Le modèle couplé THM ainsi que la loi de comportement thermomécanique LTVP ont été validés à travers un certain nombre de tests[2].

Données générales

Les puits sont supposés être forés depuis des galeries à intervalles réguliers et cela uniformément sur l'axe de la galerie. Dans le cas présenté ici, nous considérerons une équidistance entre les axes des puits de vingt mètres. Une présentation schématique de la partie modélisée est donnée sur la figure 1 (50 cm de diamètre du puits et 20 m d'entraxes de sol adjacent). Le puits de stockage peut être assimilé à un cylindre de 55m de hauteur auquel s'ajoute une bouchon en argile en tête. Un puits contient quarante conteneurs. Ces derniers sont caractérisés par la même loi de dégagement thermique puisque supposés être déposés au même instant. La périodicité du réseau de puits et la supposition qu'ils se situent tout autour du puits considéré permet de faire de modéliser cet ouvrage en axisymétrie. Ainsi, les flux de chaleur seront pris nuls sur les limites du maillage et les températures seront imposées sur la paroi intérieure. Pour tenir compte des conditions d'utilisation et d'environnement, nous avons simulé un scénario dans lequel après la mise en place des déchets, nous déterminons, dans un premier temps, la répartition du champ thermique dans le milieu avoisinant par une analyse de diffusion transitoire. Cette analyse porte sur une période de cent ans durant laquelle le forage est soumis à une sollicitation thermique. Cette dernière, qui a l'allure du dégagement thermique de conteneurs des déchets radioactifs (Fig. 2), est d'abord croissante. Après une période avoisinant les 35 ans, le milieu commence à refroidir. Nous étudions par la suite, l'effet de cette sollicitation thermique sur le comportement hydro-mécanique en phase d'exploitation.

Le massif est maillé avec 153 éléments quadratiques pour les déplacements et bilinéaires pour les pressions. La section modélisée est considérée à mi-hauteur du puits.

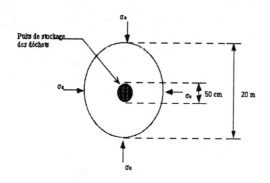

Figure 1 : Représentation schématique de la partie modélisée

Caractéristiques du matériau

Nous considérons comme matériau de référence l'argile de Mol, à l'exception de la valeur de la perméabilité. Cette argile, potentiellement favorable pour le stockage, présente un ensemble de propriétés intéressantes pour le stockage, que nous supposons homogènes dans tout le massif considéré. Les caractéristiques utilisées pour le milieu saturé sont les suivantes:

Conductivité thermique	1.69	W/m°C
Chaleur spécifique	$2.1\ 10^6$	J/°C kg
Dilatation thermique du fluide	$3.8\ 10^{-4}$	°C^{-1}
Porosité	0.39	
Compressibilité du fluide	$3\ 10^{-8}$	Pa^{-1}
Perméabilité	10^{-9}	m/s

Les caractéristiques thermomécaniques ont été déterminées par calage de courbes expérimentales [4].

Différents cas étudiés

Après la détermination de la répartition de la température dans le massif sous l'effet du dégagement thermique supposé des conteneurs de déchets, le comportement du massif peut être analysé. Dans cette phase, nous considérons différentes hypothèses. Elles portent sur l'état initial du massif (degré de surconsolidation) et la prise en compte de l'eau (approches mono ou biphasiques). Dans chacun des cas, le sol est modélisé par les trois modèles de comportement suivants:

1- thermo-élasticité linéaire avec un coefficient de dilatation constant. Cela revient à considérer que, dans la dernière équation donnée plus haut et définissant l'incrément de contrainte, le deuxième terme du second membre est nul, et que la matrice a est isotrope constante.

2- thermo-élastoplasticité (prise en compte d'une déformation thermique réversible dans le bilan des déformations élastoplastiques) avec un coefficient de dilatation constant. Ce cas est simulé avec la loi élastoplastique multimécanisme de Hujeux [5]. Dans ce modèle, la partie élastique est non linéaire avec une dépendance des modules de la pression moyenne effective. L'une des conséquences de l'ajout de la composante de dilatation thermique réversible est la modification de l'état de contraintes et la génération de déformations plastiques: c'est un modèle thermo-mécanique avec surface de charge adiabatique[3]. Dans la relation définissant l'incrément de contrainte, ce cas revient à éliminer les termes où figure la dérivation de l'expression de la surface de charge pa rapport à la température.

3- thermoélasticité/thermoplasticité (loi LTVP) avec un coefficient de dilatation du solide dépendant de la pression moyenne et de la température[2]. Dans ce troisième cas, nous considérons l'ensemble des termes de la relation définissant l'incrément de contrainte donné plus haut.

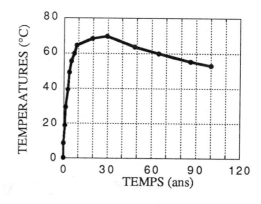

Figure 2 : Dégagement thermique supposé des conteneurs

Quelques résultats de simulations numériques

L'analyse de l'effet de l'état de consolidation a été effectuée en considérant le matériau soit normalement consolidé soit surconsolidé. Ainsi, pour chaque loi de comportement deux contraintes in situ différentes ont été considérées, (1MPa : état surconsolidé, 3MPa: état normalement consolidé) et deux valeurs de perméabilité (l'une nous rapprochant de l'état non-drainé et l'autre, fictive, d'un drainage complet proche d'un comportement monophasique). Ce dernier nous permet de distinguer les effets directement liés au couplage thermo-mécanique, des effets de la dilatation du fluide et ses conséquences sur l'état de contrainte. Par manque d'espace, les résultats ne sont présentés que pour un nœud du maillage situé à la frontière du puits, et principalement en terme de convergence de la paroi

La figure 3 présente les convergences radiales autour de la cavité pour le cas où le milieu est monophasique surconsolidé. Nous remarquons qu'en phase d'échauffement, ces convergences sont du même ordre de grandeur pour les trois types de comportement supposés. Parmi les phénomènes à retenir, nous notons ceux qui résultent du caractère irréversible du comportement. Cela est particulièrement mis en évidence en phase de refroidissement, et cela jusqu'à une date de cent ans, où le modèle thermo-plastique présente une pente différente de celles des deux autres qui indiquent une possible irréversibilité à long terme. Cependant, nous ne pouvons extrapoler cette ligne de décharge du moment que le coefficient de dilatation est non linéaire.

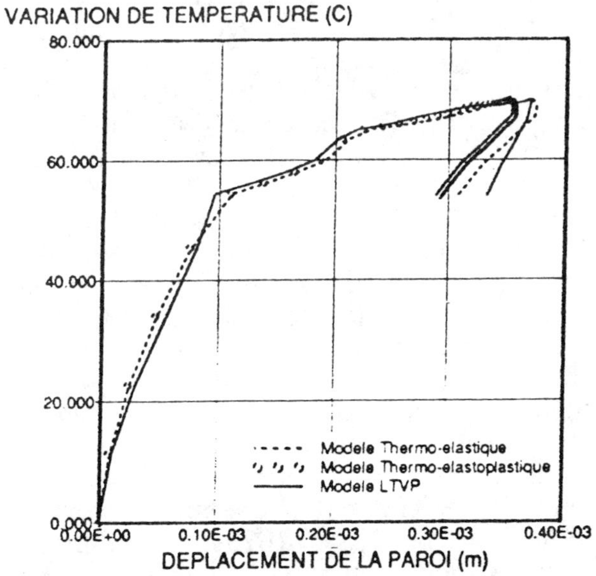

Figure 3 : Convergence des parois du puits en condition surconsolidé monophasique

Nous devons souligner que les changements de pente brusques apparaissant dans l'intervalle des températures de 55°C à 65°C sont dus aux interpolations linéaires du champ thermique et ne correspondent pas à un phénomène physique.

Les résultats obtenus suggèrent que pour cet état du matériau, la seule utilisation du modèle thermo-élastique aurait permis de déterminer la convergence en phase de chargement. Les pentes de décharge laissent entrevoir une légère différence qui peut être attribuée à la non-linéarité du coefficient de dilatation.

Lorsque l'état initial du matériau est normalement consolidé, la réponse du modèle thermo-plastique en terme de déformation du matériau est complètement différente des deux autres approches (Fig. 4). Entre ces deux dernières nous notons que le modèle thermo-élastoplastique (modèle 2) prédit une dilatation moins forte que celle du modèle thermo-élastique. La réponse du modèle thermo-plastique prévoit quant à elle un comportement essentiellement contractant, avec une irréversibilité presque totale.

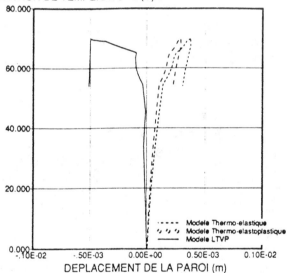

Figure 4 : Déplacement radial autour de la cavité en condition normalement consolidé monophasique

Le comportement du massif en présence d'eau est étudié avec les mêmes hypothèses que pour les cas monophasiques. La figure 5 présente le cas biphasique avec un matériau surconsolidé. Les convergences de la paroi du puits indiquent un plus grand déplacement radial pour le modèle LTVP. En effet, à partir d'une température de l'ordre de 60°C, une accélération de la convergence de la paroi est obtenue pour ce modèle. Nous remarquons que, quel que soit le modèle de comportement, la prise en compte de la phase fluide favorise un déplacement radial plus important.

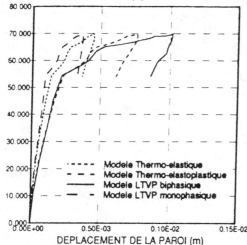

Figure 5 : Convergence du puits en terme de déplacement radial autour de la cavité pour un milieu biphasique surconsolidé

En condition non drainée, la dilatation de l'eau interstitielle fait que le modèle LTVP réagit comme le modèle thermo-élastoplastique avec une amplitude de déplacement comparable (Fig.6).

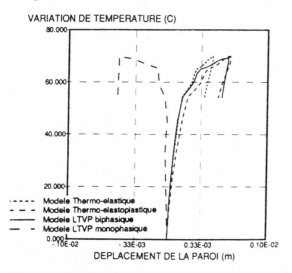

Figure 6 : Convergence du puits biphasique normalement consolidé

Conclusion

Les simulations numériques du comportement d'un puits de stockage de déchets générateurs de chaleur sont réalisées suivant deux configurations extrêmes (condition monophasique et condition biphasique non drainée). Deux états de contraintes initiales ont été pris pour le massif pour lequel trois types de comportement sont supposés. La première configuration nous permet d'étudier le comportement de l'ouvrage dans des conditions thermo-mécaniques non-homogènes. Avec l'hypothèse du comportement biphasique, nous évaluons le rôle de l'eau interstitielle dans ce type de prédictions.

De ces simulations paramétriques, nous pouvons en particulier observer des cas où les réponses thermo-plastiques et thermo-élastoplastique sont nettement différentes. Ces différences sont appréciablement notées dans les conditions normalement consolidées en monophasique. Des comparaisons avec des mesures in-situ devront nous permettre de mieux évaluer ces effets thermiques transitoires sur le comportement des puits de stockage. Des validations de ces différentes prédictions nous indiqueront les limites d'utilisation de chacun des modèles de comportement testés.

Réferences

1- Modaressi H., Laloui L. et Aubry D.(1990): "Numerical modelling of thermal consolidation". 2nd European Conference on Numerical Methods in Geotechnical Eng., pp 280-292, Santander - Espagne.

2- Laloui L.(1993): "Modélisation du Comportement Thermo-Hydro-Mécanique des milieux poreux anélastiques". Thèse de Doctorat de l'Ecole Centrale de Paris. 220 p.

3- Modaressi H. et Laloui L.(1992): "A cyclic Thermoviscoplastic Constitutive Model for Clays", Proceedings of 4th Symposium on Numerical Models in Geomechanics (NUMOG IV), Swansea, Balkema, pp 125-137.

4- Modaressi H., Foerster E. et Laloui L.(1991): "Stability analyses of coupled thermo-hydro-mechanical schemes for deformable porous media". International Conference on Numerical Methods in Thermal Problems (Stanford, USA), volume 7, pp 1073-1083. Edition Lewis, Chin et Houmsy; Pineridge Press Swansea, U.K.

5- Hujeux J.C.(1985): "Une loi de comportement pour les chargements cycliques des sols". Génie Parasismique. Edition V. Davidovici. Presses de l'E.N.P.C., pp 287-302.

Geoenvironmental Design of a Uranium Mill Tailings Facility in Northern Saskatchewan

H.K. Mittal
H.K. Mittal and Associates Ltd., Saskatoon, Saskatchewan, Canada

N. Holl
Cameco Corporation, Saskatoon, Saskatchewan, Canada

S. Donald
Golder Associates Ltd., Waterloo, Ontario, Canada

SYNOPSIS

This paper provides an overview of the geoenviromnental studies undertaken for the design of a proposed new tailings disposal facility in the mined-out Deilmann pit at the Key Lake uranium mine in northern Saskatchewan. The tailings would be deposited at the proposed facility at depths over 90 m below the restored water level in the pit. A prethickened (non-segregating) tailings slurry would be deposited underwater at a density of over 35 % solids. To minimize remixing with the ponded pit water, the tailings slurry would be discharged below the surface of the existing tailings deposit. Subaqueous tailings deposition offers significant advantages at the Deilmann facility including an early reduction of contaminant loading from the peripheral pit dewatering system and a frost free tailings deposit. Extensive engineering studies were performed to quantify the long term behaviour of the proposed facility and its impact on the surrounding environment. The results of these state-of-the-art numerical modelling studies confirmed the findings of the earlier preliminary assessments that the proposed system would result in a fully consolidated tailings mass with a coefficient of permeability less than 10^{-7} m/s allowing the timely decommissioning of the proposed facility. The downstream impacts on the groundwater and/or surface water quality and potential receptor groups are determined in the analyses to be minimal and well within regulatory requirements.

INTRODUCTION

Approximately 300,000 tonnes of tailings are produced annually at the Key Lake uranium mine operated by Cameco Corporation (Cameco). The mine site is located in northern Saskatchewan, approximately 600 km north of Saskatoon. At Key Lake, uranium is extracted by a sulphuric acid leach process and lime is used to neutralize the tailings at the mill.

Since the start of the Key Lake mill in 1983, the tailings have been deposited in an above-ground tailings management facility (TMF), located south of the mill site. The storage capacity of the existing TMF is expected to be exhausted by about 1995. A new TMF is proposed for the storage of tailings from the remaining Key Lake ore at the mined-out Deilmann pit located east of the mill site. In order to meet the latest environmental criteria (consistent with the present day regulatory requirements) comprehensive studies were undertaken in view of the observed and anticipated performance of the above-ground Key Lake TMF and other uranium mill tailings storage facilities in northern Saskatchewan. A description of the proposed design for the Deilmann TMF and the results of the supporting engineering studies are presented in this paper.

TAILINGS CHARACTERISTICS

Key Lake tailings can be described as a sandy/clayey silt with approximately 60% fines (< 75μm). Approximately 80 % of the Key Lake tailings consist of mineral solids from the acid leach circuit. The remaining 20 % of the tailings consist of chemical precipitates. The mineral solids in the tailings consist of sandy/clayey silt and the chemical precipitates of 2/3 gypsum and 1/3 metal hydroxides. Typical concentrations of critical contaminants in the tailings solids and the porewater are shown in Tables 1 and 2.

TABLE 1 -- Key Lake Tailings Solids

Material	U ppm	Ni %	As %	Ra226 Bq/g
Tailings Solids	400	1.5	1.2	260

TABLE 2 -- Tailings Pore Water

Material	U μg/L	Ni μg/L	As mg/L	Ra226 Bq/L	pH
Tailings Pore Water	8	10	6.3	148	10

PERFORMANCE OF ABOVE-GROUND TMF

The above-ground TMF is a 600 m x 600 m square facility contained on all sides by till embankments. To minimize seepage losses through the facility a bentonite modified impervious liner is provided at the base as well as in the embankments. A seepage collection system is provided at the base of the tailings above the impervious liner.

The tailings have been deposited in the TMF at a slurry density of about 30 % solids. The original design of the facility was based on the anticipation of achieving an average dry density of about 850 kg/m^3. The anticipated dry density appears to be achieved only during the summer months when desiccation of the freshly placed tailings can occur. During the initial operation much lower densities, frequently less than 500 kg/m³, were achieved during the winter months due to segregation effects and permafrost build-up in the tailings. The permafrost build-up in the tailings resulted from concurrent freezing in the winter months of tailings placed over a relatively long beach in thin layers (i.e. a mechanism similar to that used for developing outdoor skating rinks). Not all of the frozen low density layers of tailings were thawed during the ensuing summer due to insulating effects of the overlying summer tailings. This resulted in a much lower overall average density of the tailings mass of about 600 kg/m³ by 1990 with significant inclusions of frozen tailings at relatively high water contents.

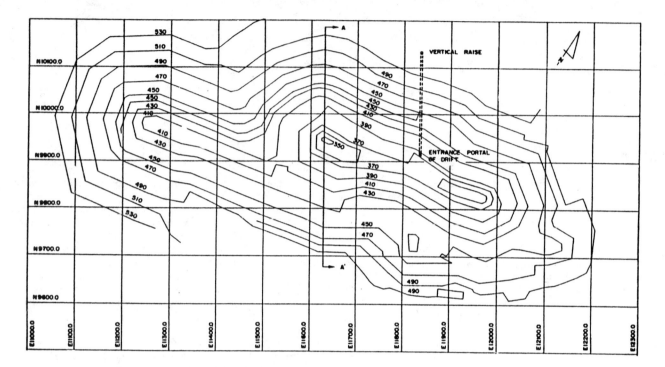

Figure 1 - Plan of Deilmann Pit

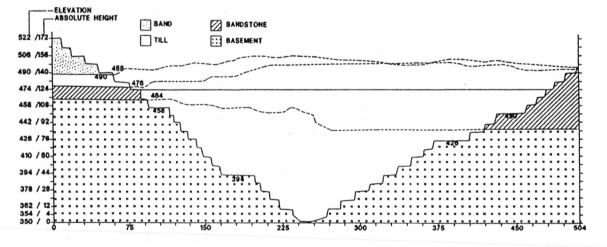

Figure 2 - Section A-A' - Deilmann Pit

In 1990, a dyke was constructed across the tailings to subdivide the facility into two cells for separate summer and winter tailings deposition allowing seasonal thawing of frozen layers. An average dry density of about 700 kg/m³ has been achieved since 1990 using the two cell design.

The results of geothermal studies indicated a period of up to 100 years for the full thawing of the existing frozen tailings. Since estimated thaw consolidation settlements of up to 3 m in the eastern part of the facility would have made any conventional cover design impractical, remedial action was initiated to thaw the low density frozen tailings. There was also concern about the release of tailings leachate due to post-closure thaw consolidation of frozen tailings.

Remedial action commenced in 1992 in which 50 m wide strips of frozen tailings are annually thawed by injecting heated water through closely spaced well points placed in the tailings. The remedial program is to be completed over a period of about 10 to 12 years. Reprocessing of the nickel rich tailings and redisposal of the reprocessed tailings in the Deilmann pit are also under consideration.

PIT HYDROGEOLOGY

As shown in Figure 1, the mined-out Deilmann pit is approximately an oval excavation with a maximum depth of about 170 m. A typical section through the deepest part of the pit is shown in Figure 2. The typical stratigraphic profile at the Deilmann pit consists, in descending order, of about 45 m of outwash sand, about 40 m of sandstone and basement rock (Precambrian gneiss) to a depth far below the pit bottom. The average hydraulic conductivities of these materials have been estimated by pumping tests at approximately 2×10^{-4} to 1×10^{-3} m/s for the outwash sand, 1×10^{-7} to 1×10^{-5} m/s for the sandstone and less than 1×10^{-7} m/s for the basement rock. The principal direction of natural groundwater flow at the Deilmann pit is from south-west to north-east at a horizontal hydraulic gradient of about 0.002.

PIT DEWATERING SYSTEM

The Deilmann pit, located in the bed of former Key Lake, is continually maintained in a dewatered state by a series of peripheral wells with intakes located at the bottom of the sandstone. In addition, horizontal drains are installed in the pit at the sandstone/basement contact. At present, approximately 14 million m³ of groundwater is discharged annually from the dewatering system. Approximately 12 % of the total dewatering discharge is unsuitable for direct discharge to the environment and is pumped to the mill for treatment and/or for use as process make-up water.

PROPOSED DEILMANN TMF

To facilitate consolidation of the tailings, the Deilmann TMF would be provided with bottom and partial side drains. As shown in Figure 3, the seepage water in the drains would be pumped, via a horizontal drift and a vertical raise behind the north pit wall, to the mill for treatment.

To minimize particle size segregation, upon deposition of the tailings solids, the tailings slurry would be prethickened to a minimum density of 35 % solids. Initially, the tailings would be subaerially deposited in the Deilmann TMF for about 4 years while supernatant and seepage water would be pumped from the drainage system. To commence subaqueous deposition of tailings, contrary to present practice at other in-pit disposal systems, the pit would be flooded when the side drains are fully covered by the tailings. The tailings would be discharged from a pipe connected to a floating barge as shown in Figure 3. To minimize mixing of the tailings slurry with

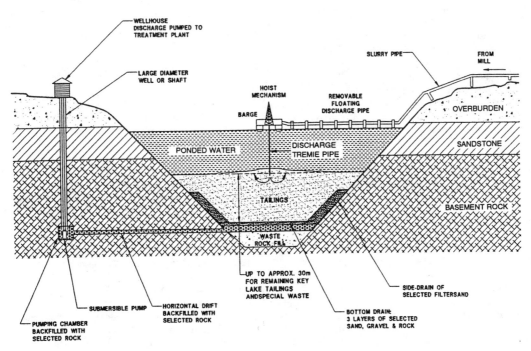

Figure 3 - Subaqueous Deposition

open water, during subaqueous deposition, the tailings discharge pipe would be kept below the surface of the deposited tailings. The discharge points would be moved as required to provide a relatively even distribution of tailings in the pit. To enhance consolidation of the tailings, pumping from the bottom and side drains would be continued throughout the operating life of the facility.

During subaqueous deposition, the level of the flood water in the pit would continue to rise eventually reaching a maximum depth of about 90 m above the tailings in about 4 years. The horizontal drains and the peripheral wells of the dewatering system would be phased out during flooding of the pit. To ensure inward flow of groundwater towards the TMF the water level in the pit would be controlled by a barge pump. The inward flow would ensure complete hydrodynamic containment during the operating life of the proposed facility.

At the end of tailings deposition, to stabilize the surface of the tailings and provide a diffusion barrier, a nominal 2 m thick cover of sand would be hydraulically placed over the surface of the tailings. Prior to decommissioning the facility, water would be continually pumped from the barge pumps on the Deilmann pond and treated in the mill until the ponded water in the pit meets regulatory requirements.

ENVIRONMENTAL CRITERIA

The environmental criteria for the proposed Deilmann TMF according to the existing regulations and/or guidelines are summarized below.
- The tailings facility must be designed in a suitable manner so as to minimize any outward flux of contaminants beyond the limits of the facility during the operating phase as well as after decommissioning.
- The facility must be suitably prepared and decommissioned within a reasonable time period after tailings deposition is completed.
- The tailings must be consolidated to a satisfactory condition prior to decommissioning of the facility.
- The post closure groundwater flux through the tailings mass must be less than 5 % of the average annual precipitation on an equivalent above-ground facility.
- The long term water quality in the Deilmann pond above the tailings mass must meet regulatory requirements. The Saskatchewan Surface Water Quality Objectives (SSWQO) as summarized in Table 5 were used as a basis for comparison.
- The net impact on surface water quality in Outlet Creek, approximately 4 km downgradient of the proposed facility, must be minimized so as to meet regulatory criteria.
- The net risk to nearby receptor groups must be minimized as determined by the pathways risk analysis in accordance with regulatory requirements.

GEOTECHNICAL OBJECTIVES

In order to meet the required environmental criteria, the following geotechnical objectives were identified for the design of the Deilmann TMF.
- In the proposed design, the tailings must be fully consolidated soon after the end of tailings deposition.
- The hydraulic conductivity of the consolidated tailings must be less than 10^{-7} m/s in order to inhibit groundwater flow through the tailings mass to less than 5 % of the average annual precipitation for an equivalent above-ground facility.
- The tailings must either be frost free at the time of decommissioning or must have achieved sufficiently high densities so that subsequent thaw consolidation would not lead to any significant expulsion of tailings leachate to the environment.

ENGINEERING STUDIES

A team of consultants was retained by Cameco Corporation to assist with the design of the Deilmann TMF. Approximately 25,000 engineering hours were spent in an intense 18 month study. Engineering analyses and numerical modelling studies were undertaken to quantify the design requirements and predict the long term behaviour of the proposed facility and its impact on the surrounding environment. The results of these studies are briefly described below.

Consolidation Analyses

Consolidation analyses were performed for the slurried placement of tailings at the proposed facility using the computer code ACCUMV based on the one-dimensional large strain, non-linear consolidation theory. The computer code ACCUMV has been developed under the direction of Professor R.L. Schiffman at the University of Colorado. Consolidation analyses were performed for the deposition of 2.4×10^6 tonnes of tailings from the remaining Key Lake ore at an annual rate of deposition of 300,000 tonnes. Analyses were also performed for a potential expansion case of approximately 8×10^6 tonnes of Key Lake type tailings at the same annual rate of deposition. The analyses were performed for three alternative design options of "no drain", a "partial side drain" and a "full side drain" along the sides of the pit. The results of the consolidation analyses are summarized in Table 3.

TABLE 3 -- Results of Consolidation Analyses

Design Option	Key Lake Tailings		Potential Expansion Case	
	Average Dry Density After Consolidation (t/m³)	Time For Full Consolidation After End of Deposition (years)	Average Dry Density After Consolidation (t/m³)	Time For Full Consolidation After End of Deposition (years)
No Drain	0.72	10	0.75	30
Full Side Drain	0.78	0	0.83	0
Partial Side Drain	0.80	0	0.82	0

The partial side drain option as shown in Figure 3 was selected for the design of the proposed facility. The predicted ranges of void ratios and hydraulic conductivities for the fully consolidated tailings are shown in Figure 4 and 5. The calculated pore pressure distributions at the end of full consolidation for the Key Lake tailings is shown in Figure 6. It is noted from the results in Figure 6 that:
- the pore pressure distribution is highly non-linear due to the decreasing coefficient of permeability with depth of the fully consolidated tailings; and
- the tailings are over-consolidated with respect to the hydrostatic porewater pressure distribution which is expected to be established in the long term after decommissioning of the facility.

Geothermal Analysis

In view of the frost related problems at the above-ground facility, geothermal analyses were undertaken to assess the likely extent of frost build-up in a subaerial in-pit tailings deposition. A modified computer code was developed to specifically deal with the concurrent freezing of a multi-layered tailings deposition during winter placement. Results from these analyses indicated that

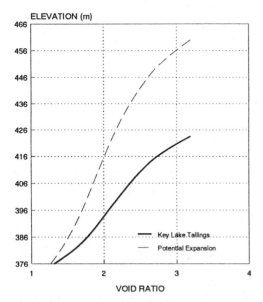

Figure 4 - Computed Void Ratio Distribution
Deilmann TMF - Part Side Drain

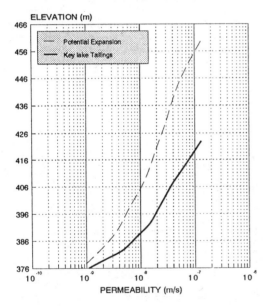

Figure 5 - Computed Permeability Distribution
Deilmann TMF - Part Side Drain

potential for frost-build up in the tailings is dependent upon a number of variables which are described below.

(i) Deposition Rate of Tailings

No permafrost build-up is predicted in the tailings where deposition rates are either very low (< 0.2 m/a) or very high (> 10 m/a). The maximum permafrost build-up is predicted in the range of deposition rates of about 1.5 to 3.0 m/a. Average annual deposition rates at the above-ground facility are approximately 1.5 to 2.0 m. The estimated deposition rates at the proposed Deilmann TMF would range from an average of more than 10 m/a for the initial 4 years to about 2 to 2.5 m/a in the later stages of the potential expansion case.

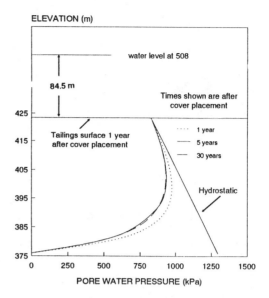

Figure 6 - Porewater Pressure Distributions
for Key Lake Tailings Deposition

(ii) Thickness of Snow Cover

Minimum permafrost build-up is predicted if evenly distributed average seasonal snow cover is maintained over the tailings surface during winter placement. Operating experience indicates, however, that there is an uneven snow cover on an active tailings beach with large areas of little or no snow cover.

(iii) Distance of Tailings From the Discharge Point

Much less frost build-up occurs in the tailings near the discharge point due to a high initial temperature (20°C) of the tailings slurry.

(iv) 3-D Heat Transfer from Rock Walls

Relatively fast thawing of the tailings is predicted near the rock walls compared to the tailings located in the central portion of the facility. To avoid the possibility of permafrost build-up in the tailings during later stages of deposition, which could complicate the decommissioning of the facility, the subaqueous method of tailings disposal is proposed for the Deilmann TMF after the first 4 years of operation.

Hydrogeological Modelling

Computerized numerical models were constructed to simulate groundwater flow and contaminant transport from the proposed Deilmann TMF and to assess the long term environmental impacts after decommissioning of the proposed facility. Three dimensional regional and local groundwater flow simulations were performed using the computer code MODFLOW. The regional model covered an area of about 60 km² around the Key Lake mine site and the local model included an area in the immediate vicinity of the Deilmann pit. Boundary conditions used in the local model were derived from the regional model. The long-term groundwater flows were examined in the hydrogeological modelling for the Key Lake tailings as well as
the potential expansion case. The analyses were performed for all three design options of "no drain", "partial side drain" and "full side drain". Results for the proposed partial side drain option for the Key Lake tailings are summarized in Figure 7.

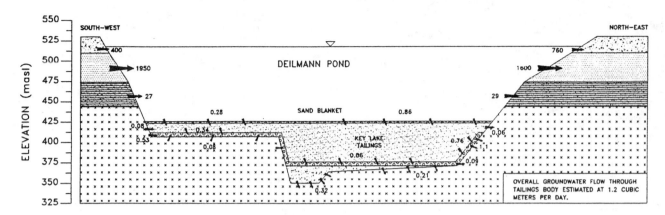

Figure 7 - Groundwater Flow Estimates For Key Lake Tailings Deposition

The approximate distribution of daily flow rates through different materials in the partial side drain design for the Key Lake tailings is summarized below.

Material	Flow Rate m³/day
Waste Rock	400
Outwash Sand	1950
Sandstone	27
Basement Rock	1.5
Tailings	1.2

Due to the high hydraulic conductivities of the overburden materials and the great depth of water (>90 m) on top of the Key Lake tailings, more than 99.9 % of the total groundwater flow will naturally bypass far above the tailings body after full restoration of the water table. Relatively small groundwater flows are calculated through the tailings body as summarized in Table 4.

TABLE 4 -- Long Term Groundwater Flows Through Tailings

Design Option	From Key Lake Tailings m³/d	From Potential Expansion Case m³/d	Environmental Criteria 5% of Avg. Precip. m³/d
No Drain	1.67	2.58	10 to 15
Partial Side Drain	1.22	1.66	10 to 15
Full Side Drain	0.01	0.02	10 to 15

As can be seen, virtually no flow occurs through the tailings in the "full side drain" case while approximately 1.2 m³/day is estimated for the proposed "partial side drain". It is noted that total groundwater flow through the tailings in the "partial side drain" case is a small fraction of that specified by the environmental criteria in Table 4.

Contaminant Transport Modelling
In combination with the 3-D groundwater flow modelling, 2-D finite-element models of groundwater flow and solute transport (AQUIFER and LTGPLAN) were used to assess long term leachate flux into the Deilmann pond and contaminant plume development downgradient of the facility.

The predicted long term contaminant concentrations in the Deilmann pond are summarized in Table 5. As can be seen, the predicted concentrations are nearly one order of magnitude lower than the criteria specified by the Saskatchewan Surface Water Quality Objectives (SSWQO). It should be noted that, due to the low hydraulic conductivity of the tailings and a low hydraulic gradient of about 0.002, the contaminant transport into the Deilmann pond is controlled by the diffusive flux through the sand cover rather than by advective transport.

TABLE 5 -- Predicted Long Term Contaminant Concentrations In Deilmann Pond

For Key Lake Tailings			
Design Option	As (mg/L)	Ni (mg/L)	Ra226 (Bq/L)
Full Side Drain	0.006	0.003	0.01
Partial Side Drain	0.004	0.003	0.01
No Side Drain	0.005	0.003	0.01
SSWQO	0.050	0.025	0.11
For Potential Expansion Case			
Design Option	As (mg/L)	Ni (mg/L)	Ra226 (Bq/L)
Full Side Drain	0.011	0.003	0.02
Partial Side Drain	0.008	0.003	0.01
No Side Drain	0.011	0.003	0.01
SSWQO	0.050	0.025	0.11

With respect to the impact on the groundwater downstream of the facility, the predicted long term concentrations of the highly mobile contaminants such as sulphates are found to be of no consequence. Also the predicted long term concentrations of other contaminants such as molybdenum and nickel in the tailings pore water are extremely low and therefore of no consequence. The transport of less mobile contaminants downgradient of the Deilmann TMF will be extremely slow due to the high retardation potential of the sandstone and basement materials combined with the relatively low hydraulic conductivities of the bedrock units and a very low hydraulic gradient (.002). As summarized in Table 6, it is predicted in the analysis that concentration of the critical contaminants such as arsenic and radium - 226 would not exceed SSWQO beyond about 200 m from the proposed facility in 10,000 years.

TABLE 6 -- Contaminant Plume In Rock Formations Downgradient of TMF

Parameter	Travel Time (Years)	Travel Distance in Sandstone (m)	Travel Distance in Basement (m)
As	6,400	80	9
Ra-226	10,000	200	16

The contaminant concentrations in Outlet Creek will be dominated by solute transport in the more permeable overlying outwash sand. As summarized in Table 7, the long term (steady state) concentrations of critical contaminants in groundwater discharge at the entrance of Outlet Creek would be a small fraction of the criteria specified in the SSWQO.

TABLE 7 -- Concentration Of Critical Contaminants In Ground Water Discharge at Outlet Creek

Case	Arsenic µg/L	Nickel µg/L	Radium 226 Bq/L
Key Lake Tailings in Deilmann TMF	4	3	0.01
SSWQO	50	25	0.11

Pathways Risk Analysis

The impact of radionuclides and metals carried from the TMF via groundwater and surface water pathways to a hypothetical future community at the closest major waterbody (Russell Lake), downgradient of the TMF, was defined by a pathways risk analysis. The risk analysis for Russell Lake, approximately 20 km downstream from the TMF, (based on the ETP model) gave the following results:
- there would not be any measurable impact on the water quality in Russell Lake, i.e., contaminants in Russell Lake will remain at baseline concentrations;
- the annual radiation dose to an individual living at Russell lake would be less than 1 µsv/a which is less than 0.05% of the average annual dose received by the general population of Canada from natural sources, and less than 0.01 % of the proposed new Canadian public dose limit;
- the toxicological impact of contaminants on an individual at Russell Lake would also remain orders of magnitude below acceptable levels in accordance with the latest EPA regulations.

CONCLUDING REMARKS

The proposed method of tailings disposal at the Deilmann TMF is shown to be technically sound. The proposed design is expected to produce a relatively uniform and fully consolidated frost free tailings mass of a relatively low average permeability.

In recent years an in-pit design with subaerial deposition and a fully drained pervious envelope at the bottom and the sides of the tailings mass has been favoured for the uranium mill tailings disposal in northern Saskatchewan. The proposed design for the Deilmann TMF includes optimization for the specific conditions at the Deilmann site. The site specific considerations of major interest in this regard are:

(i) an early reduction of the dewatering discharge which minimizes the environmental load from the dewatering system;
(ii) elimination of potential detrimental permafrost build-up in the tailings and the need for remedial thawing prior to decommissioning of the facility; and
(iii) elimination of radon emanation and dust problems from the surface of the tailings leading to safer operating conditions.

ACKNOWLEDGEMENTS

The authors are grateful to Cameco for permission to publish and the encouragement during the preparation of the paper. The design studies for the Deilmann TMF were a team effort with the following list of participants.

Cameco Corporation	General direction and overall coordination
Golder Associates Ltd.	Hydrogeological, Contaminant Transport, and Thermal Modelling
Steffen Robertson and Kirsten	Definition of Source Terms for Contaminant Transport Modelling and Pathways Risk Analysis, and Acid Drainage Potential
Beak Consultants	Pathways Risk Analysis
H.K. Mittal and Associates Ltd.	Geotechnical Design, Consolidation Modelling and Overall Review

A Model for Coupled Deformation and Non-isothermal Multiphase Flow Through Saline Media. Application to Borehole Seal Behaviour

S. Olivella, A. Gens, J. Carrera, E.E. Alonso
Geotechnical Department, Civil Engineering School, Technical University of Catalunya, Barcelona, Spain

Abstract

In salt rock environments important specific phenomena take place that are relevant for problems of waste disposal. These are related mainly to solubility, creep deformation, brine inclusions and porosity variation. Strong coupling exists between multiphase mass flow of brine and gas, medium deformation and heat transport problems. A new coupled theory has been formulated which has lead to the development of a numerical program (CODE-BRIGHT). The application to the study of borehole seal behaviour is included as an example of coupled phenomena simulation.

1. Introduction

Salt rock formations are considered possible environments for radioactive waste disposal. In this context, salt granular aggregates are used to fill mine openings and to build sealing systems with the objective of isolating the disposed wastes. In order to characterize the hydro- thermo- mechanical behaviour of the complete system composed by different seal elements, porous backfills and the host rock, it has been necessary to establish a new general formulation and to develop suitable constitutive relationships.

The governing equations for coupled deformation and non-isothermal multiphase flow of brine and gas through saline media have been obtained. The specific aspects that arise due to the particular physical properties of salt have been taken into account in the development of this new coupled formulation. These include:

– Deformation of salt (rock and aggregates) is time dependent and can be very large. Creep deformations occur caused by different mechanisms. Temperature and presence of brine strongly influence creep strain rates. Porous salt agregates incorporate also structural effects which combine with the intrinsic mechanisms of deformation.

– High solubility of salt in water. Dissolution/ precipitation can be induced by different causes: water condensation/ evaporation, temperature, liquid pressure variations and stress variations. In fact, stress in solid salt in contact with brine causes variations of concentration which in turn produce salt migration. This is the mechanism that may explain deformation of wet salts [1]. Finally, the presence of an important amount of solute in liquid strongly influences its equilibrium vapour pressure, which is generally known as the hygroscopic character of salts.

– The migration of brine inclusions [2] within grains and the possibility of their creation/ destruction at grain boundaries may be important in some cases. Brine inclusions can move when subject to temperature gradients because concentration gradients are induced inside the brine bubbles. By molecular diffusion, salt is transferred from the region where dissolution occurs to the region where precipitation occurs. Due to this mechanism, amounts of brine can be supplied to the connected pores that are significant when porosity is low and the medium relatively dry.

– Porosity variation. The void volume changes due to the processes outlined above. Its variation influences storage and transport terms. Particularly, transport properties such as hydraulic permeability, may undergo variations of several orders of magnitude.

These especific aspects of saline media, and the more general ones for porous materials, lead to the necessity of characterizing the behaviour in a coupled framework. As an example of coupled phenomena, deformation is significantly influenced by the presence of brine. Hence, porosity evolution depends on the movement of brine. In turn, hydraulic permeability is a function of porosity. This illustrates that it is important to use a coupled approach. However, due to the difficulty involved in solving coupled problems, there is a lack of analyses in which the various aspects affecting the problem are treated in a sufficiently general way.

This paper reports a new theoretical framework to deal with this type of problems. The framework developed has been transformed into a finite element code (CODE-BRIGHT) capable of solving coupled problems in these saline environments. It handles COupled DEformation, BRIne, Gas and Heat Transport problems.

2. Governing Equations

The problem is described as a porous medium composed by a body of salt solid grains (with brine inclusions), brine as the liquid phase (including not only dissolved salt but dissolved air as well) and a mixture of dry air and water vapour as the gaseous phase. Figure 1 is a schematic representation of the medium. We take into account three species: salt(h), water(w) and air (a) and three phases: solid (s), liquid (l) and gas (g).

The governing equations subject to certain initial and boundary conditions will allow to compute the state variables. These are: solid displacements, liquid phase pressure, gas phase pressure, temperature and concentration of water in solid phase. The equations can be grouped into different cathegories:
(1) definition constraints (e.g. liquid and gas degree of saturation, $S_l + S_g = 1$)
(2) equilibrium restrictions (solubility, Henry's law, psychrometric law)
(3) constitutive equations (e.g. gases law)
(4) balance equations used as constitutive (e.g. Darcy's law)
(5) macroscopic balance equations.

The macroscopic balance equations that have been derived are: stress equilibrium equations, mass balance equations (for salt, water and air) and energy balance equation (thermal equilibrium is assumed). These equations have been derived according to a macroscopic approach. Theoretical basis for this macroscopic approach can be found in [3]. Reference works for multiphase flow are those from [4], [5] and [6].

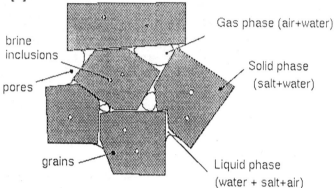

Figure 1: Schematic representation of the medium composed by salt grains, brine and gas.

2.1. Stress Equilibrium Equation

The stress equilibrium equation is obtained from momentum balance and is expressed as:

$$\nabla \cdot \boldsymbol{\sigma} + \boldsymbol{b} = \boldsymbol{0} \qquad (1)$$

where $\boldsymbol{\sigma}$ is the stress tensor and $\boldsymbol{b}$ is the body force vector.

2.2. Mass and Energy Balance Equations

A balance equation has been established for each species. This methodology is referred as the compositional approach and is preferred because phase change terms do not appear explicitly (equilibrium is assumed), while non-advective fluxes are explicit and easy to express.

The balance equations for salt, water and air species are:
Salt mass balance:

$$\frac{\partial}{\partial t}(\omega_s^h \rho_s (1-\phi) + \omega_l^h \rho_l \phi S_l) + \nabla \cdot (\boldsymbol{j}_s^h + \boldsymbol{j}_l^h) = 0 \qquad (2)$$

Water mass balance:

$$\frac{\partial}{\partial t}(\omega_s^w \rho_s(1-\phi) + \omega_l^w \rho_l \phi S_l + \omega_g^w \rho_g \phi S_g) +$$
$$+ \nabla \cdot (\boldsymbol{j}_s^w + \boldsymbol{j}_l^w + \boldsymbol{j}_g^w) = 0 \qquad (3)$$

Air mass balance:

$$\frac{\partial}{\partial t}(\omega_g^a \rho_g \phi S_g + \omega_l^a \rho_l \phi S_l) + \nabla \cdot (\boldsymbol{j}_l^a + \boldsymbol{j}_g^a) = 0 \qquad (4)$$

where ϕ is porosity, ρ is density, $\boldsymbol{j}$ is mass flux, ω is mass fraction (superscripts h, w and a refer to salt, water and air; and subscripts s, l and g refer to solid, liquid and gas respectively), and S_l, S_g are degrees of saturation of the liquid and gaseous phases. The sum of species fluxes over phases gives the total flux. Each species flux is the sum of advective and diffusive/dispersive terms.

As the problem is non-isothermal, the balance of energy is expressed as:

$$\frac{\partial}{\partial t}(E_s \rho_s(1-\phi) + E_l \rho_l \phi S_l + E_g \rho_g \phi S_g) +$$
$$+ \nabla \cdot (\boldsymbol{j}_c + \boldsymbol{j}_{Es} + \boldsymbol{j}_{El} + \boldsymbol{j}_{Eg}) = f^E \qquad (5)$$

where E is specific internal energy, $\boldsymbol{j}_c$ is the conductive heat flux, the terms $\boldsymbol{j}_E$ are energy fluxes due to mass movement and f^E is an internal production term to account for internal viscous dissipation or other causes such radioactive decay.

There is also another equation which balances the mass of water contained in brine inclusions. This equation is required because water in inclusions can not be considered in equilibrium with liquid in the connected porosity. It contains the terms of the water mass balance equation related with water in solid phase. Brine inclusions have been treated in [7].

The equations briefly presented here have been established in [8]. A detailed derivation is presented there.

2.3. Constitutive Theory

The preceding formulation requieres adequate constitutive laws for the different processes that take place. Any process needs a specific constitutive relationship. For example, Darcy's law and Fick's law are models to express advective and non-advective mass fluxes. The parameters contained in these laws (e.g. permeability, viscosity, etc) depend on the state variables.

For most of the constitutive relationships, available models have been adopted. However, mechanical behaviour of porous salt aggregates is not well known and new theoretical and experimental work is necessary. Based on the basic mechanisms of deformation and a proposed geometry for grains and pores, a new model for creep behaviour has been developed [9]. The mechanisms of deformation are fluid assisted diffusional transfer (FADT) and dislocation creep (DC). The former consists of the deformation of grains due to salt dissolution/ molecular diffusion/ precipitation due to the presence of brine in pores. The latter includes the intracrystalline deformation of grains due to dislocation climb and glide.

3. Computer Code

A computer code, CODE-BRIGHT, has been developed. Details of its characteristics and performance are beyond of the scope of this paper. The program has been verified (comparison with known solutions) for a wide range of uncoupled and coupled problems. Of special relevance for this paper, three particular uncoupled verification cases have been performed: (1) gas pulse test in a borehole, (2) constant heat flow in a borehole, and (3) instantaneous stress distribution and stress relaxation around a borehole. As those three verification cases concern the three equations to be solved in the example of application described below, the results obtained there can be regarded with a high degree of confidence.

4. Application to Borehole Seal

4.1. Problem Definition

One of the possiblilities of waste disposal in salt mines consists of placing it in vertical boreholes drilled in the floor of the galleries. Each vertical hole receives a canister and the space between the cylinder and the wall is filled with compacted crushed salt with relatively low porosity. Figure 2 is a representation of such scheme. For our purposes, the medium is considered indefinite and the scheme very long. This allows geometry simplifications and preliminary simulations. On the other hand, the porous material is assumed to be filled with dry air, initially at atmospheric pressure (0.1 MPa).

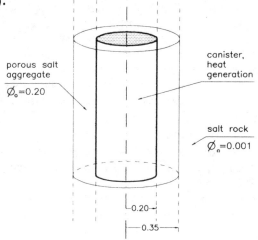

Figure 2: Schematic representation of the borehole where a canister is placed. The annular space is backfilled with compacted porous salt.

Firstly, the borehole is drilled and the stresses around it change instantaneously. Afterwards, a progressive stress relaxation occurs because of salt creep. At ambient temperature, the hole converges relatively slowly. After placing the canister, the heat output induces stress field variations, both because of the thermal dilatation of rock and of the high temperature dependence of creep parameters. The porous backfill is then strongly compressed as convergence rates increase. Both temperature increment and porosity-permeability reduction cause gas pressure build up. Permeability reduction of rock near the hole decreases the possibility of gas pressure dissipation towards the medium, which is the only permitted possibility of gas flow in the analysis.

4.2. Simulation

A preliminary simulation has been performed in order to accumulate knowledge of the behaviour of such system. The geometry is very simple and a linear grid (42 nodes) has been sufficient because the scheme is considered very long, hence, plane strain axisymmetrical conditions apply. Conditions at the far boundary are prescribed at the last node which is at a distance of 192.2 m. An initial isotropic stress state in the rock of 15 MPa is assumed.

365 days after drilling the borehole the canister is installed, the annular space backfilled and heating starts.

The parameters that have been used are respectively, for salt rock (r), and for porous salt backfill (b):

Initial porosity (ϕ_o): 0.001(r), 0.20(b)

Young's modulus (E): 25000 MPa (r)
$7000 + 95000(0.30 - \phi)$ MPa (b)

Poisson's ratio (ν): 0.30

Creep power law parameters: $n = 5.375$
$A = 4.98 \times 10^{-6} \exp(-59650/RT) \text{s}^{-1}$

Thermal linear dilatation (β): 4×10^{-5} K^{-1}

Thermal conductivity (λ): 5 J/s/m/K

Heat capacity (C_s): 920 J/Kg/K

Permeability (Kozeny's function) (k):
1.0×10^{-18} m^2 at $\phi_o = 0.001$ (r)
5.0×10^{-14} m^2 at $\phi_o = 0.20$ (b)

The parameters for creep power law correspond to a rock with zero porosity. The creep constitutive model includes the dependence on void ratio given in [9].

The simulation has been carried out for two different exponentially decaying values of the canister heat output:

$j = j_o \exp(-\alpha t) \text{J/s/m}$

where $j_o = 500$ and 1000 J/s/m, and $\alpha = $(-log 0.5)/(30 year). For the highest heat output problem, temperature is very high, the material becomes very deformable and reduction of porosity and permeability to near zero values can be obtained. A very low minimum permeability value (1.0×10^{-21} m^2) has been imposed by limiting porosity to 10^{-4}.

4.3. Results

The results of the two simulations performed are presented here. Figure 3 shows the stress relaxation around the borehole during the first year after excavation. Figure 4 is the evolution of temperature at the borehole wall. Due to exponentially decaying heat production, a maximum temperature is reached, while a constant heat output would have lead to a monotously increasing logarithmic evolution.

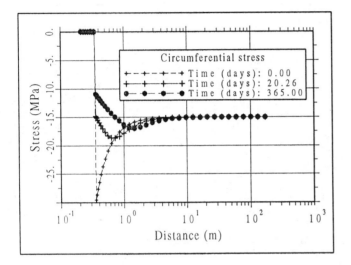

Figure 3: Circumferential stress during stress relaxation. Points correspond to element centre.

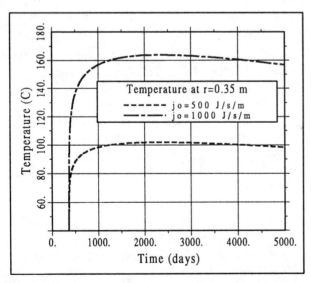

Figure 4: Temperature at borehole wall. Heating begins after 1 year from borehole drilling.

Figure 5 shows the influence of the temperature and the presence of the backfill on the stress distribution. It is interesting to see how the backfill is progressively loaded, from zero initial stress state to a stress level similar to that in the rock. For the problem with $j_o = 1000$ J/s/m of initial heat output, a similar evolution is obtained.

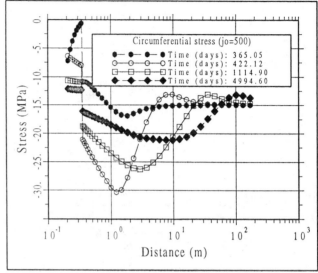

Figure 5: Circumferential stress after canister emplacement ($j_o = 500$ J/s/m).

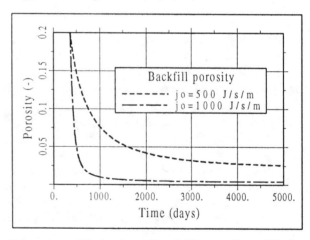

Figure 6: Variation of backfill porosity with time.

Figure 6 compares the evolution of the backfill porosity. The difference is caused by the larger creep strain rates of salt, both in the rock and in the porous backfill, at the higher temperature.

Finally, Figure 7 shows the gas pressure build up due to temperature increment and porosity reduction. The progressive decay is due to temperature decrease and to flow towards the rock.

As commented, only gas flow towards the rock has been permitted. This assumption seems reasonable in order to prevent undesirable gas migration through the gallery. This implies that the sealing system which isolates the borehole should be highly impervious.

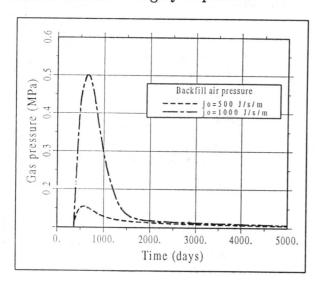

Figure 7: Gas pore pressure in the backfill.

5. Conclusions

A model for COupled DEformation, BRIne, Gas and Heat Transport (CODE-BRIGHT) in saline media has been developed. The model consists of a finite element numerical code based on a theoretical formulation, which includes governing equations for multiphase/multispecies flow in deformable media and adequate constitutive theories. Building on three simple verification exercises, a preliminary simulation for a coupled problem (deformation, gas flow, heat flow) has been carried out and the results presented. The resulting plots highlight important aspects which have to be taken into account in the design and construction of such disposal systems.

6. Acknowledgements

This work has been performed under contract with ENRESA (Empresa Nacional de Residuos) in the framework of a project partially funded by the ECC (European Community Commision).

7. References

[1] Spiers, C.J., P.M.T.M. Schutjens, R.H. Brzesowsky, C.J. Peach, J.L. Liezenberg and H.J. Zwart, (1990): "Experimental determination of constitutive parameters governing creep of rocksalt by pressure solution", Deformation Mechanisms, Rheology and Tectonics, Knipe, R.J. and Rutter, E.H., Geological Soc. Special Publ., no. 54: 215-227.

[2] Roedder, E. (1984): "The Fluids in Salt", American Mineralogist, Vol. 69: 413-439.

[3] Hassanizadeh, S.M. (1986): "Derivation of basic equations of mass transport in porous media, Part 1. Macroscopic balance laws, Part 2. Generalized Darcy's and Fick's laws ", Adv. Water Resour. 9, 196-222.

[4] Bear, J., J. Bensabat, and A. Nir (1991): "Heat anf Mass Transfer in Unsaturated Porous Media at a Hot Boundary: I. One-Dimensional Analytical Model", Transport in Porous Media 6, 281-298.

[5] Milly, P.C.D. (1982): "Moisture and Heat Transport in Hysteretic, Inhomogeneous Porous Media: A Matric Head Based Formulation and a Numerical Model", Water Resour. Res., vol. 18, N° 3: 489-498.

[6] Pollock, D.W. (1986): "Simulation of Fluid Flow and Energy Transport Processes Associated With High-Level Radioactive Waste Disposal in Unsaturated Alluvium", Water Resour. Res., Vol. 22, N° 5: 765-775.

[7] Ratigan, J.L. (1984): "A Finite Element Formulation for Brine Transport in Rock Salt", Int. J. for Num. and Analit. Meth. in Geomech., Vol. 8: 225-241.

[8] Olivella, S., J. Carrera, A. Gens, E.E. Alonso (1994): "Non-isothermal multiphase flow of brine and gas through saline media". Submitted to Transport in Porous Media, Kluver A.P.

[9] Olivella, S., A. Gens, E. E. Alonso, J. Carrera, (1993): "Behaviour of Porous Salt Aggregates. Constitutive and Field Equations for a Coupled Deformation, Brine, Gas and Heat Transport Model". Proc. of the "3rd Conf. on The Mechanical Behaviour of Salt", in press. Paris, September 14-16.

A Thermo-Elastoplastic Model with Thermal Hardening for Saturated Clay Barriers

J-C Robinet, A. Rohbaoui
Eurogeomat Consulting, Orléans, France

F. Plas
Agence Nationale pour la Gestion des Dechets Radioactifs (ANDRA), Fontenay-aux-Roses-France

Abstract

A constitutive model based on the microscopic irreversible phenomena occurring in saturated clays undergoing a thermo-mechanical loading is proposed. The first mechanism results from the modification of adsorbed water ratio during heating. Moreover, the molecular agitation increase induces an expansion of remaining adsorbed layers and constitutes the second microscopic phenomenon. Finally, for expansive clays, the internal repulsive stresses build up with temperature give rise to macroscopic irreversible strains which characterise a third irreversible mechanism. Thermo-mechanical paths on *remoulded Boom, "Bassin Parisien"* and *Pontida* clays are given with a satisfying agreement between experimental data and model predictions.

1. Introduction

Three classes of pore space are defined from the textural organisation of clay flakes:

1. the inter-lamellar space between clay flakes known by an average size from 15 to 25 Å *[14]*,
2. the inter-particle space among the stacks of flakes with 200 to 1500 Å of size in the high porosity clays *[14]* and about 10 to 25 Å in the low porosity clays *[10]*,
3. the inter-aggregate space of sizes in the range of *1.5* to *1.6* μ.

Owing to this classification, two families of water in the saturated clays are distinguished: the free water lying mainly in the inter-aggregate space and the adsorbed water which is located in the inter-particular and inter-lamellar spaces.

According to the clay settlement ratio, the rate partition of free and adsorbed water is strongly varied because of its dependence on distribution and size pores. Thus, for high level compacted clays, all the more expansive, the pore distribution is monomodal (inter-lamellar) so the water is essentially in adsorbed state. However, for high porosity clays, the pore distribution is bimodal and the free water is predominant.

At isothermal conditions, the adsorbed water behaviour is identical to the one of a plastic granular medium, so the clays can be considered as bimodal *[7]*. However, heating increases the atomic agitation of adsorbed water which can consequently move as free water. This process occurs with a decrease of adsorbed layers strength *[8]*. Besides, the nuclear magnetic resonance *(NMR)* indicates that the dielectric relaxation time T_2 of proton is proportional to the free water content. *Carlsson [4]* shows that the heating provokes an increase of free water content due to adsorbed water decrease. Therefore, during heating the saturated clay must be considered as a triphasic medium constituted by clay flakes, adsorbed water and free water.

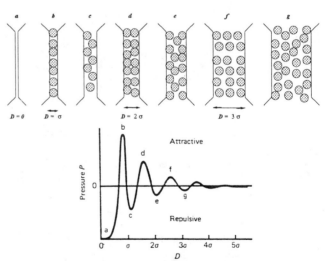

Figure 1: Detailed measurements of forces between two curved Mica surfaces (Israelachvili and al. 1978)

In expansive clays the flakes get a negative charge by means of the ionic substances (Al^{+3} exchanged by Mg^{+2} for the Montmorillonite). This charge is properly the origin of clay flakes separation in presence of polar solvent and confers to the clay an expansive character. *Israelachvili [7]* gives an illustration of the forces distribution between two mica-plates in electrolytic solution *(cf. figure 1)*.

2. Inventory of microscopic mechanisms responsible of macroscopic irreversible strains

In isothermal conditions, the clay mud consolidation shows that the micro-porosity remains unchanged. The macroscopic irreversible volumetric strains of consolidation proceed essentially from macro-porosity reduction of inter-aggregate pore spaces *(cf. figure 2)*. It results that for expansive clays the repulsive forces generated in the inter-lamellar and inter-particle spaces remain constant and hence the macroscopic irreversible strains developed are conventionally described by the plasticity rule.

The study of thermal strains is carried out on two remoulded non expansive "Bassin Parisien" and expansive Boom clay. Their characteristics are given in the *table 1*:

CLAY	B. PARISIEN (%)	BOOM (%)
SiO_2	51.60	57.58
Al_2O_3	21.54	12.93
Fe_2O_3	6.42	7.57
CaO	2.035	2.22
MgO	1.72	2.40
Na_2O	0.62	0.12
K_2O	2.25	1.96
MnO	0.015	0.01
TiO_2	1.12	0.88
P_2O_5	0.44	0.18
Lost by heating	12.24	14.15

CLAY	BASIC CEC (meq)	TOTAL CEC (meq)	BET-SS (m^2/g)
B. PARISIEN	46	145	33.5
BOOM	59	213	42

CEC = Cation Exchange Capacity   *SS = Specific Surface*

CLAY		B. PARISIEN	BOOM
γ_s	(g/cm³)	2.68	2.67
W_L	(%)	45-60	60-75
W_P	(%)	15-30	25-35
activity by oedometric test	(cg/cc)	0.21	0.27-0.32

Table 1: Physical characteristic of remoulded "Bassin Parisien" and Boom clays (Euro-géomat 1992)

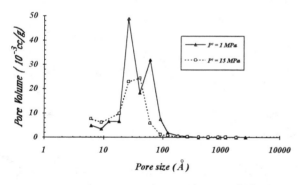

a) macro-porosity--porosimetry method

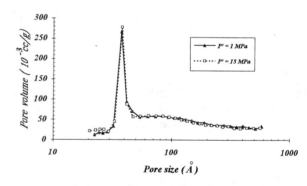

b) Micro-porosity--BET method

Figure 2:   Porosity distribution versus consolidation pressure - Boom clay (Belanteur and Al Mukhtar 1993)

The thermal loading originates two opposite microscopic mechanisms in saturated clays:

• the first is contractive by diminution of the double layers thickness owing to the transition from adsorbed to free water,
• the second mechanism is dilatant and produced by molecular agitation increase giving rise to the expansion of remaining adsorbed layers.

The first mechanism for normally consolidated states *(NCS)* affects mainly the inter-aggregate pore space producing hence a macroscopic volumetric strain. This strain displays both reversible and irreversible parts *(cf. figure 3)*. The second mechanism of adsorbed layers dilatancy have a low magnitude and produces in overconsolidated states *(OCS)* an expansion of inter-aggregate pores *(cf. figure 4)*

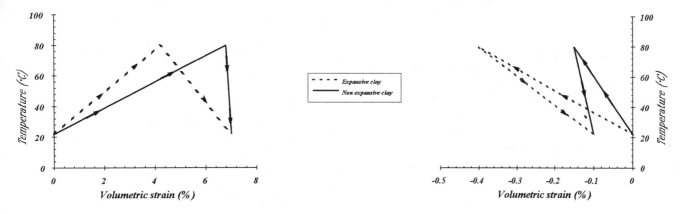

| Figure 3: Schematic thermal behavior of NC clays | Figure 4: Schematic thermal behavior of OC clays (OCR > 8) |

Sridharan [13] suggests that in the description of macroscopic expansive clays behaviour, the effective stress concept of *Terzaghi* must be extended. Hence, the medium strains are controlled by three independent variables: σ'_c contact stress, U interstitial pressure, and σ_{R-A} internal stress resulting from attractive and repulsive forces:

$$\sigma = \sigma'_c + u + \sigma_{R-A} \qquad (1)$$

Heating at constant total stress causes a contact stress drop and increases the repulsive stresses. Therefore, in the *NCS* a thermal cycle generates contractive volumetric strains *(cf. figure 3)*.

In the *OCS* with an *OCR* less than *8* a part of the thermal loading is responsible for an elastic response which characterises a modification of water adsorbed structures.

The inter-aggregate pore spaces settlement induced by mechanical consolidation is unable to produce the rearrangement of internal structure of clay. The compactive irreversible response is required when the inter-aggregate friction forces become greater than the consolidation one *(cf. figure 5)*.

In non expansive clays we remark that the settlement induced by thermal cycle decreases with the *OCR* in agreement with experimental results *[5]*. In the higher *OCS* (*OCR > 8*), the first mechanism of adsorbed layers contraction is inefficient to produce an irreversible settlement of inter-aggregate pore spaces. The macroscopic volumetric strains are then considered to be elastic. Moreover, the adsorbed layers dilatancy induces an irreversible response of inter-aggregate spaces *(cf. figure 6)*.

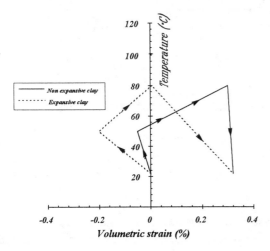

Figure 5: Schematic thermal behavior of OC clay with OCR ≤ 8

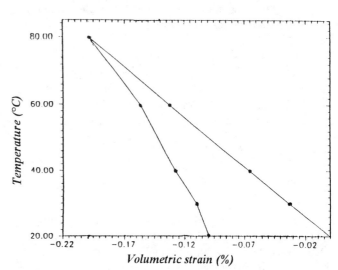

Figure 6: Thermal cycle on RBP clay with OCR = 10 (Belanteur 1993)

3. Thermo-mechanical characteristic of saturated clays

3.1 Oedometric behavior

The thermo-mechanical oedometric tests are performed on remoulded "Bassin Parisien" and Boom clays. The experiments are carried on *20, 40* and *80 °C* with *1MPa* of back-pressure. Above a total stress of *1 MPa*, the behaviour is identical to the normally consolidated clays. The consolidation slope (λ) is independent of applied stress and temperature *(cf. figure 7), [11]*.

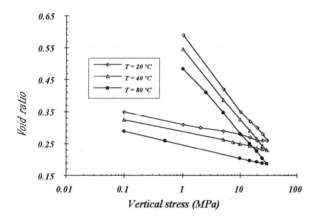

Figure 7: Thermal oedometric tests on RBP clay (Rahbaoui and Belanteur 1992)

In low consolidation stress (*< 0.1 MPa* for "Bassin Parisien"*[11]* and Boom clays, *< 0.21 MPa* for Bentonite clays *[9]*) a small dependence of consolidation coefficient on temperature is observed, but in the study range the λ-*coefficient* is taken constant. The thermal cycles at constant stress in *NCS* produce a material hardening.

Besides, in the *OCS* (*OCR < 8*) we notice a *settlement* for non expansive clay and a *low swelling* for expansive clay during the thermal cycle. The former response decreases as the *OCR* increases *[5]*. Moreover, in the high *OCS* (*OCR > 8*) the behaviour becomes dilatant for both saturated clays.

3.2 Triaxial behavior

The triaxial tests performed on remoulded Pontida clay show that the critical state concept is not affected by temperature *[1]*. At constant deviatoric stress, the heating generates an axial strain increase which can be characterised by a failure in undrained conditions if the heating is significant *[6] (cf. figure 9)*. This failure is caused by a pore pressure increase during heating. We remark also that the value of clay strength remains constant in the drained condition and falls in the undrained condition.

4. Model constitutive equation for saturated non expansive clays

The studies at microscopic scale associated with thermo-mechanical macroscopic tests show that three mechanisms are the origin of macroscopic irreversible thermal response. Only two mechanisms subsist for non expansive clays. Therefore, the macroscopic irreversible settlement of inter-aggregate pore spaces is produced by the first mechanism which is modelled by a non associative plasticity.

4.1. Yield surface of settlement for the NCS called thermo-mechanical loading collapse "TMLC"

In general, the yield surface f_{TMLC} of thermo-mechanical settlement is defined in the (Q, P', e, T) space. It is schematised by two plan projections:

a- **Yield surface projection on the (T, P') plan**

A yield surface is characterised by the same plastic void ratio. Thus, to determinate this locus requires a minimum of two isotropic paths which produce the same plastic total strains.

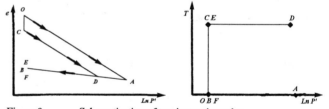

Figure 8: Schematization of two isotropic cycles

For the mechanical loading *(OAB)*, we have:

$$\varepsilon_{vm}^p = \frac{1}{\beta} Ln \frac{P_c'}{P_{co}'} \quad , \quad P_c' = P_{co}' Exp(\beta \varepsilon_{vm}^p) \qquad (2)$$

The irreversible strain due to thermal loading *(OC)* is:

$$\varepsilon_{vT}^p = f(T) \Delta T \qquad (3)$$

The mechanical plastic strain at high constant temperature *(CDE)* is:

$$\varepsilon_{vmT}^p = \frac{1}{\beta} Ln \frac{P_D'}{P_C'} \qquad (4)$$

P_{cT}' is the critical mean stress at temperature *T*,

The thermal unloading *(EF)* induces only an elastic strain.

The projection of yield surface *(cf. figure 9)* on the *T-P'* plan is:

$$P_{cT}' = P_{co}' Exp(\beta \varepsilon_{vm}^p) Exp(-\beta f(T) \Delta T) \qquad (5)$$

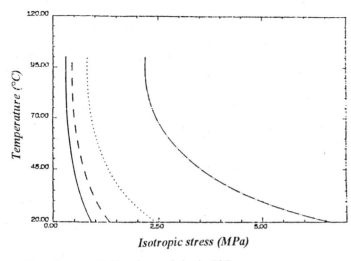

Figure 9: Yield surface evolution in (P',T) space

b- Projection of yield surface f on the Q-P' plan

This projection is defined by the modified Cam-Clay model associated with the deviatoric and isotropic mechanisms:

$$f_{TMLC}(P',Q,P'_{cT},T) = Q^2 + M^2 r^2 P'(P' - b P'_{cT}) R(\theta) = 0 \quad (6)$$

P'_{cT} and r are two hardening variables with $r = \dfrac{\varepsilon^p_{dTMLC}}{a + \varepsilon^p_{dTMLC}}$

where:
ε^p_{dTMLC} is the volumetric deviatoric strain
a and b are two model parameters depending on material.
$R(\theta)$ is a function of Mohr-Coulomb failure.

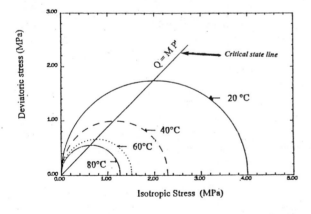

Figure 10: Schematic projection of thermomechanical yield surface

The critical state is given by the slopes M_c in compression and M_E in extension:

$$M_C = \frac{6 \sin \phi}{3 - \sin \phi} \quad , \quad M_E = \frac{6 \sin \phi}{3 + \sin \phi} \quad (7)$$

Figure 10 shows the projections of thermo-mechanical yield surfaces of settlement due to first elasto-plastic mechanism such as:

$$\dot{\varepsilon}_{ij_{TMLC}} = \dot{\varepsilon}^e_{ij_{TMLC}} + \dot{\varepsilon}^p_{ij_{TMLC}} \quad (8)$$

The elastic strains are expressed as:

$$\dot{\varepsilon}^e_{ij_{TMLC}} = \frac{\dot{s}_{ij}}{2G} + \frac{\dot{I}_1}{9K} \delta_{ij} \quad (9)$$

where:
$$K = K_a P_a \left(\frac{P}{P_a}\right)^n \quad , \quad G = G_a P_a \left(\frac{P}{P_a}\right)^n \quad (10)$$

The plastic strains are materialised by:

$$\dot{\varepsilon}^p_{ij_{TMLC}} = \lambda \frac{\partial g}{\partial \dot{\sigma}_{ij}} \quad (11)$$

where g is the potential of plastic flow:

$$g = Q^2 + M^2 P'(P' - P'_{cT}) = 0 \quad (12)$$

4.2. Yield surface of swelling called Thermal Increase

The second plastic mechanism due to adsorbed layers dilatancy during heating is independent of applied stresses. This mechanism will induce an irreversible dilatancy of micro-porosity of material and is distinguished by an associative plastic rule called *"Thermal Increase"* (TI).

The hardening is traduced by displacements of f_{TI} plan along of the temperature axis, thus the thermal hardening is expressed as follows:

$$\dot{\varepsilon}^p_{v_{TI}} = -a_T b_T Exp(-b_T T) \dot{T} \quad (13)$$

The total strains corresponding to the second irreversible mechanism of dilatancy are given by:

$$\dot{\varepsilon}_{ij_{TI}} = \dot{\varepsilon}^e_{ij_{TI}} + \dot{\varepsilon}^p_{ij_{TI}} \quad (14)$$

where: $\dot{\varepsilon}^e_{ij_{TI}} = -\dfrac{\alpha^a_w}{3} n_a \dot{T} \delta_{ij}$ is the reversible part.

α^a_w : coefficient of thermal expansion of adsorbed water

5. Comparison between a thermo-elasto-plastic model and the laboratory tests

The calibration of the model have been realised on the remoulded non expansive Pontida clay with adequate parameters.

a- Simulation of oedometric paths with thermo-mechanical loading

Figure 11 presents the oedometric paths simulations and confirms the above macroscopic description. The heating from *20°C* to *95°C* provokes an obvious settlement of material in *NCS* and induces a low swelling in *OCS*.

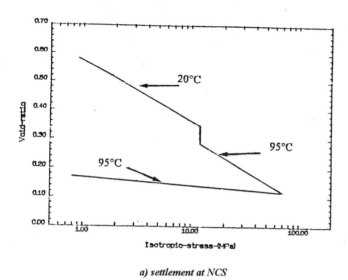

a) settlement at NCS

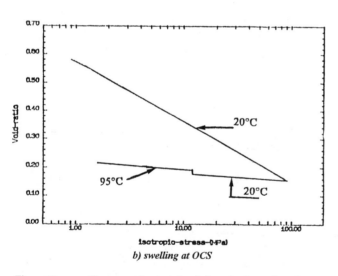

b) swelling at OCS

Figure 11: Thermomechanical simulation of oedometric path

b- Simulation of triaxial thermo-mechanical tests in drained conditions:

In the *NCS*, the two irreversible mechanisms are active:

$$\dot{\varepsilon}_{ij} = \dot{\varepsilon}^e_{ij_{TMLC}} + \dot{\varepsilon}^e_{ij_{TI}} + \dot{\varepsilon}^p_{ij_{TMLC}} + \dot{\varepsilon}^p_{ij_{TI}} \quad (15)$$

Figures 12 and *13* give the comparison between experiment and numerical triaxial simulation. At $Q = 1.2$ *MPa*, when the temperature increases from *22°C* to *100°C*, the axial strain builds up to a determinate value. After the mechanical loading, the material softens (*cf. figure 10*).

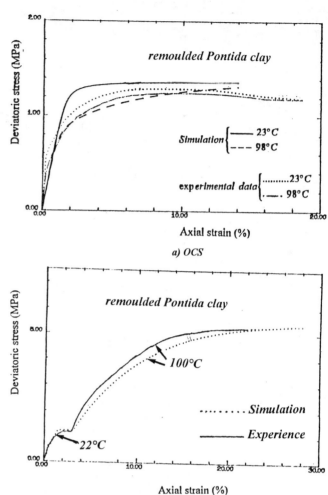

a) OCS

b) NCS

Figure 12: Comparison between experiments and simulation of RP clay in drained conditions

c- Simulation of undrained triaxial paths:

The response is determined by the volume change of free water with temperature, thus:

$$\dot{\varepsilon}_v = -\frac{\alpha}{3} n \dot{T} = \dot{\varepsilon}_1 + 2\dot{\varepsilon}_2 \quad (16)$$

The plastic strain of both mechanism are given by equations *(11)* and *(13)*.

The information about the thermal elastic strain $\dot{\varepsilon}^e_{TI}$ and the total imposed strain $\dot{\varepsilon}_{ij}$ allow the determination of the thermo-mechanical elastic strain:

$$\dot{\varepsilon}^e_{TMLC} = \dot{\varepsilon} - \dot{\varepsilon}^e_{TI} - \dot{\varepsilon}^p_{TMLC} - \dot{\varepsilon}^p_{TI} \quad (17)$$

The stresses and the interstitial pressure are given by the elasticity rule:

$$\dot{\sigma}' = C\,\dot{\varepsilon}^e_{TMLC} \quad , \quad \dot{u} = -\dot{\sigma}'_2 \quad (18)$$

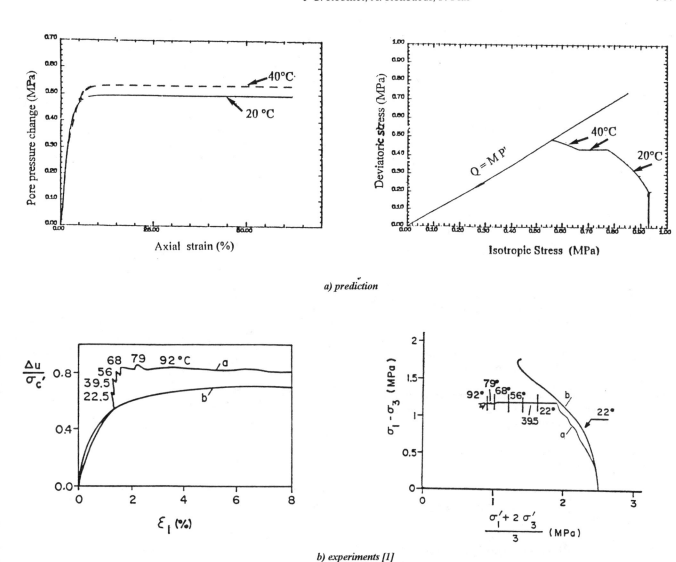

Figure 13: Comparison between experiments and simulation of RP clay in undrained conditions

CONCLUSION

The inventory of thermo-mechanical microscopic mechanisms due to transfer of adsorbed water ratio to the free water and the dilatancy of the remaining adsorbed layers are taken into account in order to establish a thermo-elasto-plastic model related to the stress-strain response aspects. The model is characterised by two yield surfaces f_{TMLC} and f_{TI}. In the *NCS*, both irreversible mechanisms are activated corresponding to the first mechanism of contraction and to the second of dilatancy. In contrast, in the overconsolidated states, only the first mechanism f_{TI} is excited.

The model is verified using the oedometric and triaxial paths. The comparison between the experiments on remoulded Pontida and "Bassin Parisien" clays and the model predictions shows that all observed phenomena are correctly described.

Shortly, our purpose is to quantify the heating/cooling effect on the internal stresses in adsorbed layers which have an important significance specially for expansive clays.

Acknowledgements

We wish to express our most sincere thanks to the *"Agence Nationale pour la Gestion des Déchets Radioactifs, ANDRA" (France)* who supports this work and to the SCK-CEN *(Belgium)* for supplying the clays.

References

[1] Baldi G., Hueckel T. and Pellegrini R. 1988. Thermal volume changes of mineral-water system in low porosity clays soils. Can. Geotech. J. V 25 n° 4 pp 807-825.

[2] Bolt G. H. 1956, Physico-chemical analysis of the compressibility of pure clays, Geotechnique 86-93.

[3] Campanella R. G. and Mitchell J. K. 1968, Influence of temperature variations on soil behavior. J. Soil Mech. and Found. Div. ASCE, 709-734.

[4] Carlsson T. 1985, NMR-Studies of pore water in Bentonite/water/electrolyte, Scien. Basic for Nuclear Waste Management IX, Ed. L. O. Werme, Stockholm 1985, Mat. Res. Symp. Proc. 609-615.

[5] Demars K. R. and Charles R. D. 1982. Soil volume changes induced by temperature cycling, Canad. Geot. J., 188-194

[6] Hueckel T. and Pellegrini R. 1989. Modeling of thermal failure of saturated clay. In numerical models in geomechanics. Eds. S. Pietruszczak and GN Pande Elsevier pp 81-90.

[7] Israelachvili J. N and Adams G.E. 1978. Measurements of forces between two mica surfaces in aqueous electrolyte solutions. J. Chem. Soc. Faraday Transact, 1 74 pp 975-1001.

[8] Paaswell R. E. 1967, Temperature effects on clay consolidation. J. Soil Mecha. and Found. Div. ASCE 9-21.

[9] Plum R. L. and Esrig M. I. 1969, Some temperature effects on soil compressibility and pore water pressure. Special Report 103, Highway Res. Board, Washington, D. C. 231-242.

[10] Pusch R., Hökmark H. and Karnland O. 1989. Microstructure and conductivity of smectite clays. A/PEA 9th Int. Clay Conf. Strasbourg

[11] Robinet J. C., Al-Mukhtar M., Belanteur N. and Rahbaoui A. 1992. Thermomecanical behaviour of high compacted clays. Coll. of René Houpert ENSG-INPL Nancy pp 335-342

[12] Robinet J. C., Al-Mukhtar M. and Rahbaoui A. 1992. Thermomechanical modeling of non active, low porosity and saturated clays. Franco-American Workshop 19-20 October 1992

[13] Sridharan A. and Venkatappa Rao G. 1973. Mechanisms controlling volume change of saturated clays and the role of the effective stress concept. Geotechnique 23 n° 3 pp 359-382

[14] Touret O. and Pons C. H. 1989, Etude de la répartition de l'eau dans les argiles saturées au forte teneur en eau, Clay Minerals 25, 217-223.

[15] Yong R. N., Chang R. K. and Warkentin B. P. 1969, Temperature effect on water retention and swelling pressure of clay soils, Special Report 103, Highway Res. Board, Washington D. C. 132-137.

Notations:

NCS : Normally consolidated states
OCS : Overconsolidated states
NC : Normally consolidated
OC : Overconsolidated
RBP : Remoulded "Bassin Parisien"
RP : Remoulded Pontida
σ : Total stress tensor
s : Deviatoric stress tensor
I : First invariant of stress tensor
σ' : Effective stress or contact stress
U : Water pressure
σ_{R-A} : Resultant stress of attractive and repulsive stresses
e : Void ratio
T : Temperature
P' : Effective mean stress
Q : Deviatoric stress
M : Slope of critical state line
ε : Strains
C : Tensor of elasticity
K : Bulk modulus
G : Shear modulus
λ : Consolidation slope / Plastic multiplier
"•" : rate symbol
α : Coefficient of thermal expansion
β : Coefficient of plastic compressibility
n : Porosity

Modélisation des Transferts de Masse dans les Argiles à Faible Porosité. Application au Stockage des Déchets Radioactifs

J.C. Robinet, M. Rhattas
Euro-Geomat, Orléans, France

F. Plas, J.M. Palut
Agence Nationale pour la Gestion des Déchets Radioactifs, Fontenay-aux-Roses, France

Abstract: The ventilation of underground repositories, meant for the storage of hazardous waste, to the creation of a unsaturated zone. We have studied, by numerical simulations, the desaturation of a clayey formation. After a quick presentation of the transfer mechanisms and of the principal results obtained on the hydrodynamic characterisation of Boom clay, two applications of transfer model are performed: the first one concerns the evaluation of a unsaturated zone; the second one deals with the desaturation of the massif in closing phase. The results showed that the time of resaturation is more important than that of desaturation.

Résumé: La ventilation des ouvrages souterrains destinés à l'entreposage de déchets toxiques, conduit à la création d'une zone désaturée. Nous avons étudié, par la voie de simulations numériques, la desaturation d'un massif argileux. Après une présentation rapide des mécanismes de transfert et des principaux résultats obtenus sur la caractérisation hydrodynamiques de l'argile de Boom, deux applications du modèle de transfert sont traitées: la première concerne l'évaluation de la zone désaturée; la deuxième application correspond à la resaturation du massif en phase de fermeture du stockage. Les résultats montrent que le temps de resaturation est plus important que celui de la desaturation.

1- Introduction

L'étude du transfert de l'humidité dans les sols touche des domaines très variés, en particulier les sciences de l'environnement telle que l'hydraulique souterraine (stockage des déchets urbains, industriels et nucléaires). Le besoin de trouver un environnement géologique convenable dans lequel les déchets peuvent être évacués avec succès et sur de longues périodes de temps, est à l'origine d'une recherche significative sur les propriétés du transfert de l'eau dans les matériaux argileux. Les ouvrages souterrains implantés en profondeur dans un milieu argileux caractérisé par une certaine perméabilité, sont le siège d'un transfert d'humidité. En effet, la ventilation des ouvrages pendant leur exploitation conduit à la création d'une zone désaturée d'extension non négligeable (Robinet et al. 1992). Ce phénomène a été, par ailleurs, observé et évalué dans un massif granitique sur une profondeur de 1,60 m (Schneebeli et al. 1991). En milieu rocheux, la désaturation peut atteindre plusieurs dizaines de mètres au-delà des parois: ainsi dans l'ancienne mine de fer de Konrad (Allemagne), la ventilation a permis d'obtenir en régime permanent un débit d'eau de $0,43 \times 10^{-3} l/s$ extrait d'une galerie de 74 m de long (Cottez et Piraud 1993).

Dans les argiles à très faible porosité, l'eau se trouve essentiellement sous forme adsorbée. Fripiat et Coll (1982) montrent que la couche d'eau adsorbée dans les argiles est de quelques dizaines d'Angströms. Pusch et Hökmark (1990) estiment que dans une montmorillonite sodique artificielle de masse volumique 1,8 g/cm^3, 60 % du volume de l'eau interstitielle serait lié. Selon Baldi et al. (1990) l'eau liée représente dans l'argile de Boom naturelle entre 24 % et 49 % du volume total de l'eau.

Notre contribution consiste en l'étude des transferts de masse isotherme dans les argiles à faible porosité en prenant en compte le transfert de l'eau adsorbée dans la phase liquide. Les caractéristiques hydrodynamiques du modèle sont déterminées par la voie expérimentale. En nous appuyant sur ces données, nous appliquons le modèle au phénomène de désaturation d'un massif argileux au voisinage des ouvrages ventilés pendant leur période d'exploitation. La validation du modèle est effectuée sur l'évolution du front de pression nulle à l'intérieur du massif argileux. Enfin, une attention particulière est portée à l'étude de la resaturation du massif en phase de fermeture du stockage.

2- Mécanismes de transfert

Dans les matériaux poreux, l'eau liée est caractérisée par de fortes liaisons avec la matrice solide qui sont à l'origine de deux phénomènes physiques souvent considérés comme statiques: l'adsorption de surface et la capillarité. Les phénomènes dynamiques de migration de l'eau sont: le transfert convectif de la vapeur, le transfert de l'eau à l'état condensé et le transfert diffusif de l'eau adsorbée. A l'échelle microscopique, la saturation d'un pore, supposé cylindrique, s'effectue en 4 étapes correspondants aux phénomènes physiques fondamentaux intervenant dans le comportement de l'eau dans les matériaux poreux (Figure 1):

① formation d'un film d'eau adsorbée sur les parois du pore par diffusion de la vapeur d'eau (adsorption mono-moléculaire)

② Transfert diffusif dans les couches adsorbées; la diffusion de la vapeur d'eau donne naissance à l'adsorption de surface. Il s'agit donc d'une diffusion surfacique

③ Transfert de l'eau dans le domaine capillaire (évaporation-condensation)

④ saturation de l'espace poral et écoulement suivant un mécanisme convectif dans la phase d'eau liquide condensée.

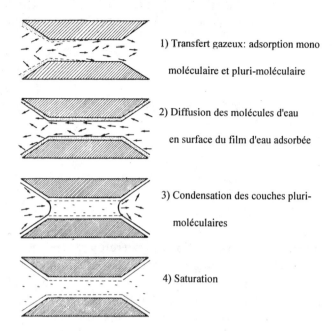

Fig. 1. — *Processus de transfert dans un pore cylindrique*

Les approches théoriques développées par Philip et Devries (1957) et Luikov (1966) sont toujours une référence en matière de transfert de masse dans les milieux poreux. On peut citer également les travaux plus récents de Quintard et Puiggali (1986) ainsi que ceux de Daian (1986) qui font apparaître, moyennant l'hypothèse des flux en parallèle, la contribution de deux paramètres: le coefficient de diffusion isotherme sous forme vapeur et un coefficient de transfert en phase liquide. Notre point de départ est le modèle proposé par Rhattas et al. (1992) sur le transfert hydrique dans les argiles à faible porosité. Ce modèle conduit, dans le cas général, à un système de deux équations décrivant, d'une part, le transfert de l'eau sous forme adsorbée, capillaire et vapeureuse et d'autre part, le transfert de l'air. Dans ce travail, nous nous limitons au cas où la phase de l'air est immobile, ce qui réduit le système à l'équation différentielle suivante:

$$\rho_w . C_w(h_w) . \frac{\partial (h_w)}{\partial t} + \nabla \left[\left(-\rho_w . K_w + \rho_g . K_{dv} \right) \nabla (h_w) \right] = 0 \qquad (1)$$

le coefficient de transfert global K_{wt} apparaît comme la somme de deux termes: i) coefficient de transfert dans la phase liquide capillaire et ii) coefficient de transfert de l'eau adsorbée. Il s'exprime par:

$$K_{wt} = K_w + K_{dv} \ , \ K_{dv} = -\left(\rho_{vs}/\rho_g\right).D_v.\frac{dh_r}{dh_w} \quad (2)$$

avec:

C_w capacité capillaire
D_v coefficient de diffusion dans la phase vapeur
h_w pression de l'eau en hauteur d'eau équivalente
h_r humidité relative
K_w coefficient de transfert dans la phase liquide
ρ_g masse volumique du mélange gazeux (air-vapeur)
ρ_{vs} masse volumique de la vapeur saturante
ρ_w masse volumique de l'eau

Nous identifions dans le coefficient K_{wt} la part de chacun des phénomènes de transfert sous forme de 2 termes:

① un terme représentant le transfert diffusif de l'eau adsorbée (K_{dv})

② un terme représentant le transfert convectif en phase condensée (K_w)

3- Caractéristiques hydrodynamiques de l'argile de Boom

La détermination des caractéristiques hydrodynamiques dans les argiles à faible porosité en condition d'écoulement non saturé, est particulièrement délicate. En ce qui concerne la mesure du potentiel de l'eau en fonction de la teneur en eau, notre choix est porté sur le contrôle de l'humidité relative par des solutions salines saturées. Cette technique permet d'exploiter les propriétés spécifiques de sorption que possèdent les sels en présence de l'eau et permet d'imposer un potentiel de l'eau dans un domaine allant de 3 à $1 \times 10^3 \ MPa$. Nous avons reporté sur les figures 2 (a) et (b) la courbe caractéristique du milieu dans les graphiques $\Theta_w - h_r$ et $\Theta_w - h_w$ où Θ_w représente la teneur en eau volumique.

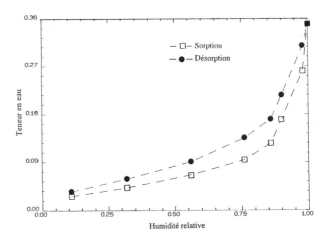

<u>Fig. 2</u> (a) — *Courbe caractéristique du milieu. Cas de l'argile de Boom*

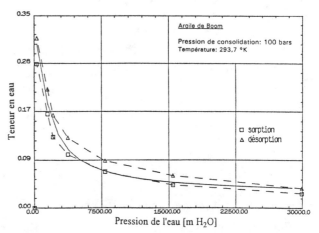

<u>Fig. 2</u> (b) — *Courbe caractéristique du milieu. Cas de l'argile de Boom*

L'étude du processus d'imbibition nous permet de connaître les variations du coefficient de transfert global en fonction de la succion. A l'issue d'un essai, on dispose de mesures de la teneur en eau massique, d'une manière discrète, aux différents points de la colonne d'argile et pour un temps d'imbibition prédéfini. La figure 3 montre les résultats des essais réalisés sur l'argile de Boom pour différents temps d'imbibition. Ces courbes présentent à un instant donné une allure comparable indiquant l'homogénéité des propriétés hydriques du matériau. On constate d'une manière global que la cinétique du transfert dans l'argile de Boom compactée est très faible.

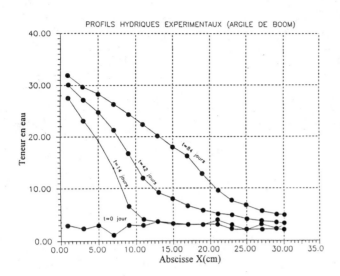

Fig. 3 — *Profils hydriques. Cas de l'argile de Boom*

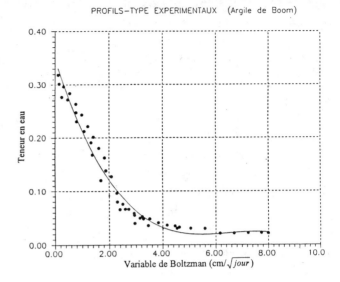

Fig. 4 — *Evolution de la teneur en eau en fonction de la variable de Boltzmann (ξ)*

L'exploitation des données de l'expérience nécessite l'adoption de l'équation de diffusion non linéaire qui s'avère la mieux adaptée pour estimer le coefficient de transfert global $D_\Theta(\Theta_w)$ Angulo (1986), Daian (1986).

$$\frac{\partial \Theta_w}{\partial t} = \frac{\partial}{\partial x}\left(D_\Theta(\Theta_w)\frac{\partial \Theta_w}{\partial x}\right) \quad (3)$$

La solution $\Theta_w(x,t)$ de l'équation (3) peut s'exprimer en fonction d'une seule variable $\xi = x.t^{-1/2}$ dite de Boltzmann (Hall 1989), (Angulo 1986) et (Daian 1986). Cela revient à regrouper tous les profils hydriques $\Theta_w(x,t)$, établis en régime transitoire, autour d'une courbe unique $\Theta_w(\xi)$ (Figure 4). Une méthode de dépouillement par bilan hydrique permet pour 4 essais d'imbition, de calculer le coefficient de diffusion à l'aide de l'expression:

$$D_\Theta(\Theta_w) = \frac{1}{2}.\frac{d\xi}{d\Theta_w}.\int_{\Theta_w}^{\Theta_{wi}} \xi.d\Theta_w \quad (4)$$

où Θ_{wi} est la teneur en eau initiale de la colonne d'argile. Sur la figure 5 nous avons tracé la courbe représentant la variation du coefficient de transfert global.

4- Application à la désaturation d'un massif argileux au voisinage des ouvrages ventilés

Il est reconnu de façon pragmatique que pendant la phase d'exploitation des ouvrages souterrains, la ventilation peut provoquer une zone désaturée. Des simulations numériques en régime transitoire sont effectués et ont pour objectif d'évaluer la zone désaturée. Nous appliquons à la section ventilée de la galerie une valeur du potentiel de l'eau symbolisant une valeur constante de l'humidité relative imposée par la ventilation. Les profils hydriques correspondant à une valeur de l'humidité relative de 50 % et pour deux valeurs du coefficient de transfert global $K_{wt} = 10^{-12} m/s$ et $K_{wt} = 10^{-13} m/s$ sont présentés sur la figure 6.

Sur une période de désaturation de 120 mois, on constate que cette accroissement affecte grandement le développement de la zone désaturée et améliore en conséquence la sécurité des ouvrages. La validation du modèle est effectué à partir de l'évolution du front de pression interstitielle induit dans le massif argileux par la ventilation (Figure 7). Les résultats relatifs à l'étude portant sur la resaturation en phase de fermeture du stockage sont présentés. Ils illustrent la distribution du champs de pression interstitielle dans une couronne d'argile ayant été initialement désaturée par ventilation pendant 1 an. La figure 8 montre que la période au bout de laquelle la resaturation du massif sera atteinte est d'environ 10 ans.

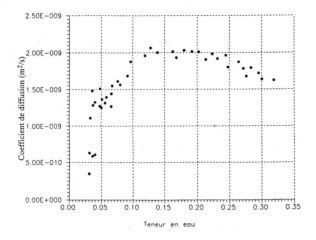

Fig. 5 — *Variation du coefficient de transfert*

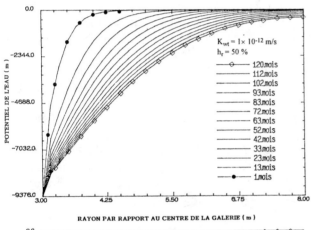

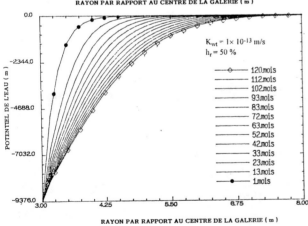

Fig. 6 — *Influence du coefficient de transfert sur la zone désaturée*

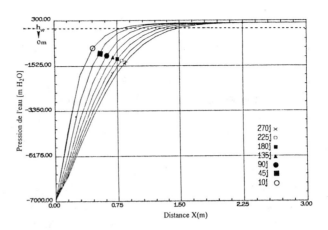

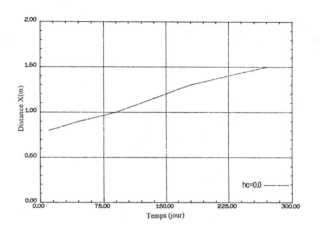

Fig. 7 — *Désaturation du massif argileux*
(a) Profils de pression de l'eau
(b) Evolution du front de pression

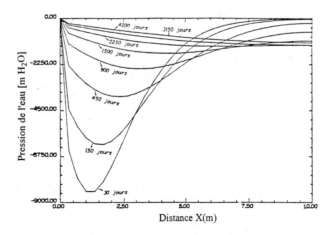

Fig. 8 — *Resaturation du massif argileux*

5- Conclusion

Un modèle de transfert de masse est appliqué à la ventilation des ouvrages souterrains profonds implantés en milieu argileux, hygroscopiques, caractérisés par une faible porosité.

La description de la saturation d'un pore cylindrique a permis de faire la synthèse des différents mécanismes mis en jeu au cours du processus de transfert. Les résultats expérimentaux concernant la caractérisation hydrodynamique de l'argile de Boom sont présentés. Ils constituent un outil indispensable pour la modélisation; en particulier, le coefficient de transfert global et la courbe caractéristique potentiel matriciel-teneur en eau. Les résultats de la simulation numérique ont permis de mettre en évidence, la présence d'une zone désaturée suite à la ventilation du massif argileux. Cette désaturation est d'autant plus importante que le coefficient de transfert global est grand. L'amplitude de la zone désaturée dépend principalement du coefficient de transfert global. On a pu observer également que la resaturation du massif argileux exige 10 fois plus de temps que la désaturation. Une extension de la modélisation faisant appel à l'usage d'un modèle de structure poreuse du matériau argileux est envisagée. Celle-ci permettrait à l'avenir, de rendre compte de façon satisfaisante des différents processus de transfert.

Références

[1] **Angulo, R.J. 1989.** Caractérisation hydrodynamique de sols déformables partiellement saturés. Etude expérimentale à l'aide de la spectrométrie Gamma double-sources. Thèse de Docteur de l'Institut National Polytechnique de Grenoble.

[2] **Baldi, G., Hueckel, T., Peano, A. and Pellegrini 1990.** Developments in modeling of thermo-hydro-geomechanical behaviour of Boom clay and clay based buffer materials. Final report for CEC, ISMES, Bergamo, Italy.

[3] **Cottez, S. et Piraud, J. 1993.** Intérêt potentiel d'un stockage souterrain de déchets en galeries accessibles à niveau. Geoconfine 93, Volume 1. pp. 493 - 498.

[4] **Daian, J.F. 1986.** Processus de condensation et de tansfert d'eau dans un matériau méso et macroporeux. Etude expérimentale du mortier de ciment". Thèse de Docteur d'état, INPG et Université de Grenoble.

[5] **Fripiat, J.J. and Coll, J. 1982.** Comportement micro-dynamique et thermo-dynamique de l'eau dans les suspensions argileuses, Symp. Adsorption Gas-sol and Liquid-sol. Interface, pp 447-449.

[6] **Hall, C. 1989.** Water sorptivity of mortars and concretes. Magazine of Concrete Research, 41, No. 147, June, 51-61.

[7] **Luikov, A.V. 1966.** Heat and mass transfer in capillary porous bodies. Pergamon. Press. Edition.

[8] **Philip, J.R et Devries, D.A. 1957.** Moisture movements in porous materials under température gradients. trans. Am. Geoph. Union, Vol. 38, n° 2, avril.

[9] **Push, R. and Hökmark, H. 1990.** Basic model of water and gas flow through smectite clay buffers, Eng. geology n°28, pp 379-389.

[10] **Quintard, M. et Puiggali, J. R. 1986.** Numerical modelling of transport processes during the drying of a granular porous medium". Heat and Technology, Volume 4, number 2.

[11] **Rhattas, M., Al Mukhtar, M. Robinet, J. C. 1992.** Mass transfer model for low porosity clays. Franco-American workshop on Environmental Geomaterials 19-20. Oct 1992; CRMD-CNRS Orléans.

[12] **Robinet, J.C, Al Mukhtar M., Rhattas M., Plas F. et Lebon, P. 1992.** Modèle de transfert de masse dans les argiles à faible porosité, Application à l'effet de la ventilation dans les galeries, Rev. Franç. Géotech. n° 61. pp. 31-43 (décembre 1992).

[13] **Schneebeli, M., Bear, T., Wydler, H. and Flühler, H. 1991.** In situ measurement of water potential and water content in unsaturated granitic rock, Proc. of NEA-OECD Workshop on gas generation and release from radioactive waste repositories, Aix en Provence pp 1-9.

Design and Construction of the Closure Plan for the Conquista Uranium Mill Tailings Impoundment

Clint Strachan
Water, Waste & Land, Inc., Fort Collins, Colorado, USA

Claude Olenick
Conoco Inc., Falls City, Texas, USA

Abstract

This paper outlines the key components associated with closure of the Conquista uranium mill tailings impoundment in south Texas, including the major design and permitting issues in receiving closure plan approval and the key factors associated with reclamation construction.

The Conquista Uranium mill was operated from 1972 to 1982. Approximately 8.75 million tons (7.95 million tonnes) of uranium mill tailings were discharged in an impoundment adjacent to the mill covering approximately 250 acres (617 ha). In 1987 Conoco initiated engineering and licensing for closure of the tailings impoundment. Closure construction started in 1989 and was completed in 1993. Closure of the tailings impoundment was permitted through the Texas state agencies under regulations consistent with U.S. Nuclear Regulatory Commission (NRC) uranium tailings regulations.

The key design components included a sloping reclaimed surface and 3.5-foot (1.07-m) thick compacted clay cover designed to reduce the average rate of radon emanation to NRC-regulated levels and to have a permeability (for reduction of infiltration) less than 10^{-7} cm/sec. The cover surface, reclaimed embankment slopes, and surrounding channels were designed to provide acceptable erosional stability, assessed by flow velocities due to runoff from the Probable Maximum Precipitation (PMP) and tractive force analyses. Cover materials were evaluated and selected for suitability based on permeability, radon attenuation, and dispersivity.

Closure plan construction started in 1989 and proceeded concurrently with evaporation of remaining tailings pond water. Key construction issues included regrading of tailings to meet closure plan subgrade elevations and placement of fill over tailings of varying grain-size distribution, shear strength, and stress history. Regrading required consideration of excess pore pressures and equipment trafficability, and utilized specialized equipment, materials, and techniques where necessary. Remaining construction activities include maintenance of vegetation and monitoring of closure plan performance prior to eventual transferral of the site to the state of Texas.

Project Background

The Conquista site is located in south Texas, approximately 60 miles (100 km) southeast of San Antonio within a region of uranium mining and milling from the late 1950's. The Susquehanna/Falls City site being reclaimed under the U.S. Department of Energy Uranium Mill Tailings Remedial Action Program (UMTRAP) is approximately two miles (3 km) west of the Conquista site.

The site is in the South Texas Plains vegetation region, characterized by gently rolling hills covered with grasses and limited brushwood. Annual precipitation averages approximately 30 inches (760 mm) and annual pan evaporation averages approximately 80 inches (2030 mm). The site is within the San Antonio River watershed, one of the primary river systems in the region draining to the southeast to the Gulf of Mexico.

The site lies on southeast-dipping Tertiary sediments, consisting primarily of poorly consolidated siltstones, claystones, and fine-grained sandstones. The formation immediately beneath the tailings impoundment, found in the immediate site area, and used for cover material is the Dubose Clay Member of the Eocene Whitsett formation. The Dubose Clay Member is composed predominantly of soft claystones of tuffaceous to montmorillonitic composition from deposition in a marginal marine barrier environment.

Tailings Impoundment Description

The Conquista mill was operated by Conoco Inc. and Pioneer Nuclear Inc. to recover uranium from several open-pit mines in the area. Approximately 8.75 million tons (7.95 million tonnes) of ore were milled from 1972 to 1982, with uranium oxide recovery by sulfuric acid leach and solvent extraction.

The tailings impoundment was initially constructed across a small valley in 1971. The embankment height was raised 30 feet (9 m) in 1979, forming the ring-dike structure shown in Figure 1. The earthen embankment forming the perimeter of the impoundment was constructed as a central-core embankment from borrow materials initially within the impoundment perimeter and subsequently from a borrow area on the east side of the impoundment.

The foundation soils beneath the impoundment and the materials used for embankment construction were Dubose Member clays, as mentioned above. The final tailings embankment height ranged from approximately 16 to 70 feet (5 to 21 m), had a 3:1 (horizontal:vertical) outside or downstream slope, and a 2:1 inside or upstream slope.

The tailings impoundment was operated by spigotting from the inside slope of the embankment along the entire perimeter of the impoundment. Ponded water was pumped from a barge inside the impoundment for reuse in the mill. The final elevation of spigotted beaches was approximately 12 feet (3.7 m) from the embankment crest elevation of 435 feet (133 m) above sea level.

Initial Reclamation

Decommissioning of the mill site began at the end of 1982, and consisted of dismantling of equipment, demolition of buildings, and cleanup of ore stockpile areas and mill site area soils. Equipment and materials that could not be decontaminated and salvaged were buried in the tailings impoundment. Initial reclamation of the tailings impoundment itself began in 1984, and consisted of covering the western 2/3 of the impoundment surface with soils from selected stockpiles and a borrow area immediately southwest of the impoundment. The nominal thickness of fill was approximately five feet (1.5 m) thick, and was constructed by pushing fill from jetties built across the impoundment (shown in Figure 1). Burial of mill debris and mill site reclamation waste materials were in selected areas along the constructed jetties.

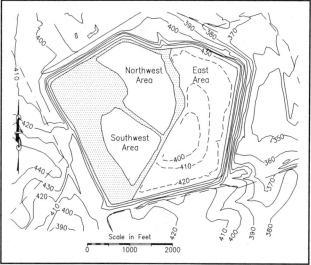

FIGURE 1 -- Site Layout at End of Initial Reclamation

Tailings pond water and precipitation during initial reclamation were managed by pumping residual tailings pond water from between the jetties to the east area of the tailings pond prior to completion of the fill surface (Figure 1). After completion of initial reclamation, runoff collected on fill in the western portion of the impoundment could be discharged. Pond water and runoff from the eastern portion of the site were contained on the tailings. Evaporation of water from the tailings was enhanced by a sprinkling system.

Reclamation Plan Description

The final reclamation design was submitted to the Texas Department of Health, Bureau of Radiation Control (TDH) in 1987 [Ref 1]. The reclamation plan was prepared to make as much use as possible of the existing tailings impoundment configuration (Figure 1) and the earthen embankments forming the impoundment perimeter. The reclamation plan consisted of a domed surface sloping to the west at a 0.5 percent slope and sloping to the east at a 1.0 percent slope (shown in Figure 2). The 3:1 outside embankment slopes were reduced to 5:1. This configuration was selected to provide a surface meeting NRC reclamation criteria while minimizing the volume of required earthworks. This surface was chosen in place of other alternatives due to the elevation and location of buried mill debris and wastes and the fill placed during initial reclamation. The reclaimed surface elevations were lowered as much as possible to minimize the required volume of random fill. The reclaimed surface of the impoundment was constructed by a combination of tailings regrading and random fill placement.

The engineered cover on top of the regraded surface consisted of a 3.5-foot (1.07-m) thick compacted clay cover and 0.5 feet (150 mm) of topsoil. The cover was designed to reduce the average rate of radon emanation to the NRC-regulated level of 20 pCi/m^2-sec and to have a permeability less than 10^{-7} cm/sec (for reduction of infiltration related to ground water protection issues).

A key design issue in meeting NRC design criteria for long-term stability is providing acceptable erosional stability. The reclamation plan at Conquista used vegetated surfaces, requiring gentle slopes and uniform surfaces to provide acceptable erosional stability. NRC criteria for acceptable erosional stability include analysis of runoff from the Probable Maximum Precipitation, and demonstration that this runoff would not cause erosion sufficient to expose tailings. The PMP selected for the analyses was 31.5 inches (800 mm) in six hours. Erosional stability analyses included comparison of calculated flow velocities with maximum non-erosive velocities, comparison of calculated shear stresses at the ground surface due to runoff with maximum non-erosive soil strengths, and evaluation for the potential for gullying.

Cover Materials

Cover materials were evaluated and selected for suitability based on permeability testing, radon attenuation testing and modeling, and dispersivity testing. Dubose clay soils excavated from borrow areas on the southwest, north, and southeast side of the impoundment met these suitability criteria and were the primary materials used for cover construction. Selected soils from initial reclamation of the western portion of the impoundment and the tailings embankment core were also used for cover materials.

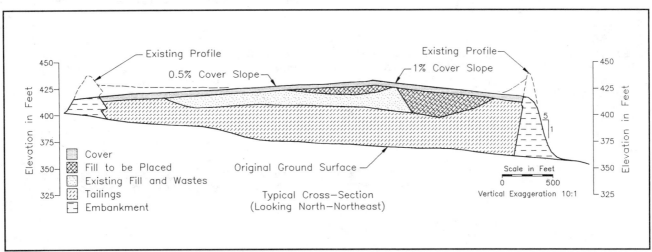

FIGURE 2 -- Typical Cross-Section through Reclaimed Impoundment

Borrow materials not meeting cover material specifications were used for random fill to reach desired slopes and elevations beneath the cover.

The cover materials were mostly high-plasticity clays, with liquid limit values primarily above 40 and plasticity index values plotting above the "A" line. Most of the Standard Proctor test optimum moisture contents ranged from 25 to 40 percent. Cover material specifications included minimum Atterberg limit and No. 200 sieve values, which allowed incorporation of minor amounts of sand in the cover.

Reclamation Plan Permitting

The closure plan was submitted to the TDH in late 1987 for termination of the Radioactive Materials License for mill operation and tailings disposal at the Conquista site. Following several rounds of meetings and written questions and responses, a set of construction drawings and specifications were prepared in 1989 that included the minor modifications to the reclamation plan from TDH review. Approval for reclamation plan construction was given by TDH in 1989.

During reclamation plan construction, several design issues and minor modifications were discussed with the TDH and included in the reclamation plan. Modifications included the configuration of the runoff drainage channels outside the perimeter of the impoundment and final cover surface elevations on the eastern side of the impoundment. The TDH also reviewed and inspected construction quality assurance testing during construction. Due to state agency reorganization in 1992 and 1993, authority for radioactive materials licensing and review of site reclamation was transferred from TDH to the Texas Water Commission and subsequently to the Texas Natural Resources Conservation Commission.

Tailings Regrading and Covering

Tailings regrading allowed the elevation of dome surface to be lowered significantly, thereby reducing the volume of random fill required to reach desired slopes and elevations. The tailings requiring regrading were generally along the perimeter of the impoundment. Tailings regrading and covering were generally in one of three shear strength and stress history conditions, each requiring different techniques for regrading.

The tailings beneath the fill on the western portion of the impoundment (from initial reclamation) had been pre-loaded with a minimum of 5 feet (1.5 m) of random fill and allowed to consolidate for approximately four years. For these tailings, regrading could be done with conventional earthmoving equipment as long as work was conducted several feet above the level of saturation of the tailings.

The tailings along the perimeter of the eastern portion of the impoundment did not have the pre-loading and consolidation history of the western tailings. Regrading was done with small dozers and high-floatation pull scrapers. This equipment allowed regrading work to take place within approximately one foot (300 mm) of the zone of saturation of the tailings.

The tailings in the center of the eastern portion of the impoundment were primarily the fine fraction of the tailings in the deepest part of the impoundment. As expected, these tailings experienced the most settlement of any tailings on site. Regrading of these tailings was not required, but covering with fill was necessary following evaporation of remaining tailings pond water. Small dozers, rubber-tracked tractors (Cat Challengers), and high-flotation pull scrapers were used to spread random fill over these tailings. One area of approximately 3 acres (7 ha) required laying geogrid over the tailings prior to fill placement. The geogrid significantly reduced the displacement of tailings due to fill placement.

Tailings Fluid Management

At the end of initial reclamation work in 1985, approximately 160 million gallons of pond water were contained in the eastern portion of the impoundment. Evaporation of this water was augmented initially by spraying along the tailings beaches to provide for evaporation from the wetted surfaces of the tailings beaches. In later stages of reclamation, evaporation cells were constructed on the regraded tailings surface. The elevation of each evaporation cell was designed to be below the elevation of the bottom of the cover. In later stages of reclamation, an additional 20-acre (49-ha) evaporation pond was constructed on part of the completed cover surface (shown in Figure 3). In early 1993, this pond was reclaimed and the cover was reconstructed. The final volume of concentrated pond water was covered with random fill in late 1993.

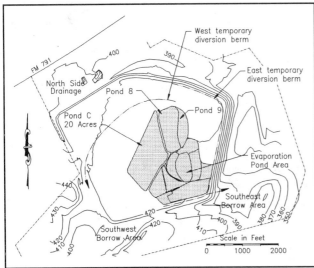

FIGURE 3 -- Evaporation Pond Layout

Cover Construction

Cover construction started in mid-1989 and was completed at the end of 1993, using both an outside earthmoving contractor and Conoco personnel and equipment. The final configuration of the impoundment is shown in Figure 4. The rate of construction varied depending on areas available for construction, weather conditions, and equipment, and was as much as 6,000 cubic yards (4620 m^3) per day. Cover construction was done with conventional earthmoving equipment, with materials loaded, hauled, and placed by scrapers; materials moisture-conditioned on the cover with disks and graders; and cover material compacted primarily with wedge-foot rollers.

Construction quality assurance testing was conducted according to NRC guidelines and Texas agency review for the cover and embankment slopes. Testing included field density tests at a minimum frequency of one test per 500 cubic yards (385 m^3) along with daily compaction and index tests. Field density measurements were made with a nuclear density gauge and correlated with sand cone tests. Undisturbed samples from initial cover placement and later stages of construction were collected and tested for confirmation of permeability values in the laboratory on site.

Establishment of Vegetation

Vegetation on the cover surface and surrounding site slopes was established by Conoco using species recommended for long-term performance and erosion protection by the U.S. Soil Conservation Service and from Conoco experience. Prior to the tailings impoundment, Conoco had several years of experience reclaiming and establishing vegetation on the uranium mine sites in the area.

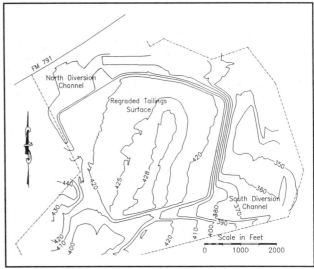

FIGURE 4 -- Reclaimed Site Layout

Establishing self-sustaining vegetation was accomplished in phases, with erosion-resistant plants established initially, and native species established as seasonal and soil moisture conditions permitted. Initially, common bermudagrass was established with transition to native species consisting of King Ranch bluestem, buffalograss, sideoats grama, and kleingrass on the top surface of the reclaimed impoundment. The embankment slopes and runoff drainage channels were initially planted with rye grass and sprigged with coastal bermudagrass. Maintenance of the vegetation is primarily by mowing or chopping to discourage growth of brushwood.

Monitoring and Performance

Monitoring of the reclaimed impoundment includes continuation of the ground water monitoring program conducted during and after mill operation. Monitoring also includes sampling of soils to ensure cleanup of uranium-contaminated areas and acceptable radon emanation from the cover, monitoring of established settlement points on the cover surface, and observation of the surface of the cover and surrounding site for excessive settlement or erosion.

Reference

[1] Water, Waste & Land, Inc. and Conquista Project, 1987. "Description of Closure Plan In Application for Termination of Radioactive Materials License No. 9-1634," submitted to Texas Department of Health, Bureau of Radiation Control, November.

Strength and Deformation of Cement-Bentonite Underground Filling Materials

Phi Oanh Tran Duc
Hazama Corporation, Japan

Hiroya Komada, Takao Endo, Michihiko Hironaga
Central Research Institute of Electric Power Industry, Japan

Koichi Taniguchi
Hazama Corporation, Japan

ABSTRACT

Cement-asphalt and cement-bentonite mixtures have been developed as filling materials for the disposal of low-to medium-level radioactive waste in underground sites. The disposal system was established underground because of their apparent low-permeability, high resistance to chemical reactions, and high deformation capacity.

This study examined the strength, deformation and permeability of cement-betonite mixtures (especially mortars of 5mm maximum aggregate grain diameter) during extended curing conditions. Investigations were performed in a laboratory and included compression, drying shrinkage, creep and permeability tests.

These tests indicated that 1) the compressive strength of the mixture depended on the water-cement ratio, regardless of the bentonite content, 2) the capacity of deformation that is valued by a modulus of elasticity and creep strain increases with the addition of bentonite, and 3) the coefficient of permeability of cement-bentonite mixture varied within the range of 10^{-8} to 10^{-12} cm/s, and showed decreasing tendencies with the addition of bentonite in 1:1 water-cement ratios.

1. INTRODUCTION

In the underground (-200 to -300 m) disposal system for low to medium-level radioactive waste, the filling material which is utilized between the rock surface and the waste storage structure plays an important role in stabilizing the structure and protecting it against the permeation from ground water.

Fig.1 illustrates the underground radioactive waste disposal system (R.W.D. System).Table 1 lists the engineering standards for the filling materials. The compressive strength is based on estimates of earth (rock) and water pressures at a depth of 300m.
The modulus of elasticity and the coefficient of permeability were determined by making comparisons with some of rock mass.

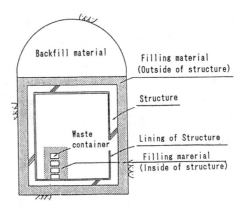

Fig. 1. The Image of R.W.D System

TABLE 1. Standard Values for Filling Material

Item	Standard Value
Compressive strength	50～100 kgf/cm^2
Modulus of elasticity	under 10^5 kgf/cm^2
Coefficient of permeability	under 10^{-7} cm/s

2. METHODOLOGY

2.1 Mix Ratio

The materials used in the investigation were Portland cement, SPV-volclay bentonite(USA), and emulsified asphalt.

Materials were prepared by mixing aggregate, cement and bentonite for 1 minute, then filling water into a forced mixer (capacity 60 liters) and mixing for 2 minutes.

Slump tests for concrete materials (20mm maximum aggregate grain diameter) and table flow tests for mortar materials were conducted to control the quality of these materials. Afterward, air content and bleeding tests were performed on the mixing materials. Mix ratios and standard values for slump/flow are shown in Table 2.

TABLE 2. Mix ratio

Water cement ratio(%)	100,130,160
Bentonite cement ratio	0,20,30,40,50,60
Standard price of slump and flow	Slump=16～21cm Flow=160～220mm

2.2 Post Curing Tests

Table 3 lists tests which were performed after the mortar materials hardened.

TABLE 3. Post Curing Tests

Name of test	Size of specimen, Method
Compression test	5cm,H10cm Strain gauges
Drying shrinkage test	H4cm,W4cm,L16cm Contact gauge (min.of reading=0.001mm)
Creep test	φ15cm,H60cm Specimen was closed up by a thin copper sheet. Compressive load is less than 40% of σc. Term of curing:1 year
Permeability test	φ20cm,H10cm Term of curing:4 weeks Hydraulic pressure: 28 kgf/cm²

Fig. 2 shows a specimen from the creep test, which was based on ASTM standards.

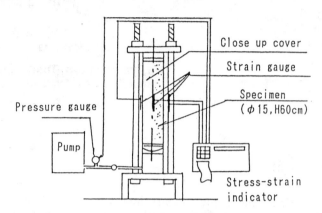

Fig. 2. Diagram of Creep Test

3. TEST RESULTS

3.1 Air Content

The air content of materials decreases with the addition of bentonite. For example, it falls below 2% when the bentonite-cement ratio (B/C) is greater than 10% (Fig. 3). This has been caused by the swelling of bentonite, which fills up the pore of material.

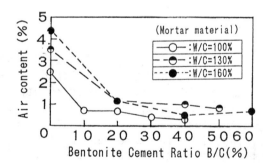

Fig. 3. Air-B/C Content Curve

3.2 Bleeding Rate

The Bleeding Rate decreases as B/C increases and the water-cement ratio (W/C) decreases. The Bleeding rate of CB materials after 1 day of mixing was under 1% (Fig. 4).
This decrease in bleeding may have been caused by the bentonite absorbing water.

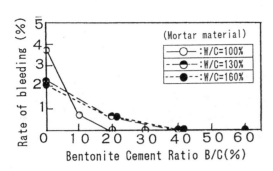

Fig. 4. Bleeding Rate-B/C Curve

Fig. 5 shows the volume change of the materials before and after bleeding. The volume decreases as B/C increases. When B/C ≧ 40%, the rate of volume decrease falls below 1%. The cause of this phenomena is considered to be the absorption of surplus water in materials by the montmorillonite layer in bentonite.

In addition, it is suggested from the viewpoint of maintaining the liquidity and workability of the materials just after mixing, the bentonite-cement ratio should be less than 0.6.

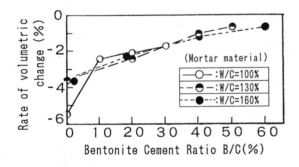

Fig. 5. Volume Change of Material

3.3 Compression Strength and Coefficient of Elasticity

The stress-train curves shown in Fig. 6 indicate that the shape of curve is affected by B/C. In other words, the maximum value of strain (i.e., strain response to compressive strength) increases as B/C increases.

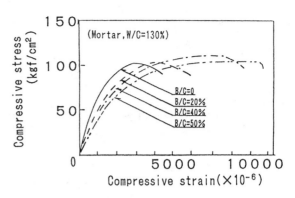

Fig. 6. Illustration of Stress-Train Curve (after 1 year)

Figs 7 and 8 show the variation of compressive strength (σc) and coefficient of elasticity (Ec) with B/C ratio respectively. These figures show that the compressive strength of the material depends on the water cement ratio, regardless of the addition of bentonite.

The compressive strength after 28 days for W/C=100% was 126 to 143 kgf/cm^2, W/C=130% was 53 to 73 kgf/cm^2, W/C=160% was 28 to 42 kgf/cm^2.

On the other hand, the modulus of elasticity was indirectly proportional to B/C. When W/C= 100 to 130% and B/C≧40%, Ec falls under 10^5 kgf/cm^2, within the range of standard values.

From the above results of compression tests, the most suitable mix proportion to satisfy the standard values of compressive strength and coefficient of elasticity is W/C=130% and B/C≧40%.

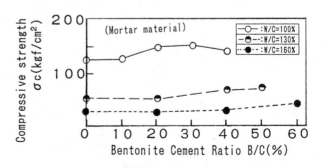

Fig. 7. Relationship Between σc and B/C

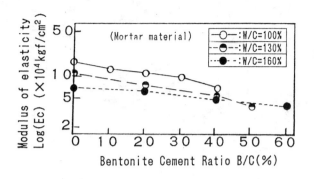

Fig. 8. Relationship Between Ec and B/C

Figs 9 and 10 show the change compressive strength and coefficient of elasticity during curing time when B/C=130%.

In most cases, the tendencies of σc and Ec during curing time are believed to be unaffected by the addition of bentonite. σc increases with the curing time, and Ec increases until 91 days, after which it remains nearly constant.

These results make it clear that bentonite has no influence on the changes of strength and deformation during time of materials.

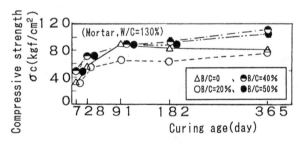

Fig. 9. Changes in σc During Curing Time

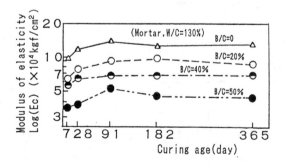

Fig. 10. Changes in Ec During Curing Time

Fig. 11 shows that $Ec/\sqrt{\sigma c}$ is inversely proportional to B/C.

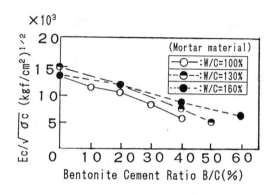

Fig.11. Relationship Between $Ec/\sqrt{\sigma c}$ and B/C

From this result, a rough relationship between σc, Ec and B/C can be expressed by the following equations.

$$B/C = (13674 - Ec/\sqrt{\sigma c})/15500 \quad (1)$$

or

$$Ec = \sqrt{\sigma c}(13674 - 15500(B/C)) \text{ kgf/cm}^2 \quad (2)$$

Based on Eq. 1 and 2, B/C can be calculated when the required σc and Ec are known.

3.4 Drying Shrinkage

Fig. 12. shows changes in the rate of drying shrinkage over time.

The rate of drying shrinkage (ratio of shrunk length to the original length of specimen) of the CB mixture after 1 year is 2 to 8×10^{-3}, 2 to 6 times greater than that of ordinary mortar (bentonite not added).

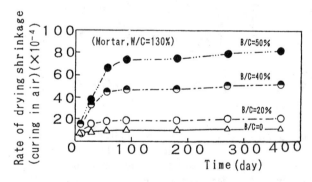

Fig. 12. Changes in Rate of Drying Shrinkage

However, shrinkage does not occur in underwater curing (see Fig. 13).

This means that when CB materials are applied in underground areas of high moisture content, the problem of drying does not occur.

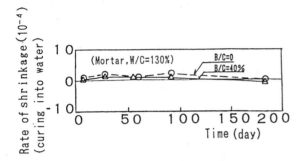

Fig. 13. Changes in Rate of Shrinkage for CB Placed under Water

3.5 Creep

Table 4 shows the results of creep tests. The creep rate F(K) in the table is used to evaluate the rate of advance of creep strain, and is calculated by the following equation.

$$\delta = 1/E_c + F(K) \times Ln(t+1) \quad (3)$$

Where
  δ = total strain
    = elastic strain + creep strain
  E = coefficient of elasticity (kgf/cm^2)
  F(K) = creep rate
  t = time (days)

TABLE 4. Results of Creep Tests

Sample	M-130-40	C-130-40	C-130-0
Mix ratio	Mortar W/C=130% B/C=40%	Concrete W/C=130% B/C=40%	Concrete W/C=130% B/C=0
Creep stress σcr (kgf/cm^2)	17.3	17.3	17.7
$\sigma cr/\sigma c$ (%)	23.7	25.2	25.4
Creep rate F(K)	7.2	4.4	1.2

The strain per unit creep stress over time are depicted in Fig. 14.

The test results are summarized as follows.
1) The creep rate F(K) of CB materials (samples M-130-40 and C-130-40) ranges from 4.4 to 7.2, which is 3.6 to 6 times greater than that of ordinary mortar (F(K) of C-130-0 = 1.2). This fact means that the rate of advance of creep strain increases with the addition of bentonite.

2) As compared F(K) of BC-mortar (M-130-40) with BC-concrete (C-130-40), F(K) of BC-mortar is greater than that of BC-concrete. This fact may have been caused by the differences of the aggregate grain size and the quantity of water contained in materials. In other words, when grain size of aggregate is small and a large amount of water is contained, the creep rate F(K) becomes large.

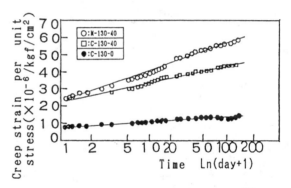

Fig. 14. Creep Strain over Time

3.6 Coefficient of Permeability

Fig. 15 shows the relationship between the coefficient of permeability (k) and B/C.

In case of W/C=100%, k decreases as B/C increases. When B/C= 20% to 40%, k=10^{-9} to 10^{-12} cm/s.

On the other hand, when W/C=130% and 160%, the effect of bentonite on coefficient of permeability was not seen clearly. However, the coefficients of permeability of BC materials in these cases (W/C=130% to 160%, B/C=20%~60%) alter from 10^{-8} to 10^{-10} cm/s, within the range of the standard k of below 10^{-7} cm/s for filling materials.

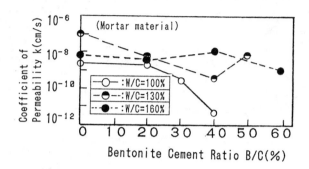

Fig. 15. Changes in k with B/C and W/C

4. CONCLUSION

We can make the following conclusions from the results of the laboratory tests.

1) The compressive strength of the material depends on the water-cement ratio, but is unaffected by the addition of bentonite.

On the other hand, the modulus of elasticity decreases as the amount of bentonite increases.

From these results, the required compressive stress and modulus of elasticity of the CB mixture can be obtained easily by selecting the proper water-cement and bentonite-cement ratios.

However, to maintain the liquidity and the workability of the materials, the bentonite-cement ratio should be less than 0.6.

2) The rate of drying shrinkage of the CB mixture after 1 year is 2 to 8×10^{-3}, 2 to 6 times greater than that of ordinary mortar (bentonite not added).

However, shrinkage does not occur in underwater curing.

3) Due to the addition of bentonite, the creep rate of the CB mixture is 4 to 7, which is 4 to 6 times greater than that of ordinary mortar.

This property is a reflection of an improved CB deformation capacity, a condition which is necessary for the filling material to protect material against failure or cracking due to deformation.

4) The coefficient of permeability of CB (W/C= 100~160%, B/C=20~60%) varies from 10^{-8} to 10^{-12} cm/s, within the range of the standard values of under 10^{-7} cm/s.

However, the effect of bentonite on the permeability of BC is not clearly seen.

Laboratory tests have helped us to understand the fundamental properties of CB materials such as compressive strength, drying shrinkage, creep strain, the coefficient of elasticity during long-term curing, and coefficient of permeability.

The resistance of the CB material against chemical reaction is currently being investigated.

Perorations

Perspectives of Sustainable Human Development in Environmental Geotechnics

Joseph D. Ben Dak
United Nations Development Programme, New York, New York, USA

Final manuscript not received in time for inclusion in the *Proceedings*. For additional information, please contact:

        Dr. Joseph D. Ben Dak
        United Nations Development Programme
        Bureau for Programme Policy & Evaluation
        1 United Nations Plaza
        New York, NY 10017
        USA
        Voice: +1-212-906-5028
        Fax: +1-212-906-5365

Comparing the Practice of Environmental Geotechnics in Various Regulatory Environments

W.F. Brumund, President
Golder Associates Corporation, Atlanta, Georgia, USA

Abstract

The paper identifies some of the important differences between Environmental Geotechnics and other areas of Civil Engineering and discusses the increasingly global aspects of the environmental market. Environmental services are largely created and driven by regulations; these vary considerably between countries and jurisdictions. The role of the consultant in the design and permit process varies greatly depending on local custom and requirements. Frequently, the consultant must help an owner understand the subtle differences in approach required in various jurisdictions and work with a client in developing environmental strategy. The US regulations are more prescriptive in nature than most other countries and the litigious attitude prevalent in the US tends to inhibit addressing some of the important environmental problems. It is doubtful if the European countries or Australia will develop as codified a set of environmental regulations as exist in the United States. Specific differences in landfill design requirements for different countries are cited to illustrate some of the different design requirements that currently exist.

Introduction

This Conference deals with the topic of "Environmental Geotechnics" yet I am not sure there is any definition or consensus as to what constitutes environmental geotechnics. Recall that the breadth of activities in the geotechnical engineering field has expanded dramatically in the past several decades. In fact, it was only 20 years ago (January 1974) that the American Society of Civil Engineers (ASCE) changed the name of the "Soil Mechanics and Foundations Division" to the "Geotechnical Engineering Division."

The practice of environmental geotechnics grew out of the environmental movement of the 1970s and was initially fueled by the promulgation of environmental regulations in the United States. The U.S. Environmental Protection Agency was founded in December 1970, and the first major piece of legislation which spawned the need for environmental geotechnics was not enacted until 1976.

The practice of environmental geotechnics is significantly different from many other areas of geotechnical engineering or civil engineering. Some markets served by civil engineers are relatively independent of regulations, some are influenced by regulations, and some markets are largely created by and driven by regulations. The field of environmental geotechnics falls largely in this last category. This creates unique and different challenges for professionals practicing in the field of environmental geotechnics. In a traditional project, engineering and functional issues normally dominate a design; regulations or statutory codes may influence a design (e.g. building code requirements) but the design methodology is determined largely by engineering logic and economy. One of the more philosophical questions and challenges for professionals in the field of environmental geotechnics is how to apply creative, innovative, and cost effective solutions to environmental problems in a highly regulated and increasingly liability prone setting.

The business area of environmental geotechnics is becoming increasingly global. As a result, there is a need to understand and be able to work in other countries in order to serve the needs of multinational and foreign-owned corporations. This paper will review the general framework and background of the United States environmental regulations primarily because these regulations spawned the business of environmental geotechnics. The paper will also outline the approach several other countries are taking in the environmental field and the role of the environmental consultant in these various settings.

Global Business Opportunities

Although the environmental movement may have started in the United States it should be recognized that there are tremendous environmental opportunities outside the USA. Today, 95% of the world's population does not live in the USA, and the non-USA population possesses about two-thirds of the world's purchasing power. The newly industrialized countries have had sustained annual growth rates of more than 8% for the past five years. China has had a growth rate of greater than 10% per year since 1980. A recent survey of 400 middle market companies in 20 different countries indicated that 86% of these firms are selling products outside of their own country and almost half of the companies have non-domestic manufacturing operations.

In 1992, 30 countries altered their laws to make foreign investment easier; none made their laws more restrictive. The trend to building global business opportunities will continue to accelerate. If companies are going to be increasingly global in their activities then the knowledge of and adherence to relevant environmental regulations is critical. This trend helps explain why it is important to understand the environmental regulations in different countries, and it provides an opportunity for the consulting community to be able to offer a service in environmental geotechnics worldwide.

Summary and Application of U.S. Environmental Regulations Affecting Environmental Geotechnics

U.S. Legislation

Beginning in the mid-1970's, the United States developed comprehensive legislation to address municipal and hazardous waste disposal for existing or proposed facilities and for closed or abandoned facilities. The most important laws that affect environmental geotechnics and the acronyms of these laws are as follows:

1. Resource Conservation Recovery Act of 1976 (RCRA)
2. Toxic Substances Control Act of 1976 (TSCA)
3. Comprehensive Environmental Response, Compensation, and Liability Act of 1980 (CERCLA or Superfund)
4. Hazardous and Solid Waste Amendments Act of 1984 (HSWA)
5. Superfund Amendments and Reauthorization Act of 1986 (SARA)

The regulations resulting from the RCRA Act were the first comprehensive federal effort to deal with the problems of hazardous and solid wastes in the United States. RCRA is a regulatory statute designed to provide cradle to grave management of hazardous wastes by imposing management requirements on the generators and transporters of hazardous materials and upon the owners and operators of treatment, storage, and disposal facilities. RCRA deals with existing and proposed facilities but does not address the problems associated with abandoned or inactive sites. The act is divided into nine parts or "Subtitles." The most often referenced parts of this act are Subtitle C which deals with hazardous wastes and Subtitle D which addresses solid or municipal waste.

With the passage of TSCA, Congress recognized that cancer can be environmentally induced. To deal with the issue of toxic chemicals, including pesticides and carcinogens, Congress provided EPA with the authority to require testing of all chemical substances, both new and old, entering the environment and to regulate them where necessary. To a large degree, the enactment of TSCA was a result of concerns over polychlorinated biphenyl's (PCB's).

Congress enacted CERCLA, also referred to as Superfund, to provide funding and enforcement authority for responding to hazardous substance spills and for cleaning up the thousands of hazardous waste sites that had been created in the United States over the previous decades. The intent was that CERCLA, along with RCRA, would provide regulations sufficient to address the clean up and remediation of closed or abandoned facilities and provide a regulatory framework for environmental control of current and future waste facilities.

Some of the more important concepts that need to be understood when reviewing the CERCLA legislation are as follows:

1. CERCLA covers all environmental media: air, surface water, groundwater, and soil. CERCLA can apply directly to any type of industrial, commercial, or even non-commercial facility irrespective of whether there are specific regulations affecting that type of facility.
2. Events that may trigger a CERCLA response or liability would be "the release or threatened release into the environment of a hazardous substance, pollutant, or contaminant." Each of these terms is broadly defined under the CERCLA regulations.
3. The cost of evaluating sites, developing the most appropriate remedial designs, and implementing remedial action at a site is extremely expensive. The CERCLA legislation authorizes the U.S. EPA to draw on two basic sources of funds to pay for waste remedies. These include the "Superfund" set up by Congress and from the "responsible parties" who have specific involvement with a site. "Potentially Responsible Parties" (PRP's) are statutorily defined as:

   a. Past and present owners or operators of a site.
   b. Parties who transported waste to a site.
   c. Parties (usually referred to as generators) who arranged for waste to be disposed or treated at a site.

4. Under CERCLA, the doctrines of "strict liability" and "joint and several liability" have been imposed on the PRP's in order to recover funds necessary to implement remedial studies and action at a given site. The Courts continue to render expansive rulings under CERCLA with the intent of increasing PRP liability (i.e. piercing the corporate veil, parent company liability). EPA and the Department of Justice have been given broad authority to seek criminal sanctions against companies as well as the individual executives of these various companies.

In 1984, Congress passed HSWA which amended the RCRA legislation. HSWA significantly expanded both the scope and coverage of the previous RCRA legislation. For example, with HSWA, Congress required EPA to more fully regulate an estimated 200,000 companies that produce only small quantities of hazardous waste (less than 1,000 kilograms per month). In addition, HSWA specified minimum technology requirements for the design of hazardous waste facilities. For example, the HSWA legislation mandated that hazardous waste landfills be designed and constructed with two synthetic liners.

The SARA amendments to the CERCLA legislation passed in 1986 provided EPA with considerably more funding to pay for the clean-up of abandoned and inactive sites in the United States. This legislation also required EPA to move more aggressively toward initiating remedial activities on sites in the United States.

EPA Guidance Documents

In addition to the regulations promulgated to enact the above legislation, the EPA has developed "Guidance Documents" which probably have more impact on the day to day activities in the practice of environmental geotechnics than do the regulations themselves. The EPA deals with the regulated community through devices known as Administrative Guidance Documents. These instructive interpretations do not have the force of law although EPA sometimes treats these documents as if they were binding. The Guidance Documents serve both as a guide to agency staff and to the regulated community by indicating the agency's views as to how it would like to see the laws implemented.

Comparison of Practice in Several Countries

United States

This paper will compare one facet of the environmental regulations in various countries, and to do so will focus specifically on the varying philosophies associated with landfill design and waste containment and the role of the Consultant in the process. The basic landfill containment philosophy in the United States is to provide and depend on the integrity of bottom liners to prevent the release of leachate and contaminants during the active life of the facility and the post-closure period. During this period of time, any leachate in the landfill must be evacuated from a primary leachate collection system. There is a prohibition on landfilling of liquids, so the predominant source of any leachate comes from incident rainfall. Since the leachate collection system must be operated well after the final closure cover is installed, there should be only "de minimus" amounts of liquid in the landfill after the post-closure period.

After closure, the reliance on minimizing or eliminating any releases from the landfill shifts from the bottom liner to the closure cover. If one can ensure that the closure cover limits the amount of infiltration into the landfill to negligible quantities, there should be little or no hydraulic gradient causing exfiltration of leachate from the landfill after closure.

Although the issue of finding superior geologic settings for landfills is addressed by the U.S. EPA in the Guidance Documents, the dominant regulatory philosophy is to depend primarily on engineered barriers as liners for landfills. The current regulations require that hazardous waste landfills have two or more synthetic or synthetic/clay composite liners in the landfill. The use of at least one synthetic or synthetic/clay composite liner is required in most municipal waste landfills in the United States today. Additionally, under the new Subtitle D regulations, leachate recycling is permitted in municipal waste landfills in order to accelerate the biodegradation of the waste. Such practice is not common and represents a philosophical change from EPA's initial strategy of minimizing all liquids within the landfill. There are limitations placed on the types and concentrations of the wastes that can be landfilled; there is a prohibition on the burial of bulk or containerized liquids for both hazardous and solid wastes. As mentioned previously, a high quality closure cover must be installed as soon as practical after completion of landfilling.

In the United States, the permit process is often adversarial, particularly during the public comment period. Consultants retained by an applicant are normally viewed as an advocate for the applicant and not an unbiased expert. The regulatory staff tasked with granting landfill permits often rely heavily on the Guidance Documents as an example of what constitutes adequate design. Significant differences in design philosophy and design details often occur between the Guidance Document suggestions and a site specific design; these technical differences and the sometimes significant cost impact of the different approaches help create the tension between regulators and applicants. In the United States, there are ten EPA regional offices and there are strong "state's rights" issues in regulating municipal solid waste. Notwithstanding the state's rights issues, the dominant federal regulations governing municipal and hazardous waste disposal help ensure consistency of approach throughout the United States.

The regulations dealing with landfills are extremely prescriptive in nature. It is doubtful if a performance standard basis of design will be adopted in the foreseeable future. That is not the case, however, when dealing with the remediation of contaminated sites. There is a desire in the U.S. today to move from the study phase to the actual clean up of sites. Given the staggering costs of clean up, there is a growing trend to try to better quantify the true impacts of soil and groundwater contamination through the use of human health and ecological risk assessment in order to evaluate the viability of various clean up options. In the future, it is anticipated that greater use will be made of decision theory, health risk analyses, and cost benefit studies to determine the most appropriate remedial option.

Canada

Unlike the United States, Canada does not have a strong set of Federal regulations regarding waste disposal. Therefore, there is considerable variability in legislation and approach in the various provinces. Recently, the Canadian Council of Ministers of the Environment (CCME) developed guidelines for managing contaminated soil and groundwater. These contaminant guidelines are used when remediating a site that is Federally owned. There is a move in various provinces to adopt the CCME guidelines as provincial standards. As in the U.S., health based risk assessments are being used more often to judge the effectiveness of contaminant cleanup options. There is considerable variability in groundwater protection standards in the various provinces; New Brunswick is more aggressive than other provinces in pushing strong groundwater protection standards, primarily because of their large reliance on groundwater in that province.

There is a fairly strong emphasis on finding sites with superior geologic containment potential across Canada. This is emphasized both by the federal government and by professors in several of the major universities. However, because some of the provinces have a dearth of readily accessible clay to serve as liners, there is strong reliance on synthetics in some of the provinces. The issue of leachate recycling in municipal waste landfills also varies provincially. Leachate recycling is permitted but not encouraged in British Columbia; whereas a significant number of the landfills in Southern Ontario permit leachate recycling. In British Columbia, a landfill operator now must post a bond or put money in an escrow account to provide for post-closure care as a condition for getting a permit.

The Consultant is often viewed by the regulators as more of an unbiased expert or almost a mediator during the permit process. The role of the Consultant, therefore, in Canada is significantly different from the role normally taken by Consultants in the U.S. during the permitting process. However, during public hearings, the Consultant is normally viewed by the public as more of an advocate for the applicant than an unbiased expert. All in all, the permit process is less adversarial in Canada than it is in the United States.

Australia

The legislation governing waste disposal in Australia is State promulgated. The charge of the recently formed Commonwealth EPA is primarily to ensure the consistency of the regulations in the various states. In the U.S. and Canada, the desire to prevent contamination of groundwater has driven much of the legislation and controls many of the decisions associated with operating landfills or cleaning up sites. In Australia, however, there is only limited reliance on the use of groundwater in the States other than Western Australia. Therefore, groundwater is not as emotive a concern as it is in North America. The primary reason for this is that in many regions, much of the groundwater is too saline to be used for drinking water or irrigation.

The regulations and consistency of practice varies widely among the states in Australia. This is partially explained by the fact that the state of Victoria has had an Environmental Protection Authority for over 20 years whereas the state of New South Wales has only had an EPA for about two years and some other states have only recently instituted Environmental Protection Agencies or are currently in the process of establishing these groups. Since the authorities are reasonably small and embryonic they rely on the Consultants more as unbiased experts and rely on common sense in issuing permits as opposed to being driven by a consistent set of regulations. In most situations, the Consultants work with the regulators in developing design or clean up standards for sites. Seldom do the regulators hire their own experts.

The state of Victoria has the most comprehensive environmental regulations and policies of any of the states in Australia. There is a clear preference to select potential sites which offer superior geologic containment when siting landfills. Currently, there are no regulations which address leachate recycling of municipal waste landfills. Victoria is starting to address the issue of post-closure care of closed facilities by requiring an operator to post a bond or provide a bank guarantee. New licenses now require an after-care period for landfill operators. Owners of old sites are being required to surrender their licenses and control is being maintained through the issue of pollution abatement notices.

For landfills, clay liners are not required but are now generally being considered if not expected. There are very few synthetically lined landfills or waste repositories in the country. Only two municipal waste landfills and one hazardous waste landfill are synthetically lined. A number of tailings and sludge lagoons are also synthetically lined. At the time of writing this paper, there were no landfills utilizing synthetics in the closure cover although one has been proposed. The currently accepted closure design practice in Victoria is to provide surface drainage and 0.5 meter of earthen material as closure cover over emplaced waste.

United Kingdom

In the UK, the environmental regulations are set at the national level. England has a Department of Environment (DOE) and has passed an Environmental Protection Act in 1990 and a Water Act in 1991. The legislation is supplemented by regulations and guidance developed by the DOE and the National Rivers Authority (NRA).

Each county in England has a Waste Regulation Authority (WRA) that is responsible for receiving, reviewing, and granting permits to operate landfills in England. In Wales and Scotland, these permits are granted by the various District Councils. There is considerable variability in the way the various WRA's interpret and enforce the environmental laws. The strictest standards and enforcement occur in the south of England.

Liners are required in all landfills. The minimum acceptable liner is one meter of clay having a coefficient of permeability of 1×10^{-9} meters per second, but the use of synthetics is extensive. Approximately 70% of the new landfills have composite synthetic/clay liners. The primary driver for liner installation is the desire to minimize the potential for methane migration away from the landfill as opposed to groundwater issues.

Currently, the NRA is developing an aquifer protection policy and it is expected that the NRA strongly recommend siting landfills at sites offering superior geologic containment potential. The responsibility of the operator for post-closure care varies somewhat from the practice in Australia. In the UK, an operator turns the permit back into the WRA at the completion of landfilling, but this can only be done after the WRA issues a certificate stating that the landfill no longer represents a significant risk to the environment. Once the operating permit is turned back into the WRA, the operator has no further responsibility for environmental damage caused by the landfill.

The role of the Consultant in the UK in working on landfill applications varies considerably. Sometimes, the Consultant's role is adversarial depending on the client and/or the WRA. In other instances, the Consultant is viewed as a professional expert in reasonably non-adversarial proceedings. Seldom do the WRA's retain their own technical experts. As with most countries whose environmental regulations have exploded in recent years, the regulators tend to rely heavily on guidance documents and other technical papers in evaluating and granting permits to operate.

Sweden

Environmental legislation in Sweden is established at the federal level and is applied in the 24 counties of Sweden. Because of the size of the country, enforcement is consistent and there is not a great deal of local flexibility. At this time, there is little use of synthetics as liners in landfills primarily because of the regulatory concerns for the safety and longevity of synthetics as liners. Leachate recycling is encouraged to accelerate biodegradation of municipal waste.

The regulations state that the typical closure cover will consist of two meters of till covering the emplaced waste. The role of the Consultant in the permit process is generally that of an unbiased expert. The regulators tend to trust the Consultant and seldom do the regulators bring in their own technical experts during the permit process.

Italy

Environmental regulations in Italy are set at the federal level; however, in practice the application of the federal legislation varies widely throughout the country. There is an increasing tendency to let the local regulators modify the application of the laws to fit local conditions. It is possible for an applicant to get a variance or an ad hoc exemption from specific legislation. Such variances are granted at the local level by the mayor of a town on an "emergency basis." These emergency variances can extend for a ten-year period.

Clay liners are commonly provided in landfills and the need for such liners is largely driven by hydrogeologic considerations. Today, the use of synthetics as liners is almost mandatory, but the use of synthetics is not common in closure covers. There is little emphasis on finding sites offering superior geologic containment potential.

In the past, leachate recycling in municipal solid waste landfills was encouraged, but that is not the case today. The federal legislation is silent on the issue of post-closure care, but it is typically the owner's responsibility.

The Consultant is normally viewed as an unbiased expert in the permit review process. The permit process is not adversarial and the regulators seldom retain their own technical experts to review a permit application. The local regulators have considerable power in granting, rejecting, or modifying a permit application.

Germany

The environmental regulations in Germany are considerably different from other European countries as well as from those in the United States. Solid waste is regulated at both the Federal and State level. Generally, the Federal regulations set the philosophy and general objectives to be achieved, but the States have the responsibility for implementing the regulations and granting operating permits. There is reasonable consistency in the application of landfill design standards in the 16 German states with the exception of Bavaria, where there is a lack of readily accessible clay in many areas.

The standard liner system consists of a synthetic/clay composite employing typically one meter of clay with a permeability of 1×10^{-9} meters per second. Concern about groundwater contamination is the primary factor necessitating the use of liners in landfills. Leachate collection systems are standard; leachate is treated and discharged as opposed to being recycled. Leachate recirculation back into emplaced waste is not permitted. Closure cover systems typically incorporate synthetics in the closure cover.

There is considerable emphasis placed on locating sites offering superior geologic containment potential and in doing extensive site characterization studies. The extensive siting and site characterization work is necessary to address the concerns usually expressed by the public who are resisting the development of any new landfills.

Most solid waste landfills are developed and owned by the local counties. The county authorities normally retain consultants to help plan their solid waste needs, locate potentially acceptable sites, characterize the site, design the landfill, and interact with the state regulators tasked with granting the operating permits. Typically the consultant is viewed by the county and state authorities as an unbiased expert in the design process.

granting agencies in various countries. Golder Associates has had more than one international client who was satisfied in dealing with professionals in one country but was frustrated with professionals in another. This frustration can sometimes be traced to the fact that the client did not recognize or appreciate the subtle differences in design philosophy, permit approach, or local custom required in various regulatory jurisdictions or countries.

Finally, while it should be stressed that the environmental market is largely influenced by the promulgation and ensuing enforcement of laws and regulations, the Consultant must also understand that, increasingly, the major market drivers for environmental services are economics and the client's concern for liability reduction or containment. If the Consultant recognizes this in assessing his client's needs, he will be able to provide a more valued service, no matter what the country.

Summary

In order to provide a consulting service in the field of environmental geotechnics in various countries, the Consultant must understand the local regulatory framework, the needs, concerns, and modus operandi of the permit granting agency and how the Consultant's role is viewed in the different countries. Consultants should recognize that owners, operators and other potential international clients may not understand the important, but sometimes subtle, differences in approach to dealing with local permit

Redevelopment of Contaminated Sites in the UK

Andrew Lord, MA(Cantab), PhD, CEng, FICE, MConsE, AFPWI
Director of Ove Arup & Partners, London, UK

In the UK, the Industrial Revolution has left a legacy of contamination, particularly in cities, leading to blight and dereliction. This will remain unless positive steps are taken for redevelopment. The success of any development relies on close co-operation between the developer, the architect, the engineer and the quantity surveyor - success being the fulfilment of the developer's requirements at an economical cost. In the case of redevelopment of derelict land the need for close co-operation is paramount, with the engineer assuming an even more important role than normally (including the essential task of co-ordination) if an economical scheme is to be achieved. Although almost any derelict land could be redeveloped for almost any purpose at a price, the cost of such development would probably be prohibitive in comparison with that for a green field site, for which very few constraints would exist. Consequently it is essential in any redevelopment of derelict land to maximise and harness its strengths as far as possible and to minimise and overcome its weaknesses such as pollution. The paper describes various schemes in the UK where the development of contaminated sites has been successfully implemented.

HISTORICAL BACKGROUND

Contamination of ground in the UK began about 200 years ago with the start of the Industrial Revolution. Urban development sprang up around industry in order to support it, so creating the 20th century industrial cities and towns. Subsequent industrial decline, often as a result of the exhaustion of natural resources, has led to urban dereliction, all too frequently on sites that are contaminated, thereby exacerbating redevelopment. But, by way of compensation, such areas were frequently the first to be mapped by cartographers and/or photographed from the air. Thus in assessing the potential for contamination on a site, the need for a desk study is paramount. Such a study would include:
- Pre-Ordnance Survey (OS) maps, including Town Plans dating back to the 17th century (and thus predating most of the now harmful industries) and Tithe Maps at 1:2500 scale recording land and property with schedules of use in the 1830s and 1840s.
- Ordnance Survey maps first published in 1804 to 1" to 1 mile scale, with larger scale mapping from the 1860s onwards.
- Non-OS maps, such as fire insurance plans and mining plans
- Geological maps dating from the 1860s and 1890s.
- Aerial photographs dating from 1941 onwards with oblique views of towns and cities from the 1920s.

ASSESSMENT OF CONTAMINATION

The current approach to the assessment of contamination in the UK is to identify the principal hazards and targets. For a limited range of contaminants, concentrations at which action should be considered ("trigger" levels) or is essential ("action" levels) have been derived for different land uses, such as
- domestic gardens, allotments
- parks, playing fields, open space.
- landscaped areas, buildings, hard cover

according to the form of the contaminants (ICRCL, 1987). These trigger values are intended to assist in the selection of the most appropriate end-use for a site and in deciding if any remedial treatment is required. A common complaint is the lack of guidance on acceptance levels of contamination and toxicity. This arises in part because the toxicity and physiological effects of contaminants are generally poorly understood almost universally. For this reason there has been a move in the UK to assess contamination in terms of the risk that it possesses, similar to the "National Classification System for Contaminated Sites" developed in Canada by CCME (1992). A recent Construction Industry Research & Information Association (CIRIA) report (RP475) on "Methane and associated hazards to construction" has sought to assess the risk posed by methane emissions by quantifying the risk of an explosion occurring.

TREATMENT OF CONTAMINATED SITES

The most abundant source of case histories in the UK is Building on Marginal and Derelict Land (I.C.E. 1987). Since then, there has been a marked reluctance to publish case histories of successful remedial treatment for contaminated land. This may be partly due to the proliferation of seminars on various aspects of contaminated land, although detailed case histories are seldom quoted, but more probably to a reluctance on the part of clients unwilling to draw attention to the existence of contamination on a site and the perceived risk from such publicity.

The basic precepts applied are:
- taking risk avoidance measures to prevent contact between contaminants and susceptible targets by adopting alternative afteruse or modified layout.
- taking risk reduction measures by removing, controlling and modifying the contamination source or migration pathways.

Risk reduction measures rely on civil engineering techniques based on the broad principles outlined by Smith (1987) and Lord (1991), namely:
1. Removal (excavation of contaminated material for treatment or disposal)
2. Containment using barriers or covers to isolate contaminated material from specified targets
3. Hydraulic control measures used in support of 1 and 2 above; as the primary means of control; or specifically for the treatment of contaminated surface or groundwater.

In the UK the use of in-situ process-based techniques for remediation, such as thermal, physical, chemical, biological, and stabilisation/ solidification, are perceived as having a much greater degree of uncertainty in their application and so have yet to find application of any significance; instead the broad thrust of treatment is to encapsulate contamination to prevent it migrating with a view to in-situ treatment when the techniques have been advanced sufficiently to remove the uncertainties while at the same time maintaining economic viability.

For many years the problem of methane emission arising from the biodegradation of organic matter in contaminated ground has been perceived as the major risk to any redevelopment of that site. But, as described subsequently, redevelopment of such sites has been safely undertaken for over a quarter of a century and this, together with the formulation of a quantifiable risk assessment, has now tended to diminish public perception of the problem. Instead the Water Resources Act (1990) together with the Environmental Protection Act (1991) have increasingly drawn attention to the risk of contaminant migration in the flow of leachate from a contaminated site, with the ensuing risk of polluting neighbouring clean sites and the groundwater. Sites in the UK which have experienced a variety of industrial uses not only contain a variety of contaminants at different depths and locations, but the infrastructure that supported them often buried, poses particular problems for redevelopment. Examples of a gasworks, former naval dockyard and power station are given.

METHANE GENERATING SITES

Factory near Nottingham

One of the earliest developments on a landfill site at which methane protection measures were adopted was for a factory near Nottingham constructed in the early 1970's. The site covered an area of 12ha and contained 10 year old landfill to a depth of 7m. The ground floor comprised a car park and primary production area each occupying approximately one third of the floor space. The remaining floor space was occupied by a despatch area and power house, with the latter extending some 1.5m below the general floor level, with lower external ground levels. It was recognised at an early stage in the development that precautions would have to be taken to protect the building from the possibility of landfill gas ingress. The protection measures employed varied between different parts of the factory:

1. **The car park**: No special protective measures were used as ventilation of the car park was required to prevent the build-up of exhaust fumes.
2. **The primary process area**: A 1.2m high ventilation space actively ventilated by directing the ventilation exhaust air from the power house through the void and to atmosphere.
3. **Power house and despatch area**: A 150mm layer of no fines hardcore with open-jointed porous earthenware pipes was installed beneath the power house and despatch area to allow the horizontal discharge of gas (see Fig. 1) venting through the inside face of a void behind the edge of a despatch docks to atmosphere by means of an extract fan.

Following construction, it was expected that the landfill would continue to settle below the suspended ground slabs opening up additional spaces. The land drains and granular fill were therefore only required for an intermediate period while the settlement of the ground occurred. Settlements following construction were observed to be about 150-300mm under the buildings.

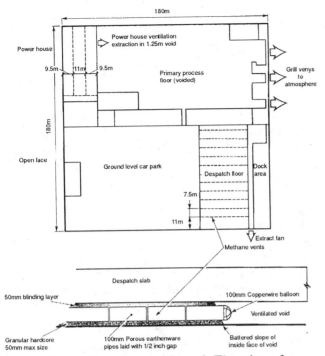

Fig. 1 Nottingham Factory: Plan & Elevation of methane ventilation details

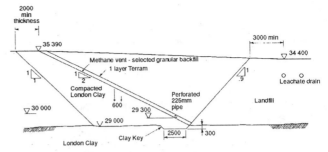

Fig. 2 Stockley Park: Details of perimeter bund

The Stockley Park Project

Stockley Park is a 36ha Business Park development which has taken place on a former landfill. The project, which involved the rehabilitation of a 140ha of derelict land formerly used for gravel workings and which had subsequently been backfilled with a mixture of domestic refuse and industrial waste, has been described by Gordon et al. (1987). Filling took place between 1916 and 1984 and raised the site level with a fill thickness of up to 13.5m. Leachate from the site was polluting the Grand Union Canal along its southern boundary.

Despite recommendations that the landfill could be left in place, commercial considerations dictated that the Business Park development was constructed on clean fill with leachate and landfill gas from adjacent areas completely excluded. All the landfill was therefore removed from the Business Park area, involving the excavation of an estimated $2.7Mm^3$ of landfill and replacement with $1.5Mm^3$ of clean fill. The excavated landfill, except special waste, was used in landscaping of the recreational area thereby minimising the removal of material off-site. Maximum use was made of naturally occurring fill material for the replacement backfill. A clay bund of 2m minimum thickness (see Fig. 2) was constructed around the entire building area to eliminate the possible entry of methane or leachate. In order to allow methane to vent to atmosphere and leachate to be removed for treatment, a 600mm layer of granular material was installed between the clay bund and the landfill. A perforated pipe was laid at the base of the granular layer to allow leachate to be removed. Beneath the car parks outside the bunds the landfill was left in place to minimise the quantity of material excavated and clean fill required.

Business Units in Buckinghamshire

A development comprising 7 two-storey business units, together with two linked warehouse units, has been constructed on a site which had been used for gravel and clay extraction since the middle of the 19th century. The gravel and clay pits had been backfilled between 1932 and 1980 to a depth of 7m over an area of approximately 2ha. Two site investigations identified areas of decomposing, oily fill on part of the site, although no more than 10% of the volume of the fill was of an organic nature. As design and site work for the development had reached an advanced stage, it was not possible to install gas monitoring boreholes, but in the majority of the trial pits, gas emissions were detected which, in areas of high concentration, did not reduce over the 30 minute monitoring period.

Three of the units were constructed over the fill with piled foundations, with suspended ground floor slabs; other units were constructed with pad footings. Large concrete obstructions in the fill were removed to a depth of about 3m. In so doing a proportion of the contaminated material was removed from the site. Remaining contamination was contained by the underlying impermeable London Clay. The following measures were adopted in the protection of the buildings against the presence of methane:

(i) natural ventilation of sub-floor voids
(ii) incorporation of a gas proof barrier into the floor
(iii) gas monitoring using:
(a) land drains below three units to allow monitoring of gas levels during construction
(b) sensors located at the void ventilation slots for long term monitoring
(iv) provision for mechanical ventilation comprising:

(a) an extraction fan and associated ducting
(b) gas monitoring sensors and fan controls.

The natural ventilation system comprised a fully suspended ground floor slab with a 350mm deep void (see Fig. 3), ventilated through slots on the upwind and downwind sides of the perimeter beams. Measures to prevent the ingress of gas vertically through the ground floor slabs involved the incorporation of a methane impermeable membrane into the floor design. At downstanding beams and at ventilation slots, the bituthene was wrapped around to give continuity to the gas proof barriers.

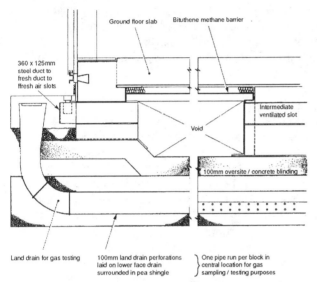

Fig. 3 Office & Warehouse Units: Gas protection details

During construction, methane levels were monitored using land drains placed under the units built on the fill. The minimum period between readings was one week but this was extended to four weeks after early results showed very low methane levels (less than 0.01%). Readings were taken continuously over a 22 month period to March 1987 and thereafter in June 1988. Continuous methane readings of less than 10ppm were measured beneath the buildings and at the ventilation slots of completed buildings. Standpipes in gas monitoring boreholes produced similar low readings; it was concluded that methane generation had virtually ceased. Therefore it was decided to discontinue the monitoring programme and not to install an active ventilation system or permanent gas monitoring.

Leckwith Athletics Stadium, Cardiff
Leckwith Stadium is built on an area of extensive landfill dating from the 1960s, overlying soft alluvium. The landfill is typically 5m thick and groundwater is at about 3m depth. Standpipes in the landfill recorded high concentrations of methane, which have persisted. The Stadium itself was founded on piles driven into the gravel stratum which underlies the alluvium; driven precast concrete piles at 4.5m centres, were also used as a foundation beneath the athletics track on account of the need to ensure very tight differential settlement tolerances. The remainder of the field sports area was dynamically compacted.

Methane venting measures consisted of a an impermeable membrane over a granular venting layer. Services ducts through the slab were sealed and confined spaces eg. cupboards under stairs, were ventilated. Gas monitoring has been carried out inside the building on a regular basis since construction was completed in 1987. Early in the life of the building, methane was detected at very low concentrations within a service manhole in the plant room. The source was traced to a poorly sealed service entry and this was repaired. Since then no gas has been detected, indicating the gas control measures are performing adequately.

Superstore and County Council offices in S. England
The 6.1ha site is within a river flood plain and is crossed by a tributary stream. The site had been used for controlled filling with domestic refuse during the 1960s and 1970s. Site investigation showed that this fill was 3½ - 4½m thick and overlay 10m thickness of alluvial clay and silt which in turn overlay chalk. The fill was capped with a layer of inert material while the alluvium contained lenses of peat particularly in the lower half. Chemical testing identified that soil contamination was generally low, but the water quality was poor; gas monitoring showed low emission rates but high concentrations of methane. In addition, methane was being generated in the alluvium under the landfill.

As the Local Authority was unwilling to permit lorry loads of refuse to be transported around local roads, designs were developed retaining the landfill insitu. The building protection system consisted of three separate levels, as illustrated in Fig. 4:
(i)   an actively vented void between the building and clay capping layer always maintained at a negative pressure relative to the building.
(ii) a gas impermeable membrane on top of the floor slab and beneath the protective screed.
(iii) a gas detector and alarm within the building.

Meteorological data studies of the area showed that passive ventilation by the wind could not be relied upon to keep gas concentrations below the building to less than 20% LEL, so an active ventilation system was adopted. The smaller office block has an air permeable,

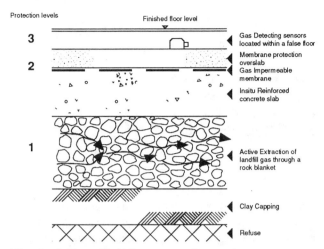

Fig. 4 Offices: Building protection system

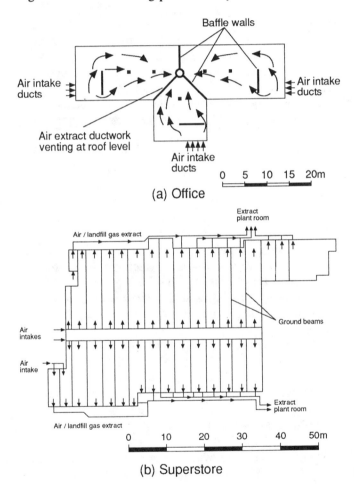

Fig. 5 Underslab ventilation systems

cobble-filled void which would reduce the cost of forming the ground floor slab. The larger superstore has a completely open void formed by a composite pre-cast/insitu concrete floor supported by ground beams and piled foundations.

Under the office block the space is divided into three equal zones (see Fig. 5). Air is drawn through each by a fan, operating continuously at low speed, connected to a centrally-placed plenum/collection chamber, entering the void blanket through ventilation openings at the three extremities of the building, with eventual discharge through the roof at high level. A back-up fan and power supply have been provided. The system is always under negative pressure so that any potential leak will allow air to be drawn in rather than allow an air/methane mixture to escape. If any sensor in the extract duct detects a methane level greater than 20% LEL, it will switch the fans to high speed; they run until the level reduces to 5% LEL or for 30 minutes, whichever is the greater. Gas-detecting sensors are also installed in the building in areas of high risk: confined spaces, penetrations through the ground floor slabs, and electrical switchrooms. The sensors are linked to a control board with alarms: visual at low levels of landfill gas and aural at high levels. In practice the volume of methane in the extracted air was very low and never exceeding 0.4% LEL.

Perimeter trenches, backfilled with cobblestones are vented at 20m centres to provide preferential paths to prevent lateral migration of landfill gases out of or into the site. Although Waste Management Paper 27 dictates the need for a gas-proof barrier, this is not justified where landfill is present outside the confines of the site. All manholes have perforated covers, other than those near the building which are vented to more distant soft landscaping, whilst in the car parks, isolated passive gas wells are provided. All light standards are protected from gas ingress. Services trenches, which are potential conduits for landfill gas, are sealed with a clay plug at entry and exit positions from the site.

Assessment

From the foregoing examples, it may be seen that sites in the UK where landfill is present and where methane is being generated have been successfully developed for a variety of end-uses. The absence of appropriate design criteria in the Building Regulations until recently has resulted in a variety of solutions to the problem, the majority involving passive ventilation but some requiring active ventilation, but in all cases two or three lines of defence have been provided, frequently including monitoring systems. Irrespective of these measures, careful attention to detail is essential in designing the structure, as illustrated by Barry (1987). Further details of the Protection of New and Existing Developments from Methane and Associated Gases in

the Ground are given in the CIRIA Project Report RP453/1. The type of development and usage is an important factor in the selection of individual protection measures. Residential housing is probably the most sensitive and more at risk from methane and gases. This is because of the difficulty in incorporating gas/control systems which:
- need to be effective for considerable length of time, either the design life of the residence or when the gas regime ceases to be hazard
- need little or no maintenance
- impose little or no infringement or constraint on the householders' use of the property including building modifications, land use, etc.
- are energy efficient and incur little or no cost to run.

The capability for assessing the need for gas protection measures has been improved within the past two or three years by the publication of the Building Regulations Approved Document (DoE, 1992) and Waste Management Paper 27 (DoE 1991).

Over the past 2 or 3 years, CIRIA has undertaken an Environmental Geotechnics Programme: Methane and Associated Hazards to Construction. One of the research projects which forms Phase III of this work provides guidance on assessing the degree of risk from gas to which a development could be exposed (O'Riordan and Milloy (1994)). Their report proposes a rational methodology for gas hazard evaluation and risk assessment that would be appropriate for a wide range of construction situations and ground gases but particularly for methane and carbon dioxide. Outlining the principles of risk assessment and applying them to the context of ground gas entering a building, the risks presented by these gases are related to other tolerable risks and specifically to those of natural gas supply mains. The proposed methodology starts from a qualitative screening that leads on to a quantitative procedure.

SITES OF FORMER GASWORKS
With the advent of natural gas supplies from the North Sea, so vanished the need for gasworks to produce gas from coal. This left a legacy of derelict sites, which were invariably polluted, close to the centre of cities and towns. Moreover the pollutants are composed of tars, phenols and hexacyanoferrates, but various factors help mitigate the problems posed:
- the layout of gasworks is generally well mapped so that the location of the buildings where the different processes were performed, and hence the wastes likely to accumulate, is known
- many of the sites are frequently underlain by naturally occurring clay deposits which inhibit downward migration of contaminants
- most contaminants have accumulated at shallow depths in known areas

Many of the sites are close to rivers or streams and hence pose a threat to pollution of water courses.

In Building on Marginal and Derelict Land, Collins et al (1987) and Mills & Clark (1987) respectively describe the redevelopment of Norwich and Wandsworth Gasworks. In both cases, very little of the contaminated material was removed: naturally occurring basal seals and, where necessary, sheet pile walls prevented migration of the contaminants.

St Helens Gasworks
St Helens Gasworks previously occupied a 4ha site on the outskirts of St Helens, Lancashire. The use of the site as a gasworks dates back to the early production of gas from coal. Prior to this the site had been used for an alkali works and prior to that deep and shallow mining had taken place under the site. This had left a legacy of mine shafts and workings, contamination, buried foundations and tanks. The proposed end use after redevelopment was for a non-food retail park.

The fundamental principle of reclamation design was to understand the groundwater regime at the site. Being in part on former railway sidings, areas of the site comprised fill above natural ground. The site was bounded on two sides by road which are up to 5m lower than the site itself. This together with the results of a desk study and site investigation led to the conclusion that groundwater was at some depth in natural ground. Thus the potential for migration of contaminants was an acceptably low risk, and so the existing contaminated material could be left on the site but with a 1m thick capping of compacted insitu contaminated material, summounted by a 500mm layer of crushed arisings. In areas of spread foundations (light steel buildings were proposed and therefore piling was unnecessary) not less than 1.5m of compacted granular fill would spread foundation loads. Foundation areas were set out to suit building platforms for the anticipated end-users. Prescribed service trenches were routed between the platforms and filled with clean granular material for future excavation and service provision.

In addition to the above, earthworks included:
- excavation to find and cap buried mine shafts appro-

ximately 4-6m beneath existing ground level
- manipulation of excess material previously tipped on the site and unsuitable as engineering fill.

Part of the temporary works involved the creation of three interconnected lagoons separated by a filter bund for coarse primary decontamination. Contaminated liquid from the site was deposited in the lagoon giving an opportunity for testing for compliance with discharge limits into the foul sewer, or if necessary, be removed from site by tanker. The primary filter removed some of the worst contaminants including suspended solids and particles.

Unchartered tanks, pipes and containers of often tarry and contaminated liquors are typical hazards of such contaminated sites. The site is traversed by a culverted river and part of the initial investigation was to survey the culvert and to report on its integrity. This revealed concrete defects in the roof which, on subsequent removal of precast planks, were found to be in the conversion of high-alumina cement concrete, so necessitating their replacement.

CHATHAM MARITIME

English Estates are redeveloping the former Chatham Naval Dockyard with a mix of commercial and residential areas and a marina based around the old dockyard basins. St Mary's Island is a 60ha area lying to the north of the dock basin as shown in Fig. 6. Naval use of this site has led to contamination of the soil with heavy metals and asbestos; remediation is being undertaken on St Mary's Island as a Ground Treatment Contract and the East Bund Contract, each with a very different technical and philosophical approach.

The level of St Mary's Island had been raised by about 2-3m (more where creeks were filled) using spoil from excavation of the dock basins, rubble and ash and other waste from brickworks located on the Island. Only the area adjacent to the basins had a significant number of buildings, most of the island being playing fields and woodland. Past uses of the site have included timber stores, sawmills and seasoning kilns, machine shops, laboratories, paint shops and stores, asbestos lagging workshops and stores, electrical substations, boiler houses and boiler makers shops, battery stores and workshops, railways, a decontamination centre, a fire school and even a cemetery for Napoleonic French prisoners of war.

The site lies over what was once the channel of the River Medway carved out of the Chalk and Thanet Sand to a depth of about 18m below current ground level (see Fig. 7). The old channel has been infilled with river gravel, typically 3m thick, and alluvial silts and clays, often with peat, which overly the gravel and extend to the underside of the fill material, typically 3-5m below current ground level. The fill, generally a silty clay, derived from the river alluvium, also contains ash, rubble and other man-made materials.

St Mary's Island ground treatment contract

A site investigation consisting of trial pits on 100m grid had been carried out prior to letting the Ground Treatment Contract. Trial pits had been excavated to visually "clean" material, usually clay free or ash of other contaminants, but this had not always been sampled and tested to confirm it to be clean. The investigation showed the presence of heavy metals above ICRCL residential end-use threshold levels but, apart from a relatively small area of the site, asbestos contamination was believed to be at less than 0.75m depth.

English Estates' requirement was for a site capable of development for a residential end-use throughout, although about half the site was to be landscaped open space. To maintain a completely flexible masterplan, it was not possible to remediate the site to different levels in different areas dependent on the final end-use. Testing in pits excavated from the base of excavated areas showed that contaminants exceeding the ICRCL Threshold Trigger Levels for residential end-use

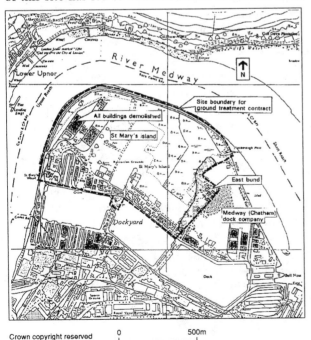

Fig. 6 Plan of St. Mary's Island, Chatham

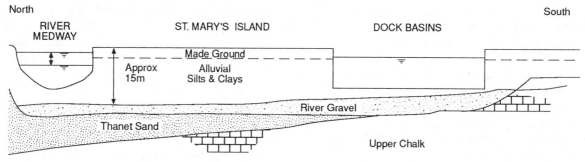

Fig. 7 Geology of St Mary's Island, Chatham

extended to greater depths than had been anticipated. Excavations of 5m, and sometimes more, were not uncommon and the final volume of soil excavated was 1.2Mm³, considerably more than that originally envisaged.

Some clean-up criteria were also found to be inappropriate in the context of the site. The limit of 5000-mg/kg for Toluene Extractable Material was often exceeded in peaty material. Initially this was excavated and removed before more specific testing showed that the relatively high levels of TEM were not caused by contaminants but by naturally occurring immobile hydrocarbons in peat. Large volumes of material with concentrations of the phytotoxic metals, copper, nickel and zinc slightly above the contract clean-up criteria were also removed (sometimes from considerable depth) despite the fact that the ambient pH of the soil was normally greater than 7. At these pH levels higher concentrations of these metals are tolerable to plants as their availability is reduced. In the later stages of the Contract the acceptable levels of copper, nickel and zinc were increased by 50% if pH>7 in line with DoE guidance on the use of sewage sludge. It should also be noted that the soil at Chatham is naturally very saline and concentrations of Boron often exceed the ICRCL Threshold Level for phytotoxicity. Copper, nickel and zinc are in general probably less of a limitation to plant growth than the natural saline environment.

Initially all material exceeding the original clean-up criteria was exported, including quite large volumes of material suitable for use in landscaping areas. This was subsequetnly stockpiled for later use in the East Bund capping works, thereby saving nearly £3.5M.

East Bund Contract
On the eastern side of St Mary's Island (see Fig. 6) a 4.5ha area of miscellaneous material, mostly contaminated soil and demolition arisings, had been placed in the mid 1980s to a height of about 8m above the surrounding marsh levels. The material's chemical composition was unacceptable in terms of the adjacent 60ha Ground Treatment contract for the new housing site. The cost of removal of the bund material was assessed to be about £15M, so that it was considerably cheaper to encapsulate the bund to USEPA and UKNRA standards. It was demonstrated that the 13m of underlying alluvium would adequately attenuate the flow of mobile contaminants into nearby water courses. The 10m high bund, landscaped with grass terraces, footpaths and wooded glades with views along the Medway River Valley, would not only provide a physical barrier with an amenity value between the housing site and an adjacent dockyard, but would also save enormous removal costs. The encapsulation comprised:
- regrading the bund to form a stable profile
- placement of LDPE membrane and engineered cover system
- construction of a 5m deep slurry/membrane cut-off wall on housing side
- construction of surface water drainage system
- landscaping

Following a groundwater survey, the contamination present on St Mary's Island (see Fig. 6) was found to be mostly in solid form. Migration of the soluble components of the main contaminants, heavy metals and hydrocarbons, had either ceased or was occurring at a sufficiently low rate that the deeper groundwaters were not significantly contaminated. This was considered to be due to the high pH and high surface area presented by the fine grained alluvial strata that underlay the site; a fairly typical salt marsh environment that, furthermore, is located close to alkaline chalk strata. In order to ensure that:
- the material left on the island did not pose a threat to the adjacent housing site
- the works would not worsen the existing groundwater quality

the bund (see Fig. 8) was to be encapsulated by:
(i) an engineered capping up to 2m thick, with sur-

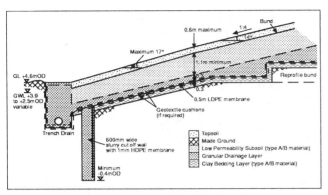

Fig. 8 Sections through East Bund, Chatham Maritime

face water run-off control
(ii) an engineered vertical barrier on the western (housing) side in case the capping were to fail

Although the ground had been investigated using boreholes and trial pits, finding all potential contaminants and pathways could not be guaranteed, so standpipes and piezometers were installed to enable long-term monitoring of groundwater quality to be carried out.

OLDHAM BROADWAY REDEVELOPMENT

The Oldham Broadway Redevelopment at Chadderton, about 3km west of the centre of Oldham and 10km north-east of Manchester typifies the problems of redevelopment of a site which has been subject to a variety of industrial uses and activities over the past 200 years. The site (see Fig. 9) is about 1350m long and tapers from 600m wide on it southern boundary to 100m wide on its northern boundary, totalling some 54 ha in extent. It is bounded on its western side by railway tracks, while the Rochdale Canal forms at least 50% of its eastern boundary. The canal was completed in 1804 and the railway line in 1839. Both were constructed on fill where they crossed the east-west flowing Wince Brook and its associated alluvium in the northern third of the site, the canal being 5-10m high and the railway up to 13m high. For development purposes, the site has been divided into two phases: Phase 1 in the southern sector occupies 34ha (excluding electricity substations) and Phase 2 in the north totalling some 14ha. These phases also reflect past uses of the site, which may also be sub-divided into Areas 1W and 2W on the west side of the canal and 1E and 2E on east side of the canal.

Historically the first part of the site to be developed was Area 2W, when the Slacks Valley Chemical Works was constructed between 1800 and 1825. This produced vitriol (sulphuric acid) and continued in existence until between 1880 and 1905, with the building demolished by 1909. From 1891, the Gibraltar Works (soap and candle) were located immediately to the south of the Vitriol Works adjacent to the Rochdale Canal, but shown as disused by 1922. Adjacent to the latter works the canal had been widened to form a basin in 1840s. The 1933 OS map shows tipping had already taken place towards the north between the railway and the canal which, by 1948, had created a level surface at the same elevation as the main railway line. This was subsequently extended over Wince Brook when the existing culverts beneath the railway line and canal were joined in 1956. Tipping was believed to be complete by 1974 when the Refuse

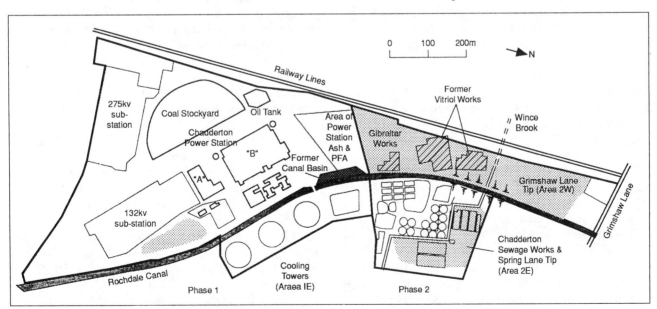

Fig. 9 Oldham Broadway: Combined history of site

Pulverising Depot on the northern boundary of the site was removed. Extensive site investigations have shown that the filled ground varies between 5 and 13m in thickness, being greatest in the vicinity of Wince Brook. As a result of foundations and infrastructure of the former Vitriol Works are now likely to be buried by up to 11m of fill.

To the east of the canal (Area 2E), Chadderton Sewage Works was first shown on the 1933 OS map. By 1975, aerial photographs show that with the exception of Wince Brook and six rectangular settling tanks, the remainder of the sewage works appeared to be unmaintained and overgrown with vegetation. In addition, fill comprising domestic refuse had been placed over the eastern part of the works. By 1979 the area containing the six rectangular tanks had been backfilled and graded level to the same elevation as that of the banks of the Wince Brook, but it is not known whether the tanks were removed prior to backfilling. This area of the site is about 10m lower than the remainder of the area while further to the east is situated the present Slacks Valley Sewage Works.

Area 1W was first developed between 1925 and 1927 with the construction of the Chadderton "A" power station whose location is shown in Fig. 9, together with sidings serving it and connected to the main railway line. The "A" station was replaced in 1953 with the larger "B" station with three large cooling towers located on the east bank of the canal (Area 1E), extensive railway sidings, a semicircular coal stock area and two large electrical substations (one of 275kV where power is delivered from the National Grid and the other of 132kV where power is connected to the distribution network). The original station was demolished sometime between 1959 and 1975, while the "B" station was sold in 1984 and subsequently demolished apart from the administrative block and the workshop. The northern part of this area was used as an ash and PFA dump, while in about 1975 two oil storage tanks surrounded by spillage containment walls were constructed to the north west of the coal store.

Constraints on site development
Major constraints to the development of Areas 1W and 1E may be summarised as follows:
- Existing electricity substations
- Underground and overhead cables connecting the substations to the distribution networks. Corridors and areas can generally be identified within which the services are located: in addition to electricity cables, a 610mm diameter foul sewer runs alongside the canal while the main foul sewer feeding to the Slacks Valley Treatment Works crosses the site between Areas 1W and 2W.
- Buried foundations and underground structures remaining from the former power station. The main building of Chadderton "B" was founded on piles, with pile caps up to 3m thick and a ground floor slab up to 750mm thick. No details are known about the foundations of the "A" station.
- Contamination - the major areas of potential ground contamination would be the coal and ash stockpiles. Minor areas of contamination could arise from oil spillages from tanks and transformers and chemical spillages from water treatment plant. All asbestos was apparently removed during the demolition of the main power station building.
- Although an existing road network serves the power station, this and the provision of services was inadequate for the development area.

In addition to the non-existent access and service provisions for the Phase 2 area of the site, the major constraints to development was the depth of landfill material in both the former Grimshaw Lane (Area 2W) and Springs Road (Area 2E) tips. Furthermore, spillages of raw materials, waste and biproducts from the now buried Vitriol Works are likely to exist beneath the landfill. Monitoring of gas levels within the Area 2W tip was undertaken for two years during which gas levels remained steady at levels considerably in excess of the published guideline levels. Therefore, for any development on the landfill area, gas control measures would be required. The landfill which was deposited on the former sewage works to the east of the canal (Springs Road Tip, Area 2E) was also found to be generating methane.

Development Masterplan
From its initial conception, Oldham Broadway has been seen as a high quality Business Park providing 1Million ft^2 of development in a landscaped environment. The proposal to construct the M66 Manchester Outer Ring Motorway across the southern part of the site (see Fig. 10) and to provide a major junction with the A662 Broadway will assist the objectives by considerably enhancing accessibility to the site.

The Masterplan for the development was drawn up to reduce the influences of the site constraints and to maximise development potential. For example, the buried structures for the Chadderton "B" power station, particularly the main building and cooling towers, would be very costly to remove prior to constructing

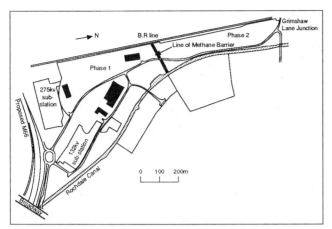

Fig. 10 Oldham Broadway: Highway layout after motorway construction with current development

new foundations. However, the presence of these buried foundations presents an opportunity for founding the new structures on top of the old foundations since it is extremely unlikely that the new structural loads would exceed those imposed by the power station. Similarly, locating new structures on the area of the coal stockyard could also be advantageous since the ground beneath would have been subject to a surcharge load from the coal itself for about 30 years, so reducing future settlement. In this way a central core to the development would be created which would then be supplemented by a high quality environment with the reclaimed Rochdale Canal forming a central water feature.

The redevelopment of the site is proceeding in stages:
- To overcome the poor access a new road network has been constructed through the site from the A662 Broadway in the southeast corner, to Grimshaw Lane in the northwest corner, where a new junction was created, as illustrated in Fig. 10. This involved the resurfacing of 400mm of existing road to the National Grid Substation and the construction of 1900m of new road including a 20m span bridge utilising the foundations from one of the cooling water culverts over the Rochdale Canal. Nearly 600m of the new road was constructed along the edge of the Grimshaw Lane Tip (Area 2W) founded partly on the original canal construction and partly on landfill.
- Some services were diverted during the course of the road construction, but many, particularly the high voltage cables, had to be left in place on account of the excessive cost of diverting them.
- The Grimshaw tip (Area 2W) was thoroughly investigated and the water table was found to be approximately 4m below the surface of the tip. Landfill gas was found to be generated but was venting naturally to the atmosphere, as the tip had not been capped. As the tip had been constructed in a natural depression which formed the valley of the Wince Brook, the railway embankment to the west and the canal embankment to the east formed natural barriers to contaminant migration. The glacial clay underlying the site also provided a natural lining to the tip.
- As landfill gas is being generated in the Phase 2, but not in the Phase 1 Area, a methane barrier was constructed across the southern boundary of the Grimshaw Lane tip (see Fig. 10) to prevent the escape of landfill gas to the Phase 1 of the development. The barrier consists of a 7m deep cut-off toeing into the underlying clay stratum and incorporating an HDPE membrane in a cement/bentonite slurry. Venting facilities, comprising a blanket of stone beneath the road on the east side and a 5m deep trench on the west side, have also been provided.
- Reclamation of the Rochdale Canal has involved the dredging of approximately 8,000m^3 of silt so as to permit reclamation of 1.6km of the abandoned canal to navigable standards; this has included repairs to the canal wall, refurbishment of the tow path and the construction of a large basin in the vicinity of the 1840s original. The silt from the canal was placed in a lagoon constructed on the Springs Lane Tip (Area 2E). Extensive sampling and chemical testing of the silt showed it to be contaminated and unsuitable for use as topsoil, and hence it will be encapsulated within the lagoon.
- Development of the Phase 1 Area is progressing as indicated in Fig. 10: two industrial units have been developed on site now for almost two years - one of these units, on the site of the former Chadderton "A" station, makes ice cream, while another, a 30,000ft^2 warehouse and 5,500ft^2 office is located close to the railway line. A 102,000ft^2 printing and warehouse facility as being developed on the "B" station site. Further development is taken place on the coal stockyard area.

CONCLUDING REMARKS
There is a growing recognition in the UK that there are no "no risk" situations and that decisions must be made as to the tolerability of risk. Political, economic and other intangible factors (e.g. fear of unknown, confidence in predictions) may decide the final outcome. Nevertheless risk assessment provides an invaluable aid to decision making and is used by the Health & Safety Executive (HSE) to assess major industrial hazards. Quantitative estimates, usually of risk of

serious injury or death, are made for the particular hazard under consideration. Risk levels are compared with risks entailed in other human activities to allow decision makers to set risks in context. HSE criteria for assessing risk are based on a three-zone approach. Above an upper limit, the risks are a major consideration and probably unacceptable. Below a lower limit, the risks are probably insignificant. Between the limits, a more detailed study may be necessary and the ALARP Principle applies i.e. risks should be reduced to a level which is As Low As Reasonably Possible.

Risk assessment is particularly appropriate to contaminated land problems in view of the uncertainties involved in determining the effects of contaminants on receptor groups (often humans), in how exposure may occur, and in the distribution and concentration of contaminants in the ground. Methods of assessment range from qualitative/semi-quantitative to full quantitative risk assessment. Semi-quantitative systems generally take the form of specified contaminant "guidance" levels (triggering particular actions e.g. Dutch A and C levels, ICRCL guidelines) or scoring systems which consider the principal factors likely to affect risk (e.g. CCME screening system). Such systems are useful for quickly screening a large number of sites, prioritising sites and indicating the likely significance of risks, but problems can arise when they are used in a prescriptive manner rather than for guidance. Where preliminary screening methods indicate risks which are too high, a quantitative risk assessment which is site specific and addresses particular end use should be carried out. This type of approach is now being more generally adopted e.g. Canada (CCME guidelines), UK (CIRIA report 475 on Methane hazards), HSE (major industrial hazards, nuclear installations).

REFERENCES

Barry, D.L. (1987) "Hazards from methane and carbon dioxide". Reclaiming contaminated land (ed. Cairney, T.) Blackie, Glasgow, Ch 11, 223-255.

Card, G.F. (1994) "Protection of new and existing development from methane and associated gases in the ground". CIRIA Report RP453/1.

CCME (1992) "National Classification System for Contaminated Sites". Report CCME EPC-CS39E, Winnipeg, Manitoba.

Collins, S.P. et al. (1987) "Rehabilitation of the Old Palace Gasworks site for the Norwich Crown and County Courts". Building on marginal and derelict land. Thomas Telford. 449-496.

DoE (1991) Waste Management Paper No. 27. The control of landfill gas.

DoE (1992) The Building Regulations Approved Document, London, HMSO.

Gordon, D.L., Lord, J.A. & Twine, D. (1987) "The Stockley Park project". Building on marginal and derelict land. Thomas Telford, 359-379.

I.C.E. (1987) "Conference on Building on Marginal and Derelict Land". Glasgow, 1986, Thomas Telford.

Interdepartmental Committee on the Redevelopment of Contaminated Land (1983) "Notes on the redevelopment of sewage works and farms". Dept. of Environment, Cent. Div. on Environmental Protection, London, 2nd edn. ICRCL 23/79.

Interdepartmental Committee on the Redevelopment of Contaminated Land (1987). "Guidance on the assessment and redevelopment of contaminated land". Dept. of Environment, Cent. Div. on Environmental Protection, London 2nd edn, ICRCL 59/83.

Lord, J.A. (1991) "Options available for problem-solving". Recycling derelict land (ed. Fleming, G.), Thomas Telford, Ch. 6, 145-195.

Mills, G. & Clark, J.C. (1987) "The redevelopment of Wandsworth Gasworks site". Building on marginal and derelict land. Thomas Telford, 497-520.

O'Riordan, N.J. & Milloy, C. (1994) "Risk assessment for methane and other gases in the ground". CIRIA Report RP 475.

Smith, M.A. (1987) "Available reclamation methods". Reclaiming contaminated land (ed. Cairney, T.), Blackie, Glasgow, Ch. 5, 114-143.

Tailings Disposal in an Equatorial Environment

William J.C. Meynink
Pells Sullivan Meynink, Brisbane, Queensland, Australia

Final manuscript not received in time for inclusion in the *Proceedings*. For additional information, please contact:

        Mr. William J.C. Meynink
        Pells Sullivan Meynink
        55 Northam Avenue Bardon
        Brisbane, Queensland 4065
        AUSTRALIA
        Voice: +61-7-217-5195
        Fax: +61-7-369-5062

Physical Barriers for Waste Containment

James K. Mitchell
*The Edward G. Cahill and John R. Cahill Professor of Civil Engineering Emeritus,
University of California, Berkeley, California, USA*

Abstract: Contained land disposal is likely to remain a major technology for safe disposal of newly generated hazardous and municipal wastes for many years to come, and in-situ containment of wastes within already contaminated ground by means of barrier walls and covers can be a viable means for environmental protection. Vertical barriers include slurry walls, grouting, sheet pile walls, deep soil-mixed walls and geomembranes installed within slurry trenches. Bottom barriers include natural soil and rock formations, grouted barriers, composite liner systems, jet grouting, and some recently devised special systems. Covers are usually constructed in layers of compacted soil, geosynthetic and drainage materials. Materials, construction methods, interactions with the surrounding environment and wastes of various types, advantages and limitations, and permittivity of different types of physical barrier systems are considered. More data on the long-term effectiveness of barriers is needed.

Introduction

There are three strategies for site remediation and waste management for existing contaminated sites and wastes that cannot be controlled by minimization at the source and incineration: clean up, stabilization, and containment. Containment in the ground using passive physical barriers is the subject of this paper. Hydrologic barriers are not considered; however, hydrologic modifications are often integrated with physical barriers, as for example, by pumping from within a walled off region.

Vertical barriers (walls), bottom barriers, and covers, as shown schematically in Fig. 1, are considered. Landfill liner systems are outside the scope of this paper. Space limitations prevent presentation of all details of materials, properties, design, construction, and performance. Therefore, principal focus is on barrier types, functions, general characteristics, construction methods, compatibility between waste and barrier, and permittivity. Many references are available concerning most details of containment barrier systems. Two recent references that provide comprehensive coverage are Mitchell and van Court [1] and Rumer and Ryan [2].

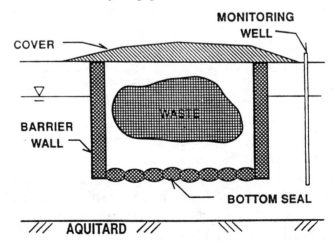

Fig. 1 General configuration of physical barriers for waste containment.

Criteria for Containment Barriers

The major function of all barrier systems is to provide isolation of wastes from the surrounding environment, thereby protecting groundwater, soil, and air from contamination. Thus, the ability of any barrier to prevent movement of waste liquids and gases through it is of paramount importance. Accordingly, the hydraulic and gas conductivities of properly manufactured and installed barrier materials are the properties of greatest concern. In the U.S.A. it is usually required that the hydraulic conductivity be less than 1×10^{-9} m/s for earth and treated earth barrier materials, with values up to 1×10^{-8} m/s allowed in some cases. Geomembrane barriers are usually specified in terms of minimum acceptable thickness and material type rather than a maximum allowable conductivity.

In reality, however, the overall effectiveness of a barrier in preventing seepage and diffusion through it depends on both its conductivity and its thickness. The hydraulic conductivity divided by the thickness is termed the *permittivity*, and this quantity provides a basis for comparing the relative effectiveness of different barrier systems.

In some cases a containment system must be designed to insure that concentrations of different contaminants at specified compliance points, for example at a property line, are less than specified values. Extensive, expensive, complex, and often uncertain field investigations, hydrologic modeling, and numerical analysis studies may be required for this purpose. Recent experiences related to landfill stability failures; e.g., the Kettleman Hills failure [Mitchell et al., 3], and to ground movements associated with the installation of slurry walls, as well as concern for the seismic stability of landfills, have introduced additional criteria that must be considered for most projects.

The design of liners and covers for hazardous and solid waste landfills in the U.S.A. is now governed by regulations developed by the U.S. Environmental Protection Agency (U.S.E.P.A.) and by the states. The U.S.E.P.A. has promulgated minimum technology guidance documents to aid in design and construction. A series of articles in *Geotechnical News* [4] provides detailed illustration of how regulatory requirements, in this case those arising from Subtitle D of the 1984 Hazardous and Solid Waste Amendments to the 1976 Resource Conservation and Recovery Act, can impact the design and construction of municipal solid waste landfills.

Vertical Barriers (Walls)

Vertical barriers are used to block lateral flows. They may be installed to contain wastes and contaminants or to redirect groundwater flow. In many cases a vertical barrier will be installed to completely surround a site; however, upgradient and down gradient barriers are also used. When possible, a vertical barrier is keyed into an underlying aquitard layer. If the aquitard is at very great depth or if the pollutants are predominantly LNAPLs (light, non-aqueous phase liquids), a *hanging wall* may be used. A hanging wall extends beneath the water table, and interior pumping may be required to maintain an inward gradient.

Types of vertical barrier walls that are now available include compacted clay, soil-bentonite slurry trench walls, cement-bentonite slurry trench walls, plastic concrete and concrete diaphragm cutoff walls, vibrating beam walls, composite geomembrane-slurry walls, deep soil-mixed and jet grouted walls, steel and geosynthetic sheetpile walls, and walls of grouted soil.

Compacted clay barriers can be constructed above the water table in trenches and may be suitable for shallow depths. An obvious advantage of this type of barrier is that the construction can be monitored, and, therefore, good quality assurance is possible.

Slurry trench walls are typically of the order of one meter thick and may be constructed to depths of several tens of meters. Excavation techniques for slurry walls are summarized in Table 1 taken from [2]. The general construction procedure for slurry walls is illustrated in Fig. 2, which shows also some of the defects that may develop in improperly constructed walls.

During excavation a mud slurry, usually of bentonite clay, is used to stabilize the open trench against collapse. The depth of slurry is maintained at a level above the water table so a net outward hydrostatic pressure acts against the trench walls. The bentonite-water slurry should have a unit weight in the range of about 10.1-11.0 kN/m^3. In some cases an attapulgite clay may be used for greater chemical stability, or a biodegradable mud might be used if a gravel filled trench is to be constructed for use as a cutoff and drain for lateral flow.

Soil-bentonite cutoff walls are made by backfilling the slurry trench with a soil-bentonite mixture. Typically the excavated material is mixed with up to 5 percent bentonite to achieve an acceptably low hydraulic conductivity, $< 1 \times 10^{-9}$ m/s if possible.

The hydraulic conductivity of a soil-bentonite wall depends on the gradation of the soil and the nature of its fines as well as on the bentonite content. Studies; e.g., Barvenik [6] and Evans [7], have shown that little further decrease in hydraulic conductivity is likely for fines contents above about 20 percent and bentonite contents above a few percent. As the sensitivity of high water content clay-water systems to changes in structure and properties as a result of changes in chemical environment usually increases with increasing clay content, it follows that the long term stability of soil-bentonite backfills should be greatest for mixes with the lower contents of fines and clay.

Cement-bentonite cutoff walls are formed by mixing cement into the bentonite slurry that stabilizes the trench. The slurry sets in place to form the low permeability barrier. This type of wall has the advantage that it can be used at sites where there is insufficient space to mix the excavated soil and bentonite for a soil-bentonite wall. It has the disadvantages that data presented by Ryan [8] show the hydraulic

Table 1. Slurry Trench Excavation Techniques

Type	Trench Width (m)	Trench Depth (m)	Comments
Standard backhoe	0.5-2.0	10	Rapid, least costly
Backhoe with extended boom	0.5-2.0	16	Rapid and low cost (extended excavation boom on unmodified backhoe)
Modified backhoe with extended boom	0.5-2.0	20	Extended excavation boom; modified power & counterweights
Clamshell	0.3-2.0	75	Attached to crane
Dragline	1.0-3.0	50	Slow, for wide, deep excavations
Rotary	0.5-2.0	75	Slow, for boulders & rock
Specialty	varies	varies	e.g. Hydrofraize (from Soletanche)

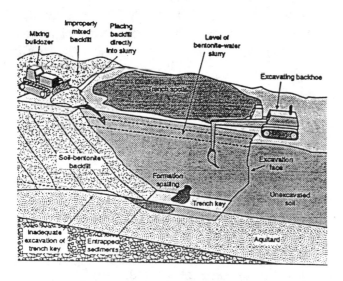

Fig. 2 Soil-bentonite slurry trench cutoff wall construction illustrating potential defects (from LaGrega, et al., 5)

conductivity to be higher than for a soil-bentonite wall and the soil excavated from the trench must be disposed.

Plastic concrete cutoff walls are similar to cement-bentonite walls except that they also contain aggregate. They are usually constructed in panels rather than continuously, as is usually done for cement-bentonite and soil-bentonite walls. The greater strength and stiffness of these walls may make them better suited for applications where ground stability and movement control are important. *Concrete diaphragm* cutoff walls in which concrete is tremied from the bottom up into a slurry trench, may also be considered when high structural strength is needed. In addition to their high cost, the difficulty of insuring good hydraulic seals between adjacent panels may be a problem.

The *vibrating beam* cutoff wall is constructed by driving an H-pile into the ground with a vibratory hammer and filling the void with grout as the beam is withdrawn. As this method produces a barrier that is relatively thin; i.e., 25 to 100 mm, it is difficult to insure a wall without discontinuities. Furthermore, if this method is used in cohesionless soils that are subject to densification by vibration, then deep beam penetration may be difficult or impossible.

Composite walls formed of vertical geomembranes inserted within a slurry trench are being used increasingly. Proper installation of a high density polyethylene (HDPE) sheet or envelope within a slurry trench may reduce the overall hydraulic conductivity of the trench-geosynthetic system very significantly - up to 4 to 5 orders of magnitude if high quality joining of the geomembrane sheets can be obtained [Jessberger, 9].

Insertion of the geomembrane sheet into the slurry trench is usually done either by mounting it on an installation frame, which is subsequently detached and withdrawn, or by hanging weights on the bottom which serve to pull the sheet down into the trench. Installation of the sheets without puncturing them or tearing them and the development of a continuous seal between adjacent sheets are potential problems in the use of this method.

Steel sheet pile cutoffs, while many times suitable for groundwater control at construction sites, are less suitable for waste containment applications owing to potential contaminant leakage through the interlocks. Special sealable joints have been developed to overcome this problem. In addition, *geosynthetic sheet piles* with sealable joints are now available [1], Gundle [10], Starr, et al. [11].

Mixed-in-place seepage barriers are constructed using deep soil mixing or jet grouting. Cement and bentonite are the most commonly used additives to produce a low hydraulic conductivity wall. Treated columns are overlapped to make a continuous wall, usually about one meter thick. Advantages of mixed-in-place walls are that there is minimal disturbance of adjacent areas, and excavation of potentially contaminated ground is avoided. Augers for deep mixing are usually mounted in groups so that several columns can be treated at one time. A typical procedure is shown schematically in Fig. 3. Jet grouting for formation of barrier walls against contaminant flow is described in [2].

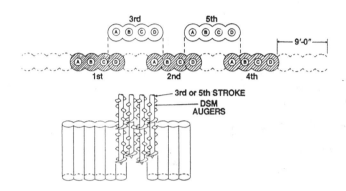

Fig. 3 Construction of a soil-mixed wall (from Geo-Con, Inc., 12)

It is pointed out in [2] that the state of stress in slurry trench cutoff walls needs to be known for proper interpretation of hydraulic conductivity data obtained in the laboratory. As bentonite slurries, soil-bentonite backfills, and cement-bentonite mixes are usually more compressible than the adjacent trench sidewalls, interface adhesion and/or friction may transfer some of the self weight of the slurry and backfill to the surrounding ground. As a result the consolidation of the backfill may be impeded, with the result that the effective vertical stress increases at a lower rate than given by the effective unit weight times the depth.

Of perhaps greater importance, especially with the increased use of slurry trench construction in developed areas, is the potential for adjacent ground movements. There do not appear to be any validated procedures for estimating the deformations that may result from digging the trench, supporting it with a slurry, and backfilling it with a highly compressible backfill. Nonetheless, deep excavation construction experience has shown that movements, with magnitudes dependent on the soil type, depth, construction procedures, and support system characteristics, are inevitable. It is prudent, therefore, to give consideration to possible ground movements when slurry cutoff walls are constructed close to existing buildings, streets, and utilities.

Injection grouting can be used to form low permeability curtains in both soil and jointed rock. Usually 2 or 3 rows of grout holes, at spacings of 1.5 to 3 meters, are required to achieve the required degree of sealing. In addition to considerations of compatibility between the grout and contaminants in the ground, the choice of grout will depend on the size of soil particles or rock fracture widths, as shown in Fig. 4. In addition to the grout types shown, microfine cements are now available that can penetrate the voids in fine sand. Different grout types are sometimes used in combination. For example, cement may be used first to fill the larger voids followed by a chemical grout to fill the smaller voids. A major concern with grouted barriers, as with virtually all below ground barrier construction, is the near impossibility of insuring adequate overlap of the grout columns and assessing the overall integrity of the grout curtain until post-construction monitoring data are available.

Bottom Barriers

Natural bottom barriers may exist if the site is underlain at reasonable depth by intact aquicludes. In such cases the side wall vertical barriers are keyed into the aquiclude to provide isolation and containment. Consolidated clay layers, tills, and many rock types may serve as

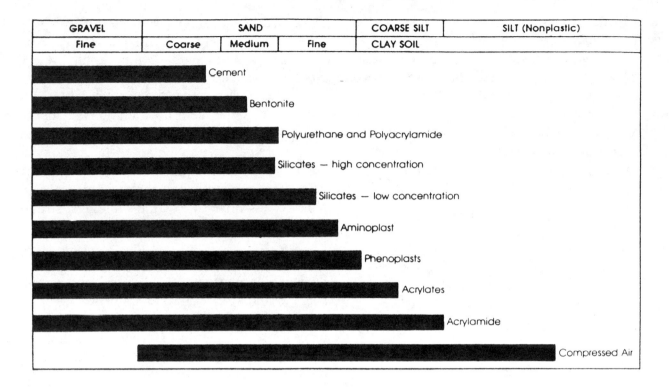

Fig. 4 Penetrability of various grouts (from Karol, 13).

suitable bottom barriers. The most difficult aspect in utilization of natural bottom barriers is demonstrating to everyone's satisfaction that the layer is continuous, that it has sufficiently low hydraulic conductivity, that it does not contain joints, fissures, or solution channels, and that it will not deform, crack, or otherwise deteriorate as a result of exposure to aggressive chemical wastes or changed loading conditions.

In coarse-grained soils *penetration grouting* can be used to build a layer of overlapping grout bulbs beneath a site. The grout is injected at pressures adequate to displace water from voids, but not so high as to cause hydraulic fracturing. The "tube-a-manchette" (sleeve pipe) technique can be used if best control over grout quantities and locations is needed. In this method the grout tubes are set in drill holes, and packers or other control measures are used to inject the grout within a desired depth range.

If such tight control is not needed, then grouting at the desired depth from open pipes may be satisfactory provided sufficient overlap between adjacent grout bulbs is obtained.

A vibration grouting technique is described by Tausch [14]. A vibratory hammer and a pipe pile are used to drive a grout valve to the desired depth. As the pipe pile is withdrawn, the soil falls in around the grout valve and a polyethylene tube connecting it to the surface, thus holding the valve in place. Grout is then injected through the polyethylene tube.

If jet grouting is used for bottom sealing, the jets are used to form short columns or disks that will overlap each other to form a seal beneath the contaminated region.

At some sites drilling through contaminated zones or waste for installation of a bottom barrier may not be allowed. In such cases *slant drilling* for injection or jet grouting might be considered, provided the overall width of the area to be treated is not too great. The concept of a bottom barrier in conjunction with two vertical walls to provide complete isolation is shown schematically in Fig. 5.

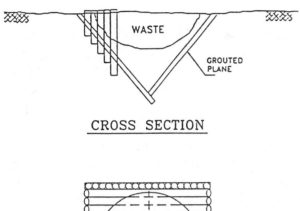

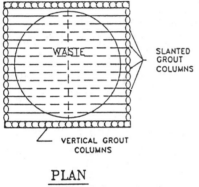

Fig. 5 Bottom barrier formed by slanted grout columns [2].

The procedures and potential applications of *directional drilling* for construction of bottom barriers are described in [2]. Directional drilling enables following a parabolic path beneath the ground. This may be useful for reaching beneath a waste pile or contaminated zone without going through it. Complete encapsulation of a waste site may be possible using the proposed Zublin System, described by Jessberger [15] and Rumer and Ryan [2]. This system involves trenching and installation of a bottom membrane liner laterally from the trench bottom from one side to the other using specially devised excavators. Both directional drilling and special systems such as this are likely to be very expensive.

Both construction quality assurance and reliable assessment of the effectiveness of bottom barrier systems are likely to be very difficult and uncertain. About the only way to determine if a bottom barrier installed below ground is *not* functioning as planned would be to detect contaminants outside the contaminated zone whose presence cannot be attributed to leakage through side wall barriers.

Covers (Surface Control Barriers)

Covers, also called *caps*, are constructed over contaminated sites or waste landfills to isolate the waste from the environment above. A cover may be designed to perform one or more of several important functions, including (1) raising the ground level to a specified elevation, (2) separating the waste from plants, animals, and the air above, (3) minimizing the infiltration of water into the waste from above, and (4) controlling the release of landfill gas from below. Detailed coverage of the details of cover systems is given in [2] and by Daniel and Koerner [16]; only a few considerations and details are included here.

Covers over waste landfills and hazardous waste sites are exposed to more severe exposure conditions; e.g., wetting and drying, freezing and thawing, precipitation and runoff, than are usual for other components of a waste containment barrier system. In addition, the activities of humankind, animals, and vegetation above and the influences of gas and settlement from below mean that durable and flexible designs are required. Furthermore, surface slopes will be required in most cases in order to insure adequate drainage of surface water, and, in the case of above-ground waste piles, to provide maximum storage volumes for a landfill. Slopes introduce stability considerations into the design of most cover systems. Settlements mean that special provisions may be required to insure continuous operation of leachate and gas collection systems.

Five possible components that might be used in a cap over a solid waste landfill in the U.S.A. to meet the requirements for final closure are shown in Fig. 6, from Daniel in [2]. Not all layers are always included.

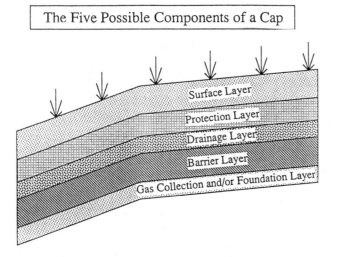

Layer	Description of Layer	Typical Materials	Typical Thickness (m)
1	Surface Layer	Topsoil; Geosynthetic Erosion Control Layer; Cobbles; Paving Material	0.15
2	Protection Layer	Soil; Recycled or Reused Waste Material; Cobbles	0.3 - 1
3	Drainage Layer	Sand or Gravel; Geonet or Geocomposite	0.3
4	Barrier Layer	Compacted Clay; Geomembrane; Geosynthetic Clay Liner; Waste Material; Asphalt	0.3 - 1
5	Gas Collection Layer and/or Foundation Layer	Sand or Gravel; Soil; Geonet or Geotextile; Recycled or Reused Waste Material	0.3

Fig. 6 Basic components of a final cap [2].

The *surface layer* serves mainly to separate the underlying components from the air above, to provide a working surface that is compatible with the planned use of the area, and to minimize temperature variations in the materials below. The *protection layer* separates the waste from burrowing animals and plant roots, protects the underlying layers from excessive wetting and drying, and provides protection against freezing below. The *drainage layer* serves to reduce the head on the underlying barrier layer, drains the surface and protection layers above, and reduces the pore pressure in the materials. The pore pressure reduction may be very significant in maintaining adequate strength in the soil materials and along layer interfaces to insure slope stability.

The function of the *barrier layer* is to impede the flow of precipitation and surface water through the cap into the waste below. New solid waste regulations in the U.S.A. require that the hydraulic conductivity of a cover be less than that of the liner system underlying the waste. The purpose of this requirement is to prevent an increase in the leachate level above the bottom liner by water infiltrating from above. Demonstrating that this requirement has been satisfied by the design and construction may be difficult in practice. Compacted clay and geomembranes have been most commonly used for barrier layer construction in the past. A compacted clay layer at least 2-ft thick with a hydraulic conductivity less than 1×10^{-9} m/s has been a typical requirement. Satisfactory long term performance of compacted clay layers cannot always be assured when there is desiccation, freezing and thawing, or excessive settlement of the waste below, any of which can cause cracking.

Partly in response to these types of concerns, *geosynthetic clay liners* (GCL) have been developed for use as barrier layer materials. GCLs are composed of geosynthetics and bentonite that are specially fabricated into layers. Four types of GCL that are currently on the market are shown and described in Fig. 7. All of these GCLs contain about 4 to 5 kg/m^2 of sodium bentonite. Their dry thickness is about 6 mm, but they expand to about 12 mm thickness on hydration. The GCLs are manufactured in panels with widths of about 5 m and lengths up 60 m and shipped in rolls. Mechanical joining of overlapped seams is not required to insure hydraulic barrier continuity; i.e., the materials are self-sealing at the overlaps provided the manufacturer's instructions are followed. Additional information concerning GCLs is given in [2].

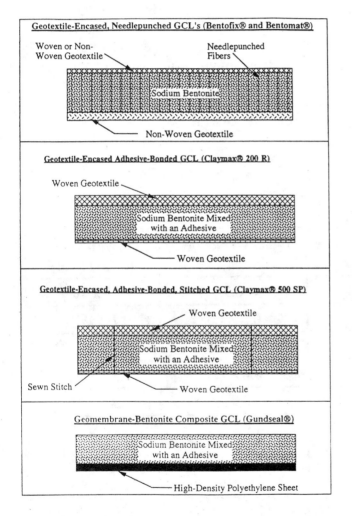

Fig. 7 Four types of geosynthetic clay liners [2].

The lowermost layer in the cover system is the *foundation and/or gas collection* layer. It must be strong enough to support the overlying layers and construction equipment. A range of materials may be used for this layer, as indicated in Fig. 6.

Chemical Compatibility

Much has been written about potential adverse interactions between chemical wastes and barrier materials that could lead to cracking or other forms of deterioration resulting in increases in hydraulic conductivity. Test methods for assessment of waste soil interactions and their effects on properties have also been studied extensively. Space limitations preclude their review here. Review articles that deal with these issues include Bodosci, et al. [17], Mitchell and Madsen [18], Quigley and Fernandez [19], and Jefferis [20].

Site specific compatibility tests on proposed barrier materials should be done under representative conditions before the final barrier design is determined whenever there is reason to believe that there could be a problem. Of particular concern are sites where there are concentrated, free phase organics, very acidic (low pH) or basic (high pH) solutions, and/or high concentrations of dissolved salts. High plasticity clays at high water content, such as bentonite slurries and soil-bentonite mixtures, are more susceptible to shrinkage and cracking when exposed to aggressive chemicals than are lower water content, densely compacted soils.

Flows Through Barriers

No barrier type or barrier material is totally impervious to chemical transport through it.

Nevertheless, the barriers that are now available can be built to meet regulatory hydraulic conductivity requirements and be sufficiently resistant to hydraulic flow by advection and diffusion that they are safe for environmental protection. Typical values for the hydraulic conductivity of different barrier materials are given in Table 2. Also shown are usual thicknesses for barriers made with the indicated materials. The permittivity, defined earlier as the ratio of hydraulic conductivity to thickness, corresponds to the flow rate for a unit difference in head across the barrier. The permittivity is a useful measure of the relative effectiveness of different barrier types in blocking advective flows.

Molecular diffusion will contribute a small amount to the total chemical flow across a barrier, with the flow direction being from the high concentration side to the low concentration side, and the rate proportional to the concentration difference. Chemical diffusion coefficients are typically in the range of 2×10^{-10} to 2×10^{-9} m^2/s for diffusion through soils and on the order of 3×10^{-14} m^2/s for diffusion through geomembranes. In materials with low hydraulic conductivities, molecular diffusion of contaminants can be important relative to advection. The relative importance of advection and diffusion to total chemical transport for soils of different hydraulic conductivity k are:

$k > 1 \times 10^{-8}$ m/s - advection dominant
$k = 1 \times 10^{-10}$ to 1×10^{-8} m/s - both are important
$k < 1 \times 10^{-10}$ m/s - diffusion dominant

Even with diffusion, however, the total chemical transport through soil and geomembrane barriers will be low for low hydraulic conductivity barrier systems, provided the head difference across the barrier is small.

TABLE 2--Hydraulic conductivities and permittivities of barrier materials

Material	Hydraulic Conductivity (m/s)	Typical Thickness (m)	Permittivity (m^3/s/m^2/m)
Compacted Clay	1×10^{-9}	0.9	1.1×10^{-9}
Soil-Bentonite	1×10^{-9}	0.5 - 1.5	$0.7-2 \times 10^{-9}$
Cement-Bentonite	1×10^{-8}	0.5 - 1.5	$0.7-2 \times 10^{-8}$
Soil Mixed Wall	1×10^{-9}	1.0 - 2.0	$0.5-1 \times 10^{-9}$
Grout Barrier	1×10^{-8}	1.0 - 2.0	$0.5-1 \times 10^{-8}$
Jet Grout Barrier	1×10^{-8}	1.0 - 2.0	$0.5-1 \times 10^{-8}$
Geomembrane	1×10^{-14}	.001 - .002	$0.5-1 \times 10^{-11}$
Geosynthetic Clay Layer	$1 \times 10^{-12} - 1 \times 10^{-10}$	0.012	$8 \times 10^{-11} - 8 \times 10^{-9}$
Steel Sheet Piling*	1×10^{-7}	0.01 - 0.015	$0.7-1 \times 10^{-5}$
Drainage Layer*	1×10^{-4}	0.3 - 0.6	$1.5-3 \times 10^{-4}$

* For Comparison

Conclusion

The most commonly used types of wall, bottom, and cover barriers for waste containment have been described. When properly designed and constructed, such systems should provide safe waste containment over time. It is likely that in-ground storage of new wastes and isolation of zones of contaminated ground will have important roles in environmental protection and enhancement for many years to come.

Unfortunately sufficient long-term performance case histories are not yet available for many types of systems, and quality assurance at the time of construction is difficult. Collection and promulgation of such information could go a long way towards changing public perception that all contaminated ground must be completely cleaned and that new wastes can never be stored below ground anywhere near anything.

References

[1] Mitchell, J.K. and van Court, W.A. (1992). Contaminant Immobilization and Containment: Barriers-Walls and Covers, Geotechnical Engineering Report No. UCB/GT/92-09, University of California, Berkeley, 53 pp.

[2] Rumer, R.R. and Ryan, M.E. (1994). Review and Evaluation of Containment Technologies for Remediation Applications. New York State Center for Hazardous Waste Management, State University of New York ar Buffalo, 156 pp.

[3] Mitchell, J.K., Chang, M., and Seed, R.B. (1993). The Kettleman Hills landfill Failure: A Retrospective View of the Failure Investigations and Lessons Learned, Proc. 3rd Int. Conf. on Case Histories in Geotechnical Engineering, St. Louis, MO, U.S.A., Vol. II, pp. 1379-1392.

[4] *Geotechnical News*, (1993) Solid Waste Landfills, Vol. 11, No 3, pp. 36-52.

[5] LaGrega, M.D., Buckingham, P.L., and Evans, J.C. (1994), *Hazardous Waste Management*, McGraw-Hill, NY, (in press)

[6] Barvenik, M.J. (1992). Design Options Using Vertical Barrier Systems, ASCE Int. Convention and Exposition, Environmental Geotech. Symposium, New York, September.

[7] Evans, J.C. (1993). Vertical Cutoff Walls, Ch. 17 of *Geotechnical Practice for Waste Disposal*, D.E. Daniel, ed., Chapman & Hall, pp. 430-454.

[8] Ryan, C.R. (1985). Slurry Cutoff Walls: Applications in the Control of Hazardous Waste, *Hydraulic Barriers in Soil and Rock*, ASTM STP 874.

[9] Jessberger, H.L., ed. (1993). Geotechnics of Landfill Design and Remedial Works - Technical Recommendations "GLR", 2d Edition, Ernst & Sohn.

[10] Gundle Lining Systems (1991). Gundwall Subsurface Polyethylene Vertical Barrier, 20, CHP 0991, Houston, TX.

[11] Starr, R.C., Cherry, J.A., and Vales, E.S. (1991). Sealable Joint Sheet Pile Cutoff Walls for Preventing and Remediating Groundwater Contamination, Technology Transfer Conference, Ontario Ministry of the Environment, Toronto, Canada.

[12] Geo-Con (1989). Technical Brief: Deep Soil Mixing, T-DSM-01-89, Pittsburgh, PA.

[13] Karol, R.H. (1990). *Chemical Grouting*, Marcel Dekker, Inc., New York, 465 pp.

[14] Tausch, N. (1991). Recent European Developments in Constructing Grouted Slabs, Grouting, Soil Improvement and Geosynthetics, ASCE Geotechnical Special Publication, No. 30, pp. 301-312.

[15] Jessberger, H.L., ed. (1993). *Sicherung von Altlasten*, A.A. Balkema, Rotterdam, pp. 85-92.

[16] Daniel, D.E. and Koerner, R.M. (1993). Cover Systems, Ch. 18 of *Geotechnical Practice for Waste Disposal*, D.E. Daniel, ed., Chapman & Hall, pp. 455-496.

[17] Bodocsi, A., Bowers, M.T., and Scherer, R.A. (1987). Waste Contamination Effects on Grout Barriers, *Geotechnical Practice for Waste Disposal '87*, ASCE Geotechnical Special Publication No. 13, pp. 306-319.

[18] Mitchell, J.K. and Madsen, F.T. (1987). Chemical Effects on Clay Hydraulic Conductivity, *Geotechnical Practice for Waste Disposal '87*, ASCE Geotechnical Special Publication No. 13, pp. 87-116.

[19] Quigley, R.M. and Fernandez, F. (1989). Clay/organic interactions and their effect on the hydraulic conductivity of barrier clays, Proc. Int. Symp. on Contaminant Transport in Groundwater, Stuttgart, Germany.

[20] Jefferis, S.A. (1992). Contaminant-Grout Interaction, *Grouting, Soil Improvement and Geosynthetics*, ASCE Geotechnical Special Publication, Vol. 2, p. 30.

The Observational Method in Environmental Geotechnics

N.R. Morgenstern
University of Alberta, Department of Civil Engineering, Edmonton, Alberta, Canada

Abstract

The observational method is an important aspect of risk management in geotechnical engineering. Recent considerations in its evolution are reviewed. It is noted that there are some circumstances in environmental geotechnics that are ideal for the application of the observational method. However there are others in which it is seriously constrained. Constraints arise from: 1) the regulatory environment, 2) the nature of decision-making related to environmental matters and, 3) the issue of longevity. These matters are amplified by discussions on the role of the observational method in mine waste management, in ground remediation, in landfill design, and in nuclear waste management.

Introduction

The overriding requirement in engineering is for the constructed (manufactured) entity or process to fulfill its intended function. That it should do so safely, economically and in an environmentally acceptable manner are also usually desirable, but not always essential, objectives. That is, the dam must store water in a safe, economical and environmentally approved manner; the foundation must support the load in a safe and economic manner; and the landfill must function in a contained manner while still being economical and in compliance with environmental regulations.

The geotechnical engineer has a long tradition of success in meeting these requirements under conditions that differ from many other types of technological endeavour. The natural materials that the geotechnical engineer must deal with are complex and do not afford the luxury of replication. Construction processes, either in-situ or associated with the construction operation itself, are performed under circumstances very different from the controlled environment of a manufacturing plant. As a result, uncertainty is a perpetual component of geotechnical design and construction. In view of the successes in the past of geotechnical design practice, there is considerable value in understanding how the geotechnical engineer deals with this uncertainty and in evaluating how this methodology might be applied to environmental design problems.

It is rare for the geotechnical engineer to rely on prediction to meet his objectives. In practice, prediction is considered to be a chimera, only worth contemplating under the most idealized circumstances. The practice of the geotechnical engineer is more modest. Risk is managed to overcome the limitations of site characterization, knowledge of material properties, other unknowns and the vagaries of construction practice. Performance is assured through design that is not driven by prediction in any direct manner and this is executed by means of the observational method.

The evolution of the observational method has had a profound influence on the practice of geotechnical engineering. It is not without its pitfalls, as will be noted below, but the observational method is widely recognized as providing a conceptual framework for geotechnical design that differs from other types of engineering design.

Some problems of environmental geotechnics are similar in kind to traditional geotechnical engineering, but many raise new issues. In some instances the observational method provides an excellent framework from which to address these problems. In others, it provides restraints on the practice of the geotechnical engineer in dealing with certain environmental problems.

It is the intent of this paper to summarize the evolution of the observational method, to assess the differences between design issues in environmental geotechnics and more traditional aspects of geotechnical engineering and to underline circumstances not only where the observational method can be applied effectively but also where there are limitations to its application.

Evolution Of The Observational Method

Peck's [22] classical paper summarizes the evolution of the observational method in its restricted geotechnical sense and it is essential to quote his summary of the method.

"In brief, the complete application of the method embodies the following ingredients:

a) Exploration sufficient to establish at least the general nature, pattern and properties of the deposits, but not necessarily in detail.

b) Assessment of the most probable conditions and the most unfavourable conceivable deviations from these conditions. In this assessment geology often plays a major role.

c) Establishment of the design based on a working hypothesis of behaviour anticipated under the most probable conditions.

d) Selection of quantities to be observed as construction proceeds and calculation of their anticipated values on the basis of the working hypothesis.

e) Calculation of values of the same quantities under the most unfavourable conditions compatible with available data concerning the subsurface conditions.

f) Selection in advance of a course of action or modification of design for every foreseeable significant deviation of the observational findings from those predicted on the basis of the working hypothesis.

g) Measurement of quantities to be observed and evaluation of actual conditions.

h) Modification of design to suit actual conditions.

The degree to which all these steps can be followed depends on the nature and complexity of the work. We can readily distinguish between projects, on the one hand, in which events have already set the stage for the observational method as being almost the only hope of success, and those, on the other hand, in which use of the method has been envisioned from the inception of the project. Applications of the first type are much the more familiar."

Peck [22] went on to provide examples of successful application but then drew attention to significant pitfalls such as:

i) Failure to anticipate unfavourable conditions: Potentially the most serious blunder in applying the observational method is the failure to select in advance appropriate courses of action for all foreseeable deviations of the real conditions, as disclosed by the observations, from those assumed in the design.

ii) The dominance on the project exercised by the concern whether or not a single potential problem can be solved.

iii) Choice of significant observations: The selection of proper quantities to observe and measure requires a feel for the significant physical phenomena governing the behaviour of the project during construction and after completion.

iv) Influence of progressive failure: The presence of brittle elements in a resisting mass may, if not appreciated, lead to failure in spite of the use of the observational method.

v) Complications in contractual relations.

One might also note that the observational method is limited when dealing with dynamic loading.

Others have added to our understanding of the concept and its application. In particular D'Appolonia [5] extended it to the operational phase of facilities and processes through his concept of "monitored decisions", see Figure 1. This extension is particularly useful when considering various problems in environmental geotechnics that arise during operational phases and require contingency planning. Muir Wood [20] characterized the observational method as a flexible design philosophy, capable of achieving very substantial benefits, provided that flexible adaptation can be accommodated in both procedural and contractual arrangements. Muir Wood [20] also notes some conflict between the implementation of certain Quality Assurance and Quality Control programs and the need to foster iterations between geotechnical design and construction.

The application of the observational method is enhanced by new developments in instrumentation and data-processing. It is also enhanced by new developments in computer modelling and simulation. Chen, Morgenstern, and Chan [2] have emphasized that the most effective use of finite element modelling and related techniques is in history matching of performance as an adjunct to the

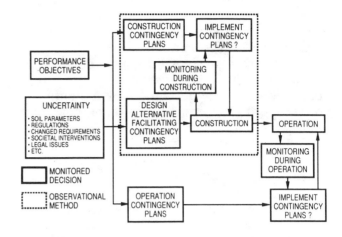

Figure 1. Flow Diagram of the Monitored-Decision Process (D'Appolonia, 1990)

observational method, as opposed to forecasting behaviour during the design phase.

There has been a considerable effort expended in recent years to develop a reliability based design approach that can accommodate different kinds of uncertainty such as uncertainty in the geological model, parameter uncertainty, cost uncertainty and other important considerations. The work of Freeze and his students [7] [13] may be the most ambitious to date. Quantitative case histories based on these methods tend to reaffirm principles of the observational method such as " . . . it was more important to know whether the field value is above or below some threshold value than it is to know its actual numerical value . . ., and . . . indicating the importance of a conceptual understanding of the site geology in estimating prior probabilities . . . ."

There is no fundamental conflict between reliability based design and the observational method, provided that Bayesian statistical updating is used. However there may be a conflict if the decision criteria controlling the design are overly simplistic. As Freeze et al., [7] state, the common risk-cost-benefit approach to design is based on the calculation of the probability of failure, P_f, or the reliability, $(1 - P_f)$. Hashimoto et al. [9] [10] have suggested that reliability does not provide a complete measure of technical performance. They introduce the concept of robustness in engineering design. A design is robust if it has the flexibility to permit adaptation to a wide range of potential conditions at little cost. When there is large uncertainty in future loads (or capacities) robustness is very desirable. There may be economic tradeoffs available between reliability and robustness.

In this terminology, it would appear that the observational method is a method that emphasizes robustness. It will be of interest in the future to see robustness incorporated into engineering design decision analysis.

Issues In Environmental Geotechnics

Morgenstern [18] reviewed the emergence of the field of environmental geotechnics and concluded that it should be primarily identified with the geotechnical aspects of waste management. It is possible to characterize the geotechnically sensitive problems according to the waste stream encountered and for geotechnical purposes it is useful to distinguish the following: 1) municipal waste, 2) industrial waste, 3) agricultural waste, 4) mining waste, and 5) nuclear waste:

1) Municipal waste: Municipal waste may be contained in landfills which are not generally regarded as hazardous. In industrialized countries, landfill siting, design, construction, operation and closure are becoming increasingly complex and costly. Geotechnical engineering has input to many aspects of landfill development and operation.

2) Industrial waste: Many industrial activities result in hazardous waste that has entered the ground. Non-hazardous waste materials, such as coal fly-ash, are also produced by industrial processes. There are geotechnical challenges associated with the increased utilization of such waste. Otherwise storage issues do not differ greatly from those encountered in the disposal of municipal waste.

Generators of hazardous waste are increasingly under pressure to reduce or eliminate their waste stream. Legislation is aggressive, particularly in the USA, requiring "cradle-to-grave" management and cleanup of past contaminated site. Complex management systems have evolved. The geotechnical engineer makes contributions to many aspects of contaminant control and site remediation.

3) Agricultural waste: Geotechnical aspects of farm waste management have not received as much attention as other waste management issues. Intensive farming can readily result in ground water contamination.

4) Mine waste: Geotechnical engineering has long made contributions to mine waste management. A distinction can be made between dry waste streams and wet waste streams [17]. The design of dry waste dumps and of tailings dams are illustrative of different geotechnical issues arising from these two streams. In some mineral processes, large amounts of water are utilized in separation and large volumes of waste result. Many other geotechnically sensitive issues arise in the mining industry such as mitigation of acid generation, environmentally safe operation of heap leaching and design for reclamation and abandonment.

5) Nuclear waste: Nuclear waste generates different public perceptions than other industrial wastes. In many countries dealing with this issue low and intermediate level reactor wastes are stored in either above ground or in-ground engineered structures. Some of the most costly research and development programs in geotechnical engineering have been mounted to address the issue of long term storage of high level waste. No nation has

yet fully resolved this issue and a variety of repository and containment strategies are under investigation.

Important differences arise in the practice of geotechnical waste management when compared with other areas of geotechnical engineering. For example, undertakings are often highly interdisciplinary and the geotechnical engineer is obliged to extend his scientific and technological range in order to function effectively in the design team. The appropriate technology is also very regional dependent. This is more so than, say, the design of a dam. These and other related factors warrant analysis, but they have little bearing on the application of the observational method. Other issues arise that do and they are the focus of this presentation.

While there are some circumstances in environmental geotechnics such as within the mining industry that are ideal for the application of the observational method, there are others in which it is seriously constrained. Constraints arise from i) the regulatory environment, ii) the nature of decision-making related to environmental matters and iii) the issue of longevity.

Environmental issues are highly regulated. The geotechnical engineer is no stranger to regulations but he has usually had little to no input into the framing of the regulations that circumscribe his environmental design practice. The regulatory structure is most highly developed in the USA and it has emphasized the use of quantitative risk analysis in the decision-making process. For example Wallace and Lincoln [26] summarize this process as implemented by the U.S. Environmental Protection Agency (EPA) in their evaluation of Superfund site remediation. They note that quantitative risk assessments are conducted for each potential remedial alternative to determine whether the remedy can reduce the health and environmental risks to acceptable levels and the subsequent decision with regard to remedial design and actions is driven by this process. This tendency of enforced quantification of what might be unknowable is in conflict with the effective application of the observational method. It promotes the false view that more study will necessarily reduce uncertainty and it encourages a separation between design and construction or implementation. Jasanoff [14] has analyzed the broad implications of this trend in the United States with fascinating conclusions:

"I have suggested that risk assessment in the United States is the product of a political process that combines large policy expectations with little trust in those called upon to formulate specific policy outcomes. Such a system threatens the credibility both of regulators and science as an institution. Pressing the evidence to produce levels of precision that it cannot support augments controversy and may feed the disenchantment with science and the scientific community already endemic in the U.S. political environment".

The observational method with its emphasis on reasonable assurance has much to commend it as an alternate basis for decision making.

The resolution of environmental issues almost always involves decision-making in the public domain. The specific process will vary from jurisdiction to jurisdiction. It may involve formalized quasi-judicial hearings to review an environmental impact statement or it may require more local informal hearings. Increasingly, the geotechnical engineer, together with other technical specialists, will find that he is not trusted. There is a disaffection with specialists on the part of the public. The geotechnical engineer must learn to earn trust on these occasions through his objectivity, the reasonableness of his claims and the ability to translate his engineering into terms and concepts appropriate for decision-making in the public domain. Under these circumstances it is important that the observational method not be construed as trial and error.

There is an understandable desire on the part of the public for waste management and related containment to have negligible impact in perpetuity. This is manifest at its most extreme in public policy related to high level nuclear waste storage. The issue of longevity creates problems for the geotechnical engineer.

It becomes increasingly necessary to understand the physical phenomenon involved in the long term to a high degree and it is necessary to have confidence in the long term behaviour of materials, whether natural or manufactured. However, this may not be enough. The use of the observational method relies on systematic performance monitoring and effecting changes in the light of this monitoring. It should be noted that the consistent application of such measures is limited to about 50-75 years of experience. There is difficulty in claiming that one can rely on such methods for a contaminating life span of hundreds of years, let alone the thousands of years that might be required in the design of nuclear waste repertories. The existence of a stable society is a necessary condition for reliance on the observational method. History suggests that even the required social stability cannot be assumed to exist too far into the future. This creates both an ethical and a technical dilemma for the geotechnical engineer. These issues will be discussed in more detail in the context of the examples of the use of the observational method in environmental applications that follow.

The Observational Method In Mine Waste Management

The mining industry often provides near ideal conditions for the application of the observational method. Excavations or construction of waste management structures proceed in an incremental manner over many years. As a result there is often ample opportunity to observe and modify design or procedures. This concept is entirely consistent with the philosophy of continuous improvement that is characteristic of, at least, the larger mining operations. Moreover the artificial contractual barriers that arise between owners, consultants and contractors and which inhibit making changes can be reduced to a minimum in the mining industry. The successful completion of the Mildred Lake Settling Basin (tailings dam) at the Syncrude Canada Ltd. oil sand mine site, Ft. McMurray, Alberta, Canada provides a vivid example. Details on the geotechnical aspects of oil sand recovery are given by Morgenstern et al. [19].

Approximately $475 \times 10^6 \, m^3$ of sand, $400 \times 10^6 \, m^3$ of sludge and $50 \times 10^6 \, m^3$ of free water will require storage for several decades within this basin. To accommodate these volumes approximately 18 km of dyke ranging from 32 to 90 m in final height have been constructed. At completion, the basin will have a surface area of $17 \, km^2$. It is currently (1994) nearing completion.

The general layout is illustrated in Figure 2. For planning purposes the dyke perimeter has been divided into 700 m long segments which are referred to as cells. The cell locations are also shown in Figure 2.

Figure 3 illustrates a typical design section for the tailings dyke. The compacted shell is constructed by utilizing hydraulic construction techniques employing dozer compaction. During the winter months when this is not feasible, the tailings stream is discharged upstream of the compacted shell to form a beach. The downstream slope angles are largely dictated by the underlying geology and associated shear strength parameters together with pore pressure response. Beneath much of the dyke is found the Cretaceous Clearwater Formation, a highly plastic clay-shale that has been weakened to its residual strength by glacial drag porcesses.

Given the complexity of the foundation and the scale of the project, it was considered at the outset that design and subsequent construction should utilize the observational method. Initial design was based on approximately average properties and instrumentation was installed in different cells. The intensity of instrumentation depended upon whether the geologic conditions were favourable or not. As shear movements were discovered monitoring was intensified. During periods of maximum construction about 125 inclinometers and 950 pneumatic piezometers would be under active observation to various degrees. Details of the instrumentation program have been provided by Fair and Handford [6].

Cell 23 proved to be particularly problematic and illustrates the use of the observational method. Construction began in 1979. The slope was originally designed and constructed at 4:1. Shear movements in the clay shale were detected in 1981 and the dyke slope was changed to 8.5:1 for all post 1983 construction, with intensive monitoring. The ultimate height in 1993 was to be 44 m (Elv. 352). Movements continued in the foundation with the addition of each 3 m lift of the dyke. One location is illustrated in Figure 4.

Traditional limit equilibrium analysis proved powerless to deal with the assessment of these movements and the forecast of ultimate

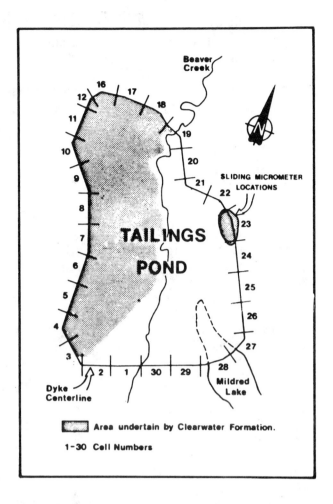

Figure 2. Layout of Syncrude Canada Ltd. Tailings Disposal Area (Fair and Handford, 1986)

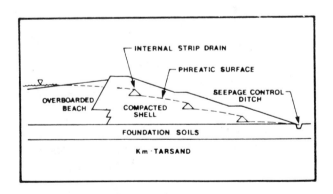

Figure 3. Typical design cross-section of Tailings Dyke (Fair and Handford, 1986)

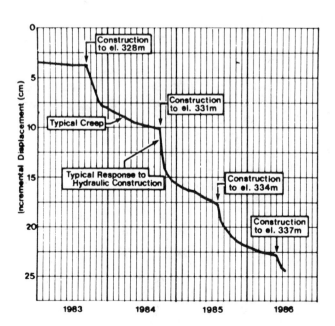

Figure 4. Performance history 319 Berm (Fair and Handford, 1986)

stability. Construction proceeded cautiously. Advanced stress analysis was initiated to study the movement mechanism and this revealed that strain in the toe zone was a key issue. Advances were made in instrumentation for strain in the toe region and the dyke was completed on the basis of strain monitoring. In this way in excess of 40 cm of slip were accommodated in the foundation of the dyke in a safe manner. There was always the additional option of more slope flattening but at substantial cost.

The observational method provided an economical resolution of a difficult geotechnical problem that standard analytical procedures could not encompass. The application of the observation method employed not only flexible planning but also advances in geological understanding, instrumentation and theoretical analysis. The relatively slow rate of construction of the dyke facilitated its application.

The Observational Method In Ground Remediation

One of the most widely used methods for remediation of contaminated groundwater is the pump-and-treat method. Contaminated groundwater is pumped to the surface and contaminants are then removed in an appropriate treatment system so that the water can be discharged or re-injected. There are many problems associated with estimating the cleanup time and hence the cost and efficiency of this technology. Kent and Mann [15] note that there have been very few documented case histories of complete cleanup of a contaminated aquifer. Many recovery operations have been terminated, but most have not recovered all of the contaminant mass. A review of several case histories indicates that removing 10-25 plume volumes may be necessary to reduce contaminants to acceptable levels. Many professionals are beginning to believe that recovery well systems may never reduce concentrations of organic contaminants to background levels.

Many of the technical problems associated with this and other remediation techniques are reasonably well understood and undoubtedly advances in scientific understanding will increase the range of problems that can be addressed adequately. However, it is important to recognize McCarty's [16] caution:

".... there needs to be a recognition that there are many sites of contamination that, if not entirely beyond our ability for rectification in an environmentally satisfactory way, may at least require many years to remediate, may involve enormous sums of money, and may create other environmental and social problems that may be equal to or greater than that posed by the contamination itself. Because of the great diversity of the problem sites, setting criteria and priorities for cleanup is not a simple task. An easy solution is not likely to be found."

The frustration expressed by all parties involved in the remediation of contaminated soils and ground water is understandable.

This frustration is compounded by the management process in which site remediation studies, design and execution unfold. In a series of penetrating studies Wallace and Lincoln [26], Brown, Lincoln and Wallace [1], and Holm [12] have analyzed the U.S. Environmental Protection Agency (EPA) procedures and find them wanting in their handling of uncertainty. They propose instead a process that by explicitly recognizing uncertainty in a proper application of the observational method offers the opportunity to reduce project time and costs as well as risks.

The current management process for Superfund site remediation is illustrated in Figure 5. It follows a traditional study-design-build approach. A remedial investigation (RI) is initiated which results in a feasibility study (FS) that compares the alternatives for remediation. The client proposes and the EPA selects an alternative (the record of decision, ROD). A consultant designs the remediation (remedial design - RD) and contractors bid on implementation

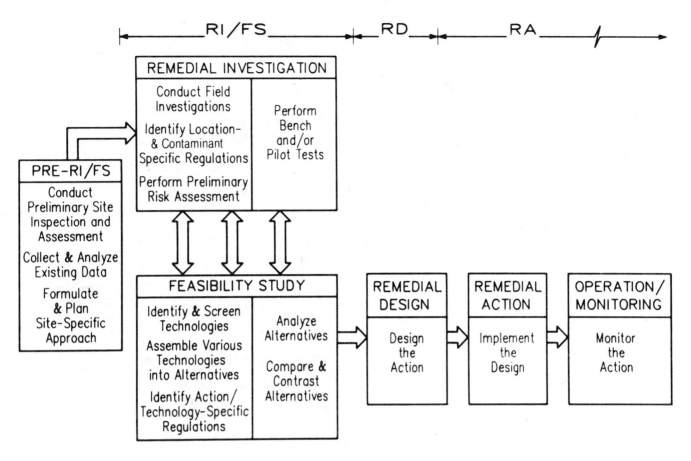

Figure 5. Schematic of the remedial process (Holm, 1993)

(remedial action - RA). Both RD and RA may involve supplemental site investigation. This procedure, common in many types of engineering, emphasizes the reduction of uncertainty early in the life of a project. However, as stressed in the preceding references, this is not an appropriate strategy for coping with the inherent uncertainty of ground remediation.

Often difficulties result in failure to advance in the process due to this uncertainty. There is an implied separation of the decision - making process into specific steps and there may be an inference that the remediation goal is to be reached at completion of implementation. Holm [12] observes that the combined effect may manifest itself as a hesitancy to establish objectives for remediation and an insistence that additional data be gathered often with the mistaken impression that uncertainty will be eliminated.

These authors advocate the incorporation of the observational method into the methodology of waste remediation by requiring the total process to be a continuum. That is, the interpretation of conditions, criteria and performance that are made in the process are working hypotheses. Reassessment continues from pre - RI/FS through implementation. Additional data needs are assessed based on their potential to disprove the working hypotheses. Decisions are finalized only after the remedy is in place and sufficient monitoring has been conducted to confirm performance. This distinction is illustrated in Figure 6.

As summarized by Brown, Lincoln and Wallace [1] the key contributions of the observational method to ground remediation

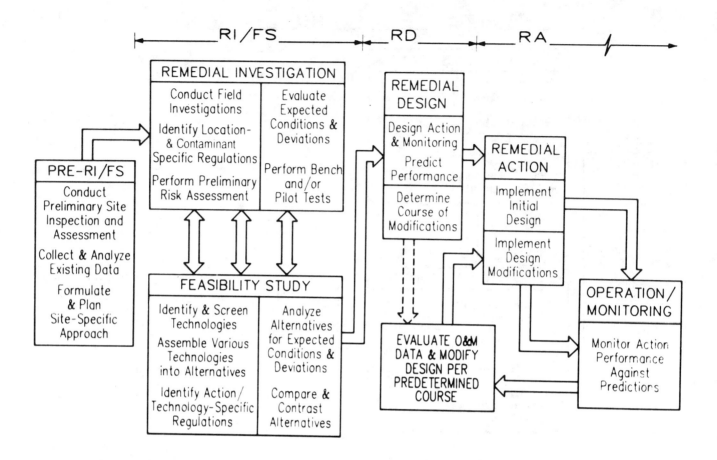

Figure 6. Schematic of the observational method (Holm, 1993)

are:

1. Remedial design based on the most probable site conditions,

2. Identification of reasonable deviations from those conditions,

3. Identification of parameters to observe so as to detect deviations during remediations,

4. Preparation of contingency plans for each deviation.

The observational method offers the potential to reduce time and cost, as well as to decrease the risks associated with remediation. This is becoming recognized with advances in pump-and-treat technology that employ dynamic adaptive well field management techniques [11].

The Observational Method in Landfill Design

The design of landfills has become increasingly complex. In the case of non-hazardous waste landfills, it is common to require a leachate collection and removal system overlying a composite liner composed of a geomembrane and compacted clay liner. For hazardous waste landfills much practice is influenced by EPA requirements of a double liner system with a leachate collection layer located above a primary geomembrane liner and a leak detection and removal system located below the primary liner and above the secondary composite liner. Even more complex systems have been proposed and constructed.

Final covers are also increasingly multi-layer systems designed to control erosion, support vegetation, inhibit infiltration and facilitate gas drainage.

The observational method is often embedded in the monitoring measures and implementation of remedial measures should they become necessary. Say that the operational life of the landfill is 20 years, it would not be difficult for the geotechnical engineer to promise acceptable performance for a 30 year post closure period. There is experience with monitoring engineered structures over this period and concern over fundamental durability of the geosynthetics utilized in landfills is minimal for this duration. However, social fairness in dealing with environmental problems increasingly requires negligible impact over a much longer period; even perpetuity! This taxes reliance on the observational method.

The issue of longevity arises in many ways. In some instances, leachate production and accumulation may be deferred for many decades. In other instances the long term reliability of landfill components may be questioned. This concern might arise from intrinsic degradation of geosynthetic materials themselves or, as is more likely, from practical considerations of clogging of leachate control measures. As emphasized by Rowe [23], there is a moral responsibility to consider the longer term consequences. In some areas there is also a regulatory responsibility to consider environmental protection in perpetuity.

An interesting example of the latter is the Ontario Ministry of the Environment guidelines associated with its Reasonable Use Policy which provides a means of quantifying long term negligible effects in a rational manner. This policy places no time constraint on the period during which the landfill is to have a negligible effect on local reasonable use of groundwater. An unacceptable impact could be expressed in terms of a limiting contamination concentration at the site boundary. As a result it becomes necessary to assess the service life of the facility and compare it with the "contaminating lifespan", which is the period of time during which the landfill will produce contaminants at levels that could have unacceptable impacts if they were discharged to the environment.

The contaminating lifespan will depend on the contaminant transport pathway, the leachate strength, the mass of waste and infiltration through the cover. It is conceivable that the contaminating lifespan, driven by diffusion processes, could be measured in hundreds of years and under these circumstances the geotechnical designer cannot in good conscience rely on the observation method. Rowe [24] has indicated methods for estimating the contaminating lifespan of landfills and he notes the need to monitor leachate levels and concentrations for the entire contaminating lifespan of the landfill. If these levels exceed trigger levels, appropriate leachate control measures would be initiated. In essence, he is advocating allowable contaminating lifespans that can be managed by the observational method, say 50-75 years.

This would be consistent with good geotechnical practice, but would introduce more active landfill control measures than are common today. For example, it would make sense to maximize leachate generation during the active phase of landfill development. While simple in concept, this is not so readily achieved. Crutcher and Mosher [4] have described the moisture addition and leachate recycle program at a large landfill and note problems that arise from interference with the gas collection system, moisture short-circuiting during recycle in the refuse, enhanced gas and odour problems and leachate treatment issues. Other measures that can affect the contaminating lifespan include inward gradient systems during the active operating phase and improving the sorption characteristics of clay barriers [8].

It is important that the geotechnical engineer

involved in landfill design recognize the constraints that long contaminating lifespans impose upon the application of the observational method in practice.

The Observational Method In Nuclear Waste Management

The development of a nuclear waste management policy has been both difficult and contentious in many countries. There are a variety of issues involved but the comments here will be restricted to scientific and technological issues. The debates over scientific evidence are most extensive in the USA and this experience will be used to highlight the role of the observational method in developing a nuclear waste management policy.

Controversies over evidence have been particularly important because of the many scientific uncertainties and problems inherent in trying to ensure that nuclear waste in a geological repository will harm neither people nor the environment for the thousands of years that the waste will remain hazardous. Colglazier [3] notes that this requirement of guaranteeing adequate safety over millennia is an unprecedented undertaking for our regulatory and scientific institutions.

There is widespread international agreement that deep geological disposal is the best option for disposing of high-level radioactive waste. While there is no reason to doubt that such a repository could be built, and not withstanding very substantial expenditures to date, a study conducted by the Board on Radioactive Waste Management, U.S. National Research Council [21] concluded that the U.S. Program, as conceived and implemented over the past decade, was unlikely to succeed. They attributed a high degree of inflexibility to the U.S. approach which was not well matched to the technical task. In particular the approach assumed that the properties and future behaviour of a geological repository can be determined and specified with a very high degree of certainty. In reality, they observed, the inherent variability of the geological environment will necessitate frequent changes in the specifications, with resultant delays, frustration and loss of public confidence.

The Board was also concerned that geological models, and indeed scientific knowledge generally, had been inappropriately applied. They noted that computer modelling techniques and geophysical analysis can and should have a key role in the assessment of long-term repository isolation. In the face of public concerns about safety, however, models were being asked to predict the detailed structure and behaviour of sites over thousands of years. The Board believes that this was scientifically unsound and would lead to bad engineering practice.

Following a study, the Board ultimately advocated a strategy based on the following premises:

i) Surprises are inevitable in the course of investigating any proposed site, and things are bound to go wrong on a minor scale in the development of a repository.

ii) If the repository design can be changed in response to new information, minor problems can be fixed without affecting safety, and major problems, if any appear, can be remedied before damage is done to the environment or the public health.

Three principles are embodied in the advocated flexible approach:

i) Start with the simplest description of what is known, so that the largest and most significant uncertainties can be identified early in the program and given priority attention.

ii) Meet problems as they emerge, instead of trying to anticipate in advance all the complexities of a natural geological environment.

iii) Define the goal broadly in ultimate performance terms, rather than immediate requirements, so that increased knowledge can be incorporated in the design at a specific site.

This approach is, for all practical purposes, the observational method and it is a recognition of the realities of geotechnical practice.

Nuclear repository design requires consideration of safety for periods well in excess of those that can be relied upon by monitoring alone. One means of extending the time scale to offer a reasonable assurance in the design is to rely on natural analogues. Fortunately, several exist that are meaningful for underground repository design.

Conclusions

The geotechnical engineer, artful in the application of the observational method, has no difficulty in accepting the dictum cited by Southwood [25]: "The things we would like to know may be unknowable".

The observational method is an effective method for coping with uncertainty in the implementation of engineering works, particularly uncertainty arising from ground conditions. As such it is a means of risk management and it is sometimes in conflict with other procedures for risk management.

The value of the observational method in conventional geotechnical engineering has been proven beyond a doubt. It has also been shown to be effective in various aspects of environmental geotechnics such as mine waste management and in-situ remediation. However restraints arise in its application from a limited understanding of the observational method, from a dominance of regulations that pre-empt its use, and from issues associated with longevity that limit its practical application.

Neither the public at large nor many regulatory agencies have a full appreciation of and a trust in the application of the observational method for waste management engineering. This is an issue that challenges the geotechnical engineer and requires on-going attention so that the observational method is not perceived as design by trial and error. Risk communication is an integral part of risk management.

References

[1] Brown, S.M., Lincoln, D.R. and Wallace, W.A., 1990. Application of the observational method to hazardous waste engineering. J. Management Eng., ASCE, Vol. 6, p. 479-500.

[2] Chen, Z., Morgenstern, N.R., and Chan, D.H., 1992. Progressive failure of the Carsington Dam: a numerical study. Canadian Geotechnical Journal, Vol. 29, p. 971-988.

[3] Colglazier, E.W., 1991. Evidential, ethical and policy disputes: admissible evidence in radioactive waste management. In Acceptable Evidence ed. by D.G. Mayo and R.D. Hollander, Oxford University Press, New York, p. 137-159.

[4] Crutcher, A.J., and Mosher, F.A., 1991. Leachate collection and moisture recirculation at the Keele Valley Landfill site. Proc. 1st Canadian Conference on Environmental Geotechnics, Canadian Geotechnical Society, Montreal, p. 57-64.

[5] D'Appolonia, E., 1990. Monitored decisions. J. Geotech. Eng., ASCE, Vol. 116, p. 4-34.

[6] Fair, A.E., and Handford, G.T., 1986. Overview of the tailings dyke instrumentation program at Syncrude Canada Ltd. In Geotechnical Stability in Surface Mining ed. by R.K. Singhal, Balkema, p. 245-254.

[7] Freeze, R.A., Massmann, J., Smith, L., Sperling, T., and James, B., 1990. Hydrogeological decision analysis, 1.A framework. Ground Water, Vol. 28, p. 738-766.

[8] Gray, D.H., 1993. Strategies and techniques for the geo-containment of toxic wastes. *In* Geoconfine '93, ed. by M. Arnould, M. Barrès and B. Come, Balkema, p. 213-218.

[9] Hashimoto, T., Stedinger, J.R., and Loucks, D.P., 1982(a). Reliability, resiliency and vulnerability criteria for water resource system performance evaluation. Water Resources Research, Vol. 18, p. 14-20.

[10] Hashimoto, T., Loucks, D.P, and Stedinger, J.R., 1982(b). Robustness of water resource systems. Water Resources Research, Vol. 18, p. 21-26.

[11] Hoffman, F., 1993. Ground-water remediation using "smart pump and treat". Ground Water, Vol. 31, p. 98-114.

[12] Holm, L.A., 1993. Strategies for remediation. *In* Geotechnical Practice for Waste Disposal ed. by D.E. Daniel, Chapman and Hall, London, p. 289-310.

[13] James, B.R., and Freeze, R.A., 1993. The worth of data in predicting aquitard continuity in hydrogeological design. Water Resources Research, Vol. 29, p. 2049-2065.

[14] Jasanoff, S., 1991. Acceptable evidence in a pluralistic society. *In* Acceptable Evidence ed. by D.G. Mayo and R.D. Hollander, Oxford University Press, New York, p. 29-47.

[15] Kent, B., and Mann, P., 1993. Recovery well systems. *In* Geotechnical Practice for Waste Disposal, ed. by D.E. Daniel, Chapman and Hall, London, p. 497-519.

[16] McCarty, P.L., 1990. Scientific limits to remediation of contaminated soils and ground water. *In* Ground Water and Soil Contamination Remediation: Toward Compatible Science, Policy and Public Perception, p. 38-52, National Research Council, National Academy Press, Washington.

[17] Morgenstern, N.R., 1985. Geotechnical aspects of environmental control. Proc. 11th International Conference Soil Mechanics Found. Eng., San Francisco, Vol. 1, p. 165-185, Balkema.

[18] Morgenstern, N.R., 1991. The emergence of environmental geotechnics. Proc. 9th Asian Regional Conference, Int. Soc. Soil Mechanics Found. Eng., Bangkok, Vol. 2, p. 19-25.

[19] Morgenstern, N.R., Fair, A.E., and McRoberts, E.C., 1988. Geotechnical engineering beyond soil mechanics - a case study. Canadian Geotechnical Journal, Vol. 24, p. 637-661.

[20] Muir Wood, A., 1990. The observational method revisited. Proc. 10th Southeast Asian Geotechnical Conference, Taipei, Vol. 2, p. 37-42.

[21] National Research Council, 1990. Rethinking High-Level Radioactive Waste Disposal. Board on Radioactive Waste Management, National Academy Press, Washington, D.C.

[22] Peck, R.B., 1969. Advantages and limitations of the observational method in applied soil mechanics. Geotechnique, Vol. 19, p. 171-187.

[23] Rowe, R.K. 1991 (a). Some considerations in the design of barrier systems. Proc. 1st Canadian Conference on Environmental Geotechnics, Canadian Geotechnical Society, Montreal, p. 157-164.

[24] Rowe, R.K. 1991 (b). Contaminant impact assessment and the contaminating lifespan of landfills. Canadian Journal of Civil Engineering, Vol. 18, p. 244-253.

[25] Southwood, T.R.E., 1985. The roles of proof and concern in the work of the Royal Commission on Environmental Pollution. Marine Pollution Bulletin, Vol. 16, p. 346-350.

[26] Wallace, W.A., and Lincoln, David, R., 199 How scientists make decisions about ground water and soil remediation. *In* Ground Water and Soil Contamination Remediation: Toward Compatible Science, Policy and Public Perception, p. 151-165, National Research Council, National Academy Press, Washington.

World Development Report on Environment and Development

Andrew Steer
The World Bank, Washington, D.C., USA

Final manuscript not received in time for inclusion in the *Proceedings*. For additional information, please contact:

> Mr. Andrew Steer
> Deputy Director, Environment Department
> The World Bank
> 1818 H Street, NW
> Washington, DC 20433

Geotechnical Studies for the Waste Isolation Pilot Plant Repository

Wendell D. Weart
Sandia National Laboratories, Albuquerque, New Mexico, USA

Final manuscript not received in time for inclusion in the *Proceedings*. For additional information, please contact:

        Dr. Wendell D. Weart
        Sandia National Laboratories
        Org. 6303
        P.O. Box 5800
        Albuquerque, NM 87185-5800
        USA
        Voice: +505-844-4855
        Fax: +505-848-0789

Appendix 1

Report of Technical Committee on Environmental Control (TC5)

Charles D. Shackelford
Secretary for TC5, Department of Civil Engineering, Colorado State University, Fort Collins, Colorado, USA

1. Membership

The membership of the International Society for Soil Mechanics and Foundation Engineering's (ISSMFE) Committee TC5 on Environmental Control is as follows:

Chairman John Nelson
          Department of Civil Engineering
          Colorado State University
          Fort Collins, Colorado 80523
          USA

Secretary C. Shackelford   USA

Members
           Y. Acar   USA
           M. Aubertin   Canada
           J. Baumann   Denmark
           J. Booker   Australia
           R. Chapuis   Canada
           R. Clark   United Kingdom
           L. de Mello   Brazil
           M. Gandais   France
           K. Jha   India
           M. Kamon   Japan
           P. Kablena   Czech Republic
           T. Lundgren   Sweden
           M. Manassero   Italy
           Z.-C. Moh   Taiwan
           A. Musso   Italy
           M. O'Connor   Canada
           J. Saarela   Finland
           B. Soyez   France
           E. Togrol   Turkey
           J. Troncosco   Chile
           M. Tumay   USA
           M. Usmen   USA
           W. Van Impe   Belgium
           J. Wates   South Africa
           W. Wolski   Poland
           Prof. Yudhbir   India
           T. Zimmie   USA

2. Terms of Reference

At the last meeting of Committee TC5 on Environmental Control in Cesme, Turkey, May 25-27, 1992, the Terms of Reference were defined as follows.

(1) To assess whether an international specialty conference on engineered barriers is appropriate and, if so, to encourage its organization in the next four to five years.

(2) To contribute a survey paper to the 13th ICSMFE as arranged by the Organizing Committee of the 13th ICSMFE.

(3) To promote cooperation and exchange of information about geotechnical aspects of environmental control with special reference to i) site investigation practice for pollution problems and ii) design, construction, and performance of engineered barriers.

The first item under the Terms of Reference identified above obviously has been achieved with the advent of this *First International Congress on Environmental Geotechnics*. The survey paper in item (2) of the Terms of Reference is presented herein as a part of this TC5 report since it is felt that *First International Congress on Environmental Geotechnics* is the most appropriate venue for the Technical Report of the ISSMFE Committee on Environmental Control. The publication of this report should help to achieve the cooperation and exchange of information about geotechnical aspects of environmental control mentioned in item (3) under the Terms of Reference.

3. Report

A summary report concerning various aspects of environmental control from the Geotechnical Engineering viewpoint has been written by members of TC5. Due to space limitations, the report contained herein represents only about 40 percent of the original draft manuscripts submitted by the participating authors. While the manuscripts submitted by the participating authors have been edited without exception resulting in some reduction in detail, the report still contains a significant amount of information, especially for the uninitiated. In addition, the numerous references should be of benefit in providing more detailed information. Alternatively, the authors can be contacted personally for clarification and/or more information. The Secretary of TC5 accepts responsibility for any errors and/or misinformation contained within the report.

The report contains overviews of the following topics.
- General Design Aspects of Containment and Drainage Systems
- Clay Liner Design, Testing, and Construction
- Design and Construction of Landfill Covers
- Slurry Wall Systems
- Groundwater Control and Pollution
- Site Remediation
- Risk Assessment/Risk Management Methodology
- References

3.1 General Aspects of Containment and Drainage Systems (W. F. Van Impe, Kijksuniversiteit-Gent, Labo voor Gronmechanica, Grotesteenweg Noord 12, 97100 Zwignaarde, Belgium)

3.1.1 Containment Systems

The first step in design of containment systems is to choose the most appropriate site based primarily on the optimal conditions from hydrological, geological, and geotechnical points of view. The second step is to perform extensive geotechnical investigations at the site (e.g., deep borings, CPT, SPT and/or PMT, phreatic level and seepage determinations, seismic testing, geophysical tests). Also, in some specific cases, geo-electric, gravimetric, geomagnetic, and/or geothermal measurements may be required, particularly in case of existing waste ponds at the same site.

In terms of the design of the physical containment system, the lining system underneath the disposal must be evaluated. This must include consideration of long-term permeability, changes in properties due to interaction with the waste material, plastic deformability, long-term brittleness, creep, and relaxation behavior. When the bearing or deformation capacity of the subsoil does not fit the requirements, soil improvement of foundation soils must be considered. In some cases, intermediate liners at different levels in the containment pond must be considered. Combined base liners are typical; several proposed designs can be found in the literature.

In terms of liner materials, there is a clear distinction between "mineral" liners (clay and clay sand mixtures) and "artificial" liners (geomembrane, asphalt liners, etc.). Based on minimum requirements with respect to hydraulic conductivity or permeability, k (e.g., $k \leq 5 \times 10^{-10}$ m/s), pure clay or silt liners frequently are unsuitable. A homogeneous mixture of natural clay with bentonite can result in a brittle mixture which would be unsuitable. A more suitable material for a "mineral" liner is a mix of silty sand with bentonite. The quality of such mixtures will be highly dependent on the grain size distribution of all constituents, the homogeneity of the mixture, the water content at mixing, and the mixture density at installation. The silty sand should be as well graded as possible to provide for the optimal interaction with the bentonite minerals which is of utmost importance in prohibiting small fissuring under temperature gradients common in waste disposal systems.

On the other hand, geosynthetic liners are almost impermeable and many combinations for composite synthetic liner systems have been proposed in the literature. Designs using geosynthetics should be based on combined mineral and synthetic linings due to the severe requirements for overlap of geosynthetic liners and the possibility for small punctures through these very thin (≤ 15 mm) materials.

In all cases, appropriate monitoring systems in and under a base liner should be included, allowing for the long-term follow-up of the system behavior and possible remediation, if necessary. However, remediation is typically difficult to perform and must be investigated thoroughly from a technological viewpoint.

An alternative to liner systems is the concept of a surrounding or encapsulating trench or retaining walls. The objective is to design a watertight barrier embedded in an impermeable natural or artificial layer at given depth. Examples would include bored pile walls, jet or VHP wall, thick bentonite walls, bentonite combined with overlapping geomembranes, and/or frozen and vitrified soil walls. Depths up to 35 m, even with bentonite membranes ≈ 100 mm thick can be reached. With large bored bentonite walls up to 3 m in thickness, depths up to 100 m can be reached.

The best material mixture for impermeable walls must be analyzed with respect to each specific case of soil conditions. Other parameters, such as uniaxial strength and sensitivity to erosion or "aging" of the wall material, also play a role in the final decision. Optimizing the wall behavior obviously also will include the knowledge of its resistance against chemical attack from the waste. Mixtures without cement are available which allow for permeability coefficients of from 10^{-11} to 10^{-12} m/sec.

The final step in the general design of containment systems is related to the top covering layer. The primary functions of the covering layer are to (1) prohibit infiltration of water into the containment system, (2) protect the environment from gas emanation, (3) protect the inner slopes of the surrounding walls against erosion and general weather conditions, and (4) allow for restoration of the area of the containment pond with vegetation. As a result, such a covering bed basically has to consist of several layers.

3.1.2 Drainage Systems

For drainage systems, it is quite common to use washed gravel materials. Gravel having well rounded grains is preferable to crushed gravel to avoid damaging or even perforating synthetic liners by angular grains. The gravel should be of good quality and contain as little as possible any minerals which could be dissolved in case of acids percolating through the pond. Minimum thicknesses of 50 cm should be required for drainage beds. Although synthetic drainage mats or grids also exist, their long-term behavior is still unproven.

In addition to horizontal gravel beds, containment systems also may be provided with drain pipes. Such pipes are generally made from perforated synthetic material. In order to maintain their efficiency, drainage pipes should allow for regular inspection by camera and should be accessible to cleanout systems.

Pipes may drain to collection pumping wells. Pumping wells must be designed to resist negative skin friction and must allow for maintenance.

3.2 Clay Liner Design, Testing, and Construction (C. D. Shackelford and L. G. de Mello, Rua Capitao Antonio Rosa 297, Sao Paulo 01443, Brazil)

3.2.1 Design and Testing

3.2.1.1 Field- Versus Laboratory-Measured Hydraulic Conductivity

Numerous case studies performed over the past decade indicate that hydraulic conductivity (k) values measured in the laboratory on compacted soil are often less than those measured in the field for the same soil. Daniel [14] examined several existing earthen liners and found that the field hydraulic conductivity values ranged from about 5 to 100,000 times greater than the hydraulic conductivity values measured in the laboratory. Daniel [15] presented a list of 10 cases for which field hydraulic conductivity values of compacted clay liners are documented and found that only one of the cases resulted in field hydraulic conductivity value less than 1×10^{-9} m/s. Daniel and Brown [19] compiled a table of 14 case studies of earthen liners constructed throughout the United States and Canada and found that only two of the 14 met the criterion of $k \leq 1 \times 10^{-9}$ m/s. Several detailed field experiments also have indicated that in situ hydraulic conductivity values on compacted clay soil can be several orders of magnitude greater than the laboratory-measured values particularly if construction quality control and construction quality assurance is poor [22, 34, 88].

Nonetheless, there have been many reports of successful construction of clay liners. Examples of good performance of well-constructed liners are reported by Reades et al. [85, 86], Chen and Yamamoto [13], Laihti et al. [56], Gordon et al. [45], Johnson et al. [54], and Trautwein and Williams [102]. Several of these reported cases represent prototype liners or test fills and, in most cases, rigorous construction quality assurance was implemented [11].

As a result of the documentation on the relationship between laboratory- and field-measured hydraulic conductivity values, the U.S. Environmental Protection Agency (EPA) now requires that laboratory-measured hydraulic conductivity values be confirmed in the field on a test pad which is constructed using the same materials and construction procedures that will be used on the actual compacted clay liner.

3.2.1.2 Compatibility Testing

The purpose of compatibility testing is to determine the ability of the waste leachate to attack or to otherwise cause an increase in the hydraulic conductivity of the clay liner material. Numerous studies have been performed to determine the effect of different chemical solutions on the laboratory-measured hydraulic conductivity of various compacted clay soils [76, 95]. The results of these studies have been conflicting in that some of the liquids have resulted in large increases in the measured hydraulic conductivity values while other test results indicate either no change or a slight decrease in hydraulic conductivity. Various reasons have been postulated for these test results, including inappropriate test procedures. However, due to the wide range of chemical compositions of waste leachates and the numerous different potential liner materials, compatibility testing is a necessary component of the evaluation of potential clay liner materials. Based on the potential for chemical attack of the clay liner materials by the waste leachate, compatibility testing should be performed on all proposed soil liner materials. Modified Atterberg limits tests and modified hydrometer tests may be used in some cases to provide a relatively rapid indication of a potential compatibility problem [95].

3.2.1.3 Contaminant Transport

Although the design of clay liners traditionally has been based on the assumption of advective (Darcian) transport, more recent studies indicate that molecular diffusion tends to be the dominant mechanism of contaminant transport in fine-grained soils [32, 43, 53]. The significance of diffusion has been illustrated by Shackelford [93], Mitchell [74], and Manassero and Shackelford [63]. Two main conclusions resulting from these analyses are (1) the best containment barrier that can be built is a diffusion-controlled barrier, and (2) the lower the hydraulic conductivity, the more non-conservative the analysis based on consideration of only hydraulic conductivity (i.e., pure advection). As a result of both field evidence and analytical modeling, a low hydraulic conductivity for a field compacted clay liner is a necessary, but not sufficient, condition to ensure adequate containment of contaminants.

3.2.2 Construction

3.2.2.1 Index Tests for Construction Materials

Several specifications on Atterberg limits for proposed soil liner materials have been recommended in the literature [3, 17, 46, 54, 75]. In general, these recommendations have been made to achieve a required hydraulic conductivity in the field (e.g. $k \leq 1 \times 10^{-9}$ m/s) by increasing the workability of the soil, thereby minimizing the existence of macropores and other potentially detrimental secondary defects in the clay liner. Specifications for grain-size characteristics of proposed liner materials also have been reported [3, 11, 17, 46, 54, 75]. In general, the goal of the specifications with respect to grain-size characteristics is to achieve as low a hydraulic conductivity as possible. This can generally be achieved by (1) increasing the amount of fine-grained material, in general, and clay material, in particular, (2) decreasing the amount of coarse-grained material, and (3) decreasing the potential for structural defects and/or macropore effects by minimizing the maximum particle/clod size which is allowed for the liner material.

3.2.2.2 Construction of Compacted Clay Liners

Recommended tests and observations for subgrade preparation, pre-construction testing of soils considered for clay liners, and testing of constructed clay liners for construction quality control and construction quality assurance are described by Gordon et al. [44, 45, 46] and Daniel [17]. Some of the more important aspects of these construction considerations are discussed in the following paragraphs.

Desiccation has been noted as one of the primary suspected causes for the performance failure of several of the clay liners reported in the literature [14, 19]. Desiccation, a potentially serious problem in arid climates, refers to the drying out of the compacted soil and the resulting formation of tension cracks leading to preferential flow paths through the clay liner. Desiccation also can cause cracking of soils in tropical climates, particularly for soils with high ($\geq 40\%$) optimum water contents [26]. In order to minimize the potential for desiccation cracking, each lift of a compacted clay liner should be wetted to above optimum water content, and each compacted lift should be proof-rolled with a smooth-drum roller in order to form a hard "skin" which lessens the potential for desiccation cracking [17]. The smooth surface should be scarified (roughened) with a disc before placement of the next lift of loose soil to aid the connection between compacted lifts of soil and minimize interlift flow paths. If a relatively long period of exposure of any lift of compacted clay is experienced, the lift should be periodically moistened and/or covered with a sheet of plastic or layer of soil.

Two criteria typically are specified for construction of compacted clay liners, viz. (1) the dry unit weight (γ_d) in the field must be greater than, or equal to, some percentage of the maximum dry unit weight (e.g., $\gamma_d \geq 0.95\gamma_{dmax}$) based on a specified compaction procedure and energy in the laboratory, and (2) the molding water content at compaction (w) must be greater than or equal to optimum water content ($w \geq w_{opt}$) based on the same specified compaction procedure and energy. Daniel and Benson [20] provide evidence that indicates that this commonly used approach does not succeed very well in distinguishing between w-γ_d points that correspond to a regulatory limit for the saturated hydraulic conductivity, k (e.g., $k \leq 10^{-9}$ m/s). This discrepancy results from the fact that the unit weight criterion for structural fills is based on consideration of strength and compressibility, not hydraulic conductivity, and on the fact that some w-γ_d points which achieve the standard criteria may actually be dry of optimum water content due to the variation between compaction energy in the laboratory versus compaction energy in the field. Based on this evidence, Daniel and Benson [20] propose an alternative criterion in which water content, not dry unit weight, is the controlling factor and the line of optimum water contents is used as a guideline for the construction specifications. This procedure avoids defining the field water content as a percentage of optimum water content. Further details on this criterion can be found in Daniel and Benson [20].

A test pad or demonstration strip has been either recommended by several investigators, or required (e.g., by the U. S. EPA), to verify that the materials and methods of construction will produce the desired results [3, 11, 15, 17, 75, 102, 104]. A test pad allows evaluation of the parameters affecting the construction and performance of the clay liner, such as the compaction water content, the actual field equipment (type and weight), the number of passes of the equipment for each lift, the lift thickness, etc. The recommended dimensions of the test pad are a width and length of at least three construction vehicles, and a thickness of at least 3 or 4 compacted lifts. In addition, in situ hydraulic conductivity testing should be performed on the constructed test pad to ensure that the required hydraulic conductivity has been achieved in the field using the same construction equipment and procedures that will be used in construction of the full-sized liner. Methods for evaluating the hydraulic conductivity of clay liner via field testing are described by Daniel [16]. The sealed double-ring infiltrometer (SDRI) has been recommended [21, 102] because the method tests a relatively large volume of soil thereby allowing for a better evaluation of the existence of any macropores or structural defects in the clay liner. However, since the SDRI measures the infiltration rate and not hydraulic conductivity, there remains some debate about the use of the SDRI to measure hydraulic conductivity. Other procedures may be acceptable in other parts of the world. For example, the procedure by Matsu et al. [68] is popular in Brazil.

A sheepsfoot compactor usually is specified for compaction of compacted clay liners because the feet on the sheepsfoot compactor will break up and destroy the existence of large clods of soil and provide good interlift bonding. However, one important consideration which often is overlooked in specifying a sheepsfoot compactor is the length of the feet; i.e., the length of feet on the sheepsfoot compactor should be as long, if not longer, than a loose lift of soil to assist in good interlift bonding [18]. Therefore, it is recommended that a maximum loose lift thickness be specified to be not thicker than the length of the feet on the sheepsfoot roller used in the actual construction of the test pad and the full-sized liner. This specification may result in compacted lift thicknesses which are less than 15 cm (6 inches). In addition, heavier compaction equipment is usually preferred to generate significant shear stresses and destroy clods [25].

3.3 Design and Construction of Landfill Covers
(J. Saarela, National Board of Waters and the Environment, PB 250, 00101 Helsinki, Finland)

A cover at a landfill site is designed to (1) exclude surface wastes from contact with disposed wastes, (2) act as a barrier against human or animal contact with the wastes, and (3) prevent release of vapors to the environment. Covers are used in most site remediation plans and usually are included as part of an integrated concept of waste containment.

A cover may perform a number of functions [69]. However, the prevention of percolation and leachate generation is predominant because contamination of surface water and groundwater supplies is the most serious environmental threat of uncontrolled hazardous waste sites.

3.3.1 Materials for Cover Systems

Soils will be used in most cover systems when suitable soils are available and are relatively cheap. However, not all local soils are suitable for all cover purposes, and either imported soils or non-soil materials may have to be used. The cover designer needs to consider soils both from the engineering and the agricultural points of view.

Permeability (hydraulic conductivity) is an important property of cover soils since the cover system is largely intended to prevent or reduce the inflow of water. The permeability of soil is less when it is unsaturated because out of a large number of possible flow channels through the soil, only a fraction are utilized. Unsaturated flow is important because the soils in a cover system will be completely saturated only a small portion of the time, if ever.

The elements that may be found in a multi-layered cover system include (top to bottom) (a) a surface (vegetation) layer, (b) filter(s), (c) biotitic barrier, (d) a drainage layer, (e) a hydraulic barrier, (f) foundation (buffer) material, (g) filter(s), (h) a gas control layer, and (i) the waste. Not all elements may be required for a given site, but it is likely that cases may exist where all of the cover elements are required. Any cover system must possess a surface layer. Unless water percolation is not a concern, a hydraulic barrier must be present. The wastes always will be present at the bottom of the system. The need for different layers depends on the type of waste, the wideness, and the placement and the afteruse of sanitary landfill site. The functions of the other layers include a gas-control layer which intercepts gases evolved by the wastes and leads them to the atmosphere via venting mechanisms, a foundation or buffer layer which isolates the hydraulic barrier from the wastes and serves as a strong base to support the rest of the system, a drainage layer which intercepts downward-percolating water and conveys it laterally out of the system, a biotic-barrier layer (largely conceptual at present) which hinders plant roots and burrowing animals from disrupting the layers below, particularly the hydraulic barrier, and filter layers which serve to separate fine from coarse materials and prevent clogging of coarse interstices by fine particles.

The designer must consider all specific aspects of the site that will affect the cover design, including excessive surface-water requiring a diversion system, topography, water table location, local availability of suitable materials, and possible utilization of synthetic materials. Non-soil materials finding applications in cover systems, beyond those previously mentioned as soil additives, include asphalt, residual or waste substances from various industries, pipes and tiles, and synthetic products such as geomembranes and geotextiles. These materials perform various functions in various parts of cover systems. While some of these materials are relatively impermeable, others are permeable and water-conducting.

Asphalt has been used in hydraulic barriers in several forms. A hot-sprayed asphalt membrane consists of a tough, high-softening-point asphalt or tar sprayed on a base surface to conform to its irregularities and form a ductile yet durable membrane. If tensile strength is important, a reinforced sprayed-asphalt membrane comprising a geotextile into which the asphalt is sprayed may be used. Careful design and installation are required to preclude delaminations.

Industrial waste materials, where abundant and cheap, may be used in cover components. The greatest quantities of potentially useful inorganic industrial waste materials are generated in the mining and metallurgical industries, including mine and pit wastes, mill tailings, and furnace slag, among others. In general, the coarse-grained waste materials may be useful in the permeable and load-bearing portions of cover systems. The finer waste materials usually are not useful in cover systems because of undesirable non-plastic properties.

Geomembranes have been used as pond and lagoon liners for two decades, but their use in cover systems is relatively new. Geomembranes are fabricated from three basic types of polymeric materials: elastomers (rubbers), thermoplastics (plastics) and combinations of elastomers and thermoplastics. Geotextiles fulfill five basic functions: filtration, drainage, separation, reinforcement and armoring. Geotextiles are made from a variety of synthetic materials of which the most common are polypropylene and polyester.

Soil-cement is a relatively impermeable, inflexible, and brittle material produced by mixing portland cement and water with soils (usually granular) to form a weak concrete. Mixing may be done either in place, using a disk-harrow, or in a batch plant. Soil-cement may be used in cover systems where strength or imperviousness is required and where the threat of cracking is not severe or can be tolerated.

3.3.2 Cover System Design

A key component of the cover design process is estimating the potential amount of leachate that may be produced. This process requires a water-balance analysis. Water-balance analysis may be performed manually or by using computer models. A manual method was described by Fenn et al. [40]. A computer model developed specifically for modeling landfill hydrology is the Hydrologic Evaluation of Landfill Performance (HELP) model. HELP was designed for interactive use and requires climatological, soil, and design data as inputs. Default climatological and soil data are available to satisfy the needs of many users. The program allows designers and regulators to rapidly screen alternative designs. Along with total leachate, HELP computes various component conditions within the system, e.g., total saturated thickness in the soil above a barrier layer.

Gas generation may be expected where hazardous and municipal wastes are disposed together. Evolution of gases may persist over many years, although the rate decreases steadily with time. The main gases produced are methane and carbon dioxide. Only traces of other gases are produced, although these trace gases may have prominent odors.

Vapors, the simple product of volatilization of liquid or solid compounds, are not likely to escape from buried wastes in hazardous quantities, unless active gas evolution from organic decay is taking place, because only the pressures generated by active gas evolution will be sufficient to force appreciable gas flows to the surface.

A layer of coarse-grained material should be present to control landfill gases. The coarsest gradation possible is desirable to inhibit the growth of a biomass of anaerobic slimes. Pipe vents, with or without exhaust blowers and set in gravel packs to provide a clear channel for the gas to flow, may be placed at regular intervals to vent the gases to the atmosphere.

Both graded cohesionless soils (sands and gravel) and geotextiles may be used as filters. Graded filters are durable, have a long history of use, and often can be made from readily available materials, although careful attention is required for both the gradation of the material selected and the installation. Formulas giving filter criteria in terms of the gradations of the materials involved are available. Geotextile filters are relatively simple to install; however, they do not yet have a service history from which to predict long-term behavior. As with granular filters, careful attention to the relation between the size of openings in the fabric and the size of soil particles to be separated is required.

The foundation (buffer) layer separates the hydraulic barrier from the wastes and serves as a loadbearing member to support the weight of the rest of the cover. Subsidence within the wastes undermining the cover system may be mitigated by placing a good foundation layer. The best materials for the foundation layer are coarse-grained, granular soils or their equivalents in the non-soil category.

The primary function of the hydraulic barrier is to divert or impede the downward percolation of any water coming into contact with it. There are three primary failure mechanisms for the hydraulic barrier: chemical, mechanical, and environmental. Chemical failure is probably the least significant. Mechanical failure especially needs to be guarded against during construction. Delayed mechanical failure might take place through undermining of the hydraulic barrier from severe subsidence of the underlying waste. Environmental failure includes both weathering effects (freezing and thawing, wetting and drying, etc.) and the attack of plant roots or burrowing animals. There simply is not enough service history at the present time to predict how well hydraulic-barrier members of cover systems will survive this threat.

Barrier materials include fine-grained, cohesive soils, amended soils, asphalt and geomembranes. Compacted natural soils have the advantage of durability. Amended soils, as produced by the addition of bentonite, have been used to seal farm ponds and may be useful in a hydraulic barrier layer. Asphalt can be used to form a hydraulic barrier, although experience with asphalt in cover systems is limited.

Geomembranes provide an impervious barrier, in contrast to other systems, but the service history of geomembranes in cover applications is not demonstrated.

The function of the drainage layer is to intercept water which has entered the cover system as infiltration and to conduct it to one or more safe disposal outlets. The drainage layer consists of a blanket of free-draining sands and gravels and a collection/transport system comprising gravel drains, pipes, or tiles. The collection system forms a network of varying complexity depending on the site. The exits from the collection system should be designed with care to prevent erosion from concentrated flows.

The need for a biotic barrier layer arises from the threat of damage to the hydraulic barrier from plant roots and burrowing animals. No biotic barrier is known to have been installed in cover system to date. Research results indicate, however, that cobblestones, brick rubble or other large particles appear to be effective both in repelling animals and in halting the progress of plant roots. Some chemical toxins also have been found to be effective under certain conditions.

Any cover must have a surface layer. In all but very arid climates, this layer will be vegetated, spontaneously if not by design. A well-designed controlled vegetative community will present erosion, expel water from the cover system through evapotranspiration and promote site aesthetics. Vegetation is usually very cost-effective. However, fertile soils are rarely available at landfill sites, and it is necessary to work with the inferior soils that may be present.

A cover section designed to impede percolation presents an environment that, in many ways, is the direct opposite of one favorable to vegetation. Because vertical drainage is poor, roots are subject to swamping. Anaerobic conditions, detrimental to healthy roots, may develop. Conversely, cover systems designed to drain and dry quickly deprive vegetation of needed reservoirs of moisture. A drought, even a short one, may kill or severely injure the vegetative community. For all these reasons, the establishment and maintenance of a stable vegetative cover are very challenging problems.

The major factors determining effectiveness of a soil for supporting vegetation are grain size, pH, organic matter and nutrient content. Many other factors play necessary, though less critical, roles. Design consideration for waste cover systems should be given to soils that are locally available. Soil tests can indicate if the soil material is acid, saline and sodic, excessively drained, poorly drained, wind erodible or contains dispersive clay. Appropriate tests can be conducted on the soils proposed for use in the vegetative layer at a soil testing laboratory. In addition, waste materials such as sewage sludge, manure or fly ash as a liming agent should be considered to improve the fertility of the available soil materials.

The thicker the vegetative layer, the more stable it will be and the better it can support desirable naturally deep-rooted plants such as legumes. The Resource Conservation and Recovery Act (RCRA, U.S. EPA) specifies a minimum surface-layer thickness of 61 cm for controlled landfill covers. This would seem to be a desirable minimum at uncontrolled sites as well. Among the factors to be considered is the need to retain enough moisture in the surface layer to sustain vegetation through dry periods.

Surfacewater management refers to all features concerned with the management or control of runoff. The goal of surfacewater management is to conduct surface water off the site (runoff) in such a way that it does not erode the cover system. With regard to the cover, the surfacewater management system, along with the vegetative layer, is the only part of the cover system that is directly affected by extreme weather conditions (e.g. violent storms).

Surfacewater management involves land grading, waterways, diversion structures, check dams and outlet structures. Land grading, which is the reshaping of the existing topography at a site, is carried out in accordance with a plan based on an engineering survey and layout. The steepness of cut and fill slopes should be controlled according to established principles. Reserve-slope benches and diversions may be used to limit maximum overland flow distances. The land grading plan must be integrated with waterways and diversion structures to form a coherent overall drainage system.

The threat of damage from frost action in colder climates may be controlled by proper design. The most serious damage is that caused by frost heaving, the formation of ice lenses within a soil. Formation of such lenses, and collapse upon thawing and draining, could disrupt the cover system. Among the factors necessary for frost heaving to occur is the presence of a nearby water supply to sustain the growth of ice lenses. By providing for proper drainage, the designer can ensure that such a water supply is not available and, thus, that frost heaving will not take place.

Good construction is equally as important as good design in providing an effective, durable cover system. Construction includes site preparation, planning and scheduling of work, selection and use of proper equipment and site closure. Effective quality control and quality assurance measures are necessary to assure that design specifications are met. Site preparation includes establishment of site security, initial clearing and grubbing, establishment of a controlled surface drainage network to prevent runoff of contaminants, establishment of support facilities, internal access roads, soil stockpiling and site precontouring. The slopes established in precontouring may or may not parallel the final contours at the site.

3.4 Slurry Wall Systems (M. Manassero, Ingegneria Geotechnica, Corsa Montevecchio 50, 10129 Torino, Italy)

The common practice, the modern trend, and the future perspectives in slurry cutoff wall systems are addressed briefly with particular consideration for (1) construction systems, (2) criteria and procedures for design, (3) suitability tests for construction materials, and (4) the effect of construction quality control on in situ performance.

3.4.1 Construction Systems

Vertical barriers (cutoff walls) used to contain contaminated waste or ground water include slurry walls (simple and composite), driven beam thin walls, steel sheeting, grout curtains, jet-grouting, compacted clay or mineral mixtures, deep soil mixing, intersecting bored piles, and freezing [37, 38, 81, 107]. The focus of this paper is on slurry walls.

The greatest advancements in excavation techniques for slurry walls are related to the development of equipment. For example, the new hydromills can reach about 100 m in depth in fully controlled conditions as far as verticality of the trench is concerned [28, 87, 100]. Also, very difficult materials (i.e. presence of boulders, cemented soil and weak rocks) can be excavated with this kind of equipment. Many contractors are setting up agile trenching machines to speed excavation and filling operations [64, 82]. These machines are able to execute conveniently narrow (up to 30 cm) and shallow (up to 5 m) trenches to be filled with the final sealing mixtures.

The composite slurry wall is probably the most outstanding advancement in the last few years as far as containment systems of polluted subsoils are concerned. Following the concept of the composite lining system for new landfills (geomembrane plus compacted clay), geomembrane sheets with special joints are introduced into the fresh slurry of the diaphragm [27, 33, 70, 90]. The latest products available in the field of geomembranes for composite slurry walls are composite sheets integrated with a monitoring system and possibly with leachate collection [12, 79]. As an alternative to the geomembrane with joint systems, a geomembrane film unrolled into the trench also has been proposed [27]. However, practical use of this technology has not been used on a large scale.

3.4.2 Design Criteria and Procedures

The critical design criteria for slurry wall containment systems include consideration for the permeability, deformability, and durability [73]. For long-term containment systems, the design approach must refer to the allowed contaminant flow rate or migration time given by national or local regulations if available; otherwise, an acceptable international standard is the only reference. In terms of permeability, reference must be made both to the basic microstructure permeability of the mixture to be adopted and to the actual full-scale in situ permeability bearing in mind that this last feature is influenced strongly by the construction procedures. Discussions about the differences between microstructure and full-scale in situ permeability of slurry wall are given elsewhere [5, 61, 65, 90, 103]. The influences of basic component mixture contents on permeability trends are given by D'Appolonia [23], Jefferis [52], Evans et al. [39], Meseck and Hollstegge [71] and Millet et al. [73]. Recently, the contribution of diffusion to the contaminant flow rate through mineral barriers has been fully recognized [84, 89, 94, 97, 98]. Design criteria for cutoff walls based on both hydraulic conductivity and diffusion are given by Gray and Weber [47], Manassero [61], and Manassero and Shackelford [63]. Simplified calculation procedures for definition of wall thickness and embedment into impervious layer based on both flow rate and migration time limits are given by Manassero and Pasqualini [64]. Moreover, the same authors provide some guidance for the evaluation of composite wall performance on the basis of geomembrane joint features.

Design criteria related to deformability are well established [50]. In any case, deformability is often a minor concern for containment systems due to generally low strains by external loads or water pressures.

Research about interaction phenomena between typical pollutants and sealing mixtures is continuing [1, 8, 15, 29, 48, 76, 90, 91, 95]. General addresses on test programs for sealing mixture definition and design are given by USEPA [103], Evans [38], and ETC8 [37].

3.4.3 Construction Materials and Suitability Tests

Backfill mixtures for slurry cutoff walls include (1) cement-bentonite (CB) self-hardening slurry, (2) soil-bentonite (SB) mixture, and (3) plastic concrete (PC) mixtures. The main characteristics of these materials in terms of hydraulic conductivity, stress-strain behavior and, in some cases, chemical compatibility can be found in Caron [10], Mastrantuono and Tornaghi [67], D'Appolonia [23], Millet and Perez [72], De Lucca and De Paoli [24], Li et al. [59], Meseck and Hollstegge [71], Schweitzer [92], and Manassero and Viola [65]. References for basic test programs to evaluate the suitability of materials are given in the design criteria. For particular tests and/or new sealing products, see Gandais and Delmas [41], Esnault [36], De Paoli et al. [27], Ryan [90], Khera and Tirumala [55], Meseck and Hollstegge [71], Carlsson and Marcusson [9], Li et al. [59], Paviani [82], Boyd et al. [7], Marbach [66], and Smith and Booker [99]. The major research effort to understand complex chemico-physical phenomena and to define the basic framework and procedures for practical professional and construction activities is in the interaction between sealing backfill materials and contaminants, long-term behavior, and the related suitability tests.

3.4.4 Construction Quality Control and In Situ Performance

Long-term monitoring of the containment structures plays a fundamental role in confirming the design hypothesis and checking the performance of the system. For example, it is rather simple to check hydraulic conductivity of the mixtures with laboratory tests using small scale samples taken in situ during construction operations or prepared and cured in the laboratory. Unfortunately, this kind of indirect control, while satisfactory for suitability analysis, is not fully reliable in terms of the actual in situ performance of the barrier. In the case of hydraulic conductivity, disagreement often arises due to accidental variation of backfill mixture composition, sandy or gravel lense inclusions, or cracks and fissures following slurry consolidation or displacement of surrounding soil. In general, in situ tests are potentially more suitable than laboratory tests to assess the actual performance of confinement barriers. However, in situ tests are more expensive and time consuming.

As a result of the above considerations, the importance of improving and advancing the field of in situ test and monitoring is evident. Good perspectives are given by some of the new trends and developments provided by Hulla et al. [49], Ryan [90], De Paoli et al. [27], Muller-Kirchenbauer et al. [78], Armbruster et al. [2], Mitchell [74], Bertero et al. [5], Torstensson [101], Chen and Yamamoto [13], Bhatia and Eldin [6], Daniel [16], Leps [58], Manassero and Viola [65], Druback and Arlotta, [33], Cavalli [12], Daniel [18], Manassero [62]. For example, the possible use of the piezocone to provide a continuous assessment of CB slurry wall hydraulic conductivity has been investigated by Manassero [62]. The assessment uses an opportune combination of pore pressure increment, total point resistance, and sleeve friction data calibrated by dissipation tests. Major research efforts should be devoted to this specific topic in the near future. From this point of view, it would be useful to form an international data bank with the purpose of collecting short- and long-term records about actual performance of the constructed walls in order to enhance a fundamental knowledge in this specific field.

3.5 Groundwater Control and Pollution (M. Tumay, G^3S Program, National Science Foundation, Room 545.17, 4201 Wilson Blvd., Arlington, Virginia 22230, USA)

3.5.1 Potential Contamination of Ground Water from Pesticides

Groundwater contamination from the use of agricultural chemicals, pesticides and herbicides, is a recognized threat to the environment. Perhaps more than any other source of pollution, agricultural chemicals can be regulated to promote safe use for increased production while promoting pollution prevention (waste minimization), the safest and most cost-effective way of protecting our environment. The central question is the effect of pesticides on the groundwater system beyond the unsaturated zone.

3.5.2 Integrated Groundwater Transport Simulation Models

Contamination from industrial chemicals is an increasing threat to the environment whether caused by agricultural pesticides and herbicides, petroleum storage tanks, landfills and other waste sites, or by statistically inevitable accidents. Processes that affect the source and migration of contaminants include advection and dispersion in response to the movement of the water mass, interaction of the bulk mass of the chemical with the water mass (dissolution, adsorption/desorption), chemical and biological processes that affect the fate of the chemical and interactions with the soil.

Numerous models for the simulation of individual processes exist, more or less sophisticated, accurate or reliable. Therefore, the issue is not which model is best (or how to improve it), but rather how to simplify the use of a model, how to control errors associated with model set-up and use, and how to support and improve the decision making ability of the user of the model. An exercise in modeling typically encompasses the following steps: extraction of geographic information from maps, areal photography or other sources; collection of meteorologic/hydrologic/chemical data; discretization and data preparation for the selected simulation model; and interpretation of the model results.

3.5.3 New Developments in Stochastic Finite Element of Random Three-Phase Geo-Media

The evaluation of the reliability of soil structures is a crucial component in conceptualization and design. The prediction of the behavior of soil structures is sensitive to even small uncertainties of pertinent design variables. Several of these variables are inherently random and can be modeled most appropriately as random processes. Such variables include the modulus of elasticity, the poisson ratio, the shear strength, hydraulic conductivities, and other constitutive parameters. Clearly, versatile numerical finite element algorithms can be used to obtain accurate mathematical predictions of the physical behavior of complex modern soil structures.

Randomness has been introduced in finite element modeling over the past decade using first and second order perturbation techniques. However, these methods are often impractical due to the scale of the required computational effort. Recent developments aim at reducing the computational burden of such stochastic analyses. In this respect, the classical deterministic finite element approach is still used to discretize the soil medium, but a spectral expansion of the nodal random variables is introduced requiring the definition of a basis in the space of random variables. The basis consists of polynomial expressions orthogonal to the Gaussian probability measure. This new approach allows the computation of the entire probability distribution functions (not just the first two moments) of the response variables in a cost-efficient way. Preliminary results show that considerable savings can be obtained for real case studies. Further work is needed to investigate the applicability of this approach to a wide range of geoenvironmental processes.

3.5.4 Simulation of Biodegradation Phenomena in Soil Media

Many organic contaminants can be degraded by microbes. These processes occur in natural and engineered soil systems, and new compounds including intermediate toxic compounds and/or stable polymerized compounds are being developed. The microbial transformation and/or degradation processes contribute towards the overall fate and transport of contaminants. However, distribution of micro-organisms and associated activities are difficult to represent in a simulation model due to the microscale. Also, when contaminants and other organic nutrients are consumed, microbial populations are affected which may lead eventually to alteration of the hydrophysical soil environment (e.g. permeabilities and structural parameters). A quantitative understanding of the interaction between microbial population dynamics and chemical transformation, as influenced by the soil environment, is essential for evaluating the effectiveness of bioremediation.

Attention must be given to enzyme kinetics, native water and soil skeleton, pollutant, nutrient availability, toxic waste, diffusion and advection, and physical/chemical soil characteristics as related to the microbial population. The need for a quantitative description of these processes is recognized by the scientific community.

3.5.5 New Developments in Landfill Siting Procedures Using Fuzzy Sets

Fuzzy set theory has become a very powerful production tool in China, Japan, and Europe, but is still not well-known. In most of the existing applications, the Fuzzy set theory has been used in feed-back control implementations. However, research is needed in the application of this theory in the area of knowledge-based systems where processing of the ambiguous and imprecise information will be handled through the use of fuzzy sets. This research should have a significant impact in advancing the current knowledge associated with landfill siting decision which frequently is based on ambiguous and imprecise data and opinions.

3.5.6 Impact of Waste Liquids Injection on Natural Flow Systems

The impact of liquid injection systems to the natural underground environment is still not well defined. Typically, industrial facilities, such as manufacturers of synthetic fibers, inject waste liquids into the aquifer and assess the water quality on the basis of simplified analytical models. More rigorous simulation models obviously are needed to assess the water quality of the injected aquifer, focusing on the quantitative analysis of the movement of the dissolved solids and calibration based on known in-situ data of Durov plots.

3.5.7 Need for In Situ Determination of Geotechnical/Chemical Properties of an Aquifer

Recently, the conventional cone penetrometer has been coupled with a ground-water sampler for environmental applications such as the determination of the extent and preferential flow pathways of a soluble hydrocarbon plume in an aquifer. Aqueous concentrations of hydrocarbons are measured in situ by a gas chromatograph with a photoionozation detector. This technique gives satisfactory results in good agreement with direct concentration measurements obtained in the laboratory using standard ground-water analytical procedures. However, further studies are needed to refine this new technology and satisfy stringent environmental requirements.

3.5.8 Geostatistics in Geoenvironmental Problems

Three of the basic problems in geostatistics are to estimate (1) the value of a parameter (hydraulic conductivity, thickness, concentration) at a given point by statistical inference based on a set of sampled data points (point estimate), (2) the mass or volume of an expansive parameter (such as concentration) over a given volume (volume estimate), and (3) the average value of a given parameter over a given volume in space or time (volumetric estimate).

The theory of Kriging consists of a clear statement of assumptions and procedures for the optimal estimation of the distribution of physical properties over space and time. The assumptions and procedures have evolved over time, culminating with the theory of General Order Increments which uses a Generalized Intrinsic Hypothesis and the notion of Generalized Covariance. Kriging has many advantages as compared with traditional techniques of interpolation because it is an exact estimator at the sampled points and quantifies the accuracy of the estimate. Further research is needed to enhance the performance of Kriging in three-dimensional geomedia taking into consideration spatial and temporal discontinuities in typical characteristics exhibited by the natural processes in ground contamination problems.

3.6 Site Remediation (R. G. Clark, CL Associates, Haywood House, Mucklow Hill, Halesowen, Birmingham, United Kingdom B62 8EL)

There have been extensive efforts in recent years to recycle contaminated land. This generally requires some form of site remediation to make the land suitable for its intended use. The techniques available range from those that are already well established to new technologies that are becoming available. The method or methods that are appropriate for a particular site will depend on the location of the site and the intended land use. Some techniques are practically and economically viable and others are not. Also, some techniques are likely to be under the control of geotechnical engineers whereas other techniques may require a lesser involvement.

In general, remediation processes can be categorized as in situ, ex situ or both. These remediation processes generally include excavation and disposal, containment (e.g., vertical and horizontal barriers, clay caps), hydrogeological methods (e.g., pumping either for disposal or for some other form of treatment), in situ systems (e.g., soil washing, electrokinetic processes, vegetation or algae based systems, lime treatment), ex-situ technologies (e.g., particle separation, volume reduction, thermal desorption, incineration, vitrification, vapor extraction, air or steam stripping), solidification and stabilization, and biological and chemical treatment. These technologies rely on either removing, neutralizing or immobilizing the contaminants or blocking the path by which contaminants could reach susceptible targets.

Conventional engineering techniques such as excavation and disposal and containment generally are based on the principle of preventing contact between the contaminated ground or groundwater and any susceptible targets. Excavation of the contaminated ground may be followed by disposal on-site or off-site (usually to a licensed landfill facility). This procedure is widely used and is likely to remain popular as long as landfill capacity is available and the costs of disposal are less than other forms of treatment.

Containment (macro encapsulation) also uses well established civil engineering techniques such as surface covering, vertical barriers (diaphragm walls, cement/bentonite cut offs, grout curtains or geomembranes) and horizontal barriers. Such systems can be used with a wide range of contaminants. However, because the contaminant remains in place, the long-term durability of the system is of prime importance. Containment is less readily accepted in some countries where it is regarded as a temporary measure.

Hydraulic systems can involve simple pumping of the contaminant from the ground. However, pumping may require large quantities of contaminated groundwater to be removed and treated. The method is usually used in conjunction with another technology and is only applicable in certain hydrogeological conditions.

The majority of the in situ treatment systems (such as soil washing and electrokinetic processes) utilized are based on fairly well established technologies. For example, soil washing generally is based on construction dewatering systems and the use of solvents to remove high molecular weight organic compounds. However, there are a number of other techniques, such as bioremediation, which are becoming popular. In certain countries, methods such as vitrification also have been used in situ. In this sense, in situ processes are processes which do not require the removal of contaminated soil or water from the ground.

Solidification and stabilization also can be carried out as in situ techniques with the aim being to immobilize the contamination rather than to destroy it. These techniques are usually accomplished by the addition of some form of cement or chemical compound. The intention generally is to use the technology in situ but the methods also can be applied ex situ. The techniques are potentially suitable for a wide range of contaminants but have been used more for inorganic contamination. The techniques imply the need for long-term monitoring because the processes are containment rather than removal or destruction.

Biological and chemical treatment can be carried out either in situ or ex situ. In the case of bioremediation, the principle is to use micro organisms to destroy the contaminants. The micro organisms are either introduced or naturally occurring. In some cases, plants or algae can be used.

Some contaminants are more amenable to bioremediation than others. Biological processes generally are used for treating organic contaminants in soils or in water. On the other hand, chemical treatments (i.e., chemical reactions used to destroy or otherwise modify the contaminants) are generally ex situ processes most commonly used for slurries or liquids. Thus far, only limited experience with respect to the use of chemical treatment has been gained and, hence, reservations over the consequential effects of certain reactions exist.

Ex situ technologies tend to involve methods which require either highly specialized plant and equipment or very controlled conditions. In this sense, ex situ is defined as either those processes where the contaminated ground and groundwater are removed from site or those processes that are carried out on or adjacent to the site on contaminated media which has been removed from the ground. In general, only very heavily contaminated soils can justify the considerably greater expense of carrying out off-site ex situ methods of treatment.

In addition, there are a number of proprietary systems which are not covered in the above summary, but which may represent completely new technologies, a variation of the above methods, or a combination of these methods.

All the relevant factors must be considered when deciding on a particular remediation technique or combination of techniques for any particular contaminated site. For example, consideration needs to be given to whether treatment should be carried out in situ, thus avoiding the transporting of contaminants and the associated environmental impact, or ex situ where complicated or refined treatments are much easier to control.

Consideration also must be given to planning requirements, health and safety issues during remediation, cost effectiveness, public opinion and current legislation in the country in question. The technical factors that are of importance include the nature and extent of contamination, site geology, geotechnical ground conditions, hydrogeology and hydrology, site size, land value, development type, time proximity to the general public, and the presence of buildings and underground structures.

The geotechnical engineer has a major role in all site remediation, and particularly in the areas of site investigation, excavation and earthworks, diaphragm walls, slurry trench walls and grout curtains, clay liners and capping, horizontal subsurface barriers (e.g., jet grouting, chemical grouting or claquage grouting), break layers (e.g. to prevent the upward migration of contaminants due to soil suction), geomembranes, extraction wells, treatment wells and monitoring wells (for water, leachate, liquid contaminant or gas), stabilization of mine workings, reinstatement and development (e.g. replacement or cleaned fill materials, slope stability, settlement, ground improvement, foundations, protection measures to buildings and foundations), and project management.

The choice of technology needs to be approached with caution regardless of whether the technology is proven or emerging. Many of the technologies do not have a sufficient track record for a sufficient range of different situations to be usable without significant trials and monitoring. In the past, the benefits of many good technologies have been overstated or the processes misused and, therefore, the credibility of the technology has come into question. A critical review is required of the various methods and systems that are presently available. Despite a high level of scientific involvement in some technologies, site remediation is a civil engineering exercise and, in many respects, the geotechnical engineer should take the lead in carrying out this review.

3.7 Risk Assessment/Risk Management Methodology (J. C. Paslawski, O'Connor Assoc. Environmental Inc., 201, 1144 - 29 Avenue N.E., Calgary, Alberta, Canada T2E 7P1)

Risk assessment provides a means of identifying and quantifying the potential impacts of industrial activities on human health and the environment. The quantification of these impacts in financial terms provides industrial operators with a rational basis for decisions governing the management of environmental issues. The consideration of health and environmental risks and the concept of risk management in the establishment of regulatory policies and guidelines ensures protection of the public while acknowledging the economic constraints of private industry. An increasing trend towards the use of risk assessment and risk management in both industry and government is anticipated.

Thus far, risk assessment has been used primarily in the development of site-specific remediation criteria, i.e. contaminant concentrations derived to meet target risk levels. However, a potentially more significant role of risk assessment lies with evaluation of the quantitative results for the development of risk management alternatives. The option to manage or reduce risks, rather than eliminate risks by complete clean-up, is becoming more attractive under the rising financial constraints of site remediation.

3.7.1 Traditional Approaches

Traditional approaches for selection of remediation criteria for contaminated sites include absolute methods, whereby fixed criteria are adopted by the governing regulatory agency, and ad hoc approaches whereby criteria are assigned on a case-by-case basis. Many jurisdictions prescribe remediation levels in terms of contaminant concentrations or other parameters that must be met during clean-up of a site. In the past, a single clean-up concentration was often specified for a given contaminant, without consideration of differing levels of human health risk and/or environmental risk between sites. More recently, human health risk has been addressed, to a limited degree, in some jurisdictions by specifying different clean-up levels for different land uses (i.e., residential versus industrial). Compliance with these remediation criteria, while satisfying regulatory requirements of the day, would not necessarily ensure a consistent level of accepted health risk between sites. Many statutory criteria are conservative, requiring clean-up measures more thorough and costly than would be justified on the basis of health risks. Other criteria, originally based on best available detection or clean-up technology, may be found in the future to be unconservative with respect to health risks as new information becomes available.

3.7.2 Health Risk-Based Methods

A superior approach to the management of risks during remediation of a contaminated site is the use of risk-based clean-up criteria. As noted previously, risk assessment can be used to determine the required contaminant concentrations to meet target levels of acceptable risk. When carried out on a site-specific basis, this would provide some assurance that risks to receptors on and near the site are within acceptable levels. When considering the human health risks associated with a given site, all likely future land uses should be considered to ensure that future redevelopment does not result in unacceptable levels of health risk and hence potential liabilities. The development of health risk-based clean-up criteria is now considered by many regulatory agencies to be an acceptable alternative to the use of prescribed numerical values.

3.7.3 Environmental Risk Assessment Methods

The process of assessing environmental risks is essentially parallel to that for human health risks. Environmental risks are not normally quantified in the same way as human health risks. Environmental risks tend to be expressed in terms of effects rather than risks. These effects may be aesthetic (visual or odor), or they may be reflected in distressed vegetation or mortality within a particular population of animals, fish or other organisms. Target environmental quality criteria, including effluent discharge limits, are frequently effects-based, and are related to plant toxicity or toxicity to certain organisms. The latter is frequently expressed in terms of the dose that is determined on the basis of bioassay experiments to be lethal to 50 percent of the exposed population.

3.8 References

1. Acar, Y.B., D'Hollosy, E. (1987). Assessment of Pore Fluid Effects Using Flexible Wall and Consolidation Permeameters. *Geotechnical Practice for Waste Disposals '87*, ASCE Geotechnical Special Publication No. 13, New York, 231-245.
2. Armbruster, H., Merkler, G.P., Troger J.H.M. (1989). The Closure of a Leakage in the Sealing System of a Dyke. *Proc. of the Twelfth ICSMFE* - Rio de Janeiro, 1472-1474.
3. Bagchi, A. (1990). Design, Construction, and Monitoring of Sanitary Landfill. John Wiley and Sons, Inc., New York, 284 pp.
4. Barnes, F. J., Rodgers, J.C. (1988). Evaluation of Hydrologic Models in the Design of Stable Landfill Covers. EPA/600/2-08/048.
5. Bertero, M., Marcellino, P., Paviani, A. (1983). Determination of In-Situ Permeability of Slurry Trench Cutoff Wall. *Proc. of the Int. Symp. Recent Developments in Lab. and Field Tests and Analysis of Geotech. Problems*, Bangkok, Balkema Publ., Rotterdam. 261-266.
6. Bhatia, S.K., El-Din, K. (1989). In Situ Measurement of the Permeability of Slurry Walls. *Proc. of the Twelfth ICSMFE* - Rio de Janeiro, 1475-1478.
7. Boyd, S.A., Lee, J.F., Mortland, M.M. (1988). Attenuating Organic Contaminant Mobility by Soil Modification. *Nature*, 26:345-347.
8. Brunelle, T.M., Dell, L.R., Meyer, C.J. (1987). Effect of Permeameter and Leachate on a Clay Liner. *Geotechnical Practice for Waste Disposals '87*, ASCE Geotechnical Special Publication No. 13, New York,

347-361.
9. Carlsson, B., Marcusson, L. (1989). Stabilized Residue. *Proc. of the Twelfth ICSMFE*, Rio de Janeiro, 1857-1860.
10. Caron, C. (1973). Un Nouveau Style de Perforation La Boue Autodurcissable. *Annales de L'Institut Technique du Batiment et des Travaux Publics* n° 311.
11. Cartwright, K., Krapac, I. G. (1990). Construction and Performance of Long-Term Earthen Liner Experiment. *Waste Containment Systems: Construction, Regulation and Performance*, ASCE Geotechnical Special Publication No. 26, New York, 135-155.
12. Cavalli, N.J. (1992). Composite Barrier Slurry Wall. *Slurry Walls: Design, Construction and Quality Control,* ASTM STP 1129, Philadelphia, 78-85.
13. Chen, H. W., Yamamoto, L. O. (1987). Permeability Tests for Hazardous Waste Management Unit Clay Liners. *Geotechnical and Geohydrological Aspects of Waste Management*, Lewis Publ., Inc., Chelsea, Michigan, 229-243.
14. Daniel, D. E. (1984). Predicting Hydraulic Conductivity of Clay Liners. *ASCE Journal of Geotechnical Engineering*, 110(4): 465-478.
15. Daniel, D. E. (1987). Earthen Liners for Land Disposal Facilities. *Geotechnical Practice for Waste Disposal '87*, ASCE Geotechnical Special Publication No. 13, New York, 21-39.
16. Daniel, D. E. (1989). In Situ Hydraulic Conductivity Tests for Compacted Clay. *ASCE Journal of Geotechnical Engineering*, 115(9): 1205-1226.
17. Daniel, D. E. (1990). Summary Review of Construction Quality Control for Earthen Liners. *Waste Containment Systems: Construction, Regulation and Performance*, ASCE Geotechnical Special Publication No. 26, New York, 175-189.
18. Daniel, D. E. (1990). Impervious Barriers for Pollutants Containments. *Seminar organized by Italian National Electricity Board*-Research Centre of Milan-Italy
19. Daniel, D. E., Brown, K. W. (1988). Landfill Liners: How Well Do They Work and What Is Their Future?" Land Disposal of Hazardous Waste, Ellis Horwood Ltd., West Essex, England, 235-244.
20. Daniel, D. E., and Benson, C. H. (1990). Water Content-Density Criteria for Compacted Soil Liners, *ASCE Journal of Geotechnical Engineering*, ASCE, 116(12):1811-1830.
21. Daniel, D. E., Trautwein, S. J. (1986). Field Permeability Test for Earthen Liners. *Use of In Situ Tests in Geotechnical Engineering*, ASCE Geotechnical Special Publication No. 6, New York, 146-160.
22. Day, S. R., Daniel, D. E. (1985). Hydraulic Conductivity of Two Prototype Clay Liners. *ASCE Journal of Geotechnical Engineering*, 111(8): 957-970.
23. D'Appolonia, D.J. (1980). Soil-Bentonite Slurry Trench Cutoffs. *J. Geotechnical Engineering Division,* ASCE, 106 (GT4): 399-.
24. De Lucca, M., De Paoli, B. (1986). The Use of Slurry Wall Cutoffs for Remediation. *Proc. ISMES Seminar on Hazardous Waste: Production, Control and Disposal,* Bergamo, Italy.
25. de Mello, L.G. (1980). Comparative Behaviors of Similar Compacted Earthwork Dams in Basalt Geology in Brazil. Symposium on Problems and Practice in Dam Engineering, Bangkok, Vol. 1.
26. de Mello, L.G., Carrier III, W.D., Lapa, R., and Carrarezi, F. (1991). Optimized Design of Uncompacted Earth Dikes - the Trombetas Case History. *IX Pan American Conf.*, SMFE, Vol. III, Vina del mar, Chile.
27. De Paoli, V., Marcellino, P., Mascardi, C., Paviani, A. (1991). Le Applicazioni dei Diaframmi nella Difesa Ambientale. *Proc.*

of the Fifteenth Conference on Soil Mechanics, Torino, Italy, November 20-22.

28. De Paoli B., Mascardi C., Stella C. (1989). Construction and Quality Control of a 100 m Deep Diaphragm Wall. *Proc. of the Twelfth ICSMFE,* Rio de Janeiro, 1479-1482.

29. Demetracopoulos, A.C., Dharmapal, A.P. (1987). Flow and Mass Transport for Hazardous Waste Liners. *Geotechnical Practice for Waste Disposal '87,* ASCE Geotechnical Special Publication No. 13, New York, 392-405.

30. Dendrou, S. (1992). *EIS/PRZM- A Computer Platform for Pesticide Root Zone Models Managing the Potential Contamination of Ground Water from Pesticides Applied to Different Soils,* Final Report submitted to Geomechanical, Geotechnical and Geo-Environmental Systems Program, National Science Foundation.

31. Dendrou, S., Dendrou B., Williams, W. (1992). *New Interface Platforms for Pesticide Transport Models,* 204th American Chemical Society (ACS) National Meeting, Washington, D.C.

32. Desaulniers, D. E., Cherry, J. A., Fritz, P. (1981). Origin, Age and Movement of Pore Water in Argillaceous Quaternary deposits at Four Sites in Southwestern Ontario. *Journal of Hydrology*, 231-257.

33. Druback, G.W., Arlotta, S.V. (1984). Subsurface Pollution Containment Using a Composite System Vertical Cutoff Barrier. *ASTM Symposium,* Denver, USA.

34. Elsbury, B. R., Daniel, D. E., Sraders, G. A., Anderson, D. C. (1990). Lessons Learned from Compacted Clay Liner. *ASCE Journal of Geotechnical Engineering*, 116(11):1641-1660.

35. <u>Environmental Geotechnology</u>, (1992), *Proceedings*, Mediterranean Conference in Environmental Geotechnology. Usmen, M. and Acar, Y.B. (eds), A.A. Balkema, Rotterdam, 594 pp.

36. Esnault, A. (1992). Adaptation et évolution des techniques de traitement de sol en matière de protection de l'environnement. *Revue Francaise de Geotechnique*, Presses de l'école nationale des Ponts et Chaussées n° 60 - 27-40.

37. ETC8 - ISSMFE (1993). <u>Geotechnics of Landfills and Contaminated Land - Technical Recommendations</u> - Ernst & Sohn - Berlin. Second Edition.

38. Evans, C.J. (1991). Geotechnics of Hazardous Waste Control Systems. *Foundation Engineering Handbook*, Van Nostrand Reinhold, New York, 750-777.

39. Evans, C.J., Stahl, E.D., Droof, E. (1987). Plastic Concrete Cutoff Walls. *Geotechnical Practice for Waste Disposal '87.* ASCE Geotechnical Special Publication No. 13, New York, 462-472.

40. Fenn, D.G., Hanley, K.J. and DeGeare, T.V. (1975). Use of the Water Balance Method for Predicting Leachate Generation from Solid Waste Disposal Sites. EPA/530/SW-168, U.S. EPA, Cincinnati, OH.

41. Gandais, M. and Delmas, F. (1989). Paroi d'etancheité pour retention de déchets toxiques à forte acidité. *Proc. of the Twelfth ICSMFE* Rio de Janeiro, 1871-1872.

42. Goldman, L. J., Greenfield, L. I., Damle, A. S., Kingsbury, G. L., Northeim, C. M., Truesdale, R. S. (1986). <u>Design, Construction, and Evaluation of Clay Liners for Waste Management Facilities</u>. Draft Technical Resource Document for Public Comment (March), EPA/530/SW-86/007F, U.S. EPA, Washington, D. C.

43. Goodall, D. C., Quigley, R.M. (1977). Pollutant Migration from Two Sanitary Landfill Sites Near Sarnia, Ontario. *Canadian Geotechnical Journal*, 14:223-236.

44. Gordon, M. E., Huebner, P. M., and Kmet, P. (1984). An Evaluation of the

Performance of Four Clay-Lined Landfills in Wisconsin," *Proc., Seventh Ann. Madison Waste Conference on Municipal and Industrial Waste,* Madison, Wisconsin, 399-460.
45. Gordon, M. E., Huebner, P. M., Miazga (1989). Hydraulic Conductivity of Three Landfill Clay Liners. *ASCE Journal of Geotechnical Engineering,* 115(8):1148-1160.
46. Gordon, M. E., Huebner, P M., Mitchell, G. R. (1990). Regulation, Construction, and Performance of Clay-Lined Landfills in Wisconsin. *Waste Containment Systems: Construction, Regulation and Performance,* Rudolph Bonaparte (ed.), ASCE Geotechnical Special Publication No. 26, New York, 14-29.
47. Gray, D.H., Weber, W.J., Jr. (1984). Diffusional Transport of Hazardous Waste Leachate Across Clay Barriers. *Proc., Seventh Ann. Madison Waste Conference, Municipal and Industrial Waste,* Madison, Wisconsin.
48. Ho, Y.A., and Pufahl, D.E. (1987). The Effects of Brine Concentration on the Properties of Fine Grained Soils. *Geotechnical Practice for Waste Disposals '87,* ASCE, Geotechnical Special Publication No. 13, New York, 547-561.
49. Hulla, J., Ravinger, R., Kovac, L., Turcek, P., Dolezalova, M., Horeni, A. (1985). Dewatering and Stability Problems of Deep Excavation. *Proc. of the Eleventh ICSMFE,* San Francisco, 2099-2102.
50. ICOLD (1985). Filling Materials for Watertight Cutoff Wall. *Bulletin 51,* Paris.
51. Inyang, H. (1992). *Advances and Traends in Geo-Environmental Engineering,* Proceedings of the U.S.-Canada Workshop on Recent Accomplishments and Future Trends in Geomechanics in the 21st Century, National Science Foundation, Washington, D.C.
52. Jefferis, S.A. (1981). Bentonite-Cement Slurries for Hydraulic Cutoffs. *Proc. of the Tenth ICSMFE,* Stockholm, June 15-19, A.A. Balkema, Rotterdam, 435-440.
53. Johnson, R. J., Cherry, J. A., Pankow, J. F. (1990). Diffusive Contaminant Transport in Natural Clay: A Field Example and Implications for Clay-Lined Waste disposal Sites. *Journal of Environmental Science and Technology,* 23:340-349.
54. Johnson, G. M., Crumbley, W. S., and Boutwell, G. P. (1990). Field Verification of Clay Liner Hydraulic Conductivity. *Waste Containment Systems: Construction, Regulation and Performance,* ASCE Geotechnical Special Publication No. 26, New York, 226-245.
55. Khera, R.P., Tirumala, R.K. (1992). Materials for Slurry Walls in Waste Chemicals. *Slurry Walls: Design, Construction and Quality Control,* ASTM STP 1129, Philadelphia, 172-180.
56. Laihti, L. R., King, K. S., Reades, D. W., Bacopoulos, A. (1987), Quality Assurance Monitoring of a Large Clay Liner. *Geotechnical Practice for Waste Disposal '87,* ASCE Geotechnical Special Publication No. 13, New York, 640-654.
57. Lechner, P. (1987). The Austrian Guidelines for Sanitary Landfills. *Proceedings of International Symposium on Process, Technology and Environmental Impact of Sanitary Landfill.* Cagliari, Sardinia, 19th-23rd, October 1987.
58. Leps, T.M. (1989). Slurry Trench Cutoffs and the In-Situ Permeability Test. *De Mello Volume* Editoria Edgard Blucher Ltda. Sao Paulo, Brazil, 259-270.
59. Li, J.C., Hwang, C.L., Yao, H.L., Lee, H.J., Lee, R.J. (1989). A Study of Slag Cement-Bentonite Slurry. *Proc. of the Twelfth ICSMFE.* Rio de Janeiro, 3:1499-1502.
60. Lutton, R.J., Regan, G.L., Jones, L.W. (1979). Design and Construction of Covers for Solid Waste Landfills. EPA-600/2-79-

165. Municipal Environmental Research Laboratory, U.S. EPA, Cincinnati, OH, 250.
61. Manassero, M. (1991). Metodologie di intervento nei terreni inquinati. *Proc. of the Fifteenth Conference on Soil Mechanics of Torino*, Italy, November 20-22.
62. Manassero, M. (1992). In Situ Quality Control of Cement-Bentonite Cutoff Wall Using Piezocone Test. *Annual Seminar of National Research Council; Geotechnical Group*. Roma, Italy.
63. Manassero, M., Shackelford, C. (1994). The Role of Diffusion in Contaminant Migration through Soil Barriers. *Italian Geotechnical Journal* (in print).
64. Manassero M., Pasqualini, E. (1992). Ground Pollutant Containment Barriers. *Proc. of the Mediterranean Conference on Environmental Geotechnology*, Balkema, Rotterdam, 195-204.
65. Manassero, M., Viola, C. (1992). Innovative Aspects of Leachate Containment with Composite Slurry Walls: A Case History. *Slurry Walls: Design, Construction and Quality Control*, ASTM STP 1129, Philadelphia, 181-193.
66. Marbach, W.D. (1988). Designed Clays That Drink Up Toxic Wastes. *Business Week*, June 13, 60.
67. Mastrantuono, C., Tornaghi, R. (1977). Fanghi autoindurenti per la realizzazione di diaframmi plastici. *Atti del Gruppo Lombardo* (Italia Nord-Ovest) dell'AGI. Milano, Italy.
68. Matsuo, S., Honmachi, Y., and Akai, K. (1953). A Field Determination of Permeability. *III Int. Conf. Soil Mech. Found. Eng.*, Zurich, 268-271.
69. McAneny, C.C., Hatheway, A.W., (1985). Design and Construction of Covers for Uncontrolled Landfill Sites. *The 6th National Conference on Management of Uncontrolled Hazardous Waste Sites*. Washington, DC.
70. Meseck, H. (1987). Dichtwande und Dichtsohlen. *Fachseminar*, June 2-3, Stadthalle Braunschweig, Technische Universitat Braunschweig.
71. Meseck, H., Hollstegge, W. (1989). Case History on the Construction of Cutoff Wall Diaphragm Wall Method. *Proc. of the Twelfth ICSMFE*, Rio de Janeiro, 3:1503-1506.
72. Millet, R.A., Perez, J.Y., (1981). Current USA Practice: Slurry Wall Specifications. *Journal of Geotechnical Engineering Division*, ASCE, 107(GT8):1041-1056.
73. Millet R.A., Perez, J.Y., Davidson R.R. (1992). USA Practice Slurry Wall Specification 10 Years Later. *Proc. of Conf. on Slurry Walls: Design, Construction and Quality Control*, ASTM-STP 1129, Philadelphia, 42-66.
74. Mitchell, J. K. (1992). Conduction Phenomena: From Theory to Geotechnical Practice. *Geotechnique*, 41(3):299-340.
75. Mitchell, J. K., Jaber, M. (1990). Factors Controlling the Long-Term Properties of Clay Liners," *Waste Containment Systems: Construction, Regulation and Performance*, ASCE Geotechnical Special Publication No. 26, New York, 84-105.
76. Mitchell, J. K., Madsen, F. T. (1987). Chemical Effects on Clay Hydraulic Conductivity. *Geotechnical Practice for Waste Disposal '87*, ASCE Geotechnical Special Publication No. 13, New York, 87-116.
77. Mitchell, J. K., Hooper, D. R., Campanella, R. G. (1965). Permeability of Compacted Clay. *Journal of the Soil Mechanics and Foundations Division*, ASCE, 91(SM4):41-65.
78. Muller-Kirchenbauer H., Friedrich, W., Rogner J. (1987). Ergebnisse der Dichtwandtestmanahme im Rahmen eines F + E-Vorhaben fur die Sanierung der Sonderabfalldeponie Gerolsheim. *Symposium, Ministerium fur Umwelt und*

Gesundheit des Landes Rheinland-Pfalz, Mainz, S. 63-77.
79. Nussbaumer, M. (1987). Beispiele fur die Herstellung von Dichtwanden im Schlitzwandverfahren. *Fachseminar,* June 2-3, Stadthalle Braunschweig, Technische Universitat Braunschweig, 21-34.
80. Olson, R. E., Daniel, D. E. (1981). Measurement of the Hydraulic Conductivity of Fine-Grained Soils. ASTM STP 746, Philadelphia, 18-64.
81. Paul, D.B., Davidson, R.R., Cavalli, N.J. (1992). *Slurry Wall: Design, Construction, and Quality Control.* ASTM, STP 1129, Philadelphia.
82. Paviani, A. (1992). Personal Communication.
83. Putti, M., Yeh, W.G. and Mulder, W.A. (1990). A Triangular Finite Volume Approach with High-Resolution Upwind Terms for the Solution of Groundwater Transport Equations. *Water Resources Research,* (26):12-.
84. Quigley, R.M., Yanful, E.K., Fernandez, F. (1987). Ion Transfer by Diffusion Through Clayey Barriers. *Geotechnical Practice for Waste Disposal '87,* New York, 137-158.
85. Reades, D. W., Lahti, L. R., Quigley, R. M., Bacopoulos, A. (1990). Detailed Case History of Clay Liner Performance. *Waste Containment Systems: Construction, Regulation and Performance,* ASCE Geotechnical Special Publication No. 26, New York, 156-174.
86. Reades, D. W., Pohland, R. J., Kelly, G., King, S. (1987), Discussion, *ASCE Journal of Geotechnical Engineering,* 113(7):809-813.
87. Ressi di Cervia A. (1992). History of Slurry Wall Construction. *Slurry Wall: Design Construction and Quality Control - ASTM STP 1129,* Philadephia, 3-15.
88. Rogowski, A. S. (1986). Hydraulic Conductivity of Compacted Clay Soils. *Proceedings,* 12th Annual Research Symposium on Land Disposal, Remedial Action, Incineration, and Treatment of Hazardous Waste, U.S. EPA, Cincinnati, OH, 29-39.
89. Rowe, R.K. (1987). Pollutant Transport Through Barriers. *Geotechnical Practice for Waste Disposal '87,* ASCE, Geotechnical Special Publication No. 13, New York, 159-181.
90. Ryan, C.R. (1987). Vertical Barriers in Soil for Pollution Containment. *Geotechnical Practice for Waste Disposal '87,* ASCE Geotechnical Special Publication No. 13, New York, 182-204.
91. Schubert, W.R. (1987). Bentonite Matting in Composite Lining Systems. *Geotechnical Practice for Waste Disposal '87,* ASCE Geotechnical Special Publication No. 13, ASCE, New York, 784-796.
92. Schweitzer, F. (1989). Strength and Permeability of Single-Phase Diaphragm Walls. *Proc. of the Twelfth ICSMFE* Rio de Janeiro, A.A. Balkema Rotterdam, 3:1515-1518.
93. Shackelford, C. D. (1988), Diffusion as a Transport Process in Fine-Grained Barrier Materials. *Geotechnical News,* 6(2):24-27.
94. Shackelford, C.D. (1990). Transit Time Design of Earthen Barriers. *Engineering Geology,* Elsevier, Amsterdam, 29:79-94.
95. Shackelford, C.D. (1994). Waste-Soil Interactions that Alter Hydraulic Conductivity. *Hydraulic Conductivity and Waste Contaminant Transport in Soils,* ASTM STP 1142, ASTM, Philadelphia (in press).
96. Shackelford, C. D., Javed, F. (1991). Large-Scale Laboratory Permeability Testing of a Compacted Clay Liner. *Geotechnical Testing Journal,* ASTM, Philadelphia, 14(2):171-179.
97. Shackelford, C.D., Daniel, D.E. (1991a). Diffusion in Saturated Soil. I: Background. *ASCE Journal of Geotechnical Engineering,*

117(3):467-484.
98. Shackelford C.D., Daniel D.E. (1991b). Diffusion in Saturated Soil. II: Results for Compacted Clay. *ASCE Journal of Geotechnical Engineering*, 117(3):485-506.
99. Smith, D.W., Booker, J.R. (1993). Decontamination of a Polluted Aquifer by Interception and Sorption. *Proc. of the International Conference: The Environment and Geotechnics: from Decontamination to Protection of the Subsoil,* Paris, Ecole Nationale des Ponts et Chaussées.
100. Stoetzer, E. (1989). The New Diaphragm Walling Technique. *Proc. of the Twelfth ICSMFE,* Rio de Janeiro.
101. Torstensson, B.A. (1984). A New System for Ground Water Monitoring. *Ground Water Monitoring Review*, 4(4):131-138.
102. Trautwein, S. J., Williams, C. E. (1990). Performance Evaluation of Earthen Liners. *Waste Containment Systems: Construction, Regulation and Performance*, ASCE Geotechnical Special Publication No. 26, New York, 30-51.
103. U.S. EPA (1984). Slurry Trench Construction for Pollution Migration Control. *Handbook EPA*-540/2-84-001, Cincinnati, OH.
104. U.S. EPA (1985). Draft, Minimum Technology Guidance on Double Liner Systems for Landfills and Surface Impoundments. EPA/530-SW-85-014, Cincinnati, OH.
105. U. S. EPA (1988). Seminars-Requirements for Hazardous Waste Landfill Design, Construction, and Closure. Center for Environmental Research Information (CERI-88-33), Cincinnati, OH.
106. Wiemer, K. (1987). Technical and operational possibilities to minimize leachate quantity. *Proceedings of International Symposium on Process, Technology and Environmental Impact of Sanitary Landfill.* October 19-23, Cagliari, Sardinia.
107. Woods, R.D. (1987). Geotechnical Practice for Waste Disposal, ASCE, Geotechnical Special Publication No. 13, New York, 864 pp.

Appendix 2

First Author Index

Al-Tabbaa	599
Alimi-Ichola	115
Alonso	847
Amann	121
Artières	605
Aubertin	427
Avci	127
Azevedo	433
Ben Dak	927
Bernhard	611
Betelev	133
Betelev	137
Black	617
Blight	417
Boldt-Leppin	141
Booker	147
Brand	1
Brumund	929
Bullock	441
Calabresi	449
Campanella	153
Chan	161
Cherry	77
Clark	167
Clark	625
Cooke	173
Coulet	631
Crawford	179
Crosnier	185
Dasika	187
Davidson	415
Davidson	455
de Campos	461
de O. e Castro	637
De Souza	467
Deaver	193
Deaver	201
Del Greco	643
do Val	651
Dusseault	209
Dusseault	855
Egyir	33
Ehrlich	657
Erickson	665
Ericson	475
Esnault	215
Fahey	481
Fair	79
Favaretti	221
Federico	671
Fleureau	227
Fourie	487
Gandais	861
Garassino	677
Gemperline	233
Genske	493
Germanov	499
Gertz	827
Gomyoh	241
Grebenets	247
Hadjigeorgiou	505
Hansen	255
Hellou	261
Hinchee	101
Hong	683
Hulla	867
Imamura	267
Inyang	273
Iturbe	689
Jang	695
Jessberger	577
Jong	279
Kamon	287
Karim	701
Kavazanjian	707
Khera	293
Kikkeri	299
Knox	41
Komine	873
Kozicki	305
Kozicki	713
Kulachkin	879
Laloui	881
Lefebvre	313
Lighthall	513
Lim	319
Lo	519
Lord	937
Manassero	103
Martin	325
Meynink	949
Mitchell	951
Mittal	887
Miyake	721
Morgenstern	963
Nnadi	331
O'Shaughnessy	729
Ogino	49
Olivella	895
Oweis	735
Papademetriou	743
Perrin	749
Potter	525
Pradhan	755

Przygocki	337
Quigley	589
Rahman	761
Rainbow	343
Rakotondramanitra	767
Rampello	531
Reddy	539
Reznik	545
Robinet	901
Robinet	909
Rogers	837
Rollings	21
Rowe	349
Saarela	773
Sakai	781
Samtani	57
Sanchez	355
Sawa	361
Schlosser	91
Shackelford	789
Shackelford	981
Shepherd	829
Sillito	795
Sjöholm	367
Smiley	801
Sohn	807
Steer	977
Stoeva	551
Strachan	915
Suthaker	557
Suzuki	65
Swarbrick	563
Szymanski	569
Taha	373
Thevanayagam	379
Tran Duc	921
Uehara	813
Unruh	819
Viergever	11
Vogler	387
Weart	979
Winiarski	393
Yamanouchi	71
Yasuhara	401
Yudhbir	409

Author Index

Author	Corresponding First Author	Author	Corresponding First Author
Aachib	Aubertin	Chan	Chan
Acar	Taha	Chang	Shackelford
Acedera	Jong	Chapuis	Aubertin
Al-Tabbaa	Al-Tabbaa	Cherry	Cherry
Alimi-Ichola	Alimi-Ichola	Chiu	Shackelford
Almeida	Ehrlich	Clark	Rainbow
Alonso	Olivella	Clark	Clark
Alonso	Alonso	Clark	Clark
Alonso	Dusseault	Cook	Erickson
Alves	Azevedo	Cooke	Cooke
Alves	de Campos	Correnoz	Artieres
Amann	Amann	Coulet	Rakotondramanitra
Aref	Hadjigeorgiou	Coulet	Perrin
Artières	Artières	Coulet	Coulet
Atwater	Dasika	Coward	Lo
Aubertin	Aubertin	Crawford	Crawford
Avci	Avci	Crosnier	Crosnier
Azevedo	de Campos	Crotty	Hansen
Azevedo	Azevedo	Curtil	Coulet
Azzouz	Rakotondramanitra	Das	Erickson
Barbosa	de O. e Castro	Dasika	Dasika
Barbosa	Ehrlich	Davidson	Davidson
Barbour	Lim	Davidson	Davidson
Batra	Jong	Davies	Campanella
Battey	Jong	de Almeida	de O. e Castro
Bell	Bullock	de Campos	Azevedo
Bell	Sillito	de Campos	de Campos
Ben Dak		de O. e Castro	de O. e Castro
Bentoumi	Alimi-Ichola	De Souza	De Souza
Bernhard	Bernhard	Deaver	Deaver
Betelev	Betelev	Deaver	Deaver
Betelev	Betelev	Del Greco	Del Greco
Bianco	Rampello	Delahaye	Alonso
Bianco	Calabresi	Delmas	Artieres
Bilak	Dusseault	Delolme	Crosnier
Black	Black	Demers	Davidson
Blight	Blight	Didier	Winiarski
Bodine	Avci	Didier	Alimi-Ichola
Boldt-Leppin	Boldt-Leppin	do Val	do Val
Booker	Booker	Domvile	Lighthall
Boudissa	Coulet	Donald	Mittal
Bowman	Kikkeri	Donovan	Rainbow
Boyd	Campanella	Dorr	Vogler
Boyd	Hadjigeorgiou	Dounias	Papademetriou
Brand	Brand	Dufournet-Bourgeois	Esnault
Brumund	Brumund	Dufournet-Bourgeois	Gandais
Bullock	Bullock	Dusseault	Davidson
Burbidge	Martin	Dusseault	Dusseault
Burnotte	Lefebvre	Dusseault	Dusseault
Bussiere	Aubertin	Egyir	Egyir
Calabresi	Rampello	Ehrlich	de O. e Castro
Calabresi	Calabresi	Ehrlich	Ehrlich
Campanella	Campanella	El Yazidi	Hellou
Carrera	Olivella	Endo	Tran Duc

Author	Corresponding First Author	Author	Corresponding First Author
Erickson	Erickson	Imai	Pradhan
Ericson	Ericson	Imamura	Imamura
Esnault	Gandais	Inyang	Inyang
Esnault	Esnault	Ishibashi	Yamanouchi
Everard	Campanella	Iturbe	Iturbe
Fahey	Fahey	Iwatani	Miyake
Fair	Fair	Jang	Jang
Favaretti	Favaretti	Jang	Hong
Federico	Federico	Jeong	Hong
Fedoseev	Grebenets	Jessberger	Jessberger
Fleureau	Fleureau	Johnston	Przygocki
Fong	Cooke	Jong	Jong
Fourie	Fourie	Kabir	Oweis
Fraser	Rowe	Kamon	Sawa
Fredlund	Lim	Kamon	Kamon
Fujiyasu	Fahey	Karim	Karim
Gale	Taha	Kataoka	Ogino
Gandais	Gandais	Katsumi	Kamon
Garassino	Garassino	Katzenbach	Vogler
Garga	O'Shaughnessy	Kavazanjian	Kavazanjian
Gemperline	Gemperline	Khera	Khera
Gens	Olivella	Khizardjian	Perrin
Gens	Alonso	Kikkeri	Kikkeri
Genske	Genske	Kim	Jang
Germanov	Germanov	Kirillov	Kulachkin
Gertz	Gertz	Kislyakov	Betelev
Giambastiani	Garassino	Kitazono	Suzuki
Gibson	Potter	Knox	Knox
Gomyoh	Gomyoh	Komada	Tran Duc
Goto	Ogino	Komine	Komine
Gourc	Bernhard	Korenkov	Kulachkin
Grebenets	Grebenets	Kosinski	Przygocki
Grovel	Sanchez	Kostov	Germanov
Guler	Avci	Kozicki	
Hadjigeorgiou	Hadjigeorgiou	Kozicki	Boldt-Leppin
Hagarty	Kikkeri	Kozicki	Kozicki
Haider	Karim	Kozicki	Kozicki
Hansen	Hansen	Kozicki	Boldt-Leppin
Hanzawa	Gomyoh	Kozicki	Kozicki
Hara	Uehara	Kraft	Unruh
Harty	Kozicki	Kulachkin	Kulachkin
Haug	Boldt-Leppin	Kulachkin	Betelev
Hayashi	Suzuki	Kumar	Yudhbi
Hellou	Hellou	Kumar	Yudhbi
Hinchee	Hinchee	Kumar	Yudhbi
Hirao	Yasuhara	Kuroda	Ogino
Hironaga	Tran Duc	Laloui	Laloui
Holl	Mittal	Lancelloti	Samtani
Hong	Hong	Le Tellier	Bernhard
Horiuchi	Yasuhara	Lebon	Dusseault
Hsu	Chan	Lee	Jang
Hulla	Hulla	Lee	Sohn
Hyodo	Yasuhara	Lee	Jang
Ilyichev	Kulachkin		

Author	Corresponding First Author	Author	Corresponding First Author
Lefebvre	Lefebvre	Plas	Robinet
Leo	Booker	Plas	Rohinet
Lighthall	Lighthall	Plaskett	Ericson
Lim	Lim	Plsko	Hulla
Lloret	Alonso	Polsky	Kulachkin
Lo	Lo	Poran	Kavazanjian
Lolaev	Grebenets	Potter	Potter
Lord	Lo	Poulin	Hadjigeorgiou
Lord	Lord	Pradhan	Pradhan
Loxham	Viergever	Pradhan	Gomyoh
MacPhie	Szymanski	Previatello	Favaretti
Madsen	Amann	Przygocki	Przygocki
Manassero	Manassero	Quigley	Quigley
Martin	Martin	Radkevich	Kulachkin
Martinenghi	Amann	Rahbaoui	Robinet
Maruyama	Suzuki	Rahman	Rahman
Maruyama	Miyake	Rainbow	Rainbow
Massey	Brand	Rakotondramanitra	Rakotondramanitra
Matasovic	Kavazanjian	Rampello	Calabresi
Matichard	Bernhard	Rampello	Rampello
McLeod	Lighthall	Reddy	Reddy
Medlock	Shepherd	Reznik	Reznik
Meynink	Meynink	Rhattas	Robinet
Miller	Shepherd	Ricard	Aubertin
Mitchell	Mitchell	Robinet	Robinet
Mittal	Mittal	Robinet	Robinet
Miyake	Miyake	Rocha	Winiarski
Modaressi	Laloui	Rogers	Rogers
Molitor	Vogler	Rollings	Rollings
Moraci	Favaretti	Rosenberg	Lefebvre
Morgenstern	Morgenstern	Rothenburg	Dusseault
Mowder	Deaver	Rowe	Rowe
Murata	Sakai	Saarela	Saarela
Naito	Sawa	Sakai	Sakai
Najjar	Knox	Salesi	Garassino
Neto	do Val	Samtani	Samtani
Nnadi	Nnadi	Sanchez	Sanchez
O'Shaughnessy	O'Shaughnessy	Santamarina	Dusseault
Oberti	Artieres	Satoh	Kavazanjian
Ogata	Komine	Savchenko	Grebenets
Oggeri	Del Greco	Sawidou	Potter
Ogino	Ogino	Sawa	Sawa
Olenick	Strachan	Scarrow	Clark
Olivella	olivella	Schlosser	Schlosser
Ouvry	Gandais	Schuh	Reddy
Oweis	Oweis	Scott	Suthaker
Palut	Robinet	Sego	Egyir
Pane	Calabresi	Shackelford	Manassero
Pane	Rampello	Shackelford	Shackelford
Pankov	Kulachkin	Shackelford	Shackelford
Papademetriou	Papademetriou	Shakoor	Black
Parisopoulos	Papademetriou	Sheeran	Fair
Parks	Crawford	Sheinin	Kulachkin
Perrin	Perrin	Shepherd	Shepherd

Author	Corresponding First Author	Author	Corresponding Firest Author
Shimomura	Imamura	Wrench	Fourie
Sillito	Sillito	Yamanouchi	Yamanouchi
Sills	de Campos	Yamshchikov	Kulachkin
Silva	Iturbe	Yang	Suzuki
Sjöholm	Sjöholm	Yasuhara	Yasuhara
Skinner	Clark	Yudhbir	Yudhbir
Smiley	Smiley	Zamiskie	Oweis
Snow	Kavazanjian	Zappi	Taha
Sohn	Sohn	Zlatanov	Stoeva
Somers	Martin		
Soyez	Schlosser		
Stavnitser	Kulachkin		
Steer	Steer		
Stoeva	Stoeva		
Strachan	Strachan		
Strandberg	Sjoholm		
Sueoka	Imamura		
Suthaker	Suthaker		
Suzuki	Suzuki		
Swarbrick	Swarbrick		
Szymanski	Szymanski		
Taha	Hulla		
Taha	Taha		
Taibi	Fleureau		
Takagi	Yamanouchi		
Tanal	Samtani		
Taniguchi	Tran Duc		
Tarumi	Sakai		
Tashima	Yamanouchi		
Thein	Genske		
Thevanayagam	Thevanayagam		
Tomohisa	Sawa		
Tran Duc	Tran Duc		
Tremblay	Aubertin		
Trentesaux	Dusseault		
Tworkowski	Deaver		
Uchiyama	Pradhan		
Uehara	Uehara		
Unruh	Unruh		
Vaciago	Martin		
Van den Berg	Kulachkin		
Viergever	Viergever		
Vogler	Vogler		
Wada	Miyake		
Walsh	AlTabbaa		
Wang	Samtani		
Wang	Thevanayagam		
Weart	Weart		
Westrate	Viergever		
Whiteside	Brand		
Wilcher	Kikkeri		
Winiarski	Winiarski		
Winkler	Ericson		
Wojnarowicz	Schlosser		

Appendix 3

COUNTRY INDEX

Country	First Author	Country	First Author
AUSTRALIA	Booker		Fleureau
	Cooke		Gandais
	Fahey		Hellou
	Meynink		Laloui
	Swarbrick		Perrin
			Rakotondramanitra
BANGLADESH	Karim		Robinet
	Rahman		Sanchez
			Schlosser
BRAZIL	Azevedo		Winiarski
	de Campos		
	de O. e Castro	GERMANY	Genske
	Ehrlich		Jessberger
	do Val		VoGler
BULGARIA	Germanov	GREECE	Papademetriou
	Stoeva	HONG KONG	Brand
CANADA	Aubertin	INDIA	Yudhbir
	Boldt-Leppin		
	Campanella	ITALY	Calabresi
	Cherry		Del Greco
	Crawford		Favaretti
	Dasika		Federico
	Davidson		Garassino
	De Souza		Manassero
	Dusseault		Rampello
	Egyir		
	Falr	JAPAN	Gomyoh
	Hadjigeorgiou		Imamura
	Kozicki		Kamon
	Lefebvre		Komine
	Lighthall		Miyake
	Lim		Ogino
	Lo		Pradhan
	Mittal		Sakai
	Morgenstern		Sawa
	O'Shaughnessy		Suzuki
	Quigley		Tran Duc
	Rowe		Uehara
	Suthaker		Yamanouchi
	Szymanski		Yasuhara
		KOREA	
FINLAND	Saarela		Hong
	Sjöholm		Jang
FRANCE	Alimi-Ichola	MEXICO	Iturbe
	Artières	NETHERLANDS	Viergever
	Bernhard		
	Coulet		
	Crosnier		
	Esnault		

Country	First Author	Country	First Author
RUSSIA	Betelev	USA	Strachan
	Grebenets		Taha
	Kulachkin		Thevanayagam
SLOVAK REPUBLIC	Hulla		Unruh
SOUTH AFRICA	Blight		Weart
	Bullock		
	Fourie		
SPAIN	Alonso		
	Olivella		
SWITZERLAND	Amann		
TURKEY	Avci		
UNITED KINGDOM	Al-Tabbaa		
	Clark		
	Lord		
	Martin		
	Potter		
	Rainbow		
USA	Ben Dak		
	Black		
	Brumund		
	Chan		
	Davidson		
	Deaver		
	Erickson		
	Ericson		
	Gemperline		
	Gertz		
	Hansen		
	Hinchee		
	Inyang		
	Jong		
	Kavazanjian		
	Khera		
	Kikkeri		
	Knox		
	Mitchell		
	Nnadi		
	Oweis		
	Przygocki		
	Reddy		
	Reznik		
	Rogers		
	Rollings		
	Samtani		
	Shackelford		
	Shepherd		
	Smiley		
	Sohn		
	Steer		